Linker Strategies In Solid-Phase Organic Synthesis

Linker Strategies In Solid-Phase Organic Synthesis

Edited by

PETER J. H. SCOTT

University of Michigan, Ann Arbor, USA

A John Wiley and Sons, Ltd., Publication

This edition first published 2009
© 2009 John Wiley & Sons Ltd
Reprinted June 2010

Registered office
John Wiley & Sons Ltd, The Atrium, Southern Gate, Chichester, West Sussex, PO19 8SQ, United Kingdom

For details of our global editorial offices, for customer services and for information about how to apply for permission to reuse the copyright material in this book please see our website at www.wiley.com.

The right of the author to be identified as the author of this work has been asserted in accordance with the Copyright, Designs and Patents Act 1988.

All rights reserved. No part of this publication may be reproduced, stored in a retrieval system, or transmitted, in any form or by any means, electronic, mechanical, photocopying, recording or otherwise, except as permitted by the UK Copyright, Designs and Patents Act 1988, without the prior permission of the publisher.

Wiley also publishes its books in a variety of electronic formats. Some content that appears in print may not be available in electronic books.

Designations used by companies to distinguish their products are often claimed as trademarks. All brand names and product names used in this book are trade names, service marks, trademarks or registered trademarks of their respective owners. The publisher is not associated with any product or vendor mentioned in this book. This publication is designed to provide accurate and authoritative information in regard to the subject matter covered. It is sold on the understanding that the publisher is not engaged in rendering professional services. If professional advice or other expert assistance is required, the services of a competent professional should be sought.

The publisher and the author make no representations or warranties with respect to the accuracy or completeness of the contents of this work and specifically disclaim all warranties, including without limitation any implied warranties of fitness for a particular purpose. This work is sold with the understanding that the publisher is not engaged in rendering professional services. The advice and strategies contained herein may not be suitable for every situation. In view of ongoing research, equipment modifications, changes in governmental regulations, and the constant flow of information relating to the use of experimental reagents, equipment, and devices, the reader is urged to review and evaluate the information provided in the package insert or instructions for each chemical, piece of equipment, reagent, or device for, among other things, any changes in the instructions or indication of usage and for added warnings and precautions. The fact that an organization or Website is referred to in this work as a citation and/or a potential source of further information does not mean that the author or the publisher endorses the information the organization or Website may provide or recommendations it may make. Further, readers should be aware that Internet Websites listed in this work may have changed or disappeared between when this work was written and when it is read. No warranty may be created or extended by any promotional statements for this work. Neither the publisher nor the author shall be liable for any damages arising herefrom.

Library of Congress Cataloging-in-Publication Data

Scott, Peter J. H.
 Linker strategies in solid-phase organic synthesis / Peter J. H Scott.
     p. cm.
 Includes bibliographical references and index.
 ISBN 978-0-470-51116-9
 1. Solid-phase synthesis. I. Title.
 QD262.S36 2009
 547′.2—dc22
                                                2009030808

A catalogue record for this book is available from the British Library.

ISBN: 978-0-470-51116-9 (H/B)

Typeset in 10/12pt Times by Laserwords Private Limited, Chennai, India
Printed and bound in the United Kingdom by CPI Antony Rowe, Chippenham, Wiltshire

Contents

Foreword	*xv*
Preface	*xix*
List of Contributors	*xxi*
About the Editor	*xxiii*
Abbreviations	*xxv*

I   INTRODUCTION — 1

1   General Overview — 3
Scott L. Dax

1.1	Introduction, background and pivotal discoveries	3
1.2	Fundamentals of conducting solid-phase organic chemistry	9
	1.2.1 Apparatus	9
	1.2.2 Typical solid supports	10
	1.2.3 Fluorous supports	12
	1.2.4 Linker strategies	12
	1.2.5 Challenges	17
	1.2.6 Linker groups	18
1.3	Concluding comments	20
1.4	Personal perspective and testimony: solid-phase Mannich chemistry	21
	References	22

II   TRADITIONAL LINKER UNITS FOR SOLID-PHASE ORGANIC SYNTHESIS — 25

2   Electrophile Cleavable Linker Units — 27
Michio Kurosu

2.1	Introduction	27
2.2	Resins for use with electrophilic linkers	28
2.3	Electrophile cleavable linkers	30
	2.3.1 Acid labile linkers	31
2.4	Conclusion	70
	References	71

3   Nucleophile Cleavable Linker Units — 77
Andrea Porcheddu and Giampaolo Giacomelli

3.1	Introduction	77

3.2	Linker units		78
3.3	Nucleophilic labile linker units		79
	3.3.1	Cleavage by saponification or basic trans-esterification	80
	3.3.2	Cleavage by aminolysis	86
	3.3.3	Cleavage by hydrazinolysis	101
	3.3.4	Cleavage by Hydroxylamines	105
	3.3.5	Cleavage by other nucleophiles	109
	3.3.6	Linker cleavage by intramolecular nucleophilic reaction	119
3.4	Conclusion		129
	References		130

4	**Cyclative Cleavage as a Solid-Phase Strategy**		**135**
	A. Ganesan		
4.1	Introduction		135
4.2	C–N bond formation		137
	4.2.1	Cyclopeptides and cyclodepsipeptides	138
	4.2.2	Heterocycles, five-membered ring formation	139
	4.2.3	Heterocycles, six- and seven-membered ring formation	142
4.3	C–O bond formation		145
4.4	C–C bond formation		146
4.5	Conclusion		148
	References		148

5	**Photolabile Linker Units**	**151**
	Christian Bochet and Sébastien Mercier	
5.1	Introduction	151
5.2	Linkers based on the ortho-nitrobenzyloxy function	151
5.3	Linkers based on the ortho-nitrobenzylamino function	158
5.4	Linkers based on the α-substituted ortho-nitrobenzyl group	161
5.5	Linkers based on the ortho-nitroveratryl group	165
5.6	Linkers based on the phenacyl group	173
5.7	Linkers based on the para-methoxyphenacyl group	176
5.8	Linkers based on the benzoin group	180
5.9	Linkers based on the pivaloyl group	184
5.10	Traceless linkers	187
5.11	Other types of photolabile linker units	187
5.12	Conclusion	188
	References	191

6	**Safety-Catch Linker Units**		**195**
	Sylvain Lebreton and Marcel Pátek		
6.1	Introduction		195
6.2	Activation of a carbonyl group by the inductive effect (I–) of an adjacent substituent		196
	6.2.1	Kenner-type safety-catch linker	196
	6.2.2	*N*-boc-activated safety-catch linker	197

	6.2.3	Sulfide/sulfone safety-catch linker	198
	6.2.4	Dpr(phoc) safety-catch linker	199
6.3	Activation by the mesomeric effect (M-) of the $-X-Y=Z$ moiety adjacent to a carbonyl group		199
	6.3.1	Carbonyl activation by oxidative aromatization	199
	6.3.2	Carbonyl activation by indole ring formation	200
	6.3.3	Benzyl/phenyl-hydrazide safety-catch linker	200
	6.3.4	Dehydration activated safety-catch linker	202
6.4	Activation by the positive mesomeric effect (M+) of the $-X-Y=Z$ moiety adjacent to a N-acyl or O-alkyl group		202
	6.4.1	Benzhydryl-based safety-catch linker	202
	6.4.2	Indole-based safety-catch linker	203
	6.4.3	Nitrobenzyl alcohol-based safety-catch linker	204
6.5	Aromatic $S_N Ar$ substitution		205
6.6	Fragmentation by β-elimination		207
6.7	Safety-catch linker for release in aqueous buffers		208
	6.7.1	Geysen safety-catch linker	208
	6.7.2	Frank safety-catch linker	210
	6.7.3	Lyttle safety-catch linker	210
	6.7.4	Multiple cleavable linkers	211
6.8	Photochemical activation		212
6.9	Miscellaneous safety-catch linkers		213
	6.9.1	Activation by reductive aromatization	213
	6.9.2	Activation via intramolecular H-bonding	214
	6.9.3	Activation by formation of an alkyne-cobalt complex	215
	6.9.4	Activation by oxidation of arylsulfide for pummerer rearrangement	216
	6.9.5	Activation by oxidative N-benzyl deprotection	217
	6.9.6	Activation by thioether alkylation	218
6.10	Conclusion		219
	References		219

7	**Enzyme Cleavable Linker Units** *Mallesham Bejugam and Sabine L. Flitsch*		**221**
7.1	Introduction		221
7.2	Enzyme cleavable linker units		222
	7.2.1	Exo linker units	222
	7.2.2	Endo linker units	225
7.3	Conclusion		237
	References		237

III	**MULTIFUNCTIONAL LINKER UNITS FOR DIVERSITY-ORIENTED SYNTHESIS**	**239**
8	**An Introduction to Diversity-Oriented Synthesis** *Richard J. Spandl, Gemma L. Thomas, Monica Diaz-Gavilan, Kieron M. G. O'Connell and David R. Spring*	**241**
8.1	Introduction	241

8.2		Exploring chemical space	243
8.3		Sources of skeletally diverse small molecules	244
8.4		Enriching chemical space using DOS	244
8.5		The subjective nature of 'Diversity'	245
8.6		Differing strategies towards similar goals	246
	8.6.1	DOS based on privileged scaffolds	246
	8.6.2	DOS from simple starting materials	248
8.7		Generating skeletal diversity	248
	8.7.1	Strategy 1: Pluripotent functional groups	249
	8.7.2	Strategy 2: Pluripotent (densely functionalised) molecules	253
	8.7.3	Strategy 3: Folding pathways	256
8.8		DOS and solid-phase organic synthesis	257
	8.8.1	An overview of linkage cleavage strategies	258
	8.8.2	Diversity linkers: A summary of the approaches used	259
8.9		Conclusion	260
		References	260

9 T1 and T2 – Versatile Triazene Linker Groups 263
Kerstin Knepper and Robert E. Ziegert

9.1		Introduction	263
9.2		The T1 linker	264
	9.2.1	The dibenzyl-type T1 resins	266
	9.2.2	The piperazinyl-type T1 resins	278
9.3		The T2 linker units	282
	9.3.1	The T2 Linker	282
	9.3.2	The T2* linker for synthesis	287
	9.3.3	The T2* scavenger resin	292
9.4		Miscellaneous triazene linkers	293
9.5		Conclusion	300
		References	300

10 Hydrazone Linker Units 303
Ryszard Lazny

10.1	Introduction	303
10.2	Hydrazone linker units	303
10.3	Conclusion	312
	References	314

11 Benzotriazole Linker Units 317
Daniel K. Whelligan

11.1		Introduction	317
11.2		Syntheses of polymer-supported benzotriazoles	318
	11.2.1	Carbon–carbon tethered benzotriazoles	318
	11.2.2	Ether tethered benzotriazoles	319
	11.2.3	Amide tethered benzotriazoles	320

		11.2.4 Ester tethered benzotriazoles	322
11.3	Polymer-supported benzotriazole linked reactions		322
	11.3.1	Mannich-type reaction and cleavage	322
	11.3.2	Enolate acylation	325
	11.3.3	Urea synthesis	325
	References		329

12 Diversity Cleavage Strategies from Phosphorus Linkers — 331
Patrick G. Steel and Tom M. Woods

12.1	Introduction	331
12.2	Diversity cleavage through olefination reactions	332
	12.2.1 Diversity cleavage through the Wittig reaction	332
	12.2.2 Diversity cleavage using the Horner–Wadsworth–Emmons reaction	336
12.3	Diversity cleavage of enol phosphonates through palladium catalysed cross-coupling reactions	338
12.4	Oxidative diversity cleavage of cyanophosphoranes	339
	References	340

13 Sulfur Linker Units — 343
Peter J. H. Scott

13.1	Introduction	343
13.2	Sulfide linker units	344
	13.2.1 Introduction	344
	13.2.2 Reductive traceless cleavage	345
	13.2.3 Multifunctional cleavage via nucleophilic substitution reactions	347
	13.2.4 Multifunctional cleavage via elimination reactions	350
13.3	Sulfonium Linker Units	351
13.4	Sulfoxide linker units	354
	13.4.1 Introduction	354
	13.4.2 Traceless cleavage	355
	13.4.3 Multifunctional cleavage using the pummerer rearrangement	355
13.5	Sulfone linker units	358
	13.5.1 Introduction	358
	13.5.2 Reductive traceless cleavage	360
	13.5.3 Multifunctional cleavage via elimination reactions	362
	13.5.4 Multifunctional cleavage via nucleophilic substitution reactions	370
13.6	Sulfonate ester linker units	373
	13.6.1 Introduction	373
	13.6.2 Alkanesulfonate ester linker units	373
	13.6.3 Perfluoralkanesulfonyl (PFS) linker units	378
	13.6.4 Tetrafluoroarylsulfonyl linker units	381
13.7	Sulfamate linker units	383
13.8	Thioester linker units	385
13.9	Conclusions	387
	References	387

14 Selenium- and Tellurium-Based Linker Units   391
Tracy Yuen Sze But and Patrick H. Toy

14.1 Introduction   391
14.2 Selenium- and tellurium-based linker group reagents and their syntheses   391
14.3 Selenium-based linker group attachment methods   398
    14.3.1 Electrophilic attachment at selenium   398
    14.3.2 Nucleophilic attachment at selenium   398
    14.3.3 Radical attachment at selenium   402
    14.3.4 Attachment at other positions   402
14.4 Selenium-based linker group cleavage methods   403
    14.4.1 Oxidative cleavage   403
    14.4.2 Nucleophilic displacement cleavage   410
    14.4.3 Homolytic cleavage   411
    14.4.4 Miscellaneous cleavage methods   415
14.5 Conclusions   415
References   416

15 Linker Units Cleaved by Radical Processes: Cleavage of Carbon-Sulfur, -Selenium, -Tellurium, -Oxygen, -Nitrogen and -Carbon Linkers   419
Giuditta Guazzelli, Marc Miller and David J. Procter

15.1 Introduction   419
15.2 Linkers cleaved using tin hydride, alkyltin and silicon hydride reagents   420
    15.2.1 Oxygen-based linkers   420
    15.2.2 Sulfur-based linkers   421
    15.2.3 Selenium-based linkers   421
    15.2.4 Tellurium-based linkers   430
15.3 Linkers cleaved by oxidative electron-transfer   431
    15.3.1 Ether and amine linkers cleaved by oxidative electron transfer   431
    15.3.2 A homobenzylic ether linker cleaved by oxidative electron transfer   439
    15.3.3 A sulfur linker cleaved by oxidative electron transfer with CAN   440
    15.3.4 Safety-catch linkers cleaved by oxidative electron transfer   443
15.4 Linkers cleaved by reductive electron-transfer   447
    15.4.1 N–O linkers cleaved using samarium(II) iodide   448
    15.4.2 Sulfonamide linkers cleaved by reductive electron-transfer   450
    15.4.3 Ether linkers cleaved using samarium(II) iodide   450
    15.4.4 Alkyl and aryl sulfide/sulfone linkers cleaved by reductive electron-transfer   453
15.5 Radical processes that indirectly trigger linker cleavage   462
    15.5.1 Nitro group reduction as a trigger for linker cleavage   462
    15.5.2 Radical carbon–carbon bond formation as a trigger for linker cleavage   462
15.6 Conclusions   465
References   465

16 Silicon and Germanium Linker Units   467
Alan C. Spivey and Christopher M. Diaper

16.1 Introduction   467
16.2 Silicon-based linkers   468

	16.2.1	The preparation of silyl resins	468
	16.2.2	Activation of Si–H and Si–Aryl resins for substrate attachment	471
	16.2.3	Silyl ether linkers	475
	16.2.4	Fragmentation-based silyl linkers	479
	16.2.5	Traceless/diversity silyl linkers	487
16.3	Germanium-based linkers		495
	16.3.1	The preparation of germyl resins	496
	16.3.2	Activation of Ge–Methyl and Ge–Aryl resins for substrate attachment	497
	16.3.3	Traceless/diversity germyl linkers	498
16.4	Conclusions		500
	References		501

17 Boron and Stannane Linker Units — 505
Peter J. H. Scott

17.1	Introduction	505
17.2	Organostannane linker units	507
	17.2.1 Introduction	507
	17.2.2 Organostannane linker units	508
17.3	Organoboron linker units	511
	17.3.1 Introduction	511
	17.3.2 Diversity cleavage through suzuki–miyaura reactions	512
	17.3.3 Alternative cleavage strategies from organoboron linkers	514
17.4	Conclusion	516
	References	516

18 Bismuth Linker Units — 519
Peter J. H. Scott

18.1	Introduction	519
18.2	Bismuth linker units	519
18.3	Conclusions	523
	References	523

19 Transition Metal Carbonyl Linker Units — 525
Susan E. Gibson and Amol A. Walke

19.1	Introduction	525
19.2	Chromium carbonyl linker units	525
19.3	Cobalt carbonyl linker units	533
19.4	Manganese carbonyl linker units	535
19.5	Conclusion	536
	References	536

20 Linkers Releasing Olefins or Cycloolefins by Ring Closing Metathesis — 537
Jan H. van Maarseveen

20.1	Introduction	537

20.2	Cycloolefins via method I		539
20.3	Terminal olefins via route II		545
20.4	Terminal and internal olefins via route III		547
20.5	Conclusion		548
	References		549

IV ALTERNATIVE LINKER STRATEGIES   551

21 Fluorous Linker Units   553
Wei Zhang

21.1	Introduction		553
	21.1.1	Polymer versus fluorous linkers	553
	21.1.2	Different types of fluorous linkers	553
	21.1.3	Methods for separation of fluorous linker attached substrates	557
21.2	Fluorous linkers for synthesis of small molecules		557
	21.2.1	Synthesis of heterocyclic compounds	557
	21.2.2	Synthesis of natural product analogs	562
	21.2.3	Fluorous mixture synthesis	563
21.3	Fluorous linkers for synthesis of biomolecules		566
	21.3.1	Synthesis of peptides	566
	21.3.2	Synthesis of oligosaccharides	566
	21.3.3	Synthesis of glycopeptides	569
	21.3.4	Synthesis of oligonucleotides	569
21.4	Other applications of fluorous linkers		570
	21.4.1	Isolation of proteomics samples	571
	21.4.2	Microarray screening	572
	21.4.3	Enzymatic synthesis	573
21.5	Conclusion		574
	References		574

22 Solid-Phase Radiochemistry   577
Brian G. Hockley, Peter J. H. Scott and Michael R. Kilbourn

22.1	Introduction		577
22.2	Solid-phase surrogates in radiochemistry		578
	22.2.1	Thin film radiochemistry	578
	22.2.2	Germanium solid-phase surrogates	580
22.3	Solid-phase radiochemistry		581
	22.3.1	Stannane linkers in radiochemistry	581
	22.3.2	Germanium linkers in radiochemistry	582
	22.3.3	Fluorous linkers in radiochemistry	583
22.4	Conclusions and perspectives		585
	References		586

V LINKER SELECTION TABLES — 589

23 Linker Selection Tables — 591
Peter J. H. Scott

23.1	Introduction	591
23.2	Linkers for alcohols, phenols and diols	592
23.3	Linkers for carboxylic acids, esters and related compounds	598
23.4	Linkers for aldehydes, ketones and related carbonyl compounds	606
23.5	Linkers for amides, ureas and related compounds	611
23.6	Linkers for amines	623
23.7	Linkers thiols, thioethers and disulfides	629
23.8	Linkers for sugars	629
23.9	Linkers liberating alkyl groups	631
23.10	Linkers for alkenes, alkynes and related compounds	635
23.11	Linkers for aryl compounds	644

Index — 657

Foreword

When I have the honor to write the Foreword to a technical book on Chemistry, I do my very best to place the subject matter of the book into a non-technical perspective. It's challenging, and often causes the reader to reflect on a *raison d'être* more expansive than structures, numbers and yields. How many of us have had the common experience of presenting a copy of our PhD thesis to Mom and Dad, who then tell us that the angular hieroglyphics and impenetrable vocabulary makes it impossible for them to comprehend the fruit of our labors – but they are very proud of us just the same? We try, typically in vain, to give them everyday things that would help explain what we have done. But this seldom succeeds. So many of the concepts we use in Chemistry have no common day counterpart.

So what of a book on 'Linkers in Solid-Phase Organic Synthesis?' Mom and Dad, this could be a tough one. A 'linker' is easy enough to understand. A linker connects one solid (not liquid; that would be a pipe) thing with another. Ropes keep young children linked to their parent, eliminating the chance for escape until the parent wishes it. A phone cable links your telephone with other wires that link to a telephone switching station (as least, that's the way it used to work).

But these, 'Solid-Phase' and 'Organic Synthesis' things... what are they? And do I need to understand them in order for me to understand exactly what the 'linker' does? To the extent creativity serves to avoid these questions, that is the extent to which you will enjoy success in your explanation. So what functions must a linker serve in this context, and are there any examples in everyday life that are similar?

OK, now we get to the part we'd better be able to explain as one chemist to another. First, what is 'solid-phase organic synthesis?' Amazingly, few of your chemist friends know of this. Why? Because this unbelievably useful, Nobel Prize winning technology seldom finds its way into any college chemistry courses. Which is crazy. Automated peptide synthesis... DNA synthesis... soon polysaccharide synthesis... these are the chemistry inventions upon which the majority of the molecular biology revolution are based. Most molecular biologists will use biomolecules if they can: (a) buy them or (b) make them in a way that is effortless. Otherwise, forget it. We chemists provided the research that led to automated synthesis machines that led molecular biologists to make landmark discoveries. Why did Chemistry lose out on the 'recognition' part of this endeavor? Allow me to opine.

Oligo peptide/DNA/saccharide syntheses rely on using a finite set of reagents (21/4/many, respectively) in sequence to make a molecular chain composed of a sequence of the chemical entities embodied in the reagents. For many years, all types of oligo synthesis required protected reagents, coupling, purification, deprotection, then start again. Then in the mid-1960s, a humble, brilliant chemist named Bruce Merrifield working at Rockefeller University had an idea: why not conduct this reaction sequence with the first reagent covalently bonded to something solid? That way, with high yielding reactions every purification step requires only washing the solid support much as you might wash your clothes; the undesirable substances (e.g., stains, oil, etc.) simply dissolve and float away while the thing you want – your shirt – can just be physically pulled from the water. Doing most of the lab work himself, Merrifield experimented with spherical polymers before settling on the now ubiquous 2–4% cross-linked polystyrene that swells with solvent yet does not crumble.

The apex of this solution-phase vs. solid-phase philosophy came in 1969, when two groups published back-to-back the chemical syntheses of ribonuclease A. 128 amino acids long, which both groups found folded spontaneously into the shapes that demonstrated catalytic activity. My biochemistry professor dismissed this achievement with the statement, 'Any cow can make ribonuclease A better than an organic chemist.' The group using the solution-phase approach (*J. Am. Chem. Soc.*, **1969**, *91*, 502) was hailed as a monumental achievement for organic synthesis... because it followed the protect-react-deprotect paradigm that organic chemists used and understood.

But what of this solid-phase method (*J. Am. Chem. Soc.*, **1969**, *91*, 501)? I only recently learned that this paper by Merrifield was not even cited by other authors for several years following its publication. Given that the solid-phase method has become the clear winning paradigm, that's an astonishing bit of history. Why did this revolutionary bit of research go unrecognized for so long? I can only offer my own viewpoint: biochemists could trivially isolate (and even crystallize) RNase A from slaughterhouse waste, and chemists did not want to do solid-phase synthesis themselves. 'Good' reactions were ones in which a solvent led to complete dissolution of all reagents – the sole exception being reactions such as hydrogenation that occur on the surface of certain insoluble transition metals. The average chemist didn't even know that the polystyrene beads Merrifield had so meticulously optimized allowed for reactions not just on the surface of the solid support, but all through the sponge-like polymer. The few chemists who saw the future – machines that would allow biochemists to make their own proteins at the touch of a few buttons – went on to become famous within small circles and, more importantly, wealthy. Merrifield won the Nobel Prize in Chemistry for his work in 1984 (http://nobelprize.org/nobel_prizes/chemistry/laureates/1984/). I wish I had known him. I'm certain he's the kind of man I would have respected. I was never introduced to his work in any of my chemistry or biochemistry classes – graduate or undergraduate (http://nobelprize.org/nobel-prizes/chemistry/laureates/1984/presentation-speech.html).

So now the need and necessary properties for a linker become clear to a chemist. They must stably (most often covalently) connect the first reagent in a sequential synthesis to a polymer (most often insoluble, but more recently can be soluble under some conditions and not under others). They must not react with the reagents used in the course of the synthesis. They must be de-linked using conditions under which the product is stable. In many instances, they will allow the synthesis site to reside far enough away from the polymer that the growing product does not suffer steric hindrance from the polymer.

Mom and Dad didn't make it this far; just not a good read. So you can't just give them this Foreword and feel you have fulfilled your obligation. Engage your mind; this is the best way to force yourself to think 'out-of-the-box.'

I've come up with two everyday examples, neither of which is perfect, but either of which is better than nothing.

1. Once upon a time, clothes were dried by hanging each garment on a line. This allowed the wind to 'wash' the excess water off of the garment and into the air. Just draping the garment on a line is frought with peril; the item can be blown off by the wind and lost (or at least rendered dirty again). And so, the clothing pin was invented. For those of you dear readers who have not seen a clothing pin, look here: http://en.wikipedia.org/wiki/Clothes-pin. The garment is temporarily immobilized on a solid support (the clothes line), and then later easily removed after the drying process is complete without damage to the garment.
2. Fishing. One cannot fish by throwing a worm into the water. This 'sacrificial first reagent' is linked to a fishing line by way of a hook (harms the fish little; harms the worm a lot). The linked first reagent is cast into the vessel of reagents (e.g., a lake). After the capture 'reaction' takes place, the fish can be removed from the water and de-linked by a deft pull of the hook. (If you have no knowledge of 'fishing', look here: http://en.wikipedia.org/wiki/Fishing ... and God bless you.)

Can you think of equally useful or even better examples? Please write and tell me. I'm at ACzarnik@unr.edu.

My thanks to the Editor of this important new book and to the Production Editors who never gave up on my promise to deliver. These days, I must wait for my muse.

Anthony W. Czarnik
University of Nevada, Reno, USA

December, 2008

Preface

With the passing of Prof. Robert Bruce Merrifield in 2006, the field of solid-phase organic synthesis (SPOS) regrettably lost its champion. Whilst there are those critics that group SPOS with combinatorial chemistry claiming both have had their day, I hope this volume serves not only to highlight the legacy that Prof. Merrifield left behind but also that SPOS has not become outdated, rather that it has matured into a powerful synthetic technique that is as relevant today as it was during its genesis. The first half of the book provides an introduction to SPOS and the early linker units, whilst the second half introduces the current state-of-the-art multifunctional linker units which are finding applications in the growing fields of diversity-oriented synthesis, chemical genetics and focused library preparation.

All that remains is for me to express my personal thanks to the editorial team at Wiley (particularly Paul Deards, Richard Davies and Jamie Summers) for the invaluable opportunity to edit this volume and for their support throughout the entire process; to all of the authors who have devoted so much of their time, enthusiasm and expertise to preparing the reviews in this book; and, finally, to my family and friends for their constant love and encouragement.

Peter J. H. Scott
Los Angeles, California

November 2008

List of Contributors

Mallesham Bejugam, The University Chemical Laboratory, University of Cambridge, Lensfield Road, Cambridge, CB2 1EW, United Kingdom

Christian Bochet, Department of Chemistry, University of Fribourg, CH-1700 Fribourg, Switzerland

Scott L. Dax, Galleon Pharmaceuticals, Inc., 213 Witmer Road, Horsham, PA 19044, USA

Christopher M. Diaper, NAEJA Pharmaceutical Inc., Edmonton, Alberta, T6E 5V2, Canada

Monica Diaz-Gavilan, Department of Chemistry, University of Cambridge, Lensfield Road, Cambridge, CB2 1EW, United Kingdom

Sabine L. Flitsch, Manchester Interdisciplinary Biocentre (MIB) and School of Chemistry, The University of Manchester, 131 Princess Street, Manchester, M1 7DN, United Kingdom

A. Ganesan, School of Chemistry, University of Southampton, Southampton, SO17 1BJ, United Kingdom

Giampaolo Giacomelli, Dipartimento di Chimica, University of Sassari, via Vienna 2, 07100-Sassari, Italy

Susan E. Gibson, Department of Chemistry, Imperial College London, London, United Kingdom

Giuditta Guazzelli, The School of Chemistry, University of Manchester, Oxford Road, Manchester, United Kingdom

Brian G. Hockley, Department of Radiology, University of Michigan, Ann Arbor, MI 48109, USA

Michael R. Kilbourn, Department of Radiology, University of Michigan, Ann Arbor, MI 48109, USA

Kerstin Knepper, Institut für Organische Chemie, Karlsruhe Institute of Technology, Fritz-Haber-Weg 6, D-76131 Karlsruhe, Germany

Michio Kuruso, Department of Microbiology, Immunology, and Pathology, Colorado State University, CO, USA

Ryszard Lazny, Instytut Chemii, University of Bialystok, ul. Hurtowa 1, 15–399 Bialystok, Poland

Sylvain Lebreton, Exploratory Chemistry, Oncology, Sanofi Aventis, Vitry Sur Seine, France

Jan H. van Maarseveen, Van't Hoff Institute for Molecular Sciences, University of Amsterdam, Nieuwe Achtergracht 129, 1018 WS Amsterdam, The Netherlands

Sébastien Mercier, Department of Chemistry, University of Fribourg, CH-1700 Fribourg, Switzerland

Marc Miller, The School of Chemistry, University of Manchester, Oxford Road, Manchester, United Kingdom

Kieron M. G. O'Connell, Department of Chemistry, University of Cambridge, Lensfield Road, Cambridge, CB2 1EW, United Kingdom

Marcel Pátek, Department of Chemistry, Sanofi-Aventis, Tucson, AZ, USA

Andrea Porcheddu, Dipartimento di Chimica, University of Sassari, via Vienna 2, 07100-Sassari, Italy

David J. Procter, The School of Chemistry, University of Manchester, Oxford Road, Manchester, United Kingdom

Peter J. H. Scott, Department of Radiology, University of Michigan Medical School, Ann Arbor, MI, USA

Richard J. Spandl, Department of Chemistry, University of Cambridge, Lensfield Road, Cambridge, CB2 1EW, United Kingdom

Alan C. Spivey, Department of Chemistry, South Kensington Campus, Imperial College, London, United Kingdom

David R. Spring, Department of Chemistry, University of Cambridge, Lensfield Road, Cambridge, CB2 1EW, United Kingdom

Patrick G. Steel, Department of Chemistry, University of Durham, Durham, United Kingdom

Gemma L. Thomas, Department of Chemistry, University of Cambridge, Lensfield Road, Cambridge, CB2 1EW, United Kingdom

Patrick H. Toy, Department of Chemistry, The University of Hong Kong, Pokfulam Road, Hong Kong, People's Republic of China

Amol A. Walke, Department of Chemistry, Imperial College London, London, United Kingdom

Daniel K. Whelligan, Cancer Research UK Centre for Cancer Therapeutics, The Institute of Cancer Research, Haddow Laboratories, Sutton, Surrey, United Kingdom

Tom M. Woods, Department of Chemistry, University of Durham, Durham, United Kingdom

Tracy Yuen Sze But, Department of Chemistry, The University of Hong Kong, Pokfulam Road, Hong Kong, People's Republic of China

Wei Zhang, Department of Chemistry, University of Massachusetts Boston, Boston, MA, USA

Robert E. Ziegert, Institut für Organische Chemie, Karlsruhe Institute of Technology, Fritz-Haber-Weg 6, D-76131 Karlsruhe, Germany

About the Editor

Dr Peter Scott received his BSc in medicinal and pharmaceutical chemistry from Loughborough University in the United Kingdom in 2001 following research into peptide nucleic acids under the guidance of Prof. Raymond Jones. Upon graduating he pursued doctoral research, designing novel multifunctional linker units for solid-phase synthesis, with Dr. Patrick Steel at the University of Durham, also in the United Kingdom, and was awarded his PhD in 2005. He then moved to the United States for postdoctoral research, initially in rhodium carbenoid chemistry at the University of Buffalo with Prof. Huw Davies (2005–2006), and subsequently in radiochemistry and PET imaging with Prof. Michael Kilbourn at the University of Michigan Medical School (2006–2007). In 2007, he joined Siemens Molecular Imaging and Biomarker Research, where he was head of radiochemistry at the LA Tech Center and involved in the design and synthesis of novel radiopharmaceuticals for use in PET imaging. In April 2009, Dr Scott returned to the University of Michigan and a faculty position in the Department of Radiology, where his research interests are developing novel tracers and technology for PET imaging, including solid-phase radiochemistry. He currently lives in Ann Arbor, Michigan, with his wife Nicole.

Abbreviations

AA	Amino acid
Ac	Acetyl
Acac	Acetylacetonate
AcOH	Acetic acid
AIBN	2,2-Azobisisobutyronitrile
Ala	Alanine
All	Allyl
BAL	Backbone amide linker
BEMP	2-*tert*-Butylimino-2-diethylamino-1,3-dimethylperhydro-1,3,2-diazaphosphorine
BHA	Benzhydrylamine
Binap	2,2′-Bis(diphenylphosphino)−1, 1′-binaphthyl
Boc	*tert*-Butoxycarbonyl
Bn	Benzyl
BOBA	4-Benzyloxybenzylamine
BOP	Benzotriazol-1-yl-oxy-tris-(dimethylamino)phosphonium hexafluorophosphate
BtOH	1-Hydroxybenzotriazole
n-BuOH	n-Butanol
t-BuOH	t-Butanol
t-BuOK	Potassium t-butoxide
Bz	Benzoyl
Boc	*tert*-Butyloxycarbonyl
CAN	Cerium(IV) ammonium nitrate
Cbz	Benzyloxycarbonyl
*m*CPBA	*meta*-Chloroperbenzoic acid
CPD	Cyclopentadiene
CPG	Control pore glass
CSA	10-Camphorsulfonic acid
Da	Dalton
DAST	Diethylaminosulfur trifluoride
DBU	1,8-Diazabicyclo[5.4.0]undec-7-ene
DCC	*N,N*′-Dicyclohexylcarbodiimide
DCM	Dichloromethane
DCU	Dicyclohexylurea
DDQ	2,3-Dichloro-5,6-dicyano-1,4-benzoquinone
DEAD	Diethylazodicarboxylate
DHQ	1,2-Dihydroquinoline
DIAD	Diisopropylazodicarboxylate

DIBAL	Diisobutylaluminium hydride
DIC	Diisopropylcarbodiimide
DIPEA	*N,N*-Diisopropylethylamine
DMAP	4-(*N,N*-Dimethyl)aminopyridine
DMB	*meta*-Dimethoxybenzene
DMDO	Dimethyldioxirane
DME	Dimethoxyethane
DMF	*N,N*-Dimethylformamide
DMG	Directing metalation group
DMP	Dess–Martin periodinane
DMPU	1,3-Dimethyl-3,4,5,6-tetrahydro-2(1*H*)-pyrimidinone
DMPU	*N,N'*-Dimethyl-*N,N'*-propylene urea
DMSO	Dimethylsulfoxide
DMTMM	4-(4,6-Dimethoxy-1,3,5-triazyn-2-yl)-4-methylmorpholinium chloride
DMTST	(dimethylthio)methylsulfonium triflate
DoM	Directed ortho-metalation
Dppe	1,1'-Bis(diphenylphosphino)ethane
Dppf	1,1'-Bis(diphenylphosphanyl)ferrocene
Dppp	Bis(diphenylphosphino) propane
DSA	Dialkoxyalkylsulfinylbenzhydrylamine
DSB	4-(2,5-Dimethyl-4-methylsulfinylphenyl)-4-hydroxybutanoic acid
DSEM	Diphenylmethylsilyl)ethoxymethyl
DTBP	2,6-Di-*tert*-butylpyridine
DVB	Divinylbenzene
EDCI	(1-Ethyl-3-(3'-dimethylaminopropyl)carbodiimide
ES-MS	Electrospray mass spectrometry
Et	Ethyl
Et$_3$N	Triethylamine
EtOAc	Ethyl acetate
EtOH	Ethanol
Fmoc	9-Fluorenylmethoxycarbonyl
FSPE	Fluorous solid-phase extraction
FT	Fourier transform
GC	Gas chromatography
HASC	α-Heteroatom substituted carbonyl
HATU	2-(1H-7-Azabenzotriazol-1-yl)-1,1,3,3-tetramethyl uronium hexafluorophosphate methanaminium
HBTU	*O*-Benzotriazole-*N,N,N',N'*-tetramethyluronium hexafluoro-phosphate
HFIP	Hexafluoroisopropanol
HMPA	Hexamethylphosphoric triamide
HMPB	4-(4-Hydroxymethyl-3-methoxyphenoxy)-butyric acid
HOA	Hydroxy octanoic acid
HOBt	*N*-Hydroxybenzotriazole
HPLC	High-performance liquid chromatography
HTPM	Hydroxytetrachlorodiphenylmethyl
HTS	High-throughput screening

hυ	Light
Ipc	Isopinocamphenyl
IR	Infra red
LAH	Lithium aluminium hydride
LDA	Lithium diisopropylamide
MALDI–TOF	Matrix-assisted laser desorption/ionisation–time of flight
MBHA	Methylbenzhydrylamine
mCPBA	*meta*-Chloroperbenzoic acid
Me	Methyl
MeCN	Acetonitrile
MeI	Methyliodide
MeOH	Methanol
MeOTf	Methyl trifluoromethanesulfonate (or Methyl triflate)
MOBHA	4-Methoxybenzhydrylamine
MOM	Methoxymethyl
MW	Microwave(s)
NBS	*N*-Bromosuccinimide
NCS	*N*-Chlorosuccinimide
NIS	*N*-Iodosuccinimide
NMM	*N*-Methylmorpholine
NMO	*N*-Methylmorpholine-*N*-oxide
NMP	*N*-Methylpyrrolidone
oNBS	*o*-Nitrobenzene
P	Protecting group
PA	Polyacrylamide
PAL	Peptide amide linker
PAM	Phenylacetamide
PEG	Polyethyleneglycol
PEGA	Polyethyleneglycol-polyacrylamide
PFS	Perfluoroalkanesulfonyl
Ph	Phenyl
Phe	Phenylalanine
PHFI	9-Phenylfluoren-9-yl
PMB	*para*-Methoxybenzyl
PNA	Peptide nucleic acid
Pnm	*p*-Nitromandelic acid
PPTS	Pyridinium *p*-toluenesulfonate
PS	Polystyrene
PTC	Phase transfer catalysis
PBu$_3$	Tributyl phosphine
Py	Pyridine
PyBOP	(Benzotriazol-1-yloxy)tripyrrolidinophosphonium hexafluorophosphate
RAMP	(R)–1- amino-2-methoxymethylpyrrolidine
RCM	Ring-closing metathesis
ROMP	Ring opening metathesis polymerization
rt	Room temperature

SAMP	(S)-1- amino-2-methoxymethylpyrrolidine
SAR	Structure-activity relationship
SCAL	Safety-catch amide linker
SCX	Strong cationic exchanger
SEM	(Trimethylsilyl)ethoxymethy
Ser	Serine
SPE	Solid-phase extraction
SPOS	Solid-phase organic synthesis
SPS	Solid-phase synthesis
SPPS	Solid-phase peptide synthesis
SPRC	Solid-phase radiochemistry
TBAF	Tetrabutylammonium fluoride
TBD	1,5,7-Triazabicyclo[4.4.0]dec-5-ene
TBDMS	*tert*-Butyldimethylsilyl
TBDPS	*tert*-Butyldiphenylsilyl
TBHP	*tert*-Butyl hydroperoxide
TBS	*tert*-Butyldimethylsilyl
TEA	Triethylamine
TES	Triethylsilyl
Tf	Trifluoromethanesulfonyl
TFA	Trifluoroacetic acid
TFAA	Trifluoroacetic anhydride
TFE	Tetrafluoroethane
TFMSA	Trifluoromethanesulfonic acid
TG	Tentagel^{TM}
THF	Tetrahydrofuran
THP	Tetrahydropyranyl
Thr	threonine
TIPS	Tri-*iso*-propylsilyl
TIS	Triisopropylsilane
TLC	Thin layer chromatography
TMEDA	N,N,N',N'-tetramethylethylenediamine
TMG	Tetramethylguanidine
TMS	Trimethylsilyl
TMSE	2-(Trimethylsilyl)ethanol
TPAP	Tetrapropylammonium perruthenate
TPP	Triphenylphosphine, PPh$_3$
TsOH	*p*-Toluenesulfonic acid
TTMSS	Tris(trimethylsilyl)silane
18-c-6	18-Crown-6 (1,4,7,10,13,16-hexaoxacyclooctadecane))

Part I
Introduction

1
General Overview

Scott L. Dax

Galleon Pharmaceuticals, Inc., USA

1.1 Introduction, background and pivotal discoveries

In 1963, Bruce Merrifield described the synthesis of short peptides making use of a new concept in which a terminal amino acid was first covalently joined to, or immobilized onto, a solid support and subsequently reacted sequentially with other amino acids to essentially 'grow' a desired (tetra)peptide (Figure 1.1).[1] This monumental work would usher in a new paradigm within organic synthesis that was markedly different from previous traditional approaches and would change synthetic chemistry in striking ways. Prior to Merrifield's insightful work, which would garner him a Nobel Prize, organic synthesis, as a discipline, was often relegated to cumbersome and inefficient solution-based reactions that necessitated precise stoichiometry and laborious purifications.

In the early 1990s, several research groups presented findings that solidified this exciting concept within organic synthesis. The Ellman laboratory reported on an expedient solid-phase method for the synthesis of 1,4-benzodiazepines.[2–4] In this work, three components, namely immobilized 2-aminophenones, aminoacids and alkylating agents, were reacted in a combinatorial manner to generate a small library of these therapeutic compounds (Figure 1.2). The reaction sequence was highly efficient, tolerant of functional group diversity and devoid of racemization. Collectively, this work demonstrated that versatile organic reactions could be conducted with one reactant immobilized on a solid support, and that this methodology allowed the different modules or components that make up a target molecule to be chemically joined in a combinatorial fashion, which could offer advantages to traditional linear synthesis.

Researchers at Parke-Davis also developed a similar approach to non-peptide chemical diversity with the introduction of their 'diversomer' technology.[5, 6] In this seminal work, a series of structurally-related compounds were synthesized in a multiple, simultaneous manner by making and using structurally diverse building blocks which were immobilized on an insoluble polystyrene-based solid support. The potential

Linker Strategies in Solid-Phase Organic Synthesis   Edited by Peter Scott
© 2009 John Wiley & Sons, Ltd

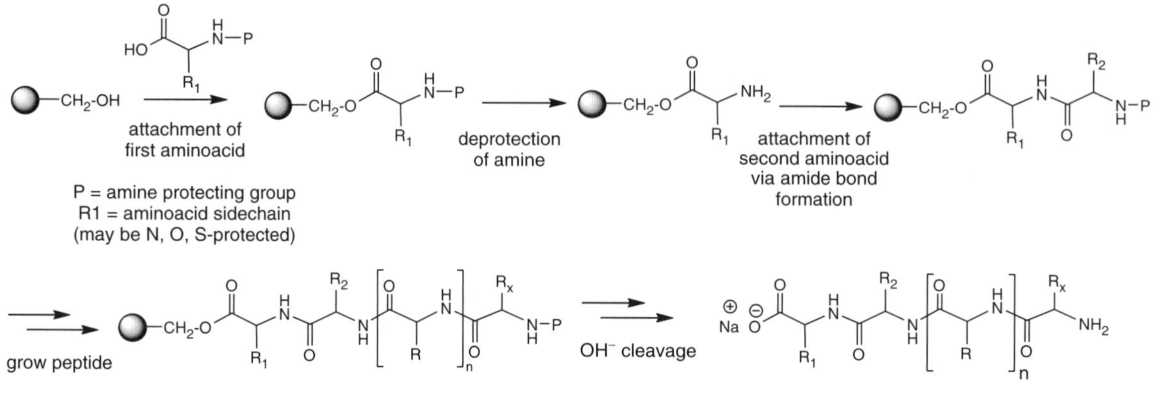

4 Introduction

Figure 1.1 Representation of Merrifield solid-phase synthesis

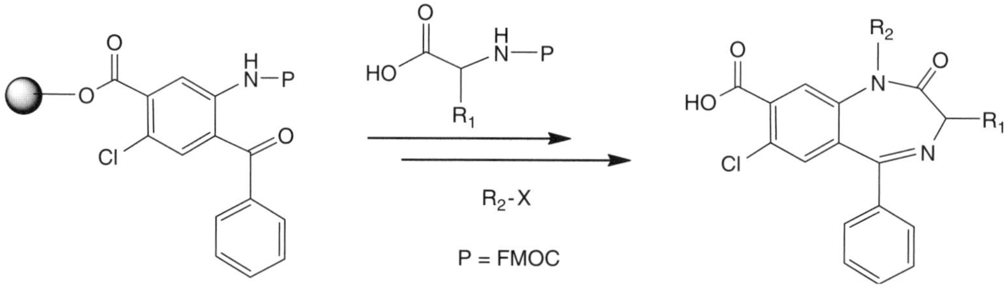

Figure 1.2 Representation of Ellman's solid-phase synthesis of 1,4-benzodiazepines

applicability to medicinal chemistry was enticing from the outset, as collections of pharmacologically-privileged structures (hydantoins (Figure 1.3) and benzodiazepines) were synthesized. An innovative feature of this work was the use of a cyclization reaction to release hydantoin final products from the solid support. This general strategy of cyclative mechanism-based cleavage has found much use and is discussed in detail in Chapter 4.

In retrospect, the Merrifield synthesis opened up seemingly countless possibilities in its heyday, as peptides and proteins, often inaccessible or at best available only in small quantities through heroic traditional chemical transformations, would eventually lie within reach through the expansion and automation of this technology. So, too, would this become the case for the Parke-Davis 'diversomer' technology and the many useful modifications that would follow.[7–15]

The Parke-Davis group (Sheila DeWitt, John Kiely, Mike Pavia, Walter Moos, Donna Reynolds Cody, Tony Czarnik and colleagues) introduced apparatus, built upon materials often found in an organic chemistry laboratory, in which to carry out 'diversomer' technology.[16–18] This was a key event because other groups swiftly constructed their own homemade versions in which to conduct parallel solid-phase synthesis. The 'diversomer' technology, arguably the underpinnings of non-peptide solid-phase organic synthesis, became widely available to the chemist and would lead to many creative applications.

Solid-phase technology has reshaped the landscape of organic synthesis and thus it is useful to provide some general definitions at the outset and to draw some contrasts. Traditional organic synthesis, a tremendously powerful science, relies upon establishing a usually homogenous solution phase to dissolve and

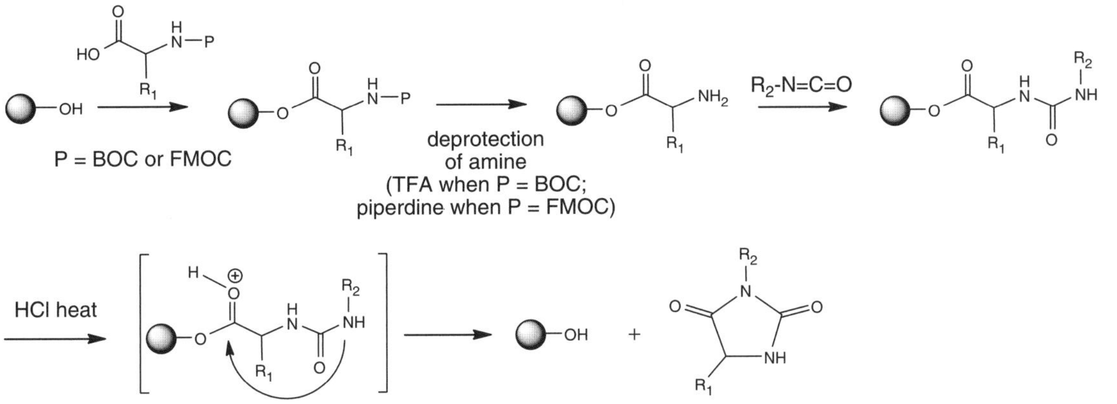

Figure 1.3 Solid-phase synthesis of hydantoins using 'Diversomer' technology

liberate reactants, so that they may collide with each other with proper orientation and sufficient energy to allow for new bond formation at the expense of weaker or less favorable chemical bonds. In contrast, a solid-phase synthesis invokes that at least one reactant exists and remains as a undissolved component in the reaction medium, thus inferring that a heterogeneous mixture is present. While every practicing chemist at one time or another may debate the homogeneity of a given solution-based reaction, it is important to realize that solid-phase chemistry intentionally makes use of the interaction of two (or more) components that exist in different phase states – one being that of a solid and the other(s) in solution or liquid form. The term solid support refers to an inert insoluble macromolecule (also called a resin) to which a much smaller organic molecule or moiety is or can be attached. This molecular fragment or moiety, that joins the two, is termed a linking group (or linker unit) because it will serve as an atomic or molecular bridge between the solid support and the starting material of the synthesis, which is referred to as the 'organic substrate'. The organic substrate is usually very similar, perhaps even identical, to any starting material that might be destined for 'traditional' solution-based synthesis. The starting material should be a carefully selected molecule that can be subjected to a series of reactions that transform its functional groups or reactive substituents, to provide the desired target molecule(s). However, as shall be seen, making a given starting material or organic substrate amenable to solid-phase technology can often require that additional functionality be present or be installed. The 'organic substrate' terminology places this component within the conglomeration that constitutes solid-phase synthesis. It is the 'organic substrate' that is of interest to the synthetic chemist, for this component will be subjected to subsequent chemical transformations to make the targeted molecules.

Returning to the topic of this textbook, the linker group is a critical structural element in this interplay of inert polymeric solid support and low molecular weight organic substrate because it dictates how the two will be joined and under what conditions the two can be eventually separated. The development of linker groups has followed many varied approaches, mostly to accommodate the chemical transformations needed to convert any given organic substrate into desired product(s), but also to push the limits of solid-phase technology. The reader will learn of the many strategies that have evolved and will encounter specific examples that illustrate the utility of linker groups in solid-phase organic synthesis. The material presented herein purposely does not cover polypeptide or oligonucleotide solid-phase synthesis, which have been tremendously successful predecessors to this broad area of science, but will rather focus upon what may be viewed as solid-phase synthesis of 'small organic' molecules. There will be an effort to showcase synthetic versatility by providing specific examples that will be rich in diverse functional group chemistry.

As background to such strategies and at the simplest level, most organic molecules, particularly those that can be used for medicinal, biological or other industry-based applications, contain heteroatoms and more complex arrays of heteroatoms referred to as functional groups, in addition to carbon and hydrogen atoms. Synthetic organic chemists have long recognized that such heteroatoms and functional groups can, in essence, mark viable (dis)connection points from which smaller molecular pieces (fragments, precursors) can be joined together to build a target molecule. Indeed, chemists have been trained to look for such features when devising a synthetic pathway, a strategy pioneered by the legendary R. B. Woodward and others.

Historically, and without solid-phase techniques, a typical 'traditional' synthesis involves a sequence of reactions that transform a starting material and various selected reagents into a desired final product. In each reaction, a starting material is dissolved in an inert solvent as is the reactant, and reaction of the two occurs in a (usually) homogenous solution. In some cases, the reactivity of one component may require that key functional groups are made inert at various stages through the use of protecting group chemistry. Separately, in other scenarios, one component may be used in (molar) excess in order to increase the efficiency of the transformation by 'driving the reaction to the right'. Regardless, side reactions occur between the starting material and the reagent, or with impurities or side products that form under the reaction conditions. Therefore, the reaction mixture must be subjected to purification and this is often accomplished by quenching the reaction and conducting some (work-up) extraction procedure to concentrate impure final product, which is then further purified by chromatographic techniques. The isolation and purification of a molecular entity from a solution-based reaction mixture can often be laborious and require much more effort than conducting the chemical reaction itself.

In contrast, solid-phase synthesis entails the immobilization of a starting material onto a polymeric solid support (resin) that, if chosen properly, is inert to reagents and subsequent reaction conditions. Solid-phase techniques can be optimized to offer advantages compared to traditional solution chemistry. Since the resin does not dissolve in solvent, the organic substrate (starting material) can be exposed to solutions containing large excesses of reagent to drive the reaction to completion. Relatively straightforward filtering and washing techniques can often be used to remove impurities and circumvent chromatographic purifications. Additionally, the use of scavenger resins to remove unwanted excess starting materials and/or by-products can be effectively employed in some instances. Figure 1.4 is a representation of solid-phase synthesis in which the components or building blocks or diversity elements (terms often used interchangeably) are depicted by the differently shaped symbols; the solid support and linker group are represented by the gray circle and oval, respectively. The X substituent is used generically to indicate the point at which the organic substrate (or starting material) is joined to the linker group. The focus of this book is on the role of this attachment within the linker group and the chemistry used to attach and liberate small-molecule products from the solid support.

Referring back to Merrifield protein synthesis, the amide linkage marks a logical (dis)connection point from which smaller pieces (amino acids) can be joined together to build a desired peptide sequence. An advantage of the Merrifield methodology is that amino acid protecting group chemistry can be accomplished through immobilization onto the solid support. Subsequent functional group activation and amide bond formation allows for the (poly)peptide to be built one amino acid at a time. Thus, the attachment of the organic substrate (amino acid or peptide) to a resin can serve as a means not only to protect a key functional group, but to also provide a 'directionality' from which to assemble the peptide. While the original work grew the peptide from the C-terminus towards the N-terminus, it is important to realize that peptides can be assembled in the 'opposite direction', namely by attaching the N-terminus to a solid support and adding amino acids in the opposite order.

In designing ligands for proteins, the underlying principle is that receptors, enzymes and channels recognize certain structural elements, such as molecular shape, size, lipophilicity/hydrophilicity and charge. Proteins, as diverse as they are as a family, however, are composed of a basis set of only about 20 common

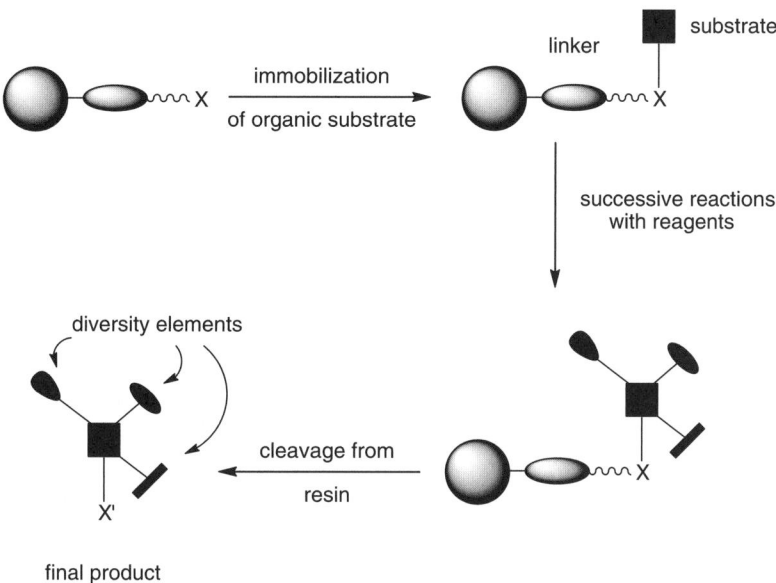

Figure 1.4 Representation of solid-phase organic synthesis (SPOS)

amino acids. Therefore, a large number of unique monomers or building blocks is apparently not essential for structural and functional diversity, but rather it is the 'space' which key amino acids can occupy once attached to a peptide backbone or a unique structural motif or scaffold that is critical. In fact, despite being comprised of only one type of chemical reaction (amide bond formation) and a limited set of building blocks (amino acids), highly specific peptidic ligands for many proteins are well known, whether endogenously formed or synthesized using some variation of the Merrifield technology.

Using solid-phase chemistry to synthesize non-peptidic organic molecules should provide markedly greater diversity since, theoretically, the vast repertoire of organic reactions should be amenable to the technology and, therefore, not limited solely to amide bond formation. It is this nearly unbound opportunity to build compound libraries by applying nearly *all* organic reactions and this 'sampling of vast chemical space' that holds the promise of discovering new small-molecule drugs. The number of building blocks should be nearly without boundary given the countless ways carbon-based molecules exist and react through the rich universe of functional groups.

But even more striking is the ability to construct organic molecules in a combinatorial manner. Very often, synthetic organic chemistry has been primarily inspired by very specific but narrow endeavors, such as syntheses of complex natural products or the construction of hand-crafted molecules to drive pharmacological evaluation and drug discovery. One powerful opportunity afforded by combinatorial chemistry was the ability to 'sample' molecular diversity and conformational space effectively by incorporating many structural variations during the course of synthetic sequences used to make compound libraries. An ideal compound library is diverse so as to include members that span a spectrum of shape, size and lipophilicity in order to probe for binding to a protein of interest. It is desirable to introduce multiple elements of diversity when possible at each step of the synthesis. In practical terms, a well conceived strategy makes use of versatile chemical transformations that not only afford structurally diverse compounds of high purity, but also can serve as source of novel intermediates that can be further elaborated to prepare additional libraries, preferably prior to liberation from the solid support. This is the 'libraries from libraries' approach that is a well-known strategy to optimize the molecular diversity one obtains in any given synthetic sequence.

8 Introduction

The idea of generating large number of molecules simultaneously was certainly not new given the groundbreaking Houghten[19] syntheses of thousands of peptides. However, having the capability to assemble molecules in a modular fashion with the control afforded by immobilization onto a solid support made such processes feasible. The technology would be used to generate a single compound per well (from a single resin – see Figure 1.4) or, in contrast, so-called 'split and pool' techniques (Figure 1.5) would be used to rapidly expand diversity. In split and pool synthesis, individual resins each containing a different organic substrate are pooled together and then subjected to solid-phase synthesis. Pools resulting from each transformation can be further pooled to generate large numbers of distinct molecules within a few reiterations (reactions).

Thus some labs chose to use this approach and generate large numbers of molecules in a parallel fashion, so giving rise to massive compound libraries. These features allow for the synthesis of hundreds or even thousands of compounds in the time it typically took to make a handful or dozen of molecules using conventional linear syntheses. This prospect was eagerly welcomed as high-throughput screening methods were routinely coming online as the industry focused on increasing R&D productivity.

An exceptional example was reported by the Schreiber group in which the synthesis of over 3000 spirooxindoles was achieved (Figure 1.6).[20] A key feature of this work was the use of a three-component reaction to install multiple elements of diversity in a single step. (Multicomponent reaction on a solid phase will be discussed later.) Remarkably, despite the complexity of the chemistry used to assemble the final products, it was determined that more than 80% of the compounds in the library had a purity of greater than

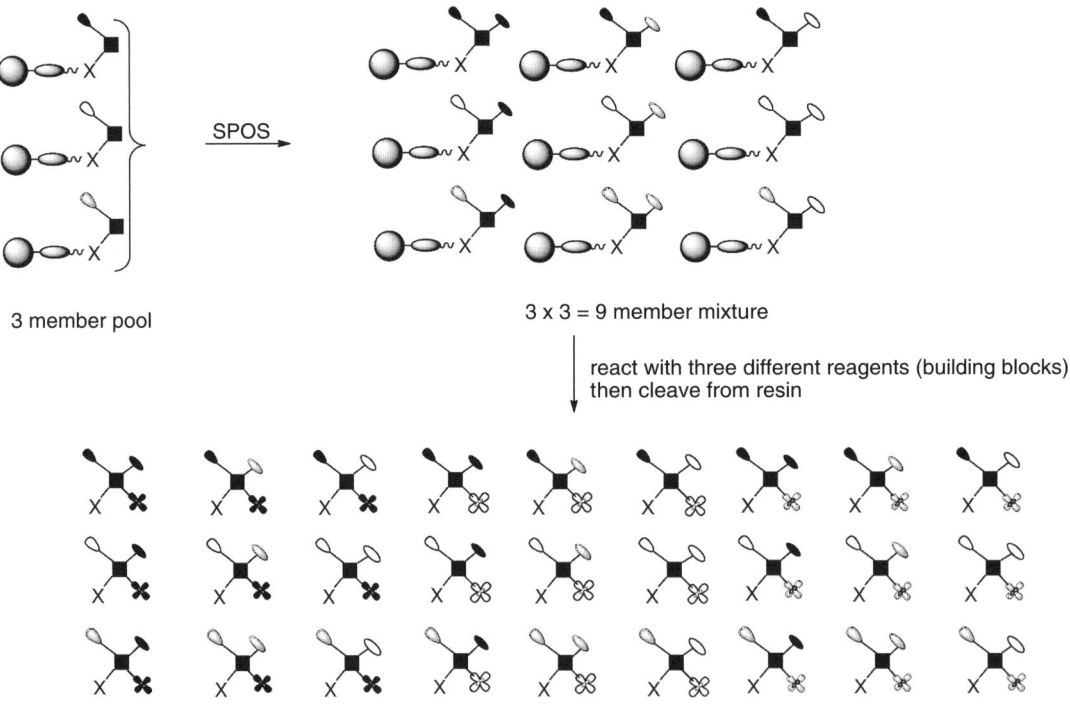

Figure 1.5 Representation of 'split and pool' methodology

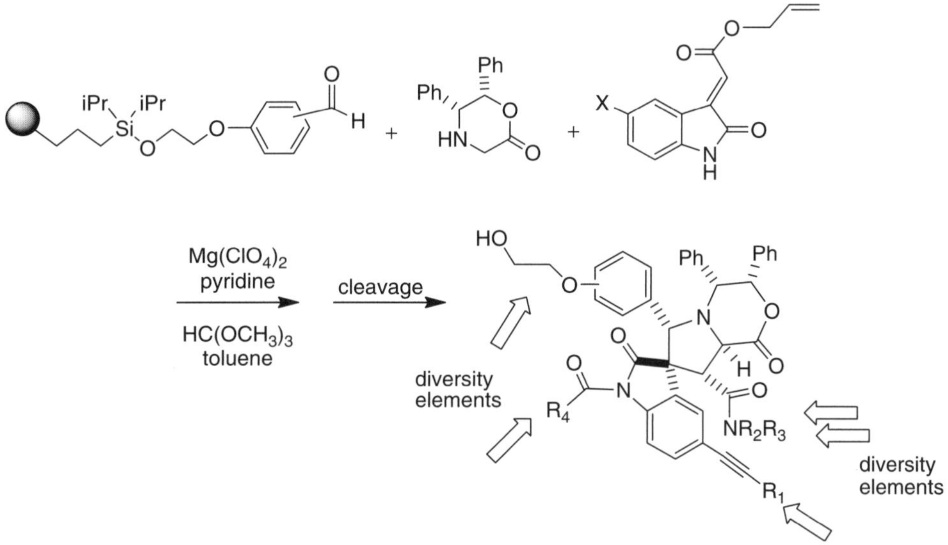

Figure 1.6 *Schreiber's split and pool three-component reaction to generate a three thousand member spirooxindole library*

80%. The Schreiber work sets a noteworthy precedence for future applications of combinatorial pool and spilt applications in the context of synthesizing libraries of challenging, high functionalized and complex small molecules.

1.2 Fundamentals of conducting solid-phase organic chemistry

1.2.1 Apparatus

There is a host of apparatus available to conduct solid-phase organic synthesis and the purpose herein is not to try to define each variation, but rather speak to the purpose and principles. The critical features for any are, firstly, the ability to mix the heterogeneous reaction (resin-bound organic substrate and reagent solution) and, secondly, to provide a means to separate solid resin from solution, preferably in a way that allows for washing of the resin. A variety of simple mixing techniques are applied, including devices to shake or invert the vessel or mechanical stirrers to agitate the mixtures or, lastly, gas bubblers to confine resin or physically move resin within the solution. Thus, in its simplest form, resins containing organic substrate can be exposed to reagent in solution in a conventional glass vessel with stirring and then the resin can be merely filtered off by some means (e.g., glass frit). The use of vacuum pressure can be applied to draw washing solutions over and through resins and to aid in drying of the resin. Many variations of this theme have been developed, from simple glass cylinders, a glass frit and a stockcock (Figure 1.7).

More sophisticated approaches have been developed. Organic substrates can be enveloped in polymeric 'bags' that are permeable to solutions of reagents in order to carry out chemical transformations. After the diversity components are attached, the 'bags' can be removed and exposed to washing solutions to remove excess reagent and by-products. The final products can then be removed from the 'tea bag'. The Houghten laboratory first developed this powerful methodology for the synthesis of peptides[19] but the methodology has been extended to a variety of small molecules.[21]

10 Introduction

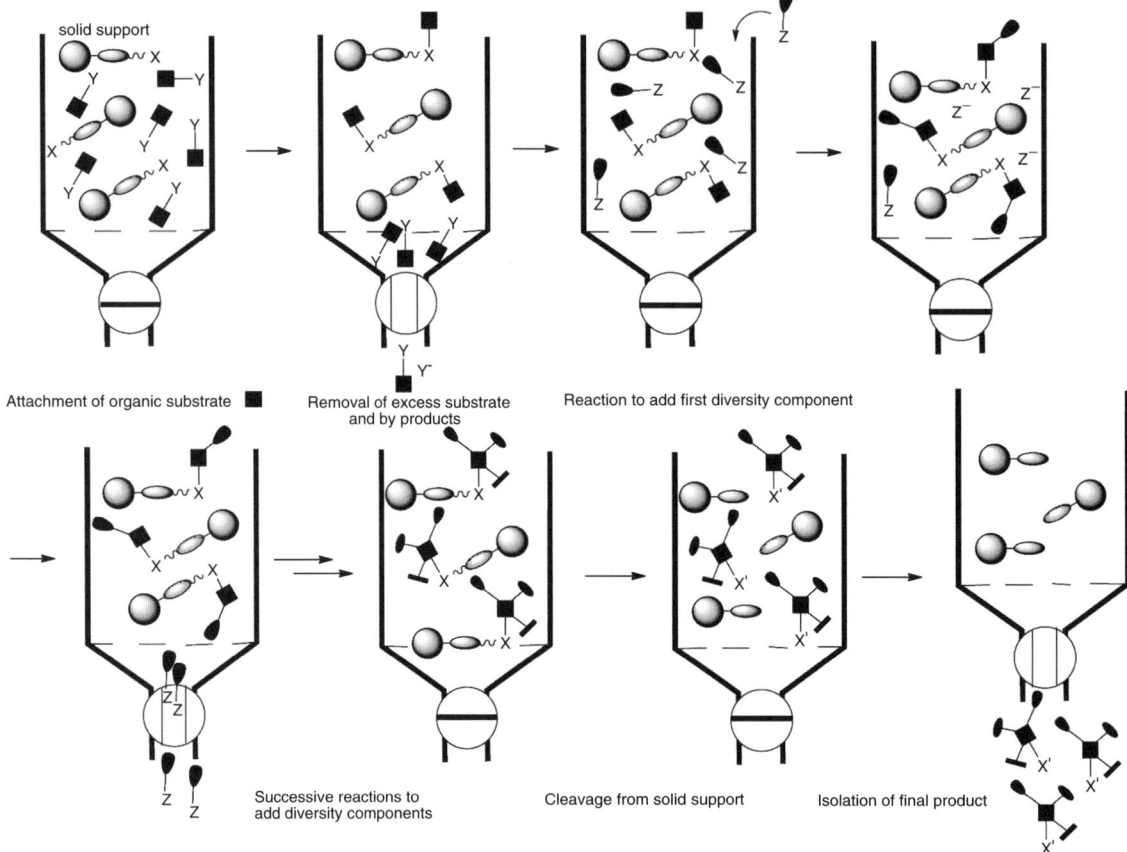

Figure 1.7 Principle operations of solid-phase organic synthesis

1.2.2 Typical solid supports

Solid supports comprise a polymeric resin that has been chemically derivatized to incorporate a functional group within the matrix that is able to undergo reaction with small organic molecules. The polymer or resin itself can be made from a number of materials. Not surprisingly, the criteria for selecting a resin revolves around its chemical and physical properties. Firstly, the material must be inert to the contemplated reaction conditions and reagents. Since the mass of the resin far exceeds that of the organic substrate, even minor competing reactions or degradation is usually problematic. In addition, the material needs to swell sufficiently in solvent to expose a large enough surface area to allow the chemical reactions to occur. Additionally, the resin should suspend well in the desired solutions to allow for efficient reaction and washing. Perhaps most importantly, the resin needs to be amenable to functionalization, so organic substrates can be covalently attached and so sequential transformations are therefore possible. The extent to which a functional group has been incorporated onto a resin is quantified and presented as a theoretical loading value, typically given in millimoles per gram.

The most common solid supports are derived from polymeric polystyrene and polyethylene glycol. Often these polymers are cross-linked with additives (divinylbenzene for example) to impart desired physical characteristics (size, swelling). Copolymers are also employed as solids supports as with the Tentagel™

family of resins, which are low cross-linked polystyrene matrices upon which polyethylene glycol is grafted. Polystyrene resins, typically of 50–400 mesh size with loading values of typically 0.5–1.5 mmol/g, have been extensively functionalized to include many common and versatile organic functional groups. Some resins are functionalized with nucleophilic moieties such as alcohols and amines whereas others contain electrophilic centers (e.g., α,β-unsaturated ketone, carbonates, etc.). More about how moieties are used to form linker groups is given in subsequent chapters. The functional groups themselves can be incorporated onto a resin directly, or via some inert tether (Table 1.1 below shows representative examples). Furthermore, these moieties can be obtained in a protected form or activated towards subsequent reaction in many cases.

Table 1.1 Resin derivatization; attachment of versatile functional groups

Functional group	Moiety presented to organic substrate
Alcohol	∼∼∼CH_2-**OH**
	∼∼∼$C_6H_4CH_2OC_6H_4CH_2$-**OH**
	∼∼∼$C_6H_4CH($**OH**$)(C_6H_5)$
	∼∼∼$C_6H_4C(C_6H_5)_2$-**OH**
Aldehyde	∼∼∼C_6H_4-**CHO**
	∼∼∼$C_6H_4CH_2O((CH_3O)_2C_6H_2)$-**CHO**
	∼∼∼$CH_2CH_2[OCH_2CH_2]_5NHC(O)(CH_2)_4$-**CHO**
(Acetal)	∼∼∼$C_6H_4CH_2CH_2NHC(O)(CH_2)_4$-**CH(OEt)$_2$**
Alkenyl	∼∼∼C_6H_4-O-C(O)-**CH** = **CH$_2$**
	∼∼∼$C_6H_4CH_2OCH_2CH_2$-SO_2-**CH** = **CH$_2$**
	∼∼∼C_6H_4-**CH** = **CH$_2$**
Amine	∼∼∼CH_2-**NH$_2$**
	∼∼∼C_6H_4-CH_2-**NH$_2$**
	∼∼∼$C_6H_4CH_2OC_6H_4CH_2$**NH$_2$**
	∼∼∼-CH_2-**N(CH$_3$)$_2$**
	∼∼∼$C_6H_4C(Ph)_2$-OC(O) $CH_2C_6H_4CH((CH_3O)_2C_6H_3)$-**NH$_2$**
Amine (benzyl)	∼∼∼$C_6H_4CH_2$-**NH**-CH(CH$_3$)CH$_2$Ph
Amine (sulfonyl)	∼∼∼NHC(O)-C_6H_4**SO$_2$-NH$_2$**
Amidine	∼∼∼$C_6H_4CH_2OC_6H_4$-**N** = **CH-N(CH$_3$)$_2$**
Carbonates	∼∼∼**OC(O)O**C_6H_4-**NO$_2$**
Carboxylic acid	∼∼∼C_6H_4-**COOH**
	∼∼∼$C_6H_4CH_2$-**COOH**
	∼∼∼$C_6H_4CH_2NHC(O)CH_2CH_2$-**COOH**
	∼∼∼CH_2CH_2 $OCH_2CH_2NHC(O)CH_2CH_2$-**COOH**
Halogenated	∼∼∼**CH$_2$-Cl**
	∼∼∼$C_6H_4CH_2OC_6H_4$-**CH$_2$-Cl**
	∼∼∼C_6H_4-**I**
Hydroxylamine	∼∼∼$C_6H_4CH_2OC_6H_4CH_2$-**NH-OH**

Examples (as will be seen in later chapters) include the common fmoc ((9H-fluoren-9-ylmethoxycarbonyl)) derivatives (not shown) used to cap amines, esters to activate carboxylic groups and *para*-nitrocarbonates used to activate carbonyl centers.

1.2.3 Fluorous supports

The development of fluorous chemistry, led by the Curran research group, has hugely impacted solid-phase organic synthesis as well as traditional solution-based chemistry.[22] Highly fluorinated organic molecules, particularly perfluorinated hydrocarbons, possess unique physical properties (e.g., high degree of hydrophobicity) that have been described as 'orthogonal' to traditional (non-fluorinated) 'organic' and 'aqueous' phases. Thus, fluorous solvents can be used to solubilize fluorous reactants and react with an organic substrate. The fluorous product can be separated from the reaction media by several methods. Liquid–liquid extraction is a straightforward procedure to isolate fluorous products from non-fluorinated by-products by using solvents with high fluorine content to separate product from non-fluorous materials. Such separations closely resemble aqueous quench, work-up and extraction protocols commonly used in organic synthesis.

Separately, fluorous tags can be used to covalently modify organic molecules so as to make the physical and chromatographic properties of the tagged molecule similar to fluorohydrocarbons, and thus allow for fluorous separation and purification. Ideally, the fluorous tag can subsequently be removed to yield the final product. Lastly, chromatographic stationary phases can be made fluorous and used to effect separation and purification, much as in the same way silica gel is routinely used with non-fluorinated solvents.[23]

Following these advances, highly fluorinated or perfluorinated hydrocarbons have also been functionalized and immobilized on a solid support to serve as linker groups (Chapter 21). All non-fluorous materials (unreacted reagents for example) are then readily removed by simple washing techniques and the final product is then cleaved from its fluorous host. Resin-bound fluorinated alkylsulfonate esters, for example, aryltriflates, can undergo palladium catalyzed transfer hydrogenolysis as a means to induce traceless cleavage or, conversely, undergo Suzuki coupling to arylate upon cleavage (Figure 1.8).[24, 25]

1.2.4 Linker strategies[26, 27]

To accommodate the burgeoning traditional chemistry that would be adapted to solid-phase organic synthesis, the composition and fate of the so-called linker group garnered great interest for several reasons. Since attachment of the organic substrate constitutes the first chemical reaction within the synthesis, the linker group needs to form a covalent bond with the organic substrate readily and efficiently. Ideally, this

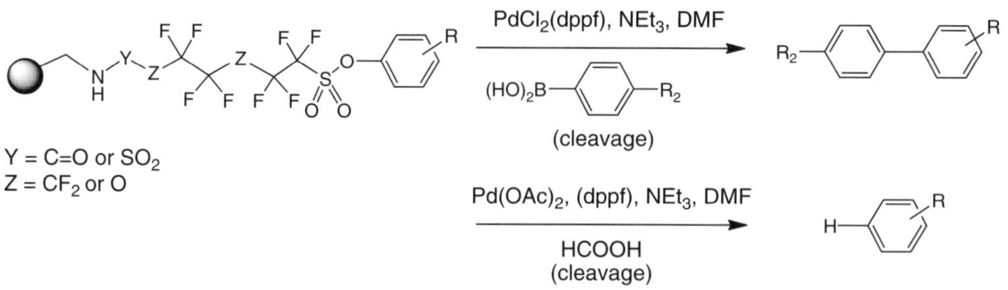

Figure 1.8 *Examples of fluorous linker groups*

process will result in a high degree of 'loading' of the organic substrate onto the resin. In cases where this does not happen, additional procedures are often needed to cap remaining active sites within the resin in order to carry out subsequent chemistry as desired. Since the linkage needs to remain intact for the duration of the synthesis, the optimal linker needs to be inert to a wide range of reaction conditions and reagents. Moreover, the linker group often needs to provide some 'molecular' distance from the macromolecular solid support, so as to allow for exposure of the organic substrate to the solution-based reagent(s). While this factor is difficult to ascertain empirically, an optimum linker group will allow for the organic substrate to dispose and orientate itself within the solvent, so as to allow for an efficient reaction with reagent and preferably an acceptable (rapid) rate of reaction. Lastly, once fully synthesized, it is vital that the organic substrate can be cleaved from the solid support with ease. Thus, specialized reaction conditions have been developed depending upon the specific linking moiety, in order to liberate pure final products. Notice the requirements of the successful linker group – it has to be robustly attached to the solid support for the duration of the synthesis yet able to react with the organic substrate, inert towards subsequent chemical transformations and, lastly, under a separate set of conditions, able to release the organic substrate.

The linking groups themselves can be divided into categories based upon their structural impact on the organic substrate molecule (Figure 1.9). Many linkers rely upon a reactive functional group that forms a robust functional group upon capturing the organic substrate. In such cases, cleavage from the resin revolves upon a reaction of that functional group and this process (e.g., hydrolysis) often leaves a heteroatom bonded to the organic substrate at site of cleavage. This heteroatom (or functional group) can be 'traced back' to the precise site of attachment to the linker group and, hence, to the solid support. Such traditional linkers can be especially valuable if they are used to install a key heteroatom or functional group at some desired

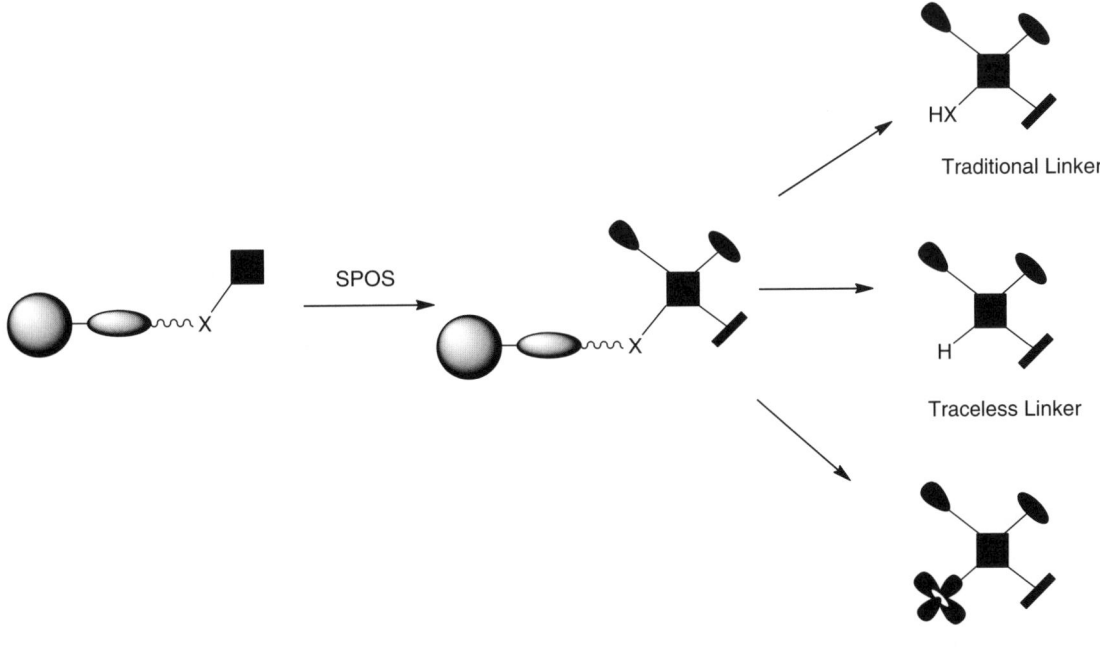

Figure 1.9 *Depiction of different linker group strategies. (Reproduced with permission from Eur. J. Org. Chem* **2006**, *2251–2267. Copyright Wiley-VCH Verlag GmbH & Co. KGaA.)*

position (e.g., a structural element needed for some pharmacological activity). However, too often, this is not the case and one settles for an extraneous, if not undesirable, functional group uniformly installed at the same position across a compound library.

For these reasons, a great amount of effort went into developing so-called 'traceless' linkers, in which a hydrogen atom replaces the heteroatom or functional group that covalently joins the organic substrate to the solid support. In doing so, there is no tell-tale sign that the organic molecule was joined to a solid support and, more importantly, there is no extraneous heteroatom or functional group that needs to be addressed (e.g., removed via additional chemical transformations). Traceless linker groups removed a major shortcoming with the early groups. After all, a chemist carrying out solution-based synthesis would rarely find reason to choose a more complex starting material containing superfluous heteroatoms; to be relegated to do otherwise in order to conduct solid-phase synthesis diminishes the application and usefulness of the technology.

The last category of linker groups is those that structurally alter the original attachment point of the organic substrate during the process of cleavage. This strategy has been another major advance to the field as a whole because it allows for an additional element of diversity to be introduced into the compound library, and at the very last step of solid-phase synthesis, namely upon cleavage. Diversity-based (or oriented) linker groups contain functional groups (or heteroatoms) that are specifically chosen for their latent reactivity towards cleavage conditions (e.g., a nucleophile) and their ability to undergo an efficient chemical transformation as a means to induce cleavage. This may be accomplished in a number of ways, including either the addition of a nucleophilic component or an electrophilic component to react with the linking group and liberate a modified organic substrate. The key distinction compared to the other types of linker groups is that diversity-oriented linkers afford liberated molecules that do not have the original heteroatom or functionality present at the site of attachment, nor are the final products 'traceless' (i.e., have a hydrogen atom at the site of attachment). The majority of linker groups discussed in this book will deal with diversity-oriented linker groups which, as will become apparent, have been cleverly conceived and crafted to install even more diversity into compound libraries.

Hundreds of linkers that have been developed, many are variations on a common theme; only a few will be highlighted here to illustrate key concepts about the technology as a whole. The reader is encouraged to consult the subsequent chapters, in which specific linking groups are discussed in detail. The early linker units, not surprisingly, followed chemistry developed for Merrifield peptide synthesis (and oligonucleotide synthesis) and attached the organic substrate to the solid support through the covalent reaction of a polar functional group with a nucleophilic or electrophilic site within the resin. Alcohols, amines and carboxylic acids serve as common linker groups. For example, resin-bound benzyl alcohol readily reacts with acids (as activated esters) to form ester linkages, or with isocyanates to form carbamate linkages. After synthetic modification, the organic product can be cleaved using rather strongly acidic conditions (Merrifield cleavage originally used hydrofluoric acid) or by the reaction of strong nucleophile such as hydroxide (Figure 1.10). Additionally, amides are accessible from ester-linked substrates by the addition of nucleophilic amines.

Unfortunately other functional groups can be labile under such harsh conditions and thus milder conditions have since been developed. The Wang resin makes use of a *para*-oxygen atom within the benzylic alcohol linker to stabilize the resultant cation and allow for cleavage under much milder conditions. Thus, Wang resin can first be activated by conversion to its *para*-nitrobenzylcarbamate to which nucleophilic amines readily add. N-methylated amine products are then accessible by a reductive cleavage from the resin while des-methylated analogs can be obtained via mild acid cleavage (Figure 1.11).[28] In a similar way, benzhydryl amines and amides, such as the Rink linker, can be used to make amides and ureas with mild cleavage from the resin (Figure 1.11).[29]

Mild base-induced cleavages, to compare to the use of strong nucleophiles, have also been developed and one such method has led to a specialized regenerative Michael acceptor (REM) resin that contains an

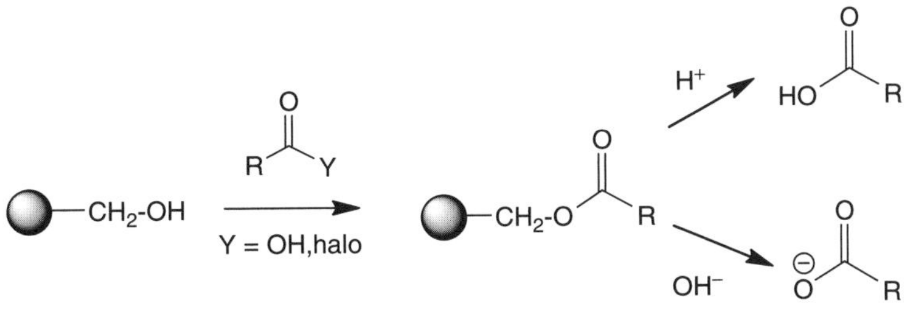

Figure 1.10 Cleavage via acid or by the reaction of a nucleophile

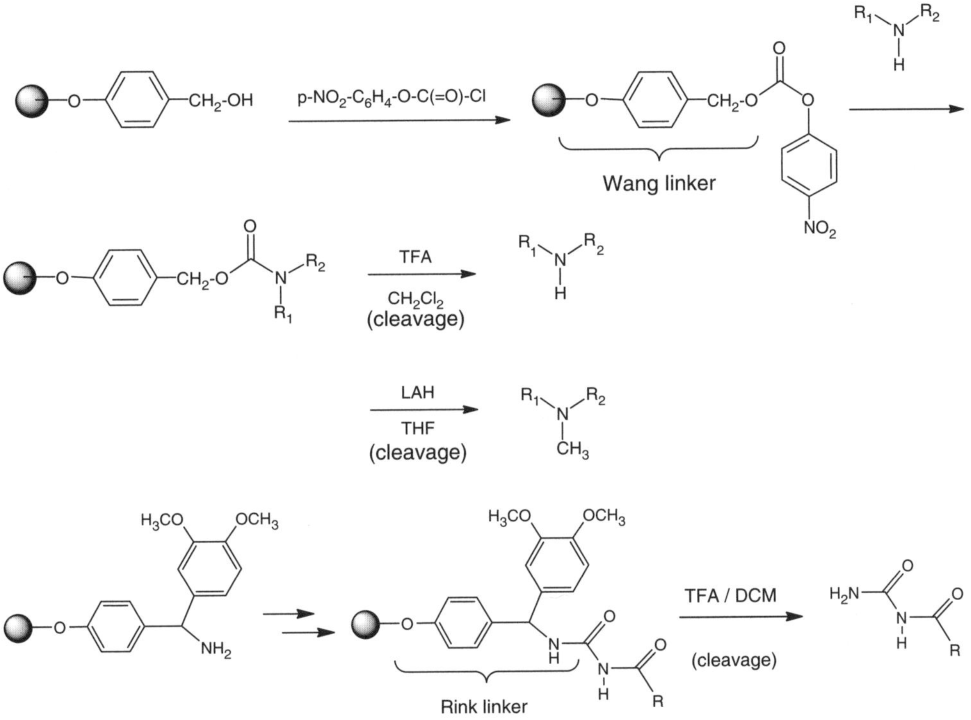

Figure 1.11 Wang and Rink linker groups allow for mild acid cleavage

acrylate group to readily accept a nucleophilic organic substrate via a Michael addition.[30] After subsequent transformations, the final product is cleavage by the action of amine base which promotes a Hoffmann elimination reaction (Figure 1.12). In this case, it is the basicity of the added amine, not its nucleophilicity, that is responsible for cleavage.

A major limitation of using strong acids or bases, or many nucleophiles, to induce cleavage of organic substrate joined to the resin via simple carboxylic acids derivatization, such as esters, is their lability. Ideally, a linker group will release an organic substrate only as called upon, despite the myriad of reaction conditions it may encounter throughout its synthetic lifetime. The development of a so-called 'safety catch' linker group introduced the concept of making use of a robust sulfonamide linker group that could be activated towards cleavage as desired, through simple and specific N-alkylation (Figure 1.13).[31, 32]

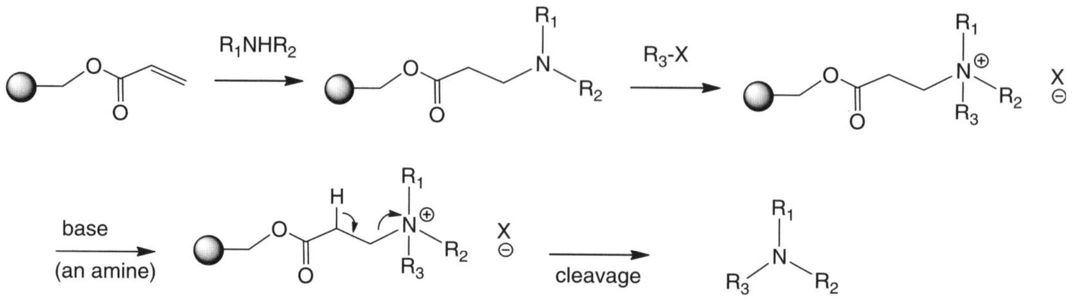

Figure 1.12   Cleavage by a base inducing an elimination reaction

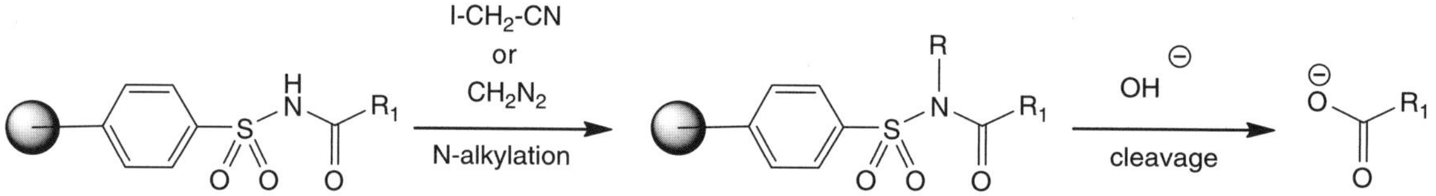

Figure 1.13   A sulfonamide 'safety-catch' linker group

This linker group is inert to most nucleophiles, including hydroxide, and this feature greatly expands the types of nucleophilic transformations that can be accomplished on the organic substrate without causing premature release.

In addition to perfecting cleavage methodology through the use of functional group cleavage (ester, sulfonamides, etc. as shown above), a great deal of attention was to develop chemistry that would remove all atomic evidence that an organic substrate, a small molecule, was indeed ever attached to a resin. Towards this, the first traceless linker to be developed and widely used makes use of the ability of silicon to undergo ipso substitution, by which hydrogen can be introduced in place of the heteroatom or functional group that originally bonded the organic substrate. The ability to introduce a hydrogen atom makes the linker traceless; the final product contains no tell-tale sign or signature that allows the deduction of which position of the organic substrate was bonded to the linker group, and therefore to the solid support. The prototypic traceless linker was pioneered by Ellman and again applied initially to the synthesis of benzodiazepines (Figure 1.14).[33]

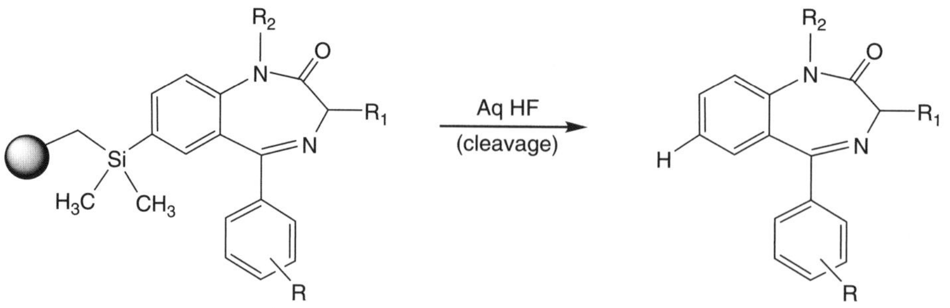

Figure 1.14   Traceless cleavage using a silicon-based linker group

Diversity-oriented (or diversity-generating) linker groups introduce another element of diversity to the organic substrate upon the cleavage process. There are numerous examples of this approach, which will be covered extensively in Part 2 of this book. Some are very direct examples in which an atom, often an electrophile, mimics hydrogen by interacting with an organometallic atom within the linker group, and induces a cleavage similar to the traceless cleavage previously discussed. Thus, germanium-based linking groups react with electrophilic halogens to directly substitute bromine, chlorine or iodine upon releasing the substrate from resin. Germanium, like silicon, undergoes ipso substitution via a β-stabilized carbocation, allowing for a halogen to be introduced (Figure 1.15).[34]

Very innovative linker groups have been developed based upon the unique properties of other elements and so almost all non-metal and many organometallic elements have found their way into solid-phase organic synthesis (SPOS). One striking example is the use of phosphonates to firstly serve as a linker group and then, secondly, to allow for arylation via palladium catalyzed Suzuki chemistry upon cleavage (Figure 1.16).[35]

The examples provided above capture some of the breadth of chemistry that has been developed in the context of linker groups. The early work was based primarily on carboxylic acid functionality to attach and then cleave the organic substrate. As milder cleavage methods were sought, specialized variants, such as benzylic alcohols and amines, were immobilized and developed as linkers. Efforts to design linker groups that were traceless ensued, along with the recognition that linker groups themselves could be used as synthetic handles to install additional diversity. Collectively, these major advances allowed for remarkable expansion of the technology as is illustrated in subsequent chapters.

1.2.5 Challenges

Despite the many advantages that solid-phase organic chemistry can afford, there remain significant challenges. Macromolecular resins and polymers remain limited in the amount, or number of substrate

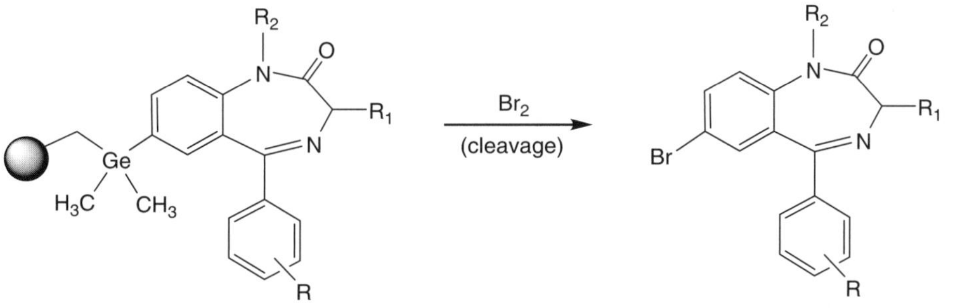

Figure 1.15 Cleavage via an electrophile: modification of a traceless approach to install a diversity element (halogen) directly

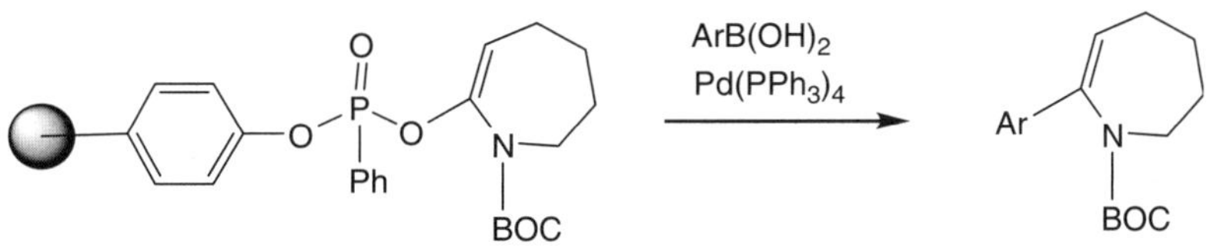

Figure 1.16 Phosphonate-based linker groups (arylation as a diversity element)

molecules that can be accommodated onto the solid support. First of all, even under optimal loading conditions, the amount of organic substrate per volume or weight, can pale in comparison to the traditional homogeneous solution-based chemistry (depending upon concentrations of the latter). Secondly solid-phase synthesis, in typically yielding milligram to gram quantities, has been relegated to basic research applications such as stocking compound collections or driving *in vitro* structure–activity relationship (SAR), rather than finding widespread use in scale-up or process operations. Additionally, even highly efficient loading reactions that covalently attach the organic substrate to the solid support, will leave reactive sites on the resin intact. These moieties can then undergo chemical transformations that can result in the formation of significant quantities of side products. For example, a resin bearing free hydroxyl groups can readily accept organic substrates via formation of an ester or ether bond but any remaining unreacted free hydroxyl groups can then act as nucleophiles themselves towards subsequent reagents (e.g., electrophiles). Conversely, while most resins are remarkably inert, the scope of chemical reactions for any given solid-phase supported organic substrate can be limited by side reactions that either cleave the substrate or covalently modify the resin itself. Since the bulk of material in any solid-phase synthesis comes from the solid-support, even minor side reactions to the macromolecule can be problematic in either consuming a reagent or by altering the physical or chemical properties of the resin.

Similarly, applying solid-phase organic synthesis in a combinatorial manner is not without limitations. The concept of conducting multiple simultaneous chemical reactions infers that all reactions will proceed with similar efficiency and, usually, at similar rates. These assumptions are necessary when generating large number of compounds unless there is a willingness to check and monitor each and every substrate and its chemical reactions. If one goal of generating a library is to install a high level of diversity, as is often the case, then structurally diverse building blocks (reagents and organic substrates) are required. The selection of building blocks may include members with wide variations in molecular size, sterics, electronic character around a key functional group, polarity and lipophilicity. Such a rich medley of molecular participants will present differing reactivities and solubilities, which can be challenging in terms of efficient production, isolation and purification of desired target molecules.

While many of the issues raised above are addressable to some degree, one fundamental intangible, central to the very heart of combinatorial chemistry, is subject to near constant debate: What constitutes diversity? While there have been attempts to define 'diversity' in terms of 'chemical space', it is often the pharmacological or biological activity that generates interest in a molecule. Such interactions, with enzymes, proteins, channels, receptors, and so on, most likely involve conformational changes, pockets of solvation and lipophilic interactions and so the way in which a chemist may view a molecule (e.g., with this backbone 'extended' and its substituents neatly disposed in two dimensions) may be quite different from how a molecule actually 'works' with respect to 'activity'. Even from a pure chemical perspective, 'diversity' may be in the 'eye of the beholder' because the sum of individual molecular fragments, however delineated, does not accurately describe a given molecule as a whole.

1.2.6 Linker groups

This textbook is dedicated to describing linker strategies in solid-phase organic synthesis. After this introduction, the Part Two of the book covers traditional linker units, namely those which leave a molecular (or atomic) footprint on the final products that can be traced back to the site of attachment to the linker group. Chapter 2, *Electrophile Cleavage Linker Units* by Michio Kurosu (Colorado State University, USA) and Chapter 3, *Nucleophile Cleavable Linker Units* by Andrea Porcheddu and Giampaolo Giacomelli (University of Sassari, Italy), discuss these large categories of linker groups and build upon a few of the landmark examples that were briefly illustrated in this chapter (Figures 1.1, 1.10–1.12). Such linkers clearly originated from Merrifield peptide synthesis and were instrumental in moving solid-phase technology ahead from peptide assembly into the realm of pure organic synthesis. Chapter 4, *Cyclative Cleavage as a Solid-Phase Strategy* by A. Ganessan (University of Southampton, UK), describes cyclization cleavage methodology, which has been a very creative undertaking, often giving rise to interesting heterocyclic compound libraries (one example is provided in Figure 1.3 in this chapter). Chapter 5, *Photolabile Linker Units* by Christian Bochet and Sébastien Mercier (University of Fribourg, Switzerland), introduces the topic of mild photolytic cleavage from the solid support, typically using rather straightforward irradiation reactions. However, one specific challenge to such linker groups arises from shadowing by the polymeric support, which can slow photolysis. Chapter 6, *Safety-Catch Linker Units* by Sylvain Lebreton and Marcel Patek (Sanofi Aventis, USA and France), covers a family of linkers that require activation through a chemical transformation prior to cleavage (Figure 1.13). This approach has been a breakthrough in circumventing side reactions often encountered with cleavages that require harsh conditions. Chapter 7, *Enzyme Cleavable Linker Units* by Mallesham Bejugam (University of Cambridge, UK) and Sabine Flitsch (University of Manchester, UK), introduces a rather new, tailored concept by describing linker units that are cleaved by enzymatic processes.

The evolution of traceless and multifunctional linker groups is covered extensively in Part Three of the book, which makes up the bulk of this book. Chapter 8, *Introduction to Diversity-Oriented Synthesis* by Richard Spandl, Gemma Thomas, Monica Diaz-Gavilan, Kieron O'Connell and David Spring (University of Cambridge, UK), discusses how linker groups can be designed to participate in well known contemporary organic reactions thereby enabling a diversity element to be introduced during the cleavage process. The adoption of such solution-phase chemistry to solid-phase technology in this context has been extraordinarily powerful as subsequent chapters reveal.

Chapter 9, *T1 and T2 – Versatile Triazene Linker Groups* by Kerstin Knepper and Robert Ziegert (Woergl, Austria), Chapter 10, *Hydrazone Linker Units* by Rysard Lazny (University of Bialystok, Poland), and Chapter 11, *Benzotriazole Linker Units* by Daniel Whelligan (Institute of Cancer Research, UK), cover linkers that revolve around nitrogen chemistry in describing triazene-based linker groups and hydrazone/benzotriazole variants, respectively. The remaining chapters turn to linker groups that use certain chemical properties that are unique to specific elements. Thus, Chapter 12, *Diversity Cleavage Strategies from Phosphorous Linkers* by Patrick Steel and Tom Woods (University of Durham, UK), illustrates Wittig and Horner–Emmons type chemistry as well as palladium mediated aryl couplings, which are extremely valuable in adding complex diversity units (e.g., Figure 1.16). Chapter 13, *Sulfur Linker Units* by Peter Scott (University of Michigan, USA), showcases various thiol-derived linkers that can be oxidatively activated for cleavage and the subsequent installation of an additional diversity unit. Intriguing organometallic-based linker groups include selenium and tellurium congeners (Chapter 14, *Selenium and Tellurium Linker Units* by Tracy Yuen Sze But and Patrick Toy (University of Hong Kong, Hong Kong)). There have been a number of cleavage methods that make use of free radical chemistry (from sulfur, oxygen and selenium linker units) and this topic is covered in Chapter 15, *Sulfur, Oxygen, Selenium and Tellurium Linker Units Cleaved by Radical Processes* by Guiditta Guazzelli, Marc Miller and David Procter (University of Manchester, UK). Chapter 16, *Silicon and Germanium Linker Units* by Alan Spivey (Imperial College, UK) and Chris Diaper (NAEJA Pharmaceutical Inc., Canada), discusses the use of these organometallic-centered groups to act as either traceless linkers, or as linkers that are readily cleaved by a simple halogen, respectively. Other rarer organometallic linker units are discussed in Chapter 17, *Boron and Stannane Linker Units*, and Chapter 18, *Bismuth Linker Units* (Peter Scott, University of Michigan, USA), which address these rather rare linkers, highlighting Stille/Suzuki-type couplings and bismuth cross-coupling reactions, respectively. Transition metal variants such as chromium-based linkers are discussed in Chapter 19, *Transition Metal Carbonyl Linker Units* by Sue Gibson and Amol Walke (Imperial College, UK). Finally in this section, Chapter 20, *Linkers releasing Olefins or Cycloolefins by Ring Closing* by Jan H. van Maarseveen (Universiteit of Amsterdam, The Netherlands), turns to the linkers with carbon–carbon unsaturated bonds and the use of metathesis chemistry as a powerful means to attach and cleave organic substrates.

Part Four of the book, *Alternative Linker Strategies*, contains the final chapters discussing emerging uses of solid-phase synthesis. Chapter 21, *Fluorous Linker Units* by Wei Zhang (Fluorous Technologies, USA), is devoted to a discussion on alternative linker strategies including fluorous technologies which is a rapidly expanding area. Chapter 22 by Brian Hockley, Peter Scott and Michael Kilbourn (University of Michigan, USA) showcases the emerging use of solid-phase synthesis in radiochemistry and the process of producing radiopharmaceuticals. Lastly, Part Five, *Linker Selection Tables*, provides comprehensive and extremely useful information that the reader can turn to when designing and using linker groups in solid-phase chemistry.

1.3 Concluding comments

In its infancy (mid-1990s), the lure of solid-phase synthesis, coupled with combinatorial techniques, was irresistible. Large pharmaceutical companies, selling perhaps at best dozens of drugs looked in dissatisfaction at their meager corporate compound libraries that numbered in the mere tens of thousands. In a desire to discover new drug leads, it was easy to fathom how many compounds were historically needed to obtain one clinical candidate (estimates run from 3000 to 10,000), yet alone a successful, revenue-generating pharmaceutical product. Therefore, it was assumed or could even be calculated that having the ability to synthesize (and screen) many tens or hundreds of thousands of compounds would give better lead compounds and therefore more marketable drugs quicker (by some measure). Remember, at the time, earlier biological break-throughs, such as polymerase chain reaction (PCR) and other molecular biology advances (cloning), were well in hand and in widespread use and, thus, biology was 'outpacing' chemistry. Chemists, as well as others in the industry, were eager to embrace and drive this new chemistry technology to new heights.

Along the way, the combi-chemists (as they were rightfully or wrongly termed) accomplished so much as they redefined productivity and structural diversity on some level, but as many have witnessed, more is not always better. Having the capability and capacity does not alone drive progress in research. Science must be built upon previous learnings and discoveries and is not subject to calculated odds. Thus, while the field of chemistry has benefitted enormously from solid-phase synthesis and combinatorial techniques, the industries that use and pay for chemistry innovation in terms of bringing products to market, have been left perhaps somewhat disappointed (i.e., pharmaceutical houses). The translation of the technology to the marketplace has yet to be realized because the expectations from the outset were misplaced. More meaningful expectations need to be set within the context of delivering to 'the bottom line' within the pharmaceutical sectors.

Few, if any technologies within the field of synthetic organic chemistry have generated more promise and potential than combinatorial chemistry. Looking back upon the past decade or two, and under the cloud of stumbling productivity within the pharmaceutical industry, it may be time to ask what this amazing technology is best used for, rather than using its power to plays odds against the discovery of a drug. Indeed, pharmaceutical sciences have advanced in wonderful ways due to solid-phase organic synthesis and combinatorial chemistry. The staggering numbers of compounds that have been synthesized attests to this fact. The increasing ways in which complex synthetic chemistry can be made more efficient, or even automated, are evident. Furthermore, analytical techniques, such as magic-angle NMR, tagging and deconvolution processes, which were once reserved for the few, are now widely used and appreciated by many in the field.

In the future, a continuation of expansion of the solid-phase synthesis/combinatorial chemistry universe will be seen because there are significant frontiers that have yet to be surmounted. Moreover, further refinement in scope will be seen, so as to better translate the chemistry from the resin in the vessel to the biology in the well on the plate, or even into the animal.

1.4 Personal perspective and testimony: solid-phase Mannich chemistry

In our labs at Johnson & Johnson, we became interested in developing small compound libraries that could be loosely targeted towards G-protein coupled receptors (GPCRs). The general premise was to install lipophilicity to allow for penetration within the transmembrane region while maintaining requisite charge site(s) to interact with conversed aspartic acid residues within specific members of the family. We desired a way to rapidly expand 'diversity' and to 'dial in' shape to our molecules and envisioned additional chemical transformations that could be carried out late in the synthetic route so as to add yet another element of diversity. With these items in mind, we turned to multicomponent reaction systems because multiple elements of diversity can be introduced in a single transformation. The Mannich reaction (Figure 1.17) is a classic three-component system in which 'hydrogen active' substrates react with imine species that arise from condensation of an amine with an aldehyde. The application of Mannich chemistry to resin-bound substrates had been meager despite the general utility of this reaction in traditional solution-based organic synthesis.[36] The Kobayashi group recognized that silyl enol ethers can serve as the 'hydrogen active' component and applied this variation to solid-phase technology.[37] Resin-bound silyl enol ethers reacted with imines that were pre-formed via the reaction of amines with aldehydes in the presence of a Lewis acid catalyst and a dehydrating agent. However, terminal alkynes, in the presence of a copper(I) salt, can also behave as the 'hydrogen active' Mannich coupling partner and without the need to pre-form the imine species.[38] From this finding, we decided to explore solid-phase Mannich chemistry and in order to fully use the power of this reaction, with respect to generating highly diversified compound libraries, we sought to separately immobilize each component onto a solid support.

We went on to demonstrate that solid-phase Mannich reactions of aldehydes, amines and alkynes indeed occur smoothly and efficiently, and is not hampered by the heterogeneity of the reaction (Figure 1.18).[39–43] This multicomponent strategy is powerful for several reasons.[44] Any single one of the components can be immobilized on an appropriate solid support, which makes the methodology very versatile from a synthetic perspective. Structurally-diverse compound libraries can be readily prepared due to the numerous amines, aldehydes and alkynes that are either commercially available or easily synthesized. The alkyne moiety itself presents an opportunity for further synthetic elaboration and provides a site for the introduction of another element of diversity. Finally, many functional groups are tolerant to the experimental conditions. In our studies, the solid-phase Mannich reactions are very efficient with isolated products typically being of high purity (>90%).

The resultant Mannich adducts contain a carbon–carbon triple bond, which is a particularly attractive feature. Alkynes are among the most versatile functional groups in organic chemistry[45] and can serve as a synthetic handle for a myriad of additional manipulations. Furthermore, simple modifications to alkynes greatly alter their shape and flexibility, changing considerably the molecular scaffold thus allowing for more conformational space to be accessed. Whereas carbon–carbon triple bonds are rigid and linear, *cis*- and *trans*-alkenes possess characteristic cupped or extended arrays, respectively, and the corresponding saturated alkane is a floppy tether. In addition to reductions, other synthetic modifications of carbon–carbon

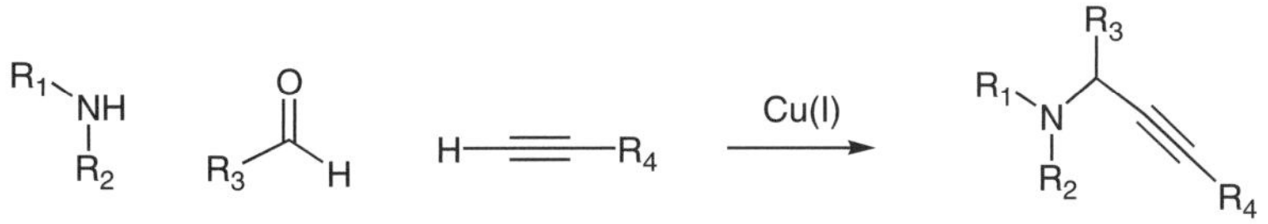

Figure 1.17 A Mannich reaction

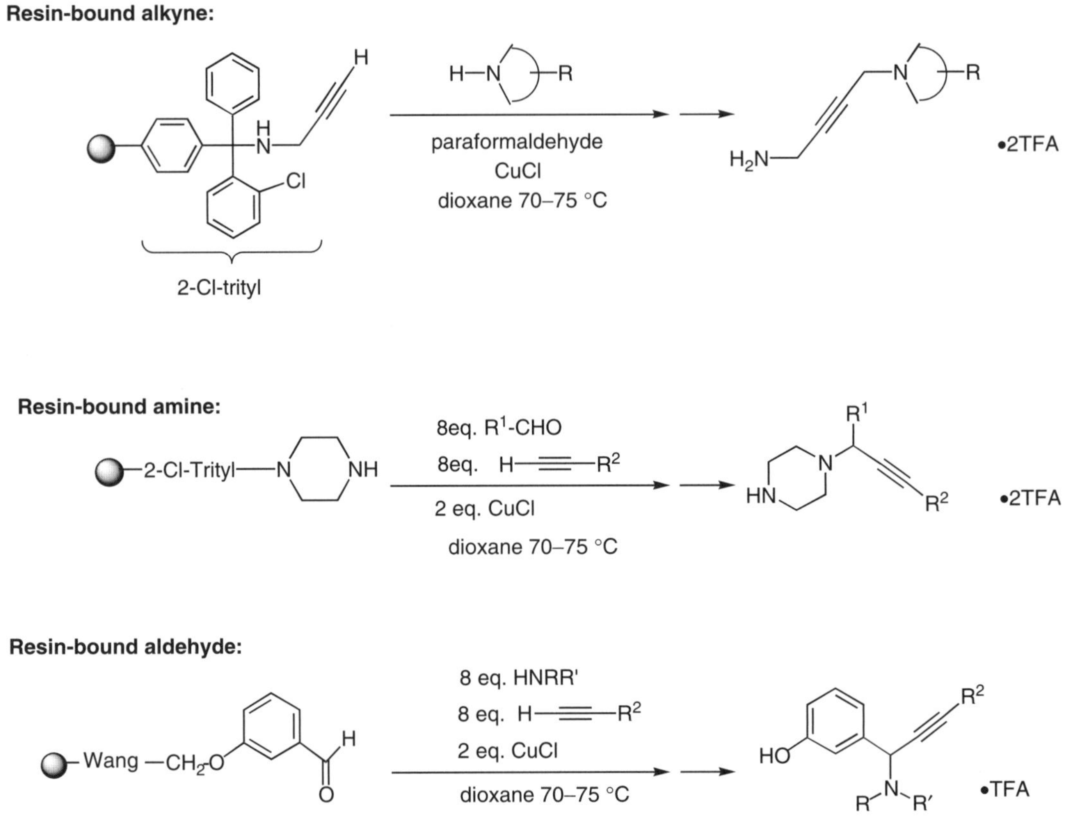

22 Introduction

Resin-bound alkyne:

Resin-bound amine:

Resin-bound aldehyde:

Figure 1.18 Solid-phase Mannich reactions

triple bonds are well-known and include hydrometalation[46] and cycloaddition[47] chemistry. Such heterocyclic systems are widely found within biologically active substances with medicinal, veterinary, and agricultural application. By developing and using this chemistry, we were able to synthesize hundreds of novel compounds and rapidly found submicromolar ligands for a variety of GPCRs thus validating our approach.

This project was not a main focus of our laboratories and two superb chemists, Mark A. Youngman and James J. McNally conducted this work literally 'working in the back of the hood'. The solid-phase technology outpaced the number of compounds made by other members of the group, again demonstrating the power of solid-phase synthesis and combinatorial approaches.

References

[1] Merrifield, R. B.; *J. Am. Chem. Soc.* **1963**, *85*, 2149.
[2] Bunin, B. A., and Ellman J. A.; *J. Am. Chem. Soc.* **1992**, *114*, 10997.
[3] Bunin, B. A., Plunkett, M. J., and Ellman J. A.; *Proc. Natl. Acad. Sci. USA* **1994**, *91*, 4708.
[4] Ellman, J. A.; (The Regents of the University of California), Solid phase and combinatorial synthesis of benzodiazepine compounds on a solid support, US Patent 5,288,514 (February 22, **1994**).
[5] DeWitt, S. H., Kiely, J. S., Stankovic, C. J., *et al.*; *Proc. Natl. Acad. Sci. USA* **1993**, *90*, 6909.
[6] DeWitt, S. H., Schroeder, M. C., Stankovic, C. J., *et al.*; *Drug Dev. Res*, **1994**, *33*, 116.

[7]   Dolle, R. E., Le Bourdonnec, B., Goodman, A. J., *et al.*; *J. Combi. Chem.* **2007**, *9*, 855.
[8]   Dolle, R. E., Le Bourdonnec, B., Morales, G. A., *et al.*; *J. Combi. Chem.* **2006**, *8*, 597.
[9]   Dolle, R. E.; *J. Combi. Chem.* **2005**, *7*, 739.
[10] Dolle, R. E.; *J. Combi. Chem.* **2004**, *6*, 623.
[11] Dolle, R. E.; *J. Combi. Chem.* **2003**, *5*, 693.
[12] Dolle, R. E.; *J. Combi. Chem.* **2002**, *4*, 369.
[13] Dolle, R. E.; *J. Combi. Chem.* **2001**, *3*, 477.
[14] Dolle, R. E.; *J. Combi. Chem.* **2000**, *2*, 383.
[15] Dolle, R. E., and Nelson Jr, K. H.; *J. Combi. Chem.* **1999**, *1*, 235.
[16] Reynolds Cody, D., Dewitt, S. H., Hodges, J. C., *et al.*; (Warner-Lambert Co., USA), Apparatus and method for multiple simultaneous synthesis of peptides and other organic compounds, WO 9408711 A1 19940428 (1994).
[17] Dewitt, S. H., Bear, B. R., Brussolo, J. S., *et al.*; A modular system for combinatorial an automated synthesis, in *Molecular Diversity and Combinatorial Chemistry: Libraries and Drug Discovery* (eds I. M. Chaiken, and K. D. Janda) developed from a Conference, Coronado, CA, January 28–February 2, 1996, **1998**, 207–218.
[18] Czarnik, A. W., DeWitt, S. H., Schroeder, M. C., *et al.*; *A practical approach to simultaneous, parallel organic synthesis*, Polymer Preprints (American Chemical Society, Division of Polymer Chemistry) **1994**, *35*, 985.
[19] Houghten, R. A.; *Proc. Natl. Acad. Sci. USA* **1985**, *82*, 5131.
[20] Lo, M. M.-C., Neumann, C. S., Nagayama, S., *et al.*; *J. Am. Chem. Soc.* **2004**, *126*, 16077.
[21] Fahad, A-O., Hruby, V. J., and Sawyer, T. K.; *Mol. Biotech.* **1998**, *9*, 205.
[22] Curran, D. P.; *J. Fluorine Chem.* **2008**, *129*, 898.
[23] Studer, A., Hadida, S., Ferritto, R., *et al.*; *Science* **1997**, *275*, 823.
[24] Pan, Y., and Holmes, C. P.; *Org. Lett.* **2001**, *3*, 2769.
[25] Pan, Y., Ruhland, B., and Holmes, C. P.; *Angew. Chem. Int. Ed.* **2001**, *40*, 4488.
[26] Scott, P. J. H., and Steel, P. G.; *Eur. J. Org. Chem.* **2006**, 2251.
[27] James, I. W.; *Tetrahedron* **1999**, *55*, 4855.
[28] Ho, C. Y., and Kukla, M. J.; *Tetrahedron Lett.* **1997**, *38*, 2799.
[29] Rink, H.; *Tetrahedron Lett.* **1987**, *28*, 3787.
[30] Morphy, J. R., Rankovic, Z., and Rees, D. C.; *Tetrahedron Lett.* **1996**, *37*, 3209.
[31] Kenner, G. W., McDermott J. R., and Sheppard, R. C.; *J. Chem. Soc. Chem. Commun.* **1971**, *12*, 636.
[32] Backes, B. J., Virgilio, A. A., and Ellman, J. A.; *J. Am. Chem. Soc.* **1996**, *118*, 3055.
[33] Plunkett, M. J., Ellman, J. A.; *J. Org. Chem.* **1995**, *60*, 6006.
[34] Plunkett, M. J., and Ellman, J. A.; *J. Org. Chem.* **1997**, *62*, 2885.
[35] Campbell, I. B., Guo, J., Jones, E., and Steel, P. G.; *Org. & Biomol. Chem.* **2004**, 2725.
[36] Tramontini, M., and Angiolini, L.; *Mannich Bases: Chemistry and Uses*, CRC Press Inc, Boca Raton **1994**.
[37] Kobayashi, S., Moriwaki, M., Akiyama, R., *et al.*; *Tetrahedron Lett.* **1996** *37*, 7783.
[38] Cook, S. C., and Dax, S. L.; *Bioorg. Med. Chem. Lett.* **1996**, *6*, 797.
[39] Youngman, M. A., and Dax, S. L.; *Tetrahedron Lett.* **1997**, *38*, 6347.
[40] McNally, J. J., Youngman, M. A., and Dax, S. L.; *Tetrahedron Lett.* **1998**, *39*, 967.
[41] Dax, S. L., and Youngman, M. A.; *J. Combi. Chem.* **2001**, *3*, 469.
[42] Dax, S. L., and McNally, J. J.; Solid-phase Mannich reactions of a resin-immobilized secondary amine, in *Solid-Phase Organic Syntheses* (ed. A. W. Czarnik), John Wiley & Sons, Inc., New York **2001**, 9–13.
[43] Dax, S. L., and Youngman, M. A.; Solid-phase Mannich reactions of a resin-immobilized alkyne, in *Solid-Phase Organic Syntheses* (ed. A. W. Czarnik), John Wiley & Sons, Inc., New York **2001**, 45–53.
[44] Dax, S. L., McNally, J. J., and Youngman, M. A.; *Curr. Med. Chem.* **1999**, *6*, 255.
[45] Patai, S. (Ed.); *The Chemistry of the Carbon–Carbon Triple Bond, Parts 1–2*, John Wiley & Sons Ltd, Chichester **1978**.
[46] Negishi, E.-I.; Reaction of Alkynes with Organometallic Reagents in *Prep. Alkenes* (ed. J. M. J. Williams), Oxford University Press, Oxford **1996**, 137.
[47] Bastide, J., and Henri-Rousseau, O.; Cycloadditions and Cyclizations Involving Triple Bonds, in *The Chemistry of the Carbon–Carbon Triple Bond, Parts 1–2* (ed. Patai, S.), John Wiley & Sons Ltd, Chichester, UK **1978**.

Part II
Traditional Linker Units for Solid-Phase Organic Synthesis

2
Electrophile Cleavable Linker Units

Michio Kurosu

Department of Microbiology, Immunology and Pathology, Colorado State University, USA

2.1 Introduction

The concept of solid-phase synthesis was first introduced by Merrifield in the late 1950s. The general advantages of solid-phase synthesis have been particularly exploited in oligopeptide synthesis and the simplicity of the methodology has led to the development of solid-phase syntheses of polynucleotides[1] and polysaccharides.[2–5] Automated methods for the syntheses of these biopolymers allow for the rapid preparation of structures of interest. Until the early 1990s, solid-phase organic synthesis was not widely used except for the field mentioned above. However, the resurgence of solid-phase organic synthesis was triggered by the introduction of combinatorial chemistry techniques.[6–9] Applications of combinatorial chemistry are very wide. For example, in pharmaceutical companies, large numbers of structurally distinct molecules have been synthesized on polymer supports in a short time and submitted for high-throughput screening. Generation of a large number of ligand molecules via solid-phase synthesis and optimization of identified molecules for certain catalytic asymmetric organic reactions were popularized in 1990s. Thus, solid-phase synthesis has greatly advanced research in organic chemistry, biochemistry, molecular biology, pharmacology and drug discovery. Productivity has been amplified beyond the levels that were routine for the last hundred years.

To date, a large number of reactions have been translated from the solution to the solid phase. However, because some reactions developed in solution have never been transferred successfully to the solid phase, the power of solid-phase synthesis is best used when it offers complementary advantages to solution-phase chemistries. As the choice of protecting groups is very important to the design of synthetic schemes for target molecules in solution, the choice of the linker, which is often described as a bifunctional protecting group, requires careful consideration when applying diverse organic reactions on the solid phase. The linker should fulfill a number of criteria: (i) the linker should be stable against a planned set of reaction conditions; (ii) the immobilization of organic molecules (the first building blocks or natural product derivatives) should be readily achieved in high yield; and (iii) synthesized molecules on polymer supports should be cleaved under mild conditions that do not degrade the products. When combinatorial chemistries are performed in order to synthesize a large number of molecules, the cleavage method should be very simple (i.e. by using a volatile cleavage reagent) and should not introduce impurities that are difficult to remove. Although recent advances in analysis and purification methods have dramatically enhanced the usefulness of solid-phase synthesis using the established linkers, continuous development of new linkers that allow the delivery of target molecules in high yield and purity is required.

The examples discussed in this chapter are electrophiles, such as H^+, cleavable linkers. These linkers represent the most widely used linkers in current solid-phase organic synthesis.

2.2 Resins for use with electrophilic linkers

As it is very important to understand characteristics of polymer resins when designing syntheses of target molecules on polymer supports, a brief description of the most commonly used resins (Figure 2.1) is useful.

In solid-phase synthesis, the solid support is generally based on a polystyrene (PS) resin. The most commonly used resin supports for solid-phase synthesis include spherical beads of lightly cross-linked polystyrene [**1**, 1–2% divinylbenzene (DVB)]. The polymer beads need to swell and increase the size of the pores within them in order to enable the desired organic reactions to be carried out on the polymer supports. A low level of cross-linker, typically 1% DVB, ensures the resins can swell in apolar solvents to allow rapid permeation and access to all active sites. The reactions on polystyrene resins are best performed in (DCM) or toluene or in mixed solvents such as dichloromethane/N,N-dimethylformamide (DCM/DMF). Although styrene-based polymers are by far the most commonly used solid supports in solid-phase organic chemistry, the need for new polymer supports with high mechanical and chemical stability, good swellability and excellent compatibility with a wide range of hydrophilic or hydrophobic solvents is increasing rapidly. In order to improve performance of organic reactions on polymer supports, a number of new polymer supports have been developed.

JandaJel^{TM}, **5**, is a member of an emerging class of lightly cross-linked polystyrene resins in which the standard cross-linker DVB has been replaced with 1,4-bis(4-vinylphenoxy)butane. Due to the more polar and flexible nature of 1,4-bis(4-vinylphenoxy)butane, JandaJel^{TM} absorbs more of the typical organic solvents (benzene, DCM, DMF, THF, etc.) than PS/DVB resin do. This increased swelling means that the molecules attached to this polymer are more readily accessible to the reaction solution compared to the molecules attached to the cross-linked PS/DVB resin used in heterogeneous polymer-assisted organic synthesis. Aminomethyl JandaJel resin is available from Sigma-Aldrich and can be functionalized with a variety of linkers (Section 2.3.1.1). Leana and Kumar developed cross-linked polystyrene-ethyleneglycol acrylate resin (**6**, CLEPSER) by introducing an acrylamide (Acr$_2$PEG), which was derived from [(NH$_2$)$_2$PEG$_{1900}$] cross-linker into a polystyrene network. The authors claimed that CLEPSER possesses a very high degree of swelling in a broad range of nonpolar and polar solvents, and shows stability against various reagents.[10] Thus, it is possible to design new cross-linked polystyrene-based polymer supports by the introduction of the cross-linkers possessing different physico-chemical characteristics.[11]

Sheppard designed polyacrylamide (PA) polymers, **8**, for peptide synthesis as it was expected that these polymers would more closely mimic the properties of the peptide chains themselves and have greatly improved solvation properties in polar, aprotic solvents (i.e. DMF, or N-methyl pyrrolidinone).

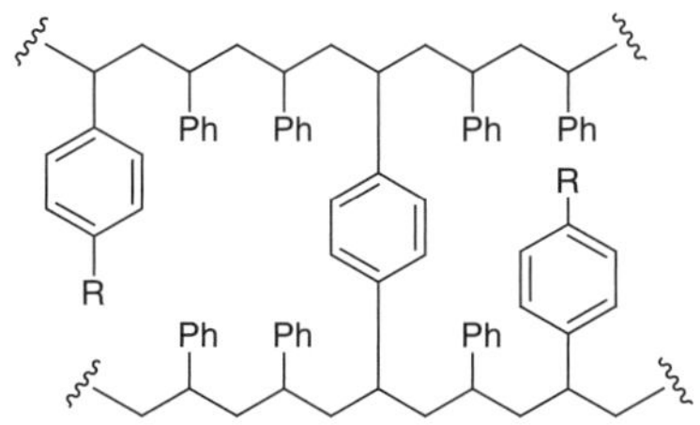

1: R = H: **Polystyrene** (**PS**) **resin**

2a: R = CH_2Cl: **Merrifield resin** (**Cl**)

2b: R = CH_2OH: **Merrifield resin** (**OH**)

2c: R = CH_2NH_2: **Merrifield resin** (**NH_2**)

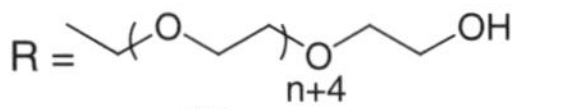

3a: **TentaGel™ resin (Rapp resin)**

3b: **HypoGel™ 400** (n = 5)

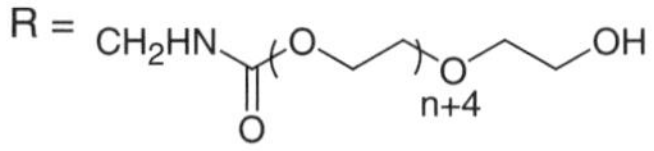

3c: **Novagel resin**

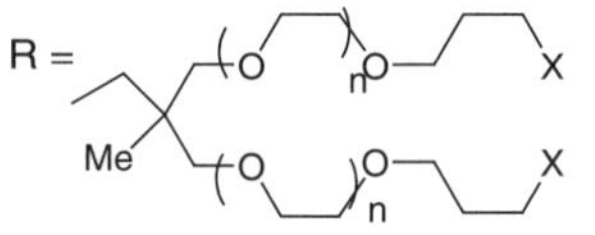

4: **ArgoGel™ resin**

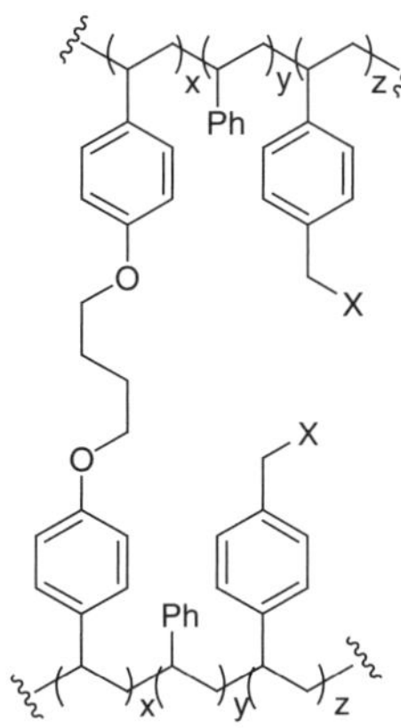

5: **JandaJel™**

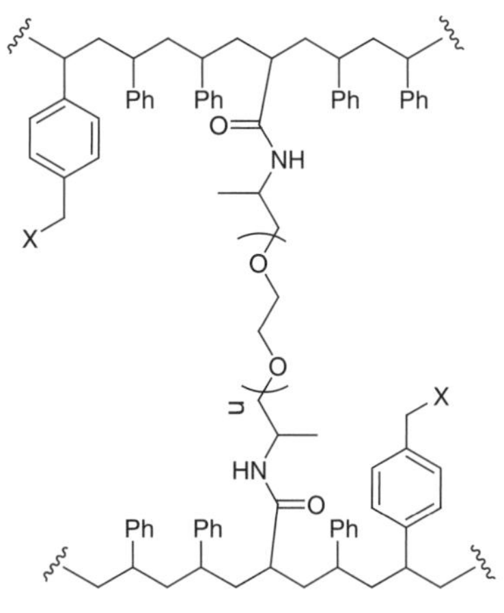

6: X = Cl or OH **CLEPSER**

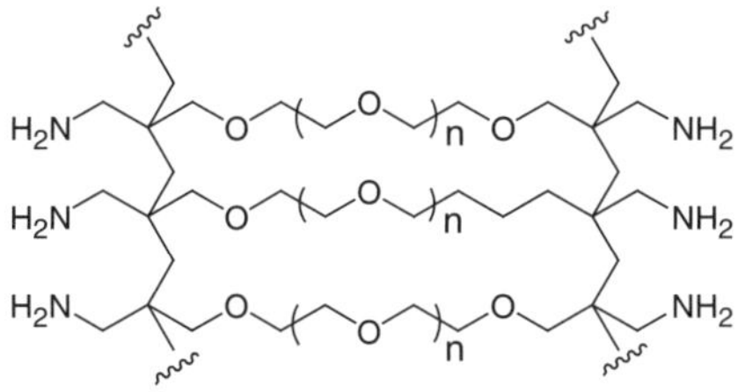

7: **ChemMatrix** (100% PEG)

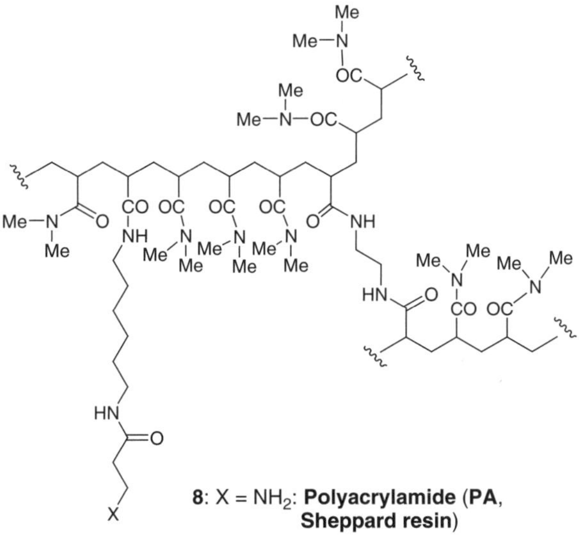

8: X = NH_2: **Polyacrylamide** (**PA**, **Sheppard resin**)

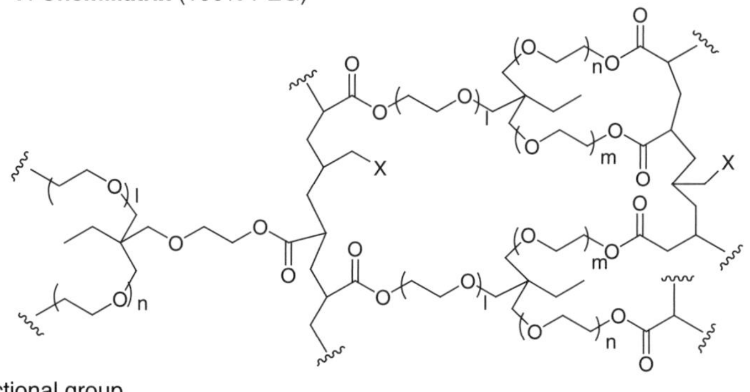

X = functional group

9: X = NH_2: **CLEAR resin**

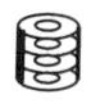

10: **SynPhase™ resin** (**Lantern**) **PS** or **PA resin**

Figure 2.1 *Chemical structures of representative polymer resins*

Poly(styrene-oxyethylene) graft copolymers, **3a**, first reported by Bayer and Rapp, are another class of widely used supports for organic synthesis. As with the polyacrylamide (PA) resins, in order to produce a polar reaction milieu that is closer to the solvents generally used in organic synthesis in solution, grafted polymer beads have been prepared. TentaGel resin (Rapp resin) is a copolymer which has polyethylene glycol (PEG) grafted on a low cross-linked polystyrene matrix. In general, polystyrene (PS) and PEG are connected through benzylic ether linkages. As the copolymer contains about 50–70% PEG, its properties are dominated by those of PEG rather than those of the polystyrene matrix.[12] HypoGel^{TM} resin, **3b**, is based on a low cross-linked (1% DVB) polystyrene matrix, oligo ethylene glycols are grafted to form a high loaded hydrophilic resin. Swelling data indicate that HypoGel is right between TentaGel and polystyrene. ArgoGel^{TM} resin, **4**, is a highly cross-linked grafted PEG-PS copolymer. The resins have very low reachable PEG impurities, and have higher loading and greater stability as a result of its bifurcated linkage.[13] NovaGel^{TM} resin is a PS-PEG resin that possesses carbamate linkages, which are claimed to be more chemically stable than benzylic ethers, to graft PS and PEG.[14]

CLEAR^{TM} resins, **9**, are cross-linked ethoxylate acrylate resins, developed by Barany and Kempe.[15] The CLEAR particles swell in a wide range of solvents including water, DCM or DMF. They are also compatible with relatively nonpolar solvents such as tetrahydrofuran (THF) or dioxane. CLEAR resin particles are highly cross-linked, structurally uniform and are generated from a large-scale suspension polymerization process (not grafted onto polystyrene). CLEAR resin is mechanically stable. Since there are no aromatic rings in CLEAR-based resins, it is claimed that there are no large blocks of hydrophobic structure that can result in unwanted aggregation with hydrophobic materials. Because CLEAR is composed of ester linkages, this resin is not recommended for use under strong basic conditions. On the other hand, polyethylene glycol-polyacrylamide (PEGA) resin is a cross-linked amide functionalized copolymer, and contains only secondary amide and ethers.[16] Thus, PEGA is stable against basic conditions. However, PEGA resins show mechanical instability and need to be maintained in a moist slurry.

ChemMatrix® resin, **7**, which is based on PEG (100%) that is cross-linked, is highly recommend for synthesizing long, complex or highly hydrophobic peptides. ChemMatrix® resin has a high chemical and thermal stability, which is compatible with microwave synthesis, and has shown a higher degree of swelling in acetonitrile, dichloromethane, DMF, *N*-methyl pyrollidone, TFA and water compared to the polystyrene-based resins. There are number of reports on the development of soluble polymer resin in polar solvents which are useful for immobilization of catalysts including enzymes.[17]

The SynPhase^{TM} lantern, **10**, is a polymer-grafted polypropylene that is cylindrical in appearance with the physical dimensions shown in Figure 2.1. Each SynPhase Lantern represents a rigid polymer support that can be individually tagged. Tagging (radiofrequency tagging or color coded tagging) offers the ability to efficiently identify, track and sort library molecules generated on the polymer support. Two surface polymer types of lanterns, polystyrene and polyacrylamide, are well-characterized on SynPhase resins.[18]

Several established linkers (i.e. carboxylic acid, carboxamide, alcohol, amine and traceless) can be conjugated onto the resins described above and some of linker resins are available from specialized vendors. The choice of resin needs careful consideration at the stage of designing synthesis; one should use the resin which ensures the integrity of the resin throughout the operations of polymer-supported organic synthesis including swelling, washing, drying and mechanical stirring (in case necessary) steps.

2.3 Electrophile cleavable linkers

The choice of linker requires careful consideration for performing multiple step organic reactions on a polymer support. The linker should be stable against a planned set of reaction conditions, but be cleaved under mild conditions that do not degrade the products. As described above, the solvation efficiency of

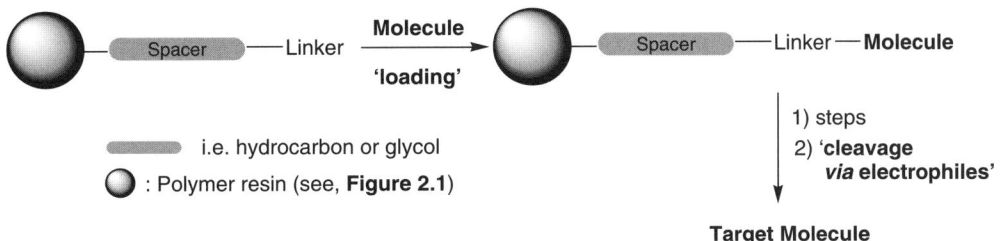

Scheme 2.1 *Immobilization onto and cleavage from polymer resin*

the polymer resin is the most important factor to be considered when deciding which resin to use for planned reactions on polymer supports. Although several studies indicate that choice of spacer (the region between the polymer resin and the linker) is also an important factor for improving loading and cleavage efficiencies,[19] and the reaction rate, detailed kinetic studies by altering physico-chemical property of spacers have not been thoroughly investigated. Thus, spacers are not emphasized in this chapter (Scheme 2.1).

Linkers are generally categorized according to the method of cleavage (i.e. acid labile linkers). In following several reviews on solid-phase organic synthesis, linkers are sub-classified based on the liberated functional group of the product after cleavage (i.e. carboxylic acid, carboxamide, alcohol, amine and ketone linkers). A great number of electrophile cleavable linkers, such as Brønstead acid (proton donor), have been developed since the late 1950s. Many acid cleavable linkers were investigated on polystyrene resin for oligopeptide and nucleotide synthesis, and some of them were later modified for other types of resins (Figure 2.1). To date, several comprehensive books and reviews on linkers for solid organic synthesis and combinatorial organic synthesis have been published. They give an overview about the current state-of-the-art of solid-phase organic synthesis with a special focus on the synthesis and strategies of compounds employing solid-supported chemistry. The chapters or contents of some of these reviews cover linker strategies (attachment/cleavage and linker type), and are still valuable to all scientists engaged in polymer-supported organic reactions, both in industry and academia. Acid labile linkers are a major class of linkers being used in solid-phase organic synthesis. Useful linker resins were developed over the long period between the early 1960s and the late 1990s. Thus, the interested reader may find some overlap between this chapter and those excellent reviews published up to 2007.[20–26] Linkers for the generation of carboxylic acids, carboxamides, alcohols, amines and carbonyls have been improved continuously. The author of this chapter overviewed acid (or electrophile) cleavable linkers.

2.3.1 Acid labile linkers

2.3.1.1 Carboxylic acid linkers

Acid labile or photolabile linkers (Chapter 5)[27] have been used in the production of large libraries of molecules through the split/pool method.[28] Acid labile linkers were originally developed for the syntheses of oligopeptides on polystyrene resin (Figure 2.2 and Table 2.1). Many of these early linkers are reviewed extensively and still in regular use. However, if the aim is to synthesize nonpeptidic molecules using conditions other than standard solid phase oligopeptide synthesis, modified versions of known carboxylic linkers that are stabilized against a planned set of reaction conditions are recommended.

Bruce Merrifield was the first to recognize the usefulness of polymer-supported organic synthesis, and is best known for the invention and development of solid-phase peptide synthesis (SPPS), which revolutionized synthetic organic chemistry.[29] Merrifield planned to attach the C-terminal amino acid to an

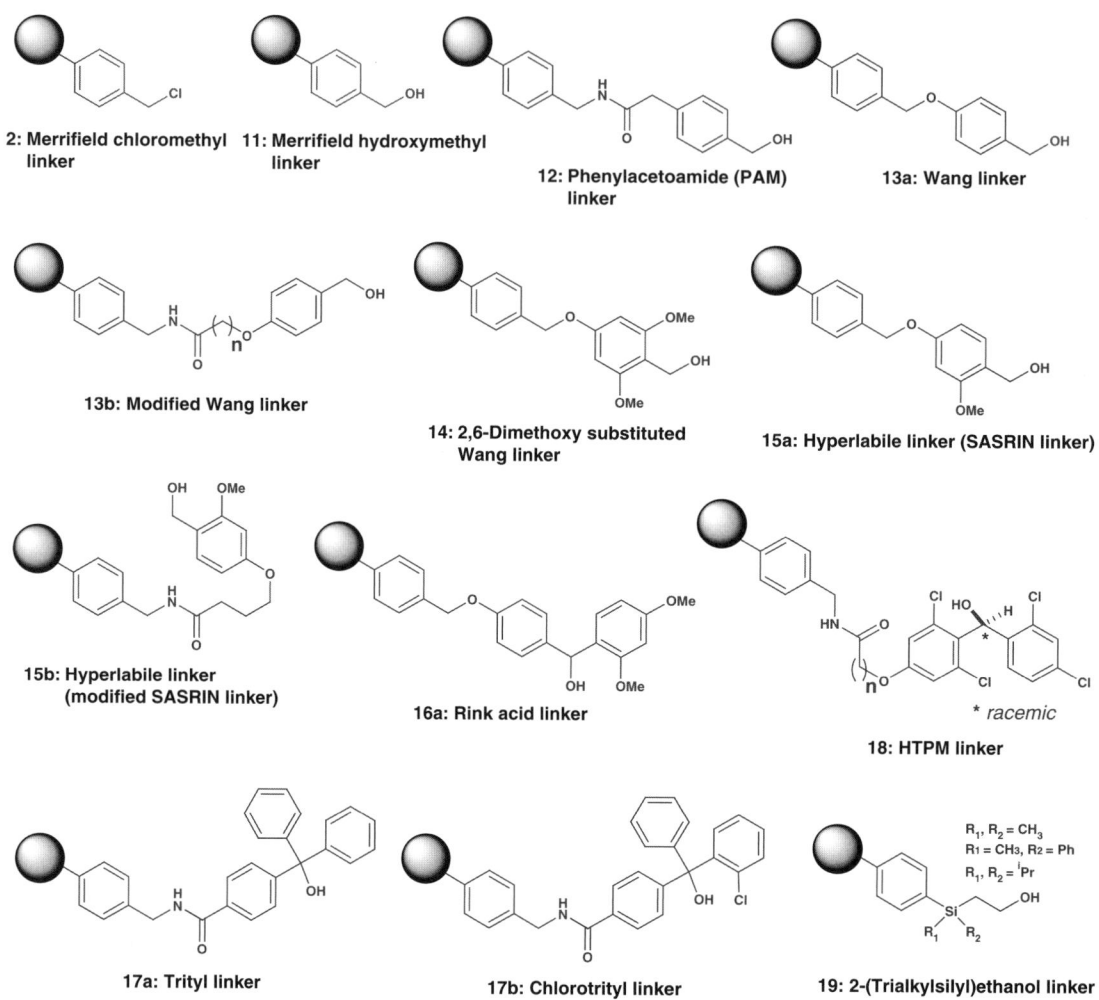

Figure 2.2 Carboxylic acid linkers

insoluble particle throughout the synthesis and to grow the peptide chain on that residue.[30] His first successful synthesis by SPPS was of the tetrapeptide, L-Leu-L-Ala-L-Gly-L-Val on nitrochloromethylphenyl polymer resin using benzyloxycarbonyl (Cbz)-protected amino acids (30% hydrogen bromide in acetic acid for the deprotection, N,N'-Dicyclohexylcarbodiimide (DCC) in DMF for the peptide-forming step, 2 N sodium hydroxide in ethanol for the peptide liberation step).[31] An important improvement was made for the solid-phase synthesis of the nonapeptide, bradykinin (L-Arg-L-Pro-L-Pro-L-Gly-L-Phe-L-Ser-L-Pro-L-Phe-L-Arg), in which the readily removal t-butoxycarbonyl (Boc)-protected amino acids (introduced in the late 1950s)[32, 33] were used in place of Cbz-protected derivatives. With this modification it was no longer necessary to use the nitrobenzyl ester resin, which is far more stable against aqueous hydrogen bromide, and the overall yield was improved significantly due to decrease in the loss of peptide from the resin.[34] Thus, Merrifield established a reliable oligopeptide synthesis on chloromethylated linker-PS/DVB resin (Merrifield linker, **2**) using the Boc strategy.[35] With the automation of this technique, Merrifield and coworkers

Table 2.1 Representative Carboxylic Acid Linkers

Linker	Suitable substrates	Attachment Conditions	Cleavage Conditions	Stable towards	Refs
Merrifield chloromethyl Linker	• Carboxylic acids	• Cs_2CO_3	HF, 0°C, (30–60 min)	>50%TFA Suitable for the Boc peptide strategy	[29–37, 56, 57]
Merrifield hydroxymethyl linker	• Carboxylic acids • DCC or DIC/DMAP	• DCC or DIC/DMAP • TPP, DIAD	HF, 0°C, (30–60 min)	>50% TFA Suitable for the Boc peptide strategy	[23, 58–63]
Phenylacetoamide (PAM) Linker	• Carboxylic acids	• DCC or DIC/DMAP • TPP, DIAD	HF, 0°C, (30–60 min)	More stable than Merrifield linker Suitable for the Boc peptide strategy	[40–42, 54–57, 64–67]
Wang linker	• Carboxylic acids • Phenols	• DCC or DIC/DMAP • TPP, DIAD	>20% TFA	Suitable for the Fmoc peptide strategy	[43, 68–70]
Modified Wang linker	• Carboxylic acids • Phenols	• DCC or DIC/DMAP • TPP, DIAD	>20% TFA	Suitable for the Fmoc peptide strategy	[44, 45, 71]
Hyperlabile linker (SASRIN or HMPB linker)	• Carboxylic acids • Phenols	• DCC or DIC/DMAP • TPP, DIAD	0.1–1% TFA	Suitable for the Fmoc peptide strategy	[44, 75, 76]
Hyperlabile linker (modified SASRIN or HMPB linker)	• Carboxylic acids • Phenols	• DCC or DIC/DMAP • TPP, DIAD	0.1–1% TFA	Suitable for the Fmoc peptide strategy	[77–85]

(*continued overleaf*)

Table 2.1 (continued)

Linker	Suitable substrates	Attachment Conditions	Cleavage Conditions	Stable towards	Refs
Rink acid linker	• Carboxylic acids	• Benzenesufonic acid/dioxane, 50 °C	0.1% TFA	Suitable for the Fmoc peptide strategy	[50, 86–88]
Chlorotrityl linker	• Carboxylic acids • Alcohols • Amines • Thiols	• activation with AcCl then nucleophile/base	0.1% TFA, 5% triisopropylsilane	Better nucleophilic base stability than benzyl ester linkers	[51, 52, 89–91]
HTPM linker (*racemic*)	• Carboxylic acids	• DCC or DIC/DMAP • TPP, DIAD	20% TFA	• 20–25% TsOH • Bases, Lewis acids and Nucleophiles	[53]
$R_1, R_2 = CH_3$ $R_1 = CH_3, R_2 = Ph$ $R_1, R_2 = {}^iPr$	• Carboxylic acids	• HATU, HOBt, ${}^iPr_2NEt/DMF$	30% AcOH/ CH_3OH, 60 °C	Suitable for the Fmoc peptide strategy	[54, 55]

demonstrated the total synthesis of an enzyme, bovine pancreatic ribonuclease A (a 124-residue molecule).[36, 37] Merrifield's work was recognized when he received the Nobel Prize in Chemistry 1984 'for his development of methodology for chemical synthesis on solid matrix'.[38]

Although the Merrifield linker, a benzyl ester, is sufficiently stable to repetitive trifluoroacetic acid (TFA) treatments for N^α-Boc deprotection, it was observed that an average of 1.4% of the growing peptide chain was lost after each deprotection step using 50% TFA. In addition, the overall yield of the target molecule has never reached close to ideal because of requirement of an inefficient cleavage of molecule from the resin via a strong acid such as hydrogen bromide or hydrogen fluoride at 0 °C for 30–60 minutes. Because of hydrogen fluoride's rapid skin penetration and the serious toxicity of the fluoride ion, use of hydrogen fluoride requires extremely careful operation. Later, trifluoromethanesulfonic acid, as an alternative cleavage reagent, was introduced to cleave ester-Merrifield linkers.[39] Loading carboxylic acid to form the ester linkage onto the Merrifield resin was generally performed via S_N2–type reactions of cesium carboxylate. Because the active surface (the chloromethyl residues) was partially hydrolyzed during the loading step, hydroxylmethyl linker, **11**, has been used in place of the chloromethyl linker, so that high yielding esterification can be achieved by using conventional esterification methods including DCC/DMAP in DCM or toluene, or Mitsunobu conditions (TPP and DIAD in THF). The miscellaneous utility of the Merrifield hydroxymethyl linker (i.e. carbonate linker to generate amines) was summarized in Bradley's extensive review.[23] On the basis of chemical characteristics of the Merrifield linker, several carboxylic acid linkers have been developed to date. The majority of acid labile linkers are a variation of benzyl or methoxybenzyl alcohol.

The phenylacetamide linker (PAM linker, **12**) provides an ester linkage that is several to 100 times more stable than the Merrifield ester against TFA.[40] This linker has been demonstrated for the synthesis of oligo to large peptides by using the Boc strategy, with negligible chain loss in each N^α-Boc deprotection with TFA. Thus, this linker is recommended for the synthesis of peptides by using the Boc strategy.[41, 42]

In 1973, Wang functionalized the chloromethylated Merrifield resin, **2**, via a Williamson type ether synthesis to provide 4-hydroxybenzyl alcohol linker resin, **13a**, whose esters can be cleaved under milder acidic conditions (>20% TFA).[43] Later, the Wang linker was modified with 2-(4-(hydroxymethyl)phenoxy)acetic acid, which can be coupled with aminomethyl resins in near quantitative yield using standard amide forming reactions to afford the modified 4-hydroxybenzyl alcohol linker, **13b**.[44] The Wang linker can be cleaved under a weaker acid concentration than the N^α-Boc group.[45] Thus, by using 2-(*p*-biphenylyl)-2-propyloxycarbonyl (N^α-Bpoc) protected amino acid building blocks, which can be deprotected with 0.5% TFA in DCM, Wang demonstrated the syntheses of oligopeptides using his linker with significantly improved overall yield (68% yield for the tetrapeptide). The acid susceptibility is rationalized based on the stability of the benzylic cation, which is stabilized by the resonance effect of the electron donating group. Thus, the additional methoxy group on the Wang linker, widely called SASRIN **15a** (the trade name) or HMPB (4-(4-Hydroxymethyl-3-methoxyphenoxy)-butyric acid) linker, dramatically decreases acid stability; polymer-bound molecules on the SASRIN linker resins can be cleaved with 1% TFA.[46] The 2,6-dimethoxy substituted Wang linker ester, **14**, is far more labile to acid; the ester linker can be cleaved with 0.1–1% TFA. Because of the experimental result that the 2,6-dimethoxy substituted Wang linker exhibited poor loading yield, due probably to the formation of a hydrogen bond between the hydroxy and methoxy groups, or simple steric reasons, the 2,6-dimethoxy substituted Wang linker has not been widely used as a carboxylic acid linker. However, the corresponding aldehyde linker has been used in generating imine molecules by immobilizing primary or secondary amines via reductive amination reactions (Section 2.1.4). Similar to the modified Wang linker, HMPB can be conveniently coupled with a wide variety of aminomethyl resins to afford **15b**. These acid-sensitive linkers have been applied to the synthesis of peptide segments, which are useful in the synthesis of target peptides or depsipeptides in solution, with the N^α-9-fluorenylmethoxycarbony (Fmoc) solid-phase technique. The Fmoc strategy differs from the Boc strategy in that N^α- and C-terminal functional groups are orthogonally protected. The Fmoc solid-phase peptide synthesis (Figure 2.3) was introduced by Carpino in 1972[47] having previosuly described Fmoc as a protecting group for amino acids in solution peptide chemistry.[48] To remove Fmoc from a growing peptide chain, basic conditions (e.g. 20% piperidine in DMF or tetra *n*-butylammonium fluoride (TBAF) in DMF) are used. Cleavage of the oligo peptide from the resin is designed to achieve by using low concentrations of TFA.[49]

In attempts to further increase acid lability of the ester linker, the trialkoxy-diphenylmethanol conjugated onto Merrifield resin **16a** was developed by Rink.[50] Disadvantages of using Rink linkers for the synthesis of carboxylic acid peptides is that the trialkoxy-diphenylmethanol shows poor nucleophilicity for esterification with carboxylic acids under standard methods. Thus, it is recommended to cap unreacted hydroxyl groups on resins with a convenient acyl group (e.g. acetyl or benzoyl group). The generated peptide bound to resins through a Rink linker can be cleaved with 0.1% TFA in dichloromethane within two minutes. Alternatively, AcOH/DCM can be used to cleave molecules from the Rink linker. The Rink amide linker has been

36 Traditional Linker Units for Solid-Phase Organic Synthesis

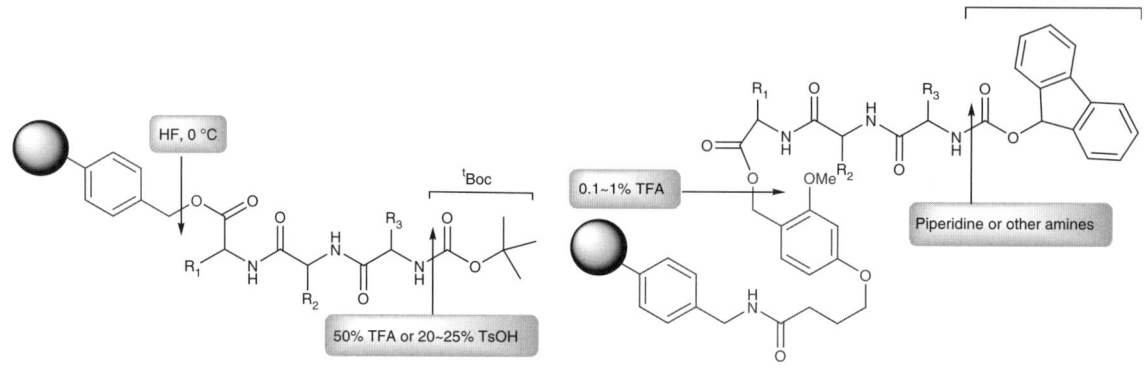

Figure 2.3 ^{t}Boc and Fmoc peptide synthesis strategies using representative linker resins

widely used to prepare peptide amide derivatives (Section 2.1.2). The utility of the Rink linker is amplified by commercially available Fmoc protected 2-(4-(amino(2,4-dimethoxyphenyl)methyl)phenoxy)acetic acid, and thus the Rink linker is formed on a wide variety of aminomethyl resins.

Trityl linkers **17a** are representative of another group of acid labile linkers frequently used for immobilizing carboxylic acids and secondary amines. Chlorotrityl linker **17b** is more stable than the trityl resin itself but 1% TFA is still strong enough to cleave synthesized molecules from the linker.[51] On the other hand, due to the steric bulkiness of trityl group, it is far more stable against nucleophilic bases than the benzyl-based linkers. The trityl linker resins are commercially available but need to be activated by treatment with acetyl chloride, thionyl chloride ($SOCl_2$) or phosphorus trichloride (PCl_3) prior to loading the starting building block. In addition to widespread availability from chemical vendors, loading carboxylic acid building blocks can be carried out using a simple protocol (mixing with the linker and base such as pyridine, DMAP or DIPEA) without racemization of the α-center of N-protected amino acids. However, a drawback is that adventitious water in solvents reacts with the trityl groups faster than carboxylates.

Thus, loading yields usually vary depending on the conditions used, and thus careful analysis of loaded molecules is recommended.[52]

In order to take advantage of abundantly available N^α-Boc protected amino acids (less expensive than Fmoc protected amino acids) and to carry out syntheses with base-sensitive building blocks, the hydroxytetrachlorodiphenylmethy (HTPM) linker **18** was recently developed by Kurosu.[53] Interestingly, the HTPM linker exhibited stability against a wide variety of bases and nucleophiles. However, the HTPM linker can be cleaved with 20% TFA. Iterative N^α-Boc deprotections in oligopeptide syntheses can be carried out using 20% TsOH. The HTPM linker/Boc strategy offers a mild method for solid-phase peptide synthesis that can avoid the use of hydrogen fluoride for the final cleavage step.

2-(Trimethylsilyl)ethanol (TMSE) has been used as convenient protecting group for carboxylic acids and anomeric positions of carbohydrates. Kim and coworkers developed the polymer-supported 2-(trialkylsilyl)ethanol linker, **19**, and this linker was examined as a C-terminal linkage to synthesize several diketopiperazines. The ester linkages could efficiently be cleaved with 30% AcOH/MeOH at 60 °C.[54] 2-(Trimethylsilyl)ethanol is a starting material for 2-(trimethylsilyl)ethoxymethyl chloride (SEM-Cl), and the SEM protecting group has been used for protections of alcohols and amines in solution chemistries. Kim and coworkers established the transformation of the polymer-supported 2-(diphenylmethylsilyl)ethanol to the corresponding polymer-supported 2-(diphenylmethylsilyl)ethoxymethyl chloride (DSEM-Cl) with paraformaldehyde and HCl(g).[55]

Typical Experimental Procedures

A vast number of peptide esters have been prepared by Merrifield solid-phase synthesis (in the original Merrifield procedure, the ester linkage was prepared from the chloromethylated resin and N-protected amino acid in the presence of triethylamine or cesium carbonate). However, these esterification reactions generally resulted in incompletion. Thus, careful analysis of remaining chlorine atoms of Merrifield resin is strongly recommended after loading reactions. Several methods have been suggested for the esterification of hydroxy resin; dicyclohexylcarbodiimide (DCC) or diisopropylcarbodiimide (DIC) mediated coupling in the presence of diaminomethylaminopyridine (DMAP) is most often used to esterify nonracemizable carboxylic acids. However, loading of Fmoc-protected amino acids requires careful control of the reaction conditions such as amount of base and reaction time due to the base labile characteristics of the Fmoc group. For loading racemizable α-chiral carboxylic acids Mitsunobu reaction using triphenyl phosphine and diethyl azocarboxylate (DEAD) or diisopropyl azocarboxylate (DIAD) in THF or DMF is a typical esterification condition.[92, 93]

Loading procedures

Procedure A (carbonyl imidazole method): Carbonyl diimidazol (CDI, 2–5 equiv.) is dissolved in dichloromethane and Fmoc-protected amino acid (2–5 equiv.) is mixed thoroughly for 10–15 minutes. This mixture is added to the hydroxy resin and the suspension is gently shaken overnight at room temperature. The resin is collected by filtration, washed with an appropriate solvent system (i.e. dichloromethane, ^{i}PrOH–water (3:1) and dichloromethane). The level of loading is determined based on the Fmoc chromophore released by the treatment with secondary amine. For example, the loading yield is determined on a 10 mg (for available hydroxy groups 1–3 mmol/g resin) of resin by cleaving off the Fmoc group with 20% pyrrolidine (1.0 ml), filtering off the resin and diluting the volume to 10 ml. The optical density of a 1:10 dilute sample is measured.

Procedure B (diisopropylcarbodiimide (DIC) method): The resin (i.e. Wang resin) is swollen in the solvent mixture of DCM/DMF (9:1) and treated with DIC (3 equiv.), carboxylic acid (5 equiv.) and DMAP (0.1–3 equiv. theoretically, catalytic amount of DMAP is enough). The reaction mixture is gently stirred (the microwave irradiation greatly enhances reaction rate).[94–96] *The loading of carboxylic acid on resin can be estimated via Fourier transform infrared (FTIR) measurement of the resin and also determined accurately by cleavage of the loaded molecules or resins.*

Cleavage procedure

The loaded polymer is treated with a cleavage cocktail such as a TFA/dichloromethane (1:1). The reaction suspension is stirred for several minutes to hours depending on the lability of the linkage and subsequently filtered. After removal of all volatiles the desired product is generated (if necessary the crude product is purified).

2.3.1.2 Carboxamide linkers

The continuing need for the synthesis of biologically interesting peptide carboxyl-terminal amides was a driving force for chemists to develop solid-phase syntheses of peptide carboxamides (Figure 2.4 and Table 2.2). The Merrifield ester or Wang ester linker can be cleaved with ammonia to generate primary carboxamides; however, aminolysis of esters on polymer supports in the absence of catalysts is very slow and generally results in unsatisfactory yields. When a prolonged treatment with amines is necessary to achieve these reactions, racemization (or epimerization) of chiral centers is a serious concern. To efficiently perform aminolysis of ester linkers, ester linkers which are more susceptible to nucleophilc attack by amines were developed. The benzophenone oxime linker can be esterified efficiently with carboxylic acids under a standard coupling condition (i.e. DCC/DMAP), and aminolysis of ester linkers can be performed in the presence of excess amines.[97] In the case of molecules which are prone to racemization, it is recommended to tune basicity with acetic acid.[23]

Pitta and Marshall developed the benzhydrylamine (BHA) linker **20**.[98] Polymer-bound molecules on the BHA linker resin exhibited less acid stability (during a standard cleavage with hydrogen fluoride

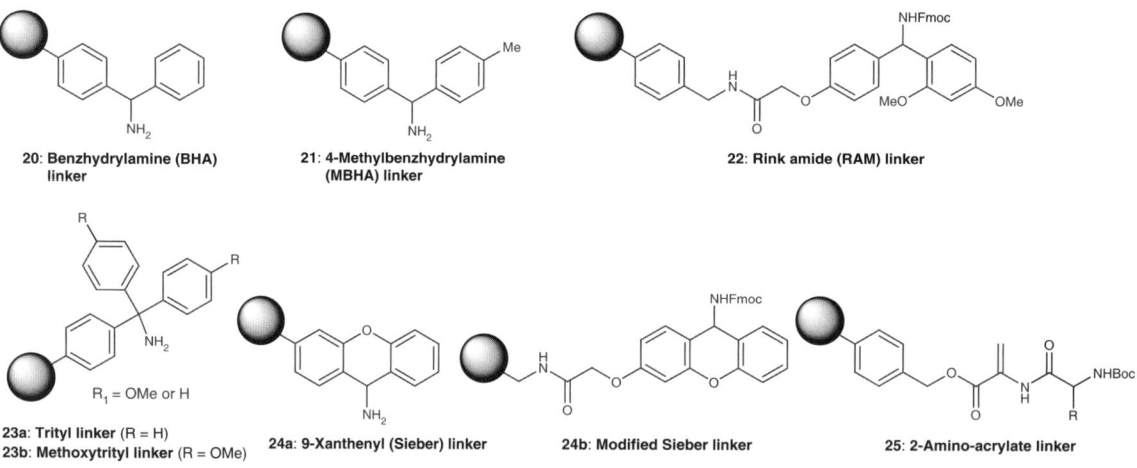

Figure 2.4 *Carboxamide linkers*

Table 2.2 Representative Carboxamide Linkers

Linker	Suitable substrates	Attachment Conditions	Cleavage Conditions	Stable towards	Refs
Benzhydrylamine (BHA) Linker	• Carboxylic acids	• Amide forming reagents	HF, 0 °C, (30–60 min)	>50%TFA Suitable for the Boc peptide strategy	[98, 105–107]
4-Methylbenzhydrylamine (MBHA) linker	• Carboxylic acids	• Amide forming reagents	HF, 0 °C, (30 min) TFA (24 h)	>50%TFA Suitable for the Boc peptide strategy	[99, 108, 109]
Rink amide (RAM) linker	• Carboxylic acids • Sulfonyl chloride	• Deprotection of Fmoc with piperidine followed by DCC, HOBt, DMF	>20% TFA	Suitable for the Fmoc peptide strategy	[110–112], Table 2.1
9-Xanthenyl (Sieber) linker	• Fmoc-protected carboxylic acids	• DCC, HOBt, DMF	2% TFA	Suitable for the Fmoc peptide strategy	[101, 113–115]
Modified Sieber linker	• Fmoc-protected amino acids	• DCC, HOBt, DMF	2% TFA	Suitable for the Fmoc peptide strategy	[116, 117]
Trityl linker	• Carboxylic acids	With or without ^{i}Pr$_2$Net	$R_1 = H$: TFA-DCM (50:50), thioanisol as a scavenger	Basic and neutral conditions	[104, 118]
2-Aminoacrylate linker	• Boc-protected amino acids	• Amide forming reagents	H$_2$O, 1N HCl/AcOH, 50 °C	Suitable for the Boc peptide strategy	[103]

at 0 °C) than those immobilized onto the Merrifield linker resin. It was reported that the first building block with sterically hindered residue (such as Val, Tyr, or Phe) immobilized on the BHA linker caused incomplete cleavage of the generated oligopeptides from the resin. The limitation in the synthesis of peptides using this linker resin is due to the less susceptible characteristic of BHA resin against hydrogen fluoride. Matsuda and Stewart demonstrated the improved synthesis of the oligopeptide carboxamides by using the 4-methylbenzhydrylamine (MBHA) linker **21**.[99] The MBHA amide liker can be cleaved with hydrogen fluoride at 0 °C within 30 minutes or anhydrous TFA for 24 h, and is the standard linker resin for the solid-phase synthesis of peptide amides using the Boc strategy. The 4-methoxybenzhydrylamine (MOBHA) linker was introduced for pursuing the same goal; however, this linker is too labile to use the Boc peptide synthetic strategy.[100] However, similar to the development of acid labile carboxylic acid linkers (Section 2.3.1.1), the BHA linkers possessing electron donating groups provide acid labile carboxamide linkers that can be applied for the Fmoc-based peptide carboxamide synthesis.

As described in Section 2.3.1.1, the Rink linker **22** is preferred for generating primary carboxamides on the solid phase. The greater acid sensitivity in this linker is a consequence of the two additional electron donating methoxy groups. Rink originally demonstrated loading of carboxamides onto the Rink acid linker resin (Figure 2.2) with benzenesulfonic acid in dioxane (50 °C for 20 h) in which these reactions involve the generation of a carbocation followed by trapping the stabilized cation intermediate with N^α-Fmoc-protected carboxamide building blocks. These reactions furnish carboxamide linker resins in 50–60% yield. Commercially available Rink carboxamide linkers are supplied in Fmoc protected form and require deprotection prior to use. The Rink amine is far more convenient to load carboxylic acids onto the resins with standard peptide forming reactions. Cleavage of the linkage of the Rink linker resin, however, requires 20–80% TFA for 1–2 h to provide primary carboxamide peptide. Thus, a carboxamide linker with greater acid lability was investigated for the base labile Fmoc strategy.

Sieber developed the 9-xanthenyl linker **24a** which exhibited greater acid instability.[101] The amide linkages of the 9-xanthenyl linker can be cleaved with 2% TFA to provide primary carboxamide-peptides. Miscellaneous carboxamide linkers which can be cleaved with low concentrations of TFA were reported in the 1990s. However, no dramatic superiority of these linkers in comparison to the Sieber and Rink linkers has been reported to date.[102]

The 2-aminoacrylate linker **25** to generate peptide carboxamides was developed by Nisula and coworkers.[103] The Boc-AA-dehydroalanines were synthesized from the corresponding Boc-AA-O-tosylserine (AA: any amino acid) by β-elimination reaction during immobilization onto chloromethylated Merrifield resin with Et₃N in DMF at 40 °C. The 2-amideacrylate linkage was stable to a Boc deprotection condition (25% TFA). Thus, using the Boc peptide strategy oligopeptides were synthesized, and cleavage of the linkage was performed with 1 N HCl in AcOH at 50 °C in the presence of stoichiometric amount of water to result in the formation of peptide primary carboxamides and the pyruvate resin.

The use of trityl linkers for solid phase synthesis of amines is well established (Section 2.3.1.4). For the synthesis of biologically active peptide amides via the Fmoc strategy the tritylamine linker **23a** needs to enhance its acid lability. Voelter and coworkers systematically synthesized the methoxy substituted tritylamine linker resins and realized that the amides of dimethoxytritylamine linker **23b** dramatically increased the acid lability. The peptide amide could be cleaved within 30 minutes from the resin **23b** using 1% TFA/dichloromethane.[104]

Typical Experimental Procedures

Amino acids can be loaded onto the amine residue of polymer resin using standard amide bond formation methods. Most of commercially available carboxamide linkers are supplied in HCl salts or Fmoc protected forms. Thus, these resins must be neutralized with a 10% $^i Pr_2 NEt/DCM$ solution or be deprotected with

20% piperidine/DMF solution prior to using. Cleavages are typically carried out with 20% TFA/DCM, and a scavenger such as water is often added.

Loading procedures[119]

The resin is swollen in the solvent mixture of DCM/DMF (1:1) or DMF and treated with DIC (diisopropylcarbodiimide) or EDCI (1-ethyl-3-(3'-dimethylaminopropyl)carbodiimide, 3 equiv.), Fmoc-protected amino acid (3 equiv.), and HOBt or HOAt (3.3 equiv.). Microwave-assisted coupling of Fmoc-protected amino acid using DIC (3 equiv.) and HOBt (3 equiv.) in DMF at 60 °C provides the desired Fmoc-peptide resin in near quantitative yield at short reaction time.[119]

2.3.1.3 Alcohol linkers

Until the late 1990s only a few examples of the attachment of alcohols to polymer supports had been reported (Table 2.3). These included using trityl, Wang and Rink linkers to immobilize alcohols. These linkers are recognized as useful linkers for the synthesis of biopolymers; however, their utility in the solid organic synthesis of many other organic molecules (nonpeptide, nonoligosaccharide and non-nucleotide) is limited by the very acid sensitive characteristics of the ether linkages. On the other hand, the benzyl ether linkage formed on Merrifield resin is very difficult to cleave, under mild conditions. A number of protecting groups for alcohols have been demonstrated for solution-phase organic synthesis and have since been developed into a variety of linkers for the immobilization of alcohols onto polymer supports. To date, besides the application of some of the carboxylic acid linkers (vide supra), phenol ether, tetrahydropyranyl (THP) and other ketal protecting groups, silyl groups, p-methoxybenzyl (PMB) groups, succinyl groups and carbonates have all been adapted into linker units for SPOS.

The benzaldehyde Merrifield linker[120] was applied to immobilize 1,2- or 1,3-diols to form O-benzylidene acetals.[121] As well as being used in protecting diols in solution organic synthesis, a wide variety of aldehyde and ketone linkers can be devised based on acetal or ketal protecting groups. Although these linkers have not been widely used in solid-phase organic synthesis, several examples of the immobilization of carbohydrates have been reported.[122] Success of hydrolysis of the acetal linkers are dependent on the structure of the diol and conditions used. TFA or HCl in DCM in the presence of a nucleophile (i.e. water or alcohol) at room temperature (for acetal linkers) or at elevated temperatures (for ketal linkers) is the standard protocol for the regeneration of alcohols. In further applications, immobilization of a series of proteins or DNA on aldehyde-terminal polymers (i.e. agarose bead or glass slide) is now a fundamental method for making affinity purification resins or microarray analysis.[123]

Although the concept is analogous to the acetal (or ketal) linkers (Section 2.3.1.5), dioxolanes possessing carboxylic acid groups can easily be conjugated onto a wide variety of carboxamide linker resins in which cleavage and deprotection can be conducted in a single operation using an appropriate acid. For example, Wang and coworkers demonstrated the synthesis of a diamino alcohol library using levulinic acid (**27**) and 10-(vinyloxy)decanoic acid derivative **31**, and the 4-methylbenzhydrylamine (MBHA) linker resin **21**.[124] As illustrated in Scheme 2.2, levulinic acid was condensed with the diol building block **26** to form the 3-(2-methyl-1,3-dioxolan-2-yl)propanoic acid derivative **28**. After a protecting group manipulation (Cbz →Fmoc), the carboxylic acid group was conjugated with the MBHA linker resin **21**. Similarly, 3-(1,3-dioxolan-2-yl)decanoic acid derivative **32** was synthesized via an acid catalyzed ethylidene acetalization of the alcohol **30** with 10-(vinyloxy)decanoic acid derivative **31**. The ethylidene acetal-carboxylic acid was conjugated onto the MBHA linker resin. In their work, the automated solid-phase synthesis was carried out on a peptide synthesizer via the Fmoc strategy. Deprotections of a library of the diamino alcohol molecules

Table 2.3 Representative Alcohol Linkers

Linker	Suitable substrates	Attachment Conditions	Cleavage Conditions	Stable towards	Refs
(levulinic acid + aminobenzhydryl resin)	• Diols	• Preformation of the ketal with levulinic acid followed by amide forming reactions	95% TFA-water	Basic conditions	[124]
Trityl linker (X = Cl or Br)	• Alcohols • Carboxylic acids • Amines	• Bases (^{i}Pr$_2$NEt or DMAP/DCM) • AgOTf, 2,6-di-*tert*-butylpyridine/DCM, or NMP	2% TFA or HCO$_2$H	Basic condition	[125–129, 154]
Rink chloride linker	• Alcohols	• ^{i}Pr$_2$NEt	5% TFA	Basic condition	[130, 155, 156]
(methylimidazolium carbonate linker, *racemic)	• Alcohols	• DMAP	20–25% TFA	Bases, nucleophiles	[53]
DHP linker	• Alcohols	• PPTS/ClC$_2$H$_4$Cl, 80 °C	TFA-water (95:5)	Bases, nucleophiles	[131, 132, 157–159]
(chlorodimethylsilyl linker)	• Alcohols	• imidazole	Acids, TBAF		[134]

(*continued overleaf*)

Table 2.3 (continued)

Linker	Suitable substrates	Attachment Conditions	Cleavage Conditions	Stable towards	Refs
(aryl-Si(iPr)₂Cl on resin)	• Alcohols	• imidazole	Acids, TBAF	Nucleophiles, bases	[135, 136, 160–162]
(aryl-propyl-SiEt₂H on resin)	• Alcohols • Carbonyls	• Activation with Rh(PPh₃)₃Cl, or 1,3,-dichloro-5,5-dimethyl-hydantoin then imidazole	HF·pyridine, TBAF, AcOH-THF-H₂O (6/6/1)	Neutral conditions, weak bases, mild nucleophiles	[137–139, 163–166]
(aryl-butyl-Si(iPr)₂OTf on resin)	• Alcohols	• 2.6-Lutidine/DCM	HF·pyridine, TBAF	Neutral conditions, weak bases, mild nucleophiles	[140–143, 167–175]
(aryl-Si(Ph)(tBu)Cl on resin)	• Alcohols	• Imidazole/DCM	HF·pyridine, TBAF	Weak acidic conditions, nucleophiles	[145]
(aryl-CH(OH)- with Ph₂SiCl₂ or iPr₂SiCl₂)	• Alcohols	• Imidazole/DCM	acids, TBAF	Organic reactions with neutral conditions	[149–151, 176]
(aryl-CH₂-O-aryl-CH₂-O-C(=NH)CCl₃ on resin)	• Alcohols	• BF₃·OEt₂	10% TFA	Mimic of MPB protecting group	[152]
(aryl-CH₂-O-aryl-CH₂-O-C(=O)-S-pyridyl on resin)	• Alcohols	• AgOTf	10% TFA	Mimic of MPB protecting group	[153]

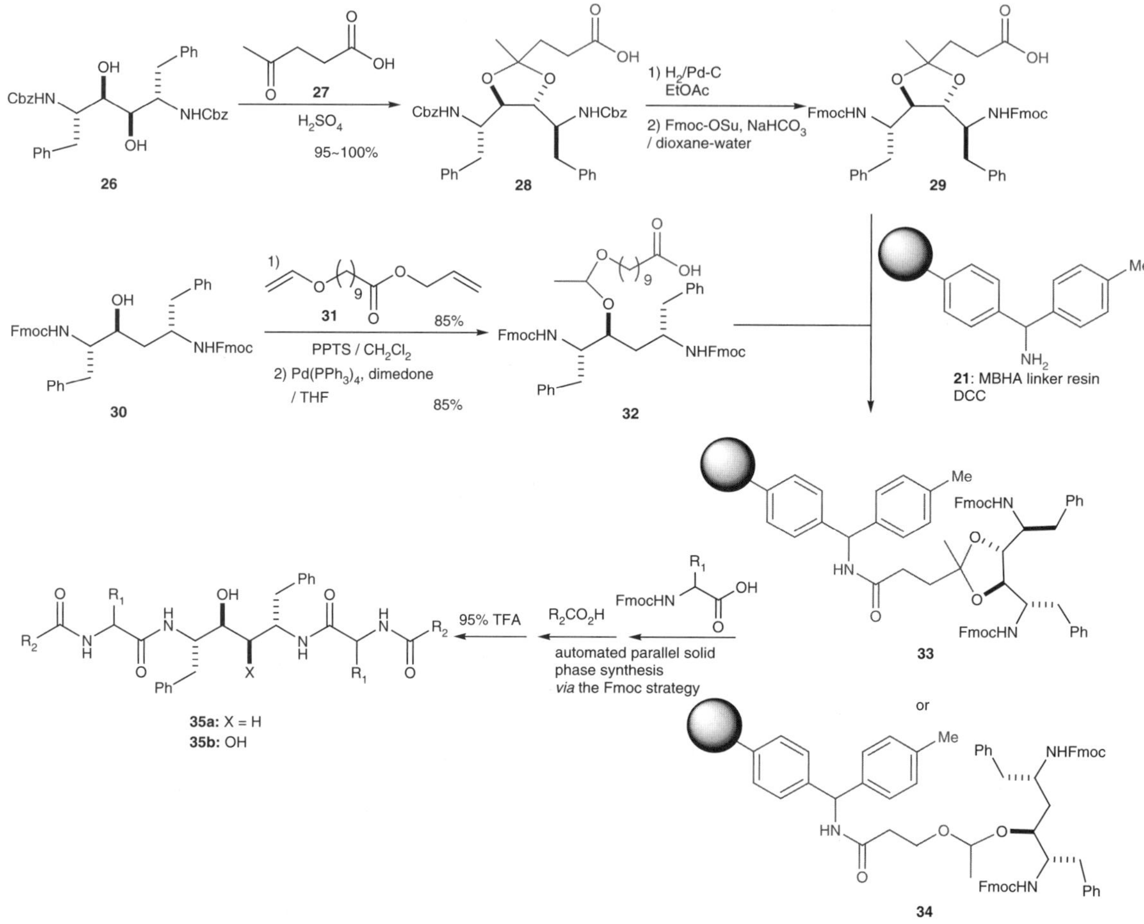

44 *Traditional Linker Units for Solid-Phase Organic Synthesis*

Scheme 2.2 *Synthesis of a diamino alcohol library*

synthesized from **33** and **34** were cleaved by the treatment with 95% aqueous TFA (for ethylidene acetals) or with 30% TFA in DCM (for 2,2-dialkyl-1,3-dioxolans).

The trityl linker **17c** is a very acid labile linker and a few applications of this linker to anchor alcohols for the synthesis of libraries of molecules have been reported.[125] For example, Chen and coworkers used the trityl linker to immobilize 1,4-butanediol (**36**) in pyridine (for 2 days at room temperature) and the remaining free alcohol of **37** was oxidized to generate the aldehyde. Horner-Emmons C2-unit extension of the aldehyde resin provided the α,β-unsaturated ketone resin **38**. Michael reactions of the resin **38** with arylthiolates **39** furnished the Michael adducts **40**. The trityl ether linkages were cleaved with formic acid to deliver the targeted formates in 7–24% overall yields (1994)[126] (Scheme 2.3). (4-Ethenylphenyl)diphenyl methanol can readily be synthesized. Thus, this monomer is used in the preparation of the trityl linker resins through a copolymerization reaction with styrene/divinylbenzene. Janda and coworkers successfully copolymerized (4-ethenylphenyl)diphenyl methanol with styrene and 1,4-bis(4-vinylphenoxy)butane to furnish *JandaJel*-trityl linker.[127] Conveniently, 4-(hydroxydiphenylmethyl)benzoic acid is commercially available and can functionalize a wide variety of aminomethylated resins to form trityl alcohol

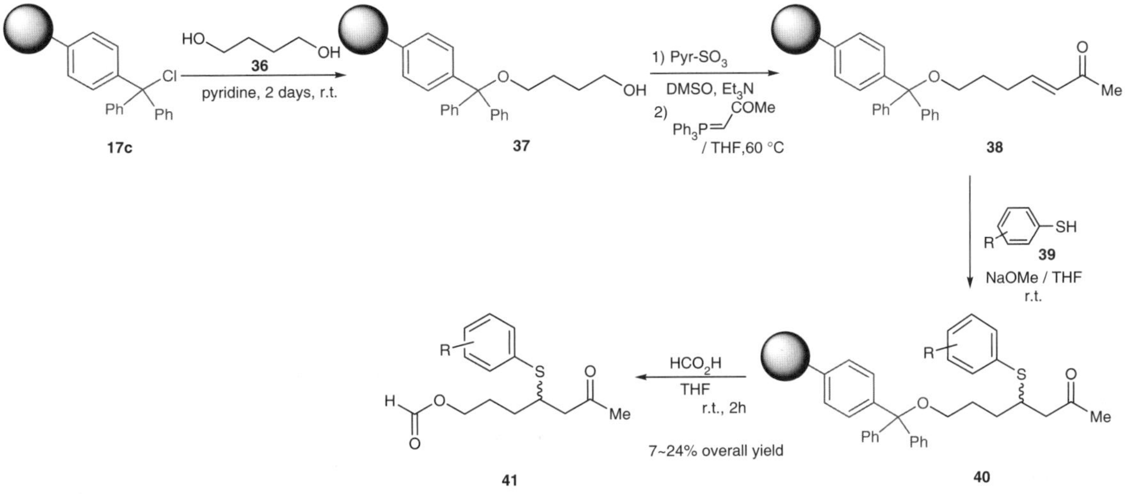

Scheme 2.3 *Solid-phase synthesis of formates*

linker resins. Recently, improved methods (using AgOTf or trityl bromide) for trityl ether formation were developed for loading alcohols on solid support.[128, 129]

Some of the carboxylic acid and carboxamide linkers can also apply to immobilize alcohols. For example the Rink carboxylic acid linker **16a** can readily be converted to the Rink chloride linker **16b** via a chlorination reaction with PPh$_3$ and hexachloroethane.[130] The Rink chloride linker **16b** has been demonstrated for loading alcohols, amines, thiols and carboxylic acids in the presence of *N*,*N*-diisopropylethylamine (DIPEA) in DCM. However, acid susceptibility of the Rink ether linkage (general cleavage conditions: 5% TFA in DCM) may limit the use of the Rink linker for performing a wide range of solid-phase organic reactions. On the other hand, the hydroxytetrachlorodiphenylmethy (HTPM) linker is stable to a variety of acids, bases and nucleophiles, but can conveniently be cleaved by solvolytic displacement reactions with 20–25% TFA. The utility of the HTPM linker **18a** for the immobilization of alcohols and phenols was demonstrated by the carbonate formation reactions of the HTPM-carbonylimidazolium salt **18b** (Scheme 2.4).[53] Thus, these examples indicate that diphenylmethanol linkers will be used in loading alcohols through the derivatization of the linker moiety to the chloride or carbamate derivative.

A hydroxy linker based on the tetrahydropyranyl (THP)-Merrifield linker **45a** was developed by Thompson and Ellmann in 1994.[131] Primary and secondary alcohols are readily reacted with dihydropyran in the presence of *p*-toluenesulfonic acid (PPTS). The hydroxyproline derivative **46** was demonstrated for loading onto and cleavage from the THP linker (**45a** →**48** in Scheme 2.5). The resulting THP protecting group is stable to bases, but can be cleaved with acids (PPTS in 1:1 butanol/ClCH$_2$CH$_2$Cl at 60 °C or 95% TFA). Basso and Ernst synthesized 2-((3,4-dihydro-2H-pyran-2-yl)methoxy)acetic acid and conjugated with aminomethyl SynPhase Lanterns (Figure 2.1). The THP linker Lanterns **45b** were used for the synthesis of a small library of hydroxyproline-based cyclic hexapeptides using the building block **49** in which intramolecular macrolactamization and simultaneous deprotections of the Boc and ^{t}Bu ester groups were performed on the polymer support. Cleavage of the linkages was carried out using 95% TFA. Overall yields for 13 steps were 42–82% (based on HPLC analyses) (**49** →**52** in Scheme 2.5).[132]

Silyl ethers are among the most frequently used protecting group for primary and secondary alcohols. The steric environment around the silicon atom and the protected molecular framework must be considered for the choice of silyl protecting group. In general, the stability of common silyl ethers against Brønstead

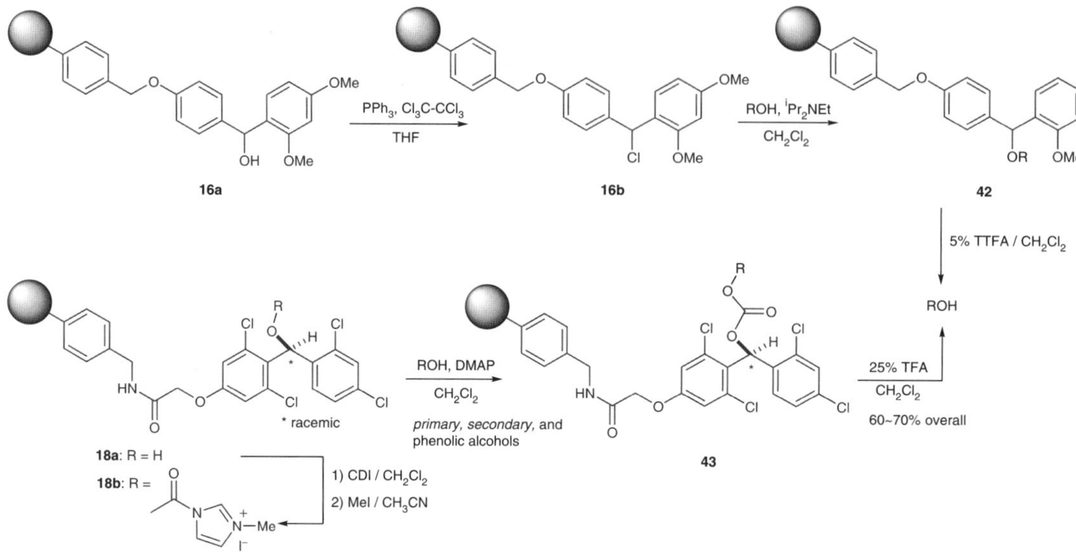

Scheme 2.4 *Loading onto and cleavage from the modified diphenylmethyl ether linker resins.*

acids is TMS < TES < TBDMS < TIPS < TBDPS and the stability against bases increases in the order TMS < TES < TBDMS ~ TBDPS < TIPS.[133] Chlorodimethyl(phenyl)silane linker **54a** was initially synthesized by Farrall and Fréchet in their studies of bromination and lithiation followed by the reaction with electrophiles.[134] They also reported the direct lithiation of cross-linked polystyrene resin with ^{n}BuLi-TMEDA in cyclohexane (65 °C, 4.5 h). Danishefsky and coworkers developed the phenyldiisopropylsilyl linker **54b** based on the reported procedures and demonstrated the synthesis of oligosaccharides on the polymer support in 1993. The acid stability of the diisopropylphenylsilyl linker **54b** over the triphenylsilyl linker **54c** was realized based on the superiority of overall isolated yield for the synthesis of the pentasaccalide **61** by using a zinc chloride ($ZnCl_2$) promoted glycosylation of the 1,2-anhydro derivatives, **55** and **59**; leakage of the molecule from the polymer support during each coupling stage was diminished by using the diisopropylphenylsilyl linker **54b**.[135, 136] The cleavage of the silyl linker can be carried out with tetrabutylammonium fluoride (TBAF), or buffered TBAF with acetic acid (40 °C, 12 h), or HF·pyridine complex (Scheme 2.6).

Silyl chloride linkers require careful consideration of storing without loss of active surface. Hydrosilanes are hydrolytically stable and have been used for dehydrogenative coupling with alcohols.[137] Hu and Porco developed polymer-supported trialkylsilane **63** to immobilize alcohols and carbonyls using Wilkinson's catalyst.[138] The synthesis of butyl diethylsilane was very reliable and amenable to a large scale reaction. Merrifield resin **2** was subjected to allylation reaction with allyl Grignard reagent and subsequent hydrosilylation of the pent-1-ene resin **62** with Rh(Ph$_3$)$_3$Cl (0.4 mol%).[139] The resin-bound silyl ethers could be formed by direct treatment with alcohol or carbonyl compounds in the presence of Rh(Ph$_3$)$_3$Cl. Cleavage of the silyl ethers was performed using HF·pyridine in THF or acetic acid/THF/water (6:6:1) to furnish (or regenerate) the corresponding alcohols. It was established that excess hydrogen fluoride can be quenched with TMSOCH$_3$ to disproportionate to volatile FSi(CH$_3$) and methanol. Alternatively, the trialkylsilane resin **63** was converted to the corresponding silyl chloride **65** with 1,3-dichloro-5,5-dimethylhydantoin (**64**) which can be used to load alcohols in the presence of imidazole in DCM (Scheme 2.7).

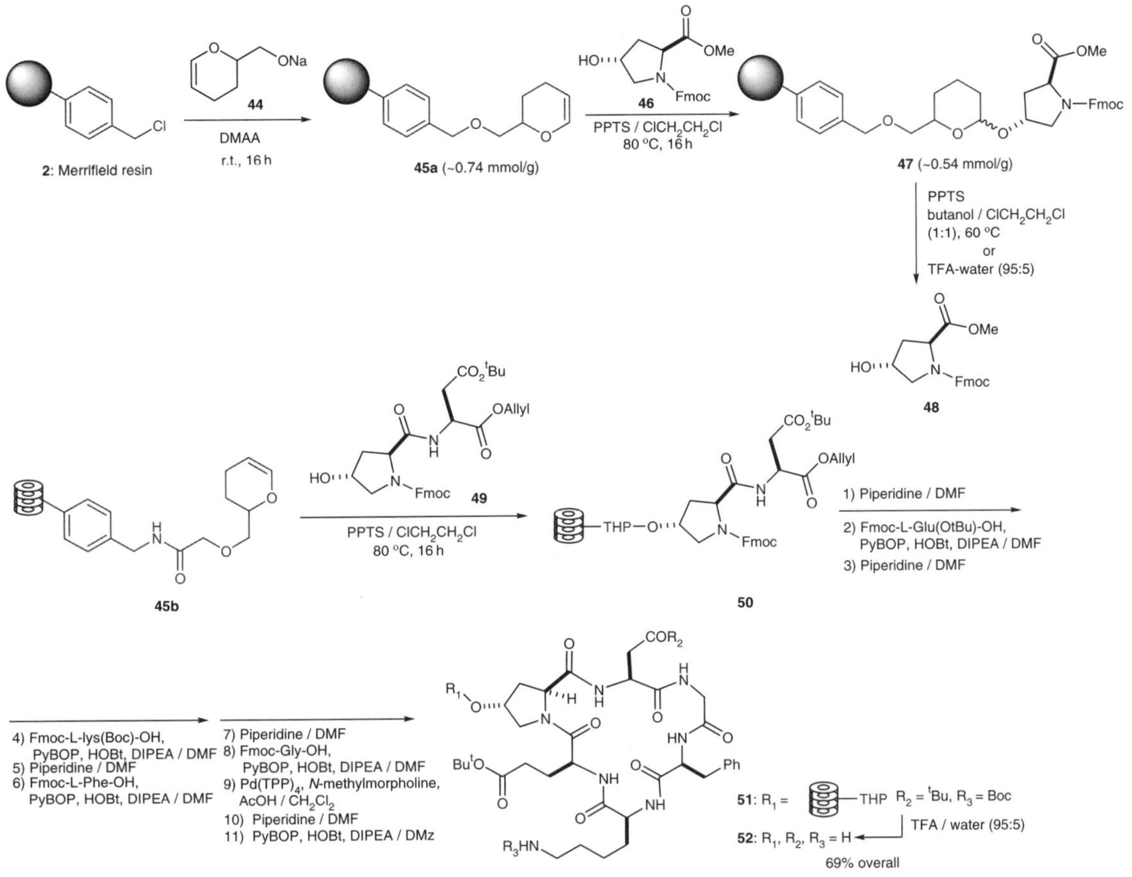

Scheme 2.5 *Synthesis of a small library of hydroxyproline-based cyclic hexapeptides*

Refinement of the modification of surface of DVB cross-linked polystyrene **1** to the alkyldiisopropylsilyl triflate linker **69** was reported in 2001.[140] Bromination of DVB cross-linked polystyrene **1** under thallium acetate (18 mol%) catalyzed electrophilic aromatic bromination in DCM (original conditions: 9 mol% Tl(OAc)$_3$ in carbon tetrachloride) was reproducibly and uniformly brominated to the bromo resin **53** (95–97% based on Br$_2$ used). The brominated DVB/polystyrene **53** was cross-coupled with the alkyl borane **67** via a Suzuki reaction (Pd(PPh$_3$)$_4$, NaOH) to furnish the latent alkyldiisopropylsilyl linker **68**. This was activated with TfOH within 20 minutes to afford the silyltriflate linker **69** [141, 142] which could immobilize a wide variety of primary and secondary alcohols with good loading yields in the presence of 2,6-lutidine (Scheme 2.8). The usefulness of the alkyldiisopropylsilyl linker has been demonstrated by the generations of large membered libraries of complex small molecules on DVB/polystyrene or Lanterns.[143] In library productions, HF·pyridine and TMSOMe (or TMSOEt) are used for cleaving and scavenging (excess hydrogen fluoride) conditions, respectively (Scheme 2.7).

By analogy to the *tert*-butyldiphenylsilyl (TBDPS) group for protecting alcohols in solution organic synthesis, the *tert*-butyldiphenylsilyl linker **73** was developed by Tan and coworkers in 2005.[144] The halogen metal exchange reaction was carried out by using an excess of ^{t}BuLi (5 equiv) and extensive washing of excess reagent. The generated phenyl lithium was reacted with *tert*-butylchloro(phenyl)silane[145] to afford

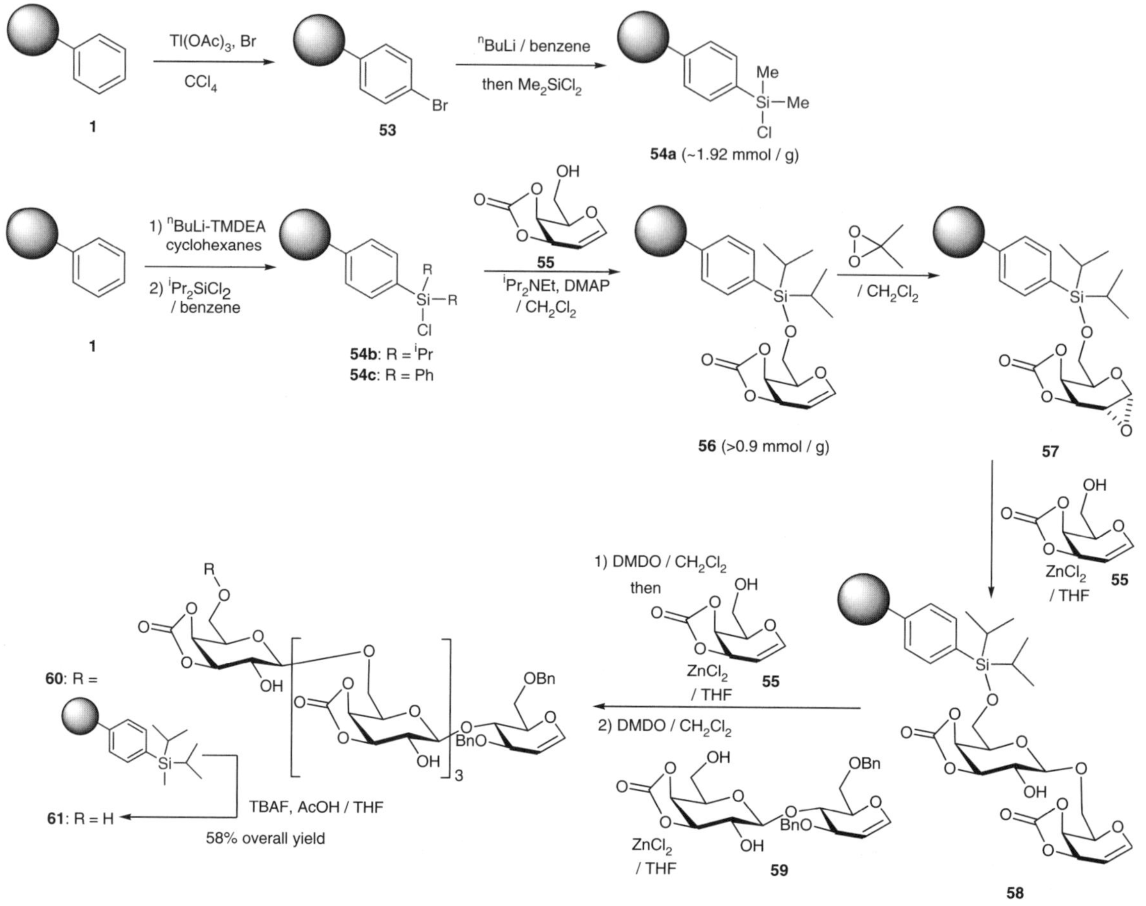

Scheme 2.6 *Synthesis of oligosaccharides on a polymer support*

the silyl hydride resin **72**. The silyl hydride group was activated via 1,3-dichloro-5,5-dimethylhydantoin (**64**) to provide the silyl chloride linker **73**; this transformation was achieved in 93–100% yield, being established by coupling of the linkers with *N*-Fmoc-β-alaninol and quantitation of subsequent release of Fmoc chromophore. Because of stereoelectronic characteristics of the *tert*-butyldiphenylsilyl (TBDPS) linker, the alcohols immobilized onto the TBDPS linker should be more stable against Brønstead and Lewis acids than the previously reported sily linkers. Cleavage of TBDPS-linked alcohols can effectively be cleaved by using a buffered HF·pyridine (HF·pyridine/pyridine/THF) at 35–55 °C for 16–70 h or TBAF or tris(dimethylamino)sulfur (triethylsilyl)difluoride (TASF).[146–148] The TBDPS linker has successfully demonstrated in several synthetic applications (i.e. acetal deprotection, Wittig and Julia couplings, and allylation reactions). Acetal deprotection and aldehyde olefination on the TBDPS linker (**73** →**77**) are illustrated in Scheme 2.9.

By introducing a bifunctional silyl group, alcohols can readily be immobilized onto hydroxy linker resins. Paterson and coworkers reported the acid labile diisopropyldialkoxysilyl (RO-Si-OR) linker for performing asymmetric aldol reactions with the α-chiral aldehydes on a solid support.[149] The hydroxy Merrifield resin **2b** was used to couple with the alkoxyldiisopropylsilylchloride to furnish the dialkoxysilyl

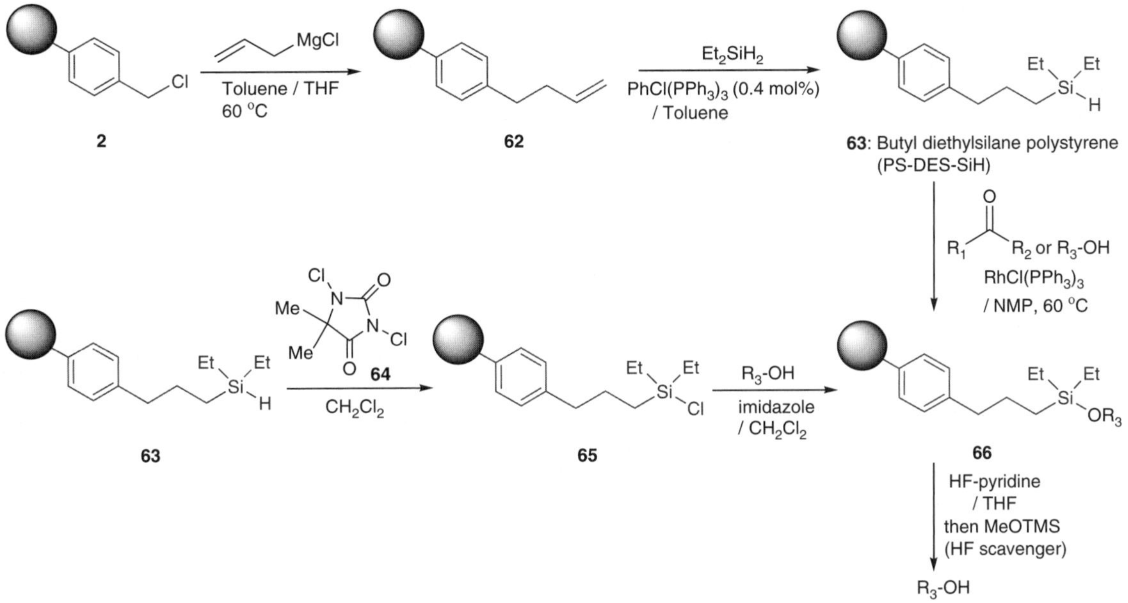

Scheme 2.7 Silyl linker groups for alcohols

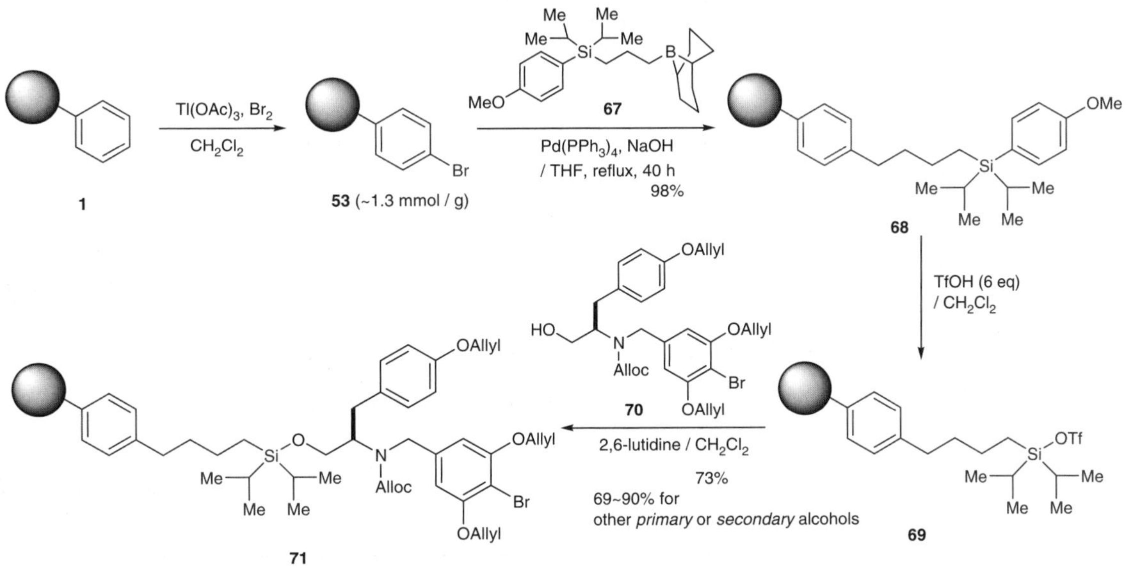

Scheme 2.8 Silyltriflate linker for immobilization of primary and secondary alcohols

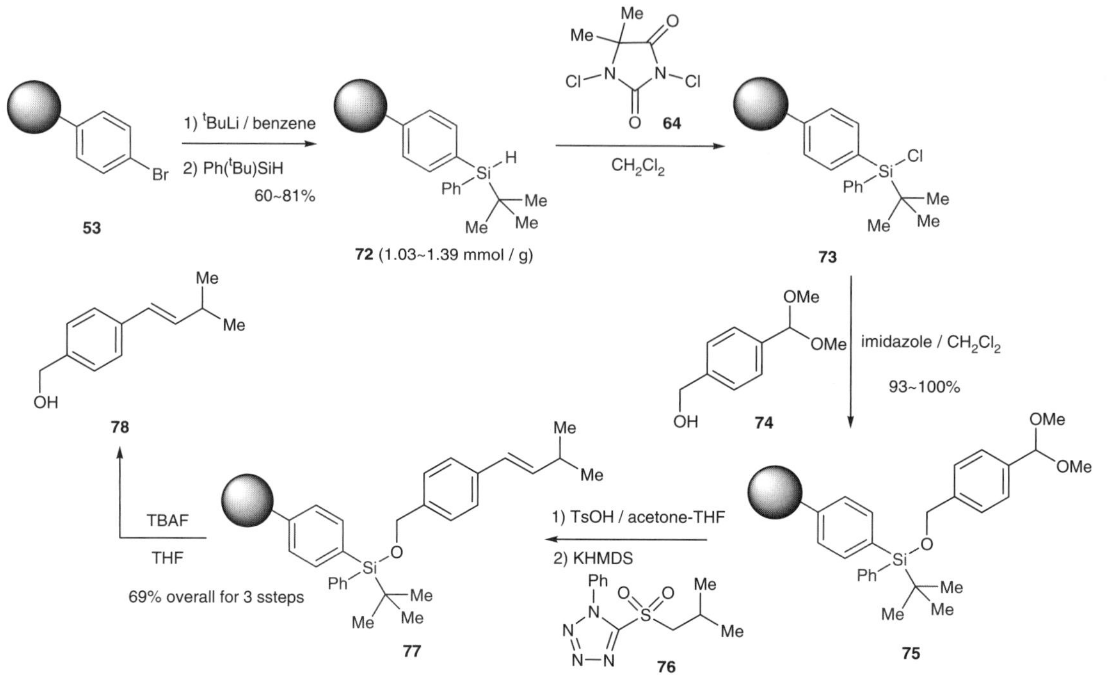

50 *Traditional Linker Units for Solid-Phase Organic Synthesis*

Scheme 2.9 *Silyl linker unit for benzyl alcohols*

resin **80**. The *p*-methoxybenzyl (PMB) ether of **80** was deprotected with DDQ and the resulting alcohol was oxidized with Dess–Martin periodinane. Reaction of the polymer-bound aldehyde **81** with the (*E*)-enol dicyclohexylborinate generated from **82** in Et_2O gave the *anti*-aldol product **83** after an oxidative work-up (hydrogen peroxide, methanol, DMF, pH 7 buffer).[150] A modified Evans–Tishchenko reaction of the aldol product **83** with a stoichiometric amount of samarium iodide (SmI_2) and acetaldehyde at 0 °C gave the polymer-bound acetate **84** with complete conversion after repeating the same reaction twice. Reductive removal of the acetate group with lithium borohydride ($LiBH_4$), acetonide formation, and cleavage of the linkage with a buffered HF·pyridine provided a mixture of the desired alcohols **85** in 47% yield for over seven steps (Scheme 2.10). Thus, stereo-controlled solid-phase synthesis of the tetraketide was demonstrated by using the silyl linker resin. Waldmann and coworkers accomplished the synthesis of the 6,6-spiroketal using the Wang linker resin (**13a** in Figure 2.2), in which asymmetric aldol reactions were used to synthesize the linear polyketide molecules. In their synthesis 2,3-dichloro-5,6-dicyano-1,4-benzoquinone (DDQ) was applied to cleave the *p*-alkoxybenzyl ether linkage (scheme not shown).[151]

According to *p*-methoxybenzyl (PMB) ether formation using an imidate method, Hanessian and coworkers modified the Wang linker **13a** to generate the trichloroacetimidate linker **86** which can immobilize primary, secondary and tertiary alcohols.[152] The formation of polymer-bound *p*-alkoxylbenzyl ethers from alcohols can be performed based on well established methods in solution organic chemistries; $BF_3 \cdot OEt_2$ or $TrClO_4$ are typically used for immobilization of alcohols. Cleavage of the polymer-bound *p*-alkoxylbenzyl ethers was achieved using 1–10% TFA in DCM (**88** → **89**). Similarly, Hannesian developed the 2-pyridylthiocarbonate linker **91** which can form a *p*-alkoxylbenzyl ether linkage with alcohols in the presence of AgOTf.[153] The *p*-alkoxylbenzyl linker resins are amenable to a wide range of organic reactions such as copper mediated alkylation, hydroxylation, oxidation, reduction, Witting reaction and

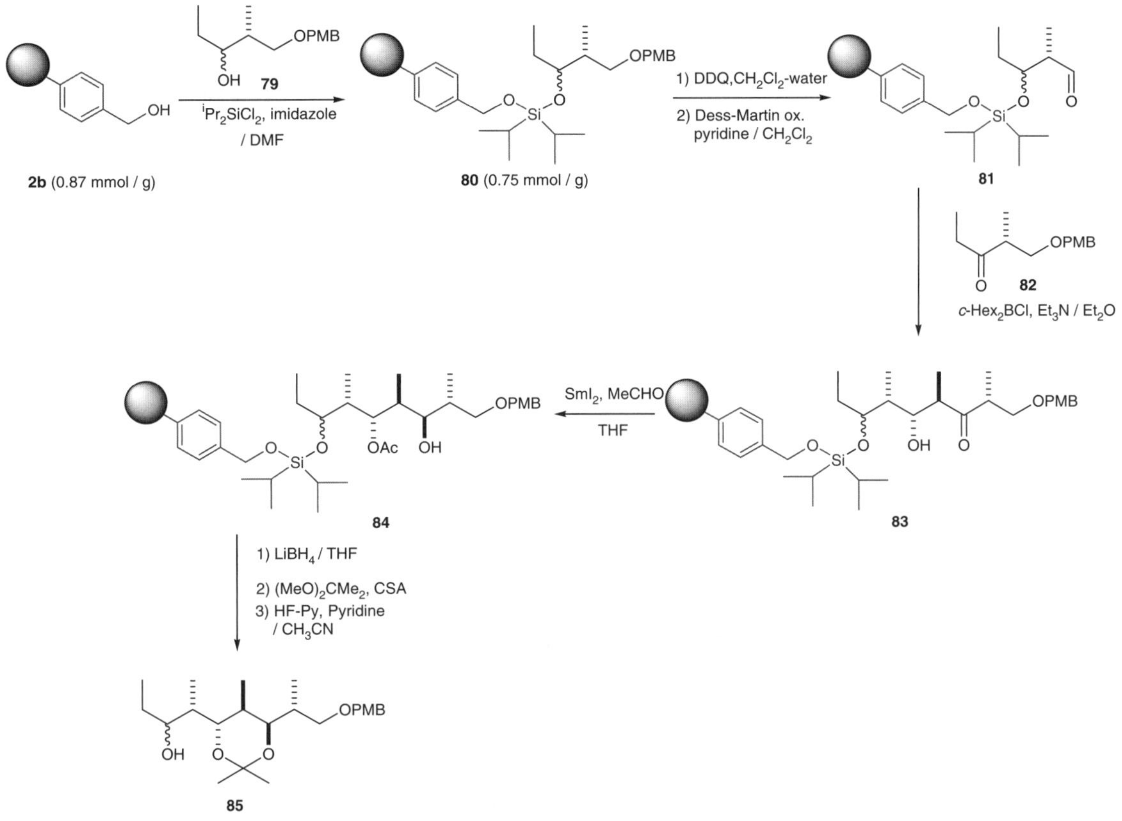

Scheme 2.10 Dialkoxysilyl linker units for alcohols

hydroxy group protection. A representative example demonstrated using the *p*-alkoxylbenzyl ether linker is illustrated in Scheme 2.11 (**91** →**95**).

2.3.1.4 Amine linkers

A number of linker units for attachment and release of amines have been developed; these are summarized in Table 2.4. For example, the bulky triphenylmethyl (trityl) group has been used to protect a variety of amines including amino acids in solution.[177, 178] Trityl linker resins have been described for immobilization of carboxylic acids (Section 2.3.1.1), carboxyamide (Section 2.3.1.2), and alcohols (Section 2.3.1.3). *N*-Trityl amines show better relative stability against acids than *O*-trityl ethers and carboxylic acid esters. In addition, the convenient loading of amino acids and the mild enough cleavage conditions make these resins very attractive for the synthesis of carboxamides and small molecules containing amines. As discussed in Section 2.3.1.1, the chlorotrityl linker is more stable against acids than the trityl resin itself. The usefulness of the 2-chlorotrityl resin was demonstrated for the generation of a focused library of peptidomimetics by Hoekstra.[179] The synthesis of peptidomimetics which contain a thiazole-constrained Leu-xxx (where xxx = any amino acid) core unit is summarized in Scheme 2.12. Saponification of the resin bound *N*-methylglycine ester **97** was achieved by using potassium trimethylsilanolate (KOTMS) in THF to provide the corresponding carboxylic acid. The thiazole containing amino acid **98** was coupled with

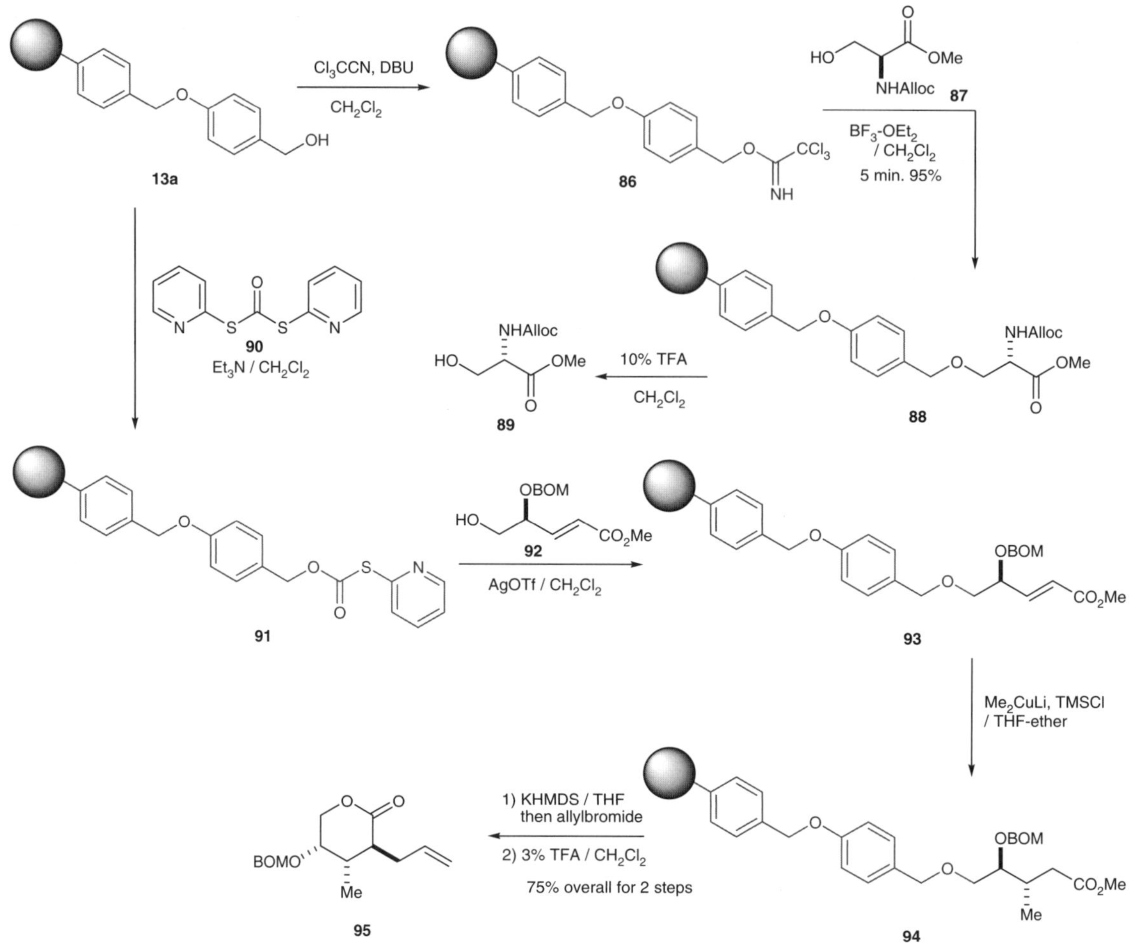

52   Traditional Linker Units for Solid-Phase Organic Synthesis

Scheme 2.11   *p-Alkoxylbenzyl linker unit*

the carboxylic acid resin by using EDCI and *N*-methylmorpholine (NMM) to furnish the pseudotripeptide **99**. The trityl linkage was cleaved after saponification by using 50% TFA to yield the desired product **101**.

The 9-phenylfluoren-9-yl group (PHFl) has been used for the protection of primary and secondary amines as an acid stable protecting group which is more acid stable than the trityl group.[180–182] Bleicher developed phenylfluorenyl linkers **103** and established cleavage conditions to regenerate amines.[183] The phenylfluorenyl linker **103** was synthesized by coupling reaction of 5-(4-(9-hydroxy-9H-fluoren-9-yl)phenoxy)pentanoic acid (**102**), which was synthesized from 9-fluorenone in four steps, and aminomethylated-PS **2c**. The alcohol resin **103** was converted to the phenylfluorenylchloride **103b** upon treatment with acetyl chloride, and **103b** can be used to link amines and carboxylic acids. As summarized in Scheme 2.13, an inverse peptide synthesis was achieved in excellent overall yield. L-Phe-OAlly was loaded in the presence of ^{i}Pr$_2$NEt. A palladium mediated deallylation and subsequent ligation with L-Phe-OAlly furnished the dipeptide resin **105**. Cleavage of the linkage was achieved by treatment with 50% TFA in the presence of Et$_3$SiH in DCM/MeOH (9:1) to yield the C-protected dipeptide with over 92% yield from **103a**.

Table 2.4 Representative Amine Linkers

Linker	Suitable substrates	Attachment Conditions	Cleavage Conditions	Stable towards	Refs
R = H or Cl	• *Primary* or *secondary* amines, or alcohols, or carboxylic acid	• With or without ^{i}Pr$_2$NEt	TFA-CH$_2$Cl$_2$ (50:50). Addition of scavenger is recommended.	Basic and neutral conditions	[179] Table 1 and 2
(fluorenyl linker)	• *Primary* or *secondary* amines, or carboxylic acid	• With or without ^{i}Pr$_2$NEt	TFA-CH$_2$Cl$_2$ (50:50) in the presence of scavenger (i.e. Et$_3$SiH)	Basic and neutral conditions	[180–183, 195]
(dimethoxybenzaldehyde linker)	• *Primary* or *secondary* amines	• Reductive amination in the presence of NaCNBH$_3$	TFA-water (95:5)	Basic and neutral conditions	[184, 185, 187–189, 196–213]
X = H or OMe (**AMEBA linker**)	• *Primary* or *secondary* amines	• Reductive amination reactions	TFA in the presence of scavenger (e.g. Et$_3$SiH)	Basic and neutral conditions	[190, 214–218]
(sulfinamide linker)	aldehydes	• Ti(OEt)$_4$, then alkylations to form sulfonamides	0.67 N HCl	Basic and neutral condition	[191]
(diazonium linker)	• *Primary* or *secondary* amines	• In the presence of *tertiary* amine (e.g. NEt$_3$)	10–50% TFA	Basic and neutral conditions	[192]
(diazonium linker)	• *Primary* or *secondary* amines	• In the presence of *tertiary* amine (e.g. NEt$_3$)	10–50% TFA	Basic and neutral conditions	[193]

(*continued overleaf*)

Table 2.4 (continued)

Linker	Suitable substrates	Attachment Conditions	Cleavage Conditions	Stable towards	Refs
(4-nitrobenzoate ester of Wang-type linker)	• Primary or secondary amines	• In the presence of $^{i}Pr_2NEt$	25% TFA	Basic and neutral conditions	[194]
(imidazole carbamate of trichloro-diphenylmethyl linker, racemic)	• Primary or secondary amines	• In the presence of DMAP	20–25% TFA	Basic conditions and a wide range of nucleophiles	[53]
(4-formylphenoxy acetamide linker)	• Diols	• $CH(OMe)_3$, TsOH	15% TFA, 1% water/ CH_2Cl_2 or acetic acid 40 °C	Basic and neutral conditions	[219]

The aminomethyl-3,5-dimethoxyphenoxy linker (widely called PAL (peptide amide linker), the highlighted moiety in **109** in Scheme 2.14), which can be prepared by reductive amination of 2,6-dimethoxy-4-alkoxybenzaldehyde linker **107** (widely called BAL (backbone amide linker)), was originally developed as an acid labile linker for the solid-phase synthesis of C-terminal peptide amide using the Fmoc strategy see, Section 2.3.1.1 and 2.2.1.2.[184, 185] Although C-terminal modified peptides, heterocycles and other small organic molecules have been synthesized through peptide bond anchoring using the PAL, syntheses of primary or secondary amine derivatives using the BAL was popularized by Albericio and Barany.[184] The BAL functional moiety can be introduced to a wide variety of aminomethylated resins by ligation reactions with 5-(4-formyl-3,5-dimethoxyphenoxy)pentanoic acid (i.e. **2c** →**107**). Primary or secondary amine building blocks can be incorporated onto the BAL resin via a reductive amination ($NaBH_3CN$ in DMF-AcOH). A representative example of synthesis of secondary amines **110** using the BAL linker is illustrated in Scheme 2.14.[186] Although the backbone linker strategy offers a direct way to prepare a variety of C-terminal modified peptides or amines, one major disadvantage of using these linkers is the requirement of neat or very high concentrations of TFA at room temperature to elevated temperature for

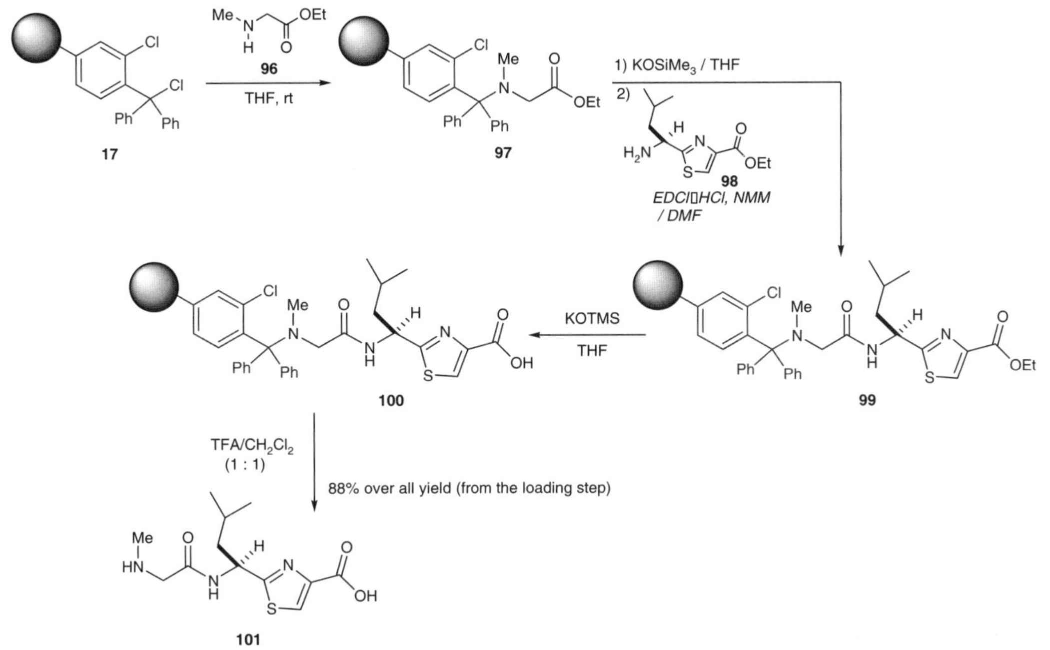

Electrophile Cleavable Linker Units

Scheme 2.12 Trityl linker unit for amines

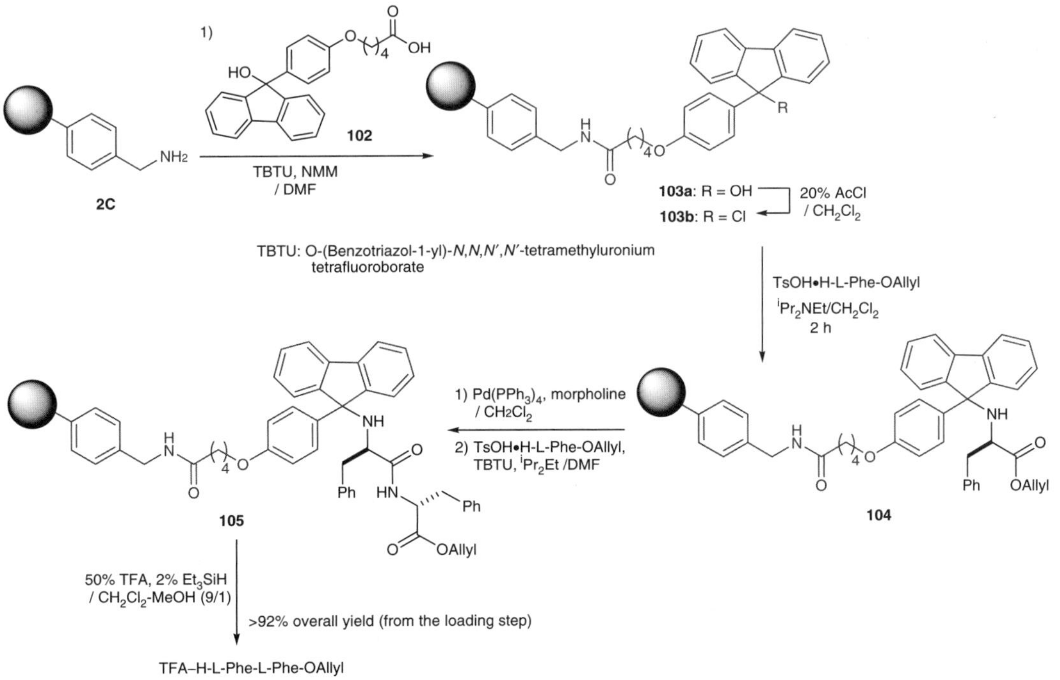

Scheme 2.13 Phenylfluorenyl linkers for amines

TBTU: O-(Benzotriazol-1-yl)-N,N,N',N'-tetramethyluronium tetrafluoroborate

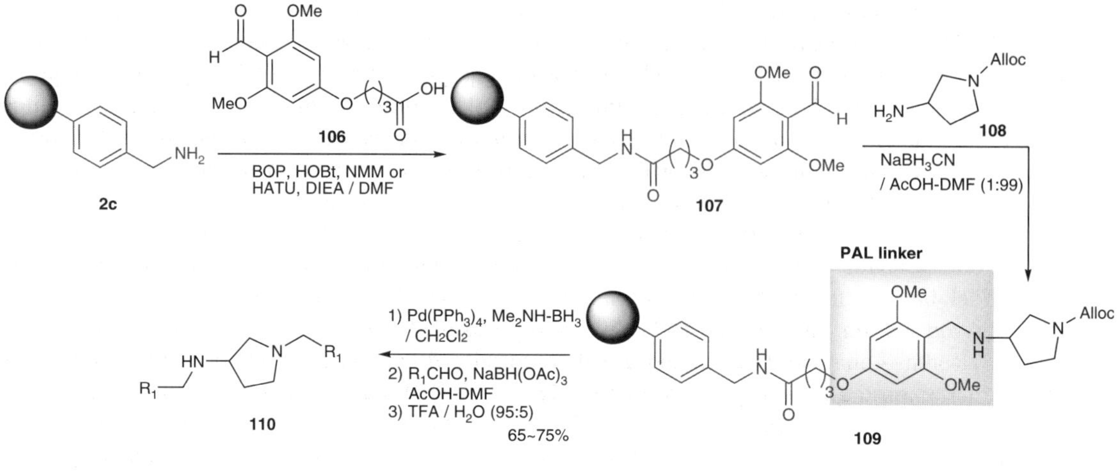

Scheme 2.14 PAL (peptide amide linker)

cleaving molecules from the polymer support.[187] Recently, Boas and coworkers developed a new class of acid labile BAL linker based on 3,4-ethylenedioxythiophene (T-BAL). The substrates could be released from the T-BAL resin by using 1–50% TFA (scheme not shown).[189, 190]

Thus, secondary amines can be loaded using the benzaldehyde functionality of polymer resin through reductive amination. Swayze reported a reliable method for the conjugation of *p*-hydroxybenzaldehyde (**111a**) and 4-hydroxy-2-methoxybenzaldehyde (**111b**) to ArgoGel™-OH **4** with the sulfonamide betaine **112**[197] in DCM. The aldehyde functionality was reductively aminated with cyclohexylamine (**114**) to form the secondary amine linker conjugate **115**. Further functionalization via amide formation reaction

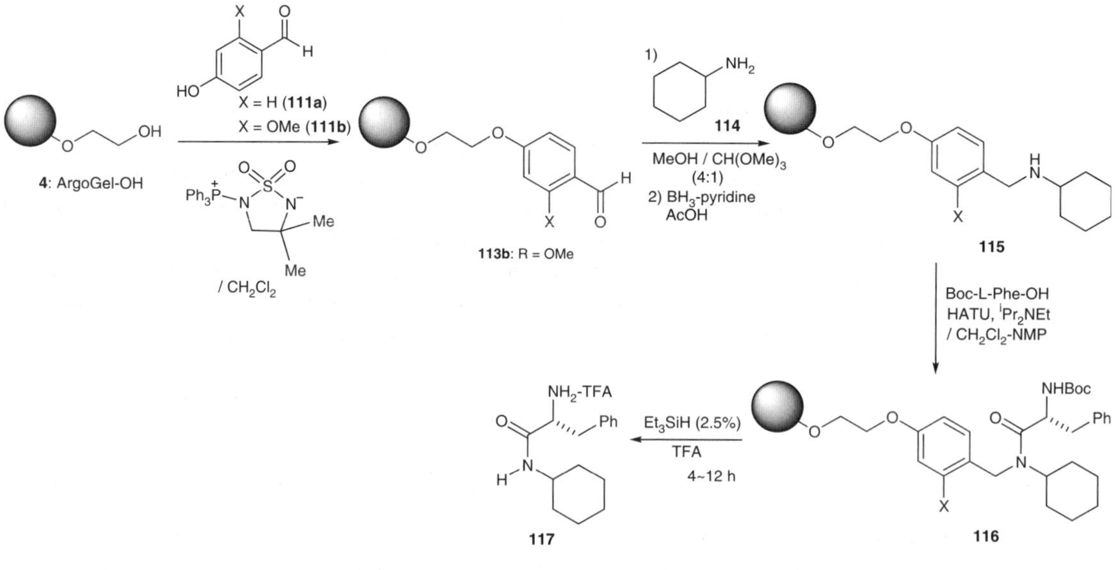

Scheme 2.15 Aldehyde linkers for loading amines via reductive amination

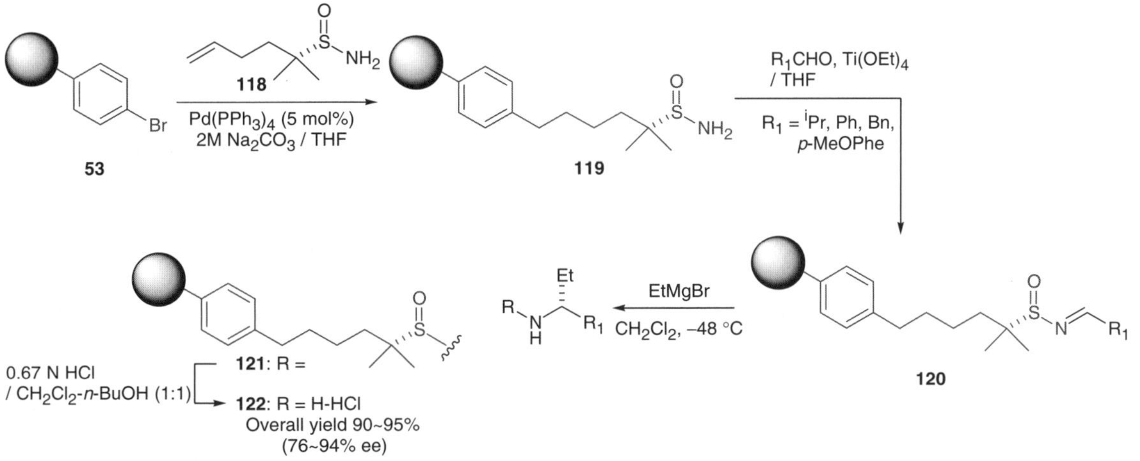

Scheme 2.16 Polystyrene-supported chiral tert-butanesulfinamide linker

with Boc-L-Phe-OH followed by cleavage of the molecule from the resin **116** with TFA in the presence of Et$_3$SiH (as a cation scavenger) provided the secondary amide **117** in good overall yield (Scheme 2.15).

Ellman and coworkers reported the synthesis of polystyrene-supported chiral *tert*-butanesulfinamide derivative **119** and demonstrated its utility in the synthesis of optically active chiral amines (2001).[191] The intermediate imines **120** were synthesized with aldehydes and Ti(OEt)$_4$. Addition of ethylmagnesium bromide to the polymer-supported imines in DCM/THF provided the desired α-branched amines **121** with 76–94% diastereomeric ratios. Cleavage of amines from the linker resins was achieved using hydrogen chloride in DCM/nBuOH to give amine hydrochlorides **122** (Scheme 2.16).

Aliphatic amine can be immobilized by using a diazonium salt. Reaction of various amines with the diazonium salt linker gave rise to the triazenes **126**, which can be cleaved by the treatment with 10–50% TFA.[192] Although the aniline linker can conveniently be prepared by grafting the functional molecules onto the Merrifield linker resin **2a**, a straightforward preparation of the aniline linker **128** was achieved via a Schmidt rearrangement of azidomethyl-polystyrene. Similarly, the triazine molecules were demonstrated to synthesize on the aniline–polystyrene resin.[193] The synthesis of piperazine amide **127** was achieved in three steps with the linker **125** (Scheme 2.17).

Carbamate linkers can be used in immobilizing amines. Carbamate linkers are prepared with acid cleavable carboxylic acid linkers (Section 2.3.1.1) and carbonylating reagents (i.e. carbonyldiimidazole (CDI), 4-nitrophenyl chloroformate (**128**), N,N'-disuccinimidyl carbonate and bis(4-nitrophenyl) carbonate). For examples, Wang linker **13a** is converted to the *p*-nitrophenyl carbonate linker resin **129** which can immobilize primary and secondary amines to form the corresponding carbamates **130**. The Wang carbamate linkers are cleaved with 25% TFA.[194] Similarly, the hydroxytetrachlorodiphenylmethy (HTPM) linker **18** was demonstrated for the immobilization of a variety of amines with carbonyldiimidazole (Scheme 2.18).[53]

2.3.1.5 Ketal (or acetal) linkers

The linkers possessing the diol and dithiane functionality were developed to immobilize aldehydes or ketones (Table 2.5). Because cleavage of ketal linkages on resins requires a harsh condition and is dependent on the structure of carbonyls, these linkers have not been widely used in solid-phase organic reactions. However, because diols and dithianes are good chelators of transition metals, the linkers possessing the bidentate

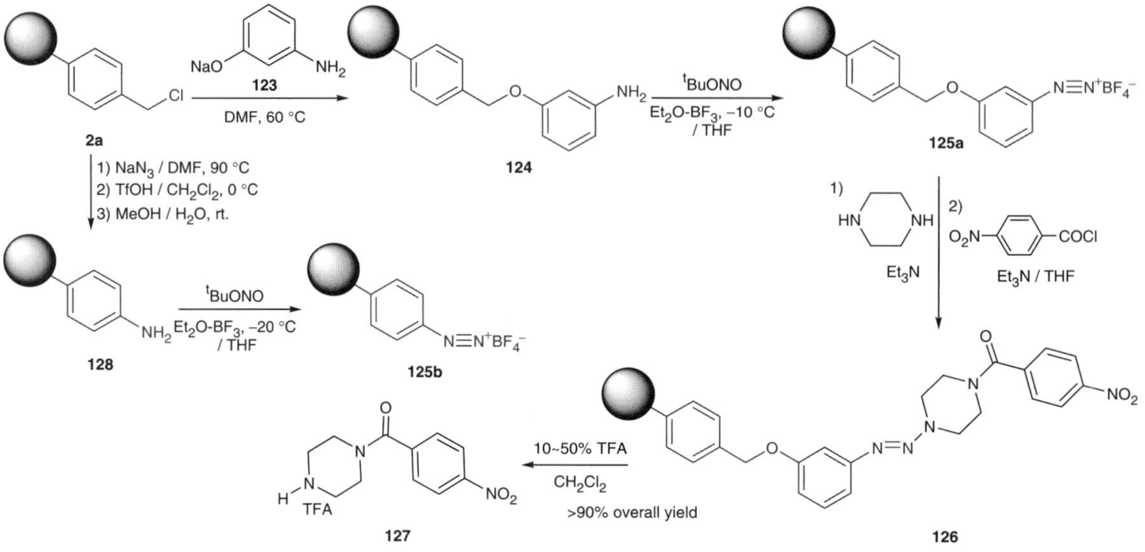

58 *Traditional Linker Units for Solid-Phase Organic Synthesis*

Scheme 2.17 *Diazonium salt linker*

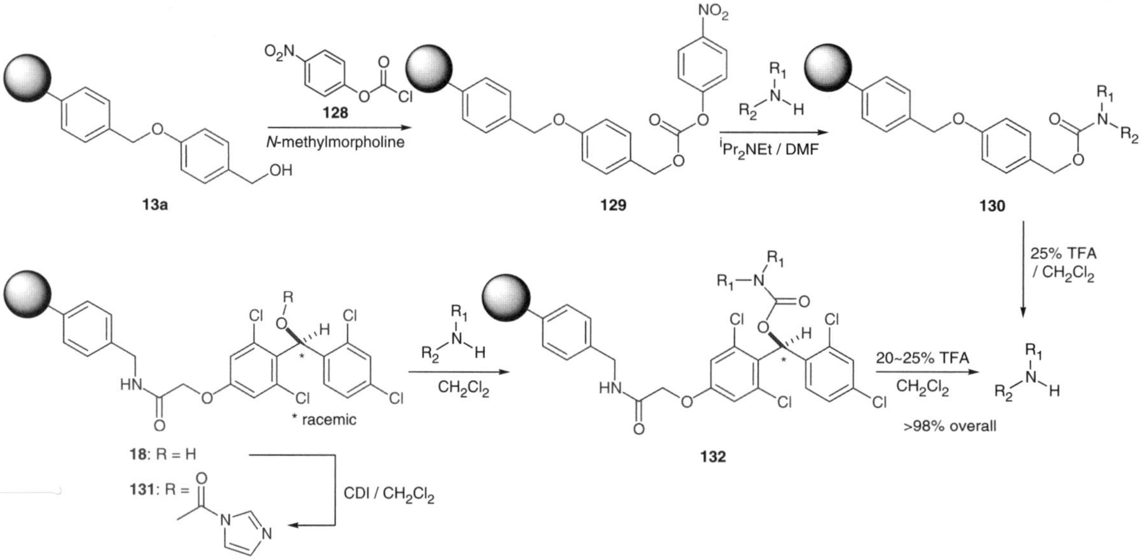

Scheme 2.18 *Carbamate linkers for immobilizing amines*

functionalities have been widely applied to immobilize reagents for organic reactions or to scavenge (or capture) organometallic by-products in solution-phase organic synthesis. Specific examples of immobilizations of ketal or acetal molecules and applications of ketal linkers have been extracted from the literature.

O-Alkyl glycosides are considered as mixed acetals. Ogawa and coworkers achieved the synthesis of oligosaccharide using the Merrifield resin **2a**.[220, 221] The *p*-hydroxybenzyl lactoside **133** was immobilized onto the Merrifield linker resin to provide the Wang linker lactoside **134**. The iterative β-selective

Table 2.5 Representative Ketal Linkers

Linker	Suitable substrates	Attachment Conditions	Cleavage Conditions	Stable towards	Refs
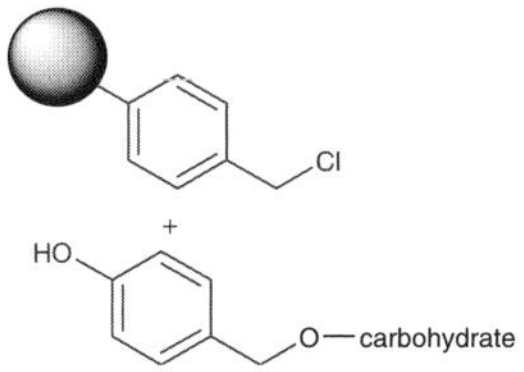 (resin-C₆H₄-CH₂Cl + HO-C₆H₄-CH₂-O-carbohydrate)	carbohydrates	Cs_2CO_3/ DMF, 50 °C	$TrBF_4$	Conditions used in carbohydrate chemistries	[220, 221], Table 2.1
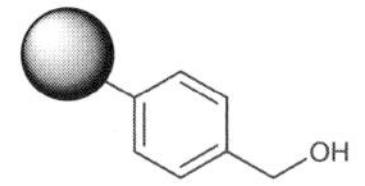 (resin-C₆H₄-CH₂OH) or other alcohol linkers	Ketal or acetal molecules	PPTS or CSA or Lewis acids	Acid catalyzed hydrolysis	Basic and neutral conditions	[222], Table 2.1

lactosylations via Schmidt glycosylations of **135** onto the lactoside resin provided the octasaccharide **136** which was cleaved with $TrBF_4$ in DCM. Subsequent purification of the crude product afforded the anomeric free octasaccharide **137** (Scheme 2.19).

Nicolaou and coworkers reported the generation of a library of sarcodictyin analogs using the hemiacetal linker **139** in which the hemiacetal moiety of the molecule was used in the formation a ketal with the alcohol linker.[222] Hydrolysis, esterification, desilylation, oxidation and esterification reaction were conducted on the polymer support. The molecules were cleaved from the resin **141** by the treatment with

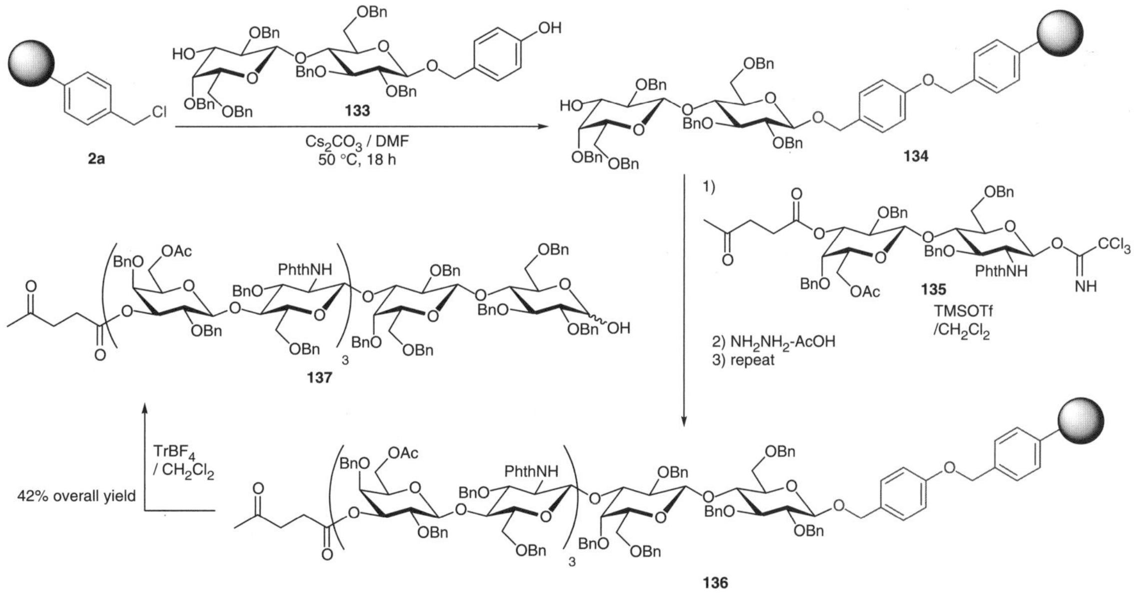

Scheme 2.19 Wang linker lactoside

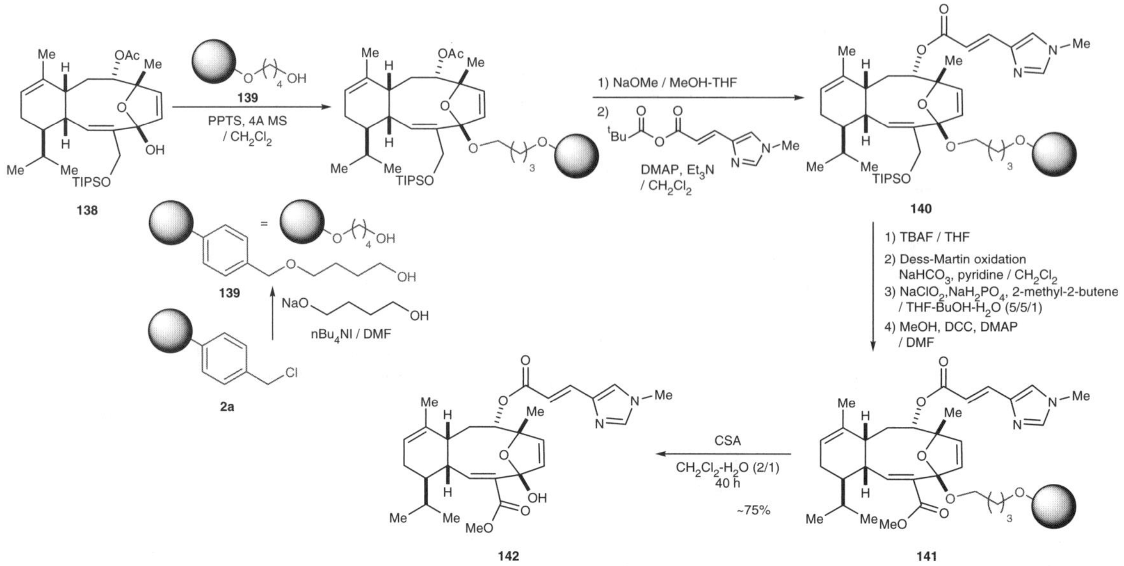

Scheme 2.20 Generation of a library of sarcodictyin analogs using a hemiacetal linker

10-camphoresulfonic acid (CSA) (3 equiv.) in DCM−H_2O (2:1) to give rise to the hemiacetal **142** in 42–48% overall yield (Scheme 2.20).

Waldmann and coworkers demonstrated an enantioselctive solid-phase synthesis of α,β-unsaturated δ-lactone **147**. The bromo-Wang linker resin **13c** was treated with the sodium 3-oxoprop-1-en-1-olate **143** to provide the polymer-bound α,β-unsaturated aldehyde (loading: 1.0–1.6 mmol/g), which was transformed into the butadine resin **144** by treatment with the Wittig reagent. An enantioselctive selective hetero Diels–Alder reaction with ethyl glyoxylate was performed via the (R)-BINOL-Ti complex, which was originally developed by Mikami.[223] The desired cycloadduct was obtained with 90–95% ee. These reactions also proceeded well for the 1-alkoxypenta-1,3-dine-polymer conjugate. To date, Diels–Alder or hetero Diels–Alder reactions on polymer supports have been investigated by a number of research groups.[224, 225] The ethyl ester of **145** was saponified and the resulting acid was converted to amides and other biologically interesting functional groups. Release of the molecule and subsequent oxidation of the lactol to the corresponding lactones **147** were effectively achieved in a single step by treatment with the Jones' reagent (chromium trioxide, sulfuric acid) (Scheme 2.21).[226]

2.3.1.6 Carbonyl linkers

Carbonyl linkers are especially important to generate peptide C-terminal aldehydes, peptidyl ketones and peptidomimetics (or amino acid mimetics). Semicarbazides, semicarbazone and oximes are hydrolytically more stable than imines. Thus, immobilized carbonyls using these linkers are amenable to mild organic reactions.

Webb and coworkers developed a semicarbazide linker **151** to generate C-terminal peptide aldehydes in 1992.[227] In the synthesis of the linker, the hydrazinecarboxamide **148** was first condensed with the amino aldehyde **149** to provide the corresponding semicarbazone **150**, which was then coupled with the

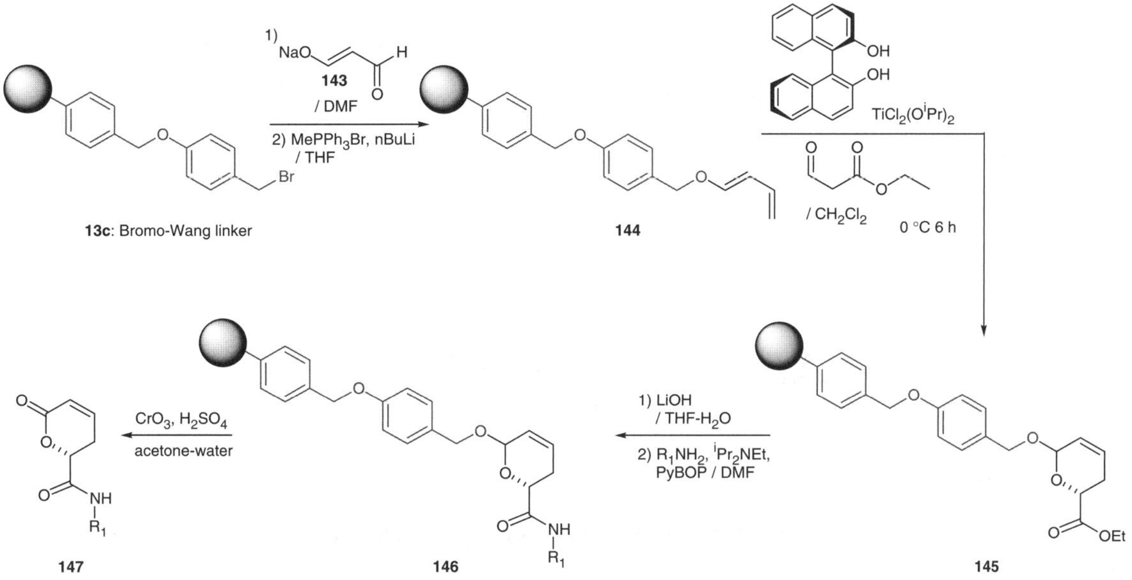

Scheme 2.21 *Enantioselctive solid-phase synthesis of α, β-unsaturated δ-lactones*

MBHA linker **21**. Standard automated Boc protocols were applied to the synthesis of oligopeptides. Peptide C-terminal aldehydes could be generated by hydrolysis of the semicarbazone linkage of **152** (Scheme 2.22). Webb's semicarbazide linker can be applied to immobilize trifluoromethyl ketones; Paupart and coworkers delivered a library of peptidyl trifluoromethyl ketones by using this linker.[228]

Ellman and coworkers devised the hydrazinecarboxylate linker to immobilize chloromethyl ketones and generated the amidomethyl ketones (i.e **159**) (1999).[229] The hydrazino group was introduced onto the ArgoGel-OH **4** (Figure 2.1) by using carbonyldiimidazole (**4** →**154**). The hydrazone formation with the α-chloromethyl ketone **155** was completed within four hours at room temperature. No racemization at the α-chiral center was observed during the hydrazone formation. A representative sequence for the synthesis of the amidomethyl molecule (**156** →**159**) is illustrated in Scheme 2.23.

Ley and coworkers demonstrated the synthesis of substituted spiroketals via the ethylene glycol linker **162**.[230] Polyethylene glycol (PEG) resin **160**[231, 232] was used as a soluble polymeric support. They developed a new PEG-5000 supported ketal linker **162** for immobilizing the ketones. One end of the PEG (the other end capped as a methyl ether) was alkylated with allyl bromide to afford **161**. The terminal olefin was then osmylated under the Upjohn's conditions to provide the ketal–PEG linker **162**. Loading the ketones onto the ketal linker **162** was performed via Dowex 1 × 2 − 100 in benzene at reflux in which generated water was trapped with a Dean–Stark trap. Under these conditions undeca-1,10-dien-6-one (**163**) could be immobilized in 91% yield. The terminal double bonds of **164** were epoxidized with *m*-chloroperbenzoic acid (MCPBA) followed by opening of the bisepoxide with sodium benzenethiolate afforded the desired thio ether **165** in 95% yield. Cleavage of the molecule and concomitant spiroketalization were conducted with the same acidic ion exchange used for the formation of the ketal in acetone to give rise to a mixture of spiroketals **166** and **167** in greater than 70% yield (Scheme 2.24).

Barrett and coworkers reported solid-phase synthesis of a series of isoxazoles using the vinyl ether resin (2001).[233] They previously demonstrated that esters supported on the Merrifield resin **168** (or Wang

Table 2.6 Representative Carbonyl Linkers

Linker	Suitable substrates	Attachment Conditions	Cleavage Conditions	Stable towards	Refs
PEG-OH	• carbonyls	Dowex1x2-100/benzene, reflux	Dowex1x2-100/wet acetone	Bases, nucleophiles	[231]
Webb linker	• aminals	Peptide forming reactions (e.g. BOP)	THF-H_2O-HCHO	Boc peptide strategy	[228, 229]
ArgoGel-OH	• α-chloro-carbonyls	• in THF	THF-H_2O-CH_3CHO-TFE	Neutral conditions	[230]
	• ketone oximes	^{t}BuOK/DMF	THF-water (4/1)	Boc peptide strategy	[236]

linker resin) underwent methylation reaction using the Tebbe reagent **169** to produce polymer-supported vinyl ethers.[234] The Suzuki–Miyaura coupling reactions on the vinyl ether resin **170** were achieved using Pd(OAc)$_2$, 1,1′-bis(diphenylphosphanyl)ferrocene (dppf) and potassium carbonate (K$_2$CO$_3$) at 90 °C in DMF to provide the biaryl vinyl ether resin **172**. The polymer-supported biaryl vinyl ethers were further treated with 2-chloro-2-(hydroxyimino)acetate and Et$_3$N to furnish the polymer-supported isoxazoline **173** through the [3+2] reactions. Treatment of the cycloadduct **173** with 5% TFA provided the corresponding oxazole **174** in good yield (Scheme 2.25). More than 20 examples were listed in this article.

Lepore and Wiley observed that a variety of ketone oximes reacted with the 2-fluorobenzonitrile derivatives to provide the corresponding aryloxime adducts (2003).[235] 4-Hydroxyl-2-fluorobenzonitrile (**175**) could readily be coupled with the Merrifield resin **2a** to afford the 2-fluorobenzonitrile linker **176a**. Alternatively, the 4-cyano-3-fluorobenzamide linker **176b** was prepared by coupling of the carboxypolystyrene (1% cross-linked) **177** and **178**. A variety of ketone oximes were immobilized onto the both linker resins. Although a limited number of transformation reactions were demonstrated with the aryloxime resins (i.e. **180**), their stable chemical characteristics would warrant their applicability to a wide range of organic reactions. The ketone **182** could be regenerated by a hydrolytic reaction of **181** with aqueous TFA at 55 °C (Scheme 2.26).

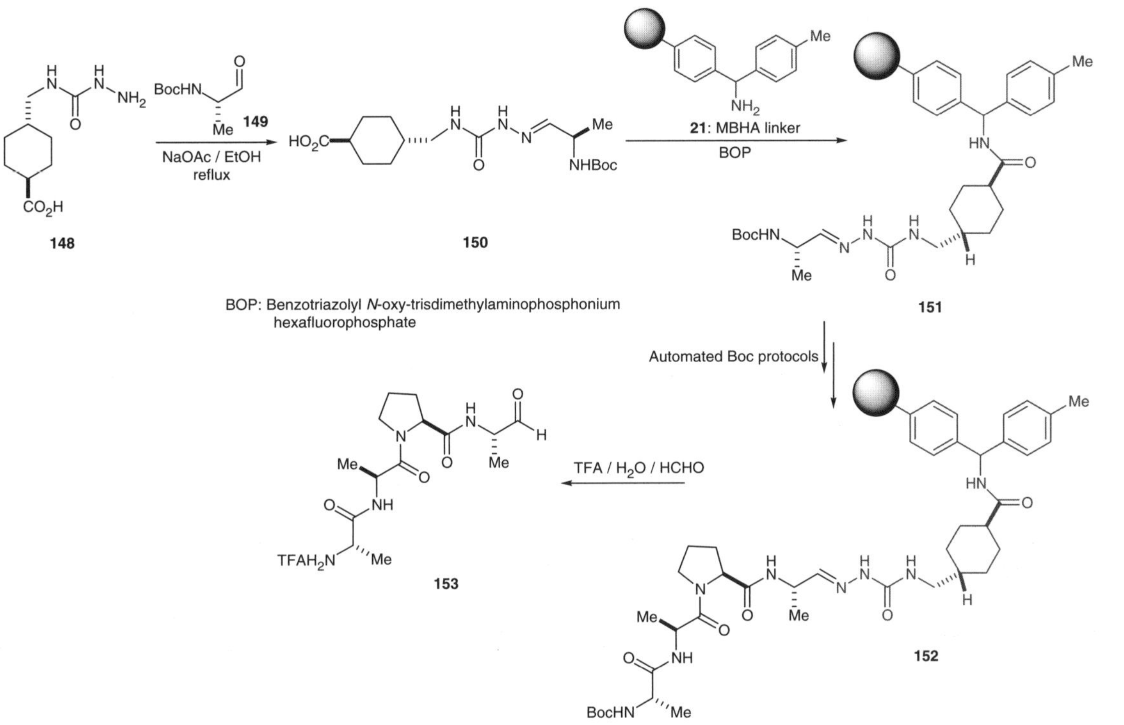

Electrophile Cleavable Linker Units 63

Scheme 2.22 A semicarbazide linker for generation of C-terminal peptide aldehydes

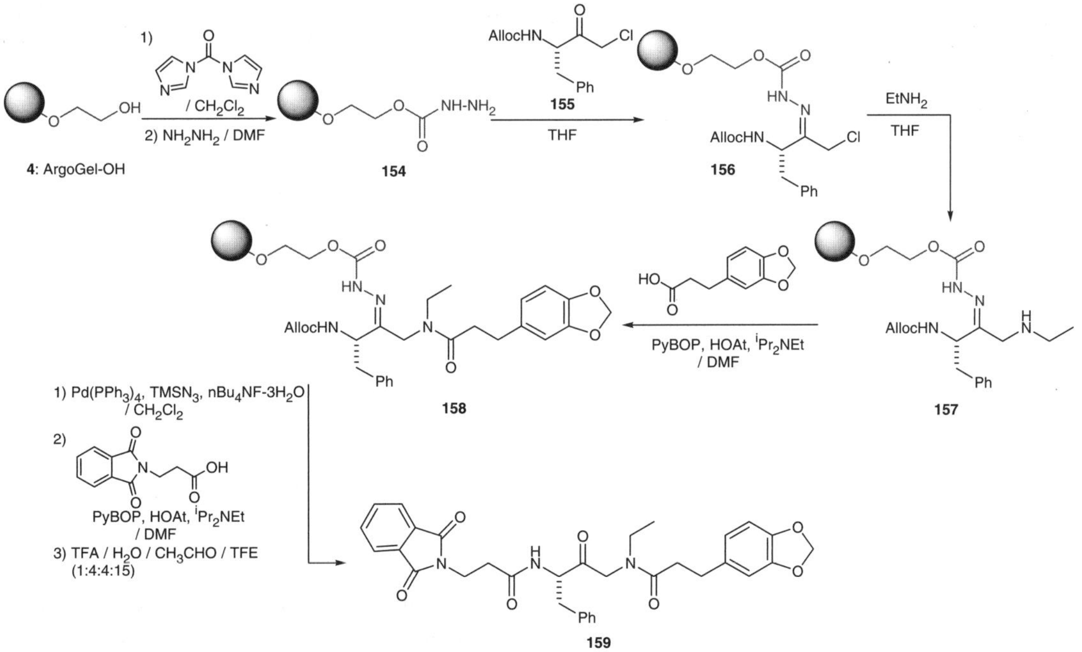

BOP: Benzotriazolyl *N*-oxy-trisdimethylaminophosphonium hexafluorophosphate

Scheme 2.23 A hydrazinecarboxylate linker for immobilizing chloromethyl ketones

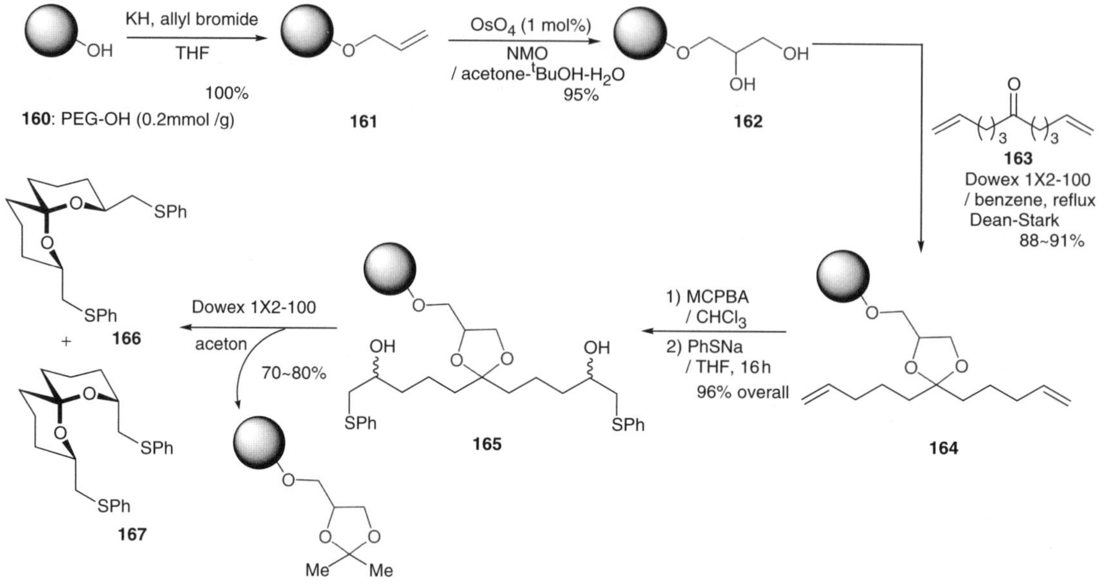

Scheme 2.24 Synthesis of substituted spiroketals using an ethylene glycol linker

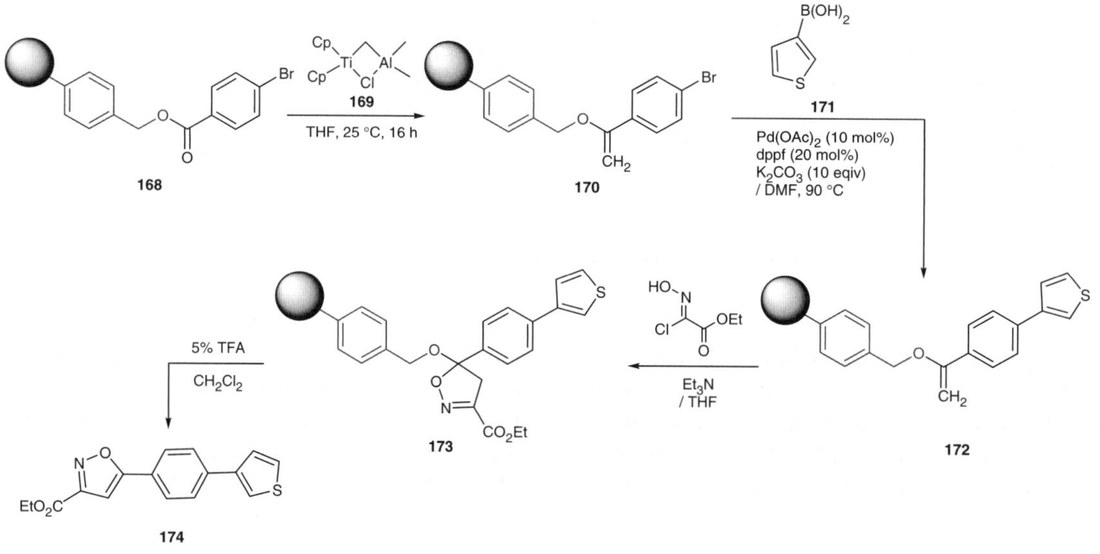

Scheme 2.25 Synthesis of a series of isoxazoles using a vinyl ether resin

2.3.1.7 Solid phase asymmetric synthesis using acid (or electrophile) cleavable linkers

To date, several linkers have been used to immobilize chiral auxiliaries, chiral controllers, and ligands for transition metals that enable immobilized molecules for use multiple times. In addition, many useful reagents have been immobilized onto convenient linker resins to facilitate purifications of desired molecules in solution phase organic reactions.[237–238] Moreover, suitably functionalized polymer supports

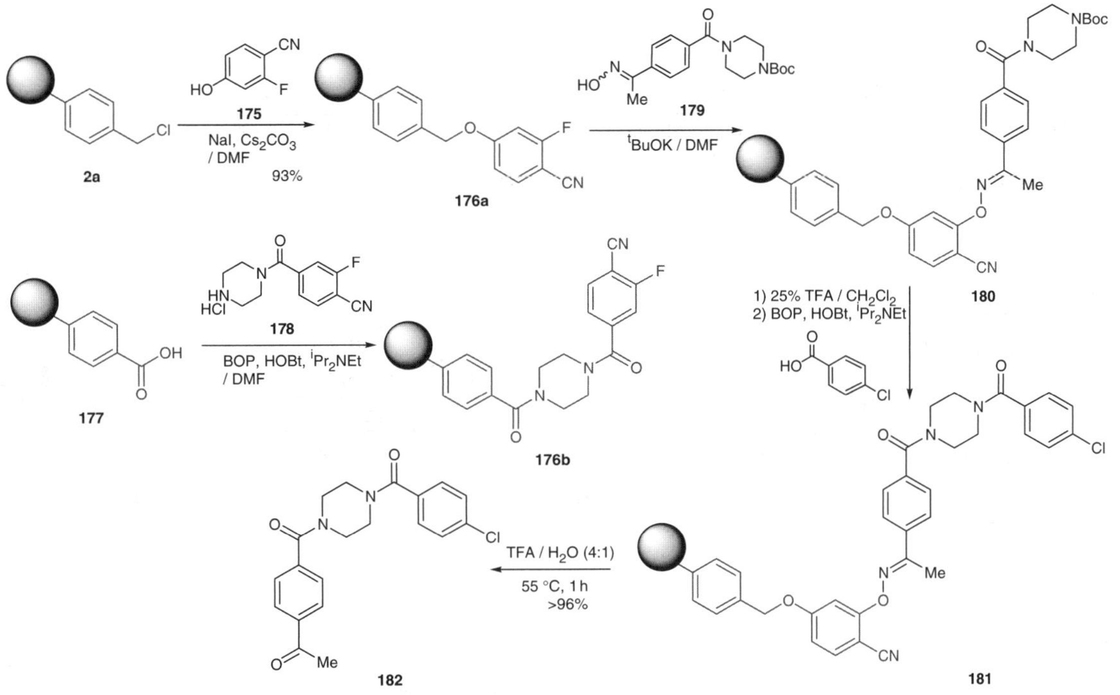

Scheme 2.26 2-Fluorobenzonitrile linker units

to selectively capture the target molecules away from other molecules including generated by-products (the catch-and-release technologies) are routinely used, especially in the field of pharmaceutical research.[239] In these cases, the linkages between molecules and resins are not expected to be cleaved under the reaction and regenerating (or washing) conditions. There is an increasing demand for polymer-supported reagents in organic synthesis[240–246] and a number of useful polymer-supported reagents are available from chemical vendors and manufacturers. On the other hand, a limited number of studies regarding asymmetric reactions using acid or electrophile cleavable linkers have been reported.[247]

Kawana and Emoto reported an asymmetric alkylation of the polymer-bound phenylglyoxylate.[248] The phenylglyoxylate group was introduced to 1,2-O-cyclohexylidene-α-D-xylofuranose **183** immobilized onto the trityl linker resin **17**. Methyl Grignard addition to the polymer-bound phenylglyoxylate **185** provided (R)-2-hydroxy-2-phenylpropanoic acid (**187**) in 77% with 65% ee after cleavage of the ester linkage (Scheme 2.27).

Worster and coworkers reported asymmetric alkylations of the immobilized imine on polymer resin. Coupling of (S)-2-phthalamido-1-propanol **188** with the Merrifield resin **2a** followed by deprotection of the phthaloyl group of **189** with hydrazine provided the polymer-bound (S)-2-aminopropan-1-ol **190**. This polymer-bound chiral amine was condensed with cyclohexanone to furnish the chiral polymer-bound imine **191**. Methylation with lithium diisopropylamide (LDA) and methyl iodide followed by hydrolysis of the imine linkage yielded (S)-methylcyclohexanone (**193**) in 80% yield with 95% ee (Scheme 2.28).[249]

Similarly, an optically active C_2-symmetric pyrrolidinone-based linker **195** was exploited to prepare optically active 3,5-disubstituted-γ-butyrolactones.[250] The N-propionated C_2-symmetric pyrrolidine-2,5-diyldimethanol **194** was loaded onto the Merrifield resin **2a** and the free alcohol was protected as the benzyl

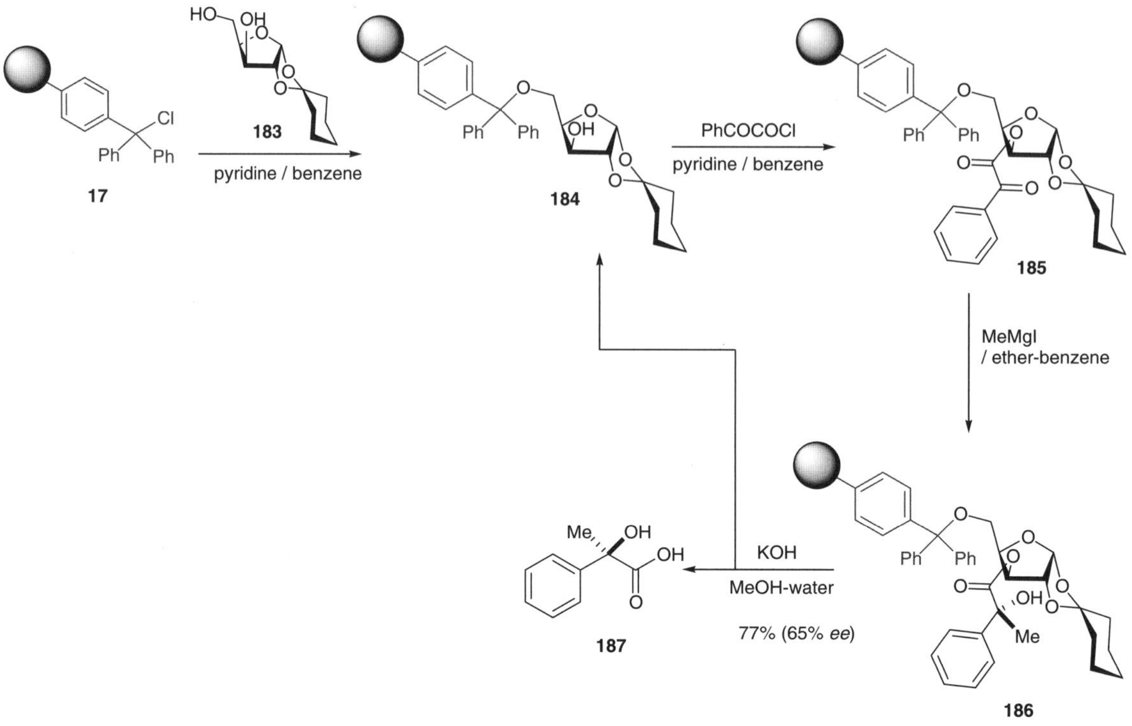

Scheme 2.27 Asymmetric synthesis of (R)-2-hydroxy-2-phenylpropanoic acid

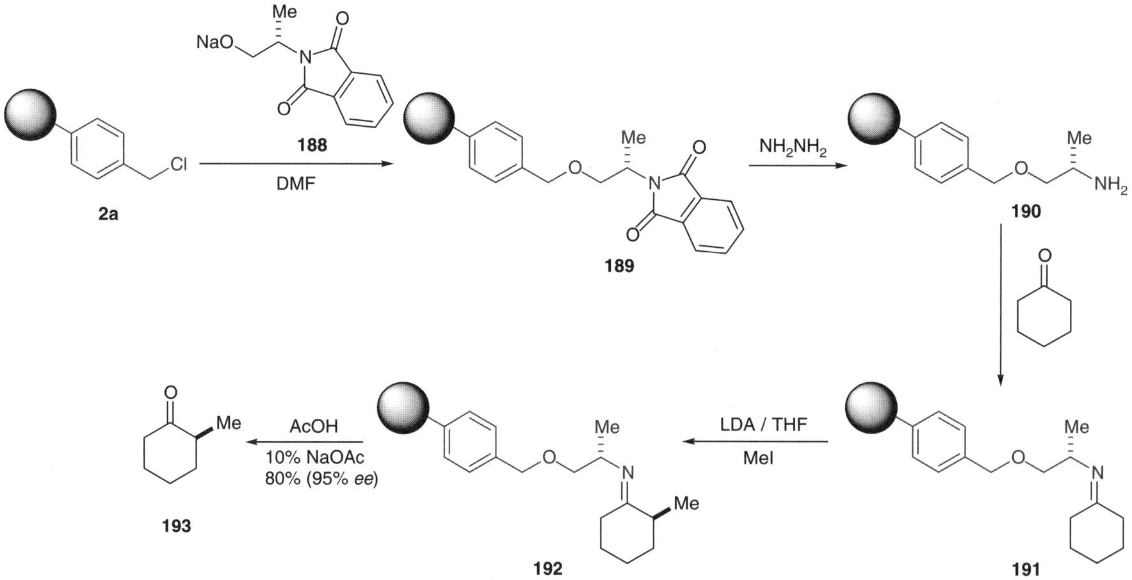

Scheme 2.28 Asymmetric alkylations of immobilized imines

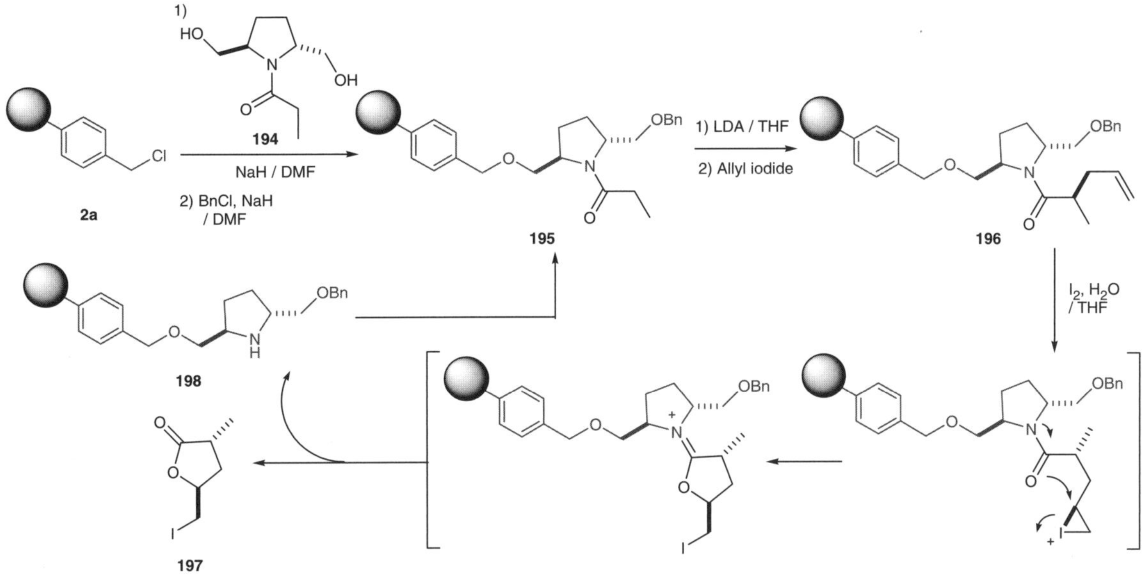

Scheme 2.29 *Preparation of optically active 3,5-disubstituted-γ-butyrolactones*

ether. Allylation of the propioamide resin with LDA furnished the polymer-bound (*R*)-2-methylpent-4-enamide **196** with 87% de. The cleavage of the linkage was performed using an electrophile such as iodine in aqueous THF to liberate (3*R*, 5*S*)-5-(iodomethyl)-3-methyldihydrofuran-2(3H)-one (**197**) and the resin **198**. The recovered resin **198** could reuse to operate the same reaction after the propionation reaction (Scheme 2.29).

Asymmetric syntheses of the α-chiral carboxylic acids and their esters using the polymer-bound oxazoline were reported by McManus and coworkers.[251] *Trans*-(4*S*, 5*S*)-2-ethyl-4-(hydroxymethyl)-5-phenyl-2-oxazoline (**199**) was attached to the Merrifield resin **2a** to give the polymer-bound oxazoline **200**. Benzylation of the optically active oxazoline **195** followed by an acid catalyzed solvolysis of the resulting product **201** provided (*S*)-ethyl 2-methyl-3-phenylpropoanoate (**202**) in 43–48% with 56% ee. Chemical and optical yields were similar to those observed in solution chemistry (Scheme 2.30).

Kunz and coworkers reported the asymmetric synthesis of unnatural amino acids using the immobilized galactosylamine as an auxiliary.[252] Later, Kunz extended the utility of the immobilized galactosylamine for stereoselective synthesis of chiral piperidine derivatives on a solid phase.[253] The silicon linker **54b** developed by Fréchet (Section 2.3.1.3) was stable against the conditions used in their synthesis on a solid phase. The primary alcohol **203** was loaded onto the diisopropylphenylsilyl linker **54b** in quantitative yield. The anomeric azide was reduced to the corresponding amine and the subsequent imine formation reaction with benzaldehyde afforded the β-galactosyl imine **205**. The formal hetero Diels–Alder products were synthesized via zinc chloride (ZnCl$_2$) catalyzed tandem Mannich–Michael reactions with Danishefsky's diene **206**. The resin-bound dihydropiperidinone **207** was cleaved with tetrabutylammonium fluoride (TBAF) buffered with acetic acid to afford **208** (Scheme 2.31).

The solid-phase synthesis of unnatural α-amino acids using the benzophenone imines of resin-bound glycinates was reported by O'Donnell and coworkers in 1999.[254, 255] O'Donnell originally reported alkylation reactions of the benzophenone imines of Gly-L-Leu-Merrifield and Gly-L-Leu-Wang linker

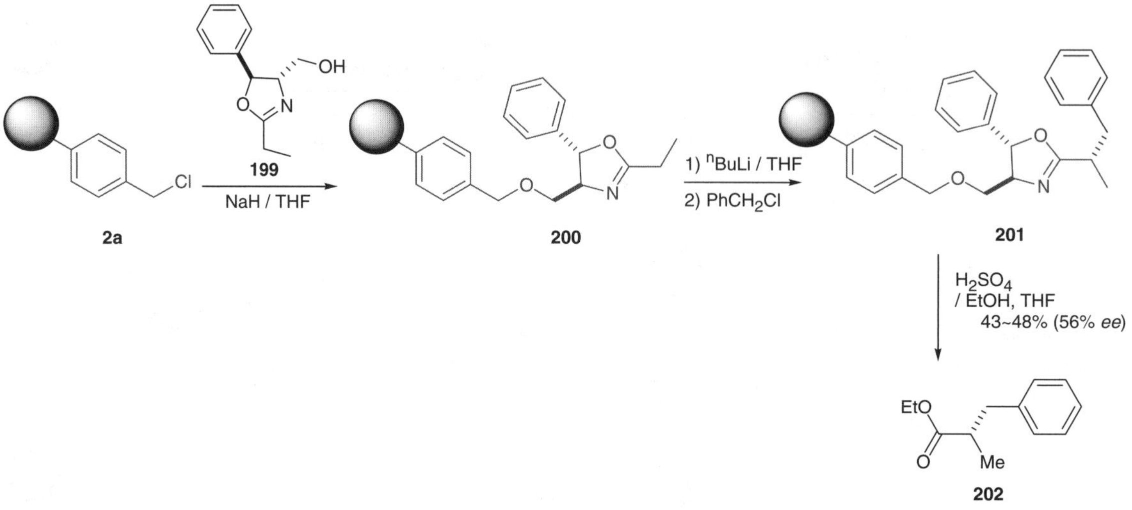

Scheme 2.30 Asymmetric syntheses of α-chiral carboxylic acids and esters

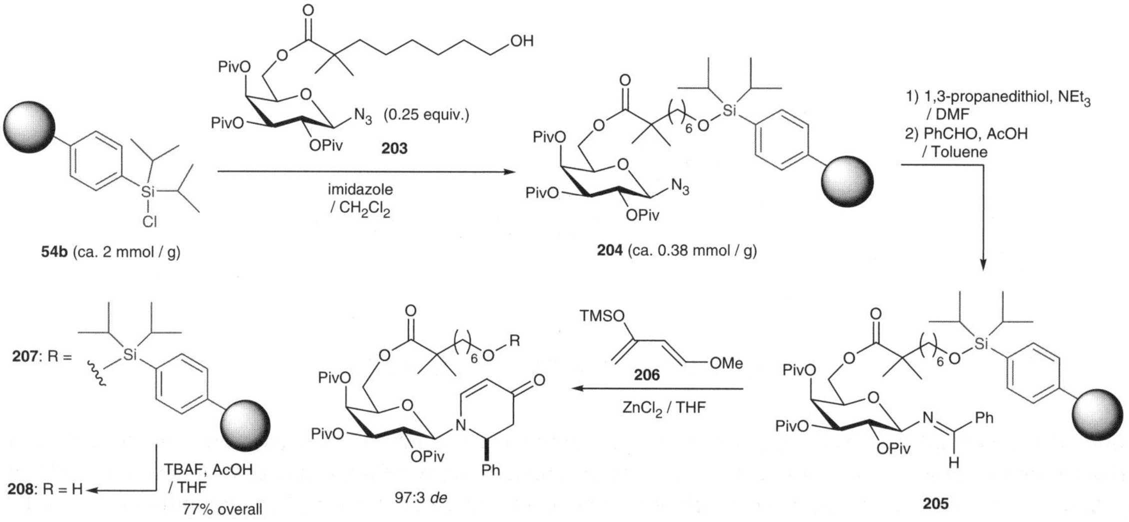

Scheme 2.31 Asymmetric synthesis of unnatural amino acids using an immobilized galactosylamine auxiliary

resins. As illustrated in Scheme 2.32, simultaneous deprotonation and alkylation of the α-position of the benzophenone imine of Gly-L-Leu-Wang linker resin **210** with benzyl bromide was accomplished using the solvent soluble nonionic phosphazene base, 2-*tert*-butylimino-2-diethylamino-1,3-dimethylperhydro-1,3,2-diazaphosphorine (BEMP) (**212**), to yield the benzophenone imine of (D,L)-Phe-L-Leu resin **213**. The imine moiety of **212** was selectively hydrolyzed with aqueous NH$_2$OH·HCl in THF and the dipeptide was cleaved with 95% TFA to furnish the dipeptide **214**.[256] The solid-phase enantioselctive alkylation of resin-bound glycinate was achieved using the cinchona-derived reagent **216**,[257] in the presence of

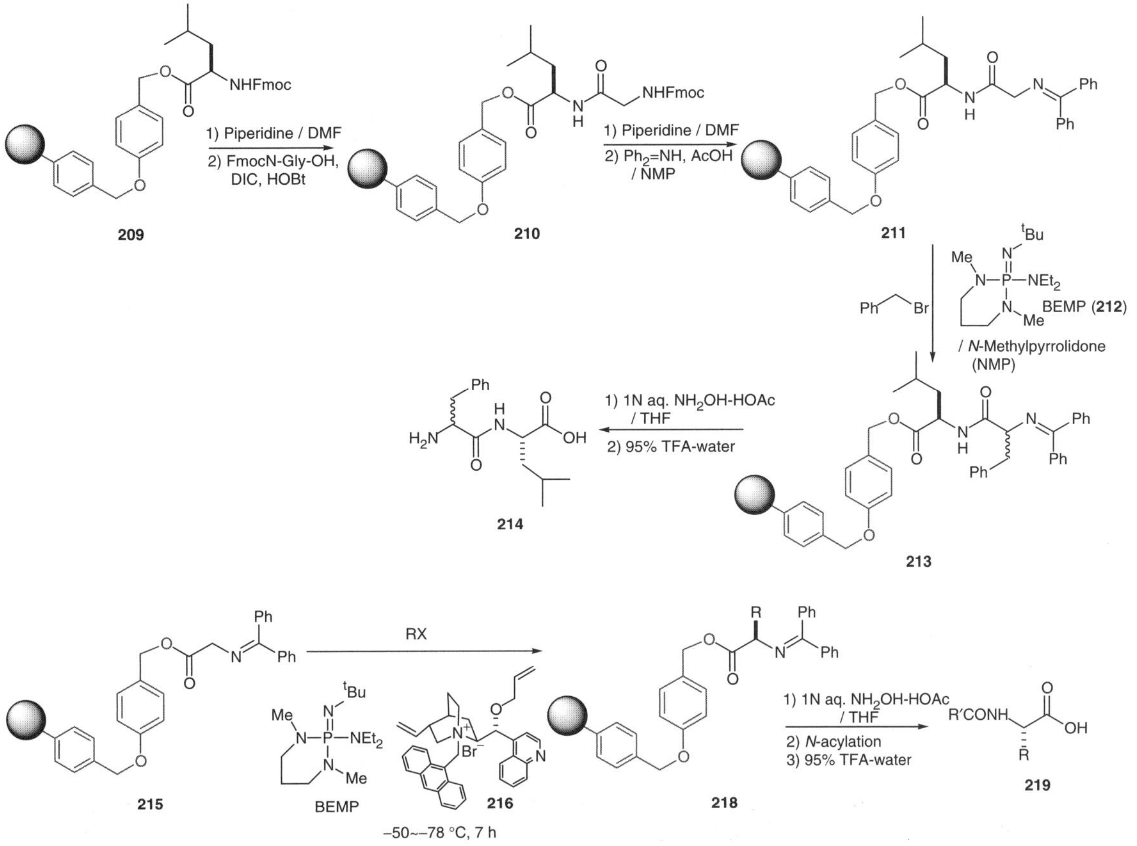

Scheme 2.32 *Solid-phase synthesis of unnatural α-amino acids using the benzophenone imines of resin-bound glycinates*

BEMP to furnish the α-alkylated glycine derivative **219** with 76% ee (for benzylation) and 89% *ee* (for methallylation). The synthesis of optically active α-amino acids using a simple and easy procedure remains an important subject in organic synthesis. The use of a Schiff base for the synthesis of α-amino acids by phase transfer catalysis (PTC) has received considerable attention since 1978. As solid-phase synthesis already involves a biphasic environment, use of PTC reaction conditions with heterogeneous base is problematic. Thus, discovery of the base, which is compatible to both polymer-supported alkylation reaction and base sensitive electrophiles (e.g. alkyl halides), is very important.[258]

A limited number of catalytic asymmetric reactions have been applied to substrates bound to polymer supports. Davies reported catalytic asymmetric cyclopropanations of alkenes on the polymer support (2001).[259] A resin-bound olefin was prepared with the butyl diethylsilane linker **63**. The silane group was activated with 1,3-dichloro-5,5-dimethylhydantoin to give the chlorosilane linker **64** (Scheme 2.7 in Section 2.3.1.3), and 2-(4-(1-phenylvinyl)phenoxy)ethanol **220** was loaded (0.83–1.0 mmol/g). The Rh$_2$(*S*-DOSP)$_4$ (**222**) catalyzed cyclopropanation was carried out with methyl phenyldiazoacetae (**223**).[260] Using as low as 0.01 mol% of the rhodium catalyst, a mixture of the tetrasubstituted cyclopropanes **225** and **226** were isolated, after cleavage of the linker of **222** with HF·pyridine, with almost identical stereoselectivities to those obtained in the corresponding reaction in solution (Scheme 2.33). Thus, it was experimentally

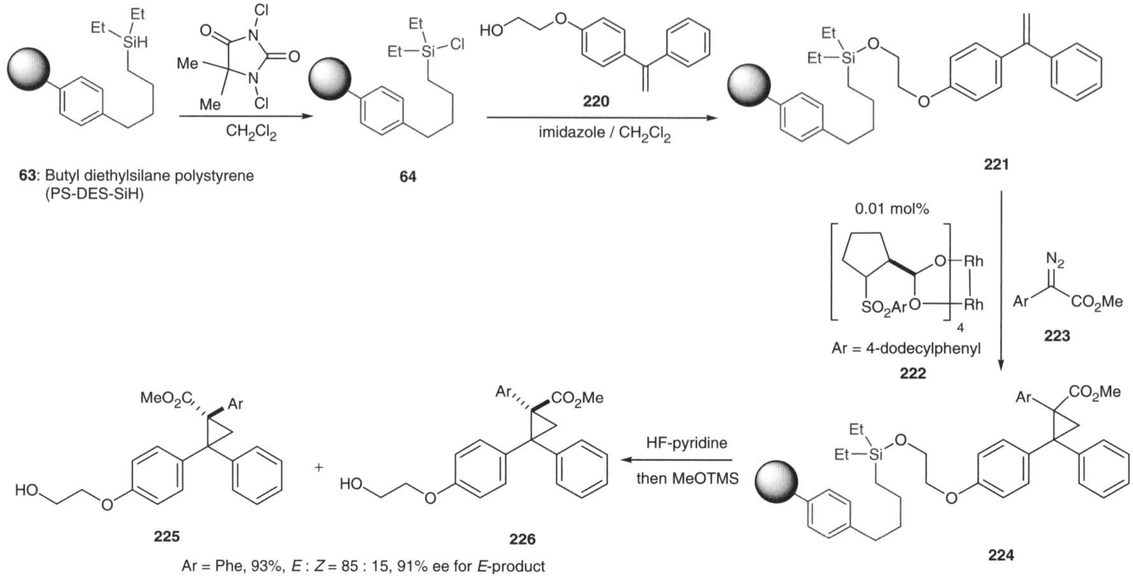

70 *Traditional Linker Units for Solid-Phase Organic Synthesis*

Scheme 2.33 *Catalytic asymmetric cyclopropanations of polymer-supported alkenes*

proved that catalytic asymmetric organic reactions developed in the solution phase can be transferred to polymer-supported reactions by using homogeneous catalysts which do not interfere with polymer supports.

An asymmetric syntheses of optically active chlorohydrins was achieved using polymer-bound allylic sulfoximines.[261] *S*-Methyl-*S*-phenyl sufoximine **227** was loaded onto the Merrifield resin **2a** in 84% yield. Lithiation of the polymer-bound sulfoximine **228** with ^{n}BuLi followed by alkylation with 3-methylbutanal furnished the polymer-bound lithium alkoxide, which upon treatment with $ClCO_2CH_3$ and, subsequently, DBU afforded the vinylic sulfoxime. The isomerization of the vinylic sulfoxime with DBU provided the allylic sulfoximine **230**. This was subjected to the lithiation, transmetallation with the titanium complex and alkylation with benzaldehyde to yield the polymer-bound homoallylic alcohol **231**.[262] Cleavage of the linker was achieved by treatment with an electrophile, $ClCO_2CH(Cl)CH_3$ in DCM resulted in the formation of the chlorohydrins **232** and **233** as a 2:1 mixture of *anti* and *syn* products with > 53% and 70% ee, respectively. These chlorohydrins could be converted to the corresponding alkenyloxiranes with DBU (Scheme 2.34).

2.4 Conclusion

In summary, electrophile cleavable linkers have been summarized in this chapter. A significant number of applications of acid cleavable linkers have been reported to date. To efficiently generate target oligo to long peptides by using the Boc or Fmoc peptide strategy, a wide variety of linkers have been developed. Knowledge of the physico-chemical properties of these linkers obtained from peptide chemistries is crucial to develop new acid labile linkers to immobilize different class of molecules (other than amino acids). The development of new methods for functional group protection and deprotection continues in solution-phase chemistry. In addition, because many useful protecting groups developed for solution chemistry have not yet been translated for use as a corresponding solid-phase version (linker unit), it is certain that efforts will continue in developing new acid labile linker units.

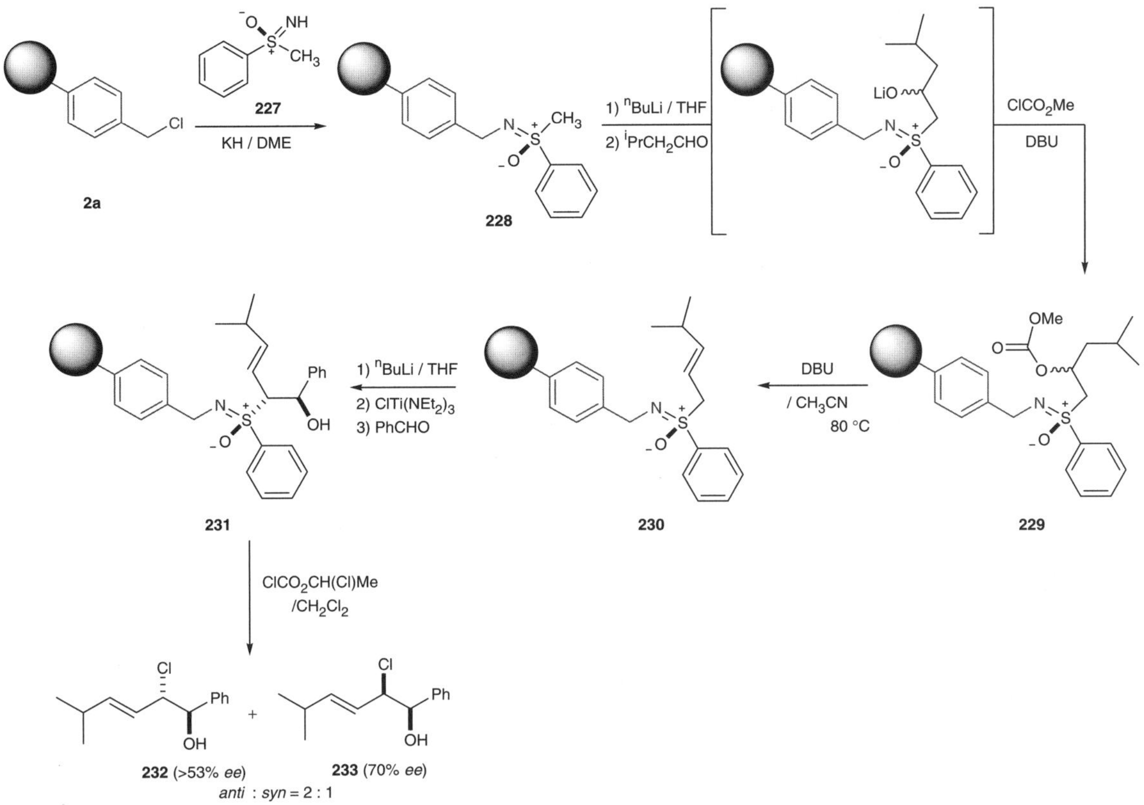

Scheme 2.34 Asymmetric syntheses of optically active chlorohydrins using polymer-supported allylic sulfoximines

References

[1] Reese, B. C., Rao, V. M., Serafinowska, H. T., et al.; *Nucleosides and Nucleotides* **1991**, *10*, 81.
[2] Frechet, J. M. J., and Schuerch, C.; *J. Am. Chem. Soc.* **1971**, *93*, 492.
[3] Liang, R., Yan, L., Loebach, J., et al.; *Science* **1996**, *274*, 1520.
[4] Planate, O. J., Palmacci, E. R., and Seeberger, P. H.; *Science* **2001**, *291*, 1523.
[5] Roberge, J. Y.; *Synthesis* **2000**, 579.
[6] Bunin, B. A., and Ellman, J. A.; *J. Am. Chem. Soc.* **1992**, *114*, 10997.
[7] DeWitt, H., Kiely, J. S., Stankovic, C. J., et al.; *Proc. Natl. Acad. Sci. USA.* **1993**, *90*, 6909.
[8] Krchnak, V., and Holladay, M. W.; *Chem. Rev.* **2002**, *102*, 61.
[9] Dolle, R. E.; *J. Comb. Chem.* **2002**, *4*, 369–418.
[10] Leena, S., and Kumar, K. S.; *J. Peptide Res.* **2001**, *58*, 117.
[11] Roice, M., and Pillai, V. N. R.; *J. Poly. Science, Part A: Poly. Chem.* **2005**, *43*, 4382.
[12] Kates, S. A., McGuinness, B. F., Blackburn, C., et al.; *Biopolymers* **1999**, *47*, 365.
[13] Gooding, O. W., Baudart, S., Deegan, T. L., et al.; *J. Comb. Chem.* **1999**, *1*, 113.
[14] For swelling studies of polymer resins, see Cavalli, G., Shooter, A. G., Pears, D. A., and Steinke, J. H. G.; *J. Comb. Chem.* **2003**, *5*, 637.
[15] Kempe, M., and Barany, G.; *J. Am. Chem. Soc.* **1996**. *118*, 7083.
[16] Auzanneau, F-I., Meldal, M., and Bock, K.; *J. Peptide Science* **2004**, *1*, 31.
[17] Garcra-Martrn, F., White, P., Steinauer, R., et al.; *Peptide Science* **2006**, *84*, 566.

[18] Parsons, J. G., Sheehan, C. S., Wu, Z., et al.; *Method Enzymol.* **2003**, *369*, 39.
[19] Wang, Y., and Lee, Y-S.; *Kongop Hwahak* **1993**, *4*, 132–143.
[20] Thompson, L. A., and Ellman, J. A.; *Chem. Rev.* **1996**, *96*, 555.
[21] James, I. W.; *Tetrahedron* **1999**, *55*, 4855.
[22] Krchnak, V., and Holladay, M. W.; *Chem. Rev.* **2002**, *102*, 61.
[23] Guiller, F., Orain, D., and Bradley, M.; *Chem. Rev.* **2000**, *100*, 2091.
[24] Bergbreiter, D. E.; *Curr. Opin. Drug Disc. Develop.* **2001**, *4*, 736.
[25] Ley, S. V., Baxendale, I. R., Longbottom, D. A., and Myers, R. M.; *Drug Disc. Develop.* **2007**, *2* 51–89.
[26] Dolle, R. E., Le Bourdonnec, B., Goodman, A. J., et al.; *J. Comb. Chem.* **2008**, *10*, 753.
[27] Bochet, C. G.; *J. Chem. Soc. Perkin Trans. 1* **2002**, *2*, 125.
[28] Sun, Y., Chan, B. C., Ramnarayanan, R., et al.; *J. Comb. Chem.* **2002**, *4*, 569.
[29] Mitchell, A. R.; *Peptide Science* **2008**, *90*, 175.
[30] Merrifield R. B.; *Fed Proc Am Soc Exp Biol.* **1962**, *21*, 412 (Abstract).
[31] Merrifield, R. B.; *J. Am. Chem. Soc.* **1963**, *85*, 2149.
[32] Carpino, L. A.; *J. Am. Chem. Soc.* **1960**, *82*, 2725 and references therein.
[33] Carpino, L. A., Parameswaran, K. N., Kirkley, R. K., et al.; *J. Org. Chem.* **1970**, *35*, 3291.
[34] Merrifield, R. B.; *J. Am. Chem. Soc.* **1964**, *86*, 304.
[35] Merrifield, R. B.; *Biochemistry* **1964**, *3*, 1385.
[36] Gutte, B., and Merrifield, R. B.; *J. Biol. Chem.* **1971**, *246*, 1922.
[37] Merrifield, R. B.; *Pure Appl. Chem.* **1971**, *50*, 643.
[38] Kresge, N., Simoni, R. D., and Hill, R. L.; *J. Biol. Chem.* **2006**, *281*, e21.
[39] One should consider that MsOH is a high boiling point reagent. Extensive co-evaporation and high vacuum equipment are necessary to remove MsOH out of the products.
[40] Mitchell, A. R., Erickson, B. W., Ryabtsev, M. N., et al.; *J. Am. Chem. Soc.* **1976**, *98*, 7357.
[41] Furlan, R. L. E., Mata, E. G., Mascaretti, O. A., et al.; *Tetrahedron* **1998**, *54*, 13023.
[42] Seitz, O.; *Angew. Chem. Int. Ed.* **1998**, *37*, 3109.
[43] Wang, S-S.; *J. Am. Chem. Soc.* **1973**, *95*, 1328.
[44] Sheppard, R. C., and Williams, B. J.; *Int. J. Peptide Protein Res.* **1982**, *20*, 451.
[45] Carey, R. I., Bordas, L. W., Slaughter, R. A., et al.; *J. Peptide Res.* **1997**, *49*, 570.
[46] Mergler, M., Tanner, R., Gosteli, J., and Grogg, P.; *Tetrahedron Lett.* **1988**, *29*, 4005.
[47] Carpino, L. A., Han, G. Y.; *J. Org. Chem.* **1972**, *37*, 3404.
[48] Carpino, L. A., Han, G. Y.; *J. Am. Chem. Soc.* **1970**, *92*, 5748.
[49] Dick, F.; Acid Cleavage/Deprotection in Fmoc/tBu Solid-phase Peptide Synthesis, in *Peptide Synthesis Protocols* (*Methods in Molecular Biology*), volume 35 (eds Pennington, M. W., and Dunn, B. M.) **1994**, Humana Press, Totowa, NJ, 63–72.
[50] Rink, H.; *Tetrahedron Lett.* **1987**, *28*, 3787.
[51] Barlos, K., Gatos, D., Kallitsis, J., et al.; *Tetrahedron Lett.* **1989**, *30*, 3943.
[52] For a direct analysis of molecules on polymer resins, see. Blas, J., Rivera-Sagredo, A., Ferritto, R., and Espinosa, J. F.; *Magnetic Resonance Chem.* **2004**, *42*, 950.
[53] Kurosu, M., Biswas, K., and Crick, D. C.; *Org. Lett.* **2007**, *9*, 1141.
[54] Wang, B., Chen, L., and Kim, K.; *Tetrahedron Lett.* **2001**, *42*, 1463.
[55] Kim, K., and Wang, B.; *Chem. Comm.* **2001**, *21*, 2268.
[56] Leger, R., Yen, R., She, M. W., et al.; *Tetrahedron Lett.* **1998**, *39*, 4171.
[57] Park, K-H., Ehrler, J., Spoerri, H., and Kurth, M. J.; *J. Comb. Chem.* **2001**, *3*, 171.
[58] Barany, G., and Merrifield, R. B.; in *The Peptide: Analysis, Synthesis, Biology* (eds E. Gross and J. Meienhofer), Academic Press, New York, **1980**, *2*, 1–255.
[59] Mitchell, A. R., Kent, S. B. H., Engelhard, M., and Merrifield, R. B.; *J. Org. Chem.* **1978**, *43*, 2845.
[60] Mojsov, S., Mitchell, A. R., and Merrifield, R. B.; *J. Org. Chem.* **1980**, *45*, 555.
[61] Bayer, E., Dengler, M., and Hemmasi, B.; *Int. J. Pept. Protein Res.* **1985**, *25*, 178.
[62] Sivanandaiah, K. M., Babu, V. V., Suresh, G., and Beechanahalli, P.; *Tetrahedron Lett.* **1996**, *37*, 5989.
[63] Gouilleux, L., Fehrentz, J-A., Winternitz, F., and Martinez, J.; *Tetrahedron Lett.* **1996**, *37*, 7031.

[64] Giralt, E., Andreu, D., Pons, M., and Pedroso, E.; *Tetrahedron* **1981**, *37*, 2007.
[65] Takahashi, S., Hibino, T., and Sawada, S.; *Bull. Chem. Soc. Jap.* **1988**, *61*, 2467.
[66] Seitz, O.; *Angew. Chem. Int. Ed.* **1998**, *37*, 3109.
[67] Rosenthal, K., Erlandsson, M., and Unden, A.; *Tetrahedron Lett.* **1999**, *40*, 377.
[68] Grandas, A., Pedroso, E., Giralt, E., et al.; *Tetrahedron* **1986**, *42*, 6703.
[69] Boggian, D. B., and Mata, E. G.; *Synthesis* **2006**, *20*, 3397.
[70] Stanger, K. J., and Krchnak, V.; *J. Comb. Chem.* **2006**, *8*, 652.
[71] Albericio, F., Pons, M., Pedroso, E., and Giralt, E.; *J. Org. Chem.* **1989**, *54*, 360.
[72] Rahman, S. A.; *Egy. J. Chem.* **1996**, *39*, 99.
[73] Lattmann, E., Billington, D. C., Poyner, D. R., et al.; *Pharm. Pharmacolog. Lett.* **2001**, *11*, 18.
[74] Dublanchet, A-C., Lusinchi, M., and Zard, S. Z.; *Tetrahedron* **2002**, *58*, 5715.
[75] Katritzky, A. R., Toader, D., Watson, K., and Kiely, J. S.; *Tetrahedron Lett.* **1997**, *38*, 7849.
[76] Mergler, M., Gosteli, J., Grogg, P. N. R., and Tanner, R.; *Chimia* **1999**, *53*, 29.
[77] Krchňáka, V., and Slough, G. A.; *Tetrahedron Lett.* **2004**, *45*, 5237.
[78] McMurray, J. S., and Lewis, C. A.; *Tetrahedron Lett.* **1993**, *34*, 8059.
[79] Riniker, B., Floersheimer, A., Fretz, H., et al.; *Tetrahedron* **1993**, *49*, 9307.
[80] Wells, N. J., Davies, M., and Bradley, M.; *J. Org. Chem.* **1998**, *63*, 6430.
[81] Roice, M., Suma, G., Kumar, K. S., and Pillai, V. N. R.; *J. Protein Chem.* **2001**, *20*, 25.
[82] Grotenbreg, G. M., Buizert, A. E. M., Llamas-Saiz, A. L., et al.; *J. Am. Chem. Soc.* **2006**, *128*, 7559.
[83] Ay, B., Landgraf, K., Streitz, M., Fuhrmann, S., et al.; *Bioorg. Med. Chem. Lett.* **2008**, *18*, 4038.
[84] Riniker, B., Floersheimer, A., Fretz, H., et al.; *Tetrahedron* **1993**, *49*, 9307.
[85] Colombo, A., De la Figuera, N., Fernandez, J. C., et al.; *Org. Lett.* **2007**, *9*, 4319.
[86] Bernatowicz, M. S., Daniels, S. B., and Koster, H.; *Tetrahedron Lett.* **1989**, *30*, 4645.
[87] Varadi, G., Toth, G. K., and Penke, B.; *Int. J. Pept. Protein Res.* **1994**, *43*, 29.
[88] Brummer, O., Clapham, B., and Janda, K. D.; *Tetrahedron Lett.* **2001**, *42*, 2257.
[89] Takahashi, T., Nagamiya, H., Doi T., et al.; *J. Comb. Chem.* **2003**, *5*, 414.
[90] Henkel, B., Zhang, L., Goldammer, C., and Bayer, E.; *Zeitsch. Natur. B: Chem. Science* **1996**, *51*, 1339.
[91] Choi, M. K. W., and Toy, P. H.; *Tetrahedron* **2004**, *60*, 2903.
[92] Krchňá, V., Cabel, D., Weichsel, A., and Flegelová, Z.; *Lett. Pept. Science* **1994**, *1*, 277.
[93] Richter, L. S., and Gadek, T. R.; *Tetrahedron Lett.* **1994**, *35*, 4705.
[94] Al-Obeidi, F., Austin R. E., Okonya, J. F., and Bond, D. R. S.; *Mini Rev. Med. Chem.* **2003**, *3*, 449.
[95] Wathey, B., Tierney, J., Lidström, P., and Westman, J.; *Drug Disc. Today* **2002**, *7*, 373.
[96] Bowman, M. D., Jacobson, M. M., Pujanauski, B. G., and Blackwell, H. E.; *Tetrahedron* **2006**, *62*, 4715.
[97] DeGrado, W. F., and Kaiser, E. T.; *J. Org. Chem.* **1980**, *45*, 1295.
[98] Pietta, P. G., and Marshall, G. R.; *J. Chem. Soc. D.* **1970**, 650.
[99] Matsuda, G R., and Steward, J. M.; *Peptides* **1981**, *2*, 45.
[100] Pitta, P. G., Cavallo, P. F., Takahashi, K., and Marshall, G. R.; *J. Org. Chem.* **1974**, *39*, 44.
[101] Sieber, P. A.; *Tetrahedron Lett.* **1987**, *28*, 2107.
[102] Bernatowicz, M. S., Daniels, S. B., and Koster, H.; *Tetrahedron Lett.* **1989**, *30*, 4645.
[103] Gross, E., Noda, K., and Nisula, B.; *Angew. Chem. Int. Ed.* **1973**, *12*, 664.
[104] Meisenbach, M., and Voelter, W.; *Chem. Lett.* **1997**, *26*, 1265.
[105] Swistok, J., Tilley, J. W., Danho, W., et al.; *Tetrahedron Lett.* **1989**, *30*, 5045.
[106] Christensen, M., Schou, O., and Pedersen, V. S.; *Acta Chem. Scand., Series B: Org. Chem. Biochem.* **1981**, *B35*, 573.
[107] Newlander, K. A., Chenera, B., Veber. D. F., et al.; *J. Org. Chem.* **1997**, *62*, 6726.
[108] Fehrentz, J-A., Paris, M., Heitz, A., et al.; *Tetrahedron Lett.* **1995**, *36*, 7871.
[109] Alsina, J., Giralt, E., and Albericio, F.; *Tetrahedron Lett.* **1996**, *37*, 4195.
[110] Breipohl, G., Knolle, J., and Stueber, W.; *Int. J. Pept. Protein Res.* **1989**, *34*, 262.
[111] Calmes, M., Daunis, J., David, D., and Jacquier, R.; *Int. J. Pept. Protein Res.* **1994**, *44*, 58.
[112] Teno, N., and Takai, M.; *Pept. Chem.* **1995**, *32*, 229.
[113] Meisenbach, M., Echner, H., and Voelter, W.; *Chem. Comm.* **1997**, *9*, 849.

[114] Malkinson, J. P., and Falconer, R. A.; *Tetrahedron Lett.* **2002**, *43*, 9549.
[115] Han, Y., Bontems, S. L., Hegyes, P., et al.; *J. Org. Chem.* **1996**, *61*, 6326.
[116] Chan, W. C., White, P. D., Beythien, J., and Steinauer, R.; *Chem. Comm.* **1995**, *5*, 589.
[117] Somlai, C., Nyerges, L., Penke, B., et al.; *Chem. Chem. Zeitung* **1994**, *336*, 429.
[118] Meisenbach, M., Gruebler, G., Paulus, G., and Voelter, W.; *Peptides* **1995**, 265.
[119] Fara, M. A., Draz-Mocón, J. J., and Bradley, M.; *Tetrahedron Lett.* **2006**, *47*, 1011.
[120] Fréchet, J. M., and Schuerch, C.; *J. Am. Chem. Soc.* **1971**, *93*, 492.
[121] Hanessian, S., Ogawa, T., Guidon, Y., et al.; *Carbohydr. Res.* **1974**, *38*, C15.
[122] Fréchet, J. M., and Pellé, G.; *J. Chem. Soc. Chem. Commun.* **1975**, 225.
[123] Kulesh, D. A., Clive, D. R., Zarlenga, D. S., and Greene, J. J.; *Proc. Natl. Acad. Sci. USA* **1987**, *84*, 8453.
[124] Wang, G. T., Li, S., Wideburg, N., et al.; *J. Med. Chem.* **1995**, *38*, 2995.
[125] Borhan, B., Wilson, J. A., Gasch, M. J., et al.; *J. Org. Chem.* **1995**, *60*, 7375.
[126] Chen, C., Randal, L. A. A., Miller, R. B., et al.; *J. Am. Chem. Soc.* **1994**, *116*, 2661.
[127] Manzotti, R., Reger, T. S., and Janda, K. D.; *Tetrahedron Lett.* **2000**, *41*, 8417.
[128] Lundquist, J. T., Satterfield, A. D., and Pelletier, J. C.; *Org. Lett.* **2006**, *8*, 3915.
[129] Cresteya, F., Ottesena, L. K., Jaroszewskia, J. W., and Franzyk, H., *Tetrahedron Lett.* **2008**, *49*, 5890.
[130] Garigipati, R. S.; *Tetrahedron Lett.* **1997**, *38*, 6807.
[131] Thompson, L. A., and Ellman, J. A.; *Tetrahedron Lett.* **1994**, *35*, 9333.
[132] Basso, A., and Ernst, B.; *Tetrahedron Lett.* **2001**, *42*, 6687.
[133] Greene, T. W., and Wuts, P. G. M.; Protection for the Hydroxyl Group, Including 1,2- and 1,3-Diols in *Protective Groups in Organic Synthesis* (eds Greene, T. W., and Wuts, P. G. M.), 3rd edn, John Wiley and Sons, Inc., New York; **1999**, 113–148.
[134] Farrall, M. J., and Fréchet, M. J.; *J. Org. Chem.* **1976**, *41*, 3877.
[135] Danishefsky, S. J., McClure, K. F., Randolph, J. T., and Ruggeri, R. B. A.; *Science* **1993**, *260*, 1307.
[136] Randolph, J. T., McClure, K. F., and Danishefsky, S. J.; *J. Am. Chem. Soc.* **1995**, *117*, 5712.
[137] Lukevics, E., and Dzintara, M.; *J. Organomet. Chem.* **1985**, *295*, 265.
[138] Hu, Y., and Porco, J. A. Jr.; *Tetrahedron Lett.* **1998**, *39*, 2711.
[139] Hu, Y., Porco, J. A. Jr, Labadie, J. W., and Gooding, O. W.; *J. Org. Chem.* **1998**, *63*, 4518.
[140] Tallarico, J. A., Depew, K. M., Pelish, H. E., et al.; *J. Comb. Chem.* **2001**, *3*, 312.
[141] Smith, E. M.; *Tetrahedron Lett.* **1999**, *40*, 3285.
[142] Poco, J. A., and Hu, Y.; *Tetrahedron Lett.* **1999**, *40*, 3289.
[143] Kurosu, M., Porter, J. R., and Foley, M. A.; *Tetrahedron Lett.* **2004**, *45*, 145.
[144] DiBlas, C. M., Macks, D. E., and Tan, D. S.; *Org. Lett.* **2005**, *7*, 1777.
[145] Fleming, I., Roberts, R. S., and Smith, S. C.; *J. Chem. Soc. Perkin Trans.* **1998**, *1*, 1209.
[146] Noyori, R., Nishida, I., and Sakata, J.; *J. Am. Chem. Soc.* **1983**, *105*, 1598.
[147] Kurosu, M., Marcin, L. R., Grinsteiner, T. J., and Kishi, Y.; *J. Am. Chem. Soc.* **1998**, *120*, 6627.
[148] Scheidt, K. A., Chen, H., Follows, B. C., et al.; *J. Org. Chem.* **1998**, *63*, 6436.
[149] Paterson, I., and Temal-Laib, T.; *Org. Lett.* **2002**, *4*, 2473.
[150] Paterson, I., Goodman, J. M., and Isaki, M.; *Tetrahedron Lett.* **1989**, *30*, 7121.
[151] Barun, O., Sommer, S., and Wadmann, H.; *Angew. Chem. Int. Ed.* **2004**, *43*, 3195.
[152] Hanessian, S., and Xie, F.; *Tetrahedron Lett.* **1998**, *39*, 733.
[153] Hannesian, S., Ma, J., and Wang, W.; *Tetrahedron Lett.* **1999**, *40*, 4631.
[154] Patel, V. F., Hardin, J. N., Starling, J. J., and Mastro, J. M.; *Bioorg. Med. Chem. Lett.* **1995**, *5*, 507.
[155] Beer D., Bhalay G., Dunstan A., et al.; *Bioorg. Med. Chem. Lett.* **2002**, *12*, 1973.
[156] Brown, D. S., Revill, J. M., and Shute, R. E.; *Tetrahedron Lett.* **1998**, *39*, 8533.
[157] Ma, S., Duan, D., and Wang, Y.; *J. Comb. Chem.* **2002**, *4*, 239.
[158] Ramaseshan, M., Ellingboe, J. W., Dory, Y. L., and Deslongchamps, P.; *Tetrahedron Lett.* **2000**, *41*, 4743.
[159] Wang, X., Choe, Y., Craik, C. S., and Ellman, J. A.; *Bioorg. Med. Chem. Lett.* **2002**, *12*, 2201.
[160] Nakamura, K., Ishii, A., Ito, Y., and Nakahara, Y.; *Tetrahedron* **1999**, *55*, 11253.
[161] Lipshutz, B. H., and Shin, Y-J.; *Tetrahedron Lett.* **2001**, *42*, 5629.
[162] Lindsley, C. W., Hodges, J. C., Filzen, G. F., et al.; *J. Comb. Chem.* **2000**, *2*, 550.

[163] Doi, T., Sugiki, M., Yamada, H., *et al.*; *Tetrahedron Lett.* **1999**, *40*, 2141.
[164] Doi, T., Hijikuro, I., and Takahashi, T.; *J. Am. Chem. Soc.* **1999**, *12*, 6749.
[165] Lindsley, C. W., Chan, L. K., Goess, B. C., *et al.*; *J. Am. Chem. Soc.* **2000**, *122*, 422.
[166] Suginome, M., Iwanami, T., and Ito, Y.; *J. Am. Chem. Soc.* **2001**, *123*, 4356.
[167] Spring, D. R., Krishnan, S., and Schreiber, S. L.; *J. Am. Chem. Soc.* **2000**, *122*, 5656.
[168] Pelish, H. E., Westwood, N. J., Feng, Y., *et al.*; *J. Am. Chem. Soc.* **2001**, *123*, 6740.
[169] Blackwell, H. E., Perez, L., and Schreiber, S. L.; *Angew. Chem., Int. Ed.* **2001**, *40*, 3421.
[170] Blackwell, H. E., Perez, L., Stavenger, R. A., *et al.*; *Chem. Biol.* **2001**, *8*, 1167.
[171] Kwon, O., Park, Seung B., and Schreiber, S. L.; *J. Am. Chem. Soc.* **2002**, *124*, 13402.
[172] Wei, Q., Harran, S., and Harran, P. G.; *Tetrahedron* **2003**, *59*, 8947.
[173] Taylor, S. J., Taylor, A. M., and Schreiber, S. L.; *Angew. Chem. Int. Ed.* **2004**, *43*, 1681.
[174] Burke, M. D., Berger, E. M., and Schreiber, S. L.; *J. Am. Chem. Soc.* **2004**, *126*, 14095.
[175] Mukherjee, S., Robinson, C. A., Howe, A. G., *et al.*; *Bioorg. Med. Chem. Lett.* **2007**, *17*, 6651.
[176] Meloni, M. M., Brown, R. C. D., White, P. D., and Armour, D.; *Tetrahedron Lett.* **2002**, *43*, 6023.
[177] Bol, K. M., and Liskamp, R. M. J.; *Tetrahedron* **1992**, *48*, 6425.
[178] Sliedregt, K. M., Schoute, A., Kroon, J., and Liskamp, R. M. J.; *Tetrahedron Lett.* **1996**, *37*, 4237.
[179] Hoekstra, W. J., Greco, M. N., Yabut, S. C., *et al.*; *Tetrahedron Lett.* **1997**, *38*, 2629.
[180] Christie, B. D., and Rapoport, H.; *J. Org. Chem.* **1985**, *50*, 1239.
[181] Gerspacherm M., and Rapoport, H.; *J. Org. Chem.* **1990**, *56*, 3700.
[182] Gosselin, F., Betsbrugge, J. V., Hatam, M., and Lubell, W. D.; *J. Org. Chem.* **1999**, *64*, 2486.
[183] Bleicher, K. H., Lutz, C., and Wüthrich, Y.; *Tetrahedron Lett.* **2000**, *41*, 9037.
[184] Jansen, K. J., Alsina, J., Songster, M. F., *et al.*; *J. Am. Chem. Soc.* **1998**, *120*, 5441.
[185] Alsina, J., Jensen, K. J., Albericio, F., and Barany, G.; *Chemistry Eur. J.* **1999**, *5*, 2787.
[186] Forns, P., Sevilla, S., Erra, M., *et al.*; *Tetrahedron Lett.* **2003**, *44*, 6907.
[187] Albericio, F., Kneib-Cordonier, N., Biancalana, S., *et al.*; *J. Org. Chem.* **1990**, *55*, 3730.
[188] Jessing, M., Brandt, M., Jensen, K. J., *et al.*; *J. Org. Chem.* **2006**, *71*, 6734.
[189] Hammershoej, P., Jessing, M., Madsen, A. O., *et al.*; *Int. J. Pept. Res. Ther.* **2007**, *13*, 209.
[190] Castro, J. L., and Matassa, V. G.; *J. Org. Chem.* **1994**, *59*, 2289.
[191] Dragoli, D. R., Burdett, M. T., and Ellman, J. A.; *J. Am. Chem. Soc.* **2001**, *123*, 10127.
[192] Bräse, S., Köbberling, J., Enders, D., *et al.*; *Tetrahedron Lett.* **1999**, *40*, 2105.
[193] Areniyadis, S., Wagner, A., and Mioskowski, C.; *Tetrahedron Lett.* **2002**, *43*, 9717.
[194] Ho, C. Y., and Kukla, M. J.; *Tetrahedon Lett.* **1992**, *38*, 2799.
[195] Gosselin, F., Betsbrugge, J. V., Hatam, M., and Lubell, W. D.; *J. Org. Chem.* **1999**, *64*, 2486.
[196] Boojamra, C. G., Burow, K. M., Lorin, K. M., *et al.*; *J. Org. Chem.* **1997**, *62*, 1240.
[197] Swayze, E. E.; *Tetrahedron Lett.* **1997**, *38*, 8465.
[198] Bourne, G. T., Meutermans, W. D. F., and Smythe, M. L.; *Tetrahedron Lett.* **1999**, *40*, 7271.
[199] Alsina, J., Jensen, K. J., Albericio, F., and Barany, G.; *Chem. Eur. J.* **1999**, *5*, 2787.
[200] Guillaumie, F., Kappel, J. C., Kelly, N. M., *et al.*; *Tetrahedron Lett.* **2000**, *41*, 6131.
[201] Jin, J., Graybill, T. L., Wang, M. A., *et al.*; *J. Comb. Chem.* **2001**, *3*, 97.
[202] Alsina, J., Jensen, K. J., Songster, M. F., *et al.*; *Solid-Phase Org. Syn.* **2001**, *1*, 121.
[203] Jamieson, C., Congreve, M. S., Hewitt, P. R., *et al.*; *J. Comb. Chem.* **2001**, *3*, 397.
[204] Boas, U., Brask, J., Christensen, J. B., and Jensen, K. J.; *J. Comb. Chem.* **2002**, *4*, 223.
[205] Shannon, S. K., Peacock, M. J., Kates, S. A., and Barany, G.; *J. Comb. Chem.* **2003**, *5*, 860.
[206] Pittelkow, M., Boas, U., Jessing, M., *et al.*; *Org. Biomol. Chem.* **2005**, *3*, 508.
[207] Gross, C. M., Lelievre, D., Woodward, C. K., and Barany, G.; *J. Pept. Res.* **2005**, *65*, 395.
[208] Olsen, C. A., Witt, M., Franzyk, H., and Jaroszewski, J. W.; *Tetrahedron Lett.* **2006**, *48*, 405.
[209] Springer, J., de Cuba, K. R., Calvet-Vitale, S., *et al.*; *Eur. J. Org. Chem.* **2008**, *15*, 2592.
[210] Gu, W., and Silverman, R. B.; *Org. Lett.* **2003**, *5*, 415.
[211] Alsina, J., Yokum, T. S., Albericio, F., and Barany, G.; *J. Org. Chem.* **1999**, *64*, 8761.
[212] Bilodeau, M. T., and Cunningham, A. M.; *J. Org. Chem.* **1998**, *63*, 2800.
[213] Harikrishnan, L. S., and Hollis Showalter, H. D.; *Synlett* **2000**, *9*, 1339.

[214] Fivush, A. M., and Willson, T. M.; *Tetrahedron Lett.* **1997**, *38*, 7151.
[215] Ouyang, X., Tamayo, N., and Kiselyov, A. S.; *Tetrahedron* **1999**, *55*, 2827.
[216] Aoki, Y., and Kobayashi, S.; *J. Comb. Chem.* **1999**, *1*, 371.
[217] Jeon, M-K., Kim, D-S., La, H. J., and Gong, Y-D.; *Tetrahedron Lett.* **2005**, *46*, 4979.
[218] Hasegawa, M., Ohno, H., Tanaka, H., *et al.*; *Bioorg. Med. Chem. Lett.* **2006**, *16*, 158.
[219] Maletic, M., Antonic, J., Leeman, A., *et al.*; *Bioorg. Med. Chem. Lett.* **2003**, *13*, 1125.
[220] Shimizu, H., Ito, Y., Kanie, O., and Ogawa, T.; *Bioorg. Med. Chem. Lett.* **1996**, *6*, 2841.
[221] Manabe, S., Ito, Y., and Ogawa, T.; *Synlett.* **1998**, 628.
[222] Nicolaou, K. C., Winssinger, N., Vourloumis, W. D., *et al.*; *J. Am. Chem. Soc.* **1998**, *120*, 10814.
[223] Mikami, K., Motoyama, Y., and Terada, M.; *J. Am. Chem. Soc.* **1994**, *116*, 2812.
[224] Yli-Kauhalouma, J.; *Tetrahedron* **2001**, *57*, 7053.
[225] Stavenger, R. A., and Schreiber, S. L.; *Angew. Chem. Int. Ed.* **2001**, *40*, 3417.
[226] Leßmann, T., Leuenberger, M. G., Menninger, S., *et al.*; *Chem. Biol.* **2007**, *14*, 443.
[227] Murphy, A. M., Dagnino, R. Jr, Vallar, P. L., *et al.*; *J. Am. Chem. Soc.* **1992**, *114*, 3156.
[228] Poupart, M-A., Fazal, G., Goulet, S., and Mar, l. T.; *J. Org. Chem.* **1999**, *64*, 1356.
[229] Lee, A., Huang, L., and Ellman, J. A.; *J. Am. Chem. Soc.* **1999**, *121*, 9907.
[230] Haag, R., Leach, A. G., Ley, S. V., *et al.*; *Syn. Comm.* **2001**, *31*, 2965.
[231] Bayer, E., and Mutter, M.; *Nature* **1972**, *237*, 512.
[232] Bayer, E., Mutter, M., Uhmann, R., *et al.*; *J. Am. Chem. Soc.* **1974**, *96*, 7333.
[233] Barrett, A. G. M., Procopio, P. A., and Voigtmann, U.; *Org. Lett.* **2001**, *3*, 3165.
[234] Ball, C. P., Barrett, A. G. M., Commercon, A., *et al.*; *J. Chem. Soc. Chem. Commun.* **1998**, *18*, 2019.
[235] Lepore, S. D., and Wiley, M. R.; *Org. Lett.* **2003**, *5*, 7.
[236] Gordon, K., and Balasubramanian, S.; *J. Chem. Technol. Biotecnol.* **1999**, *74*, 835.
[237] Clapham, B., Reger, T. S., and Janda, K. D.; *Tetrahedron* **2001**, *57*, 4637.
[238] Lazny, R., Nodxewska, A., and Zabicka, B.; *Wiadomsci Chemiczne* **2006**, *60*, 191.
[239] Solinas, A., and Taddei, M.; *Synthesis* **2007**, *16*, 2409.
[240] Allin, S. M., and Shuttleworth, S. J.; *Teterahedron Lett.* **1996**, *37*, 8023.
[241] Purandare, A. V., and Natarajan, S.; *Tetrahedron Lett.* **1997**, *38*, 8777.
[242] Phoon, C. W., and Abell, C.; *Tetrahedron Lett.* **1998**, *39*, 2655.
[243] Reggelin, M., Brening, V., and Welcker, R.; *Tetrahedron Lett.* **1998**, *39*, 4801.
[244] Suginome, M., Iwanami, T., and Ito, Y.; *J. Am. Chem. Soc.* **2001**, *123*, 4356.
[245] Huchison, P. C., Heightman, T. D., and Procter, D. J.; *Org. Lett.* **2002**, *4*, 4583.
[246] Wang, X., Yin, L., Yang, T., and Wang, Y.; *Tetrahedron: Asymmetry* **2007**, *18*, 108.
[247] Chung, C. W. Y., and Toy, P. H.; *Tetrahedron: Asymmetry* **2004**, *15*, 387.
[248] Kawana, M., and Emoto, S.; *Tetrahedron Lett.* **1972**, *18*, 4855.
[249] Worster, P. M., McArthur, C. R., and Leznoff, C. C.; *Angew. Chem. Int. Ed.* **1979**, *18*, 221.
[250] Moon, H., Schrore, N. E., and Kurth, M. J.; *Tetrahedron Lett.* **1994**, *35*, 8915.
[251] Colwell, A. R., Duckwall, L. R., Brooks, R., and McManus, S. P.; *J. Org. Chem.* **1981**, *46*, 3097.
[252] Oertel. K., Zech, G., and Kunz, H.; *Angew. Chem. Int. Ed.* **2000**, *39*, 1431.
[253] Zech, G., and Kunz, H.; *Angew. Chem. Int. Ed.* **2003**, *42*, 787.
[254] O'Donnell, M. J., Delgado, F., and Pottorf, R. S.; *Tetrahedron* **1999**, *55*, 6347.
[255] O'Donnell, M. J., and Delgado, F.; *Tetrahedron* **2001**, *57*, 6641.
[256] O'Donnell, M. J., Zhou, C., and Scott, W. L.; *J. Am. Chem. Soc.* **1996**, *118*, 6070.
[257] Corey, E. J., Xu, F., and Noe, M. C.; *J. Am. Chem. Soc.* **1997**, *119*, 12414.
[258] Chinchilla, R., Mazón, P., and Nájera, C.; *Tetrahedron: Asymmetry* **2000**, *11*, 3277.
[259] Nagashima, T.; and Davies, H. M. L.; *J. Am. Chem. Soc.* **2001**, *123*, 2695.
[260] Davies, H. M. L., Nagashima, T., and Klino, J. L. III.; *Org. Lett.* **2000**, *2*, 823 and references cited therein.
[261] Gais, H-J., Babu, G. S., Guenter, M., and Das, P.; *Eur. J. Org. Chem.* **2004**, *7*, 1464.
[262] Hachtel, J., and Gais, H-J.; *Eur. J. Org. Chem.* **2000**, *8*, 1457.

3
Nucleophile Cleavable Linker Units

Andrea Porcheddu and Giampaolo Giacomelli

Department of Chemistry, University of Sassari, Italy

3.1 Introduction

Many synthetic compounds exhibit various pharmacological properties and any modification of the core structures, such as replacement of functional substituent groups, provides a high degree of molecular diversity, which has been demonstrated to be useful in the search for new therapeutic agents.

Literature reports have focused on the basic principles and the fundamental reactions for the synthesis of biologically active compounds, and most of these reactions refer to the homogeneous solution phase. However, it is now standard practice to conduct many synthetic operations on a solid phase where the substrate (or sometimes the reagent) is covalently bound to an insoluble polymeric support. This is particularly important for creating large 'libraries' of compounds.

Moreover, recently, the synthesis of new compounds has received special attention within combinatorial chemistry, which has grown-up to become a key tool in many aspects of chemistry in general and in drug discovery process in particular. Combinatorial compounds are produced either by solution-phase synthesis or by synthesizing compounds bound covalently to polymeric supports.

One of the most persistent debates in the field of combinatorial chemistry involves the relative advantages and disadvantages of solid-phase versus solution-phase combinatorial synthesis. Solid-phase synthesis allows multi-step reactions to be carried out and reactions to be forced to completion because an excess reagent can be added. The great advantage of solid-supported reactions is, however, the manipulation and purification of products; a sequence of reactions can be performed with the by-products and excess reagents being washed away, without the need for chromatography or isolation of intermediates. The final products are then cleaved from the polymer in a state of relatively high purity. Many of the reactions, both functional group transformations and scaffold constructions, with a wide range of reagents, have already been

78 Traditional Linker Units for Solid-Phase Organic Synthesis

adapted to the solid state and combinatorial chemistry, and the list is growing rapidly. But a much wider range of organic reactions is available for solution-phase synthesis, the technology used traditionally by most synthetic organic chemists, and products in solution can be more easily identified and characterized.

Since the beginning, combinatorial libraries have been developed using primarily parallel synthesis and specific techniques, that is, resin-based chemistry. The rapid generation of small-molecule libraries can be carried out effectively by using combinatorial or simultaneous-parallel synthesis on solid supports. The combination of solid-phase organic synthesis (SPOS) and the development of high-throughput screening (HTS) has greatly increased the number of substances that are being tested and has emerged as a valuable tool in the search for novel lead structures. Furthermore, SPOS offers the opportunity to synthesize drug-like molecules via novel routes, which may be difficult or impossible using traditional solution methods, and allows the possibility to rapidly synthesize these molecules without tedious and time-consuming purifications.

In brief, solid-phase synthesis is no longer just a tool to prepare large numbers of compounds through coupling chemistry. More and more it is becoming a domain on its own right, which gives rise to important research efforts and numerous publications. Synthetic approaches have evolved from the direct, though not trivial, transposition of existing homogeneous phase schemes to the design of original methods that exploit the best advantages of SPOS. At the same time, specific linkers and anchoring strategies have been developed to better fit all synthetic approaches.

3.2 Linker units

The arrival of combinatorial techniques, first for peptides and nucleotides then for small molecules, has brought about a major renewal in the study of solid-phase organic synthesis. Large numbers of reactions have been converted from solution phase to solid phase and more continue to be developed.

The attachment of the molecule to the solid phase is critical to this process. Effectiveness in anchoring and removing a small organic molecule from the polymeric resin depends on the right choice of the linker group. This fragment is key in the development of a synthetic strategy. This can be accomplished through a cleavable linker. A linker acts as an attachment point to the polymer and should provide an efficient chemical means of attachment and cleavage (Scheme 3.1). Reactivity of the linker affects the scope of the chemistry that can be employed during the library synthesis. Many linkers have their roots in protecting group chemistry.

The attachment point of the linker to the solid support should be chemically stable during the synthesis and cleavage; yields for its loading and cleavage should be as quantitative as possible. A linker is really

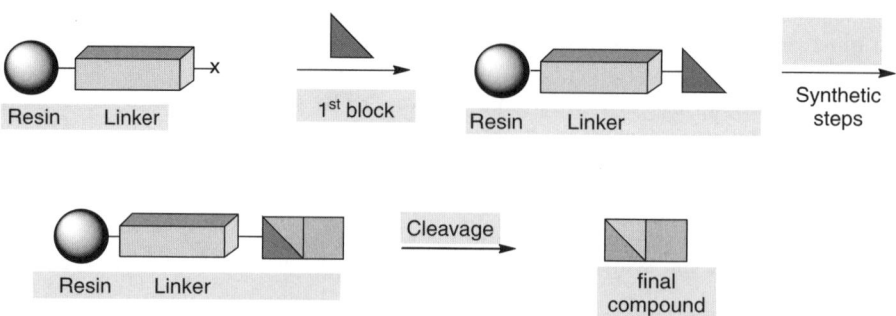

Scheme 3.1 *Step-wise solid-phase synthesis*

a bifunctional protecting group that is attached to the molecule being synthesized through a bond labile to the cleavage conditions and to the solid-phase polymer through a more stable bond. This definition gives a clear image for the most of the linkers used. Indeed, linkers achieve similar functions to protecting groups and many of the linkers developed in recent years are based on the protecting groups frequently used in solution-phase synthesis. There are, however, many other linkers that are not based on common protecting group methods, such as the increasing range of traceless linkers, and those that rely on cleavage or cyclization. These linkers fall into the broader definition as a connection between the molecule being synthesized and the solid-phase polymer that is cleaved to release the desired molecule. This cleavage may involve a range of techniques and give a range of functional groups. The functional group obtained may be dependent not only on the linker but also on the method of cleavage.

The ideal linker would match many important criteria. First of all it would be cheap and readily available and the attachment of starting material would be readily accomplished in high yield. The linker would be stable to the reaction conditions used in the synthesis and cleavage would be not wasteful so that the final product is not damaged.

As linkers are often applied to the synthesis of large libraries of compounds, a further important criterion for the ideal linker is that the cleavage method should be easily worked up in large numbers and should not introduce impurities that are difficult to remove. However, many linkers do not realize all of these criteria. The attachment process may be tricky, although the procedure of attachment in the solution phase may be a valid starting point. All linkers must go on in the course of the synthesis, although the choice of synthetic route may determine the choice of linker, and vice versa. Cleavage conditions are often severe, for example trifluoroacetic acid (TFA), and not all molecules will survive these conditions. It is important to consider this when planning a library synthesis.

Generally, linkers, as immobilized protecting groups, have to be classified into two types: (i) *Integral linkers*, in which the linker is formed directly from part of the solid support core, and (ii) *Grafted linkers*, in which the linker is attached to the resin core. For example, the original chloromethylated polystyrene Merrifield resin must be considered as an integral linker. This allowed Merrifield to anchor N-protected amino acids onto the resin by formation of immobilized benzyl esters. However, the majority of linkers used in solid-phase synthesis are of the grafted type, as the disadvantage with any integral linker is the control of synthesis. In fact, this occurs directly on the resin, its outcome being so influenced by the whole range of steric and electronic effects. The exact degree of loading and functionalization can be hard to control. The grafted linkers can be loaded onto the resin and then derivatized or preloaded prior to linking. For example, the p-alkoxybenzyl alcohol Wang linker was initially prepared by reacting 4-hydroxybenzyl alcohol with Merrifield resin in the presence of sodium methoxide ($NaOCH_3$).[1]

Many families of solid phase linkers are currently used in SPOS and are classified according to conditions used for the cleavage. The most frequently kind of linkers used in SPOS are **electrophilic** (Chapter 2) and **nucleophilic labile** (Chapter 3), **cyclative** (Chapter 4), **photolabile** (Chapter 5), **safety catch** (Chapter 6) and **traceless/multifunctional** (Part Two) linkers. Many historically important resins (Merrifield, Wang, Sasrin, Sieber, Rink resins) have linkers that are cleaved under acidic conditions. In particular, acidic conditions were intended to prevent racemization of amino acids during solid phase peptide synthesis.

3.3 Nucleophilic labile linker units

The wide use of electrophilic labile linkers in SPOS has reduced the development of nucleophilic labile linkers. However, they are popular in oligonucleotide SPOS. Although the term 'base labile' is more used, two types of cleavage are included under this heading. The more common type actually involves nucleophilic addition/elimination, usually on an ester, whereas the less common type involves a base

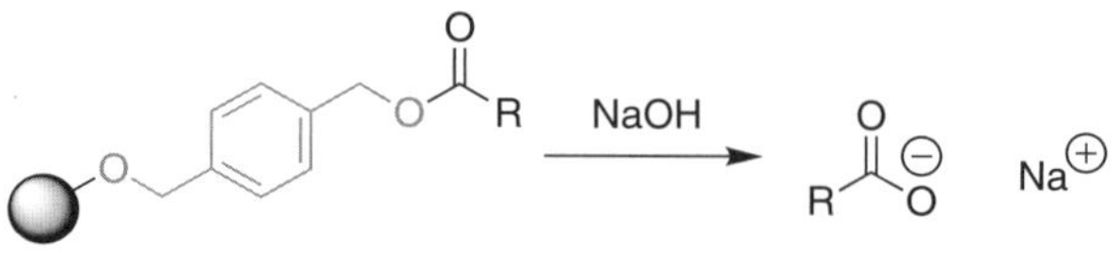

Scheme 3.2 Base-mediated cleavage

catalysed reaction, such as elimination or cyclization. In its simplest form, nucleophilic cleavage can be considered as an extension of the base mediated cleavage employed by Merrifield (Scheme 3.2).[2] Cleavage may also be achieved using sodium methoxide in methanol to give the methyl carboxylic ester, or by an unhindered amine such as methylamine to give the amide.

All of these cleavage reagents are bases, although the reactivity is determined primarily by their nucleophilicity. However, one the most frequent drawbacks associated with these kind of linkers is the possibility of unwanted cleavage, particularly with ester-linked substrates. This drawback can be solved by using safety-catch linkers, in which a specific activation must be done before the cleavage.[3]

Even in this case there may be a disadvantage because often the polar functional group that is introduced remains attached to the final compound and can greatly influence the possible biological activity of the target molecule if not removed. A possible solution to this problem can involve the use of cyclo-release linker units: these are linked to the support but the cleavage leaves the polar functional group attached to the support. In every case, the use of this kind of linker is limited, as many target substrates are not cyclic or do not have a suitable ring size.

This drawback has led to the development of traceless linkers that are, for definition, those which leave a hydrogen residue upon cleavage. The disadvantage is that no functional group can be introduced in the cleavage steps.

3.3.1 Cleavage by saponification or basic trans-esterification

The simplest linkers from which alcohols or carboxylic acids can be cleaved are carboxylic acid and hydroxyalkyl resins, respectively. Looking at the commonly applied basic cleavage conditions, it is evident that post cleavage work-up procedures are necessary to obtain pure products. However, the advantage of this kind of anchorage is the high stability under many reaction conditions.[4]

Saponification and the release of peptide acids has been used since the introduction of solid-phase chemistry by Merrifield in 1963 (Scheme 3.3)[2] and represents the classical preparation of peptide acids, the method being completely orthogonal to *tert*-butoxycarbonyl (Boc) and benzyloxycarbonyl (Cbz) protecting groups. Mild conditions such as five equivalents of potassium carbonate (K_2CO_3) in methanol for 48 hours at room temperature or tetrabutylammonium hydroxide in tetahydrofuran (THF) are often used. These work-up conditions are necessary to prevent any racemizing reaction occurring during cleavage. Moreover, the use of this method resulted in limited by-products, even if with not complete yields. Better results were achieved using a mixture of five equivalents of lithium hydroxide in water/methanol/THF (1:2:5) under reflux for 1–2 days.[5]

A recent application of hydroxymethylated polystyrene resin included the synthesis of a dihydropyridone scaffold **4**.[6] In this communication, commercially available hydroxymethylated polystyrene resin was converted into the chloroformate solid support **1**, to which a premixed solution of 4-methoxypyridine and the desired Grignard reagent in anhydrous THF was added to afford the polymer-bound dihydropyridone **3** (Scheme 3.3). After a few minutes the reaction was quenched with 3 N aqueous HCl/THF (1:1) and

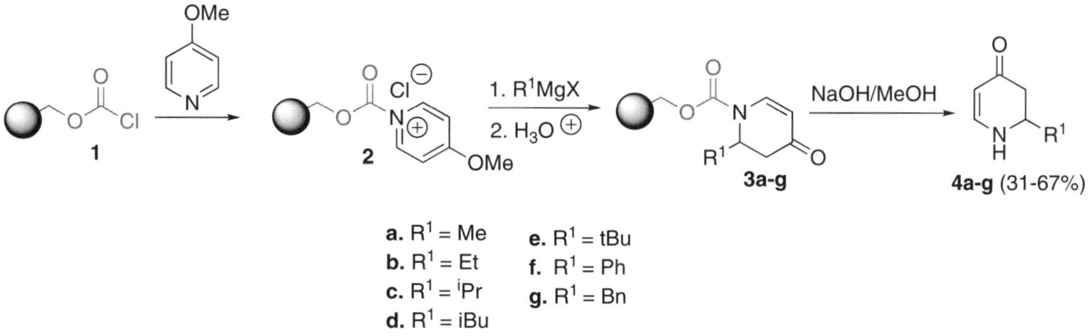

Scheme 3.3 Synthesis of 2-alkyl-2,3-dihydropyridin-4(1H)-ones

a. R^1 = Me
b. R^1 = Et
c. R^1 = ^{i}Pr
d. R^1 = iBu
e. R^1 = tBu
f. R^1 = Ph
g. R^1 = Bn

subsequently washed with solvent and dried. Cleavage was accomplished by adding a catalytic amount to one equivalent of a 4.37 M solution of sodium hydroxide in methanol to a suspension of the resin in THF. The advantage of this procedure is that any **2**, which does not react with the Grignard reagent, collapses to parent resin upon workup.

Typical Experimental Procedures

Synthesis of 4a (Scheme 3.3)[6]

*To a suspension of chloroformate resin **1** (0.5 g, 0.41 mmol, based on loading of 0.81 mmol/g) in THF (6 ml) in a 15 ml polypropylene tube fitted with a frit was added a pre-mixed solution of 4-methoxypyridine (1.53 mg, 0.48 mmol) and a solution of methylmagnesium chloride (3.0 M in THF solution, 0.27 ml, 0.81 mmol). The resulting reaction mixture was vigorously mixed then immediately filtered. The resin was then washed with a 1:1 mixture of 3 M aqueous HCl:THF solution (5 × 5 ml), 3 M HCl (3 × 5 ml), methanol (3 × 5 ml), DMF (3 × 5 ml) and dichloromethane (CH_2Cl_2) (3 × 8 ml) then air dried to afford the resin bound **3a**. Analysis of a small sample of beads (3 mg) by FTIR confirmed the dihydropyridone scaffold **3a** on solid support. THF (8 ml) was added to a 4.37 M solution of sodium methoxide/methanol (0.093 ml, 0.41 mmol) and the reaction mixture was agitated for 1 h. Ammonium chloride (5 equiv. 100 mg) was then added and agitation was continued overnight. The mixture was filtered and the resin was washed THF (3 × 5 ml). The filtrates were passed through a plug of celite (celite was washed with THF). The filtrate was collected, concentrated and dried under reduced pressure to give **4a** as colourless oil (65% yield, 98% purity HPLC).*

A major aspect of solid-phase chemistry is synthesizing heterocyclic compounds from various methods. In particular, low molecular weight heterocyclic compounds are used in pharmaceutical lead compounds because they are highly functionalizable scaffolds. For example, the solid-phase synthesis of indoles is mainly focused on two different approaches, namely, the palladium catalysed approaches and the Fischer indole synthesis. Recently, Kondo *et al.* have developed a new method for the solid-phase indole syntheses via enaminoester.[7] Enaminoesters **8** were prepared in a solid phase, rhodium catalysed $N-H$ insertion into immobilized carbenoids, which were generated from the immobilized α-diazophosphonoacetate **6** followed by a Horner–Emmons reaction (Scheme 3.4). Intramolecular palladium catalysed cyclization of the α- or β-(2-halophenyl)amino-substituted α,β-unsaturated esters **8** was effective in the solid-phase synthesis of indole 2- and 3-carboxylates **9** with various functional groups on the benzene ring. Moreover,

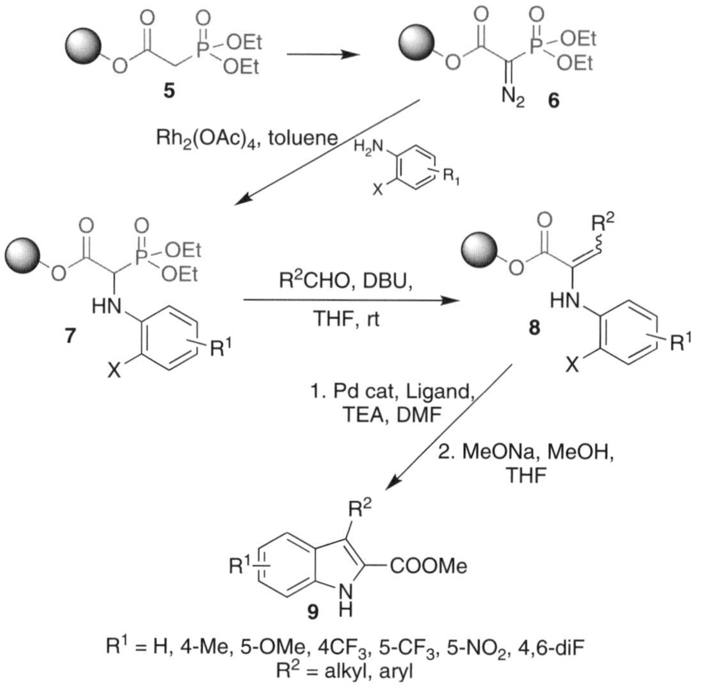

82 *Traditional Linker Units for Solid-Phase Organic Synthesis*

R^1 = H, 4-Me, 5-OMe, 4CF$_3$, 5-CF$_3$, 5-NO$_2$, 4,6-diF
R^2 = alkyl, aryl

Scheme 3.4 *Solid-phase synthesis of indole-2-carboxylates*

the intramolecular palladium catalysed amination route was a useful strategy and the palladium catalysed tandem C, N-arylation also provided a new one-pot method for solid-phase indole synthesis.

More recently, cleavage by basic trans-esterification with excess NaOCH$_3$ in a 1:4 mixture of methanol/THF was conducted for the solid-phase synthesis of 1,2,3-trisubstituted indoles **14** via iodocyclization (from **12** to **13**, Scheme 3.5).[8] The benzoic acid **10** was anchored to commercially available Wang chlorinated resin (3.0 equiv. of dicaesium carbonate (Cs$_2$CO$_3$), 0.5 equiv. of potassium iodide (KI), 1.5 equiv. of acid for Cl residue, DMF, 80 °C), providing resin **11** with a loading of 0.67 mmol/g. The Sonogashira reaction of **11** under standard reaction conditions [catalyst PdCl$_2$(PPh$_3$)$_2$, co-catalyst copper iodide (CuI), HNEt$_2$] afforded the immobilized alkynylaniline **12a** in excellent yield and purity.[9] The resin-bound anilines **11** and **12a** have been verified by cleaving a small sample of the resin with 50% TFA/DCM (dichloromethane) (1 h, room temperature); however, in later work, aniline **11** and all Sonogashira products **12** have been verified by NaOCH$_3$ cleavage. The cyclization of **12a** to the polymer-bound 3-iodoindole **13a** was carried out in DCM using a threefold excess of iodine at room temperature for 24 hours. The resulting polymer-bound indole **12a** was cleaved from the resin using 50% TFA/DCM (1 h, room temperature).

Unfortunately, the desired 3-iodoindole **14** was isolated in only a 40% yield with poor purity. On the contrary, when the polymer-bond indole **13a** was cleaved from the resin by basic trans-esterification with excess NaOCH$_3$ in a 1:4 mixture of methanol/THF, the corresponding methyl indole-5-carboxylate **14a** was obtained in a 99% yield with excellent purity.

Brase and coworkers have developed for the first time Bartoli indole synthesis on solid supports.[10] Starting from Merrifield resin, they performed immobilization of five nitro benzoic acids **15** (Scheme 3.6). Addition of four different alkenyl Grignard reagents and basic cleavage leads to substituted methyl indole carboxylates **17** in excellent purities. Features of this reaction are the stability of halide groups, ester

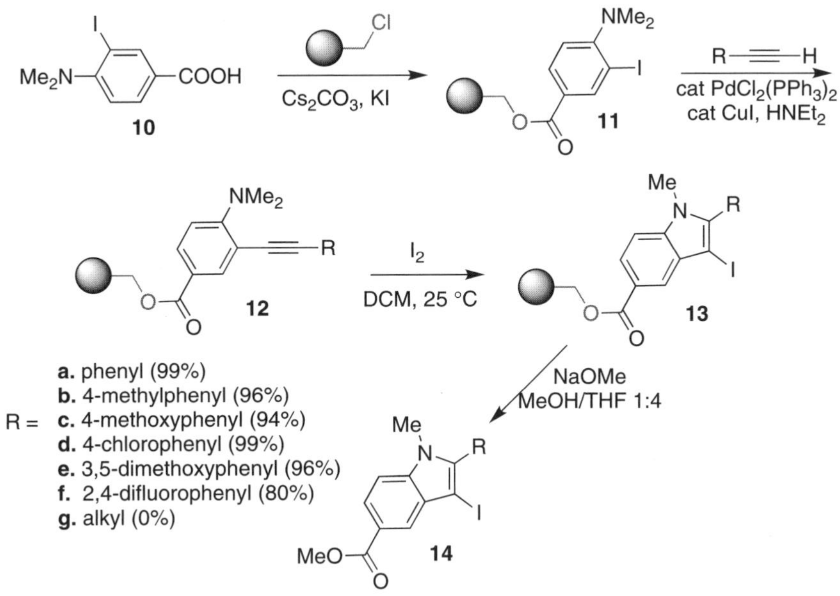

Scheme 3.5 Solid-phase synthesis of 1,2-disubstituted 3-iodoindoles

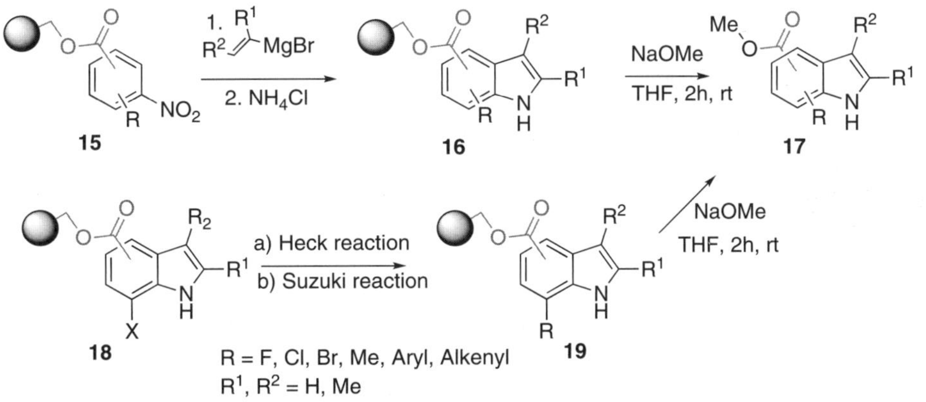

R =
a. phenyl (99%)
b. 4-methylphenyl (96%)
c. 4-methoxyphenyl (94%)
d. 4-chlorophenyl (99%)
e. 3,5-dimethoxyphenyl (96%)
f. 2,4-difluorophenyl (80%)
g. alkyl (0%)

Scheme 3.6 Solid-phase synthesis of methyl indole carboxylates

R = F, Cl, Br, Me, Aryl, Alkenyl
R^1, R^2 = H, Me

moieties and tolerance of O, O-unsubstituted nitro resins. Heck and Sonogashira reactions are also possible with immobilized indoles **18**.

Nicolaou *et al.* have developed four novel classes of potent FXR activators originating from natural product-like libraries.[11] Initial screening of a 10,000-membered, diversity-orientated library of benzopyran containing small molecules for farnesoid X receptor (FXR) activation using a cell-based reporter assay led to the identification of several lead compounds possessing low micromolar activity.

With the initial lead compound identified and validated, the stage was set for the systematic optimization of the three regions of the lead structure. These compounds were systematically optimized employing parallel solution-phase synthesis and solid-phase synthesis to provide four classes of compounds that

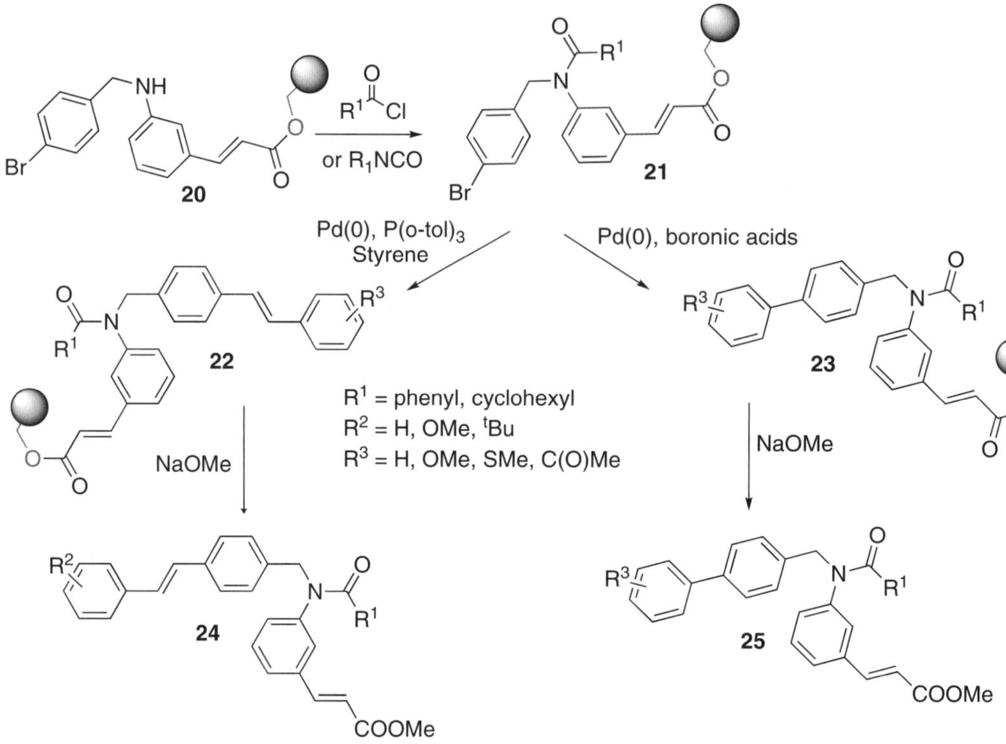

84 Traditional Linker Units for Solid-Phase Organic Synthesis

Scheme 3.7 *Synthesis of libraries of substituted cinnamates as FRX agonists*

potently activate FXR. They have employed the Heck coupling [Pd(dba)$_3$P(o-Tol)$_3$, TEA] with substituted styrenes **22** and the Suzuki coupling [Pd(PPh$_3$)$_4$, Cs$_2$CO$_3$] with boronic acids **23** as the key step for the generation of focused libraries of biaryl and stilbene cinnamates **24** and **25** (Scheme 3.7). Two series of compounds, bearing stilbene or biaryl moieties, contain members that are the most potent FXR agonists reported to date in cell-based assays.

Hetero Diels–Alder reactions (HDAR) represent one of the more efficient and valuable tools for the construction of six-membered heterocycles which, in turn, probably are the most common cyclic moiety encountered in bio-organic and medicinal chemistry.[12] The development of [4+2] cycloadditions on a solid phase represents an appreciated target either for a possible simplification of the synthetic procedure or for their potential application in combinatorial chemistry.

In a very interesting paper, Menichetti *et al.* have reported the successful extension on a solid phase of the *N*-thiophthalimide mediated generation of α,α′-dioxothiones. Using the easily available Wang or OH-modified Merrifield resins as solid supports, and avoiding the introduction of any linker, dioxothiones were obtained under very mild reaction conditions and reacted in HDAR either as electron-poor dienes or dienophiles. Using highly reactive small dienophiles the isolation from the solid phase of the oxathiin cycloadducts was more simple and convenient than the parallel solution-phase procedure.

Into details, the β-ketoester function was linked to a Wang resin by direct trans-esterification with *tert*-butylacetoacetate[13] in toluene at 110 °C, to obtain immobilised ketoester **26** (Scheme 3.8). The reaction of a DCM pre-swelled suspension of resin **26** with 1.4 equiv. of phthalimidesulfenyl chloride (PhtS-Cl) at room temperature, allowed the transformation into the α,α′-dioxothiophthalimide modified resin **27** (Scheme 3.8). Eventually, the reaction of **27** with 1 equiv. of pyridine in trichloro methane (CHCl$_3$) at

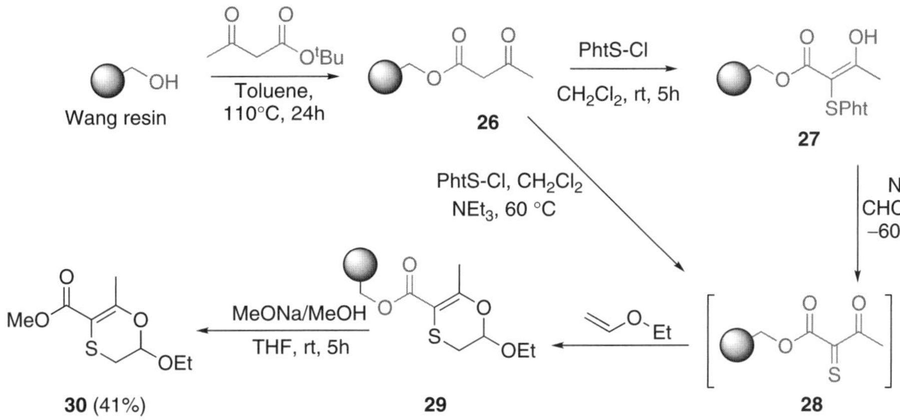

Scheme 3.8 *Solid-phase synthesis of methyl 6-ethoxy-2-methyl-5,6-dihydro-1,4-oxathiine-3-carboxylate*

room temperature afforded the solid-supported α,α'-dioxothione **28** that, in the presence of 5 equiv. of ethyl vinyl ether, underwent an inverse electron demand HDAR to give the resin-linked oxathiin **29**.

Studies into the biological activities of nucleosides and their phosphorylated derivatives have been a fundamental and fruitful field of research since the 1940s.[14] In a recent paper, Agrofoglio *et al.* described the solid-phase synthesis of various substituted pyrimidine nucleosides starting from 2'-deoxyuridine **31**, which had been attached through a nucleophilic labile linker **32** to polystyrene resins. Thus, starting from 2'-deoxyuridine **31**, the nucleoside was attached via the 5'-OH moiety to a base labile linker **32** (easily obtained in seven steps from commercially available *p*-bromomethyl benzoic acid, Scheme 3.9). Following acetylation of the 3'-hydroxyl group and removal of the *tert*-butyl ester group of **33**, the carboxylic acid **34** was coupled to commercially available resin **35** using the *N*-hydroxybenzotriazole activated ester and subsequent capping of residual amine groups borne by the resin. Cleaving **36** via a basic trans-esterification with excess CH$_3$ONa in a mixture of methanol/dioxane performed analysis of the substitution level of the resin, leading to a highly pure form of 2'-deoxy-D-uridine. An iodo group was selectively introduced at the

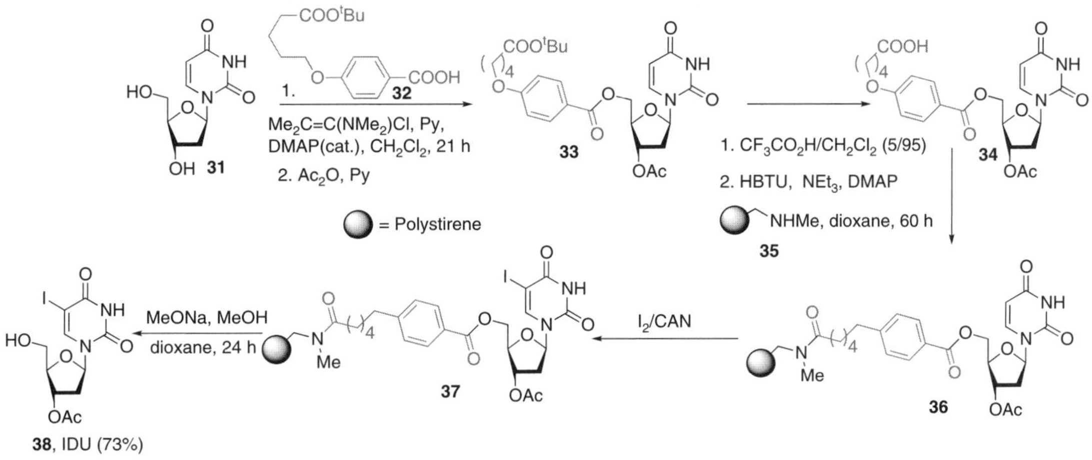

Scheme 3.9 *Solid-phase synthesis of 5-iodo-2'-deoxyuridine*

C-5 position of the heterocyclic moiety using iodine/cerium(IV) ammonium nitrate (CAN), and cleavage of resulting **37** afforded the well-known antiviral **38**, 5-iodo-2'-deoxyuridine (IDU), reported by Prusoff et al.[15] in the early 1960s.

The utility of the palladium(0) cross-coupling to functionalized pyrimidine nucleosides has been expanded to include reactions of resin-supported 5-iodo-2'-deoxyuridine **37** under Sonogashira, Stille, Heck and Suzuki conditions. Upon cleavage with CH_3ONa, a library of 5-substituted pyrimidine nucleosides **39a-e, 40a-c, 41a-e** (Scheme 3.10) was obtained in good (under Sonogashira and Stille conditions) to moderate (under Heck or Suzuki conditions) yields and high purity. Except for the Suzuki-type reactions, the presented methods exhibit a significant improvement and facilitate the synthetic procedure with respect to purification and yields (**42a-b**, Scheme 3.10).

3.3.2 Cleavage by aminolysis

Solid supported active esters have been employed as intermediates for amide preparation on a solid phase for over forty years, starting by Merrifield's pioneering work on solid-phase peptide synthesis.[2] Early examples of polymer-bound active esters include both the o-nitrophenol systems[16] and the polymer-bound 1-hydroxybenzotriazole derivatives[17] developed by Patchornik and coworkers. More recently, Salvino et al.,[18] have prepared tetrafluorophenol resins, which facilitates the use of ^{19}F NMR to quantitate loading (Scheme 3.11). This new resin provides a useful tool for acylation and a novel activated polymeric sulfonate ester to generate sulfonamides. This activated resin reacts with a wide scope of N-nucleophiles including primary and secondary amines and anilines. The quality of the resin **43** and the resulting polymeric activated resins **44** and **45** may be quantitatively determined in a nondestructive analysis by ^{19}F NMR spectroscopy. The authors have observed that loading diverse acids to generate polymeric activated resins requires optimization of the loading conditions for each class of acid to load.

Chang and coworkers[19] have prepared isomeric varieties of the nitrophenol linker (**47a–c**) employing different solid supports (polystyrene, TentaGel, macroporous, PEGA and silica gel) for facile amide and sulfonamide library synthesis (Scheme 3.12). The broad choice of resin materials available will allow the reaction to occur successfully in solvents ranging from nonpolar organic solvents to aqueous media.

Typical Experimental Procedures

Representative preparation of nitrophenol resin 47a–c (PS, TG, MP, PEGA-2) (Scheme 3.12)[19]

*In a 50 ml polystyrene cartridge, 4-hydroxy-3-nitrobenzoic acid (**46a**, 1 g, 5.5 mmol), HOBt (1 g, 7.4 mmol), and diisopropylcarbodiimide (DIC) (1 mL, 6.4 mmol) were added to an amino polystyrene resin (1 g, 1.2 mmol) in DMF (15 ml). After overnight shaking, the reaction mixture was washed with DMF (20 ml, 5 times), DCM and methanol (20 ml, 5 times alternatively). To remove any undesirable side product, DMF (5 ml) and piperidine (0.5 ml) were added to the cartridge and allowed to shake for 1.5 h. The resin was filtered and washed with DMF (20 ml, 5 times). The resulting piperidine salt was removed via the addition of a 10% HCl solution (in DMF, 20 ml) and was allowed to shake for 1.5 h. The resin was then filtered; washed with DMF, methanol and DCM (20 ml, 5 times each); and dried by nitrogen gas flow.*

Representative synthesis of amide: N-pyridin-2-ylmethylbenzamide (Scheme 3.12)[19]

*A mixture of benzoate ester resin (PS-**48a**, 40 mg, 0.048 mmol) in THF (1 ml) was treated with 2-(aminomethyl) pyridine (2.6 µL, 0.024 mmol) and stirred at room temperature overnight. The reaction*

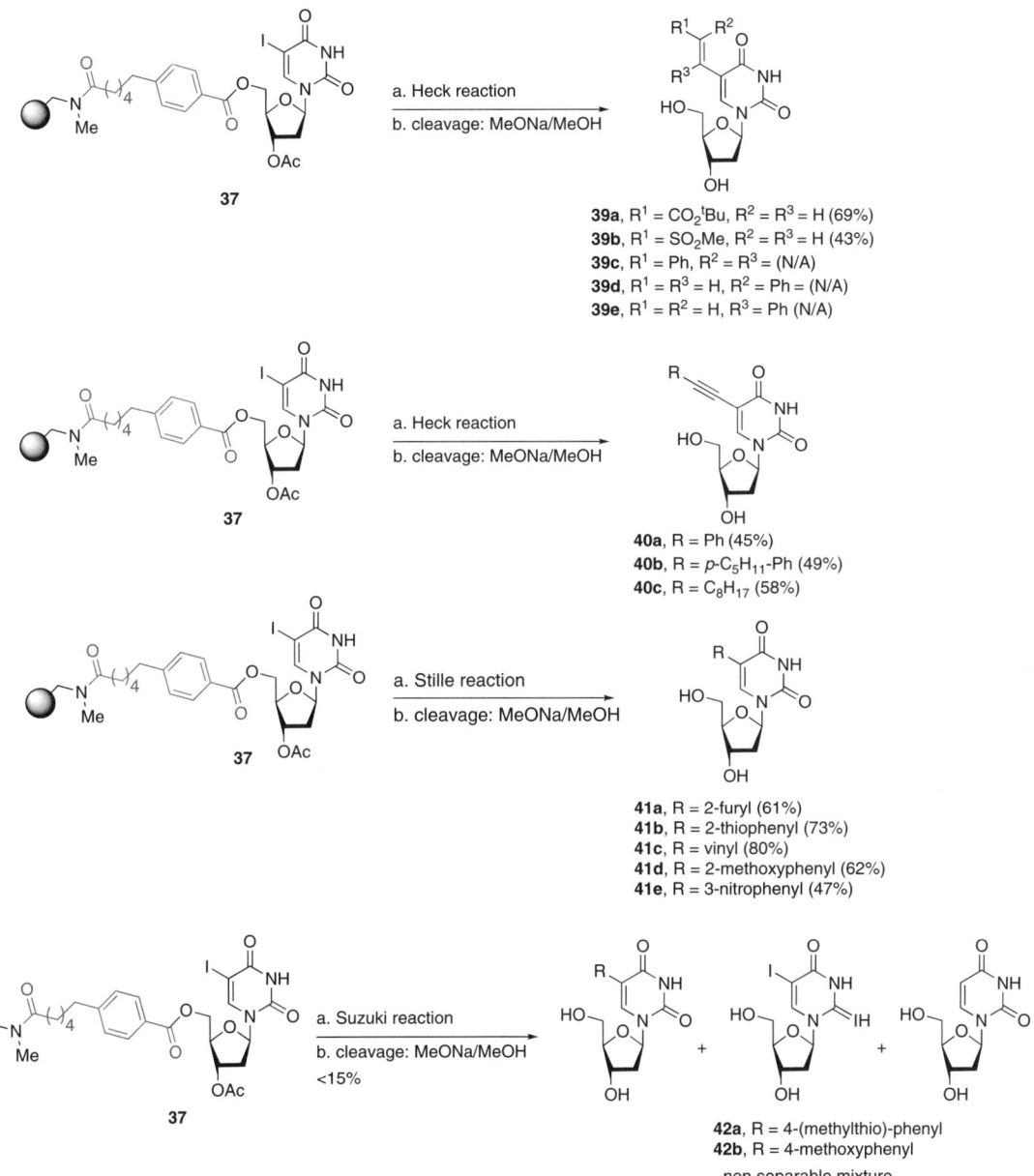

Scheme 3.10 Combinatorial synthesis of a library of 5-substituted pyrimidine nucleosides

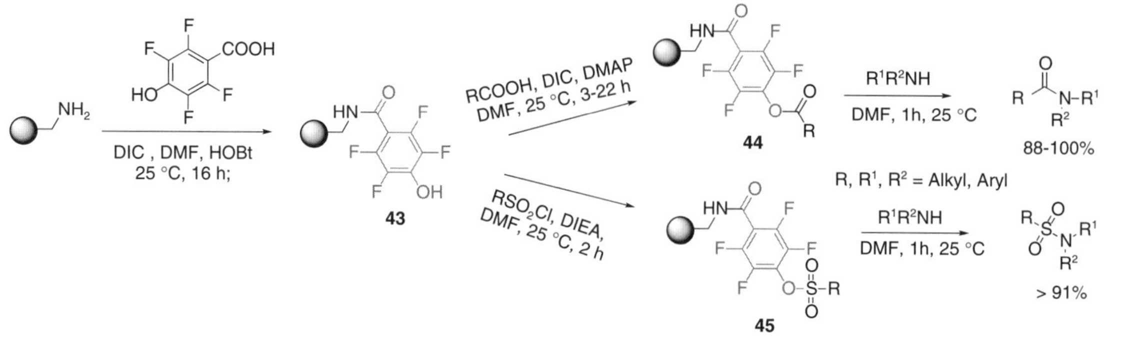

Scheme 3.11 *Aminolysis cleavage in the preparation of libraries of amides and sulfonamides with tetraphenol resins*

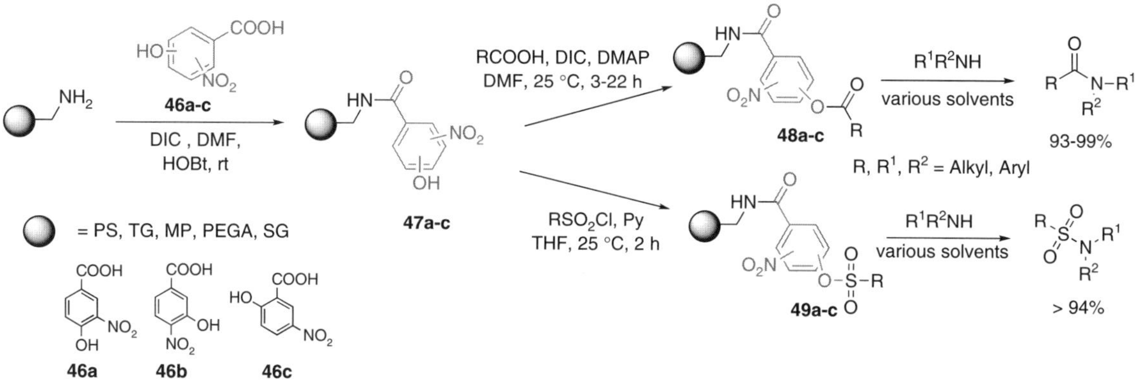

Scheme 3.12 *Aminolysis cleavage in the preparation of libraries of amides and sulfonamides with nitrophenyl resins*

mixture was filtered and washed with THF (1 ml, 3 times). The combined filtrate was evaporated and analyzed by liquid chromatography/mass spectrometry (LC/MS).

Afterwards, a modified polymer-supported 1-hydroxybenzotriazole derivative **52** (Scheme 3.13) with high reactivity developed by Tartar and coworkers.[20] In detail, an aminomethylated polystyrene was functionalized by reaction with 3-nitro-4-chlorobenzenesulfonyl chloride **50** followed by treatment with hydrazine hydrate to give the polymeric *N*-benzyl-1-hydroxybenzotriazole-6-sulfonamide **52**. The polymeric reagent **52** was shown to be highly efficient for the synthesis of amides. The efficiency of **52** could be attributed to its high acidity, conferred by the sulfonyl moiety. The procedure for amide synthesis involves the formation of an activated ester **53** on the derivatized polymer followed, in a second step, by treatment with an amine to generate the amide in solution. Simple filtration allowed the separation of the product from the polymeric reagent, which in this case played the role of leaving group. An optimization study of this two-step procedure was performed. As amides are obtained in solution free of reaction by-products, this method can be used in an automated procedure to recover them directly into a 96-well plate, ready to be used in high throughput screening assays.

Masala and Taddei[21] have loaded chlorotriazine functioning as an active ester on different types of NH$_2$-functionalized resins to give the new supported reagent **55**. The best results, in term of yields products,

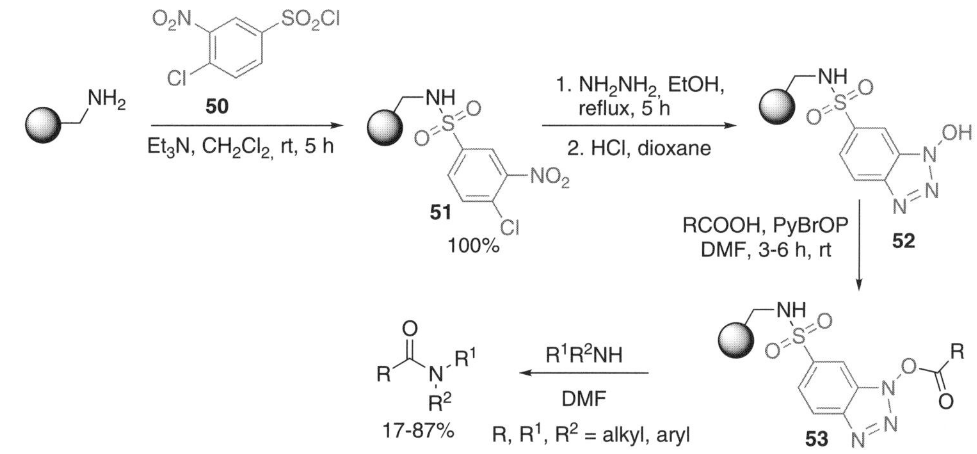

Scheme 3.13 *N-Benzyl-1-hydroxybenzotriazole-6-sulfonamide in the preparation of libraries of amides*

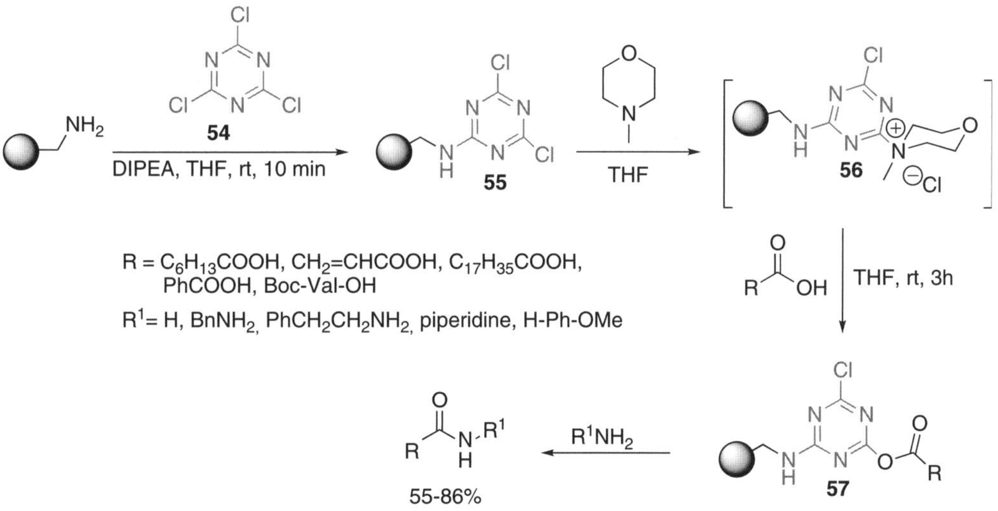

Scheme 3.14 *Chlorotriazine as a linker in the solid-phase synthesis of amides*

were obtained using the chlorotriazine linked to a polystyrene-poly(ethylene glycol) resin. This reagent was employed for the solution-phase synthesis of different amides and dipeptides, as showed in Scheme 3.14.

Typical Experimental Procedures

General procedure to prepare a library of amides (Scheme 3.14)[21]

*A Wang type resin loaded with Fmoc-Gly (1.0 mmol/g resin) was used as starting material. After deprotection of the Fmoc under standard conditions, the NH_2 on the resin was treated with a THF solution of cyanuric chloride **54** (5 equiv.) in the presence of DIPEA. The reaction occurred within 10 minutes, as showed by a negative ninhydrine test. Alternatively, the resin was mixed with THF and the loaded chlorotriazine* *activated with a THF solution of NMM (1.5 equiv. with respect to the theoretical loading). After few minutes, a solution of the carboxylic acid (2 equiv. with respect to the loading) in THF was added and the mixture stirred for three hours under a nitrogen stream in a vial equipped with a sintered glass plate. The excess of the acid and the salts eventually formed were washed away with THF and DMF. The resin 55, loaded with the appropriate carboxylic acid (57), was treated with a THF solution of the amine (1.2 equiv.) in the presence of 1 equiv. of N-methylmorpholine (NMM). thin layer chromatography (TLC) monitored the conversion of the reaction. At the end (three hours was generally a long enough reaction time for a high conversion) the product was recovered by filtration on the sintered glass plate, followed by elimination of the excess of amine by acidic workup and evaporation of the solvent.*

Nevertheless, apart from the Kenner sulfonamide system,[22] the employment of active esters as linkers for solid-phase synthesis has been rather limited. This is not surprising as the esters have been chosen for their high intrinsic reactivity and are designed to function as activating reagents, rather than for more robust attachment of acids.

In a very interesting paper, Graden et al.[23] have prepared and investigated the reactivity of three different polymer-bound cyclohexadienoic acid active esters 60, 62 and 63 modified by complexation with iron tricarbonyl in order to evaluate their potential use as linker systems for solid phase chemistry (Scheme 3.15). Polymer-bound ortho-nitrophenol resin 58 was synthesized by coupling of 46a to aminomethyl polystyrene, using a protocol adapted from Berteina and De Mesmaeker,[24] and was reacted with an excess of iron tricarbonyl cyclohexadienoic acid 59 in the presence of DIC and 4-(N,N-dimethyl)aminopyridine (DMAP) to yield the corresponding polymer-bound ester 60 (Scheme 3.15). IR analyses have showed vibrations at 1974 and 2050 cm^{-1} for the Fe(CO)$_3$ moiety and at 1722 cm^{-1} for the ester carbonyl group. The 4-hydroxy-2,3,5,6-tetrafluorobenzamidomethyl polystyrene (PS-TFP) and N-hydroxybenzotriazole polystyrene (PS-HOBt) linkers, both commercially available, were esterified using the same protocol to form 62 and 63.

All three activated esters were effective and gave the expected amides in all cases. Although the PS-HOBt linkage gave the highest yields, the reaction rates were too rapid for its use as a linker. However, the PS-HOBt activated ester 58 would be the best choice for direct amide formation using iron tricarbonyl cyclohexadienoic acids. The PS-NP (nitrophenyl) (60) and PS-TFP (62) systems showed similar reactivity towards amine nucleophiles, although the final isolated yields for the PS-TFP system were generally higher. The fact that PS-TFP is also commercially available makes the active ester 62 more suitable as a linker system. Nevertheless, amine nucleophiles react very rapidly with cationic complexes[25] making this a useful approach even for less sterically hindered amines.

Lewis acid assisted nucleophilic cleavage of resin bound esters has been shown to be an effective method for the preparation of secondary and tertiary amides (Scheme 3.16). Morphy et al.[26] have examined a variety of Lewis acids reagents known to produce good results for the conversion of esters to amides in solution, including BBr$_3$, RhCl$_3$ and RuCl$_3$, ZrCl$_4$ and AlCl$_3$, Me$_3$Al[(Me$_3$Si)$_2$N]$_2$Sn. They have established that the chlorides of aluminium and zirconium are particularly effective reagents for the synthesis of a diverse range of amides using Wang resin. The precise mechanism of action of these reagents in the aminolysis step is unknown. However, their ability to act as Lewis acids is probably an important factor. An array of amino amides was prepared by a procedure which involves short reaction times and generally mild conditions and which is appropriate for automated synthesis. Separation of the metal salts from the amide products was easily accomplished by solid-phase extraction, providing crude products with a high level of purity.

Linkers that enable release of the target molecule from the solid support under mild conditions and are robust towards a variety of reaction conditions are particularly desirable, as are those which allow the introduction of structural variability in the cleavage step. In this context, Brown et al.[27] have report the

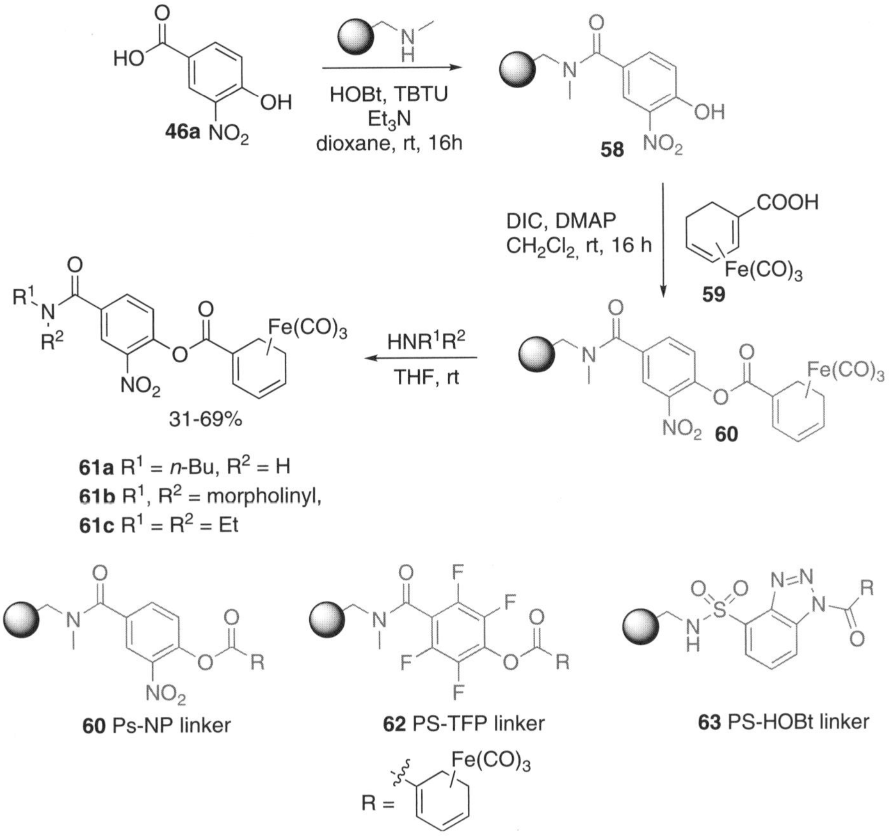

Scheme 3.15 *Complexed iron tricarbonyl linkers for solid-phase synthesis*

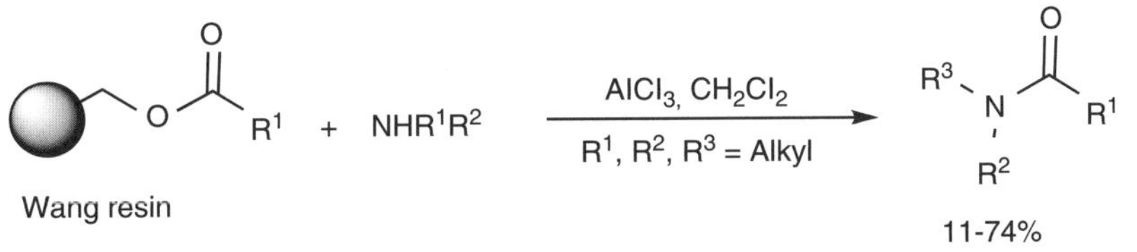

Scheme 3.16 *Lewis acid assisted cleavage in the synthesis of amides*

development of a complementary strategy based on the cleavage of allyl phenyl ethers **66** (Scheme 3.17),[28] which are stable to nucleophiles in the absence of palladium, but readily cleaved by amines in the presence of catalytic palladium(0). The allylic amine structural motif is present in a large number of biologically active molecules, and the methodology described may present a means of preparing arrays of allylic amines for screening. The solid supported allyl ethers **66a–d** were prepared employing a Mitsunobu coupling reaction to form the ether bond (Scheme 3.17).[28] The presence of the second allylic ether **66** was supported by signals at 138.4 (CH), 116.4 (CH$_2$) and 81.1 (CH) ppm in the gel phase ^{13}C NMR spectrum.

To investigate the scope of the palladium catalysed nucleophilic cleavage process further, the reactions of four different amines with four resin-bound allylic ethers were investigated (Scheme 3.17). With the

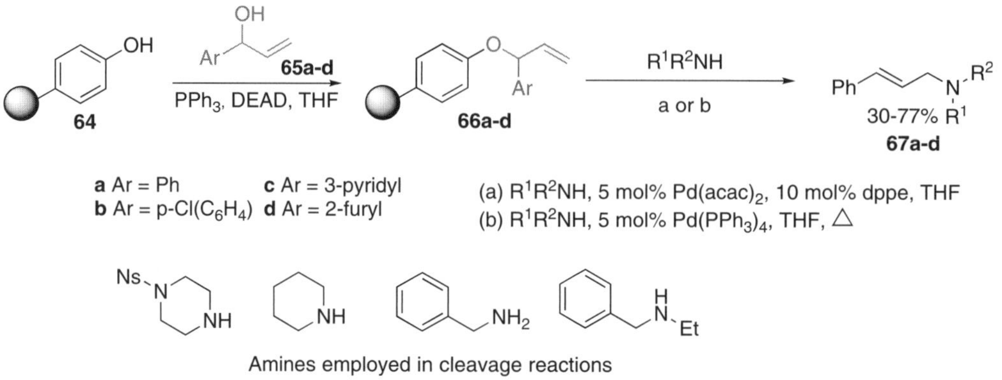

Scheme 3.17 Cleavage of bounded allyl phenyl ethers

exception of the 2-furyl-substituted allylic ether **66d**, all reactions afforded the desired products in acceptable yield. The intermolecular substitution reactions of allylic acetates with primary amines has been reported not to be a useful process due to the formation of by-products arising from diallylation and direct nucleophilic attack on the ester.[29] In this case the latter side reaction was not an issue.

Typical Experimental Procedures

Typical procedure for Mitsunobu coupling (Scheme 3.17)[28]

*THF (25 ml) was added to a mixture of phenoxypolystyrene resin (**64**) (2.04 g, 3.06 mmol), triphenylphosphine (5.24 g, 20 mmol) and allylic alcohol **65b** (3.36 g, 20 mmol) under an inert atmosphere. The suspension was cooled to 0 °C and stirred for 10 minutes. diethylazodicarboxylate (DEAD) (3.2 ml, 3.48 g, 20 mmol) was added drop-wise and the reaction was allowed to warm to room temperature; the resulting orange suspension was stirred at room temperature for 12 hours. The resin was collected by filtration, washed with DCM, DMF, methanol, DMF and DCM (100 ml each) and dried under vacuum (0.7 mmHg, 50 °C) for 2 hours. This afforded the product **66b** as a beige solid (2.62 g).*

Typical procedure for palladium catalysed nucleophilic cleavage (Scheme 3.17)[28]

*THF (3 ml) was added to a mixture of resin **66b** (149.6 mg, 0.19 mmol), Pd(acac)$_2$ (2.8 mg, 9.3 mol), 1,2-bis(diphenylphosphino)ethane (dppe) (7.4 mg, 18.6 mol) and piperidine (15.8 mg, 0.19 mmol) under an inert atmosphere. The reaction was heated at reflux with gentle stirring for 1.5 hours. After cooling to room temperature, the resin was collected by filtration, washed with DCM (50 ml) and the filtrate concentrated under vacuum to give the crude product as an orange oil (45 mg). Purification by column chromatography eluting with DCM/methanol (19:1) afforded the allylic amine **67d** as a colourless oil (31.5 mg, 0.13 mmol, 72%).*

In 2002, Austin et al.,[30] discovered a novel, self-promoted and chemoselective resin cleavage process while adapting for solid-phase synthesis a diastereofacially selective 1,3-dipolar cycloaddition reaction of isomunchones with vinyl ethers. This solid-phase asymmetric linker (**69**), structurally analogous to the most successful solution phase auxiliary design,[31] was constructed starting with the benzhydrylamine (BHA) resin **68**.[32]

Coupling of the BHA **68** resin with both enantiomers of *R*-hydroxyvaline furnished the immobilized auxiliaries (*S*)-**69** and (*R*)-**69** (Scheme 3.18). These, in turn, were functionalized with *N*-acylmalonamic acid and subjected to diazotransfer, providing diazoimide resins (*S*)-**70** and (*R*)-**70**. Rhodium(II) catalysed nitrogen extrusion, followed by cycloaddition in the presence of vinyl ethers, completed the solid phase assembly. The high solubility and high turnover ratio[33] of the $Rh_2(pfbm)_4$ catalyst enabled a smooth adjustment of the solution phase protocol to the solid phase synthesis. After extensive washing of the resin, the disconnection of the bicyclic structures (Scheme 3.19) was accomplished with 1 M methylamine in methanol, at 23 °C, and with sonication.

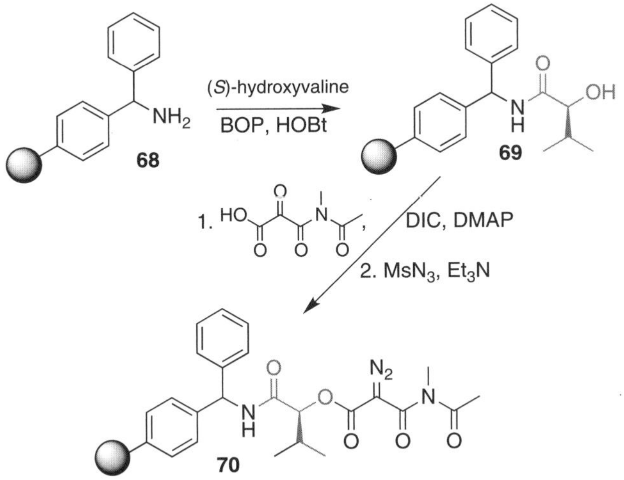

Scheme 3.18 Synthetic steps for preparation of chiral diazoimide resins

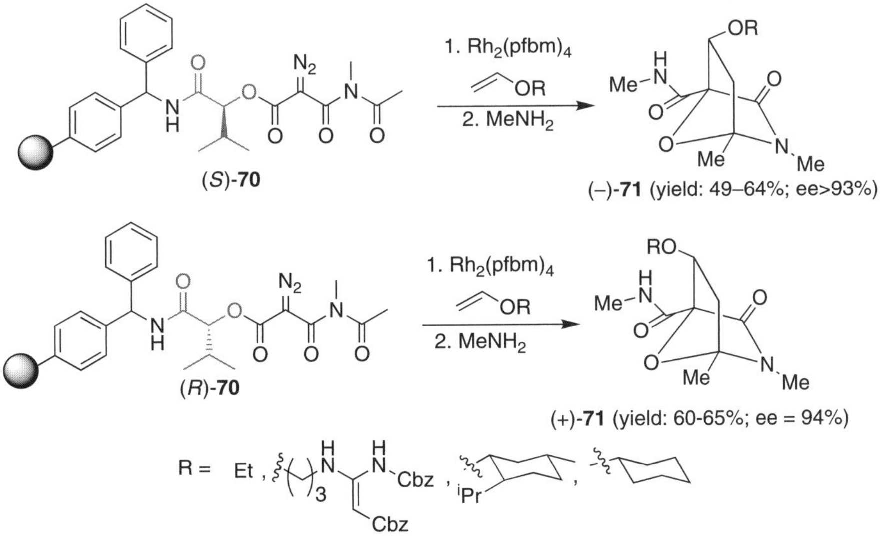

Scheme 3.19 Use of chiral diazoimide resins for the synthesis of (+)-5-alkoxy-N,1,2-trimethyl-3-oxo-7-oxa-2-azabicyclo[2.2.1]heptane-4-carboxamide

The induction of stereoselectivity of the auxiliary can be best illustrated by the chiral (-)-menthyl-based dipolarophile. The cycloaddition proceeded with a high degree of selectively in both the 'matched' (> 95% de) and even the 'mismatched' case (90% de). In every case, the ^{1}H NMR and HPLC of the liberated crude material indicated a remarkable purity of the aminolysis-derived products with the desired bicyclic species being the sole products of the multistep synthetic sequence.

Even pyrazole and triazole compounds can give privileged scaffolds for generation of drug-like compounds to drug discovery. In this context, Scheme 3.20 reports an elegant SPOS synthesis of 5-amino-1-(substituted thiocarbamoyl)-pyrazole (**77**) and 1,2,4-triazole (**75**) derivatives based on the cyclization of polymer-bound dithiocarbazate with various electrophiles, such as 3-ethoxyacrylonitriles and cyanocarboimidates.[34] The polymer-bound dithiocarbazate **77**, key intermediate for subsequent heterocyclic diversification (**74** and **76**), has been prepared by nucleophilic reaction with carbon disulfide and Fmoc-hydrazine on the Merrifield resin (Scheme 3.20). Further nucleophilic substitution on these polymer-bound 5-amino-1-dithiocarboxypyrazoles and 1,2,4-triazoles with various amines under thermal cleavage condition produces the desired 5-amino-1- (substituted thiocarbamoyl)pyrazoles **77** and 1, 2, 4-triazoles **75** (Scheme 3.20).

Ureas are important targets in combinatorial chemistry since this functional group is widespread in many biologically relevant molecules.[35] Thus, it is not surprising if several solid-phase strategies that provide

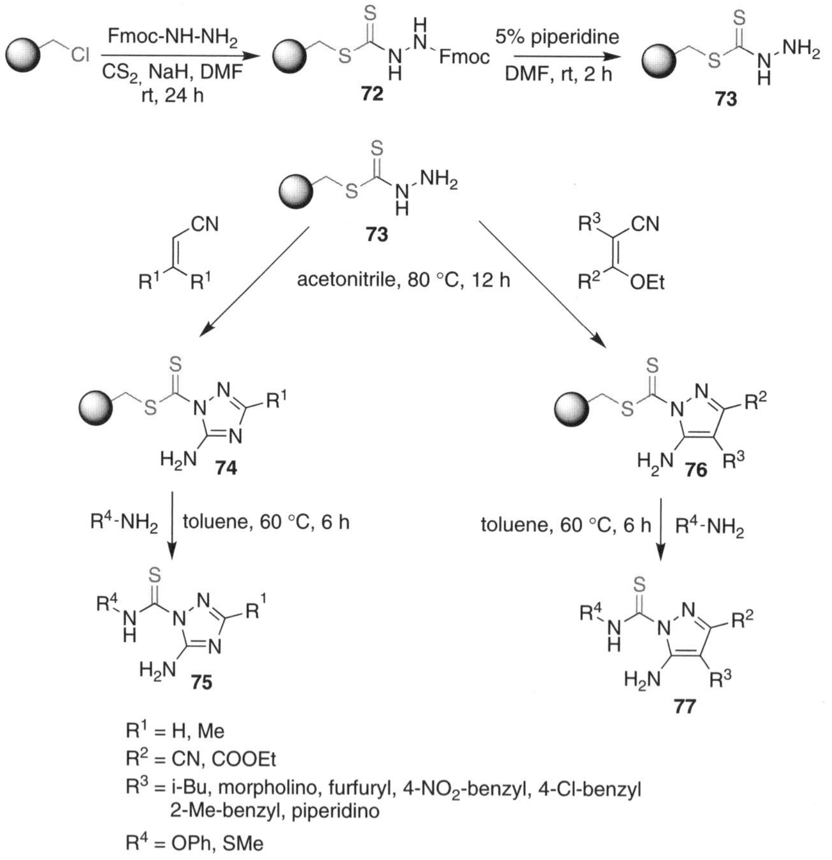

Scheme 3.20 *Combinatorial synthesis of 5-amino-1- (substituted thiocarbamoyl)pyrazoles and 1,2,4-triazoles*

arrays of urea compounds have been published in the literature.[36] Unfortunately, all these strategies have the shortcoming that only a maximum of tri-substituted urea products are produced. Janda et al.[37] have successfully developed a solid-phase method that enables the synthesis of tetra-substituted ureas using a diversity-building cleavage reaction (Scheme 3.21). The keystone for the urea library is the preparation of polymer-bound carbamates **79** and **80**. Polymer-bound *p*-nitrophenyl carbonate **78**, synthesized starting from a hydroxymethyl resin by reaction with *p*-nitrophenyl chloroformate, was treated with a series of primary amines (R^1NH_2) to provide the primary amine polymer-bound carbamates **79**. When this reaction was performed with alkyl- and benzylamines, the reaction proceeded readily at room temperature using Hunigs base; however, the lesser reactive anilines required more forcing conditions, namely sodium bis(trimethylsilyl)amide. The progress of the last reaction was checked by IR spectroscopy monitoring the disappearance of the carbamate N–H stretch. When alkyl iodides, benzyl bromides and allyl bromides were employed as the alkylating reagent, the reaction proved to be straightforward and complete conversion to the desired carbamates **80**, could be achieved using lithium *tert*-butoxide base at room temperature.

These polymer-bound primary and secondary amine carbamates were then treated under 'smart' diversity-building cleavage conditions using a series of aluminum amide complexes to form the corresponding urea cleavage products. Firstly, the aluminum amide reagent was prepared by addition of $Al(CH_3)_3$ (2.5 equiv.) to the amine (R^3R^4NH) **82** (5 equiv.). This mixture was then added to the polymer-bound carbamates **79** and **80** and the reaction was heated between 50 and 110 °C for the appropriate period before quenching the reaction with water in THF. After quenching, the reaction mixture was passed through a strong-acid ion exchange resin that removed both the excessively used amine starting materials **82** and the aluminum salts, providing pure urea products **83**. When weakly basic amines **82**, such as diphenylamine or indole, were used the ion exchange resin is not able to remove these amines from the crude products, and it is necessary to purify final urea products by preparative TLC.

The Marshall linker[38] has been widely used to synthesize compounds that can be cleaved by primary and secondary amines to afford the corresponding amides.[39] While the original report on the Marshall linker involved the oxidation of the linker before cleavage, recently the efficient release of the resin-bound compounds using nucleophiles from the unoxidized linker has been reported.[40] In this context, Yan *et al.* have studied the kinetics of cleavage reactions of seven resin-bound thiophenol esters **85a–g** with three amines **86a–c** by single-bead FTIR (Scheme 3.22).[41]

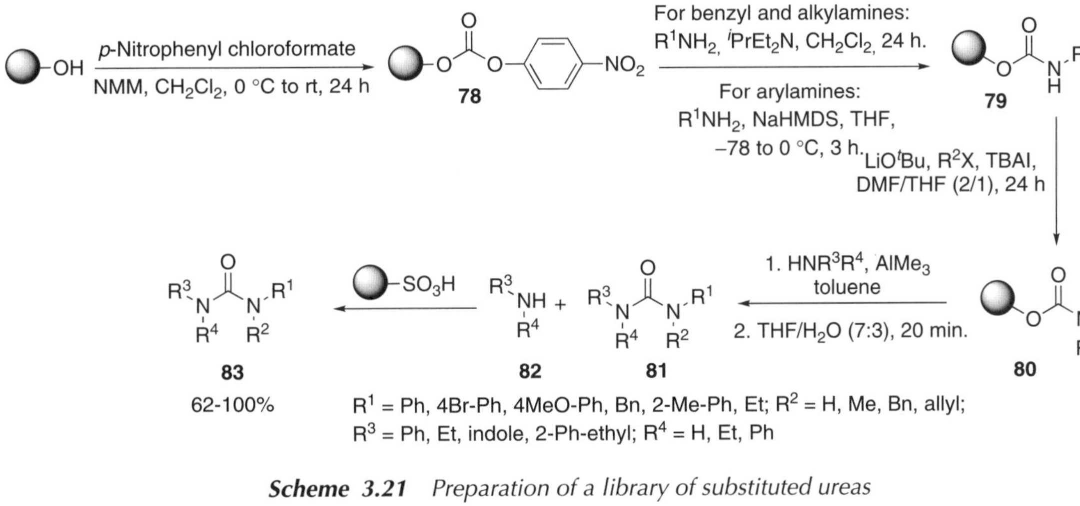

Scheme 3.21 *Preparation of a library of substituted ureas*

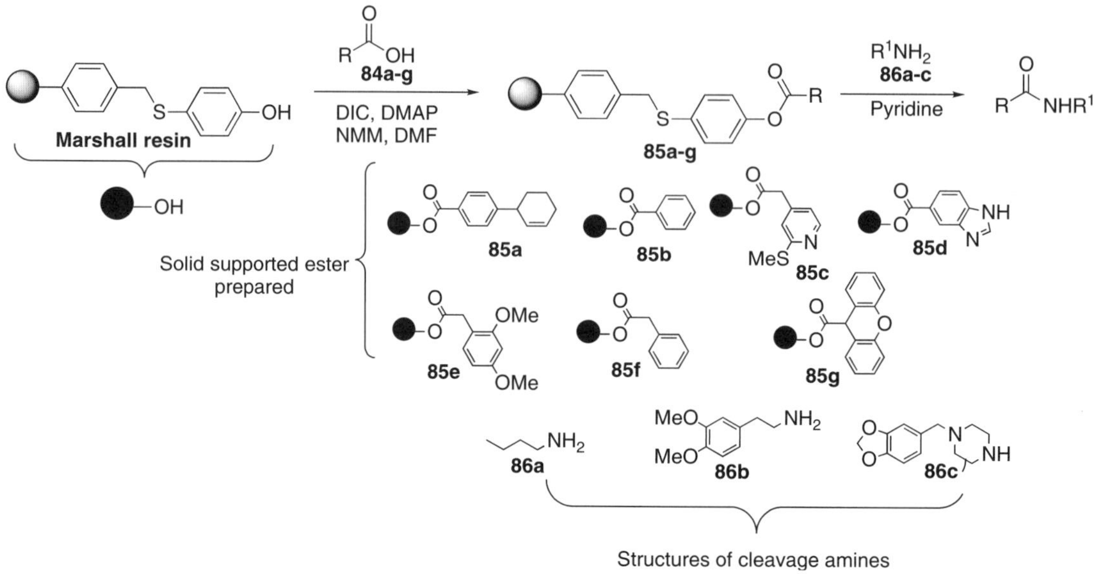

96 Traditional Linker Units for Solid-Phase Organic Synthesis

Scheme 3.22 The 'Marshall linker' in the synthesis of amides

The reaction rate of these thiophenol esters could be divided into three groups based on their chemical structures: alkyl thiophenol esters (resins **85e–g**), aromatic thiophenol esters (resins **85a–c**) and 5-benzimidazolecarboxylic thiophenol ester (resin **85d**). The calculated rate constants of alkyl thiophenol esters were almost 13 times higher than those of aromatic thiophenol esters. The 5-benzimidazolecarboxylic thiophenol ester was even more reactive than alkyl thiophenol esters. The reactivity of the same resin-bound thiophenol esters with three amines varied as follows: *n*-butylamine > 3,4-dimethoxyphenethylamine > 1-piperonylpiperamine. The investigation of reactivity on temperature has shown a twofold increase per 10 °C rise in reaction temperature.

Since the hydroxyl group on the thiophenol is regenerated after cleavage, the used resin is expected to be functional again. To test the reusability of the used resins, Marshall resin was subjected to reaction with five different activated esters **88a–e** made *in situ* from acids **87a–e** and then the phenol ester products were cleaved with *n*-butylamine (Scheme 3.23).[42]

The results obtained from three reaction–cleavage cycles are summarized in Table 3.1. Loading reported here was determined by a quantitative HPLC. The first column (cycle 1) contains results from the reaction performed on resins directly purchased from the manufacturer, with no pre-treatment except a solvent washing step. It appears that the reactivity of resins improved in subsequent experiments. The significance of the results from Table 3.1 is the achievement of relatively consistent resin reactivity and the loading capacity from cycle to cycle. The results demonstrate the potential for resin recycling applicable for resins without being chemically altered during the synthesis. For example, oxidized Marshall resins (to sulfone and sulfoxide) cannot be recycled.

Recently, methods for constructing heterocyclic small molecules on a solid support have attracted considerable attention.[43] In this context, Boldi *et al.*[44] have carried out a library of alkoxyprolines **91** using a nucleophile-cleavable solid support. Boc-hydroxyproline scaffolds **89** were coupled to the Marshall linker with DIC and catalytic DMAP. Complete coupling to the solid support was confirmed by a negative colorimetric test for phenols (FeCl$_3$/pyridine) (Scheme 3.24).[40] After, the Boc protecting group was removed with trifluoroacetic acid in DCM with anisole present as a cation scavenger. The proline

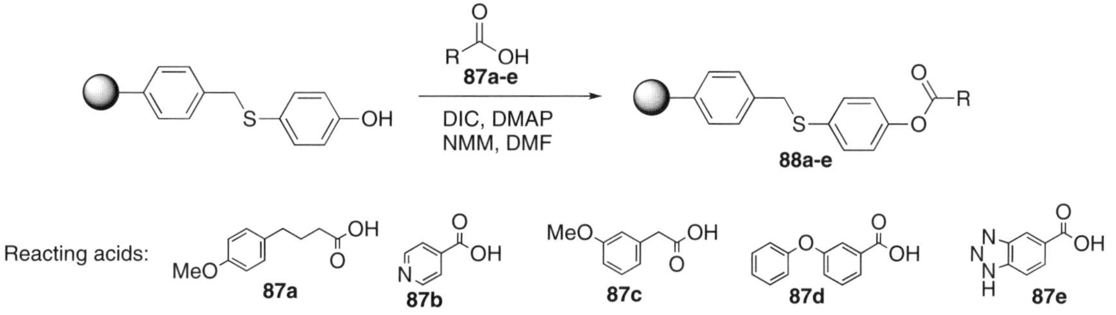

Scheme 3.23 Loading of 'Marshall resin' with carboxylic acids

Table 3.1 Loading value of 87a–e from three reaction cycles

Reacting acids	New resin 1.03 mmol/g	After cycle 1 0.99 mmol/g	After cycle 2 1.11 mmol/g
87a	0.99	1.12	1.10
87b	0.97	1.06	0.96
87c	0.69	1.08	1.05
87d	0.73	1.12	1.10
87e	0.77	0.85	0.85

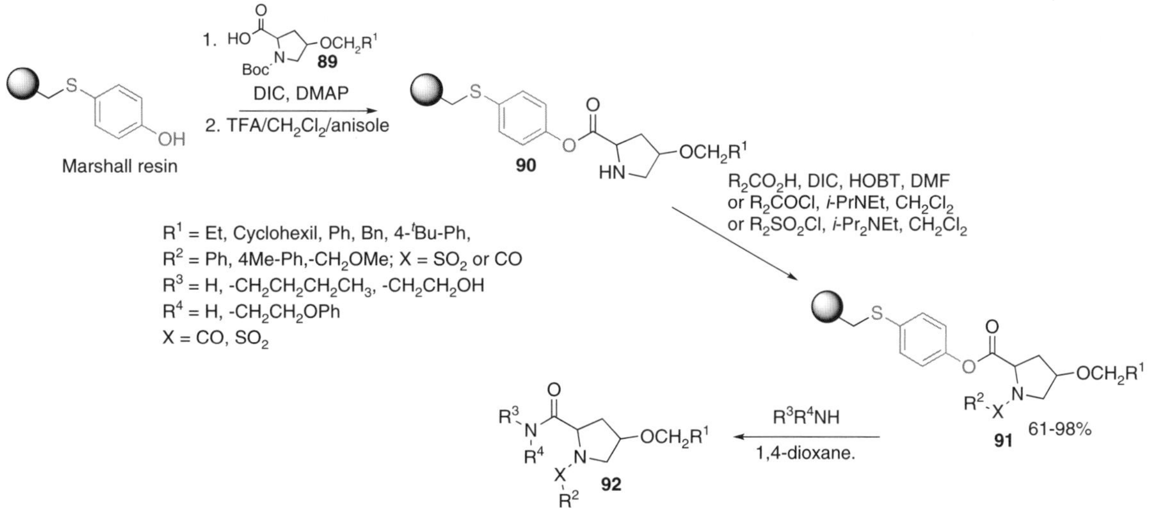

R^1 = Et, Cyclohexil, Ph, Bn, 4-^{t}Bu-Ph,
R^2 = Ph, 4Me-Ph,-CH$_2$OMe; X = SO$_2$ or CO
R^3 = H, -CH$_2$CH$_2$CH$_2$CH$_3$, -CH$_2$CH$_2$OH
R^4 = H, -CH$_2$CH$_2$OPh
X = CO, SO$_2$

Scheme 3.24 Use of 'Marshall resin' for alkoxyproline derivatives

98 Traditional Linker Units for Solid-Phase Organic Synthesis

nitrogen was acylated with carboxylic acids (DIC/HOBt in DMF), acid chlorides or sulfonyl chlorides (in the presence of ^{i}Pr$_2$NEt in DCM). Treatment of esters **91** with various primary and secondary amines in 1,4-dioxane gave the alkoxyproline products **92**. A reaction time of more than 24 hours was generally required for cleavage of products from the solid support. Cleavage with various secondary amines was much slower (36–48 h) than with primary amines (24 h). Pyridine was also found to be a suitable solvent for cleavage, but 1,4-dioxane offered the advantage of solvent removal via lyophilization. Excess amine was removed by supported liquid–liquid extraction (SLE) using 4:1 DCM/THF as the extraction solvent and 2 N hydrochloric acid as the priming buffer on the Hydromatrix column.

More recently Kamal et al.[45] have developed a versatile method for the solid-phase synthesis of imidazo[1,2-a]pyridine-8-carboxamides **95** (Scheme 3.25). The use of the phenol–sulfide linker (Marshall's linker)[38] with its chemical versatility in the cleavage step adds to the overall diversity of the library. The reaction conditions used in this protocol are mild and the compounds are obtained in good yields and purity. Further, this procedure also demonstrated the application of high-throughput purification (SLE) methods, which have been employed in conjunction with react and release solid-phase techniques to improve the purity of resulting libraries. This method has potential for use in the generation of a large number of imidazo[1,2-a]pyridine-based compounds using an automated synthesizer.

Liquid crystals are widely used in optoelectric devices and electron-transporting materials. Considerable synthetic effort and time are required to develop new liquid crystals. Therefore, Hioki et al.,[46] have focused their attention on the synthesis of rod-shaped azomethine derivatives **99**, which are typical liquid crystals following a combinatorial approach (Scheme 3.26). The loading of the substrates on a solid support and cleavage from the solid support were performed by an imine synthesis (from **96** to **97**) and by imine-exchanged process (from **98** to **99**). This linker has the advantage of being able to release the final product under mild conditions.

Typical Experimental Procedures

Synthesis of 97 on solid support 96 (Scheme 3.26)[46]

*Eighty four pieces of the solid supported aniline **96** (loading: 38 μmol × 84, 3.19 mmol) were reacted with 4-carboxybenzaldehyde (2.54 g, 16.9 mmol, 5.3 equiv.) in DMF solution (80 ml) at room temperature for*

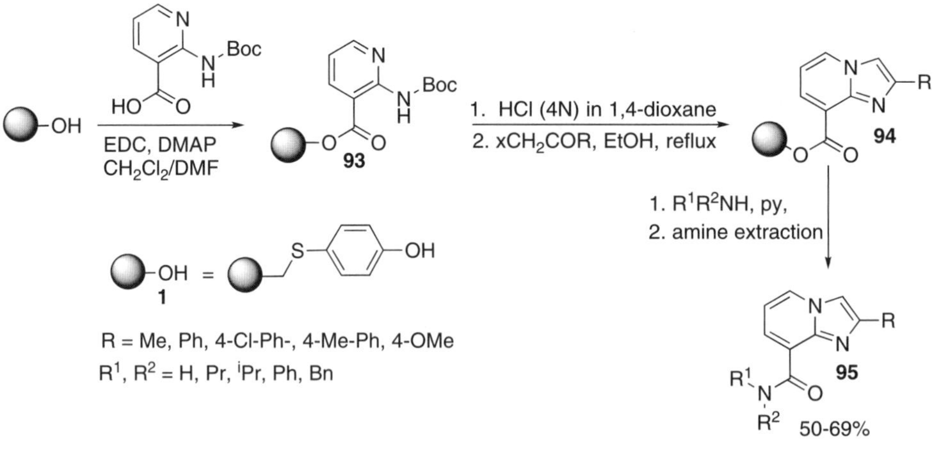

Scheme 3.25 *Preparation of a library of imidazo[1,2-a]pyridine-8-carboxamides*

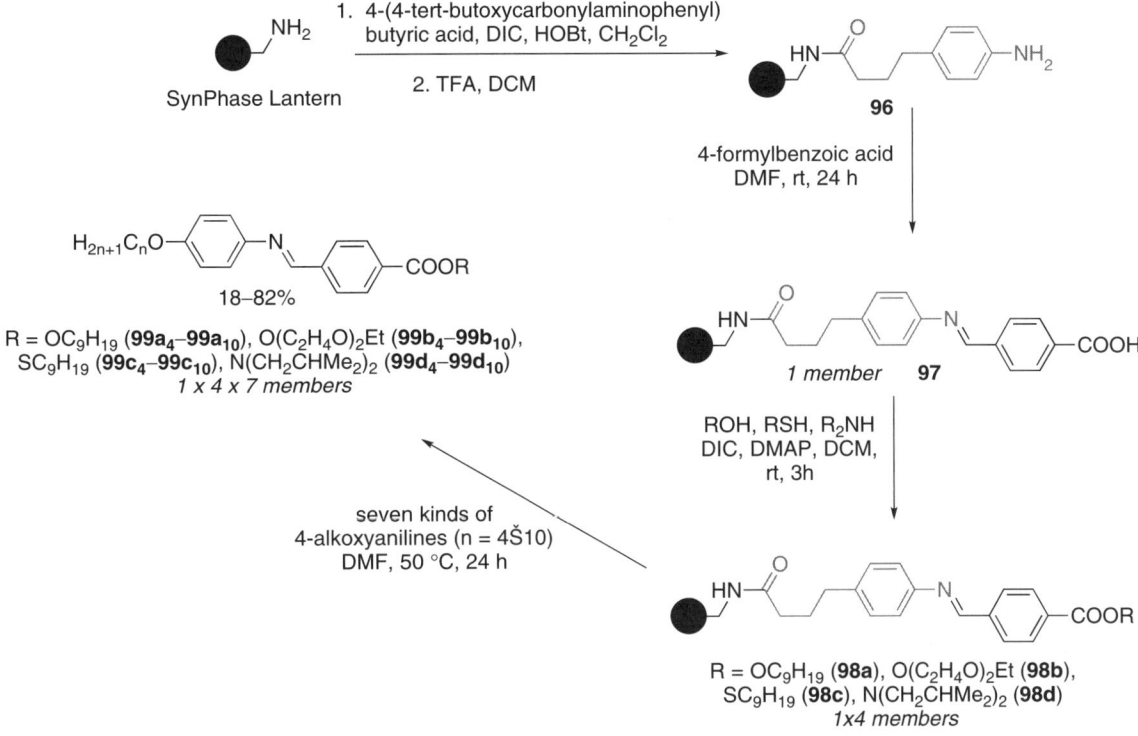

Scheme 3.26 *Preparation of a library of rod-shaped azomethine derivatives*

24 h. The solution was removed by decantation and the resulting resins were washed with DMF (3 × 0.5 min) and DCM (3 × 0.5 min).

Synthesis of 98a on solid support 97 with n-nonanol (Scheme 3.26)[46]

*Twenty eight pieces of the solid supported azomethine **97** (loading: 38 μmol × 28, 1.06 mmol) were reacted with 4-N,N-dimethylaminopyridine (32.5 mg, 0.266 mmol, 0.25 equiv.), n-nonanol (0.74 ml, 4.2 mmol, 4 equiv.) and 1,3-diisopropyl-carbodiimide (1.3 ml, 8.5 mmol, 8 equiv.) in DCM at room temperature for 3 h. The solution was removed by decanting and the resulting resins were washed with DMF (3 × 0.5 min) and DCM (3 × 0.5 min). Cleavage from the solid support: Four pieces of the solid supported ester **98a** (loading: 38 μmol × 4, 152 μmol) were reacted with 4-n-octyloxyaniline (168.2 mg, 0.76 mmol, 5 equiv.) in DMF (5 ml) at 50 °C for 3 h. The resulting resins were washed with DMF (3 × 3 min). The combined DMF solution was evaporated and purified by HPLC (hexane/ethyl acetate = 9/1) to give **98a₈** in 66% yield (48.3 mg, 107 μmol) as pale yellow solid. All 28-library members have been fully characterized by analytical and spectroscopic means.*

The precence of the *O*-alkyl hydroxylamine function in natural and synthetic products displaying potent pharmacological activities[47] has motivated a medicinal interest as a non basic substitute for biologically active amines[48] or as part of bioisosteric groups.[49] More recently, it has been demonstrated that its incorporation into peptides can induce and stabilize turns or helical structures.[50] In this context, a

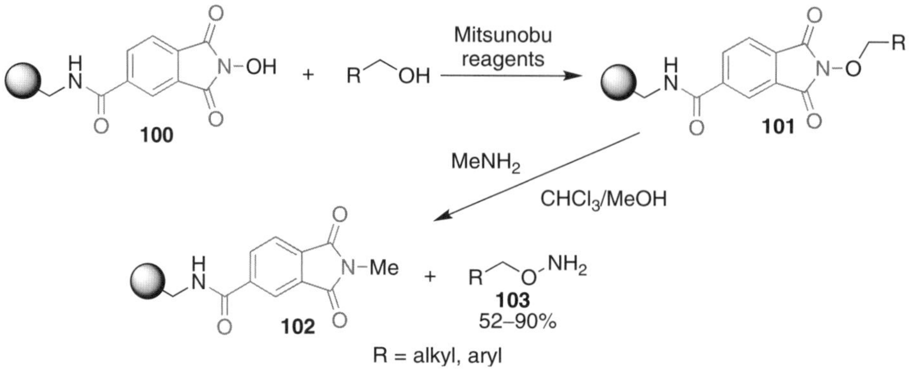

Scheme 3.27 Parallel synthesis of O-alkyl hydroxylamines

solid-phase approach based on the alkylation by alcohols of a new supported N-hydroxyphthalimide reagent **100** using a Mitsunobu reaction followed by methylaminolysis has been recently reported;[51] it was used to synthesize in parallel a large diversity of O-alkyl hydroxylamines **103** (Scheme 3.27).

Typical Experimental Procedures

Preparation of N-hydroxyphthalimide resin 100 (Scheme 3.27)[51]

*To a solution of 2-hydroxy-1,3-dioxoisoindoline-5-carboxylic acid (1.82 g, 4 equiv.) in DCM (200 ml) were added PyBOP (4.48 g, 5 equiv.) and DIPEA (3.2 ml, 10 equiv.). The mixture was stirred for one hour and the dark orange solution was added to a pre-swollen (CH_2Cl_2) aminomethylated polystyrene resin (0.92 mmol/g, 2.0 g, 1.84 mmol). The suspension was stirred mechanically for 24 hours at room temperature and monitored using the Kaiser test. After filtration, the resin **100** was washed with DCM (3 × 50 ml), DMF (3 × 50 ml), methanol (3 × 50 ml), DCM (50 ml), methanol (50 ml), and DCM (50 ml) and dried under reduced pressure (2.38 g).*

General procedures for synthesis of alkoxyamines 101 with supported reagent 100 (Scheme 3.27)[51]

*Resin **100** (0.80 mmol/g, 50 mg, 39 μmol) was loaded into a 6 ml SPE cartridge. The resin was suspended in DCM (5 ml). The SPE cartridge was capped, loaded horizontally onto an orbital shaker and agitated for one hour. The resin was drained and suspended in DCM (5 ml). Alcohol (5 equiv.) and triphenyl phosphine (52 mg, 5 equiv.) were added, mixed by inverting the cartridge until dissolution and DIAD (39.8 μL, 5 equiv.) and imidazole (14 mg, 5 equiv.) were added. The cartridge was capped and agitated as previously for 24 hours at room temperature. The resin was drained and washed with DCM (3 × 5 ml), DMF (3 × 5 ml), methanol (3 × 5 ml), DCM (5 ml), methanol (5 ml) and DCM (5 ml). This resin was then subjected to the cleavage step by methylaminolysis.*

General procedure for cleavage (Scheme 3.27)[51]

The resin was suspended in trichloromethane/methanol (9/1, 5 ml), the SPE cartridge was capped, loaded horizontally onto an orbital shaker and agitated for one hour. The resin was drained, suspended in 5 ml of a 0.05 M methylamine solution in trichloromethane/methanol (9/1, 5 ml) and agitated for two hours at room temperature. The solution was collected by filtration, the resin was washed with trichloromethane (1 ml) and the filtrate collected and evaporated under reduced pressure (higher than 20 mbar) to afford the O-alkyl hydroxylamine. Yields were determined by NMR titration.

3.3.3 Cleavage by hydrazinolysis

Although several carboxy linkers are available, as a rule the majority are either acid (Wang[52]) or *hyper* acid (hypersensitive acid labile (HAL),[53] Super Acid Sensitive ResIN (SASRIN™),[54] trityl[55]) labile. Linkers such as hydroxymethylbenzoic acid (HMB)[56] and glycolamido[57] are stable to acid, but release carboxylic acids on treatment with sodium hydroxide or Bu_4NOH. Effective silicon based linkers which are stable to base only, yet cleaved by fluoride ion, have been described by Chao[58] and Ramage,[59] but they limit the use of silyl protecting groups in solid-phase synthesis (SPS), particularly for carbohydrate manipulation. Other carboxy linkers, requiring mild photochemical,[60] enzymatic[61] or palladium(0)[62] cleavage conditions to release the product are also available and have specific stabilities and applications. Further methodology, which has additional advantages, now offers the opportunity to extend the range of compound types amenable to SPS.

4-{*N*-[1-(4,4-Dimethyl-2,6-dioxocyclohexylidene)-3-methylbutyl]amino}benzyl ester (*O*Dmab) **105**, which is employed as a carboxy protecting group, displays inherent stability towards the base deprotection conditions required for Fmoc/ᵗBu solid-phase peptide synthesis (SPPS) and the acid conditions required for side chain deprotection. Usually, the ester cleavage is achieved by hydrazinolysis, followed by elimination of the resulting *p*-aminobenzyl group via a 1,6-electron shift (**106**, Scheme 3.28).

In this context, Kellam *et al.*[63] have successfully developed a 1-(4,4-dimethyl-2,6-dioxocyclohexylidene)ethyl (Dde)-derived carboxy protecting group, as a carboxy functional linker, stable to acidic and basic environments, yet easily cleaved in 2% v/v hydrazine monohydrate.

The synthetic approach followed two distinct routes and is outlined in Scheme 3.29. Route **A** was undertaken primarily in solution to highlight any potential problems in synthesis, since the reactions could be monitored using standard chromatographic techniques. This route, however, generated a linker with the first amino acid (e.g. Fmoc-Leu-OH) pre-attached (**110**). Therefore, to construct a 'generic linker' route **B** was developed in parallel to yield **112** with free benzylic–OH functionality available for the attachment of any carboxylic acid. In route **B** the free acid **111** was immobilised onto the solid support (complete

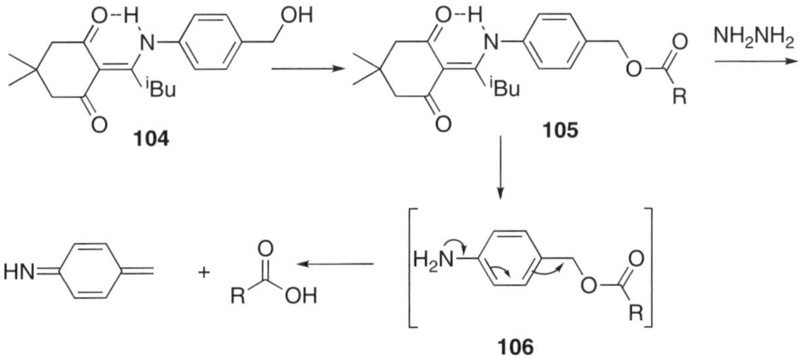

Scheme 3.28 *Hydrazinolysis of O Dmab*

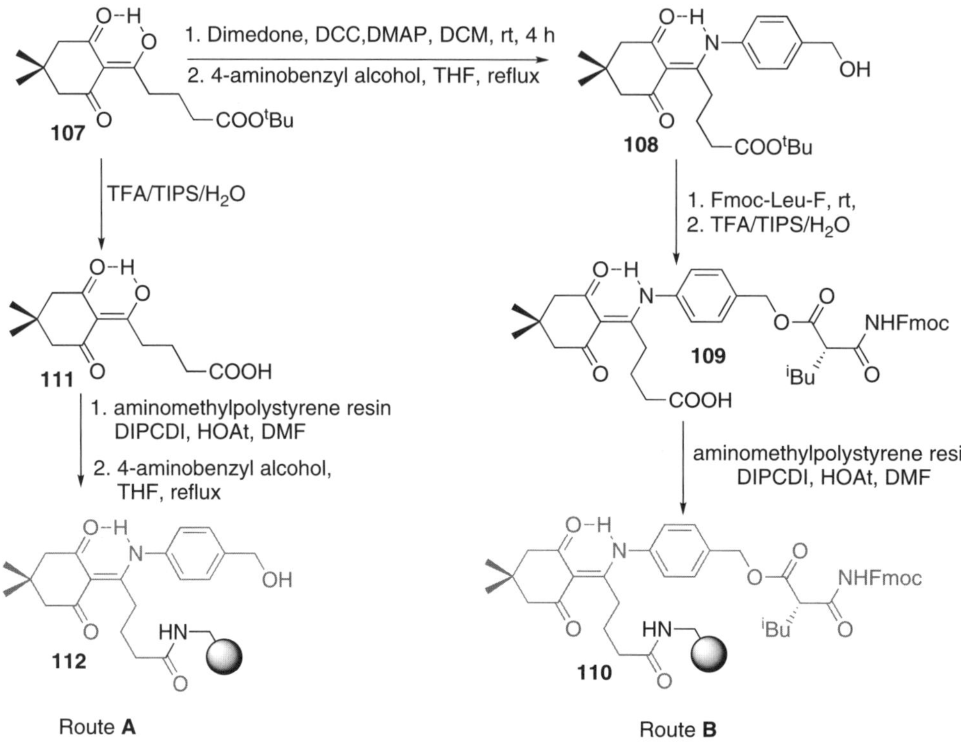

Scheme 3.29 *Preparation of Dde based linkers*

acylation was confirmed by a negative 2,4,6-trinitrobenzene sulfonic acid (TNBS) test), which was then condensed with 4-aminobenzyl alcohol to yield the desired linker **112**. Coupling of the first residue was achieved via either the corresponding amino acid fluoride or the pre-formed symmetrical anhydride, the latter resulting in higher loading (> 80%) as confirmed by Fmoc loading tests.[64]

The stability of the linker to standard acid and base conditions employed in Fmoc:^{t}Bu SPPS has been demonstrated and its utility illustrated by the construction of model peptides, Leu-enkephalin and Human Angiotensin II.

Recently, Toth *et al.*[65] have described the application of a Dde (*N*-1-(4,4-dimethyl-2,6-dioxocyclohexylidene)ethyl ester) based linkers for the solid-phase synthesis of oligosaccharides (Scheme 3.30). Dimedone **113** was reacted with glutaric anhydride **114** in the presence of triethylamine and dimethylaminopyridine in anhydrous dichloromethane to give vinylogous acid **111** (80% yield after crystallisation), which was subsequently coupled to 4-methylbenzhydrylamine (MBHA) resin (0.7 mmol/g) to give compound **115** (with higher than 99.5% coupling efficiency according to quantitative ninhydrin test). Resin linker conjugate **115** was then heated at reflux in THF/ethanol (1/1, v/v) with *p*-aminobenzyl alcohol (fivefold excess) to afford resin-linker-converter conjugate **116**. Alternatively, glutaric acid was coupled to MBHA resin (0.42 mmol/g), in presence of 1.2 equiv. of *N*, *N'*-dicyclohexylcarbodiimide (DCC), then the intermediate was stirred with dimedone in the presence of DCC resulting in compound **115**, which was reacted with *p*-aminobenzyl alcohol, resulting in compound **112**. The first sugar can be attached to the resin linker via a vinylogous amide bond, or by ether linkage using a *p*-aminobenzyl alcohol converter. The final oligosaccharide products (**118**) can be cleaved from the resin by hydrazine, ammonia or primary amines, but the linker is stable under the conditions of oligosaccharide synthesis.

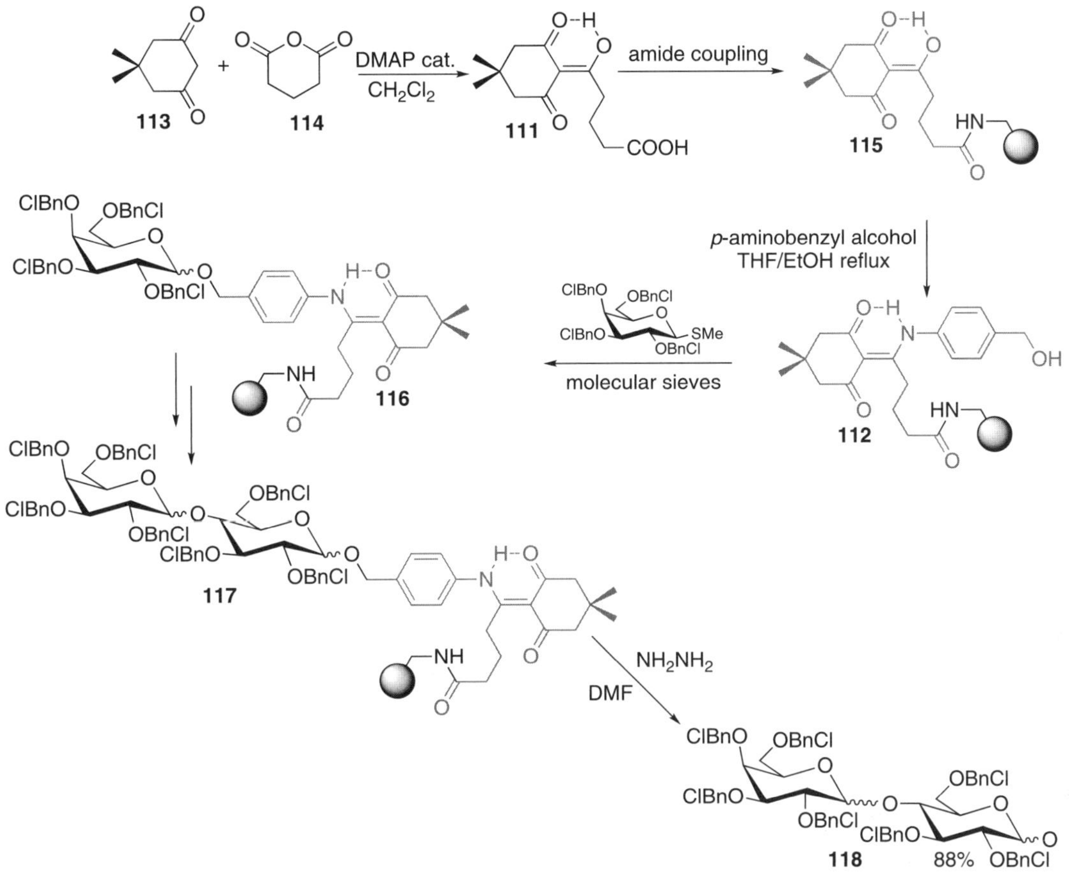

Scheme 3.30 *Use of Dde-based linkers for solid-phase synthesis of oligosaccharides*

Traditionally, access to structurally complex carbohydrates has been very arduous.[66] Unfortunately, solid-phase synthesis of oligosaccharides can be problematic due to low coupling yields, inadequate stereocontrol and the necessary employment of low loading resin to achieve significant coupling yields.[67] Although some recent progresses in solid-phase synthesis have made the preparation of oligosaccharides less tedious, a high level of technical expertise is still indispensable to achieve the desired structures.

Recently, in a seminal paper, Seeberger *et al.*[68] have developed an automated chemical synthesis of complex oligosaccharides on a solid-phase synthesizer taking into consideration several key issues such as: (i) an instrument capable of performing repetitive chemical manipulations at variable temperatures; (ii) the design of an overall synthetic strategy with either the reducing or the non-reducing end of the growing carbohydrate chain attached to the support;[69] (iii) selection of a polymer and linker that are inert to all reaction conditions during the synthesis but cleaved efficiently when desired; (iv) protecting group strategies consistent with the complexity of the target oligosaccharide; and (v) stereospecific and high yielding glycosylation. A branched dodecasaccharide was synthesized through the use of glycosyl phosphate building blocks and an octenediol functionalized resin.

The synthesis of the branched β-(1–3)/β-(1–6) glucan structure was accomplished using two glycosyl phosphate building blocks (Scheme 3.31). A levulinoyl ester (**120** and **121**) served as 6-*O* temporary

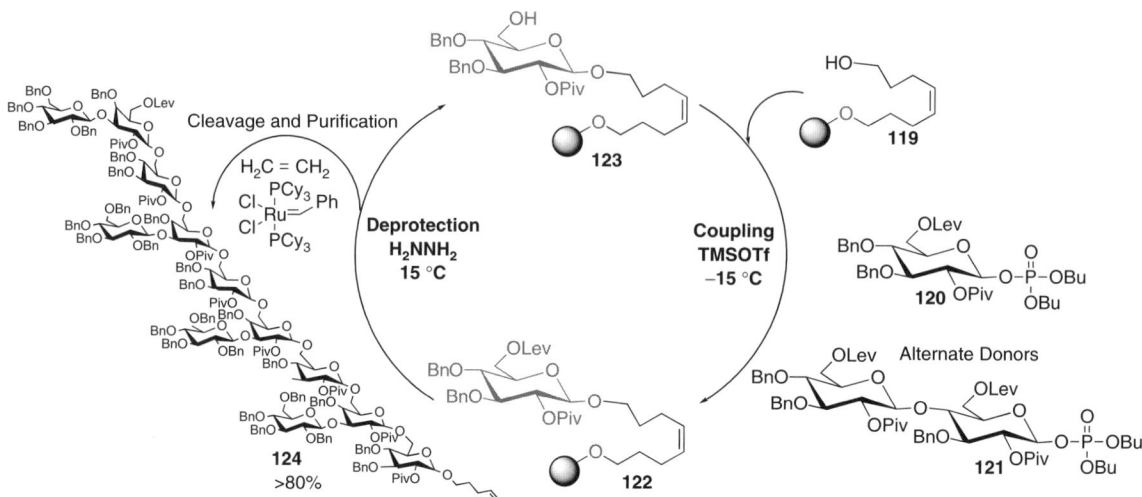

Scheme 3.31 *Synthesis of branched β-(1−3)/β-(1−6) glucan structures*

protecting group and the 2-*O*-pivaloyl group ensured complete trans-selectivity in the glycosylation reactions. Deprotection of the levulinoyl ester was achieved with a hydrazine solution in pyridine/acetic acid while the phosphate building block was activated with TMSOTf. This linear synthesis used a disaccharide (**121**), such that alternating elongation with monosaccharide building blocks resulted in a branched structure. The rapid automated assembly of this complex carbohydrate established the principle of automated synthesis and addressed all of the challenges. Still, a merely linear structure was established containing exclusively trans glycosidic linkage.

1,3-Diketones are key intermediates in heterocyclic chemistry, being highly considered for production of libraries.[70] Early strategies derived from the well established deprotonation/alkylation chemistry, with diversity introduced on a grafted 1,3-diketone backbone by combining various electrophiles.[71] Noteworthy is that the diketone's two oxygen atoms are eliminated during the cyclodehydration reaction. In a seminal paper, Wagner *et al*.[72] have therefore considered a *de novo* approach using these two atoms as anchors onto the resin (Scheme 3.32). 1,3-oxazinium salts bearing two 1,3-electrophilic centres behave similarly to 1,3-diketones upon treatment with an excess of hydrazines or other bis nucleophiles. Hence, if the oxazinium species **129** were bound to a resin through the 2-position, the trans heterocyclization reaction would result in a versatile multifunctional cleavage. Resin-bound 4*H*-1,3-oxazines **129** are synthesized by the stepwise condensation of an amide resin **125**, an aldehyde, and an alkyne (Scheme 3.32). Upon 2,3-dichloro-5,6-dicyano-1,4-benzoquinone (DDQ) activation, oxazines **128** are converted into oxazinium salts. When treated with hydrazines, these resin-bound α-diketone equivalents yield pyrazoles through a functionalizing release process. This multicomponent capture strategy, tedious to handle in classical synthesis in solution, is well suited to solid-supported chemistry. It facilitates the handling of sensitive and unstable intermediates, such as *N*-α-methoxyalkylamides **126** and 1,3-oxazinium salts **129**.

Typical Experimental Procedures

Typical procedure for the preparation of supported 4H-1,3-oxazines (Scheme 3.32)[72]

*The resin-bound N-(R-methoxyalkyl)-amide **126** (100 mg, 1 equiv.) was allowed to swell in 1 ml of DCM and cooled to 0 °C. The alkyne (5 equiv.) was then added along with $BF_3 \cdot Et_2O$ (3 equiv.). The slurry was*

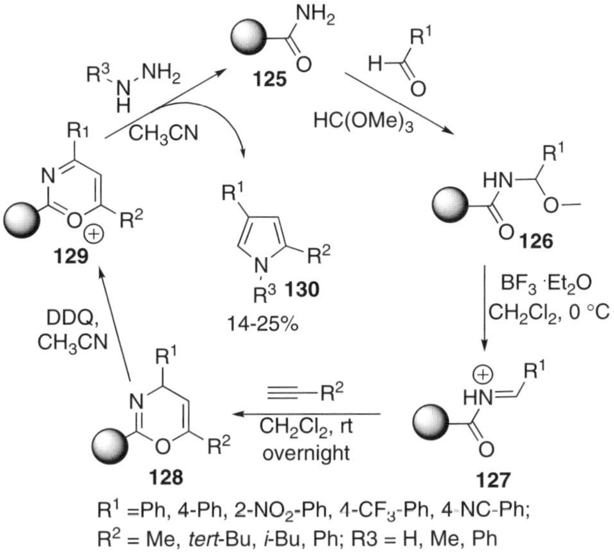

Scheme 3.32 *Supported 4H-1,3-oxazines in the synthesis of pyrazoles*

shaken at 0 °C for one hour, then allowed to warm to room temperature and to react for another 12 hours. The resin was filtered and successively washed with DCM (3 ml), DCM/triethylamine (3 ml), and then five times alternatively with DCM and methanol (3 ml).

Typical procedure for the functional cleavage (Scheme 3.32)[72]

To a suspension of a resin-bound 4H-1,3-oxazine (100 mg, 1 equiv.) in acetonitrile (2 ml) was added DDQ (3 equiv.). The slurry was warmed to 60 °C and allowed to react for 12 hours. The resin was filtered and successively washed several times with acetonitrile (3 ml), DCM (3 ml) and ethyl acetate (3 ml) and, finally, dried under vacuum. The activated resin was then allowed to swell in acetonitrile (1 ml). Nucleophile was added (20 equiv.) and the suspension was allowed to react at 60 °C for 12 hours. The resin was cooled, filtered and washed five times alternatively with DCM and methanol (3 ml). The combined filtrates were evaporated to dryness to yield a yellow-brown solid. The resulting material was eluted through a cartridge of silica gel with ethyl acetate/hexane 1:1 to filter off the by-products (DDQ residues).

3.3.4 Cleavage by Hydroxylamines

Although hydroxamic acids may be obtained by direct cleavage of resin-bound esters with hydroxylamine derivatives, it has been reported that sometimes this approach does not give reproducible results.[73] Therefore, solid-phase synthesis of hydroxamic acid derivatives generally involves either immobilization of the hydroxylamine group through a special *N*- or *O*-linkage or formation of a protected resin-bound hydroxamate.

In a very interesting paper, Weigel *et al.*,[74] have described a facile approach for the synthesis of peptide hydroxamic acids based on cleavage of resin-bound thioesters. Peptide chains were assembled on polystyrene resins bearing the 3-mercaptopropionyl or 2-mercaptoacetyl moiety and TentaGel derivatized

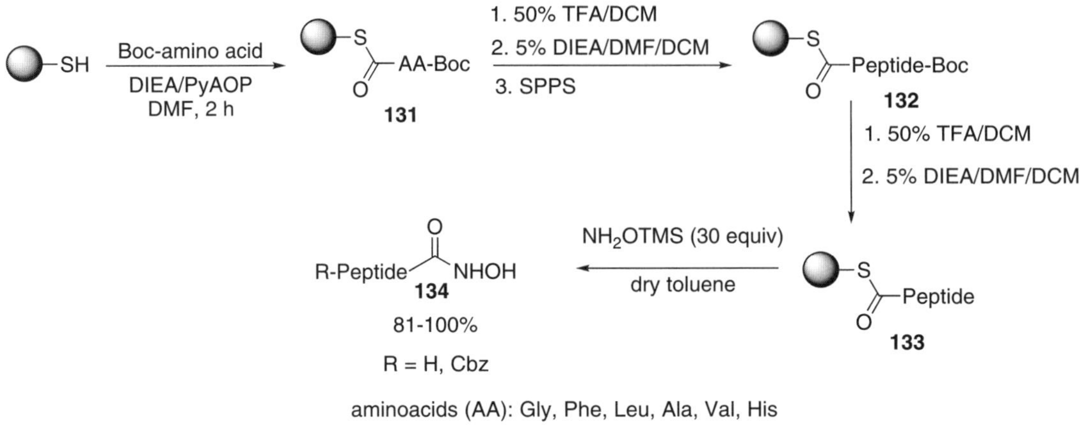

106 Traditional Linker Units for Solid-Phase Organic Synthesis

Scheme 3.33 Synthesis of peptide hydroxamic acids

with 3-mercaptopropionic acid (Scheme 3.33).[75] After attachment of the first Boc-amino acid to a alkylthiol resin (**131**), assembly of the peptide sequence was achieved by standard solid-phase peptide synthesis methodology using preactivated Boc-amino acids (Boc-amino acid/PyAOP/DIEA, 1/1/1.5). Removal of the N-terminal Boc protecting group and subsequent neutralization with 5% diisopropylcarbodiimide (DIPEA) in DMF resulted in peptide thioester resin **133**. Making use of O-trimethylsilyl hydroxyl-amine instead of aqueous hydroxylamine, cleavage of the thioester linkage in the presence of esters is selective (**134**).

Hydroxamic acids have been obtained by displacement of supported esters using aqueous hydroxylamine in THF[76] or in methanol.[77] Although several routes have been published for the preparation of hydroxamic acids on solid phase,[78] these generally involve the preparation of a special linker to which hydroxylamine is attached. Dankward's approach[76] obviates the need for special linkers or protecting groups, by displacing the desired hydroxamic acid from the resin directly using hydroxylamine, as illustrated in Scheme 3.34. Polymer-supported Cbz protected amino acids esters **135** were displaced from the resin with aqueous hydroxylamine to provide the corresponding hydroxamic acids **136**.

Concerning this last aspect, a convenient two-step procedure for the parallel synthesis of low molecular weight hydroxamic acids from carboxylic acids and hydroxylamine with the use of polymer-supported 1-hydroxybenzotriazole **137** (Scheme 3.35) has been reported recently.[79] It involves the formation of a

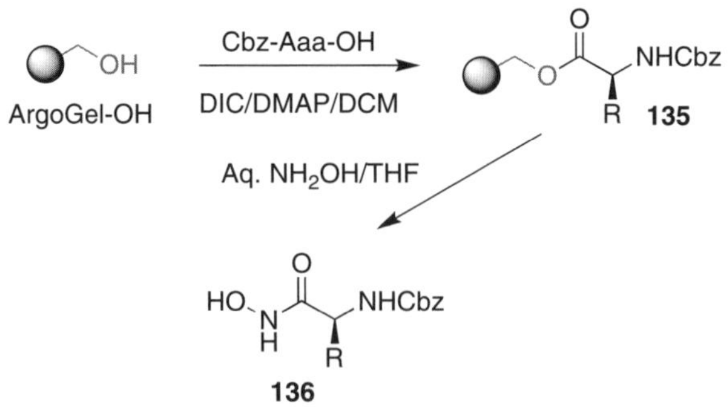

Scheme 3.34 Synthesis of 2-aminohydroxamic acids

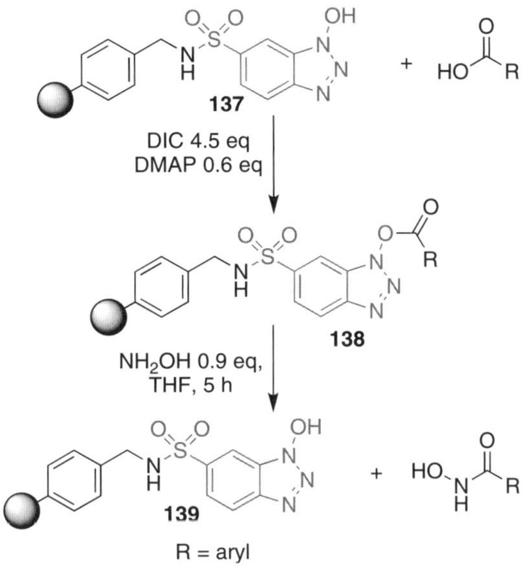

Scheme 3.35 Solid-phase synthesis of hydroxamic acids

polymer-bound HOBt active ester **137** and subsequent reaction with O-protected or free hydroxylamine (Scheme 3.35). The use of free hydroxylamine leads to increased yields while maintaining high purities. Recycling of the exhausted resin to produce the same or a different hydroxamic acid has been achieved by a three-step protocol, which is easily amenable to automation and cost economical.

Typical Experimental Procedures

Synthesis of hydroxamic acids (Scheme 3.35)[79]

*In a typical experiment, the PS-HOBt **137** (150 mg, 1.41 mmol/g. loading, 0.2 mmol) was pre-swollen with DMF. A solution of DMAP in 1:1 DMF/DCM (2.42 ml, 0.05 M) and a solution of carboxylic acid in DMF (853 µl, 0.38 M) were then added to the resin and the mixture was agitated for two minutes. A solution of DIC in DCM (565 µl, 1.65 M) was then added to the resin and the resulting mixture was agitated for three hours. The resin was collected by filtration, washed successively with DMF, DCM, DMF and THF (3 × 5 ml aliquots of each) and dried under nitrogen. To the resin-bound esters was added a solution of O-tertbutyldimethylsilylhydroxylamine in THF (3 ml, 0.5–0.8 equiv. relative to the loading of the PS-HOBt resin) or a solution of anhydrous hydroxylamine (3 ml, prepared from hydroxylamine hydrochloride and sodium methoxide) in 2.3:1 THF/methanol, and the mixture was agitated for five hours. The solution was filtered, the resin washed with THF (3 ml) and the combined organic solutions evaporated under reduced pressure. The cleavage of the tert-butyldimethylsilyl protecting group was carried out by dissolution in 95:5 TFA/water (3 ml) and stirring for 15 hours. The solution was then concentrated with nitrogen and the resulting solid was dried in vacuo.*

2,5-dimethylpyrrole can be considered a good protecting group for protecting/masking terminal amines on the solid phase. For this reason, Botta *et al*.[80] have synthesized and used a polymer-supported diketone **140** to fully protect/mask primary amines by the formation of a pyrrole ring (**141**) (Scheme 3.36).

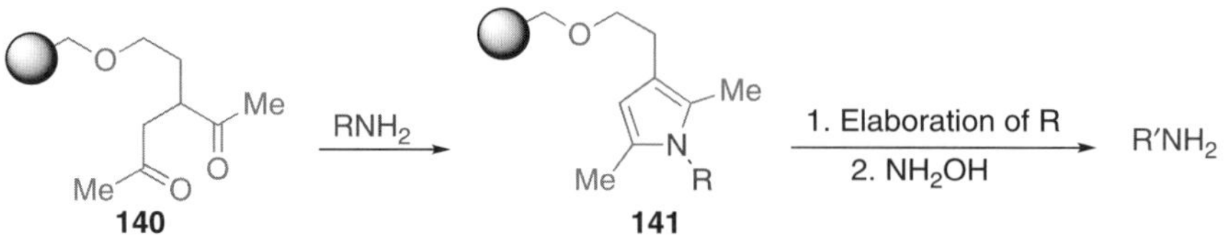

Scheme 3.36 *Synthesis of amines by a polymer-supported 1,4-diketone*

TentaGel-NH$_2$ (TG), a standard type of resin used for peptide synthesis, solid-phase organic synthesis and combinatorial chemistry, was chosen as the solid support since it swells well even in aqueous solvent; moreover, as the reaction occurs at the end of poly(ethylene glycol) (PEG) spacer and there is no cross-linking between them, the access of reagents to the reaction site is easier than with polystyrene resin (PS) and, hence, the reaction rate on TG is usually higher than on PS.

The linker was anchored to the solid support by the formation of a benzyl ether **143**, a functionality which is usually compatible with the conditions required by a variety of synthetically useful transformations, particularly basic conditions. Tentagel resin (Scheme 3.37) was acylated with 4-chloromethylbenzoic acid in the presence of EDC/HOBt and the complete transformation into **142** was confirmed by a negative colorimetric Kaiser test,[81] which is specific for the detection of primary amino groups. The chloromethyl polymer **142**, treated with CH$_3$COOK, gave **143**, which by trans-esterification with benzyl alcohol in the presence of Bu$_4$-NBr, afforded the alcohol **144**, which represents the key intermediate for the activation of the polymer in the form of trichloracetimidate (TCA). The presence of the hydroyl group on the resin was visualized with a colorimetric test.[82] The whole sequence, starting from the commercially available polymer to the alcohol **144**, could be performed in only 30 minutes by the use of MW irradiation. The TCA derivative **145** (appearance of a strong C=N stretching band at 1664 cm^{-1} in the IR spectrum) was obtained by reacting a suspension of resin **144** in DCM with trichloroacetonitrile in the presence of 1,8-diazabicyclo[5.4.0]undec-7-ene (DBU). Treating **145** with AcCl/Et$_3$N showed the complete conversion of the polymer-bound benzyloxy group to benzyl trichloroacetimidate: no AcO absorption band was observed

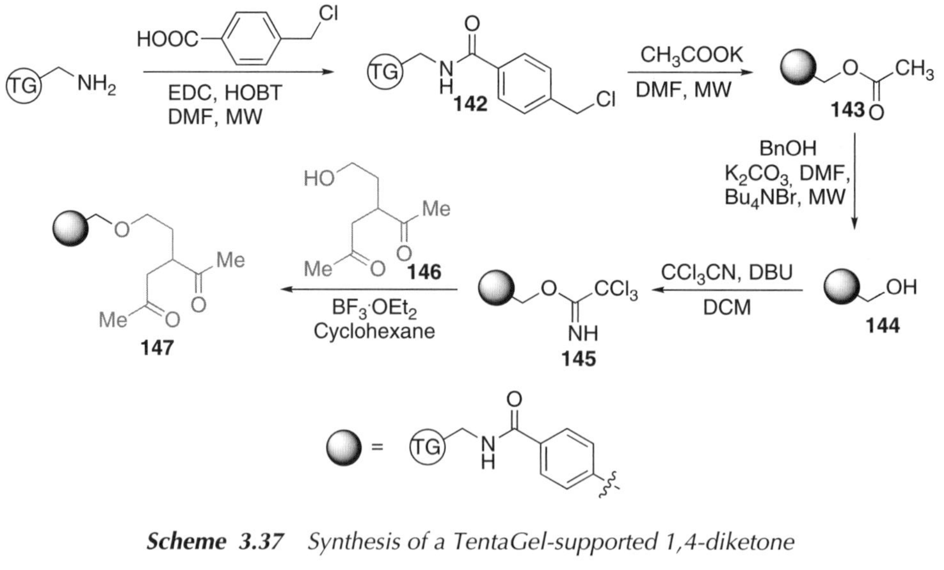

Scheme 3.37 *Synthesis of a TentaGel-supported 1,4-diketone*

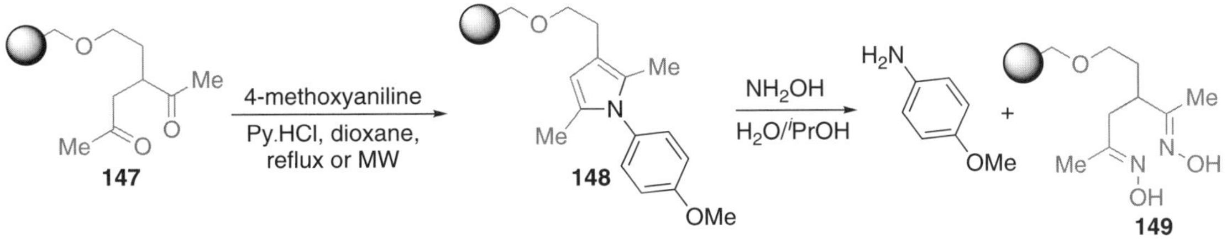

Scheme 3.38 *Microwave mediated formation of a polymer-bound pyrrole derivative*

in the IR spectrum of the product. Finally, polymer **145** was used for *O*-benzylation of the alcohol **146** in the presence of BF′$_3$OEt$_2$: the reaction was monitored by IR spectroscopy, observing the disappearance of the band of the C=N stretching and the appearance of the carbonyl band of the diketone **147** at 1713 cm^{-1}.

To prove its ability to react with amines, **147** was treated with *p*-anisidine in the presence of catalytic amount of pyridine hydrochloride in refluxing dioxane (Scheme 3.38). The formation of the polymer-bound pyrrole derivative **148** proved to be complete after two cycles of reaction as shown by IR analysis. This result was further confirmed by opening of the pyrrole ring using hydroxylamine hydrochloride and triethylamine in a refluxing water/^{i}PrOH mixture. After washing of the resin **149** and alkalinization of the filtrate, *p*-anisidine was obtained in pure form and quantitative yield, demonstrating that along the whole solid-phase sequence the loading of each intermediate had been quantitative. Alternatively, the formation of the pyrrole ring was successfully achieved under microwave irradiation using a few drops of DMF as solvent and an excess of amine. With this approach, the solid-supported 2,5-dimethylpyrrole was obtained in only five minutes, thus confirming the importance of microwaves in speeding the reaction rate on solid phase.

Typical Experimental Procedures

General procedure for the preparation of polymer-bound 2,5-dimethylpyrroles 147 (Scheme 3.38)[80]

Method A: To a suspension of resin **147** (2.23 g, 0.09 mmol) in 10 ml of dioxane the appropriate amine (0.27 mmol) was added, followed by pyridine hydrochloride (0.005 mg, 0.04 mmol). The reaction mixture was refluxed for 24 hours, the resin was filtered, washed successively with DCM, THF and DCM and then dried in vacuo.

Method B: Polymer **147** (0.22 g, 0.09 mmol) was swollen in a few drops of DMF, the appropriate amine was added and the mixture was irradiated at 400 W for 5 minutes.

Various reactions could be performed on this system, which then can be cleaved with full restoration of the amine functionality. The same authors have also investigated the possibility of preparing compounds having a guanidine core. The resin can also be recycled at least once after a treatment with *tert*-butyl hydroperoxide (TBHP)[83] in acetone, without loss of purity of the final compound.

3.3.5 Cleavage by other nucleophiles

The search for synthetic efficiency has dominated the recent evolution of synthetic methods. This is proved by its influence on the development of polymer linkers. Recently, the design of polymer linkers as a rather

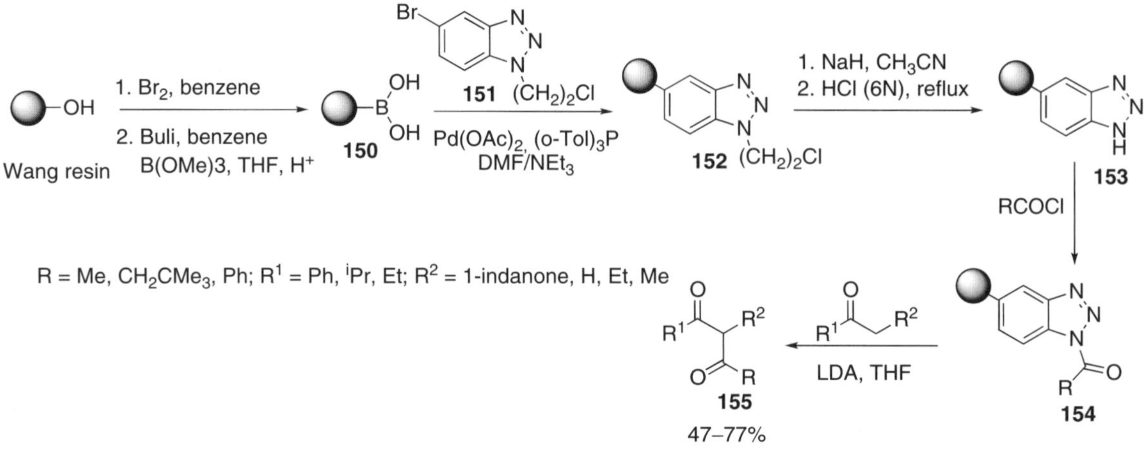

110 *Traditional Linker Units for Solid-Phase Organic Synthesis*

Scheme 3.39 *Immobilized benzotriazole linker for the synthesis of 1,3-diketones*

elaborate protecting groups[84] has been exemplified by the advent of *'leaving group linkers'*.[85] Such linkers increase the synthetic efficiency by allowing a wider variety of bond formations at the linkage site.

An efficient application of this concept was illustrated by the preparation of a immobilized benzotriazole linker (**132**) to exploit as a leaving group (Scheme 3.39).[86] The benzotriazole moiety has been directly attached to a polymer support via a carbon–carbon bond employing the classical Suzuki coupling reaction. Upon treatment of benzotriazole with bromochloroethane in the presence of sodium hydride followed by direct bromination, the desired 5-bromo compound **151** was isolated together with traces of the 7-bromo isomer in 58% yield. Compound **151** was heated with polymer-bound boronic acid in the presence of a palladium catalyst at 90 °C for 24 hours. Subsequent deprotection afforded **153** and the loading was determined by CHN analysis as 1.3 mmol/g.

This solid supported heterocyclic leaving group was tested in enolate C-acylation sequence (Scheme 3.39). Polymer support **153** was first transformed into the corresponding polymer azolides **154** which were subsequently reacted with ketone lithio enolates to yield 47–77% of the corresponding 1,3-diketones **155**. NMR spectra analysis indicated that such acylation is regiospecific and yielded only products of C-acylation.

Typical Experimental Procedures

General procedure for the preparation of polymer-supported azolides 154 (Scheme 3.39)[86]

*The 1-(2-chloroethyl)-1H-1,2,3-benzotriazole protected resin **152** (3.6 g, 1.6 mmol/g) was swollen in anhydrous acetonitrile (100 ml) under nitrogen, then sodium hydride (546 mg, 21.6 mmol) was added and the mixture was heated under reflux for 10 hours. After cooling to room temperature, 6 N hydrochloric acid (36 ml) was added drop-wise and the mixture was heated again at 80 °C overnight. The resulting mixture was cooled again to room temperature followed by the addition of a solution of sodium acetate (19.4 g, 0.24 mol) in water (33 ml). The mixture was stirred at room temperature for two hours and the resulting resin was then filtered off and washed successively with water (200 ml), water–methanol 50% (100 ml), methanol (100 ml), methanol/DCM (100 ml), DCM (100 ml) and ethyl ether (100 ml). Loading: 1.3 mmol/g (%N, 5.49). Lithium bis(trimethylsilyl)amide (3 ml, 3 mmol, 1 N solution in hexanes) and benzotriazole resin **153** respectively (2 mmol) pre-swelled in 1-methyl-2-pyrrolidinone (50 ml) were stirred for one hour at room temperature* and the solution of the corresponding acyl chloride (3 mmol) in dry THF (10 ml) was added drop-wise. The slurry was stirred at room temperature for six hours and then at 60 °C for two hours, cooled and washed successively with DMF (50 ml), DMF–methanol 50% (100 ml), methanol (100 ml), water–methanol 50% (100 ml), methanol (100 ml) and ethyl ether (100 ml).

General procedure for the C-acylation of ketones with polymer-supported azolides 154 (Scheme 3.39)[86]

A 50 ml round-bottom flask with septum inlet was charged with a solution of lithium diisopropylamide (LDA) (1 ml, 0.15 M solution in hexanes) in dry THF (10 ml) under nitrogen and cooled to −78 °C, and a solution of the corresponding ketone (1.5 mmol) in dry THF (15 ml) was added drop-wise. After stirring for one hour at this temperature, the resulting mixture was added drop-wise to the slurry of resin **154** (0.5 mmol) in dry THF (50 ml) by cannula and the mixture was stirred for another 12 hours while the temperature was allowed to rise to 20 °C. The resin was separated and washed with THF (20 ml) and the organic layer was washed with 10% ammonium chloride (2 × 50 ml) and water (50 ml) and then dried (anhydrous magnesium sulfate). Chromatography on silica gel with hexanes/ethyl acetate (4:1) as an eluent gave diketones **12**, which were analyzed by LC/MS.

A well-sustained multi-step synthetic protocol has been designed for the PEG-functionalized aromatic acid amide to generate a molecular library of 2-alkylthio bis-benzimidazoles **159** (Scheme 3.40). An attempted synthesis of benzimidazole-2-thiol in dichloromethane has led to S-chloromethyl methyl sulfides, mimicking bacterial enzymatic systems. Regioselective S-alkylation was brought about under controlled conditions using a mild base at room temperature. The polymer-free compounds, 2-sulfanylated bis-benzimidazoles **159**, were obtained in high yields and high purities through a mild cleavage by using potassium cyanide (Scheme 3.40).[87]

Typical Experimental Procedures

General procedures for the synthesis of benzimidazolyl benzimidazolones 159 (Scheme 3.40)[87]

To the solution of polymer immobilized conjugate **156** (0.33 mmol, 1.0 equiv.) in 1,2-dichloroethane, trifluoroacetic acid (0.5 ml) and magnesium sulfate (0.5 g) were added. The mixture was refluxed for 12 hours and the progress of the reaction was monitored by ^{1}H NMR. After completion of the reaction, magnesium sulfate was removed by filtration through a layer of Celite. Then the solvent was evaporated and the product was precipitated by slow addition of an excess of cold ether. The precipitated benzimidazole conjugate **157** was filtered through a fritted funnel and washed with ether. The resulting conjugate **157** was treated with various amines (2.31 mmol, 7.0 equiv.) in dichloromethane at ambient temperature for 12 hours. When the reaction was finished, the solvent was changed from dichloromethane to methanol because most of the excess amine that would hinder the zinc reduction can be removed by precipitation from methanol and because a small amount of conjugate product may be lost during ether washing. Conjugate product in methanol was completely reduced by zinc (3.3 mmol, 30 equiv.) and ammonium formate (3.75 mmol, 15 equiv.) for 30 minutes at ambient temperature. Zinc was removed by centrifugation and filtration, and dichloromethane was added to precipitate ammonium formate. After filtration through Celite, polymer-immobilized diamine was obtained. Polymer immobilized diamine (0.33 mmol, 1 equiv.), triethylamine (0.78 mmol, 2.6 equiv.) and thiocarbonyldiimidazole (0.78 mmol, 2.6 equiv.) were mixed in 1,2-dichloroethane and the reaction mixtures

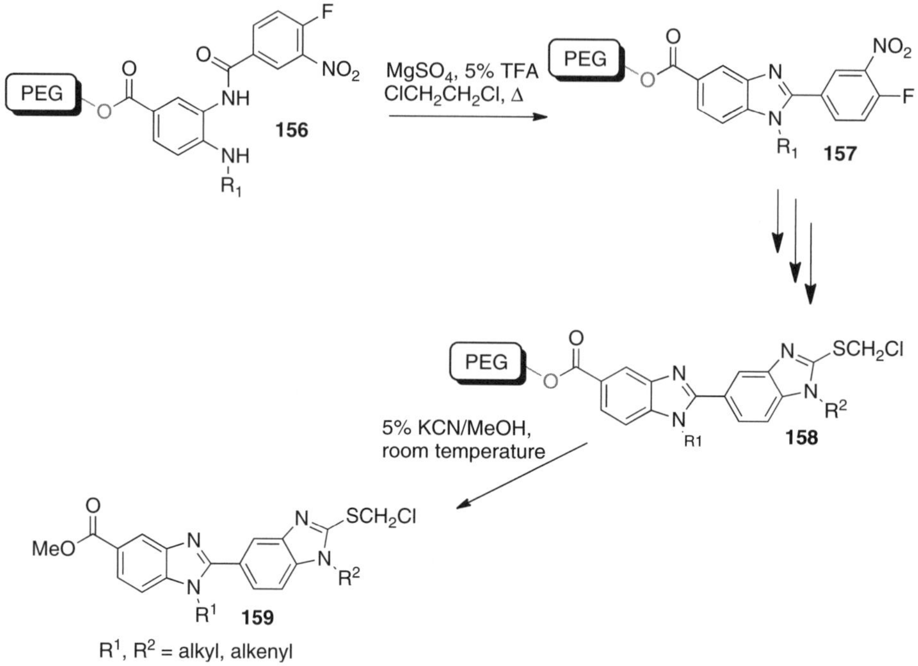

Scheme 3.40 *Preparation of a library of 2-alkylthio bis-benzimidazoles*

were refluxed for 12 hours. The corresponding cyclised conjugated was isolated by a similar ether precipitation and washing procedure. Added to the conjugated product (0.33 mmol, 1 equiv.) in dichloromethane were triethylamine (1.0 mmol, 3.3 equiv.) and various alkyl halides (1 mmol, 3.3 equiv.) at room temperature, and the reaction mixture was reacted for 14 hours at ambient temperature. Conjugated product was purified by precipitation and washing with ether. The target compound was obtained by treating conjugate **158** with potassium cyanide in methanol for three days. The crude products **159** was obtained and subsequently checked by HPLC after precipitation. The crude product was then purified by column chromatography (DCM/methanol).

The tropane ring system is found in numerous naturally occurring alkaloids.[88] Although many synthetic methods are available for the preparation of this structural component in solution phase,[89] their extension to solid-phase synthesis and chemical library production has been very limited.[90] In an very interesting revisitation of the classical Robinson tropinone synthesis, a solid-phase version has been developed,[91] which involved 1,3-Dipolar cycloaddition of 3-oxidopyridinium betaine to activated olefins on a solid-phase leading to resin-bound 8-azabicyclo[3.2.1]octenones which undergo further transformations such as 1,4-addition of nucleophiles. This sequence began with the bromomethyl linker **160** obtained by bromination of Wang resin (CBr$_4$, PPh$_3$, DMF).[92] Reaction of **160** with 3-hydroxypyridine in refluxing propanol for 30 hours gave the polymer-bound 3-hydroxypyridinium bromide, which was treated with sodium methoxide (CH$_3$ONa) in propanol to afford **161** (Scheme 3.41). This resin-bound betaine, readily isolated by washing with propanol and drying under vacuum, was immediately subjected to 1,3-dipolar cycloaddition conditions. Heating **161** in the presence of excess phenyl vinyl sulfone (6 equiv.) in THF at refluxing temperature for 16 hours led to resin-bound cycloadduct **162** [IR (potassium bromide): max 1691 cm^{-1}].

Unfortunately, the benzyl linker in **162** has been found to be very stable towards a variety of cleavage conditions. First attempts using acidic (TFA, pure or as a solution in DCM, for one hour to three days) or oxidative [2,3-dichloro-5,6-dicyano-1,4-benzoquinone (DDQ) in benzene or in $CH_2Cl_2-H_2O$, ceric ammonium nitrate (CAN) in CH_3CN-H_2O] conditions were disappointing. Under these conditions the cleavage was incomplete releasing the tropane derivative in low yield. After extensive experimentations, the best results have been found when **162** was treated with acid chlorides in acetonitrile in the presence of potassium iodide.[93] Thus heating **162** with acetyl chloride (7.5 equiv.) and potassium iodide (5 equiv.) in refluxing acetonitrile for six hours furnished a separable mixture of amides **163a** (6-*exo* sulfone) and **164a** (7-*exo* sulfone) (87:13 ratio) in 51% overall yield on the basis of the initial loading level of resin. The use of potassium iodide in this reaction was crucial, otherwise the cleavage was ineffective. When acetyl chloride was replaced by benzoyl chloride, only the 6-*exo* isomer **163b** can be isolated in 54% overall yield. The cleavage using acryloyl chloride led to a separable mixture of amides **163c** and **164c** in a similar ratio as with acetyl chloride in 58% overall yield (Scheme 3.41).

In an interesting paper, Kurth and coworkers[94] have described the synthesis of a library of 1,3-diols with potential antioxidant activity by the use of a two-step aldol reduction sequence (Scheme 3.42). The common precursor **166** is readily prepared from Merrifield resin and can readily be converted to a zinc enolate that react smoothly with a range of aldehydes. DIBAL reduction of the intermediate aldol products **167** effects concomitant removal from the resin giving diols such as **168** (26% overall as a 7:5 threo/erythreo mixture). It is noteworthy that polymer-supported reactions to give **166** and **167** can be monitored by potassium bromide pellet FT-IR analysis of the polymer. In this manner, making use of a set of 27 different aldehyde, they have prepared a library of 27 diols.

Martinez[95] and Armstrong[96] have reported two independent examples of the use of a linker based on the Weinreb amide for the synthesis of aldehydes and ketones. The synthesis of the linker was performed as showed in the Scheme 3.43. *O*-methoxy hydroxylamine was reacted with benzyl acrylate and the *tert*-butoxycarbonyl group protected the resulting alkylated methoxy-amine **169**. After deprotection of compound **170** by hydrogenolysis, the linker **171** was couplet to the solid support (i.e. MBHA) with an activating agent to yield the substituted resin **172**. After deprotection of the *N*-terminal Boc, elongation by classical solid phase synthesis (Boc or Fmoc strategies) was possible. Reductive cleavage of resin-bound peptide **173** using lithium aluminium hydride (LiAlH$_4$) provided a 40% yield of the peptide aldehyde

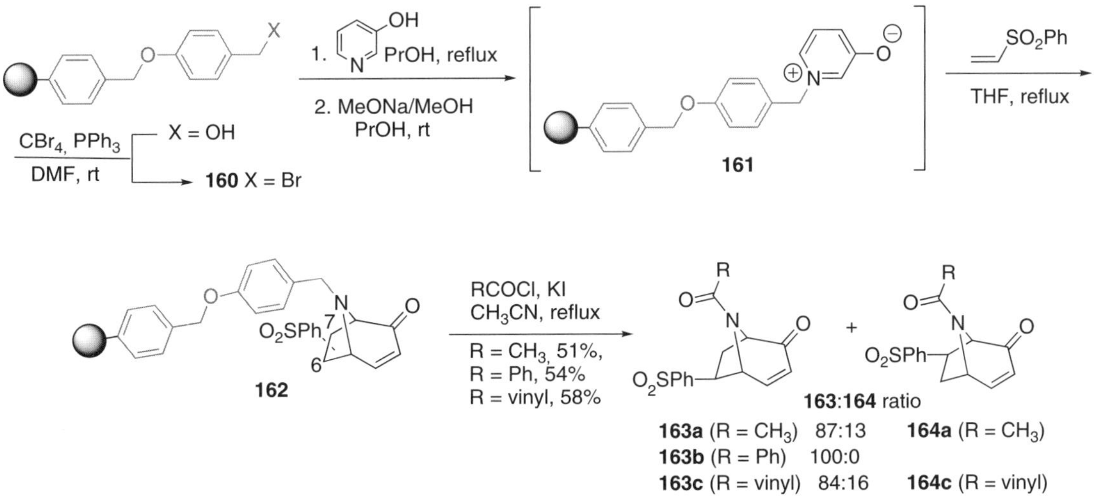

Scheme 3.41 *Solid-phase synthesis of 8-acyl-8-azabicyclo[3.2.1]oct-3-en-2-ones*

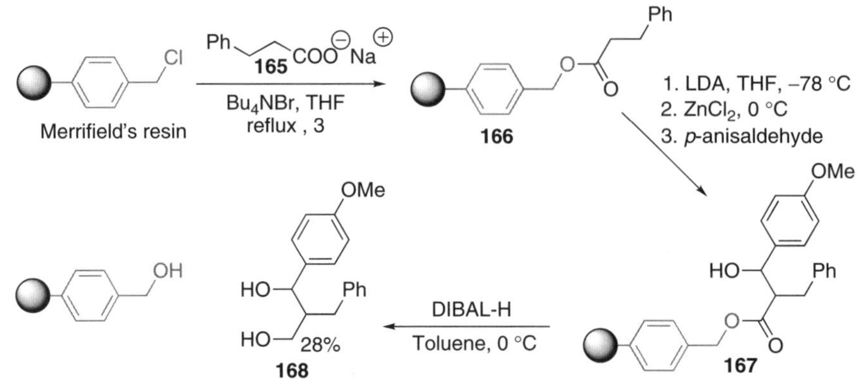

114 Traditional Linker Units for Solid-Phase Organic Synthesis

Scheme 3.42 Solid-phase synthesis of 2-benzyl-1-(4-methoxyphenyl)propane-1,3-diol

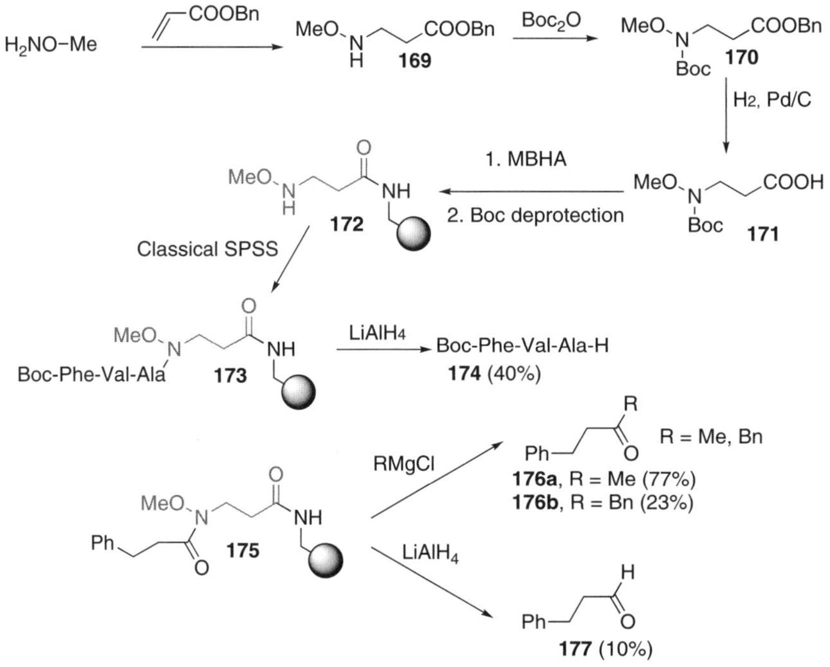

Scheme 3.43 Weinreb-type linkers for the synthesis of carbonyl compounds

Boc-Phe-Val-Ala-H **174** after purification. Similarly, treatment of **175** with CH_3MgCl, BnMgCl, PhMgCl or $LiAlH_4$, provided the corresponding ketones **176a–b** or aldehyde **177** in 77%, 23% and 10% yields respectively.

Typical Experimental Procedures

General procedure for the synthesis of linker 173 (Scheme 3.43)[95, 96]

9.65 g (60 mmoles) of benzyl acrylate were added to a solution containing the methoxy-aminechlorhydrate (60 mmol, 5 g) in acetonitrile (50 ml) in the presence of 50 ml of diisopropylethylamine (DIPEA). After

48 hours at 52 °C the oily residue was dissolved with ethyl acetate, washed with saturated solutions of sodium bicarbonate and sodium chloride. After drying with sodium sulfate, the solution was concentrated and gave an oil (crude yield: 11 g = 88%). The compound was purified on a silica gel chromatography (ethyl acetate/hexane: 5/5) to yield the pure 3-amino-N-methoxy-propionic acid benzyl ester (7.49 g = 60%). 33 mmol (7.3 g) of Boc_2O were then added to a solution of dioxane/water (2/1) containing 33 mmol (7.0 g) of the precedent compound in the presence of sodium hydroxide 1 M in order to maintain the pH between 12 and 13. After three hours the solvent was evaporated, the mixture dissolved in ethyl acetate and washed with potassium bisulfate, aqueous solution (5%). After drying and concentration, an oily residue is obtained (yield: 10 g = 98%). This 3-amino N-tert-butyloxycarbonyl-N-methoxypropionic acid benzyl ester was hydrogenolysed with Pd/C in ethanol 95% in three hours. After filtration of the catalyst and evaporation of the solvent, the linker is quantitatively obtained (6.8 g) and ready to be coupled to the resin.

Typical reduction protocol (Scheme 3.43)[95]

1.5 g of peptidyl-resin (0.66 mmole) were suspended in anhydrous THF and placed in an ice bath. Lithium aluminium hydride (114 mg, 5 molar equivalents) was added and the reaction was stirred for 30 minutes. The reaction was then hydrolysed with a potassium bisulfate, aqueous solution (5%). The resulting mixture was filtrated in order to eliminate the resin, which was washed twice with dichloromethane. The liquid phases were gathered diluted with dichloromethane and washed with a potassium bisulfate, aqueous solution, saturated solutions of sodium bicarbonate and sodium chloride. After drying and evaporation a white powder was obtained (crude yield: 180 mg = 71%,). This aldehydic peptide was then purified either on a silica gel chromatography with solvents containing 0.1% pyridine or by reversed phase HPLC in gradient mode using acetonitrile/water/TFA: x/y/0.1% as solvent system.

Representative procedure for the reaction of CH_3MgCl with Weinreb-type amide 175 (Scheme 3.43)[96]

To a suspension of solid supported amide 175 (0.015 mmol) in THF (0.6 ml) was added CH_3MgCl (70 ml in THF, 0.22 mmol). The flask was glass-stoppered and parafilmed and the mixture was stirred for 13 hours at room temperature. The mixture was then quenched with 5% HCl/ethanol (0.5 ml) and stirred for 30 minutes at room temperature. The resin was filtered and washed alternatively with 3×0.5 ml ethanol and 3×1 ml dichloromethane. After the filtrate was extracted with dichloromethane and water, the crude product was clean for 4-cyclohexyl-2-butanone (1.8 mg, 78% yield).

Among the few methods available for the generation of tertiary alcohols in solution, addition of nucleophiles to esters is the most straightforward. Accordingly Chandrasekhar et al.[97] have developed a new protocol, wherein the ester bound polymer can be cleaved by addition of a Grignard reagent to give tertiary alcohols in excellent yields and high purity (Scheme 3.44). In this reaction the use of different Grignard cleaving agents can contribute to diversity.

Four polymer supported esters **178** have been prepared from 10-undecenoic, 3,4,5-trimethoxybenzoic, pyridine-2-chloro-3-carboxylic and 3,4-dichlorobenzoic acids. Each acid was neutralized with sodium bicarbonate and azeotroped with benzene to give a solid sodium salt, which was dissolved in DMF and treated with Merrifield resin that had been preloaed in DMF for 30 minutes. The mixture was stirred for 16 hours at 80 °C, allowed to cool to room temperature, then filtered, washed with DMF, water, THF, methanol, DCM and ether, and then dried under vacuum to yield the resin bound ester **178**. The acid loaded resins

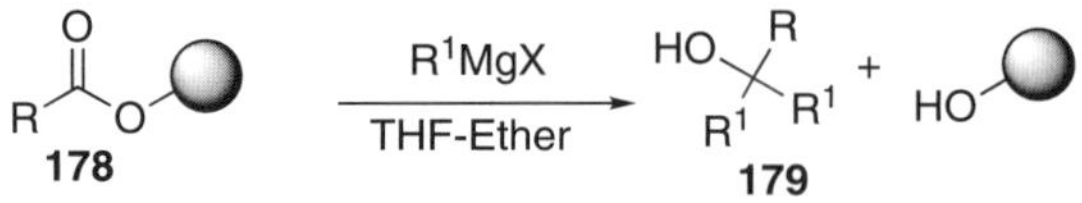

116 Traditional Linker Units for Solid-Phase Organic Synthesis

Scheme 3.44 *Tertiary alcohols from polymer-supported esters*

also showed a strong carbonyl stretch near 1735 cm^{-1}, corresponding to the ester carbonyl frequency. The polymer bound acids were exposed to different Grignard reagents, in threefold molar excess equivalents to get the corresponding tertiary alcohol **179** in good yield by simple filtration through a pad of silica gel–ammonium chloride (1/1) mixture.

One of the still unresolved problems in parallel synthesis is the availability of a general and rapid method for the transformation of a primary amine into the corresponding secondary amine without the issue of polyalkylation. Following the Fukuyama method,[98] which is based on the alkylation of *o*-nitrobenzenesulfonamides, followed by removal of the sulfonyl group, Taddei *et al.*[99] have developed a simple protocol which can be easily applied to parallel synthesis making use of supported reagents and scavengers. This approach could be considered as the first example where a solid supported nucleophile cleavages the amino protecting group catching all the side reaction products (Scheme 3.45). The monomethylation of benzylamine was selected as a model reaction to assess the influence of several bases.

The conventional method for the preparation of sulfonamides requires the use of a slight excess of the nosylchloride in DCM and the use of triethylamine as the base. These conditions were tried at first using a 1:1:3 ratio of the chloride, the amine and triethylamine. After the mixture was stirred for four hours at room temperature and the solid formed was filtered, evaporation of the solvent gave **180** in very good yields. Unfortunately, the crude product **180** was contaminated with triethylammonium hydrochloride, which could be purified by selective strong cationic exchanger (SCX) extraction at the end. The present reaction provides an efficient alternative, in particular, for monomethylation of primary amines (see the preparation of **181** in Scheme 3.45). As the last step of the reaction is compatible with the condition of reductive amination, the primary amine can be transformed into a tertiary amine with three different substituents in good yields with

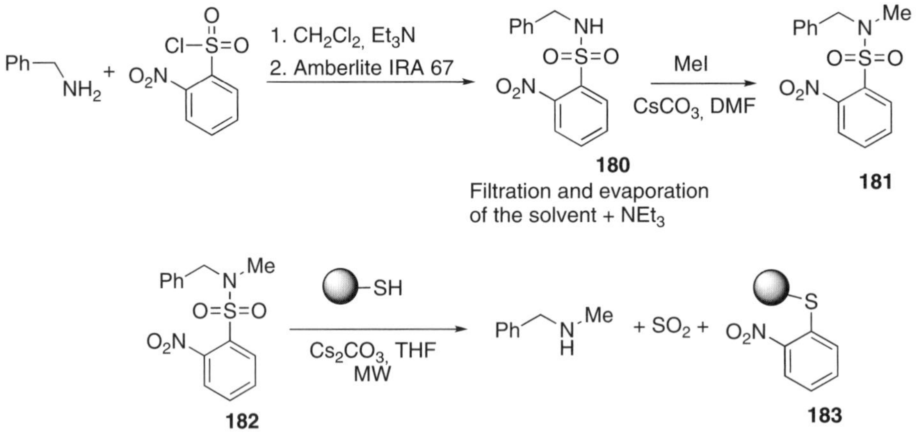

Scheme 3.45 *Polymer-assisted transformation of primary amines into secondary amines*

a simple To verify the robustness of the method, a small representative array of secondary amines have been prepared. Moreover, taking advantage of the possibility to use different supported reagents in the same pot, they, also prepared, starting from primary amines, a series of differently substituted tertiary amines.

Typical Experimental Procedures

General procedure for classical oil-bath-assisted alkylation (Scheme 3.45)[99]

*Benzylamine (0.068 ml, 0.63 mmol) was dissolved in dry DCM (3 ml), followed by the addition of Et_3N (0.19 ml, 1.9 mmol) and 2-nitrobenzenesulfonyl chloride (0.139 g, 0.63 mmol). The mixture was stirred at room temperature for six hours. The mixture was filtered; DCM (6 ml) was added, and Amberlite IRA-67 (1.0 g, 2 mmol, previously washed with water, methanol, THF and DCM) was added to the solution. The mixture was shaken for two hours. The resin was filtered off, and the solvent was evaporated under vacuum to give crude sulfonamide **180** (0.18 g, 98% yield) which was analysed by ^1H NMR and HPLC-MS analysis. The crude was dissolved into dry DMF (5 ml); to this solution, caesium carbonate (0.2 g, 0.616 mmol) and methyl iodide (0.088 ml, 1.39 mmol) were added and the mixture stirred at room temperature for 12 hours. The mixture was filtered; the solvent was evaporated, and the residue was dissolved into dry trichloromethane. After one hour of shaking in trichloromethane, the solution was filtered through a sintered glass disc (maximum pore size of 16–40 μm) and the solvent was evaporated to give pure compound **181** (0.18 g, 95% yield), which was analyzed by ^1H NMR and HPLC-MS analysis. The crude was dissolved into dry THF (5 ml) and, to this solution, Cs_2CO_3 (0.38 g, 1.2 mmol) was added, followed by PS-thiophenol resin (0.4 g of a 1.41 mmol/g loading resin, 0.56 mmol). The mixture was shaken for 12 hours at room temperature. The solid was filtered off and Cs_2CO_3 (0.38 g, 1.2 mmol) and PS-Thiophenol resin (0.4 g of a 1.41 mmol/g loading resin, 0.56 mmol) were added to the solution. The mixture was shaken for 12 hours and filtered, and then the solvent evaporated to give compound **182** (0.067 g, 96% yield), which was analyzed by ^1H NMR and GC/MS analysis, in comparison with an authentic sample.*

General procedure for microwave-assisted alkylation (Scheme 3.45)[99]

2-Phenylethylamine (0.050 g, 0.413 mmol) was reacted with nosyl chloride as previously described. The crude sulfonamide was dissolved in DMF (5 ml) into a microwave reaction tube and Cs_2CO_3 (0.268 g, 0.826 mmol) added, followed by iodocyclopentane (0.201 g, 1.032 mmol). The tube was sealed, inserted inside a microwave cavity (monomode irradiation) and heated at 100 °C and 200 psi for two minutes, followed by two minutes of rest and an additional two minutes of irradiation. The tube was cooled, the solvent diluted with DCM (10 ml) and the solid filtered off. The solvent was evaporated and the crude was dissolved into dry THF (8 ml). Cs_2CO_3 (0.268 g, 0.826 mmol) was added, followed by PS-thiophenol (0.5 g of a 1.41 mmol/g loading resin, 0.75 mmol), and the mixture was shaken for 12 hours at room temperature. The solid was filtered away and to this solution Cs_2CO_3 (0.266 g, 0.826 mmol) and PS-thiophenol resin (0.5 g of a 1.41 mmol/g loading resin, 0.75 mmol) were added. The mixture was shaken for 12 hours, filtered, and the solvent was evaporated to give N-Cyclopentyl-2-phenylethyl amine (0.070 g, 90% yield) as an oil which was analyzed by ^1H NMR and ^{13}C NMR in comparison with literature data.

The Nicholas reaction, developed in the 1970s by Kenneth Nicholas,[100] involves the treatment of an alkynol or alkylyl ether with a Lewis acid and the reaction of the cation thus formed with different nucleophiles. The reaction could be used for both the formation of carbon–carbon as well as carbon heteroatom bonds complementing the numerous palladium-based methodologies for carbon–carbon cross-coupling.[101] Recently, Cassel et al.[102] have published a short report on the Nicholas reaction on the

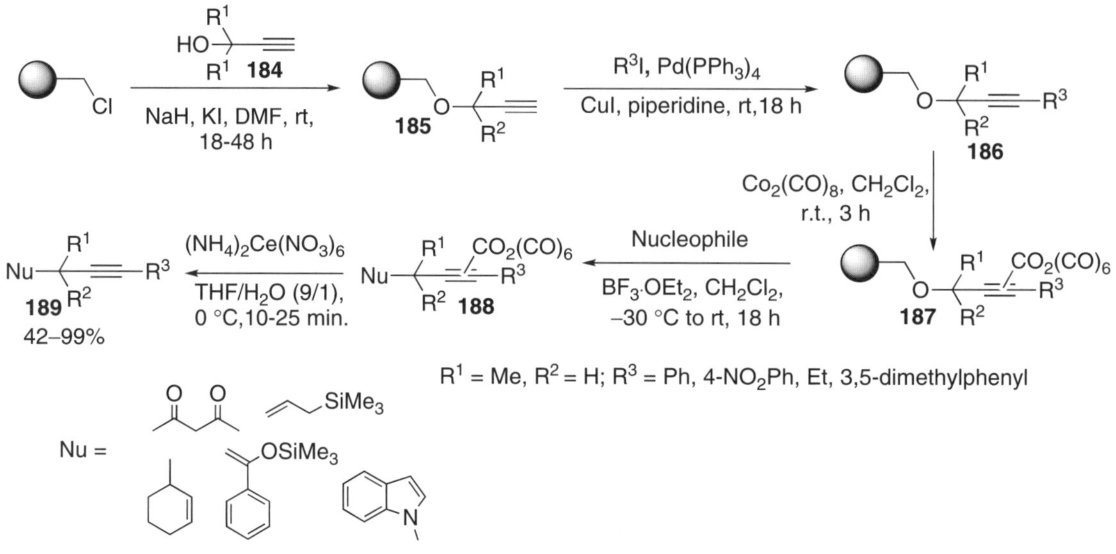

Scheme 3.46 *The Nicholas reaction on solid phase*

solid phase, involving the reaction of a polymer-bound alkynol **185** with aryl halides via a Sonogashira coupling, followed by treatment with dicobalt octacarbonyl and subsequent reaction with boron trifluoride in conjunction with oxygen or carbon nucleophiles (Scheme 3.46).

In a subsequent paper,[103] the same authors have optimized the procedure performing both the attachment of the alkynol to the solid support and the Sonogashira coupling under microwave irradiation. The use of polymer-bound scavengers or a catch-and-release approach for the purification allows the preparation of target compounds of high purity. Carbon, oxygen, nitrogen and sulfur nucleophiles could be used in the Nicholas step, while reactions with hydride and fluoride did not give any product.

Dehydrogenation of aryl hydrazines by a variety of oxidants has long been known to produce arenes and nitrogen via a transient aryl diazene **191** (Scheme 3.47).[104] Coupling a carboxylic acid to an aryl hydrazine forms an aryl hydrazide linker **190**. This may be cleaved oxidatively by a range of reagents including Fehling's solution, Tollen's reagent, *N*-bromosuccinimide/pyridine and periodic acid (Scheme 3.47).

Lowe *et al.*[105] have developed an acid/base stable aryl hydrazide linker (**192**, Scheme 3.48), which is readily coupled to solid-phase resins (**193**). Cleavage is specific and facile, requiring a copper(II) catalyst, base and a nucleophile to proceed. The conditions are compatible with all 20 proteinogenic amino acids and quantitative cleavage is achieved within two hours at 20 °C to give peptides with C-terminal acid, amide or ester functionalities. Aryl hydrazides also offer scope as simple 'traceless' linkers.

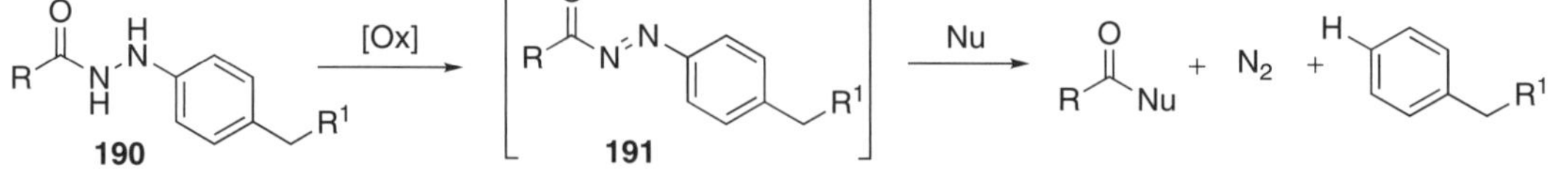

Scheme 3.47 *Dehydrogenation of aryl hydrazines to arenes*

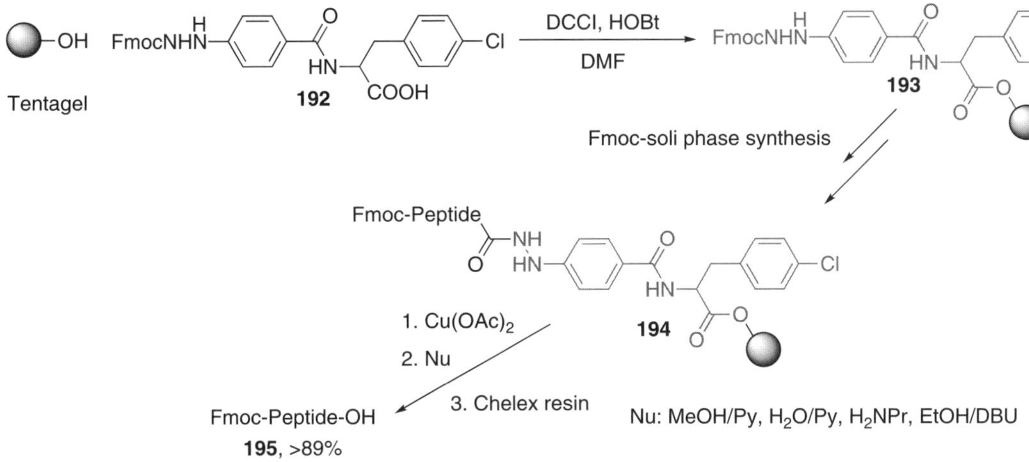

Scheme 3.48 Stable aryl hydrazide linker in the preparation of peptides

3.3.6 Linker cleavage by intramolecular nucleophilic reaction

A promising SPOS method of cleavage that became very common is the so-called cyclative cleavage. It produces an advanced open intermediate that undergoes cyclization with concomitant release of the final cyclized product. This method is, therefore, particularly used in the synthesis of heterocycles. If the bond being broken in an intramolecular reaction is involved in attaching the molecule to the solid phase (i.e. part of the linker) then it is possible to achieve both cyclization and cleavage in the one step, that is cyclative cleavage. Induction of cleavage by cyclisation during the final step of a synthesis has the major advantage that reaction products, which are not capable of cyclisation, for instance, due to an incomplete transformation earlier in the reaction sequence, remain attached to the solid phase. High levels of purity can therefore result, as only complete cyclised product has the ability to be released into solution.

One of the examples of nucleophilic cyclative cleavage to appear in the literature described the production of heterocycles. In the synthesis of oxazolidinones, Buchstaller[106] used a cyclisation cleavage strategy (Scheme 3.49).[107] Reaction of isocyanates **195** with Wang resin provided resin-bound carbamates **196**

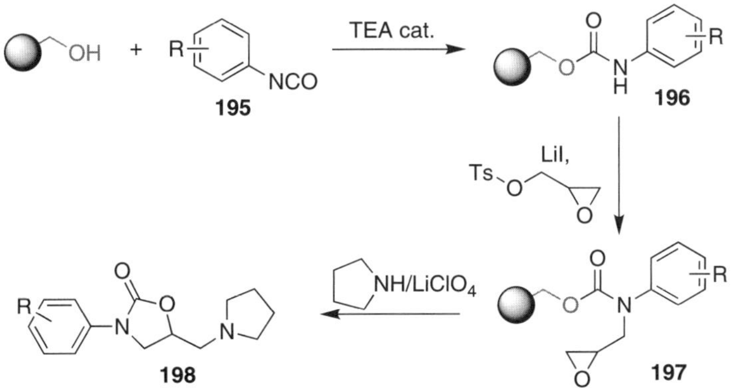

Scheme 3.49 Synthesis of 3-aryl-5-(pyrrolidin-1-ylmethyl)oxazolidin-2-ones

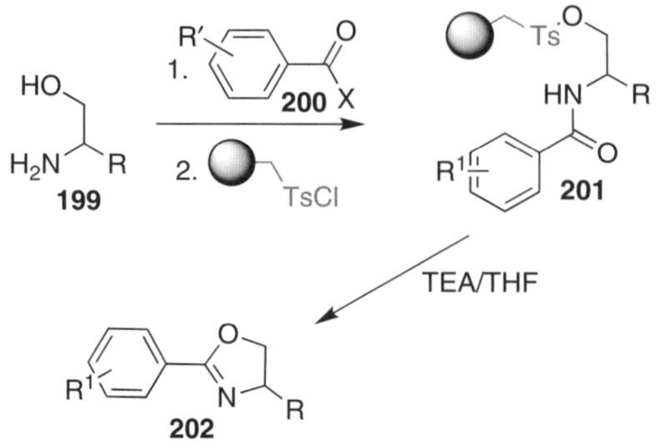

Scheme 3.50 *Synthesis of 4-alkyl-2-aryl-4,5-dihydrooxazoles*

which were alkylated with glycidyltosylate to the corresponding epoxide **197**. Nucleophilic opening of the epoxide with pyrrolidine and subsequent cyclisation provided oxazolidinones **198** in high yield and purity, with any by-products formed during the reaction remaining bound to the solid phase.

The method exemplified into Scheme 3.50 shows another significant example. Polymer-bound tosyl chloride was used to capture hydroxyamides from the reaction mixtures in which they were formed. The resulting support-bound amide/sulfonates **201** undergo ring-forming cleavage from the polymer on treatment with weak base (such as pyridine or TEA), forming oxazolines and oxazines in generally good yield and high purity.[108]

Typical Experimental Procedures

General procedure for the oxazoline synthesis from hydroxyamides 201 (Scheme 3.50)[108]

*Alcohol **199** (0.323 mmol, 87 mg, R = COOCH$_3$, R^1 = OCH$_3$) was dissolved in 3.5 ml of freshly distilled pyridine. The solution was slowly added to a peptide synthesis vessel containing 0.4 g (0.97 mmol) of (polystyrene)tosyl chloride resin. After allowing the resin to swell for three minutes, the vessel was sealed and placed in a freezer (−15 °C) for 48 hours. The resin was washed twice with DCM/methanol (1:1), three times with DMSO, and four times with DCM/methanol (1:1). Resin **201** was shaken at room temperature for 24 hours in THF/pyridine (10/1). The resin was filtered and washed twice with DCM/methanol (1:1). The filtrate was evaporated to give 50 mg (66%) of a yellow solid. ^{1}H NMR (CDCl$_3$) δ 7.92 (2H, d, J = 9 Hz), 6.90 (2H, d, J = 9 Hz), 4.92 (1H, dd, J = 7.8, 10.5 Hz), 4.66 (1H, dd, J = 8.1, 8.7 Hz), 4.56 (1H, dd, J = 8.7, 10.5 Hz), 3.84 (3H, s), 3.81 (3H, s). GC/MS (M + H): 236.*

Since polymer-bound tosylates are shown to be stable to a wide variety of reaction conditions including Suzuki couplings, reductive aminations and acylations,[109] it seems to be possible to use a resin-bound tosylate in a multi-step solid-phase synthesis ending with an acylation and ring-forming cleavage reaction to make other three-, five- and six-membered heterocycles such as aziridines, epoxides, tetrahydrothiophenes and tetrahydrofurans.

A traceless approach for the solid-phase synthesis of 6-amino-1,3,5-triazine-2,4-diones **207** is an additional example (Scheme 3.51).[110] Starting from *p*-nitrophenyl carbonate resin **78**, *S*-methylisothiouronium

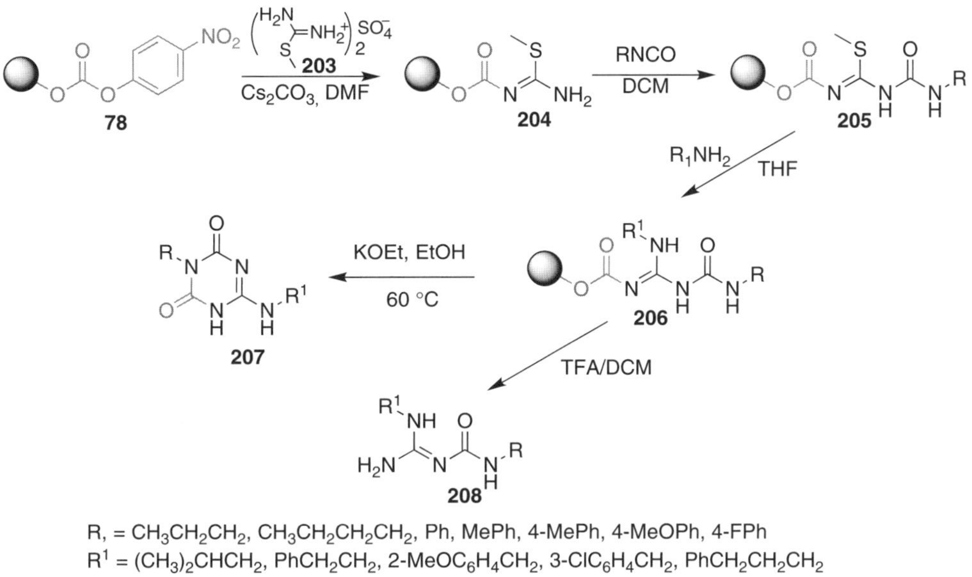

Scheme 3.51 *Combinatorial synthesis of 6-amino-1, 3, 5-triazine-2,4-diones and guanidines*

sulfate **203** was coupled to the resin in the presence of caesium carbonate in dimethyl formamide.[111] The resin-bound *S*-methylisothiourea **204** was reacted with an isocyanate to yield the corresponding iminourea **205**. Reaction of this resin-bound compound with an amine led to the displacement of the methylthio group to give the resin-bound guanidine **206**. The 6-amino-1, 3, 5-triazine-2,4-diones **207** were obtained via intramolecular cyclization with concomitant cleavage from the resin using potassium ethoxide as a base at 60 °C in good yield.

The method reported from the authors is appealing because, carrying out an acidic cleavage, a library of guanidines **208** can be obtained instead.

The same *p*-nitrophenyl carbonate resin **78** was previously used for the solid-phase synthesis of quinazoline-2,4-diones[112] via *N*-terminal amino group linkage to the solid support and a base catalysed cyclisation/cleavage strategy used for the synthesis of dipeptides and hydantoins **211** too (Scheme 3.52).[113]

A more recent example, concerning the synthesis of pyrazolones, involved a hydrazide linker **211** (Scheme 3.53). Treatment of the compound bound to the resin with alcoholic base carried out to the formation of pyrazolones **212**.[114]

An alkyl *N*-methyl-*N*-polystyreneamino-2-isocyanoacrylate **214** was instead used for the synthesis of 1-substituted imidazole-4-carboxylates **215** using the cyclative methodology.[115] The reactions were performed under microwave irradiation affording the title compounds in both high yields and chemical purity directly to solution, from the solid phase support (Scheme 3.54).

The commercial *N*-methyl aminomethylated polystyrene was charged with alkyl isocyanoacetate **213** and *N*-formyl imidazole diethylacetal under acid catalysed conditions (10% 10-camphorsulfonic acid (CSA)) in DMF, under microwave irradiation in order to shorten the reaction time. Treatment of the *N*-methyl-*N*-polystyreneamino-2-isocyanoacrylate **214** so formed with an amine afforded the isomerically pure alkyl 1-substituted-4-imidazolecarboxylate **215**, restoring the starting amino polyester resin. The cyclization step was carried quantitatively by refluxing the resin in *n*-butanol in the presence of either amines or amino acid esters.

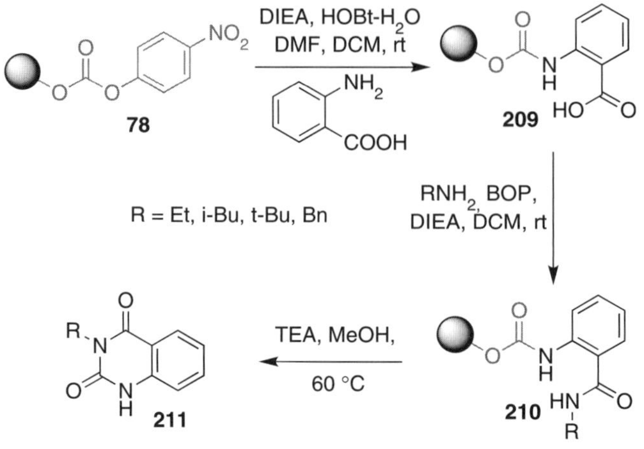

Scheme 3.52 Base catalyzed cyclisation–cleavage strategy for the synthesis of hydantoins

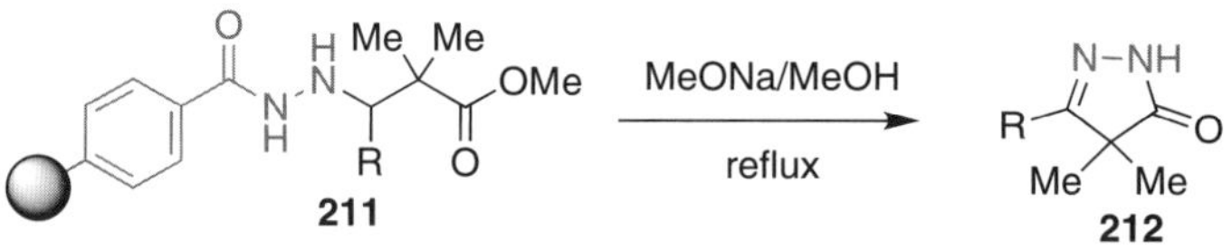

Scheme 3.53 Formation of pyrazolones through a hydrazide linker

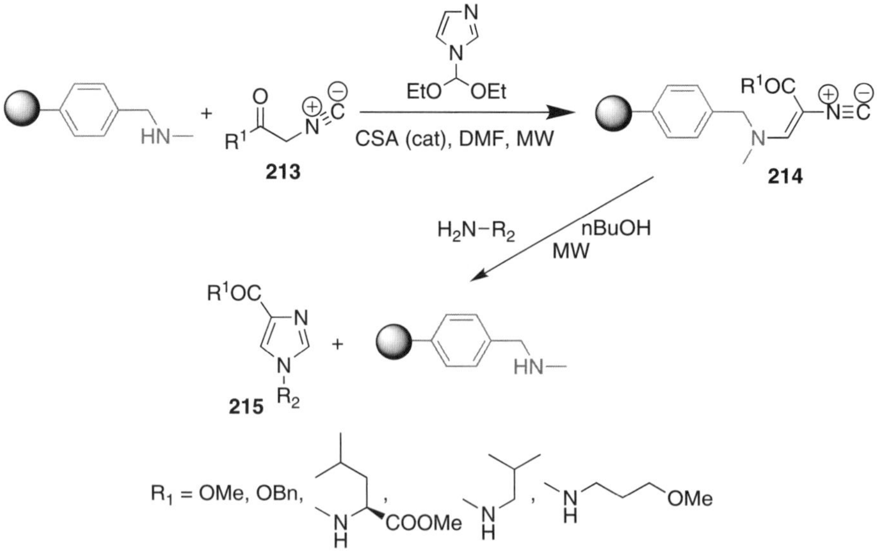

Scheme 3.54 Synthesis of 1-substituted imidazole-4-carboxylates

Typical Experimental Procedures
General procedure for microwave-assisted imidazole synthesis (Scheme 3.54)[115]

To a suspension of N-methyl aminomethylated polystyrene resin (0.10 g, 0.138 mmol) swollen in DMF (2 ml), was added a solution of N-formylimidazole dimethylacetal (0.138 g, 0.83 mmol), methyl isocyanoacetate (0.816 g, 0.83 mmol) and camphorsulfonic acid (13.8 mg, 10% w/w) in DMF (8 ml). The resulting mixture was irradiated to 80 °C for 20 minutes in a sealed tube (CEM designed 10 ml pressure-rated reaction vial) in a self-tuning single mode CEM Discover Focused Synthesizer. The mixture was cooled rapidly to room temperature by passing compressed air through the microwave cavity for three minutes. The reaction progress was monitored by the colorimetric chloranil test (negative). After cooling to room temperature, the resin sample was collected by filtration using a sintered glass funnel. The resin was thoroughly washed with alternative portions of DMF (3 × 10 ml), hexane (3 × 10 ml), THF (3 × 10 ml), hexane (3 × 10 ml) and n-butanol (3 × 10 ml). The resin sample was dried under reduced pressure. The IR spectrum of this resin sample was compared with the amine starting resin (potassium bromide pellet) and showed a characteristic absorption band at 2106 cm^{-1}, indicating successful solid support capture of the isocyanide building block. The resin **214** was suspended in dry n-butanol (5 ml) in a round-bottomed flask with a reflux condenser. Benzylamine (0.1 ml, 1.38 mmol) was added and the resulting mixture was exposed to microwave irradiation at 114 °C for two cycles of 30 minutes in a self-tuning, single mode, CEM Discover Focused Synthesizer. The mixture was cooled rapidly to room temperature by passing compressed air through the microwave cavity for three minutes. The reaction progress was monitored by the colorimetric chloranil test (positive). After cooling to room temperature, the resin sample was collected by filtration using a sintered glass funnel and successively worked up as above. After evaporating the solvent to dryness under reduced pressure, pure methyl 1-benzyl-1H-imidazole-4-carboxylate **215** (28.9 mg, 97% yield, 99% purity) was obtained as a colourless oil.

Porcheddu et al. reported an interesting nucleophilic cyclo-release in the synthesis of a library of 2,4,5-substituted pyrimidines from resin-bound β-enaminones and a series of guanidines (Scheme 3.55).[116]

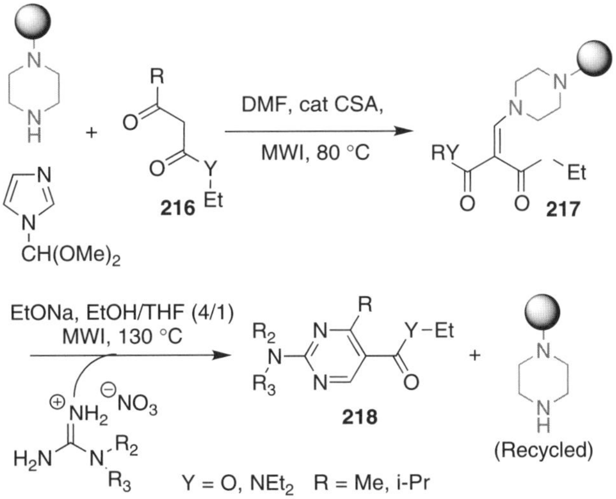

Scheme 3.55 Synthesis of a library of 2,4,5-substituted pyrimidines

Through a cyclocondensation reaction, the solid supported enaminone **217** was reacted with several guanidines under heating or microwave irradiation (MWI) affording the corresponding pyrimidines **218** in good yield and chemical purity directly on solution. After this final step, the support could be effectively recycled.

Resin-bound enaminone synthesis can be effectively performed in high yield within 30 minutes using a self-tunable microwave synthesizer at 80 °C. The reaction was carried out in an open vessel to allow the removal of the formed methanol from the equilibrium. After some experiments, it was discovered that the desired pyrimidines **218** could be efficiently released from the resin by controlled single-mode microwave irradiation of supported enaminone in presence of guanidine, working in a sealed tube for 10 minutes at 130 °C and using ethanol/THF (4/1) as solvent.

The same authors have described the synthesis of libraries of substituted pyrazoles and isoxazoles **221** via *in situ* generation of polymer-bound enaminones **220** (Scheme 3.56).[117] The synthetic protocol used a new solid-supported reagent, commercial available aniline cellulose **219**, as new low cost and versatile biopolymer, under very mild condition and through a microwave strategy. This support allowed carrying out reactions in polar solvents under conventional heating or microwave irradiation without degradation of the polymer. The key step was the reaction between cellulose-bound enaminone and hydroxylamine or hydrazines to afford the target heterocycles in high yield directly in solution (Scheme 3.56). At the end, the support could be recycled.

According to this scheme, enaminones were identified as convenient starting materials as they can react with different bidentate nucleophiles to give pyrazoles and isoxazoles. Thus, in order to design a versatile rapid synthesis of a library delivered in solution, the β-enamino ketone was anchored on cellulose support.

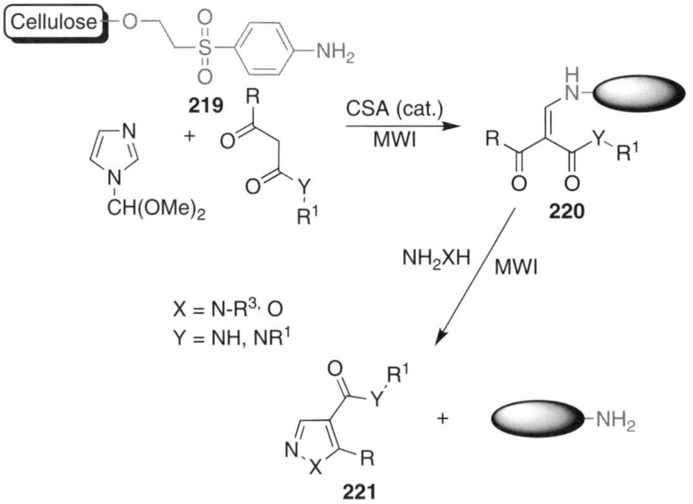

220 a-c: Y = O; R = Me; R$_1$ = PhCH$_2$, Et, Me
220 d-j: Y = O; R = iPr; R$_1$ = Et, Allyl, Ph, PhCH$_2$CH$_2$, iPr, -CH$_2$CH$_2$OH, -CH$_2$CH$_2$CN
220 k-m: Y = O, R = tBuCH$_2$; R$_1$ = PhCH$_2$CH$_2$, -CH$_2$CH$_2$OH, -CH$_2$CH$_2$CN
220 n-r: Y = N; R = iPr; R$_1$ = Ph, PhCH$_2$, PhCH$_2$CH$_2$, cyclopentyl, cyclohexyl

Scheme 3.56 *Synthesis of libraries of substituted pyrazoles and isoxazoles*

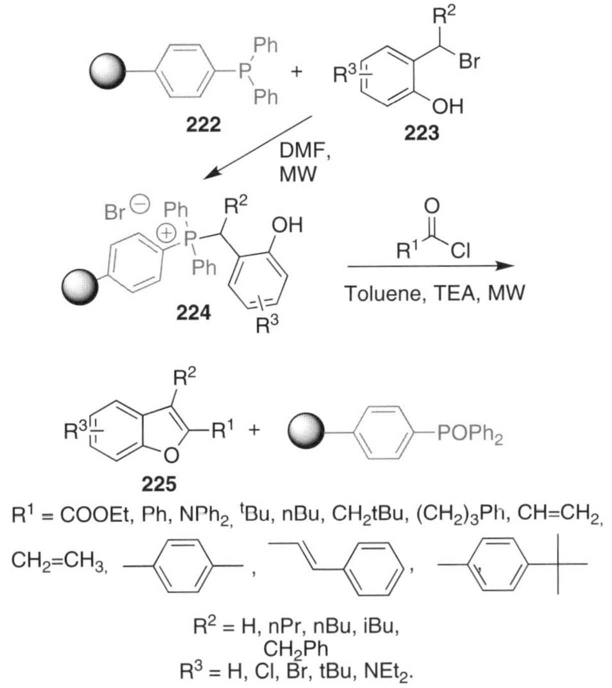

Scheme 3.57 *Microwave mediated synthesis of substituted benzofurans*

Consequently, the NH$_2$ group on cellulose was an attractive point to realize a *'cyclative release'* approach and deliver the target libraries with minimal purification steps. A modified Bredereck's reagent[118] was prepared to convert the amino group of cellulose into the formamide acetal[119] (Scheme 3.56) that was further employed to immobilize a β-ketoester or amide. The cellulose was treated with an excess of formyl imidazole under acid catalysed conditions (camphorsulfonic acid: CSA) in DMF. The mixture afforded the functionalized support, which was treated with an excess of a β-dicarbonyl compounds in DMF to give the corresponding solid-supported β-enaminodiones that can be isolated by simple filtration.

Very recently, a microwave enhanced procedure for the synthesis of substituted benzofurans **225** starting from 2-(1-hydroxyalkyl)-phenols **223** and using triphenylphosphine polystirene resin **222** was reported.[120] Thus, triphenylphosphine polystirene resin was charged with 2-(bromoalkyl)-phenols in dry DMF (Scheme 3.57).

Typical Experimental Procedures

General procedure for microwave-assisted benzofuran synthesis (Scheme 3.57)[120]

*To a suspension of triphenylphosphine polystirene resin **222** (0.10 g, 0.160 mmol) swollen in dry DMF (1 ml) 2-(bromomethyl)phenol (0.12 g, 0.640 mmol) was added. The resulting mixture was irradiated to 85 °C for 15 minutes in a sealed tube (CEM designed 10 ml pressure rated reaction vial) in a self-tuning, single-mode CEM Discover Focused Synthesizer. The mixture was cooled rapidly to room temperature by passing compressed air through the microwave cavity for three minutes; then 2-(bromomethyl)-phenol (0.12 g, 0.640 mmol) was added again, and the mixture was irradiated to 85 °C for 15 minutes as above.*

*After it was cooled to room temperature, the resin sample was collected by filtration using a sintered glass funnel. The resin was thoroughly washed with alternative portions of DMF (3 × 10 ml), hexane (3 × 10 ml), THF (3 × 10 ml), hexane (3 × 10 ml) and DCM (3 × 10 ml). The resin **224** was dried under reduced pressure and then suspended in dry toluene (1 ml). Benzoyl chloride (0.09 ml, 0.80 mmol) and TEA (0.33 ml, 2.4 mmol) were added and the resulting mixture was irradiated to 110 °C for 60 minutes in a sealed tube. After it was cooled to room temperature, the resin was removed by filtration using a sintered glass funnel and successively washed alternatively with toluene (3 × 10 ml) and THF (3 × 10 ml). All the organic layers were combined, washed with aqueous sodium bicarbonate (2 × 10 ml) and water (2 × 10 ml), and concentrated in vacuum. The crude 2-phenylbenzofuran **225** (0.11 g, 65% yield) was further purified by chromatography (ethyl acetate/petroleum ether, 1:9) to obtained a white solid (global yield 61%), melting point: 119–122 °C.*

The formation of the functionalized support had to be conducted under microwave irradiation in order to shorten the reaction time. Treatment of the resin with an acyl chloride and TEA afforded the isomerically pure benzofuran derivative **225**, which can be recovered from the solution, leaving the triphenylphosphine oxide polystirene resin as a solid residue.[121] The cyclization step was carried quantitatively in dry toluene. The benzofurans were isolated in good to high yields and purities by simple work-up. The procedure was applied to chiral α-alkyl-2-benzofuranmethanamines too.[120]

Although in the past decades there has been an increasing number of reactions performed on the solid phase, there has been little success in the area of directed ortho-metalation (DoM) of arenes.[122] The lack of success of solid-phase DoM is most likely due to an incompatibility between the alkyl lithium bases used in the reaction and the polystyrene resins most commonly used in solid-phase synthesis. The right linker would have to both withstand the alkyl lithium bases and be free of directing metalation groups (DMGs). Therefore, the Garibay's team have chosen to attach benzoic acids or acid chlorides to a standard aminomethylated polystyrene resin (Scheme 3.58) and use the formed secondary benzamide as the directing metalation group.[123] The corresponding secondary benzamides **227** are then ortho-lithiated with excess *n*-butyl lithium, forming a dark dianion. Addition of ketones or aldehydes to the dianion **228** forms an alcohol **229**. When it is heated in toluene at 90 °C, a cyclative cleavage occurs, yielding the desired phthalide **230** without the need of any further cleavage reagents. Resin-bound benzamides possessing meta

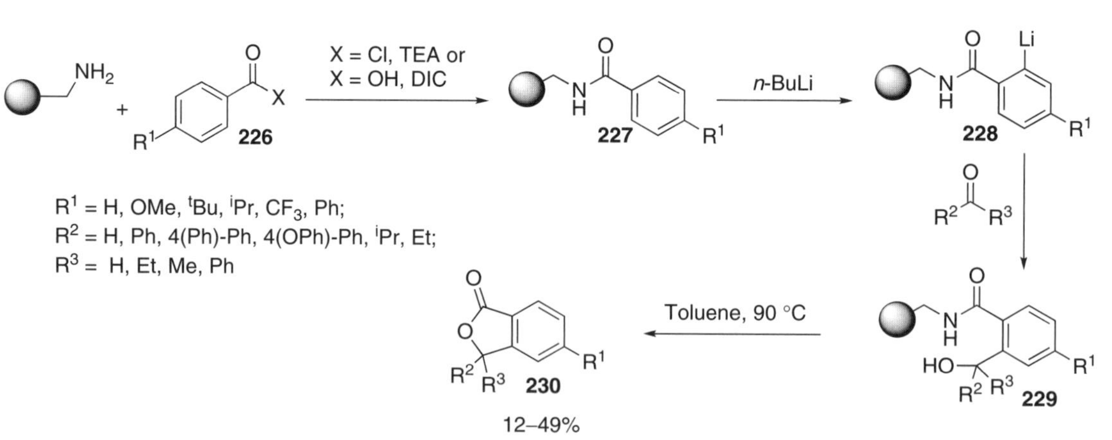

Scheme 3.58 *Parallel synthesis of 3,3,5-trialkylisobenzofuran-1(3H)-ones*

substitutions were observed to be much less reactive than those reported in solution reactions. The method was further investigated by the parallel synthesis of 100 well-defined phthalides **230**. Solid-phase methods for the synthesis of phthalides via lithium–iodine exchange and magnesium–iodine exchange were also shown to be feasible.

Among the heterocyclic templates, minor attention has been paid to the synthesis of 2,6-diketopiperazines (2,6-DKPs), although their framework is present in the topoisomerase II inhibitor ICRF-154[124] and related compounds,[125] having promising anticancer efficacy.[126] At Menarini Ricerche, Altamura's group have extensively researched into a general method to prepare 1,3-disubstituted 2,6-diketopiperazines **236** (2,6-DKPs) by SPOS (Scheme 3.59). A Wang resin carboxylic ester **235** has been used as acylating agent under solid-phase conditions, allowing the cyclization to take place with simultaneous cleavage of the product from the resin ('cyclocleavage'). The synthetic method worked well with several couples of amino acids, independently from their configuration, and was used for the parallel synthesis of a series of fully characterized compounds. The use of iterative conditions in the solid phase (repeated addition of fresh solvent and potassium carbonate to the resin after filtering out the product-containing solution) allowed the diastereoisomeric content to be kept below the detection limit by HPLC and ^{1}H NMR (200 MHz). The more interesting comparison between the homogeneous and solid-phase approach is that solid-phase methodology can be automated more easily than solution-phase methodology, and considering the global reaction time to prepare the final products, the SPOS was found to be four times faster than the homogeneous-phase method.

The benzopyrone ring system, present in several natural products, interacts with various enzymes and receptors of pharmacological significance.[127] In a recent communication, Brueggemeier et al.[128] have described an one-pot conversion of bis-silylated salicylic acids into alkynyl ketones via an acid chlorination and subsequent Sonogashira coupling with terminal alkynes (Scheme 3.60). The alkynyl ketones **238** were treated with solid-supported secondary amines to form enaminones **239**, which underwent a facile cyclization and elimination of support bound secondary amine to provide the benzopyrone nucleus **240** (Scheme 3.60).

In detail, a piperazinyl resin was reacted in separate reactions with a 10-fold excess of alkynones **238** (Scheme 3.60). In all the cases, support-bound enaminones **239** were detected by the presence of the carbonyl absorption band using IR. Subsequent cyclization and concentration of filtrates produced

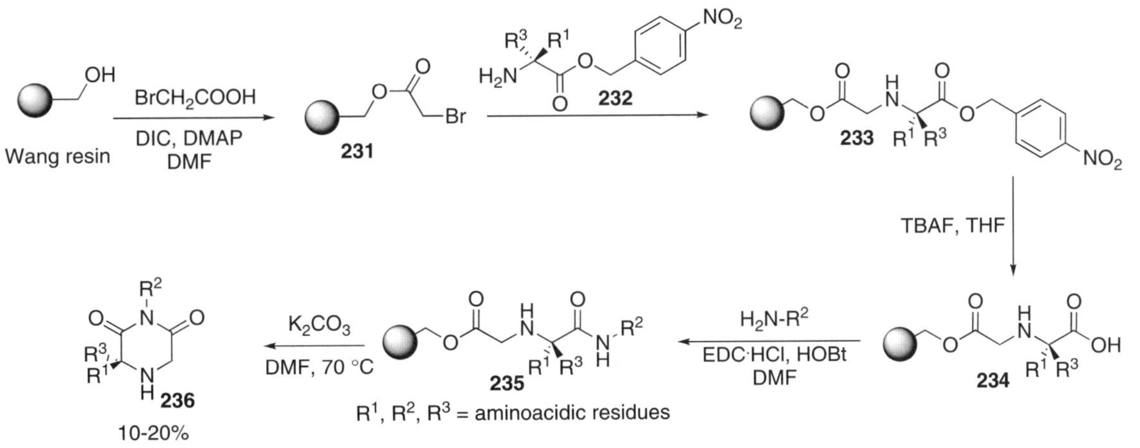

Scheme 3.59 *Synthesis of 1,3-disubstituted 2,6-diketopiperazines*

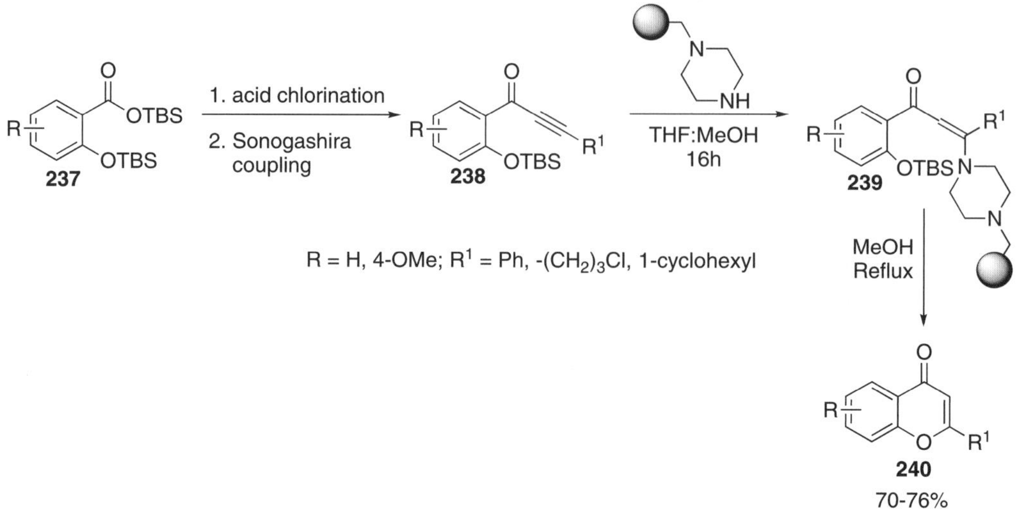

Scheme 3.60 Preparation of benzopyrone derivatives

the crude benzopyrones in approximately 70–80% yield, based upon TLC analysis showing only the particular benzopyrone as the major product and a small baseline impurity. Flash chromatography on silica gel afforded purified benzopyrones **240** in final yields of 70–76%.

A convenient solid-phase synthesis approach to novel spiropyrans **244** through the use of a succinate linker **241** using commercially available, inexpensive starting materials was recently reported (Scheme 3.61).[129] In this case, potassium *tert*-butoxide in THF was used for cleavage that produced compounds under excellent yields and purities.

Typical Experimental Procedures
Preparation of 1,3,3-trimethyl-indoline-5-yl-succinic amide Wang ester 242 (Scheme 3.61)[129]

Method A: Under nitrogen atmosphere, 5-amino-1,3,3-trimethyl indoline (1.94 g, 10.3 mmol) in anhydrous THF (5 ml) was added dropwise of 5 ml THF solution of succinic anhydride (1.0 g, 10 mmol) over one hour. The mixture was stirred at room temperature for seven hours. Wang resin (theoretical loading 1.28 mmol/g, 5 g, 6.4 mmol), diisopropyl carbodiimide (DIC, 1.26 g, 10 mmol) and DMAP (61.1 mg, 0.5 mmol) were added. The suspension was shaken at room temperature for 24 hours. Filtered through glass sinter, washed with THF (4 × 15 ml), DMF (4 × 15 ml), dichloromethane (4 × 10 ml) and dried in vacuo. 7.82 g Of beads were obtained. 50.9 mg Of beads were swelled in 1 ml THF and released with 0.01 M THF solution of potassium tert-butoxide (3 × 0.2 ml, 15 min). The combined THF solution was shaken with finely powdered sodium dihydrophosphate and anhydrous sodium sulfate, filtered, and washed with THF. Solvent was removed in vacuo. Chloroform was added to the residue and then removed in vacuo to remove trace of THF. The residue was dried in vacuo, 1.0 ml of 0.01 M hexamethyldisiloxane in $CDCl_3$ was added. From the integral of proton of $N-CH_3$ or succinyl verse that of internal standard in ^1H NMR spectra, the loading was estimated to be 96%.

Method B: Wang resin (theoretical loading 1.28 mmol/g, 3.0 g, 3.84 mmol), succinic anhydride (2.0 g, 20 mmol) and DMAP (47 mg, 0.38 mmol) were refluxed in THF for 48 hours. The mixture was cooled

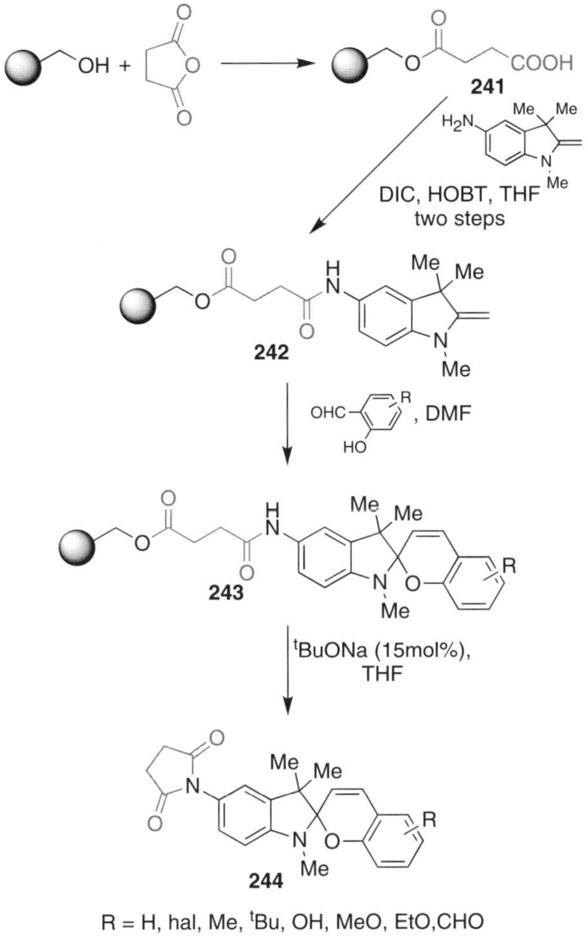

Scheme 3.61 *Solid-phase synthesis to spiropyrans*

down, filtered through glass sinter and washed with THF (3 × 10 ml), DMF (3 × 10 ml), dichloromethane (2 × 10 ml), methanol (3 × 10 ml) and dichloromethane (2 × 10 ml), dried in vacuo. 3.524 g of white resin were obtained. The suspension of beads (1.37 g, 1.5 mmol), 5-amino-1,3,3-trimethyl indoline (3.76 g, 2.0 mmol), HOBt (12% water, 322 mg, 2.1 mmol) and DIC (265 mg, 2.1 mmol) in 15 ml THF were shaken overnight at room temperature under nitrogen, filtered through glass sinter, washed with THF (3 × 5 ml), DMF (3 × 5 ml), water (2 × 5 ml), DMF (2 × 3 ml), THF (3 × 3 ml), dichloromethane (3 × 3 ml), and dried in vacuo. 1.725 g purple beads were obtained.

3.4 Conclusion

The last few years of drug discovery have been heavily influenced by solid-phase synthesis, which has allowed the generation of compound collections with large numbers of drug-like molecules. Synthetic approaches have evolved from the direct, though not trivial, transposition of existing homogeneous phase schemes to the design of original methods that exploit the best advantages of SPOS. As a result, new linkers and cleavage strategies are constantly under investigation. Linkers, which enable release of the target molecule from the solid support under mild conditions or are robust towards a variety of reaction conditions, are particularly desirable, as are those which allow the introduction of structural variability in the cleavage step. However, in many cases, many of these linkers exhibit considerable substrate dependence and, consequently, there is an ongoing need for new and more versatile examples to be developed. In the field of nucleophile labile linkages many new linkers and methodologies are presented. A promising strategy is nucleophilic cleavage with simultaneous introduction of additional diversity such as with different amines. It is hoped that more examples for such strategies will emerge in the future, especially with the consequent use of scavenger resins to remove excess reagents.

References

[1] Wang, S.-S.; *J. Am. Chem. Soc*. **1973**, *95*, 1328.
[2] Merrifield, R. B.; *J. Am. Chem. Soc*. **1963**, *85*, 2149.
[3] Guillier, F., Orain, D., and Bradley, M.; *Chem. Rev*. **2000**, *100*, 2091.
[4] Eggenweiler, H.-M.; *Drug Discovery Today*, **1998**, *3*, 552.
[5] Chamoin, S., Houldsworth, S., and Snieckus, V.; *Tetrahedron Lett*. **1998**, *39*, 4175.
[6] Chen, C.; McDonald, I. A., and Munoz, B.; *Tetrahedron Lett*., **1998**, *39*, 217.
[7] Yamazaki, K., Nakamura, Y., and Kondo, Y.; *J. Org. Chem*., **2003**, *68*, 6011.
[8] Yao, T., Yue, D., and Larock R. C.; *J. Comb. Chem*. **2005**, *7*, 809.
[9] Koradin, C., Dohle, W., Rodriguez, A. L., *et al*.; *Tetrahedron* **2003**, *59*, 1571.
[10] Knepper, K., and Brase, S.; *Org. Lett*., **2003**, *16*, 2829.
[11] Nicolaou, K. C., Evans, R. M., Roecker, A. J., *et al*.; *Org. Biomol. Chem*., **2003**, *1*, 908.
[12] Nicolaou, K. C., Snyder, S. A., Montagnon, T., and Vassilikogiannakis, G.; *Angew. Chem., Int. Ed*., **2002**, *41*, 1668, and reference cited therein.
[13] Strohmeier, G. A., and Kappe, O. C.; *J. Comb. Chem*., **2002**, *4*, 154.
[14] Agrofoglio, L. A., and Challand, S. R.; *Acyclic, Carbocyclic and L-Nucleosides*; Kluwer Academic Publishers, Dordrecht, The Netherlands, **1998**.
[15] Prusoff, W. H.; *Biochim. Biophys. Acta* **1959**, *32*, 295.
[16] Packham, D. I.; *J. Chem. Soc*., **1964**, 2617.
[17] Kalir, R., Warshawsky, A., Fridkin, M., and Patchornik, A.; *Eur. J. Biochem*., **1975**, *59*, 55.
[18] Salvino, J. M., Kumar, N. V., Orton, E., *et al*.; *J. Comb. Chem*., **2000**, *2*, 691.
[19] Lee, J. W., Louie, Y. Q., Walsh, D. P., and Chang, Y. T.; *J. Comb. Chem*., **2003**, *5*, 330.
[20] Pop, I. E., Deprez, B. P., and Tartar, A. L.; *J. Org. Chem*., **1997**, *62*, 2594.
[21] Masala, S., and Taddei, M.; *Org. Lett*., **1999**, *1*, 1355.
[22] Kenner, G. W., McDermot, J. R., and Sheppard, R. C.; *J. Chem. Soc., Chem. Commun*., **1971**, 636.
[23] Eriksson, J., Olsson, T., Kann, N., and Graden, H.; *Tetrahedron Lett*., **2006**, *47*, 635.
[24] Berteina, S., and De Mesmaeker, A.; *Tetrahedron Lett*., **1998**, *39*, 5759.
[25] Bandara, B. M. R., Birch, A. J., and Kelly, L. F.; *J. Org. Chem*., **1984**, *49*, 2496.
[26] Barn, D. R.; Morphy, J. R.; and Rees, D. C.; *Tetrahedron Lett*., **1996**, *37*(18), 3213.
[27] Fisher, M., and Brown, R. C. D.; *Tetrahedron Lett*., **2001**, *42*, 8227.
[28] Chang, S., and Grubbs, R. H.; *J. Org. Chem*., **1998**, *63*, 864.
[29] Tsuji, J.; *Organic Synthesis with Palladium Compounds*; Springer-Verlag, Berlin, Germany, **1980**.
[30] Savinov, S. N., and Austin, D. J.; *Org. Lett*. **2002**, *4*, 1419.
[31] Savinov, S. N., and Austin, D. J.; *Org. Lett*. **2002**, *4*, 1415.
[32] Heavner, G. A., Doyle, D. L., and Riexinger, D.; *Tetrahedron Lett*. **1985**, *26*, 4583.
[33] Savinov, S. N., and Austin, D. J.; *Chem. Commun*. **1999**, 1813.
[34] Hwang, J. Y., Choi, H., Lee, D., *et al*.; *J. Comb. Chem*. **2005**, *7*, 136.
[35] Patil, S. P., Vananthakumar, G.-R., and Suresh Babu, V. V.; *J. Org. Chem*. **2003**, *68*, 7274.

[36] Acharya, A. N., Nefzi, A., Ostrech, J. M., and Houghten, R. A.; *J. Comb. Chem.* **2001**, *3*, 189.
[37] Lee, S.-H., Matsushita, H., Koch, G., *et al.*; *J. Comb. Chem.* **2004**, *6*, 822.
[38] Marshall, D. L., and Liener, I. E.; *J. Org. Chem.* **1970**, *35*, 867.
[39] Johnson, C. R., Zhang, B., Fantauzzi, P., *et al.*; *Tetrahedron* **1998**, *54*, 4097.
[40] Breitenbucher, J. G., Johnson, C. R., Haight, M., and Phelan, J. C.; *Tetrahedron Lett.* **1998**, *39*, 1295.
[41] Fang, L., Demee, M., Sierra, T., *et al.*; *J. Comb. Chem.* **2002**, *4*, 362.
[42] Irving, M. M., Kshirsagar, T., Figliozzi, G. M., and Yan, B.; *J. Comb. Chem.* **2001**, *3*, 407.
[43] Bunin, B. A., Dener, J. M., Livingston, D. A.; Application of Combinatorial and Parallel Synthesis to Medicinal Chemistry, in *Annual Reports in Medicinal Chemistry* (ed. Doherty, A.); Academic Press: San Diego, CA, **1999**; Vol. *34*, Chapter 27, 267–286.
[44] Boldi, A. M., Dener, J. M., and Hopkins, T. P.; *J. Comb. Chem.* **2001**, *3*, 367
[45] Kamal, A., Devaiah, V., Reddy, K. L., *et al.*; *J. Comb. Chem.* **2007**, *9*, 267
[46] Hioki, H., Fukutaka, M., Takahashi, H., *et al.*; *Tetrahedron Lett.*, **2004**, *45*, 7591.
[47] a) Best, W. M., Macdonald, J. M., Skelton, B. W., *et al.*; *Can. J. Chem.* **2002**, *80*, 857; b) Liu, W., Peterson, P. E., Carter, R. J., *et al.*; *Biochemistry* **2004**, *43*, 10896.
[48] Kuksa, V., Buchan, R., Kong, P., and Lin, P. K. T.; *Synthesis* **2000**, 1189.
[49] Lima, L. M., and Barreiro, E. J.; *Curr. Med. Chem.* **2005**, *12*, 23.
[50] Yang, D., Zhang, Y.-H., Li, B., and Zhang, D.-W.; *J. Org. Chem.* **2004**, *69*, 7577.
[51] Maillard, L. T., Benohoud, M., Durand, P., and Badet, B.; *J. Org. Chem.* **2005**, *70*, 6303.
[52] Wang, S.; *J. Am. Chem. Soc.* **1973**, *95*, 1328.
[53] Albericio, F., and Barany, G.; *Tetrahedron Lett.* **1991**, *32*, 1015.
[54] Mergler, M., Tanner, R., Gosteli, J., and Grogg, P.; *Tetrahedron Lett.* **1988**, *29*, 4005
[55] Barlos, K., Gatos, D., Kallitsis, I., *et al.*; *Tetrahedron Lett.*, **1989**, *30*, 3943.
[56] Valerio, R. M., Benstead, M., Bray, A. M., *et al.*; *J. Anal. Biochem.* **1991**, *197*, 168.
[57] Baleux, F., Calas, B., and Mery, J.; *Int. J. Peptide Protein Res.* **1986**, *28*, 22.
[58] Chao, H.-G., Bernatowicz, M. S., Reiss, P. D., *et al.*; *J. Am. Chem. Soc.* **1994**, *116*, 1746.
[59] a) Ramage, R., Barton, C. A., Bielecki, S., and Thomas, D. W.; *Tetrahedron Lett.* **1987**, *28*, 4105. b) Ramage, R., Barton, C. A., Bielecki, S., *et al.*; *Tetrahedron* **1992**, *48*, 499.
[60] Peukert, S., and Giese, B.; *J. Org. Chem.* **1998**, *63*, 9045.
[61] Sauerbrei, B., Jungmann, V., and Waldmann, H.; *Angew. Chem., Int. Ed. Engl.* **1998**, *37*, 1143.
[62] Kunz, H., and Dombo, B.; *Angew. Chem., Int. Ed. Engl.* **1988**, *27*, 711.
[63] Chhabra, S. R., Parekh, H., Khan, A. N., *et al.*; *Tetrahedron Lett.* **2001**, *42*, 2189.
[64] Atherton, E., Logan, C. J., and Sheppard, R. C.; *J. Chem. Soc., Perkin Trans. 1* **1981**, 538.
[65] Drinnan, N., West, M. L., Broadhurst, M., *et al.*; *Tetrahedron Lett.*, **2001**, *42*, 1159.
[66] G. J. Boons (Ed.); *Carbohydrate Chemistry*, Blackie, London, **1998**.
[67] Adolfi, M., Barone, G., De Napoli, L., *et al.*; *Tetrahedron Lett.* **1998**, *39*, 1953.
[68] a) Plante, O. J., Palmacci E. R., and Seeberger, P. H.; *Science*, **2001**, *291*, 1523. (b) Seeberger, P. H.; *Chem. Soc. Rev.*, **2008**, *37*, 19, and references cited therein.
[69] Ito, Y., and Manabe, S.; *Curr. Opin. Chem. Biol.*, **1998**, *2*, 701.
[70] a) Shen, D.-M., Shu, M., and Chapman, K. T.; *Org. Lett.* **2000**, *2*, 2789. b) Marzinzik, A. L., and Felder, E. R.; *Tetrahedron Lett.* **1996**, *37*, 1003. c) Stauffer, S. R., and Katznellenbogen, J. A.; *J. Comb. Chem.* **2000**, *2*, 318. d) Nicolaou, K. C., Cao, G.-Q., and Pfefferkorn, J. A.; *Angew. Chem., Int. Ed.* **2000**, *39*, 739.
[71] a) Wilson, R. D., Watson, S. P., and Richards, S. A.; *Tetrahedron Lett.* **1998**, *39*, 2827. b) Shankar, B. B., Yang, D. Y., Giron, S., and Ganguly, A. K.; *Tetrahedron Lett.* **1998**, *39*, 2447. c) Haap, W. J., Kaiser, D., Walk, T. B., and Jung, G.; *Tetrahedron* **1998**, *54*, 3705.
[72] Vanier, C., Wagner, A., and Mioskowski, C.; *J. Comb. Chem.* **2004**, *6*, 846.
[73] Floyd, C. D., Lewis, C. N., Patel, S. R., and Whittaker, M.; *Tetrahedron Lett.* **1996**, *37*, 8045.
[74] Zhang, W., Zhang, L., Li, X., *et al.*; *J. Comb. Chem.* **2001**, *3*, 151.
[75] Zhang, L., and Tam, J. P.; *J. Am. Chem. Soc.* **1999**, *121*, 3311.
[76] Dankwardt, S. M.; *Synlett* **1998**, 761.
[77] Beyerman, H. C., and Maassen van den Brink-Zimmermannová, H.; *Recl. Trav. Chim.* **1968**, *87*, 1196.

[78] Yang, K., and Lou, B.; *Mini Rev. Med. Chem.* **2003**, *3*, 349.
[79] Devocelle, M., McLoughlin, B. M., Sharkey, C. T., et al.; *Org. Biomol. Chem.* **2003**, *1*, 850.
[80] Paladino, A., Mugnaini, C., Botta, M., and Corelli, F.; *Org. Lett.*, **2005**, *7*, 565.
[81] Kaiser E., Colescott R. L., Bossinger C. D., and Cook P. I.; *Anal. Biochem.* **1970**, *34*, 595.
[82] Attardi, M. E., Falchi, A., and Taddei, M.; *Tetrahedron Lett.* **2000**, *41*, 7395
[83] Joseph, R., Sudalai, A., and Ravindranathan, T.; *Tetrahedron Lett.* **1994**, *35*, 5493.
[84] James, I. W.; *Tetrahedron*, **1999**, *55*, 4855.
[85] Halm, C., Evarts, J., and Kurth, M. J. A.; *Tetrahedron Lett.* **1997**, *38*, 7709.
[86] Katritzky, A. R., Pastor, A., Voronkov, M., and Tymoshenko, D.; *J. Comb. Chem.* **2001**, *3*, 167.
[87] Chang, C-M., Kulkarni, M. V., Chen, C-H., et al.; *J. Comb. Chem.* **2008**, *10*, 466.
[88] For reviews on tropane alkaloids see: Fodor, G. R.; *Nat. Prod. Rep.* **1994**, *11*, 603 and references cited therein.
[89] Brossi, A. (Ed.); *The Alkaloids*; Academic Press: New York, **1988**; Vol. *333*.
[90] a) Koh, J. S., and Ellman, J. A.; *J. Org. Chem.* **1996**, *61*, 4494. b) Jonsson, D., Molin, H., and Unden, A.; *Tetrahedron Lett.* **1998**, *39*, 1059.
[91] Caix-Haumesser, S., Hanna, I., Lallemand, J.-Y., and Peyronel, J.-F.; *Tetrahedron Lett.*, **2001**, *42*, 3721.
[92] Morales, G. A., Corbett, J. W., and DeGrado, W. F.; *J. Org. Chem.* **1998**, *63*, 1172.
[93] Coskun, N., and Tirli, F.; *Synth. Commun.* **1997**, *27*, 1.
[94] Kurth, M. J., Ahlerg Randall, L. A., Chen, C., et al.; *J. Org. Chem.*, **1994**, *59*, 5862.
[95] Fehrentz, J.-A., Paris, M., Heitz, A., et al.; *Tetrahedron Lett.* **1995**, *36*, 7871.
[96] Dinh, T. Q., and Armstrong, R. W.; *Tetrahedron Lett.* **1996**, *37*, 1161.
[97] Chandrasekhar, S., Padmaja, M. B., and Raza, A.; *J. Comb. Chem.* **2000**, *2*, 246.
[98] Kan, T., and Fukuyama, T.; *Chem. Commun.* **2004**, 353.
[99] Cardullo, F., Donati, D., Fusillo, V., et al.; *J. Comb. Chem.*, **2006**, *8*, 834.
[100] Nicholas, K. M.; *Acc. Chem. Res.* **1987**, *20*, 207.
[101] Sammelson, R. E., and Kurth, M. J.; *Chem. Rev.* **2001**, *101*, 137.
[102] Cassel, J. A., Leue, S., Gachkova, N. I., and Kann, N. C.; *J. Org. Chem.* **2002**, *67*, 9460.
[103] Cassel, J. A., Leue, S., Gachkova, N. I., and Kann, N. C.; *J. Comb. Chem.* **2005**, *7*, 449.
[104] Newbold, B. T.; Oxidation and Synthetic Uses of Hydrazo, Azo and Azoxy Compounds, in *The Chemistry of the hydrazo, azo and azoxy group (Part 1)*, (ed. Patai, S.); John Wiley & Sons Ltd, Chichester, **1975**, 542.
[105] Millington, C. R., Quarrell, R., and Lowe, G.; *Tetrahedron Lett.*, **1998**, *39*, 7201.
[106] Buchstaller, H.-P.; *J. Comb. Chem.* **2003**, *5*, 789.
[107] Gordon, K., and Balasubramanian S.; *J Chem Technol Biotechnol* **1999**, *74*, 835.
[108] Pirrung, M. C., and Tumey, L. T.; *J. Comb. Chem.* **2000**, *2*, 675
[109] Baxter, E. W., Rueter, J. K., Nortey, S. O., and Reitz, A. B.; *Tetrahedron Lett.* **1998**, *39*, 979.
[110] Yu, Y., Ostresh, J. M., and Houghten, R. A.; *J. Comb. Chem.* **2004**, *6*, 83.
[111] Dodd, D, S., and Zhao, Y.; *Tetrahedron Lett.* **2001**, *42*, 1259.
[112] Gouilleux, L., Fehrentz, J-A., Winternitz, F., and Martinez, J.; *Tetrahedron Lett.* **1996**, *37*, 7031.
[113] Letsinger, R. L., Komet, M. J., Mahadevan, V., and Jerina, D. M.; *J. Am. Chem. Soc.* **1964**, *86*, 5163.
[114] Kobayashi, S., Furuta, T., Sugita, K., et al.; *Tetrahedron Lett.* **1999**, *40*, 1341.
[115] De Luca, L., Giacomelli, G., and Porcheddu, A.; *J. Comb. Chem.* **2005**, *7*, 905.
[116] Porcheddu, A., Giacomelli, G., De Luca, L., and Ruda, A. M.; *J. Comb. Chem.* **2004**, *6*, 105.
[117] De Luca, L., Giacomelli, G., Porcheddu, A., et al.; *J. Comb. Chem.* **2003**, *5*, 465.
[118] Brown, R. S., and Curtis, N. J.; *J. Org. Chem.* **1980**, *45*, 4038.
[119] Bienaymé, H.; *Tetrahedron Lett.* **1998**, *39*, 4255.
[120] De Luca L., Giacomelli, G., and Nieddu, G.; *J. Comb. Chem.* **2008**, *10*, 517.
[121] Regeneration of the resin may be achieved via treatment with a clear solution of LiAlH$_4$ in THF, according to literature: Sieber, F., Wentworth Jr, P., Toker, J. D., et al.; *J. Org. Chem.* **1999**, *64*, 5188 in 80% yield. However, the regenerated resin is less efficient than the original and the yields are lowered.
[122] Snieckus, V.; *Chem. Rev.* **1990**, *90*, 879.
[123] Beak, P., Tse, A., Hawkins, J., et al.; *Tetrahedron* **1983**, *39*, 1983.
[124] Creighton, A. M., Hellmann, K., and Whitecross S.; *Nature* **1969**, *222*, 384.

[125] Singh, S. B.; *Tetrahedron Lett*. **1995**, *36*, 2009.
[126] Andoh, T., and Ishida, R.; *Biochim. Biophys. Acta* **1998**, *1400*, 155.
[127] Harborne, J. B. (Ed.); *The Flavonoids Advances in Research Since 1986*; Chapman & Hall, London, **1994**.
[128] Bhat, A. S., Whetstone, J. L., and Brueggemeier, R. W.; *J. Comb. Chem*. **2000**, *2*, 597.
[129] Zhao, W., and Carreira, E. M.; *Org. Lett*. **2005**, *7*, 1609.

4
Cyclative Cleavage as a Solid-Phase Strategy

A. Ganesan

School of Chemistry, University of Southampton, United Kingdom

4.1 Introduction

Solid-phase synthesis was first introduced by Bruce Merrifield as a means for the rapid construction of peptides without purification of intermediates. Initially applied towards linear peptides, the advantages of the method soon led to its extension for cyclic peptides. In designing the solid-phase synthesis of cyclic peptides, three general strategies have been adopted:

i) *Off-resin cyclization.* The linear peptide **1** is prepared in the normal way on solid phase, usually by backbone attachment of the C-terminus to the resin and chain extension. The peptide **2** is then cleaved from the resin and cyclized in a separate solution-phase step. An early example from Merrifield is the total synthesis of the antibiotic valinomycin **3** (Scheme 4.1).[1]
ii) *On-resin cyclization, side chain attachment.* An amino acid is attached to the resin via a functional group on the side chain. Peptide synthesis then leads to a linear peptide **4**. Upon unmasking of N- and C-terminal protecting groups, the peptide is cyclized on solid phase, followed by cleavage from the resin. Merrifield was again a pioneer, as illustrated by his synthesis of a cyclic hexapeptide **5** via immobilization of the imidazole ring of a histidine residue (Scheme 4.2).[2]
iii) *On-resin cyclization, backbone attachment.* The peptide is attached via a backbone carbonyl group and extended to the linear peptide. This undergoes intramolecular nucleophilic attack by the N-terminal amine, resulting in cyclization and cleavage from the resin in the same step. The first example appears to be from Marshall, who reported the synthesis of cyclic di-, tri- and tetrapeptides **7** (Scheme 4.3).[3]

The three methods have fundamental differences in terms of practice. The first, off-resin cyclization, ignores the cyclic nature of the target and reduces it to solid-phase synthesis of a linear intermediate

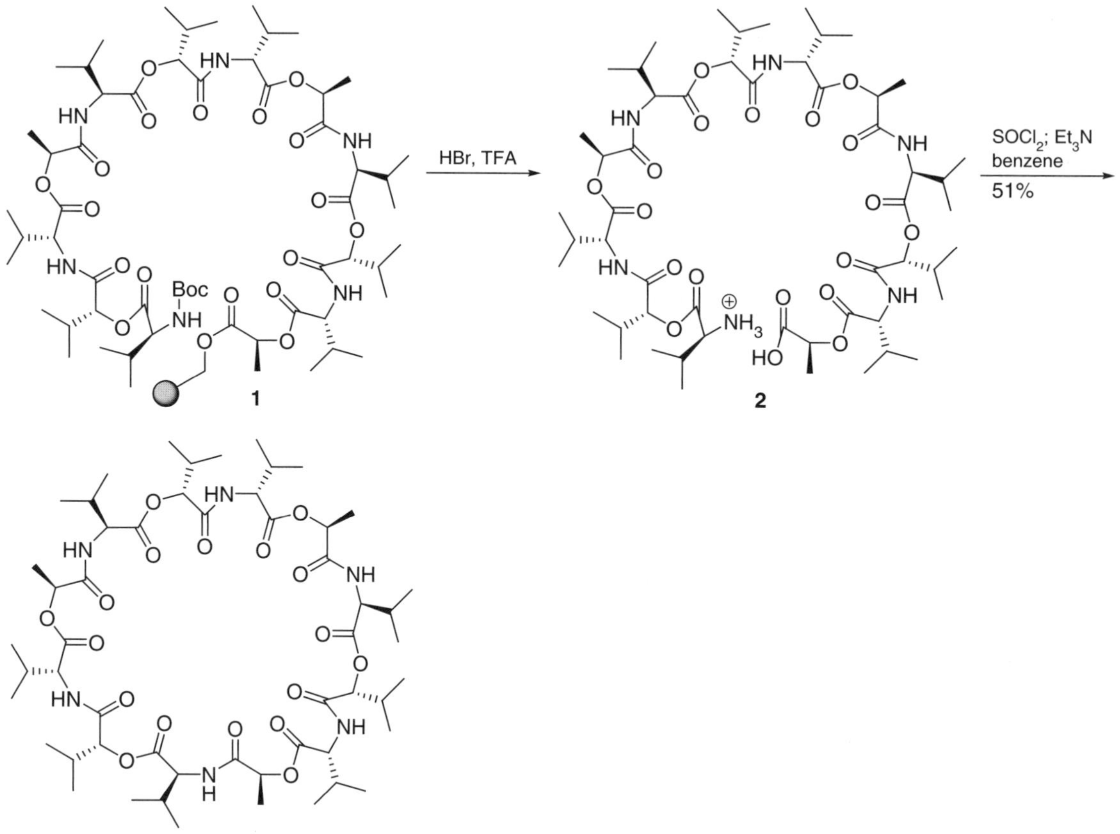

136 *Traditional Linker Units for Solid-Phase Organic Synthesis*

Scheme 4.1 Solid-phase synthesis of valinomycin (*3*)

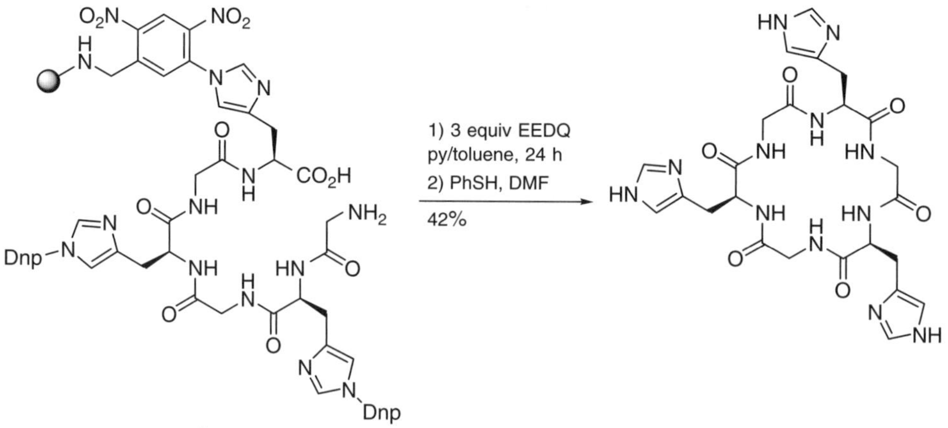

Scheme 4.2 Solid-phase synthesis of a cyclic hexapeptide (*5*)

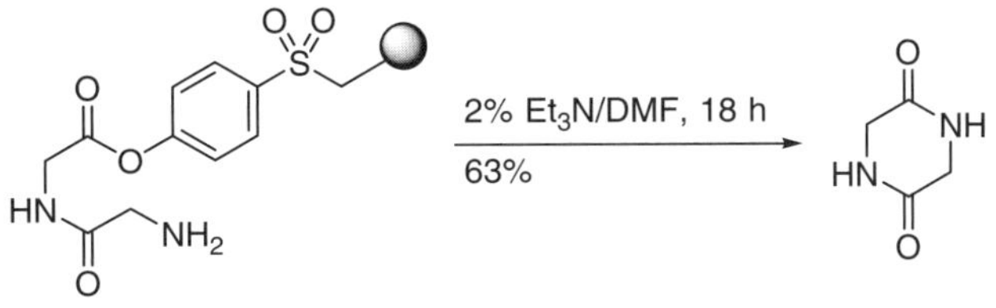

Scheme 4.3 Cyclative cleavage to yield cyclic dipeptides (**7**)

by standard methods. However, the key cyclization is then carried out in solution phase, as is the final deprotection of any masked side chains. For these two operations the efficiencies of solid-phase synthesis are lost and the product needs to be purified from residual starting material, reagents and by-products. Furthermore, if macrocycle formation is involved, there is the potential formation of mixtures of oligomers. The second method, on-resin cyclization with side chain attachment, avoids the purification bottleneck as the whole synthesis is performed on solid phase. Solid-phase macrocyclization tends to minimize dimer or oligomer formation, although the reasons are not fully understood and go beyond the pseudo-dilution effect of the resin. Nevertheless, this method is only applicable for side chains with a functional group suitable for immobilization. If the cyclization is not quantitative, the product will be obtained together with uncyclized materials upon resin cleavage. Finally, on-resin cyclization with backbone attachment is aesthetically pleasing as there are no limitations on the residue selected for immobilization. On the other hand, both peptide synthesis and the cyclization involve the same chemistry of amide bond formation. The backbone attachment must remain inert throughout the cycles of peptide coupling, and become reactive only for the cyclization step. This was a sufficient hindrance that the method was rarely used, although Marshall had already demonstrated one solution using a 'safety-catch' linker. In his example, the phenolic linker with a thioether in the *para* position is stable to peptide coupling, after which it is oxidized to the sulfone, generating a much more labile linkage that undergoes intramolecular cyclization.

With the advent of combinatorial chemistry, interest in combining cyclization and cleavage was rejuvenated both for peptides and non-peptidic small molecules. For the preparation of compound libraries, this strategy had three important benefits. Firstly, a separate solution-phase cyclization with its attendant purification issues was not involved. Secondly, it avoids the presence of a common dangling functional group in all library members, this being the original site of resin attachment and cleavage. Thirdly, only the desired linear precursor can undergo the cyclization reaction. By-products or failed intermediates earlier in the solid-phase sequence would be unreactive, and remain attached to the resin. In principle, the supernatant should contain only the product and the reagents used for cyclization, thus facilitating compound purification. The cyclization–cleavage strategy was first called 'cyclative cleavage' by DeWitt and Czarnik and this nomenclature continues to be popular.[4] As the technique has been recently and extensively reviewed,[5] the ensuing sections illustrate key concepts with newer and representative examples rather than a comprehensive listing.

4.2 C–N bond formation

Amide and urea bond formation is by far the most common disconnection for cyclative cleavage, due to the strength of the resulting bond. For five- and six-membered ring formation, cyclization is often spontaneous, and it remains reasonably efficient for macrocyclic oligomers.

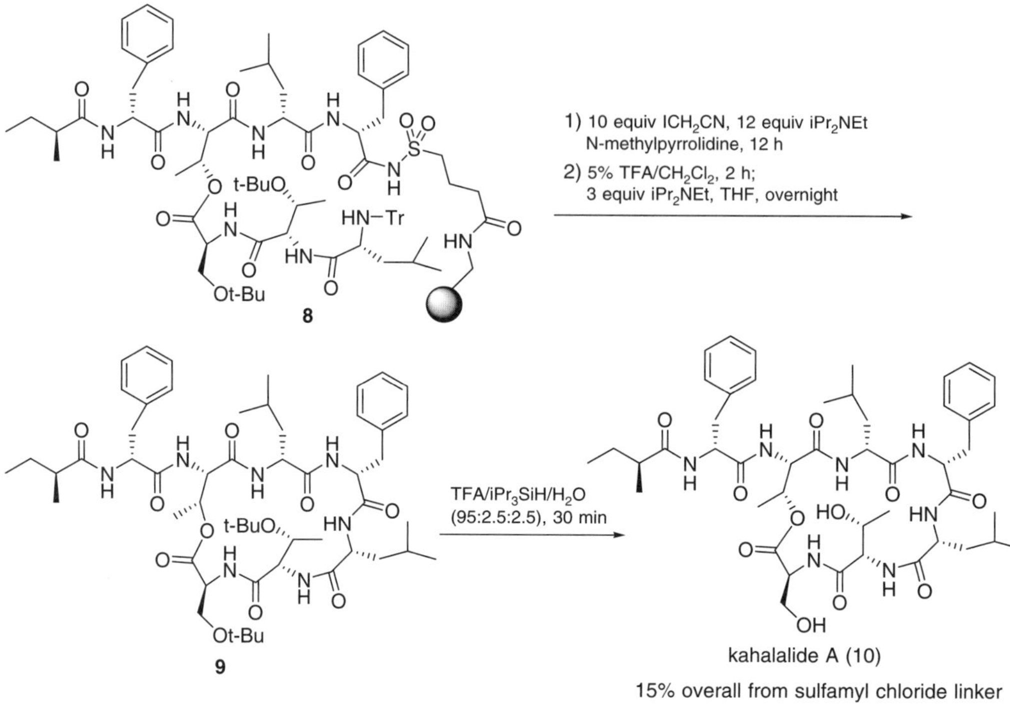

138   Traditional Linker Units for Solid-Phase Organic Synthesis

Scheme 4.4   *Solid-phase total synthesis of Kahalalide A (**10**)*

4.2.1   Cyclopeptides and cyclodepsipeptides

Kenner's acylsulfonamide 'safety-catch' linker is a popular recent option for macrocyclative peptide cleavage. The linker is stable to acidic, basic and nucleophilic conditions, whereas sulfonamide alkylation activates the linker towards nucleophilic cleavage. The original Kenner procedures were improved by Ellman for combinatorial library generation.[6] Since Yang and Moriello at Merck first demonstrated cyclative cleavage with the Kenner linker,[7] it has been widely used for cyclopeptide and cyclodepsipeptide synthesis, aided by the commercial availability of the linker.

Our total synthesis of kahalalide A (**10**) is representative of the procedure.[8] The linear depsipeptide **8** (Scheme 4.4) bearing an N-trityl protecting group was activated by alkylation with iodoacetonitrile. Trityl deprotection and neutralization then triggered macrocyclative cleavage to release 9 from the resin, and the total synthesis was completed by removal of the *t*-butyl protecting groups on serine (Ser) and threonine (Thr) residues. Other examples using the sulfonamide linker include the synthesis of polymyxins,[9] streptogramins,[10] phakellistatin 12[11] and phakellistatin 13,[12] while Clark has successfully cyclized a 22-residue peptide to generate a 66-membered ring.[13]

Typical Experimental Procedures

Kenner safety-catch activation and cyclative cleavage to cyclopeptide

The sulfonamide resin (prepared from 500 mg of initial resin) was activated by iodoacetonitrile (0.4 ml) and Hünig's base (1.73 ml) in 2.5 ml of N-methylpyrrolidinone for 12 hours. This activation step was repeated under the same conditions. After resin washing (N-methylpyrrolidinone, DCM), the N-trityl group was removed by 5% TFA in dichloromethane (DCM) treatment for two minutes. The deprotection was repeated for two hours. The linear peptidyl resin was then washed and suspended overnight in THF with addition of Hünig's base (0.29 ml). The resin is washed, and the combined filtrates concentrated to yield the cyclatively cleaved peptide.

In their studies on tyrocidine antibiotics,[14] Qin showed that macrocyclative cleavage does not require side chain protection of the phenol in tyrosine or the alcohol in hydroxyproline, presumably due to their reduced nucleophilicity compared to the N-terminal amine. The side chain amine of an ornithine residue next to the C-terminus also did not require protection. While this may appear surprising, cyclization was likely to be hindered due to the unfavorable 9-membered ring size thus formed.

Cyclative cleavage with the sulfonamide linker has given mixed results with tetrapeptides. Silverman reported an uneventful total synthesis of microsporin,[15] while the synthesis of somatostatin mimics with beta-amino acids proceeded poorly. The cyclative cleavage gave the product as a mixture of epimers in 6.5% yield, whereas the same peptide was made pure in 18–28% yield by side chain attachment and on-resin cyclization (Section 4.1, strategy 2).[16] Meanwhile, the cyclative cleavage failed completely in the attempted total synthesis of the cyclic tetradepsipeptide serratamolide.[17] None of the product was detected by MS or HPLC, and off-resin cyclization of a linear peptide (Section 4.1, strategy 1) was the best approach.

New linkers continue to be developed for cyclative cleavage of peptides. A thioester linkage **11** was used in the preparation of motilin antagonists **12**, with macrolactamization promoted by silver(I) (Scheme 4.5).[18]

Another example takes advantage of the relative lability of catechol monoesters compared with their alkyl derivatives.[19] The *t*-butyl catechol ester **13** (Scheme 4.6), prepared as a mixture of two regioisomers, is stable while its deprotection **14** leads to approximately three orders of magnitude increase in reactivity and concomitant cyclative cleavage. As proof of concept, penta- to hepta-cyclopeptides were prepared.

4.2.2 Heterocycles, five-membered ring formation

Among the five-membered ring heterocycles accessible by cyclative cleavage, hydantoins **17** have received the most attention. They are readily prepared by reacting immobilized amino acid esters **15** with iso-cyanates and cyclization of the intermediate urea **16**. In a study by Parrot, both steps were accelerated by

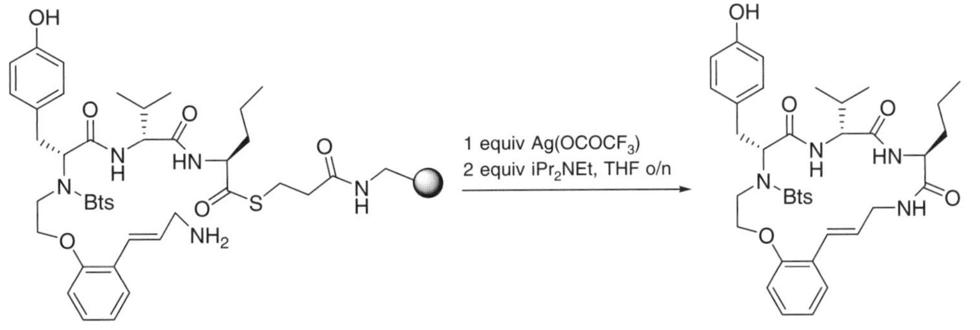

Scheme 4.5 *Preparation of motilin antagonists (**12**)*

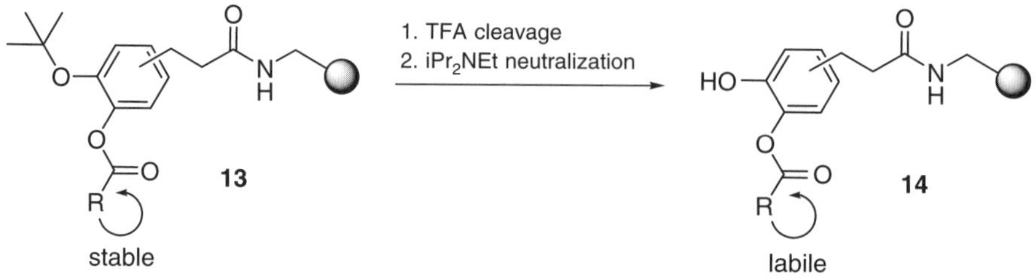

Scheme 4.6 Cyclative cleavage using catechol ester linkers

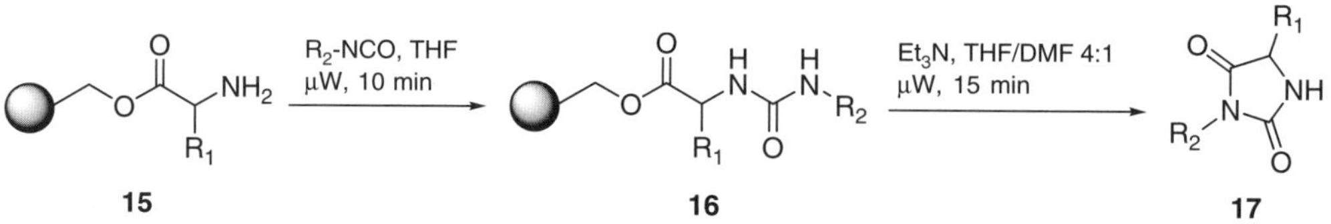

Scheme 4.7 Cyclative cleavage to provide hydantoins (17)

microwave assistance (Scheme 4.7) and a variety of resins found to give good results.[20] A more complex example features an N-acyliminium Pictet–Spengler reaction on solid phase preceding urea formation and cyclization (Scheme 4.8).[21]

Typical Experimental Procedures

N-acyliminium Pictet-Spengler reaction and cyclative cleavage

L-Trp Wang resin (200 mg, loading of 0.54 mmol/g) was shaken overnight with an aldehyde (10 equiv.) in 1:1 DCM/trimethoxy methane followed by resin washing and drying. The resin was then swollen in 1:1 DCM/THF, with the addition of pyridine (14 equiv.), DMAP (1 equiv.) and p-nitrophenyl chloroformate (10 equiv.). After overnight reaction, the resin was washed, filtered and dried. The resin was then swollen in DMF and reacted overnight with an amine (8 equiv.) and triethylamine (8 equiv.) at 90 °C for 24 hours. The resin was filtered and washed, and the combined filtrate purified to yield the tetrahydro-β-carboline-hydantoin product.

The standard solution-phase synthesis of phthalimides **23** via ring opening of phthalic anhydrides was readily translated to solid-phase cyclative cleavage (Scheme 4.9).[22] The intermediate amide-ester **22** underwent cyclization upon heating, and the authors mention this can be carried out under solvent-free conditions.

The diversion of a solid-phase intermediate into different pathways depending on the reagents is an efficient route to the generation of multiple scaffolds. In Kappe's bicyclic dihydropyrimidinone synthesis, solid-phase Biginelli condensation led to intermediate **24** with a pendant chloride (Scheme 4.10).[23] Microwave-assisted reaction with amines displaced the halide, followed by cyclative cleavage to the fused pyrrolidinones **25**. In the same vein, reaction with hydrazines generated fused pyridazinones, while heating of the chloride on its own led to cyclative cleavage by the ester to yield butyrolactones **26**.

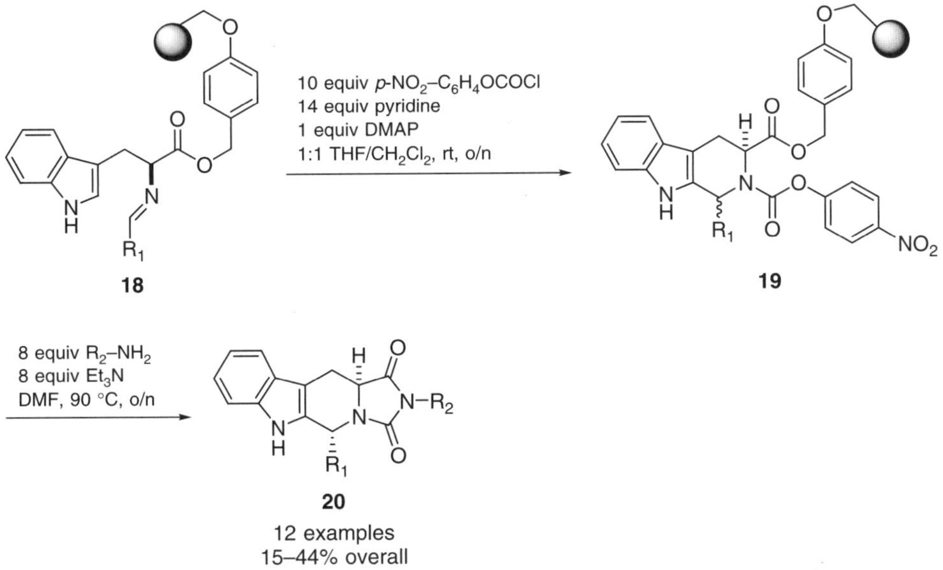

Scheme 4.8 *Urea formation and cyclative cleavage*

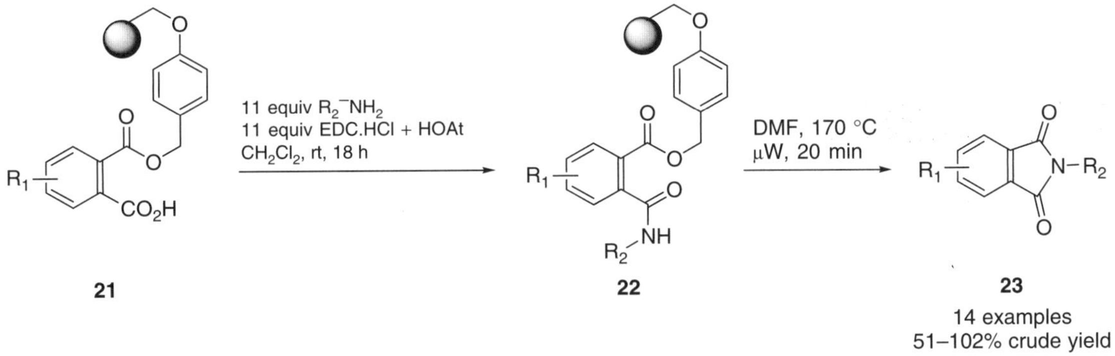

Scheme 4.9 *Solid-phase synthesis of phthalimides (23)*

Bräse showed (Scheme 4.11) that immobilized 2-formylbenzoic acids **27** react with amines to undergo imine formation **28**, followed by further amine addition to induce intramolecular cyclization to isoindolinones **29**.[24] The intermediate imine could also be reacted with a second alcohol or amine nucleophile to introduce a third element of diversity. Alternatively, the aldehyde was treated with an organometallic reagent, with the intermediate alkoxide cyclizing to the butyrolactone **30**. Grignard, organozinc and allylsilane reagents were all successfully employed.

Scott and O'Donnell prepared aldehydes **31** on solid phase by ozonolysis of allyl groups introduced by glycine imine alkylation (Scheme 4.12).[25] Reductive alkylation of the aldehyde led to pyrrolidinones **32** by cyclative cleavage. More complex templates were also obtained by linking imine formation to Cope rearrangement and Pictet–Spengler cyclization, while direct reduction of the aldehyde furnished butyrolactone products **33**.

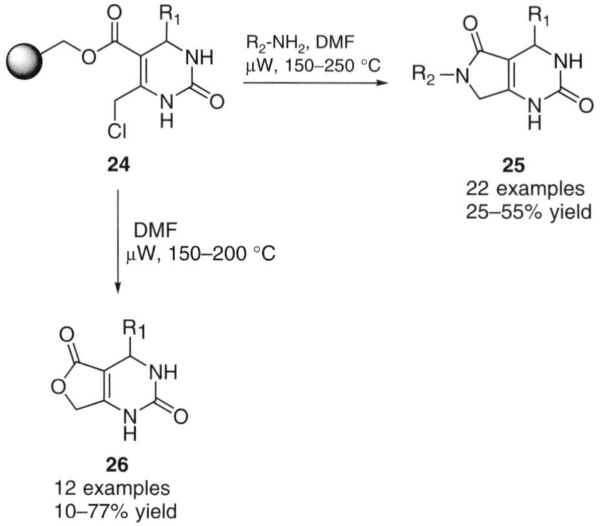

Scheme 4.10 Cyclative cleavage strategies providing pyrrolidinones (**25**) and butyrolactones (**26**)

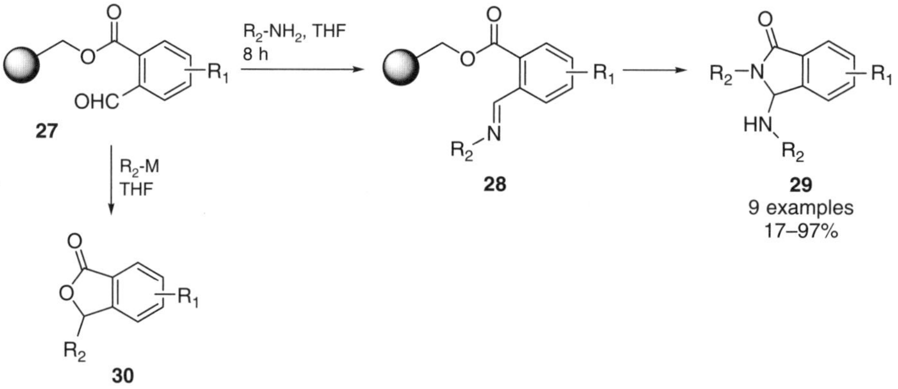

Scheme 4.11 Cyclative cleavage to provide isoindolinones (**29**) or butyrolactone (**30**)

4.2.3 Heterocycles, six- and seven-membered ring formation

Piperazines and variants thereof are the most common six-membered ring heterocycles prepared by cyclative cleavage.[26] A classic example is the total synthesis of demethoxyfumitremorgin C **37** from immobilized tryptophan by the three-step sequence of imine formation to give **35**, N-acyliminium Pictet–Spengler reaction to generate **36** and, finally, cyclative cleavage yielding **37** (Scheme 4.13).[27]

The method was applied to the synthesis of a series of analogues by varying the aldehyde and acid chloride. In a recent medicinal chemistry application, a library of spirodiketopiperazines was prepared as potential CCR5 antagonists.[28] Solid-phase Ugi multicomponent reaction with immobilized isonitrile **38**, followed by *tert*-Butoxycarbonyl (Boc) deprotection and cyclative cleavage led to the target scaffold **40** (Scheme 4.14). Both the polystyrene resin and Rink linker were compatible with the chemistry.

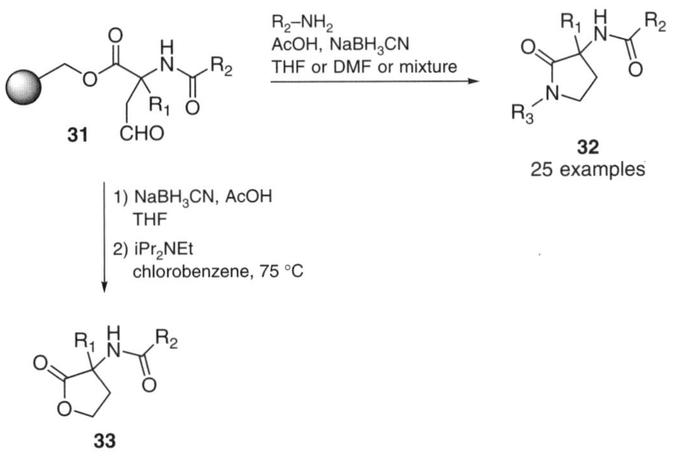

Scheme 4.12 Solid-phase synthesis of pyrrolidinones (**32**) and butyrolactones (**33**)

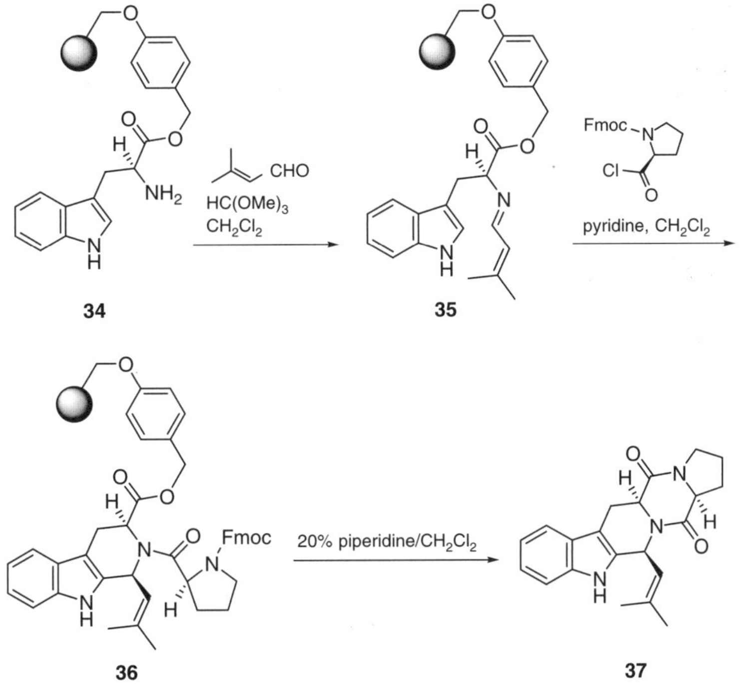

Scheme 4.13 Total synthesis of demethoxyfumitremorgin C (**37**)

Typical Experimental Procedures

Fmoc deprotection and cyclative cleavage of diketopiperazine

The resin (circa 1 g) was swollen by addition of DCM (CH_2Cl_2) (1.6 ml), followed by addition of piperidine (0.4 ml). After shaking for 30 minutes, the resin was filtered and washed. The combined filtrates were

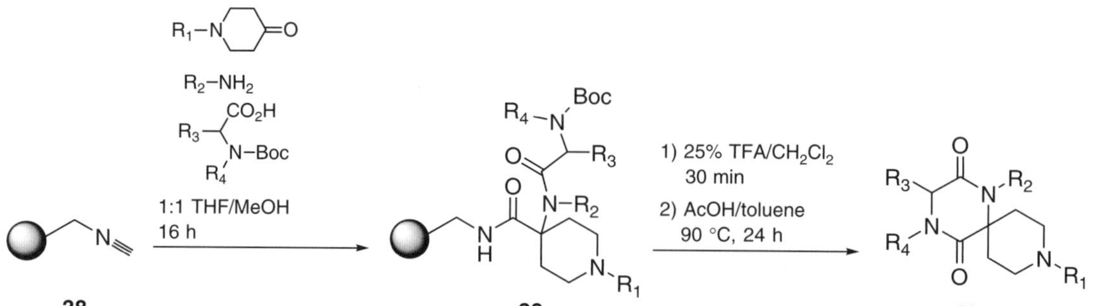

Scheme 4.14 Preparation of a library of spirodiketopiperazines (**40**)

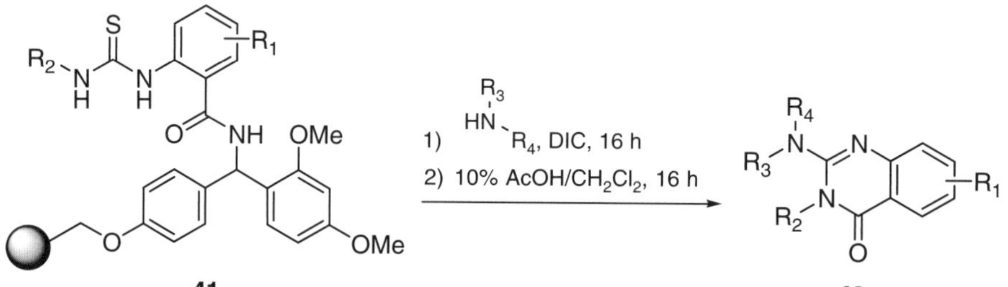

Scheme 4.15 Cyclative cleavage strategies providing quinazolinones (**42**)

concentrated and the residue purified to give the tetrahydro-β-carboline diketopiperazine as a mixture of cis and trans diastereomers.

In a quinazolinone (**42**) synthesis (Scheme 4.15), anthranilic acids were immobilized on the Rink linker and converted to thioureas **41**.[29] Reaction with secondary amines provided guanidines, and treatment with mild acid promoted cyclative cleavage to the heterocycle **42**.

A second guanidine synthesis (Scheme 4.16) involved derivatization of Baylis–Hillman adducts **43** on solid phase by conjugate addition with diamines, followed by reaction with cyanogen bromide to provide **44**.[30] Heating the guanidines in the presence of triethylamine triggered cyclative cleavage to yield fused pyrimidinones **45**.

A seven-membered ring cyclization began with epoxidation of immobilized cinnamates to give resin-bound epoxides **46**. Epoxide ring opening by *ortho*-aminothiophenol gave the linear intermediate **47**, which was cyclized to the benzothiazepinone **48** upon heating (Scheme 4.17).[31] A pyrrolo-benzodiazepinone **50** synthesis began with elaboration of *trans*-hydroxyproline and derivatives immobilized on Merrifield resin.[32] Among the conditions explored for cyclative cleavage to the tricyclic scaffold (Scheme 4.18), Ellman's lithiated oxaolidinone gave the best results. Library screening led to the identification of micromolar hits with antimycobacterial activity.

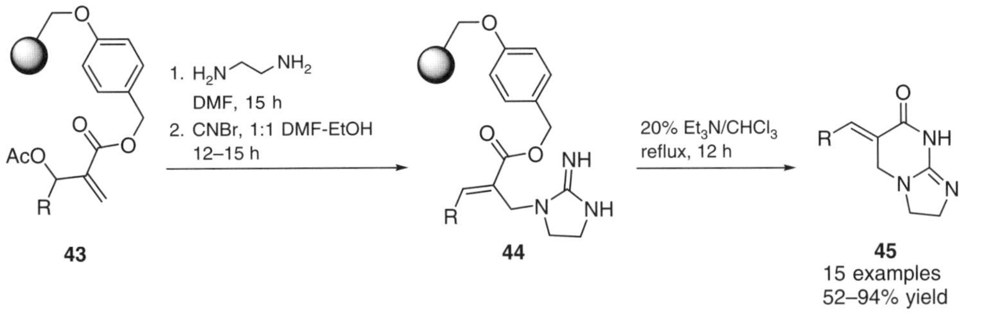

Scheme 4.16 *Cyclative cleavage to yield fused pyrimidinones (45)*

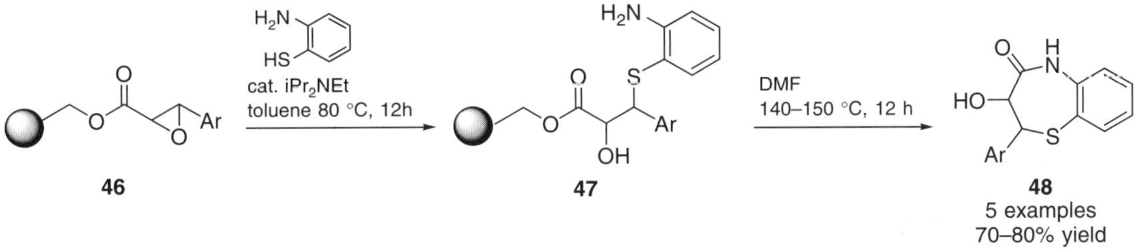

Scheme 4.17 *Cyclative cleavage to provide a benzothiazepinone (48)*

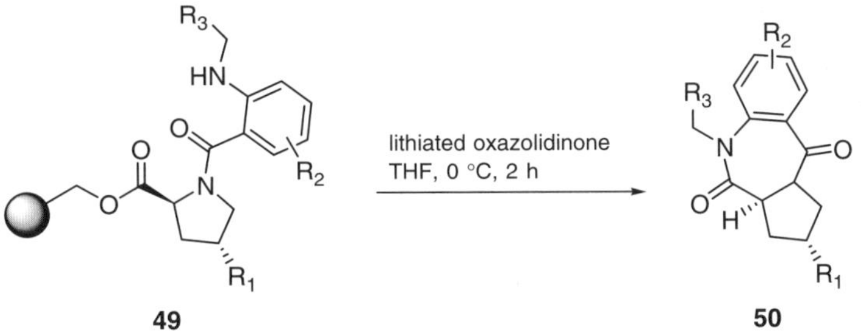

Scheme 4.18 *Cyclative cleavage yielding pyrrolo-benzodiazepinones (50)*

4.3 C–O bond formation

Lactone formation is the most common method for cyclative cleavage via C–O bond formation, and some examples featured in Section 4.2 (Schemes 4.10–4.12). For example, lithiation of immobilized 5-carboxyindoles **51** followed by addition to aldehydes and cyclative cleavage represents another route to phthalides **53**.[33] The lithiation, however, occurred at both C-4 and C-6 of the indole ring, leading to two regioisomeric products with the major one depicted in Scheme 4.19.

A cyclic depsipeptide (**55**) synthesis (Scheme 4.20) used catalytic dibutyltin oxide mediated activation of the terminal alcohol **54** to induce cyclative cleavage from the resin.[34] Ring sizes of 15–27 were achieved by varying the nature of the peptide tether and hydroxy acid.

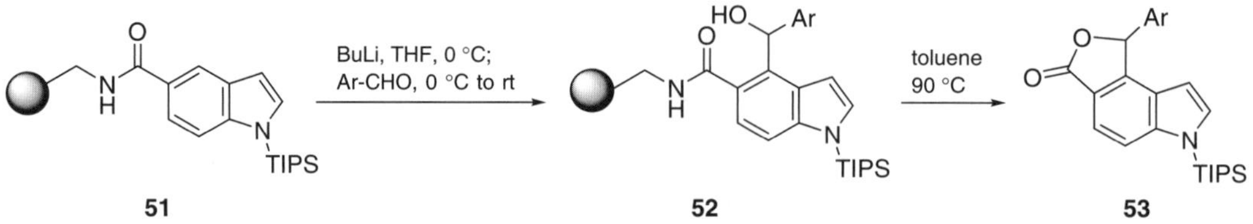

Scheme 4.19 Solid-phase route to phthalides (**53**)

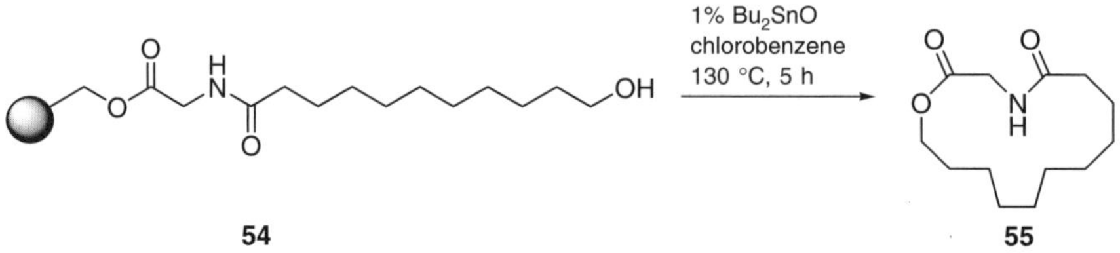

Scheme 4.20 Cyclative cleavage of cyclic depsipeptides (**55**)

Typical Experimental Procedures

Depsipeptide synthesis by macrolactonization

The resin (1.8 mmol) was suspended in chlorobenzene (55 ml). Di-n-butyltin oxide (2 mol%, 10 mg) was added and the mixture was heated under reflux for 24 hours. The filtrate and resin washings were combined and concentrated to give the cyclodepsipeptide.

4.4 C–C bond formation

Intramolecular Claisen-type processes are the most common carbanionic means for solid-phase cyclative cleavage. For example, acylated amino acids **56** immobilized on the Wang resin were treated with hydroxide ion, resulting in Lacey-Dieckmann cyclization and release of the tetramic acid **57** (Scheme 4.21).[35]

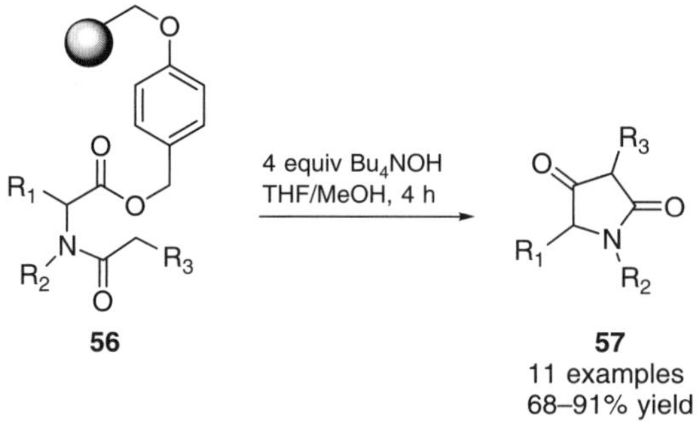

Scheme 4.21 Lacey–Dieckmann cyclization and release of tetramic acids (**57**)

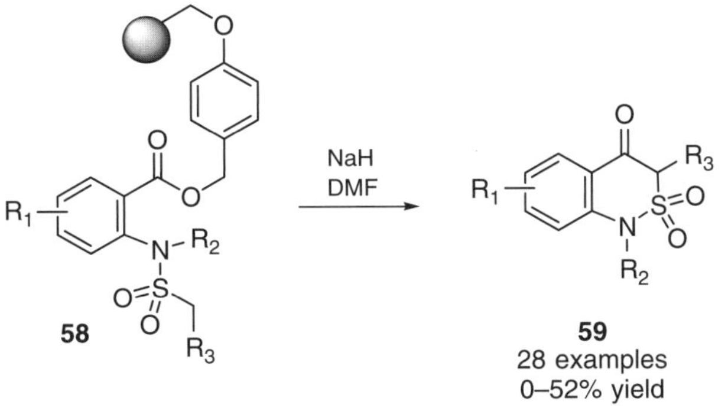

Scheme 4.22 *Jeon's benzothiazinone synthesis*

Following this first report, other groups have published similar examples.[36] The use of a sulfonamide stabilized carbanion featured in Jeon's benzothiazinone synthesis.[37] Immobilized anthranilic acids were sulfonylated and the nitrogen alkylated under Mitsunobu conditions yielding polymer-supported sulfonamide **58**, followed by deprotonation to induce cyclative cleavage generating **59** (Scheme 4.22).

Typical Experimental Procedures

Amino acid acylation and Lacey–Dieckmann cyclative cleavage to tetramic acid

The resin (300 mg) was suspended (DCM, 10 ml), followed by the addition of hydroxybenzotriazole hydrate (15 equiv.) and carboxylic acid (15 equiv.). The reaction mixture was cooled to 0 °C, and diisopropylcarbodiimide (20 equiv.) added slowly. After warming to room temperature, the resin was agitated for 18 hours, filtered, washed and dried. Cyclative cleavage was accomplished by resuspending the resin in THF (10 ml) and tetrabutylammonium hydroxide (0.8 ml of 1M solution in methanol) and agitation for 6 hours. The product was filtered off, the resin washed and the combined filtrates concentrated. The residue was stirred vigorously for one hour in THF (10 ml) and Amberlyst A-15 resin (0.7 g), followed by filtration and washing (THF, 3 × 10 ml) of the ion exchange resin. The combined filtrates were concentrated and the Amberlyst A-15 treatment repeated. The residue was then washed with hexane (2 × 5 ml) to yield the pure tetramic acid.

Carbon–carbon bond formation via an enamine featured in a solid-phase adaptation of the Friedlander quinoline synthesis. Arylimines **60** immobilized on TentaGel resin were reacted with ketones under Borsche's conditions, resulting in cyclative cleavage of aromatic quinolines **61** (Scheme 4.23).[38] As a by-product, this regenerates the amine resin, which can then be recycled.

Organometallic reactions offer one route to carbon–carbon bond formation under mild conditions. Among these, ring-closing metathesis (RCM) has been applied to cyclative cleavage by a number of groups. This concept, as both a cyclative and multifunctional cleavage strategy, has been recently reviewed[39] and is discussed at length in Chapter 20 and therefore need not be repeated in this chapter. However, as a recent representative example, peptide **62** was linked to an olefinic tether which on treatment with Grubbs II catalyst underwent RCM with the linker to release the macrocyclic product **63** (Scheme 4.24).[18]

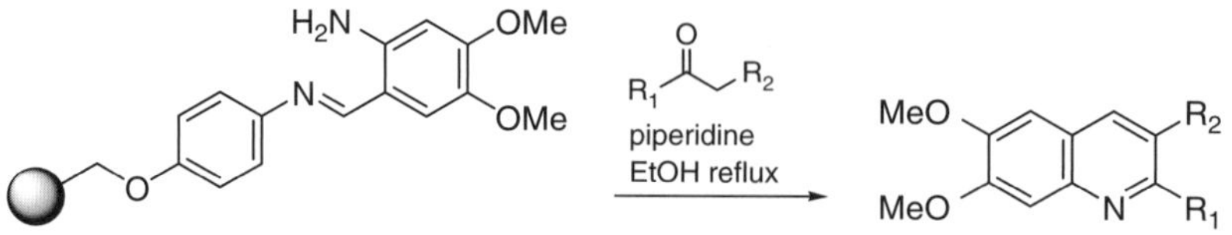

148 *Traditional Linker Units for Solid-Phase Organic Synthesis*

Scheme 4.23 *Cyclative cleavage of aromatic quinolines (**61**)*

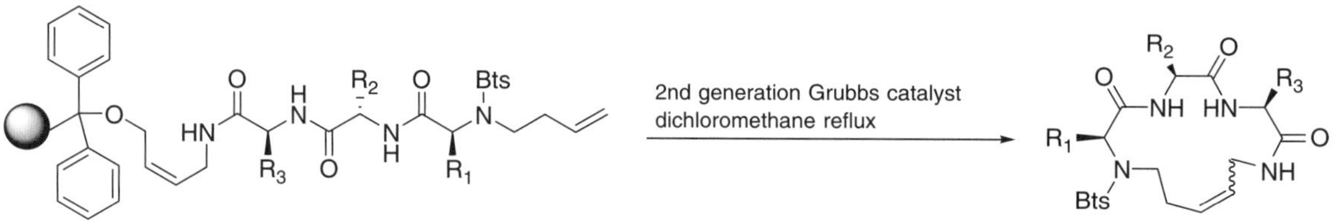

Scheme 4.24 *Ring closing metathesis (RCM) cyclative cleavage strategy*

4.5 Conclusion

Cyclative cleavage is a powerful and versatile strategy that is almost as old as solid-phase synthesis itself. Over the years, it has been applied to a diverse array of small molecules and oligomers. Given its advantages in product purification and the absence of dangling functional groups used for resin attachment, cyclative cleavage should always be considered when designing routes to cyclic target molecules.

References

[1] Gisin, B. F., Merrifield, R. B., and Tosteson, D. C.; *J. Am. Chem. Soc.* **1969**, *91*, 2691.
[2] Isied, S. S., Kuehn, C. G., Lyon, J. M., and Merrifield, R. B.; *J. Am. Chem. Soc.* **1982**, *104*, 2632.
[3] Flanigan, E., and Marshall, G. R.; *Tetrahedron Lett.* **1970**, *11*, 2403.
[4] DeWitt, S. H., and Czarnik, A. W.; *Acc. Chem. Res.* **1996**, *29*, 114.
[5] a) Park, K-H., and Kurth, M. J.; *Drugs Fut.* **2000**, *25*, 1265. b) Seitz, O.; *Nach. Chem. Tech.* **2001**, *49*, 912. c) Ganesan, A., in *Combinatorial Chemistry, Part B*; (eds Morales, G. A. and Bunin, B. A.); Methods in Enzymology, Volume 369; Elsevier, San Diego, **2003**; pp 415. d) Pernerstorfer, J., in *Combinatorial Chemistry*, 2nd edn, (eds Bannwarth, W. and Hinzen, B.); Wiley-VCH Verlag GmbH: Weinheim, Germany, **2006**; pp 111.
[6] Heidler, P., and Link, A.; *Bioorg. Med. Chem.* **2005**, *13*, 585.
[7] Yang, L., and Morriello, G.; *Tetrahedron Lett.* **1999**, *40*, 8197.
[8] Bourel-Bonnet, L., Rao, K. V., Hamann, M. T., and Ganesan, A.; *J. Med. Chem.* **2005**, *48*, 1330.
[9] de Visser, P. C., Kriek, N. M. A. J., van Hooft, P. A. V., et al.; *J. Pept. Res.* **2003**, *61*, 298.
[10] Mukhtar. T. A., Koteva, K. P., and Wright, G. D.; *Chem. Biol.* **2005**, *12*, 229.
[11] Ali, L., Musharraf, S. G., and Shaheen, F.; *J. Nat. Prod.* **2008**, *71*, 1059.
[12] Shaheen, F., Ali, L., Musharraf, S. G., et al.; *Bioorg. Med. Chem. Lett.* **2009**, in press.
[13] Clark, T. D., Sastry, M., Brown, C., and Wagner, G.; *Tetrahedron* **2006**, *62*, 9533.

[14] a) Qin, C., Bu, X., Wu, X., and Guo, Z.; *J. Comb. Chem.* **2003**, *5*, 353. b) Qin, C., Zhong, X., Ng, N. L., et al.; *Tetrahedron Lett.* **2004**, *45*, 217. c) Ding, Y., Qin, C., Guo, Z., et al.; *Chem. Biodiversity* **2007**, *4*, 2827.
[15] Gua, W., Cueto, M., Jensen, P. R., et al.; *Tetrahedron* **2007**, *63*, 6535.
[16] Seebach, D., Dubost, E., Mathad, R. I., et al.; *Helv. Chim. Acta* **2008**, *91*, 1736.
[17] Teixidó, M., Caba, J. M., Prades, R., et al.; *Int. J. Pept. Res. Ther.* **2007**, *13*, 313.
[18] Marsault, E., Hoveyda, H. R., Peterson, M. L., et al.; *J. Med. Chem.* **2006**, *49*, 7190.
[19] Ravn, J., Bourne, G. T., and Smythe, M. L.; *J. Pept. Sci.* **2005**, *11*, 572.
[20] Colacino, E., Lamaty, F., Martinez, J., and Parrot, I.; *Tetrahedron Lett.* **2007**, *48*, 5317.
[21] Bonnet, D., and Ganesan, A.; *J. Comb. Chem.* **2002**, *4*, 546.
[22] Martin, B., Sekljic, H., and Chassaing, C.; *Org. Lett.* **2003**, *5*, 1851.
[23] Prez, R., Beryozkina, T., Zbruyev, O. I., et al.; *J. Comb. Chem.* **2002**, *4*, 501.
[24] Knepper, K., Ziegert, R. A., and Bräse, S.; *Tetrahedron* **2004**, *60*, 8591.
[25] Scott, W. L., Martynow, J. G., Huffman, J. C., and O'Donnell, M. J.; *J. Am. Chem. Soc.* **2007**, *129*, 7077.
[26] O'Neill, J. C., and Blackwell, H. E.; *Comb. Chem. High Throughput Screening* **2007**, *10*, 857.
[27] Wang, H., and Ganesan, A.; *Org. Lett.* **1999**, *1*, 1647.
[28] Habashita, H., Kokubo, M., Hamano, S., et al.; *J. Med. Chem.* **2006**, *49*, 4140.
[29] Kesarwani, A. P., Srivastava, G. K., Rastogi, S. K., and Kundu, B.; *Tetrahedron Lett.* **2002**, *43*, 5579.
[30] Pathak, R., Roy, A. K., Kanojia, S., and Batra, S.; *Tetrahedron Lett.* **2005**, *46*, 5289.
[31] Kumar, H. M. S., Chakravarthy, P. P., Rao, M. S., et al.; *Chem. Lett.* **2004**, *33*, 888.
[32] Kamal, A., Reddy, K. L., Devaiah, V., et al.; *J. Comb. Chem.* **2007**, *9*, 29.
[33] Tois, J., and Koskinen, A.; *Tetrahedron Lett.* **2003**, *44*, 2093.
[34] Cook, A., Hodge, P., Manzini, B., and Ruddick, C. L.; *Tetrahedron Lett.* **2007**, *48*, 6496.
[35] Kulkarni, B. A., and Ganesan, A.; *Tetrahedron Lett.* **1998**, *39*, 4369.
[36] a) Mathews, J., and Rivero, R. A.; *J. Org. Chem.* **1998**, *63*, 4808. b) Weber, L., Iaiza, P., Biringer, G., and Barbier, P. T.; *Synlett* **1998**, 1156. c) Romoff, T., Ma, L., Wang, Y., and Campbell, D. A.; *Synlett* **1998**, 1341. d) Liu Z., Ruan X., and Huang X.; *Bioorg. Med. Chem. Lett.* **2003**, *13*, 2505. e) Evans, K. A., Chai, D., Graybill, T. L., et al.; *Bioorg. Med. Chem. Lett.* **2006**, *16*, 2205.
[37] Jeon, M.-K., Kim, M.-S., Kwon, J.-J., et al.; *Tetrahedron* **2008**, *64*, 9060.
[38] Patteux, C., Levacher, V., and Dupas, G.; *Org. Lett.* **2003**, *5*, 3061.
[39] Piscopio, A. D., and Robinson, J. E.; *Curr. Opin. Chem. Biol.*, **2004**, *8*, 245.

5
Photolabile Linker Units

Christian Bochet and Sébastien Mercier

University of Fribourg, Department of Chemistry, Switzerland

5.1 Introduction

During solid-phase organic synthesis (SPOS), the reactive molecule can be bound directly to the solid support; that would be the case in functionalized resins. However, a suitable functional group for attaching the reactive molecule is sometimes not available. It is then necessary to insert a linker between the resin and the molecule of interest (Scheme 5.1).[1–4]

This entity should fulfill conditions that are quite similar to those required for protecting groups. Thus, a linker needs to be easily attached both to the resin *and* to the substrate of interest, and it should be completely inert towards the various reaction conditions to which the molecule will be subjected. If the linker does not survive under even one of these conditions, the synthetic scheme will be ruined. On the other hand, the linker needs to be cleaved under very specific conditions that, in turn, will not affect the rest of the molecule. In this perspective, *photolabile linkers* represent attractive solutions to these selectivity problems, for several reasons. Firstly, photolabile linkers are mild, since they require only light as a reagent for their cleavage, which limits the chemical interferences with other parts of the molecule. Then, such linkers can be chosen such that they are stable under numerous chemical conditions, as are the photolabile protecting groups from which they are frequently derived. In this chapter, the various photolabile linkers used in SPOS are discussed. Indeed, as will be seen, most of them are derived from existing photolabile protecting groups.[5]

5.2 Linkers based on the ortho-nitrobenzyloxy function

Historically, photolabile linkers were introduced in 1973.[6] Rich and Gurwara were the first to include in a linker the *o*-nitrobenzyloxy group just popularized by Patchornik and Woodward.[7]

Linker Strategies in Solid-Phase Organic Synthesis Edited by Peter Scott
© 2009 John Wiley & Sons, Ltd

152 Traditional Linker Units for Solid-Phase Organic Synthesis

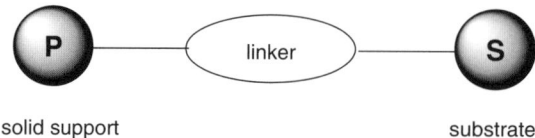

Scheme 5.1 *Role of a linker*

In this specific case, the solid support is a styrene-divinylbenzene copolymer. The first functionalization step is the Friedel–Crafts-type chloromethylation of the polymer aromatic rings (as described by Merrifield in his pioneering work on SPOS[8, 9]), followed by nitration in the *ortho* position. The resin is now ready for the attachment of an amino acid at the C-terminus as an ester, and classical peptidic synthesis can be carried out. Release of the peptide is triggered by irradiation at 350 nm, with yields ranging from 52 to 72% (Scheme 5.2).

Many variations of this *ortho*-nitrobenzyl linker were developed later, always anchoring the polymer *para* to the benzylic methylene group.

Two years after the initial work, Rich and Gurwara showed that this linker was not appropriate for peptides composed of more than four amino acids. The reason invoked to explain this was a reduced swelling capacity of the resin in organic solvents. Indeed, the enhanced polarity caused by the presence of additional nitro groups in the resin (and not in the linker part) seems to hinder the swelling. To circumvent this problem, they prepared the linker separately and *then* coupled it to the resin, thus avoiding over-nitration.[10] The synthesis started with the benzylic radical bromination of *para*-toluic acid with *N*-bromosuccinimide, followed by nitration, leading to the 3-nitro-4-bromomethylbenzoic acid. On the resin side, the polymer is chloromethylated, followed by benzylic amination with gaseous ammonia. Both fragments are then coupled with *N-N'*-dicyclohexylcarbodiimide (DCC), leading to the resin **2** (Scheme 5.3).

Now any desired peptide can be synthesized by successive peptidic couplings, and finally released by photolysis at 350 nm with an overall yield of 64%, based on the glycine-derived resin **3**.

This linker was used by Tam *et al.* in the preparation of multi-detachable resins, allowing the release of the target molecule under a wider range of conditions (Scheme 5.4).[11, 12]

Another application is the three dimensional orthogonal protection in SPOS. In the 1980s, Barany *et al.* designed a strategy allowing the independent deprotection of three different protecting groups, one of

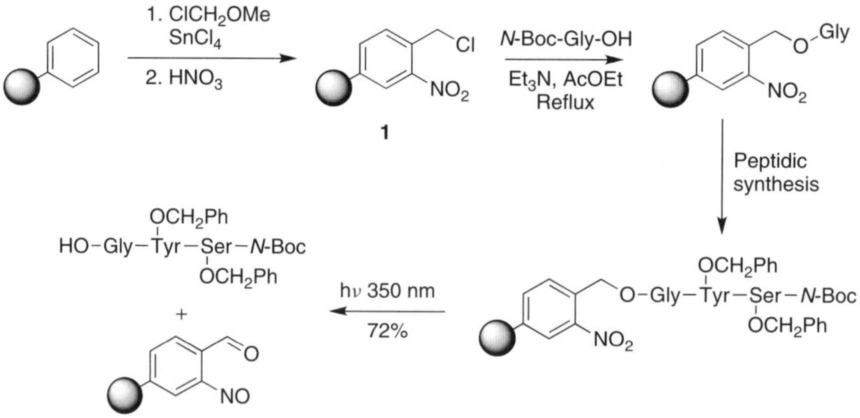

Scheme 5.2 *First nitrobenzylalcohol-derived resin*

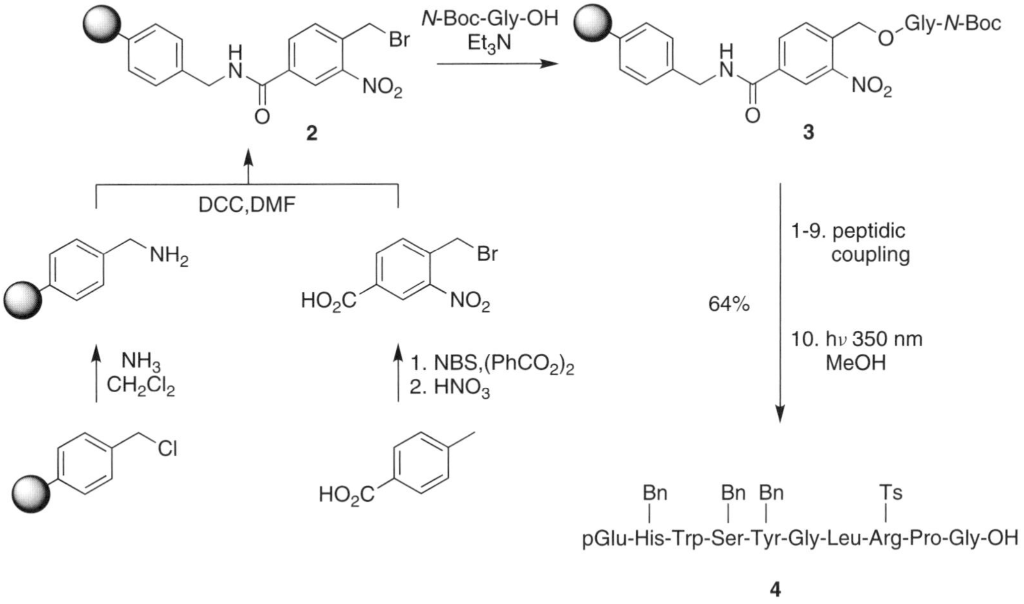

Photolabile Linker Units 153

Scheme 5.3 First resin with a nitrobenzylalcohol derivative

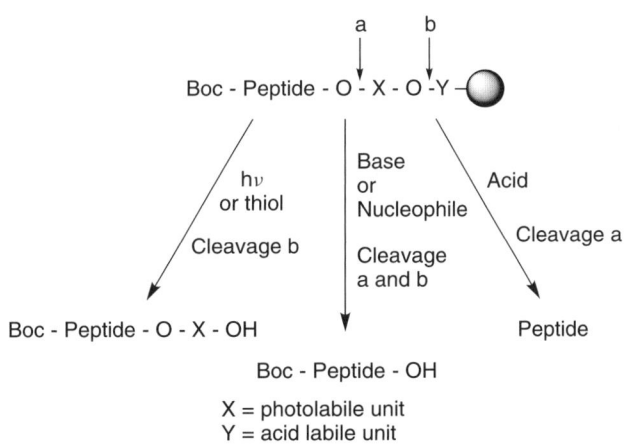

Scheme 5.4 Multidetachable linkers

them being used as a linker to the resin. In addition to the photolabile resin **2**, the *tert*-butyl group (acid labile) and the dithiasuccinoyl group (thiol labile) were used. The two latter are photochemically inert (Scheme 5.5).[13]

Numerous variations of the type of bond between the linker and the resin were used. For example, in the work of Giralt *et al.*, an additional phenyl group is added at the benzylic site (benzhydrylamine resin **5**),[14, 15] or a glycine unit acting as an internal standard is inserted (linker **6**, Scheme 5.6).[16, 17]

An interesting example is the use of similar linkers in solution-phase synthesis, in combination with a polyethylene glycol-type (PEG) polymer.[18–20] Indeed, solid-phase synthesis may be associated with

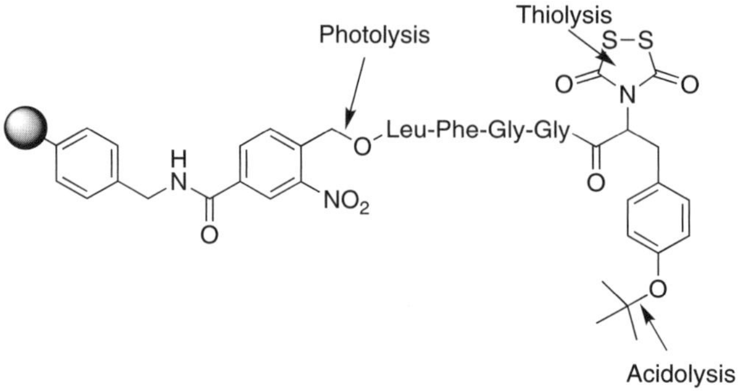

Scheme 5.5 Orthogonally protected resin-bound peptide

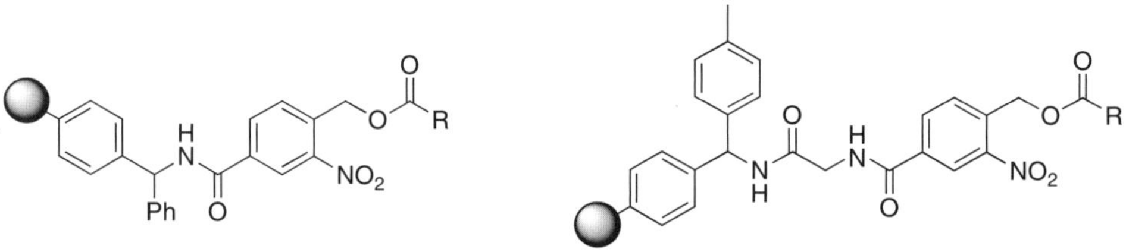

Scheme 5.6 Additional spacers between the resin and the photolabile unit

certain drawbacks, such as a limited choice of solvents, while solution-phase synthesis does not have these limitations. Tjoeng and Hodges were among the first to use, in the 1970s, photolabile linkers based on the o-nitrobenzyl moiety in solution.[21–23] Thus, simple DCC-promoted coupling between PEG and 3-nitro-4-(bromomethyl)benzoic acid gave the support **7**, which can then be directly used in peptide synthesis. After photolysis of the linker, the remaining support is removed by simple membrane filtration.[18] Pillai also studied this type of linker, in particular the various functional groups which can connect the photolabile part and the PEG.[24] The best results were obtained with the PEG-NH$_2$ type support **8**, prepared in three steps from PEG in a similar manner as **7** (Scheme 5.7).

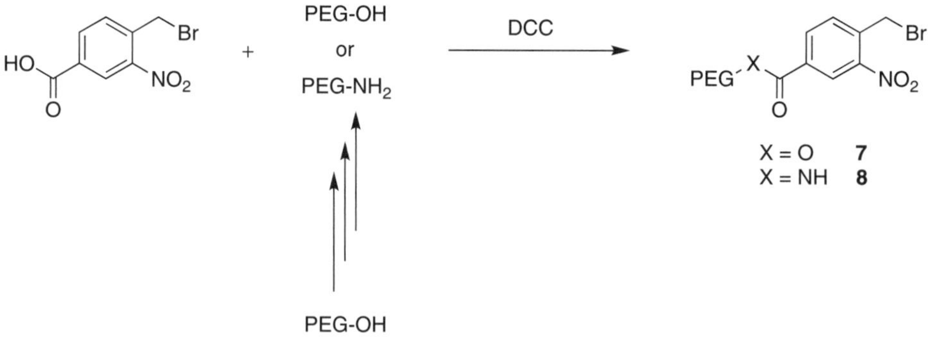

Scheme 5.7 Linkers for soluble polymers

Pillai also used another type of solid support, combining the advantages of solid-phase and solution-phase synthesis. This support consists of a PEG unit bound to a polystyrene bead (also named TentaGel) on one side and a *o*-nitrobenzyloxy unit on the other side.[25]

More recently, Greenberg showed the first application of such photolabile linkers to the liberation of oligonucleotides,[26, 27] using the support **11** based on the *o*-nitrobenzyloxy group, using LCAA-CPG (Long Chain Alkyl Amine Controlled Pore Glass) as the solid part. Its preparation requires two steps: first the coupling of 5′-O-dimethoxytrityl-3′-O-succinatothymidine **9** and the trichlorophenyl ester of 4-hydroxymethyl-3-nitro benzoic acid **10** with DCC, then the attachment of the ester with the LCAA-CPG in the presence of hydroxybenzotriazole hydrate (Scheme 5.8).[28] The advantage of such a method is the possibility of cleaving an oligonucleotide from its support under very mild conditions, at wavelengths harmless to the biopoplymer. Ohlmeyer exploited this feature in the preparation of ligand libraries.[29]

All the previous examples show that the *o*-nitrobenzyloxy-based linkers can be used with carboxylic acid derivatives, with the exception of the last case, where a succinic acid unit binds an alcohol to the linker. However, these linkers can also be used directly with alcohols and amines.

Alcohol-containing groups can also be anchored to *o*-nitrobenzylic linkers through carbonates, as shown by Albericio,[30] or as ethers, either in multi-cleavable resins (Tam and Merrifield)[31] or in oligosaccharide synthesis as shown by Zchavi (Scheme 5.9).[32] In this strategy, the solid support is an aminoethyl polyacrylamide (either acetylated or not) bound to the *o*-nitrobenzyl through an amide using *N*-ethyl-*N*′-(3-dimethylaminopropyl)carbodiimide (EDC) hydrochloride (**12**). It is then bound to carbohydrate **13** in the presence of 2,6-dimethylpyridinium perchlorate; disaccharide **14** is finally obtained after hydrolysis of the benzoyl groups. Further enzymatic glycosylation assembles the full oligosaccharide, which is then liberated photochemically at 350 nm.

Nicolaou also used this kind of linker for the synthesis of oligosaccharides using functionalized polystyrene beads containing phenols, then coupled to a carbohydrate with the anomeric hydroxyl protected with a 5-(3-iodopropoxy)-2-nitrobenzyl group using (dimethylthio)methylsulfonium triflate (DMTST). Iterative glycolysation/activation are followed by the photochemical liberation of the oligosaccharide as a mixture of two anomers (Scheme 5.10).[33]

One year later, this strategy was improved by using an additional spacer between the linker and the oligosaccharide, thus preventing epimerization at the anomeric site during the photolysis. Furthermore, treatment of the resin with trimethylsilylthiophenol and zinc iodide generated a glycosyl donor immediately ready for the next steps, contrary to the previous scheme where reactivation of the photoliberated product was necessary (Scheme 5.11).[34] This time, the Merrifield resin was used.

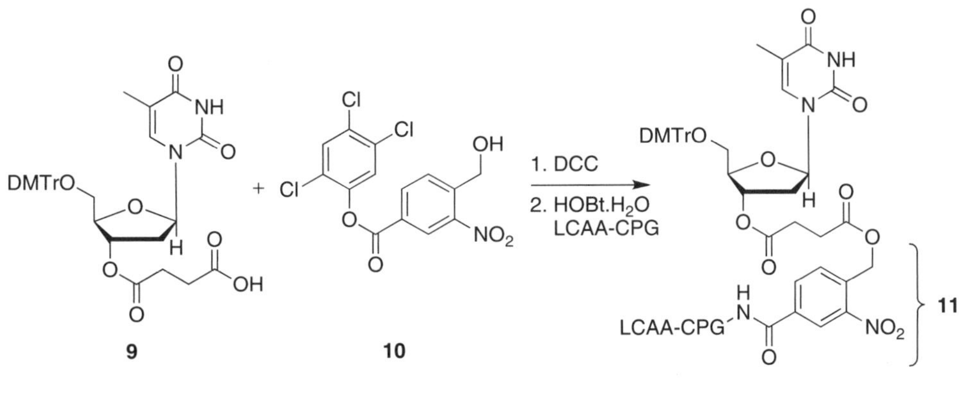

Scheme 5.8 Photolabile linker for oligonucleotide synthesis

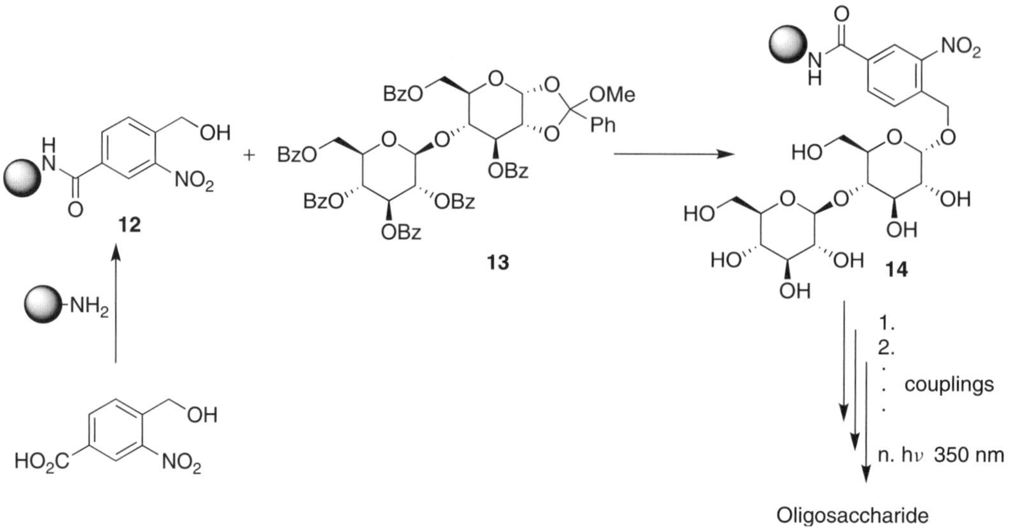

Scheme 5.9 *Photolabile linker for oligosaccharide synthesis*

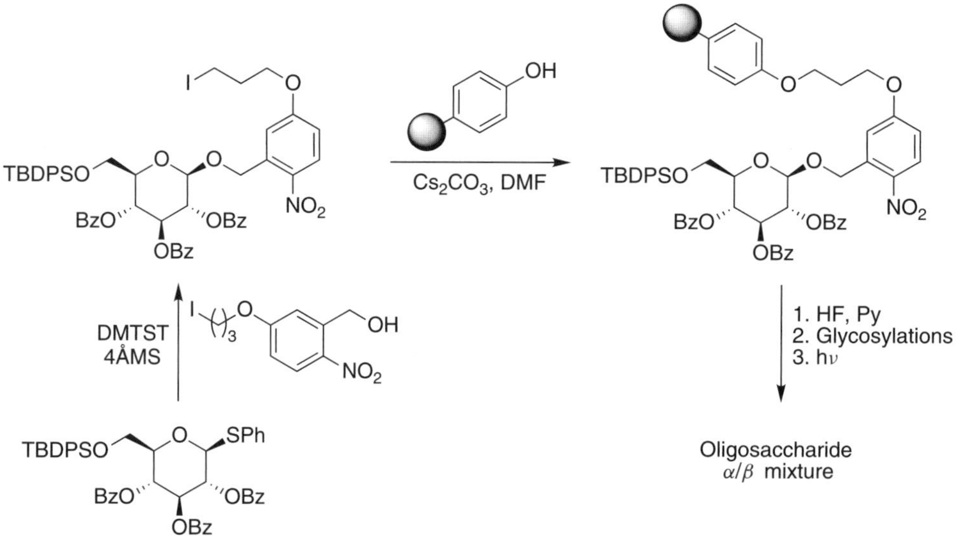

Scheme 5.10 *Photolabile linker for oligosaccharide synthesis with a three-carbon spacer*

Furthermore, it is possible to attach amines through a carbamate function, as shown by Armstrong in a Passerini reaction-based combinatorial library.[35] The linker **16** is prepared by the reaction between the trichlorophenyl ester of 4-hydroxymethyl-3-nitro benzoic acid **10** and the isocyanate **15**, which is then coupled with a methyl-benzhydrylamine glycine resin (MBHA-Gly). The Passerini reaction is then carried out and the desired molecules are released photochemically (Scheme 5.12).

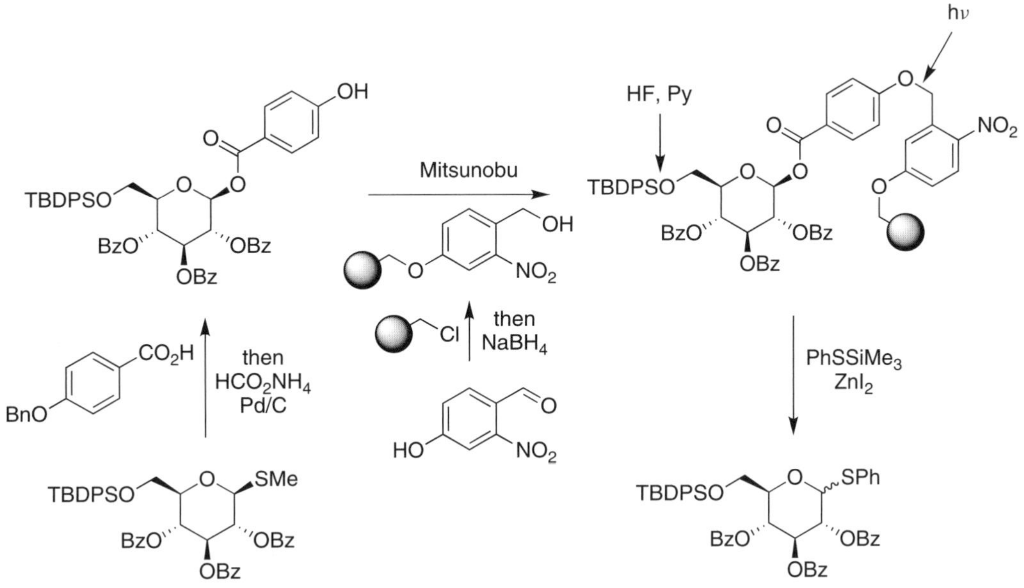

Scheme 5.11 Photolabile linker for oligosaccharide synthesis with an additional spacer

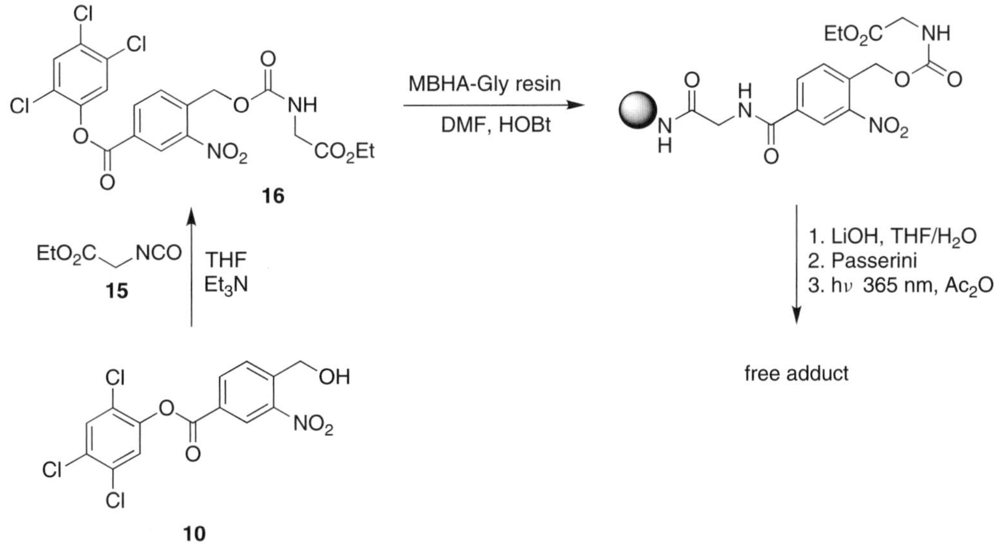

Scheme 5.12 Photolabile linker attached to amino acids N-terminus

Typical Experimental Procedures

Wang, S.-S., *J. Org. Chem.* **1976**, *41*, 3258-61.

Bromomethyl 1

Chloromethyl resin (10 g, 7 mmol) was suspended in DMF (100 ml) and stirred with 11 ml of n-propylamine for 70 hours. The resin was washed with DMF, THF and methanol to provide 10.1 g of amino resin. It was washed several times with dichloromethane (DCM) and suspended in 150 ml of dichloromethane when 2.35 g of 3-nitro-4-bromomethylbenzoic acid (9 mmol) and 2.0 g of DCC (9.7 mmol) were added. After stirring for two hours the resin was collected and washed as usual yielding 11.5 g of desired product **1** (N, 0.86; Br, 4.88). The resin absorbed strongly at 1600 cm^{-1} in the infra red (IR) spectrum.

Boc-gly resin 2

Boc-Gly-OH (0.7 g, 4 mmol) was dissolved in 8 ml of i-PrOH plus 2 ml of water and the mixture titrated to pH 7.0 with 20% cesium carbonate (Cs_2CO_3). The solution was evaporated to dryness, re-evaporated twice with DMF (40 °C), and then stirred in DMF (25 ml) with 6 g of **1** (3.68 mmol) for 24 hours. The resin was then washed as usual and dried to give 6.2 g of material. Amino acid analysis indicated that there was 0.49 mmol of glycine per gram of resin. There was virtually no residual bromide left (0.13%). There were strong absorption bands at 1750, 1710 and 1600 cm^{-1}.

Deprotection

The protected tetrapeptide resin (7.0 g, 2.14 mmol) was suspended in 250 ml of DMF that had been treated with argon gas (2 ml/s) for 15 minutes inside a jacketed Pyrex tube (3 × 30 cm). The suspension was further flushed with argon for an additional 60 minutes with gentle magnetic stirring. The reaction mixture was then tightly stoppered and irradiated at 350 nm (16 × 24 W) in a Rayonet photochemical reaction chamber for 72 hours with efficient water cooling (20 °C). The released peptide **3** was separated by suction filtration and the solvent removed at 40 °C under reduced pressure to give 3.5 g of clear oil, which solidified immediately on treatment with ethyl acetate. It was crystallized from THF and water: yield 0.8 g (49%).

5.3 Linkers based on the ortho-nitrobenzylamino function

The *o*-nitrobenzyloxy functional group is the most popular linker used in solid-phase synthesis; its photolysis releases a carboxylic acid or an alcohol. However, many biologically active peptides possess a terminal primary amide. Hence, Rich and Gurwara developed a variation of this linker, based on the *o*-nitrobenzylamino group, which would release an amide upon photolysis.[36] The preparation of such a linker is very similar to the formation of **2**; it is worth noting that the substitution of

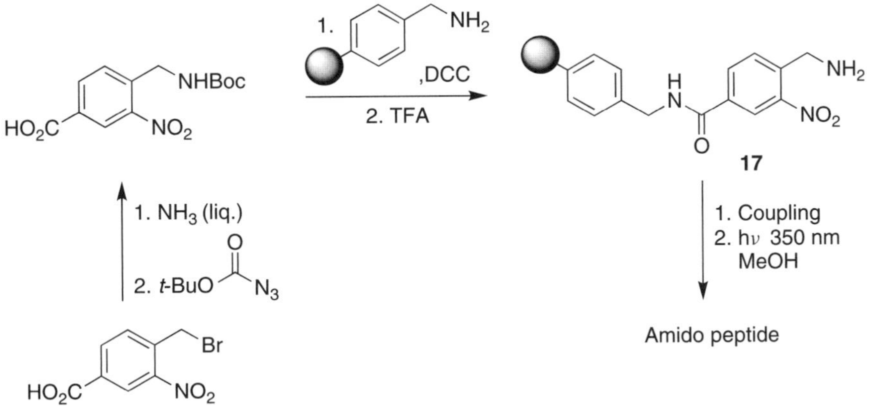

Scheme 5.13 Photolabile linker attached to amino acids C-terminus as secondary amide

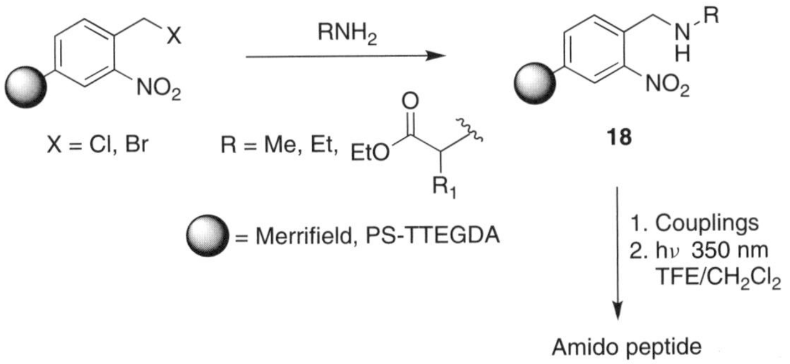

Scheme 5.14 Photolabile linker attached to amino acids C-terminus as tertiary amide

the bromide with an amino group occurs before the anchoring of the linker to the resin. 3-Nitro-4-bromomethylbenzoic acid is aminated in liquid ammonia, immediately followed by the protection of the primary amine with the *tert*-butoxycarbonyl (Boc) group. The linker is attached to a Merrifield resin by a N,N'-dicyclohexylcarbodiimide (DCC) mediated amidation, and the primary amine **17** is restored in acidic medium (Scheme 5.13). Alternative synthetic routes were proposed by Barany, starting with 4-aminobenzoic acid, with comparable overall yields.[37]

This type of linker was used with various supports, among which PEG as discussed earlier. [24, 38] It is also possible to release secondary amides if the linker is appropriately substituted. Thus, Pillai derivatized the brominated linker precursor with a primary amine (instead of ammonia), as in the family of supports **18** (Scheme 5.14).[39–41]

In these cases, the support is either a Merrifield resin or a polystyrene/tetraethyleneglycol diacrylate copolymer (TTEGDA).[40, 41] This support has the advantage of an excellent hydrophobic/hydrophilic ratio and superior swelling properties, while remaining chemically and mechanically quite robust. The final photolysis is carried out at 350 nm in the presence of trifluoroethanol and dichloromethane, with overall yields between 62% and 83%.

Typical Experimental Procedures

Kumar, K. S.; Pillai, V. N. R., *Tetrahedron* **1999**, 55, 10437-10446.

3-Nitro-4-bromomethylbenzamide PS-TTEGDA resin 1

3-Nitro-4-bromomethylbenzoic acid (2.09 g, 0.8 mmol) and N-hydroxybenzotriazole (HOBt) (2.16 g, 1.6 mmol) were dissolved in DCM (25 ml) and shaken with a DCC (1.65 g, 0.8 mmol) solution in DCM (10 ml). After one hour, dicyclohexylurea (DCU) was filtered off and the HOBt active ester formed was shaken with pre-swelled aminomethyl resin (1 g, 0.2 mmol NH/g) in DCM. After one hour the resin was filtered, washed with DCM (6 × 20 ml) and a second coupling was performed with HOBt active ester. The resin was collected by filtration, washed with DCM, DMF and methanol (4 × 15 ml each). Bromine constant of the resin as determined by Volhardt's method = 0.2 mmol/g. IR (potassium bromide): 1650, 1540, 1340 cm^{-1}.

3-Nitro-4-aminomethylbenzamidomethyl PS-TTEGDA resin 2

3-Nitro-4-bromomethylbenzamide PS-TTEGDA resin 1 (5 g, 1 mmol Br) was suspended in DMF. Potassium phthalimide (3.5 g, 2 mmol) was added and the reaction mixture was stirred at 110–120 °C. After 12 hours, the reaction mixture was filtered, washed with DMF (3 × 10 ml), dioxane (3 × 10 ml), ethanol (3 × 10 ml), methanol (3 × 10 ml) and dried in vacuum. The dried resin was suspended in ethanol (100 ml) and refluxed with hydrazine hydrate (0.3 ml, 9 mmol) for eight hours. The resin was collected by filtration and washed with hot ethanol (3 × 10 ml) and methanol (3 × 10 ml). The resin was then dried in vacuum. Amino capacity of the resin = 0.2 mmol/g, as determined by the picric acid method.

General procedure for solid-phase peptide synthesis (Scheme 5.14)[41]

Synthesis was carried out manually in a silanized glass reaction vessel of 15 ml volume. The resin bound N-alkyl derivative of the C-terminal amino acid was swollen in DCM. HOBt active esters of Boc-amino acids were prepared by reacting 2.5 equiv. acid (relative to resin capacity) of DCC, HOBt and Boc-amino acid for one hour at room temperature. The precipitated DCU was filtered off and the active ester solution was added to the pre-swelled resin in DCM. The contents were shaken for 50 minutes; the resin was filtered and washed with DCM (6 × 10 ml). Each coupling was carried out for a second time with N-methylpyrrolidone (NMP) as solvent to achieve 100% coupling. Each coupling step was monitored by semiquantitative ninhydrin test to ensure the completion of coupling. After incorporating each Boc-amino acid, the resin was treated with 30% trifluoroacetic acid (TFA) in DCM for 30 minutes to remove the Boc protection. The resin was filtered and washed with DCM (6 × 10 ml). The deprotected resin was neutralized by adding 5% DIPEA in DCM for 10 minutes and then washed with DCM (6 × 10 ml). The coupling, deprotection and washing were repeated until the desired sequence was achieved. Boc protection of the N-terminal amino acid was not removed.

General procedure for photolytic cleavage (Scheme 5.14)[41]

The peptide resin (1 g) was suspended in a mixture of tetrafluoroethane (TFE) and DCM (1:3) (100 ml) and placed in the inner jacket of a water-cooled Pyrex immersion-type photochemical reactor. The reaction vessel was silylated before photolysis of peptide resin to prevent resin adhering to the walls of the vessel. This is done by rinsing the reaction vessel three or four times with a 10% solution of Me_3SiCl in DCM followed by washing with absolute ethanol and drying. Nitrogen gas was bubbled through the suspension for about one hour to remove the dissolved oxygen gas. The suspension was irradiated with a Philips HPK 125 W medium pressure mercury-quartz lamp placed inside the inner jacket for about 10–12 hours. A solution of copper sulfate ($CuSO_4$) was circulated through the outer jacket of the photochemical reactor to filter out light wavelengths below 320 nm. After photolysis, the spent resin was filtered, washed with ethanol (3 × 20 ml), methanol (3 × 20 ml) and DCM (3 × 20 ml). The combined filtrate and washing were evaporated on a rotary evaporator under reduced pressure. The residue was collected and purified by column chromatography on a sephadex G-10 column using 5% acetic acid in water as the eluent.

5.4 Linkers based on the α-substituted ortho-nitrobenzyl group

An unpleasant feature of *ortho*-nitrobenzyl-based linker is that the side product of their photolysis is an *o*-nitrosobenzaldehyde, which is prone to polymerization and becoming colored. This has the consequence of limiting the accessibility of light to the relevant reaction site. Adding an extra methyl group at the α position results in the formation of a much less reactive acetophenone as a side product. Thus, Pillai prepared support **19** by a Friedel–Crafts reaction between the aromatic groups of polystyrene cross-linked with divinylbenzene and acetyl chloride, followed by reduction with sodium borohydride, substitution of the resulting alcohol with hydrogen bromide and *ortho* nitration with nitric acid (Scheme 5.15).[42]

Unfortunately, as in the case of the unsubstituted resins, the swelling properties are degraded by overnitration during the preparation of the linker. Thus, the photoliberation of peptides longer than five amino acids gives poor yields, despite better intrinsic performances of the substituted linker.[10] A solution to this problem is to introduce in the α position not a methyl group, but the support itself which then consists of an *ortho*-nitrobenzhydryl resin that is more robust towards acidic conditions (necessary to cleave the Boc group) than a conventional benzhydryl.[43, 44] Its preparation is similar to **19**, namely a Friedel–Crafts acylation of a cross-linked polystyrene/divinylbenzene, followed by reduction. Both the alcohol **20** and the amine **21** can be obtained, depending on whether one needs a peptide[43] or an amidopeptide (Scheme 5.16).[44] Good yields were obtained with peptides up to 10 amino acids long.

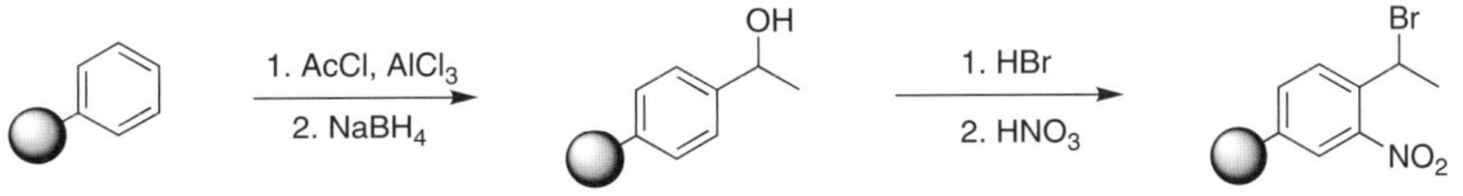

Scheme 5.15 *α-Substituted nitrobenzylalcohol-derived resin*

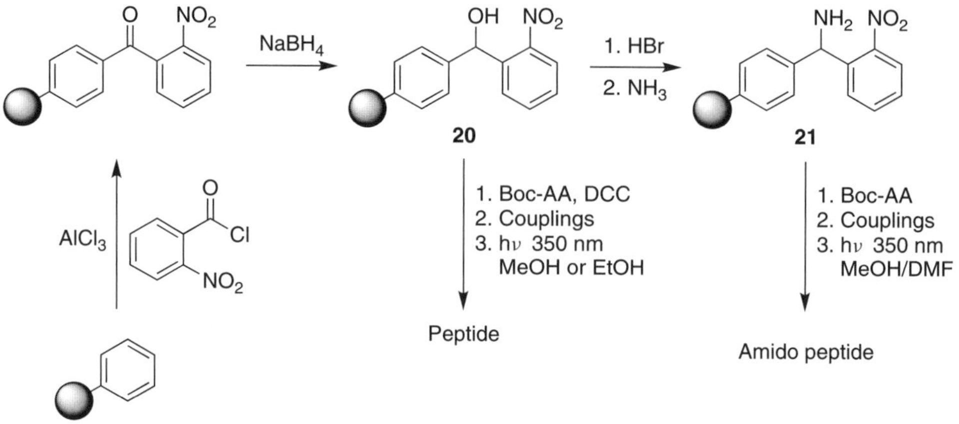

Scheme 5.16 Ortho-*nitrobenzyhdryl resin*

In a related approach, Brown, Wagner and Geysen developed the 3-amino-3-(2-nitrophenyl)propionyl linker (ANP).[45] It has the advantages of a high liberation yield and a shorter photolytic half-life. Amino acids can be released as amides, and the synthesis of tripeptides was exemplified. A single bead can be photolyzed and analyzed by Electrospray Ionization-Mass Spectrometry (ESI-MS), making the deconvolution of product libraries very straightforward. The linker is conveniently prepared starting from the imine *o*-nitrobenzaldehyde, followed by condensation/decarboxylation with malonic acid and protection with the 9-fluorenylmethoxycarbonyl (Fmoc) group. The carboxylic acid terminus is then coupled to an amino resin (in this specific case a TentaGel), and final deprotection gave the ANP resin **22** (Scheme 5.17).

Schreiber proposed a variant of this resin, with two methyl groups in the α position of the amide, for an increased stability towards acidic or basic conditions.[46]

An ANP-based linker was also used for oligosaccharide synthesis, with the amino roup replaced by an alcohol, and a longer spacer between the photosensitive part and the support, thus avoiding β-elimination problems and a possible lactonization of the alcohol.[47] The preparation of this linker starts with the addition of an organozinc valerate ester to *o*-nitrobenzaldehyde, followed by silylation of the resulting alcohol and saponification of the ester. The acid is coupled to an amino resin and the free alcohol is released by a fluoride. Resin **23** is attached to the anomeric center of a carboxydrate, and standard glycosylation/deprotection steps are carried out, before the final photolysis at 365 nm in THF, with a 42% overall yield (average of 87% per step) (Scheme 5.18).

Harran designed an alternative route, allowing access to both hydroxyl- and amino-containing versions.[48] In this case, the support is an aminopropylsiloxane-grafted glass plate. It has the advantage of high resistance to chemical conditions and does not require proper swelling. Supports **24** and **25**

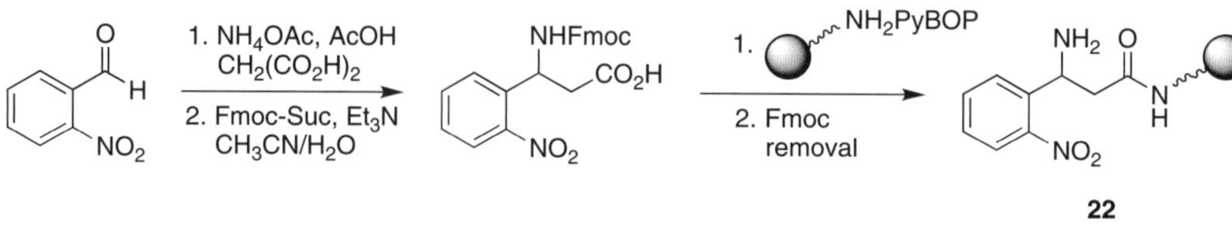

Scheme 5.17 ANP resin

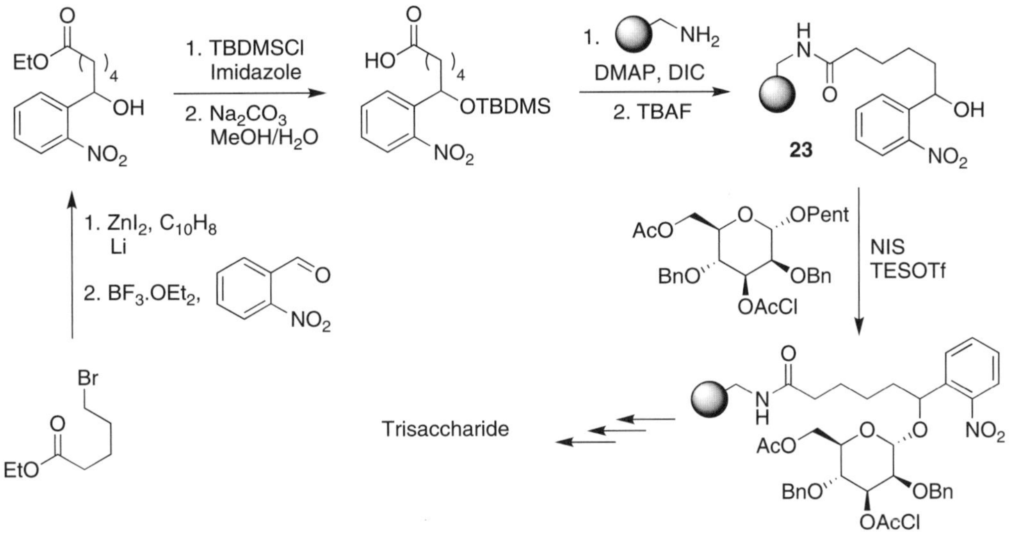

Scheme 5.18 *Alcohol-based ANP modified linker for oligosaccharide synthesis*

are prepared by nitronium tetrafluoroborate (NO_2BF_4) mediated nitration of 2-phenylcyclohexanone, followed by either a Baeyer–Villiger oxidation or a Schmidt rearrangement, and saponification of the lactone/lactame (Scheme 5.19).

These linkers are then attached to the support by standard amidation. After further suitable synthetic steps, the final photolytic release is best carried out in dioxane.

Finally, the recent work of Karuso should be mentioned, in preparing an α-methylated *o*-nitrobenzyl linker bearing two PEG chains, terminated by amino groups; this confers to the linker a high

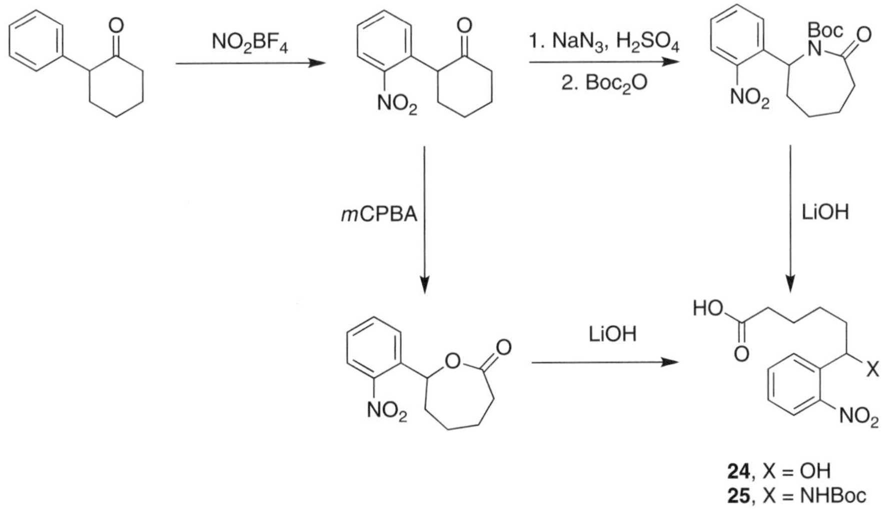

Scheme 5.19 *Improved synthetic route to ANP variants*

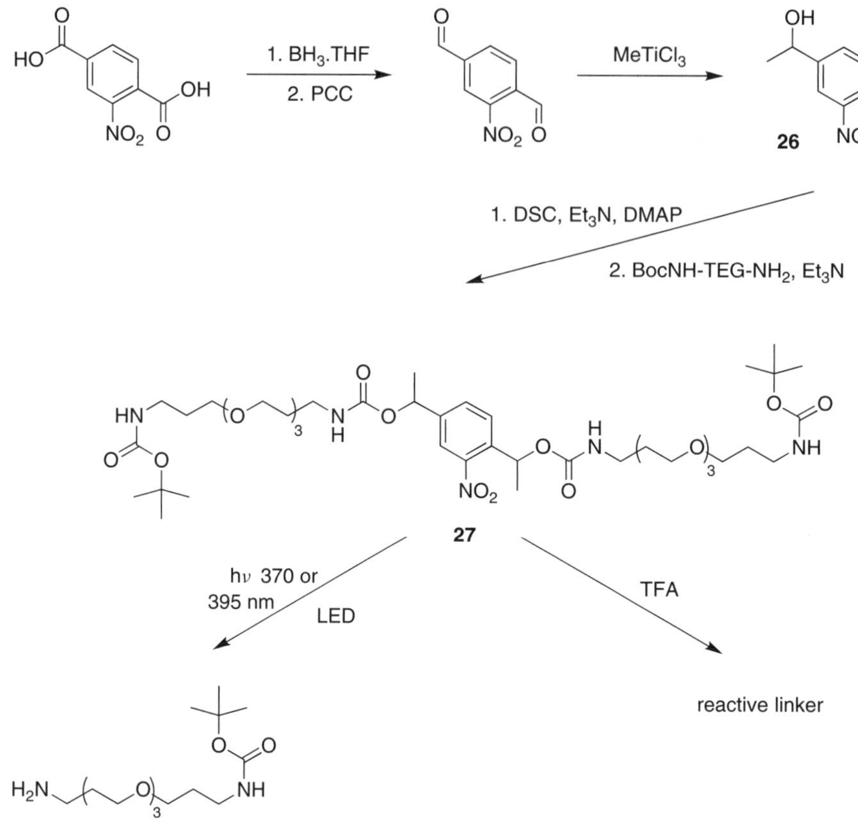

Scheme 5.20 Bifunctional linker

hydrophilicity.[49] It is prepared starting from *o*-nitroterephthalic acid, which is converted into the corresponding dialdehyde. Addition of two methyl groups with methyltitanium dichloride is followed by activation with *N*,*N*′-disuccinimidyl carbonate and addition of tetraethyleneglycol chains leading to **27**. The terminal amines can be released with trifluoroacetic acid (Scheme 5.20).

The photolysis of **27** can be carried out at long wavelengths (up to 395 nm), and is among the first examples of the use of UV-emitting LEDs in photochemistry.

Typical Experimental Procedures

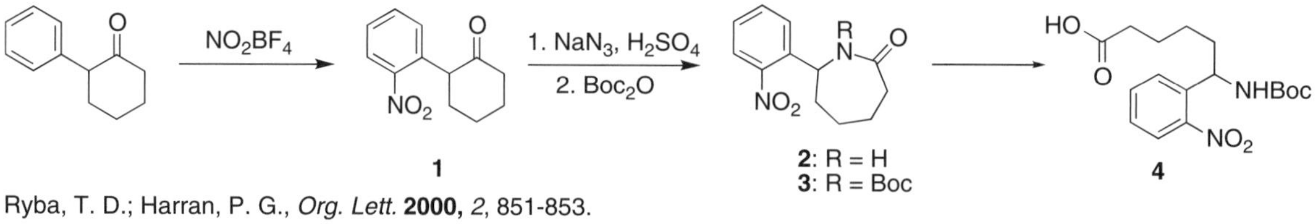

Ryba, T. D.; Harran, P. G., *Org. Lett.* **2000**, *2*, 851-853.

2-(2'-nitrophenyl)cyclohexanone 1

To a stirred solution of 85% nitronium tetrafluoroborate (Aldrich Chem. Co., 1.92 g, 12.3 mmol) in anhydrous nitromethane (25.5 ml) is added solid 2-phenylcyclohexanone (2.0 g, 11.2 mmol) at 0 °C. The resultant indigo solution is stirred at 0 °C for 30 minutes, warmed to room temperature over 30 minutes, and quenched with saturated sodium bicarbonate. The mixture is partitioned between ethyl acetate and water, the organic layer washed with brine, dried over sodium sulfate and concentrated in vacuo. Flash chromatography on silica gel (10% ethyl acetate/hexanes) affords pure **5** (1.14 g, 46%) along with a more polar para isomer (300 mg, 12%). Rf = 0.30 (30% ethyl acetate/hexanes). Melting point = 67–69 °C.

7-(2'-nitrophenyl)caprolactam 2

To a stirred solution of **1** (1.50 g, 6.8 mmol) in toluene (45 ml) at 0 °C is added concentrated sulfuric acid (7.7 ml, 150 mmol). Sodium azide (1.33 g, 20.5 mmol) is added in portions and the resultant biphasic mixture stirred vigorously at 0 °C for 30 minutes. After warming to room temperature over 20 minutes, the mixture is diluted with ethyl acetate, neutralized with saturated sodium bicarbonate, washed with water and brine, dried over sodium sulfate and concentrated in vacuo. Flash chromatography on silica gel (50% ethyl acetate/hexanes) affords recovered **1** (150 mg) and **2** (1.59 g, 65–73% based on recovered **1**) as a pale yellow solid. Rf = 0.27 (ethyl acetate). Melting point = 180–182 °C.

N-tert-butoxycarbonyl-7-(2'-nitrophenyl)caprolactam 3

To a solution of amide **2** (1.28 g, 5.44 mmol) in dry THF (22 ml) is added solid 4-(N,N-dimethyl)aminopyridine (DMAP) (997 mg, 8.16 mmol) followed by di-tert-butyldicarbonate (3.57 g, 16.3 mmol) via syringe. The result homogenous solution is stirred at room temperature for three hours, diluted with EtOAc, washed with water and brine, dried over sodium sulfate, and concentrated in vacuo. Flash chromatography on silica gel (30% ethyl acetate/hexanes) affords **3** (1.66 g, 91%) as a light yellow solid. Rf = 0.38 (30% ethyl acetate/hexanes). Melting point = 100–105 °C.

6-(2'nitrophenyl)-6-(N-tert-butoxycarbonyl)amino caproic acid 4

Solid lithium hydroxide monohydrate (646 mg, 8.62 mmol) is added to a solution of **3** (1.31 g, 3.92 mmol) in 2:1 THF/water (6 ml). The resultant solution is warmed to 45 °C and stirred for four hours. After cooling to room temperature, the mixture is partitioned between ether and water and the organic layer separated. The aqueous layer is neutralized with 10% aqueous hydrochloric acid and extracted with ethyl acetate (3×). The combined extracts are washed with water, brine, dried over sodium sulfate, and concentrated in vacuo. Recrystallization from hot water affords **4** (1.10 g, 82%) as white needles. Rf = 0.20 (50% ethyl acetate/hexanes). Melting point = 129–135 °C.

5.5 Linkers based on the ortho-nitroveratryl group

The *o*-nitroveratryl group (4,5-dimethoxy-2-nitrobenzyl) was known as a photolabile group since the 1970s, but it is only in the mid 1990s that it was included in photolabile linkers, independently by Greenberg and Holmes. Greenberg used it to circumvent the problems associated with the unsubstituted analogue, that is the long photolysis times (leading to degradation of the released molecule) and the low yields of liberation

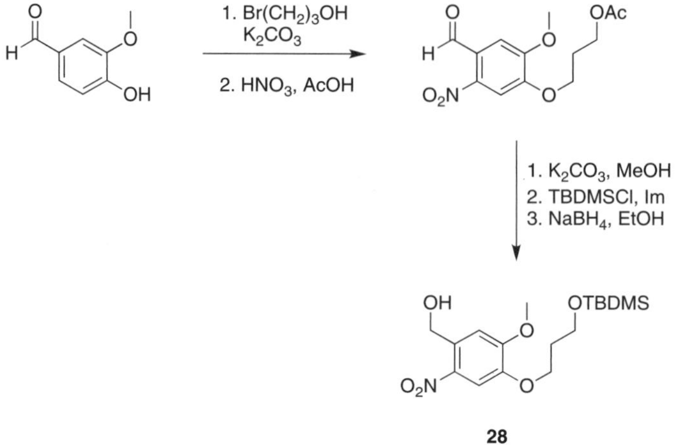

Scheme 5.21 *Nitroveratryl-based linker*

encountered in his previous work.[50] The preparation of linker **28** starts from readily available vanillin, etherified with bromopropanol and nitrated at the *ortho* position. A few functional group interconversions lead to the linker **28** with an overall yield of 35% (Scheme 5.21).

The linker is then bound to a LCAA-CPG (Long Chain Alkyl Amine Controlled Pore Glass) resin in a few synthetic steps, starting with the esterification of **28** with **30**, followed by conversion of the propyl side chain up to the stage of an activated ester, ready for the direct coupling of the amino-subsituted LCAA-CPG (Scheme 5.22).

The yield in the release, after the assembly of the oligonucleotide, strongly depends on the overall photolysis time. While 30–60 minutes of irradiation time are often sufficient to give yields greater than

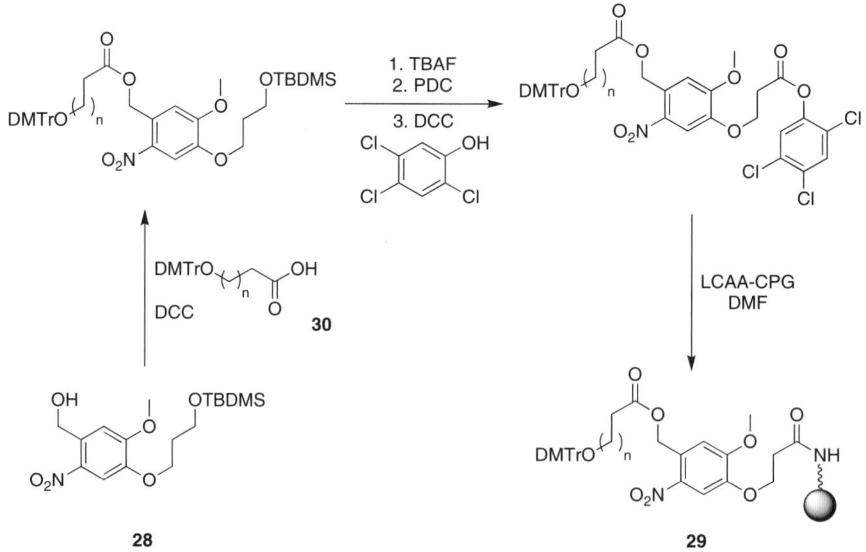

Scheme 5.22 *Photosensitive nitroveratryl-based linker on LCAA–CPG*

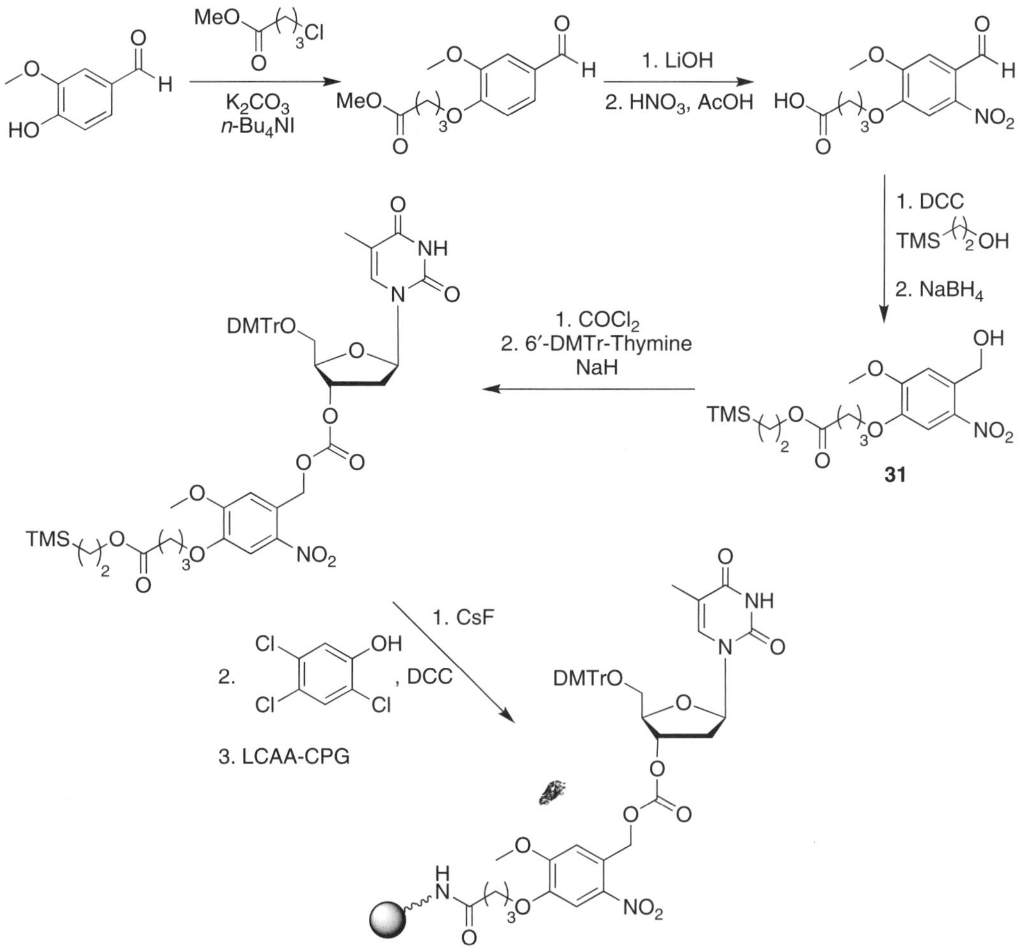

Scheme 5.23 *Improved version of nitrovertryl-based linker using a carbonate*

70%, longer times lower the overall yields. It was shown later that the photolysis was more efficient if the linker is directly attached to the hydroxy group of the nucleotide via a carbonate.[51] This linker is also prepared starting from vanillin, in a nine-step sequence and an overall yield of 10%. (Scheme 5.23). This linker was later used by other groups.[52]

Almost simultaneously, Holmes proposed a *o*-nitroveratryl-based linker, but as an amine and bearing a methyl group in the α-position. Its first use with a model compound, 4-thiazolidinone, showed indeed an increased photoreactivity.[53] Later, a systematic study of substituents was performed, using simplified units as models (a benzyl group in place of the resin and an acetate in place of the substrate). (Scheme 5.24).[54]

For the liberation of amides, the linker **37** was prepared (Scheme 5.25).

Kinetic studies showed that the rate of cleavage is 7 to 20 times higher when the *o*-nitrobenzyl ring is substituted with alkoxy substituents (except in buffered media). Furthermore, the α-substitution with an alkyl group also increases the rate by a factor of three. An effect of the spacer chain lengths was also noticed, with a slightly increased rate for longer chains, which was explained in terms of a better electrodonor ability of longer chains (the amide being electroattracting). It was also noted that, in some solvents, amides were

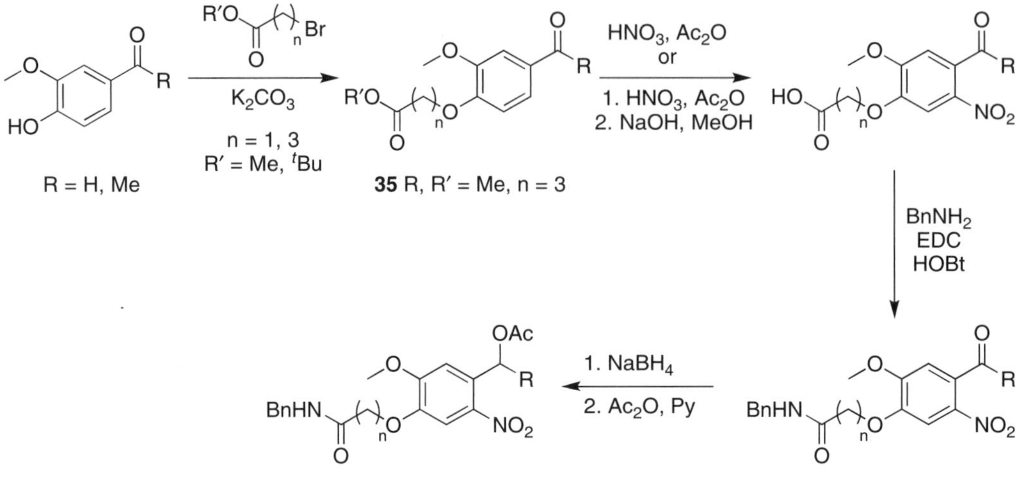

168 Traditional Linker Units for Solid-Phase Organic Synthesis

Scheme 5.24 Alcohol-containing α-methylated nitroveratryl linker

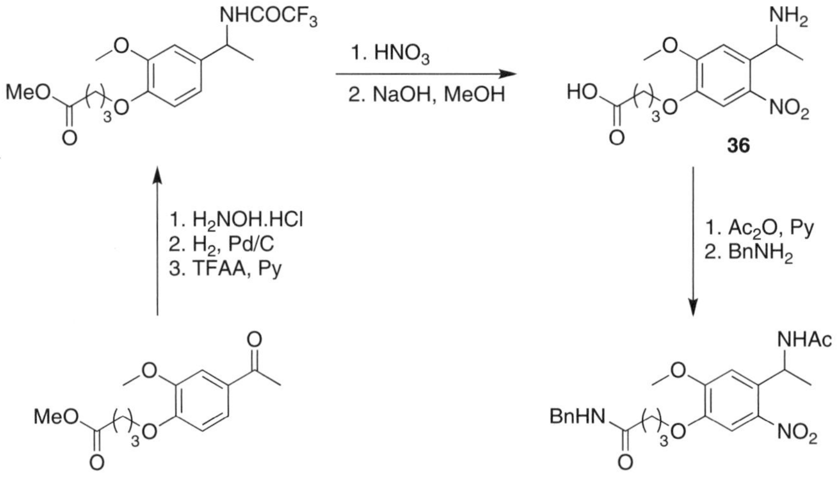

Scheme 5.25 Amine-containing α-methylated nitroveratryl linker

released between three and eight times more rapidly than esters. The general conclusion of this study is that both the substitution of the aromatic rings with alkoxides and the benzylic position with alkyl groups is significantly beneficial to the photochemical reactivity.

Numerous applications based on similar linkers have been used since then. For example, Austin showed that these linkers could be linked to the resin though an ester bond, without detrimental effects on the photolysis.[55] McKeown also used such linkers for analytical purposes.[56] Indeed, with two orthogonal linkers successively attached to a TentaGel bead (one photolabile, one acid labile), it is possible to either release a primary amine (ideally suited for ESI-MS characterization; unambiguous discrimination from background noise in very small quantities can be achieved by isotopically labeling one of the nitrogen atom of the pentanediamine spacer) or the desired carboxylic acid (Scheme 5.26).

More recently, Madder used the linker **36** in the preparation of serine protease analogues.[57] After protection of the amine with a Fmoc group, attachment to the TentaGel resin and deprotection of the amine, this support can anchor up to three peptidic chains with a suitable spacer. After the chemical steps, the tripodal peptide can be released photochemically and analyzed (Scheme 5.27).

The linker **36** was also used by Blackwell in the multicomponent assembly of combinatorial diketopiperazine libraries with the Ugi reaction.[58] The support was a modified cellulose, very simple to use, both in the synthesis and the purification steps; it is also compatible with most of the screening methods

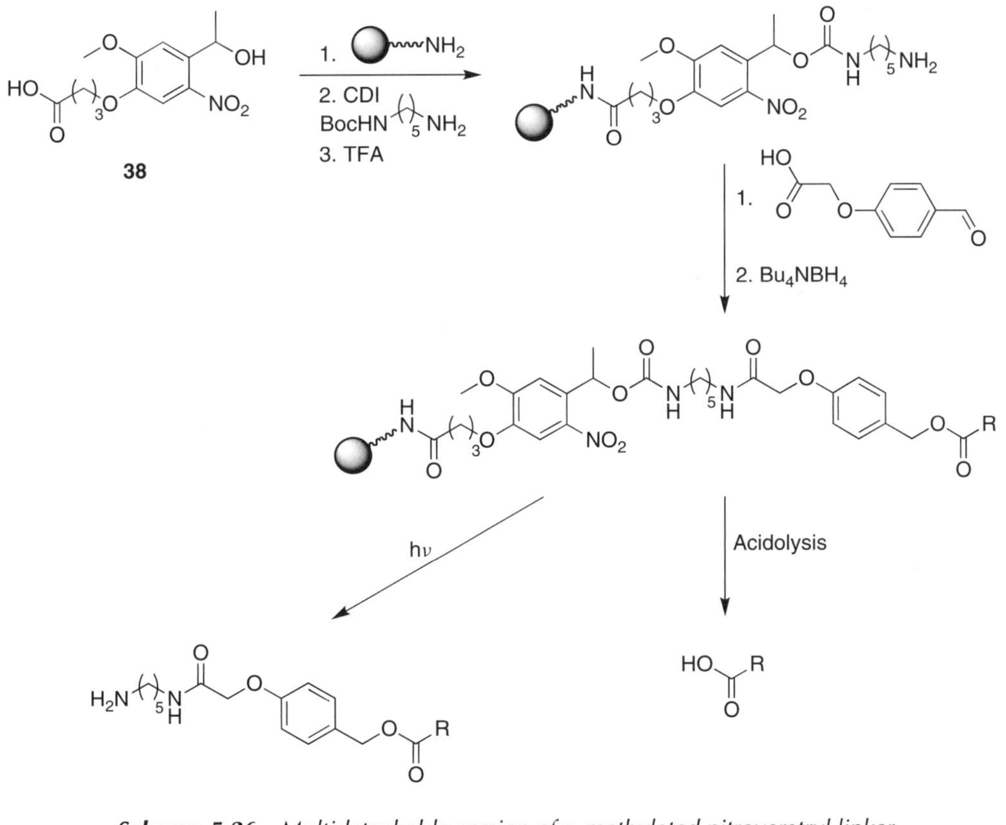

Scheme 5.26 Multidetachable version of α-methylated nitroveratryl linker

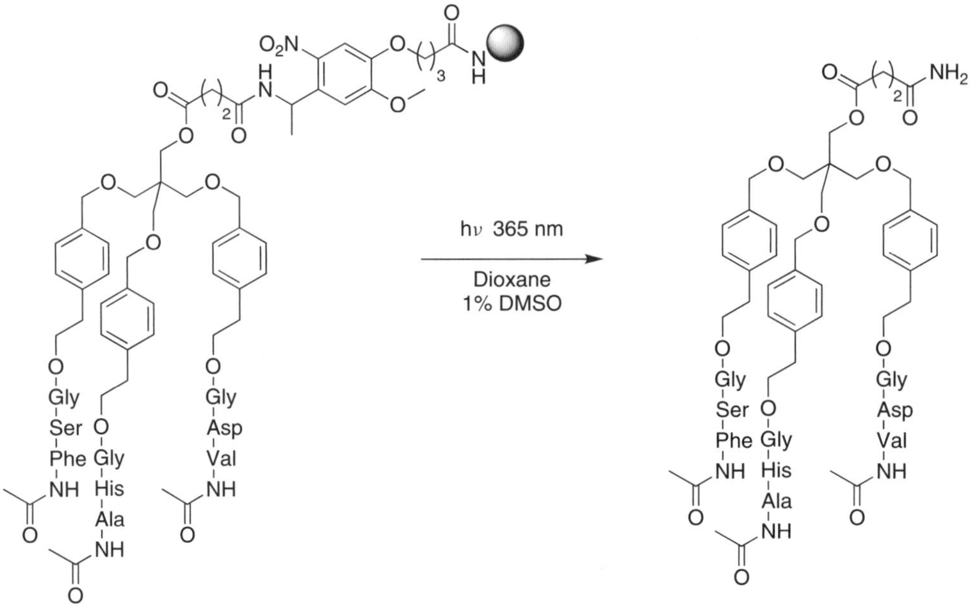

Scheme 5.27 *Tripodal linker*

used in biology and with the acidic conditions used in this example.[59] Thus, the Fmoc protected **36** is bound to the support in the presence of N,N'-diisopropylcarbodiimide (DIC), N-hydroxysuccinimide (HOSu) and diisopropylethylamine (DIPEA) and microwave irradiation. After hydrolysis of the Fmoc group, the reactive amine (N-Fmoc-4-nitro-L-phenylalanine) is attached (with standard methods) and submitted to the Ugi reaction. Methanolysis is followed by the photochemical release (at 366 nm) in very good yields, and the hydrolysis of the Fmoc group leads directly to the cyclized diketopiperazines **40** (Scheme 5.28).

The same linker **36** was used by Gennari for the preparation of libraries of vinylic sulfonamidopeptides with a TentaGel support.[60] The results of Holmes were confirmed, showing the superior properties of the *o*-nitroveratryl-based linkers compared to the original *o*-nitrobenzyl versions. The photolysis was carried out in methanol at 354 nm in very good yields (77–86%). For the release of secondary peptides, an additional alkylation of the sulfonamide can be effected before each coupling (Scheme 5.29).

Still using the linker **36**, libraries of 3,4-disubstituted β-lactams was proposed by Gallop.[61] The free amine of the linker is condensed with an aldehyde, and the resulting imine undergoes a [2+2] cycloaddition with a ketene (prepared *in situ* by deprotonation of the corresponding acid chloride). The β-lactams are then released photochemically at 365 nm in dimethylsulfoxide (DMSO) in good to excellent yields (71–90%, Scheme 5.30).

A related linker was used in the preparation of a library of melanocortin subtype-4 receptor (MCR4) containing an amide function.[62] The support has two HMBA sites (hydroxymethylbenzoic acids), one of them bound to the photolabile linker, the other to a phenylpropionic acid. The photolysis, carried out at 365 nm, in aqueous acetonitrile, is followed by a basic treatment, cleaving the two HMBA units; the 3-phenylpropionic acid thus released is used as an internal standard to determine the ratio of released/non-released amide by HPLC (Scheme 5.31).

This approach was generalized by using a series of amines other than phenylethylamine, and substituting the chloride with a piperazine.

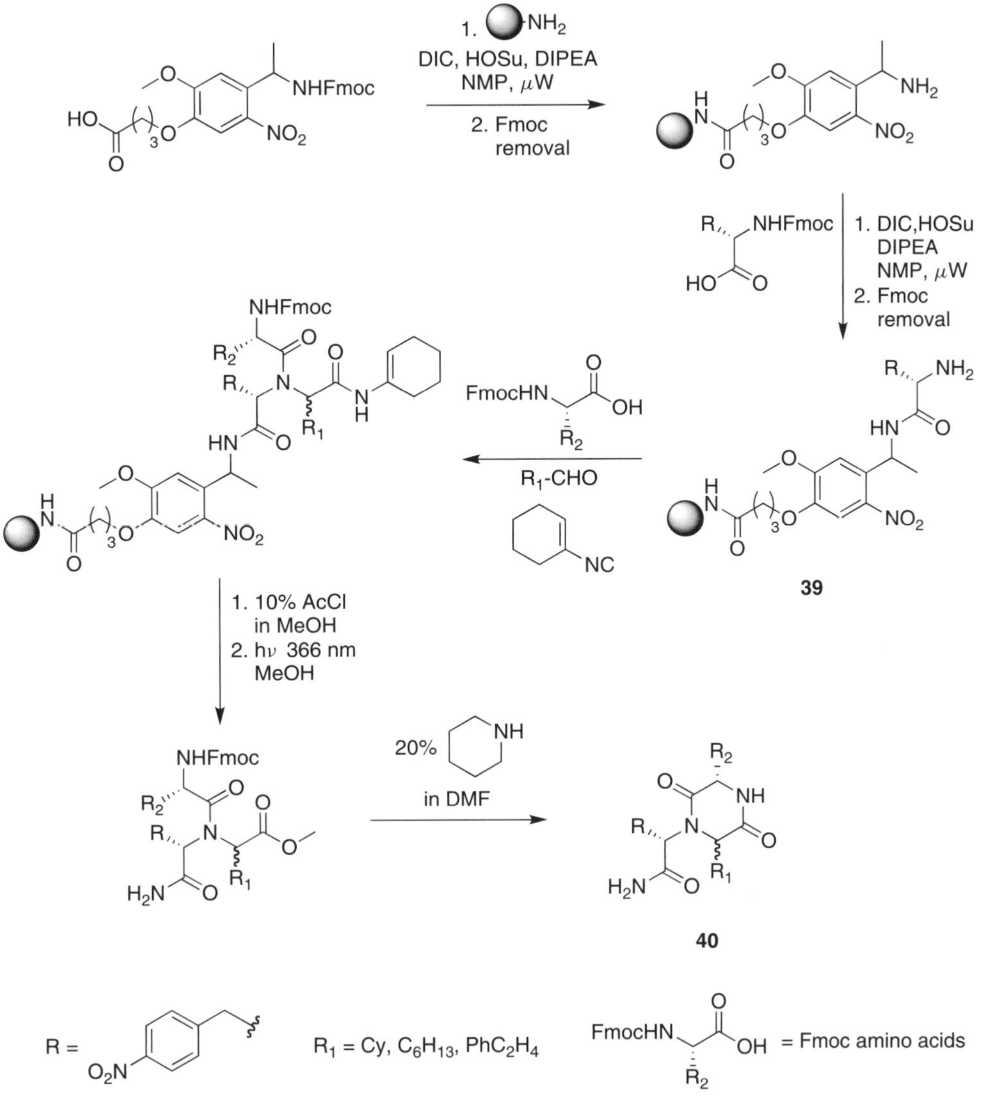

Scheme 5.28 α-Methylated nitroveratryl linker in multicomponent reactions

Typical Experimental Procedures

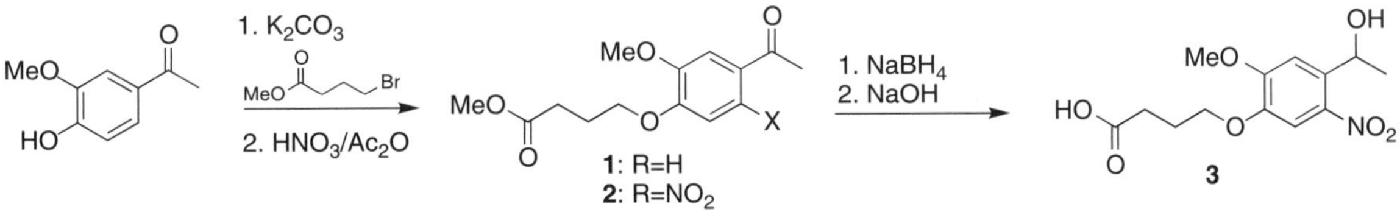

Holmes, C. P., *J. Org. Chem.* **1997**, *62*, 2370-2380.

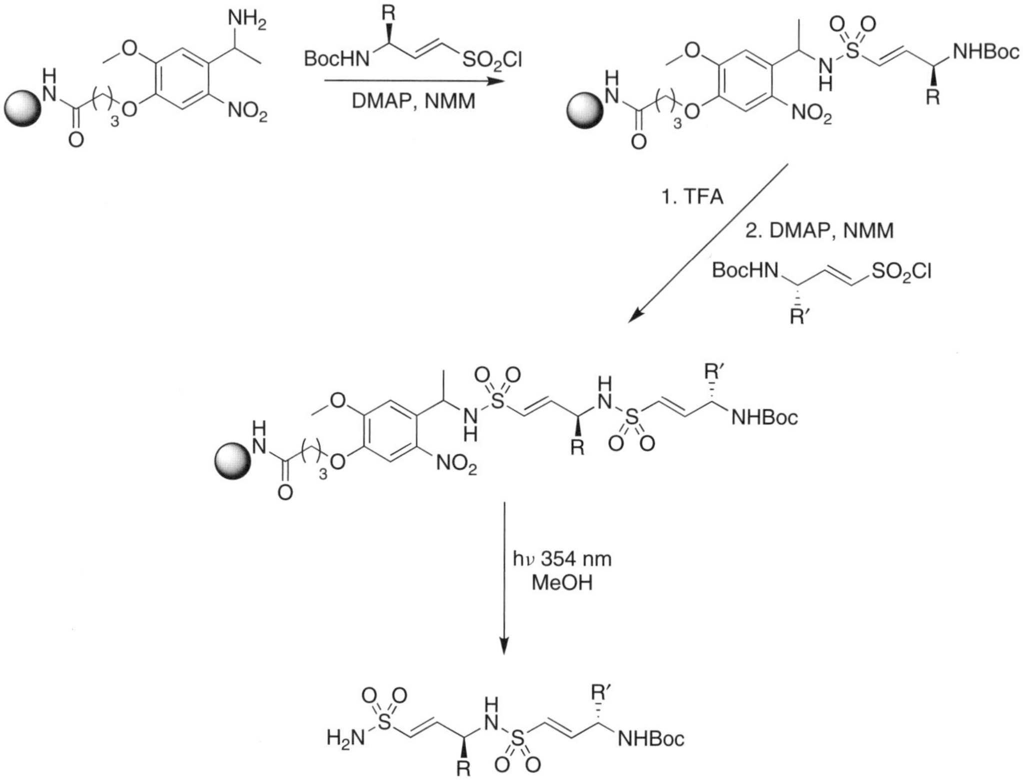

172 Traditional Linker Units for Solid-Phase Organic Synthesis

Scheme 5.29 *α-Methylated nitroveratryl linker bound as sulfonamide*

Methyl 4-(4-Acetyl-2-methoxyphenoxy)butanoate 1

*A slurry of acetovanillone (41.00 g, 246.7 mmol), methyl 4-bromobutyrate (49.63 g, 274.1 mmol), and potassium carbonate (51.1 g, 370 mmol) in 200 ml of DMF was stirred at room temperature for 16 hours. Water was added to the reaction mixture until all the salts were dissolved, and the solution was partitioned between ethyl acetate and saturated sodium chloride. The organic phase was dried (magnesium sulfate), filtered and evaporated to dryness to afford 67.90 g (100% crude yield) of product keto-ester **1** as a colorless oil which slowly solidified: melting point = 48–49 °C.*

Methyl 4-(4-Acetyl-2-methoxy-5-nitrophenoxy)butanoate 2

*A solution of keto-ester **1** (5.00 g, 18.8 mmol) in 15 ml of acetic anhydride was slowly added to a solution of 100 ml of 70% nitric acid and 20 ml of acetic anhydride at 0 °C. After stirring for 2.5 hours the reaction mixture was poured into water and chilled to 4 °C overnight. The precipitate was collected by filtration and washed extensively with water. Recrystallization from methanol/water afforded the nitro-ester **2** as yellow crystals (3.87 g, 66% yield): melting point = 112–113 °C.*

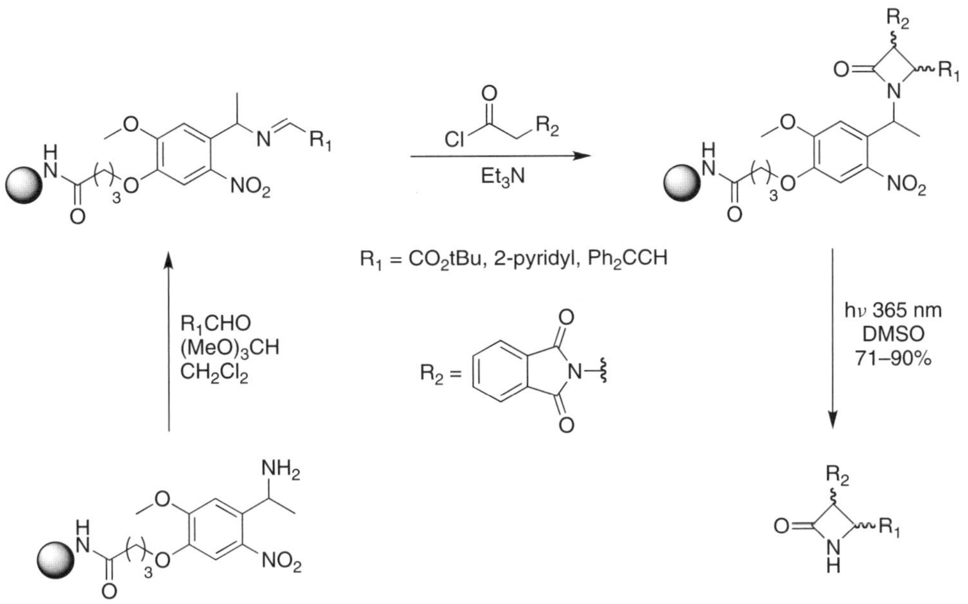

Scheme 5.30 α-Methylated nitroveratryl linker in β-lactam synthesis

4-(4-(1-Hydroxyethyl)-2-methoxy-5-nitrophenoxy)butanoic Acid 3

*To a solution of ester **2** (24.19 g, 77.72 mmol) in 400 ml of THF and 200 ml of methanol was added sodium borohydride (NaBH$_4$) (3.61 g, 95.4 mmol) at room temperature. An additional 1 g (26.4 mol) of sodium borohydride was added after stirring for one hour. The reaction mixture was stirred for 15 hours at room temperature, by which time TLC indicated that complete reduction had taken place. 1 N sodium hydroxide (200 ml, 0.200 mmol) and water (100 ml) were added and the reaction stirred for seven hours. The reaction mixture was concentrated under reduced pressure to approximately 400 ml, carefully acidified with 6 N hydrochloric acid, and then extracted with ethyl acetate. The combined organic phase was dried (magnesium sulfate) and evaporated to give a yellow residue. Recrystallization from ethyl acetate/hexane afforded photolinker **3** (16.61 g, 71% yield) as a pale yellow solid: melting point = 164–168 °C (decomposition).*

5.6 Linkers based on the phenacyl group

Already during the early exploration of photolabile linkers, the phenacyl group was introduced by Wang.[63] This type of linker is very easily prepared, by simple Friedel–Crafts acylation of polystyrene-divinylbenzene copolymer. In this case, it can be considered more as a functionalized resin. A first amino acid is coupled by displacement of the bromide with its cesium salt; the rest of the peptide is grown in a classical manner. The final release was effected by photolysis at 350 nm in DMF during 72 hours, in a 70% overall yield (Scheme 5.32). It was observed that the photolysis is two to five times slower if the first amino acid is alanine, leucine, threonine or isoleucine.

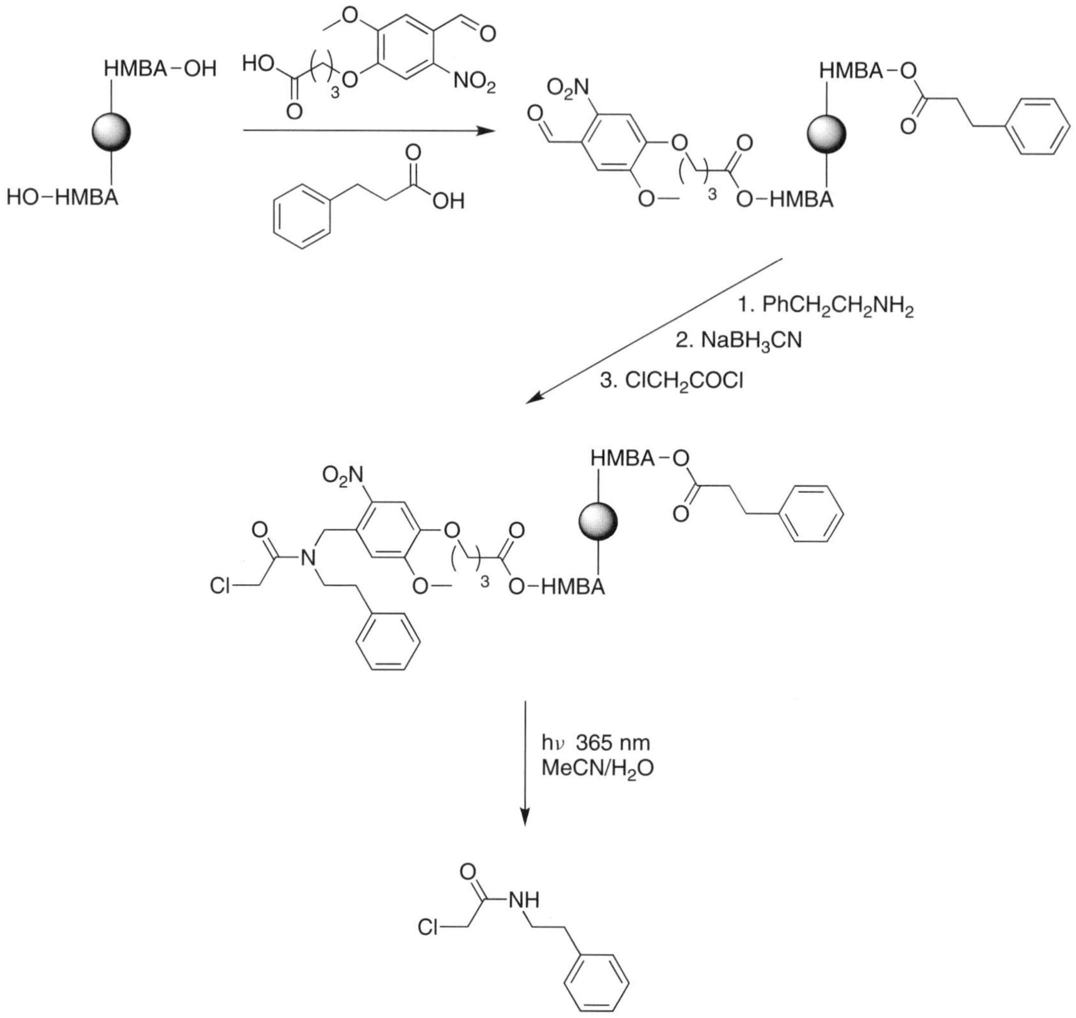

Scheme 5.31 Analytical marker on the resin

As for the *o*-nitrobenzyl resins, an α-methylphenacyl linker was used by Tam in multi-detachable resins;[11, 12] it can be cleaved both photochemically and under basic or nucleophilic conditions (Scheme 5.33). An additional acid labile linker is inserted, as an ester or an ether.[31]

Tjoeng proposed a very similar support, but this time as a real resin plus linker system; its advantage is a better control of the substitution level of the resin.[64] Thus, phenylacetic acid is esterified, then substituted in the *para* position by a Friedel–Crafts acylation. Saponification and activation is followed by the attachment to an amino resin (polystyrene/divinylbenzene) leading to the support **43**, ready for peptide synthesis. Final release is carried out in 27 hours at 350 nm in DMF using a Rayonet reactor, with isolated yields up to 78% (Scheme 5.34).

This linker has also been used with a PEG-type support, bound through an ester.[65] The preparation of the latter linker is very similar to the former, the attachment with the PEG being carried out with DCC. Final photorelease after the peptide synthesis gives excellent yield.

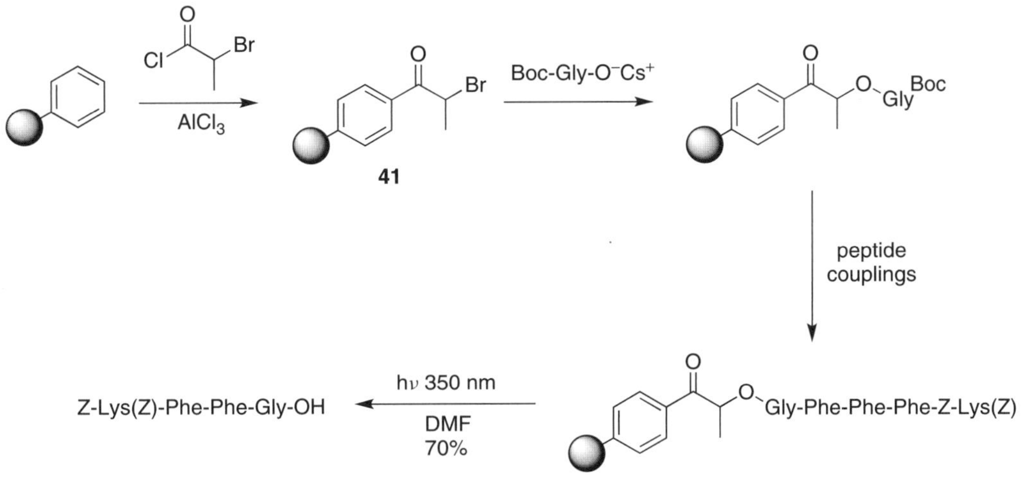

Scheme 5.32 Phenacyl photolabile resin

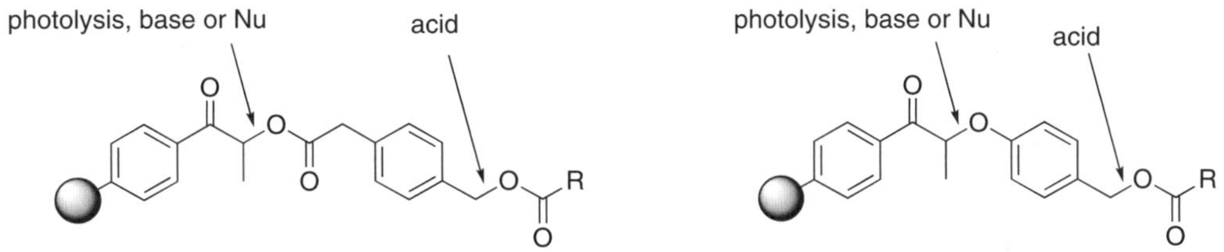

Scheme 5.33 α-Methylphenacyl linkers

Typical Experimental Procedures

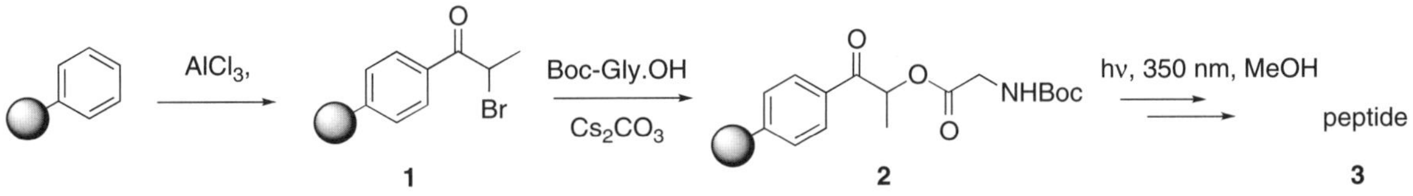

Wang, S.-S., *J. Org. Chem.* **1976**, *41*, 3258-61.

Phenacyl resin 1

2-Bromopropionyl chloride (50 g, 243 mmol) was added slowly to a suspension of aluminum trichloride (39 g) in 250 ml of dichloromethane with gentle stirring. The solid dissolved after a brief period, forming a light brown solution. It was cautiously added to a suspension of Bio-Beads S-X2 (216 g, 200–400 mesh) in 2.2 l of dichloromethane during a period of approximately 30 minutes. The mixture was stirred for an additional 17 hours. The acylated resin thus obtained was collected and washed successively with dichloromethane, nitrobenzene and THF. The slightly brownish resin was stirred in a mixture of THF/water (6 l, 2:1) for 30 minutes and collected by filtration. The operation was repeated twice more and the resin

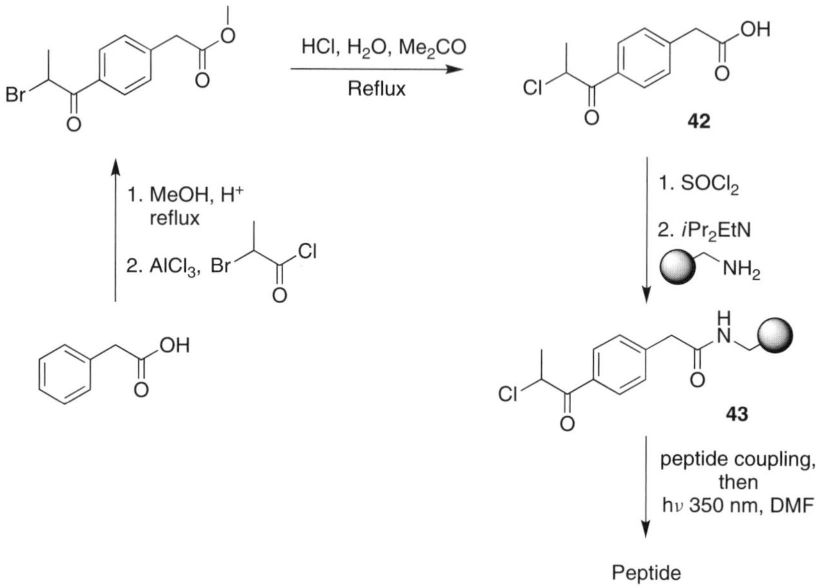

176   Traditional Linker Units for Solid-Phase Organic Synthesis

Scheme 5.34 *α-Methylphenacyl linkers with an additional spacer*

again washed with water, THF and then methanol to give 248.3 g of light buff colored material: IR (KBr) 1685 cm^{-1}; Br, 7.50 (0.94 mmol/g); Cl, 0.06.

N-Boc-glycine-phenacyl resin 2

*Boc-Gly-OH (4.38 g, 25 mmol) was dissolved in a mixture of 40 ml of ethanol and 10 ml of water. The solution was titrated to pH 7.0 with 20% cesium carbonate. The mixture was evaporated to dryness (35 °C) and the residual solid was evaporated twice with fresh DMF. Boc-Gly-O$^-$ Cs$^+$ thus obtained was stirred with 20 g of **1** (18.8 mmol) in 80 ml of DMF for 17 hours. The esterified resin was then collected and washed successively with DMF, DMF–water, water, THF–water, THF and methanol to give 20.5 g of the desired product **2**. Amino acid analysis indicated the presence of 0.57 mmol Gly/g; Br, 0.11% (0.014 mmol/g); IR (potassium bromide) 1750, 1725, 1685 cm^{-1}. Similarly prepared were the resin analogues of Boc-Ala-OH (0.60 mmol Ala/g; 0.08% Br); Boc-Leu-OH (0.58 mmol Leu/g; 0.13% Br) and Boc-Thr(Bzl)-OH (0.62 mmol Thr/g; 0.07% Br).*

Photolytic removal from the resin

As described in Section 5.2.

5.7 Linkers based on the para-methoxyphenacyl group

Originally developed in 1973 by Sheehan as a photolabile moiety, the *p*-methoxyphenacyl group was introduced into a linker by Mutter in 1985.[66] It is very similar to the phenacyl group described above, but the presence of an electron releasing methoxy group in the *para* position induces a bathochromic shift, making the photolysis at 350 nm much more efficient, thus shortening the irradiation times. The release of a single amino acid can be performed in various solvents (ethanol, DMF, dioxane) with yields up to 80% after 10 hours. Furthermore, this linker is acid stable. Its preparation starts with phenoxyacetic acid, which is derivatized in a similar fashion as shown above. The first amino acid is introduced either in the presence of potassium fluoride, or via its cesium salt. Final photolysis at 350 nm releases the peptide; in the case of a pentapeptide, the overall yield was 71% (Scheme 5.35). This linker was used in longer peptides (nine amino acids), as shown by Gauthier, with a phenylacetamide (PAM)-type resin.[67] In this case, the overall yield after photolysis was higher than 85%.

Based on this photolabile unit, Belshaw developed a safety-catch linker, bearing two *meta*-methoxy groups.[68] The strategy is to mask the ketone function as a dimethyl acetal, thus making the linker photostable; the photoreactivity can be restored by acidic hydrolysis which re-forms the ketone (Scheme 5.36). Such protection is essential, because another photolabile group, 3′-nitrophenylpropyoxycarbonyl (NPPOC), is used as a protecting group. It was used in light-driven oligonucleotide synthesis. The preparation of such a linker starts with 3′,5′-dimethoxy-4′-hydroxyacetophenone, etherifying the 4′-hydroxyl group with monosilylated ethyleneglycol under Mitsunobu conditions. Acetalization and oxidation of the methyl group with an hypervalent iodine compound were followed by the protection of the primary alcohol with the photolabile NPPOC chloroformiate and deprotection activation of the spacer for the coupling to the support, in this case a glass slide (Scheme 5.37).

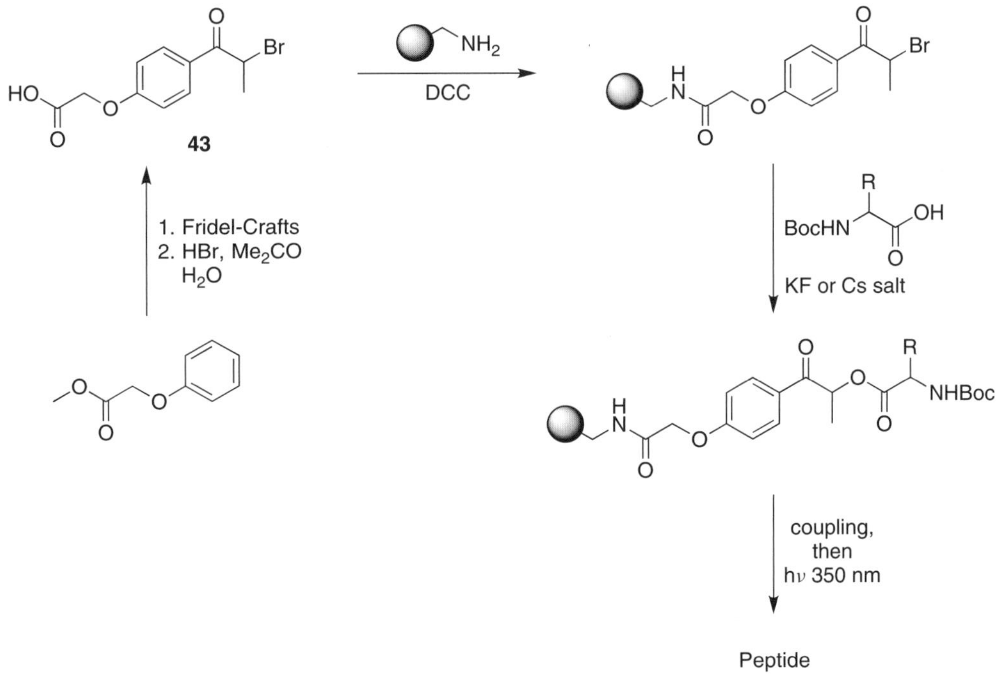

Scheme 5.35 Para-*methoxy-α-methylphenacyl linkers*

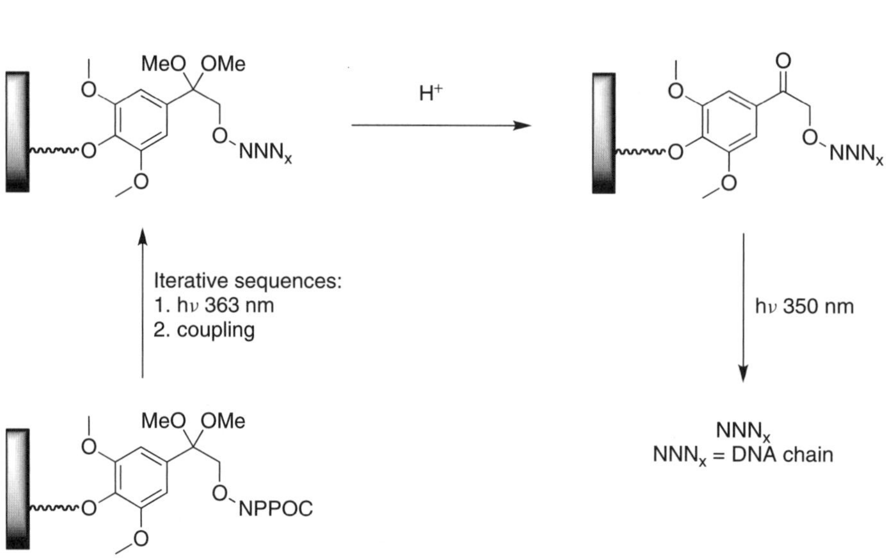

Scheme 5.36 Trimethoxyphenacyl safety-catch linker

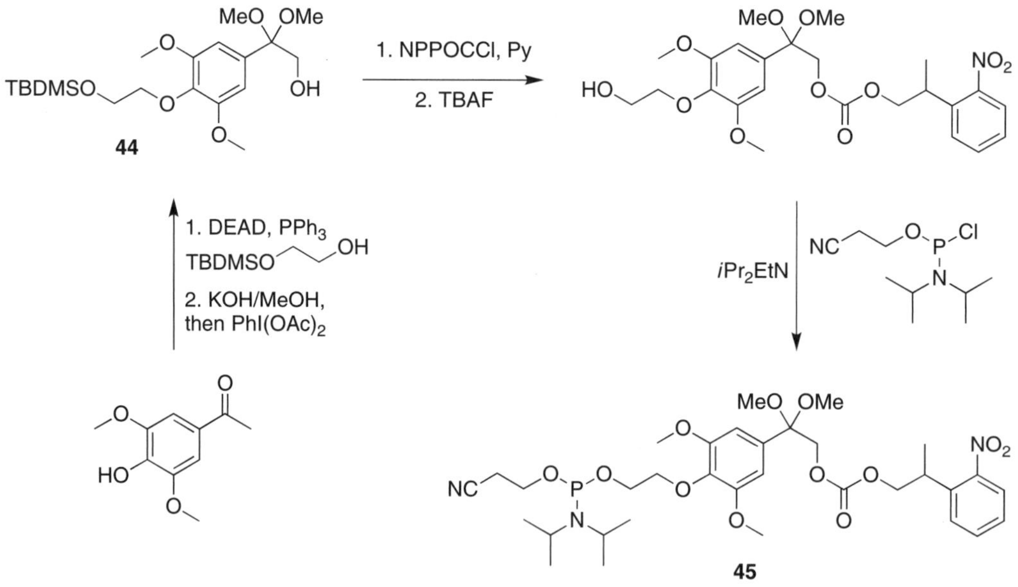

Scheme 5.37 Use of the safety-catch feature in successive photolytic deprotections

Typical Experimental Procedures

Flickinger, S. T.; Patel, M.; Binkowski, B. F.; Lowe, A. M.; Li, M.-H.; Kim, C.; Cerrina, F.; Belshaw, P. J., *Org. Lett.* **2006**, *8*, 2357-2360.

1-{4-[2-(tert-Butyl-dimethyl-silanyloxy)-ethoxy]-3,5-dimethoxy-phenyl}-ethanone 2

7.71 g **1** *(2 equiv., 44 mmol) was dried by azeotroping three times with freshly distilled toluene. Then a flame-dried flask with a stirring bar, under a nitrogen atmosphere, was charged with 4.288 g* **1** *(1 equiv, 21.9 mmol),* **6**, *and 6.89 g triphenyl phosphine (1.2 equiv., 26 mmol). These were dissolved in 60 ml of freshly distilled THF, and cooled to 0 °C. 4.14 ml of diethylazodicarboxylate (DEAD) (1.2 equiv., 26 mmol) was then added dropwise. The reaction mixture was slowly allowed to warm to room temperature. After 4.5 hours the reaction mixture was concentrated in vacuo to yield a yellow solid, which was taken up in ethyl acetate. This was then washed sequentially with 10% acetic acid, brine, saturated sodium bicarbonate (twice) and brine. The organic layer was dried over anhydrous sodium sulfate, filtered and concentrated to yield 12.91 g of oil mixed with a white solid (presumably O = triphenyl phosphine). This was purified by flash column chromatography, using a solvent gradient starting with hexanes up to 30% ethyl acetate in hexanes, to yield 6.72 g (87%) of an oil with a slight yellow tint. Caution: This compound is heat sensitive. It degraded on us several times while concentrating in vacuo with heating.*

2-{4-[2-(tert-Butyl-dimethyl-silanyloxy)-ethoxy]-3,5-dimethoxy-phenyl}-2,2-dimethoxy-ethanol 3

The procedure was based on Moriarty's methodology.[69] *Specifically, the procedure given by Luis and Andrés*[70] *was followed, with the following exceptions. Firstly, the reaction was run under anhydrous conditions, and* **2** *was dried by azeotroping three times with freshly distilled toluene. In addition, the reaction was not treated with hydrochloric acid after stirring overnight, so that the dimethyl ketal could be isolated instead of the ketone. Reaction with 1.0 g of* **2** *(1 equiv., 2.82 mmol) yielded 0.9853 g of white solid* **3** *(84%). Caution: This compound is heat sensitive. It degraded on us several times while concentrating in vacuo with heating.*

Carbonic acid 2-[4-(2-hydroxy-ethoxy)-3,5-dimethoxy-phenyl]-2,2-dimethoxy-ethyl ester 2-(2-nitro-phenyl)-propyl ester 4

6.48 g (1 equiv., 16 mmol) of **3** *was co-evaporated three times with freshly distilled pyridine. This was then added to a flame-dried flask, with a stirring bar and dropping funnel, in 85 ml of freshly distilled pyridine.*

This solution was cooled to −15 °C with an ice/methanol bath. 7.6 g (2 equiv., 31 mmol) of NPPoc-Cl (prepared from the corresponding alcohol according to the procedure described elsewhere).[71] 40 ml of freshly distilled DCM was then added drop-wise with stirring. The reaction was left to stir overnight while slowly warming to room temperature. The reaction was then concentrated in vacuo and re-dissolved in EtOAc. This was then washed with a saturated sodium bicarbonate solution followed by brine. The organic layer was dried over anhydrous sodium sulfate, filtered and concentrated. The crude material was dried by azeotroping three times with freshly distilled toluene. The flask was then flushed with nitrogen and 10 ml of freshly distilled THF was then added, followed by 16 ml (1 equiv., 16 mmol) of tetrabutylammonium fluoride (TBAF). After stirring for 15 minutes the reaction was diluted with saturated sodium bicarbonate solution, followed by extraction with EtOAc. The organic layer was then washed with brine, dried over anhydrous sodium sulfate, filtered and concentrated to yield 15.59 g of dark colored oil. This was purified by flash column chromatography, using a solvent gradient starting with 50% ethyl acetate in hexanes up to 65% ethyl acetate in hexanes, to yield 7.07 g (89% over three steps) of dark orange oil **4**.

Carbonic acid 2-(4-{2-[(2-cyano-ethoxy)-diisopropylamino-phosphanyloxy]-ethoxy}-3, 5-dimethoxy-phenyl)-2,2-dimethoxy-ethyl ester 2-(2-nitro-phenyl)-propyl ester 5

The protocol used was essentially that described in Gait.[72] All DCM used in the preparation of **5** was freshly distilled and passed through a short plug of basic alumina before use. 1 g of **4** (1 equiv., 1.96 mmol) was placed in a flame-dried flask with a stirring bar and dried by co-evaporation two times with freshly distilled DCM:pyridine (9:1). The flask was then sealed with a rubber septum, pierced with a needle and placed in a vacuum desiccator overnight. The following morning the desiccator was released from vacuum under an inert atmosphere (preferably argon), the needle removed and an inert gas line added. Then 684 μl of DIPEA (4 equiv., 3.92 mmol) was added to the flask followed by 2 ml of DCM. Meanwhile, the phosphitylating reagent was allowed to warm to room temperature for 30 minutes in a desiccator. A dry syringe was flushed with the phosphitylating reagent and then 348 mg (1.5 equiv., 1.47 mmol; added as solution in DCM) of the phosphitylating reagent was charged to the flask containing the reaction mixture. The reaction was stirred at room temperature for one hour, after which time it was diluted with a small amount of hexanes, and loaded directly onto a flash column. The flash column was prepared by wet packing the silica in ethyl acetate:hexanes:TEA (18:72:10). The crude material was then purified using a solvent gradient starting from ethyl acetate:hexanes:TEA (18:72:10) up to ethyl acetate:hexanes:TEA (45:45:10). 1.20 g (86%) of yellow oil **5** was recovered. Pure **5** was then dissolved in 60 ml of freshly distilled acetonitrile which had been filtered through a plug of basic alumina. This was divided into four equal volumes of 15 ml each and aliquoted into oven dried bottles with 4Å MS for use on the AGS instrument. Bottles not currently used were sealed with parafilm and stored in a desiccator at −20 °C.

5.8 Linkers based on the benzoin group

Benzoins are well-known photolabile protecting groups, and a safety-catch version based on the strategy discussed above was used by Chan.[73] In this case, it is literally a linker, not to bind a residue to a support, but really to bind together two parts of a compound. The acid **46** was thus prepared in a few simple steps, the most notable feature being the dithioacetal, acting as the masked ketone, its hydrolysis being the safety-catch trigger (Scheme 5.38).

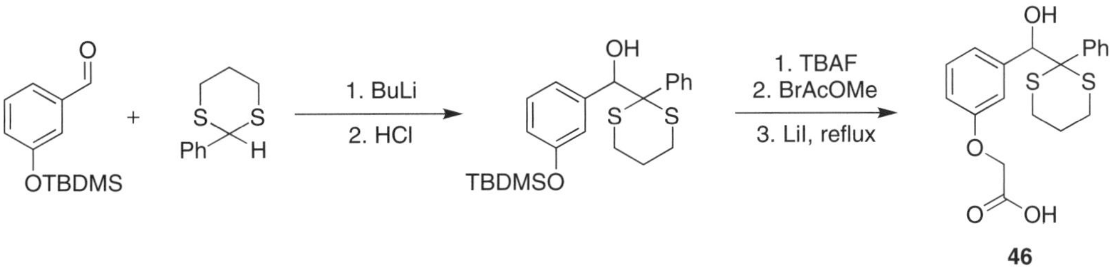

Scheme 5.38 *Safety-catch benzoin linker precursor*

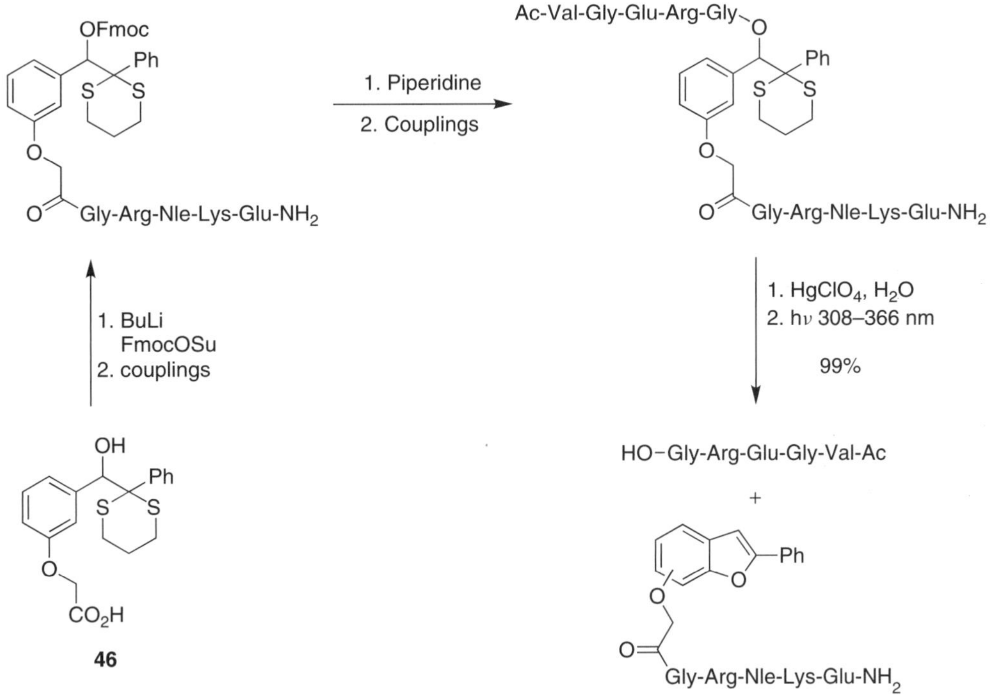

Scheme 5.39 *Use of the benzoin safety-catch linker in solution-phase synthesis*

In a first phase, the alcohol function of **46** is protected and amino acids are coupled to the carboxylic terminus. Deprotection and further amino acids couplings grows a second peptidic chain. Hydrolysis of the dithioacetal with mercuric perchlorate (activation towards photolysis) and irradiation between 308 and 366 nm for a very short time (10 minutes) liberated the pentapeptide and the two benzofurane side products (Scheme 5.39).

Later, Balasubramanian exploited this type of linker in solid-phase syntheses, coupling the 3-hydroxybenzaldehyde to a Merrifield resin, assembling the dithioacetal protected benzoic unit in the

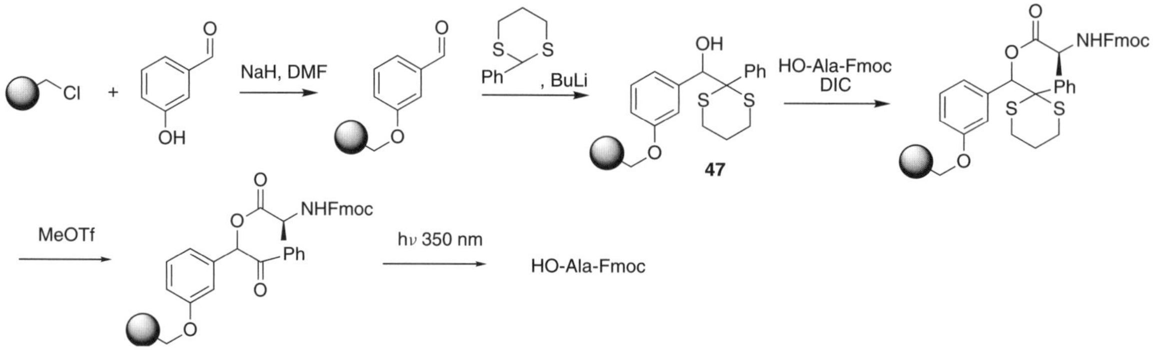

Scheme 5.40 Use of the benzoin safety-catch linker in solid-phase synthesis

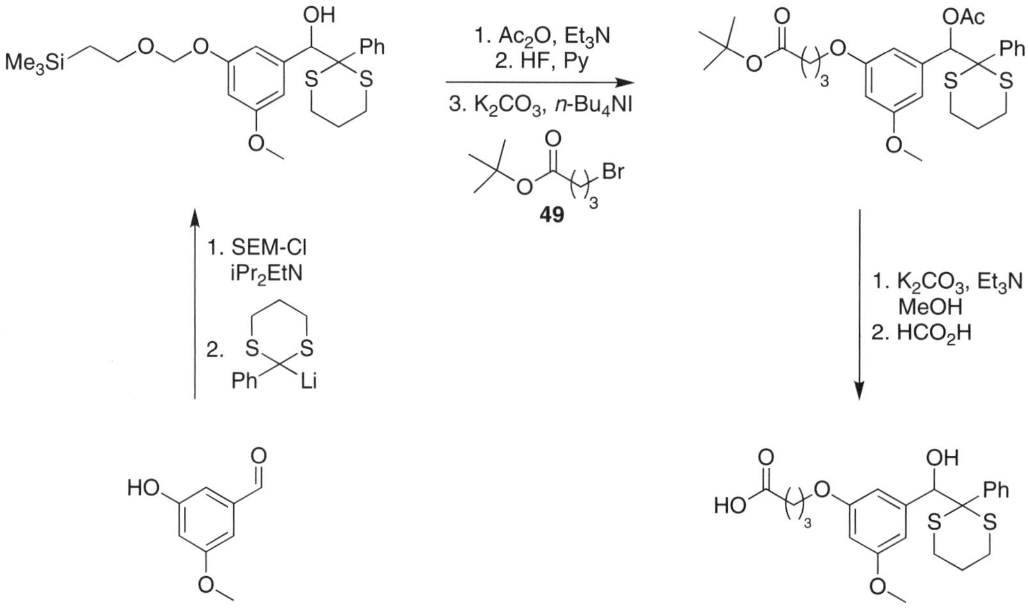

Scheme 5.41 Safety-catch benzoin linker with an additional spacer

same way as above.[74, 75] Final photolysis liberated the substrates in good to excellent yields (75–97%), with irradiation times of two hours at 350 nm in a THF/methanol mixture (Scheme 5.40). An additional methoxy group was added in the *meta* position and similar performances were recorded.

A slightly improved version was presented a few years later, with a TentaGel support and a longer spacer to prevent attacks on the α carbon (Scheme 5.41). [76]

Typical Experimental Procedures

Lee, H. B.; Balasubramanian, S., *J. Org. Chem.* **1999**, *64*, 3454-3460.

Resin-bound benzaldehyde 1

Sodium hydride (0.24 g, 6 mmol, 3 equiv.) was added in one portion to a stirred solution of 3-hydroxybenzaldehyde (0.70 g, 5.7 mmol, 285 mol %) in anhydrous DMF (10 ml) under argon at 0 °C, and stirring was continued for one hour at room temperature. This solution was syringed to a suspension of Merrifield resin (2 g, 2 mmol, 1 equiv.) in anhydrous DMF (10 ml) and the mixture shaken under argon for 12 hours at 25 °C. The resin was filtered, washed and dried under vacuum to give the resin-bound aldehyde **1**: IR 1697 cm^{-1}.

Resin-bound dithiane protected benzoin 2

To a stirred solution of 2-phenyl-1,3-dithiane (0.78 g, 4 mmol, 4 equiv.) in anhydrous THF (4 ml) under argon at −78 °C was added n-butyllithium (1.76 ml, 3.5 mmol, 3.5 equiv.), and stirring was continued for 10 minutes at 0 °C. This solution was transferred via a cannula to a suspension of aldehyde resin **1** (1.0 g, 1 mmol, 1 equiv.) cooled to 0 °C in THF (6 ml) and the mixture was gently stirred at 0 °C. The aldehyde IR peak was completely lost after stirring for one hour. The resin suspension was quenched with dilute 1 N hydrochloric acid (5 ml) and the resin was filtered, washed and dried under vacuum to give the resin-bound dithiane protected benzoin **2**: loading 0.80 mmol/g (99%); IR 3583, 3445 cm^{-1}.

General fmoc protected alanine loading procedure 3

Diisopropylcarbodiimide (94 μl, 0.6 mmol, 3 equiv.) followed by diisopropylethylamine (104 μl, 0.6 mmol, 3 equiv.) was syringed into a solution of dithiane protected benzoin resin **2** (200 mg, 0.2 mmol, 1 equiv.), Fmoc protected alanine (168 mg, 0.6 mmol. 3 equiv.), N.N-dimethylaminopyridine (10 mg) and HOBt (10 mg) in anhydrous DMF (2 ml) under argon at room temperature, and the mixture was gently stirred for five hours. The resin was filtered, washed and dried under vacuum to give the resin-bound Fmoc protected amino acid **3**. Loading was determined by Fmoc analysis of **3**.

General procedure for dithiane deprotection of 3 using methyl triflate

Methyl triflate (136 μl, 0.96 mmol, 10 equiv.) was added dropwise to a suspension of the Fmoc-alanine loaded resin **8a** (120 mg, 0.1 mmol, 100 mol %) in anhydrous DCM (1 ml), and the mixture was gently stirred at room temperature. The dithiane ^{13}C NMR signals were virtually lost after 20 minutes. The resin was filtered, washed and dried under vacuum to give the resin-bound dithiane deprotected benzoin **4**. Found: N, 1.0; S, 0.9 (79%, 0.54 mmol/g of **4**). Subjecting **4** to the same deprotecting conditions yielded a further 5–10% reduction in S content.

General photolysis procedure (5)

Dithiane deprotected resin **4** (20 mg) and Fmoc-serine (internal standard, 3 mg) in 3:1 distilled THF/methanol (12 ml) in a quartz test tube was purged with a stream of nitrogen gas for 10 minutes. The suspension was UV irradiated for two ours with constant nitrogen bubbling to yield the benzofuranyl resin and Fmoc protected alanine in solution. Twenty-five μl of solution were removed at various time intervals and analyzed by HPLC using UV detection at 254 nm. A ^1H NMR spectrum of the cleaved Fmoc protected alanine was identical to an authentic sample.

5.9 Linkers based on the pivaloyl group

Giese developed a new type of linker based on the Norrish-type I photoreactivity of *tert*-butylketones. It is used not only in the solid-phase synthesis of peptides but also in adducts arising from Stille, Suzuki or epoxidation reactions, with very high photolysis yields.[77] The linker **51** is prepared in seven steps from the 1,3-dixydroxyacetone dimer **52** with a 40% overall yield (Scheme 5.42).

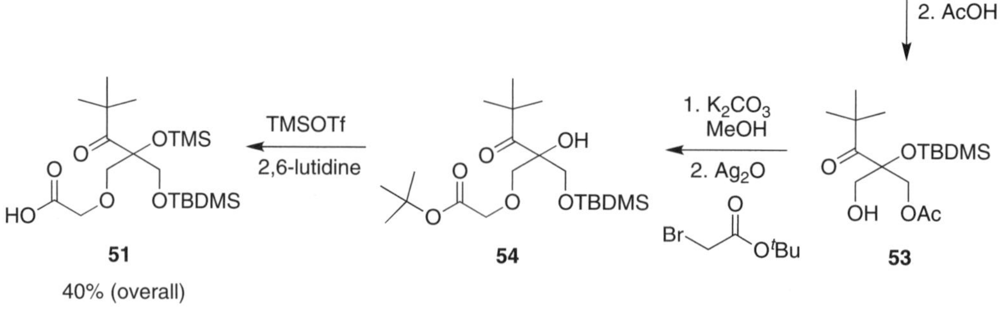

Scheme 5.42 *Pivaloyl-based photolabile protecting group*

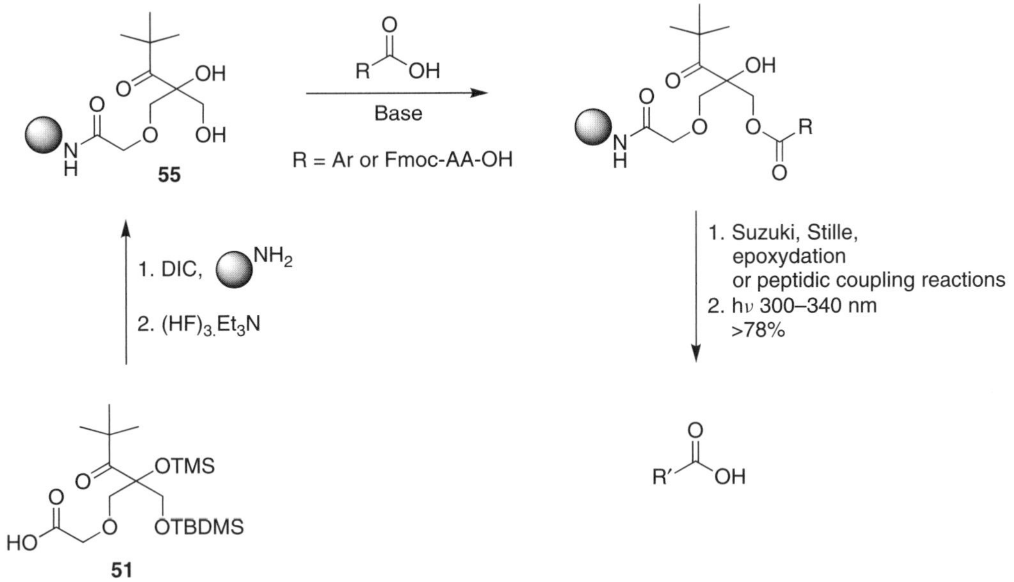

Scheme 5.43 *Pivaloyl-based photolabile linker for carboxylic acids*

The acid **51** is then attached to a TentaGel support by activation of the acid with diisopropylcarbodiimide (DIC) and the residues are then coupled to the free primary alcohol (without notable interference of the quite unreactive tertiary alcohol). Final release can be carried out at 300–340 nm in high yields in many solvents, except in acetonitrile (Scheme 5.43). This linker is stable under numerous acidic and basic conditions basiques (HCl 1 M, TFA, *p*-TsOH 80 °C, AcOH/H_2O 60 °C, BF_3.OEt_2, $^iPr_2EtN$ 60 °C, DBU 80 °C, KHMDS −78 °C). To a certain extent, it is also photostable at longer wavelengths (420 nm), and this property was exploited in a two-wavelength-based peptide synthesis.[78]

Two years later, an ether-based modified version was proposed, able to release directly alcohols.[79] Thus, the protected ethers **60** are prepared by nucleophilic attack of an epoxide precursor by the relevant alcohol, which is then released by photolysis at 300 nm for three hours, with yields ranging from 60 to 80% (Scheme 5.44).

This linker is stable in acidic consitions (TFA, *p*-TsOH, BF_3.OEt_2), basic conditions (^tBuOK, $^iPr_2EtN$) and reducing conditions ($LiAlH_4$).

Typical Experimental Procedures

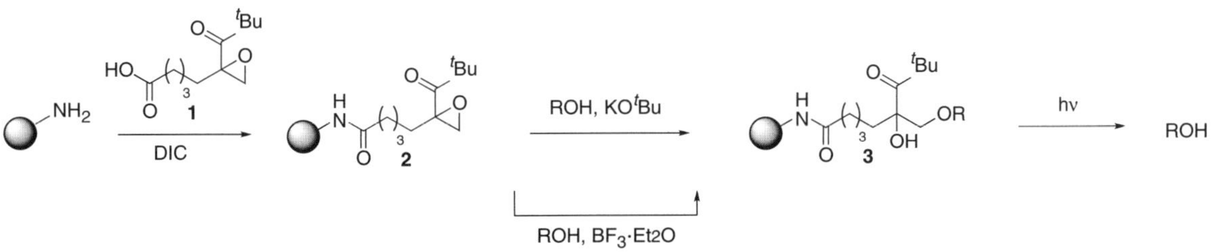

Glatthar, R.; Giese, B. *Org. Lett.* **2000**, *2*, 2315-2317.

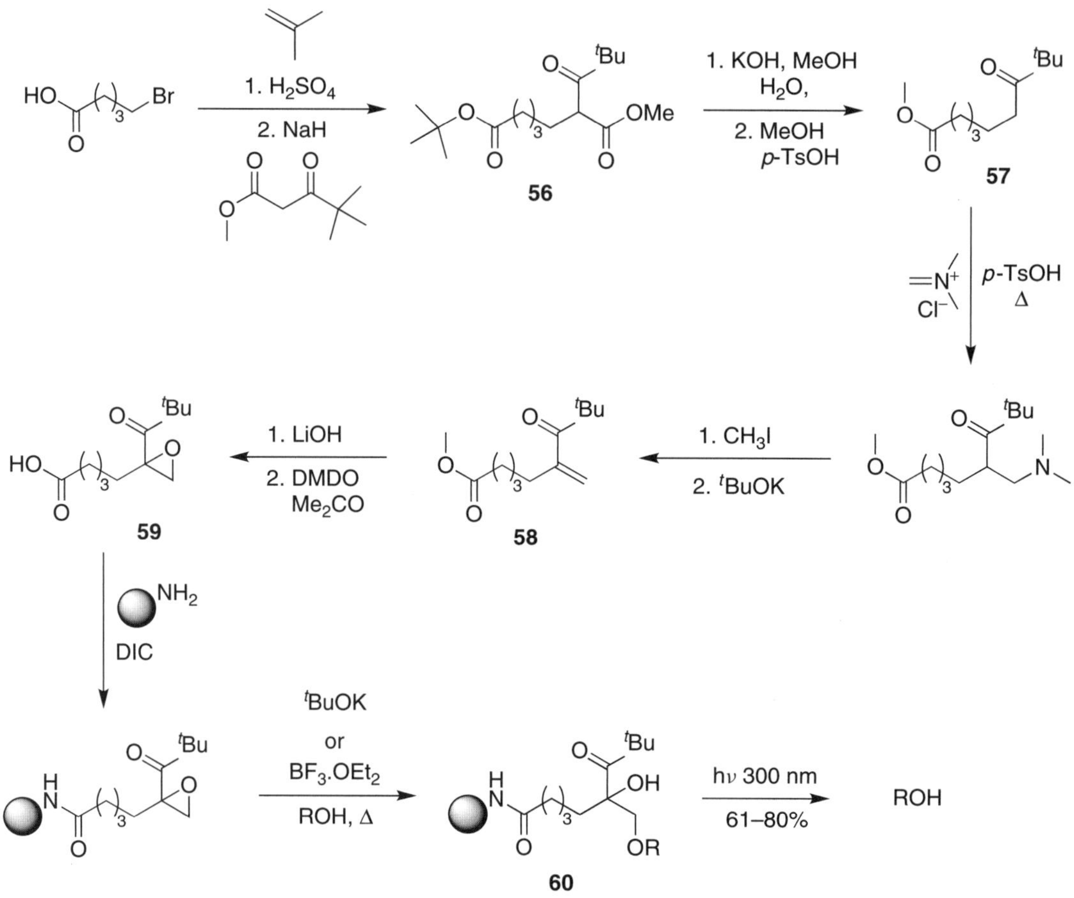

Scheme 5.44 Pivaloyl-based photolabile linker for alcohols

Photolinker-solid support 2

*TentaGel S NH$_2$ (1.00 g, 0.28 mmol/g loading) or polystyrene NHMe (1% DVB) (215 mg, 1.3 mmol/g loading) was suspended in dry DCM (5 ml) and photolinker **1** (96.0 mg, 420 mmol, 1.5 equiv.), DMAP (9 mg, 0.07 mmol, 0.25 equiv.) and diisopropylcarbodiimide (DIC, 110 µl, 0.7 mmol, 2.5 equiv.) were added. The resin was shaken for 18 hours at room temperature and washed with DCM, DMD, DCM and dried to yield the photolabile support **2**. Kaiser test revealed complete reaction.*

General procedure for base induced epoxide opening

*To a suspension of photolinker resin **2** (~100 mg, PS or TentaGel) in THF (5 ml) were added the alcohol (20 equiv.) and potassium-tert-butoxide (10 equiv.) and the mixture was shaken for 15 hours at 60 °C. Washing*

(THF, 5% HOAc in methanol, DMF, DCM, THF) after cooling to room temperature and drying afforded the ether-photolinker resin **3**.

General procedure for lewis acid induced epoxide opening

*To a suspension of photolinker resin **2** (~100 mg, PS or TentaGel) in toluene (3 ml) were added the alcohol (6 equiv.) and $BF_3 \cdot Et_2O$ (100 µl) and the mixture was shaken for 2.5 days at 100 °C. Washing (toluene, DMF, THF, DCM) after cooling to room temperature and drying afforded the ether-photolinker **3**.*

General photolysis conditions

*Photolyses in quartz glass cells (equipped with stirring bar) were conducted with 4–10 mg of resin **3**, suspended in 1.5–3 ml of solvent in the beam of a 500 W mercury high-pressure arc lamp fitted with a water filter and a cut-off filter (295 nm) or a photochemical reactor from Rayonet with up to 16 × 21 W lamps of a spectral energy distribution from 370 to 250 nm, with the maximum at 300 nm. The cells were maintained at 20 °C and irradiated horizontally with gentle mixing of the beads by means of a magnetic stirrer. After photolysis, a defined amount of standard (decane for GC analysis or naphthalene for RP-HPLC analysis) was added. After mixing, an aliquot was removed and analyzed by GC or reversed-phase HPLC with detection at 254 nm.*

5.10 Traceless linkers

Traceless linkers release a C–H bond at the former attachment point. The photolabile linker developed by Routledge belongs to this category.[80] It is based on an early observation by Barton that thiohydroxamic esters are photochemically decomposed into alkanes in the presence of hydrogen donors.[81] As an example, 1,3-dimethylindole is attached as N-methylindol-3-acetic acid by standard carboxyl activation, then released photolytically at 350 nm (Scheme 5.45). A drawback of this method is the requirement of a hydrogen donor in stoichiometric amounts, in this example a quite toxic tin hydride.

Sucholeiki proposed a thioether variant of the *p*-methoxyphenacyl, a traceless platform for metal catalyzed cross-coupling reactions.[82, 83] Photolysis liberates the corresponding toluene, in yields ranging from 21 to 58%, depending on the nature of the side products in the mixture (Scheme 5.46). The obvious drawback is the relatively modest yield but, as in the previous case, it is worth pointing out that these are the only examples of traceless linkers, and that yields are not of the highest importance in the context of combinatorial synthesis.

5.11 Other types of photolabile linker units

Nicolaou developed a linker exploiting the photoreactivity of acylated 7-nitroindolines.[84] This family of heterocycles displays an increased electrophilicity of the carboxyl group upon irradiation. Thus, indole derivative **63** was acetylated and reduced to an indoline, nitrated at the 7-position and attached to a resin through a short spacer (Scheme 5.47).

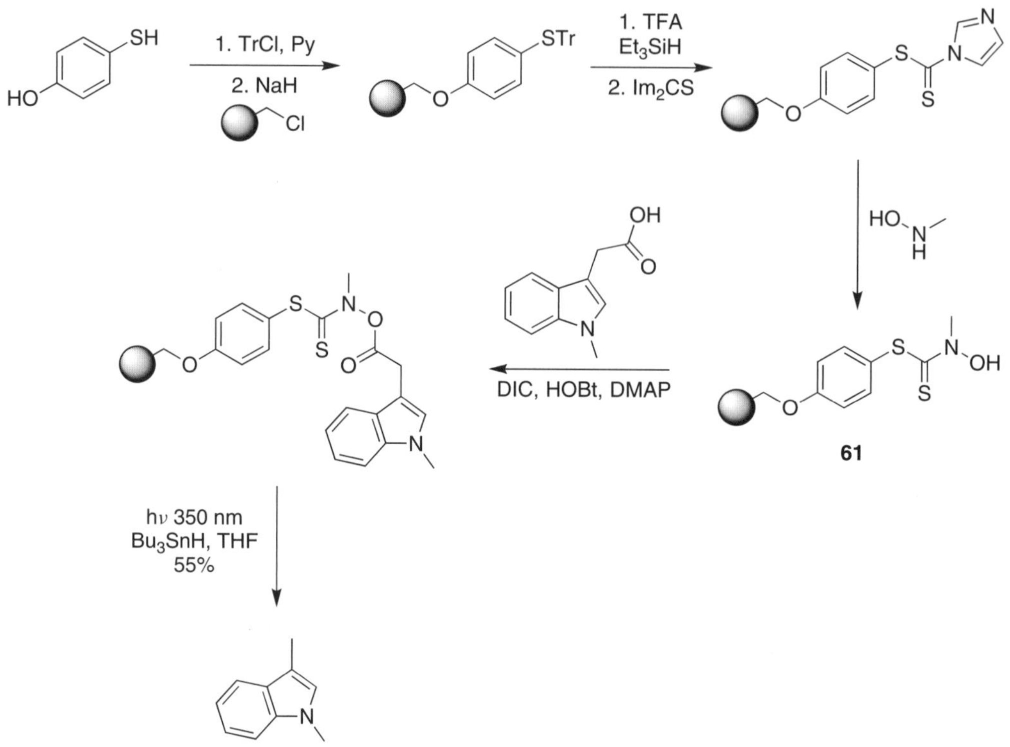

Scheme 5.45 Thiohydroxamic acid-derived traceless linker

The linked resin **65** is then attached to the substrate, which can be modified with all the features of solid-phase synthesis. Photolysis at wavelengths longer than 290 nm in the presence of a nucleophile allows not only the release from the resin, but also the connection to an additional unit (Scheme 5.48).

A unique feature is that, if a peptide is grown on such a support and contains a nucleophilic terminal, the final photolysis product can be trapped intramolecularly, leading to a cyclic peptide. The photolysis yields range between 67% and 95%.

Another original linker was developed by Enders for the release of secondary amines.[85] This phenyl-triazene linker is prepared by binding 3-hydroxyaniline to a Merrifield resin, and diazotation of the aniline nitrogen. Addition of an amine to the diazonium salt leads to a triazene **67**, which fragments upon photolysis at 355 nm, in excellent yields, typically higher than 77% (Scheme 5.49).

Finally, a very original linker was introduced by Kutateladze, allowing the binding of two units, which can be again released photolytically.[86, 87] It is based on the photosensitized fragmentation of 2-hydroxy-1,1-dithioacetals. For example, the linker **69** is twice lithiated and added to a calixarene aldehyde. The linker can then be detached photochemically in the presence of benzophenone (Scheme 5.50).

5.12 Conclusion

It is always easy, but sometimes delicate, to comment on the past from the comfortable standpoint of an observer 40 years after the initial discovery. However, it is unquestionable that the initial work by the

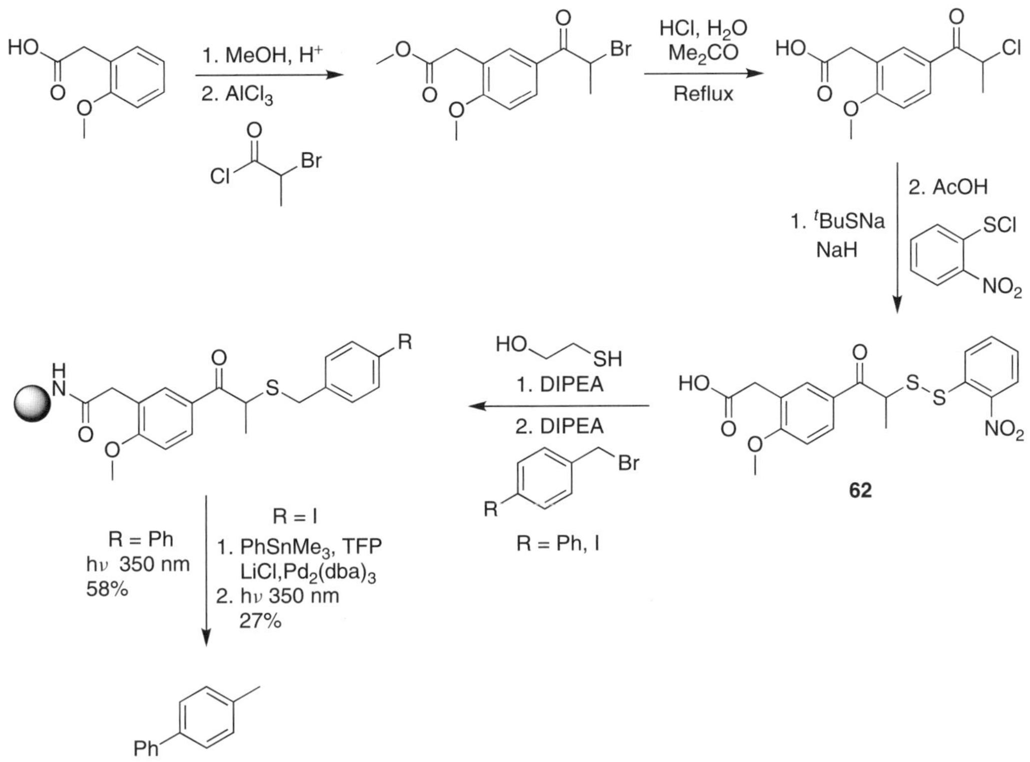

Scheme 5.46 Thioether-based para-methoxyphenacyl traceless linker

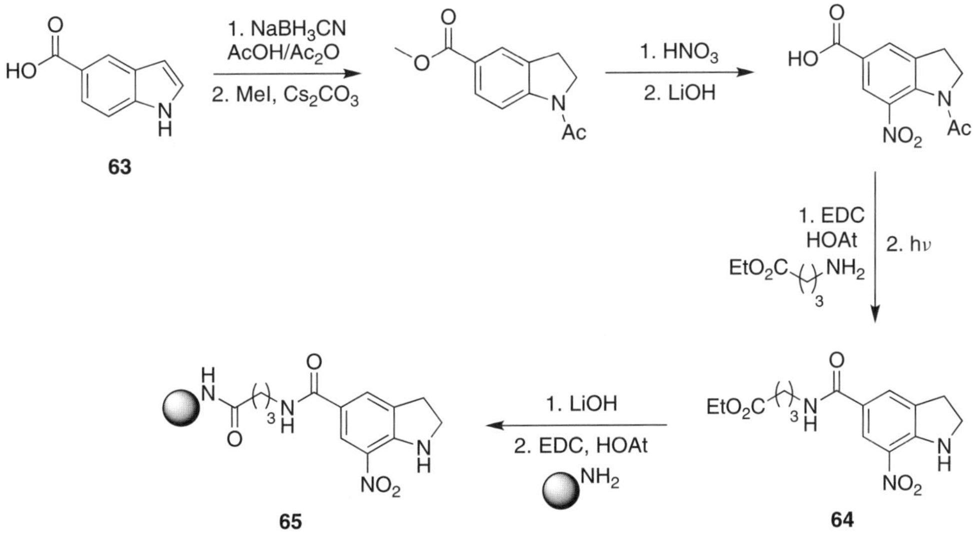

Scheme 5.47 Preparation of a nitroindoline-based linker

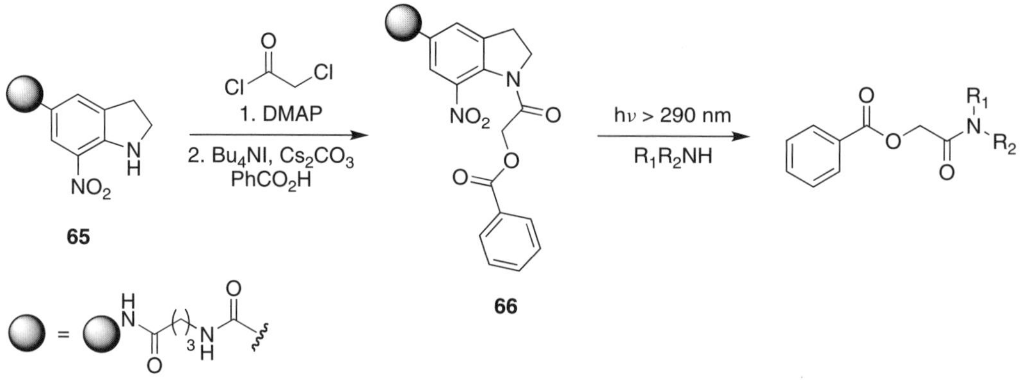

Scheme 5.48 *Use of a nitroindoline-based linker*

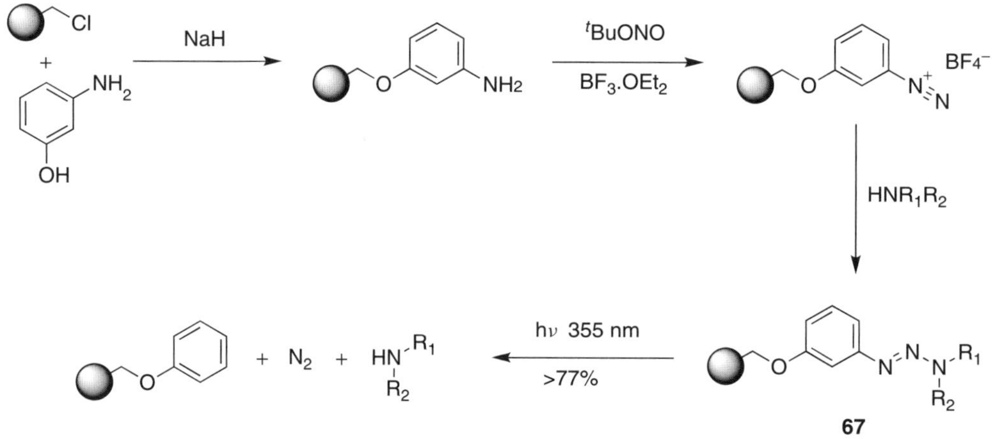

Scheme 5.49 *Triazene-based linker*

pioneers in solid-phase synthesis has undergone a formidable expansion. Peptide synthesis, oligonucleotide synthesis, combinatorial synthesis, high-throughput parallel synthesis have successively fueled an intense research activity in the field; each of these steps has necessitated the development of additional linkers, with tailored properties for specific requirements. One of the driving forces in these developments was the search for efficiency and possibility of automating the processes to the maximum extent. In this perspective, photolabile linkers have become more and more important, because avoiding the addition of a reagent yet still being able to perform a controlled chemical reaction is, by essence, one of the most efficient strategies. As seen from this short overview of available linkers, very significant progress has been made, but it is obvious that the road is still very long to reach the ultimate goal: a linker which gives the organic residue bound to the support a reactivity close to the solution phase, which is cleaved selectively in a quantitative yield, which is compatible with all functional groups, which does not release side products and which is cost effective.

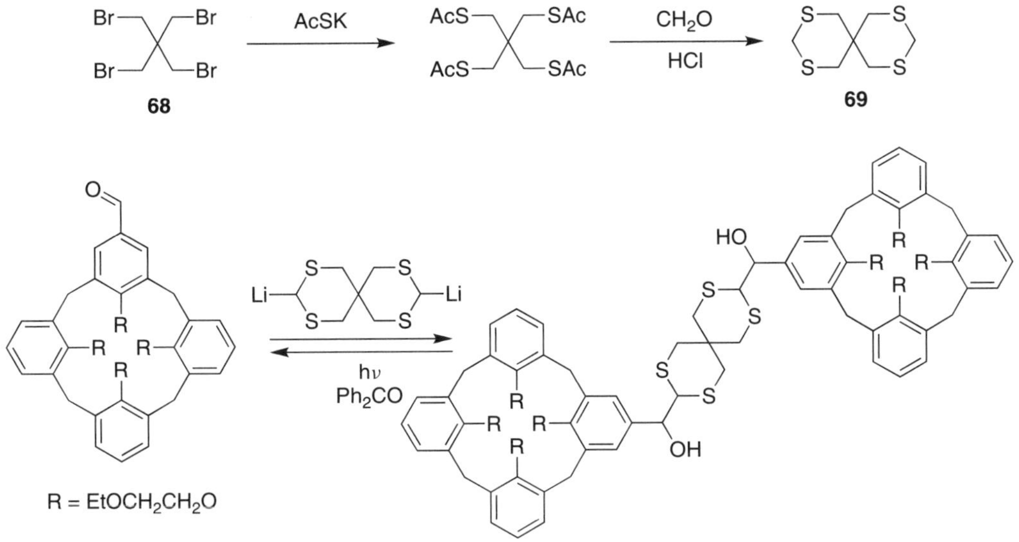

Scheme 5.50 *Bifunctional dithioacetal-based linker*

References

[1] Seneci, P.; *Solid-Phase Synthesis and Combinatorial Technologies* **2000** John Wiley & Sons, Inc., New York.
[2] Zaragoza, F.; *Organic Synthesis on Solid Phase: Supports, Links, Reactions* **2000**, Wiley-VCH Verlag GmbH, Weinheim, Germany.
[3] Terrett, N.; *Combinatorial Chemistry* **1998**, Oxford University Press Inc., New York.
[4] Scott, P. J. H., and Steel, P. G.; *Eur. J. Org. Chem.* **2006**, 2251.
[5] Bochet, C. G.; *J. Chem. Soc., Perkin Trans 1* **2002**, 125.
[6] Rich, D. H., and Gurwara, S. K.; *J. Chem. Soc., Chem. Commun.* **1973**, 610.
[7] Patchornik, A., Amit, B., and Woodward, R. B.; *J. Am. Chem. Soc.* **1970**, *92*, 6333.
[8] Merrifield, R. B.; *J. Am. Chem. Soc.* **1963**, *85*, 2149.
[9] Merrifield, R. B.; *Biochemistry* **1964**, *3*, 1385.
[10] Rich, D. H., and Gurwara, S. K.; *J. Am. Chem. Soc.* **1975**, *97*, 1575.
[11] Tam, J. P., Tjoeng, F. S., and Merrifield, R. B.; *J. Am. Chem. Soc.* **1980**, *102*, 6117.
[12] Voss, C., Dimarchi, R., Whitney, D. B., *et al.*; *Int. J. Pept. Protein Res.* **1983**, *22*, 204.
[13] Barany, G., and Albericio, F.; *J. Am. Chem. Soc.* **1985**, *107*, 4936.
[14] Giralt, E., Albericio, F., Pedroso, E., *et al.*; *Tetrahedron* **1982**, *38*, 1193.
[15] Giralt, E., Eritja, R., Pedroso, E., *et al.*; *Tetrahedron* **1986**, *42*, 691.
[16] Lloyd-Williams, P., Gairi, M., Albericio, F., and Giralt, E.; *Tetrahedron* **1991**, *47*, 9867.
[17] Lloyd-Williams, P., Gairi, M., Albericio, F., and Giralt, E.; *Tetrahedron* **1993**, *49*, 10069.
[18] Bayer, E., and Mutter, M.; *Nature* **1972**, *237*, 512.
[19] Mutter, M., and Bayer, E.; *Angew. Chem.* **1974**, *86*, 101.
[20] Stueber, W., Hemmasi, B., and Bayer, E.; *Int. J. Pept. Protein Res.* **1983**, *22*, 277.
[21] Tjoeng, F. S., Staines, W., St.-Pierre, S., and Hodges, R. S.; *Biochim. Biophys. Acta* **1977**, *490*, 489.
[22] Tjoeng, F. S., Tong, E. K., and Hodges, R. S.; *J. Org. Chem.* **1978**, *43*, 4190.

[23] Tjoeng, F.-S., and Hodges, R. S.; *Tetrahedron Lett.* **1979**, 1273.
[24] Pillai, V. N. R., Mutter, M., Bayer, E., and Gatfield, I.; *J. Org. Chem.* **1980**, *45*, 5364.
[25] Pillai, V. N. R., Renil, M., and Haridasan, V. K.; *Indian J. Chem., Sect. B* **1991**, *30B*, 205.
[26] Greenberg, M. M.; *Tetrahedron Lett.* **1993**, *34*, 251.
[27] Matray, T. J., and Greenberg, M. M.; *J. Am. Chem. Soc.* **1994**, *116*, 6931.
[28] Greenberg, M. M., and Gilmore, J. L.; *J. Org. Chem.* **1994**, *59*, 746.
[29] Baldwin, J. J., Burbaum, J. J., Henderson, I., and Ohlmeyer, M. H. J.; *J. Am. Chem. Soc.* **1995**, *117*, 5588.
[30] Alsina, J., Chiva, C., Ortiz, M., et al.; *Tetrahedron Lett.* **1997**, *38*, 883.
[31] Tam, J. P., Dimarchi, R. D., and Merrifield, R. B.; *Int. J. Pept. Protein Res.* **1980**, *16*, 412.
[32] Zehavi, U., Sadeh, S., and Herchman, M.; *Carbohydr. Res.* **1983**, *124*, 23.
[33] Nicolaou, K. C., Winssinger, N., Pastor, J., and DeRoose, F.; *J. Am. Chem. Soc.* **1997**, *119*, 449.
[34] Nicolaou, K. C., Watanabe, N., Li, J., et al.; *Angew. Chem., Int. Ed. Engl.* **1998**, *37*, 1559.
[35] Armstrong, R. W., Combs, A. P., Tempest, P. A., et al.; *Acc. Chem. Res.* **1996**, *29*, 123.
[36] Rich, D. H., and Gurwara, S. K.; *Tetrahedron Lett.* **1975**, 301.
[37] Hammer, R. P., Albericio, F., Gera, L., and Barany, G.; *Int. J. Pept. Protein Res.* **1990**, *36*, 31.
[38] Pillai, V. N. R., Mutter, M., and Bayer, E.; *Tetrahedron Lett.* **1979**, 3409.
[39] Ajayaghosh, A., and Pillai, V. N. R.; *J. Org. Chem.* **1990**, *55*, 2826.
[40] Renil, M., and Pillai, V. N. R.; *Tetrahedron Lett.* **1994**, *35*, 3809.
[41] Kumar, K. S., and Pillai, V. N. R.; *Tetrahedron* **1999**, *55*, 10437.
[42] Ajayaghosh, A., and Pillai, V. N. R.; *Tetrahedron* **1988**, *44*, 6661.
[43] Ajayaghosh, A., and Pillai, V. N. R.; *J. Org. Chem.* **1987**, *52*, 5714.
[44] Ajayaghosh, A., and Pillai, V. N. R.; *Tetrahedron Lett.* **1995**, *36*, 777.
[45] Brown, B. B., Wagner, D. S., and Geysen, H. M.; *Mol. Div.* **1995**, *1*, 4.
[46] Sternson, S. M., and Schreiber, S. L.; *Tetrahedron Lett.* **1998**, *39*, 7451.
[47] Rodebaugh, R., Fraser-Reid, B., and Geysen, H. M.; *Tetrahedron Lett.* **1997**, *38*, 7653.
[48] Ryba, T. D., and Harran, P. G.; *Org. Lett.* **2000**, *2*, 851.
[49] Piggott, A. M., and Karuso, P.; *Tetrahedron Lett.* **2005**, *46*, 8241.
[50] Yoo, D. J., and Greenberg, M. M.; *J. Org. Chem.* **1995**, *60*, 3358.
[51] Venkatesan, H., and Greenberg, M. M.; *J. Org. Chem.* **1996**, *61*, 525.
[52] Poijaervi, P., Heinonen, P., Virta, P., and Loennberg, H.; *Bioconjugate Chem.* **2005**, *16*, 1564.
[53] Holmes, C. P., and Jones, D. G.; *J. Org. Chem.* **1995**, *60*, 2318.
[54] Holmes, C. P.; *J. Org. Chem.* **1997**, *62*, 2370.
[55] Whitehouse, D. L., Savinov, S. N., and Austin, D. J.; *Tetrahedron Lett.* **1997**, *38*, 7851.
[56] McKeown, S. C., Watson, S. P., Carr, R. A. E., and Marshall, P.; *Tetrahedron Lett.* **1999**, *40*, 2407.
[57] Gea, A., Farcy, N., Roque i Rossell, M., et al.; *Eur. J. Org. Chem.* **2006**, 4135.
[58] Lin, Q., and Blackwell, H. E.; *Chem. Comm.* **2006**, 2884.
[59] Bowman, M. D., Jeske, R. C., and Blackwell, H. E.; *Org. Lett.* **2004**, *6*, 2019.
[60] Gennari, C., Longari, C., Ressel, S., et al.; *Eur. J. Org. Chem.* **1998**, 2437.
[61] Ruhland, B., Bhandari, A., Gordon, E. M., and Gallop, M. A.; *J. Am. Chem. Soc.* **1996**, *118*, 253.
[62] Minkwitz, R., and Meldal, M.; *QSAR & Combinatorial Science* **2005**, *24*, 343.
[63] Wang, S.-S.; *J. Org. Chem.* **1976**, *41*, 32583261.
[64] Tjoeng, F. S., and Heavner, G. A.; *J. Org. Chem.* **1983**, *48*, 355.
[65] Tjoeng, F. S., and Heavner, G. A.; *Tetrahedron Lett.* **1982**, *23*, 4439.
[66] Bellof, D.; and Mutter, M., *Chimia* **1985**, *39*, 317320.
[67] Abraham, N. A., Fazal, G., Ferland, J. M., et al.; *Tetrahedron Lett.* **1991**, *32*, 577.
[68] Flickinger, S. T., Patel, M., Binkowski, B. F., et al.; *Org. Lett.* **2006**, *8*, 2357.
[69] Moriarty, R. M., and Hou, K.; *Tetrahedron Lett.*, **1984**, 691.
[70] Luis, J. G., and Andrés, L. S.; *J. Chem. Research (S)*, **1999**, 220.
[71] McGall, G. H., Barone, A. D., Diggelmann, M., et al.; *J. Am. Chem. Soc.* **1997**, 5081.

[72] Atkinson, T., and Smith, M.; Solid-Phase Synthesis of Oligodeoxyribonucleotides by the Phosphite-Triester Method, in *Oligonucleotide Synthesis* (ed. Gait, M. J); Practical Approach Series; IRL Press, Oxford, **1984**; 3583 (revised **1985**, 41–45).
[73] Rock, R. S., and Chan, S. I.; *J. Org. Chem.* **1996**, *61*, 1526.
[74] Routledge, A., Abell, C., and Balasubramanian, S.; *Tetrahedron Lett.* **1997**, *38*, 1227.
[75] Lee, H. B., and Balasubramanian, S.; *J. Org. Chem.* **1999**, *64*, 3454.
[76] Cano, M., Ladlow, M., and Balasubramanian, S.; *J. Org. Chem.* **2002**, *67*, 129.
[77] Peukert, S., and Giese, B.; *J. Org. Chem.* **1998**, *63*, 9045.
[78] Kessler, M., Glatthar, R., Giese, B., and Bochet, C. G.; *Org. Lett.* **2003**, *5*, 1179.
[79] Glatthar, R., and Giese, B.; *Org. Lett.* **2000**, *2*, 2315.
[80] Horton, J. R., Stamp, L. M., and Routledge, A.; *Tetrahedron Lett.* **2000**, *41*, 9181.
[81] Barton, D. H. R., Crich, D., and Potier, P.; *Tetrahedron Lett.* **1985**, *26*, 5943.
[82] Sucholeiki, I.; *Tetrahedron Lett.* **1994**, *35*, 7307.
[83] Forman, F. W., and Sucholeiki, I.; *J. Org. Chem.* **1995**, *60*, 523.
[84] Nicolaou, K. C., Safina, B. S., and Winssinger, N.; *Synlett.* **2001**, 900.
[85] Enders, D., Rijksen, C., Bremus-Koebberling, E., *et al.*; *Tetrahedron Lett.* **2004**, *45*, 2839.
[86] Wan, Y., Mitkin, O., Barnhurst, L., *et al.*; *Org. Lett.* **2000**, *2*, 3817.
[87] Kurchan, A. N., Mitkin, O. D., and Kuteladze, A. G.; *J. Photochem. Photobiol., A* **2005**, *171*, 121.

6
Safety-Catch Linker Units

Sylvain Lebreton and Marcel Pátek

Sanofi-Aventis, Combinatorial Technologies Center, USA

6.1 Introduction

Ever since the dawn of solid-phase chemistry, the choice of a linker together with synthetic strategy has proven to be a key aspect of many robust solid phase and combinatorial library protocols. Not surprisingly, linker design has witnessed significant attention from the synthetic community being continuously paralleled by the steady growth of new linker units.[1] The most important and general features of linkers used in solid-phase synthesis is the efficient attachment of organic molecules to the solid support, their stability and inertness towards a multitude of reaction conditions, and, finally, their cleavage upon conditions that insure structural integrity of the final product. To cope with the ever increasing demands on the design and synthesis of compound libraries, availability of a wide repertoire of linkers is of great importance.

One strategy that has quickly gained the attention of solid-phase chemists is the so-called 'safety-catch' concept. Safety-catch linkers offer many advantages over the traditional types of linkers. Factors, such as stability during the synthesis, negligible or no premature release of the attached compound and propensity toward chemoselective modification that activates the linker for the final cleavage are the main characteristics of this class of linkers. In a general sense, the safety-catch principle involves a process based upon conversion of a relatively stable form of a linker to a labile, isolable and cleavable one.[2]

Based on the overall cleavage process, two types of safety-catch units can be distinguished. For the safety-catch type I, the stable and labile forms of the linker are orthogonal with respect to cleavage mechanism, that is the cleavage mechanism for both forms is different. For the safety-catch type II, the stable and labile forms of these linkers show a kinetic 'fine-tuning' effect, that is the cleavage mechanisms of the stable and labile forms are the same and their stability depends solely upon the 'strength' of the

Linker Strategies in Solid-Phase Organic Synthesis Edited by Peter Scott
© 2009 John Wiley & Sons, Ltd

cleavage agent (e.g., aqueous sodium hydroxide or amines at room temperature and higher temperature for phenol-sulfide/sulfone linker **5** and **6**). Since such a categorization has only limited use in practice, arrangement according to the cleavage mechanism is used in this review.

6.2 Activation of a carbonyl group by the inductive effect (I–) of an adjacent substituent

6.2.1 Kenner-type safety-catch linker

Among the first carboxy anchors used to demonstrate the safety-catch principle was the acylsulfonamide resin proposed by Kenner and coworkers.[3, 4] Release of the product is feasible upon activation of the carbonyl group via the negative inductive effect of an adjacent substituent. Activation is achieved through an N-alkylation that converts the stable linker into a labile one. Acylsulfonamides are stable to strong anhydrous acids, including hydrobromic acid/acetic acid, as well as to strongly nucleophilic reagents and aqueous alkali as the basic reagents ionize the acidic NH group (pKa ~ 2.5). Upon activation, for example with diazomethane, the corresponding N-acyl-N-methylsulfonamide becomes labile toward nucleophiles due to the increased electrophilicity of the amide group. Kenner demonstrated the use of this safety-catch linker by preparing amides, hydrazides or free carboxylic acids upon treatment of resin **2** with ammonia, hydrazine or alkali hydroxide, respectively (Figure 6.1). However, the original preparation and use of the Kenner safety-catch linker served mostly as a proof of concept only as it suffered from poor loading efficiencies and low reactivity toward less nucleophilic amines (e.g., anilines).

To overcome this limitation, Ellman hypothesized that an N-alkyl group with electron withdrawing properties would enhance the reactivity of the activated intermediate.[5] Dramatic increase in the reactivity of linker **3** toward a larger variety of amines and other nucleophiles was indeed observed after introduction of the electron withdrawing cyanomethyl group. Nonetheless, it was found that carboxylic acids bearing an α-electron withdrawing substituents, such as N-Boc (*tert*-butoxycarbonyl) and N-Fmoc (9-fluorenylmethoxycarbonyl) protected amino acids or benzoic acid, suffered from incomplete alkylation, probably due to an attenuated nucleophilicity of the N-acyl-N-arylsulfonamide anion. This limitation has been overcome by using the aliphatic sulfonamide linker **4**. The overall improvement from the original aryl sulfonamide linker **1** was demonstrated by higher cleavage yields in comparative experiments. He and Kiesling later reported an alternative alkylative activation via palladium(0) catalyzed allylation of

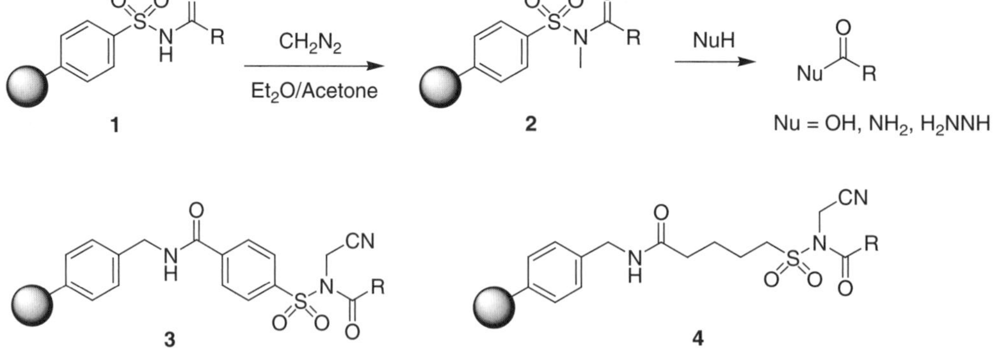

Figure 6.1 Kenner-type safety-catch linker

the *N*-acyl-sulfonamide nitrogen to generate a highly labile intermediate against different nucleophiles.[6] Their procedure allowed for activation under neutral and mild conditions, therefore significantly reducing the risk of racemization of protected amino acids.

The efficacy of this linker was further demonstrated in the synthesis of cyclic peptides via a head-to-tail cyclization–cleavage sequence involving the reaction of a terminal amine with the activated acyl sulfonamide.[7, 8] A reverse-Kenner safety-catch linker has also been used for the synthesis of sulfonamides.[9]

Typical Experimental Procedures

General activation/cleavage procedure

To a 25 ml round bottom flask was added acylsulfonamide resin (500 mg, 0.17 mmol), dimethylsulfoxide (DMSO) or NMP (4 ml), N,N-diisopropylethylamine (DIPEA) (136 µl, 0.85 mmol, and bromoacetonitrile or iodoacetonitrile (4.0 mmol). After stirring the mixture 24 hours, the resin was filtered and washed with DMSO (5 × 5 ml) and tetrahydrofuran (THF) (3 × 5 ml). To the resin containing flask was added THF (3 ml) and amine (3 mmol), and the mixture was stirred for 12 hours, after which the resin was filtered and washed with dichloromethane (CH_2Cl_2) (3 × 5 ml). Some amine may require heating at 65 °C for 20 hours. The filtrate and the washes were combined, washed with 1 M hydrochloric acid (2 × 20 ml), dried with calcium sulfate and concentrated with rotary evaporation.

6.2.2 *N*-boc-activated safety-catch linker

Resin-bound isonitrile **5** was developed for application in the Ugi multicomponent reaction in order to avoid tedious parallel purifications associated with a liquid-phase approach (Figure 6.2). As an added benefit, an excess of reagents can be used to drive the reaction to completion.[10] Upon reaction with the aldehyde, amine and carboxylic acid components, the linker was activated by introduction of a *N*-Boc group to yield resin **6**, which allowed for a facile cleavage using methoxide or hydroxide to yield the methyl esters or carboxylic acids, respectively. Once cleaved, cyclization under different sets of conditions led to diverse arrays of 1,4-benzodiazepine-2,5-diones, ketopiperazines and diketopiperazines, generally in good purity.

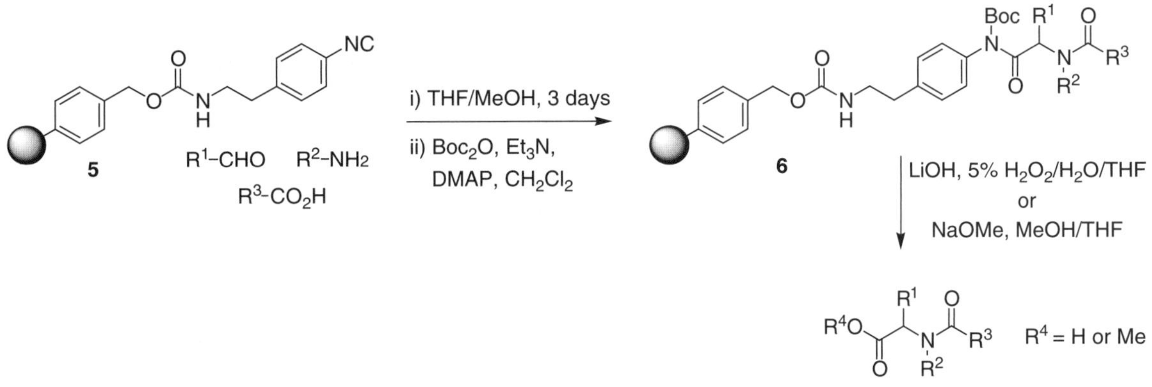

Figure 6.2 *N-Boc activated safety-catch linker*

Typical Experimental Procedures

General activation/cleavage procedure

Upon completion of the Ugi reaction, the washed and dried resin was treated with Boc_2O (10 equiv.), Et_3N (10 equiv.) and 4-(N,N-dimethyl)aminopyridine (DMAP) in dichloromethane (DCM) for 15 hours. Cleavage was achieved by adding sodium methoxide (5 mg) in THF:methanol (1:1) and shaking the resin for 20 hours. Solvent was then evaporated to yield the desired methyl esters.

6.2.3 Sulfide/sulfone safety-catch linker

The attractive feature of the sulfide/sulfone linker is its facile conversion from the arylsulfide to the activated arylsulfone, which then facilitates nucleophilic attack at the ester carbonyl group. Marshall used this particular strategy to anchor amino acids onto the thiophenol linker to give resin **7** (Figure 6.3).[11] Displacement of the ester with an amino group coming from another amino acid residue was accomplished after oxidation of the sulfide with hydrogen peroxide, thus allowing formation of the new amide bond. The utility of this strategy was also illustrated in the preparation of cyclic peptides.[12] One obvious limitation that prevents a wider applicability of this linker is related to the oxidation step and the corresponding oxidative conditions. Unfortunately, sensitive amino acids such as tryptophan, cysteine, cystine and methionine have to be excluded to avoid lower yields and impurities. Adaptation of the phenol–sulfide linker **7** for the synthesis of tetrahydro-β-carboline-3-carboxamides and piperazine-2-carboxamides suggested that the final cleavage step could be effectively carried out even without prior activation of the linker, although longer reaction times were required and excess of amines were used to ensure high cleavage yields.[13]

Dressman demonstrated the synthesis of pure ureas by reacting 0.1 equivalent of amine with a carbamate linker similar to linker **7** and also found that even with the less electrophilic carbamate the oxidation step could be omitted providing that triethylamine was added to the cleavage mixture.[14] Similarly, it was shown (still without prior activation) that cleavage was more efficient in pyridine than in N,N-dimethylformamide (DMF) and that an excess of amine could be used to achieve good yields in less than 24 hours.

Typical Experimental Procedures

General activation/cleavage procedure

The peptide-bound resin (prepared from 1 g of about 0.91 mmol/g 4-hydroxythiophenyl resin) was treated with 2 ml of 30% hydrogen peroxide in 20 ml of acetic acid and stirred for 12 hours at room temperature. After filtering and washing with ethanol, removal of the peptide from the resin was accomplished by stirring for 24 hours with 0.75 mmol of glycine (as the sodium salt) in a DMF/water mixture. The mixture was then filtered and the filtrate was evaporated to dryness.

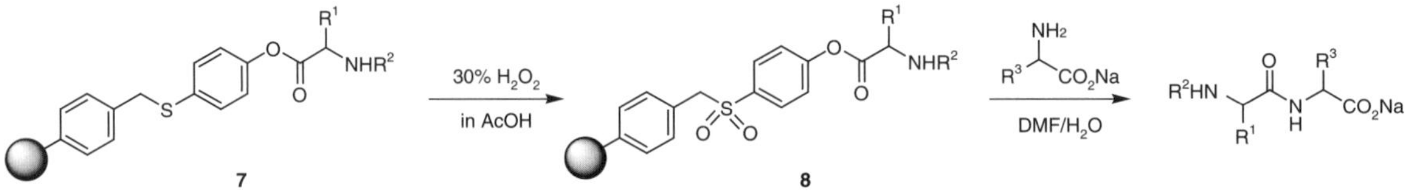

Figure 6.3 Marshall safety-catch linker

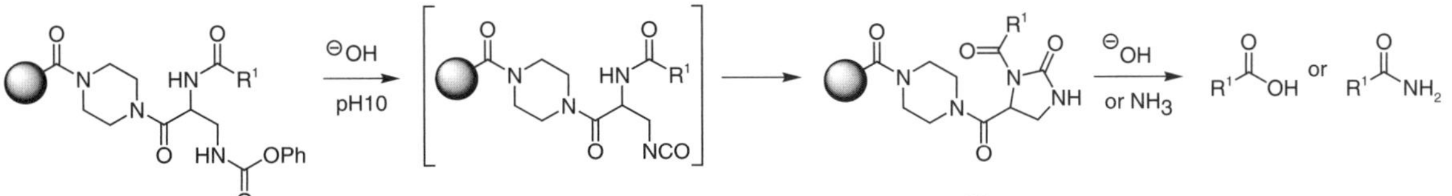

Figure 6.4 Dpr(Phoc) safety-catch linker

6.2.4 Dpr(phoc) safety-catch linker

The base-activated linker reported by Pascal is derived from β-aminophenyloxycarbonyl-2,3-diaminopropionic acid, which exhibits high stability under neutral and acidic conditions.[15] Activation was achieved with a dilute base (e.g., 0.03M sodium hydroxide) or with saturated ammonia solutions by selective intramolecular cyclization of the generated electrophilic isocyanate **10** to the adjacent secondary amide group (Figure 6.4). The *N*-acyl-2-imidazolidinone intermediate **11** formed by this cyclization can now be cleaved from the support by treatment with dilute sodium hydroxide. Alternatively, use of an aqueous buffer at pH 10 or ammonolysis (*i*PrOH saturated with ammonia) yielded peptide acids and peptide amides, respectively.

6.3 Activation by the mesomeric effect (M-) of the −X−Y = Z moiety adjacent to a carbonyl group

In contrast to derivatives of the type R−CO−X discussed in the previous section (X being an electron withdrawing substituent), yet another method of carbonyl activation is discussed here in which the negative mesomeric effect (M-) of the −X−Y = Z moiety plays a crucial role.

6.3.1 Carbonyl activation by oxidative aromatization

Mioskowski reported the preparation of resin-bound 1,2-dihydroquinoline **12** (DHQ) for the generation of amides or carboxylic acids (Figure 6.5).[16] The *N*-acylated form of the DHQ resin was found to be stable under basic, acidic and mild reducing agents. Activation of the linker was performed by the oxidative aromatization with either 2,3-dichloro-5,6-dicyanobenzoquinone (DDQ), cerium(IV) ammonium nitrate (CAN) or $Ph_3C^+ BF_4^-$, leading to the activated acylquinolinium intermediate **13**. Final release of the amide or acid was achieved via nucleophilic displacement of the quinolinium moiety using the corresponding nucleophiles.

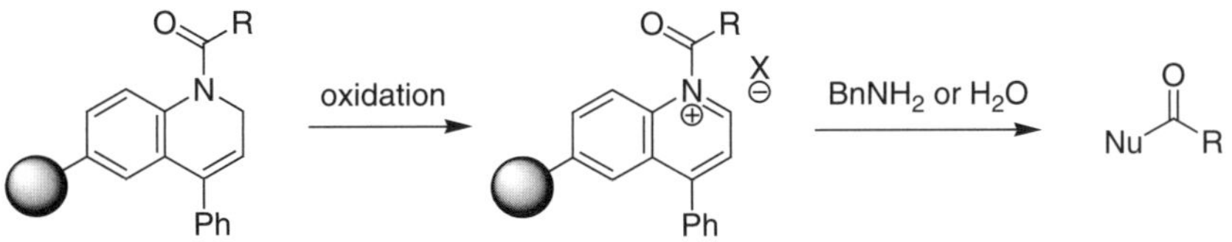

Figure 6.5 Safety-catch linker activation by oxidative aromatization

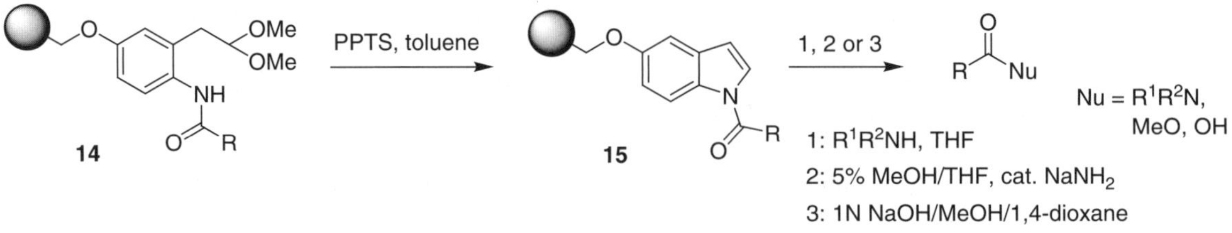

Figure 6.6 *Safety-catch linker activation by indole ring formation and carbonyl activation*

Typical Experimental Procedures

General activation/cleavage procedure

Resin 12 was treated with 3 equiv. of either DDQ in DCM/acetonitrile for 10 hours or $CPh_3^+BF_4^-$ in THF for 3 hours at room temperature. The resulting resin was filtered and treated with an excess of benzylamine in DCM. The resin was then filtered and thoroughly washed with methanol and DCM. The combined organic washes were evaporated to afford a mixture containing the desired amide along with the excess of benzylamine. The pure product was obtained after quick purification by silica flash chromatography.

6.3.2 Carbonyl activation by indole ring formation

A safety-catch linker based on carbonyl activation via an indole formation has been described by Abell et al.[17] Inspired by the acid protecting group reported by Fukuyama used in the synthesis of amides and esters,[18] resin-bound dimethyl acetal linker **14** was prepared and found to be stable at room temperature (Figure 6.6). Acylation with 4-iodobenzoyl chloride or 5-phenylvaleric acid followed by acid catalyzed ring closure yielded the activated *N*-acyl indole **15**. Reactivity of the indole amides was then studied under a variety of cleavage conditions. The corresponding amides, esters and acids were obtained in excellent yields and purities. However, only traces of the corresponding products were detected when the same conditions were applied to the inactivated dimethyl acetal resin **14**.

6.3.3 Benzyl/phenyl-hydrazide safety-catch linker

Wieland et al. have developed a safety-catch linker **16** (Figure 6.7) based on the benzyl hydrazide functionality.[19] Unreactive *N*-acyl hydrazine linker was oxidized with *N*-bromosuccinimide (NBS) to the corresponding reactive diazene intermediate **17**. The acyldiazo intermediate reacts with nucleophiles to yield the functionalized acyl derivatives while liberating nitrogen. In an attempt to reduce side reactions due to aliphatic azo group, the 4-carboxyphenylhydrazide safety-catch linker **18** was introduced. Activation by NBS oxidation followed by reaction with benzylamine afforded the benzylamide product in 62% yield. Similarly, Semenov described the use of the phenylhydrazide linker **19** where the aryl group was attached to the resin through a sulfonamide group.[20] Upon air oxidation in a mixture of DMF, pyridine, aqueous acetic acid and copper sulfate, decomposition of the resulting phenylazo compound took place *in situ*. In one example, a protected tripeptide was obtained in 83% HPLC purity after 16 hours. Incompatibility of the arylhydrazine linker and Mitsunobu reaction conditions, presumably due to the reactive hydrazine moiety being susceptible to *N*-alkylation, led Ladlow to design a modification of this safety-catch linker.[21] By temporary protecting the nitrogen directly attached to the aryl ring with the acid labile 2,4-dimethoxybenzyl group (DMB), a library of ketopiperazines bearing up to four points of diversity was

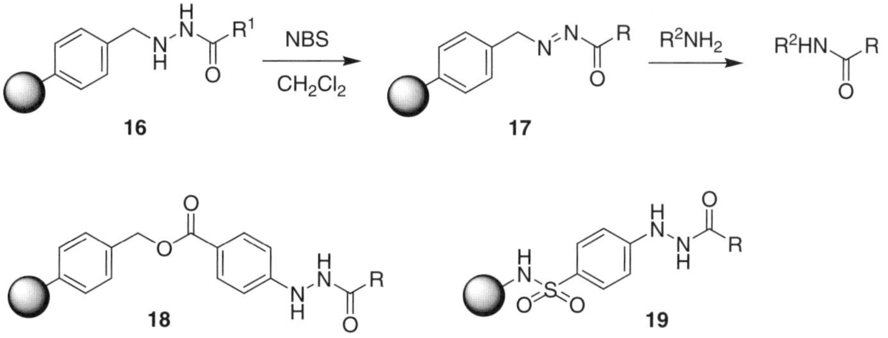

Figure 6.7 Hydrazide-based safety-catch linkers

prepared from resin **20** by means of Mitsunobu *N*-alkylation between the R^1-functionalized amine and the corresponding ethanolamine bearing the R^2, R^3 and R^4 groups (Figure 6.8).

A reverse strategy involving the hydrazide group attached to the resin through the acyl functionality was reported by Lowe.[22] Oxidation, followed by nucleophilic displacement then led to the release of aryl derivatives into the cleavage solution. This route was further validated by Waldmann *et al.*[23] by first preparing 4-iodophenylhydrazide resin **21** (Figure 6.9). Substitutions at the 4-iodo position were achieved via Heck, Suzuki, Sonogashira and Stille cross-coupling reactions. The corresponding resin-bound aryl hydrazides were then oxidized either with copper(II) acetate in *n*-propylamine or by NBS in dichloromethane and converted to the corresponding acyl diazenes. Release of the final product was accomplished by the reaction with a nucleophile either present in the oxidative reaction mixture or added after the oxidation step. This displacement resulted in the formation of a resin-bound ester or amide and the release of the desired coupling products in high yields.

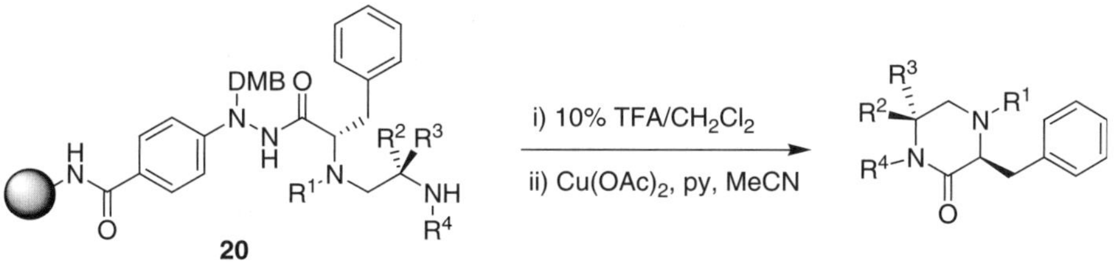

Figure 6.8 DMB protected hydrazide-based safety-catch linker for Mitsunobu derivatization

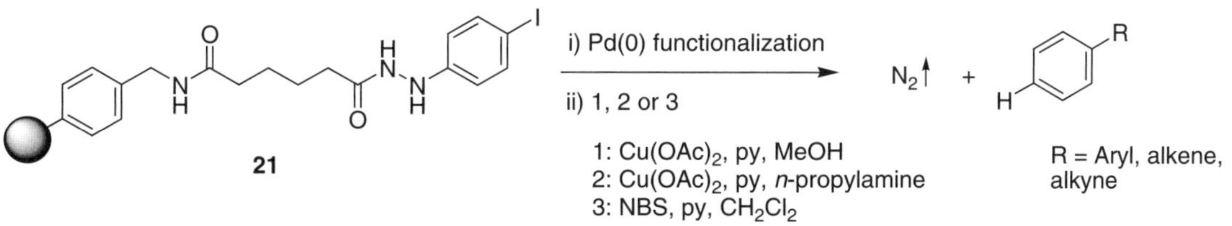

Figure 6.9 Reversed hydrazide-type safety-catch linker for the preparation of aryl derivatives

Typical Experimental Procedures

General activation/cleavage procedure

Upon DMB deprotection of resin **20** (190 mg, maximum loading 0.31 mmol/g) with a solution of TFA in DCM (10% v/v; 2 ml) for 10 minutes and subsequent washings (DMF; DCM), the resulting hydrazide resin was treated with a solution of copper(II) acetate (10.8 mg, 0.06 mmol) and pyridine (48 μl, 0.60 mmol) in acetonitrile (10 ml). After stirring for 2 hours at room temperature, the resin was filtered, washed (DCM) and the resulting combined organic phases evaporated to dryness under reduced pressure to give an oil, which was purified by preparative HPLC.

6.3.4 Dehydration activated safety-catch linker

The safety-catch 2,2-diphenyl-2-hydroxyethyl ester **22** (Figure 6.10) introduced by Wieland uses an acid catalyzed dehydration to convert the ester into the reactive enol ester **23**, which was then reacted with amines to afford the corresponding secondary amides.[24] Unfortunately, the generality of this approach is limited by the incompatibility of additional acid-labile groups with the acidic treatment required for activation.

6.4 Activation by the positive mesomeric effect (M+) of the − X−Y = Z moiety adjacent to a N-acyl or O-alkyl group

6.4.1 Benzhydryl-based safety-catch linker

Guided by the analogy of the safety-catch p-(methylthio)benzyl and p-(methylsulfinyl)benzyl protecting groups developed by Samanem and Brandeis,[25] Pátek and Lebl introduced the safety-catch amide linker (SCAL) for the synthesis of peptide amides on solid support (Figure 6.11).[26] In its oxidized form, the resin-bound linker **24** is extremely stable towards strong acids and bases. Reductive activation of the SCAL linker can be accomplished by treatment with 1 M PPh_3/Me_3SiCl/CH_2Cl_2, followed by acidolytic cleavage from the support. Alternatively, the reduction and cleavage can be carried out simultaneously by using a one-pot reductive acidolysis procedure with 1 M Me_3SiBr/thioanisole/TFA, yielding the released compounds in high purities. This linker strategy was later used for the assembly of large polypeptides and OBOC libraries.[27]

The 4-(2,5-dimethyl-4-methylsulfinylphenyl)-4-hydroxybutanoic acid (DSB) linker **26** (Figure 6.12) described by Kiso allowed for Boc-SPPS synthesis of γ-endorphin containing 17 amino acid residues.[28]

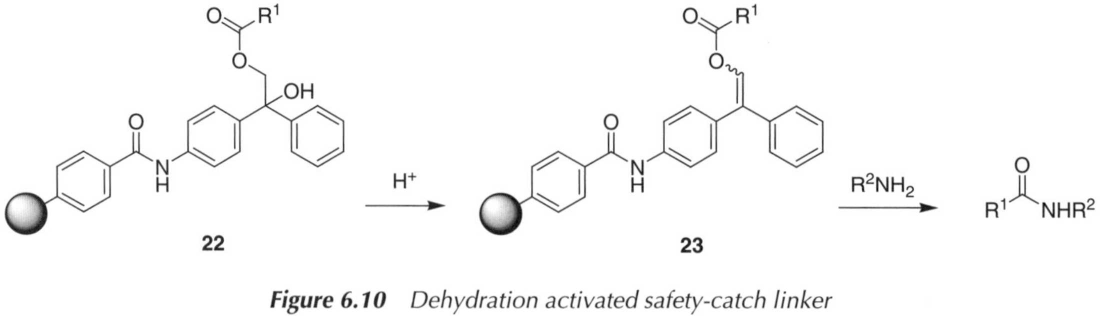

Figure 6.10 Dehydration activated safety-catch linker

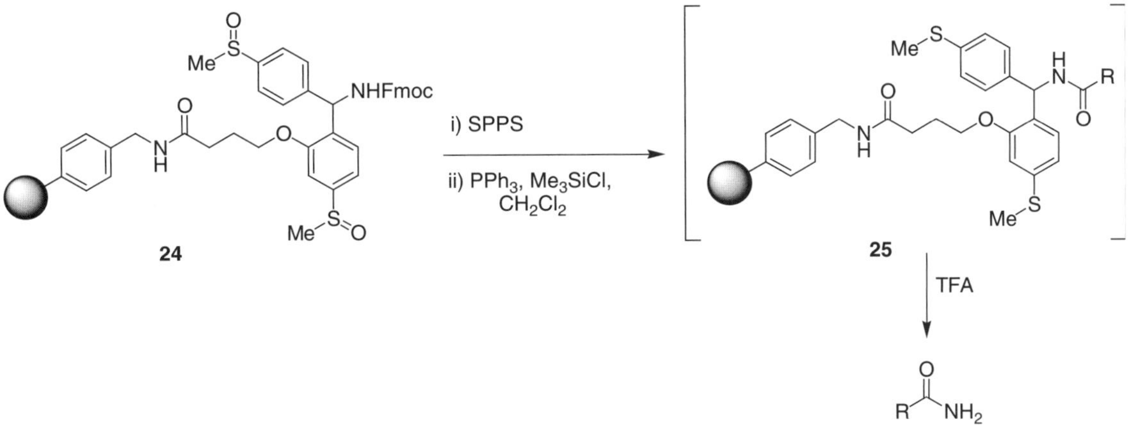

Figure 6.11 Benzhydryl-based SCAL safety-catch linker

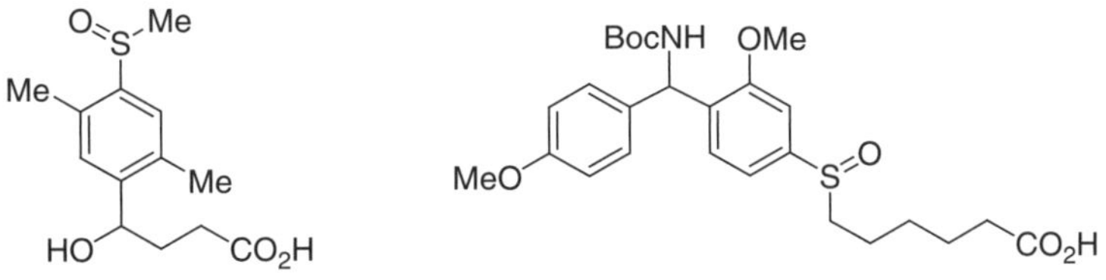

Figure 6.12 Stucture of DSB and DSA safety-catch linkers

The stability of the linker was examined by exposure of the resin-bound Boc-Leu-DSB to TFA/anisole for 24 hours. As a result, only 3% of Leu was cleaved whereas 94% was cleaved under reductive acidolysis conditions. A similar dialkoxyalkylsulfinylbenzhydrylamine (DSA) linker **27** (Figure 6.12) was described, which displayed the same reactivity toward reductive acidolysis although with slightly lower stability toward strongly acidic conditions.[29]

Typical Experimental Procedures

General activation/cleavage procedure

After peptide synthesis was completed, reductive activation of the SCAL supported resin was accomplished by repetitive treatment with either 1 M $PPh_3/Me_3SiCl/CH_2Cl_2$ or 20% $(EtO)_2P(S)SH$/DMPU (3 × 20min), followed by washing with DCM and DMF. Acidolytic cleavage is then carried out by treatment with TFA/water (95:5) or TFA/DCM/water/Bu^i_3SiH (85:10:2.5:2.5) for 1 hour.

6.4.2 Indole-based safety-catch linker

The acidic lability of indole-based systems described by Estep and coworkers[30] guided Ley *et al.* to suggest an indole-based safety-catch linker that could be used for the release of secondary amides, sulfonamides and carbamates.[31] The required *N*-tosyl protected indole linker **28** (Figure 6.13) was synthesized from

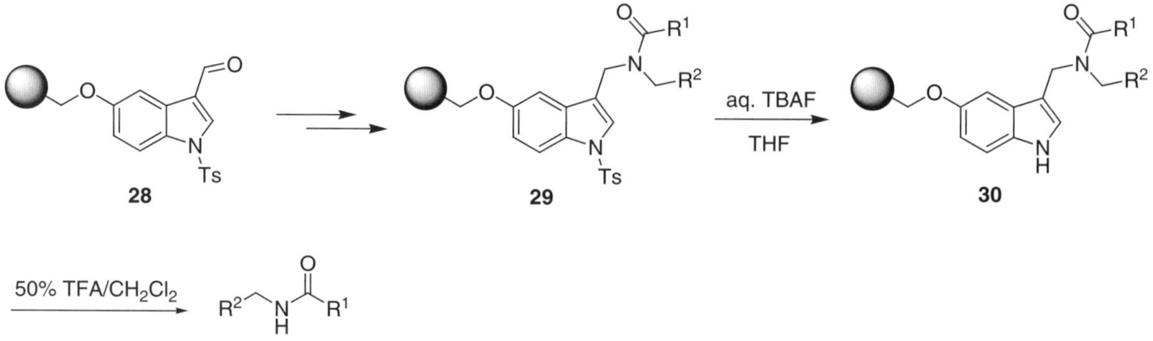

204 Traditional Linker Units for Solid-Phase Organic Synthesis

Figure 6.13 *Use of N-tosyl protected indole-based safety-catch linker*

5-benzyloxyindole-3-carboxaldehyde. The aldehyde functionality was used to attach amines via reductive alkylation. Subsequent *N*-acylation yielded intermediate **29**. Using a model construct, Ley showed that such a linker assembly was stable in 50% TFA/DCM solution. Deprotection of the tosyl group was accomplished with tetrabutylammonium fluoride (TBAF) to give activated intermediate **30**. The corresponding amide was obtained by a quantitative acid catalyzed release from the resin **30**.

Typical Experimental Procedures

General activation/cleavage procedure

*Treatment of resin **29** with aqueous TBAF in THF at 50 °C for 17 hours resulted in quantitative N-detosylation. After washing the resin with THF and DCM, the amide was cleaved from the resin by treatment with 50% TFA in DCM for five minutes.*

6.4.3 Nitrobenzyl alcohol-based safety-catch linker

Takahashi *et al.* described the use of 3-nitrobenzyl safety-catch linker for the solid-phase synthesis of a library of *N*-alkylated naltrindoles.[32] Resin-bound naltrexone **31** (Figure 6.14) whose nitro group enhances its acid stability was prepared by the Mitsunobu alkylation of the phenolic hydroxl group. The construct was then found to be stable to 50% TFA/DCM. The indole ring was formed in the next step by the Fischer indole synthesis involving arylhydrazines. Following *N*-alkylation with a set of benzyl bromides (R²) provided intermediate **32**. Activation of the linker was achieved by reduction of the nitro group with tin(II) chloride dihydrate ($SnCl_2 \cdot 2H_2O$) and chemoselective protection of the resulting amine with tosyl chloride. In the final step, treatment of resin **33** with 50% TFA/DCM afforded the desired *N*-benzyl naltrindole derivatives along with a significant amount of the *N, O*-dialkylated compounds.

In a recent report, *p*-nitromandelic acid (Pnm) was used as a safety-catch linker for the solid-phase synthesis of peptide and depsipeptide acids.[33] Resin-bound Pnm linker **34** (Figure 6.15) was functionalized via standard Boc-SPPS procedures to prepare a number of 5-membered depsipeptides. Removal of the side chain protection was performed with trifluoromethanesulfonic acid (TFMSA) in the presence of thioanisole, prior to linker activation by reduction of the nitro group with tin(II) chloride in a 1.6 mM hydrochloric acid/dioxane. The premature loss of peptide resulting from the activation step was found to be in the range of 1–24%. Finally, the treatment of the aniline resin **36** with 5% TFA/dioxane under microwave irradiation at 50 °C afforded the desired products.

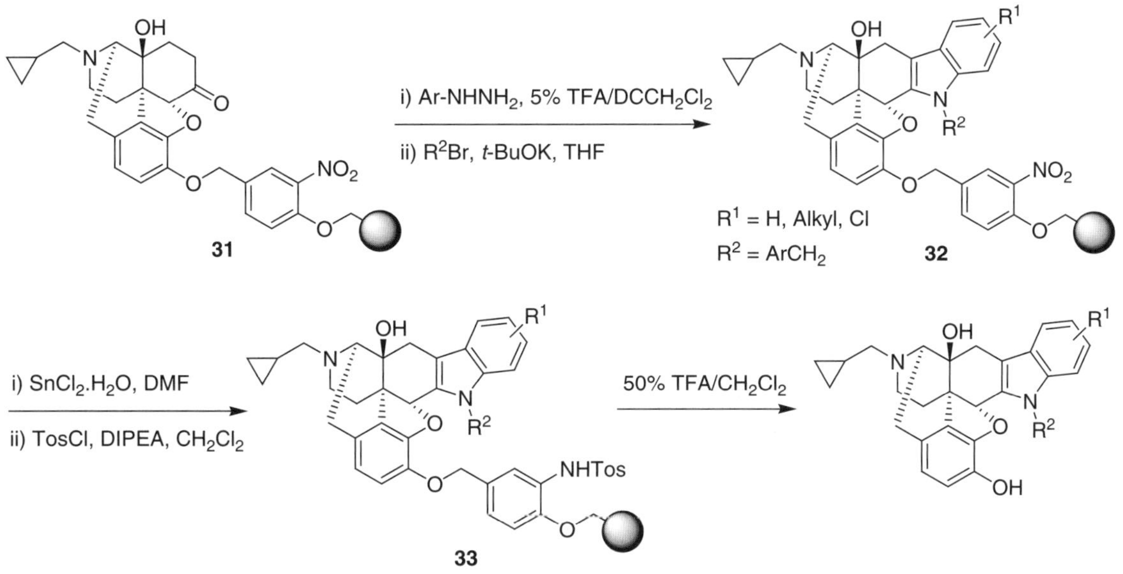

Figure 6.14 Preparation of a library of N-alkylated naltrindoles using a nitrobenzyl alcohol safety-catch linker

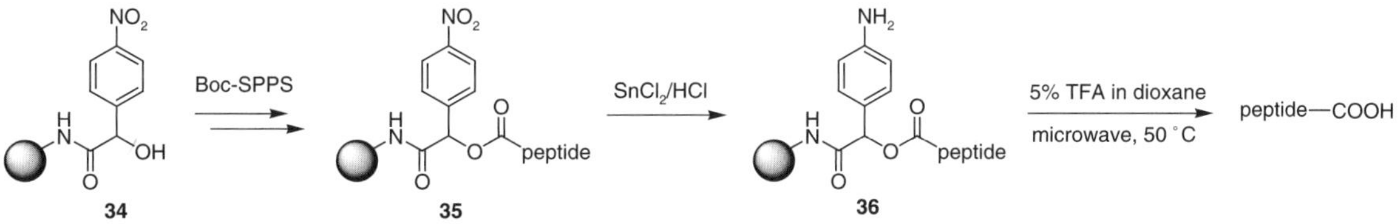

Figure 6.15 p-Nitromandelic-based safety-catch linker

Typical Experimental Procedures

General activation/cleavage procedure

*Resin **35** was treated with 6 M tin(II) chloride, 1.6 mM HCl/dioxane in DMF for one hour at room temperature. The resin was then thoroughly washed with DMF and dioxane. Peptide release was accomplished by treatment with 5% TFA in dioxane for one hour at 50 °C under microwave heating.*

6.5 Aromatic $S_N Ar$ substitution

Nucleophilic aromatic substitution has also been widely used in the preparation of heterocycles. This strategy relies on the use of a 2-thiopyrimidine skeleton, which remains inert throughout the synthesis. Upon oxidative activation to the sulfone, 2-sulfonylpyrimidine linker becomes activated toward nucleophilic *ipso*-substitution. This approach was first reported by a group from Hoffmann-La Roche in the preparation of 2,4,6-trisubstituted pyrimidines (Figure 6.16).[34] Resin-bound pyrimidine-carboxylic acid **37** was subjected

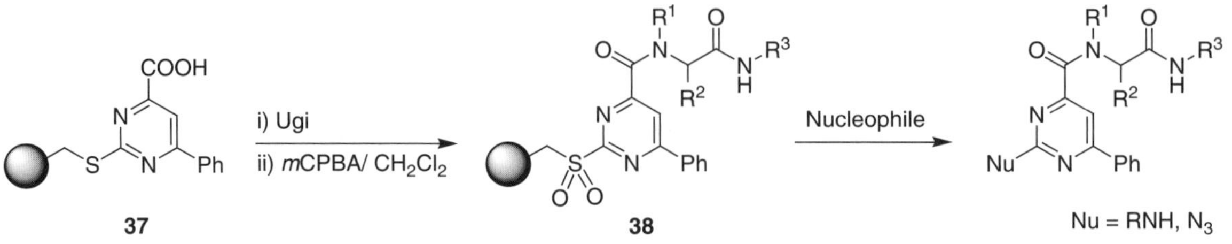

Figure 6.16 2-Thiopyrimidine-based safety-catch linker

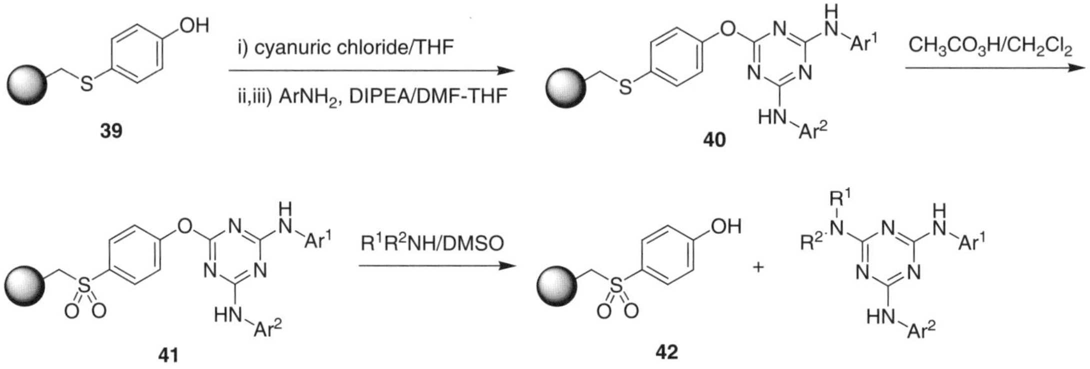

Figure 6.17 Marshall-type safety-catch linker for the preparation of 2,4,6-trisubstituted triazine derivatives

to the Ugi four-component condensation to form the carboxamides. Oxidation to the corresponding sulfone **38** followed by the treatment with different nucleophiles afforded the desired products in purities typically greater than 90%. Many examples illustrating this approach have been reported for the preparation of substituted pyrimidines,[35] purines[36] and benzoxazoles.[37]

Similarly, Bradley reported the use of Marshall-type safety-catch linker **39** for the preparation of triazine-based antibiotics (Figure 6.17).[38] Reaction of **39** with cyanuric chloride followed by successive aromatic nucleophilic substitutions with various anilines, provided the resin-bound triazine **40** that was further oxidized to the corresponding sulfone **41**. Finally, the treatment with secondary amines afforded the desired triazine derivatives in purities above 80%. The resin-bound phenol **42** can be recycled for another reaction batch.

Typical Experimental Procedures

General activation/cleavage procedure

Upon completion of the Ugi reaction starting from resin **37**, *the resin was washed with dioxane, DMF, DCM, i-PrOH, DCM, i-PrOH, DCM, and pentane and dried. DCM (3 ml/mmol of resin) and 3 equiv. of meta–chloroperbenzoic acid (mCPBA) were added. After shaking the resin for 15 hours, excess reagents were flushed and the resulting resin-bound sulfone was successively washed with three cycles of DCM, i-PrOH, and pentane. After drying, dioxane (3 ml/mmol of resin) and the amine (1 equiv.) were added. The mixture was vortexed at room temperature for six hours. The resin was washed with DCM, the filtrate evaporated and filtered through a small silica-gel cartridge.*

6.6 Fragmentation by β-elimination

In order to introduce a basic functionality to become an inherent feature of exploratory libraries and new chemotypes, Rees *et al.* have developed the REM (<u>R</u>egenerated resin after initial functionalization via a <u>M</u>ichael addition) linker with the aim to generate tertiary amines.[39] Michael addition of secondary amines onto the acrylate resin **43** gives resin-bound tertiary amine **44** (Figure 6.18), which can be further quaternized with a number or reactive halides. This step also introduces an additional point of diversity. DIPEA promoted Hofmann elimination carried on at room temperature allows release of the tertiary amine and regeneration of the acrylate resin **43**. The linker was shown to have high chemical compatibility with many synthetic protocols while consistently yielding the tertiary amines in high purity even over five cycles. However, purification by solid-phase extraction (SPE) was often required at the end of the synthesis. Murphy overcame this limitation by adding a resin-bound tertiary amine, which not only acts as the basic reagent to promote the required elimination but also traps the salts. Such modification resulted in high purity of products with no need for further purification.[40]

This two-resin approach was also chosen for the introduction of acrylamide electrophiles onto small-molecule library members.[41] In this example, it is the Michael acceptor that was released from the resin (Figure 6.19). The secondary amine in linker **46** was protected as an *ortho*-nitrobenzene sulfonamide (oNBS). Following the removal of the Boc protecting group, a variety of carboxylic acids were coupled

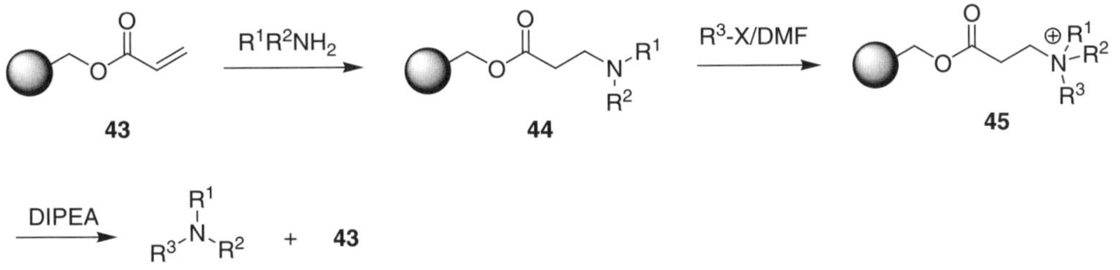

Figure 6.18 REM safety-catch linker

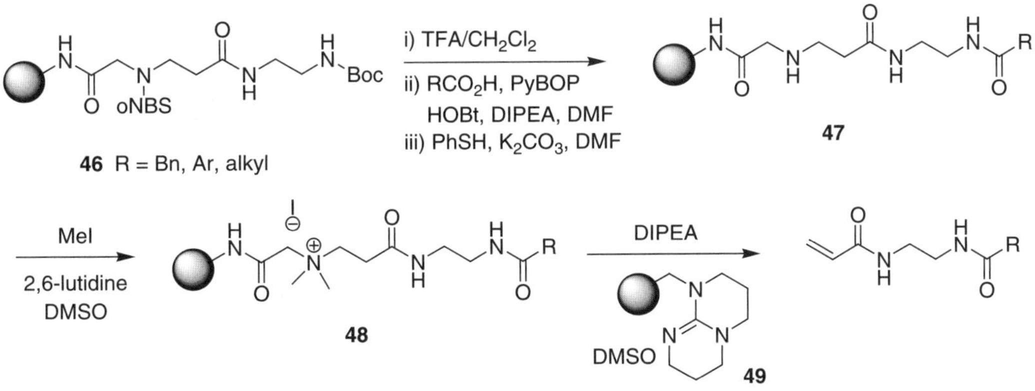

Figure 6.19 REM safety-catch linker for the release of acrylamides

208 Traditional Linker Units for Solid-Phase Organic Synthesis

onto the linker. Treatment with thiophenol and potassium carbonate removed the oNBS protecting group to give the intermediate **47**. The resulting free amine was then quaternized with a large excess of methyl iodide in the presence of 2,6-lutidine. In the final step, the acrylamides were released by the treatment of resin **48** with 10 equivalents of solid-phase bound 1,5,7-Triazabicyclo[4.4.0]dec-5-ene (TBD) **49** in the presence of two equivalents of DIPEA. All products were isolated in good yields and purities after filtration and evaporation.

In a close analogy to the REM linkers, a variety of solid-supported vinyl sulfone linkers have been successfully used for the alkylation of secondary amines.[42] In another example of β-elimination assisted cleavage, dehydroalanine derivatives have been synthesized starting from the resin-bound substituted cysteine **50** (Figure 6.20). The linker was shown to be stable to acidic conditions. Upon oxidation with *m*CPBA, the dehydroalanines were released from the activated linker by treatment with 1,8-Diazabicyclo[5.4.0]undec-7-ene (DBU). Further workup yielded the expected products with high purities.[43]

Typical Experimental Procedures

General activation/cleavage procedure

REM resin **43** (1.0 g, 0.74 mmol) was suspended in a solution of the amine (7.4 mmol) in DMF (7 ml) and agitated for 2.5 hours. The resin was drained and washed with DMF and then briefly dried under reduced pressure. The resin was resuspended in a solution of the alkyl iodide/bromide (3.7 mmol) in DMF (7 ml) and agitated for 2.5 hours. The resin was then drained and washed with DMF and DCM, then briefly dried under reduced pressure. The resin was resuspended in a solution of DIPEA (1.48 mmol) in DCM (7 ml) and agitated for 2.5 hours. Finally, the resin was drained and washed with DCM. The filtrate was collected and evaporated. Purification was achieved by an aqueous extraction.

6.7 Safety-catch linker for release in aqueous buffers

6.7.1 Geysen safety-catch linker

Linkers that would allow for the synthesis of releasable peptides on solid phase are quite attractive, as they would facilitate simultaneous preparation and screening of biological activity for a plurality of individual compounds. Such a concept was first illustrated by Geysen *et al.* by synthesis of safety-catch linker **51** (Figure 6.21).[44] Upon N^α-deprotection, resin **52** bearing an ammonium salt is generated and was shown to be stable to the protocols designed to remove contaminants from the resin-bound peptide. Cleavage was

Figure 6.20 *Preparation of dehydroalanine derivatives via β-elimination of cysteine-based safety-catch linker*

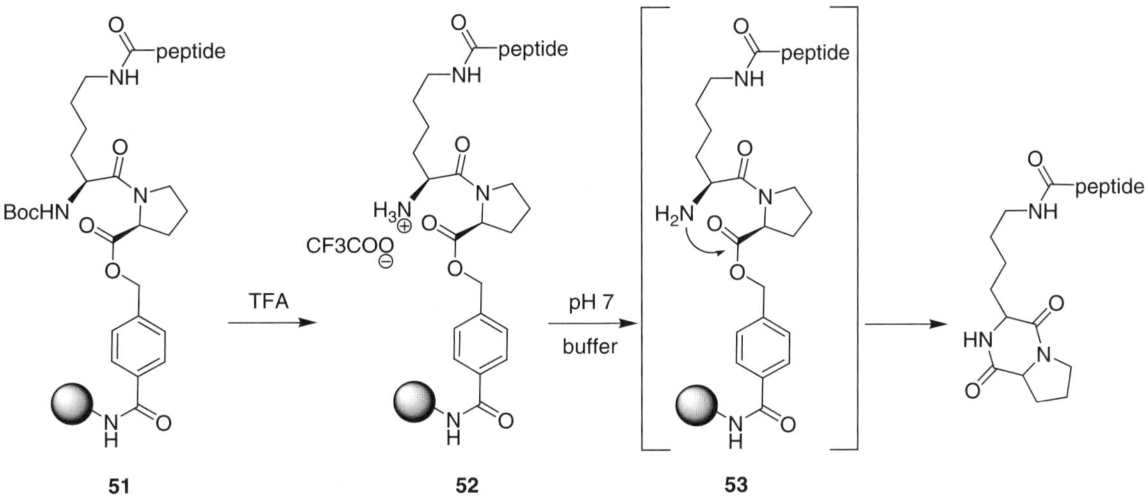

Figure 6.21 Geysen safety-catch linker

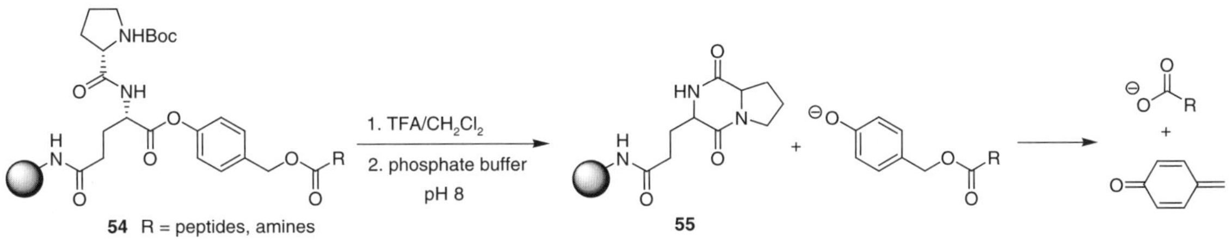

Figure 6.22 Bradley Safety-catch linker

initiated by the treatment with 0.1 M phosphate buffer at pH 7, which results in the fast intramolecular cyclization and release of the peptide bearing a *C*-terminal diketopiperazine moiety. Using a similar acid-activated safety-catch linker, Bradley *et al.* reported linker **54**, which undergoes spontaneous cleavage in aqueous buffers once activated (Figure 6.22). In contrast to the Geysen approach, the diketopiperazine formed during cyclization remains on the support while the 4-hydroxybenzyl ester intermediate is released into the solution. Subsequent 1,6-elimination provides either the peptide[45] or an amine[46] together with the corresponding quinone methide.

Typical Experimental Procedures

General activation/cleavage procedure

*To the preswollen resin **54** in DCM was added a solution of TFA/DCM (1:1) and the resin was shaken for 45 minutes. The resin was then washed with DCM, DMF, DCM, methanol and water. A solution of sodium phosphate buffer (pH 8, 0.1 M) was added (1 ml per 10 mg of resin). The resin was shaken at room temperature for four hours, filtered and washed.*

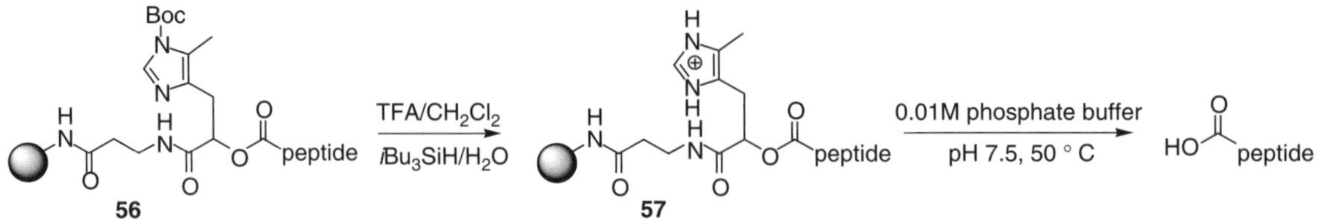

Figure 6.23 Frank safety-catch linker

6.7.2 Frank safety-catch linker

The central feature of safety-catch linker **56** developed by Frank *et al.* is the rate enhancement of the ester bond hydrolysis by an intramolecular catalysis enabled by a proper positioning of the imidazole moiety.[47] During the peptide assembly, the catalytic effect of imidazole is attenuated by the Boc group placed onto the position N-1 of the imidazole ring (Figure 6.23). Final deprotection in acidic media removes the Boc group and other acid labile groups to yield resin **57**. The deprotected imidazole ring remains 'turned off' at this stage due to the ring protonation. Upon neutralization (pH $\sim$ 7), the desired peptide product is released into the aqueous buffer and is readily available for a bioassay.

Typical Experimental Procedures

General activation/cleavage procedure

*Resin **56** was treated with TFA/DCM (1:1) containing 2.5% triisobutylsilane and 2% water. After successive extractions (3 × 10 min) with methanol/water (1:1) containing 0.1% hydrochloric acid and 1 M acetic acid, cleavage in 0.01 M phosphate buffer (pH 7.5) at 50 °C for 25 minutes gave the desired peptide acids.*

6.7.3 Lyttle safety-catch linker

Lyttle devised a safety-catch linker based on the 3-amino-1,2-dihydroxypropane group for the synthesis and release of oligonucleotide from a control pore glass (CPG) support (Figure 6.24).[48] Upon completion of the solid-phase synthesis, linker **58** is activated by palladium(0)-assisted removal of the allyloxycarbonyl

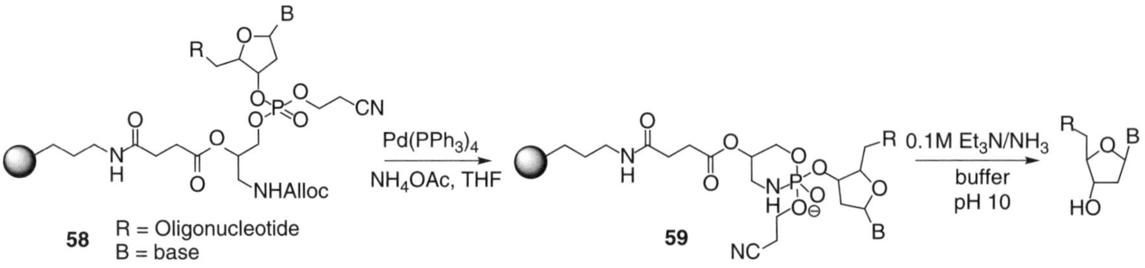

Figure 6.24 Lyttle safety-catch linker for the preparation of oligonucleotides

group. The primary amine formed in the previous step can further react with the phosphate ester to yield the intermediate **59**. The 3′-hydroxyl nucleotide is released by treatment with 0.1 M triethylammonium acetate buffer and ammonia. Although the 2-hydroxypropionitrile could also be expelled upon nucleophilic attack of the amine, the resulting cyclic intermediate is expected to be further hydrolyzed in the basic aqueous buffer to eliminate the 3′-hydroxyl nucleotide.

Typical Experimental Procedures

General activation/cleavage procedure

*A mixture of 25 mg tetrakistriphenylphosphine palladium(0), 50 mg ammonium acetate hydrate and 100 mg triphenyl phosphine in 1 ml THF was heated to 50 °C for two minutes. About 200 ml of the yellow solution was taken up in a 1 ml syringe and about half of this was passed into the oligonucleotide synthesis column containing support bound nucleotide **58**. The column with syringe attached was placed in a previously warmed 13 × 100 mm test tube and heated in an aluminum hot block at 50 °C. After 10 minutes, the rest of the solution was forced through the column, and after five minutes the column was removed from the tube and washed with 5 ml acetonitrile. Next, a solution of 1 ml 0.1 N TEA (pH 8.5) was mixed with 40 µl of 3% aqueous ammonia and 0.5 ml of this solution was taken up in a syringe. Over two hours this solution was pushed through the column in small increments, with the effluent collected in an eppendorf tube. The column was then further rinsed with 0.5 ml 50% acetonitrile in water. The combined effluent was evaporated in vacuo. The residue was then subjected to concentrated ammonia for five hours at 55 °C and evaporated for subsequent purification or analysis.*

6.7.4 Multiple cleavable linkers

Linker **60**, developed by Kocis and Lebl,[49] consists of a specific arrangement of three carboxymethyl units in N-acylated iminodiacetic acid (Figure 6.25). Compound synthesized on one bead is attached to two identical arms. This multiple cleavable linker allows for release of only a 50% fraction of the synthesized compound from the individual solid support particle. Owing to these unique structural features, testing the compound for biological activity, isolation of the relevant particle and confirmation of the activity after the release of the second portion of the compound can be done for any bead of the library. The first cleavage follows the same strategy outlined in Section 6.1. Upon treatment with TFA followed by Boc removal, neutralization in an aqueous buffer triggers the cyclization via intramolecular formation of resin-bound diketopiperazine **62**. The first portion of the released compound can be directly used in the biological assay. The second portion can be cleaved later by alkaline ester hydrolysis. A third branch that is attached to the non-cleavable part of the linker can be used for compound identification.

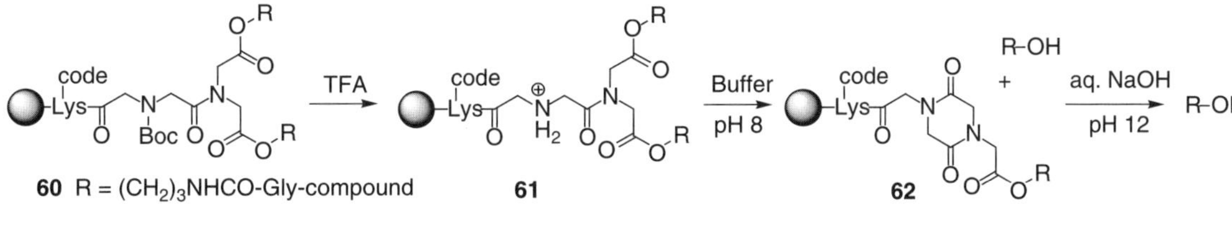

Figure 6.25 *Multiple cleavable linker*

Typical Experimental Procedures

General activation/cleavage procedure

For the first release after Boc cleavage with TFA, library beads were transferred into pH 4.5 buffer containing 1.0% carboxymethylcellulose, shaken and rapidly pipetted into the upper chambers of a 96-well filtration manifold. About 500 beads were deposited in each filtration well. After vacuum filtration of transfer buffer, first-stage peptide release was accomplished by dispensing the appropriate buffer or tissue culture medium (neutral pH) into each well; the plates were then incubated overnight to cleave the peptide by diketopiperazine formation and the peptides were vacuum filtered into microassay plates, used for bioassay. After first-stage assay, 'positive' wells were identified, and the positive beads were recovered from the corresponding well(s) of the filtration master plate. Recovered beads were transferred individually to single microwells in 96-well filtration plates. Cleavage of the ester linker was accomplished by addition of 0.2% sodium hydroxide and overnight incubation followed by pH adjustment or, alternatively, by overnight incubation in ammonia vapors. Thereafter, the peptide in buffer was filtered for bioassay. The individual peptide bead for each positive well in second-stage assay was then recovered for microsequencing.

6.8 Photochemical activation

Based on the dithiane protected 3-alkoxy benzoin tether used by Chan et al.,[50] Balasubramanian et al. reported the analogous use of the dithiane protected benzoin linker for the photolytical release of carboxylic acids and alcohols (Figure 6.26).[51] After formation of Fmoc protected β-alanine ester to give resin **63**, the dithiane safety-catch group could then be removed by the treatment with either mercury(II), periodic acid or methyl triflate. The resulting O-acyl benzoin **64** was photolyzed at 350 nm. The release of Fmoc-β-alanine reached a near quantitative yield after two hours. Further studies carried out using other carboxylic acid substrates showed that loading efficiency was sensitive to steric bulk of the acid. Also, the authors showed that the photochemical reaction was little influenced by the choice of resin as similar half-lives

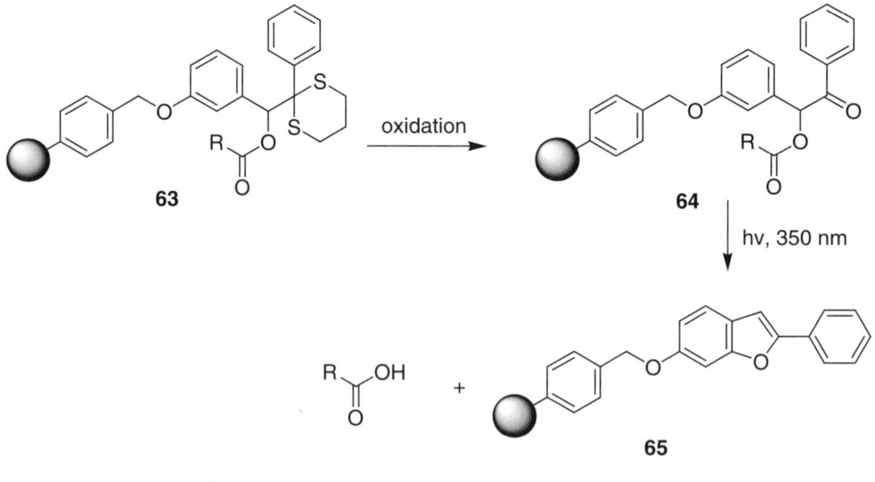

Figure 6.26 Benzoin-based safety-catch linker for photochemical release

were obtained using polystyrene or TentaGel resins. However, lowering the loading capacity of the resin was found to improve rates of cleavage. On the other hand, substitutions at the benzoin system failed to improve the cleavage efficiency compared to that of the linker **63**. [52]

Typical Experimental Procedures

General activation/cleavage procedure

Methyl triflate (0.96 mmol) was added dropwise to a suspension of the resin **63** *(~ 0.1 mmol) in anhydrous DCM (1 ml) and the mixture was gently stirred at room temperature for 30 minutes. The resin was filtered, washed and dried under vacuum to give the resin-bound dithiane deprotected benzoin* **64**. *The resin (20 mg) and Fmoc-serine (internal standard, 3 mg) in 3:1 distilled THF/methanol (12 ml) in a quartz test tube was purged with a stream of nitrogen gas for 10 minutes. The suspension was UV irradiated for two hours with constant nitrogen bubbling to yield the benzofuranyl resin* **65** *and the acid in solution.*

6.9 Miscellaneous safety-catch linkers

This section encompasses safety-catch linkers where the mechanism of activation/cleavage is either different from the ones cited above or where activation/cleavage is a complex event involving several mechanisms.

6.9.1 Activation by reductive aromatization

Taking advantage of a lactonization reaction facilitated by a trimethyl lock, a redox-sensitive linker for the solid-phase synthesis of *C*-terminal peptides was developed by Wang *et al.*[53] In this example, quinone-based resin-bound linker **66** was prepared (Figure 6.27). Ethanolamine was used as a two-carbon spacer to introduce the carboxyl group of the first amino acid. Three model peptides (tripeptide, tetrapeptide and pentapeptide) were synthesized using the standard Boc-SPPS procedures. Chemical stability of the quinone moiety was demonstrated. In the next step, the deprotection of the Boc group was achieved with 25% TFA in dichloromethane/indole. Finally, the linker was activated by the treatment with sodium hydrosulfite

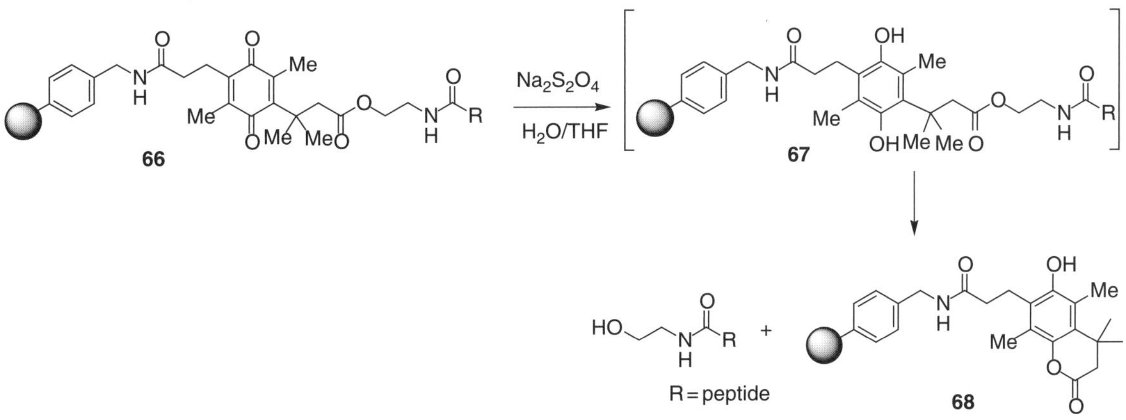

Figure 6.27 *Safety-catch quinone-based linker activation by reductive aromatization*

in water/THF to yield the hydroquinone intermediate **67**. Reaction progress could be monitored by IR spectroscopy and by the disappearance of yellow color of the resin due to the quinine reduction. Complete lactonization was found to occur after 2.5 hours at room temperature with peptides formed in a high yields and purities.

Typical Experimental Procedures

General activation/cleavage procedure

*After washing with DCM, methanol and THF, resin **66** (~ 1 g) was treated with sodium hydrosulfite (850 mg, 85% pure) in water (5 ml) and THF (8 ml) for 2.5 hours at room temperature. After filtration and washing with THF, methanol and DCM, the resin was recovered. The combined solvents were evaporated. Ethyl acetate (40 ml) and water (5 ml) were added into the residue, and this mixture was stirred for 20 minutes. The organic layer was separated, and the aqueous layer was extracted with ethyl acetate. The combined organic layers were dried over magnesium sulfate. After filtration and evaporation, the crude product was obtained.*

6.9.2 Activation via intramolecular H-bonding

One shortcoming of the Boc-solid phase synthesis of cyclic peptide lies in the final acid-induced deprotection of side chains and concomitant cleavage from the linker. Cyclic products formed in this cyclative step have to be extracted from the deprotection/cleavage mixture. In an attempt to overcome this limitation, Bourne *et al.* have developed the catechol-derived safety-catch linker **69** (Figure 6.28). During peptide synthesis, one of the phenolic groups is masked with a benzyl group thus making the linker deactivated for aminolysis.[54] Other groups have reported that the reactivity of hydroxyphenyl esters toward amines in their O-alkylated form are approximately three orders of magnitude lower than in their OH free form.[55] Moreover, it was reported that aminolysis of the o-hydroxyphenyl ester of protected amino acids or peptides occurs without racemization.[56] The authors reported that in order to further decrease the reactivity

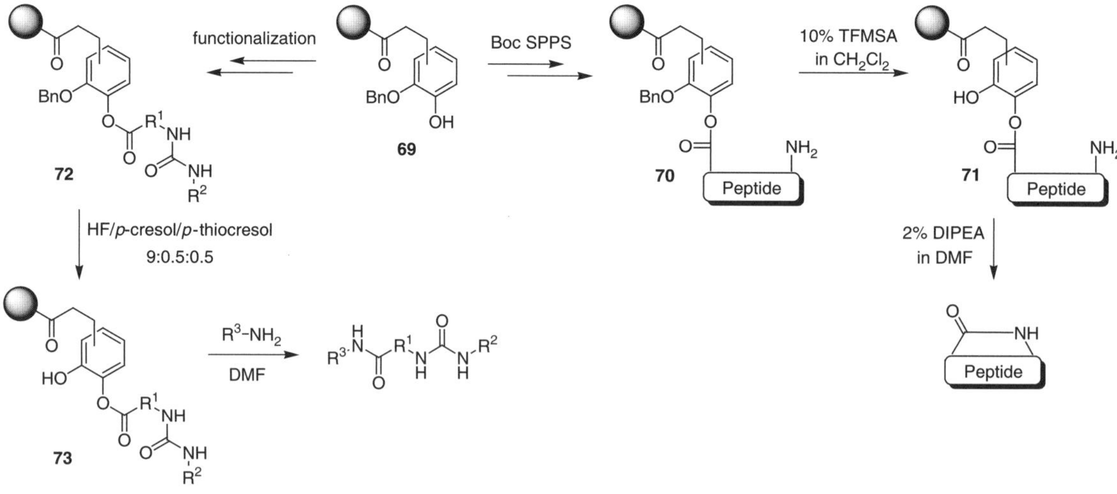

Figure 6.28 Catechol derived safety-catch linker

of the ester group during the peptide synthesis, the catechol linker had to be anchored to the support via an alkyl chain rather than a carboxylic group. Also, a tripeptide spacer was introduced between the support and the linker for a more effective peptide assembly. Several linear peptides were then assembled in a stepwise fashion using *in situ* neutralization protocols and *O*-benzotriazole-*N*, *N*, *N'*,*N'*-tetramethyluronium hexafluoro-phosphate (HBTU) activation to yield the intermediate **70**. Hydrofluoric aid or hydrobromic acid/TFA was used for the activation/deprotection of the resin **71**. Final cyclization or cleavage was accomplished by treatment with 2% DIPEA solution in DMF. Choice of the amino acid protective groups and cleavage conditions proved to be critical to attain good purity of the products. Optimized conditions where HF was replaced by 10% TFMSA/DCM were used in the automated parallel synthesis of 432 cyclic hexa- and pentapeptides.

Using the same safety-catch linker strategy, Smythe *et al.* have described the preparation of a library of peptidyl privileged structures that contain a urea moiety.[57] Activation of the resin **72** with HF was followed by treatment with excess of amine to provide the corresponding products in moderate yields and good quality (Figure 6.28). More recently, the same group reported the additional application of this linker to the synthesis of a biphenyl-based library.[58] As in the previous examples, the linker was first activated with a strong acid. The final aminolysis provided expected products in high purity and good overall yields.

Typical Experimental Procedures

General activation/cleavage procedure

*O-Benzyl group on resin **72** was deprotected by treatment with 9:0.5:0.5 HF/p-cresol/p-thiocresol for one hour at 0 °C. The resin was then washed with anhydrous ether and dried under nitrogen. The resin was suspended in amine (4 equiv.) in either DMF or DMSO and the treatment continued for one day prior to filtration and washing with DMF or DMSO. For amines that were used as the acid salts, additional DIPEA (1 equiv.) was added for neutralization in the cleavage mixture. An additional amine (4 equiv.) in DMF or DMSO was then added to the drained resin and left for an additional day. After the second filtration and washing, the combined filtrate was removed under reduced pressure and dissolved in 55:45:0.05 water/acetonitrile/TFA and lyophilized before purification by preparative RP-HPLC.*

6.9.3 Activation by formation of an alkyne-cobalt complex

Safety-catch linkers based on a propargyl alcohol have been reported for the synthesis of carboxylic acids and amines.[59] These linkers are known to be stable under strong acidic conditions, but can be activated toward acidic cleavage upon treatment with dicobalt octacarbonyl. Arylalkyne resin **74** is thus transformed into the alkyne-cobalt complex **75** (Figure 6.29). Due to the strong stabilization of a cationic charge at the α-position, the linker can now be easily cleaved by acids to provide the carboxylic acid and the resin-bound alkyne-cobalt complex **76**. While high purities of products were observed, the attachment protocol for amines suffered from low yields and requires further improvement. The solid phase synthesis of oligosaccharides using this linker strategy was reported by Kusumoto *et al.*[60]

Typical Experimental Procedures

General activation/cleavage procedure

*Resin **74** (∼ 100 mg) was shaken with a mixture of $Co_2(CO)_8$ (14.4 mg, 42.0 μmol) in DCM (3.0 ml) at room temperature for one hour. After the reaction mixture was filtered, the resin was washed with DMF*

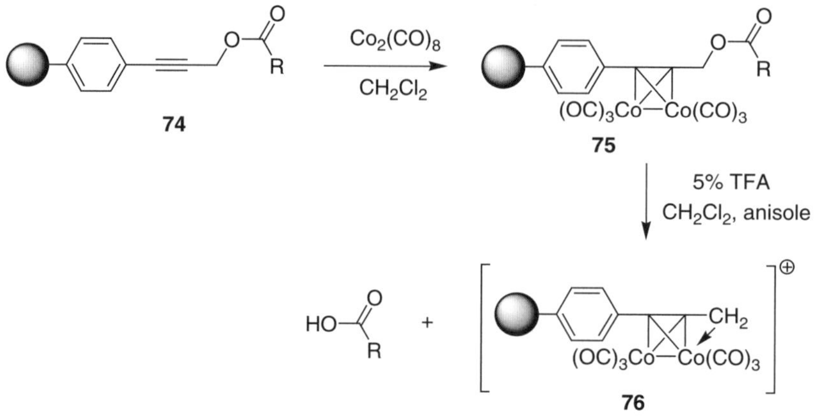

216   Traditional Linker Units for Solid-Phase Organic Synthesis

Figure 6.29 *Activation of propargyl derived safety-catch linker by formation of an alkyne–cobalt complex*

and DCM. The resin was shaken with TFA (0.5 ml) in DCM (4.0 ml) and water (0.5 ml) at room temperature for 12 hours and then filtered. The resin was then washed with ethyl acetate. The organic layers were combined, washed with saturated sodium bicarbonate solution and brine, dried over magnesium sulfate, and concentrated in vacuo. The residue was purified with preparative silica gel TLC.

6.9.4   Activation by oxidation of arylsulfide for pummerer rearrangement

The utility of this safety-catch linker strategy for the preparation of a variety of aldehydes and alcohols was reported by Li *et al.* (Figure 6.30).[61] Resin-bound aryl sulfide **78** was prepared by S-alkylation under several alkylating conditions. The sulfide linkage was then oxidized by treatment with ^{t}BuOOH/10-camphorsulfonic acid (CSA). Subsequent Pummerer rearrangement was initiated by treatment of the resin **79** with trifluoroacetic anhydride (TFAA) leading to trifluoroacetoxythioacetal intermediate **80**. Thioacetals can then be cleaved either by treatment with Et$_3$N in ethanol to give the aldehyde or by reductive release using sodium tetrahydroborate/triethylamine in ethanol to yield the corresponding alcohols.

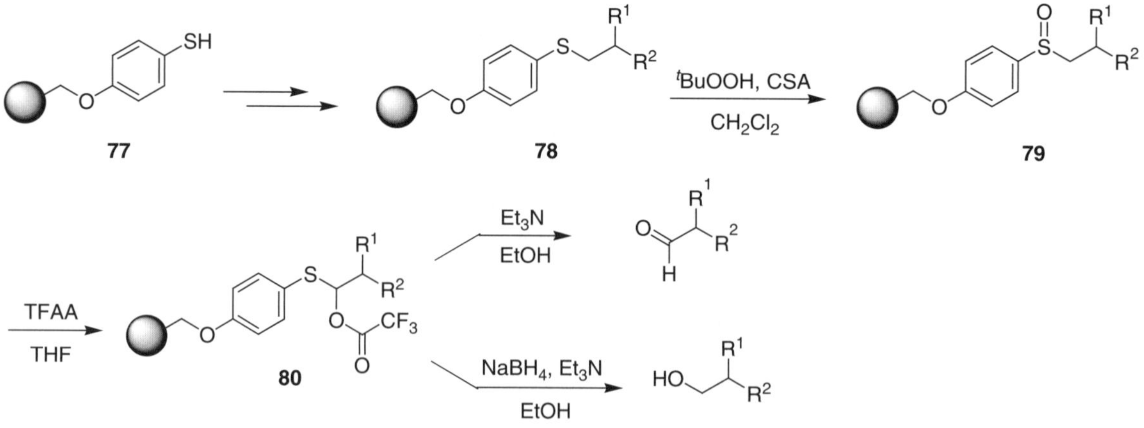

Figure 6.30 *Arylsulfide-based safety-catch linker activated by Pummerer rearrangement*

Typical Experimental Procedures

General activation/cleavage procedure

Resin 78 was swollen in DCM (1.6 ml) and was then treated with t-BuOOH (1.264 mmol) and CSA (0.158 mmol). The solution was agitated at room temperature for 12 hours. The resin was then flushed and washed with DMF and DCM, and dried by passing nitrogen through it. A suspension of this resin in THF (0.6 ml) was added slowly to a solution of trichloroacetic anhydride (0.06 ml) in THF (1.44 ml) at ice-bath temperature. The solution was slowly warmed to room temperature and allowed to stir for 45 minutes at room temperature. The reaction mixture was then flushed and washed with DMF and DCM. The resin was then suspended in THF (2 ml) and was treated with TEA (0.158 mmol) and ethanol (0.158 mmol). After the solution was agitated for two hours at room temperature, sodium borohydride (0.316 mmol) was added and the suspension was agitated for another two hours. After this time, the resin was filtered and the filtrate was concentrated in vacuo. The resulting residue was purified by column chromatography.

6.9.5 Activation by oxidative N-benzyl deprotection

Davies recently reported the preparation of safety-catch linker **81** where N-benzyl tertiary amine can be chemoselectively deprotected for a subsequent cyclative cleavage (Figure 6.31). Installation of a *gem*-dimethyl substitution α- to the carbonyl protects the carbonyl ester against nucleophiles while favoring 5-exo-trig cyclization due to the Thorpe–Ingold effect.[62] The linker applicability was demonstrated in the synthesis of a number of phenol derivatives. At the end of the solid-phase synthesis, resin-bound ester **82** is debenzylated by treatment with CAN to yield resin **83**, which further undergoes a base promoted cyclization to afford the desired phenols in high purity. Lactam **84** formed in this cyclative cleavage stays attached to the solid support.

Typical Experimental Procedures

General activation/cleavage procedure

CAN (5.0 equiv.) was added to a stirred suspension of resin 82 (1.0 equiv.) in THF/water (8:1). After stirring for 16 hours the resin was transferred to a sintered funnel and washed repeatedly with THF, water, THF and DCM. The resin was then treated with DCM/triethyl amine (5:1) and the filtrate was concentrated in vacuo. The resulting residue was filtered through a plug of silica (ethyl acetate eluent) and the solvent removed in vacuo.

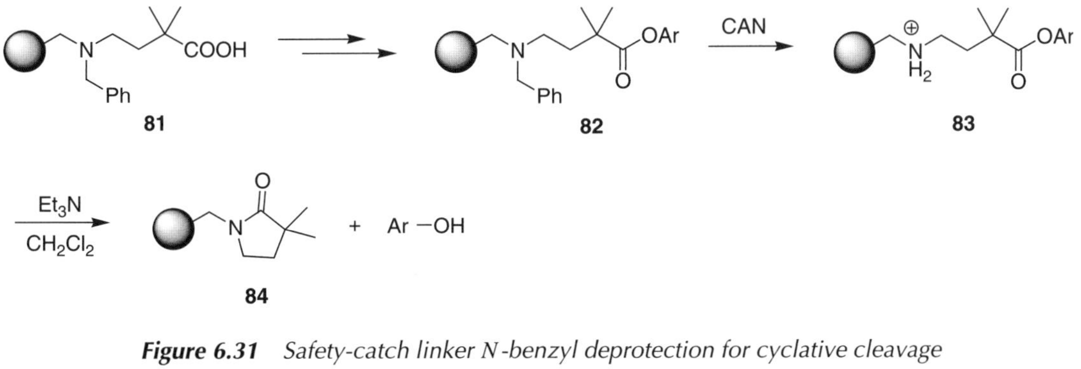

Figure 6.31 *Safety-catch linker N-benzyl deprotection for cyclative cleavage*

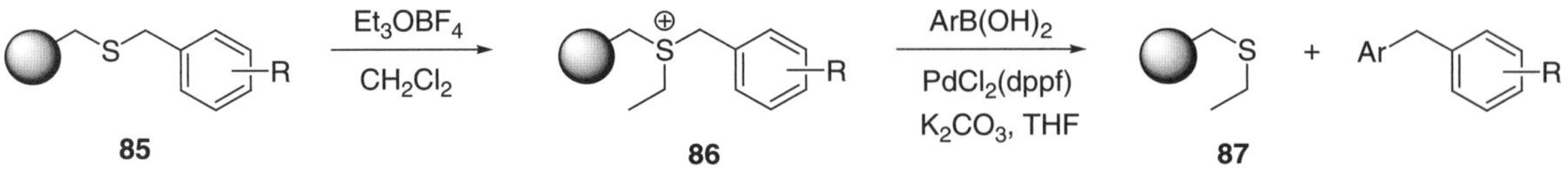

Figure 6.32 Alkylative activation of benzyl thioether linker for palladium catalyzed functionalization/cleavage

6.9.6 Activation by thioether alkylation

In an effort to develop a safety-catch linker that would allow for carbon–carbon bond formation and functionalization within the cleavage step, Wagner and Mioshowski reported the use of derivatized benzyl thioethers for the synthesis of biarylmethane derivatives.[63] Following the alkylation of a resin-bound thiol with a diverse set of benzyl bromides, benzyl thioether resin **85** (Figure 6.32) was activated by alkylation at the sulfur atom with triethyloxonium tetrafluoroborate leading to the corresponding sulfonium salt **86**. This sulfonium salt serves as a substrate for palladium catalyzed cross-coupling reactions with diverse boronic acids to yield the desired products together with the thioether resin **87**. A variety of benzylthioethers bearing miscellaneous functional groups, both electron rich and electron deficient were found to be stable to the activation conditions. Also, the unactivated sulfide linkage was reported to be stable to the palladium catalyzed cross-coupling conditions. However, formation of a diaryl compounds derived from homo-coupling of the boronic acid was found to be a significant side reaction. Since the palladium catalyst was present in the released product, purification by silica flash chromatography was necessary.

Gennari *et al.* showed that resin-bound stabilized sulfur ylides can be used for the synthesis of cyclopropanated macrocyclic lactones.[64] Reacting resin-bound mercaptoacetic acid with ω-hydroxy vinylketone gave resin **88** (Figure 6.33). Upon treatment with methyl trifluoromethanesulfonate (methyl triflate), the corresponding activated sulfonium salt **89** was formed. Treatment with DBU afforded the stabilized sulfonium ylide, which led to an intramolecular cyclopropanation via a Michael addition–elimination and cleavage. Release from the solid support provided the resin-bound methyl thioether **90**. The macrocyclic product was isolated in good yield and purity exclusively as the *trans* isomer. The authors also showed that such sulfur ylide can react with aldehydes to release the corresponding epoxides.

Typical Experimental Procedures

General activation/cleavage procedure

*The sulfide resin **85** (200 mg, 1.0 equiv.) was swollen in DCM (2 ml) and then cooled down to 0 °C. Triethyloxonium tetrafluoroborate (5 equiv.) was added and the resulting slurry was stirred for eight hours at room*

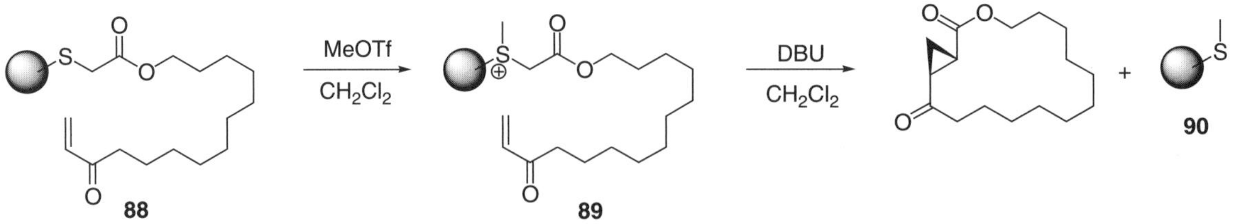

Figure 6.33 Activation of thioether acetate safety-catch linker by stabilized ylide formation

temperature. The activated resin was washed with DCM and methanol and finally dried under vacuum. To this resin under argon was added $Cl_2Pd(dppf)$ (0.2 equiv.), aryl boronic acid (2 equiv.), potassium carbonate (3 equiv.) and THF (2 ml). The suspension was degassed with argon and the reaction was carried out at 60 °C for 14 hours under argon atmosphere. After cooling, the resin was filtered off and washed successively with THF, DCM and methanol. The combined filtrates were evaporated to dryness to yield a residue which was filtered over a bed of silica.

6.10 Conclusion

As the nature of the compounds synthesized on solid supports progresses from peptides to more complex carbo- and heterocycles, there has been a significant increase in demands on the new linkers that are capable of delivering these complex structures. As such, the safety-catch approach has been quickly adapted into protocols used for the synthesis of combinatorial libraries. The usefulness of safety-catch principles has been successfully demonstrated on a multitude of examples. Nevertheless, it is the complexity of the linker synthesis together with limited commercial availability of the corresponding precursors that still remains a significant shortcoming, thus limiting a much broader use of this class of linkers. At the same time, owing to the many attractive features of the safety-catch linkers, it is expected that research efforts in this area will continue and that many new linkers will become commercially available.

References

[1] (a) Guillier, F., Orain, D., and Bradley, M.; *Chem. Rev.* **2000**, *100*, 2091. (b) James, I. W.; *Tetrahedron* **1999**, *55*, 4855.
[2] Pátek, M., and Lebl, M.; *Biopolymers (Peptide Science)* **1998**, *47*, 353.
[3] Kenner, G. W., and Dermott, J. R.; *J. Chem. Soc., Chem. Comm.* **1971**, 636.
[4] For review on the *N*-acylsulfonamide linker, see the following: Heider, P., and Link, A.; *Bioorg. Med. Chem.* **2005**, *13*, 585.
[5] Backes, B. J., Virgilio, A. A., and Ellman, J. A.; *J. Am. Chem. Soc.* **1996**, *118*, 3055.
[6] He, Y., Wilkins, J. P., and Kiessling, L. L.; *Org. Lett.* **2006**, *8*, 2483.
[7] Bourel-Bonnet, L., Rao, K. V., Hamann, M. T., and Ganesan, A.; *J. Med. Chem.* **2005**, *48*, 1330.
[8] Diaz-Moscoso, A., Benito, J. M., Mellet, C. O., and Garcia Fernandez, J. M.; *J. Comb. Chem.* **2007**, *9*, 339.
[9] Maclean, D., Hale, R., and Chen, M.; *Org. Lett.* **2001**, *3*, 2977.
[10] Hulme, C., Peng, J., Morton, G., et al.; *Tetrahedron Lett.* **1998**, *39*, 7227.
[11] Marshall, D. L., and Liener, I. E.; *J. Org. Chem.* **1970**, *35*, 867.
[12] Flanigan, E., and Marshall, G. R.; *Tetrahedron Lett.* **1970**, *11*, 2403.
[13] (a) Fantauzzi, P. P., and Yager, K. M.; *Tetrahedron Lett.* **1998**, *39*, 1291. (b) Breitenbucher, J.G., Johnson, C. R., Haight, M., and Phelan, J. C.; *Tetrahedron Lett.* **1998**, *39*, 1295.
[14] Dressman, B.A., Singh, U., and Kaldor, S. W.; *Tetrahedron Lett.* **1998**, *39*, 3631.
[15] (a) Sola, R., Saguer, P., David, M. L., and Pascal, R.; *J. Chem. Soc. Chem. Comm.* **1993**, 1786. (b) Sola, R., Mery, J., and Pascal, R.; *Tetrahedron Lett.* **1996**, *37*, 9195.
[16] Arseniyadis, S., Wagner, A., and Mioskowski, C.; *Tetrahedron Lett.* **2004**, *45*, 2251.
[17] Todd, M.H., Oliver, S. F., and Abell, C.; *Org. Lett.* **1999**, *1*, 1149.
[18] Arai, E., Tokuyama, H., Linsell, M. S., and Fukuyama, T.; *Tetrahedron Lett.* **1998**, *39*, 71.
[19] Wieland, T., Lewalter, J., and Birr, C.; *Liebigs Ann. Chem.* **1970**, *740*, 31.
[20] Semenov, A.N., and Gordeev, K. Y.; *Int. J. Peptide Protein Res.* **1995**, *45*, 303.
[21] Berst, F., Holmes, A. B., and Ladlow, M.; *Org. Biomol. Chem.* **2003**, *1*, 1711.
[22] Millington, C.R., Quarrell, R., and Lowe, G.; *Tetrahedron Lett.* **1998**, *39*, 7201.
[23] Stieber, F., Grether, U., and Waldmann, H.; *Angew. Chem. Int. Ed.* **1999**, *38*, 1073.

[24] Wieland, T., Birr, C., and Fleckenstein, P.; *Liebigs Ann. Chem.* **1972**, *756*, 14.
[25] Samanem, J.M., and Brandeis, E.; *J. Org. Chem.* **1988**, *53*, 561.
[26] Pátek, M., and Lebl, M.; *Tetrahedron Lett.* **1991**, *31*, 3891.
[27] Brik, A., Keinan, E., and Dawson, P. E.; *J. Org. Chem.* **2000**, *65*, 3829.
[28] Kiso, Y., Fukui, T., Tanaka, S., *et al.*; *Tetrahedron Lett.* **1994**, *35*, 3571.
[29] Kimura, T., Fukui, T., Tanaka, S., *et al.*; *Chem. Pharm. Bull.* **1997**, *45*, 18.
[30] Estep, K. G., Neipp, C. E., Stephens-Stramiello, L. M., *et al.*; *J. Org. Chem.* **1998**, *63*, 5300.
[31] Scicinski, J. J., and Congreve, M. S., Ley, S. V.; *J. Comb. Chem.* **2004**, *6*, 375.
[32] Ohno, H., Tanaka, H., and Takahashi, T.; *Synlett* **2004**, *3*, 508.
[33] Isidro-Llobet, A., Alvarez, M., Burger, K., *et al.*; *Org. Lett.* **2007**, *9*, 1429.
[34] Obrecht, D., Abrecht, C., Grieder, A., and Villalgordo, J. M.; *Helv. Chim. Acta* **1997**, *80*, 65.
[35] (a) Gayo, L. M., and Suto, M. J.; *Tetrahedron Lett.* **1997**, *38*, 211. (b) Kumar, A., Sinha, S., and Chauhan, P. M. S.; *Bioorg. Med. Chem. Lett.* **2002**, *12*, 667. (c) Font, D., Heras, M., and Villalgordo, J. M.; *J. Comb. Chem.* **2003**, *5*, 311.
[36] Ding, S., Gray, N. S., Ding, Q., *et al.*; *J. Comb. Chem.* **2002**, *4*, 183.
[37] Hwang, J.Y., and Gong, Y.-D.; *J. Comb. Chem.* **2006**, *8*, 297.
[38] Lebreton, S., Newcombe, N., and Bradley, M.; *Tetrahedron* **2003**, *59*, 10213.
[39] (a) Morphy, J. R., Rankovic, Z., and Rees, D. C., *Tetrahedron Lett.* **1996**, *37*, 3209. (b) Brown, A. R., Rees, D. C., Rankovic, Z., and Morphy, J. R.; *J. Am. Chem. Soc.* **1997**, *119*, 3288.
[40] Ouyang, X., Armstrong, R. W., and Murphy, M. M.; *J. Org. Chem.* **1998**, *63*, 1027.
[41] Ciolli, C.J., Kalagher, S., and Belshaw, P. J.; *Org. Lett.* **2004**, *6*, 1891.
[42] (a) Heinonen, P., and Lonnberg, H.; *Tetrahedron Lett.* **1997**, *38*, 8569. (b) Kroll, F. E. K., Morphy, R., Rees, D., and Gani, D.; *Tetrahedron Lett.* **1997**, *38*, 8573.
[43] Yamada, M., Miyajima, T., and Horikawa, H.; *Tetrahedron Lett.* **1998**, *39*, 289.
[44] Bray, A. M., Maeji, N. J., and Geysen, H. M.; *Tetrahedron Lett.* **1990**, *31*, 5811.
[45] Atrash, B., and Bradley, M., *J. Chem. Soc., Chem. Commun.* **1997**, 1397.
[46] Orain, D., and Bradley, M.; *Molecular Diversity* **2000**, *5*, 25.
[47] Hoffmann, S., and Frank, R.; *Tetrahedron Lett.* **1994**, *35*, 7763.
[48] Lyttle, M. H., Hudson, D., and Cook, R. M.; *Nucleic Acids Research* **1996**, *24*, 2793.
[49] Kočiš, P., Krchňák, V., and Lebl, M., *Tetrahedron Lett.* **1993**, *34*, 7251.
[50] Rock, R.S., and Chan, S. I.; *J. Org. Chem.* **1996**, *61*, 1526.
[51] Routledge, A., Abell, C., and Balasubramanian, S.; *Tetrahedron Lett.* **1997**, *38*, 1227.
[52] Lee, H.B., and Balasubramanian, S.; *J. Org. Chem.* **1999**, *64*, 3454.
[53] Zheng, A., Shan, D., and Wang, B.; *J. Org. Chem.* **1999**, *64*, 156.
[54] Bourne, G. T., Golding, S. W., McGeary, R. P., *et al.*; *J. Org. Chem.* **2001**, *66*, 7706.
[55] (a) Senatore, L., Ciuffarin, E., Isola, M., and Vichi, M.; *J. Am. Chem.Soc.* **1976**, *98*, 5306. (b) Ciuffarin, E., and Lois A.; *J. Org. Chem.* **1983**, *48*, 1047.
[56] Jones, J. H.; *J. Chem. Soc., Chem. Commun.* **1969**, 1436.
[57] Horton, D. A., Severinsen, R., Kofod-Hansen, M., *et al.*; *J. Comb. Chem.* **2005**, *7*, 421.
[58] Severinsen, R., Bourne, G. T., Tran, T. T., *et al. Comb. Chem.* **2008**, *10*, 557.
[59] Fürst, M., and Rück-Braun, K.; *Synlett* **2002**, *12*, 1991.
[60] Izumi, M., Fukase, K., and Kusumoto, S.; *Synlett* **2002**, *12*, 1409.
[61] Tai, C.-H., Wu, H.-C., and Li, W.-R.; *Org. Lett.* **2004**, *6*, 2905.
[62] Davies, S.G., Mortimer, D. A. B., Mulvaney, A. W., *et al.*; *Org. Biomol. Chem.* **2008**, *6*, 1625.
[63] Vanier, C., Lorge, F., Wagner, A., and Mioskowski, C.; *Angew. Chem. Int. Ed.* **2000**, *39*, 1679.
[64] La Porta, E., Piarulli, U., Cardullo, F., *et al.*; *Tetrahedron Lett.* **2002**, *43*, 761.

7
Enzyme Cleavable Linker Units

Mallesham Bejugam[1] and Sabine L. Flitsch[2]

[1]*Medicinal Chemistry Division, Aurigene Discovery Technologies, Bangalore, India*
[2]*The University of Manchester, Manchester Interdisciplinary Biocentre (MIB) and School of Chemistry, United Kingdom*

7.1 Introduction

Solid-phase synthesis has been widely used over the last two decades in the field of organic chemistry, especially in peptide, carbohydrate, oligonucleotide and peptide nucleic acid (PNA) synthesis.[1] Solid-phase methodology has several advantages over the conventional solution phase synthesis, including: (i) excess reagents can be used to drive the reaction to completion; (ii) no requirement for reaction work-up; and (iii) purification is achieved by simple filtrations, which can be easily automated. Any solid phase method involves the use of a covalently attached linker (spacer) group between the solid support and the chemical entity. These linkers should be stable under the reaction conditions used and have to be cleaved on completion of the synthesis in high yield, without affecting the structure of the target compound.

The most widely used linkers in peptide and nucleic acid synthesis have been cleaved under strongly acidic or basic conditions, respectively. However, such cleavage conditions limit the application of the linkers to other compound libraries, especially sugars, which contain acid labile glycosidic linkages. Alternative linkers, which are stable to commonly used reaction conditions but can be cleaved under mild neutral methods, have, therefore, been sought. Attractive propositions are enzyme-cleavable linkers, since enzyme-catalyzed reactions proceed under mild conditions[2] (pH 5–8, ambient temperature (25–37 °C) and high chemo, regio and stereoselectivity). Some excellent reviews have been published by Waldman *et al.* on enzyme cleavable linkers in solid phase synthesis.[3]

Linker Strategies in Solid-Phase Organic Synthesis Edited by Peter Scott
© 2009 John Wiley & Sons, Ltd

7.2 Enzyme cleavable linker units

More than a decade ago Flitsch et al. proposed that linkers can be broadly classified into two sub classes, such as exo and endo linkers, and such linkers can be cleaved by exo and endo enzymes, respectively.[4] Herein, the synthesis and cleavage strategies for a number of diverse exo and endo enzyme cleavable linkers and their application to the solid-phase synthesis of different classes of compounds are described.

7.2.1 Exo linker units

Exo linkers contain an enzyme recognition motif (C) to provide enzyme mediated hydrolysis, a site to attach a desired target molecule (A) and an additional spacer (B) attachment site to link an optional spacer (Figure 7.1).

A number of exo linkers have been developed by the groups of Waldman et al.[3c, 5, 6] and Flitsch et al.[7] Waldman et al. have reported a new enzyme labile safety-catch linker strategy for combinatorial synthesis on a soluble polymeric support POE 6000 (linear polyethyleneglycol).[3c, 5] The target molecule was cleaved by penicillin G acylase, which hydrolyze phenylacetamide of **1** with high selectivity. The enzyme releases an intermediate amine **2**, which cyclises, with liberation of the target molecule **3** (Scheme 7.1). The linker itself is attached to an amino-functionalized carrier via a urethane. It allows for the attachment of alcohols and amines as carboxylic acid esters and amides, respectively (X = O or NH).

The linker **5** was synthesized from homovanillic acid methyl ester **4** in a number of steps (Scheme 7.2). The polymeric support selected for attachment to the linker was a soluble linear polyethyleneglycol (POE 6000) functionalized at both ends with an amino group and with an average molecular weight of 6000 Da.[8, 9]

The suitability of the solid support linker conjugate **6** for combinatorial synthesis and also the possibility of its applicability were investigated for a variety of different transformations, such as Heck,[10] Suzuki,[11] Sonogashira,[12] Mitsunobu and Diels–Alder reactions.[3c, 5]

Typical Experimental Procedures

General procedure for the enzymatic cleavage of alcohols from POE 6000[3c]

A defined amount of polymer-bound substrate was dissolved in methanol and 0.2 M aqueous sodium phosphate buffer (pH 7.0). A suspension of penicillin G acylase (70 mg/ml, 560 units/ml in 0.1 M aqueous potassium phosphate buffer (pH 7.5) was added to the reaction mixture and was shaken at 37 °C. The similar amount of enzyme was added after 12, 24 and 36 hours. After 48 hours the reaction mixture was extracted with diethyl ether (6 × 20 ml). The combined organic layers were dried over magnesium sulfate and the solvent was removed under reduced pressure to afford cleaved alcohol. After that the aqueous

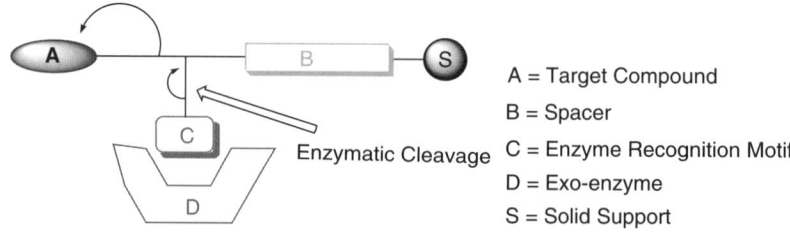

Figure 7.1 Schematic representation of exo-enzyme mediated cleavage of exo linker

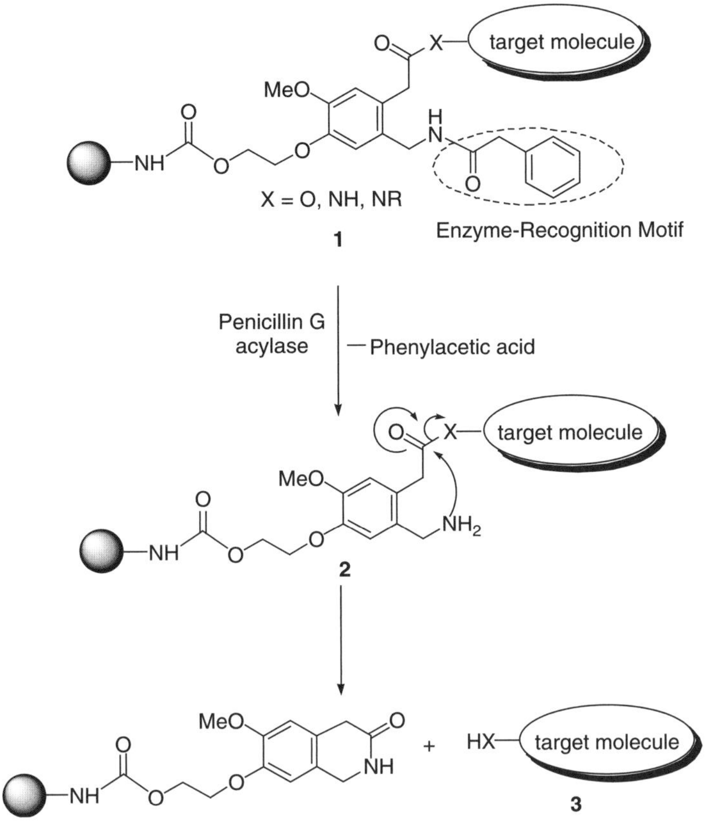

Scheme 7.1 *Principle for development of an enzyme-cleavable safety-catch linker group (X = O or NH)*[3c, 5]

phase was extracted with dichloromethane (DCM) (4 × 20 ml), and combined organic layers were dried over magnesium sulfate. The solvent was removed under reduced pressure. The polymer obtained thereby was dried in vacuo. The products were analyzed by NMR and HRMS.

In a different strategy Waldman *et al.* have reported linker system **7**, which contains an acyl group, such as acetate, that can be cleaved by lipases or esterases (Scheme 7.3).[6a] Phenolate **8** is generated, which fragments to give quinone methide **9** and then releases the desired compound **10**. The quinone methide **9** remains bound to the polymeric support and is trapped by an additional nucleophile. In this way amines (X = NH), alcohols (X = O) and carboxylic acids (X = CR_2) can be cleaved from the polymeric support.

The polymer support chosen for the subsequent transformation on the solid phase was TentaGelS-NH_2, a co-polymer of polystyrene and polyethyleneglycol unit that is terminally functionalized with –NH_2 groups.[13] Its polar surface and good solvation in aqueous systems have made it a popular choice of resin in a wide range of solvents, including water.

The linker **12** was synthesized from 5-methyl salicylic acid **11** in three steps and subsequently coupled to a TentagelS-NH_2 solid support via an amide bond, as proof-of-concept, an amino acid (Leu-O^{t}Bu) was coupled in and hydrolysis of **13** by lipase under very mild conditions generated the protected amino acid

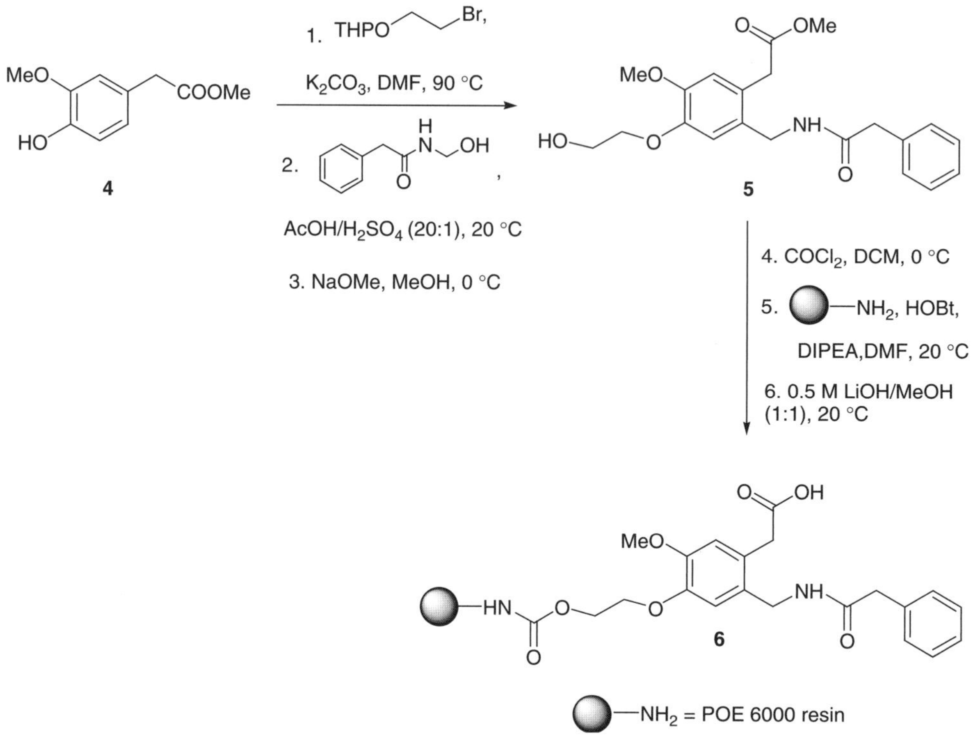

Scheme 7.2 Synthesis of the linker **5** and coupling to polymer support (POE 6000) [3c, 5] [THP = Trihydropyran; DIPEA = Diisopropylethyl amine]

14 in excellent yield (Scheme 7.4). The linker was also used for the synthesis of tetrahydro-β carbolines by Pictet–Spengler reaction.[14]

Typical Experimental Procedures[6a]

In a typical experiment a defined amount of polymer bound substrate was treated with lipase RB001-01 (Recombinant Biocatalysis, Diversa, San Diego, USA) at pH 5.8 in morpholino ethane sulfonic acid (MES) buffer/methanol mixture (60/40) and the mixture was shaken at 40 °C. The desired tetrahydro β-carbolines were isolated in 70–80% yield.

Flitsch *et al.* have introduced a novel linker system for the attachment of alcohol to the solid support and subsequent release from the solid support by penicillin G amidase.[7] Penicillin G amidase is known to catalyze the hydrolysis of a wide range of amines protected as the corresponding phenylacetyl derivatives. To incorporate the enzyme recognition site, the linker **15** was designed as shown in Scheme 7.5, in which – OR represents the alcohol group.

It was expected that cleavage would be initiated by hydrolysis of the phenyl acetamide moiety, generating the hemiaminol **16**, which should easily fragment in aqueous system releasing the alcohol.

The linker **17** was obtained by a synthetic route developed by Katritzky *et al.* (Scheme 7.6).[15] Amine linker **17** was then coupled to a variety of solid supports and gave **18** in excellent yields. A number of

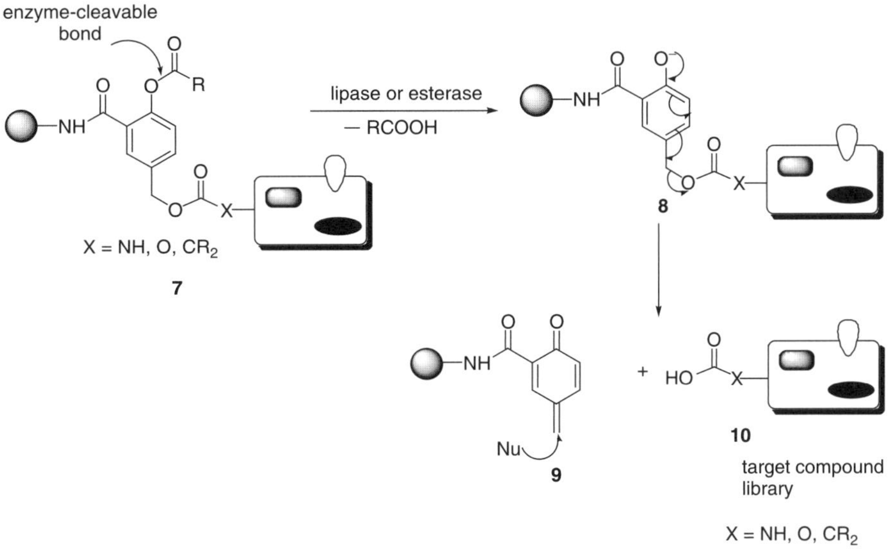

Scheme 7.3 *Design principle of an enzyme cleavable 4-acyloxy-benzyloxy linker system*[6a]

TentaGel and PEGA resins (PEGA is a co-polymer of polyacrylamide and functionalized polyethyleneglycol) were investigated because of their compatibility with aqueous reaction conditions. The thioethyl group of **18** was activated by treating with N-iodosuccinimide (NIS) followed by displacement with alcohol to afford the substrate **19**. Subsequently, **19** was treated with penicillin G amidase to afford alcohol **20** in good yield. This strategy was applied to a range of alcohols, such as sugars and amino alcohols (Figure 7.2).

Typical Experimental Procedures[7]

In a typical experiment penicillin amidase (700 units) in 0.1 M potassium phosphate buffer (pH 7.5; 0.4 ml) was added to the resin (2 mg) and the mixture was shaken at room temperature for 16 hours. The solution was removed, treated with 2 M hydrochloric acid (0.5 ml) and extracted with DCM (2 × 5 ml). The combined organic layers were concentrated under reduced pressure to afford a residue, which was dissolved in 0.3 ml of methanol and analysed by HPLC.

7.2.2 Endo linker units

Endo linkers contains a linker, in which the desired target molecule (A), an enzyme recognition motif that provides a site for enzyme mediated hydrolysis (C), and an optional spacer (B) are attached to the solid support (S) in a linear fashion (Figure 7.3). A disadvantage of endo linker is the requirement for a recognition motif, which in the majority of cases remains in the compounds released, thus limiting the type of libraries that can be generated. Nevertheless, a number of endo linkers have been reported over the last few years.

Elmore *et al.* have described the first enzyme cleavable linker system containing a phosphodiester group for peptide synthesis on Pepsyn K, a polyacrylamide resin.[16] The polymer-bound peptide, a collagenase substrate, was generated from **21** using 9-fluorenylmethoxycarbonyl (Fmoc) solid-phase peptide synthesis. In the last step the phosphodiester **22** was cleaved with a phosphodiesterase to obtain a desired peptide Fmoc-Ala-Pro-Gly-Leu-Ala-Gly-OH in excellent yield. The peptides, β-casomorphin and Leu-enkephalin were also synthesised in high yields using this methodology (Scheme 7.7).

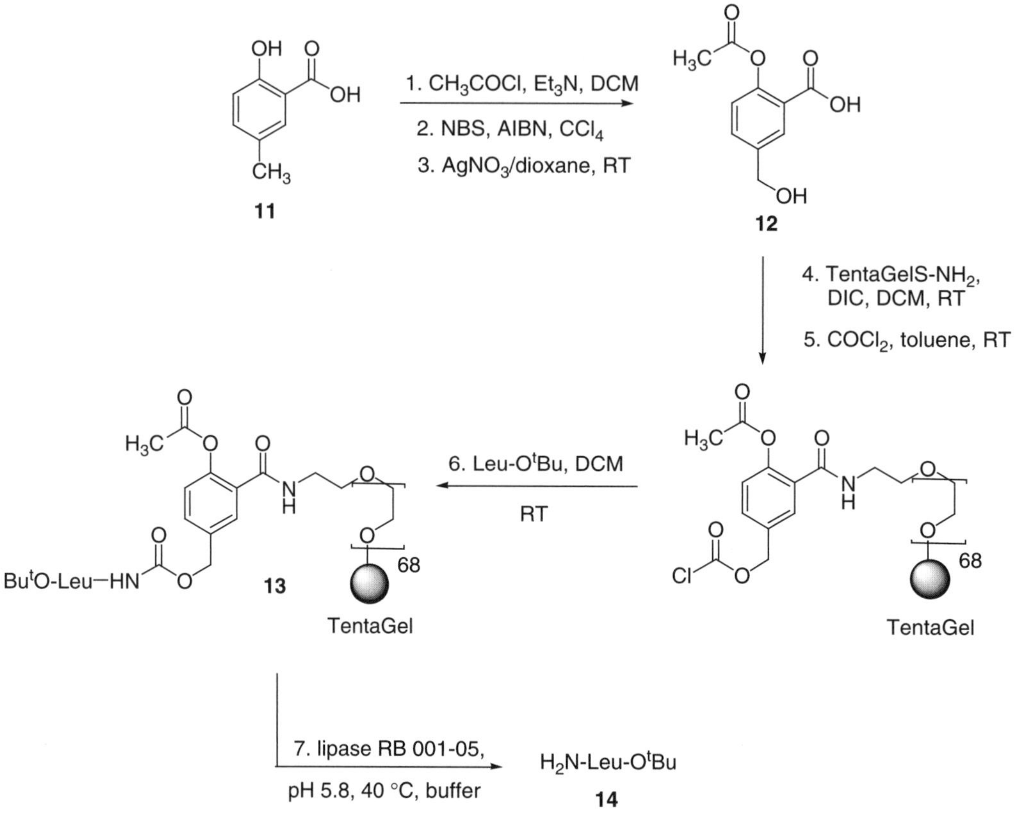

Scheme 7.4 Synthesis of substrate **13** on a TentaGel resin and cleavage by lipase[6a]

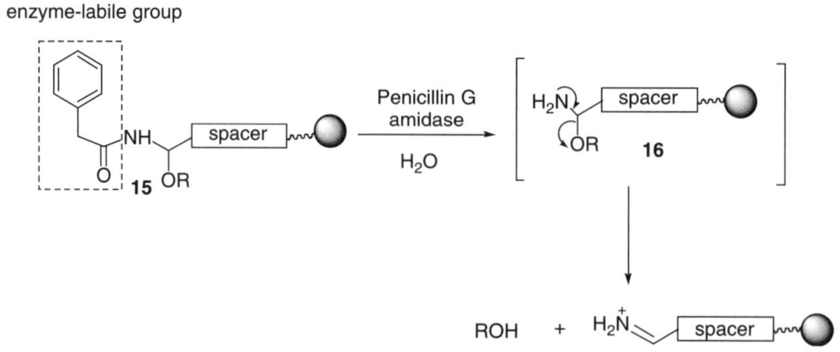

Scheme 7.5 Design of an enzyme-cleavable linker **15** and cleavage by penicillin G amidase[7]

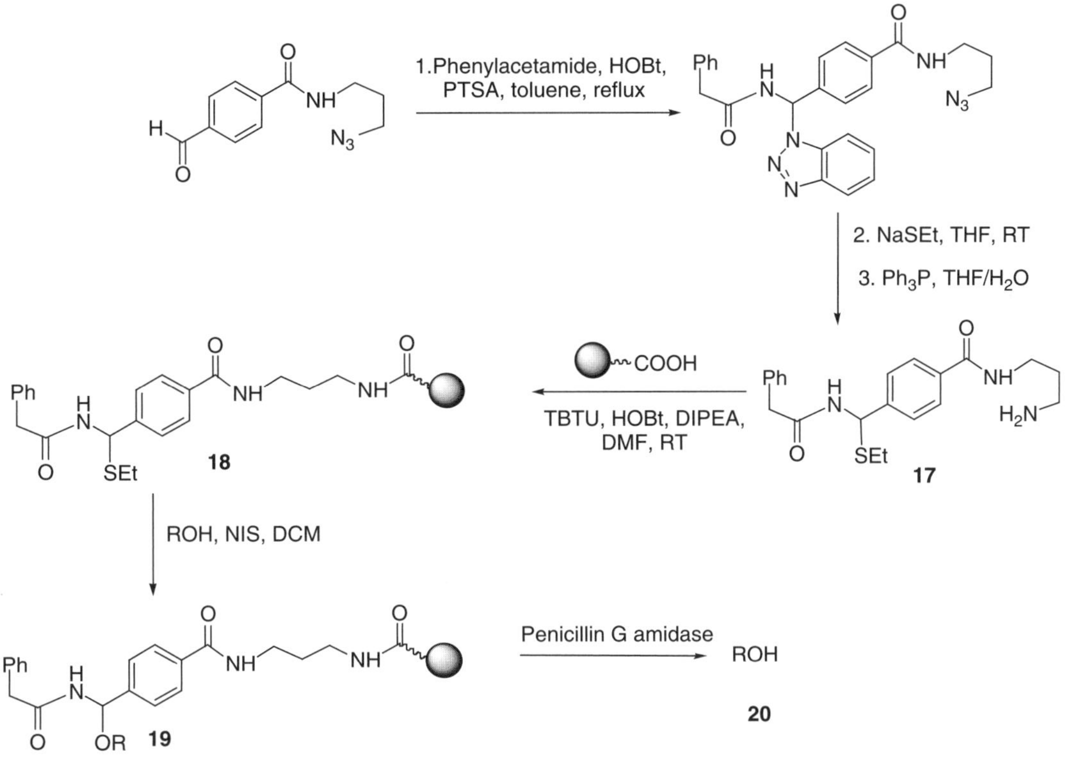

*Scheme 7.6 Synthesis of substrate **19** on polymer support and hydrolysis by penicillin G amidase*[7]

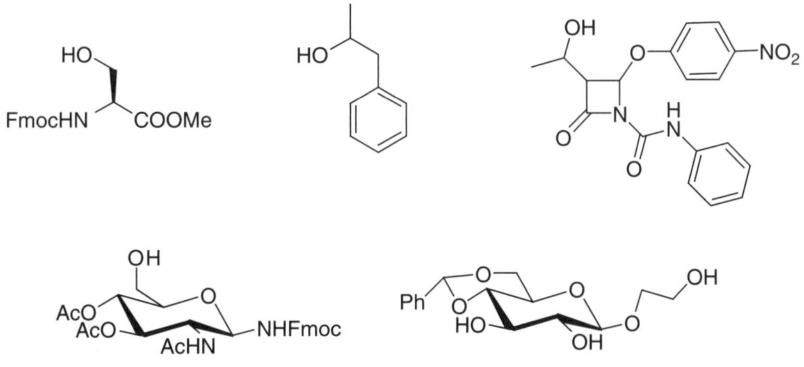

Figure 7.2 Library of alcohols cleaved from solid support by penicillin G-amidase[7]

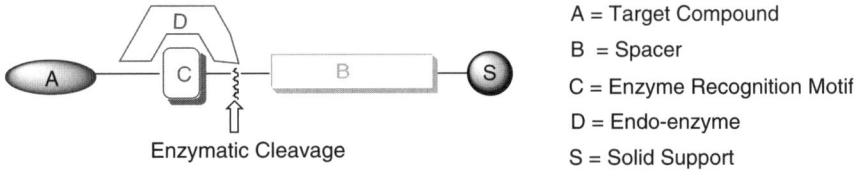

A = Target Compound
B = Spacer
C = Enzyme Recognition Motif
D = Endo-enzyme
S = Solid Support

Figure 7.3 Schematic representation of endo-enzyme mediated cleavage of endo linker

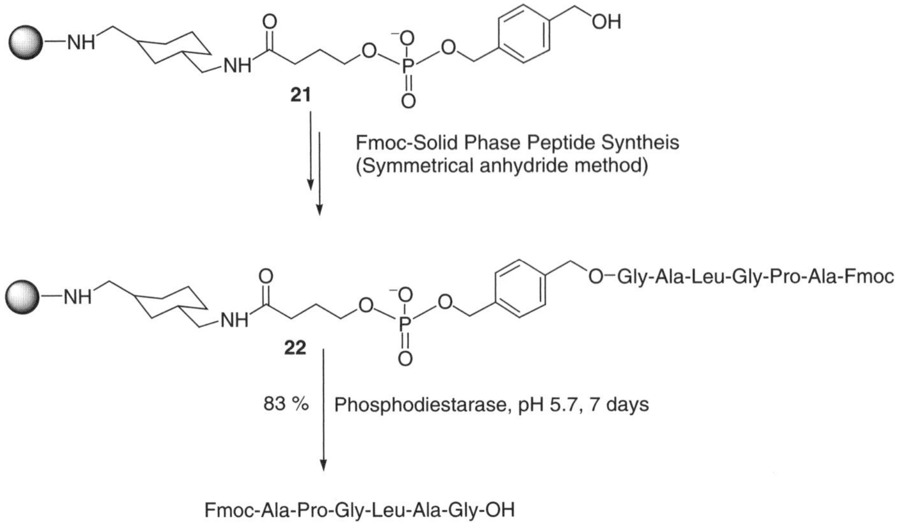

228 Traditional Linker Units for Solid-Phase Organic Synthesis

Scheme 7.7 *Synthesis of collagenase substrate **22** on solid support and cleavage by phosphodiesterase*[16]

Wong *et al.* have introduced a new linker strategy for the chemo-enzymatic synthesis of glycopeptides **24** on a silica-based solid support.[17] The key element in this strategy is to use an endoprotease recognition sequence to the solid support, so that the final glycopeptide can be released using proteases. The synthesis of glycopeptide substrate **23** was achieved by a chemo-enzymatic synthesis on aminopropyl silica as a solid support. A selective cleavage site was implemented for the release of intermediates and final products of the synthesis using a α-chymotrypsin sensitive phenylalanyl ester bond (Scheme 7.8).[18] The polymer-bound glycopeptide **23** was subjected to α-chymotrypsin mediated hydrolysis to obtain the desired glycopeptide **24** in excellent yield.

Typical Experimental Procedures[17]

*In a typical experiment the polymer bound substrate **23** (50 mg) was suspended in 0.2 ml of water and was shaken in the presence of α-chymotrypsin (Sigma, 2 mg/ml) at pH 7.0 for one hour. The filtrate was concentrated in vacuo using a SpeedVac Concentrator (SAVANT). The glycopeptide **24** was further derivatised and purified by semi preparative HPLC and characterized by NMR and HRMS.*

Nishimura *et al.* have reported a novel enzymatic method for the synthesis of neo-glycoconjuagte **29**, employing the α-chymotrypsin-sensitive, water soluble GlcNAc-polymer linker system **28** (Scheme 7.9).[19] In the first step, oxazoline derivative **25** was coupled to 6-(*N*-benzyloxycarbonyl-L-phenylalanyl)-amino-1-hexanol followed by *N*-deprotection of the Phenylalanine (Phe). Subsequent condensation with 6-acrylamidocaproic acid gave polymerisable GlcNAc derivative **26**. The intermediate **26** was subjected to Zemplen hydrolysis conditions followed by co-polymerisation of the sugar monomer with acrylamide in the presence of ammonium persulfate (APS) and *N,N,N′*,N′-tetramethylethylenediamine (TMEDA) gave the polymer **27** in good yield. The polymer **27** was subjected to galactosylation and sialylation, using corresponding glycosyl transferases to afford polymer bound substrate **28**. The polymer-bound substrate **28** was subjected to α-chymotrypsin mediated hydrolysis at 40 °C for 24 hours to afford neo-glycoconjugate **29** in excellent yield (Scheme 7.9).

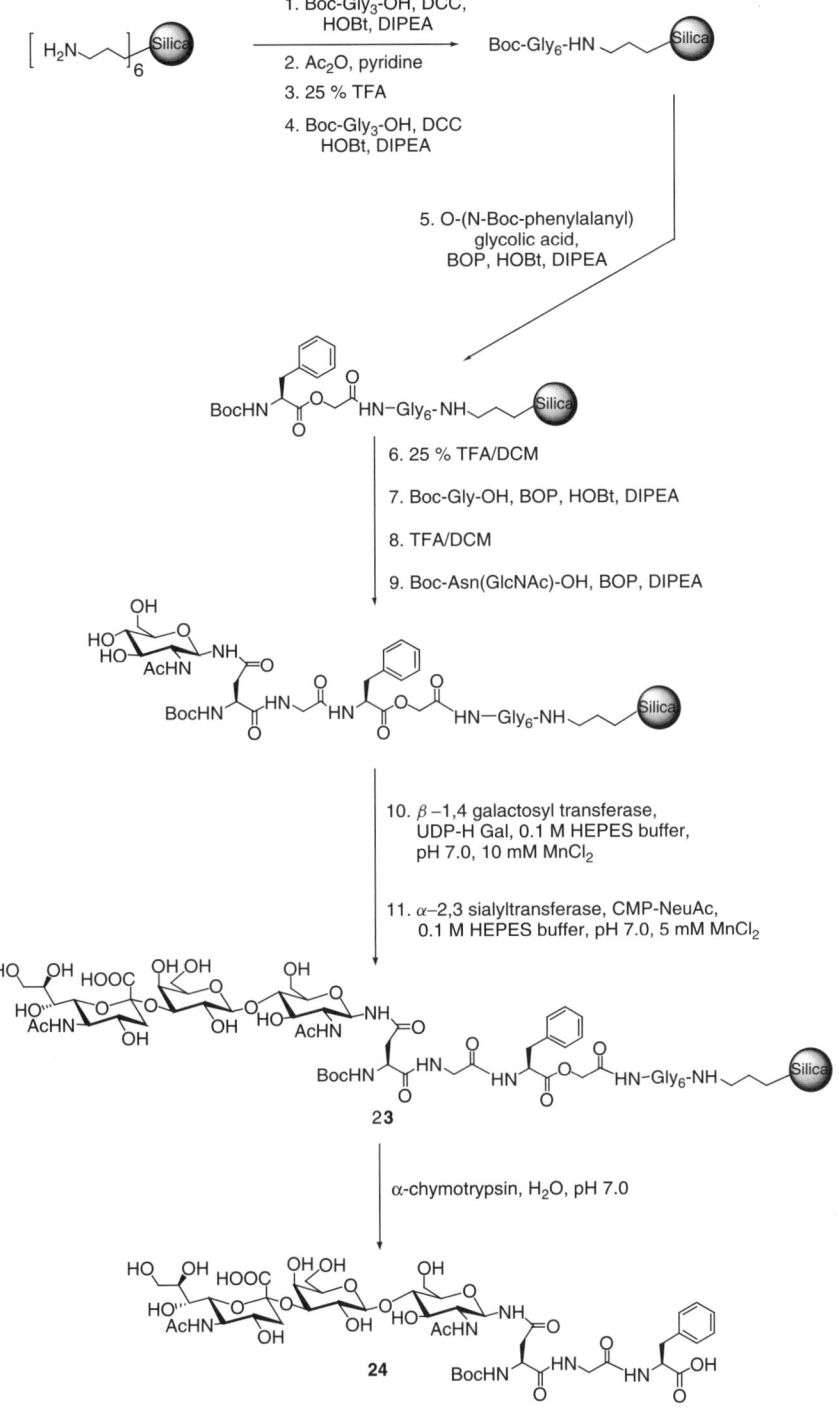

Scheme 7.8 Synthesis of glycopeptide substrate **23** by chemo-enzymatic method and cleavage by α-chymotrypsin[17]

Typical Experimental Procedures[19a]

*In a typical experiment, glycopolymer **28** (33 mg, 20 μmol of sialyl-LacNAc) and α-chymotrypsin (1 mg, 100 units) were incubated in 0.08 M Tris-HCl buffer (1 ml, pH 7.8) at 40 °C for 24 hours. The reaction mixture was directly purified by chromatography on a Sephadex G-25 column (1.5 × 40 cm) eluted with 0.05 M ammonium acetate solution. The sugar–containing fractions were collected and lyophilised to give desired compound **29** in excellent yield. The product was characterised by NMR spectroscopy.*

Nishimura *et al.* have introduced a new enzymatic synthesis of a glycosphingolipid, ganglioside **33**, on a water soluble polymer support having a specific linker system that can be recognised and cleaved by ceramide glycanase.[20] The synthesis of polymerisable lactose derivative **31** contain a ceramide glycanase-sensitive linker, as shown in Scheme 7.10. The lactosyl ceramide (LacCer) mimetic glycopolymer **31** was obtained by co-polymerisation of monomer precursor **30** with acrylamide. The solid support **31** was subjected to sialylation reaction using sialyltransferase to obtain polymer bound ganglioside **32**. Finally, ganglioside-bound substrate **32** was treated with leech ceramide glycanase contain excess ceramide (0.01 units), ceramide (4.85 equiv.), Triton CF54 (one drop), 50 mM sodium citrate buffer (pH 6.0) at 37 °C for 17 hours to afford ganglioside **33** in good yield (Scheme 7.10).

Typical Experimental Procedures[20]

*To a mixture of glycopolymer **32** (22.0 mg, about 16.2 μmol of sialyllactose residue) and ceramide (C_{18}) (50 mg, 78.7 μmol) in 50 mM sodium citrate buffer (pH 6.0, 1 ml) was added Triton CF-54 (one drop), and the mixture was sonicated for one minute in an ultrasonic water bath. The reaction was initiated by the addition of ceramide glycanase from leach (0.01 units) and incubated at 37 °C for 17 hours. The reaction mixture was directly chromatographed on Sephadex LH-20 column and eluted with chloroform/methanol/water (60:30:4.6 (v/v) to give ganglioside **33** in good yield (11.6 mg, 61%). The product was characterised by NMR and elemental analysis.*

Hoheisel *et al.* have synthesised arrays of 1000 peptide nucleic acid (PNA) oligomers with the length up to 16 nucleotides, of various different sequences on polymer membranes using a robot.[21] In this strategy, the PNA chain was attached to the peptide spacer glutamic acid-(γ-tert-butyl ester)-(ε-aminohexanoic acid)-(ε-amino hexanoic acid) [Glu(O^{t}Bu)-εAhx-εAhx] via a trypsin cleavable Glu-Lys bond. Subsequently, the Glu(O^{t}Bu)- εAhx- εAhx was coupled to an amino functionalised membrane using Fmoc chemistry. In the next stage, the membranes were mounted in the ASP 222 Automated SPOT Robot and a grid of the desired format was spotted at each position. The free amino groups outside the spotted areas were capped by acylation. Further chain elongation was achieved with Fmoc protected PNA monomers for the synthesis of desired PNA oligomers. Finally, the PNA oligomers were released from the membrane by treatment with bovine trypsin solution in ammonium bicarbonate at 37 °C for three hours. The cleaved PNA oligomers were analysed by matrix-assisted laser desorption ionisation mass spectrometry (MALDI-MS) (Figure 7.4).[21]

In a different strategy, Meldal and coworkers introduced the novel highly permeable, polar polymer support SPOCC (Scheme 7.11).[22] The feasibility of enzyme mediated transformation of SPOCC-1500 resin was examined with a decapeptide protease substrate that was synthesised on this solid support. In the first step, the hydroxyl-functionalized SPOCC resin was acylated with Fmoc-Gly-OH using 1-(2-mesitylenesulfonyl)-3-nitro-1H-1,2,4-triazole (MSNT) activation conditions to afford **34**. Subsequently, the Fmoc group was removed and a fully protected, preformed nonapeptide **35** was coupled to resin

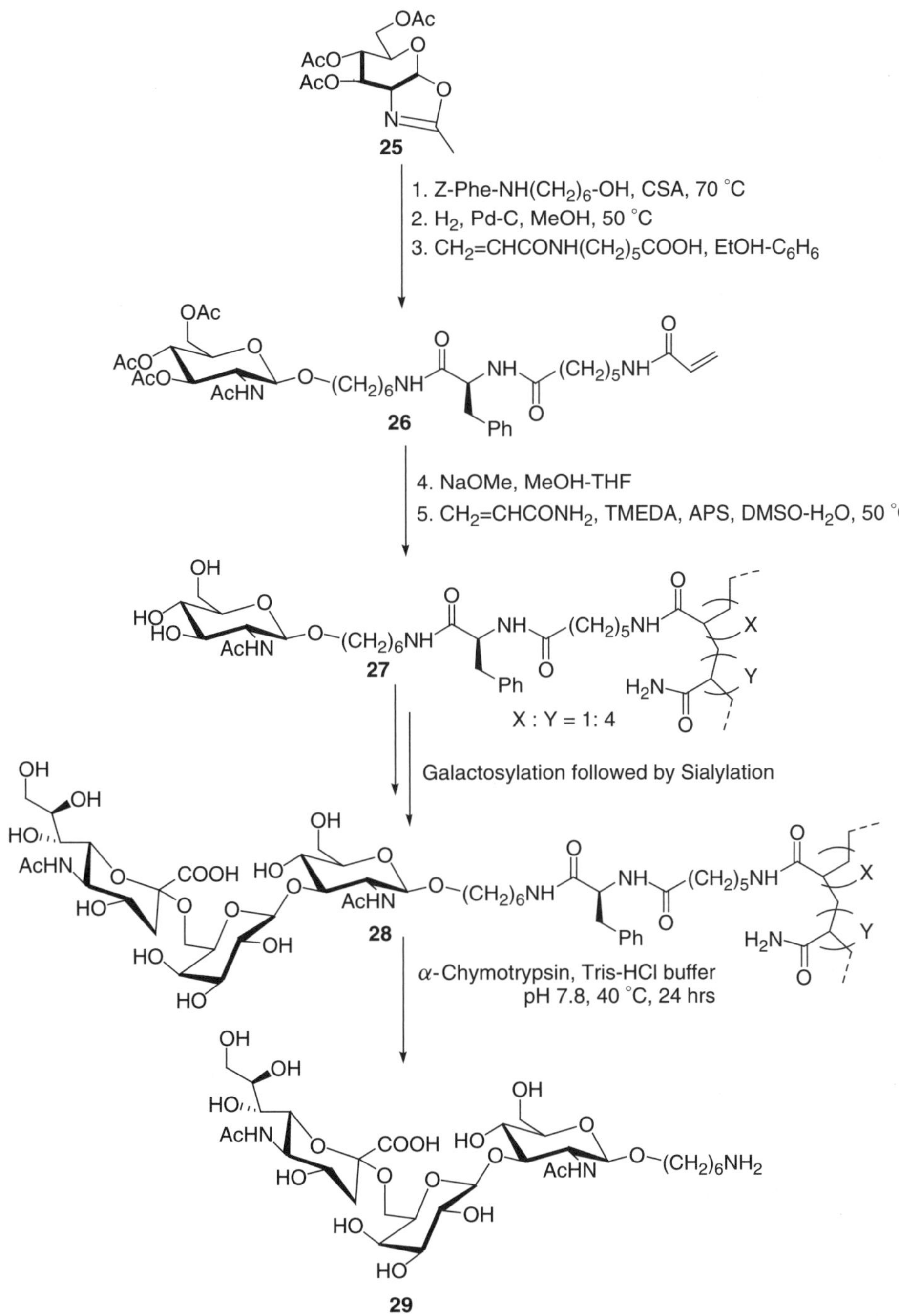

Scheme 7.9 *Chemo-enzymatic synthesis of neoglycoconjugate substrate* **28** *and α-chymotrypsin mediated cleavage from polymer support*[19]

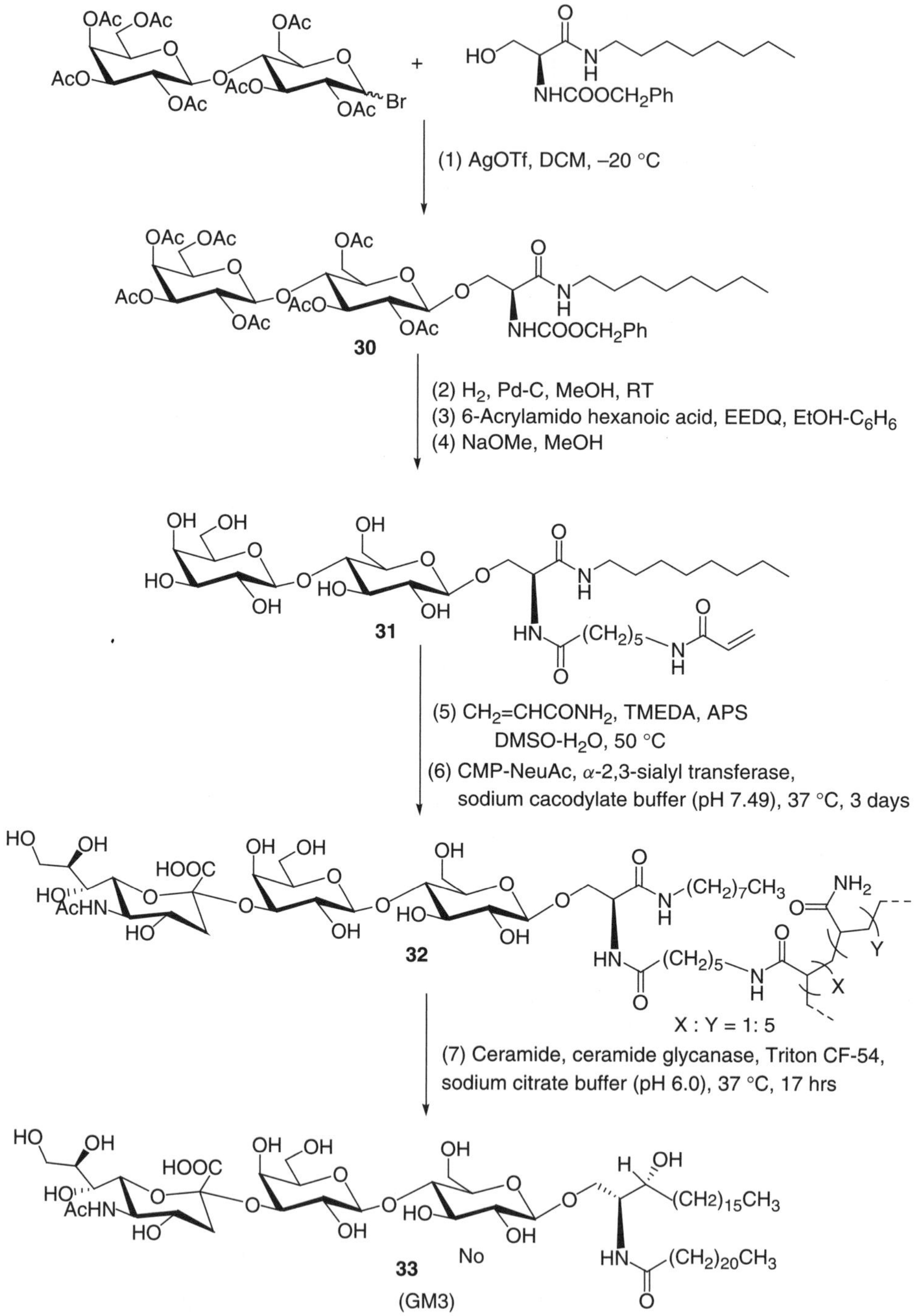

Scheme 7.10 Chemo-enzymatic synthesis of ganglioside substrate **32** on solid support and cleavage by ceramide glycanase[20]

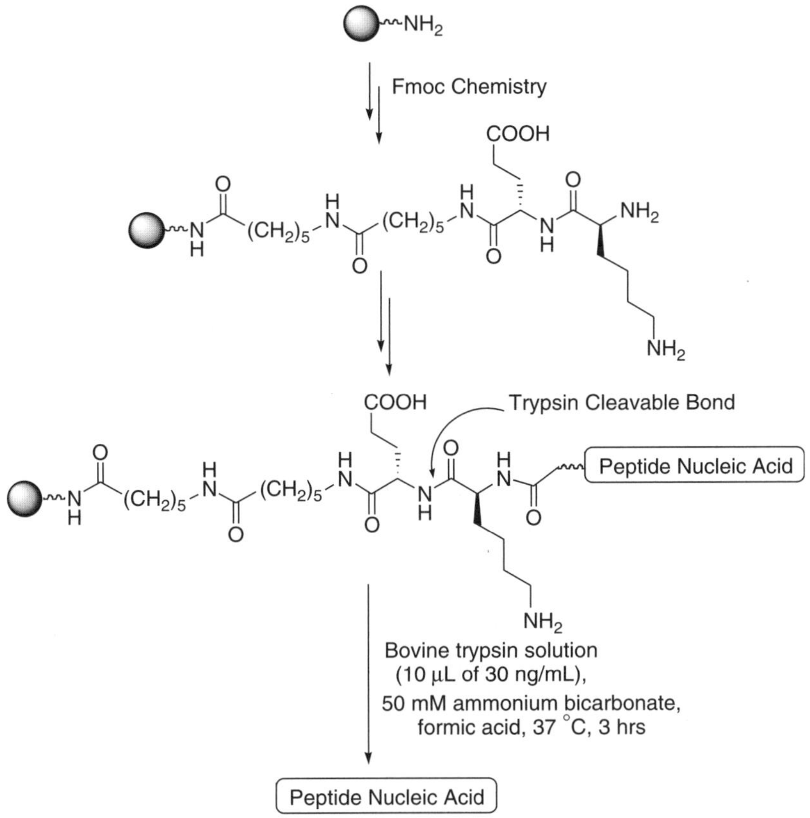

Figure 7.4 *Trypsin mediated cleavage of Glu-Lys bond in peptide nucleic acid (PNA) synthesis*[21]

to afford resin-bound protease substrate **36**. The substrate containing resin **36** was treated with 27 kDa protease subtilisin (0.01 μM in 0.01 M pH 7.0 phosphate buffer for 30 minutes) (Scheme 7.11). To elucidate the nature of the enzymatic reaction, Edman degradation, peptide cleavage and HPLC analysis was conducted. These results revealed that the digestion of the starting substrate was completed. The residual fragments cleaved from the resin were purified by HPLC and analysed by MALDI-TOF mass spectrometry.

Typical Experimental Procedures[22]

*In a typical experiment resin **36** (2 mg) was treated with subtilisin BNP' (Novo Nordisk, Bagsvaerd, Denmark, 10^{-7} M = 100 nM of the 27 kDa protein) in pH 7.0 phosphate buffer (50 mM of sodium dihydrogen phosphate (NaH_2PO_4) in water). After 15 minutes strong fluorescence was observed under UV irradiation. After three hours the resin was washed with water, DMF, THF and DCM and dried. One portion of the enzyme treated resin (1 mg) was cleaved with sodium hydroxide (50 μl of 0.1 M solution for two hours) and analysed by HPLC and MALDI-MS. The other portion of the resin (1 mg) was subjected to Edman degradation. The HPLC chromatogram indicated complete cleavage of the starting peptide substrate and yielded one peptide product. HPLC: retention time 21 minutes; ES-MS: calculated ($M = C_{26}H_{42}N_{10}O_7$) 606.7 Da and found m/z 607.6 Da; Edman degradation (three cycles): A, Abz, R and K.*

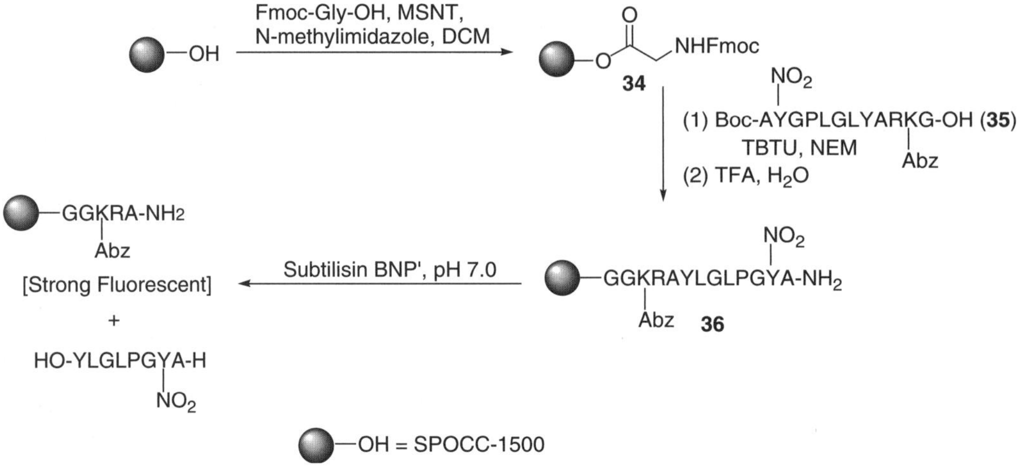

Scheme 7.11 Synthesis of peptide substrate **36** on solid support SPOCC and cleavage by subtilisin[22]

The majority of the above described linker systems have been specially designed and were synthesized by a number of steps to incorporate enzyme recognition motifs. The need for such motifs has limited the application of enzyme cleavable linkers. Recently, Flitsch *et al.* reported an unexpected enzymatic cleavage of an ester linker that would commonly employed in solid-phase peptide synthesis, such as Wang linker (hydroxymethylphenoxyacetic acid, HMPA) (Scheme 7.12).[23] Cleavage was not limited to peptides containing *C*-terminal aromatic amino acids normally associated with α-chymotrypsin recognition motifs but can be used for a wide range of peptide sequences. The interesting lability of the commonly used Wang linker to chymotrypsin catalyzed hydrolysis had not been observed before and was further investigated. First of all, the *C*-terminal amino acid residue attached to Wang linker was replaced by amino acids with diverse functionalities to investigate the tolerance of chymotrypsin at the 'pseudo P_1' site of the ester. A series of resin bound Fmoc-L-amino acids incorporating the acid labile Wang linker were synthesized by reacting HMPA-PEGA with pre-activated solution of Fmoc-amino acid (10 equiv.) and DIC (5equiv.) in DMF. Subsequently, the side chain protecting groups were removed under appropriate conditions to afford substrates **37–51**. The resin bound substrates **37–51** were incubated with α-chymotrypsin (2 mg/ml of 0.1 M potassium phosphate buffer pH 8.0) for 16 hours to afford the cleaved products **52–60** (Scheme 7.12, Table 7.1).

The cleavage reaction mixture was analyzed by LC-MS using reverse phase conditions. The LC-MS results are shown in Table 7.1.

A great advantage of biocatalysis over chemical synthesis is that reactions can often be highly stereoselective. Thus, chymotrypsin is known to be highly specific for L-amino acids. Such stereospecificity was also observed for the present linker cleavage; whereas the L-amino acid in **37** was efficiently cleaved from the resin, its D-enantiomer **44** did not yield any **59** under the same conditions (Entry 8, Table 7.1). This result confirms that the observed cleavage of the linker is indeed catalyzed by the enzyme. Given that the *C*-terminal amino acid is often prone to racemization during peptide synthesis or cleavage, α-chymotrypsin catalyzed cleavage is an attractive option for the simultaneous cleavage and kinetic resolution of *C*-terminal epimers.

Wang linker is popular in Fmoc solid-phase peptide synthesis because of its acid lability. However, acid lability did not seem to be important for α-chymotrypsin hydrolysis, since the base labile HMBA

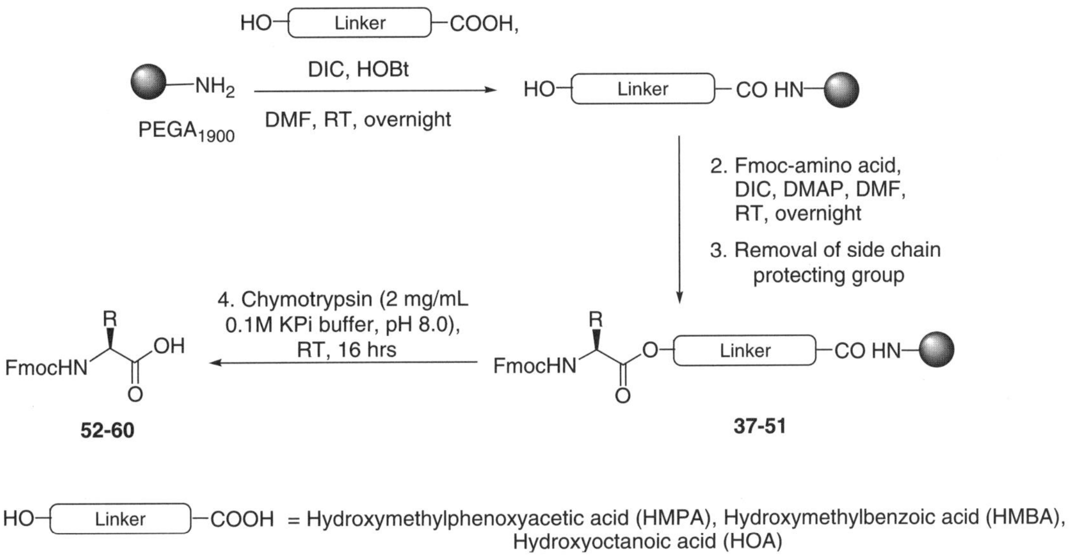

Scheme 7.12 Synthesis of substrates **37–51** and cleavage by α-chymotrypsin[23]

Table 7.1 Chymotrypsin mediated hydrolysis of substrates **37–51**[23]

Entry	Linker	Substrate	R	Product	Yield (%)
1	HMPA	37	CH_2Ph	52	100
2	HMPA	38	CH_3	53	100
3	HMPA	39	$CH_2CH(CH_3)_2$	54	100
4	HMPA	40	CH_2OH	55	60
5	HMPA	41	CH_2COOH	56	100
6	HMPA	42	$(CH_2)_4 NH_2$	57	100
7	HMPA	43	H	58	24
8	HMPA	44	D-CH_2Ph	59	0
9	HMPA	45	$CH_2CO_2(2-{}^iPrPh)$	60	25
10	HMBA	46	CH_2Ph	52	100
11	HMBA	47	CH_3	53	100
12	HMBA	48	CH_2COOH	56	100
13	HOA	49	CH_2Ph	52	100
14	HOA	50	H	58	35
15	HOA	51	CH_2COOH	56	100

(hydroxymethylbenzoic acid) linker and a neutral open chain linker HOA (hydroxy octanoic acid) was also readily cleaved by the enzyme (Table 7.1).[23]

An interesting application for enzyme cleavable linkers is in glycopeptide synthesis because of the additional complexity that arises from carbohydrate side chains in solid-phase peptide synthesis, in particular problems of acid lability. The applicability of the enzyme labile linker to the highly conserved *N*-glycan motif (Asn-X-Ser/Thr tripeptide codon where X = any amino acid except Pro)[8, 9] was investigated

(Scheme 7.13). The synthesis of glycopeptide **62** began from attachment of HMPA linker to PEGA$_{1900}$ (0.15 mmol/g (dry), 10% wt in methanol) by activating with DIC, HOBt in DMF to yield **61**. The first amino acid, Fmoc-Ser(Trt)-OH was coupled to **61** via the symmetrical anhydride approach and the trityl side chain protecting group was selected in order to enable its deprotection in the presence of the TFA labile HMPA linker. The glycopeptide **62** was then assembled using standard Fmoc solid-phase glycopeptide synthesis using the glycoamino acid, Fmoc-Asn(GlcNAc)-OH.[24] The side chain protecting group was removed with TFA/TIS/DCM (1:5:94) to afford resin bound glycopeptide substrate **62**. Then the substrate **62** was subjected to the enzymatic cleavage by using α-chymotrypsin (2 mg/ml of 0.1 M of potassium phosphate buffer pH 8.0) for 16 hours. The cleavage reaction mixture was analysed by LC-MS using reverse phase conditions and found that the desired glycopeptide **63** was obtained in quantitative yield (Scheme 7.13).[23]

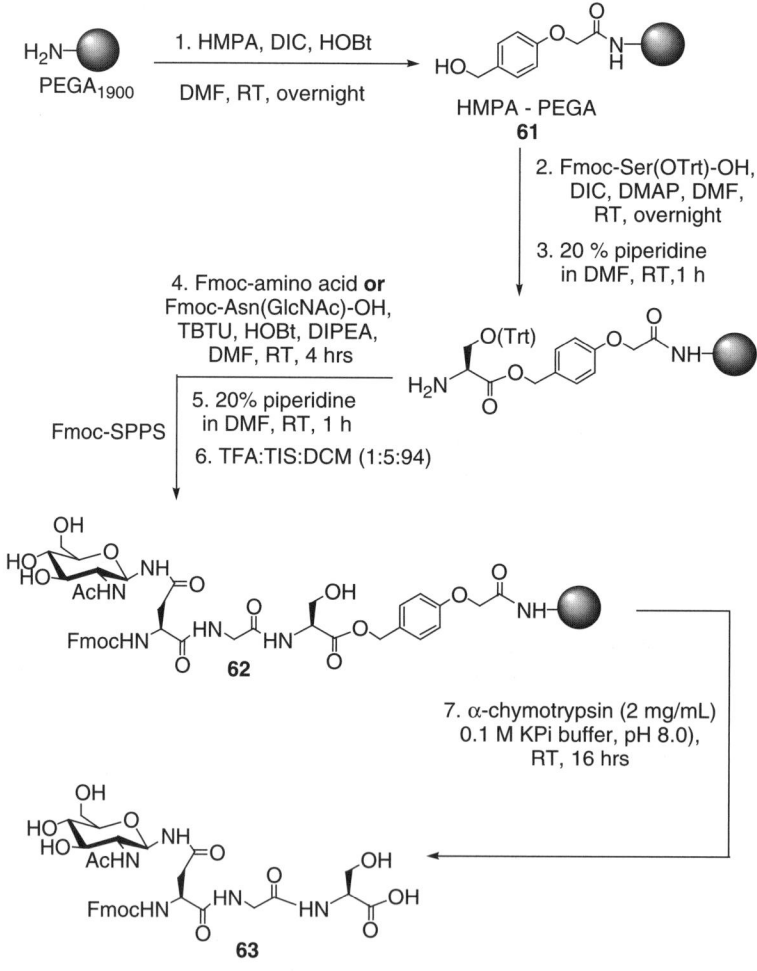

Scheme 7.13 *Fmoc-SPPS of glycopeptide substrate **62** and cleavage by α-chymotrypsin*[23]

Typical Experimental Procedures[23]

*The resin bound glycopeptide **62** (75 mg corresponding to 7.5 mg dry weight, 0.001 mmol) was washed with potassium phosphate buffer (0.1 M, pH 8.0) and was treated with α-chymotrypsin (2 mg/ml) in potassium phosphate buffer (1 ml, 0.1 M, pH 8) for 16 hours. The resin was filtered, washed with buffer and analysed by LC-MS at 214 nm. The glycopeptide **63** was eluted with a retention time of 17.2 minutes. The glycopeptide was analysed by LC-MS. MS (ESI): Calculated mass 701.19 $[M]^+$ and was found at 702.2 $[M+H]^+$.*

7.3 Conclusion

In this chapter the successful application of a few exo and endo enzyme cleavable linker strategies for solid-supported synthesis has been reviewed. The main challenge to combinatorial chemists is to discover mild and efficient conditions for the cleavage of the target compound libraries from the polymer support, preferably at room temperature and neutral conditions without affecting the structure of the target molecule. In this way the development of new enzyme cleavable linkers are valuable, since enzymatic transformations proceed under very mild conditions at room temperature with high chemo, regio and stereoselectivity. These enzyme cleavable linkers allow the synthesis of diverse classes of compounds, including, oligosaccharides, peptides, glycopeptides, peptide nucleic acids (PNA), oligonucleotides, various natural products and biologically relevant molecules on variety of solid supports.

References

[1] (a) Seeberger, P. H.; *Chem. Soc. Rev*. **2008**, *37*, 19. (b) Banfi, L., Guanti, G., Riva, R., and Basso, A.; *Curr. Opin. Drug. Discov. Devel*. **2007**, *10*, 704. (c) Tomada, H., and Doi, T.; *Acc. Chem. Res*. **2008**, *41*, 32. (d) Feliu, L., Vera-Lugue, P., Abericio, F., and Alvarez, M.; *J. Combi. Chem*. **2007**, *9*, 521. (e) Zatsepin, T. S., and Oretskaya, T. S.; *Chemistry and Biodiversity*. **2004**, *1*, 1401. (f) Jung, N., Encinas, A., and Bräse, S.; *Curr. Opin. Drug. Discov. Devel*. **2006**, *9*, 713. (g) Cironi, P., Alvarez, M., and Albericio, F.; *Mini. Rev. Med. Chem*. **2006**, *6*, 11. (h) Ganesan, A.; *Mini. Rev. Med. Chem*. **2006**, *6*, 3. (i) Seeberger, P. H., and Werz, D. B.; *Nat. Rev. Drug Discov*. **2005**, *4*, 751. (j) Marshall, W. S., and Kaiser, R. J.; *Curr. Opin. Chem. Biol*. **2004**, *8*, 211. (k) Kundu, B.; *Curr. Opin. Drug. Discov. Devel*. **2003**, *6*, 815. (l) Hojo, H., and Nakahara, Y.; *Curr. Protein. Pept. Sci*. **2000**, *1*, 23. (m) Nielsen, J.; *Curr. Opin. Chem. Biol*. **2002**, *6*, 297.

[2] Drauz, K., and Waldmann, H. (eds); *Enzyme Catalysis in Organic Synthesis*, Wiley-VCH Verlag GmbH, Weinheim, Germany, **1995**.

[3] (a) Reents, R., Jeyaraj, D. A., and Waldmann, H.; *Drug Discovery Today*. **2002**, *7*, 71. (b) Reents, R., Jeyaraj, D. A., and Waldmann, H.; *Adv. Synth. Catal*. **2001**, *343*, 501. (c) Grether, U., and Waldmann, H.; *Chem. Eur. J*. **2001**, *7*, 959.

[4] (a) Vagner, J., Barany, G., Lam, K. S., et al.; *Proc. Natl. Acad. Sci. USA* **1996**, *93*, 8194. (b) Turner, N. J.; *Current Organic Chemistry*. **1997**, *1*, 21.

[5] Grether, U., and Waldmann, H.; *Angew. Chem. Int. Ed*. **2000**, *39*, 1629.

[6] (a) Sauerbrei, B., Jungmann, V., and Waldmann, H.; *Angew. Chem. Int. Ed. Engl*. **1998**, *37*, 1143. (b) Waldmann, H., and Nägele, E.; *Angew. Chem. Int. Ed. Engl*. **1995**, *34*, 2259. (c) Waldmann, H., Schelhaas, M., Nägele, E., et al.; *Angew. Chem. Int. Ed. Engl*. **1997**, *36*, 2238. (d) Pohl, T., and Waldmann, H.; *J. Am. Chem. Soc*. **1997**, *119*, 6702.

[7] Böhm, G., Dowden, J., Rice, D. C., et al.; *Tetrahedron Lett*. **1998**, *39*, 3819.

[8] Bayer, E., and Mutter, M.; *Nature* **1972**, *237*, 512.

[9] Gravert, D. J., and Janda, K. D.; *Chem. Rev*. **1997**, *97*, 489.

[10] Hiroshige, M., Hauske, J. R., and Zhou, P.; *Tetrahedron Lett*. **1995**, *36*, 4567.

[11] Piettre, S. R., and Baltzer, S.; *Tetrahedron Lett*. **1997**, *38*, 1197.

[12] Berteina, S., Wenderborn, S., Brill, W. K. D., and Mesmaeker, A. D.; *Synlett* **1998**, 676.
[13] Rapp, W., in *Combinatorial peptide and Nonpeptide Libraries* (ed. Jung, G), Wiley-VCH Verlag GmbH, Weinheim, Germany, **1996**, 425.
[14] For Pictet-Spengler reactions on solid support see: (a) Kalijuste, K., and Unden, A.; *Tetrahedron Lett.* **1995**, *36*, 9211. (b) Mohan, R., Chou, Y.-L., and Morissey, M. M.; *Tetrahedron Lett.* **1996**, *37*, 3963. (c) Hutchins, S., and Chapman, K.; *Tetrahedron Lett.* **1996**, *37*, 4865. (d) Yang, L., and Guo, L.; *Tetrahedron Lett.* **1996**, *37*, 5041. (e) Mayer, J. P., Bankaitis-Davis, D., Zhang, J., *et al.*; *Tetrahedron Lett.* **1996**, *37*, 5633.
[15] (a) Katritzky, A. R., Lan, X., and Fan, W. Q.; *Synthesis* **1994**, 445. (b) Katritzky, A. R., and Drewniak, M.; *J. Chem. Soc., Perkin Trans 1*. **1988**, 2339.
[16] Elmore, D. T., Guthrie, D. J. S., Wallace, A. D., and Bates, S. R. E.; *J. Chem. Soc. Chem. Commun.* **1992**, 1033.
[17] Schuster, M., Wang, P., Paulson, J. C., and Wong, C. H.; *J. Am. Chem. Soc.* **1994**, *116*, 1135.
[18] α-Chymotrypsin is highly specific for substrates containing aromatic amino acid residues in P1 site. Moreover, the specificity of magnitude higher than (expressed K_{cat}/K_m) for an ester is several orders of magnitude higher than that for the corresponding amide bond: Liu, T. Y., and Elliot, S. D.; in *The Enzymes*, Boyer, P. D. (ed.); Academic Press, New York, **1971**, *3*, 609.
[19] (a) Yamada, K., and Nishimura, S. I.; *Tetrahedron Lett.* **1995**, *36*, 9493. (b) Yamada, K., Fujita, E., and Nishimura, S. I.; *Carbohydr. Res.* **1997**, *305*, 443.
[20] Nishimura, S. I., and Yamada, K.; *J. Am. Chem. Soc.* **1997**, *119*, 10555.
[21] Weiler, J., Gausepohl, H., Hauser, N., *et al.*; *Nucleic Acids Res.* **1997**, *25*, 2792.
[22] Rademann, J., Grøtli, M., Meldal, M., and Bock, K.; *J. Am. Chem. Soc.* **1999**, *121*, 5459.
[23] Maltman, B. A., Bejugam, M., and Flitsch, S. L.; *Org. Biomol. Chem.* **2005**, *3*, 2505.
[24] Bejugam, M., and Flitsch, S. L.; *Org. Lett.* **2004**, *6*, 4001.

Part III
Multifunctional Linker Units for Diversity-Oriented Synthesis

8
An Introduction to Diversity-Oriented Synthesis

Richard J. Spandl, Gemma L. Thomas, Monica Diaz-Gavilan, Kieron M. G. O'Connell and David R. Spring

Department of Chemistry, University of Cambridge, United Kingdom

8.1 Introduction

Diversity-oriented synthesis (DOS) aims to synthesize collections of molecules that represent the variety of charges, polarities, bonding interactions and architectures that can potentially be recognised by nature's biomolecules.[1–3] The structural variety present in DOS libraries confers both physico-chemical and biological diversity to the compound collection.[4] DOS libraries can be used in screening experiments to identify novel biologically active small molecule modulators.[5–8] Although there are other potential applications, it is in this context that DOS will be discussed.

Since the publication of the completed human genome sequence in 2004,[9] computational efforts towards annotation have been less than trivial.[10] Although the long established techniques of traditional genetics have a part to play, the more recently developed technique of *chemical genetics* is also predicted to be of use.[11–14] Chemical genetics, instead of using gene knock-outs directly on the level of the DNA, employs chemical methods to perturb the corresponding gene products (proteins). Small molecules can therefore be used to have a modulating effect on proteins and, consequently, the biological system under investigation can be dissected. As with traditional genetic approaches, chemical genetics can be performed in either a forward sense (i.e. induce a phenotype and identify the protein target) or reverse sense (i.e. perturb a protein and observe the phenotype) (Figure 8.1).

Although chemical genetics has been exploited successfully on an ad hoc basis,[13] its use is restricted by a lack of generality. To illustrate this point, of the 25,000 human genes proposed to encode proteins which

---

Linker Strategies in Solid-Phase Organic Synthesis   Edited by Peter Scott
© 2009 John Wiley & Sons, Ltd

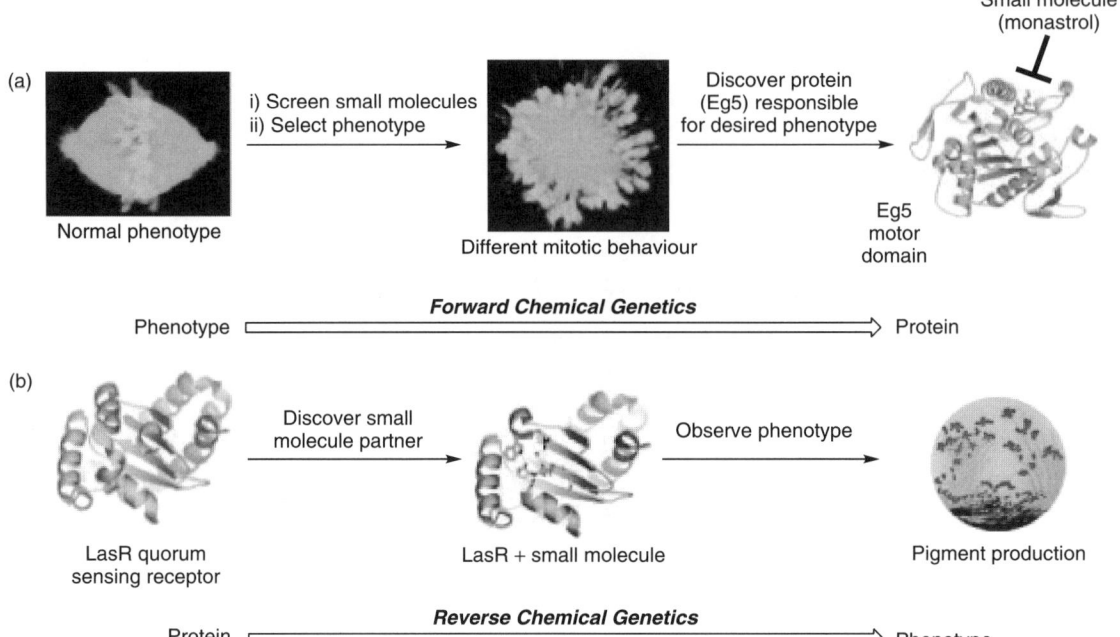

Figure 8.1 (a) In forward chemical genetics a library is screened and small molecules that induce a desired phenotype (e.g. different mitotic behaviour) are identified. Further investigation allows the protein responsible for this change (i.e. the protein partner of the small molecule identified from the initial screen) to be discovered. (b) In reverse chemical genetics the small molecules partner of the protein under investigation (e.g. LasR) is first discovered. The phenotype induced by the action of this pre-selected small molecule is then observed

will bind drug-like molecules (the 'druggable genome' i.e. 10% of the total),[15] only about 1000 have known small molecule partners. Furthermore, of these interactions, fewer can be considered as specific with regards to their effects on other proteins. Promiscuity of this nature will be detrimental to a chemical genetics experiment and is more likely when a protein with large families of homologues (e.g. proteases or kinases) is being investigated. With this in mind, it has been proposed that using complex small molecules aids interaction specificity;[16] there is, however, debate in the literature about this point.[17, 18]

The data above demonstrate that the lack of specific small molecule modulators is a major limiting factor in increasing the generality of chemical genetic approaches. Access to skeletally diverse libraries would therefore (potentially) be advantageous, especially if the most ambitious aim of chemical genetics is to be realised; that is the identification of small molecule partners for every known protein ('*chemical genomics*').[19]

Although there are a number of potential sources of small molecules (discussed below), *de novo* library synthesis using DOS may be an alternative. DOS describes the deliberate, simultaneous and efficient synthesis of multiple targets that are not only diverse in the appendages they display, but also in their molecular architectures.[2] Thus, the aim of DOS is to synthesize skeletally diverse collections of small molecules that interrogate areas of *chemical space*, in the natural product-like/drug-like region (in a broad sense), that have not previously been explored.

8.2 Exploring chemical space

With the aim of many screening projects being to identify novel biologically active small molecules, success may be aided by screening biologically (functionally) diverse compound libraries. It is the skeletal diversity of a library that has been shown to correlate to its biological diversity.[4] As a method of assessing library quality, library diversity has been calculated using chemical space analysis.[20, 21]

Chemical space encompasses all possible chemical entities and can be defined and analysed computationally.[22] The location of a molecule in chemical space is a function of the abstract representation created by an analysis of the compound's associated *chemical descriptors*. Chemical descriptors can be used to describe not only the bulk properties of a molecule,[23] but also its 3D arrangement in space (i.e. topological features).[24] Once the chemical descriptors (of which there will be many) have been defined and calculated for each library member, this information can be condensed using principle component analysis (PCA). This allows for the construction of visually gratifying 2D or 3D displays that are accessible to human interpretation. In these representations each molecule is plotted at a discrete point in chemical space (more correctly called multidimensional descriptor space) and the relative proportion of chemical space covered by different compound collections can be compared. As a result library diversity can be analysed.[4, 20, 22, 25]

As an example of chemical space representation, cyclooxygenase-1 inhibitors, defined and analysed as described above, were plotted on a background of pharmacologically active compounds (MDL Drug Data Repository) (Figure 8.2). Examination of these results demonstrates that these inhibitors are diverse in structure and not clustered in a tightly defined region.

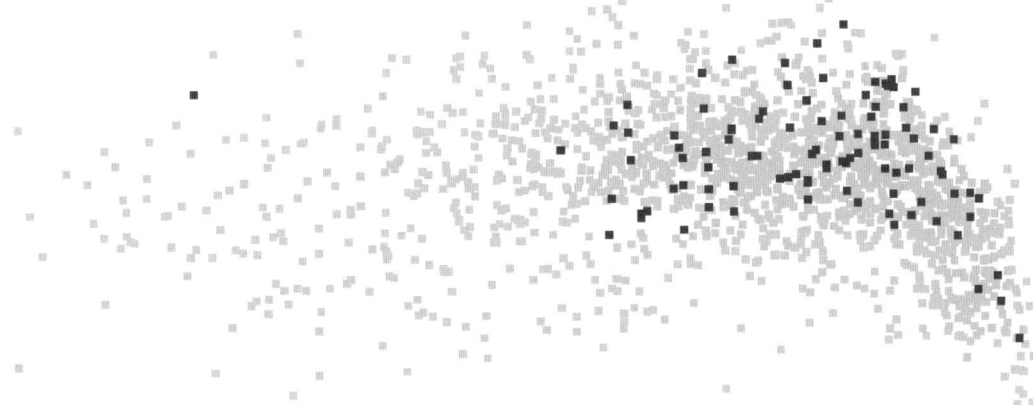

Figure 8.2 *A chemical space analysis of cyclooxygenase-1 inhibitors (blue squares) and MDDR compounds (grey squares). This visual representation shows that the cyclooxygenase-1 inhibitors populate a diffuse region of chemical space on the background of the MDDR compounds*

Since a variety of chemical descriptors can be used in the construction of chemical space representations, there may be ambiguity in these assessments of diversity. For example, in our experience, using certain chemical descriptors, a library of amides synthesized (hypothetically) from a wide variety carboxylic acids and amines appeared extremely diverse on analysis. Since this diversity is the result of the building blocks used and not the bond forming reaction itself, this highlights the problem of programming human intuition (i.e. diversity assessment) into computers. Although diversity assessment may be useful, it can also be misleading. Like-with-like comparisons are often thwarted as it is difficult to say exactly what constitutes diversity and also to know how it should be calculated.

When the target molecule in a screening programme is unknown, that is in forward chemical genetics, screening skeletally diverse collections of compounds may be useful. Although achieving skeletal diversity per se is rarely the 'end-game', its incorporation may increase hit identification.

8.3 Sources of skeletally diverse small molecules

Although DOS can be used to prepare skeletally diverse small molecules libraries, there are alternatives to this recently developed, and sometimes challenging, approach. For example, natural products, combinatorial libraries and proprietary compound collections may provide small molecule collections that offer both complexity and diversity. There are, however, potential drawbacks.

Natural products are diverse in structure and have specific modulating effects on biomolecules. Their isolation, purification and characterisation can, however, be a very complicated process.[26, 27] Furthermore, the chemical derivatisation of natural products, as a result of the complex multi-step syntheses often required, can be extremely challenging. Such chemical derivatisation is important in *focused-library synthesis*; this technique gives access to novel compounds based around the original natural product scaffold and can be used in lead-optimisation or structure–activity relationship (SAR) studies.

Combinatorial libraries, synthesized using traditional combinatorial chemistry, although they offer complexity and quantity, tend to contain mainly flat molecules with fewer chiral centres than natural products or drugs. Analysis of the mean number of chiral centres, calculated from the analysis of various databases in terms of average per molecule, demonstrates this; natural products, drugs and combinatorial library members were found to contain 6.2, 3.3 and 0.4 chiral centres respectively.[28] Likewise, a comparison of the average molecular weights (414:340:393) and number of rings (4.1:2.6:3.2) also highlights both differences and similarities between these compound types.

Combinatorial libraries tend to be synthesized using a 'one-synthesis/one-skeleton'[29] approach and, as a result, potentially exhibit limited skeletal diversity. This potential limitation can be off-set by employing many chemists to perform many syntheses; the proprietary compound collections of pharmaceutical companies are, therefore, diverse in structure. As a result of being biased by previous drug discovery programmes[30] or by meeting pre-defined criteria, for example the Lipinski rule of five (RO5),[31] these compound collections may, however, have their drawbacks. In relation to the latter point, there has been debate about the value of restricting chemists to synthesising compound that meet the RO5 criteria.[32, 33]

8.4 Enriching chemical space using DOS

Since natural products and known drugs occupy only a small proportion of bioactive chemical space,[16, 22] exploring previously uncharted regions, as is the aim in DOS, may be advantageous. When bioactive molecules are sought, although DOS aims to interrogate 'novel' regions of chemical space, the compounds prepared should still be 'natural product-like' or 'drug-like' in terms of structure. The term 'diversity' must, therefore, be taken in context.

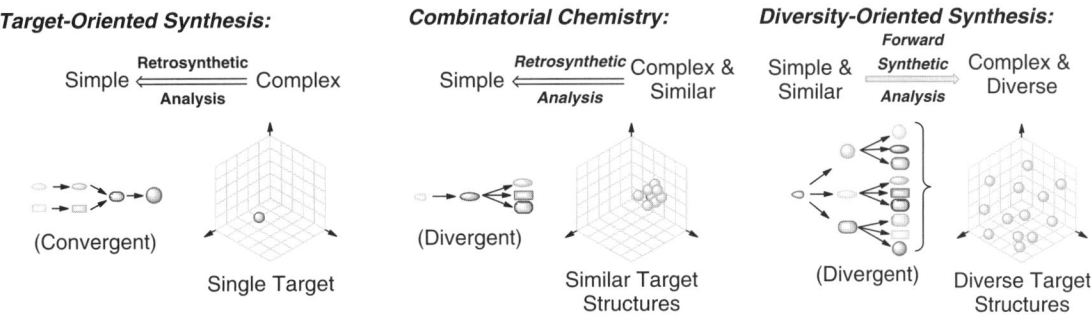

Figure 8.3 *In target-oriented synthesis and traditional combinatorial chemistry, retro-synthetic analysis is used; the target molecule or molecules occupy a discrete point in, or more densely populate a region of, chemical space. In contrast, diversity-oriented synthesis uses forward synthetic analysis to plan, and subsequently produce, compound collections that interrogate a diffuse region of chemical space*

A comparison with *target-oriented synthesis* (TOS) and traditional combinatorial chemistry (focused library synthesis) serves to highlight the DOS approach to populating chemical space. TOS and focused library synthesis attempt to synthesize compounds at a discrete point in, or clustered within a certain region of, chemical space. In these approaches *retro-synthetic analysis* is used to theoretically deconstruct complex target molecules into simple starting materials. In DOS, a *'forward synthetic'* strategy must be envisaged to facilitate the transformation of simple and similar starting materials into an array of complex and diverse products (Figure 8.3).

To be an effective strategy for library synthesis, DOS, as is the case in TOS and focused library synthesis, requires highly efficient reactions in terms of both yield and stereoselectivity.

8.5 The subjective nature of 'Diversity'

Libraries synthesized using DOS are not 'truly' diverse, nor are they designed to be. It may be better, therefore, instead of describing DOS libraries as 'non-focused', to describe them as *'soft-focused'*. Although the aim of DOS is to populate diverse regions of chemical space, the libraries constituents are, nevertheless, required to interact with biomolecules (in the case of a biological modulator or probe).

As an extension of this, the terms 'diversity' and DOS can be confusing since both are used freely in the literature. For example, the racemic synthesis of a chiral target molecule could be classified as a DOS. Also, when any compound collection is synthesized, since the constituents are not identical, there is some degree of diversity incorporated. Although these are extreme cases, they highlight the subjectivity of diversity.

To address these issues the *'molecular diversity spectrum'* (Figure 8.4) can be envisaged.[1] In one extreme of the spectrum would be where maximal chemical space coverage has been achieved and, in the other extreme, would be a TOS.[1] Although quantification of this spectrum would be difficult (see above), it should be the goal of a DOS to synthesize, in a qualitative sense, collections as near to the right hand side of this spectrum as possible.

The four principle types of diversity that can be incorporated into a compound collection are highlighted on the molecular diversity spectrum, these are: i) appendage (building block) diversity, ii) functional group diversity, iii) stereochemical diversity, and iv) skeletal diversity.[29, 34] Whereas the first three types of diversity can be introduced by using reagent controlled (stereoselective) transformations, the most

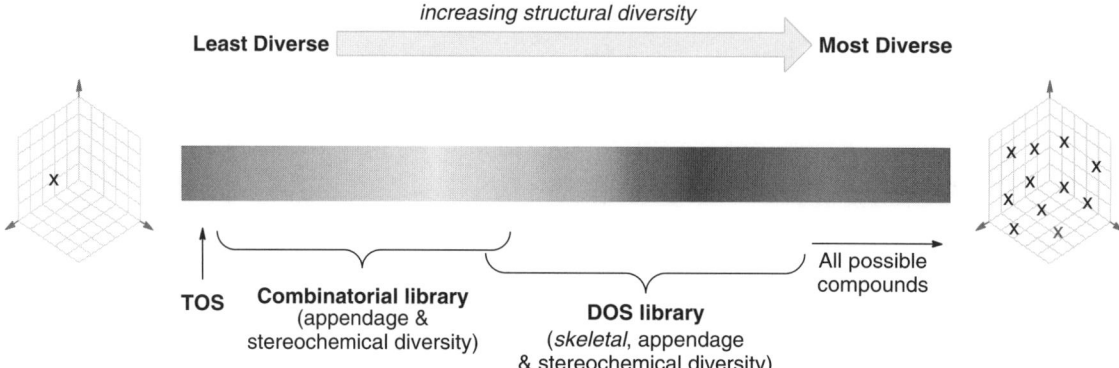

Figure 8.4 The 'molecular diversity spectrum'; a qualitative description of TOS, combinatorial chemistry and DOS in terms of structural diversity. The left-hand side of the spectrum represents minimal skeletal variety within a compound collection whereas the right-hand side represent the maximal diversity (theoretically) achievable. Combinatorial chemistry libraries and DOS libraries are placed on this axis going from left (least diverse) to right (most diverse)[1]

challenging facet of DOS is the incorporation of skeletal diversity. It is skeletal diversity that is of critical importance in a DOS project. Thus, in contrast to traditional combinatorial chemistry where diversity primarily arises by building block variation, DOS places emphasis on accessing different molecular skeletons.

8.6 Differing strategies towards similar goals

A successful DOS should allow not only for combinatorial variation in the building blocks used, but should also allow for these appendages to be displayed on different three dimensional scaffolds. In a general sense, most of the DOS strategies reported in the literature are either based on privileged scaffolds or designed from simple starting materials. These strategies, that is 'DOS based on privileged scaffolds' and 'DOS from simple starting materials', are different in their approaches and, to some extent, their end-goals.

8.6.1 DOS based on privileged scaffolds

As a result of the evolutionary pressure for specific ligand–biomolecule interactions, compounds bearing privileged scaffolds are predisposed to exhibit bioactivity. Privileged structures are defined as those which display high affinity binding to multiple protein classes; the term was first used to describe the benzodiazepines and the benzazepines by Evans and coworkers.[35] In some instances, basing libraries on privileged scaffolds may be advantageous.[36–40] In contrast to combinatorial natural product-like libraries, where limited diversification is explored,[41] DOS libraries of this nature aim to incorporate privileged motifs into structurally diverse (skeletally and stereochemically) architectures. The aim of these investigations is to identify compounds with novel bioactivities that are distinct from the natural product (or class of natural products) in which the privileged motifs are commonly found.[42]

This approach was exploited by Park and coworkers in their synthesis of a natural product-like DOS library based around the benzopyran motif **1**.[38] The benzopyran motif is common in many natural products (4636 of the compounds from the '*dictionary of natural products*' are benzyopyraniod)[43] and the synthesis of bioactive benzopyran containing compounds has been extensively studied. Using a branching strategy

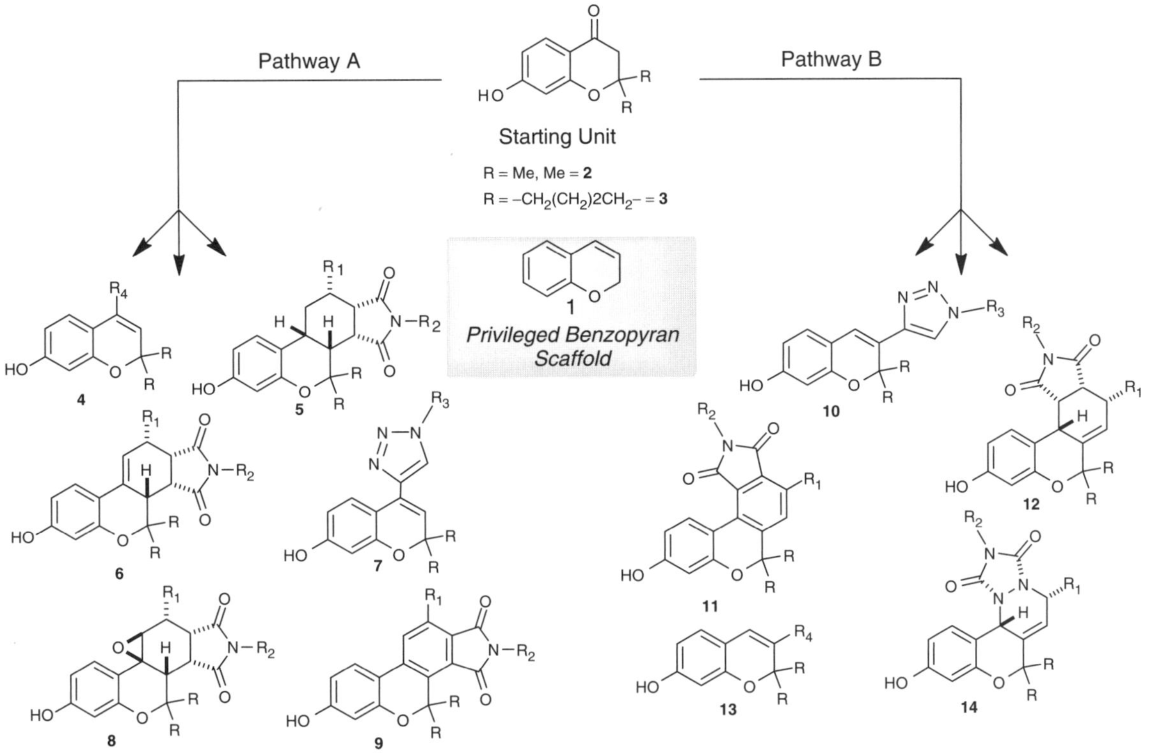

Scheme 8.1 *The scaffolds of a DOS library based around the privileged benzopyran scaffold. A range of IC_{50} values were reported (biological diversity) when this library was screened against a human cancer cell line*[38]

(also referred to as a library-from-library approach), Park and coworkers were able to synthesize 22 novel molecular architectures, each one containing the benzopyran sub-structure. From the starting materials **2** and **3**, the scaffolds **4–14** could be accessed using one of two major branching pathways (pathway A or B) (Scheme 8.1).

The reactions used in library generation were atom economical and proceeded in good yields and with good diastereoselectivities. To investigate the effect on bioactivity of using unbiased and diverse natural product-like libraries, *in vitro* cytotoxicity studies were performed against a human cancer cell line. The compounds screened showed a wide range of IC_{50} (half maximal inhibitory concentration) values and these variations were shown to correlate to differences in the molecular skeletons and not in the appendages attached. Thus, skeletal diversity led to functional (biological) diversity.

An alternative but related approach to discovering novel biologically active small molecules, known as '*biologically-oriented synthesis*' (BIOS), is noteworthy. This approach, pioneered by Waldmann and coworkers, 'builds on the inspiration given by nature through natural products'.[44] Since nature uses only a very small fraction of chemical and biological space, this allows small molecule modulators and functional proteins to be classified and grouped. For example, in the case of small molecules, the structural classification of natural products (SCONP) is used to examine the relationships between the different biologically relevant scaffolds and also to group these skeletons in a hierarchical tree.[45] Likewise, using the technique of protein structure similarity clustering (PSSC), proteins can be classified on the basis of their fold topology and also their inhibitory profiles.[46]

By merging these concepts, biologically pre-validated scaffolds can be chosen by identifying links with the target protein through architectural commonalties with other known inhibitors.[44] This process, which is the essence of BIOS and has only been discussed here briefly, allows small focused libraries to be synthesized. Indeed, BIOS has shown some initial success.[47]

Although DOS libraries based on privileged scaffolds may offer advantages when the target protein is known, there is a need to explore the chemical space occupied by neither natural products nor drugs.[1, 4] As DOS libraries from simple starting materials contain no pre-encoded bias (i.e. no pre-selected motifs), these can give access to more diverse regions of chemical space than the approaches discussed above.

8.6.2 DOS from simple starting materials

The process described by Schreiber in much of his pioneering work in DOS, that is generating skeletally diverse libraries from simple starting materials, will be the subject of the remainder of this chapter.[48, 49] As a general strategy to incorporate diversity, the use of branching pathways to access distinct molecular scaffolds will be discussed.[3]

Branching pathways commonly make use of reactions that increase structural complexity, that is complexity generating reactions. Although the complexity of a compound collection bears no relation to its diversity, it has been reported to confer specificity (see above).[16, 50] Tandem processes, where the product of one complexity generating reaction is the substrate for the next, are of immense value in DOS as both structural complexity and diversity are increased efficiently.[29] This allows the complex 3D scaffolds required to be generated. Although structural complexity is relatively straightforward to achieve, accessing skeletal diversity efficiently is a more challenging goal.

8.7 Generating skeletal diversity

When synthesising a library, skeletal diversity can be incorporated using either: branching pathways, where a common starting material is transformed into distinct skeletons using different reagents (the 'reagent based approach'); or folding pathways, where different starting materials, containing pre-encoded skeletal information, are subjected to a common set of conditions and converted to different scaffolds (the 'substrate based approach') (Figure 8.5).[34]

Successful DOS processes use these approaches to achieve skeletal diversity in a number of ways; a review of the literature suggests three general strategies. These are: i) the use of a pluripotent functional

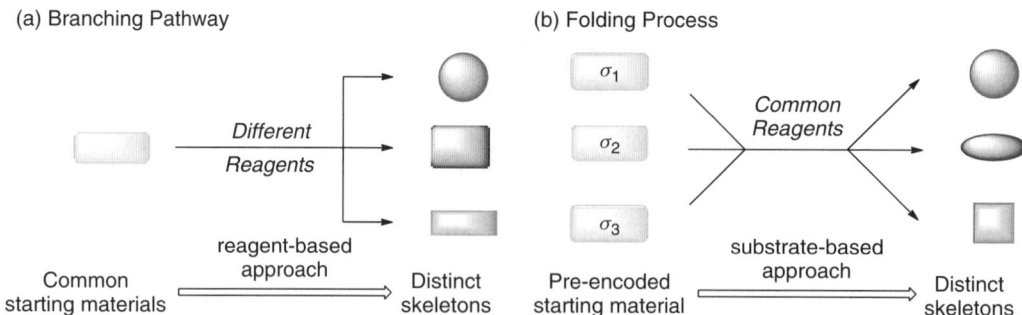

Figure 8.5 The two general strategies to accessing skeletal diversity in a compound library

group, where the same part of a molecule is subjected to different transformations induced by different reagents; ii) the use of a densely functionalised molecule, where different functionalities in the same molecule are transformed by different reagents; and iii) the use of a folding process, where different structurally encoding elements, contained in different substrates, are subjected to the same reaction conditions.

Schreiber and coworkers have recently described a 'build/couple/pair' strategy which combines approaches ii and iii.[51] After firstly 'building' the required chiral starting units, possibly from chiral pool reagents, 'coupling' reactions afford a densely functionalised molecule; it is common for multicomponent coupling reactions (MCRs) to be used. Different parts of the densely functionalised molecule, in functional group specific reactions, are then 'paired' to generate different molecular skeletons.

8.7.1 Strategy 1: Pluripotent functional groups

After a suitable functional group, which can participate in a rich variety of chemical transformations, has been chosen, other factors must also be considered for this approach to be successful. The synthesis of the starting material should allow for building block diversity to be incorporated and, more importantly, the novel scaffolds generated in the initial branching pathways should have the potential to be diversified further (preferably in subsequent branching reactions). To this end, the high reactivity and mechanistic flexibility of the diazoacetate moiety was exploited in the generation of a small molecule library by Wyatt *et al.*[52]

The fluorous-tagged diazoacetate **15** was chosen as it could participate in a variety of complexity generating carbon–carbon bond forming reactions (the products of which could be diversified further). Also, the polyfluorocarbon tag present in **15** allowed for generic purification of the library compounds by fluorous solid-phase extraction (SPE), reverse fluorous SPE or liquid–liquid extraction. The starting material **15** therefore not only gave access to multiple branching pathways, but facilitated high-throughput purification and hence increased the efficiency of the DOS process.

From the starting unit **15**, three major branching pathways were initially used (step 1): the three-membered ring forming reaction, yielding the cyclopropanes **16** and **17**; the α-deprotonation/electrophilic quenching reaction, yielding the 1,3-keto esters **19** and **20**; and the 1,3-dipolar cycloaddition reaction, yielding the heterocycles **21** and **22** (Scheme 8.2). These newly formed branch point products **16**, **17**, **19**, and **20** were then converted to a second generation of compounds **18**, **23–29** in sequential complexity generating reactions (step 2). These tandem processes thus served to increase complexity and diversity; a collection of 223 small molecules, based around 20 discrete molecular frameworks, was synthesized.

A skeletally diverse small molecule library was synthesized, also using a similar approach to that above, by Thomas *et al.* (Scheme 8.3).[53] The solid-supported phosphonate **31** was chosen as the starting unit as it could be readily synthesized from **30** (in an E-selective Horner Wadsworth Emmons reaction, step 1) and as it allowed access to three divergent reaction pathways.

The reactions from **31**, which were catalytic and enantioselective thus allowing stereochemical diversity to be incorporated, were: the 1,3-dipolar cycloaddition branching pathway to give **32** (performed using the silver acetate/(S)-QUINAP system);[54] the dihydroxylation branching pathway to give **33** (performed using a modified Sharpless asymmetric dihydroxylation (AD) reaction protocol);[55] and, the Diels–Alder branching pathway to give **34** (asymmetric catalysis was possible as the nature of **31** allowed two-point catalyst binding, that is using Evans methodology)[56] (step 2). Subsequent transformations enabled a second generation of compounds (step 3) and a third generation of compounds (step 4) to be synthesized; these processes allowed both the appendage and the skeletal diversity to be increased and afforded 242 compound based around 18 distinct scaffolds.

Similar to the fluorous-tagged technology used by Wyatt *et al.*, the solid supported nature of these compounds (Scheme 8.3) allowed for high-throughput synthesis. The imidazolidinone portion allowed the appropriate chemical entity to be attached to a novel solid support resin.[57] Simple protocols (such as

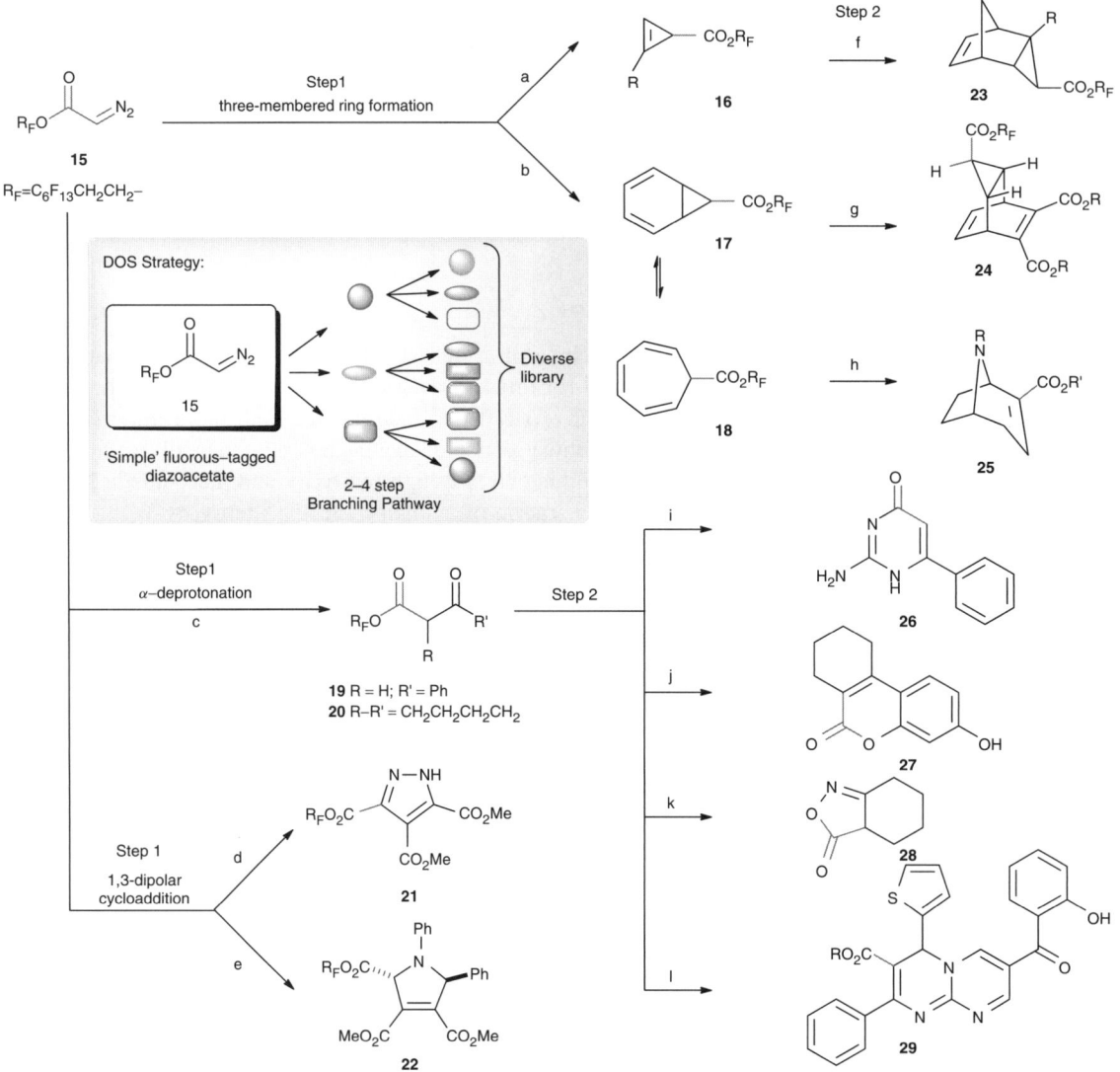

Scheme 8.2 Diversity-oriented synthesis of 223 small molecules based on 30 discrete frameworks. Steps 1: (a) RCCH, $Rh_2(OAc)_4$, [BuCCH, 57%]; (b) C_6H_6, $Rh_2(O_2CCF_3)_4$, 70%; (c) LDA, RCOR' then $Rh_2(OAc)_4$; 36: 49% (90%); 37: 68% (97%); (d) DMAD, 84% (88%); (e) PhCHO, $PhNH_2$, then DMAD, $Rh_2(OAc)_4$, d.r. = 20:1, 51% (80%). Steps 2: (f) C_5H_6, 92%; (g) dienophile [dimethyl acetylenedicarboxylate, 59%]; (h) RNH_2, NaOH then CH_3OH, H_2SO_4, [CH_3NH_2, 35%]; (i) Guanidine carbonate 62% (96%); (j) Resorcinol, H_2SO_4, 74% (95%); (k) NH_2OH, 77% (89%); (l) Thiophene-2-carboxaldehyde, guanidine carbonate, then 3-formylchromone, 43% (98%). Yields and purity (in brackets) of the product example following generic purification using (reverse) fluorous solid phase extraction or precipitation shown. Purity determined by HPLC, LCMS or ^{1}H NMR. DMAD = dimethyl acetylenedicarboxylate[52]

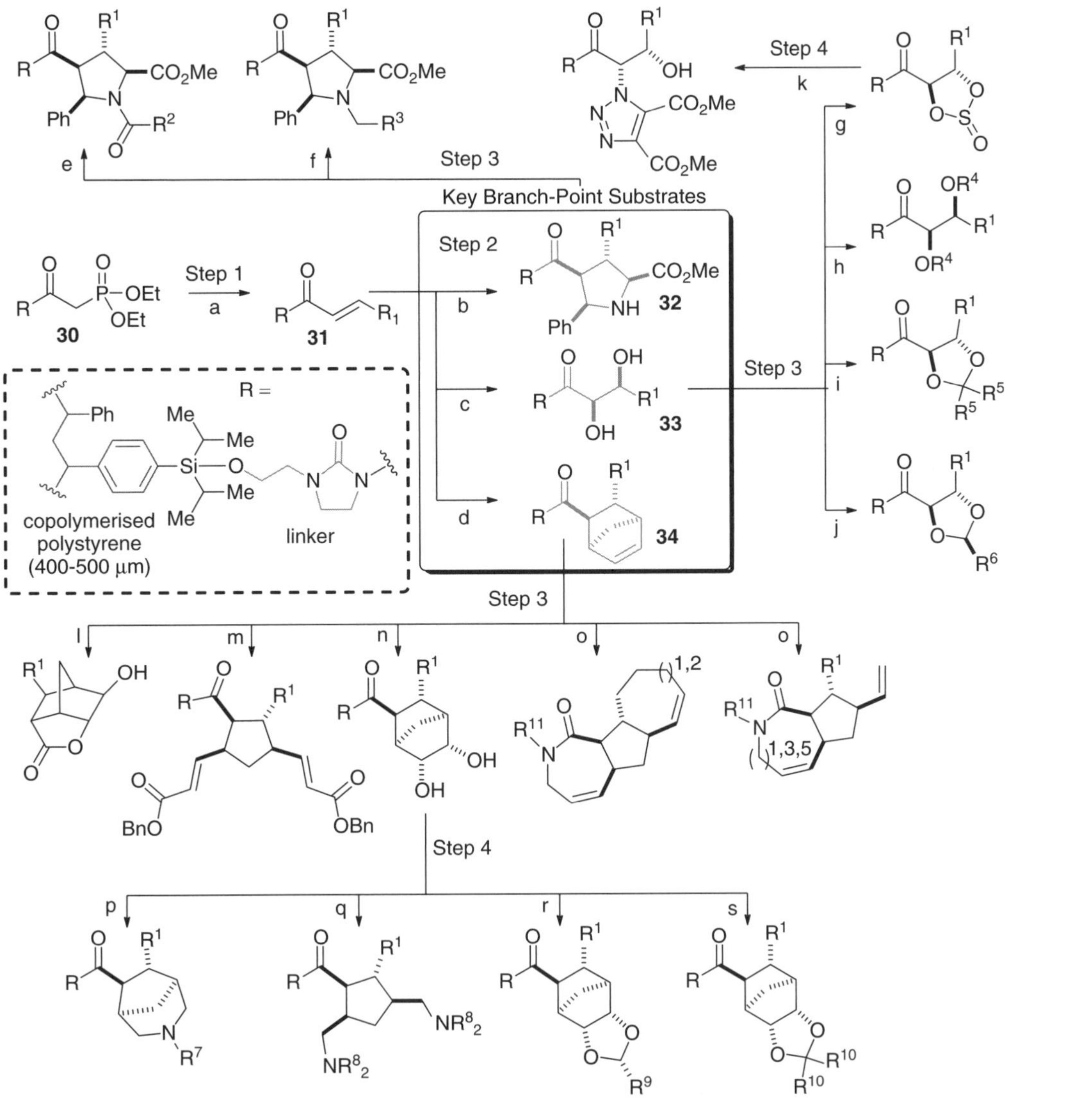

Scheme 8.3 Diversity-oriented synthesis of 242 compounds based of 18 discrete molecular frameworks. Conditions: a) LiBr, 1,8-diazabicyclo[5.4.0]undec-7-ene, R^1CHO, CH_3CN; b) (R)-QUINAP, AgOAc, i-Pr_2NEt, THF, $-78\,°C \rightarrow 25\,°C$; c) AD-mix, (DHQD)PHAL, THF:H_2O (1:1); d) chiral bis(oxazoline), Cu(OTf)$_2$, 3 Å MS, CH_2Cl_2, C_5H_6; e) R^2COCl, DMAP, pyridine, CH_2Cl_2; f) R^3CHO, BH_3·pyridine, CH_3OH; g) SOCl$_2$, pyridine, CH_2Cl_2, 40 °C; h) R^4Br, Ag$_2$O, CH_2Cl_2, 40 °C; i) R^5C(O)R^5, TsOH, DMF, 65 °C; j) R^6CHO, TsOH, DMF, 65 °C; k) NaN$_3$, DMF, 100 °C then dimethyl acetylenedicarboxylate, PhMe, 65 °C; l) mCPBA, CH_2Cl_2 then CH_3OH, 65 °C; m) CH_2=CHCO$_2$Bn, PhMe, 120 °C, Grubbs II, CH_2=CH_2; n) OsO$_4$, NMO, $CH_3C(O)CH_3$:H_2O (10:1); o) RNH$_2$, Me$_2$AlCl, PhCH$_3$ 120 °C; then NaH, R^{11}X, DMF, THF; then PhMe, 120 °C, Grubbs II, CH_2=CH_2; p) NaIO$_4$, THF:H_2O (1:1); then R^7NH$_2$, NaB(OAc)$_3$H, CH_2Cl_2; q) NaIO$_4$, THF:H_2O (1:1); then R^8NHR^8, NaB(OAc)$_3$H, CH_2Cl_2; r) R^9CHO, DMF, TsOH, 60 °C; s) R^{10}C(O)R^{10}, DMF, TsOH, 60 °C. DMF = N,N-dimethylformamide; THF = tetrahydrofuran; DMAP = N,N-dimethylaminopyridine; (DHQD)PHAL = hydroquinidine 1,4-phthalazinediyl diether; mCPBA = meta-chloroperbenzoic acid; Ts = para-toluenesulfonyl; Grubbs II = 1,3-(bis(mesityl)-2-imidazolidinylidene) dichloro (phenylmethylene) (tricyclohexylphosphine) ruthenium; NMO = 4-methylmorpholine-N-oxide[53]

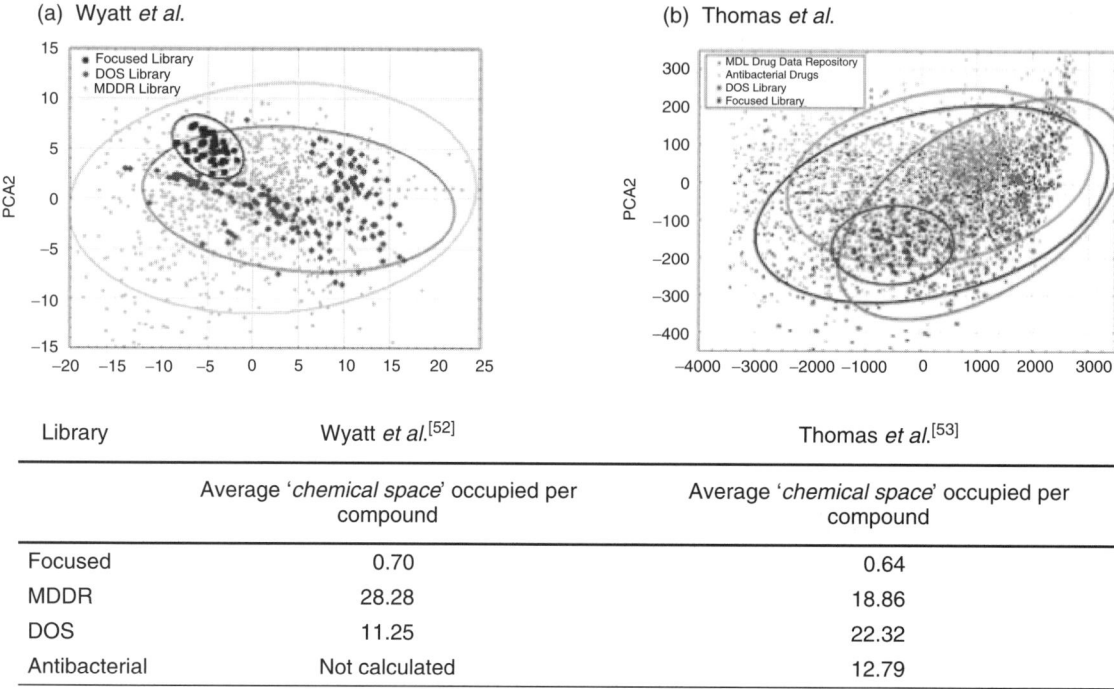

Figure 8.6 Visual representation of the diversity of different chemical collections in physicochemical and topological space using molecular operating environment (MOE) descriptors followed by principal component analysis (PCA). The DOS libraries synthesized are depicted by red squares (a: Wyatt et al.; b: Thomas et al.) For comparison, a focused library (blue squares), the MDL Drug Data Repository (MDDR; black dots), and antibacterial drugs (grey dots) (B only) are depicted. Analysis of the average 'chemical space' occupied per compound is shown in the table below[52, 53]

amide, acid and alcohol synthesis) could be used to cleave the target compounds, at the exocyclic carbonyl of the imidazolidinone, from the immobilised portion of the molecule.

The diversities of the libraries synthesized by Wyatt *et al.* and Thomas *et al.* were assessed using selected chemical descriptors and PCA (Figure 8.6a and 8.6b respectively). The diversities of these libraries were then compared to: MDDR compounds (molecular weight cut-off 650); two focused libraries (synthesized using traditional combinatorial chemistry); and (in the case of Figure 8.6b only), the 3762 compounds marked as 'antibacterial' in the MDL Drug Data Repository (MDDR) database. Both DOS libraries, as expected, were found to occupy a greater area of chemical space than the focused libraries. More significant however, using the datasets chosen, was that the compound collection produced by Thomas *et al.* was shown to be even more diverse than the MDDR library in terms of (relative) diversity units; that is 22 for the DOS library, 19 for MDDR, 13 for the antibacterials and 0.6 for the focused library (Figure 8.6b).

The more diverse compound collection synthesized by Thomas *et al.* was screened for antibacterial activity. Screens were performed against two United Kingdom epidemic strains of methicillin-resistant *Staphylococcus aureus* (EMRSA 15 and EMRSA 16); these strains are responsible for the majority of MRSA infections in the UK.[58] The most active compound **35**, which was called gemmacin, showed a broad range of activity against Gram-positive bacteria (Table 8.1). Target identification suggested that gemmacin **35** acted as a cell membrane disruptor. Thus, a unique scaffold, which could potentially be useful in the antibacterial development, had been identified using DOS.

Table 8.1 Structure and activity of gemmacin **35** with growth inhibitory activity (MIC$_{50}$) against three strains of S. aureus. For comparison the MIC$_{50}$ values for erythromycin and oxacillin are also shown. ND = not determined. MSSA = methicillin-susceptible S. aureus[53]

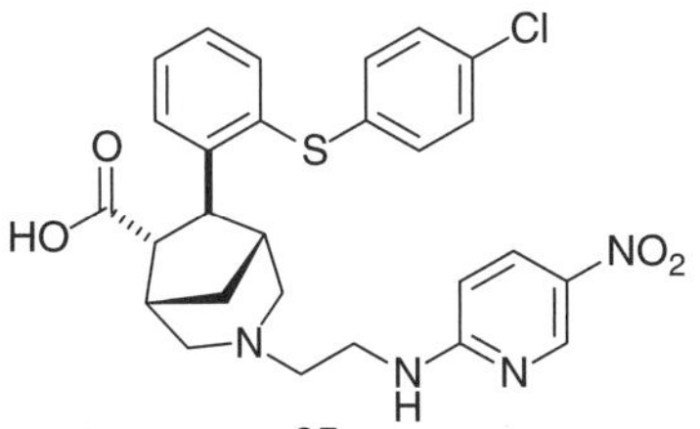

	MIC$_{50}$ (μg ml^{-1})		
	MSSA	EMRSA15	EMRSA16
(±)-gemmacin	2	16	32
(−)-gemmacin	ND	8	16
(+)-gemmacin	ND	16	32
erythromycin	0.5	> 64	> 64
oxacillin	0.5	> 32	> 32

Multicomponent coupling reactions (MCRs) feature regularly in DOS libraries as they serve to increase structural complexity; the products are, however, of identical molecular architecture. To exploit MCRs most successfully in DOS, two approaches can be used, either: use the MCR to produce a densely functionalised molecule that can then be diversified further (see Strategy 2);[59] or by incorporating a 'folding process' into the MCR (see Strategy 3).[60]

8.7.2 Strategy 2: Pluripotent (densely functionalised) molecules

Using the Petasis three-component coupling of the lactol **36**, the amino acid **37** and the boronic acid **38**, Schreiber and coworkers synthesized the β-amino alcohol **40** in good yield and diastereoselectivity (Scheme 8.4) via amine propargylation of the intermediate compound **39**. The densely functionalised compound **40** was used as the starting point for a DOS.[59]

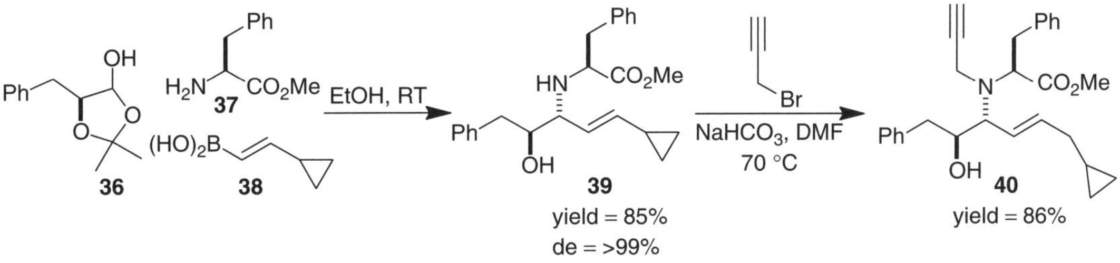

Scheme 8.4 The synthesis of the β-amino alcohol **40**. A Petasis three-component coupling reaction was used to synthesize the intermediate **39** which was then converted to **40**; this compound was used as the starting point for the DOS

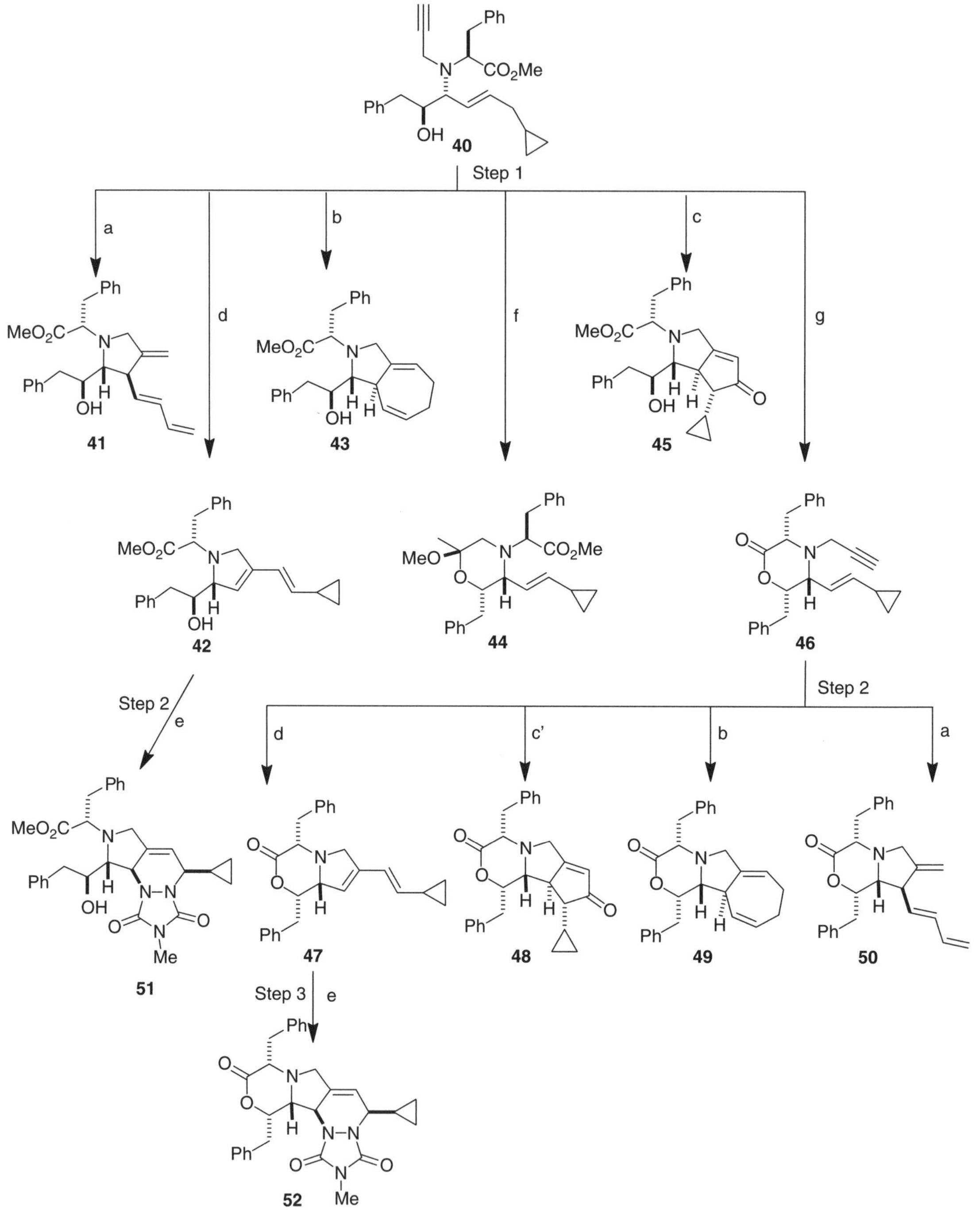

Scheme 8.5 An example of the use of densely functionalised molecules in a DOS. Conditions: a) [Pd(PPh$_3$)$_2$(OAc)$_2$] (10 mol%), benzene, 80 °C; b) [CpRu(CH$_3$CN)$_3$PF$_6$] (10 mol%), acetone, rt; c) [Co$_2$(CO)$_8$], trimethylamine N-oxide, ammonium chloride, benzene, rt; c') [Co$_2$(CO)$_8$], trimethylamine N-oxide, benzene, rt d) Hoveyda-Grubbs second-generation catalyst (10 mol%), DCM, reflux; e) 4-methyl-1,2,4-triazoline-3,5-dione, DCM, rt; f) NaAuCl$_4$ (10 mol%), methanol, rt; g) sodium hydride, toluene, rt[59]

A DOS strategy was exploited whereby the different combinations of the moieties of **40**, both polar and nonpolar, were paired in functional group specific reactions. This allowed diverse molecular skeletons to be generated using either: cycloisomerization reactions (**41** and **43**); an enyne metathesis reaction (**42**); a gold mediated alkyne addition reaction (**44**); a Pauson-Khand reaction (**45**); or a lactonisation reaction (**46**) (Scheme 8.5). In addition to this first generation of compounds, the lactone **46**, where the unsaturated functionalities remained unchanged, could be converted to a second generation of compounds **47–50** using identical pairing reactions to those used previously. The 1,3-dienes generated via the enyne metathesis reaction (i.e. **42** and **47**) were diversified further in a tandem Diels–Alder process to give the adducts **51** and **52**.[59]

The stereochemical outcome of the Petasis reaction was controlled by the lactol, and both diastereoisomers of **40** could be produced; stereochemical diversity could therefore be incorporated into any library synthesized using this approach. Although a small molecule library was not synthesized, different building blocks (i.e. different amino acids) were investigated in the initial MCR and subsequent pairing reactions. This is an elegant example of how both structural diversity and complexity can be achieved using a densely functionalised starter unit.

A similar pairing strategy has also been reported more recently by Porco and coworkers.[61] Using an enantioselective 1,4-addition to a nitro compound, the densely functionalised starter **53** (containing a combination of alkyne, alkene, nitro, and ester functional groups) could be synthesized readily. Different pairing reactions of the functional groups present in **53** led to the formation of different scaffolds: pairing the alkyne and alkene groups (route a) gave **54**; pairing the nitro and alkyne groups (route b) gave **55**; and pairing the ester and nitro groups (route c) gave **56** (**Scheme 8.6**).

Polycyclic scaffolds could be accessed by step-wise pairing reactions. For example, the starting material **57** was first converted to the 1,3-diene **58** via an enyne metathesis reaction. The diene **58** was transformed, in the same pot, to the Diels–Alder adduct **59**, which was then converted, in a pairing reaction of the nitro and ester group, to the polycyclic compound **60** (Scheme 8.7).

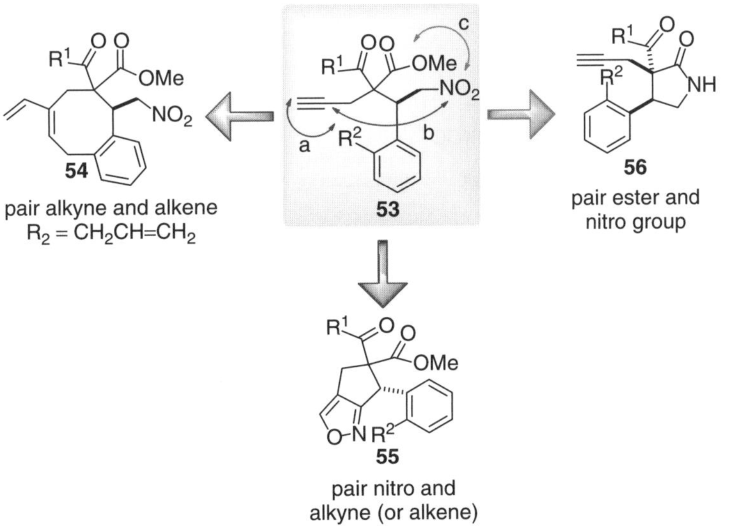

Scheme 8.6 *After the synthesis of **53**, the different scaffolds **54–56** were formed by pairing the functionalities present in the starting material*[61]

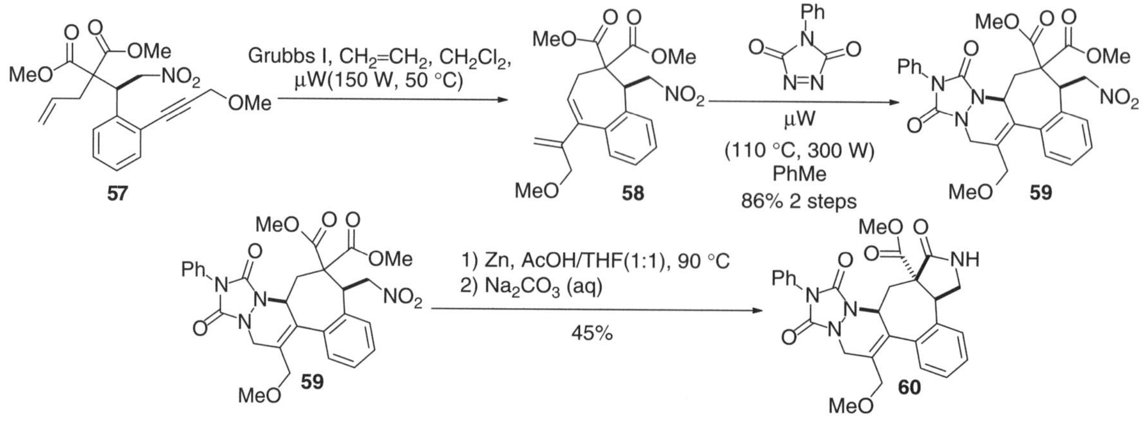

Scheme 8.7 *Polycyclic scaffolds such as **60** could be synthesized using stepwise coupling reactions*[61]

8.7.3 Strategy 3: Folding pathways

Garcia-Tellado and coworkers reported the use of an ABB' three-component coupling reaction in the synthesis of diverse scaffolds.[60] Pre-encoded information in the starting materials **61** and **64** (i.e. their comparative acidities and electrophilicities), in addition to the properties of the catalyst **62**, resulted in chemodifferentiation. The starting alkyne **61** and the α-dicarbonyl compounds **64** (which were either acidic or non-acidic α-keto esters or α-keto amides) were converted to the architectures **65–67** (Scheme 8.8).

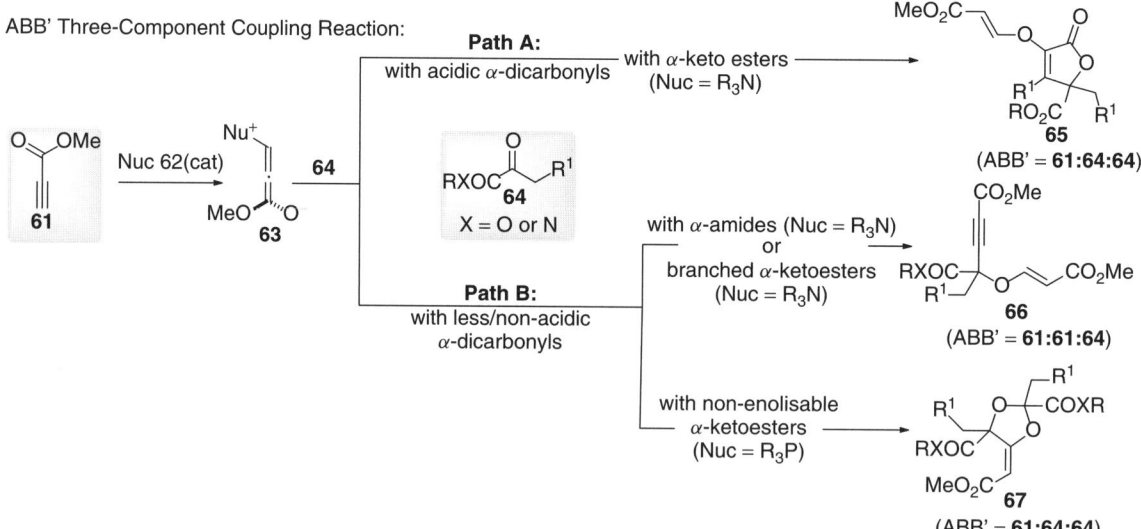

Scheme 8.8 *This ABB' MCR allowed the synthesis of scaffolds **65–67** from the alkyne **61** and the α-dicarbonyl **64** in the presence of the catalyst **62**. The chemodifferentiation, which allowed this folding process to occur, was a result of the nature of the starting materials and the catalyst*[60]

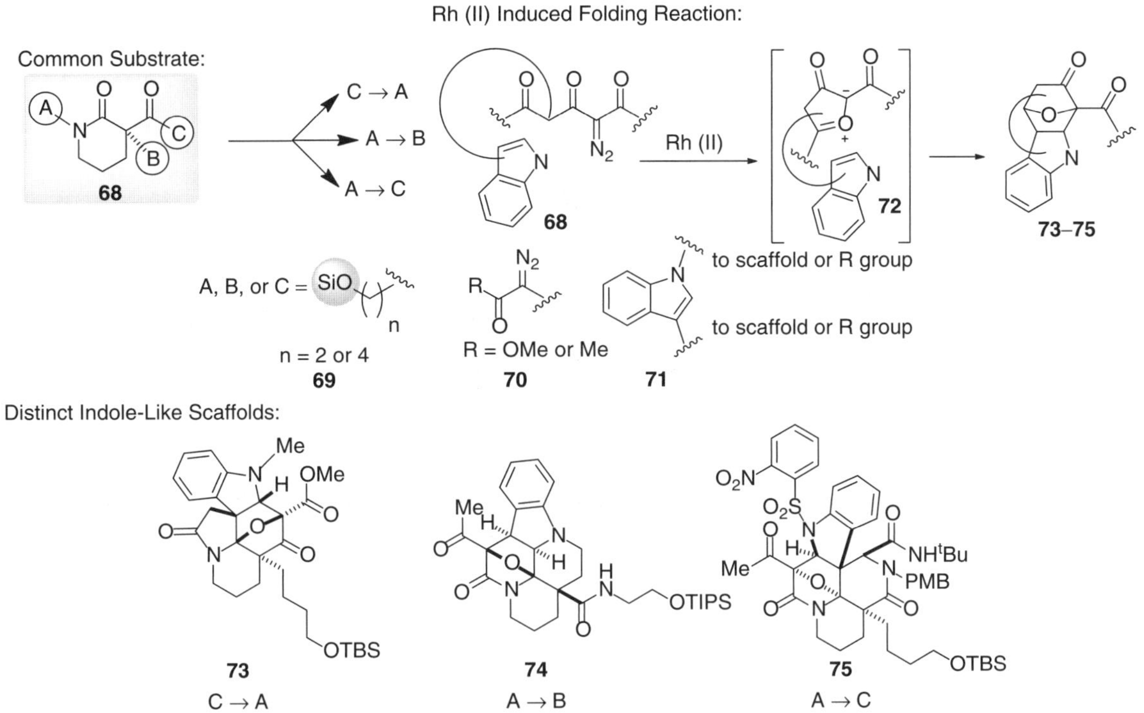

Scheme 8.9 By elaborating on the substrate scaffold **67**, with various combination of **68–70**, a rhodium(II) induced cyclisation allowed the formation of indole-like scaffolds **68–70** in this folding pathway[63]

The use of alternative ABB' MCRs has also been reviewed.[62] It is more common, however, for MCRs to feature as discrete steps in DOS pathways and be used to generate starter units. These starter units are often designed to participate in folding pathways.[48, 63, 64]

An excellent example of this was reported by Oguri and Schreiber who synthesized the starting scaffolds **68** with various combinations of key moieties at the three reactive sites A, B and C. Attached to these sites was either a silyl either linker **69**, an α-diazo ketocarbonyl group **70** or an indole moiety **71**. The relative arrangements of these groups in **68** encoded the skeletal information required to synthesize the distinct indole-like scaffolds **73–75** (Scheme 8.9).[63]

The folding pathway involved a rhodium(II) induced cyclisation in which the α-diazo ketocarbonyl group of **68** reacted to form the carbonyl ylide **72**. The newly formed ylide **72** then participated in an intramolecular 1,3-dipolar cycloaddition reaction with the 2–3 double bond of the pendant indole group to give **73–75** (Scheme 8.9).

8.8 DOS and solid-phase organic synthesis

The ability to prepare compound libraries using traditional combinatorial chemistry has been aided by the use of solid-phase organic synthesis (SPOS). Since the pioneering work by Merrifield in relation to peptide synthesis,[65] a wide range of solution phase reactions can now be performed on the solid support.[66] In addition to being amenable to automation, SPOS allows simple purification protocols, which frequently

258 *Multifunctional Linker Units for Diversity-Oriented Synthesis*

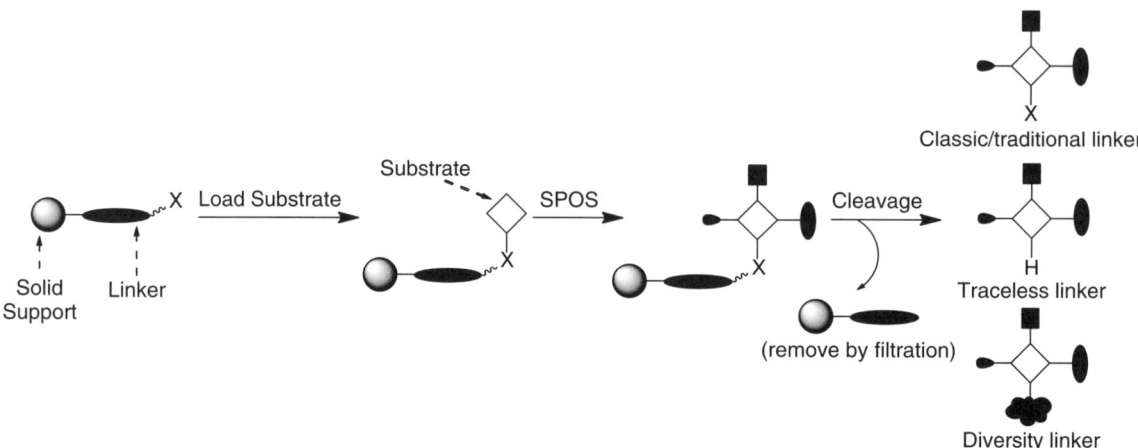

Scheme 8.10 *An overview of SPOS; a substrate molecule is attached, via a selectively cleavable linker, to a solid support. The substrate is carried through sequential rounds of chemical transformations (SPOS) before being cleaved from the resin. Linkers are classified as either: classic/traditional linkers; traceless linkers; or diversity linkers. Although the first two types are generally cleaved using a common 'cleavage cocktail', the latter can be used to incorporate structural diversity into the product*[68]

involve only a filtration to be exploited (Scheme 8.10). As a result, compound collections have been generated in a high-throughput fashion, therefore increasing the efficiency of library synthesis. Indeed, as stressed above, the problem is not the quantity of the small molecules that can be generated but the quality of the library members in terms of structural diversity.

Whereas SPOS has frequently been exploited in traditional combinatorial chemistry, the two technologies are not explicitly linked. More recently, as a result of its ever increasing applicability and generality, SPOS has been applied to the synthesis of structurally diverse compound libraries (e.g. Scheme 8.3).[7, 50, 53] Whereas the development of novel reactions on solid support is an important area of research in its own right, the major focus of this section is the use 'diversity linkers' in the generation of DOS libraries.

8.8.1 An overview of linkage cleavage strategies

To mirror the diverse array of conditions employed in organic transformations[67] over the last 20 years more than 200 linkers have been developed.[68] To be useful a linker must: contain a functional group through which the substrate to be attached; be 'long' enough to prevent any unwanted resin–substrate interactions; and be stable (especially the linker–substrate bond) to the reaction conditions employed. This latter property confers selective cleavability.

Linkers are broadly classified depending on the functionality which remains at the site of cleavage (Scheme 8.10). Whereas with classic/traditional linkers the functional group used to attach the substrate to the resin is still present after cleavage, with traceless linkers only a hydrogen atom remains.[67, 68] In contrast to these approaches, diversity linkers allow further structural variation to be incorporated and can be of value in DOS.

Although the focus of this section is not classic and traceless linkers, some considerations are noteworthy. Since classic linkers are usually acid or base labile (e.g. **76** to **77**, Scheme 8.11), which ensures the target compound can be removed under mild conditions, this can be problematic as cleavage may occur before it is required. To overcome this problem so-called safety-catch linkers were developed; these

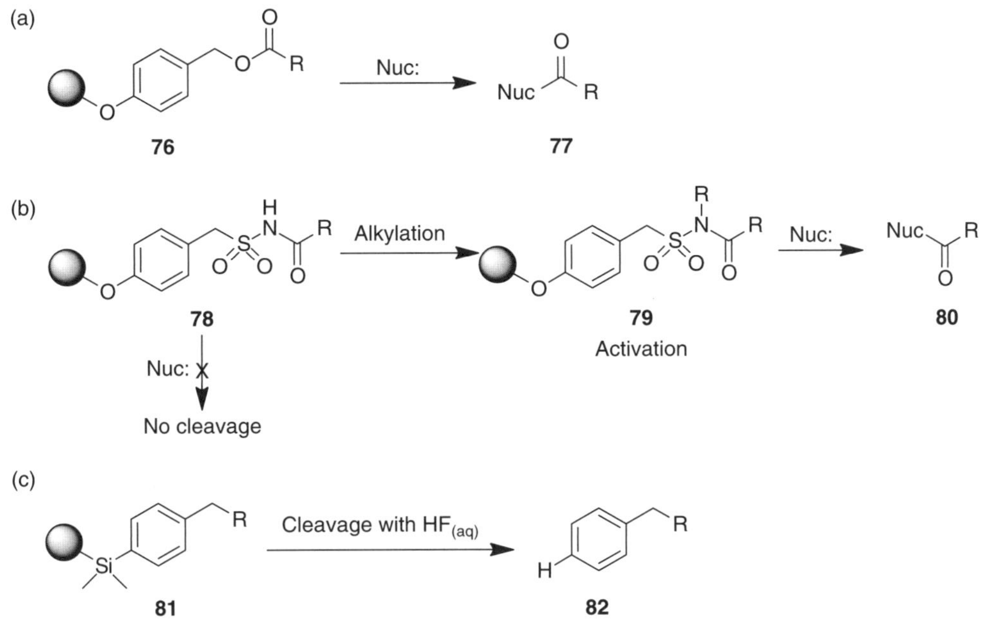

Scheme 8.11 a) An example of the use of a classic linker. Cleavage in base allows the removal of the product from the resin; the polar functional group remains. b) An extension of this strategy is the use of a safety-catch linker, in this example the nitrogen must be alkylated before base induced cleavage can occur. c) An example of the use of a traceless linker. Using the silicon-based linker, after acid induced cleavage, only a hydrogen remains at the initial site of attachment

require specific activation (i.e. **78** to **79**) before cleavage can occur (i.e. **79** to **80**). Furthermore, in the context of a chemical genetics study where this cleavage is method exploited, all of the compounds screened would contain a common polar functionality. Not only would this reduce the diversity of the collection, but the functional group itself may adversely effect bioactivity. Although this second point can be off-set by the use of traceless linkers (i.e. **81** to **82**), further chemical manipulation at this site is then problematic.

Scott and Steel have identified three sub-classes of diversity linkers in an excellent review: i) those where diversity is incorporated by the nucleophilic component; ii) those where diversity is incorporated by an electrophilic component; and ii) those which do not fit neatly into either of these classifications.[68] Herein, examples are highlighted to demonstrate key principles as a way of introducing this topic.

8.8.2 Diversity linkers: A summary of the approaches used

Although the linker used by Thomas *et al.* (Scheme 8.3) could be removed to yield acids, esters and a variety of amides by varying the nucleophilic component in the cleavage reaction, examples where electophilic intermediate is initially generated allow greater diversity to be incorporated. For example, the triazene linker **83** has been used to initially generate an elecrophilic aryldiazonium salt **84**; this was intercepted by a diverse range of nucleophiles to generate compounds **85–90** (Scheme 8.12).[69, 70]

In a similar vein, diversity can be incorporated using an electrophilic component. Generally, these linkers contain a suitably labile bond or are converted to reactive intermediates that are subsequently intercepted by a diverse range of electrophiles. As an alternative to these 'nucleophilic' and 'electrophilic'

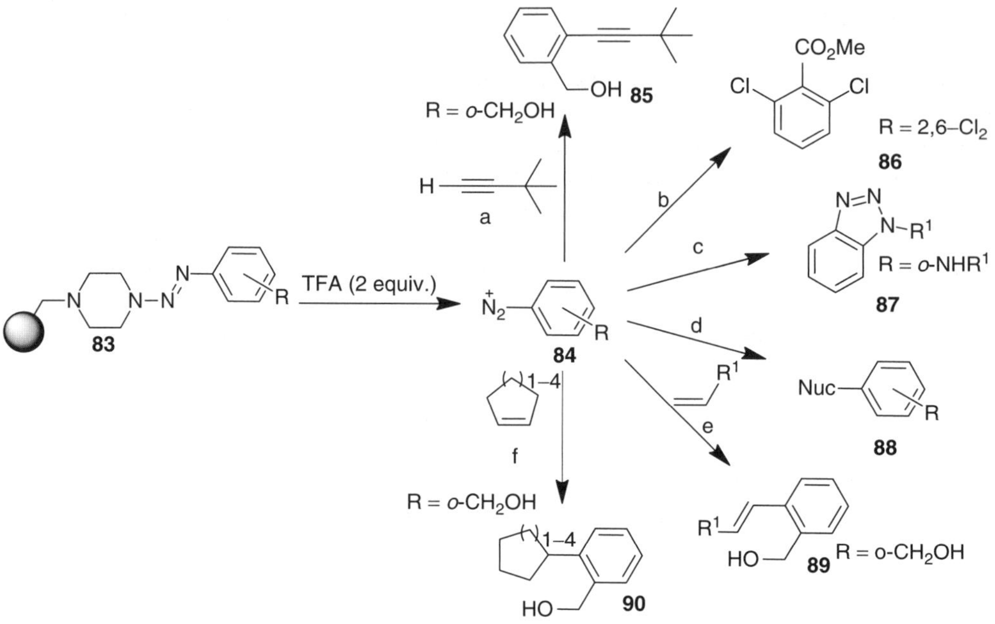

Scheme 8.12 *The generation of a highly reactive electrophilic intermediate that is then intercepted by a range of nucleophiles*[69, 70]

approaches, Diels–Alder reactions, ring closing metathesis reactions, radical cleavage reactions and modified Friedlander reactions have also been exploited. Again these have allowed diversity to be incorporated in the cleavage step in a library synthesis.

Although many diversity linkers provide a useful method of incorporating appendage diversity, approaches such as 'cyclo-release cleavage', where different skeletons can be produced under different conditions, may be more useful still.

8.9 Conclusion

Although a recently developed concept, over the last few years, novel and imaginative strategies have been used to prepare DOS libraries. Many of these compound collections have been successfully exploited in chemical genetics to identify modulators for biological systems.[5–8] DOS, to populate diverse regions of chemical space, still represents a potentially rewarding challenge for chemists.

References

[1] Spandl, R. J., Spring, D. R., and Bender, A.; *Org. Biomol. Chem.* **2008**, DOI: 10.1039/B719372F.
[2] Spring, D. R.; *Org. Biomol. Chem.* **2003**, *1*, 3867.
[3] Thomas, G. L., Wyatt, E. E., and Spring, D. R.; *Curr. Opin. Drug Discovery Dev.* **2006**, *9*, 700.
[4] Haggarty, S. J.; *Curr. Opin. Chem. Biol.* **2005**, *9*, 296.
[5] Kim, Y. K., Arai, M. A., Arai, T., et al.; *J. Am. Chem. Soc.* **2004**, *126*, 14740.
[6] Koehler, A. N., Shamji, A. F., and Schreiber, S. L.; *J. Am. Chem. Soc.* **2003**, *125*, 8420.
[7] Spring, D. R., Krishnan, S., Blackwell, H. E., and Schreiber, S. L.; *J. Am. Chem. Soc.* **2002**, *124*, 1354.

[8] Kuruvilla, F. G., Shamji, A. F., Sternson, S. M., *et al.*; *Nature* **2002**, *416*, 653.
[9] Collins, F. S., Lander, E. S., Rogers, J., and Waterston, R. H.; *Nature* **2004**, *431*, 931.
[10] Brent, M. R.; *Nat. Rev. Genet*. **2008**, *9*, 62.
[11] Spring, D. R.; *Chem. Soc. Rev*. **2005**, *34*, 472.
[12] Schreiber, S. L.; *Bioorg. Med. Chem*. **1998**, *6*, 1127.
[13] Schreiber, S. L.; *Chem. Eng. News* **2003**, *81*, 51.
[14] Walsh, D. P., and Chang, Y. T.; *Chem. Rev*. **2006**, *106*, 2476.
[15] Hopkins, A. L., and Groom, C. R.; *Nat. Rev. Drug Discovery* **2002**, *1*, 727.
[16] Lipinski, C., and Hopkins, A.; *Nature* **2004**, *432*, 855.
[17] Hann, M. M., Leach, A. R., and Harper, G.; *J. Chem. Inf. Comput. Sci*. **2001**, *41*, 856.
[18] Schuffenhauer, A., Brown, N., Selzer, P., *et al.*; *J. Chem. Inf. Mod*. **2006**, *46*, 525.
[19] MacBeath, G.; *Genome Biol*. **2001**, *2*, 2005.1.
[20] Fergus, S., Bender, A., and Spring, D. R.; *Curr. Opin. Chem. Biol*. **2005**, *9*, 304.
[21] Perez, J. J.; *Chem. Soc. Rev*. **2005**, *34*, 143.
[22] Dobson, C. M.; *Nature* **2004**, *432*, 824.
[23] Downs, G. M., Willett, P., and Fisanick, W.; *J. Chem. Inf. Comput. Sci*. **1994**, *34*, 1094.
[24] Estrada, E., and Uriarte, E.; *Curr. Med. Chem*. **2001**, *8*, 1573.
[25] Fitzgerald, S. H., Sabat, M., and Geysen, M.; *J. Comb. Chem*. **2007**, *9*, 724.
[26] Butler, M. S.; *Nat. Prod. Rep*. **2008**, *22*, 162.
[27] Lam, K. S.; *Trends Microbiol*. **2007**, *15*, 279.
[28] Feher, M., and Schmidt, J. M.; *J. Chem. Inf. Comput. Sci*. **2003**, *43*, 218.
[29] Burke, M. D., and Schreiber, S. L.; *Angew. Chem. Int. Ed*. **2004**, *43*, 46.
[30] Valler, M. J., and Green, D.; *Drug Discov. Today* **2000**, *5*, 286.
[31] Lipinski, C. A., Lombardo, F., Dominy, B. W., and Feeney, P. J.; *Adv. Drug Delivery Rev*. **1997**, *23*, 3.
[32] Keller, T. H., Pichota, A., and Yin, Z.; *Curr. Opin. Chem. Biol*. **2006**, *10*, 357.
[33] Abad-Zapatero, C.; *Drug Discov. Today* **2007**, *12*, 995.
[34] Burke, M. D., Berger, E. M., and Schreiber, S. L.; *J. Am. Chem. Soc*. **2004**, *126*, 14095.
[35] Evans, B. E., Rittle, K. E., Bock, M. G., *et al.*; *J. Med. Chem*. **1988**, *31*, 2235.
[36] Reayi, A., and Arya, P.; *Curr. Opin. Chem. Biol*. **2005**, *9*, 240.
[37] Clardy, J., and Walsh, C., *Nature* **2004**, *432*, 829.
[38] Ko, S. K., Jang, H. J., Kim, E., and Park, S. B.; *Chem. Commun*. **2006**, 2962.
[39] Zhou, C. X., Dubrovsky, A. V., and Larock, R. C.; *J. Org. Chem*. **2006**, *71*, 1626.
[40] Messer, R., Fuhrer, C. A., Iner, R. H., and Haner, R.; *Curr. Opin. Chem. Biol*. **2005**, *9*, 259.
[41] Boldi, A. M., and Dragoli, D. R.; Chemistry on the Interface of Natural Products and Combinatorial Chemistry, in *Combinatorial Synthesis of Natural Product-Based Libraries* (Ed.: A. M. Boldi), Taylor and Francis/CRC Press, London, **2006**.
[42] Goess, B. C., Hannoush, R. N., Chan, L. K., *et al.*; *J. Am. Chem. Soc*. **2006**, *128*, 5391.
[43] Buckingham, J.; *Dictionary of Natural Products (CD ROM)*, Taylor and Francis/CRC Press, London, **2005**.
[44] Kaiser, M., Wetzel, S., Kumar, K., and Waldmann, H.; *Cell Mol. Life Sci* **2008**, 1.
[45] Koch, M. A., Schuffenhauer, A., Scheck, M., *et al.*; *Proc. Nat. Acad. Sci. U.S.A*. **2005**, *102*, 17272.
[46] Dekker, F. J., Koch, M. A., and Waldmann, H.; *Curr. Opin. Chem. Biol*. **2005**, *9*, 232.
[47] Noren-Muller, A., Reis-Correa, I., Prinz, H., *et al.*; *Proc. Nat. Acad. Sci. USA* **2006**, *103*, 10606.
[48] Burke, M. D., Berger, E. M., and Schreiber, S. L.; *Science* **2003**, *302*, 613.
[49] Schreiber, S. L.; *Science* **2000**, *287*, 1964.
[50] Kumar, N., Kiuchi, M., Tallarico, J. A., and Schreiber, S. L.; *Org. Lett*. **2005**, *7*, 2535.
[51] Nielsen, T. E., and Schreiber, S. L.; *Angew. Chem. Int. Ed*. **2008**, *47*, 48.
[52] Wyatt, E. E., Fergus, S., Galloway, W., *et al.*; *Chem. Commun*. **2006**, 3296.
[53] Thomas, G. L., Spandl, R. J., Glansdorp, F. G., *et al.*; *Angew. Chem. Int. Ed*. **2008**, DOI: 10.1002/anie.200705415.
[54] Chen, C., Li, X. D., and Schreiber, S. L.; *J. Am. Chem. Soc*. **2003**, *125*, 10174.
[55] Kolb, H. C., Vannieuwenhze, M. S., and Sharpless, K. B.; *Chem. Rev*. **1994**, *94*, 2483.

[56] Johnson, J. S., and Evans, D. A.; *Acc. Chem. Res*. **2000**, *33*, 325.
[57] Thomas, G. L., Ladlow, M., and Spring, D. R.; *Org. Biomol. Chem*. **2004**, *2*, 1679.
[58] Johnson, A. P., Aucken, H. M., Cavendish, S., *et al*.; *J. Antimicrob. Chemother*. **2001**, *48*, 143.
[59] Kumagai, N., Muncipinto, G., and Schreiber, S. L.; *Angew. Chem. Int. Ed*. **2006**, *45*, 3635.
[60] Tejedor, D., Santos-Exposito, A., and Garcia-Tellado, F.; *Chem. Eur. J*. **2007**, *13*, 1201.
[61] Comer, E., Rohan, E., Deng, L., and Porco, J. A.; *Org. Lett*. **2007**, *9*, 2123.
[62] Tejedor, D., and Garcia-Tellado, F.; *Chem. Soc. Rev*. **2007**, *36*, 484.
[63] Oguri, H., and Schreiber, S. L.; *Org. Lett*. **2005**, *7*, 47.
[64] Sello, J. K., Andreana, P. R., Lee, D. S., and Schreiber, S. L.; *Org. Lett*. **2003**, *5*, 4125.
[65] Merrifield, R. B.; *J. Am. Chem. Soc*. **1963**, *85*, 2149.
[66] Dolle, R. E.; *J. Comb. Chem*. **2005**, *7*, 739.
[67] Jung, N., Weihn, M., and Brase, S.; *Top. Curr. Chem*. **2007**, *278*, 1.
[68] Scott, P. J. H., and Steel, P. G.; *Eur. J. Org. Chem*. **2006**, 2251.
[69] Brase, S., and Schroen, M.; *Angew. Chem. Int. Ed*. **1999**, *38*, 1071.
[70] Brase, S.; *Acc. Chem. Res*. **2004**, *37*, 805.

9
T1 and T2 – Versatile Triazene Linker Groups

Kerstin Knepper and Robert E. Ziegert

Institute for Organic Chemistry, Karlsruhe Institute of Technology, Germany

9.1 Introduction

For almost thirty years, pharmacologically and biologically active compounds containing a triazene moiety have been investigated as lead structures within medicinal chemistry.[1, 2] In modern organic chemistry triazene is, furthermore, used as protecting group,[3] as well as in the syntheses of polymers[1, 4] and oligomers.[5] As a result of its multifuntionality and its stability, the triazene group has become of great interest in combinatorial chemistry and solid-phase synthesis.[4, 6] Even under harsh reaction conditions like metal hydride reductions, reactions with organometallic reagents and so on, the triazene moiety turned out to be very stable and flexible.[1, 4, 5,7–9] The introduction of the triazene group in the Ullmann reaction as directing group was developed as a variation of the biaryl synthesis (Scheme 9.1)[8] by Nicolaou *et al.* during the total synthesis of Vancomycin (**3**).[9] The triazene unit was first introduced as linker group in solid-phase chemistry by Moore *et al.* (Scheme 9.2)[6] and Tour *et al.*[6b] The group of Moore used the novel 3-propyl-3-(benzyl-supported) triazene linkage for the solid-phase synthesis of phenylacetylene oligomers.[10]

In this chapter, applications in the syntheses of a broad range of compounds of versatile triazene resins are described. At the end of each section, an experimental procedure is given which provides general protocols of resin formations and cleavages as well as selected examples of transformations on the solid support.

Linker Strategies in Solid-Phase Organic Synthesis   Edited by Peter Scott
© 2009 John Wiley & Sons, Ltd

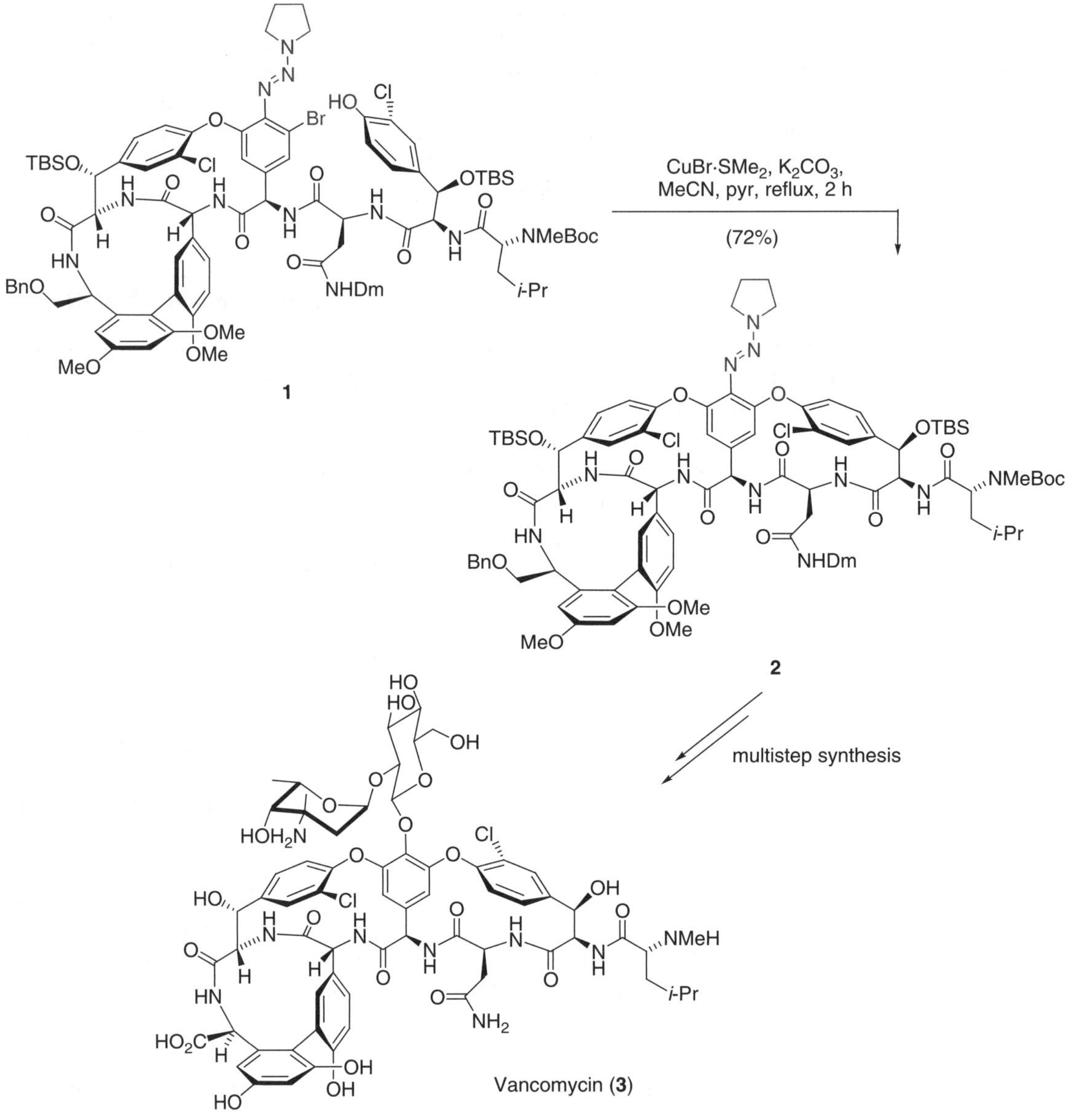

264 *Multifunctional Linker Units for Diversity-Oriented Synthesis*

Scheme 9.1 *Biaryl synthesis as the key step in the total synthesis of Vancomycin 3 by Nicolaou et al.*[8, 9]

9.2 The T1 linker

Inspired by work of Moore,[6a] Tour *et al.*[6b] and Nicolaou[8, 9] the group of Bräse developed two species of linkers based on the triazene functionality which were named the T1 and T2 linkers.[11]

As shown in Figure 9.1, the T1 resins in general are divided in the dibenzyl-type T1 resin (**7**) and the piperazinyl-type T1 resin (**8**).

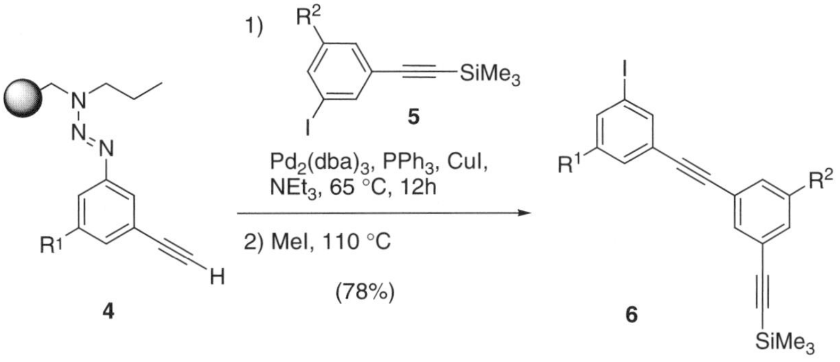

T1 and T2 – Versatile Triazene Linker Groups

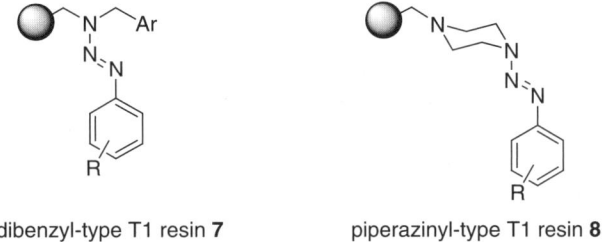

Scheme 9.2 The triazene group as linker in solid-phase chemistry introduced by Moore et al.[6]

dibenzyl-type T1 resin **7**   piperazinyl-type T1 resin **8**

Figure 9.1 Two types of triazene linkers based on the dibenzyl or piperazinyl structure[11]

In 1998, the synthesis of the piperazinyl-type T1 resin **8** was reported by Bräse et al.[12] Resin **8** can be prepared by the treatment of Merrifield resin (**9**) with piperazine and subsequent conversion with a diazonium salt **11**. Also published by Bräse et al. in 1999, was the coupling of a diazonium salt **11** to an immobilized benzylamine **12** synthesized from Merrifield resin (**9**) resulting in the dibenzyl-type T1 resin **7**. Both methods are described in general in Scheme 9.3.

The two linker systems were introduced in various reactions to show their broad applicability. The use triazene linkers in solid-phase synthesis is described in more details in the following ections.

Typical Experimental Procedures

General protocol for the synthesis of triazene resins (scheme 9.3)[12]

To a suspension of Merrifield resin (0.72 mmol/g chloride) in dimethylformamide an excess of benzyl amine is added. The mixture is stirred with an overhead stirrer for 48 hours at 60 °C. After cooling to room temperature the benzyl amine resin is washed following the general washing procedure and dried under high vacuum.

The amine (2.5 mmol) is placed in a round bottom flask and stirred in 150 ml tetrahydrofuran for 15 minutes at room temperature. The boron trifluoride diethyl etherate (37.0 mmol) is added and the mixture is stirred for 15 minutes at room temperature and then cooled down to −15 °C. The isoamylnitrite (3.5 mmol) is added and the mixture is stirred at this temperature for two hours.

In a three necked round bottom flask with an overhead stirrer the benzyl amine resin or the piperazinyl resin (5 g, 1.0 mmol/g) is swollen in a mixture of 100 ml diethyl ether and 13.5 ml pyridine and cooled down

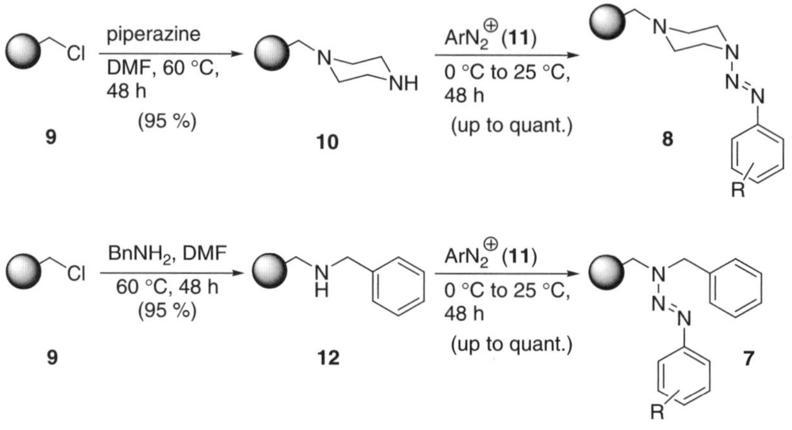

Scheme 9.3 Standard synthesis of the T1 resins **7** and **8**[12]

to −25 °C. Then the diazonium salt is added with 100 ml acetonitrile and the mixture is slowly warmed up to room temperature. The resin is washed according the standard washing procedure and dried under reduced pressure.

General washing procedure – an example

The resin is washed following three sequences:

First sequence (three times): methanol – tetrahydrofuran (THF) – pentane – dichloromethane

Second sequence (once): methanol – N,N-dimethylformamide (DMF) – pentane – THF

Third sequence (twice): pentane – dichloromethane – pentane.

9.2.1 The dibenzyl-type T1 resins

When the dibenzyl-type T1 resin **13** was described the applicability of this system was shown in principle by conversion in a Heck reaction with *t*-butyl acrylate (Scheme 9.4). After acidic cleavage the cinnamonic acid ester **15** was obtained in a good yield end high purity.

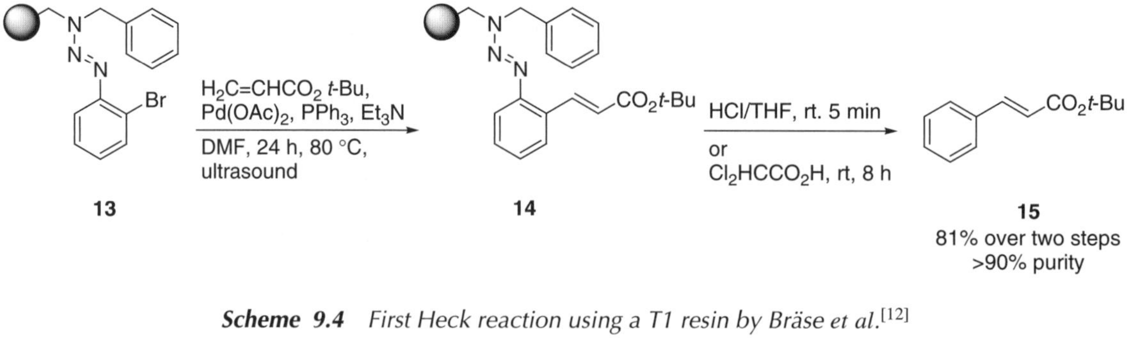

Scheme 9.4 First Heck reaction using a T1 resin by Bräse et al.[12]

T1 and T2 – Versatile Triazene Linker Groups 267

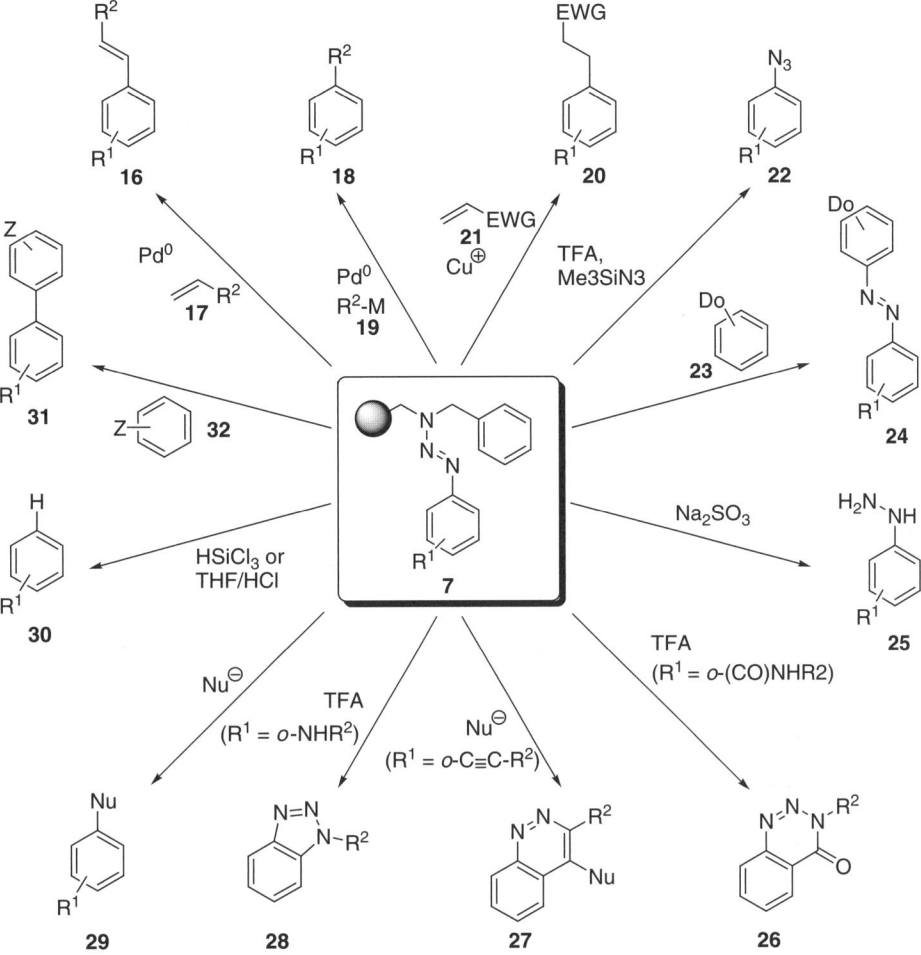

Scheme 9.5 A broad range of applications of the dibenzyl-type T1 resins 7[11]

After the first introduction of the dibenzyl-type T1 system it was applied in a range of synthetic methods as described in Scheme 9.5. The overview contains products formed directly by cleavage as well as products by *in situ* formation of a diazonium salt and subsequent transformation.[11, 13] The syntheses of nitrogen heterocycles[14–18] or metal mediated coupling reactions[14, 17, 19, 20] are possible just as well as the synthesis of azides.[15, 21] In addition to the cleavage reaction that forms azides, conditions for a traceless cleavage were developed.[22] In the present chapter the mentioned applications of the dibenzyl-type T1 resins **7** are introduced in more detail.

9.2.1.1 Cleavage strategies

A traceless cleavage was reported by Bräse *et al.* using trichlorosilane as a mild and effective reagent, obtaining products in excellent yields and purities (Scheme 9.6).[22] Compared to previously used methods,[12, 23] the reaction conditions of 32 °C and 15 minutes, as well as the high yields and purities

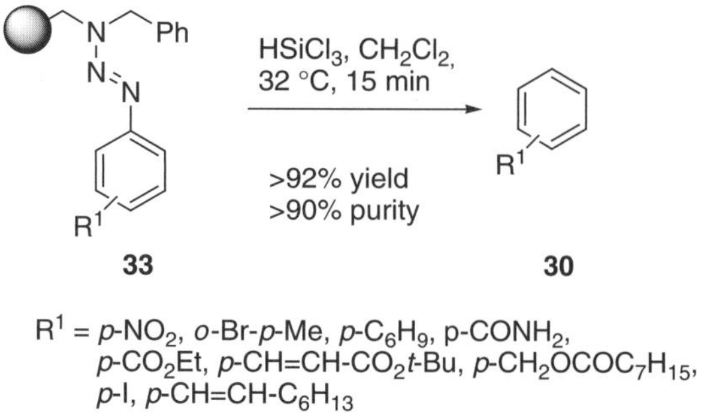

Scheme 9.6 *Traceless cleavage using trichlorosilane by Bräse et al.*[22]

made this method suitable for automated synthesis. Using cleavage conditions with THF, hydrochloric acid and ultrasound undesired by-products like polymeric impurities due to ring opening of THF were observed.[12, 22] Triethylsilane as reagent leads to the formation of triethylsilylarenes as by-products.[12, 22] The use of trichlorosilane holds one more advantage. Under metal free and radical free reaction conditions it is applicable in the presence of functional groups like double[24] or triple[25] bonds, as well as ketones,[25] esters, nitro groups and halides.[22]

A base-induced fragmentation of N,N-dibenzyl-N'-aryltriazene **34** was reported by the group of Bräse.[26] As the piperazinyl triazene system **8** is stable towards n-butyl lithium,[12] as demonstrated below, the dibenzyl-type T1 linker is labile under the same conditions due to the enhanced acidity of the bencylic position. By addition of n-butyl lithium to triazene resin **34** the color of the resin turns from yellow to deep red (Scheme 9.7). After washing with methanol the traceless cleaved product **35** was found and a yellow resin was obtained as residue. This cleavage reaction is a new pathway for the basic cleavage of this type compounds.

A cleavage to form azides **39** was developed in 2002 starting with 4-fluoro-3-nitro aniline (**36**).[15a] The conversion of fluoro resins **37** with primary or secondary amines leads to nitro resin **38**, which forms nitro azides **39** by cleavage with trimethylsilyl azide under mild conditions (Scheme 9.8). Nine azides **39** were synthesized in high purities. Using the same protocol, an azide library with 19 members was obtained in good yields and excellent purities.[15b] The main advantage using trimethylsilyl azide as azide source is the fact that this reagent is less dangerous to handle compared to, for example, sodium azide (explosive!).[15b, 27]

Azides like those described above can be modified further by post-cleavage conversion (Scheme 9.9).[15b] Bräse et al. showed that the thermal treatment of azides **42b** leads to the formation of benzofuroxane **43** in quantitative yield. The conversion with alkenes like norbornene leads to the formation of triazolines

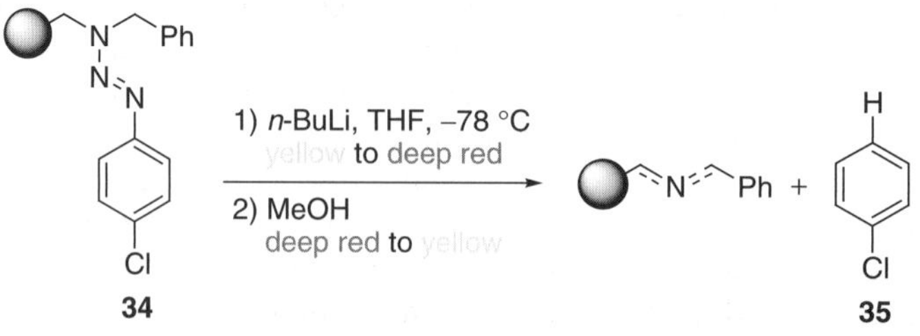

Scheme 9.7 *A base-induced fragmentation of N,N-dibenzyl-N'-aryltriazene 34 reported by Bräse et al.*[26]

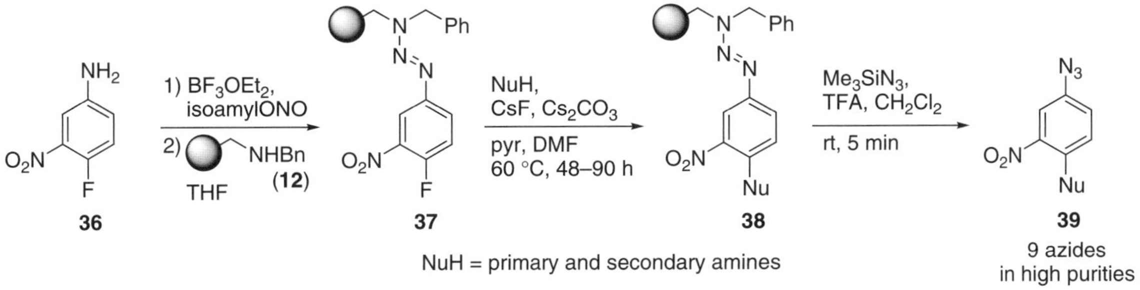

T1 and T2 – Versatile Triazene Linker Groups **269**

Scheme 9.8 *Azide synthesis by Bräse et al.*[15]

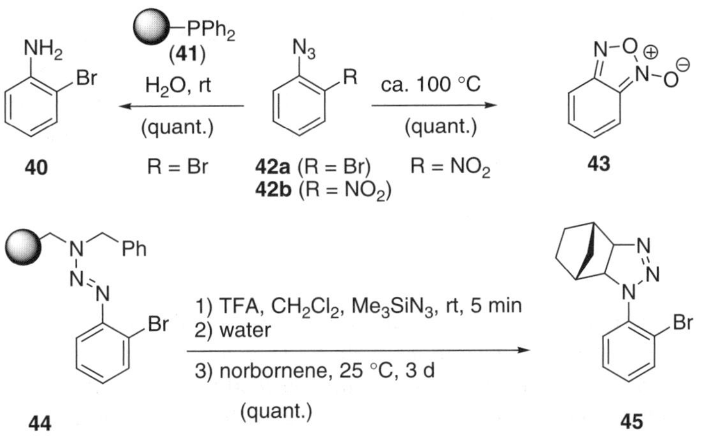

Scheme 9.9 *Post-cleavage reactions of azides by Bräse et al.*[15b]

45 likewise in quantitative yield. Furthermore, the reduction of azide **42a** to the corresponding aniline **40** with an immobilized phosphine **41** is possible.[15b, 28]

A cross-coupling product **16** can be achieved by generation of a diazonium salt **11** by treatment with trifluoro acetic acid in methanol and *in situ* coupling with alkenes **17** by the addition of the palladium catalyst (Scheme 9.10).[11, 13] The products were obtained in high yields and purities. This one-pot cleavage condition yields in salt-free products, and the formed benzyl amine resin **12** acts additionally as scavenger

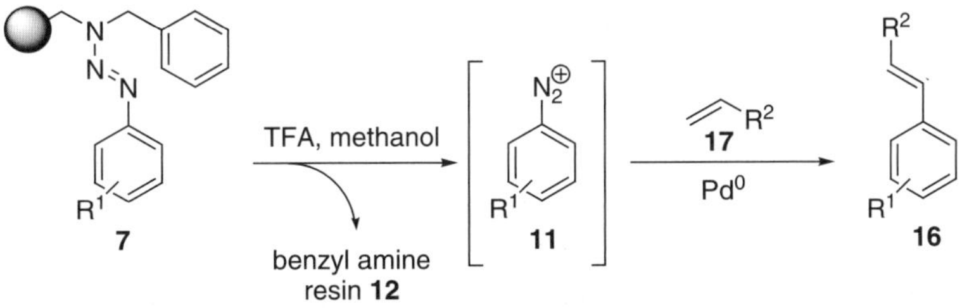

Scheme 9.10 *Cleavage/cross-coupling sequence by Bräse et al.*[11, 13]

resin catching the excess of trifluoro acetic acid. A further advantage is the use of palladium on charcoal as catalyst, which decreases palladium contamination in the products.

The strategy described above is also transferable to palladium(0) mediated coupling reactions using organometal reagents **19** to yield in product **18** as well as copper(I) mediated coupling reactions using electron-deficient alkenes **21** to yield product **20** (Scheme 9.5).[11, 13]

Further unpublished cleavage methods shown in Scheme 9.5 are the Gomberg–Bachmann radical reaction to form biaryl products **31**, the formation of diazo compounds **24** using donor-substituted benzenes **23** and the reductive cleavage to form aryl hydrazines **25**.[11, 13]

Cyclative cleavages leading to different benzoannelated heterocycles are described in Section 9.2.1.2 below.

9.2.1.2 T1 resins in solid-phase synthesis

In 1999, Bräse et al. used a Richter-type cleavage protocol for the synthesis of substituted cinnolines on a solid support.[14] After a palladium mediated cross-coupling of acetylene **47** to the T1 resin **46** the cleavage with hydrochloric acid or hydroiodic acid in acetone led to the cinnolines **49** in moderate to excellent yields and good to excellent purities (Scheme 9.11).

Enders and Schunk reported a solid-phase synthesis of monocyclic β-lactam derivatives **51** and **53** in 2002.[18] Starting from the solid-supported benzyl amine **12**, the ester **50** was synthesized in two steps. The resulting resin **50** was obtained with good to high loadings (up to 97%). The purities after cleavage to **52** were up to ≥99% (Scheme 9.12). The subsequent conversion of **50** to the immobilized lactam **51** and cleavage to the corresponding diazonium salt, which was transferred into lactam **53**, succeeded in reasonable to high yields, excellent purities and moderate to high diastereoselectivities. Using this strategy, a small library of 16 members was synthesized.

The synthesis of 1-alkyl-5-nitro-1H-benzotriazoles **55** was described by Bräse et al. in 2002.[15] Based on the experience from the synthesis of benzotriazoles starting by diazotation of diamines,[29] the benzotriazoles **55** were prepared by treatment of the nitro resins **54** with trifluoroacetic acid (Scheme 9.13). The cyclative cleavage occurred at room temperature within minutes and provided the benzotriazoles **55** in high yields and purities.

The synthesis of 3-substituted 6-aryl-3H-benzo[a][1−3]triazinones **60** and **63** using polymer-bound triazenes was developed by Bräse et al. in 2004 (Scheme 9.14).[16] Starting from various carboxyanilins **56** the resins **57** were prepared. Amide coupling and subsequent cyclative cleavage yielded the benzotriazinones **60** in reasonable to high overall yields. In addition, it was shown that the Suzuki cross-coupling

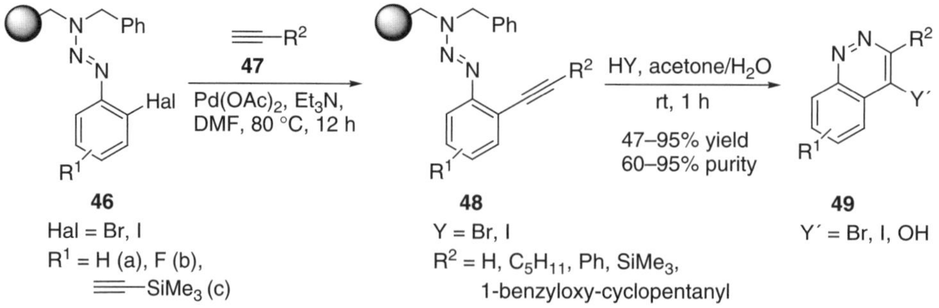

Scheme 9.11 Richter-type cleavage protocol used in the synthesis of cinnoline **49** by Bräse et al.[14]

R^2	conditions b)*
H	THF/DMF 5:2, rt, 12 h
OMe	MeOH, 2% TFA, 60 °C, 1.5 h
OCD_3	d_4-MeOH, rt, 48 h

(* optimized conditions)

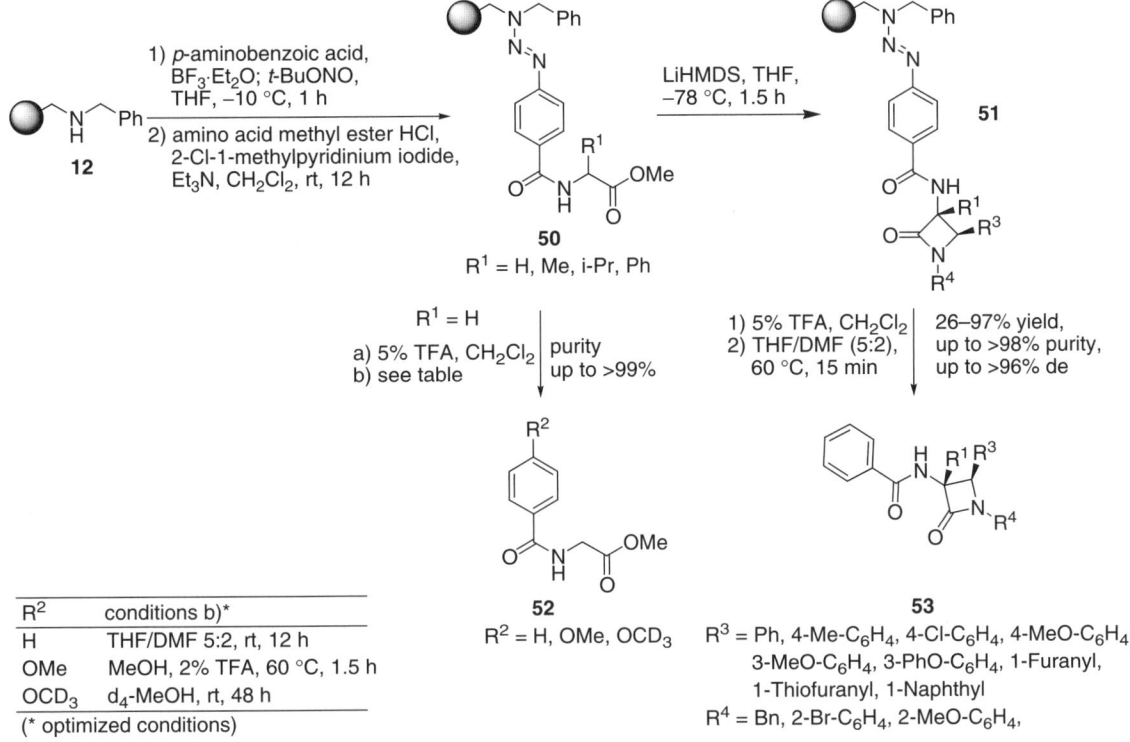

Scheme 9.12 *Solid-phase synthesis of β-lactam derivatives* ***53*** *by Enders and Schunk*[18]

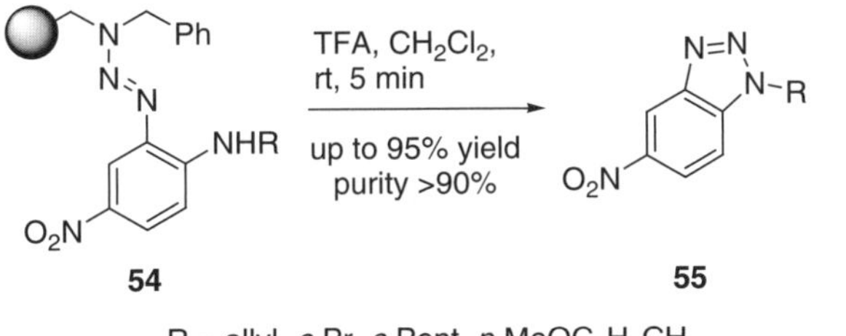

Scheme 9.13 *Synthesis of 1-alkyl-5-nitro-1H-benzotriazoles* ***55*** *using a cyclative cleavage by Bräse et al.*[15]

reaction is practicable using a iodinated resin **60** ($R^1 = I$). After amidation with phenylalanine and cyclative cleavage, the triazinones **63** were obtained in a moderate to good overall yield. Following this strategy a library of 42 triazinones **60** and **63** was synthesized.

Nicolaou and coworkers developed a variation of the Ullmann reaction for the synthesis of diarylethers using a triazene unit for activation.[8, 9] In 2004, Bräse *et al.* applied this strategy in the solid-phase synthesis of diarylethers **67**.[19] Resin **64** was formed under the previously described conditions starting from benzyl amine resin **12** and various *ortho*-bromo amines. As shown in Scheme 9.15, the *ortho*-bromo triazene **64** was converted under slightly modified conditions described by Nicolaou to the immobilized diarylether **66**. After cleavage with trimethylsilyl azide (Me_3SiN_3), a library of 32 diarylether azides **67**

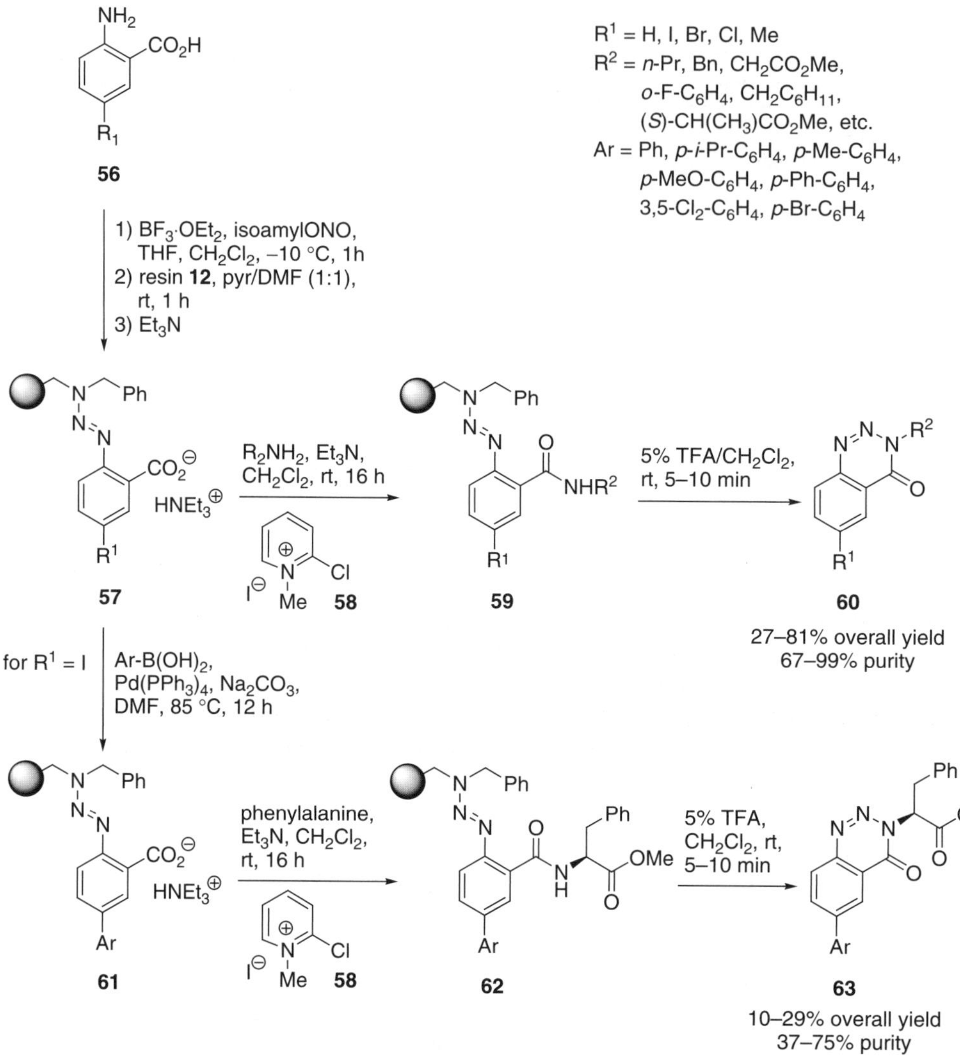

Scheme 9.14 Solid-phase synthesis of 6-aryl-3H-benzo[a][1–3]triazinones **60** and **63** by Bräse et al.[16]

was obtained in reasonable to good yields and high purities. Detailed information about the substitution pattern of the phenols **65** and the triazene resins **66** is given elsewhere.[19] In addition to the cleavage reaction leading to azides **67**, the resin **66** was treated with trifluoroacetic acid leading to the formation of diazonium salts **68**.

Next to the above mentioned monobromo resins **64**, a dibromo triazene resin **69** was introduced to the solid-supported Ullman reaction as shown in Scheme 9.16. It could be shown that the Ullman coupling only occurs at the 2-bromo position due to the directing property of the triazene/copper complex. Therefore, the 4-bromo atom is available for further coupling reactions (i.e. Suzuki cross-coupling) as demonstrated by Bräse and coworkers.[19]

Based on these experiences, Bräse and coworkers developed a solid-phase synthesis of 5-oxa-10,11-diaza-dibenzo[a,d]cycloheptenes **75** (Scheme 9.17).[17] Even today, there are only a few syntheses of

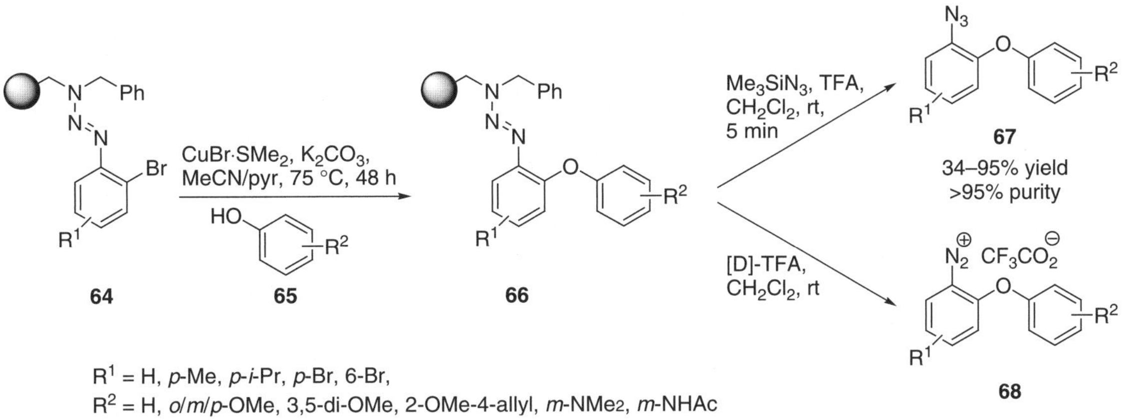

Scheme 9.15 Solid-phase synthesis using the Nicolaou variant of the Ullmann reaction by Bräse et al.[19]

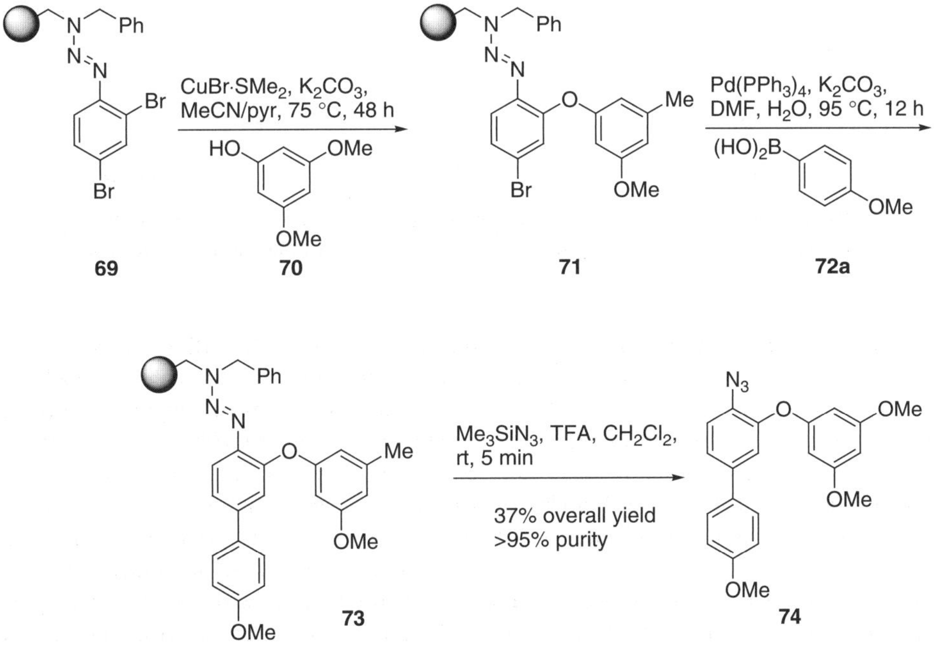

R[1] = H, p-Me, p-i-Pr, p-Br, 6-Br,
R[2] = H, o/m/p-OMe, 3,5-di-OMe, 2-OMe-4-allyl, m-NMe2, m-NHAc

Scheme 9.16 Subsequent Ullmann reaction and Suzuki cross-coupling on solid phase by Bräse et al.[19]

5-oxa-10,11-diaza-dibenzo[a,d]cycloheptenes described. The cleavage of immobilized diarylethers **71** with trifluoroacetic acid in dichloromethane without an azide source lead to the formation of the benzoannelated N,O-cycloheptenes **75** in moderate to high yields and excellent purities. A 14-membered library of cycloheptenes **75** was prepared using this strategy.

Based on the successful cross-coupling reactions described above an extensive examination on the azide functionality in palladium catalyzed reactions on solid supports was carried out by the group of Bräse in 2006.[19] The influence of the azide function on cross-coupling reactions like the Heck, the Suzuki and

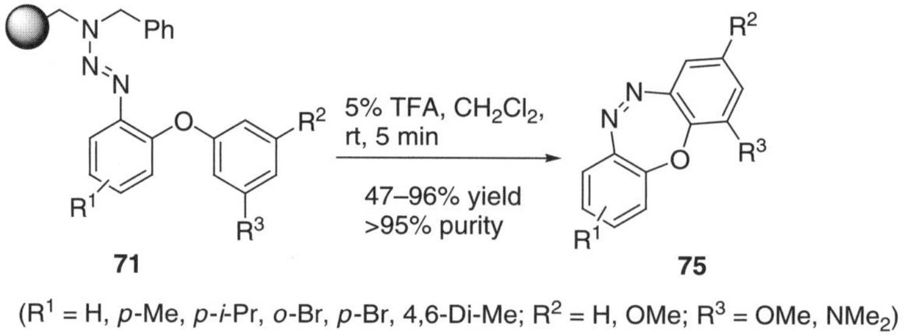

Scheme 9.17 *Solid-phase synthesis of 5-oxa-10,11-diaza-dibenzo[a,d]cycloheptenes 75 by Bräse et al.*[17]

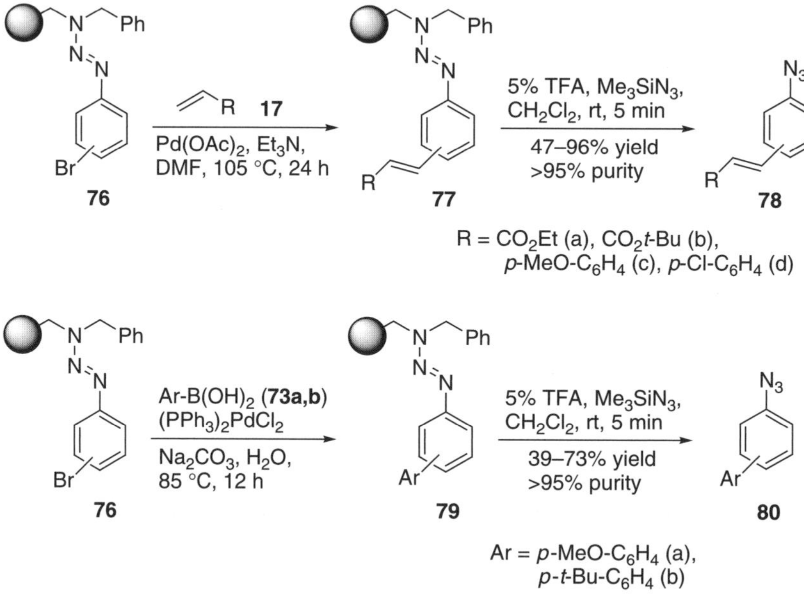

Scheme 9.18 *Heck and Suzuki cross-coupling reactions on T1 resins by Bräse et al.*[19]

the Sonogashira reactions compared to the ester linker was investigated (Scheme 9.18 and Scheme 9.19). It was shown that Heck and Suzuki cross-coupling reactions can be applied using resins with a triazene linker as well as when using resins with an ester linker. The effectiveness depends more on the coupled substrates than on the linker.

The Sonogashira cross-coupling reactions on a solid support using an ester linker were already described by Collini and Ellingboe in 1997.[30] Using triazenes as linker, for the Sonogashira cross-coupling reaction the results showed an important influence of the linker.[19] The brominated triazene resins **76** did not undergo a Sonogashira cross-coupling using the standard reaction conditions [(PPh$_3$)$_2$PdCl$_2$, copper(I) iodide, triethylamine, THF] and after cleavage only the starting material could be observed.

However, using copper-free conditions for the Sonogashira reaction, the conversion of the brominated triazene resins **76** and the terminal alkyne **81** yielded in an inseparable product mixture of the expected product **83** and the rearrangement product **84**. The reaction was not optimized and therefore no yield and purity were determined. The fact that triazene resins did not react in a copper containing medium

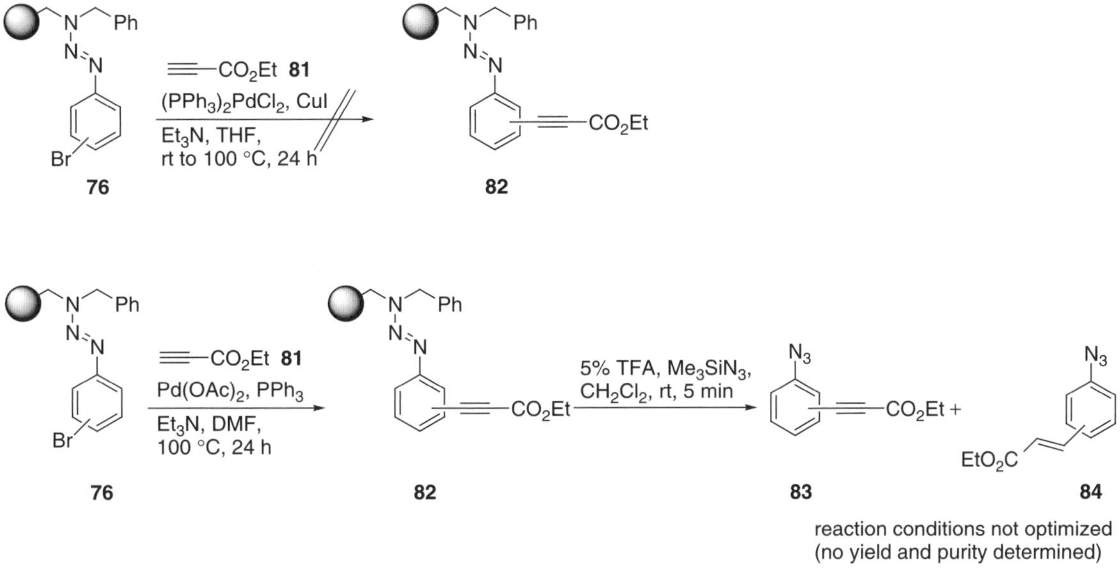

Scheme 9.19 Sonogashira cross-coupling reactions using triazene resins by Bräse et al.[19]

could be explained by the complexation of the copper by the triazene unit, since this would inactivate the copper salt and prevent the cross-coupling taking place. This is in contrast to the Nicolaou variant of the Ullmann reaction, where a complexation of the copper by a triazene leads to the activation of the coupling reaction.[8, 9]

In 2005, an efficient solid-supported synthesis of highly functionalized 1,4-benzodiazepin-5-ones **89** using an intermolecular aza-Wittig reaction was reported by Gil and Bräse (Scheme 9.20).[21] As intermediates the azides **88** were prepared starting from various anthranilic acids **56** and **85** in reasonable to good

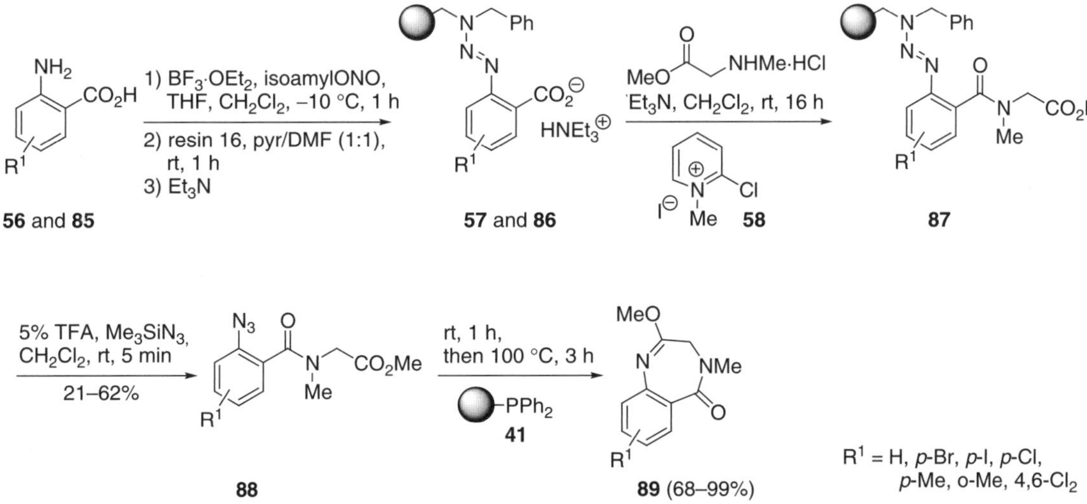

Scheme 9.20 Solid-supported synthesis of 1,4-benzodiazepin-5-ones 89 by Gil and Bräse[21]

yields. The conversion of azides **88** with the immobilized phosphine **41** in an aza-Wittig reaction led to the 1,4-benzodiazepin-5-ones **89** in yields up to 99%.

Using the same strategy, Gil and Bräse were able to synthesize the alkaloide deoxyvasicinone (**90**) and related compounds on solid phase in 99% yield (Figure 9.2).[21]

In 2007, the reduction of nitroarenes immobilized on a T1 resin were reported by the group of Bräse.[31] Different reduction reagents (sodium dithionite/potassium carbonate ($Na_2S_2O_4/K_2CO_3$), sodium sulfide/potassium carbonate (Na_2S/K_2CO_3) and Bartra reagent $[HNEt_3]^+ [Sn(SPh)_3]^-$) were examined. Under optimized reaction conditions the azido anilins **93** were obtained in up 99% yields and in excellent purities after cleavage (Scheme 9.21).

The Hartwig–Buchwald amination reaction on a solid support using triazene T1 resins was reported by Zimmermann and Bräse in 2007.[32] The immobilized *ortho*-halo aryl resins **94** or **95** were converted with 22 different amines to the corresponding immobilized amines **96** and **97** using $Pd_2(dba)_3$ or $Pd(OAc)_2$ as palladium source and *rac*-BINAP or BiPheP(Cy)2 as ligands (Scheme 9.22). After six days at 100 °C in toluene the products were obtained with turnover values up to 99% (determined through analysis of cleavage products). It was shown that the inverse variant of the Hartwig–Buchwald reaction was also conductible using immobilized anilines **98** and eleven aryl halides (turnover up to 99%).

Having the immobilized Hartwig–Buchwald products **96**, **97** and **99** in hand, resins **101** were converted to the 1*H*-benzotriazoles **103** by cleavage with TFA in dichloromethane.[32] The diazo salts **102** directly cyclized to the benzoannelated heterocycles **103** in reasonable to good yields and high purities. A library of 37 benzotriazoles **103** was produced.

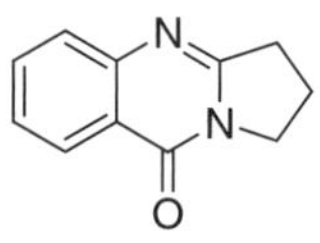

90

Figure 9.2 Structure of the alkaloid deoxyvasicinone **90**

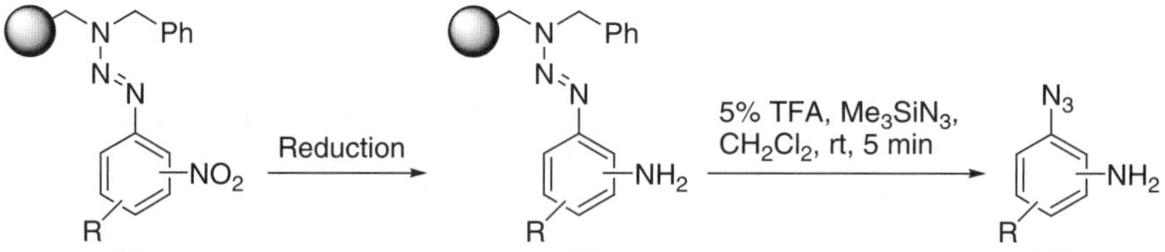

R = H, *m*-Me, *p*-Me, *m*-CO$_2$Me, *p*-MeO,
p-EtO, *p*-Cl, *p*-F, *p*-CF$_3$, 2,3-di-Me,
2,3-di-Me-4-Br, *o*-NH(CH$_2$)$_3$OMe)

Reduction reagents (for further reactin conditions see reference [30]):
a) $Na_2S_2O_4/K_2CO_3$, yield after cleavage up to 87%
b) Na_2S/K_2CO_3, yield after cleavage up to 99%
c) Bartra reagent $[HNEt_3]^+[Sn(SPh)_3]^-$, yield after cleavage up to >95%

Scheme 9.21 Reduction of nitroarenes immobilized on a T1-resin as reported by Bräse et al.[31]

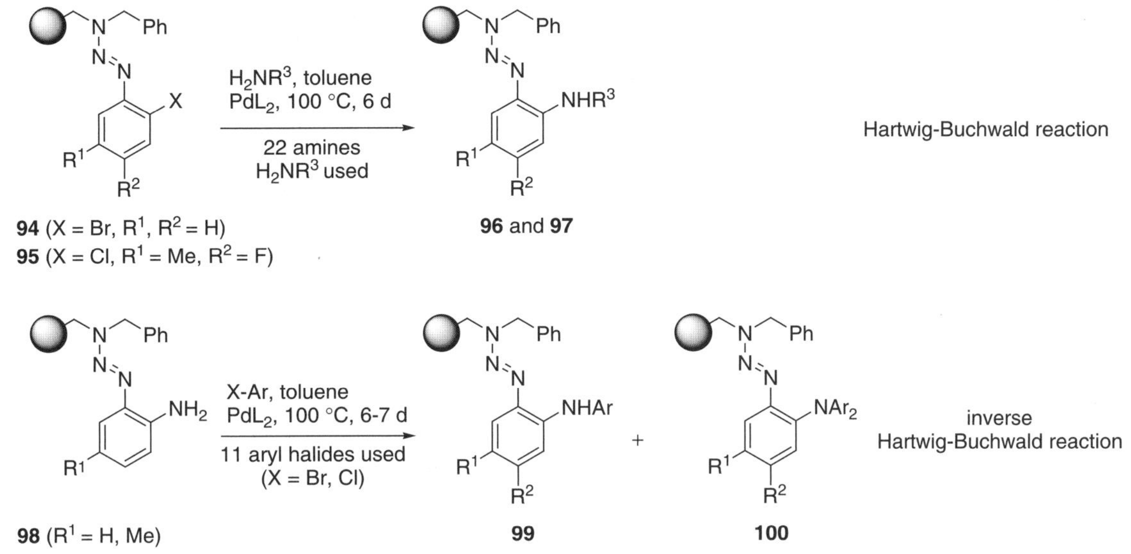

PdL_2: $Pd_2(dba)_3$ or $Pd(OAc)_2$ as Pd-source and
rac-BINAP or BipheP$(Cy)_2$ as ligands

Scheme 9.22 *(inverse) Hartwig–Buchwald reaction on solid phase reported by Zimmermann and Bräse*[32]

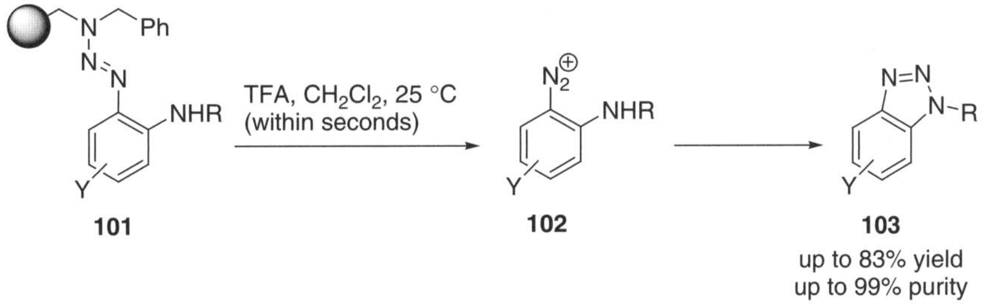

Scheme 9.23 *Solid-phase synthesis of* 1H*-benzotriazoles 103 reported by Zimmermann and Bräse*[32]

Very recently, the group of Bräse reported the solid-phase synthesis of 1*H*-benzotriazoles by reductive amination of *ortho*-anilins **101** (R = H). Subsequent cyclative cleavage, as shown in Scheme 9.23,[33] provided a library containing 14 1*H*-benzotriazoles in purities up to 99%.

Typical Experimental Procedures

Traceless cleavage with trichloromethylsilane (scheme 9.6)[22]

To a suspension of the triazene resin (1 equiv.) in dichloromethane trichlorosilane (4 equiv.)[34] *is added. The mixture is shaken for 15 minutes at 40 °C or for 60 minutes at room temperature. After being cooled to room temperature, silica gel is added the mixture is filtered and the solvent is removed under reduced pressure.*

Cleavage to azides (scheme 9.8)[15]

The triazene resin (500 mg, 0.9 mmol/g) is expanded in dichloromethane (10 ml) and mixed with azidotrimethylsilane (0.2 ml) and trifluoro acetic acid (0.15 ml).[35] The mixture is shaken for 30 minutes, then the residue is filtered off. The solvent is removed under reduced pressure.

Palladium catalyzed suzuki cross-coupling reaction (scheme 9.18)[19]

Tetrakistriphenylphosphinepalladium (0.10 equiv.) is added to a suspension of the triazene aryl halide resin (1 equiv.) under an argon atmosphere in dimethylformamide. The slurry is stirred for five minutes and then benzeneboronic acid (2 equiv.) and a solution of sodium carbonate (2 M, 2.5 equiv.) is added. The mixture is shaken for twelve hours at 85 °C and then cooled to room temperature. After diluting with ammonium acetate solution (25% in water) the mixture is shaken for additional five minutes. After filtration the resin is washed following the general washing procedure and dried under reduced pressure.

Palladium catalyzed heck cross-coupling reaction (scheme 9.18)[19]

The triazene halide resin (1 equiv.), triethylamine (1 equiv.), palladium acetate (0.10 equiv.), triphenylphospine (0.50 equiv.) and the appropriate alkene (10 equiv.) in dimethylformamide (10 ml/g resin) are placed in an argon-flashed vial. The vial is shaken at 105 °C for 24 hours. After cooling to room temperature, the resin is washed with water followed by the general washing procedure. The resin is dried under high vacuum. In the case of dibrominated resins, the amounts of all starting materials except of the resin are doubled.

Palladium catalyzed sonogashira cross-coupling reaction, copper free (scheme 9.18)[19]

The triazene halide resin (1 equiv.), triethylamine (1 equiv.), palladium acetate (0.10 equiv.), triphenylphospine (0.50 equiv.) and the appropriate alkine (10 equiv.) in dimethylformamide (10 ml/g resin) are placed in an argon-flashed vial. The vial is shaken at 80 °C for 24 hours. After cooling to room temperature, the resin is washed with water followed by the general washing procedure. The resin is dried under high vacuum.

9.2.2 The piperazinyl-type T1 resins

As shown in Scheme 9.3, piperazinyl-type T1 resins are prepared under the same conditions as the dibenzyl-type T1 resins. The piperazinyl-type T1 resins are applicable in a variety of different reactions under different conditions, which are shown in Scheme 9.24. Additional applications of the piperazinyl-type T1 resins **105** and **108** were demonstrated as they were applied in palladium catalyzed coupling reactions, Diels–Alder reactions, reductive amination reactions, asymmetric dihydroxylation or just a metal halide exchange followed by addition to phthalic acid diethyl ester. The yields (29–78%) were reasonable to good, but the purities (>90%) were very high in all reactions mentioned above.

Furthermore an efficient cleavage–cross-coupling strategy was developed by Bräse and Schroen in 1999.[36] An overview is shown in Scheme 9.25. The palladium mediated cross-coupling reactions with various alkenes, alkines or dienes were performed on solid phase using the triazene linker under standard

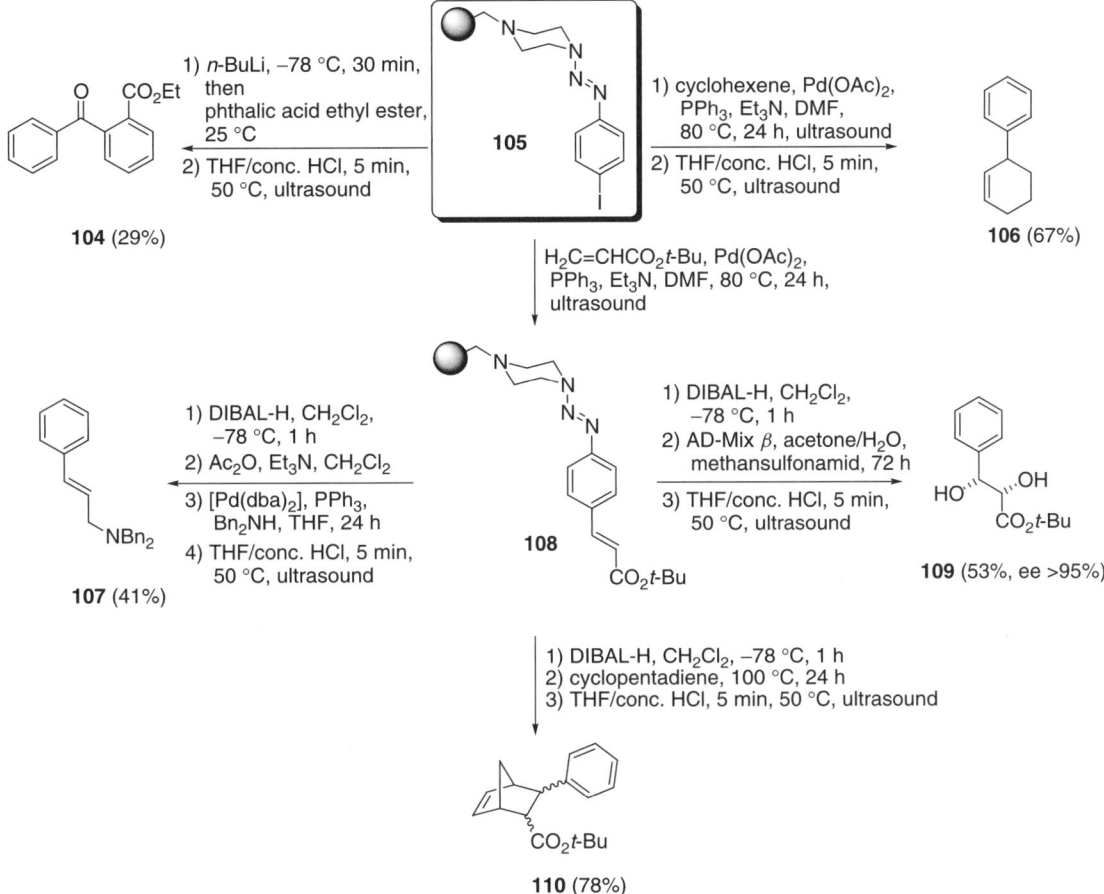

Scheme 9.24 *Various reactions using piperazinyl-type T1 resin* **105** *or* **108** *reported by Bräse et al.*[12]

conditions. The reaction conditions applied for cleavage were slightly modified by using trifluoroacetic acid as acid and methanol as solvent. The cleavage is completed within two hours in moderate to high yields and excellent purities. The possibility to recycle piperazine resins was also demonstrated.

A more complex application of the piperazinyl-type T1 resin **103** was reported by de Meijere *et al.* in 1999.[37] A Heck-Heck-Diels–Alder-Heck reaction sequence on a solid support starting with the iodated resin **103** lead to the formation of product **128** (Scheme 9.26). The bicyclopropylidene **124** was converted in a Heck reaction with phenyliodate to the intermediate **125** which was transferred without isolation to resin **126** by treatment with resin **123**. The product **128** was obtained in an overall yield of 51%.

In 2005, Bräse and coworkers reported an efficient solid phase synthesis of benzo[1−3]thiadiazoles **131**.[38] The piperazinyl-type T1 resins **129a** and **129b** were metalated by halide–lithium exchange and subsequently treated with a sulfur source like S_8 or triisopropylsilylthiol (TIPSSH) to yield the immobilized thiol **130** (Scheme 9.27). Cleavage with 5% trifluoroacetic acid in dichloromethane gave the benzo[1−3]thiadiazoles **128** in reasonable to good yields. The corresponding benzoannelated selenium heterocycles can be achieved by using a selenium source instead of sulfur.

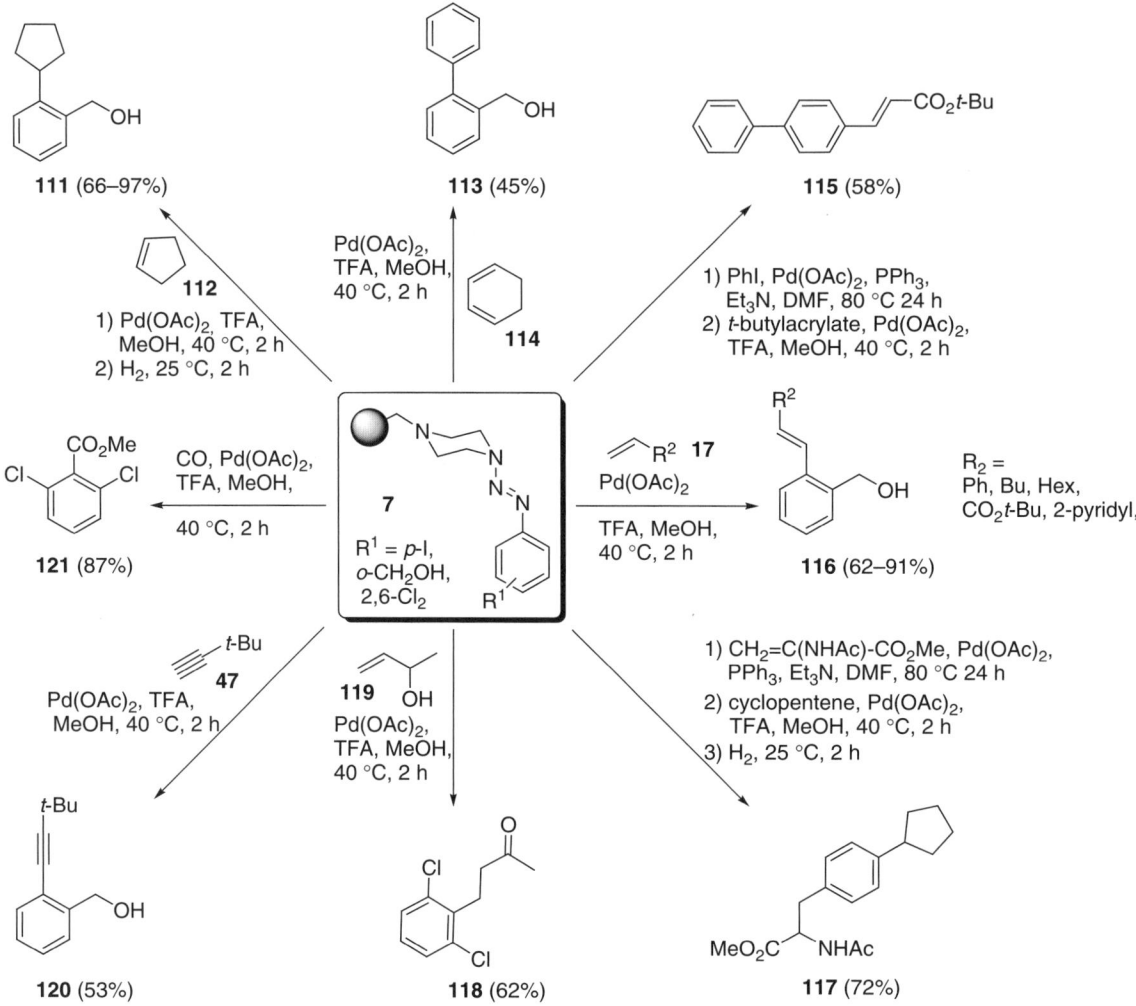

Scheme 9.25 Cleavage–cross-coupling strategies reported by Bräse and Schroen[36]

The above mentioned strategy was transferred to the solid-phase synthesis of $4H$-[1–3]-triazolo[5,1-c][1,4]benzothiazine (**135**) as shown in Scheme 9.28.[38] The immobilized thiol **130** was coupled with propargyl bromide to yield the immobilized alkyne **134**. The desired tricycle **135** was obtained by cyclative cleavage using trifluoroacetic acid and trimethylsilyl azide in 14% overall yield (four steps).

Typical Experimental Procedures

Traceless cleavage (scheme 9.24)[12]

The piperazinyl T1 resin is suspended in a mixture of THF and concentrated hydrochloric acid (10:1). The mixture is heated up to 50 °C and treated with ultrasound for five minutes. After filtration the solvent is evaporated. The residue is dried in vacuum to give the product in good yield and purity.

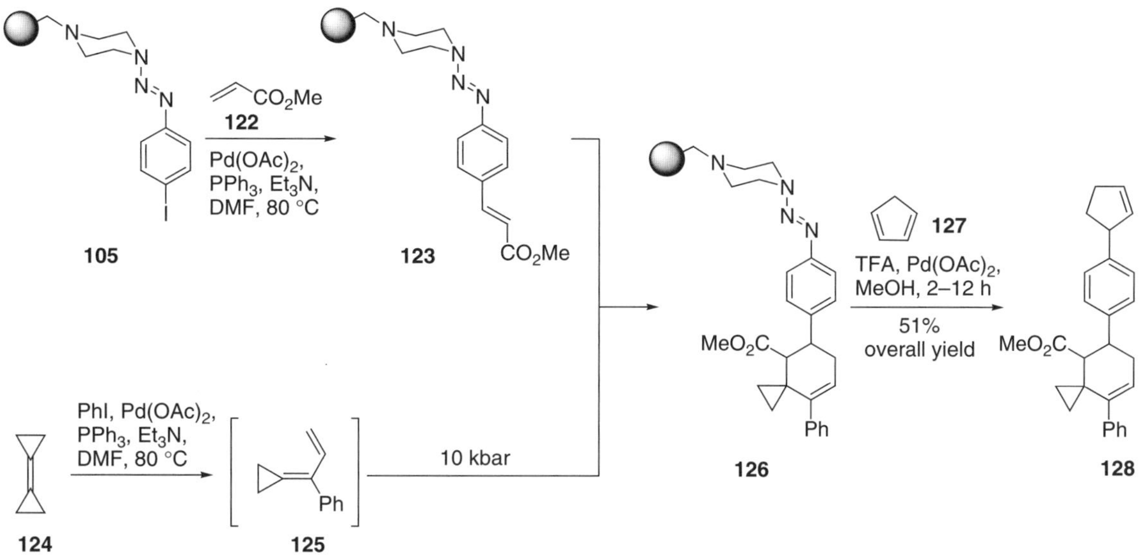

Scheme 9.26 Heck-Heck-Diels–Alder-Heck reaction sequence on solid phase by de Meijere et al.[37]

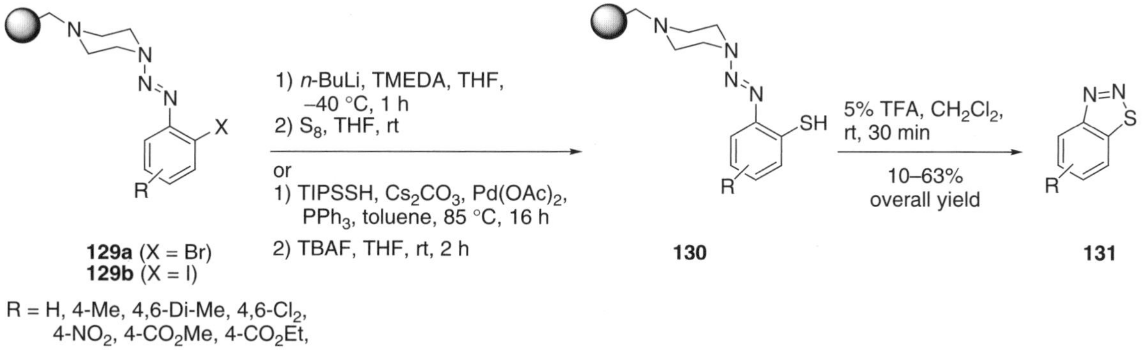

Scheme 9.27 Solid-phase synthesis of benzo[1–3]thiadiazoles 131 by Bräse et al.[38]

Cleavage by cross-coupling reactions (scheme 9.25)[36]

The piperazinyl T1 resin and an alkene/alkyne/diene as well as palladium acetate or Pd/C are suspended in methanol. After addition of two equivalents of TFA the mixture is shaken at 40 °C for 2–12 hours. After filtration, the solvent and the excess of TFA are evaporated. The residue is dried in vacuum to give the product in good yield and purity.

Heck-Heck-Diels–Alder-Heck reaction sequence on solid phase (scheme 9.26)[13, 37]

The iodinated piperazinyl T1-resin **105** (1.0 g, 0.63 mmol) is suspended in DMF (10 ml/g resin). After addition of palladium acetate (0.1 mmol/g resin), triphenylphospine (0.3 mmol/g resin) and triethylamine

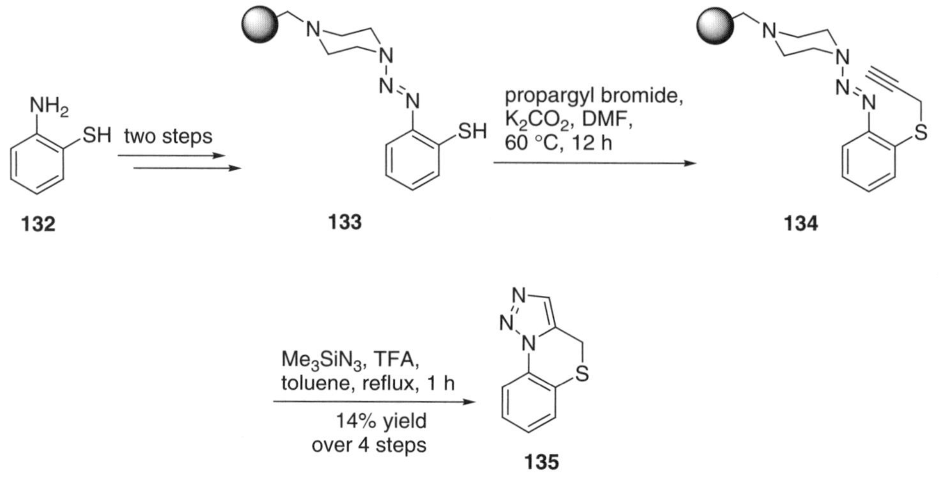

Scheme 9.28 *Solid-phase synthesis of 4H-[1–3]-triazolo[5,1-c][1, 4]benzothiazine (135) by Bräse et al.*[38]

(5.0 mmol/g resin), methylacrylate (10 mmol/g resin) is added. The mixture is heated to 80 °C for 24 hours. After cooling to room temperature, the resulting resin **123** is washed according to the standard washing procedure and dried in vacuum.

Resin **123** is suspended in DMF (10 ml/g resin). After addition of palladium acetate (0.1 mmol/g resin), triphenylphospine (0.3 mmol/g resin), triethylamine (5.0 mmol/g resin) and bicyclopropylidene (20 mmol/g resin), the mixture is heated to 80 °C for 24 hours. After cooling to room temperature, the resulting spiro-resin **126** is washed according to the standard washing procedure and dried in vacuum.

Spiro-resin **126** (0.3 g) is suspended in methanol (40 ml/g resin) and palladium acetate (0.05–0.06 mmol/g resin) is added. The mixture is cooled to 0 °C and trifluoro acetic acid (4–5 ml/g resin) is added drop-wise followed by cyclopentene (7–8 mmol/g resin). The mixture is heated to 50 °C for 12 hours and filtered after cooling to room temperature. The filtrate is evaporated and the residue is dried in vacuum to give the desired product.

9.3 The T2 linker units

An overview of the T2-type triazene resins **136–138** and the T2-type diazonium resin **139–141** is shown in Figure 9.3.

9.3.1 The T2 Linker

The T2 linkers were introduced as robust linkers for amine synthesis on solid phase (Scheme 9.29).[39] Starting from Merrifield resin (**9**) the diazonium resin **139** was prepared under the above mentioned conditions. Coupling of a secondary amine **143** and conversion of the immobilized amine **146** followed by cleavage lead to amines **147** in high yields and purities. Furthermore, it was demonstrated that the immobilized diazonium resin **139** is recyclable.

The synthesis of ureas **150** and **151** as well as amides **153** and **154** using the T2 linker was demonstrated by Bräse *et al.* in 2000.[40] The triazene resins **145** and **148** were treated with various isocyanates for twelve hours at room temperature yielding the immobilized ureas **149** (Scheme 9.30). After cleavage under mild

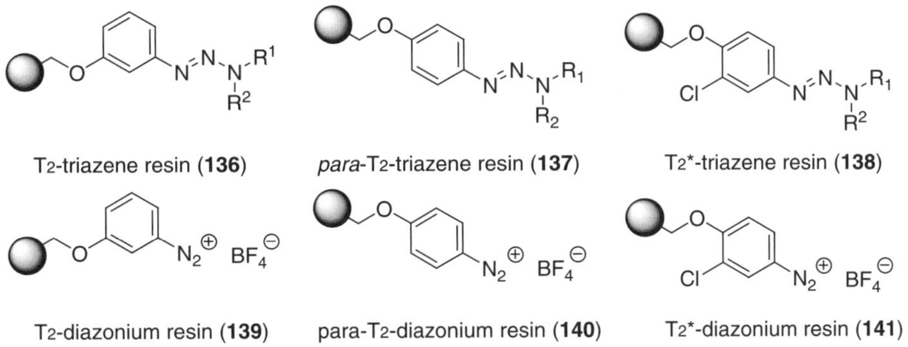

Figure 9.3 The T2 linkers

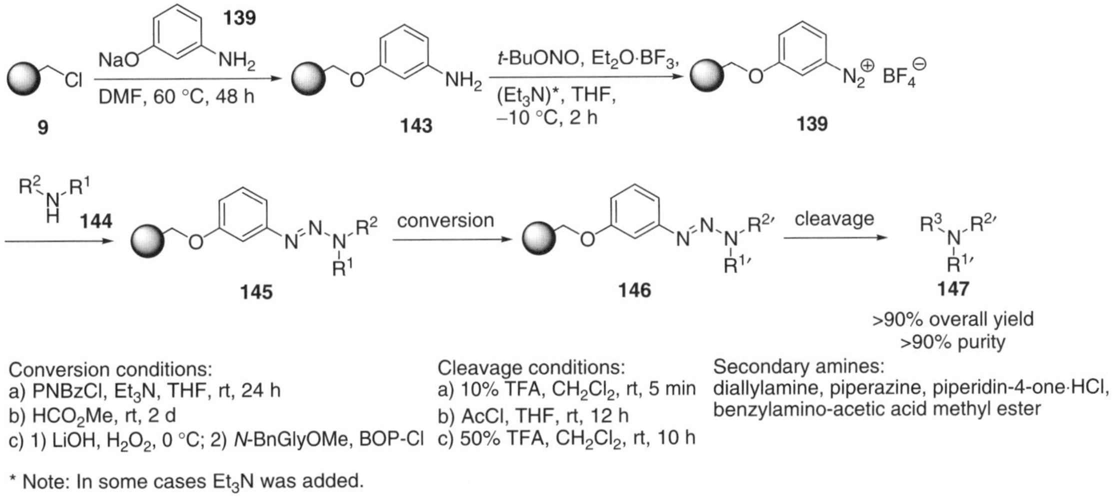

Scheme 9.29 Amines modified on solid phase using the T2-linker as reported by Bräse et al.[39]

conditions with trimethylsilyl chloride the ureas **150** were obtained in reasonable to good yields and in excellent purities. It was shown that various functional groups on the amines as well as on the isocyanates (i.e. hydroxyl, ester, amide or olefin functionalities as well as further heteroatom functionalities) were tolerated.

As shown in Scheme 9.31, it is possible to increase further structural diversity using this methodology for urea synthesis.[40] The modification of the immobilized urea **151** was carried out by N-alkylation as well as by (asymmetric) dihydroxylation or by a ozonolysis with a subsequent Wittig olefination. Depending on the nitrogen substituents the products were formed in moderate to good yields (60–>80%). The purities proved to be from moderate to very good (60–>99%).

The immobilized triazene **146** and **148** was converted with various acid chlorides providing amides **155** after cleavage in up to 75% yield (Scheme 9.32). Both solid-phase urea synthesis and solid-phase amide synthesis were successfully applied in automated synthesis using a Bohdan Neptune station equipped with a prototype solid-phase unit and a Bohdan Miniblock system. The possibility for automatization emphasizes the effectiveness and the ease of construction of this strategy. The mild reaction conditions are important advantages of the diazenyl linkers compared to other benzylamine linkers.

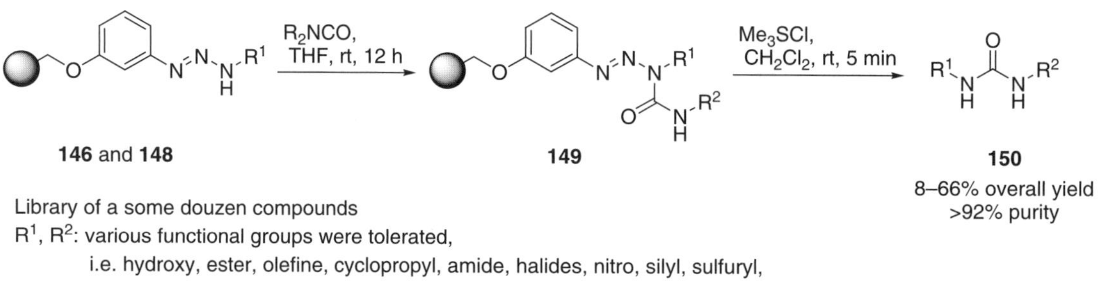
R2NCO,
THF, rt, 12 h
Me3SCl,
CH2Cl2, rt, 5 min
146 and 148
149
150
8–66% overall yield
>92% purity
Library of a some douzen compounds
R1, R2: various functional groups were tolerated,
i.e. hydroxy, ester, olefine, cyclopropyl, amide, halides, nitro, silyl, sulfuryl,

Scheme 9.30 *Solid-phase urea synthesis by Bräse et al.*[40]

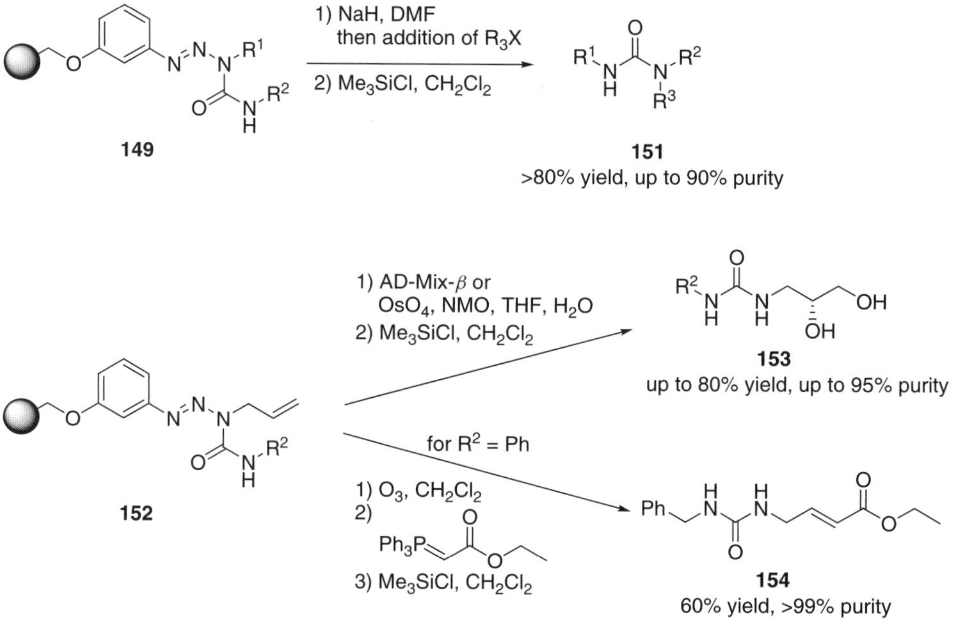
1) NaH, DMF
then addition of R3X
2) Me3SiCl, CH2Cl2
149
151
>80% yield, up to 90% purity
1) AD-Mix-β or
OsO4, NMO, THF, H2O
2) Me3SiCl, CH2Cl2
153
up to 80% yield, up to 95% purity
for R2 = Ph
1) O3, CH2Cl2
2)
3) Me3SiCl, CH2Cl2
152
154
60% yield, >99% purity

Scheme 9.31 *Further diversity of solid-phase urea synthesis by Bräse et al.*[40]

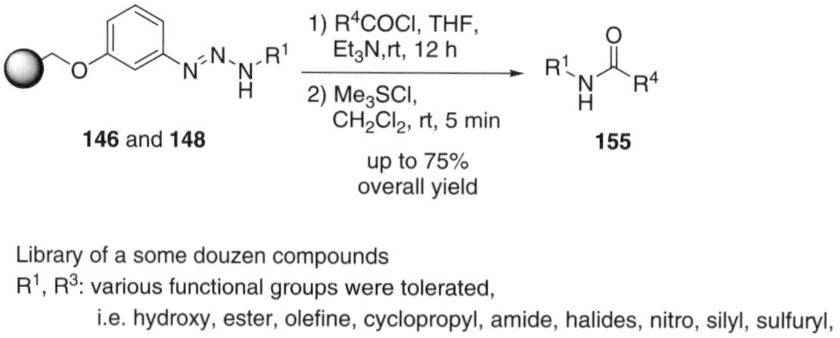
1) R4COCl, THF,
Et3N,rt, 12 h
2) Me3SCl,
CH2Cl2, rt, 5 min
up to 75%
overall yield
146 and 148
155
Library of a some douzen compounds
R1, R3: various functional groups were tolerated,
i.e. hydroxy, ester, olefine, cyclopropyl, amide, halides, nitro, silyl, sulfuryl,

Scheme 9.32 *Solid-phase urea synthesis by Bräse et al.*[40]

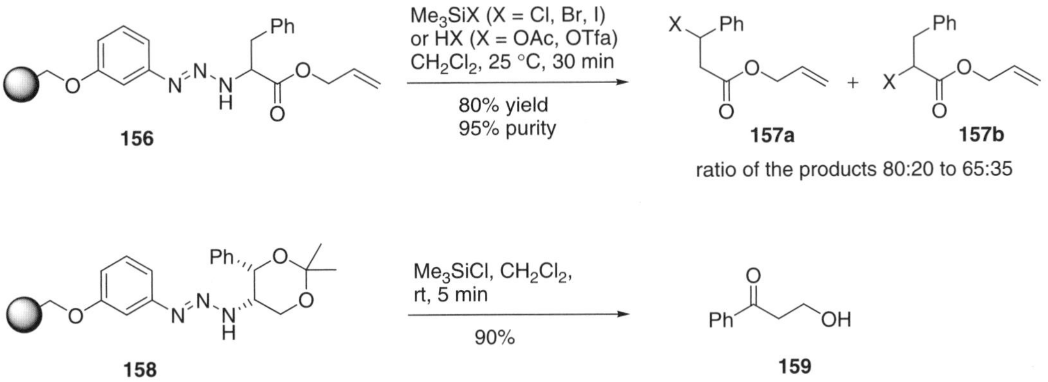

Scheme 9.33 *Cleavage from T2 resins **156** and **158** by treatment with electrophiles by Bräse et al.*[41]

In 2001, a method to release substrates from the solid support by treatment with electrophiles was demonstrated by Bräse and coworkers.[41] The immobilized phenylalanine ester **156** was treated with electrophiles like trimethyl halides as well as acetate and triflate (Scheme 9.33). The substituted alanines **157a** and **157b** were obtained as mixture in good yields and 95% purity in ratios of 80:20 to 65:35 (**157a/157b**). The cleavage of immobilized dioxane **158** with trimethylsilyl chloride led to the formation of hydroxyketone **159** in 90% yield.

A mechanism proposed by Bräse *et al.* explains the formation of the different products as shown in Scheme 9.34. The addition of the trimethylsilyl group to the aniline nitrogen giving **160** is followed by the fragmentation of the triazene to give the immobilized silyl aniline **161** and the diazonium species **162**.[42]

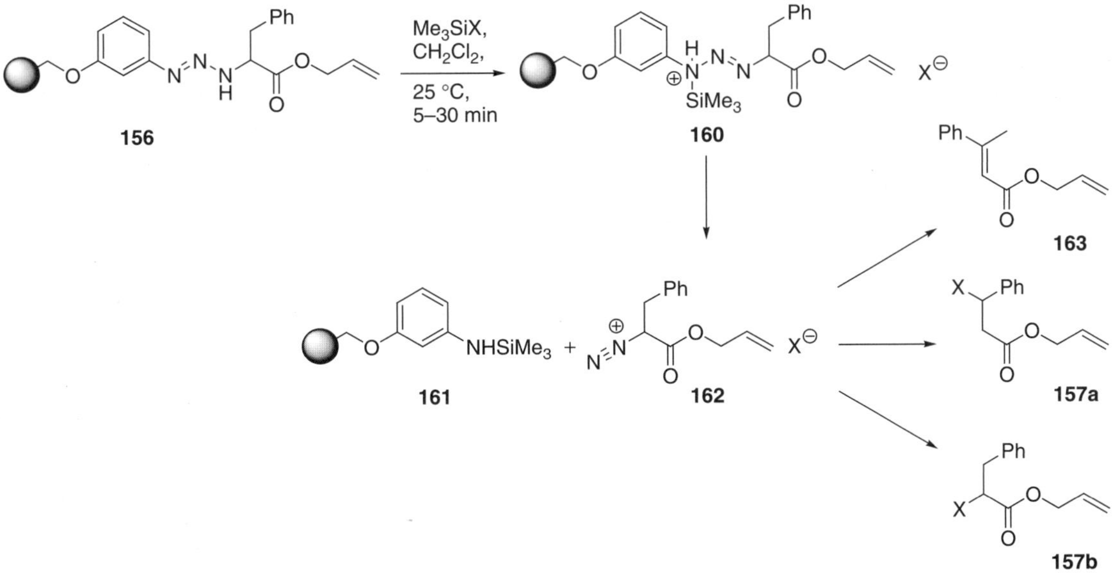

Scheme 9.34 *Proposed mechanism for of the electrophilic cleavage from T2 resins*[41]

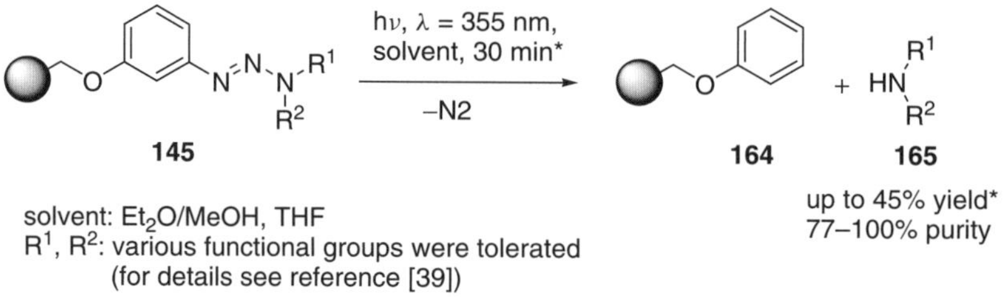

Scheme 9.35 Photolytic cleavage by Enders et al. (* For practical reasons the reaction times were limited to 30 minutes resulting only in moderate yields)[43]

The diazonium compound decomposes to the alkyl halides **157a** and **157b** as well as the α,β-unsaturated allyl ester **163**.

A photolytic cleavage protocol was developed by Enders and coworkers in 2004.[43] Resins **145** were irradiated with 3ω Nd-YAG laser at 355 nm (Scheme 9.35). For practical reasons the irradiation time was limited to 30 minutes Therefore, the corresponding cleavage products **165** were obtained in only moderate yields but in good to excellent purities.

Typical Experimental Procedures

Synthesis of diazonium T2 resins (scheme 9.29)[39]

To sodium hydride (5 equiv.) in DMF (100 ml/g resin) Merrifield resin (1 equiv., loading 0.64 mmol/g) is added. To this suspension 3-aminophenol (5 equiv.) is added portion-wise (caution: hydrogen evolution). The mixture is shaken or stirred with a mechanical stirrer at room temperature for 20 hours. After filtration and washing of the resin with THF, diethyl ether and methanol (three times with each approximately 20 ml for each solvent per gram of resin) the resin is dried in vacuum.
The resulting amine resin is suspended in dry THF and cooled to $-20\,°C$. After 20 minutes, boron trifluoride etherate ($BF_3 \cdot Et_2O$) (8–9 mmol/g resin) and subsequently after 5 minutes tert-butyl nitrite (t-BuONO) (7–8 mmol/g resin) are added. After 30 minutes the mixture is filtered and the resin is washed with chilled THF (four times with 15 ml/g resin each). The resin is further washed with THF, diethyl ether and methanol (three times with each approximately 20 ml for each solvent per gram of resin) and dried in vacuum.

Preparation of the amine T2 resins (scheme 9.29)[39]

The diazonium T2 resin is suspended in dry THF (15 ml/g resin) and treated with the amine (5 equiv.). After one hour at room temperature, a solution of methanol in THF (50%) is added to quench the reaction. After filtration and washing of the resin with THF, diethyl ether and methanol (three times with each approximately 20 ml for each solvent per gram of resin) the resin is dried in vacuum.

Preparation of the amide and urea t2 resins (scheme 9.30)[39]

The diazonium T2 resin is suspended in dry THF (10 ml/g resin) and treated with triethylamine (6 equiv. for amide resins and 1 equiv. for urea resins) and 4 equiv. of acid chloride (for amide resins) or 4 equiv. of isocyanates (for urea resins). After one hour at room temperature, a solution of methanol in THF (50%) is added to quench the reaction. After filtration and washing of the resin with THF, diethyl ether and methanol (three times with each approximately 20 ml for each solvent per gram of resin) the resin is dried in vacuum.

Cleavage protocols (scheme 9.29 and Scheme 9.30)[39]

The resin is treated three times with the following cleavage solutions and filtered:

(a) amine: 10–50% TFA in dichloromethane or acetyl chloride in THF
(b) amide: 5% TFA in dichloromethane
(c) urea: 10% trimethylsilyl chloride in dichloromethane.

After five minutes at room temperature whereby the resin turns red the combined filtrates are concentrated in vacuum.

9.3.2 The T2* linker for synthesis

Slight variation in the substitution pattern and by chlorination of the benzene ring of the T2 resin gives the T2* linker (Scheme 9.36).[44] The synthesis of the diazonium resin **141** as well as the coupling of the primary amine are carried out under standard conditions.

The alkylation with alkyl bromides as well as the acylation with acid chlorides, isocyanates or isothiocyanates yielded in the corresponding products in quantitative yields and excellent purities. All reactions were performed under mild conditions as shown in Scheme 9.37.

In 2000, Dahmen and Bräse demonstrated the solid-phase synthesis of highly diverse guanidines **178**.[45] Starting from triazene T2* resins **138** the corresponding immobilized thiourea **176** was prepared. By subsequent treatment with an amine (e.g. ammonia, ally amine or piperidine) and either silver nitrate or mercury(II) oxide followed by cleavage with trifluoroacetic acid the trisubstituted guanidines **178** were obtained in moderate to quantitative yields and in excellent purities (Scheme 9.38).

A solid-phase synthesis of alkyl sulfonates using the T2* linker was reported by Enders and Bräse (Scheme 9.39).[46] Using triazene resin **138** as an immobilized alkylating reagent various sulfonic acids **179-H** or sulfonates **179-Na** were converted to yield twelve alkyl sulfonates **180** in moderate to good yields and in high purities.

An efficient polymer-bound synthesis of sulfonic esters **180**, phosphoric esters **182a** and phosphinic esters **182b** was reported by Enders and Bräse *et al.* using the T2* linker in 2003[47] based on the results from their above described synthesis of alkyl sulfonates.[46] Starting with various sulfonic acids **179-H** or sulfonates **179-Na** as well as with phosphor-based acids **181a** and **181b**, the treatment with the polymer-bound triazene **138** led to the formation of the esters **180**, **182a** or **182b** in moderate to excellent yields

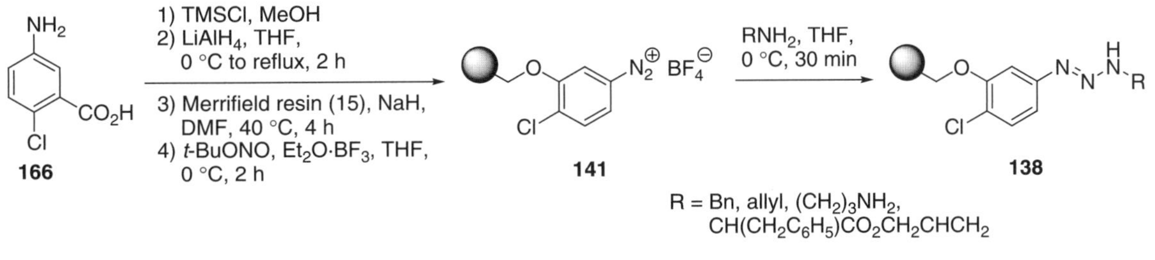

Scheme 9.36 *Synthesis of a T2* resin **138** by Dahmen and Bräse*[44]

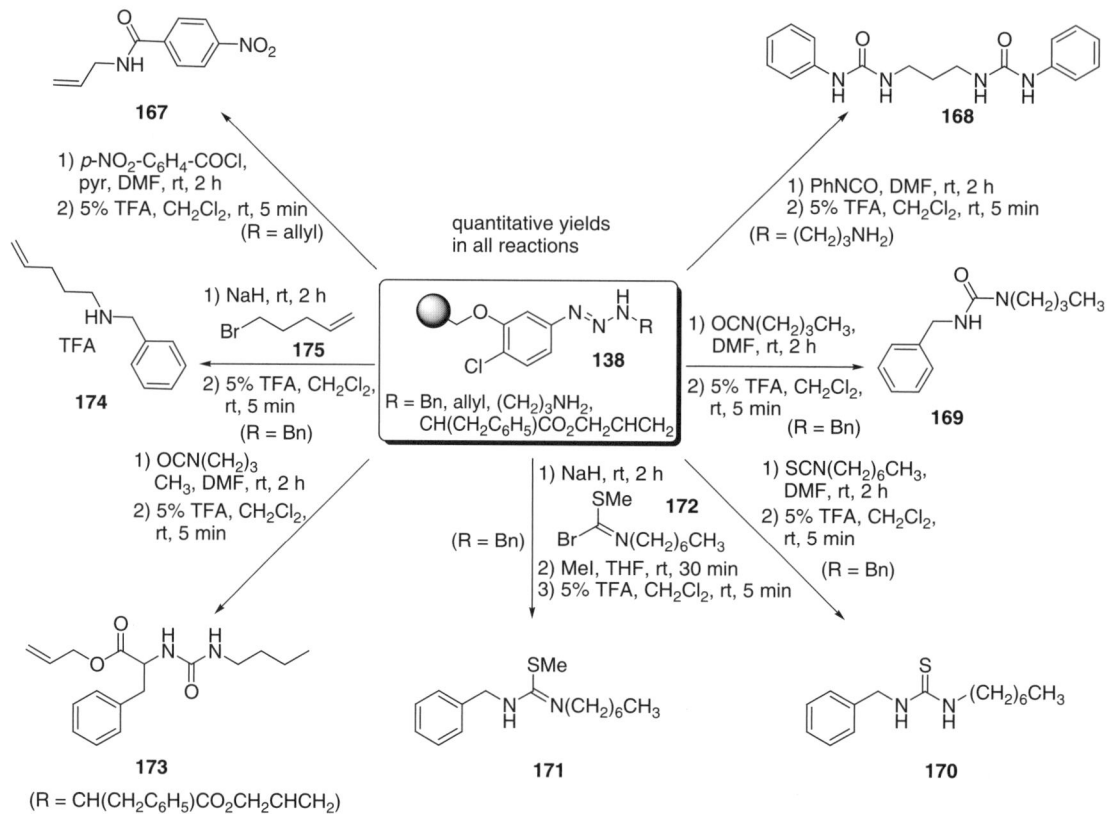

Scheme 9.37 Application of T2* resins **138** by Dahmen and Bräse[44]

and in high purities (Scheme 9.40). This method is also applicable for the synthesis of chiral sulfonic acids using enantioenriched α-substituted sulfonates. Products were obtained in 51–77% yield, above 90% pure and with an enantioselectivity greater than 98% ee.

A proposed mechanism is shown in Scheme 9.41. The protonation of the triazene resin **183** would, via defragmentation, lead to the diazonium ions **186a** and **186b**, respectively, which directly react with the sulfonates to give the desired product **183**. The reaction of phenethyl amine using this strategy results in a β-elimination of the phenethyl diazonium ion resulting in the formation of styrene. Compared to methyldiazonium ions, benzyldiazonium ions are highly unstable or nonexistent, so that a phenyl group can not be transferred using this strategy.[48]

In 2002, the T2* system was investigated in terms of its properties as modular ligand for catalysis by Bräse et al.[49] Following the above described protocol the immobilized triazene ligands **187a–h** were synthesized (Figure 9.4). Ligands **188b** and **188c** were used to complex various transition metals by treatment with cobalt(II) acetalacetonate (Co(acac)$_2$), copper(II) triflate (Cu(OTf)$_2$), iron(II) chloride, palladium acetate (Pd(OAc)$_2$), titanium isopropoxide (Ti(Oi-Pr)$_4$) and zirconium(II) acetylacetonate (Zr(acac)$_2$). The general structure of these complexes is shown in Figure 9.5. It was shown that all complexes were characteristically colored.

As the diethyl zinc addition to benzaldehyde (**192**) is a very well examined reaction and a broad range of chiral systems for the asymmetric approach has been described (e.g. Scheme 9.42),[50] this reaction

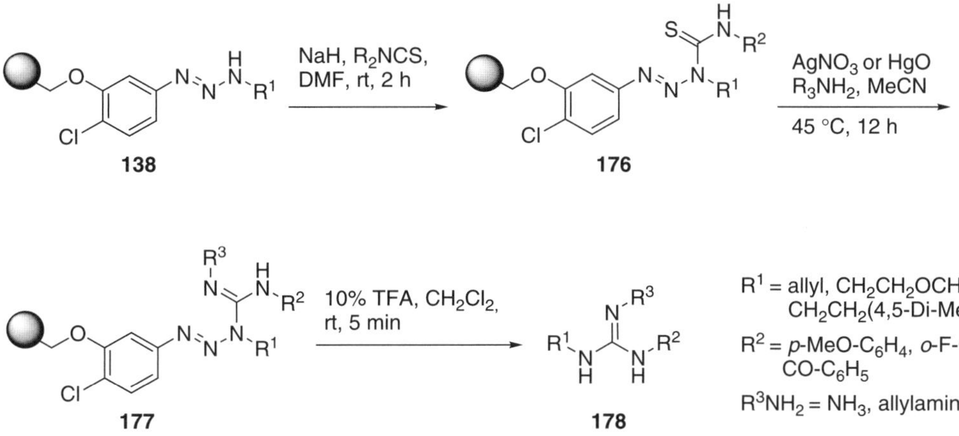

R^1 = allyl, $CH_2CH_2OCH_3$, $CH_{2\text{-}4}$-pyridinyl, CH_2CH_2(4,5-Di-MeO-C_6H_4)
R^2 = *p*-MeO-C_6H_4, *o*-F-C_6H_4, $(CH_2)_6CH_3$, CO-C_6H_5
R^3NH_2 = NH_3, allylamine, piperidine

41% to quant. overall yield
>90% purity

Scheme 9.38 *Solid-phase guanidine synthesis by Dahmen and Bräse*[45]

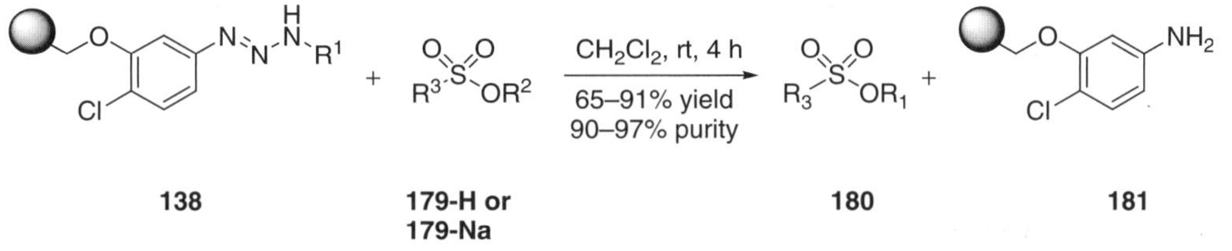

conditions:
for **154-H**: sulfonic acid, resin **138**, CH_2Cl_2, rt, 4 h
for **154-Na**: sulfonate, resin **138**, Lewatit SP 120 H form, CH_2Cl_2, rt, 4 h

R^1 = Me, *c*-Hex, Bn
R^2 = various sulfonic acids and sulfonates applicable
(for further details see reference [42])

Scheme 9.39 *Solid-phase synthesis of alkyl sulfonates **144** by the groups of Enders and Bräse*[46]

was first chosen to determine the catalytic activity of the immobilized ligands **188a** to **188h** shown in Figure 9.4.[49b] To compare the results, also the chiral amines **187a–h** (R = H, Figure 9.4), Merrifield-based resin **189a–h** without a linker and a soluble *para*-chloro-triazene-benzene system **190a–h** were introduced in the 1,2-addition reaction. As expected, the solution phase reaction using the chiral amines gave moderate to high yields (up to quantitative) and only low selectivities (up to 15%). Different solvents did not lead to higher enantioselectivity. The reactivity using the new (immobilized) ligands **188a–h** as well as the test systems **189a–h** and **190a–h** were proven to be comparable in yield (up to 98%). The selectivity of the test ligands **189a–h** and **190a–h** exhibited reasonable to moderate enantioslectivities (up to 71% *ee*), whereas the immobilized triazene ligands **188a–h** did not show significant selectivities (up to 10% *ee*). Further details on reaction conditions, yields, *ee* values and so on are given elsewhere.[49b]

Furthermore, resin **188h** was used in the syntheses of immobilized metal triazenes **194** and **195** (Figure 9.6). Complex **184** is the analogue of bis(triphenylphosphine)palladium(II) dichloride

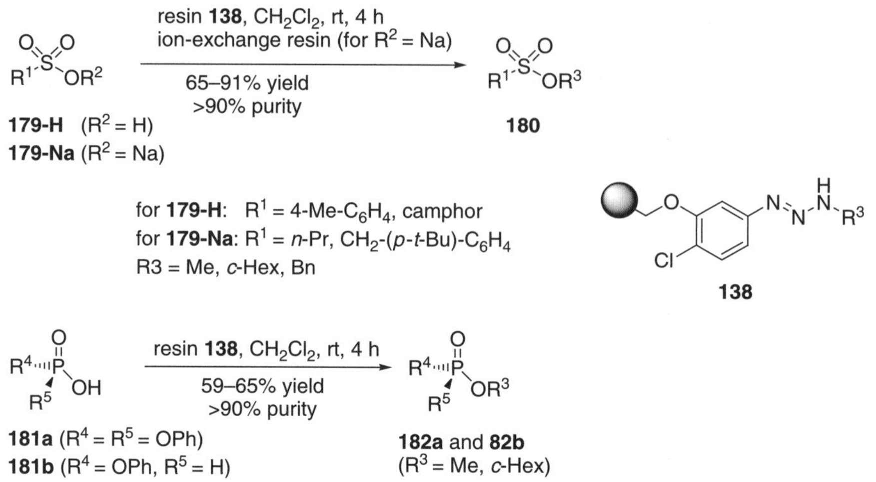

Scheme 9.40 Solid-phase synthesis of sulfonic, phosphoric and phosphinic esters by Enders and Bräse et al.[47]

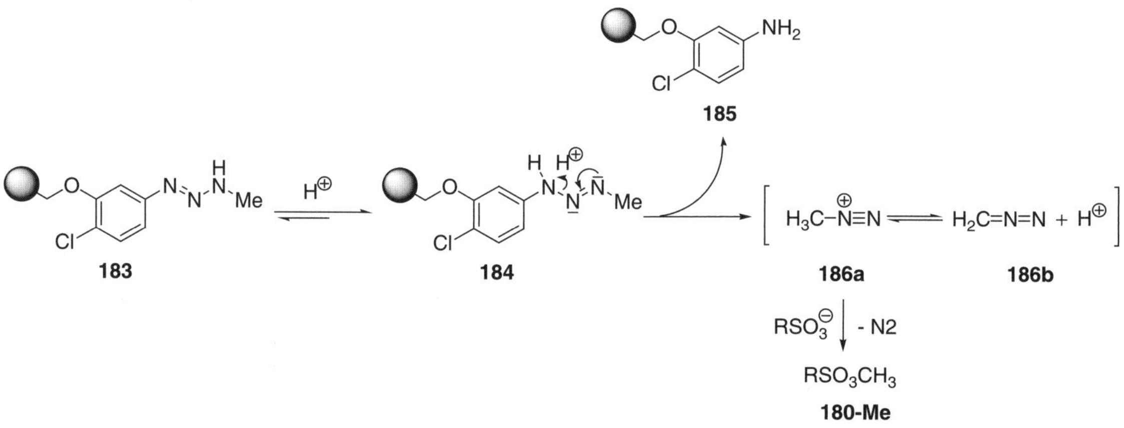

Scheme 9.41 Proposed mechanism for solid-phase synthesis of alkyl sulfonates **180**-Me[46]

(PdCl$_2$(PPh$_3$)$_2$), an established catalyst used in Sonogashira and Suzuki cross-coupling reactions.[49b] Whereas the analogue to complex **195**, RuCl$_2$(PPh$_3$)$_4$, is established for the use in transfer hydrogenation reactions. The immobilized palladium catalyst **194** was successfully used as solid supported reagent in a Suzuki cross-coupling reaction and a Sonogashira cross-coupling reaction (each 71% yield). Hydrogenation reactions using the immobilized ruthenium catalyst **195** turned out not to be efficient.

Typical Experimental Procedures

Preparation of the diazonium T2* resin (scheme 9.36)[44]

To sodium hydride (5 equiv.) in DMF (10–15 ml/g resin), Merrifield resin (1 equiv., loading 0.64 mmol/g) is added. To this suspension 2-chloro-5-amino-benzyl alcohol (5 equiv.) is added. The mixture is shaken

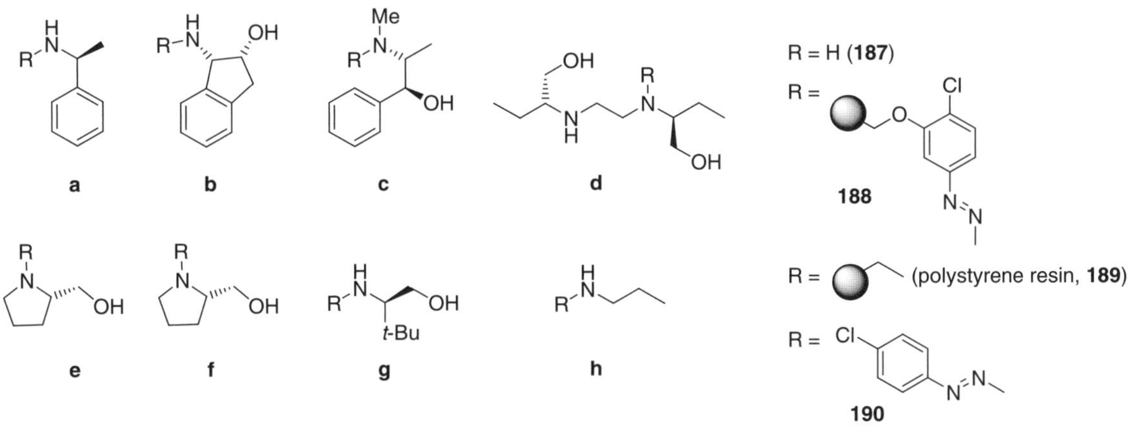

Figure 9.4 T2* system as modular ligand for catalysis by Bräse et al.[49]

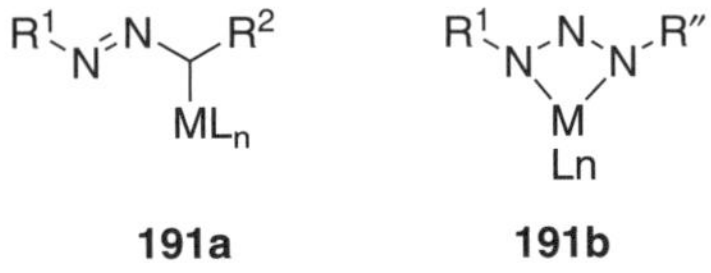

Figure 9.5 Binding modes in metal triazenes

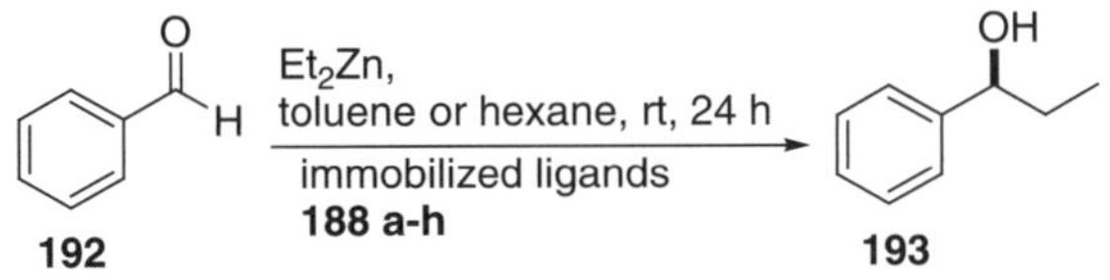

Scheme 9.42 Diethyl zinc addition to benzaldehyde **192** using immobilized chiral ligands **188a–h** by Bräse et al.[49b]

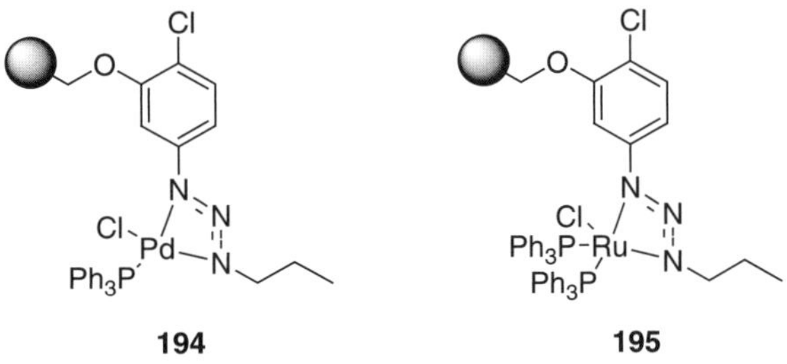

Figure 9.6 Immobilized palladium and ruthenium catalysts **194** and **195** by Bräse et al.[49]

or stirred with a mechanical stirrer at room temperature for 20 hours. After filtration and washing of the resin with THF, diethyl ether and methanol (three times with each approximately 20 ml for each solvent per gram of resin) the resin is dried in vacuum. The resulting amine resin is suspended in dry THF and cooled to 0 °C. After 20 minutes, $BF_3 \cdot Et_2O$ (8–9 mmol/g resin) and subsequently after 5 minutes t-BuONO (7–8 mmol/g resin) are added. After 30 minutes the mixture is filtered and the resin is washed with chilled THF (four times with 15 ml/g resin each). The resin is further washed with THF, diethyl ether and methanol (three times with each approximately 20 ml for each solvent per gram of resin) and dried in vacuum.

Cleavage protocol

The resin is treated with 5% TFA in dichloromethane at room temperature for five minutes. After filtration, the solvent is evaporated and the product is dried in vacuum.

9.3.3 The T2* scavenger resin

In addition to the use of T2* resins as solid support in organic synthesis, Dahmen and Bräse showed the application of this type of resin as scavengers.[44] As shown in Scheme 9.43 amines and anilines can be caught by the diazonium T2* resin **138** to give quantitative pure phenole **197**. In addition to the diethylaminomethylpolystyrene (**196**), a basic ion exchanger can be used as well. If potassium hydroxide is used as base also phenols can be quenched by scavenger **138** very effectively.

The above described electrophilic cleavage of triazenes (Scheme 9.34) can also be transferred to the T2* system (Scheme 9.44).[41] The triazene resins **200, 202** and **204** were treated with trimethylsilyl chloride or trifluoro acetic acid as electrophiles. The cleavage products **201, 203** and **205** were obtained in yields of 85–90%. The forming of product **203** can be explained by the proposed mechanism in Scheme 9.34. A comparable rearrangement was previously reported for the diazotation of α-amino acids.[51]

Typcial Experimental Procedures

Protocol for use as scavenger resin (scheme 9.43)[44]

To a mixture of the scavenger resin (1 equiv., loading 1.1 mmol/g) and diethylaminomethylpolystyrol (1 equiv., loading 2.33 mmol/g) an equimolar solution of the nucleophile and benzyl alcohol in THF are added.

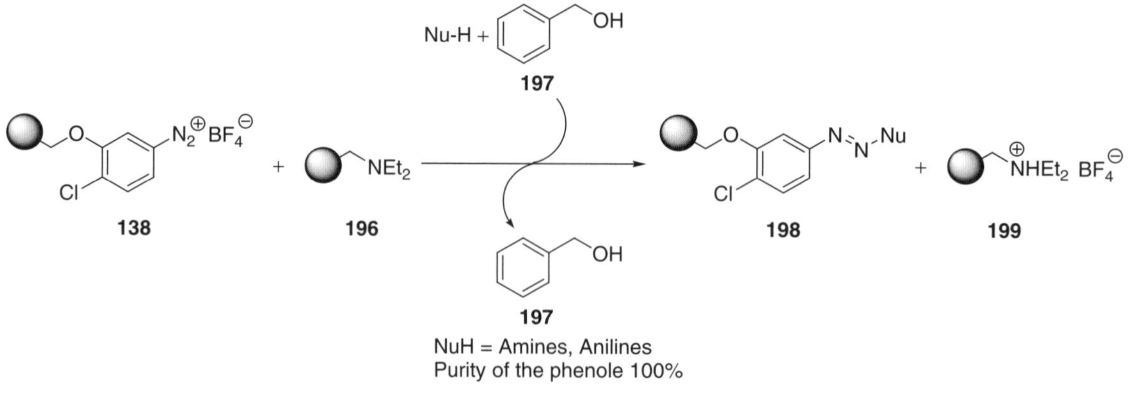

Scheme 9.43 *Diazonium T2* resin **138** as scavenger for amines or anilines by Dahmen and Bräse*[44]

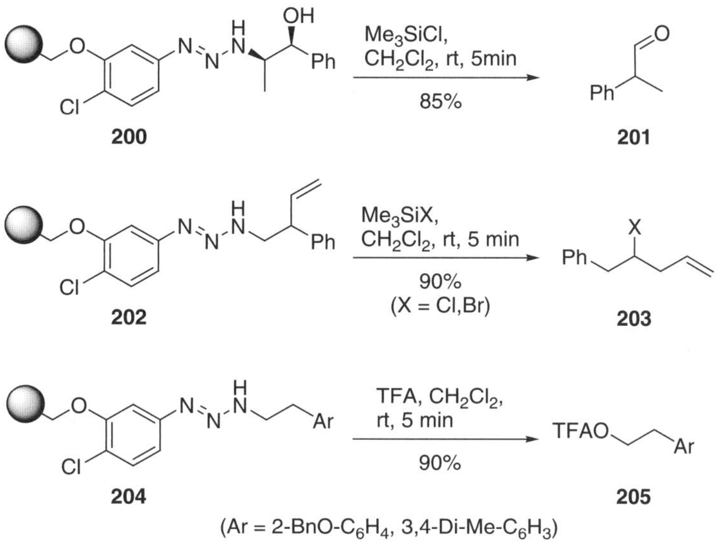

T1 and T2 – Versatile Triazene Linker Groups 293

Scheme 9.44 Cleavage of T2* resins by treatment with electrophiles by Bräse et al.[41]

The mixture is shaken for one hour at room temperature and then filtered. The benzyl alcohol is recovered by evaporation of the solvent. The nucleophile can be released after washing of the resin by cleavage with 5% TFA in dichloromethane as described above.

9.4 Miscellaneous triazene linkers

In this section, triazene linkers which could not be categorized as the T1 or T2 linkers are described. Nevertheless, they are structural comparable to the above shown linkers.

A rapid solid-phase synthesis of oligo(1,4-phenylene-ethynylene)s like **213**, **216** or **219** was reported by Tour *et al.* in 1997 (Scheme 9.45).[52] Starting with Merrifield resin (**9**) the hydroxy triazene **206** was immobilized on a solid support. A palladium catalyzed cross-coupling of aryl iodide **208** and subsequent cleavage with methyl iodide lead to the formation of dimer **210**. After the deprotection of the acetylene with *tert*-butyl amino fluoride, the coupling of dimer **210** to the immobilized dimeric acetylene **209** gave the tetrameric acetylene **212**. Repeating this sequence and final cleavage yielded the oligomer **219**. Every cleavage step gave yields of about 80%. Following this strategy the substitution pattern of these oligomers can be verified by using different monomers.[52b]

The use of alkylating polymers bearing a triazene group as transferring functionality for the synthesis of carbonic acid esters **223** was demonstrated by the group of Rademann in 2001.[53] The triazene resins **221** were prepared following established protocols for attachment, reduction and triazene formation starting with Merrifield resin (**9**) as shown in Scheme 9.46. The carbonic acid esters **223** were formed by treatment of the corresponding acids **222** (ten different carbonic acids) with the alkylation triazene polymers **221** in up to quantitative yields and excellent purities.

A straightforward preparation of amino-polystyrene resin starting from Merrifield resin (**9**) was described by Mioskowski and coworkers.[54] The benzyl amine resin **10** was synthesized from Merrifield resin in three steps. After diazotation, the amines were coupled. After cleavage with trifluoroacetic acid the amines **225** were recovered in high yields and purities (Scheme 9.47).

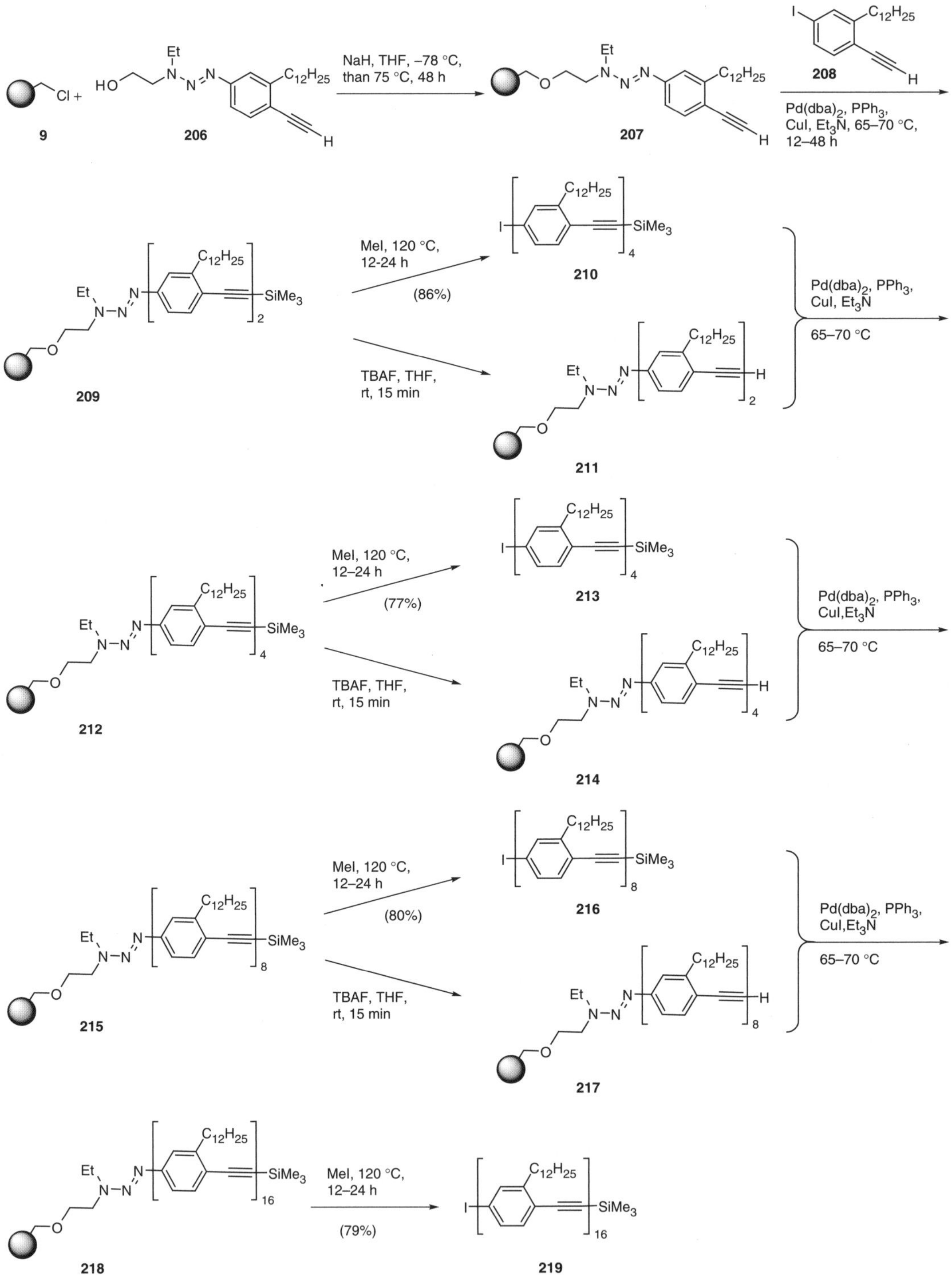

Scheme 9.45 Solid-phase synthesis of oligo(1,4-phenylene-ethynylene)s reported by Tour et al.[52]

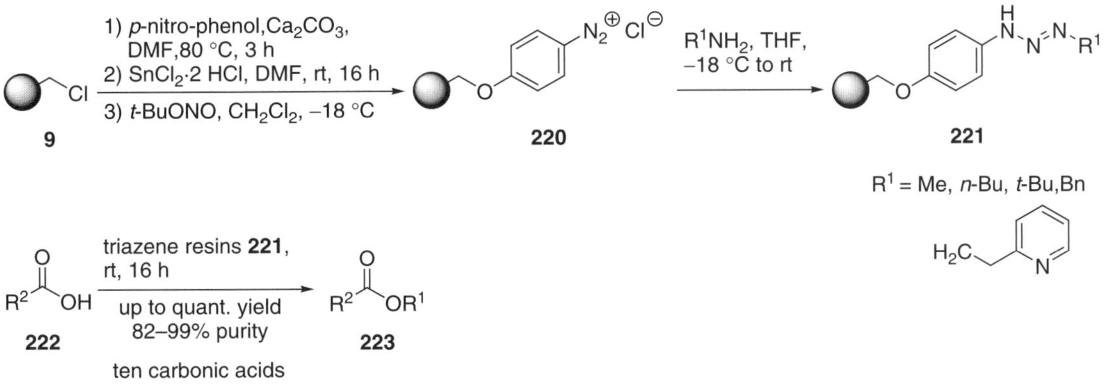

Scheme 9.46 Alkylating polymers for the synthesis of acid esters **223** by Rademann et al.[53]

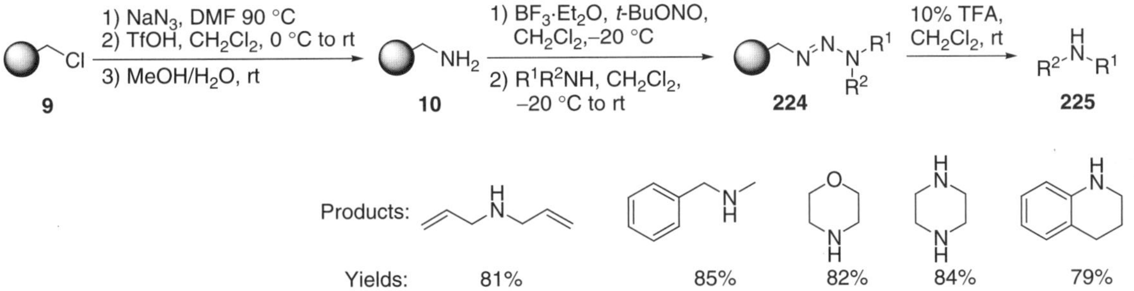

Scheme 9.47 Preparation of amino-polystyrene resin by Mioskowski et al.[54]

In 2003, aldol reactions and Grignard reactions of immobilized nortropinone using polymeric supports with spacer-modified triazene linkers were reported by Lazny and Nodzweska (Scheme 9.48).[55] Nortripinone was immobilized on different supports having no or poly propylene glycol spacers with *meta* and *para* substitution patterns. The nortripinone resins **226** were than converted in a Grignard reaction with ethyl magnesium bromide or in an aldol reaction with benzaldehyde to yield the corresponding addition products after cleavage with trifluoroacetic acid. The Grignard products were obtained in 37–71% yields. The yields and the ratio of the aldol reactions depend from the substitution pattern as well as of the spacer of the resin. The mono-aldol product was obtained in up 77% yield and the di-aldol product was obtained in up to 33% yield.

Having these results in hand, this strategy was expanded to further reaction types like oxidation reactions or *O*-alkylation reactions (Scheme 9.49).[56] As described above, nortropine was immobilized on a polymeric support. A following oxidation with Dess–Martin periodinane (DMP) or an *O*-alkylation was performed given the corresponding cycloheptenone **231** (25%) or alcoxides **232** (quantitative and 64% yield) in reasonable to high yields.

Utesch and Diederich reported a Sonogashira reaction as well as a Cadiot–Chodkiewicz-type reaction for the synthesis of poly(triacetylene)-derived oligomers **238** on the solid phase using a triazene linker (Scheme 9.50).[57] The disilyl alcohols **234** were immobilized on the triazene resin **233** by a Sonogashira cross-coupling reaction. The subsequent cleavage with methyl iodide led to the formation of product **236** in high yield. The deprotection of the acetylene function followed by a Cadiot–Chodkiewicz-type reaction

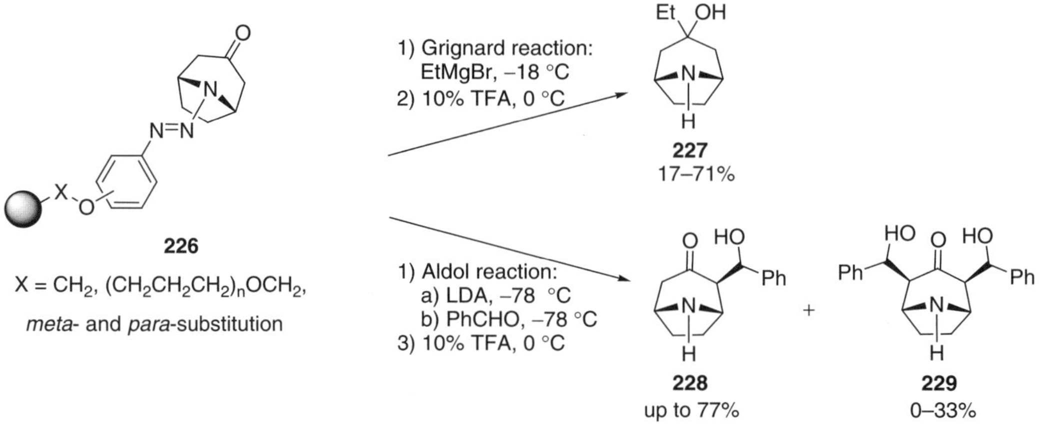

Scheme 9.48 Solid-phase Grignard and aldol reactions of nortropinone **226** by Lazny and Nodzewska[55]

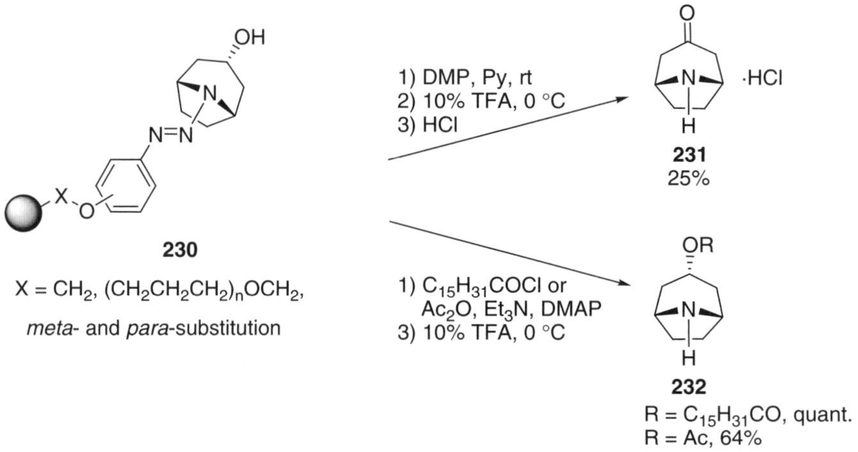

Scheme 9.49 Oxidation and esterification on a solid support reported by Lazny et al.[56]

of the brominated silyl compound **237** gave a product mixture of compounds **238a, 238b** and **239** in moderate yields. Following this strategy further oligomers could be synthesized in moderate to good yields (36–82%), depending on the number of oligomeric units. This strategy was extended by Diederich et al. also to the synthesis of oligo(phenylentriacetylene)s in reasonable to good overall yields.[57b]

In 2005, Schroen and Bräse developed polymer-bound diazonium salts containing a triazene unit for the synthesis of diazoacetic esters.[58] Treatment of amino acid esters **240** with diazonium resin **241** and subsequent cleavage in triethylamine lead to the formation of diazoacetic esters **243** in overall yields of up to 38% (based on the loading of the resin) and excellent purities (Scheme 9.51). This strategy was also successful using the depeptide HLeuLeuOMe, which gives access to labeled peptides. Further investigations of the substitution pattern of the diazonium resin **241** confirmed the supposition that the nitrogen transfer works only with electron withdrawing groups at the triazene functionality.

The extension of this strategy to glycine esters failed. After attachment to the diazonium resin **241** and basic cleavage the expected product was not obtained. Infra red investigations and elemental analysis of the formed resin indicated the formation of triazoles **244** by Dimroth cyclization (Figure 9.7).[27, 58]

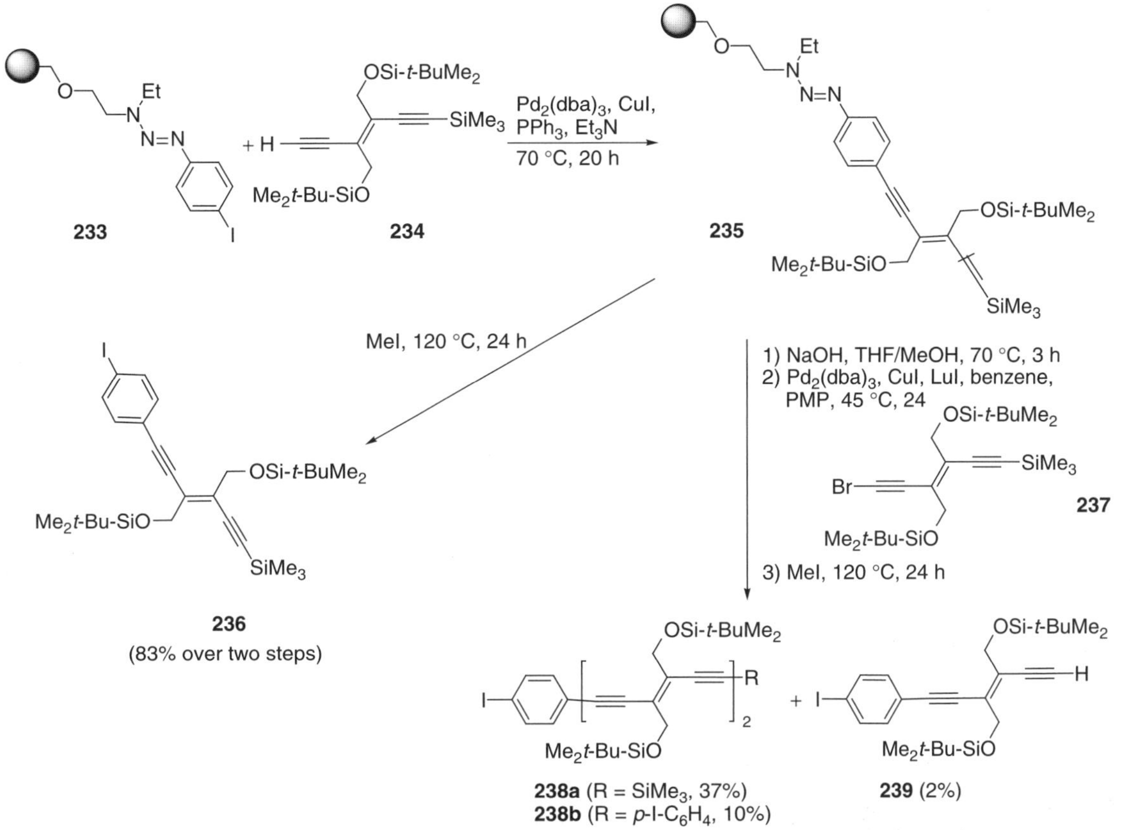

Scheme 9.50 Solid-phase synthesis of poly(triacetylene) derived oligomers reported by Utesch and Diederich (PMP = 1,2,2,5,5,pentamethylpiperidine)[57]

Recently, a solid-phase synthesis of *meta*-phenylene ethynylene heterosequence oligomers **251** based on their pioneering works was developed by Moore and coworkers.[59] Starting from resin **245**, an iterative reaction sequence of coupling terminal alkynes **246** and a subsequent cross-coupling of iodo arene **248** including an *in situ* deprotection of the alkyne functionality with tetra-*tert*-butyl ammonium fluoride lead to the formation of oligomer **249**. Repeating this sequence oligomers of up to 18 monomers (R = H and CH$_3$) were obtained in moderate to high yields and good purities. As this reaction sequence reduces the time of the synthesis of phenylene ethynylene oligomers, both cross-coupling reactions are completed within two hours each, further investigations of different sequence variations can be done. The second coupling reaction includes an *in situ* deprotection step which is a further advantage of this method.

Typical Experimental Procedures

Coupling of immobilized terminal alkynes with aryl iodides (scheme 9.44)[52]

In dry triethylamine 4 mol% of Pd(dba)$_2$, triphenylphosphine (20 mol%) and copper iodide (4 mol%) are dissolved and stirred at 70 °C for two hours. The supernatant of this solution (6 ml/g resin) is added to the immobilized terminal alkyne (1 equiv.) and the aryl iodide (2 equiv.) via cannula. The reaction mixture is shaken at 65 °C for 12–48 hours. After filtration the resin is washed with sequentially (30 ml/g resin)

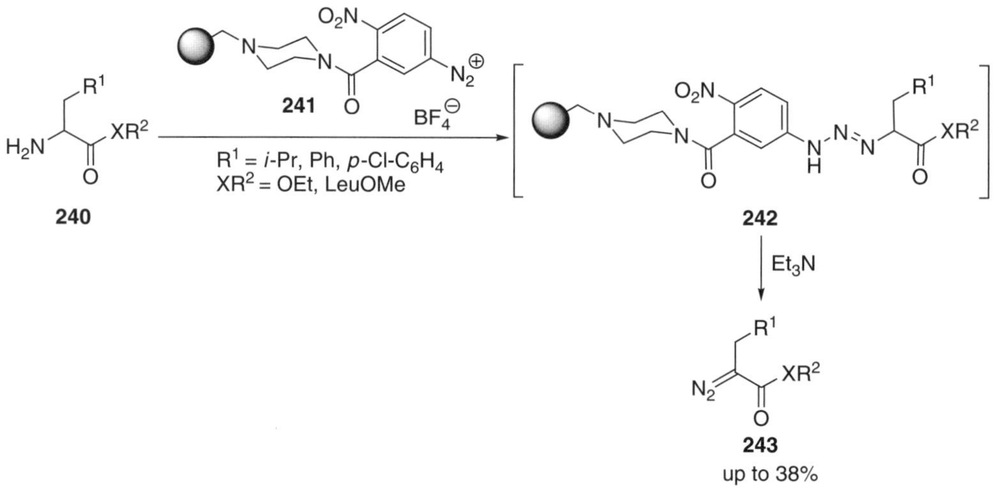

Scheme 9.51 Solid-phase synthesis of diazoacetic esters **243** as reported by Schroen and Bräse[58]

DCM, DMF, 0.05 M solution of sodium diethyl dithiocarbamate in 99:1 DMF/diisopropylamine, DMF, dichloromethane and methanol. After washing the resin is dried in vacuum at 60 °C.

Desilylation of immobilized terminal alkynes (scheme 9.44)[52]

A solution of TBAF (2 equiv., 1 molar in THF) is added to a suspension of the silyl resin (1 equiv.) in THF (9 ml/g resin). The mixture is shaken for 15 minutes and then filtered. The resin is washed with THF and methanol (30 ml/g resin each) and dried in vacuum at 60 °C.

Cleavage with methyl iodide (scheme 9.44)[52]

A suspension of the resin and methyl iodide (7 ml/g resin) is stirred at 120 °C for 12–24 hours. After filtration the resin is washed with hot dichloromethane. The combined filtrates are then passed through a plug of silica gel with DCM. The solvent is evaporated and the product is dried in vacuum.

Solid-phase synthesis of diazoacetic esters (scheme 9.51)[58]

Sodium carbonate (1.3 mmol) is dissolved in water (3 ml). Dichloromethane (8 ml) is added and the mixture is cooled to 0 °C. The amino acid ester (2.5 mmol) are added carefully and the mixture is stirred for 10 minutes The diazonium resin is added and the mixture is shaken for two hours. The resin is filtered off and washed subsequently with water, dichloromethane and THF and is the dried in vacuum. This resin is given into THF (10 ml) and triethylamine (1 ml) is added. The mixture is shaken for 12 hours and then the resin is filtered off. The solvent of the filtrate is evaporated and the remaining product is dried under vacuum.

Oxidation with dess–martin periodinane (DMP, scheme 9.49)[56]

To a suspension of the immobilized alcohol (loading 0.811 mmol/g) in dry dichloromethane (8 ml/g resin) five equivalents of Dess–Martin periodane and pyridine (0.7 ml/resin) are added and the mixture is shaken

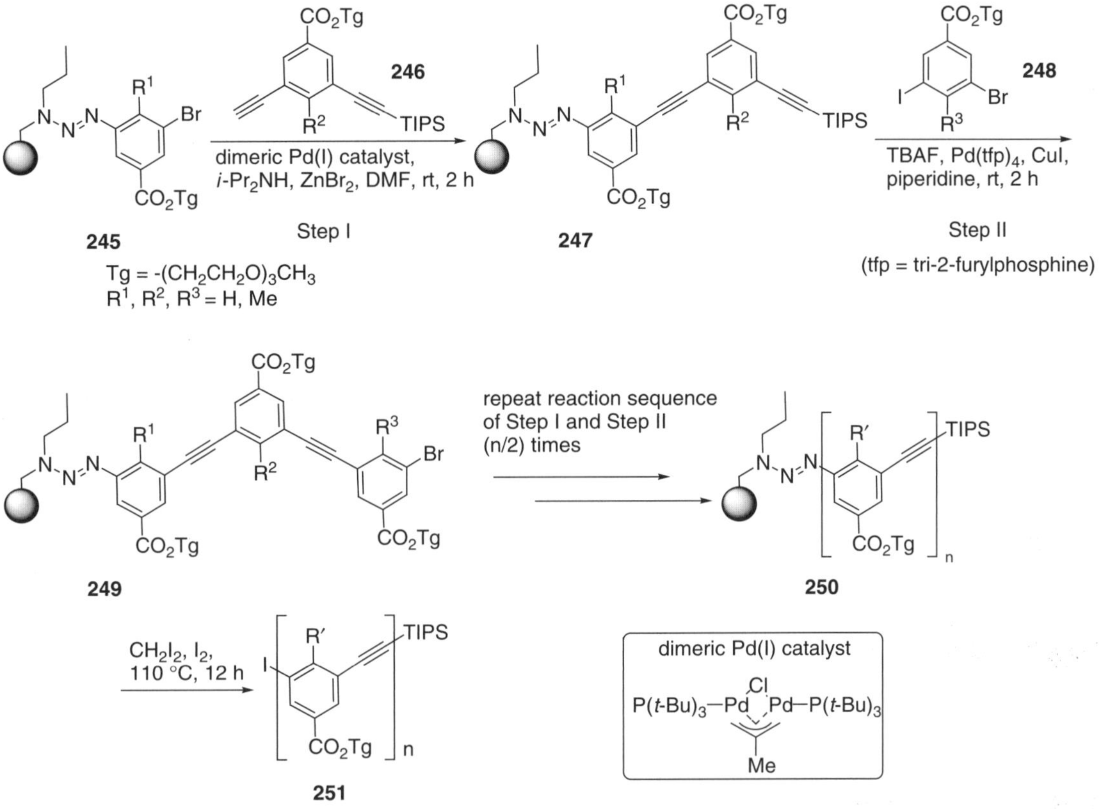

Scheme 9.52 *Iterative solid-phase heterosequence synthesis as reported by Moore et al.*[59]

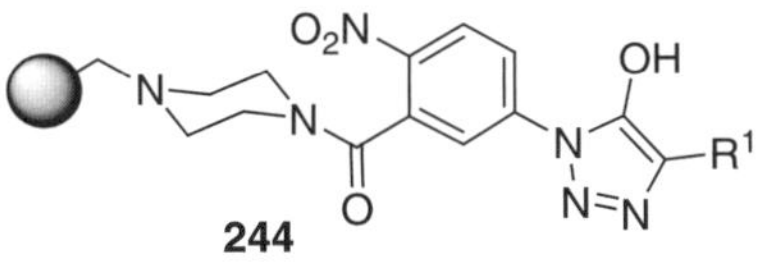

Figure 9.7 *Triazoles 244 formed by Dimroth cyclization*[27, 58]

for 72 hours at room temperature. The resin is washed according the following sequence: with methanol, a mixture of 2 M potassium hydroxide with DMF (1:4), a mixture of 2 M potassium hydroxide with methanol (1:4), 2 M potassium hydroxide, a mixture of alkaline aqueous sodium thiosulfate with DMF (1:1), a mixture of alkaline aqueous sodium thiosulfate with methanol (1:1), methanol, THF, methanol, dichloromethane and methanol. Cleavage with 10% TFA in dichloromethane, treatment with concentrated hydrochloric acid and evaporation of volatiles yields in the amino hydrochloride.

Esterification (scheme 9.49)[56]

The immobilized resin (loading 0.811 mmol/g) is suspended in dichloromethane (8–9 ml/g resin) and treated with triethylamine (6 equiv.), acid chloride or acid anhydride (5 equiv.) and a catalytic amount of DMAP.

After shaking for 12 hours at room temperature, the resin is washed according the following sequence: methanol, methanol/diethyl amine (1:1), methanol, water, methanol, DCM, methanol, dichloromethane and methanol. Cleavage with 10% TFA in dichloromethane, treatment with concentrated hydrochloric acid and evaporation of volatiles yields in the amino hydrochloride.

9.5 Conclusion

The triazene group has become of great interest as a biologically and pharmacologically active group in medicinal chemistry. The triazene moiety can be activating or directing or be used as protecting group in organic synthesis. The triazene functional group is very stable even under harsh reaction conditions. These facts and the variety of cleavage methods make the triazene linker to a useful tool in solid-phase organic synthesis. As demonstrated in this chapter, the triazene moiety is a linker system applicable to a broad range of reactions like, for example, various addition reactions, palladium mediated cross-coupling reactions or organometallic reactions. This type of linker has been also used in the syntheses of a large number of heterocycles. Furthermore, the T2 triazene resins can be used as an immobilized reagent, for example as alkylation reagent for carbonic acids, sulfonic acids or phosphoric acids. Furthermore, the T2* resin was introduced as scavenger resins for amines and the use of polymer-supported triazenes as palladium or ruthenium catalysts in cross-coupling reactions or hydrogenation reactions has been shown. As a result it could be said that the triazene moiety is one of the most manifold applicable functional groups both in liquid-phase and solid-phase organic synthesis.

References

[1] Kimball, D. B., and Haley, M. M.; *Angew. Chem.* **2002**, *114*, 3484; *Angew. Chem. Int. Ed.* **2002**, *41*, 3338.

[2] a) Connors, T. A., Goddard, P. M., Merai, K., *et al.*; *Biochem. Pharmacol.* **1976**, *25*, 241. b) Wilman, D. E. V.; *Cancer Treatment Rev.* **1988**, *15*, 69. c) Rouzer, C. A., Sabourin, M., Skinner, T. L., *et al.*; *Chem. Res. Toxicol.* **1996**, *9*, 172.

[3] a) Nicolaou, K. C., Boddy, C. N. C., Li, H., *et al.*; *Chem. Eur. J.* **1999**, *5*, 2602. b) Lazny, R., Poplawski, J., Köbberling, J., *et al.*; *Synlett* **1999**, 1304. c) Lazny, R., Sienkiewicz, M., and Brase, S.; *Tetrahedron* **2001**, *57*, 5825.

[4] Jones II, L., Schumm, J. S., and Tour, J. M.; *J. Org. Chem.* **1997**, *62*, 1388.

[5] a) Moore, J. S.; *Acc. Chem. Res.* **1997**, *30*, 402. b) Liu, S., Zavalij, P. Y., Lam, Y.-F., and Isaacs, L.; *J. Am. Chem. Soc.* **2007**, *129*, 11232.

[6] a) Young, J. K., Nelson, J. C., and Moore, J. S.; *J. Am. Chem. Soc.* **1994**, *116*, 10841. b) Jones, L., Schumm, J. S., and Tour, J. M.; *J. Org. Chem.* **1997**, *62*, 1388. c) Scott, P. J. H., and Steel, P. G., *Eur. J. Org. Chem.* **2006**, 2251. d) Laborde, M. A., and Mata, E. G., *Mini-Reviews in Med. Chem.* **2006**, *6*, 109.

[7] Gross, M. L., Blank, D. H., and Welch, W. M.; *J. Org. Chem.* **1993**, *58*, 2104.

[8] Nicolaou, K. C., Boddy, C. N. C., Natarajan, S., *et al.*; *J. Am. Chem. Soc.* **1997**, *119*, 3421.

[9] a) Nicolaou, K. C., Boddy, C. N. C., Bräse, S., and Winssinger, N.; *Angew. Chem.* **1999**, *111*, 2230; *Angew. Chem. Int. Ed.* **1999**, *38*, 2097. b) Nicolaou, K. C., Koumbis, A. E., Takayanagi, M., *et al.*; *Chem. Eur. J.* **1999**, *5*, 2622.

[10] Nelson, J. C., Young, J. K., and Moore, J. S.; *J. Org. Chem.* **1996**, *61*, 8160.

[11] a) Bräse, S., Dahmen, S., and Lormann, M. E. P.; *Methods Enzym.* **2003**, *369*, 127. b) Bräse, S.; *Acc. Chem. Res.* **2004**, *37*, 805.

[12] Bräse, S., Enders, D., Köbberling, J., and Avemaria, F.; *Angew. Chem.* **1998**, *110*, 3614; *Angew. Chem. Int. Ed.* **1998**, *37*, 3413.

[13] Bräse, S.; *personal communication* May **2008**. The cleavage methods by palladium or copper mediated cross-coupling reactions with *in situ* formed diazonium salts as as well as the the Gomberg-Bachmann reaction, the formation of hydrazins and diazo compounds are unpublished results of the group of S. Bräse.

[14] Bräse, S., Dahmen, S., and Heuts, J.; *Tetrahedron Lett.* **1999**, *40*, 6201.
[15] a) Lormann, M. E. P.; Walker, C. H., Es-Sayed, M., and Bräse, S.; *Chem. Commun.* **2002**, 1296. b) Avemaria, F., Zimmermann, V., and Bräse, S.; *Synlett* **2004**, 1163.
[16] Gil, C., Schwögler, A., and Bräse, S.; *J. Comb. Chem.* **2004**, *6*, 38.
[17] Knepper, K., Themann, A., and Bräse, S.; *J. Comb. Chem.* **2005**, *7*, 799.
[18] a) Schunk, S., and Enders, D.; *Org. Lett.* **2000**, *2*, 907. b) Schunk, S., and Enders, D.; *J. Org. Chem.* **2002**, *67*, 8034.
[19] Knepper, K., Vanderheiden, S., and Bräse, S.; *Eur. J. Org. Chem.* **2006**, 1886.
[20] Knepper, K., Lormann, M. E. P., and Bräse, S.; *J. Comb. Chem.* **2004**, *6*, 460.
[21] Gil, C., and Bräse, S.; *Chem. Eur. J.* **2005**, *11*, 2680.
[22] Lormann, M. E. P., Dahmen, S., and Bräse, S.; *Tetrahedron Lett.* **2000**, *41*, 3813.
[23] Nicolaou, K. C., Winssinger, N., Pastor, J., and Murphy, F.; *Angew. Chem.* **1998**, *110*, 2677; *Angew. Chem. Int. Ed.* **1998**, *37*, 2534.
[24] Uozumi, Y., Tsuji, H., and Hayashi, T.; *J. Org. Chem.* **1998**, *63*, 6137.
[25] Lefranc, H., Szymoniak, J., Delas, C., and Moise, C., *Tetrahedron Lett.* **1999**, *40*, 1123.
[26] Lormann, M. E. P., Dahmen, S., Avemaria, F., et al.; *Synlett* **2002**, 915.
[27] Bräse, S., Gil, C., Knepper, K., and Zimmermann, V.; *Angew. Chem.* **2005**, *117*, 5320; *Ang. Chem. Int. Ed.* **2005**, *44*, 5188.
[28] Lipschutz, B. H., and Blomgren, P. A.; *Org. Lett.* **2001**, *3*, 1869.
[29] Muir, J. C., Pattenden, G., and Ye, T.; *Tetrahedron Lett.* **1998**, *39*, 2861.
[30] Collini, M. D., and Ellingboe, J. W.; *Tetrahedron Lett.* **1997**, *38*, 7963.
[31] Zimmermann, V., Avemaria, F., and Bräse, S., *J. Comb. Chem.* **2007**, *9*, 200.
[32] Zimmermann, V., and Bräse, S.; *J. Comb. Chem.* **2007**, *9*, 1114.
[33] Zimmermann, V., Müller, R., and Bräse, S.; *Synlett* **2008**, 278.
[34] Caution: Trichlorosilane is corrosive and low boiling material.
[35] Caution: This mixture might produce potentially explosive HN_3.
[36] Bräse, S., Schroen, M.; *Angew. Chem.* **1999**, *111*, 1139; *Angew. Chem. Int. Ed.* **1999**, *38*, 1071.
[37] de Meijere, A., Nüske, H., Es-Sayed, M., et al.; *Angew. Chem.* **1999**, *111*, 3881; *Angew. Chem. Int. Ed.* **1999**, *38*, 3669.
[38] Kreis, M., Schroen, M., Niesing, C., et al.; *Org. Biomol. Chem.* **2005**, *3*, 1835.
[39] Bräse, S., Köbberling, J., Enders, D., et al.; *Tetrahedron Lett.* **1999**, *40*, 2105.
[40] Bräse, S., Dahmen, S., and Pfefferkorn, M.; *J. Comb. Chem.* **2000**, *2*, 710.
[41] Pilot, C., Dahmen, S., Lauterwasser, F., and Bräse, S.; *Tetrahedron Lett.* **2001**, *42*, 9179.
[42] Bräse, S., and Dahmen, S.; *Chem. Eur. J.* **2000**, *6*, 1899.
[43] Enders, D., Rijksen, C., Bremus-Köbberling, E., et al.; *Tetrahedron Lett.* **2004**, *45*, 2839.
[44] Dahmen, S., and Bräse, S.; *Angew. Chem.* **2000**, *112*, 3827; *Angew. Chem. Int. Ed.* **2000**, *39*, 3681.
[45] Dahmen, S., and Bräse, S.; *Org. Lett.* **2000**, *2*, 3563.
[46] Vignola, N., Dahmen, S., Enders, D., and Bräse, S.; *Tetrahedron Lett.* **2001**, *42*, 7833.
[47] Vignola, N., Dahmen, S., Enders, D., and Bräse, S.; *J. Comb. Chem.* **2003**, *5*, 138.
[48] Glaser, R., and Farmer, D.; *Chem. Eur. J.* **1997**, *3*, 1244.
[49] a) Bräse, S., Dahmen, S., Lauterwasser, F., et al.; *Bioorg. Med. Chem. Lett.* **2002**, *12*, 1845. b) Bräse, S., Dahmen, S., Lauterwasser, F., et al.; *Bioorg. Med. Chem. Lett.* **2002**, *12*, 1849.
[50] a) for liquid phase refer to: Pu, L., and Yu, H.-B.; *Chem. Rev.* **2001**, *101*, 757; b) for immobilized ligands refer to: Bräse, S., Lauterwasser, F., and Ziegert, R. E.; *Adv. Synth. Catal.* **2003**, *345*, 869.
[51] Hamman, S., and Beguin, C. G.; *Tetrahedron Lett.* **1983**, *24*, 57.
[52] a) Jones II., L., Schumm, J. S., and Tour, J. M.; *J. Org. Chem.* **1997**, *62*, 1388. b) Hwang, J.-J., and Tour, J. M.; *Tetrahedron* **2002**, *58*, 10387.
[53] Rademann, J., Smerdka, J., and Jung, G., et al.; *Angew. Chem.* **2001**, *113*, 390; *Angew. Chem. Int. Ed.* **2001**, *40*, 381.
[54] Arseniyadis, S., Wagner, A., and Mioskowski, C.; *Tetrahedron Lett.* **2002**, *43*, 9717.
[55] Lazny, R., and Nodzewska, A.; *Tetrahedron Lett.* **2003**, *44*, 2441.

[56] Lazny, R., Nodzewska, A., and Klosowski, P.; *Tetrahedron* **2004**, *60*, 121.
[57] a) Utesch, N. F., and Diederich, F.; *Org. Biomol. Chem.* **2003**, *1*, 237. b) Utesch, N. F., Diederich, F., Boudon, C., *et al.*; *Helv. Chim. Ac.* **2004**, *87*, 698.
[58] Schroen, M., and Bräse, S.; *Tetrahedron* **2005**, *61*, 12186.
[59] Elliott, E. L., Ray, C. R., Kraft, S., *et al.*; *J. Org. Chem.* **2006**, *71*, 5282.

10
Hydrazone Linker Units

Ryszard Lazny

Institute of Chemistry, University of Bialystok, Poland

10.1 Introduction

Hydrazones are well known and synthetically useful derivatives of aldehydes and ketones. They are usually easily formed from carbonyl compounds and hydrazines in a reversible reaction.[1] However, some hindered ketones, for example tropinone, may require more forcing conditions to drive the formation of hydrazones to completion. Methods of hydrazone cleavage to reform the parent carbonyl compounds are numerous and diverse.[2] The reversibility of formation of hydrazones and the extensive chemistry[1] of different types of hydrazones including (di)alkylhydrazones, acylhydrazones (hydrazides) and sulfonylhydrazones (Figure 10.1) make hydrazone units potentially attractive candidates for multifunctional linkers[3] in solid-phase synthesis. Hydrazone linkers can, in principle, be used for reversible binding of either carbonyl compounds to solid supports functionalized with hydrazine groups or for binding of hydrazines to supports exhibiting ketone or aldehyde carbonyls. Up to the end of 2008, only the first way was exploited in solid-phase synthesis. Applications of three types of hydrazone structural units for linking molecules to solid supports have been reported in the literature: sulfonylhydrazones, acylhydrazones and dialkylhydrazones. Moreover, the acylhydrazones are used for ligation of peptide hydrizides in biochemistry.[4] Hydrazones are also known as reacting species in equilibriums in dynamic combinatorial chemistry.[5]

10.2 Hydrazone linker units

The first use of a hydrazone as a linking unit between polymeric support and an organic molecule can be traced back to the work of Kamogawa[6] who prepared sufonylhydrazine on both macroreticular and gel-type (swelling) co-polymers of styrene–divinylbenzene. The sulfonylhydrazine anchoring groups were

---

Linker Strategies in Solid-Phase Organic Synthesis   Edited by Peter Scott
© 2009 John Wiley & Sons, Ltd

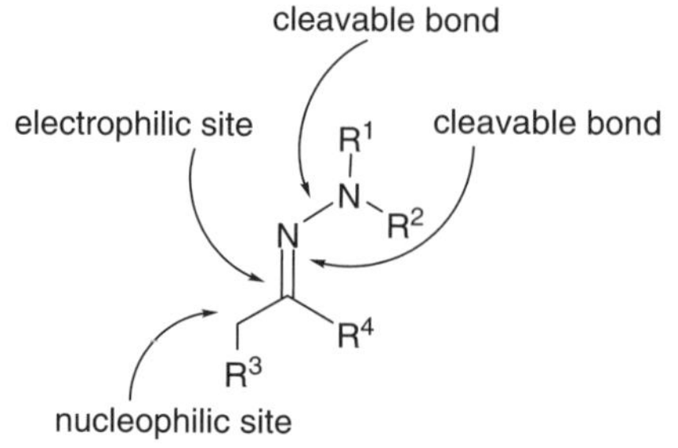

304 *Multifunctional Linker Units for Diversity-Oriented Synthesis*

N,N-dialkylhydrazone: R^1, R^2 = alkyl; sulfonylhydrazone: R^1 = SO_2Ar, R^2 = H;
acylhydrazone: R^1 = COR, R^2 = H; carbazone: R^1 = CONH-, R^2 = H; R^3, R^4 = alkyl, aryl, or H

Figure 10.1 *Reactivity profile of the hydrazone group*

introduced to the polymeric support either by chlorosulfonation followed by hydrazinolysis or by co-polymerization with *p*-vinylsulfonylhydrazine. The polymer supported sulfonylhydrazine reacted with several ketones and aldehydes (e.g. cyclohexanone, benzaldehyde, cyclodecanone). The formed hydrazones were subjected to known reaction conditions: reduction (sodium borohydride ($NaBH_4$) or lithium aluminum hydride ($LiAlH_4$), 9–77% yield), elimination ($NaOCH_2CH_2OH$, glycol, 8–27% yield) and conversion to nitriles (potassium cyanide, methanol, 0–22% yield) with modest success (Scheme 10.1).

A hydrazone linker unit as a part of a semicarbazone structure (NH–C(O)–**NHN=C**) was used for linking protected amino aldehydes in the solid-phase synthesis of peptide anologues such as peptide C-terminal aldehydes by Webb (Scheme 10.2, R_2 = H).[7] Cleavage of semicarbazones, which are resistant to dry acids and mild bases, could be achieved with aqueous acid/formaldehyde. According to the authors the approach was successfully applied in automated solid-phase synthesis of over 100 different peptide aldehydes. The required hydrazone (semicarbazone) of *N*-Boc amino aldehydes was prepared in 6 steps in solution from *tert*-butylcarbazate (*tert*-BuO-C(O)NHNH$_2$) and *trans*-4-(aminomethyl)-cyclohexanecarboxylic acid benzyl ester with the aid of carbonyl diimidazole as the coupling reagent, without preparative chromatography, in overall yield greater than 50% (so called pre-loading of the hydrazone linker). The *tert*-butoxycarbonyl (Boc) protected aldehyde (argininal, valinal, alaninal or phenylalaninal) loaded linker was anchored very

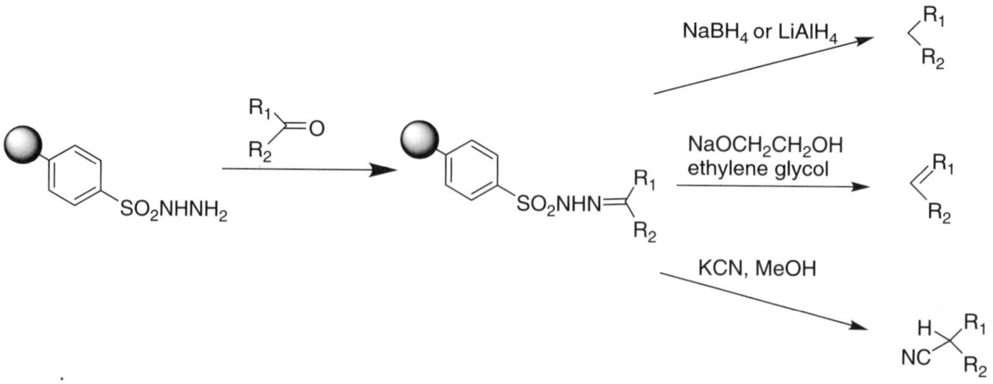

Scheme 10.1 *Early applications of polymer-supported sulfonylhydrazone linkers in synthesis*

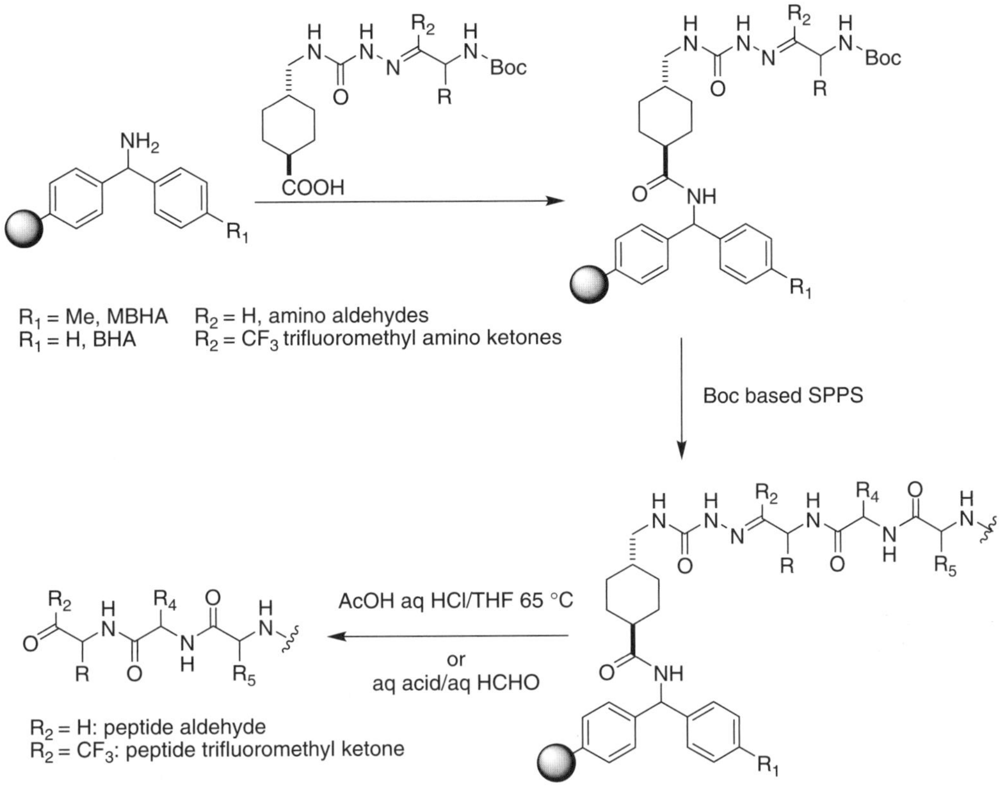

Scheme 10.2 *Automated solid-phase synthesis of peptide aldehydes and peptide ketones based on hydrazone (semicarbazone) linkers*

efficiently (99.9% yield) to a commercial methylbenzhydrylamine (MBHA) resin. The loaded resin was suitable for standard automated Boc based solid-phase peptide synthesis (SPPS) protocols. Cleavage of the final peptide aldehydes was effected with dilute aqueous acid/formaldehyde mixture. Preparation of peptide aldehydes containing amino acid residues (Thr, Ser, Tyr, Asp, Glu, Arginal) with functional groups protected by benzyl ether, benzyl ester or N^g-nitro group was also possible but required post-cleavage hydrogenolysis to free the protected functions.

The same semicarbazone linker strategy was used by Poupard for immobilization of α-amino trifluoromethyl ketones on benzhydrylamine (BHA) resin in a solid-phase synthesis of peptidyl trifluoromethyl ketones (Scheme 10.2, $R_2 = CF_3$).[8]

The direct attachment of halomethyl ketones on a solid support via a hydrazone linker (Scheme 10.3) was applied by Ellman for the synthesis of cysteine protease inhibitors.[9] From the tested hydrazines: N-alkyl, N-aryl, N,N-dialkylhydrazines and hydrazines with electron withdrawing groups such as 2,4-dinitrophenyl- and acylhydrazine, only the latter formed stable hydrazones with halomethyl ketones. Because it was found that hydrazones of methyl carbazate ($CH_3OC(O)NHNH_2$) could be completely hydrolyzed (with 1:4:4:15 mixture of trifluoroacetic acid (TFA)/water/acetaldehyde (CH_3CHO)/THF or trifluoroethanol) to carbonyl compounds the supported analogue of methyl carbazate was chosen. The supported carbazate **1** was prepared from ArgoGel-OH via activation with carbonyldiimidazol followed by reaction with anhydrous hydrazine. N-Protected α-bromo- or α-chloromethyl α'-amino ketones **2** could be

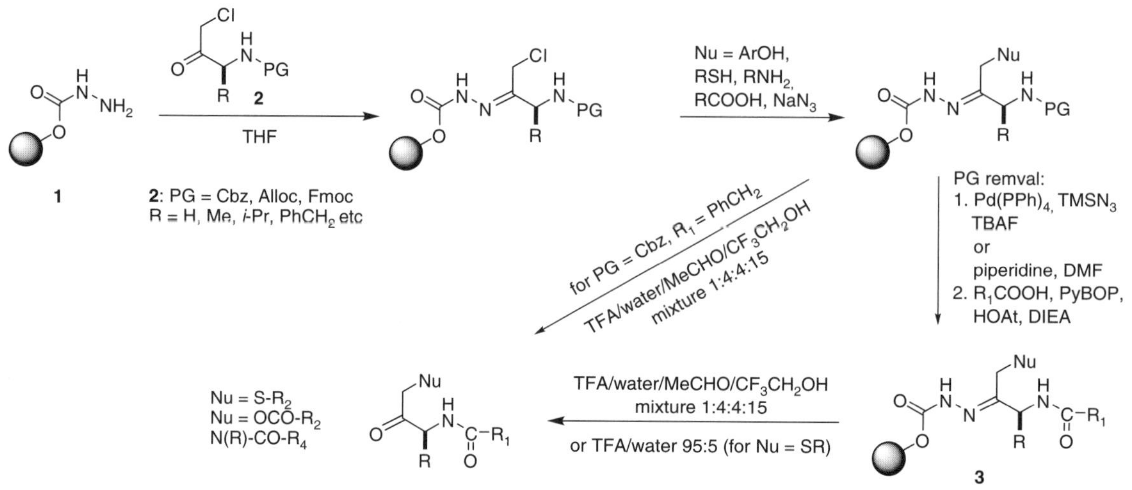

Scheme 10.3 *Ellman's approach to solid-phase synthesis of peptide ketones based on carbazate hydrazone linkers*

loaded on such hydrazine support with the latter giving better yields. Therefore, the immobilized α-chloro ketones were subjected to the nucleophilic displacement of chlorine by carboxylates, thiols, phenols, azide and primary amines. The reactions caused no detectable racemization (less than 1%) on the stereocenters α to the hydrazone group. After deprotection of the Alloc protected α'-amino group the free amine could be acylated and the all site functionalized ketones could be cleaved from the support **3**.

The developed methodolgy for immobilization and functionalization of α-chloromethyl α'-amino ketones by halide displacement was used later by Ellman (Scheme 10.4) for the synthesis of cyclic ketone inhibitors of cysteine protease[10] as well as by Lai, Payne and their coworkers[11] for parallel synthesis of substituted imidazoles (Scheme 10.4).

Elman showed that the method of immobilization on the carbazate support made from ArgoGel-OH was also applicable to 9-fluorenylmethoxycarbonyl (Fmoc) protected α-chloromethyl α'-amino ketones

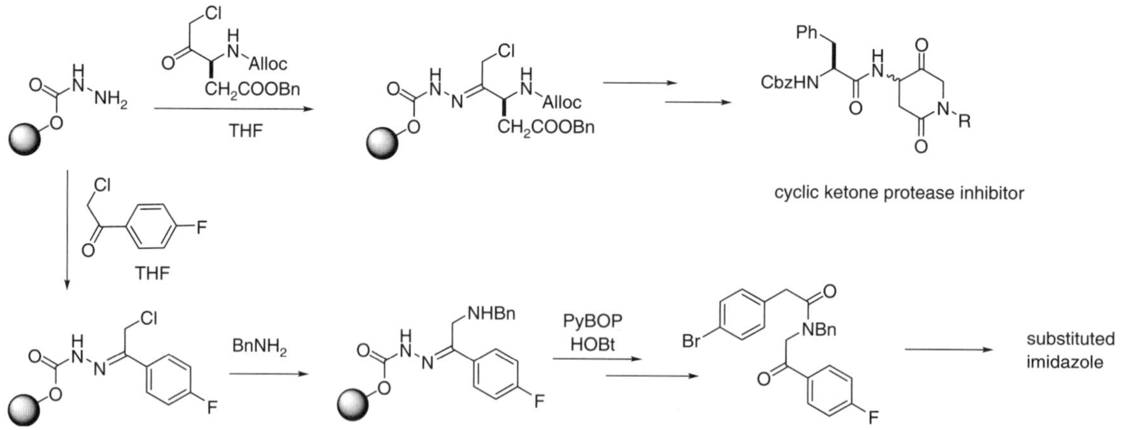

Scheme 10.4 *Syntheses from α-chloro ketones immobilized via Ellman's carbazate derived hydrazone linker*

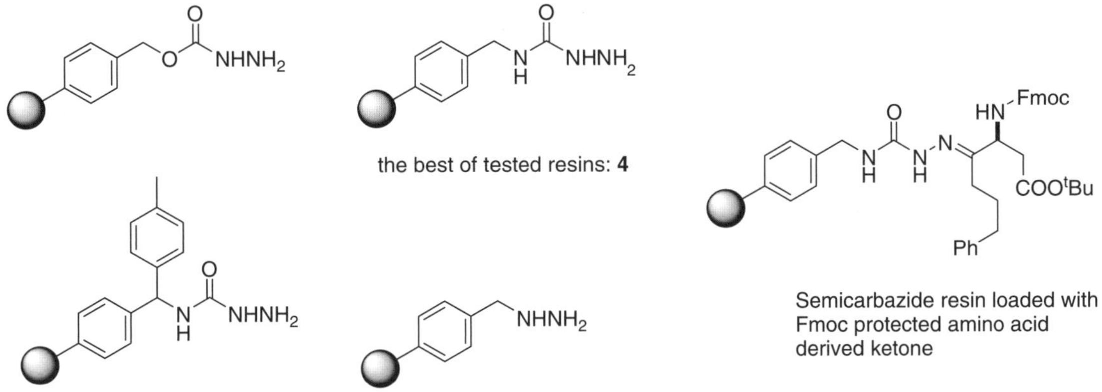

Scheme 10.5 *Hydrazine derived resins investigated by Albericio*

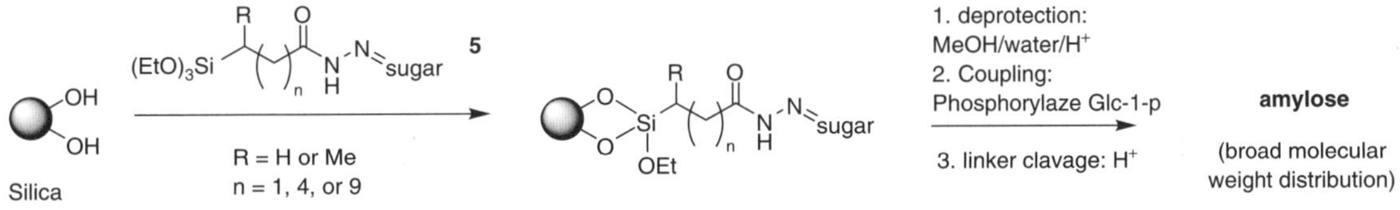

Scheme 10.6 *Preparation and reaction of a sugar-derived hydrazone-bound aldehyde on silica support*

by preparing a 2016-membered library of mercaptomethyl ketones – potential cathepsin B inhibitors.[12] In this case a simplified cleavage mixture was used – TFA/water 95:5 (Scheme 10.3). More recently, Albericio investigated the preparation, loading of ketones and performance of hydrazine derived resins (Scheme 10.5). The best in terms of yield of preparation and purity of cleaved peptide ketone product was semicarbazide polymer **4** derived from aminomethylpolystyrene resin.[13]

Breitinger applied ω-unsaturated acid hydrazides for the preparation of hydrazone linkers pre-loaded with linear maltoheptaose in solution.[14] The unsaturated bonds of the pre-loaded linkers were hydrosililated with H-Si(OEt)$_3$ and the resulting triethoxysilanes **5** were reacted with free hydroxl groups of porous silica support (Scheme 10.6). The resulting silica-supported maltoheptaose was used in enzyme (potato phosphorylase EC 2.4.1.1) catalysed synthesis of amylose. The hydrazone linker was stable under conditions of the synthesis, that is coupling and deprotection but was cleaved by citrate buffer at pH 4.0 at 60 °C in four hours. Better loadings and better results of synthesis were obtained for the linkers with longer spacers n = 4 and n = 9 respectively (Scheme 10.6).

Wessjohann synthesized benzylhydrazine functionalized support based on the polymeric matrix of poly(ethylene glycol) acrylamide resin (PEGA).[15] The supported benzylhydrazine could be used for chemoselective immobilization or for scavenging of aldehydes. The aldehydes could be released from the polymer by treatment with a mixture of acetone/THF/concentrated aqueous hydrochloric acid or released and transformed into another aldehyde hydrazone in one step (Scheme 10.7). The supported monoprotected dialdehyde, such as *p*-benzenedicarbaldehyde, was immobilized and selectively reacted with ylides, giving products hard to get chemoselectively by classical solution phase methods. The Heck reaction of a supported *p*-iodobenzaldehyde was also demonstrated (Scheme 10.7).

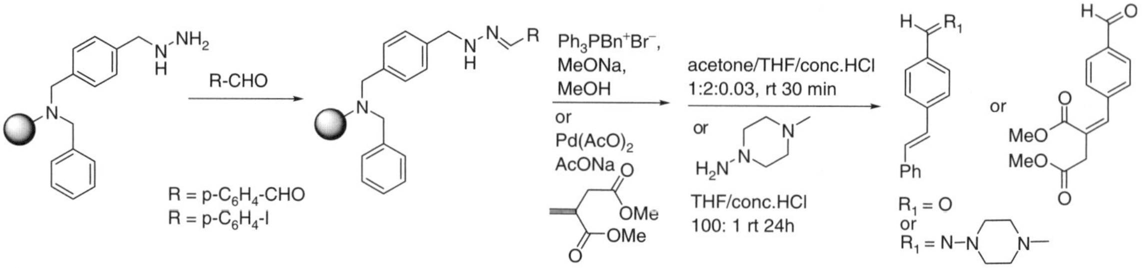

Scheme 10.7 The Wittig or the Heck reactions of polyfunctional aldehydes bound via hydrazone linkers on PEGA resin

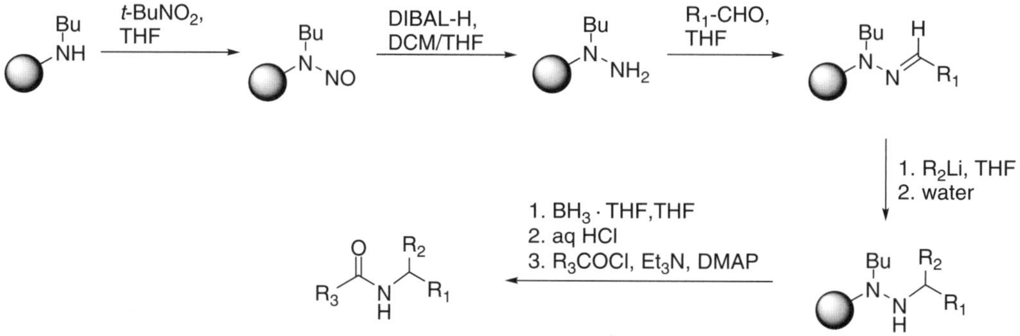

Scheme 10.8 Nucleophilic 1,2-addition to hydrazones on solid support and cleavage of the formed hydrazines to amines (isolated as amides)

Polymer-supported hydrazine synthesized from Merrifield polymer was used for immobilization of aldehydes by Enders and Bräse (Scheme 10.8).[16] Reaction of n-butylamine with chloromethylpolystyrene gave a polymer with secondary amine functionalities that could react with alkyl nitrite to form N-nitrosoamine. Reduction of supported nitrosoamine with diisobutylaluminum hydride (DIBAL-H) furnished hydrazine which was used to bind aldehydes via a hydrazone linker. Application of known in solution transformations,[17] 1,2-addition of organolithium reagents, followed by reductive cleavage of the hydrazine N–N bond with borane-THF complex provided several secondary amines (Scheme 10.8). The amines were isolated in the form of the corresponding amides for practical reasons.

Solid-phase asymmetric synthesis (SPAS) is a new and unexplored concept.[18] Known approaches to polymer-supported chiral hydrazines are based on chiral pool and as it is for their achiral counterparts discern into two kinds. The secondary amines are loaded on polymer, nitrosated and the N-nitrosoamines reduced do hydrazines[19,20] or the hydrazines are synthesized in solution and then loaded in a protected form on the polymer (Scheme 10.9).[20–22] Enders' group developed chiral hydrazone linkers applicable to asymmetric synthesis of chiral secondary amines.[19] By analogy to the highly developed chemistry of (S)-1-amino-2-methoxymethylpyrrolidine/(R)-1-amino-2-methoxymethylpyrrolidine (SAMP/RAMP) hydrazones in solution,[23] two polymer supported chiral hydrazines were synthesized

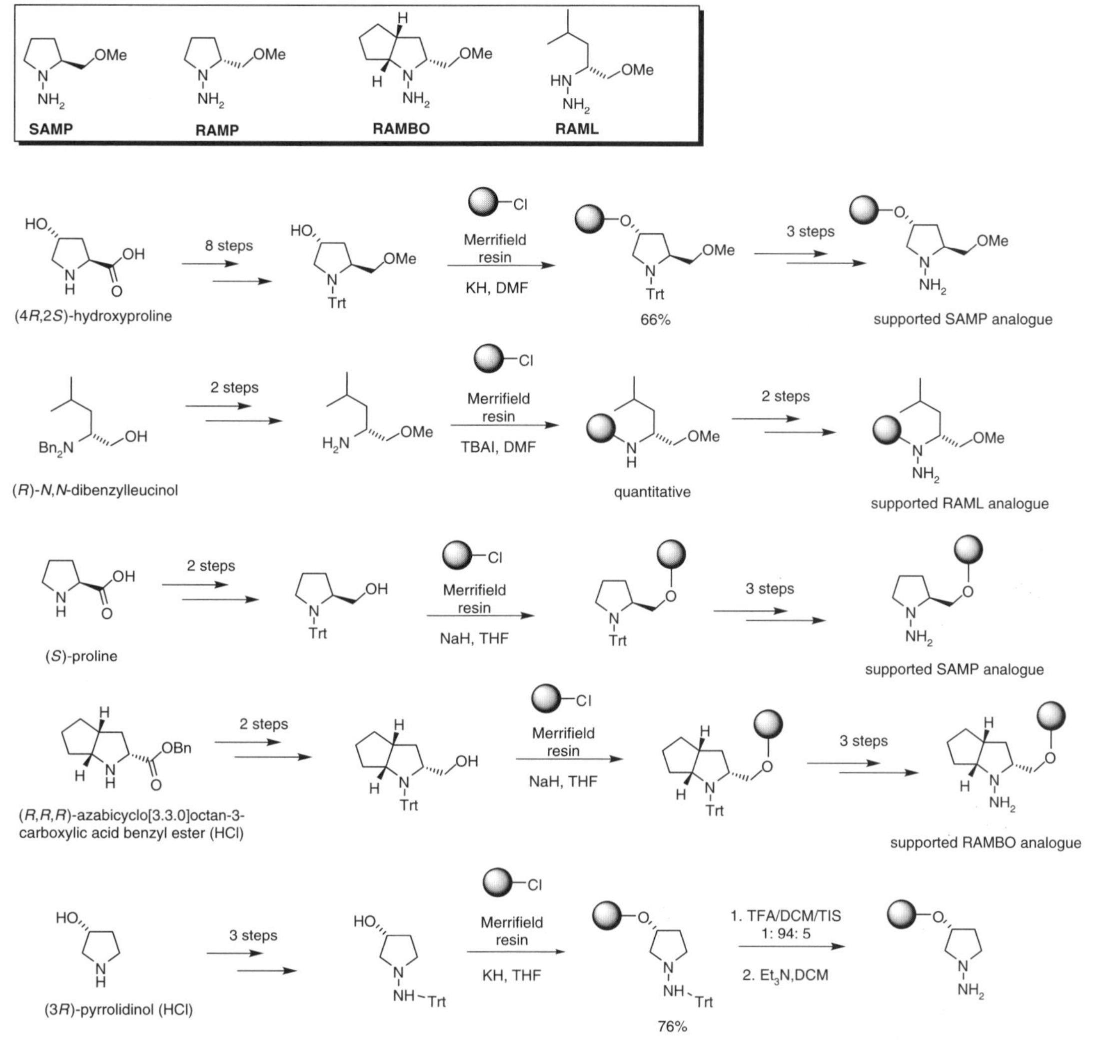

Scheme 10.9 Approaches to synthesis of chiral hydrazines useful for formation of chiral hydrazone linkers for solid-phase asymmetric synthesis (SPAS)

from (4R,2S)-hydroxyproline (*trans*-4-hydroxy-*L*-proline) and (*R-N,N*-dibenzylleucinol (Scheme 10.9). The supported hydrazines were reacted with 3-phenylpropionaldehyde and the resulting hydrazones were used as acceptors in nucleophilic 1,2-addition of *n*-BuLi, *t*-BuLi, *n*-HexLi and PhLi (Scheme 10.10). Subsequent use of the previously elaborated cleavage and work-up protocols[16] provided primary amines protected as amides in yields from 24 to 53% and moderate to high ee's (50–86%). Analogous results of the reaction with *n*-BuLi and *n*-HexLi were described by Fessner's PhD student who also investigated syntheses of supported SAMP and RAMBO analogues.[20] Rademann's PhD student synthesized immobilized hydrazine derived from (3*R*)-pyrrolidinol and used it in the nucleophilic addition of formaldehyde hydrazone to protected phenylalaninal.[21]

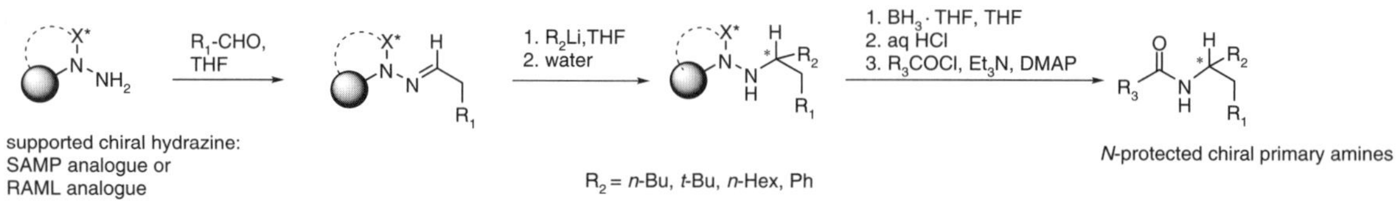

310 Multifunctional Linker Units for Diversity-Oriented Synthesis

Scheme 10.10 Asymmetric nuclephilic 1,2-addition with chiral hydrazines auxiliaries (chiral hydrazone linkers) for solid-phase asymmetric synthesis (SPAS)

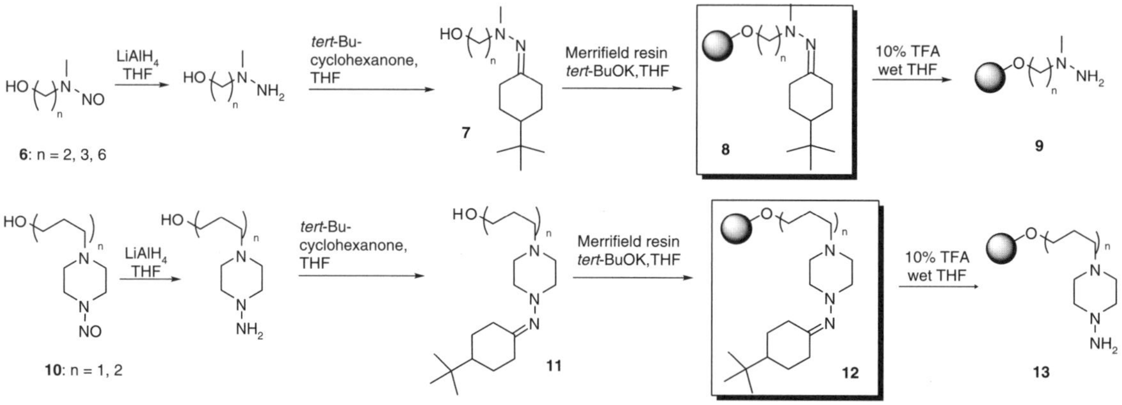

Scheme 10.11 Syntheses of hydrazines on polymeric support through immobilization of hydrazines protected as tert-butylcyclohexanone hydrazones

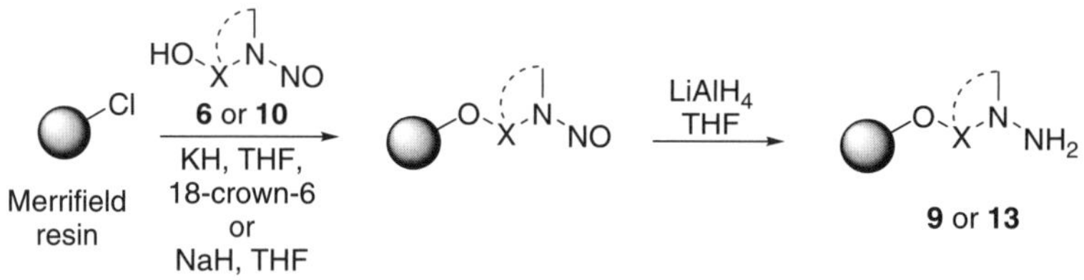

Scheme 10.12 Direct synthesis of hydrazines on polymeric support

Another approach to synthesis of polymer supported hydrazines which binds both aldehydes and ketones via hydrazone linkers was developed by Lazny (Scheme 10.11).[24]

Supported N,N-dialkylhydrazines derived from N-methyl-N-(ω-hydroxyalkyl)amin and N-(ω-hydroxylalkyl)piperidine[24] were prepared on Merrifield-type polymer, either by solution synthesis of hydrazines from nitrosoamines **6** and **10** followed by loading on the support in a protected form of 4-($tert$-butyl)cyclohexanone hydrazones **7** and **11** (Scheme 10.11),[25] or by loading of nitrosoamines **6** and **10** and their reduction on polymer (Scheme 10.12).[26] The former method was found to furnish cleaner and more uniform materials. The hydrazone protected hydrazines were stable and could be stored. The polymeric supports required activation with 10% TFA solution in wet tetrahydrofuran (THF) before loading of a ketone or an aldehyde.

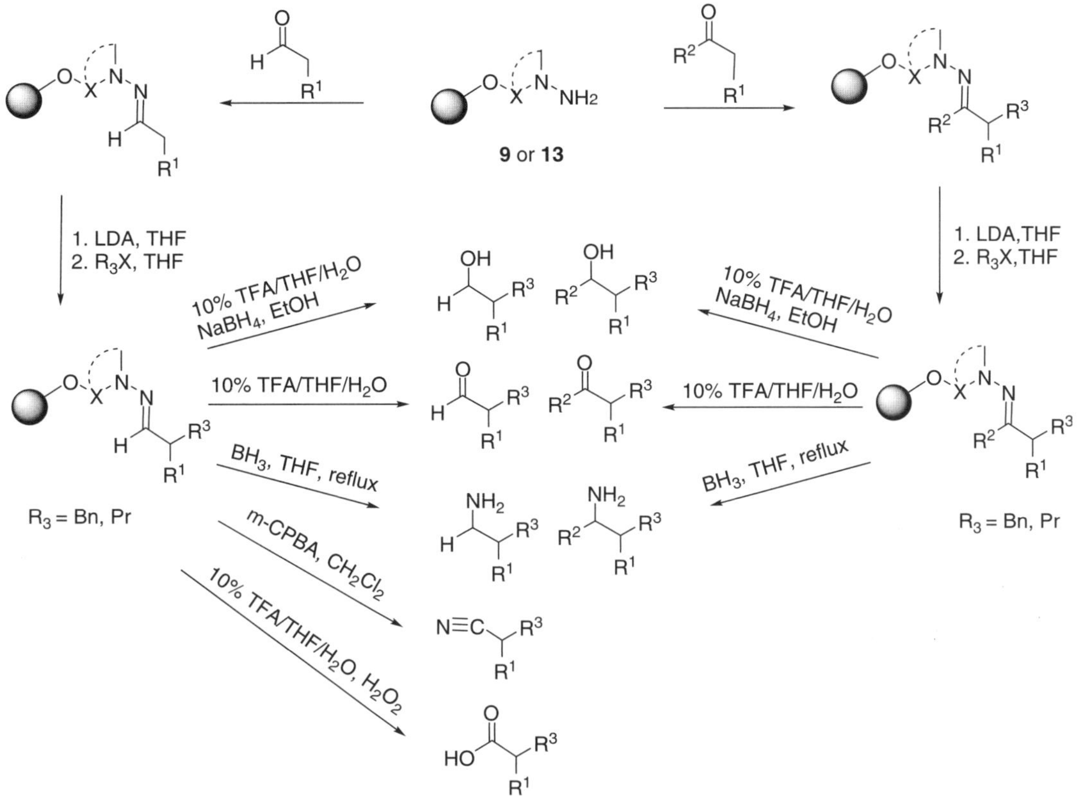

Scheme 10.13 *Solid-phase alkylation of ketones and aldehydes. Hydrazones as multifunctional linkers*

The supported hydrazines **9** and **13** reacted with several ketones and aldehydes giving hydrazones (Scheme 10.13). The specific feature of such dialkyl hydrazone linkers is their effect on reactivity of carbons in α-positions to the bound carbonyl groups. The high reactivity of lithiated hydrazones towards electrophiles is well known and used in organic synthesis in solution.[1,23] The hydrazones of the immobilized ketones and aldehydes could be lithiated with lithium diisopropylamide (LDA) and alkylated with alkyl halides.[24–26] The alkylation allows for build-up of the carbon skeleton of carbonyl compounds anchored on solid support. In addition, diverse available methods for cleavage of N,N-dialkylhydrazones provide opportunities for multifunctional cleavage of such hydrazone linkers. Hydrolytic cleavage under acidic conditions (TFA/THF/water) afforded carbonyl compounds – aldehydes and ketones. Cleavage/oxidative work-up (peroxide) or oxidative cleavage (peracids) of aldehyde hydrazones provided access to acids or nitriles respectively. On the other hand, cleavage/reductive work-up (sodium borohydride) or reductive cleavage with borane-THF complex furnished alcohols or amines respectively (Scheme 10.13).[25]

Asymmetric alkylation of ketones or aldehydes, which is a known and useful methodology in solution, is waiting to be more widely implemented in SPAS. Asymmetric (overall enantioselective) alkylation of ketones via supported chiral hydrazone linkers acting as chiral auxiliaries is theoretically feasible. Proof of concept studies were conducted independently by Enders', Fessner's and Lazny's research groups and showed that fairly high selectivities could be obtained. Enders' group used polymer-supported SAMP-analogue derived from hydroxy proline previously used for 1,2-addition. The obtained alkylated ketones

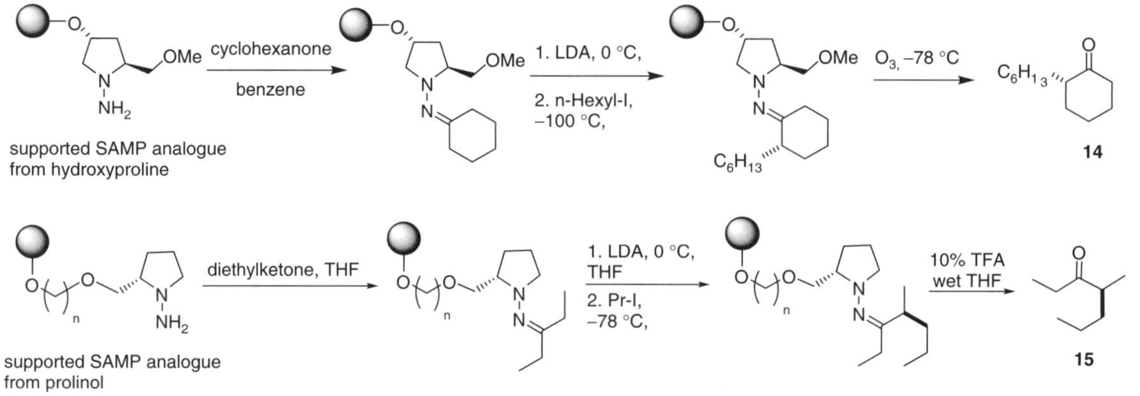

312 Multifunctional Linker Units for Diversity-Oriented Synthesis

Scheme 10.14 *Approaches to asymmetric alkylation of ketones bound to supports via chiral hydrazone linker auxiliaries*

had enantiomeric purities in the range of 55–79% at −72 °C (lowering the reaction temperature to −100 °C gave in the best case of **14** 86% ee compared to 91% ee achievable with SAMP in solution).[27] Lazny's group prepared SAMP-analogues anchored to solid support via the SAMP oxymethylene group, both with and without a spacer.[22] The enantiomeric purities of the ketones alkylated at −78 °C ranged from 10–73% ee and were the best for the product **15** and the chiral auxiliary connected to polymer matrix through a spacer (Scheme 10.14 n = 6). The same alkylation with the auxiliary anchored to the polymer matrix directly (no spacer) was less stereoselective[22] or required lower temperature (73% ee at −100 °C). [20]

10.3 Conclusion

The reviewed literature shows that hydrazone units derived either from supported dialkylhydrazines or acylhydrazines (semicarbazides, carbazates) can serve as linkers for ketones, aldehydes, amino ketones, ketone/aldehyde peptides and open chain carbohydrates. Hydrazones have a rich chemistry which partially has been implemented in SPOS, mostly for preparation of a few classes of compounds (amines, alcohols, nitriles, acids) and for some reactions of hydrazone bound carbonyl compounds (α-alkylation of aldehydes and ketones, nucleophilic displacement of α-chlorine). Solid-phase alkylation of a ketone bound via dialkylhydrazone was shown to be applicable to a synthesis of a known natural product.[28] In addition, polymer-supported hydrazones can also be used as reactants, for example acyl hydrazones in synthesis of substituted pyrazolones,[29] carbazates in synthesis of substituted pyrazolidine-3,5-diones[30] or dialkylhydrazones in synthesis of nitriles.[31] The potential of hydrazones as linkers for solid-phase asymmetric synthesis (SPAS) only just begins to unravel.

Typical Experimental Procedures

Typical procedure for loading of the chloromethyl ketone derived from an amino acid (e.g. PhAla derivative Fmoc-2, R = PhCH$_2$) on carbazate resin 1 (Scheme 10.3) [9, 12]

*A solution of chloromethyl ketone **Fmoc-2** (4 equiv., 0.2 M) in THF was added to the carbazate-derivatized resin presolvated with THF. The capped vessel was placed in a 50 °C oil bath and heated for three hours.*

The resin was transferred as a slurry into a polypropylene cartridge and subsequently rinsed with five portions of THF. The support-bound ketones were used immediately in the chloride displacement step.

General method for cleavage from the carbazate resin 3 (Scheme 10.3) [9, 12]

A mixture of trifluoroacetic acid/water (95:5, 2 ml, purged with nitrogen for five minutes) in a nitrogen glovebag or a mixture of TFA/water/acetaldehyde/trifluoroethanol (1:4:4:15, 5 ml) was added to the derivatized resin 3 (0.2 g). The resin was gently rocked for one hour or four hours, and the solution was removed. The resin was washed with five portions of THF. The acidic washings were combined and concentrated. Toluene was added to form an azeotrope with the cleavage solution (residual water and TFA) and the volatiles were evaporated. Analytically pure samples were obtained by purification by column chromatography or recrystallization.

General method for the preparation of semicarbazide resins 4 (Scheme 10.5)[13]

A solution of 1,1'-carbonyldiimidazole (5 equiv.) in DMF (3 ml) was added to a previously DMF-washed amino polystyrene resin (100 mg, 1.1 mmol/g) and the mixture was shaken at 25 °C for three hours. The resins were washed with DMF (3×), a solution of Boc-NH-NH$_2$ (5 equiv.) in DMF (3 ml) was added, and the mixture was shaken for three hours at 25 °C. Washing with DMF (3×) and DCM (3×) gave the Boc protected resins ready for storage. Before the resins can be used, the Boc group must be removed by treatment with a TFA–DCM (1:1) solution for one hour at 25 °C.

General method for binding and cleavage of ketones from hydrazine-based resin 4 (Scheme 10.5)[13]

A ketone (3 equiv.) in DCM (3 ml) was added to resin 4 and the mixture was shaken for 16 hours at 25 °C. Then the resin was washed with DCM (3×), DMF (3×) and DCM (3×). Cleavage of the ketone was done by washing the resin with TFA-water (4:1) solution, and evaporation of the combined washings.

Typical procedure for preparation of a hydrazone linker polymer (Anchoring of hydrazone protected hydrazines 7 and 11 on the Merrifield gel)[25]

2-[2-(4-tert-Butylcyclohexylidene)-1-methylhydrazino]ethoxymethylpolystyrene (8, n = 2, Scheme 10.11)

To a mechanically agitated solution of hydrazone 7 n = 2 (2.53 g, 11.2 mmol, 5 equiv.) in dry THF (20 ml) was added potassium tert-butoxide (1.258 g, 11.2 mmol, 5 equiv.). After the solids dissolved, Merrifield gel (2.042 g, 2.25 mmol, Novabiochem, 1% PS-DVB, 200–400 mesh, 1.1 mmol/g) was added. The resulting suspension was intermittently stirred at room temperature for five hours followed by heating at 60 °C for 24 hours. Then the polymer was washed under argon with help of a sintered glass filter with: THF (2 × 15 ml), methanol (3 × 15 ml), THF (2 × 5 ml), methanol (3 × 5 ml),), water/DMF mixture (2 × 2 ml + 8 ml), water (2 × 10 ml), water/DMF mixture (2 × 2 ml + 8 ml), methanol (2 × 5 ml), THF (2 × 5 ml), methanol (3 × 5 ml), DCM (3 × 5 ml), methanol (2 × 5 ml), DCM (2 × 5 ml), methanol (2 × 5 ml). The residual solvent was evaporated under vacuum and the gel was dried to a constant mass under high vacuum to give a yellowish powder 8. Solutions from the first three washings which contain excess of hydrazone used can be combined and saved for recovery of 7.

Typical procedure for anchoring of an aldehyde or a ketone via the hydrazone linker[25]

The polymer **8** (1 g, 0.70 mmol of protected hydrazine) was washed under argon with a mixture of TFA:water:THF (1:1:8, 3 × 10 minutes), followed by THF (2 × 5 ml), methanol (2 × 5 ml), THF (2 × 5 ml), methanol (2 × 5 ml), DCM (2 × 5 ml), methanol (2 × 5 ml), 10% triethylamine in DCM, (3 × 5 ml), methanol (2 × 5 ml), DCM (3 × 5 ml) and dry THF (4 × 2 ml). Then a solution of a carbonyl compound (7 mmol, 10 equiv.) in dry THF (2.5 ml) was added. The suspention was heated under reflux for 24 hours in the presence of molecular sieves 4 Å. The molecular sieves were removed with assistance of a polypropylene mesh and the gel was washed with THF (2 × 5 ml), methanol (2 × 5 ml), THF (2 × 5 ml), methanol (2 × 5 ml), DCM (2 × 5 ml), methanol (2 × 5 ml). The resulting gel was dried to a constant mass under high vacuum.

Typical procedure for acidolytic cleavage of the hydrazone linker (Cleavage from resin type 8 (Scheme 10.13))[26]

The polymeric support with an aldehyde or a ketone immobilized via the hydrazone linker was treated with a mixture of TFA/water/THF (1:1:8) for 15 minutes and washed with THF, methanol, THF, methanol, DCM, methanol. The combined acidic solutions were washed with a 20% aqueous potassium carbonate solution. The aqueous phase was back-washed three times with DCM. The combined organic extracts were dried with magnesium sulfate and concentrated under low vacuum to give the aldehyde or the ketone.

References

[1] S. Kim, J.-Y. Yoon, in Science of Synthesis, Vol. 27 (Ed.: A. Padwa), Georg Thieme, Stuttgard; New York, 2004, pp. 671.
[2] Enders, D., Wortmann, L., and Peters, R.; *Acc. Chem. Res*. **2000**, *33*, 157.
[3] Jung N., Wiehn M., and Bräse S.; *Top. Curr. Chem*. **2007**, *278*, 1.
[4] Bourel-Bonnet, L., Pecheur, E.-I., Grandjean, C., et al.; *Bioconjugate Chem*. **2005**, *16*, 450.
[5] Corbett, P. T., Leclaire, J., Vial, L., et al.; *Chem. Rev*. **2006**, *106*, 3652.
[6] Kamogawa, H., Kanzawa, A., Kadoya, M., et al.; *Bull. Chem. Soc. Jpn*. **1983**, *56*, 762.
[7] Nicolaou, K. C., Winssinger, N., Pastor, J., and Murphy, F.; *Angew. Chem. Int. Ed*. **1998**, *37*, 2534.
[8] Poupart, M. A., Fazal, G., Goulet, S., and Mar, L. T.; *J. Org. Chem*. **1999**, *64*, 1356.
[9] Lee, A., Huang, L., and Ellman, J. A.; *J. Am. Chem. Soc*. **1999**, *121*, 9907.
[10] Huang, L., and Ellman, J. A.; *Bioorg. Med. Chem. Lett*. **2002**, *12*, 2993.
[11] Cobb, J. M., Grimster, N., Khan, N., et al.; *Tetrahedron Lett*. **2002**, *43*, 7557.
[12] Wood, W. J. L., Huang, L., and Ellman, J. A.; *J. Comb. Chem*. **2003**, *5*, 869.
[13] Vázquez, J., and Albericio, F.; *Tetrahedron Lett*. **2006**, *47*, 1657.
[14] Breitinger, H.-G.; *Tetrahedron Lett*. **2002**, *43*, 6127.
[15] M. Zhu, E. Ruijter, and L. A. Wessjohann, *Org. Lett*. **2004**, *6*, 3921.
[16] Kirchhoff, J. H., Bräse, S., and Enders, D.; *J. Comb. Chem*. **2001**, *3*, 71.
[17] Enders, D., Lochtman, R., Meiers, M., et al.; *Synlett* **1998**, 1182.
[18] Leßmann, T., and Waldmann, H.; *Chem. Commun*. **2006**, *32*, 3380.
[19] Enders, D., Kirchhoff, J. H., Köbberling, J., and Peiffer, T. H.; *Org. Lett*. **2001**, *3*, 1241.
[20] Schooren, J.; **2003**, *Festphasengebundene chirale Hydrazine als Auxiliare in der Asymmetrischen Syntheses*, Ph.D. Thesis, Technical University, Darmstadt.
[21] Weik, S.; **2004**, *Entwicklung polymerer Carbanionen-Äquivalente zur Anwendung in milden C-C-Verknüpfungen und deren Einsatz in der Synthese von Aspartylprotease-Inhibitoren des Norstatin-Typss*, Ph.D. Thesis, Eberhard-Karls University, Tübingen.

[22] R. Lazny, A. Nodzewska, B. Zabicka *J. Comb. Chem.* **2008**, *10*, 986–991.
[23] Job, A., Janeck, C. F., Bettray, W., *et al.*; *Tetrahedron* **2002**, *58*, 2253.
[24] Lazny, R., and Michalak, M.; *Synlett* **2002**, 1931.
[25] Lazny, R., Nodzewska, A., Sienkiewicz, M., and Wolosewicz, K.; *J. Comb. Chem.* **2005**, *7*, 109.
[26] Lazny, R., Nodzewska, A., and Wolosewicz, K.; *Synthesis* **2003**, 2858.
[27] J. Köbberling, 2001, *Neue Ankergruppen für die enantioselektive kombinatorische Festphasensyntheses*, Ph.D. Thesis, RWTH-Aachen University, Aachen.
[28] Lazny, R., Nodzewska, A., and Sienkiewicz, M.; *Polish J. Chem.* **2006**, *80*, 655.
[29] Kobayashi, S., Furuta, T., Sugita, K., *et al.*; *Tetrahedron Lett.* **1999**, *40*, 1341.
[30] He, R., and Lam, Y.; *Org. Biomol. Chem.* **2008**, *6*, 2182.
[31] Baxendale, I. R., Ley, S. V., and Sneddon, H. F.; *Synlett* **2002**, 775.

11
Benzotriazole Linker Units

Daniel K. Whelligan

Cancer Research UK Centre for Cancer Therapeutics, The Institute of Cancer Research, United Kingdom

11.1 Introduction

Benzotriazole is a very useful auxiliary in organic synthesis for activating molecules toward nucleophilic attack and acting as a leaving group when necessary as demonstrated in the mechanism for the Mannich-type reaction in Scheme 11.1. This is because it can act as an electron donor and an electron acceptor, so, when attached to a carbon atom with another heteroatom, it can either push electrons to make the heteroatom leave – forming an iminium ion (such as step A) – or receive electrons from the σ-bond to become a leaving group itself, leaving behind, for example, an iminium cation (such as step B).[1] In resin-bound chemistry, these properties have been exploited several times in the Mannich-type reaction, as has benzotriazole's ability to form acylbenzotriazoles – stable acid chloride equivalents which can be further manipulated prior to cleavage of the benzotriazole group. A further property that makes benzotriazole a useful auxiliary is its ability to stabilize a negative charge on a carbon atom α to nitrogen. However, this has not yet been used in resin-bound chemistry. Overall, benzotriazole is easily introduced, stable to various reaction conditions and readily removed, making it well suited to adaptation into a resin-bound linker.

Resin-bound 1-hydroxybenzotriazole (HOBt) has been known since 1975[2] and is used to form solid-supported activated esters which react readily with amines and amino acids to form amides and peptides, respectively.[3] The first synthesis of resin-bound 1-*H*-benzotriazole was therefore based on that of resin-bound HOBt and was consequently bound by a CH_2 tether. Several other chemical groups have since been used to tether benzotriazole and the first section of this chapter, on the various syntheses, is divided according to the particular tether created.

The second section of this chapter is concerned with the three types of reaction for which resin-bound benzotriazoles have thus far been used. In all of these, the benzotriazole is the linker as well as the

---

Linker Strategies in Solid-Phase Organic Synthesis Edited by Peter Scott
© 2009 John Wiley & Sons, Ltd

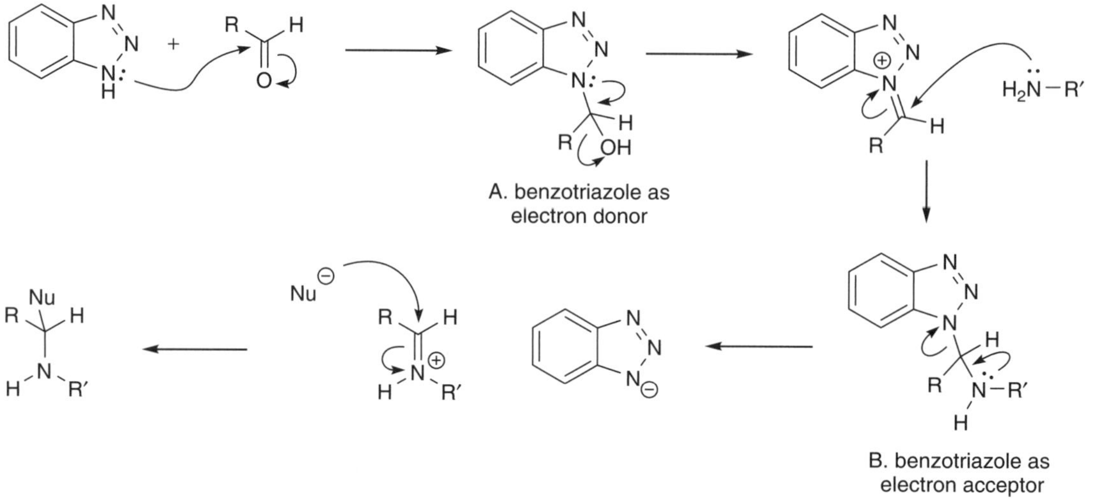

Scheme 11.1 *Benzotriazole mediated Mannich-type reaction mechanism*

activating/leaving group of an intermediate which may or may not be isolated. Once the final product is formed, the resin-bound benzotriazole will have been cleaved leaving no trace of its original presence, hence acting as a traceless linker.

11.2 Syntheses of polymer-supported benzotriazoles

11.2.1 Carbon–carbon tethered benzotriazoles

Showalter began research into benzotriazole linkers by repeating the known synthesis[2] of resin-bound 1-hydroxybenzotriazole **1** on polystyrene and then converting it to benzotriazole **2** by reductive cleavage of the 1-OH group (Scheme 11.2).[4]

This gave a polymer-supported benzotriazole with high loading capacity and a robust and slightly electron-releasing CH_2 tether. Unfortunately, it showed extremely poor swelling properties in a range of solvents. Consequently, characterization could only be obtained indirectly by infra red ($1739\,cm^{-1}$) and it was thought subsequent reactions would have been hindered.

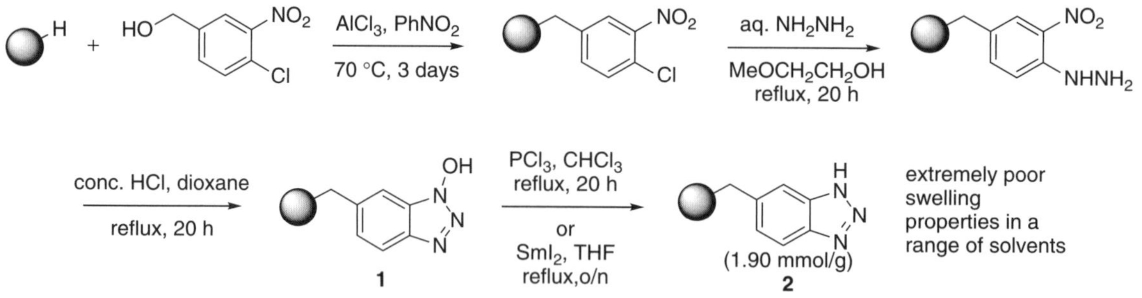

Scheme 11.2 *Preparation of polystyrene-supported, CH_2-linked benzotriazole*

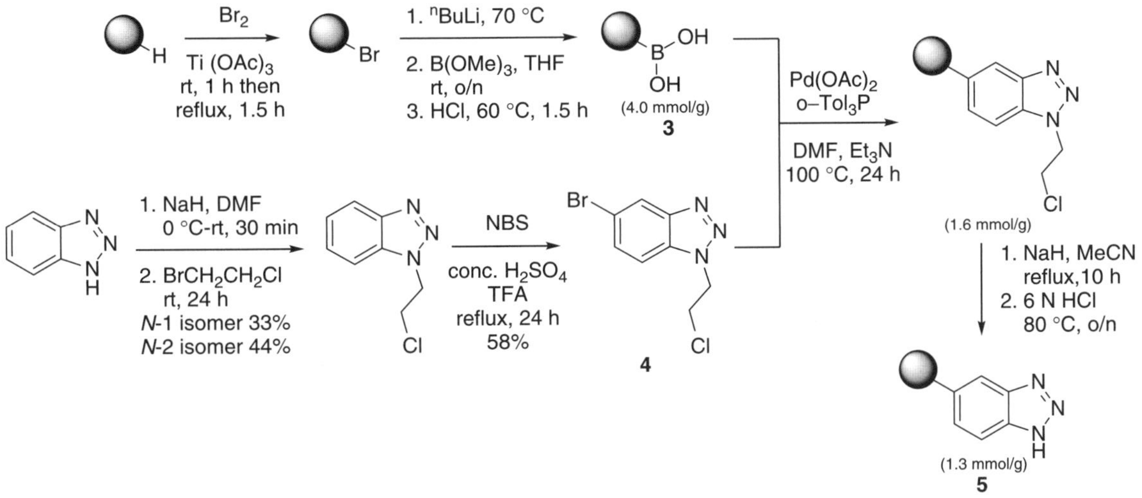

Scheme 11.3 Synthesis of benzotriazoles tethered directly to polystyrene

However, Katritzky's resin-bound benzotriazoles **5**, with an even shorter, direct carbon–carbon bond tether to polystyrene, behaved well in the reactions they promoted (Section 11.3). These were synthesized by a Suzuki coupling reaction of polystyrene boronic acid **3**[5] with the protected 5-bromobenzotriazole **4** (Scheme 11.3).[6]

11.2.2 Ether tethered benzotriazoles

To improve swelling properties, Showalter sought more flexible tethers than their original CH$_2$ group in resin **2** (Scheme 11.2), thus ethers CH$_2$OCH$_2$ and CH$_2$O(CH$_2$)$_6$OCH$_2$ were investigated.[4] The corresponding polymer-supported benzotriazoles **6** and **8** were formed by *O*-alkylation with Merrifield resin (Scheme 11.4), a step which required trityl-protection of the benzotriazole as a 2:1:2 mixture of *N*-trityl isomers.

As hypothesized, the longer tether lengths of **6** and **8** resulted in increased swelling of the resins with **8** being the most accessible to solvation in a variety of solvents. This resin also gave the highest yields in the solid-supported reaction chosen (Section 11.3).

Fang was able to access bromide **7** in the same manner as above but also showed it could be synthesized in one step by bromination of 5-methylbenzotriazole **9** (Scheme 11.5).[7] It was coupled with the sodium salt of monomethoxypolyethylene glycol (CH$_3$O-PEG-OH) of molecular weight ~5000 and deprotected to give polymer **10**. This polymer-supported resin had a rather low loading of ≤0.15 mmol/g but the advantage of being soluble during reactions after which cold ether could be added to precipitate it.

Paio *et al*. generated *tert*-butoxycarbonyl (Boc) protected 5-hydroxybenzotriazole **12** in three steps from diamine **11** and used it to make aryl–ether linkages to resins hydroxymethylpolystyrene (HMB-PS) and chloromethyl polystyrene (CM-PS) (Scheme 11.6).[8]

Katritzky *et al*. also synthesized an aryl–ether tethered benzotriazole **16**, on Merrifield resin, but avoided the use of any protecting groups by having the diamine precursor **15** bound to the resin before conversion to the benzotriazole by diazotisation[9] (Scheme 11.7).[10] Resin **16** was, therefore, in hand after only three steps via one of the two routes shown below.

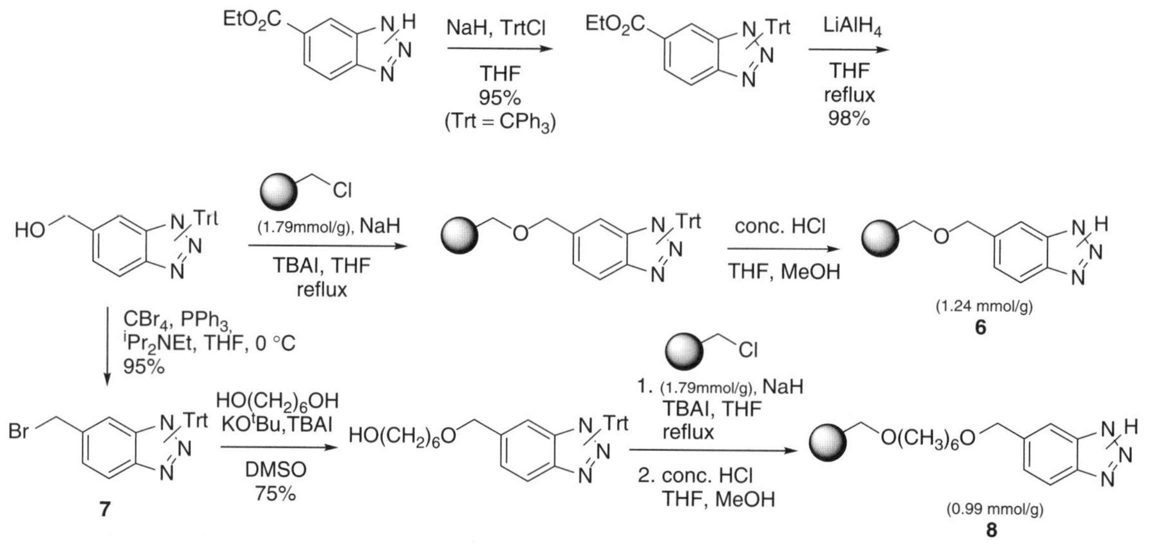

Scheme 11.4 Preparation of polystyrene-supported, ether tethered benzotriazoles

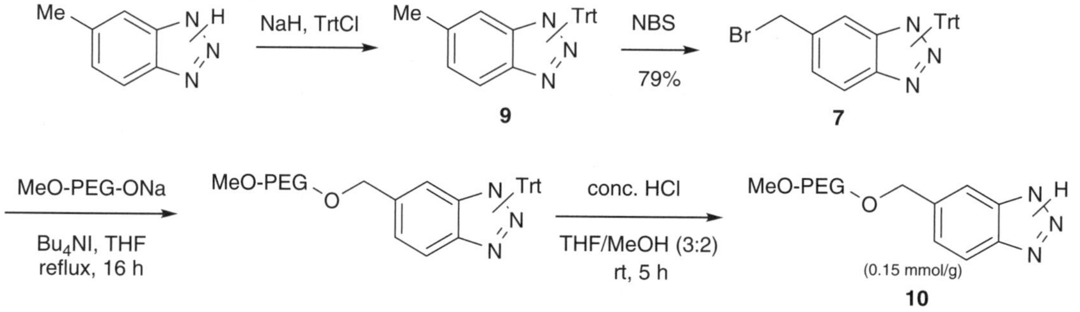

Scheme 11.5 Alternative route to bromomethylbenzotriazole 7 and coupling

11.2.3 Amide tethered benzotriazoles

Piao quickly obtained amide tethered polymer-supported benzotriazoles by Boc-protection, diisopropylcarbodiimide (DIC) mediated coupling and deprotection of commercially available carboxylic acid **17** with resins aminomethyl polystyrene (AM-PS) and Argogel-amino (AG-NH$_2$) (Scheme 11.8).[8] Later, this synthesis was further shortened to one step by development of an amide-coupling which did not require protection of the benzotriazole. Thus **17** was directly coupled to the amino resins using HOBt and DIC in N,N-dimethylformamide (DMF).[11]

Tertiary amide tethered benzotriazole **21** was prepared by Katritzky in a two-step process with Merrifield resin (Scheme 11.9).[6] The chloride was displaced by piperazine and the resulting resin-bound piperazine **20** coupled with benzotriazole-5-carboxylic acid **17** using N,N'-dicyclohexylcarbodiimide (DCC). An alternative coupling involved heating the resin-bound piperazine with the ethyl ester of benzotriazole-5-carboxylic acid under reflux for 24 hours in DMF, but this gave a lower yield of 43%. Resin **21** has not yet been tested in any reactions as alternative resins **5** and **16**, developed concurrently, had higher loadings.

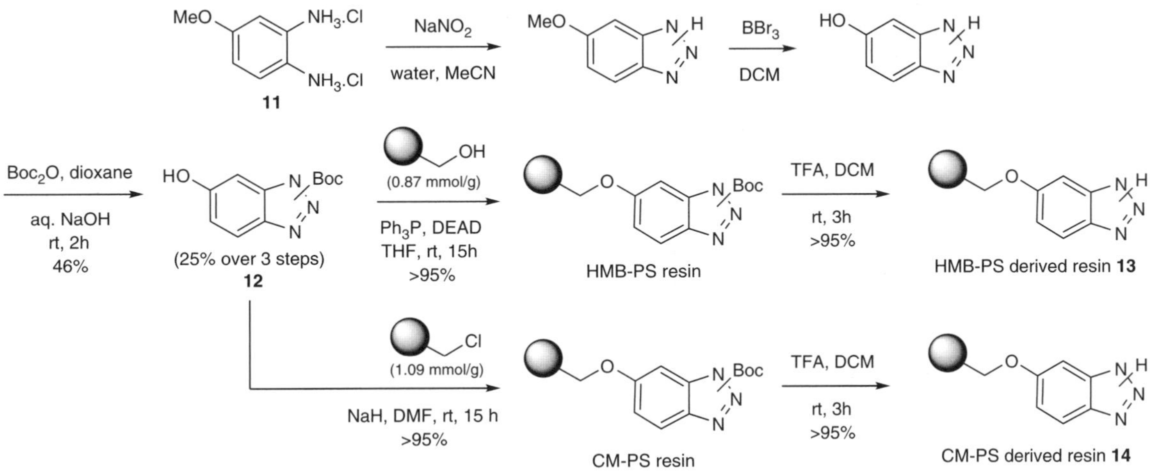

Scheme 11.6 Paio's ether tethered, resin-bound benzotriazoles

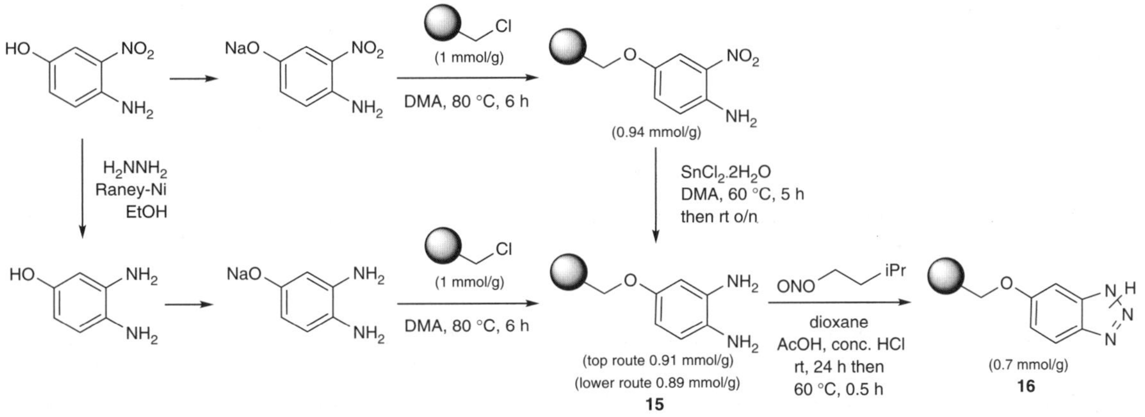

Scheme 11.7 Katritzky's synthesis of ether tethered, resin-bound benzotriazole

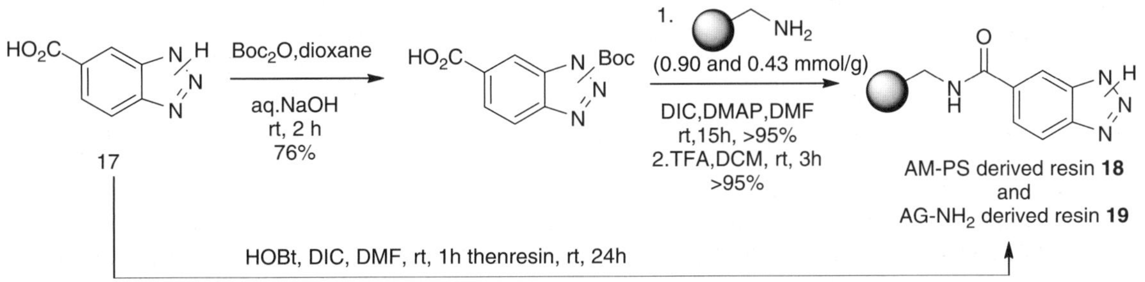

Scheme 11.8 Synthesis of amide tethered benzotriazoles

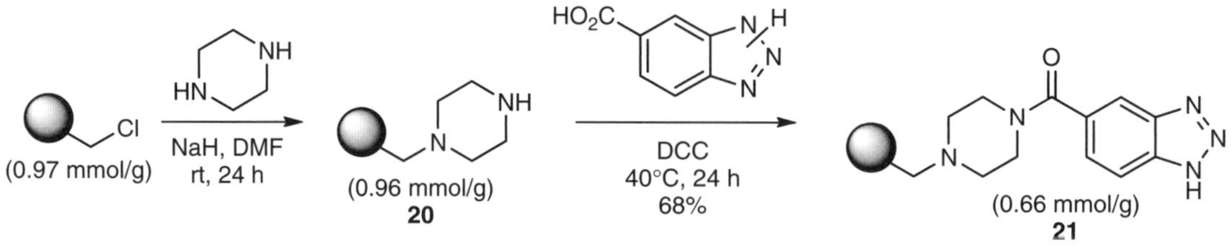

322   Multifunctional Linker Units for Diversity-Oriented Synthesis

Scheme 11.9 *Synthesis of tertiary amide tethered benzotriazole*

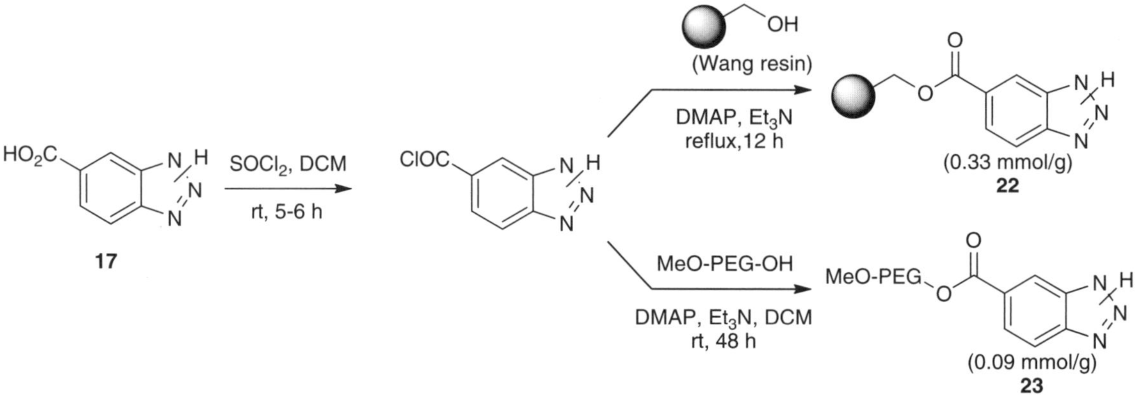

Scheme 11.10 *Synthesis of ester tethered benzotriazoles*

11.2.4 Ester tethered benzotriazoles

Fang converted acid **17** to the acid chloride and reacted it with both Wang resin and monomethoxypolyethylene glycol (CH$_3$O-PEG-OH), of molecular weight ~5000, to give ester tethered benzotriazoles **22** and **23**, respectively (Scheme 11.10).[7] CH$_3$O-PEG-bound benzotriazole **23** could be used in solution as a polymeric matrix and could therefore be analyzed by TLC and NMR during reactions then precipitated from solution afterwards by addition of ether.

In summary, a variety of syntheses have been used to tether benzotriazole, by alkyl, ether, amide and ester linkages, to polystyrene, Merrifield resin, monomethoxypolyethylene glycol (molecular weight = 5000), hydroxymethylpolystyrene (HMB-PS), chloromethyl polystyrene (CM-PS), aminomethyl polystyrene (AM-PS), Argogel-amino (AG-NH$_2$) and Wang resin.

11.3 Polymer-supported benzotriazole linked reactions

11.3.1 Mannich-type reaction and cleavage

The use of benzotriazoles in a Mannich-type reaction with aldehydes and amines to form substituted amines has been well documented by Katritzky.[1] In the resin-supported version, an aldehyde and an amine were added to afford the resin-bound aminal **24** (Scheme 11.11).[4] At this stage, the resin could be washed and dried before the product was further reacted and simultaneously cleaved from the support with a Grignard

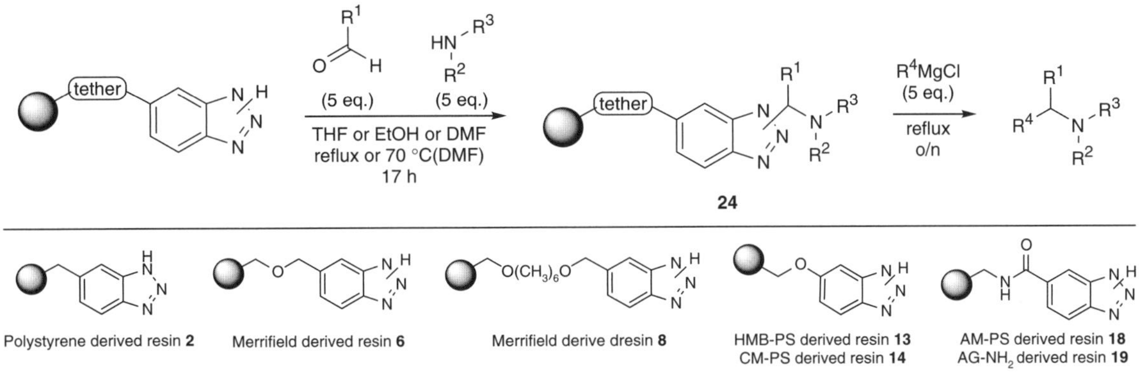

Scheme 11.11 Polymer-supported benzotriazole Mannich-type reaction

reagent. In general, resins **6** and **8**, with their longer, more flexible ether tethers, gave slightly higher yields than methylene tethered resin **2** (except in the case using *p*-methoxybenzaldehyde). However, the amount of amine produced from **2** was almost the same due to its very high loading. Strangely, primary aliphatic amines seemed to react poorly or not at all with aldehydes, whereas the secondary amine morpholine and less nucleophilic 2-aminopyridine reacted with aromatic and aliphatic aldehydes fairly well, giving, after cleavage with benzyl- or phenylmagnesium bromide, products of largely excellent purity in yields of 20–77%. These results could be attributed to the imines formed from primary amines and aldehydes being more stable than the benzotriazole aminals. These imines would then have been washed from the mixture prior to reaction with a Grignard reagent and so very little, if any, desired products would be observed.

Piao's polymer-supported, ether tethered benzotriazoles **13** and **14** and amide tethered benzotriazoles **18** and **19** were used to form substituted amines in the same manner as above but, along with Grignard reagents, sodium borohydride (NaBH$_4$) was also shown to be effective in cleaving the products from benzotriazole (Scheme 11.12).[8] An initial evaluation of the four resins using this cleavage method showed AM-PS derived resin **18** to give the highest yield, making it doubly attractive due to its synthesis from commercially available acid **17**. As can be seen, this amide resin was amenable to cleavage with both sodium borohydride and Grignard reagent (both allyl- and phenylmagnesium halide) although 30 equivalents of the latter were required. Resin **18** was then used in optimization studies to reveal the conditions shown in Scheme 11.12 as the best for adduct formation and cleavage. Work-up/purification included treatment with methanol at

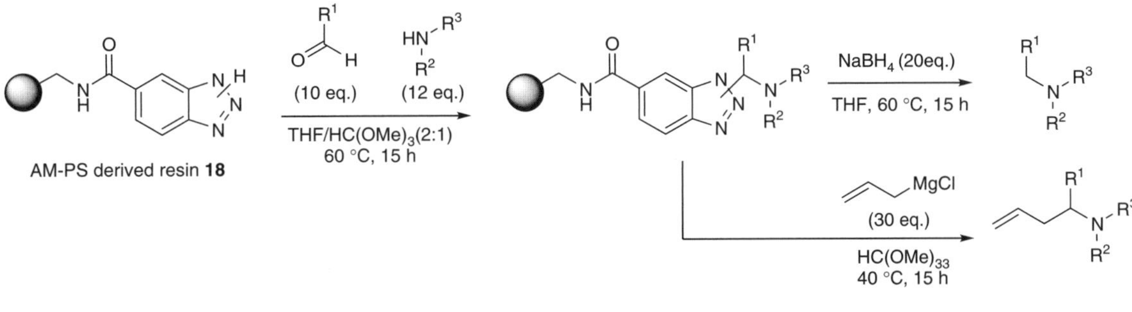

Scheme 11.12 Amide tethered benzotriazole **18** in Mannich-type reaction

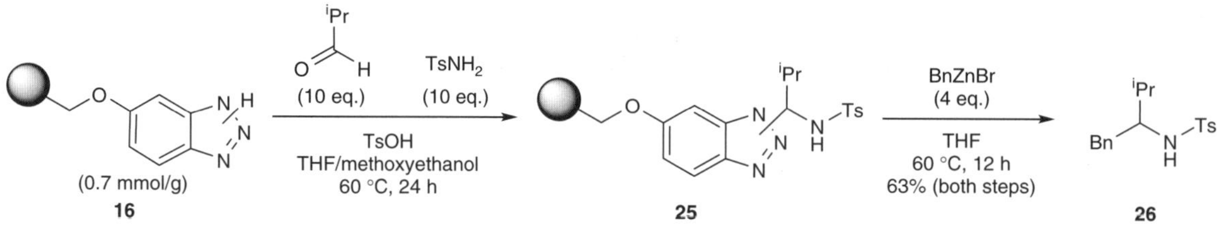

Scheme 11.13 Use of toluenesulfonamide and cleavage with a Reformatsky reagent

60 °C for one hour to break amine–borate complexes, dichloromethane (DCM) extraction from aqueous sodium carbonate solution then trapping/elution with ammonia in methanol from a strong cationic exchange resin column. The yields from the whole process (two steps and purification) ranged from 30 to 50%.

The authors went on to use this process in the formation of a library from a range of aldehydes with dibenzylamine as well as from a range of amines with 2-napthaldehyde and both cleavage conditions. Where successful, yields varied between 25 and 65% with purities of 13–100%. Of the aldehydes, only 5-bromo-2-furancarboxaldehyde failed to give a product, due to its polymerization in the reaction conditions, but of the amines, there were several failures and it was concluded that each might require its own optimized set of conditions. Katritzky noted this also; he formed aminals from amines (10 equiv.), polymer-supported ether tethered benzotriazole 16 (resin shown in Scheme 11.13) and formaldehyde or isobutyraldehyde (10 equiv.) at 60 °C in a mixture of THF and methoxyethanol (for increased swelling). However, he found the less reactive benzaldehyde required toluene at reflux and *p*-toluenesulfonic acid (TsOH) as a catalyst.[10] The aminals were cleaved using Grignard reagents *n*-butylmagnesium bromide or cyclopentylmagnesium chloride (4 equiv.). Yields were very good, ranging from 76 to 89%.

It was also shown that tosyl aminal 25 could be formed from toluenesulfonamide. This was then cleaved by reaction with a Reformatsky reagent to give tosylamine 26 (Scheme 11.13).[10]

Of further interest in this work was the observation that model benzotriazole 27 was formed as a mixture of regioisomers **a** and **b** (Figure 11.1) shown by two aminal carbon signals at 85.7 and 92.5 ppm in gel-phase ^{13}C NMR. Solid-supported benzotriazole 28, on the other hand, existed as only the 2-benzotriazole isomer with an aminal carbon signal at 92.3 ppm. This is explained as resulting from 'steric effects in the globular structure of cross-linked (1% divinylbenzene) polystyrene resin'.

Fang was interested in forming tetrahydroquinolines (THQs) using resin-bound benzotriazole after having already devised a solution-phase method employing the unbound reagent as a promoter for the coupling of two equivalents of arylamine with two equivalents of aldehyde.[12] Herein, the polymer-supported benzotriazoles acted only as catalysts, though the mechanism necessarily involved resin-bound aminal intermediates 29 and could be considered a one-pot resin capture and release reaction (Scheme 11.14).[7]

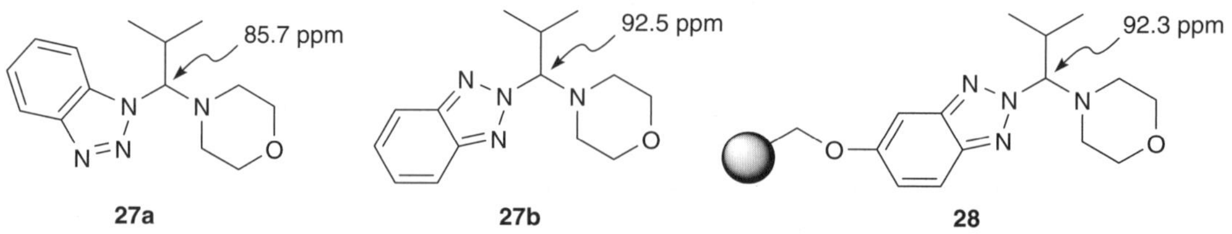

Figure 11.1 Isomers of benzotriazole aminals

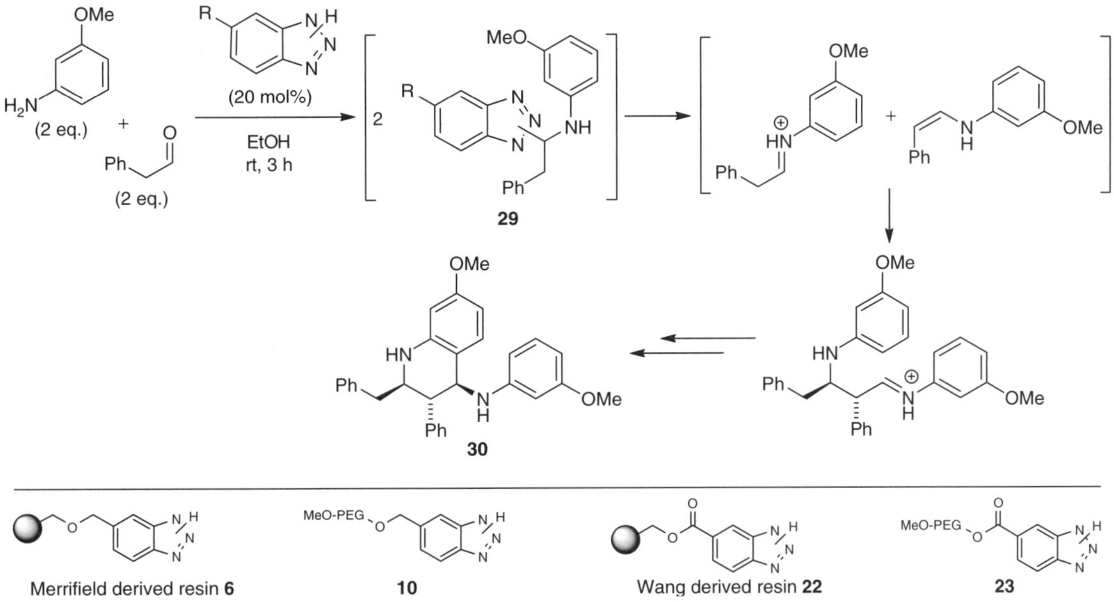

Scheme 11.14 Synthesis of tetrahydroquinolines

Ether tethered benzotriazoles **6** and **10** were tested alongside their free counterparts benzotriazole and 5-(benzyloxymethyl)benzotriazole. Ester tethered benzotriazoles **22** and **23** were also tested alongside ethylbenzotriazole-5-carboxylate. The solution-phase benzotriazoles gave higher yields than their equivalent resin-bound analogues but Merrifield resin derived ether-linked benzotriazole **6** gave 84% yield of the THQ **30** (the 2,3-trans-2,4-cis isomer only) and so almost equalled its free counterpart, which gave 90% product (97:3 2,3-*trans*-2,4-*cis*:2,3-*cis*-2,4-*cis*). Resin **6** was therefore used in the subsequent screening of a range of arylamines with phenylacetaldehyde, all giving yields of 82–89% and isomer ratios from 75:25 to 100:0.

11.3.2 Enolate acylation

Benzotriazole has been described as a 'tame halogen' substituent because its acylated analogues are stable but the benzotriazole still acts as a good leaving group.[1] It is for this reason that acylbenzotriazoles offer an alternative to unstable or unformable acid chlorides.[13]

Resin-supported carbon-carbon tethered benzotriazole **5** and ether tethered **16** were used in the acylation of enolates **32** (Scheme 11.15).[6] The advantage of using azolides **31** for this purpose is that less *O*-acylation takes place and, in this case, none was seen at all. In the range of azolides and enolates tested, ether tethered benzotriazole resin **16** gave the diketones **33** in 18–41% yield while carbon-carbon linked benzotriazole **5** gave them in 47–77% yield. This dramatic difference was not due to the swelling properties of either resin as they were both similar. It was instead attributed to the higher stability of **5**, which is less likely to undergo side reactions, such as deprotonation at the benzylic position, than ether tethered benzotriazoles.

11.3.3 Urea synthesis

Reaction of polymer-bound, amide tethered benzotriazoles **18** and **19** with phosgene afforded carbamoyl chlorides **34** which could be reacted *in situ* with anilines to give stable solid-supported benzotriazole aniline

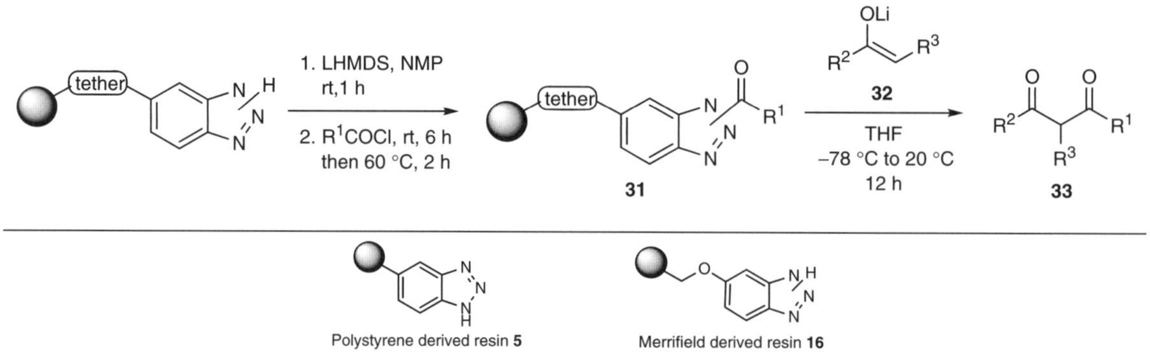

Scheme 11.15 Acylation of enolates with resin-bound benzotriazole azolides

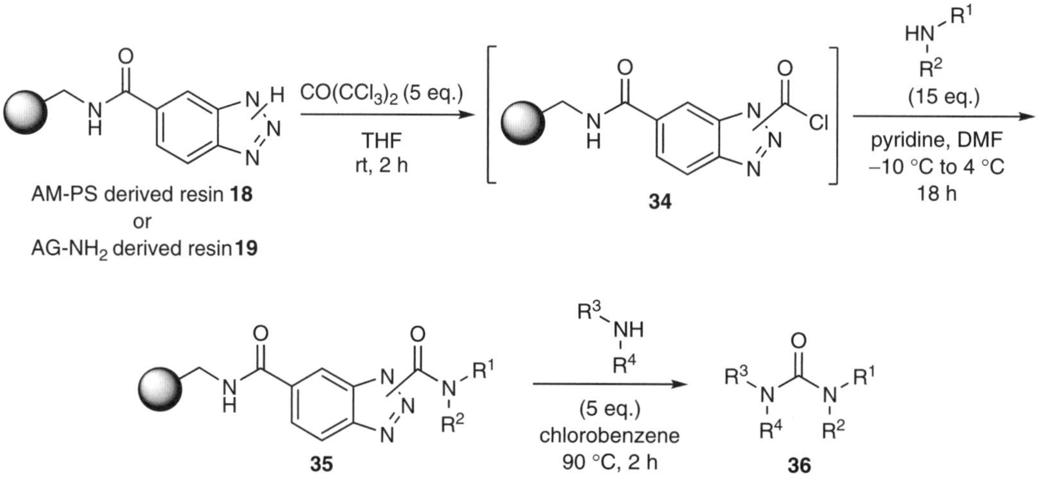

Scheme 11.16 Synthesis of unsymmetrical ureas

ureas **35** (Scheme 11.16).[11] Further reaction with another amine, this time at high temperature, facilitated cleavage from the resin to give unsymmetrical ureas **36**. Preliminary conditions showed AG-NH$_2$ derived resin **19** to be unstable to the high temperature conditions of the cleavage step, releasing polyoxyethylene residues and often rendering product ureas impure. Resin **18**, on the other hand, gave yields of 46–76% for the reaction of the carbamoyl chloride with *p*-methylaniline or *N*-methylaniline followed by benzylamine or piperidine. Less successful was the use of diphenylaniline in the first step, where lower final yields of 10–33% were obtained due to its steric hindrance and/or low reactivity. Also, use of 3,4-dichloroaniline in the second step gave low yields of 14–34%. The ureas were purified to 78–96% purity by trapping excess amines on a strong cationic exchanger (SCX).

Full optimization of the process was undertaken prior to its automation for library synthesis. The best conditions, shown in Scheme 11.16, were used with an ACT-496 synthesizer with which eight initial anilines were combined with 20 amines. 156 out of a possible 160 ureas were obtained, 138 of which were of >80% purity.

In summary, the resin-bound benzotriazoles described in the first section have been used as nucleophilic organocatalysts to promote a Mannich-type reaction between aldehydes and amines as well as a leaving group in resin-bound equivalents to acid chlorides for acylation of enolates and the synthesis of unsymmetrical ureas. In two cases, the techniques have lent themselves to automation for the formation of libraries of compounds – a technique very useful in the drug discovery process.

In conclusion, the resin-bound benzotriazoles have been shown to be readily synthesizable and applicable to a selection of the useful reactions for which free benzotriazole is used. Yields in these reactions range from poor to very good, often varying greatly with the R groups of the substrates used.

Typical Experimental Procedures

Preparation of C–C tethered benzotriazole resin 5 (Scheme 11.3)[6]

To a suspension of sodium hydride (500 mg, 20 mmol) in anhydrous DMF (20 ml) at 0 °C under nitrogen was added a solution of benzotriazole (2.38 g, 20 mmol) in DMF (10 ml) drop-wise via syringe. The resulting mixture was stirred at room temperature for 30 minutes then 1-chloro-2-bromoethane (20 mmol) was added and the mixture stirred at room temperature for 24 hours. The reaction was quenched with water (30 ml), extracted with diethyl ether (2 × 100 ml) and the organic fractions were washed with saturated sodium bicarbonate solution (2 × 50 ml) and water successively. The organic extracts were dried over magnesium sulfate, filtered and evaporated to dryness to give an oil which was purified by column chromatography on silica gel (hexanes/ethyl acetate 3:1) to give 1-(2-chloroethyl)-1H-1,2,3-benzotriazole as white needles (33%) and 2-(2-chloroethyl)-2H-1,2,3-benzotriazole as white needles (44%).

*A mixture of the former isomer (5 mmol), NBS (7.5 mmol) and concentrated sulfuric acid (1.25 mmol) was refluxed at 70 °C in TFA (3 ml) for 20 hours. The mixture was then allowed to cool to room temperature, diluted with DCM (50 ml) and washed with water (3 × 30 ml). The organic extracts were dried over magnesium sulfate, filtered and evaporated to dryness to give an oil which was purified by column chromatography on silica gel (hexanes/ethyl acetate 3:1) to give 5-bromo-1-(2-chloroethyl)-1H-1,2,3-benzotriazole **4** as white needles (58%).*

A mixture of phenylboronic acid resin 3[5] *(3 g, 4 mmol/g), bromide **4** (3.38 g, 13.2 mmol), palladium acetate (135 mg, 0.6 mmol), tris-o-tolylphosphine (365 mg, 1.2 mmol) and triethylamine (8.3 ml, 60 mmol) was heated in anhydrous DMF at 100 °C for 24 hours. The mixture was allowed to cool to room temperature and the resulting resin was filtered off, washed successively with DMF (50 ml), DMF-water 50% (100 ml), water (100 ml), 1 N hydrochloric acid (100 ml), water-methanol 50% (100 ml), methanol (100 ml), methanol-DCM 50% (100 ml), DCM (100 ml) and diethyl ether (100 ml). Loading 1.6 mmol/g.*

Preparation of ether tethered benzotriazole resin 16 (Scheme 11.7)[10]

To a suspension of Merrifield resin (1.58 g, 1 mmol/g) in DMA (30 ml) was added sodium 3-nitro-4-aminophenolate (0.7 g, 3.9 mmol) and the mixture was stirred at 80 °C for six hours. After cooling, the resin was filtered, washed sequentially with DMA, DMA-water 50%, THF-water 50%, THF and DCM then dried to give 1.75 g of resin-bound 2-nitroaniline. Loading 0.94 mmol/g.

*A suspension of this resin (0.4 g) and SnCl$_2$.2H$_2$O (1 g, 4.4 mmol) in DMA (20 ml) was stirred at 60 °C for five hours and then at room temperature overnight. After diluting with water (5 ml), the resin was filtered, washed sequentially with DMA, DMA-water 50%, THF-water 50%, THF and DCM then dried to give 0.38 g of resin-bound 1,2-phenylenediamine **15**. Loading 0.91 mmol/g.*

*To a suspension of **15** (0.2 g) in a mixture of dioxane (20 ml), acetic acid (5 ml) and concentrated hydrochloric acid (3.5 ml) was added a solution of isoamyl nitrite (2 ml, 15 mmol) in dioxane (10 ml). The* reaction mixture was stirred at room temperature for 24 hours and then heated to 50–60 °C for 0.5 hour. The resin was filtered, washed sequentially with dioxane, THF and DCM then dried to give 0.19 g of benzotriazole resin **16**. Loading 0.7 mmol/g.

Typical procedure for Mannich-type reaction and cleavage (Scheme 11.11)[10]

Benzotriazole resin **16** (0.2 g, 0.7 mmol/g), isobutyraldehyde (1.4 mmol) and morpholine (1.4 mmol) were heated to 60 °C in THF-methoxymethanol 50% for 12 hours. The resin was filtered and washed with THF (5 × 10 ml) and diethyl ether (5 × 15 ml) then dried to give resin-bound intermediate aminal **24a**. This was suspended in THF and n-butylmagnesium bromide (0.56 mmol) was added. The mixture was heated at reflux for four hours. After addition to ice-cold water and saturated ammonium chloride solution, the resin was filtered and washed with THF. The combined filtrates were extracted with diethyl ether (30 ml), the organic layer separated, washed with water and dried over anyhydrous sodium sulfate. The solvent was removed in vacuo to afford 4-(1-isopropylpentyl)morpholine (0.023 g, 82%).

Typical procedure for the acylation of enolates with resin-bound benzotriazole azolides 31a (Scheme 11.15)[6]

Preparation of resin-bound azolide: LHMDS (3 ml, 3 mmol, 1 N solution in hexanes) and benzotriazole resin **5** (2 mmol) pre-swelled in 1-methyl-2-pyrrolidinone (50 ml) were stirred for one hour at room temperature and then a solution of benzoyl chloride (3 mmol) in dry THF (10 ml) was added drop-wise. The slurry was stirred at room temperature for six hours and then at 60 °C for two hours. It was cooled and washed successively with DMF (50 ml), DMF-methanol 50% (100 ml), methanol (100 ml), water-methanol 50% (100 ml), methanol (100 ml) and diethyl ether (100 ml).

Acylation: A 50 ml round-bottom flask with septum inlet was charged with a solution of LDA (1 ml, 0.15 M solution in hexanes) in dry THF (10 ml) under nitrogen and cooled to −78 °C. A solution of acetophenone (1.5 mmol) in dry THF (15 ml) was added drop-wise. After stirring for one hour at this temperature, the resulting mixture was added dropwise to a slurry of the above resin-bound azolide **31a** (0.5 mmol) in dry THF (50 ml) by cannula. The mixture was stirred for another 12 hours while the temperature was allowed to rise to 20 °C. The resin was separated and washed with THF (20 ml). The organic layer was washed with 10% ammonium chloride solution (2 × 50 ml) and water (50 ml) then dried over magnesium sulfate. Chromatography on silica (hexanes/ethyl acetate 4:1) gave 1,3-diphenylpropane-1,3-dione **33a** (77%).

Typical procedure for the synthesis of unsymmetrical ureas 36 (Scheme 11.16)[11]

Preparation of resin-bound 1(2)-(4-toluidinocarbonyl)benzotriazole-5-carboxylic acid **35a**: AM-PS derived benzotriazole resin **18** (100 mg, 0.80 mmol/g, 0.08 mmol) was swollen with dry THF (1.0 ml) under nitrogen. Triphosgene (40 mg, 0.133 mmol) was rapidly added and the mixture shaken for two hours. After the mixture was cooled to −10 °C, pyridine (100 μl, 1.24 mmol) was added upon which precipitation of salts was immediately noticed. A solution of p-methylphenylamine (129 mg, 1.20 mmol) in dry DMF (0.70 ml) was slowly added to the reaction mixture which turned orange-red. After being shaken vigorously for a few minutes, the mixture was allowed to reach 4 °C and to react for 18 hours. It was then filtered, the resin rinsed with diethyl ether and DCM for four repeating cycles and dried under vacuum.

Urea synthesis: Piperidine (21 μl, 0.216 mmol) was added to the above described resin **35a** (60 mg, 0.043 mmol) which had been previously swollen with dry chlorobenzene (0.65 ml). The mixture was heated at

*90 °C and shaken for 18 hours. After cooling to rt, it was taken up in methanol (0.7 ml), poured into an SPE cartridge (1.0 g of SCX, 0.75 mequiv./g) and eluted with DCM (1.0 ml) and methanol (3.0 ml). After concentration at reduced pressure, N-(4-methylphenyl)tetrahydro-1(2H)-pyridinecarboxamide **36a** was obtained as a white solid (7.5 mg, 80%).*

References

[1] Katritzky, A. R., Lan, X., Yang, J. Z., and Denisko, O. V.; *Chem. Rev*. **1998**, *98*, 409.
[2] Kalir, R., Warshawsky, A., Fridkin, M., and Patchornik, A.; *Eur. J. Biochem*. **1975**, *59*, 55.
[3] Devocelle, M., McLoughlin, B. M., Sharkey, C. T., *et al*.; *Org. Biomol. Chem*. **2003**, *1*, 850.
[4] Schiemann, K., and Schowalter, H. D. H.; *J. Org. Chem*. **1999**, *64*, 4972.
[5] Farrall, M. J., and Fréchet, J. M. J.; *J. Org. Chem*. **1976**, *41*, 3877.
[6] Katritzky, A. R., Pastor, A., Voronkov, M., and Tymoshenko, D.; *J. Comb. Chem*. **2001**, *3*, 167.
[7] Talukdar, S., Chen, R.-J., Chen, C.-T., etal.; *J. Comb. Chem*. **2001**, *3*, 341.
[8] Paio, A., Zaramella, A., Ferritto, R., etal.; *J. Comb. Chem*. **1999**, *1*, 317.
[9] Kulagowski, J. J., Moody, C. J., and Rees, C. W.; *J. Chem. Soc. Perkin Trans. 1* **1985**, 2733.
[10] Katritzky, A. R., Belyakov, S. A., and Tymoshenko, D. O.; *J. Comb. Chem*. **1999**, *1*, 173.
[11] Paio, A., Crespo, R. F., Seneci, P., and Ciraco, M.; *J. Comb. Chem*. **2001**, *3*, 354.
[12] Talukdar, S., Chen, C.-T., and Fang, J.-M.; *J. Org. Chem*. **2000**, *65*, 3148.
[13] Katritzky, A. R., Suzuki, K., and Wang, Z.; *Synlett* **2005**, 1656.

12

Diversity Cleavage Strategies from Phosphorus Linkers

Patrick G. Steel and Tom M. Woods

Department of Chemistry, University of Durham, United Kingdom

12.1 Introduction

In addition to their many roles as ligands for transition metal catalysts, phosphorus reagents find many roles in organic chemistry directly participating in many important synthetic transformations. These include both carbon–carbon bond forming processes, notably the Wittig reaction[1, 2] and closely related olefination processes,[2, 3] and the many functional group interconversions promoted by phosphorus (III) reagents, including the Mitsunobu[4] and Michaelis Arbuzov reactions,[2, 5] as well as halogenation reactions of alcohols,[2, 6] thiols[2, 7] and carboxyacids.[2, 8]

As with solution-phase chemistry, the principal role of immobilised/polymer supported phosphorus chemistry is as heterogeneous supports for metal-based catalyst systems. However, largely reflecting the difficulties in dealing with phosphine oxide by-products, there have also been considerable efforts to use polymer-supported phosphorus reagents in other synthetic transformations. The first examples of immobilised phosphorus species appeared in the early 1960s and were simple polystyrene supported triphenylphosphine derivatives.[9] Since their advent, the majority (although not all) of strategies using immobilised phosphorus reagents have been based on these simple polystyrene species. There have been reports describing applications in the halogenation of alcohols, thiols and carboxylic acids,[10] the Mitsunobu[11] and Staudinger reactions[12] and, more recently as scavenger reagents in solid-phase organic synthesis strategies.[13] In addition to these supported reagent strategies, in which the phosphorus component is distinct from the substrate, there are situations in which a phosphorus atom forms an integral component of the linker unit. In most of these cases cleavage of the linker can lead to a range of substituents

Linker Strategies in Solid-Phase Organic Synthesis Edited by Peter Scott
© 2009 John Wiley & Sons, Ltd

being incorporated into the final product. The examples of this strategy are dominated by Wittig and Horner–Wadsworth–Emmons olefination chemistry. Recently, however, both transition metal catalysed cross-coupling cleavage from immobilised phosphonates and oxidative cleavage of acylcyanophosphoranes have been reported. In this chapter these diversity-oriented cleavage strategies are discussed in more detail.

12.2 Diversity cleavage through olefination reactions

The Wittig reaction is arguably the most important method of making alkenes in organic synthesis. The reaction proceeds via the formation of a phosphonium salt, which when deprotonated leads to a nucleophilic ylide or phosphorane species capable of attacking carbonyl compounds. Subsequent elimination to afford the alkene product is driven by the formation of a strong P=O bond. The selectivity of the new alkene formed from the reaction is dependent on the substituent on the carbanion of the ylide, with anion stabilising groups (stabilised ylides) leading to E-alkenes whilst unstabilised ylides afford Z-alkenes. Ylides can be split into various types according to the substituents on the phosphorus atom. Wittig reactions involve trialkyl or triaryl phosphoranes, Horner–Wittig reagents use diarylphosphine oxides, and the Horner–Wadsworth–Emmons (HWE) reaction is carried out with phosphonates. Both the Wittig and the HWE reactions have been carried out in a solid phase fashion using immobilised phosphorus reagents and are discussed in more detail below.

12.2.1 Diversity cleavage through the Wittig reaction

Isolable, immobilised phosphonium salts **2** and their respective phosphoranes **3** can be easily accessed from polymer-supported phosphines **1** on treatment with an appropriate alkyl halide followed by a base. When reacted with carbonyl compounds these reagents give access to a range of olefinic products **4**. Whilst there are many examples in which the use of such reagents lead to a supported alkene, these lie outside the scope of this chapter and only those procedures in which the phosphorus atom acts as a linker and the product is no longer bound to the resin are considered (Scheme 12.1).

A particular advantage of using such immobilised Wittig reagents is the relatively simple work-up procedures. In particular, the phosphine oxide by-products **5**, which can be difficult to separate from the products following conventional solution-phase Wittig reactions, remain polymer bound and are easily separated from the desired olefins by filtration. Moreover, the resins can be recycled and reused if desired. Direct reduction of the phosphine oxide **5** to the phosphine **1** can be effected by treatment with trichlorosilane in refluxing benzene.[10, 14, 15] Recycling can also be achieved indirectly via conversion to the supported phosphine dichloride and then subsequent reduction (Scheme 12.2).[16, 17]

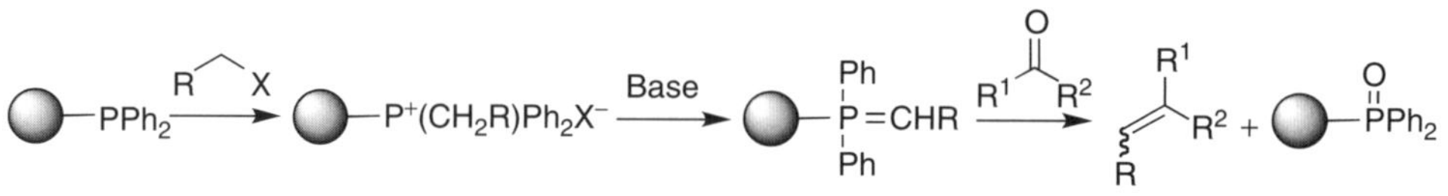

Scheme 12.1 *Diversity cleavage via the Wittig reaction*

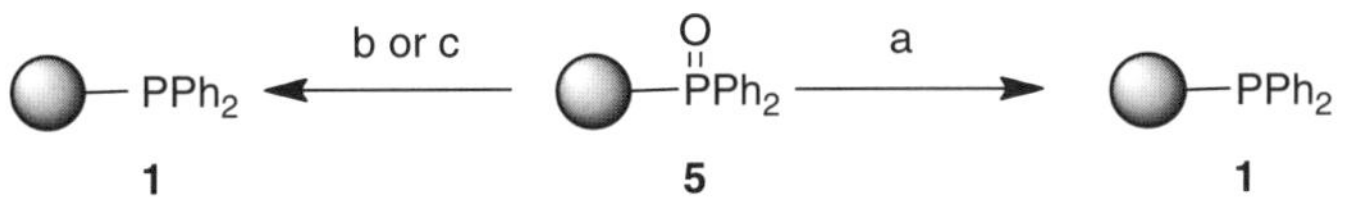

Diversity Cleavage Strategies from Phosphorus Linkers 333

a) i) HSiCl$_3$, C$_6$H$_6$, NEt$_3$, 0 °C. ii) reflux, 7days. b) i) COCl$_2$. ii) elemental phosphorus.
c) i) (COCl)$_2$. ii) DIBAL.

Scheme 12.2 *Recycling strategies for supported phosphine resins*

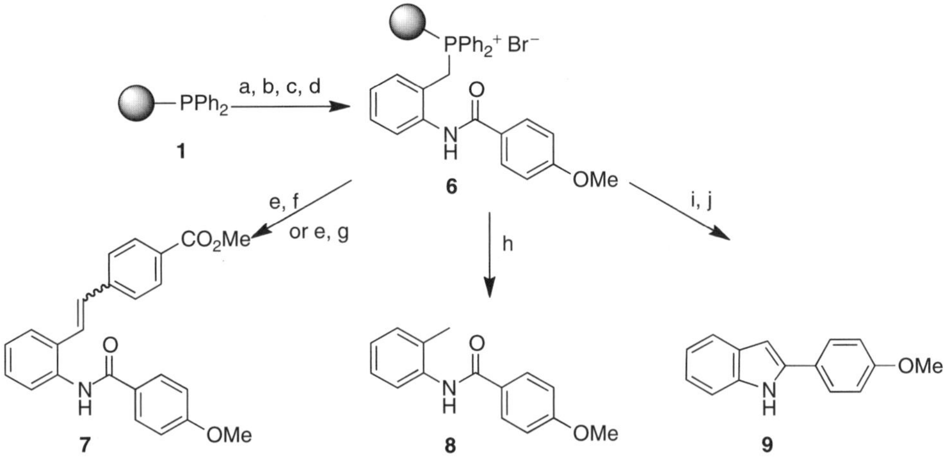

a) 2-Nitrobenzyl bromide, DMF, 70 °C, 48 h. b) Na$_2$S$_2$O$_4$, EtOH, reflux, 90 mins. c) HBr, MeOH, dioxane.
d) 4-Methoxybenzoyl chloride, py, DCM, 5 h. e) Methyl 4-formylbenzoate (2eq), NaOMe (2eq), MeOH, reflux, 2 h.
f) Girard's reagent T (Carboxymethyl) trimethylammonium Chloride Hydrazide) (3eq), AcOH, 18 h. g) Aminomethyl resin, AcOH, MeOH/dioxane, 18 h. h) NaOMe, MeOH, reflux, 4 h. i) Toluene, DMF, distill. j) KO^{t}Bu, reflux, 45 mins.

Scheme 12.3 *Diversity cleavage from supported phosphonium salts*

The earliest example of a solid-phase Wittig reaction was reported in 1971. Preparation of a diphenylbenzyl-phosphonium salt was achieved in almost quantitative yield from cross-linked polystyrene diphenylphosphine on treatment with benzylchloride. The resin was then treated with stoichiometric amounts of either potassium *t*-butoxide or sodium hydride followed by benzaldehyde affording a mixture of *cis* and *trans*-stilbenes in 40% and 60% yields, respectively (Table 12.1, entries 1 and 2).[18] The following year, McKinley and coworkers further demonstrated this polystyrene based methodology describing the reaction of three immobilised phosphoranes with a variety of carbonyl compounds (Table 12.1, entries 3–11).[19] Since then there have been several examples of this simple form of diversity cleavage (Table 12.1, entries 12–19). In general, when reacted with approximately stoichiometric quantities of carbonyl compounds these phosphoranes afforded the alkene products in good yields, which are comparable to the equivalent solution-phase reactions. Similarly, the stereochemistry of these polymeric Wittig reactions appears to mirror the corresponding solution-phase reactions, although increasing the degree of cross-linking in the polymer support has been reported to increase the *E*-selectivity.[15] In the context of stereochemical diversity, Heitz demonstrated

Table 12.1 Selected examples of polystyrene-supported phosphoranes in the Wittig reaction

Entry	R^1	X	Base	R^2R^3CO	Product	Yield (%)	Ref
1	Ph	Cl	NaH	PhCHO	PhCH=CHPh	60	[18]
2	Ph	Cl	KO^{t}Bu	PhCHO	PhCH=CHPh	40	[18]
3	H	I	Me$_2$SO$^-$Na$^+$	Cyclohexanone	Methylenecyclohexanone	90^a	[19]
4	H	I	Me$_2$SO$^-$Na$^+$	Ph$_2$CO	Ph$_2$C=CH$_2$	94^a	[19]
5	H	I	Me$_2$SO$^-$Na$^+$	PhCHO	PhCH=CH$_2$	82^a	[19]
6	H	I	Me$_2$SO$^-$Na$^+$	PhCOMe	PhC(Me)=CH$_2$	76^a	[19]
7	H	I	Me$_2$SO$^-$Na$^+$	Me(CH$_2$)$_4$CHO	Me(CH$_2$)$_4$CH=CH$_2$	58^a	[19]
8	Me	I	Me$_2$SO$^-$Na$^+$	Cyclohexanone	ethylenecyclohexanone	90^a	[19]
9	Me	I	Me$_2$SO$^-$Na$^+$	Ph$_2$CO	Ph$_2$C=CHMe	93^a	[19]
10	Me	I	Me$_2$SO$^-$Na$^+$	PhCHO	PhCH=CHMe	nr^b	[19]
11	Ph	Br	Me$_2$SO$^-$Na$^+$	PhCHO	PhCH=CHPh	72^a	[19]
12	Bu	Br	BuLi	PhCHO	BuCH=C(Me)Ph	nr^c	[14]
13	Bu	Br	BuLi	PhCHO	BuCH=C(Me)Ph	nr^d	[14]
14	n-Bu	Br	n-BuLi	PhCOMe	C$_3$H$_7$CH=C(Me)Ph	100^e	[20]
15	Geranyl	Br	n-BuLi	p-Cl-C$_6$H$_4$CHO	Me$_2$CCH(CH$_2$)$_2$(CH$_2$)$_2$=Ph-Cl	11^e/48^f	[20]
16	Ph	Cl	NaOH$_{(aq)}$^{g}	PhCHO	PhCH=CHPh	92	[21]
17	CH$_2$CH	Br	NaOH$_{(aq)}$^{h}	p-Cl-C$_6$H$_4$CHO	C$_3$H$_4$CH=CHPhCl	78	[21]
18	Ph	Br	NaOMe	PhCHO	PhCH=CHPh	93^i	[15]
19	H	I	Me$_2$SO$^-$Na$^+$	PhCH=CHCHO	PhCH=CHCH=CH$_2$	95^i	[15]
20	Ph	Br	NaHMDS	C$_7$H$_{15}$CHO	C$_7$H$_{15}$CH=CHPh	87	[22]
21	CN	Br	NaHMDS	p-MeO-C$_6$H$_4$CHO	p-MeO-C$_6$H$_4$CH=CHCN	80	[22]

^{a}Yields are based on the amount of carbonyl compound consumed.
^{b}Isolated yield not given, only the geometric isomer ratio; cis:trans 84%:16%.
^{c}Isolated yield not given, only the geometric isomer ratio; cis:trans 14%:86%.
^{d}Isolated yield not given, only the geometric isomer ratio; cis:trans 97.5%:2.5%.
^{e}Using a polystyrene cross-linked with 2% divinylbenzene.
^{f}Using a polystyrene cross-linked with 0.5% divinylbenzene.
^{g}The use of a phase transfer catalyst was required.
^{h}No phase transfer catalyst was employed.
^{i}Yield calculated by GLC analysis.

that selective formation of either the Z or E alkene from a common phosphonium salt is possible through simple modifications of the reaction conditions.[14]

Typical Experimental Procedures
Polystyrene-supported triphenylphosphonium salts[22]

*To commercial diphenylphosphine derivatised polystyrene **1** (1–2 g, 3–6 mmol phosphine) suspended in dry DMF (10–20 ml) was added 2–4 equiv. of the alkyl halide. The slurry was stirred for 48 hours at 50–70 °C. The support was then filtered under argon and extensively washed with dry toluene (40 ml), dry dichloromethane (40 ml) and dry diethyl ether (60 ml). The grey to brown powders were dried in vacuo. According to the weight increase the loading was estimated to be in the range of 1.8–3.0 mmol phosphonium salt per gramme support.*

Cleavage from polystyrene-supported triphenylphosphonium salts using a Wittig reaction (Table 12.1, Entry 20) [22]

A sintered Teflon frit reaction tube was charged with the polymer-supported phosphonium salt (300 mg, 0.54–0.9 mmol phosphonium salt). The support was placed under an argon atmosphere and washed with dry THF (8 ml). Addition of a 1 M solution of sodium bis(trimethylsilyl)amide (NaHMDS) in THF (1.8 ml) at room temperature effected the generation of the ylide within 30 minutes. Excess base was carefully removed by extensive washing of the reddish brown to black support with dry THF (25 ml). The resultant supported phosphorane was resuspended in dry THF (3 ml) and the carbonyl compound (0.25–0.3 mmol) was added at room temperature. According to GLC analysis the olefination reaction was generally complete within 20–40 minutes (conversion >95%). The olefin containing solution was collected together with additional support washings (15 ml of dry THF) and filtered through a small amount of silica gel (0.5–1 g). Evaporation in vacuo furnished the crude olefins.

2-Phenyl-2-hexene[14]

(a) *cis- selective (Table 12.1, Entry 13): Polymer-supported triphenyl-n-butylphosphonium bromide[20] (20 g, 0.0337 mol triphenylphosphine groups) was suspended in dry dioxane (50 ml) and treated with n-butyllithium in hexane (20%, 15 ml) under nitrogen at room temperature. The mixture was kept turbulent by a stream of nitrogen introduced from below for 30 minutes and was then washed with THF (1.5 l) freshly distilled from butyllithium/lithium. After this washing-out procedure freshly distilled acetophenone was added as a solution in dioxane and 2-phenyl-2-hexene was obtained in quantitative yield. A cis: trans ratio of 97.5:2.5 was determined by GC.*

(b) *trans- selective (Table 12.1, Entry 12): Polymer-supported triphenyl-n-butylphosphonium bromide (20 g, 0.0337 mol triphenylphosphine groups) in dry THF (50 ml) under nitrogen was treated with a solution of butyllithium (0.0337 mol) in n-hexane (10.8 ml) to give the polymer-supported triphenylphosphonium-ylid. Stirring was continued throughout the reaction. After 20 min the reaction mixture was cooled to −78 °C and acetophenone (4.0 g, 0.0337 mol) added. After a further 15 minutes a further portion of butyllithium in n-hexane (0.0337 mol) was added to form the betaine/lithium bromide adduct. The reaction mixture was warmed up and then cooled again to −30 °C, treated with finely powdered anhydrous lithium perchlorate (3.6 g), and kept for 60 minutes at −30 °C. At −78 °C ether saturated with HCl (at −20 °C, 100 ml) was added. Then at room temperature potassium t-butoxide (5.2 g) was added and after two hours 2-phenyl-2-hexene was obtained in 59% yield. A cis:trans-ratio of 14:86 was determined by GC.*

Whilst generation of diversity through the use of different aldehydes in the cleavage step has been demonstrated, the majority of examples use only unfunctionalised alkyl halides to form the phosphorane. Consequently, the potential for this mode of diversity cleavage to be employed in a truly combinatorial fashion remains to be fully established. The possibilities for this are illustrated in a report by Hughes which describes the diversity cleavage of *ortho*-aminotolylphosphonium salts **6** via both inter and intramolecular strategies leading to products **7** and **9**.[23] Moreover, this report also illustrated that phosphonium salts can provide greater diversity on cleavage than just alkenes through Wittig chemistry, with reaction with sodium methoxide/methanol affording alkane **8**, (hydrolysis of the carbon–phosphorus bond is often seen as a minor side reaction in the Wittig reaction[21]), Scheme 12.3.

12.2.2 Diversity cleavage using the Horner–Wadsworth–Emmons reaction

The olefination of carbonyls using a phosphonate ester containing an electron withdrawing group at the α-position to the nucleophilic carbanion is known as the Horner–Wadsworth–Emmons reaction. Deprotonation of such a phosphonate ester leads to stabilised enolate type anions which, when reacted with aldehydes and ketones, lead predominantly to *E*-alkenes and the formation of α,β-unsaturated carbonyl compounds. It should come as no surprise given its similarity to the Wittig reaction, that the solid-phase HWE reaction is also well established as a diversity generating step. The HWE reagent may be supported in several ways including through the carboxylate ester, through the phosphonate linkage and via anion capture on an ion exchange resin. The first of these leads to a supported alkene and thus falls outside the scope of this review.

Provided the pK_a of the phosphonate α-C-H is in the range $\sim$6–9 it may easily be retained on an anion exchange resin **11**. Release from the resin is then achieved by reaction with a range of aldehydes and ketones to give the alkenes in good yields **13**, Scheme 12.4. As with the solution phase version the alkenes are formed with high *E* selectivity.[24]

Whilst the above catch and release version is amenable to continuous mode operation, there are limits to the ability to incorporate further diversity in the substrate. More classical diversity cleavage can be achieved with the phosphonate immobilised through the phosphorus atom. This methodology has seen a number of applications both in inter[25] and intramolecular fashion,[26] (Scheme 12.5, equations a and b). Whilst the HWE reaction is normally *E*-selective, use of either perfluoroalkyl or aryl phosphonate esters leads to high *Z*-selectivity. Combining these strategies with the solid phase approach provides a diversity synthesis of *Z*-selective α,β-unsaturated esters as demonstrated by Taylor,[27] (Scheme 12.5, equation c).

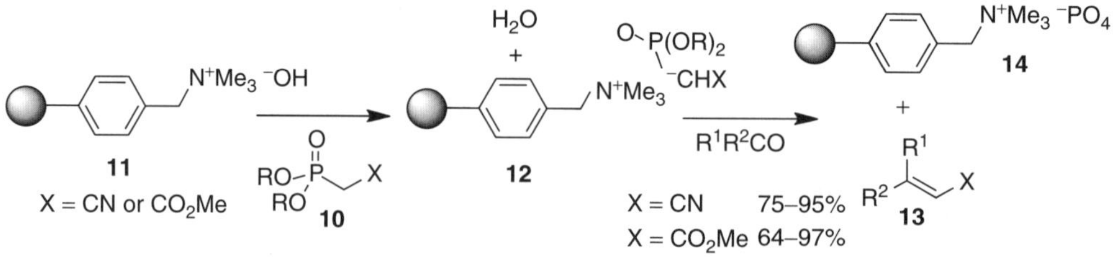

Scheme 12.4 *Phosphonate anion catch and release strategies*

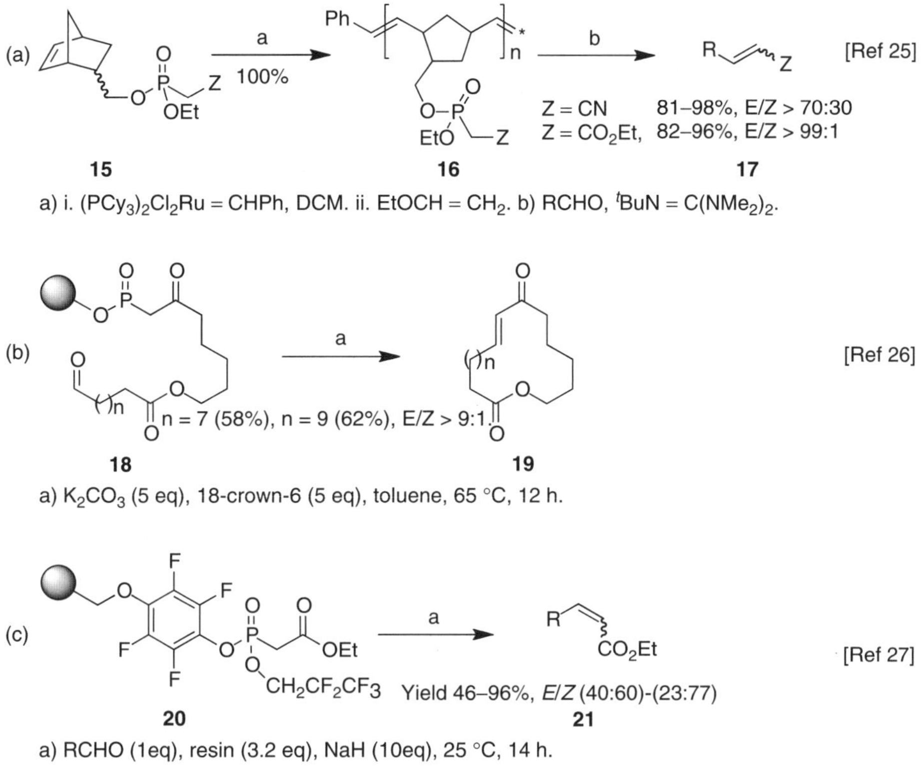

Diversity Cleavage Strategies from Phosphorus Linkers 337

a) i. (PCy$_3$)$_2$Cl$_2$Ru = CHPh, DCM. ii. EtOCH = CH$_2$. b) RCHO, ^{t}BuN = C(NMe$_2$)$_2$.

a) K$_2$CO$_3$ (5 eq), 18-crown-6 (5 eq), toluene, 65 °C, 12 h.

a) RCHO (1eq), resin (3.2 eq), NaH (10eq), 25 °C, 14 h.

Scheme 12.5 Diversity cleavage via polymer-supported HWE reactions

Typical Experimental Procedures

E- Ethyl 4-nitrocinnamate 17 (Scheme 12.5a)[25]

4-Nitrobenzaldehyde (0.045 mmol) and tert-butyltetramethylguanidine (0.09 mmol) were added to the ROMPGEL phosphonoacetate **16** (Z = CO$_2$Et, 0.09 mmol) in acetonitrile (1.5 ml). After stirring for four hours, the mixture was filtered through a 200 mg silica cartridge (Alltech Associates, cat. no. 209150), the silica was washed with ethyl acetate (10 ml), and the combined washings were evaporated in a stream of nitrogen to afford ethyl 4-nitrocinnamate (94%).

Macrolactone 19 (Scheme 12.5b)[26]

To a suspension of ketophosphonate resin **18** (1.0 equiv.) in toluene (50 ml/mmol) was added potassium carbonate (5.0 equiv.) and 18-Crown-6 (1,4,7,10,13,16-hexaoxacyclooctadecane) (5.0 equiv.). The mixture was stirred at 65 °C for 12 hours. Completion of the reaction was determined by IR analysis of the resin, monitoring the aldehyde carbonyl band. The resin was filtered and washed with hexanes (50 ml/mmol). The filtrate was poured onto a pad of silica gel, which was washed with hexanes (4 × 5 ml). Finally, the products were eluted from the silica gel with a mixture of diethyl ether (Et$_2$O)-hexanes (1:1).

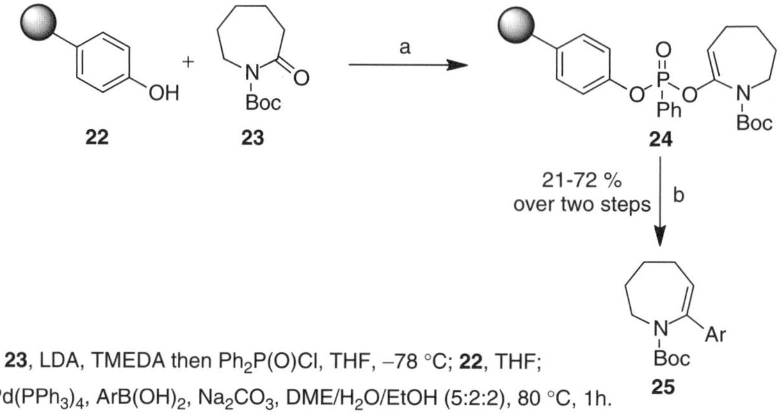

Scheme 12.6 *Diversity cleavage from polymer-supported enol phosphonates*

a) i. **23**, LDA, TMEDA then Ph$_2$P(O)Cl, THF, −78 °C; **22**, THF;
b) Pd(PPh$_3$)$_4$, ArB(OH)$_2$, Na$_2$CO$_3$, DME/H$_2$O/EtOH (5:2:2), 80 °C, 1h.

Z- Ethyl 4-nitrocinnamate 21 (Scheme 12.5c) [27]

To fluorinated phosphonate resin **20** (1.669 g, 0.766 mmol), pre-swollen in THF (20 ml), at 0 °C was added pre-washed sodium hydride (22 mg, 0.917 mmol). The mixture was stirred gently (<200 rpm) for 45 minutes at 0 °C, then solid 4-nitrobenzaldehyde (46 mg, 0.304 mmol) was added. The reaction mixture was stirred gently at 0 °C for three hours, then one hour at room temperature. The resin was filtered, washed with diethyl ether and the organic phase was concentrated in vacuo to afford ethyl 4-nitrocinnamate (57 mg, 85%, E/Z = 27:73) as a yellow oil.

12.3 Diversity cleavage of enol phosphonates through palladium catalysed cross-coupling reactions

Enol phosphonates have long been known to be effective electrophilic partners in solution-phase cross-coupling strategies with a number of different organometallic reagents.[28] In 2004, Steel and coworkers demonstrated that a similar protocol could be applied in a simple solid-phase diversity linker strategy. It was shown that polystyrene-bound lactam enol phosphonates **24** can be cleaved from the resin under Suzuki cross-coupling conditions affording the desired aryl enamines **25** in moderate to good yields (21–72%, Scheme 12.6).[29] The polystyrene bound reagent was found to be stable and storable for several months showing no loss in activity in cross-coupling reactions over this time. Moreover, these resins may be easily monitored using ^{31}P NMR spectroscopy using the characteristic, if somewhat broad, signal at δ ∼ 12 ppm for the desired enol phosphonate **14** as a convenient marker.

Typical Experimental Procedures[29]

Enol phosphonate resin 24 (Scheme 12.6)

To a cold (−78 °C) solution of N-Boc-caprolactam **23** (14 mmol, 4 equiv.) and N,N,N′,N′-tetramethylethylenediamine (16.8 mmol, 4.8 equiv.) in dry THF (50 ml) was added a solution of *lithium diisopropylamide (LDA) (16.8 mmol, 4.8 equiv.). This mixture was stirred at $-78\,°C$ for 40 minutes and phenylphosphonic dichloride (12.6 mmol, 3.6 equiv.) was added drop-wise. The reaction mixture was stirred for 0.5 hour at $-78\,°C$ and then at room temperature for one hour. It was then transferred via a canula into a suspension of phenol on polystyrene resin 22 (3.5 mmol/g, 1.0 g, 3.5 mmol) and dry TEA (7 mmol, 2 equiv.) in THF (10 ml). The suspension was stirred at room temperature for 18 hours. The reaction mixture was then filtered. The beads were washed with methanol (3 × 20 ml), DMF (3 × 20 ml), THF (3 × 20 ml), and dry ether (3 × 20 ml) and then dried to give the title resin.*

Typical Suzuki-Miyaura reaction with resin 24. Preparation of 7-(4'-methylphenyl)-2,3,4,5-tetrahydro-azepine-1-carboxylic acid tert-butyl ester (Scheme 12.6)

A suspension of resin 24 (1 g, 1.61 mmol), tolylboronic acid (4.83 mmol, 3 equiv.) and sodium carbonate (6.44 mmol, 4 equiv.) in dimethoxyethane (DME) (20 ml), water (8 ml), ethanol (8 ml) was degassed with argon for 30 minutes. Pd(PPh$_3$)$_4$ (0.16 mmol, 0.1 equiv.) was added and the reaction mixture was refluxed for one hour and then cooled to room temperature. The resin was filtered and washed with THF and ether. The combined filtrate was concentrated and extracted with ethyl acetate/brine. The organic phase was dried (magnesium sulfate), filtered and evaporated. Flash chromatography on silica (1–3% ethyl acetate/petrol ether) gave the title compound as a white solid as a 4:1 mixture of two rotomers.

12.4 Oxidative diversity cleavage of cyanophosphoranes

Whilst phosphoranes are most commonly employed in the Wittig reaction, other modes of cleavage can be employed to introduce other functionality. In addition to the base induced traceless cleavage of phosphoranes described earlier, Rademann has described an oxidative diversity cleavage strategy from immobilised acyl cyanophosphoranes 27, (Scheme 12.7).[30] The stable, storable cyanophosphorane 26 was synthesised from polystyrene-supported diphenylphosphine and, functioning as a polymer-supported acylanion equivalent, could be coupled with a variety (R^1) of activated Fmoc protected amino acids giving rise to a new carbon–carbon bond. Standard peptide chemistry allows further diversity to be introduced at this stage. Subsequent oxidative diversity cleavage from the resin was achieved by treating a suspension of the resin with ozone or dimethyldioxirane and the desired nucleophile to afford α-keto esters, α-keto amides and α-keto thioesters 28.

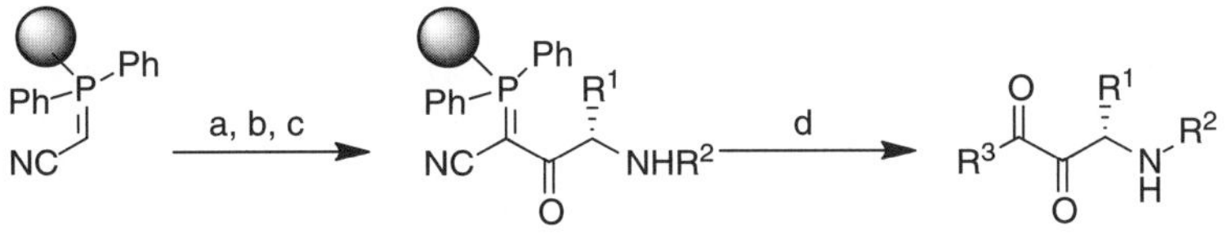

a) FmocNHCHR^{1}CO$_2$H, EDC, DMAP, DCM. b) Piperidine, DMF. c) Acylations (R^2). d) Nucleophile (R^{3}OH or R^{3}NH$_2$), DCM, O$_3$, –78 °C then r.t. for 4h or d) Nucleophile (R^{3}OH or R^{3}NH$_2$), DCM, DMDO, r.t. 4h. (11 examples 30–65% yield overall from 26).

Scheme 12.7 *Oxidative diversity cleavage from supported cyanophosphoranes*

Typical Experimental Procedures[30]

Cyanophosphorane polystyrene 26

Triphenylphosphane polystyrene (0.5 g, 1.2 mmol/g, 0.6 mmol, 1% divinylbenzene, 100–200 mesh) was suspended in dry toluene (3 ml). After addition of bromoacetonitrile (199 µl, 3 mmol), the vial was sealed and heated by microwave irradiation (15 minutes, 150 °C). Following cooling to room temperature the vial was opened. The resin was collected by filtration, washed with dry toluene, dichloromethane (DCM) and diethyl ether (3 × each) and dried. Elemental analysis (%) calculated: N 1.05; found: 1.02. The obtained triphenylphosphonium bromide resin was suspended in dry DCM (5 ml) and triethylamine (418 µl, 3 mmol) was added. The mixture was agitated (two hours, room temperature), the solid collected by filtration, washed (methanol, THF, DCM), and dried in vacuo.

Preparation of Fmoc protected acyl cyanophosphorane polystyrene 27 (Scheme 12.7)

EDCI (101 mg, 0.525 mmol), DMAP (10 mg, 0.079 mmol), and Fmoc protected amino acid (0.525 mmol) were dissolved in dry DCM (3 ml) and stirred for 5 minutes. Cyanophosphorane polystyrene 26 (100 mg, 1.05 mmol/g, 0.105 mmol) was added and the reaction mixture was agitated (12 hours, room temperature). The resin was collected by filtration, washed (methanol, DMF, THF, DCM and diethyl ether) and dried in vacuo. Coupling yields were measured by spectroscopic Fmoc determination.

Typical procedure for oxidative cleavage of resins yielding α-keto methyl esters 28 (Scheme 12.7)

Fmoc protected acyl cyanophosphorane polystyrene 27 was suspended in a mixture of dry DCM and the respective alcohol (2:1) under a nitrogen atmosphere. At −78 °C the suspension was purged with ozone until a blue-green color remains (5–10 minutes). Excess ozone was removed by a stream of nitrogen. After warming to room temperature the mixture was stirred (four hours) and filtered. The resin was washed (DCM) and the solvents were removed in vacuo to afford the title ester.

References

[1] (a) Wittig, G.; *Angew. Chem. Int. Ed.* **1956**, *68*, 505. (b) Wittig, G.; *Experientia.* **1956**, *12*, 41.
[2] For a review see Cadogan J. I. G. (Ed.); *Organophosphorous Reagents in Organic Synthesis*; Academic Press, New York, **1979**.
[3] (a) Wadsworth, W. S., and Emmons, W. D.; *J. Am. Chem. Soc.* **1961**, *83*, 1733. (b) Horner, L., Hoffmann, H., Wippel, H. G., and Klahre, G.; *Chem. Ber.* **1959**, *92*, 2499.
[4] For reviews see (a) Mitsunobu, O.; *Synthesis.* **1981**, 1. (b) Hughes, D. L.; *Org. React.* **1992**, *42*, 335. (c) Hughes, D. L.; *Org. Prep. Proced. Int.* **1996**, *28*, 127.
[5] (a) Michaelis, A., and Kahne, R.; *Ber.* **1898**, *31*, 1048. (b) Arbuzov, A. E.; *J. Russ. Phys. Chem. Soc.* **1906**, *38*, 687.
[6] (a) Lee, J. B., and Nolan, T. J.; *Can. J. Chem.* **1966**, *44*, 1331. (b) Hooz, J., and Gilani, S. S. H.; *Can. J. Chem.* **1968**, *46*, 86. (c) Downie, I. M., Holmes, J. B., and Lee, J. B.; *Chem. Ind. (London).* **1966**, 900.
[7] Weiss, R. G., and Snyder, E. I.; *Chem. Commun.* **1968**, 1358.
[8] Lee, J. B.; *J. Am. Chem. Soc.* **1966**, *88*, 3440.
[9] (a) Senear, A. E., Valient, W., and Wirth, J.; *J. Org. Chem.* **1960**, 2001. (b) Marcus, R., and Rabinowitz, R.; *J. Org. Chem.* **1961**, 4157.

[10] (a) Harrison, C. R., Hodge, P., Hunt, B. J., et al.; *J. Org. Chem.* **1983**, *48*, 3721. (b) Regen, S. L.,and Lee, D. P.; *J. Org. Chem.* **1975**, *40*, 1669.
[11] (a) Amos, R. A., Emblidge, R. W., and Havens, N.; *J. Org. Chem.* **1983**, *48*, 3598. (b) Tunoori, A. R., Dutta, D., and Georg, G. I.; *Tetrahedron Lett.* **1998**, *39*, 8751. (c) Lizarzaburu, M. E., and Shuttleworth, S. J.; *Tetrahedron. Lett.* **2002**, *43*, 2157.
[12] Wentworth Jr, P., Vandersteen, A. M., and Janda, K. D.; *Chem. Commun.* **1997**, 759.
[13] (a) Guinó, M., and Hii, K. K. M.; *Tetrahedron Lett.* **2005**, *46*, 6911. (b) Westhus, M., Gonthier, E., Brohm, D., and Breinbauer, R.; *Tetrahedron Lett.* **2004**, *45*, 3141.
[14] Heitz, W., and Michels, R.; *Liebigs. Ann. Chem.* **1973**, 227.
[15] Bernard, M., and Ford, W. T.; *J. Org. Chem.* **1983**, *48*, 326.
[16] Pommer, H., and Nurrenbach, A.; *Pure Appl. Chem.* **1975**, *43*, 527.
[17] Kobayashi, S., Suzuki, M., and Saegusa, T.; *Polym. Bull.* **1981**, *4*, 315.
[18] Camps, F., Castells, J., Font, J., and Vela, F.; *Tetrahedron Lett.* **1971**, 1715.
[19] McKinley, S. V., Rakshys, and Jun. J. W.; *J. Chem. Soc., Chem. Comm.* **1972**, 134.
[20] Heitz, W., and Michels, R.; *Angew. Chem. Int. Ed. Engl.* **1972**, *11*, 298.
[21] Clarke, S. D., Harrison, C. R., and Hodge, P.; *Tetrahedron Lett.* **1980**, *21*, 1375.
[22] Bolli, M. H., and Ley, S. V.; *J. Chem. Soc., Perkin Trans. 1.* **1998**, 2243.
[23] Hughes I.; *Tetrahedron Lett.* **1996**, *37*, 7595.
[24] Cainelli, G., Contento, M., Manescalchi, F., and Regnoli, R.; *J. Chem. Soc., Perkin Trans. 1.* **1980**, 2516.
[25] Barrett, A. G. M., Cramp, S. M., Roberts, R. S., and Zecri, F. J.; *Org. Lett.* **1999**, *1*, 579.
[26] Nicolaou, K. C., Pastor, J., Winssinger, N., and Murphy, F.; *J. Am. Chem. Soc.* **1998**, *120*, 5132.
[27] Martina, S. L. X., and Taylor, R. J. K.; *Tetrahedron Lett.* **2004**, *45*, 3279.
[28] (a) Nicolaou, K. C., Shi, G.-Q., Gunzer, J. L., et al.; *J. Am. Chem. Soc.* **1997**, *119*, 5467. (b) Nicolaou, K. C., Shi, G.-Q., Namoto, K., and Bernal, F.; *Chem. Commun.* **1998**, 1757. (c) Lepifre, F., Buon, C., Rabot, R., et al.; *Tetrahedron Lett.* **1999**, *40*, 6373. (d) Lepifre, F., Clavier, S., Bouyssou, P., and Coudert, G.; *Tetrahedron* **2001**, *57*, 6969. (e) Galbo, F. L., Occhiato, E. G., Guarna, A., and Faggi, C.; *J. Org. Chem.* **2003**, *68*, 6360.
[29] Campbell, I. B., Guo, J., Jones, E., and Steel, P. G.; *Org. Biomol. Chem.* **2004**, *2*, 2725.
[30] Rademann, J., and Weik, S.; *Angew. Chem. Int. Ed.* **2003**, *42*, 2491.

13
Sulfur Linker Units

Peter J. H. Scott

Department of Radiology, University of Michigan, USA

13.1 Introduction

Sulfur-based linker units have been widely used in solid-phase organic synthesis (SPOS) and have been the subject of a number of recent reviews, typically being grouped with selenium linker units (Chapters 14 and 15) due to similarities in cleavage strategies.[1–3] In addition, they have also been covered in more general reviews of multifunctional linker units.[4, 5] A range of sulfur-based linker units has been developed allowing for multiple different cleavage strategies, many of which are discussed elsewhere in this book and therefore will not be duplicated here. For example, sulfur-based safety-catch linker units (Chapter 6),[6, 7] photolabile linker units (Chapter 5),[8–10] and cyclo-release linker units (Chapter 4)[11–13] have all been reported. This chapter will focus exclusively upon multifunctional linker units in which the reactivity of the sulfur is exploited during cleavage. Other sulfur containing linker units[14–16] in which the sulfur is inert and not intrinsically involved in the cleavage step are discussed elsewhere in this volume and are also beyond the scope of the present chapter.

Sulfur-based linker units have been reported which use sulfur in a range of oxidation states – the sulfide, sulfoxide and sulfone linker units, which are discussed in Sections 13.2, 13.4 and 13.5, respectively. Following the discovery that sulfonate esters and perfluoroalkanesulfonate esters are viable substrates for transition metal catalyzed cross-coupling reactions, a number of linker motifs designed around these groups have also been reported and these are highlighted in Section 13.6. Other less common sulfur-linker units based around sulfonium ions (Section 13.3), sulfamates (Section 13.7) and thioesters (Section 13.8) are also discussed.

Linker Strategies in Solid-Phase Organic Synthesis Edited by Peter Scott
© 2009 John Wiley & Sons, Ltd

13.2 Sulfide linker units

13.2.1 Introduction

The sulfide or thioether linkage represents the simplest type of sulfur linker and a number of methods have been reported for the introduction of such linker units onto solid supports. The most obvious starting point is polystyrene thiol resin, which is commercially available or can be prepared.[17-21] For example, in a recent report by Wagner and Mioskowski, polystyrene **1** resin was treated with activated sulfoxide **2** in the presence of triflic anhydride to give sulfonium trifluoromethanesulfonate resin **3**. Subsequent deprotection and dealkylation on treatment with potassium *t*-butoxide provided the corresponding arylthiol resin **4** (Scheme 13.1). Note that thiol resins can be difficult to handle due to the formation of di- and polysulfide bridges. Ellman's reagent offers a simple colorimetric test to ensure thiol resins have not degraded prior to use.[22, 23]

Many sulfide linker units permit traceless (Section 13.2.2) and multifunctional cleavage of substrates by simple nucleophilic substitution (Section 13.2.3) or elimination reactions (Section 13.2.4). More sophisticated strategies allow solid-phase protocols to be carried out on the sulfide linked substrate. Following completion of the synthesis, activation (frequently oxidation to the corresponding sulfoxide or sulfone, as discussed in Sections 13.4 and 13.5, respectively) of the linker unit is required before cleavage can be realized offering safety-catch strategies (see also Chapter 6).

Typical Experimental Procedures

Polystyrene sulfonium resin 3 (scheme 13.1)[17]

*Triflic anhydride (5.27 ml, 31.2 mmol) was added drop-wise at −40 °C to a stirred suspension of polystyrene resin **1** (10 g) and 3-(2-methoxycarbonyl-ethanesulfinyl)-propionic acid methyl ester **2** (5.78 g, 26.0 mmol) in dry dichlormethane (DCM) (120 ml) under argon. The resulting mixture was maintained at −40 °C and stirred for 20 minutes. After this time, the reaction was warmed to 0 °C, then to room temperature and stirred for an additional 12 hours. The resulting orange resin was filtered, washed with DCM (3 × 150 ml),*

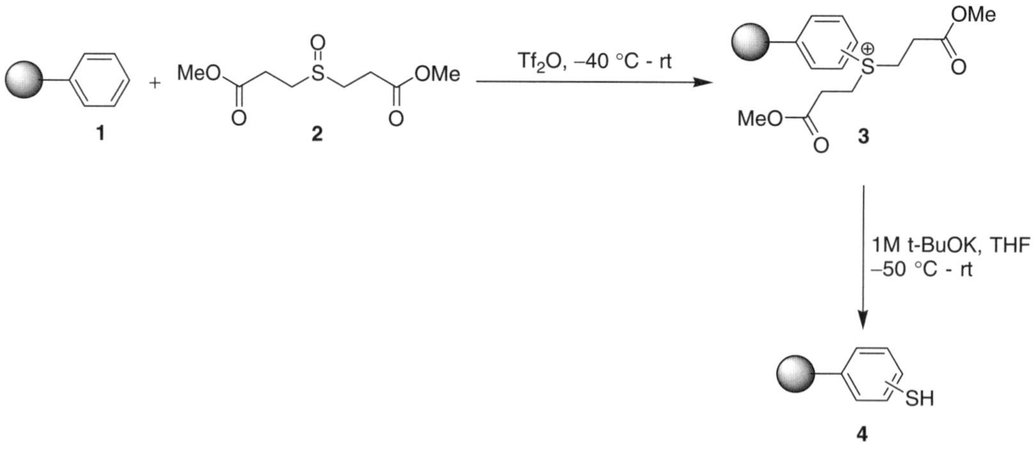

Scheme 13.1 *Preparation of arylthiol resin (4)*

methanol (3 × 150 ml) and diethylether (3 × 150 ml) and dried under vacuum to give polystyrene sulfonium resin **3** (15.4 g, loading: 1.09 mmol/g).

Polystyrene thiol resin 4 (scheme 13.1)[17]

*To a stirred suspension of sulfonium resin **3** (10 g) in dry tetrahydrofuran (THF) (90 ml) under argon at −50 °C was added potassium t-butoxide (t-BuOK) (1 M in THF, 75.0 ml, 75.0 mmol) drop-wise. The reaction was stirred for one hour at −50 °C and then warmed to room temperature and stirred for an additional 24 hours. The orange resin was then filtered, washed with THF (2 × 200 ml), 1 M hydrochloric acid (2 × 200 ml), water (2 × 200 ml), 0.8 M tributyl phosphine (PBu_3) in a 1:1 mixture of THF/water (200 ml), THF (3 × 200 ml), methanol (3 × 200 ml), DCM (3 × 200 ml) and diethyl ether (3 × 200 ml) and dried under vacuum o/n to provide polystyrene thiol resin **3** (7.16 g, loading: 1.34 mmol/g).*

13.2.2 Reductive traceless cleavage

Jung has developed a number of methods for the traceless cleavage of aliphatic substrates attached through sulfide linker units.[21, 24] Polyethylene glycol supports were employed and, in this case, the linker unit **7** was prepared in advance by loading a thiol aniline derivative **6** (Scheme 13.2). The free amine was used as a point of attachment to a PEG-CO_2H support **5** using traditional peptide coupling techniques. Alkyl bromides **8** were attached to the free resin-bound thiol to provide resin-bound thioethers **9**. Two strategies for traceless cleavage have been reported for these linker units. The first was a radical cleavage strategy (see also Chapter 15) in which resin-bound thioethers were treated with tributyltin hydride in refluxing benzene. Whilst this did cleave the desired product **10**, a sluggish reaction provided mediocre yields (40%) and

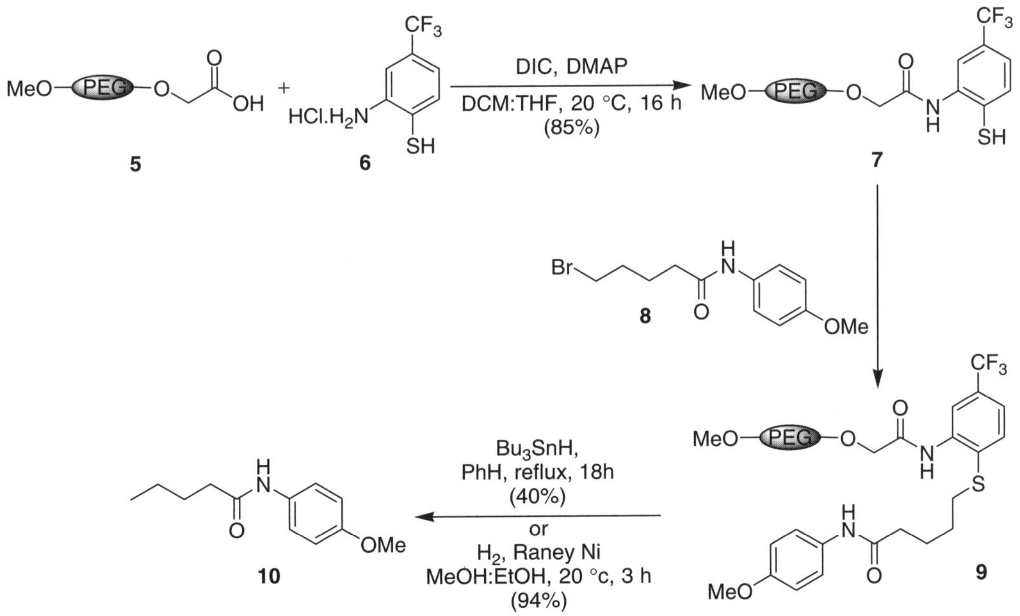

Scheme 13.2 *Traceless cleavage from sulfide linkers*

the final product had to be purified from tin by-products. Such purifications are frequently not trivial and are not ideal for realizing the quick and efficient preparation of compound libraries. Therefore, alternative cleavage strategies were considered and Jung turned to a Raney nickel mediated desulfurization reaction. In contrast to its radical counterpart, the nickel catalyzed process was complete in three hours and after precipitation of the PEG support and subsequent filtration, pure product was obtained in 94% yield after concentration of the resulting filtrate. In further applications, Chauhan has used this Raney nickel cleavage strategy in the preparation of a library of aminopyrimido[4,5-d]pyrimidines.[25] The only drawback of this approach is that functional groups that can be present in resin-bound substrates are limited due to the harsh reducing conditions.

Samarium(II) iodide (SmI_2) is a truly multifunctional reagent in organic chemistry, capable of effecting a wide range of synthetic transformations, and has been the subject of a number of reviews.[26–30] In recent work, Procter has exploited the versatility of samarium(II) iodide in traceless cleavage from a number of new linker units for SPOS (see also Chapter 15).[20, 31] One such linker is the α-Hetero-Atom Substituted Carbonyl (HASC) linker **12** illustrated in Scheme 13.3. This was generated by loading N-methyl-N-phenyl α-bromoacetamide **11** onto thiol resin **4**. The thioether linkage was then oxidized to the corresponding sulfoxide **12** using the recently developed hydrogen peroxide/hexafluoroisopropanol (HFIP) selective oxidation.[32, 33] This was necessary to achieve the subsequent Pummerer cyclization, which provided resin-bound oxindole **13** whilst simultaneously reducing the linker unit back to the thioether. Traceless cleavage was then achieved by treating with samarium(II) iodide in the presence of 1,3-dimethyl-3,4,5,6-tetrahydro-2(1 H)-pyrimidinone (DMPU) to provide oxindole **14**.

Typical Experimental Procedures

Cleavage with raney nickel (scheme 13.2)[21]

Activated Raney® nickel (50% slurry in water; 10 mg, 0.085 mmol) and absolute ethanol (1.5 ml) were added to a solution of CH_3O-PEG sulfide 9 (130 mg, 0.024 mmol) in absolute methanol (3 ml). The resulting black suspension was placed under hydrogen (34 psi) and shaken for 16 hours. After this time, the crude

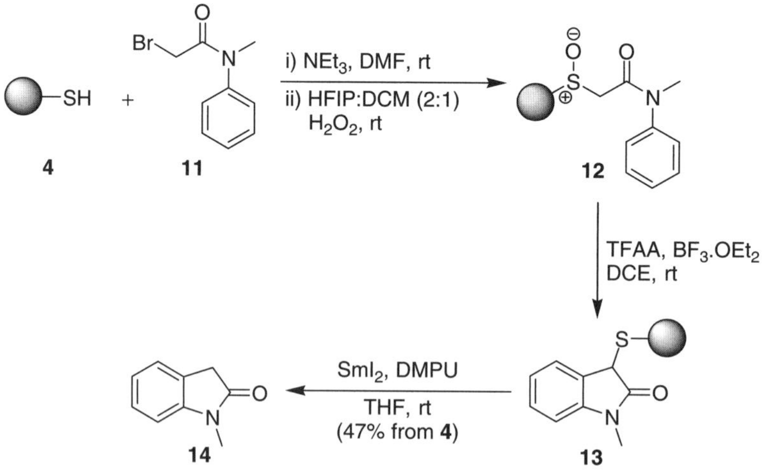

Scheme 13.3 *Samarium(II) iodide mediated traceless cleavage*

reaction mixture was filtered through Celite® and rinsed with DCM (20 ml). The filtrate was concentrated, re-dissolved in DCM (10 ml) and then triturated with diethyl ether (100 ml). The resulting precipitate (CH_3O-PEG by-products) was removed by filtration and the filtrate was concentrated to provide spectroscopically pure amide **10** (5 mg, 0.024 mmol, 99%).

Samarium(II) iodide mediated cleavage (scheme 13.3)[20]

Oxindole resin **13** (0.158 mmol) was suspended in THF (4 ml). DMPU (0.153 ml, 1.26 mmol), and samarium(II) iodide (SmI_2) (0.1 M in THF, 4.74 ml, 0.474 mmol) were added and the reaction was stirred at room temperature for 18 hours. After this time the resin was removed by filtration and the resulting filtrate was concentrated under vacuum. The concentrate was filtered through a short pad of silica gel followed by washing with 30% ethyl acetate in hexane. Further concentration under vacuum provided oxindole **14** (11 mg, 0.074 mmol, 47% over four steps).

13.2.3 Multifunctional cleavage via nucleophilic substitution reactions

An early example of diversity cleavage from a sulfide linker was demonstrated by Crosby in 1977.[34] Treatment of methyl sulfide resin **15** with *n*-butyl lithium (*n*-BuLi) provided the intermediate resin-bound lithium salt **16** onto which was loaded alkyl groups via the corresponding alkyl iodides to provide supported alkylthioethers **17**. Simple multifunctional cleavage was then demonstrated by treating with sodium iodide and iodomethane in DMF to provide alkyl iodides **18** (Scheme 13.4).

Similar cleavage protocols have been developed by Kunz[35–39] and Schmidt[40, 41] in which libraries of resin-bound glucose derivatives **19** can be cleaved from sulfide linkers as the corresponding bromides **20** by treating with *N*-bromo succinimide or elemental bromine in the presence of 2,6-di-*tert*-butylpyridine (DTBP) (Scheme 13.5).[35] In this case, sugars can be isolated as the bromide or further substituted by addition of an alcohol in a Lemieux-type glycosylation at the anomeric center. For example, Schmidt cleaved resin-bound disaccharide **21** with *N*-bromosuccinimide (NBS) in THF and methanol to provide methoxy-substituted disaccharide **22**. Importantly, none of the β-anomeric by-product was detected.[40]

Building on this concept, cleavage strategies continue to be reported in which substrates are detached from resins on treatment with a range of nucleophiles. With less reactive nucleophiles, cleavage can be enhanced by prior activation of the sulfide and strategies for this are discussed later in this

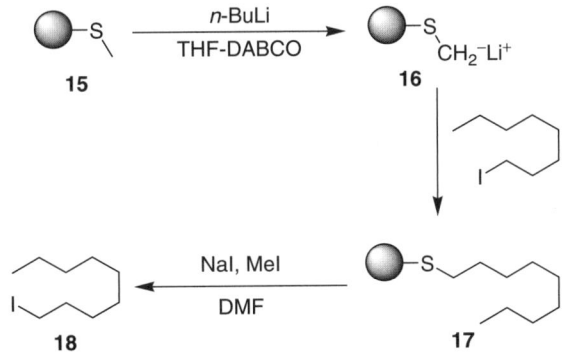

Scheme 13.4 *Crosby's early cleavage of alkyl halides*

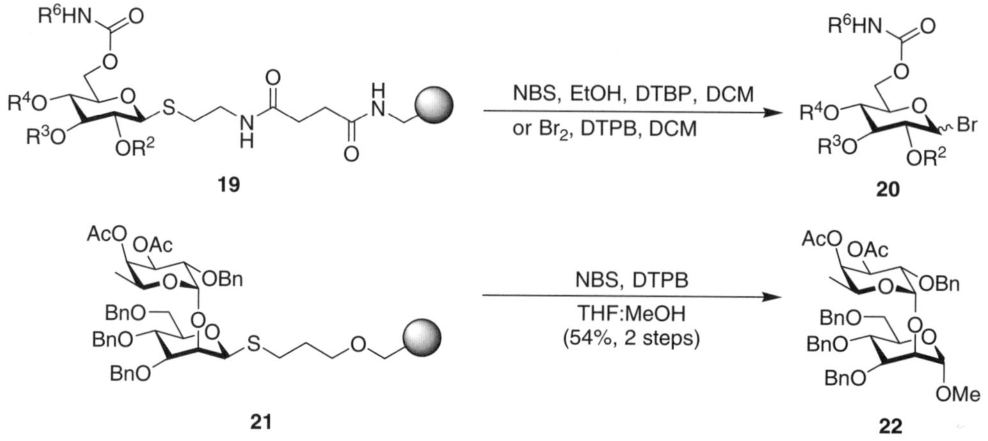

Scheme 13.5 *Nucleophilic cleavage of sugars*

chapter, for example generation of sulfonium ions (Section 13.3), sulfoxides (Section 13.4) or sulfones (Section 13.5). Zoller used a literature method to cleave amino acids loaded on Wang resin through a sulfide linker as the corresponding disulfides.[12, 42] Treatment of resin-bound amino acid **23** with dimethyl(methylthio)sulfonium tetrafluoroborate cleaved disulfide **24** without any effect on the stereochemistry of the amino acid (Scheme 13.6). The concept was also developed into a cyclative cleavage strategy by intramolecular generation of the dimethyl(methylthio)sulfonium moiety in resin-bound amino acid **25** (using *N*-chlorosuccinimide (NCS) and dimethylsulfide) to give cyclic disulfide **26** (see also Chapter 4).

The only example of cleavage with carbon nucleophiles to date was developed by Hennequin and is shown in Scheme 13.7.[43] Quinazolines **27** were treated with phosphorus pentasulfide to give thioquinazolines, which were loaded onto simple Merrifield resin to give resin-bound thioethers **28**. Multifunctional cleavage was then realized by treatment with sodium salts of oxindoles **29** to provide a library of oxindole quinazolines **30a–g**.

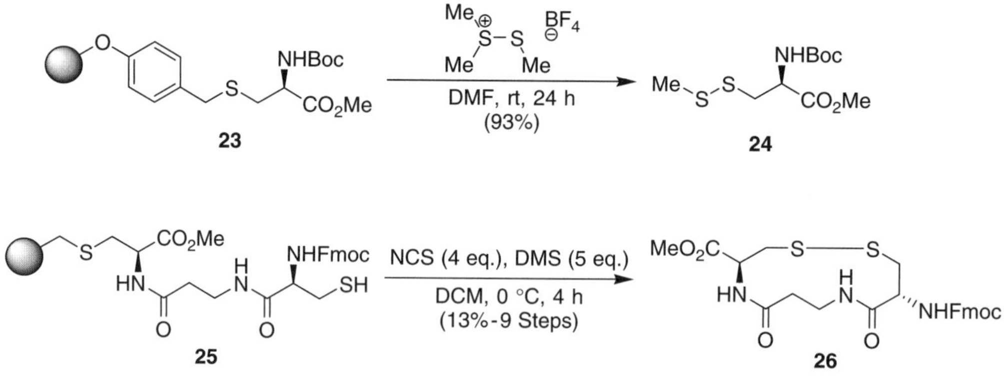

Scheme 13.6 *Cleavage of substrates as disulfides*

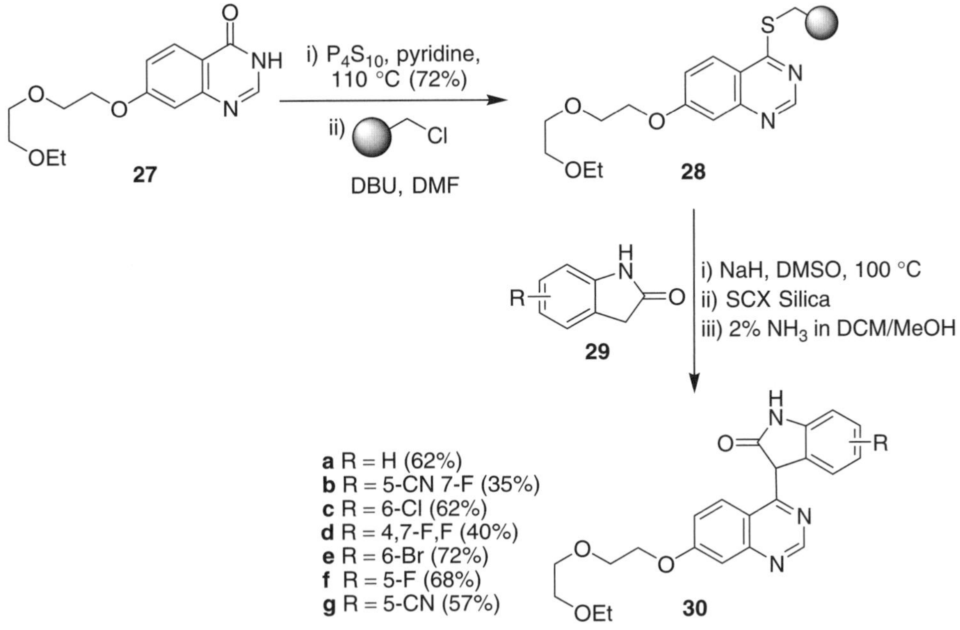

Scheme 13.7 Synthesis of a library of oxindole quinazolines (**30a–g**)

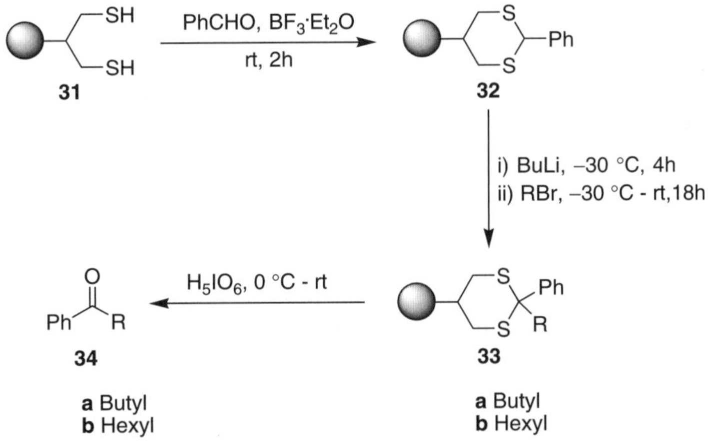

a R = H (62%)
b R = 5-CN 7-F (35%)
c R = 6-Cl (62%)
d R = 4,7-F,F (40%)
e R = 6-Br (72%)
f R = 5-F (68%)
g R = 5-CN (57%)

Scheme 13.8 The 1,3-propanedithiol linker unit

An analogue of simple sulfide linkers is the 1,3-propanedithiol linker unit **31**, onto which can be loaded aldehydes to provide resin-bound thioacetals **32** (Scheme 13.8); a number of groups have made use of this approach.[44–46] Diversity can be introduced by deprotonating with *n*-butyl lithium and quenching with electrophiles such as alkyl halides. Cleavage of the corresponding ketones has been demonstrated using either [bis-(trifluroacetoxy)iodo]benzene[44] or anyhydrous periodic acid.[46, 47]

Typical Experimental Procedures

Cleavage via bromination and Lemieux oxidation (scheme 13.5)[40]

1-O-Methyl-(di-3,4-O-acetyl-2-O-benzyl-α-L-fucipyranosyl)-tri-3,4,6-O-benzyl-α-D-mannopyranoside 22

Resin-bound disaccharide **21** *(0.01 mmol) was suspended in THF:methanol (10:1) (2.5 ml). After 15 minutes to allow swelling, NBS (9 mg, 4 equiv.) and DTBP were added and the reaction was shaken for 1.5 hours. The resin was filtered and washed with THF and DCM. Triethylsilane (10 μl) was added to quench any unreacted NBS and the combined filtrates were concentrated under vacuum and the crude material was further purified by flash chromatography (toluene:ethyl acetate, 6:1) and subsequent HPLC (hexane:ethyl acetate, 3:1) to provide the title disaccharide (8.5 mg, 54% over two steps).*

Nucleophilic cleavage with oxindoles (scheme 13.7)[43]

Oxindole Quinazolines 30a–g

A solution of the oxindole **29** *(10.9 mmol) in dimethylsulfoxide (DMSO) (9 ml) was added to a suspension of sodium hydride (10.98 mmol) in DMSO (9 ml). After stirring for 30 minutes, this mixture was added to a suspension of the resin-bound quinazoline* **28** *(200 mg) in DMSO (2 ml) and the reaction heated at 100 °C for 3.5 hours under nitrogen. After this time, the reaction was cooled and quenched with saturated. ammonium chloride (10 ml) followed by water (1 ml). The mixture was extracted with ethyl acetate and the organic layer was washed with water until pH = 6 and then dried (magnesium sulfate). The organic layer was filtered and poured onto an SCX (IST 1.9 g) column preconditioned with methanol. The column was washed successively with ethyl acetate (2 ×); DCM (2 ×) and, finally DCM:methanol (1:1) (2 ×). 1% ammonia in DCM:methanol (1:1) was then passed through the column to provide oxindole quinazolines* **30a – g** *(35–68%).*

13.2.4 Multifunctional cleavage via elimination reactions

A library of 2,4-diaminothiazoles **40** was prepared by Baer and Masquelin using substrates bonded to a resin through a sulfide linker (Scheme 13.9).[48] Supported thioureido–thiourea **35** was treated with α-bromo ketones **36** in the presence of resin-bound base **37** to give **38**, which spontaneously cyclized to give dihydrothiazole **39** bonded to the support through a thioether. Excess polymer-supported base subsequently promoted elimination to give 2,4-diaminothiazoles **40**. The simultaneous use of two resins was found to be highly advantageous, giving clean high yielding reactions allowing for simple isolation of the product after completion of the reaction by filtration of the supports.

Typical Experimental Procedures

General procedure for the preparation of 2,4-diaminothiazoles 40 (scheme 13.9)[48]

Resin-bound thiouronium salt **35** *and diethylaminomethyl polystyrene* **37** *were suspended in dry DMF (10 ml) under argon at room temperature and treated with the α-bromoketone* **36** *(1.3 mmol). The reaction mixture was then stirred at room temperature for 18 hours, after which time the resin was filtered and washed once with DMF (10 ml). The solvent was removed from the combined filtrates to yield* **40** *in high purity.*

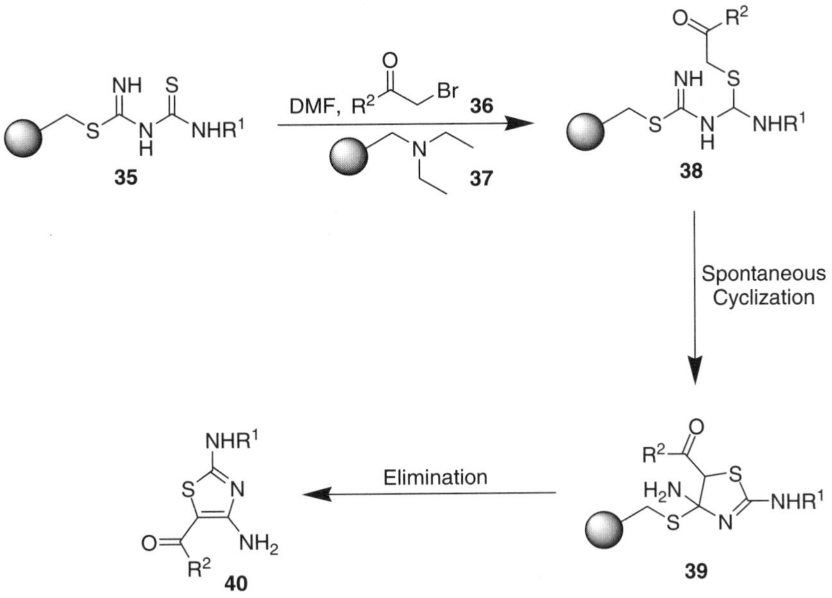

Scheme 13.9 *Preparation of a library of 2,4-diaminothiazoles (**40**)*

13.3 Sulfonium Linker Units

Substrates bound to resin through thioether linkers can be activated for cleavage by converting them into more reactive intermediates. This can be achieved by oxidation to sulfoxides (Section 13.4) or sulfones (Section 13.5) or by alkylation to the corresponding sulfonium ions (Scheme 13.10). This is typically achieved by treating with triethyloxonium tetrafluoroborate (Et_3OBF_4) or methyl trifluoromethanesulfonate (MeOTf) leading to alkylation (and activation) of the sulfur atom; such an approach has been used recently by Gennari.[49] Thioether **41** was built up on ArgoGel®-SH using standard SPOS techniques. Cleavage

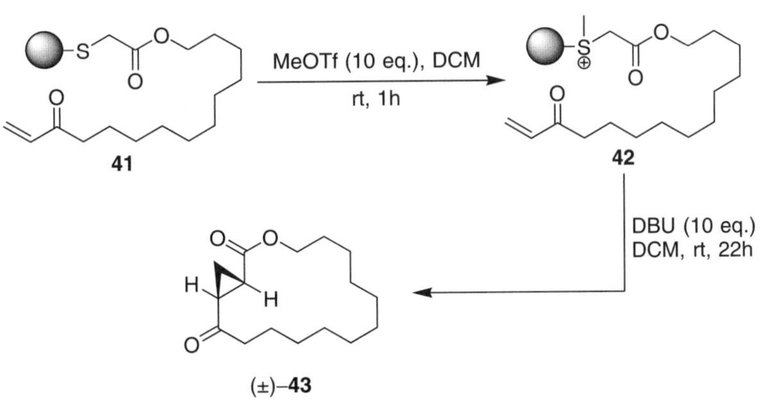

Scheme 13.10 *Cyclative cleavage via an intramolecular cyclopropanation*

352 Multifunctional Linker Units for Diversity-Oriented Synthesis

was then achieved by initially activating the sulfur with MeOTf to provide the sulfonium ion **42**. This sulfonium ion was then treated with 1,8-diazabicyclo[5.4.0]undec-7-ene (DBU) and the resulting sulfur ylide underwent an intramolecular cyclopropanation via a Michael reaction and subsequent elimination, which simultaneously cleaved the carbon–sulfur bond. This left the thiol functionality bound to the polymer support and provided the macrocycle **43** exclusively as the trans isomer. In this case, the linker is also functioning as a cyclo-release diversity linker unit (Chapter 4).

However, diversity can also be introduced through an intermolecular reaction. For example, if sulfonium **45** (generated from thioether **44**) is treated with an aldehyde and DBU, then substrates are cleaved from the resin as epoxides **46–47** (Scheme 13.11), clearly highlighting the difference between cyclo-release and true diversity linker units.

In related work, Wagner and Mioskowski loaded substituted benzylbromides onto thiol resins to generate polymer-supported thioethers **49**.[50] Activation of the sulfur with Et_4OBF_4 gave the corresponding sulfonium ions **50** and then multifunctional cleavage was demonstrated in an unusual palladium catalyzed cross-coupling reaction. Treatment of resin-bound sulfonium salts **50** with boronic acids in the presence of catalytic Pd(dppf)Cl$_2$ and potassium carbonate cleaved products generating a library of biarylmethanes **51** (Table 13.1).

Typical Experimental Procedures

Cleavage of cyclopropanes via sulfur ylides (scheme 13.10)[49]

(1S,16S)-3-oxa-bicyclo[14.1.0]heptadecane-2,15-dione 43

*Resin-bound vinyl ketone **41** (1.0 equiv.) was suspended in DCM and treated with methyl triflate to activate the thioether. The reaction was shaken at room temperature for one hour, after which time the resin was*

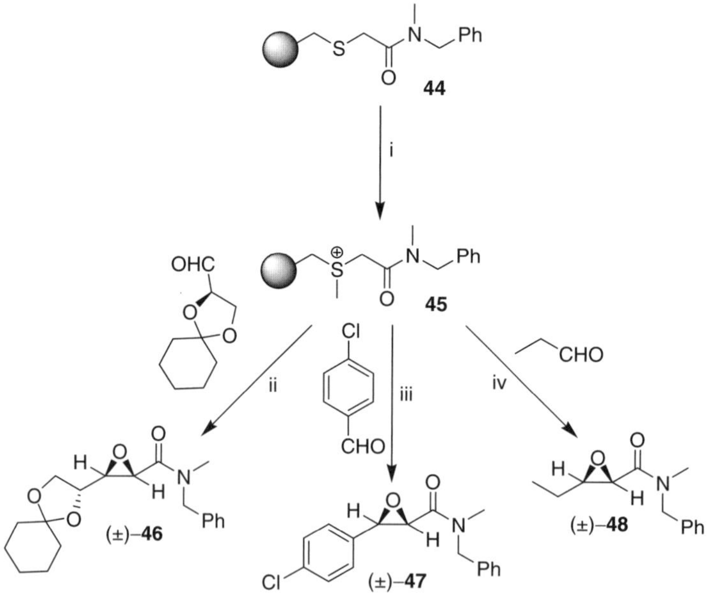

Scheme 13.11 *Multifunctional cleavage via epoxidation i) MeOTf, DCM, rt, 1 h; ii) DBU, MeCN, rt, 1.5 h; iii) DBU, DCM, rt, 3 h; iv) DBU, DCM, rt, 1.5 h.*

Table 13.1 Cleavage from a sulfonium linker using palladium catalyzed cross-coupling reactions[50]

	Resin	Boronic Acid	Product	Yield
1	resin-S-CH₂-C₆H₄-CH₃ (p-tolyl)	PhB(OH)₂	PhCH₂-C₆H₄-CH₃	46
2	resin-S-CH₂-C₆H₄-CH₃ (p-tolyl)	4-Me-3-NO₂-C₆H₃-B(OH)₂	4-Me-3-NO₂-C₆H₃-CH₂-C₆H₄-CH₃	99
3	resin-S-CH₂-C₆H₄-Ph (p-phenyl)	2-thienyl-B(OH)₂	2-thienyl-CH₂-C₆H₄-Ph	80
4	resin-S-CH₂-C₆H₄-OMe (m-OMe)	2,6-(OMe)₂-C₆H₃-B(OH)₂	2,6-(OMe)₂-C₆H₃-CH₂-C₆H₄-OMe	62
5	resin-S-CH₂-(3,4-methylenedioxyphenyl)	2-thienyl-B(OH)₂	2-thienyl-CH₂-(3,4-methylenedioxyphenyl)	24
6	resin-S-CH₂-C₆H₄-CO₂Et	4-Me-3-NO₂-C₆H₃-B(OH)₂	4-Me-3-NO₂-C₆H₃-CH₂-C₆H₄-CO₂H	99[a]

(continued overleaf)

Table 13.1 (continued)

| 7 | [resin]-S-CH₂-C₆H₄-C(O)NH₂ | PhB(OH)₂ | Ph-CH₂-C₆H₄-C(O)NH₂ | 80 |
| 8 | [resin]-S-CH₂-C₆H₄-CN | 2,6-(MeO)₂C₆H₃B(OH)₂ | 2,6-(MeO)₂C₆H₃-CH₂-C₆H₄-CN | 63 |

^a The ester was hydrolyzed to the corresponding acid with lithium hydroxide

filtered and washed to provide supported sulfonium salt **42**. DBU (10 equiv.) was added and the reaction was shaken at room temperature for 22 hours to generate the stabilized sulfur ylide, which spontaneously cyclized to give cyclopropane **43** exclusively as the trans diastereoisomer.

Palladium catalyzed cleavage to give biaryl methanes 51 (table 13.1)[50]

A mixture of resin-bound **50** (1 equiv.), boronic acid (2 equiv.), potassium carbonate (3 equiv.) and $PdCl_2 dppf$ (0.2 equiv.) in THF (2 ml) was degassed with argon. The reaction was then heated to 60 °C and stirred for 14 hours. After this time, the reaction was cooled, filtered and washed with THF (2 × 5 ml); then alternatively with DCM (5 × 5 ml) and methanol (5 × 5 ml). The combined filtrates were concentrated and the resulting red solid was filtered over a bed of silica and washed with DCM and methanol to remove palladium residues and provide biaryl methanes **51**.

13.4 Sulfoxide linker units

13.4.1 Introduction

Sulfoxide linkers can be loaded onto a resin directly[51, 52] but more typically are generated by oxidation of the corresponding thioether linkers described in Section 13.2 above.[53–55] This potential for selective activation allows sulfoxide linkers to be employed as safety-catch linkers in solid-phase synthesis (Chapter 6). Such oxidative transformations can be achieved using common reagents for oxidation such as *meta*-chloroperbenzoic acid (*m*CPBA), sodium periodate ($NaIO_4$) or ozone. However, these transformations tend to be unselective and typically result in complex mixtures of sulfoxides and sulfones.[32, 33] Note that over oxidation to the sulfone and subsequent multifunctional cleavage is a viable strategy and this is discussed further in Section 13.5 below. To control oxidation so as to generate exclusively the sulfoxide intermediate, more specialized methods are required and reported strategies include acid mediated oxaziridine oxygen transfer,[53] hydrogen peroxide/hexafluoroisopropanol (HFIP) oxidation,[20, 32, 33, 54] or *t*BuOOH/10-camphorsulfonic acid (CSA) oxidation.[55] Such sulfoxide linker units have been used in both traceless and multifunctional cleavage strategies.

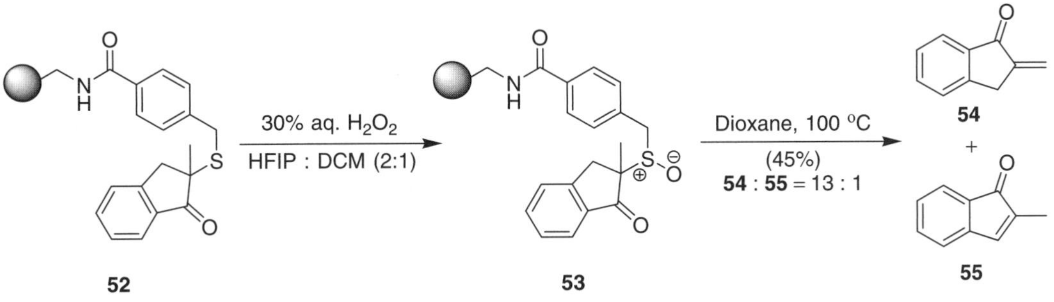

Scheme 13.12 *Traceless cleavage from sulfoxide linkers*

13.4.2 Traceless cleavage

Traceless cleavage from sulfoxide linker units has been demonstrated by simply heating resin-bound substrates. For example, Bradley was able to oxidize resin-bound sulfide **52** to the corresponding sulfoxide **53** using hydrogen peroxide/HFIP (Scheme 13.12).[54] In this case it is suspected that HFIP not only acts as an acid catalyst but also coordinates to the resulting sulfoxide preventing over oxidation to the sulfone seen with other acids. Sulfoxides **53** were highly activated towards elimination, and traceless cleavage was demonstrated by heating to 100 °C in dioxane. The desired indenones were cleaved as a 13:1 mixture of exo **54**/endo **55**. Note the analogous selenide linkers behave similarly but offer the advantage that cleavage can be achieved at room temperature (Chapters 14 and 15).

In related work, Toru loaded pre-formed sulfoxide linker unit **56** onto modified Merrifield resin to give resin-bound sulfoxide **57**.[51] Treatment with LDA and subsequent conjugate addition to methyl cinnamate **58** gave supported sulfoxide **59**. Traceless cleavage could then be realized by suspending **59** in benzene and refluxing for one hour to provide optically active ester **60** in reasonable yield (51%) and high enantiomeric excess (90% ee) as shown in Scheme 13.13. In further work, Toru demonstrated the use of tetrabutylammonium fluoride (TBAF) as an alternative cleavage strategy. TBAF effected desilylsulfination to give (*R*)-**61** in 56% yield and 90% ee.

Somewhat related to the sulfoxide linkers is the sulfinamide linker **62** reported by Ellman. Aldehydes were condensed with the linker unit in the presence of titanium(IV) ethoxide (Ti(OEt)$_4$) (Lewis acid and water scavenger) and subsequently treated with ethyl magnesium bromide to give resin-bound *tert*-butanesulfinamide **63**.[52] Mild acidic traceless cleavage (0.67 M hydrochloric acid) provided amines **64** as the corresponding hydrochloride salts in near quantitative yield over three steps (Scheme 13.14).

Typical Experimental Procedures

Thermal eliminative traceless cleavage (scheme 13.12)[54]

*A suspension of resin **53** in dioxane was heated to 100 °C. After reaction, the resin was filtered and concentration of the resulting filtrate provided the indenones **54** and **55** as an exo/endo mixture (exo/endo indenones = 13 : 1 by reverse phase HPLC) (45% yield, >95% HPLC purity).*

13.4.3 Multifunctional cleavage using the pummerer rearrangment

Multifunctional cleavage from sulfoxide linkers has also been demonstrated to generate alcohols and aldehydes using Pummerer chemistry.[53, 55] Typically the Pummerer rearrangement is achieved by treating sulfoxides with sodium acetate and acetic anhydride at elevated temperatures. However, such conditions

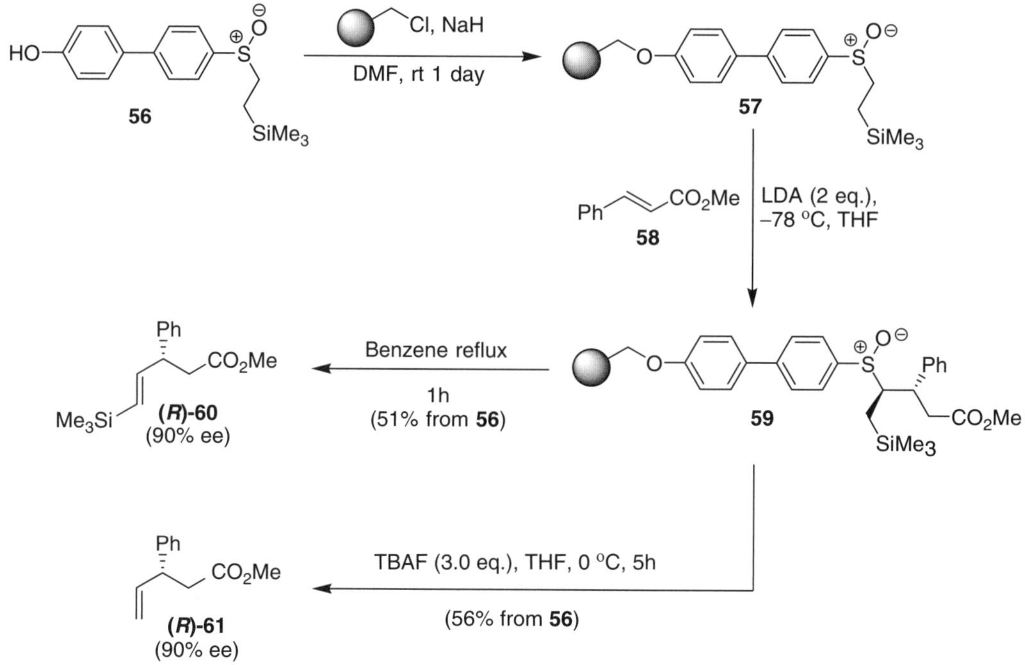

356 *Multifunctional Linker Units for Diversity-Oriented Synthesis*

Scheme 13.13 *Multifunctional-cleavage from sulfoxide linkers*

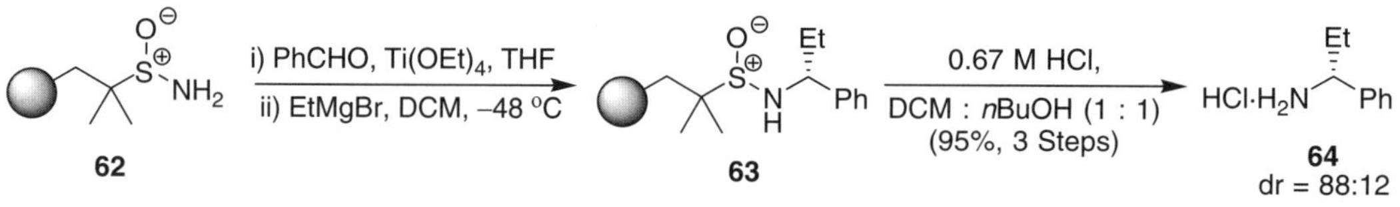

Scheme 13.14 *Ellman's sulfinamide linker unit*

are expected to be detrimental to the polystyrene resins employed in SPOS and so the more reactive trifluoroacetic anhydride (TFAA) is typically employed. For example, Li reported loading standard electrophiles such as alkyl halides, tosylates and epoxides onto a simple thiol resin by pre-treatment with sodium hydride. Alternatively, alcohols could also be loaded using Mitsunobu conditions. The resulting resin bound thioethers were then activated by oxidizing to the sulfoxide with *tert*-butyl hydroperoxide (tBuOOH) and 10-camphorsulfonic acid (CSA). Conversion to the sulfoxide was monitored using magic angle spinning gel-phase NMR. With sulfoxides **65** in hand, cleavage strategies using the Pummerer rearrangement were established. The Pummerer rearrangement was achieved using trifluoroacetic anhydride to provide resin-bound trifluoroacetoxythioacetals **66**. Cleavage was then carried out by treatment with triethylamine to give a library of aldehydes **67** (a selection of which are shown in Table 13.2). Typically, in an additional step, aldehydes are then reduced to the corresponding alcohols by treating with sodium borohydride. However, Li reported a one-pot cleavage protocol in which triethylamine and sodium borohydride (lithium aluminum hydride (LiAlH$_4$) was also tried but proved too harsh) were added together to provide alcohols **68** directly (Table 13.2).[55]

Table 13.2 Cleavage using a Pummerer rearrangement

	Sulfoxide	Products			
		Aldehyde	Yield (%)	Alcohol	Yield (%)
1	(p-nitrobenzyl sulfoxide on resin)	4-O$_2$N-C$_6$H$_4$-CHO	80	4-O$_2$N-C$_6$H$_4$-CH$_2$OH	77
2	(phthalimidoethyl sulfoxide on resin)	phthalimido-CH$_2$-CHO	87	phthalimido-CH$_2$-CH$_2$OH	80
3	(furfuryl sulfoxide on resin)	furfural	63	furfuryl alcohol	60
4	(3,4,5-tris(tetradecyloxy)benzyl sulfoxide on resin)	3,4,5-tris(OC$_{14}$H$_{29}$)benzaldehyde	85	3,4,5-tris(OC$_{14}$H$_{29}$)benzyl alcohol	80
5	(4-methoxycarbonylbenzyl sulfoxide on resin)	4-MeO$_2$C-C$_6$H$_4$-CHO	83	4-MeO$_2$C-C$_6$H$_4$-CH$_2$OH	73
6	(3-nitro-4-chlorobenzyl sulfoxide on resin)	3-NO$_2$-4-Cl-C$_6$H$_3$-CHO	78	3-NO$_2$-4-Cl-C$_6$H$_3$-CH$_2$OH	75

Typical Experimental Procedures

Aldehydes 67 (table 13.2)[55]

Resin-bound thioethers were suspended in DCM (1.6 ml). t-BuOOH (0.25 ml, 1.26 mmol) and CSA (36 mg, 0.16 mmol) were added and the reaction was shaken at room temperature for 12 hours. After this time resin-bound sulfoxide **65** was filtered, washed with DMF (3 × 2 ml) and DCM (3 × 2 ml), and dried under a nitrogen stream. Sulfoxide resin **65** was then re-swelled in THF (0.6 ml) and cooled to 0 °C. A chilled (0 °C) solution of trifluoroacetic anhydride (0.06 ml) in THF (1.44 ml) was added drop-wise and the reaction was warmed to room temperature and stirred for 45 minutes. The resin was filtered again and washed with DMF (3 × 2 ml) and DCM (3 × 2 ml) to give resin-bound trifluoroacetoxythioacetals **66** which were re-suspended in THF (2 ml). Triethylamine (0.02 ml, 0.16 mmol) and ethanol (0.009 ml, 0.16 mmol) were added and the mixture was shaken for two hours at room temperature. After this time, filtration of the resin and purification of the resulting filtrate by flash chromatography gave aldehydes **67** (48–87% yield).

Alcohols 68 (table 13.2)[55]

The general procedure described above for the preparation of aldehydes was followed to provide trifluoroacetoxythioacetals **66** which were suspended in THF (2 ml). Triethylamine (0.02 ml, 0.16 mmol) and ethanol (0.009 ml, 0.16 mmol) were added and the mixture was shaken for two hours at room temperature. After this time, sodium borohydride (11 mg, 0.32 mmol) was added and the suspension was agitated for a further two hours. The resin was then filtered and purification of the concentrated filtrate by flash chromatography provided alcohols **68** (52–82% yield).

13.5 Sulfone linker units

13.5.1 Introduction

The simplest sulfone linker can be obtained as resin-bound lithium phenyl sulfinate **71** simply by bubbling sulfur dioxide through a suspension of lithiated polystyrene resin **70** as shown in Scheme 13.15 and reported by Kurth.[56] Related to this is the lithiation of methyl sulfone resin **73**, typically prepared by quenching lithium phenyl sulfinate **71** with methyl iodide.[57, 58] Both approaches offer scope for loading electrophiles to generate polymer-supported sulfone derivatives which may enhance reactivity and permit the introduction of diversity through some interesting multifunctional cleavage strategies discussed below. Resin-bound vinyl sulfones **77** have also been used as linker units, the double bond offering a point of attachment for loading nucleophilic substrates (Scheme 13.15).[59, 60] Sulfonyl alcohol **75** is converted to either the bromide or mesylate **76** and then treatment with base eliminates to give vinyl sulfone **77**. The whole process is easily monitored by IR spectroscopy.

Whilst all of these strategies are viable means of preparing sulfone linker units, their use has been limited and by far the most common approach to obtaining sulfone linkers is through oxidation of the corresponding thioether linker units (Scheme 13.16). Oxidation of thioether-based linker units to enhance cleavage was introduced above in the context of sulfoxide linker units but the same activation strategy is also viable when sulfides are oxidized to the corresponding sulfones. This capacity to enhance the leaving group ability of a thioether by oxidation has been much exploited in safety-catch linkers (Chapter 6) but has also been widely employed in a diversity approach. The oxidation of sulfides to sulfoxides described above requires specialized reagents in order to avoid complex mixtures of sulfoxides and sulfones. In contrast,

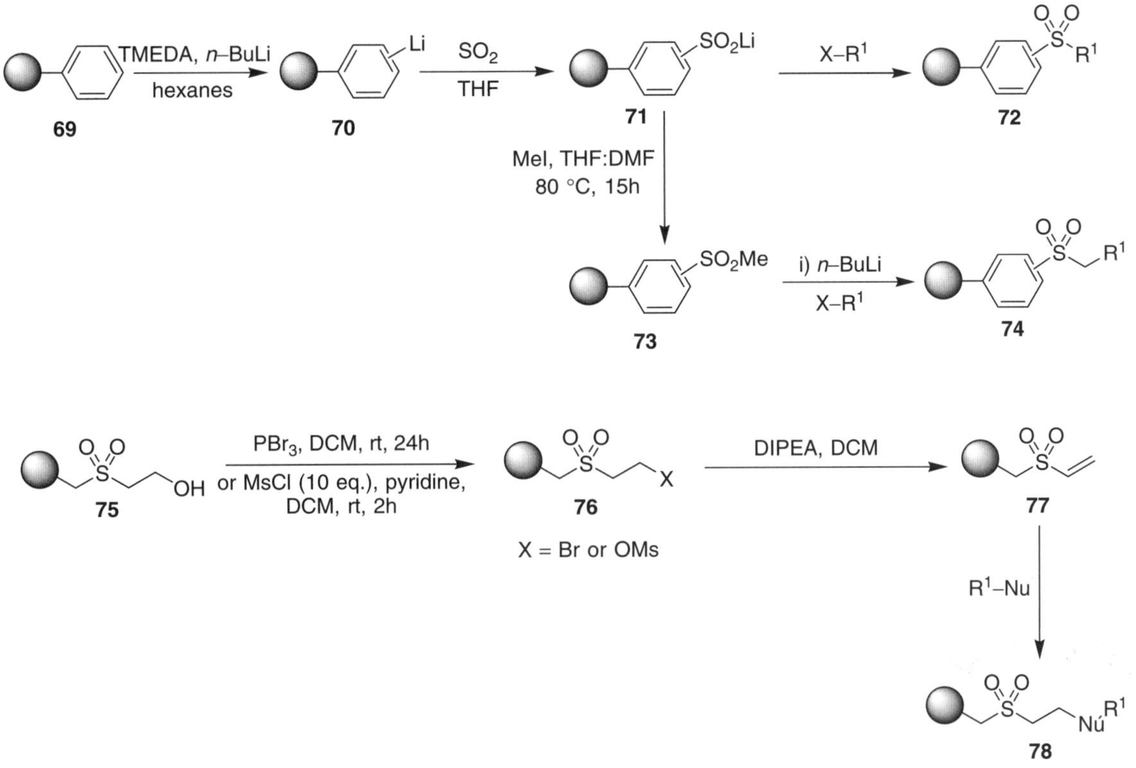

Scheme 13.15 Sulfone linker units

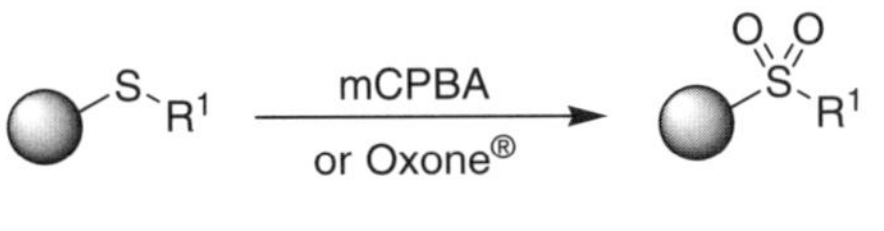

Scheme 13.16 Generation of sulfone linkers by oxidation of sulfide linkers

complete oxidation to the corresponding sulfones is relatively straightforward and can be achieved using traditional oxidizing agents such as *meta*-chloroperbenzoic acid (mCPBA),[7] Oxone® (KHSO$_5$),[20] or sodium periodate (NaIO$_4$).[61]

In this chapter, no distinction is made about how the sulfone linker units are prepared, as the multifunctional cleavage strategies discussed herein are equally viable for all substrates attached to polymer supports through sulfone linker units, regardless of how they are generated.

Typical Experimental Procedures

Polymer-supported lithium phenyl sulfinate 71 (scheme 13.15)[56]

*Polystyrene **69** (40 g) was swollen in freshly distilled cyclohexane (400 ml) and placed under an argon atmosphere. TMEDA (60 ml) was added and the reaction was cooled to 0 °C. nBuLi (1.6 M in hexanes, 250 ml) was added and the reaction was stirred overnight and then washed with THF (3×) to provide* *lithiated polystyrene **70** as an orange resin. Resin **70** was re-suspended in THF and cooled to −78 °C. Sulfur dioxide (g) was then bubbled through the reaction for one hour after which time the reaction was quenched with water (one hour). The resulting polymer-supported lithium phenyl sulfinate **71** was washed with THF, THF:water (80:20), THF and ether, and the resin was then dried in a vacuum oven at 30 °C overnight.*

Oxidation of resin-bound thioethers to resin-bound sulfones (scheme 13.16)[7]

Resin-bound thioethers were suspended in DCM. mCPBA (10 equiv.) was added and the reaction was shaken overnight. After this time, the resin was filtered and washed with DCM (3×), DMF (2×) and again with DCM (3×). Drying under vacuum provided resin-bound sulfones.

13.5.2 Reductive traceless cleavage

Substrates supported through sulfone linkers can be cleaved in a traceless fashion through a dissolving metal reduction, typically achieved using 5% sodium/mercury. This cleavage strategy has been extensively explored and refined by Janda using PEG-based soluble polymer supports (Scheme 13.17).[62–65] Initially substrates (**80**) are loaded onto the corresponding PEG–thiol support **79** and, after synthesis is complete, the thioether linkage of **81** is oxidized to the sulfone **82** using $KHSO_5$. The carbon–sulfur bond is then cleaved in a traceless fashion by treating with 5% sodium/mercury in the presence of disodium hydrogen phosphate (Na_2HPO_4) to give **83**. Following cleavage, trituration of the PEG-support facilitates its removal from the reaction mixture by simple filtration.

Traceless cleavage from Procter's α-hetero-atom substituted carbonyl linker employing samarium(II) iodide was introduced in Section 13.2 (and Chapter 15) for the thioether analog (Scheme 13.3).[20] This traceless cleavage strategy is also viable for substrates bound to supports through the corresponding sulfone

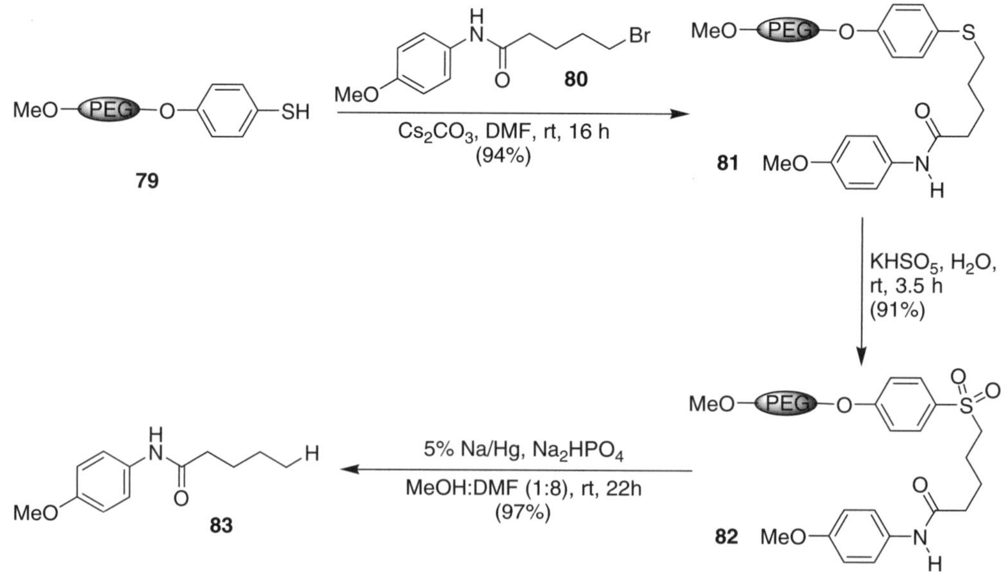

Scheme 13.17 *Reductive traceless cleavage from sulfone linkers*

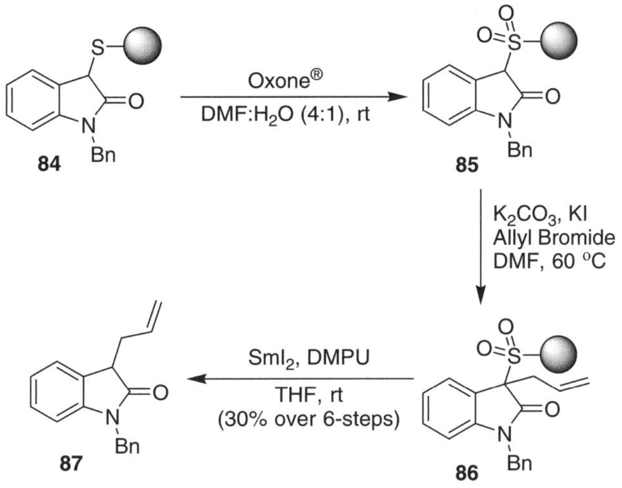

Scheme 13.18 *Procter's samarium(II) iodide mediated cleavage from sulfone linkers*

linker (Scheme 13.18). Thus, Procter oxidized resin-bound thioether **84** to the sulfone **85** using Oxone®. Further diversity was introduced by alkylation α to the sulfone using allyl bromide to give allylated sulfone **86** and then traceless cleavage with samarium(II) iodide/DMPU provided oxindole **87** in 30% yield over 6 steps on resin.

Typical Experimental Procedures

Traceless cleavage by dissolving metal reduction (scheme 13.17)[65]

*To a mixture of 5% sodium/mercury (860 mg) and disodium hydrogen phosphate (27 mg, 0.19 mmol) at room temperature under an argon atmosphere was added a solution of the polymeric sulfone **82** (200 mg, 0.0374 mmol) in anhydrous DMF (4 ml) followed by absolute methanol (0.5 ml). The reaction was then stirred for 22 hours after which time DCM (35 ml) was added. The resulting mixture was filtered through celite and washed with further DCM. The combined filtrate and washings were concentrated, redissolved in DCM (5 ml), and triturated with anhydrous diethyl ether (200 ml). The resulting suspension was filtered to remove the polymer support and washed with further diethyl ether. The combined filtrate and washings were then evaporated to dryness, redissolved in diethyl ether (30 ml) and filtered through silica gel. The filtrate was evaporated, and dried under vacuum to give the cleaved product **84** (7.5 mg, 97%).*

Traceless cleavage using samarium(II) iodide (scheme 13.18)[20]

*Oxindole resin **86** (0.158 mmol) was suspended in THF (4 ml). DMPU (0.153 ml, 1.26 mmol), and samarium(II) iodide (0.1 M in THF, 4.74 ml, 0.474 mmol) were added and the reaction was stirred at room temperature for 18 hours. After this time the resin was removed by filtration and the resulting filtrate was concentrated under vacuum. The concentrate was filtered through a short pad of silica gel followed by washing with 30% ethyl acetate in hexane. Further concentration under vacuum provided substituted oxindole **87** (30% over six steps).*

13.5.3 Multifunctional cleavage via elimination reactions

Base mediated elimination is a widely used multifunctional cleavage strategy from sulfone linker units due to the presence of acidic protons (especially those linkers with a carbonyl in the β-position). As highlighted by Procter and shown in Figure 13.1, it can proceed either via elimination of a group β to the sulfur to leave a resin-bound vinyl sulfone (type 1 elimination) or by elimination of a group attached to the sulfur to release olefinic compounds (type 2 or β-elimination).[3]

Type 1 elimination, reported by Tesser in 1976,[66, 67] was originally developed for early solid-phase peptide synthesis (SPPS) and recent reports show that such cleavage strategies continue to find extensive use in this area (Table 13.3, Entries 1 and 2).[68–70] Treatment of resin-bound sulfones **88** with a base cleaves the product and leaves supported vinyl sulfones **89**. In theory nucleophiles can be attached to the vinyl sulfones, as described in Section 13.5.1 above, allowing for recycling of the linker unit. For example, resin-bound tetrapeptides were built up on a sulfone linker unit reported by Katti using standard *tert*-butoxycarbonyl (Boc) and 9-fluorenylmethoxycarbonyl (Fmoc) SPPS. Eliminative cleavage could then be achieved using sodium hydroxide to provide pentapeptides (Table 13.3, Entry 1). Building on these concepts, such cleavage strategies have also been adapted for standard SPOS (Entries 3–7).[59–61, 66, 67, 71–73] Quaternary amino salts are generated on-bead, activating the sulfone for cleavage. Treatment with a base (typically DIPEA) then eliminates the corresponding tertiary amines such as substituted amines (Entry 3),[60] tetrahydroisoquinolines (Entry 4),[59, 60] piperidines (Entry 5)[72] and benzimidazoles (Entry 6).[71] As Table 13.3 shows, such cleavage strategies have typically concentrated on the preparation of amines. However, Schwyzer also used such an approach to generate substituted dinucleoside monophosphates (Entry 7).[61]

Whilst Type 1 elimination is traceless and offers scope for introduction of diversity immediately prior to cleavage, it has yet to be used in a truly multifunctional cleavage approach. In contrast, β-elimination (Type 2) to cleave olefinic species from sulfone linkers is a commonly employed multifunctional cleavage strategy, particularly for those compounds possessing carbonyl groups in the β-position.[11, 58, 70, 74–80] For example, such an approach was used by Yamada who prepared resin-bound cysteine derivatives **90** attached through a thioether linkage. After oxidation to the sulfone **91** with mCPBA, treatment with DBU resulted in β-elimination of enamine derivatives **92** with concomitant cleavage of the carbon–sulfur bond (Scheme 13.19).[75]

In Yamada's work, the resulting alkenes (**92**) were the desired target molecules. However, the alkene functionality can also be used as a point of further derivatization in order to introduce additional diversity into compound libraries during cleavage. Such an approach was demonstrated in cleavage of substrates from the bifunctional diversity/cyclative cleavable linker reported by Barco et al. (Scheme 13.20).[11] Resin-bound phosphonium salt **93** was treated with an array of aldehydes to introduce the first point of diversity through a solid-phase Wittig reaction and generate a small library of polymer-supported α,β-unsaturated ketones **94**. Cleavage was then achieved by treating with benzylamine to provide a series of piperidones **95**. The authors suggest that initially benzylamine underwent a Michael addition to the existing vinyl ketone, after which a second equivalent of benzylamine induced β-elimination of the sulfonyl group with concomitant generation of a second enone. Following cleavage, a second intramolecular Michael addition

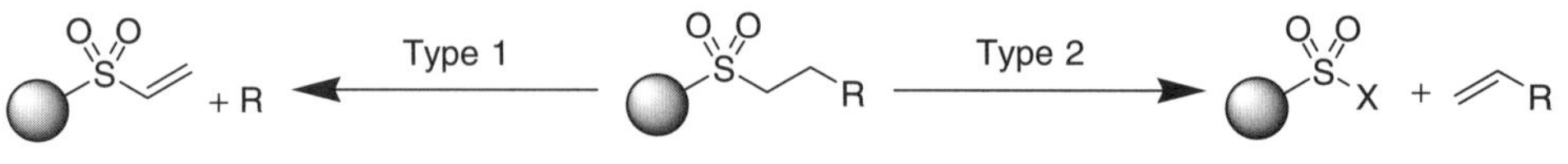

Figure 13.1 Eliminative cleavage strategies

Table 13.3 Type 1 elimination

	Resin	Cleavage Conditions	Product	Yield (%)	Ref
1	⬤–SO₂–CH₂CH₂–O–Leu-Phe-Gly-Tyr-Boc	NaOH	HO-Leu-Phe-Gly-Tyr-Boc	60	[68]
2	⬤–SO₂–CH₂CH₂–O–C(=X)–CH₂–NH–C(=O)–NHPh	4M NaOH then HCl	HO–C(=X)–CH₂–NH–C(=O)–NHPh	X = O, S[a]	[70]
3	⬤–CH₂–O–C₆H₄–SO₂–CH₂CH₂–N⁺R¹R²R³ Br⁻	DIPEA (5 equiv.)	R¹R²R³N	65–83	[60]
4	⬤–SO₂–CH₂CH₂–N⁺(R³)-tetrahydroisoquinoline(R¹,R²)	DIPEA	R³N-tetrahydroisoquinoline(R¹,R²)	25–100	[59]
5	⬤–SO₂–CH₂CH₂–N⁺(R³)-piperidine-4-N(R¹)C(=O)R² Cl⁻	Me₂NH	R³N-piperidine-4-N(R¹)C(=O)R²	8–41	[72]
6	⬤–SO₂–CH₂CH₂–N-benzimidazolium(R¹,R²)–CH₂C₆H₄R³ Br⁻	Et₃N : DCM (1 : 19)	benzimidazole(R¹,R²)–CH₂C₆H₄R³	10–75	[71]
7	⬤–NHC(=O)–C₆H₄–SO₂–CH₂CH₂–O–P(=O)(O-C₆H₄Cl)–O–(MMTO-thymidine)	Pyridine, TMG	RO–P(=O)(O-C₆H₄Cl)–O–(MMTO-thymidine)	97	[61]

[a] Yields not reported

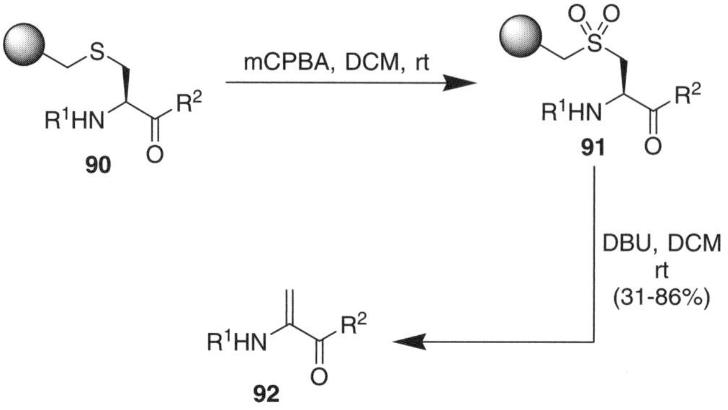

Scheme 13.19 *Type 2 elimination*

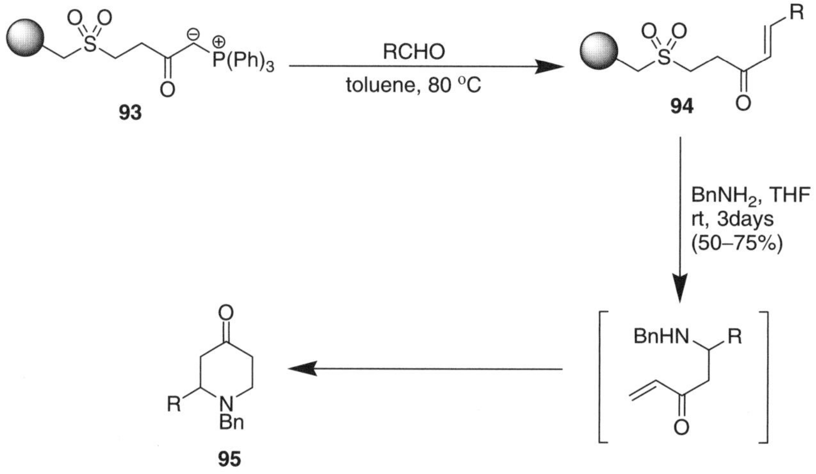

Scheme 13.20 *Synthesis of piperidones (**95**)*

of the amine to the resulting enone provided the piperidones. Whilst the authors have only demonstrated cleavage using benzylamine to date, presumably a range of amines could be employed to introduce a second point of diversity into the piperidone library during cleavage.

The carbonyl group β to the sulfone linker can also be generated prior to cleavage in a discreet step[74] or *in situ* during the cleavage step.[77, 79, 80] The former approach was used by Lam who oxidized resin-bound imidazo[1,2-a]pyridine derivatives **96** to the corresponding ketones **97** using Jones (or Dess–Martin) oxidation conditions. Simple treatment with sodium hydroxide resulted in β-elimination of the imidazo[1,2-a]pyridine **98** (Scheme 13.21).[74]

Building on this concept of multifunctional cleavage to generate simple olefins, Lam has adapted the oxidation/elimination strategy for the solid-phase syntheses of a wide assortment of heterocyclic compounds (Table 13.4). For example, demonstrating the ability to access libraries of diverse compounds from a common intermediate, Lam loaded benzyl bromide onto sodium benzenesulfinate resin **99** to provide sulfone **100** (confirmed by sulfone stretches at 1151 and 1316 cm^{-1} in the IR spectrum of **100**).[81] The sulfone was alkylated with epoxides to provide resin-bound alcohols **101** followed by oxidation to provide

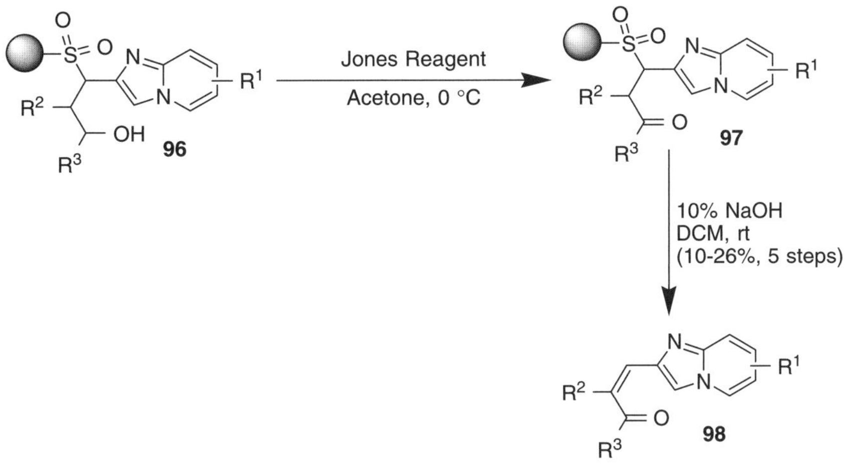

Scheme 13.21 Lam's synthesis of imidazo[1,2-a]pyridines (**98**)

ketones **102**. The oxidation was monitored by FTIR spectroscopy for the appearance of a C = O signal in the spectrum at 1687 cm^{-1}. Elimination cyclization cleavage with a range of reagents gave rapid access to very diverse classes of compound as shown in Table 13.4. For example, treatment with substituted hydrazines provided pyrazolines (cleavage under a nitrogen atmosphere, Entry 1) or a 4.5:1 mixture of pyrazoles:pyrazolines (cleaved in air, Entry 2).[81] Alternatively, cleavage could be carried out using hydroxylamines to furnish isoxazolines (Entry 3);[81] o-phenylene diamine to give benzo[b][1, 4]diazepines (Entry 4);[82] pyrimidines (Entry 5);[82]; (Entry 6)[82] or 3,4,6-trisubstituted-2-pyridones (Entry 7).[83] In the special case where R^1 is a ketone, treatment with hydrazine resulted in simultaneous condensation and cleavage to provide substituted pyridazines (Entry 8).

Typical Experimental Procedures

Pyrazolines (table 13.4, Entry 1)[81]

*The appropriate polymer-supported sulfone **102** (0.2 g) was suspended in methanol (20 ml) and nitrogen was then bubbled through for one hour. After this time, potassium hydroxide (0.30 g, 4.5 mmol) and phenyl hydrazine (0.32 g, 3.0 mmol) were added and the reaction was refluxed under nitrogen for an additional 12 hours. The resin was then removed by filtration and washed successively with DCM (2 × 10 ml), water (2 × 10 ml), DCM (2 × 10 ml), water (2 × 10 ml), DCM (2 × 10 ml) and water (2 × 10 ml). The filtrates were combined and the organic layer was separated and washed with brine (20 ml) and concentrated. The crude residue was purified by flash chromatography to provide pyrazolines (Table 13.4, Entry 2, 25–45%).*

Similarly, a three-component condensation of an amide (e.g. benzamide), aldehyde (e.g. benzaldehyde) and resin-bound sulfinic acid **103** (prepared from acidification of resin-bound sodium sulfinate) gave amide resin **104** containing two points of diversity. This reaction was monitored by IR spectroscopy, as the spectrum of the product showed an amide stretch at 1650 cm^{-1}. Treatment of **104** with excess triethylamine and an aldehyde in the presence of thiazolium catalyst generated α-ketoamide *in situ* whilst concomitantly cleaving it from the resin and introducing a third point of diversity into the target library.

Table 13.4 Multifunctional cleavage of heterocycles from sulfone linkers

	Cleavage Conditions	Product	Yield (%)	Ref
1	PhNHNH$_2$, KOH, CH$_3$OH reflux, N$_2$	4,5-dihydropyrazole with R^1, R^2, R^3, R^4	25–45	[81]
2	PhNHNH$_2$, KOH, CH$_3$OH reflux, air	4,5-dihydropyrazole : pyrazole 1 : 4.5	83[a]	[81]
3	NH$_2$OH·HCl, KOH, CH$_3$OH reflux, N$_2$	4,5-dihydroisoxazole	10–33	[81]
4	1,2-phenylenediamine (o-C$_6$H$_4$(NH$_2$)$_2$)	2,3-dihydro-1H-1,5-benzodiazepine	35	[82]
5	amidine H$_2$N–C(=NH)–R^4	pyrimidine with R^1, R^2, R^3, R^4	20–53	[82]

Table 13.4 (continued)

Entry	Reagent	Product	Yield (%)	Ref.
6	H$_2$N-C(=X)-NHR^4; X = O, S	6-membered ring: X=C with R^4-N and NH, bearing R^1, R^2, R^3; X = O, S	7–42	[82]
7	H$_2$N-C(=O)-CH$_2$-X; X = CN, CO$_2$H, CONH$_2$	2-pyridone with HN, C=O, X, R^1, R^2, R^3 substituents; X = CN, CO$_2$H, CONH$_2$	14–49	[83]
8	NH$_2$NH$_2$	Pyridazine with R^1, R^2, R^3	9–46	[78]

a Proof of concept solution phase reaction

Final cyclization of the α-ketoamide with amines provided imidazoles containing a fourth point of diversity making this SPOS strategy particularly amenable for library synthesis (Table 13.5, Entry 1).[84] In additional studies, Lam showed that quenching the α-ketoamide with Lawesson's reagent provided thiazoles (Entry 2) whilst treatment with triphenylphosphine and iodine furnished oxazoles (Entry 3).[84]

An approach for the solid-phase synthesis of oxazoles was also reported by Ganesan (Scheme 13.22).[85] Polymer-supported sulfonylmethyl isocyanide was prepared by loading N-(p-tosylsulfonylmethyl)formamide onto thiol resin **105**. The resulting thioether **106** was oxidized to the sulfone **107** with m-CPBA and then the formamide was dehydrated using PPh$_3$/carbon tetrachloride (found to be more convenient in this case than the accepted phosphoryl chloride (POCl$_3$) method) to provide the sulfonylmethyl isocyanide **108**. Cleavage to generate a small family of oxazoles **109** was then carried out by treating with benzaldehydes in the presence of base. A number of bases were screened and tetrabutylammonium hydroxide was found to be optimum, giving the highest yields of products (~50%).

Generation of the requisite carbonyl group *in situ* has been developed by Kurth et al.[77, 79, 80] In this case, resin-bound alcohols are subjected to the Swern oxidation. This generates ketone intermediates but, due to the basic conditions of the Swern reaction, immediate sulfinate elimination and release of α,β-unsaturated ketones occurs (Table 13.6). The scope of Kurth's simultaneous oxidation/elimination strategy has been successfully demonstrated in solid-phase syntheses of libraries of enones (Entry 1),[77] isoxazolocyclobutanones (Entry 2)[80] and cyclopentenones (Entry 3).[79] In the latter case, Kurth reported suitable alternatives to the Swern conditions included SO$_3$/pyridine/triethylamine/DMSO or TPAP/NMO/DCM.

Table 13.5 Cleavage of imidazoles, thiazoles and oxazoles

	Quenching Reagent	Product	Yield (%)	Ref
1	R^4NH_2, EtOH, AcOH, reflux	imidazole with R^1, R^3, R^2, R^4 substituents	24–40	[84]
2	Lawesson's reagent	thiazole with R^2, R^3, R^2 substituents	19–27	[84]
3	PPh_3, I_2	oxazole with R^1, R^3, R^2 substituents	22–32	[84]

Typical Experimental Procedures

Isoxazolocyclobutanones (table 13.6, entry 2)[80]

DMSO (8 equiv., 3.5 mmol, 0.27 g) in DCM (2 ml) was added to a solution of oxalyl chloride (4 equiv., 1.8 mmol, 0.22 g) in DCM (5 ml) at −60 °C. The reaction was stirred for five minutes and then added to a cooled (−60 °C) suspension of polymer-supported sulfone (0.69 g, 0.44 mmol) in DCM (10 ml). The reaction was stirred for 30 minutes after which time triethylamine (10 equiv., 4.4 mmol, 0.45 g) was added and the reaction was stirred for a further 20 minutes and then warmed to room temperature. The reaction was filtered and the resin washed successively with DCM (2 × 10 ml), methanol (10 ml), DCM (2 × 10 ml), methanol (10 ml), DCM (2 × 10 ml). All of the washings were combined with the original filtrate and concentrated. The resulting residue was finally purified by flash chromatography (20% ethyl acetate in hexanes) to give isoxazolocyclobutanones (34–38%).

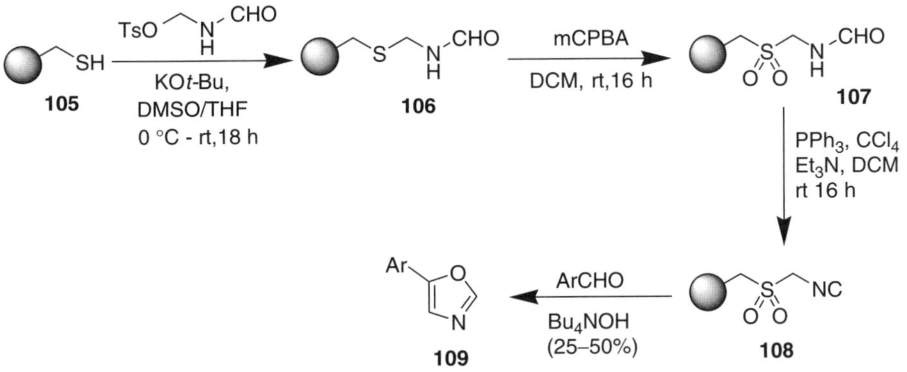

Scheme 13.22 Ganesan's oxazole synthesis

Table 13.6 Kurth's simultaneous oxidation/elimination cleavage strategy

	Resin	Cleavage Conditions	Product	Yield (%)	Ref
1	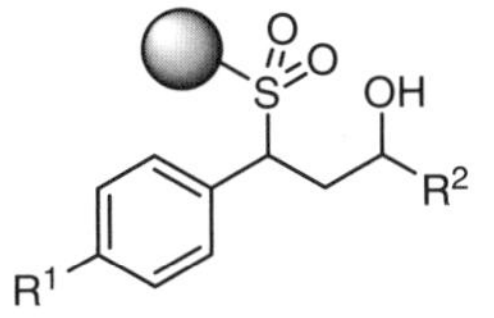 (resin-SO₂-CH(Ar-R¹)-CH₂-CH(OH)-R²)	Swern	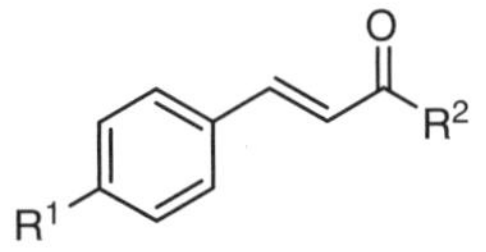 (R¹-Ar-CH=CH-C(O)-R²)	82–90	[77]
2	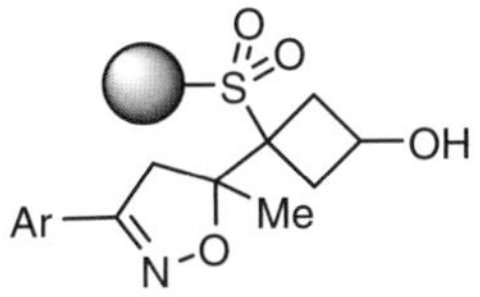 (resin-SO₂-cyclobutane-OH with isoxazoline Ar, Me)	Swern	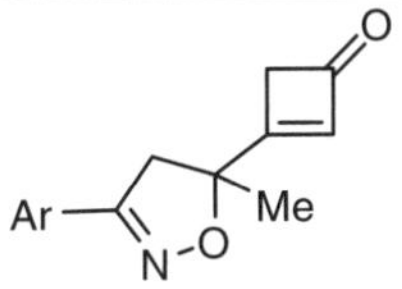 (cyclobutenone with isoxazoline Ar, Me)	34–38	[80]
3	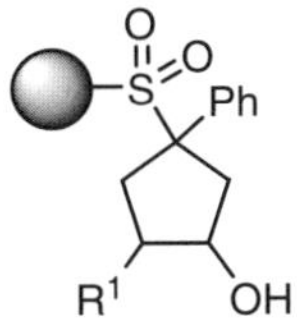 (resin-SO₂-C(Ph)-cyclopentane-OH, R¹)	Swern or SO₃/pyridine/Et3N or TPAP/NMO	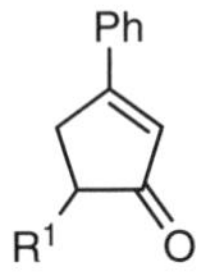 (Ph-cyclopentenone, R¹)	20–40	[79]

Finally, related to β-elimination is the modified Julia-Lythgoe olefination which in modern adaptations can be achieved with samarium(II) iodide rather than the traditional sodium/mercury reagent. De Clereq has adapted this elimination reaction for multifunctional cleavage in solid-phase synthesis (Scheme 13.23).[76] Thiophenol resin **110** was alkylated with primary tosylates to provide resin-bound thioethers which, after standard oxidation with *m*CPBA, yielded sulfone **111**. The sulfone was deprotonated with *n*-BuLi and the α-sulfonyl carbanion was treated with an aldehyde. The resulting alkoxide was quenched with acetic anhydride or benzoyl chloride to provide the corresponding acetate **112** or benzoate respectively. Cleavage

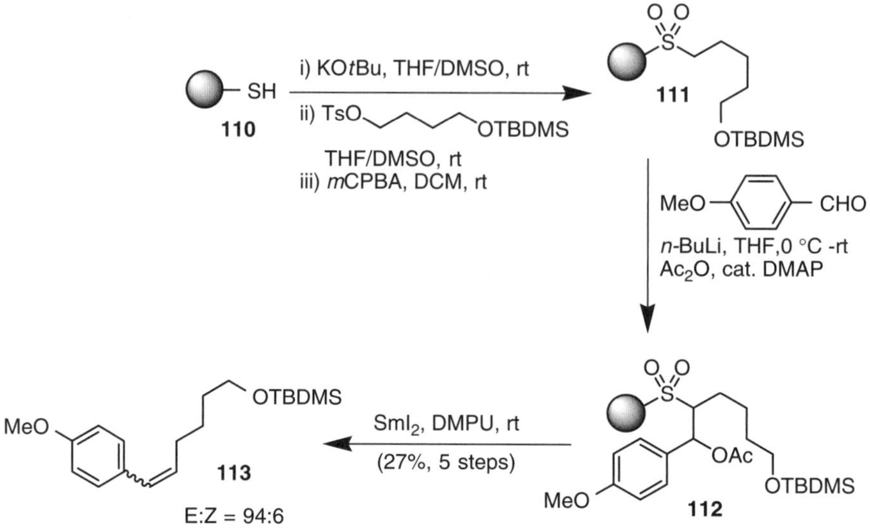

Scheme 13.23 *Cleavage using the Julia–Lythgoe reaction*

via the Julia–Lythgoe type reaction using hexamethylphosphoric triamide (HMPA) as co-solvent provided mixtures of E:Z alkenes **113** with relatively low stereoselectivity (E:Z ranged from 50:50 to 60:40). However, there was a noticeable improvement when the co-solvent was switched to DMPU (E:Z = 94 : 6).

Typical Experimental Procedures

Cleavage using the julia–lythgoe olefination (scheme 13.23)[76]

*Resin **112** (0.65 g, 0.767 mmol, loading = 1.18 mmol/g) was placed under an argon atmosphere and 0.1 M solution of samarium(II) iodide in THF (100 ml) was added. After a few minutes, freshly distilled DMPU (5 ml, 28.7 mmol) was added and the resulting blue–green solution was stirred at room temperature for 48 hours. After this time the reaction was milky white and turbid and to this was added a few drops of 2 M hydrochloric acid. The mixture was then filtered and the filtrate washed with saturated sodium thiosulfate ($Na_2S_2O_3$) solution. The crude reaction mixture was extracted with diethyl ether and the organic phase was washed with water and brine and finally dried. The solvent was removed under reduced pressure and the crude residue was purified by flash chromatography (iso-octane) to provide the alkenes **113** (27% over five steps).*

13.5.4 Multifunctional cleavage via nucleophilic substitution reactions

There are a number of examples of simple diversity cleavage from sulfone linkers via nucleophilic substitution with arrays of amines.[6, 7, 13,86–89] Typically this is an intermolecular process (although intramolecular cyclative cleavage strategies are also possible[13]) as was the case when Gayo and Suto reported an early safety-catch linker and its use in the preparation of pyrimidines (Table 13.7, Entry 1).[7] Table 13.7 highlights multifunctional cleavage from sulfone linkers using amines. A range of libraries have been prepared including pyrimidines (Entries 1 and 2), triazines (Entry 3), purines (Entry 4) and pteridinediones (Entry 5), all of which possess a diverse range of amine functionalities at the site of attachment to the resin. In

Table 13.7 Cleavage from sulfone linkers using amines

	Resin	Cleavage Conditions	Product	Yield (%)	Ref
1	pyrimidine with CF$_3$, CO$_2$Et substituents linked via -CH$_2$-SO$_2$- at 2-position	R^{1}NH$_2$, DCM	2-(R^{1}HN)-4-CF$_3$-5-CO$_2$Et-pyrimidine	50–93	[7]
2	pyrimidine with C(O)NR^{1}R^2 and R^3 substituents linked via -CH$_2$-SO$_2$- at 2-position	pyrrolidine, dioxane	2-pyrrolidinyl-pyrimidine with C(O)NR^{1}R^2 and R^3	46–65	[87]
3	triazine with R^1, R^2 substituents linked via -SO$_2$-	R^{3}NH$_2$, CH$_3$CN	R^{3}NH-triazine with R^1, R^2	90–99 purity	[6]
4	9-isopropylpurine with R^1 at 2-position, linked via -SO$_2$-CH$_2$- at 6-position	R^{2}NH$_2$	6-NHR^2-2-R^1-9-isopropylpurine	10–25	[86]
5	pteridinone (6,7-dimethyl) linked via -SO$_2$- at 2-position	NuH (R^{1}NH$_2$ or NaN$_3$)	2-Nu-6,7-dimethyl-pteridinone	34–42	[88, 89]

the latter case of pteridinediones (Entry 5), cleavage with sodium azide was also possible to leave an azide at the cleavage site.

Other sulfones are also amenable to nucleophile mediated diversity cleavage. For example, Kurth demonstrated that allyl sulfone resins **114** can be cleaved with various organocuprates in an S_N2' process to give cyclobutanols **115**, albeit with lower efficiency than the corresponding solution-phase reactions.[56] To improve this cleavage strategy, Kurth developed an alternative, more efficient process which exploits the allylic system **116** in a palladium catalyzed (or molybdenum catalyzed although with lower yields) process. This is much more versatile than the original organocuprate approach as diversity can be introduced with a much broader range of general nucleophiles to give larger libraries of protected cyclobutanols **117**.[90] In related work, similar palladium mediated cleavage processes have been described by Blechert[91] and Brown[92] with the ester-linked substrates. For example, Blechert demonstrated diversity cleavage by treating ester **118** with dimethyl malonate in the presence of palladium to give **119** (Scheme 13.24).[91]

Typical Experimental Procedures

Multifunctional cleavage from sulfone linkers using amines (table 13.7)[6]

The resin-bound sulfone was suspended in anhydrous acetonitrile (2 ml) and to this suspension was added PS-DIPEA resin (2 mg) and the amine (12.8 mol). The reaction was purged with nitrogen and heated at 90 °C for eight hours. After this time, the resin was filtered and washed with DCM (3 × 1 ml). The combined filtrates were concentrated to give the product, typically without additional purification.

Palladium mediated multifunctional cleavage (scheme 13.24)[90]

Diethyl (2-(3-benzyloxycyclobutylidene)ethyl)methylmalonate (117)

*Resin-bound sulfone **116** (3.0 g) was suspended in THF and treated with diethyl methylmalonate (2.38 g) and Pd(PPh$_3$)$_4$ (170 mg, 0.15 mmol, 5 mol%). The resulting mixture was refluxed for two days after which*

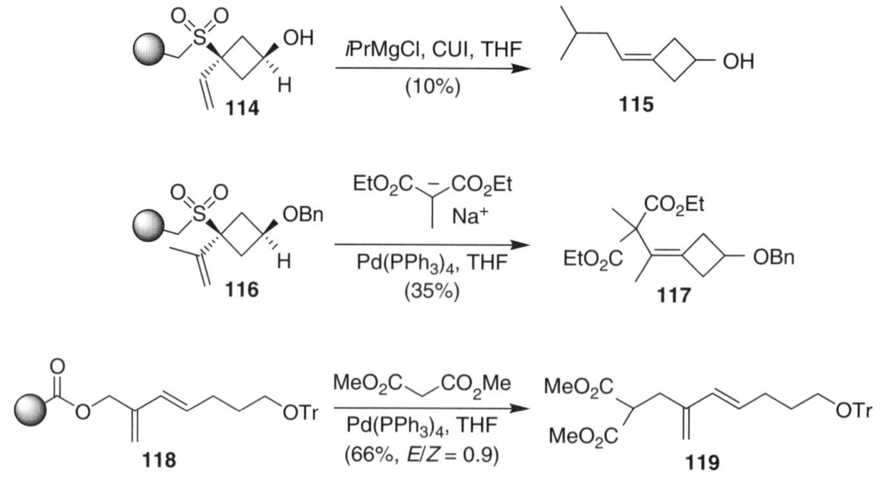

Scheme 13.24 *Cleavage using Grignard reagents or palladium catalyzed cross-coupling reactions*

*time the resin was filtered and the filtrate was neutralized with saturated aqueous ammonium chloride. The aqueous layer was extracted with diethyl ether and the combined organic extracts were dried (magnesium sulfate) and concentrated. Purification of the crude material by flash chromatography (25% ethyl acetate in hexanes) gave the title compound **117** (130 mg, 68%).*

13.6 Sulfonate ester linker units

13.6.1 Introduction

The nucleophilic displacement of alkanesulfonate groups, such as mesylates and tosylates, and their more reactive perfluoroalkanesulfonate counterparts, such as triflates and nonaflates, are classical reactions in organic synthesis and these groups therefore represent desirable linker motifs. Their stability under many common reaction conditions including Grignard reactions, Wittig reactions, $NaBH_4$-reductions, reductive aminations and treatment with common electrophiles and acylations enhances their importance as linker units. Both classes of leaving group (and simple analogs thereof) have been developed into multifunctional linker units and are discussed in this section.

13.6.2 Alkanesulfonate ester linker units

Simple alkanesulfonate leaving groups such as mesylates and tosylates have been adapted for solid-phase synthesis and a number of cleavage strategies have been developed. Alcohols and phenols are typically loaded onto supported-sulfonyl chlorides such as **123** to provide resin-bound sulfonate esters (Scheme 13.25). Typically cleavage occurs via O-alkyl or O-aryl bond breakage, although a few examples of aryl-S bond cleavage have also been reported.

13.6.2.1 Cleavage by Nucleophilic Substitution

In the case of aliphatic compounds bound to sulfonate ester linkers, multifunctional cleavage strategies have primarily focused upon substitution by treatment with nucleophiles. For example, Roush used this approach in the preparation of oligosaccharides (Scheme 13.25).[93] The resin employed was generated from simple Merrifield resin using Widlanski's reported procedure.[94] Merrifield resin **120** was treated with the lithium salt of isopropyl methanesulfonate **121** to give a resin-bound isopropyl sulfonate **122**. Treatment of this resin with sodium iodide in refluxing acetone converted the ester to the sulfonic acid and finally chlorination with thionyl chloride and triphenyl phosphine provided sulfonyl chloride resin **123**. Sugar molecules **124** were attached to the sulfonate linker through the primary hydroxyl group. Following solid-phase oligosaccharide synthesis to give resin-bound trisaccharide **126**, Roush demonstrated simple diversity cleavage using a range of nucleophiles. Treatment with sodium iodide, sodium acetate or sodium azide generated 6-iodo-**127a**, 6-acetoxy- **127b** or 6-azido-sugars **127c**, respectively, ready for further substitution if required (Scheme 13.25). In related work, Takahashi demonstrated similar multifunctional cleavage using Multipin™ systems.[95]

A number of groups have demonstrated similar multifunctional cleavage by treating resin-bound sulfonate esters with nucleophiles.[96–99] The nucleophile can also be intramolecular and in such cases allows for cyclative cleavage strategies to be developed (Chapter 4).[100] The true extent of diversity that can be introduced into compound libraries during cleavage via substitution with simple nucleophiles has been demonstrated by Nicolaou.[98, 99] A range of cyclic alkenes **128** were loaded onto immobilized sulfonic acid **129** (Table 13.8) in a one-pot procedure. Initial epoxidation of the alkene with dimethyldioxirane (DMDO) and loading of the epoxide onto the resin provided a resin-bound analog of an α-hydroxy tosy-

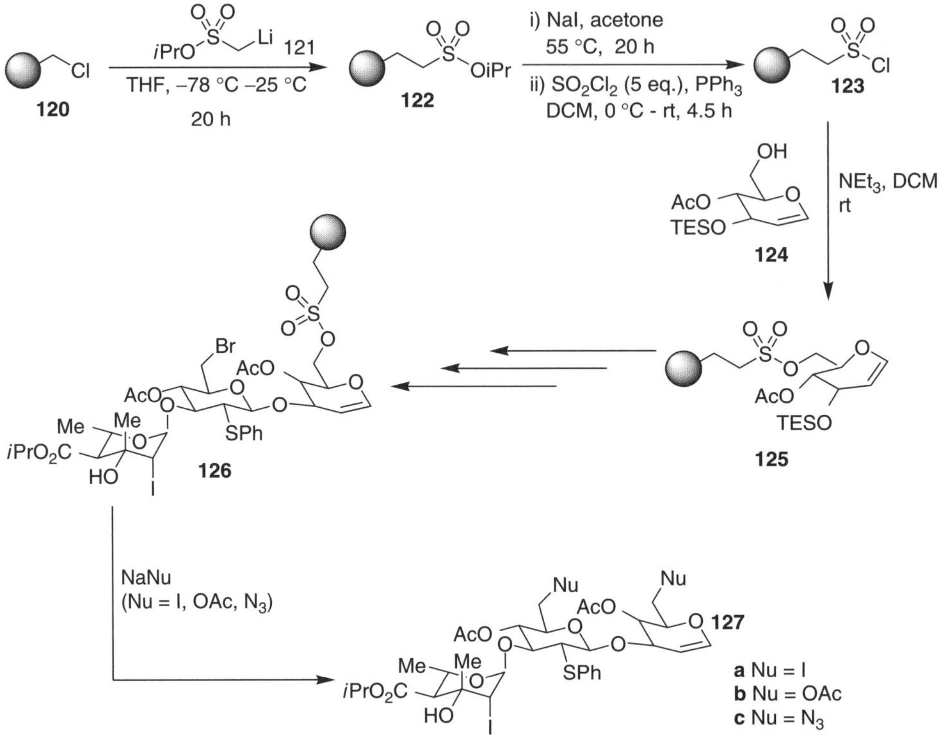

Scheme 13.25 Roush's oligosaccharide preparation

late. Subsequent oxidation with Dess–Martin periodinane (DMP) gave resin-bound α-sulfonated ketones **130** that were found to be highly stable both in solution and on solid phase. Multifunctional cleavage provided a diverse library of products, a selection of which is summarized in Table 13.8. For example, cleavage was demonstrated by treatment with a range of simple nucleophiles such as thiophenol (Entry 1), methanolic potassium carbonate (Entry 2) or morpholine (Entry 3) to give α-thiophenoxy ketones, α-methoxy ketones or α-morpholine ketones, respectively. Cleaving with 2-iodobenzoic acid (Entry 4) provides an iodobenzene, which permits further derivatization if required through, for example, Suzuki chemistry. More sophisticated multifunctional cleavage was also demonstrated, as the ketone functionality which was stabilizing the sulfonate linker also activated it upon attack by a nucleophile, and this was used in the preparation of a range of heterocyclic compounds. Thus, treatment with 2-substituted ethanols cleaved fused bicycles (Entry 5) whilst treatment with thioacetamide provided a fused thiazole (Entry 6). Tetrasubstituted pyrroles (Entry 7) were obtained upon cleavage using ethyl acetoacetate in the presence of amines whilst treatment with benzene-1,2-diamine or 2-aminothiophenol (Entry 8) provided tricyclic species. Finally, photo cleavage was also demonstrated to give the α-β-unsaturated ketone (Entry 9).

Typical Experimental Procedures

Loading substrates onto resin 129 (table 13.8)[98]

*Cyclododecene **128** (2–3 equiv.) was dissolved in DCM (0.1 M) and treated with DMDO (4 equiv.). This mixture was then added to resin **129** and the reaction was stirred at room temperature for four hours.*

Table 13.8 The Nicolaou sulfonate ester linker

	Cleavage Conditions	Product	Yield (%)
1	HSPh	cyclic ketone with SPh	95
2	K$_2$CO$_3$, CH$_3$OH	cyclic ketone with OMe	95
3	morpholine	cyclic ketone with morpholine	95
4	triethylamine, 2-iodobenzoic acid (CO$_2$H)	cyclic ketone with 2-iodobenzoate ester	78
5	H$_n$X—CH$_2$CH$_2$—OH	bicyclic product with X	X = NH 60 X = S 88 X = O 63
6	PPTS, thioacetamide (CH$_3$C(S)NH$_2$)	2-methylthiazole-fused cycle	83

(continued overleaf)

Table 13.9 (continued)

7	PPTS, 2-XH$_n$-phenyl-SH	fused tricyclic X/S heterocycle	X = NH 87 X = S 50
8	(CH$_3$)$_2$NH, diethyl malonate (EtO$_2$C-CH$_2$-CO$_2$Et)	N-methyl pyrrole fused to macrocycle with EtO$_2$C and methyl substituents	83
9	hν	macrocyclic enone	84

After this time, sodium bicarbonate (6 equiv.) and DMP (2 equiv.) were added and the reaction was stirred for 12 hours at room temperature. The resin was then filtered, washed with THF (2 × 500 ml), methanol (2 × 500 ml), DCM (2 × 500 ml) and diethyl ether (500 ml), and finally dried under vacuum to provide resin-bound α-sulfonated ketone **130**.

General procedure for cleavage of heterocyclic species[98]

(Table 13.8, Entries 5 and 7)

Resin-bound α-sulfonated ketone **130** was suspended in toluene or benzene and to this was added the appropriate bis-nucleophilic species (10 equiv.) and pyridinium p-toluenesulfonate (PPTS) (cat.). The reaction was then refluxed with a Dean–Stark apparatus after which the resin was filtered and the filtrate underwent standard aqueous work-up followed by purification by flash chromatography to provide the heterocyclic species (50–88% yield).

13.6.2.2 Cleavage by cross-coupling reactions

Whilst cleavage of aliphatic sulfonate esters using nucleophilic substitution is quite common, the corresponding cleavage of aryl sulfonate esters is comparatively rare. In contrast, the discovery that aryl sulfonate groups are viable substrates for transition metal catalyzed cross-coupling reactions has led to the development of a range of alternative multifunctional cleavage strategies. For example, in early work Wustrow cleaved aryl groups from resin-bound sulfonate **131** using a palladium mediated reduction with formic acid to leave a hydrogen residue at the cleavage site **132** (Scheme 13.26).[101] This concept has been expanded upon by Tsukamoto who found that benzenesulfonyloxy linker units are fairly unreactive towards palladium catalyzed reactions. However, after investigation of a range of traceless cleavage protocols, optimum conditions were found to be 20 mol% palladium(II) acetate (Pd(OAc)$_2$), 20 mol% PPF-*t*-Bu,

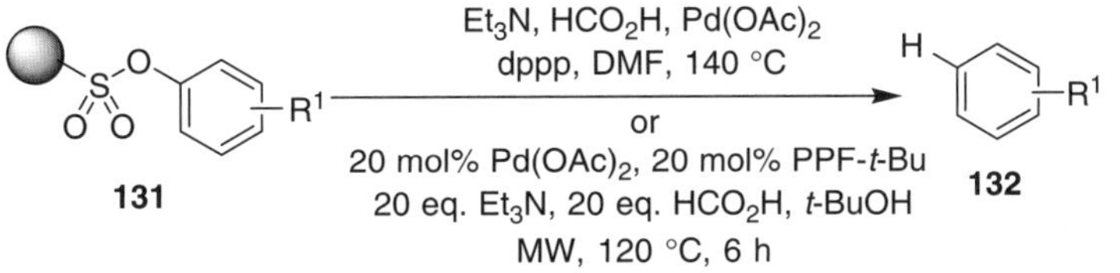

Scheme 13.26 *Traceless cleavage using palladium mediated transfer hydrogenation*

20 equiv. formic acid and 20 equiv. triethylamine in *t*-butanol at 120 °C. Cleavage of a range of aryl species was achieved in good yields.

These optimized conditions have also been adapted for multifunctional cleavage from **133** to provide a small library of substituted aromatic compounds (Scheme 13.27).[102] For example, palladium or nickel mediated cross-coupling reactions with boronic acids provided substituted phenyls (**134** and **135**), reaction with anilines under Hartwig–Buchwald conditions provided aniline derivatives **137** and palladium mediated α-arylation of cycloheptanone provided **136**.

In related work, Park has investigated cleavage of aryl sulfonate esters using nickel catalyzed cross-coupling reactions (Scheme 13.28). However, in this work the linker was attached to the resin through the O–S bond, and cleavage was achieved by activation of the S–aryl bond.[103] Arenesulfonyl chlorides **138** were loaded onto hydroxyethylmethyl resin **139** to give arenesulfonate resins **140**. In the case of bromo derivatives, the arenesulfonate linker could then be further modified, if required, using palladium catalyzed cross-coupling reactions to give resin-bound biphenyls **143**. Multifunctional cleavage with aryl Grignard reagents **141** was demonstrated by using more active nickel-based catalysts to afford biphenyls **142** or terphenyls **144**.

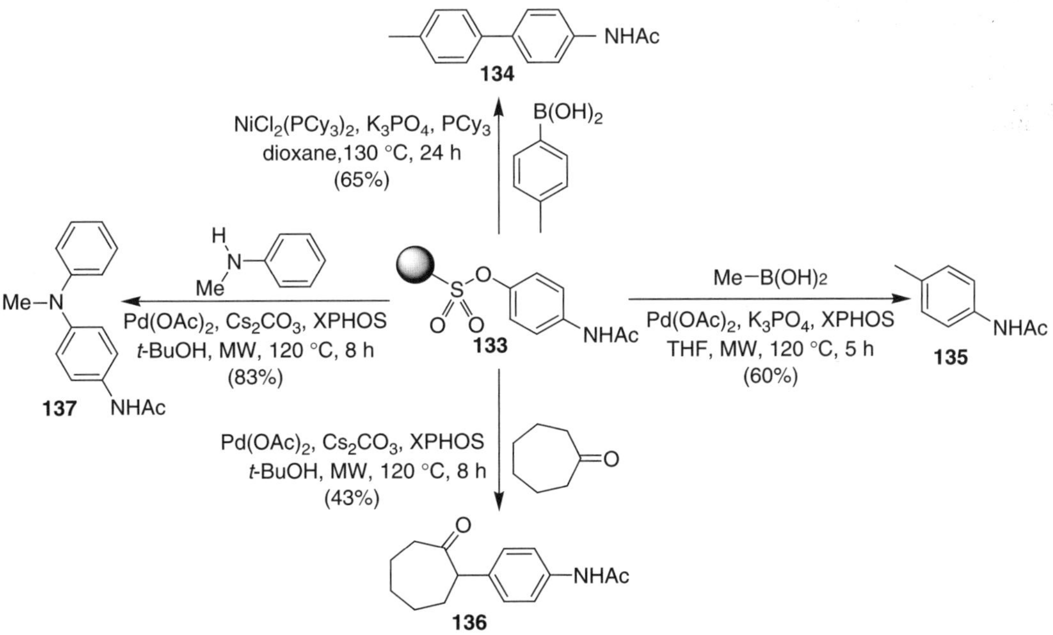

Scheme 13.27 *Palladium catalyzed multifunctional cleavage*

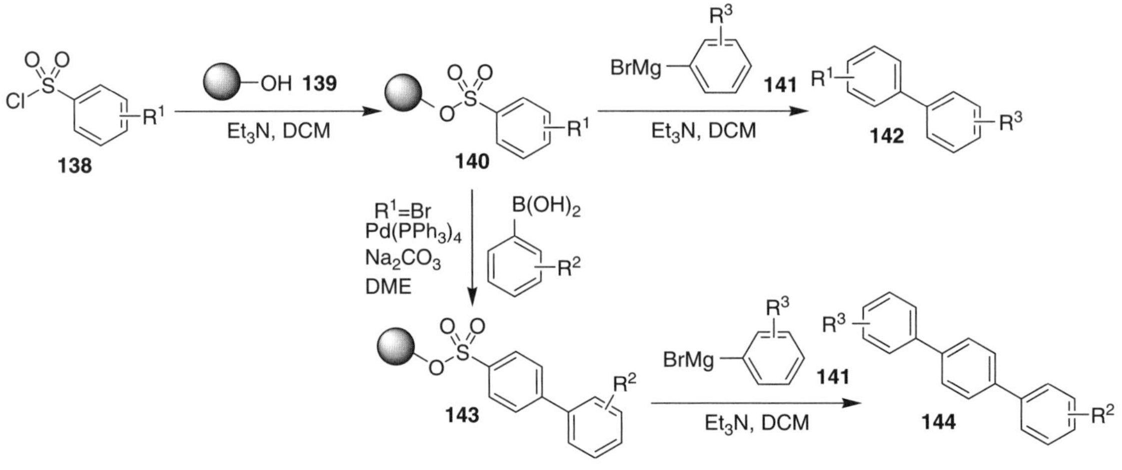

Scheme 13.28 Clevage using Grignard reagents

Typical Experimental Procedures

Traceless cleavage with formic acid (scheme 13.26)[102]

*Resin-bound sulfonate **131** (1 equiv.), palladium(II) acetate (20 mol%) and PPF-t-Bu (20 mol%) were placed in a test tube. To this was added solvent (t-BuOH, 0.33 M), triethylamine (20 equiv.) and formic acid (20 equiv.) under argon. The reaction was then heated to 120 °C under microwave irradiation for six hours. After this time, the mixture was concentrated in vacuo and the resulting residue was purified by flash chromatography to give substituted benzenes **132** (53–95% yield).*

Cleavage to generate by biphenyls (scheme 13.28)[103]

*Aryl Grignard reagents **141** (2.61 mmol) were added at room temperature to a suspension of supported arenesulfonate **140** (0.40 g, 0.26 mmol) and dppfNiCl$_2$ (53.4 mg, 0.078 mmol) in THF (8.0 ml). The reaction was then refluxed for 24 hours after which time it was cooled to room temperature, and additional **141** (1.31 mmol) was added. The mixture was then heated at reflux for a further 24 hours after which it was cooled and filtered through a sintered glass filter with diethyl ether. The filtrate was washed successively with 1% aqueaous hydrochloric acid, water and brine, dried (magnesium sulfate) and concentrated. Purification of the crude material by flash chromatography provided biphenyls **142** (64–81% yield).*

13.6.3 Perfluoralkanesulfonyl (PFS) linker units

13.6.3.1 Introduction

Perfluoroalkanesulfonates find widespread use as leaving groups in organic synthesis. Triflates and non-aflates are among the best leaving groups known and studies have shown that they are some 10^4–10^5 times more reactive than their mesylate and tosylate counterparts.[104–106] This excellent leaving group ability makes perfluoroalkanesulfonate esters labile towards a wide range of common nucleophiles such as alkoxides, amines and halides and therefore potential multifunctional linker units.

Whilst alkyl sulfonates have been used in solid-phase chemistry as described in Section 13.6.2, the corresponding alkyl perfluoroalkanesulfonates are very labile towards weaker nucleophiles and are prone to solvolysis, making them unsuitable for use as linker units that will be exposed to a wide range of SPOS reaction conditions. In contrast, their aryl and vinyl perfluoroalkanesulfonate counterparts exhibit high intrinsic stability and do not undergo solvolysis. Following the discovery that they are viable electrophilic components for palladium catalyzed cross-coupling reactions, their use in organic synthesis has greatly increased. This, in addition to the lability of aryl and vinyl triflates towards stronger nucleophiles in standard nucleophilic substitution reactions, makes them ideal multifunctional linker candidates.

13.6.3.2 Cleavage from perfluoroalkanesulfonyl linker units

Perfluoroalkanesulfonyl linkers that are based on this concept were first reported by Pan and Holmes in 2001.[107, 108] Perfluoroalkanesulfonyl linker **146** was derived in five steps from the commercially available iodosulfonyl fluoride **145**. Phenols were loaded using potassium carbonate to give resin-bound perfluoroalkanesulfonate esters **147** (Scheme 13.29).

With a resin-bound perfluoroalkanesulfonate ester linker in hand, traceless cleavage was achieved using an analogous palladium mediated reaction with formic acid to that described above for sulfonate ester linkers. The concept was demonstrated in the solid-supported synthesis of meclizine **149** and the palladium catalyzed transfer hydrogenation reaction left a hydrogen residue at the cleavage site.[107] In further applications, diversity cleavage was also developed using Suzuki cross-coupling reactions to release the desired target molecules as biphenyls **151** (Scheme 13.30).[108]

Concurrently with the efforts of Pan and Holmes, a perfluoroalkanesulfonyl linker unit was also developed by Steel and coworkers starting from 1,4-diiodooctafluorobutane (**152**).[5] Radical addition of sodium dithionite to 1,4-diiodooctafluorobutane gave the bis(sodium) sulfinate salt **153**, and subsequent treatment with molecular chlorine yielded the bis(sulfonyl) chloride **154**. This bis(sulfonyl) chloride was loaded onto amino TentaGel® and gave the solid-supported perfluoroalkanesulfonyl chloride **155** (Scheme 13.31). Phenols were loaded onto this linker unit to give resin-bound analogues of the aryl triflates **156**, and cleavage of the substrates to give biphenyls **157** was demonstrated using the Suzuki cross-coupling reaction.

Typical Experimental Procedures

General procedure for loading phenols (147) (scheme 13.29)[107, 108]

*A mixture of phenol (0.68 mmol), potassium carbonate (100 mg, 0.72 mmol), resin-bound sulfonyl fluoride linker **146** (80 mg, 0.034 mmol) and DMF (1.0 ml) was shaken at room temperature overnight. The resin*

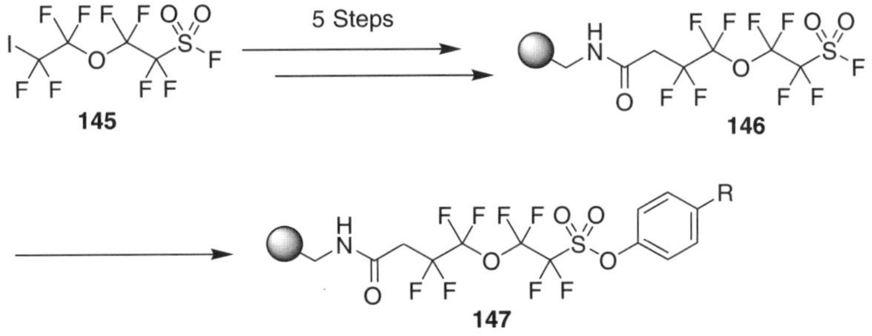

Scheme 13.29 *The Pan and Holmes perfluoroalkanesulfonyl linker unit*

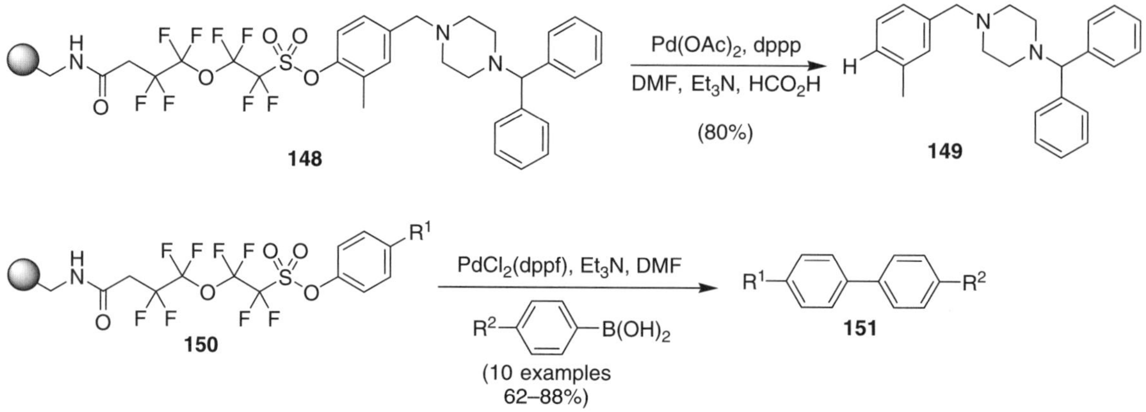

Scheme 13.30 Multifunctional cleavage from perfluoroalkanesulfonyl linkers

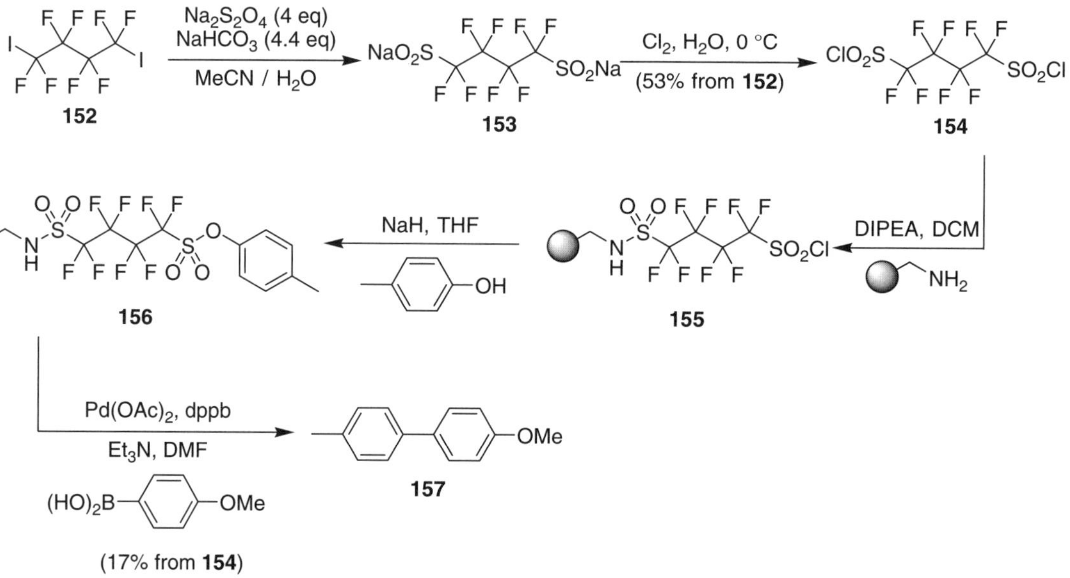

Scheme 13.31 Scott and Steel's perfluoroalkanesulfonyl linker

was then washed successively with water, DMF and DCM, and dried under vacuum overnight to give resin-bound perfluoroalkanesulfonate esters **147**.

General procedure for traceless cleavage

Meclizine 149 (scheme 13.30)[107, 108]

To the dry resin **148** was added palladium(II) acetate (8.0 mg, 0.036 mmol), 1,3-bis(diphenylphosphino) propane (dppp, 17.0 mg, 0.041 mmol), DMF (1.4 ml) and a mixture of formic acid (0.2 ml) and triethylamine (0.8 ml). The reaction was then shaken at 85 °C for two hours after which time the polymer beads were *filtered and washed with diethyl ether. The combined organic phase was washed with aqueous sodium carbonate solution and water, and evaporated to dryness. The residue was dissolved in diethyl ether and eluted through a short column of alumina to removed inorganic residues. The crude material was purified by preparative TLC to provide meclizine* **149** *(80% yield, >98% purity).*

General procedure for diversity cleavage using suzuki reactions (scheme 13.30)[107, 108]

A mixture of polymer-supported aryl perfluoroalkylsulfonate **150** *(200 mg, 0.07 mmol), [PdCl$_2$(dppf)] (7.2 mg), boronic acid (0.26 mmol) and triethylamine (88 μl, 0.62 mmol) in DMF (1.5–2.0 ml) was placed in a tube under an nitrogen atmosphere. The tube was sealed and heated at 90 °C for eight ours. After this time, the resin was filtered and washed with diethyl ether. The filtrate was washed with 10% aqueous sodium carbonate and water, and evaporated to dryness. The residue was re-dissolved in diethyl ether and filtered through a plug of silica gel to remove inorganic materials. The crude material was purified by preparative TLC to give biphenyls* **151** *(62–89% yield, >98% purity).*

13.6.4 Tetrafluoroarylsulfonyl linker units

Whilst the perfluoroalkanesulfonyl linker units function efficiently, methods for their preparation are not trivial, and consequently they have seen limited use. A simpler alternative to these perfluoroalkanesulfonyl linker units is the fluoroarylsulfonate linker **161**, reported independently by both Cammidge[109] and Ganesan[110] in 2004. Bis-acid chloride **159** was prepared in three steps from pentafluorobenzoic acid (**158**). Initially pentafluorobenzoic acid was treated with sodium hydrogen sulfide to give the benzene thiol. Oxidation to the sulfonic acid with hydrogen peroxide and acetic acid provided the bis-acid and then treatment with POCl$_3$ and phosphorus pentachloride (PCl$_5$) gave the desired bis-acid chloride **159**. This was loaded onto hydroxymethyl TentaGel® **160** to give resin-bound sulfonyl chloride **161** (Scheme 13.32). Cammidge loaded 4-hydroxybiphenyl-4′-carbonitrile (**162**) to give the resin-bound sulfonate ester **163**.

Traceless and diversity cleavage was demonstrated by both groups by using several different strategies. For example, traceless cleavage was demonstrated using formic acid to leave a hydrogen residue at the cleavage site of **164**. Alternatively, the diversity cleavage was efficiently achieved using Negishi and

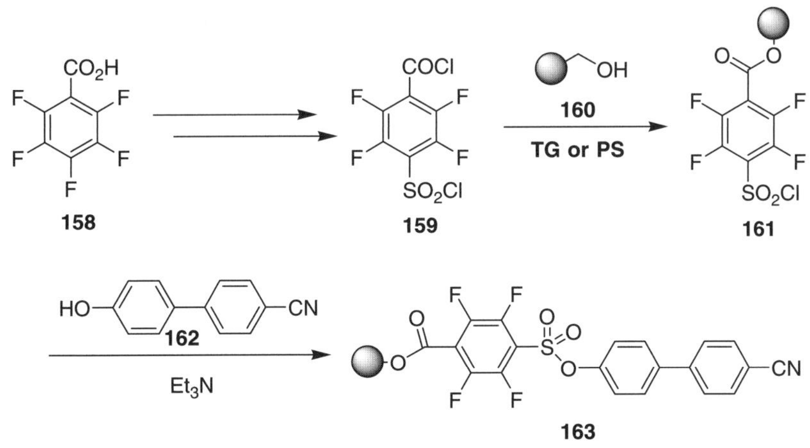

Scheme 13.32 Tetrafluoroarylsulfonyl linker units

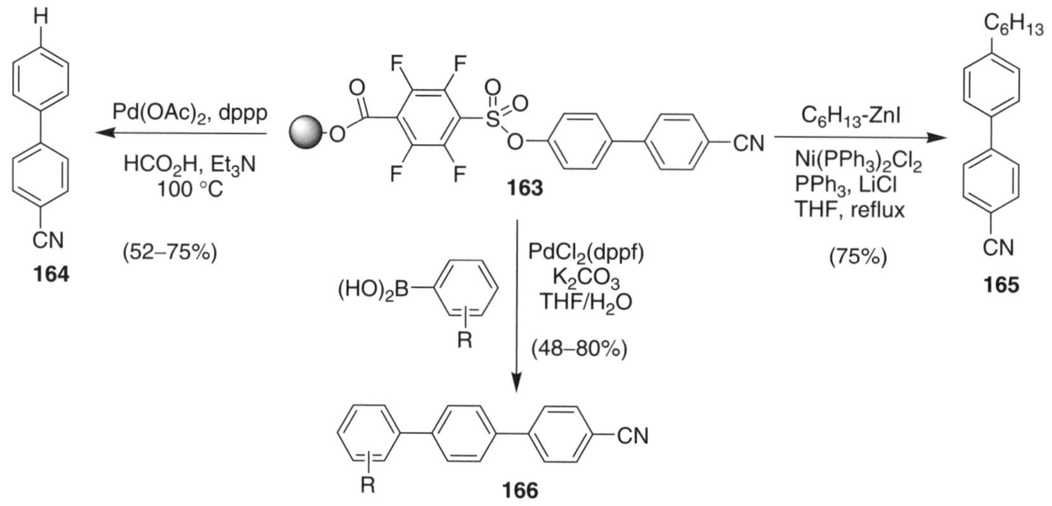

Scheme 13.33 Cleavage from tetrafluoroarylsulfonyl linker units

Suzuki reactions to leave alkyl groups and aryl groups at the cleavage site of the molecule in **165** and **166**, respectively (Scheme 13.33). This is a powerful diversity linker unit, which should see considerable exploitation in future SPOS.

Typical Experimental Procedures

Loading of substrates onto tetrafluoroarylsulfonyl linker unit (scheme 13.32)[109]

*A suspension of the resin-bound sulfonyl chloride **161** (0.28 mmol/g, 1 g, 0.28 mmol) in dry DCM (200 ml) was added to a solution of 4-hydroxybiphenyl-4′-carbonitrile **162** (0.23 g, 1.11 mmol) and triethylamine (0.17 ml, 1.20 mmol) in DCM (20 ml) at room temperature. The mixture was stirred at room temperature for 48 hours after which time the resin was filtered, washed with DCM (3 × 100 ml) and dried under vacuum to afford the resin-bound sulfonate ester **163** as a pale yellow polymer (0.97 g).*

Typical traceless cleavage protocol using formic acid (scheme 13.33)[109]

4-Cyanobiphenyl 164

*To a suspension of the resin-bound sulfonate ester **163** (0.265 mmol/g, 0.5 g, 0.13 mmol) in anhyrous. DMF (30 ml) was sequentially added triethylamine (0.334 ml, 2.4 mmol), formic acid (0.092 ml, 2.4 mmol), dppp (0.0136 g, 0.033 mmol) and Pd(OAc)$_2$ (6.7 mg, 0.03 mmol) at room temperature under an argon atmosphere. The reaction was heated at 100 °C for 16 hours and then allowed to cool to room temperature. The resin was filtered and washed with DCM (2 × 100 ml). Aqueous hydrochloric acid (5%, 50 ml) was added to the filtrate and the layers separated. The aqueous layer was extracted with DCM (2 × 20 ml) and the combined organic layers were washed with water (3 × 50 ml), dried, filtered and concentrated under reduced pressure. The crude residue was purified by preparative TLC (eluting with 9:1 petroleum ether/ether) to give 4-cyanobiphenyl **164** (0.015 g, 63%).*

Typical diversity cleavage protocol using Suzuki reactions (scheme 13.33)[109]

3″-Methoxy-[1,1′,4′,1″]terphenyl-4-carbonitrile

*Resin-bound sulfonate **4** (0.61 mmol/g, 1 g, 0.61 mmol) was suspended in THF (50 ml) and to this was added 3-methoxybenzeneboronic acid (0.46 g, 3.02 mmol), potassium carbonate (1.25 g, 9.07 mmol) and $PdCl_2(dppf) \cdot DCM$ (0.20 g, 0.27 mmol). Water (5 ml) was then added and the mixture was heated at reflux under argon for 48 hours. After this time, the resin was filtered and washed with THF (2 × 60 ml). Diethyl ether (50 ml) and water (30 ml) were added to the filtrate and the layers separated. The aqueous layer was extracted with more diethyl ether (2 × 50 ml) and the combined organic layers were washed with hydrochloric acid (1 M, 50 ml), brine (50 ml), dried, filtered and evaporated. The crude product was purified by preparative TLC (eluting with 2:1 DCM/petroleum ether) to give the 3″-methoxy-[1,1′,4′,1″]terphenyl-4-carbonitrile **166** as a colorless solid (0.11 g, 63%).*

Typical diversity cleavage protocol using Negishi conditions (scheme 13.33)[109]

4′-Hexylbiphenyl-4-carbonitrile

*Dry THF (20 ml) was added to a flame-dried flask charged with $(PPh_3)_2NiCl_2$ (69.1 mg, 0.11 mmol) and PPh_3 (0.0554 g, 0.21 mmol) at room temperature under an argon atmosphere. The reaction mixture was stirred for five minutes and subsequently n-BuLi (13.6 mg, 0.085 ml 2.5 M in hexane, 0.21 mmol), resin **163** (0.64 mmol/g, 0.3 g, 0.19 mmol) and lithium chloride (135 mg, 3.17 mmol) were added under a strong stream of argon at −78 °C. n-Hexylzinc iodide (2.90 ml of a 1.10 M solution in THF, 3.17 mmol) was then added and the reaction warmed to room temperature and stirred for 48 hours. The resin was filtered and washed with THF (60 ml). Aqueous hydrochloric acid (5%, 20 ml) was added to the filtrate and the layers separated. The aqueous layer was extracted with ethyl acetate (2 × 30 ml) and the combined organic layers were washed successively with hydrochloric acid (5%, 20 ml) and sodium hydroxide (5%, 20 ml), dried (sodium sulfate), filtered and evaporated under reduced pressure. The crude solid product was purified by preparative chromatography (eluting with DCM) to give 4′-hexylbiphenyl-4-carbonitrile **165** (0.0329 g, 65%).*

13.7 Sulfamate linker units

Sulfamate linkers are rare, but offer scope for multifunctional cleavage due to their ability to function as multidetachable linker units. For example, cleavage of substrates from linker **170** can be achieved either by breaking the sulfur–oxygen bond or the carbon–nitrogen bond (Scheme 13.34).[111] The linker unit was reported by Ciobanu and Poirier in 2003 and prepared by loading steroidal sulfamates **167** onto simple trityl chloride resin **168**. After solid-phase modifications, cleavage with piperazine (10 equiv.) at 45 °C cleaved the sulfur–oxygen bond providing the phenol **171**, whilst treatment with 5% TFA at room temperature broke the carbon–nitrogen bond to release the sulfamate **172**.

Typical Experimental Procedures

Loading steroidal sulfamates on trityl chloride resin (scheme 13.34)[111]

*Trityl chloride resin (**168**, Novabiochem, 2.05 mmol/g theoretical loading) (1 g) was swollen for five minutes in dry DCM (7 ml) and DIPEA (3.10 ml) under argon. Sulfamate **167** (1.34 g) was then added in portions*

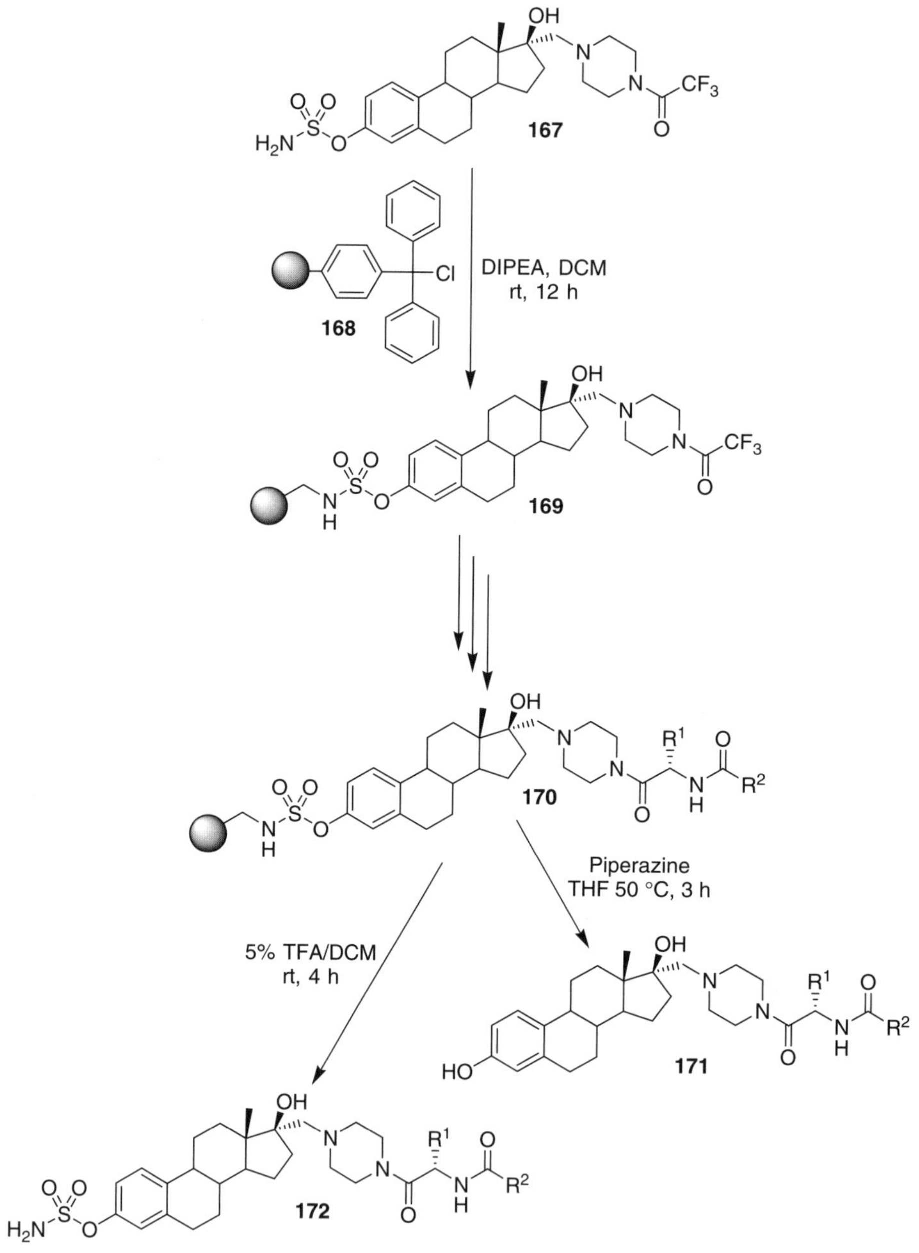

Scheme 13.34 Sulfamate linker units

*followed by additional dry DCM (3 ml). The resulting mixture was shaken for 24 hours at room temperature after which time the resin was filtered and washed three times with DCM, methanol and further DCM. Drying overnight under vacuum afforded resin (1.95 g, 91% by mass increase). The filtrate could be collected and purified by flash chromatography on alumina to recover and recycle unreacted sulfamate **167**.*

Generation of sulfamates 172 by acidic cleavage (scheme 13.34)[111]

*To resin-bound sulfamate **170** (95–126 mg) under an argon atmosphere was added a solution of 5% TFA in DCM (1.5 ml). The reaction was shaken for four hours at room temperature followed by filtration. The filtrate was collected and the solvent was evaporated in a Speedvac apparatus, to provide the sulfamate product **172** (as TFA salts, 18–65 mg, 18–66%, 90% HPLC purity).*

Generation of phenols 171 by nucleophilic cleavage (scheme 13.34)[111]

*Cleavage was carried out in 5-ml vials with Teflon caps and small magnetic stirring bars. Vials were charged with sulfamate resin **170** (60–82 mg) and piperazine (40 mg; 0.5 mmol). Freshly distilled THF (0.5 ml) was added and vials were sealed with Teflon caps and heated for three hours at 45 °C with gentle stirring. The reactions were then cooled to room temperature, filtered over a cotton pad in a small Pasteur pipet and the resins washed with ethyl acetate (3 × 2 ml). Filtrates were collected and treated with water (5 ml) and ethyl acetate (5 ml). After further extraction with ethyl acetate, the organic layers were washed with further water (2 ×). The water was removed with a pipet and the organic layers were evaporated to dryness in a Speedvac apparatus at 40 °C. Additional drying for 48 hours under vacuum provided the phenol derivatives **171** (15–27 mg; 33–54%, 90% HPLC purity).*

13.8 Thioester linker units

Thioesters are carboxylic acid derivatives that find extensive use as precursors to a range of compounds, including alcohols, aldehydes, ketones, acids, esters, lactones, amides and lactams. Easy access to such a wide range of functionalities makes thioesters ideal intermediates for diversity-oriented library synthesis and also strong candidates as multifunctional linker units. Despite this, there are limited examples of such linker units, due to difficulties in condensing thiols with sensitive resin-bound acid chlorides. Nevertheless, solutions to this problem have been reported and their potential has been recognized and exploited by a number of groups.[19, 112–114] To avoid the use of resin-bound acid chlorides, Kobayashi either loaded potassium thioacetate salts **174** onto Merrifield resin **173** or reacted solution-phase acid chlorides **176** with thiol resin **175** to provide supported thioester **177** as shown in Scheme 13.35. Reductive cleavage with lithium borohydride gave alcohols **178**.

Another solution to the problem of generating resin-bound thioesters has been developed by Bradley and Harrowven. Heating thioamide **180** with traditional Merrifield resin at 100 °C in DMF in the presence of sodium iodide provided resin-bound thioesters **181** (Scheme 13.36). The high temperature and inclusion of sodium iodide were necessary to achieve reasonable loading (∼40%) of the linker unit onto the resin. With such thioesters in hand, Bradley demonstrated multifunctional cleavage to provide a small library of benzo[d][1,3]dioxoles. Cleavage with lithium borohydride provided alcohol **182**, whilst treatment with Grignard reagents or organocuprates provided secondary alcohols **183** and ketones **184** respectively.

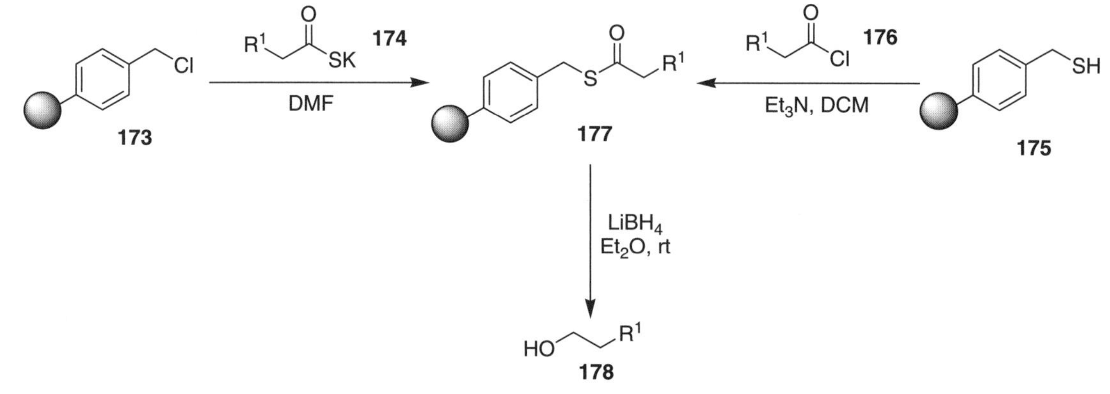

Scheme 13.35 Thioester linker units

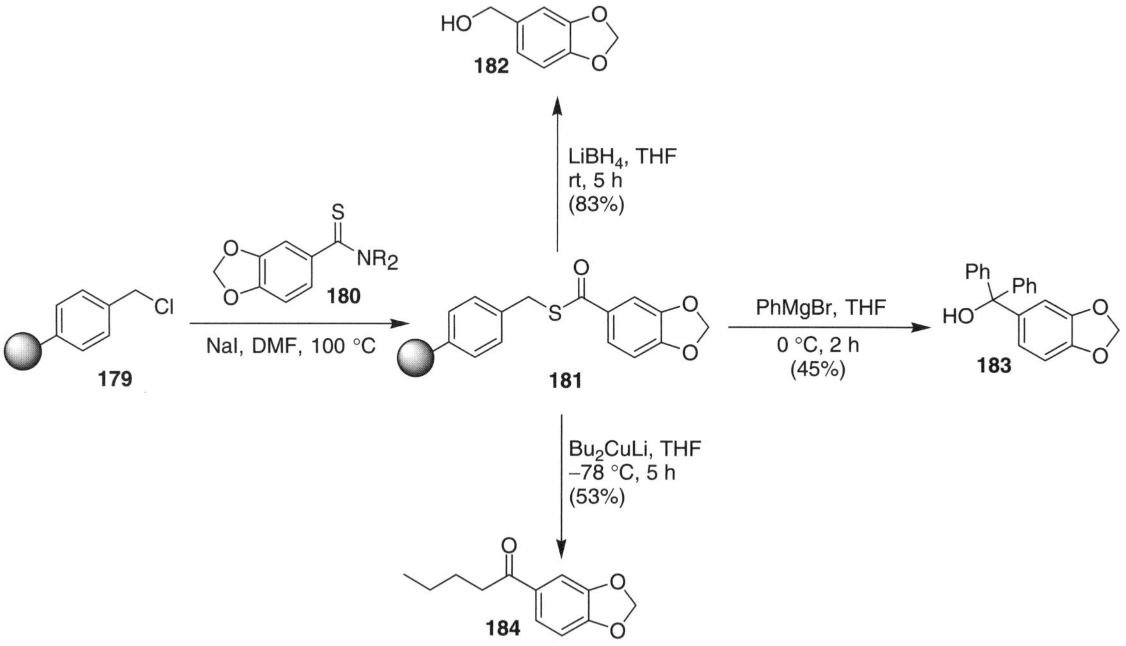

Scheme 13.36 Cleavage from thioester linker units

Typical Experimental Procedures

Cleavage of alcohols with lithium borohydride (scheme 13.35)[19]

*Resin-bound thioester **177** was suspended in diethyl ether (4 ml) and treated with lithium borohydride (21.8 mg, 1.0 mmol). The reaction was stirred at room temperature for 12 hours after which time phosphate buffer (pH = 7) and then 1 M sodium hydroxide were added. The aqueous layer was extracted with in diethyl ether and the combined organic layers were dried (sodium sulfate), filtered and concentrated.*

*Purification of the crude material by preparative TLC afforded alcohols **178**. The recovered thiol resin **175** could be recycled in further experiments.*

13.9 Conclusions

Due to the unique reactivity profile of sulfur, large numbers of sulfur linker units have been developed which make many useful solution-phase reactions available to the solid-phase organic chemist. In addition to photo cleavage and cyclative cleavage strategies, exploiting the activation of sulfur (through alkylation or oxidation) offers safety-catch possibilities. For example, substrates bound to many thioether linkers only become labile when the thioether is alkylated to the sulfonium ion or oxidized to the sulfoxide or sulfone.

Developing these concepts further has led to many multifunctional linker units based around sulfur in its varying forms. Such diversity cleavage strategies have given sulfur linker units a central role in solid-phase library synthesis and solid-phase preparations of many different classes of compound (including sugars, steroids, heterocyclic compounds, natural products, etc.), as highlighted in this chapter, have been accomplished. As many of the linker units herein offer multiple cleavage strategies (e.g. cross-coupling reactions, traceless cleavage, nucleophilic substitution, etc.), it becomes operationally simple to explore diverse chemical space from a common intermediate. Application of the linker units discussed herein to new solid-phase syntheses and, in addition, the development of the next generation of sulfur-linker units are both eagerly anticipated.

References

[1]   Zaragoza F.; *Angew. Chem. Int. Ed*. 2000, *39*, 2077.
[2]   Kan J. T. W., and Toy P. H.; *J. Sulf. Chem.* **2005**, *26*, 509.
[3]   McAllister L. A., McCormick R. A., and Procter D. J.; *Tetrahedron* **2005**, *61*, 11527.
[4]   Jung N., Wiehn M., and Bräse S.; *Top. Curr. Chem.* **2007**, *278*, 1.
[5]   Scott P. J. H., and Steel P. G.; *Eur. J. Org. Chem.* **2006**, 2251.
[6]   Khersonsky S. M., and Chang Y.-T.; *J. Comb. Chem.* **2004**, *6*, 474.
[7]   Gayo L. M., and Suto M. J.; *Tetrahedron Lett.* **1997**, *38*, 211.
[8]   Horton J. R., Stamp L. M., and Routledge A.; *Tetrahedron Lett.* **2000**, *41*, 9181.
[9]   Sucholeiki I.; *Tetrahedron Lett.* **1994**, *35*, 7307.
[10] Forman, F. W., and Sucholeiki, I.; *J. Org. Chem.* **1995**, *60*, 523.
[11] Barco A., Benetti S., De Risi C., *et al.*; *Tetrahedron Lett.* **1998**, *39*, 7591.
[12] Zoller T., Ducep J.-B., Tahtaoui C., and Hibert M.; *Tetrahedron Lett.* **2000**, *41*, 9989.
[13] Raghavendra M. S., and Lam Y.; *Tetrahedron Lett.* **2004**, *45*, 6129.
[14] Marshall, D. L., and Liener, I. E.; *J. Org. Chem.* **1970**, *35*, 867.
[15] Adamcsyk, M., Fishpaugh, J. R., and Mattingly, P. G.; *Tetrahedron Lett.* **1999**, *40*, 463.
[16] Adamcsyk, M., Fishpaugh, J. R., and Mattingly, P. G.; *Bioorg. Med. Chem. Lett.* **1999**, *9*, 217.
[17] Becht J.-M., Wagner A., and Mioskowski C.; *Tetrahedron Lett.* **2004**, *45*, 7031.
[18] Farrall M. J., and Fréchet M. J.; *J. Org. Chem.*, **1976**, *41*, 3877.
[19] Kobayashi S., Hachiya I., Suzuki S., and Moriwaki M.; *Tetrahedron Lett.* **1996**, *37*, 2809.
[20] McAllister L. A., Brand S., Gentile R. de, and Procter D. J.; *Chem. Commun.* **2003**, 2380.
[21] Jung K. W., Zhao X.-Y., and Janda K. D.; *Tetrahedron* **1997**, *53*, 6645.
[22] Virgilio, A. A., and Ellman J. A.; *J. Am. Chem. Soc.* **1994**, *116*, 11580.
[23] Kay, C., Lorthioir, O. E., Parr, N. J., *et al.*; *Biotechnol. Bioeng.* **2000**, *71*, 110.
[24] Jung K. W., Zhao X.-Y., and Janda K. D.; *Tetrahedron Lett.* **1996**, *37*, 6491.
[25] Srivastava S. K., Haq W., and and Chauhan P. M. S.; *Bioorg. Med. Chem. Lett.* **1999**, *9*, 965.
[26] Steel P. G.; *J. Chem. Soc., Perkin Trans. 1* **2001**, 2727.

[27] Dahlen, A., and Hilmersson, G.; *Eur. J. Inorg. Chem.* **2004**, 3393.
[28] Edmonds D. J., Johnston D., and Procter D. J.; *Chem. Rev.* **2004**, *104*, 3371.
[29] Gopalaiah K., and Kagan H. B.; *New J. Chem.*, **2008**, *32*, 607.
[30] Kagan H. B.; *Tetrahedron* **2003**, *59*, 10351.
[31] McKerlie, F., Procter, D. J., and Wynne, G.; *Chem. Commun.* **2002**, 584.
[32] Bégué J.-P., M'Bida A., Bonnet-Delpon D., et al.; *Synthesis* **1996**, 399.
[33] Ravikumar K. S., Bégué J.-P., and Bonnet-Delpon D.; *Tetrahedron Lett.* **1998**, *39*, 3141.
[34] Crosby G. A., and Kato M.; *J. Am. Chem. Soc.* **1977**, *99*, 278.
[35] Opatz T., Kallus C., Wunberg T., et al.; *Eur. J. Org. Chem.* **2003**, 1527.
[36] Opatz T., Kallus C., Wunberg T., et al.; *Carbohydr. Res.* **2002**, *37*, 2089.
[37] Opatz T., Kallus C., Wunberg T., and Kunz H.; *Tetrahedron* **2004**, *60*, 8613.
[38] Kallus C., Opatz T., Wunberg T., et al.; *Tetrahedron Lett.* **1999**, *40*, 7783.
[39] Wunberg T., Kallus C., Opatz T., et al.; *Angew. Chem. Int. Ed.* **1998**, *37*, 2503.
[40] Rademann J., and Schmidt R.; *J. Org. Chem.*, **1997**, *62*, 3650.
[41] Rademann J., and Schmidt R.; *Tetrahedron Lett.* **1996**, *37*, 3989.
[42] Bishop P., Jones C., and Chmielewski J.; *Tetrahedron Lett.* **1993**, *34*, 4469.
[43] Hennequin L. F, and Le Blanc S. P.; *Tetrahedron Lett.* **1999**, *40*, 3881.
[44] Huwe C. M., and Künzer H.; *Tetrahedron Lett.* **1999**, *40*, 683.
[45] Bertini V., Lucchesini F., Pocci M., et al.; *Synlett* **2003**, 1201.
[46] Bertini V., Pocci M., Lucchesini F., et al.; *Synlett* **2003**, 864.
[47] Bertini V., Lucchesini F., Pocci M., and De Munno, A.; *Tetrahedron Lett.* **1998**, *39*, 9263.
[48] Baer R., and Masquelin T.; *J. Comb. Chem.* **2001**, *3*, 16.
[49] La Porta E., Piarulli U., Cardullo F., et al.; *Tetrahedron Lett.* **2002**, *43*, 761.
[50] Vanier C., Lorgé F., Wagner A., and Mioskowski C.; *Angew. Chem. Int. Ed.* **2000**, *39*, 1679.
[51] Nakamura S., Uchiyama Y., Ishikawa S., et al.; *Tetrahedron Lett.* **2002**, *43*, 2381.
[52] Dragoli D. R., Burdett M. T., and Ellman J. A.; *J. Am. Chem. Soc.* **2001**, *123*, 10127.
[53] Rolland C., Hanquet, G., Ducep J.-B., and Solladié G.; *Tetrahedron Lett.* **2001**, *42*, 7563.
[54] Russell H. E., Luke R. W. A., and Bradley M.; *Tetrahedron Lett.* **2000**, *41*, 5287.
[55] Tai C.-H., Wu H.-C., and Li W.-R.; *Org. Lett.* **2004**, *6*, 2905.
[56] Halm C., Evarts J., and Kurth M. J.; *Tetrahedron Lett.* **1997**, *38*, 7709.
[57] Hwang S. H., and Kurth M. J.; *J. Org. Chem.*, **2002**, *67*, 6564.
[58] Sheng S.-R., Huang P.-G., Zhou W., et al.; *Synlett* **2004**, 2603.
[59] Heinonen P., and Lönnberg H.; *Tetrahedron Lett.* **1997**, *38*, 8569.
[60] Kroll F. E. K., Morphy R., Rees D., and Gani D.; *Tetrahedron Lett.* **1997**, *38*, 8573.
[61] Schwyzer, R., Felder, E., and Failli, P.; *Helv. Chim. Acta* **1984**, *67*, 1316.
[62] Zhao X. Y., Jung, K. W., and Janda K. D.; *Tetrahedron Lett.* **1997**, *38*, 977.
[63] Zhao X. Y., Metz W. A., Sieber F., and Janda K. D.; *Tetrahedron Lett.* **1998**, *39*, 8433.
[64] Zhao X. Y., and Janda K. D.; *Tetrahedron Lett.* **1997**, *38*, 5437.
[65] Zhao X. Y., and Janda K. D.; *Bioorg. Med. Chem. Lett.* **1998**, *8*, 2439.
[66] Buis J. T. W. A. R. M., Tesser G. I., and Nivard R. J. F.; *Tetrahedron* **1976**, *32*, 2321.
[67] Tesser G. I., Buis J. T. W. A. R. M., Wolters E. T. M., and Bothé-Helmes G. A. M.; *Tetrahedron* **1972**, *32*, 1069.
[68] Katti S. B., Misra P. K., Haq W., and Mathur K. B.; *Chem. Commun.* **1992**, 843.
[69] Canne L. E., Winston R. L., and Kent S. B. H.; *Tetrahedron Lett.* **1997**, *38*, 3361.
[70] Huang W., Cheng S., and Sun W.; *Tetrahedron Lett.* **2001**, *42*, 1973.
[71] Tumelty D., Cao K., and Holmes C. P.; *Org. Lett.* **2001**, *3*, 83.
[72] Wade W. S., Yang F., and Sowin T. J.; *J. Comb. Chem.* **2000**, *2*, 266.
[73] Garcia-Echeverria C.; *Tetrahedron Lett.* **1997**, *38*, 8933.
[74] Chen Y., Lam Y., and Lai Y.-H.; *Org. Lett.* **2002**, *4*, 3935.
[75] Yamada M., Miyajima T., and Horikawa H.; *Tetrahedron Lett.* **1998**, *39*, 289.
[76] D'herde J. N. P., and De Clercq P. J.; *Tetrahedron Lett.* **2003**, *44*, 6657.

[77] Cheng W.-C., Lin C.-C., and Kurth M. J.; *Tetrahedron Lett.* **2002**, *43*, 2967.
[78] Chen Y., Lam Y., and Lee Y.-S.; *Chem. Lett.* **2001**, 274.
[79] Cheng W.-C., and Kurth M. J.; *J. Org. Chem.*, **2002**, *67*, 4387.
[80] Cheng W.-C., Wong M., Olmstead M. M., and Kurth M. J.; *Org. Lett.* **2002**, *4*, 741.
[81] Chen Y., Lam Y., and Lai Y.-H.; *Org. Lett.* **2003**, *5*, 1067.
[82] Kong K.-H., Chen Y., Ma X., et al.; *J. Comb. Chem.* **2004**, *6*, 928.
[83] Li W., Chen Y., and Lam Y.; *Tetrahedron Lett.* **2004**, *45*, 6545.
[84] Li W., and Lam Y.; *J. Comb. Chem.* **2005**, *7*, 644.
[85] Kulkarni B. A., and Ganesan A.; *Tetrahedron Lett.* **1999**, *40*, 5633.
[86] Brun V., Legraverend M., and Grierson D. S.; *Tetrahedron* **2002**, *58*, 7911.
[87] Chucholowski A., Masquelin T., Obrecht D., et al.; *Chimia* **1996**, *50*, 525.
[88] Gibson C. L., La Rosa S., and Suckling C. J.; *Org. Biomol. Chem.*, **2003**, *1*, 1909.
[89] Gibson C. L., La Rosa S., and Suckling C. J.; *Tetrahedron Lett.* **2002**, *44*, 1267.
[90] Cheng W.-C., Halm C., Evarts J., et al.; *J. Org. Chem.*, **1999**, *64*, 8557.
[91] Schürer S. C., and Blechert S.; *Synlett* **1998**, 166.
[92] Brown R. C. D., and Fisher M.; *Chem. Commun.* **1999**, 1547.
[93] Hunt J. A., and Roush W. R.; *J. Am. Chem. Soc.* **1996**, *118*, 9998.
[94] Huang J, and Widlanski T. S; *Tetrahedron Lett.* **1992**, *33*, 2657.
[95] Takahashi T., Tomida S., Inoue H., and Doi T.; *Synlett* **1998**, 1261.
[96] Rueter J. K., Nortey S. O., Baxter E. W., et al.; *Tetrahedron Lett.* **1998**, *39*, 975.
[97] Baxter E. W., Rueter J. K., Nortey S. O., and Reitz A. B.; *Tetrahedron Lett.* **1998**, *39*, 979.
[98] Nicolaou K. C., Baran P. S., and Zhong Y.-L.; *J. Am. Chem. Soc.* **2000**, *122*, 10246.
[99] Nicolaou K. C., Montagnon T., Ulven T., et al.; *J. Am. Chem. Soc.* **2002**, *124*, 5718.
[100] Holte P. T., Thijs L., and Zwanenburg B.; *Tetrahedron Lett.* **1998**, *39*, 7407.
[101] Jin S., Holub D. P., and Wustrow D. J.; *Tetrahedron Lett.* **1998**, *39*, 3651.
[102] Tsukamoto H., Suzuki R., and Kondo Y.; *J. Comb. Chem.* **2006**, *8*, 289.
[103] Cho C.-H., Park H., Park M.-A., et al.; *Eur. J. Org. Chem.* **2005**, 3177.
[104] Hansen, R. L.; *J. Org. Chem.* **1965**, *29*, 4322.
[105] Streitwieser, A., Wilkins, C. J., and Kiehlmann, E.; *J. Am. Chem. Soc.* **1968**, *90*, 1598.
[106] Su, T. M., Sliwinski, W. F., and Schleyer, P. V. R.; *J. Am. Chem. Soc.* **1969**, *91*, 5386.
[107] Pan Y., and Holmes C. P.; *Org. Lett.* **2001**, *3*, 2769.
[108] Pan Y., Ruhland B., and Holmes C. P; *Angew. Chem. Int. Ed.* **2001**, *40*, 4488.
[109] Cammidge A. N, and Ngaini Z.; *Chem. Commun.* **2004**, 1914.
[110] Revell J. D, and Ganesan A.; *Chem. Commun.* **2004**, 1916.
[111] Ciobanu, L. C., and Poirier, D.; *J. Comb. Chem.* **2003**, *5*, 429.
[112] May P. J., Bradley M., Harrowven D. C., and Pallin D.; *Tetrahedron Lett.* **2000**, *41*, 1627.
[113] Kobayashi S., Wakabayashi T., and Yasuda M.; *J. Org. Chem.*, **1998**, *63*, 4868.
[114] Bolshan, Y., Tomaszewski, M. J., and Santhakumaar V.; *Tetrahedron Lett.* **2007**, *48*, 4925.

14

Selenium- and Tellurium-Based Linker Units

Tracy Yuen Sze But and Patrick H. Toy

Department of Chemistry, The University of Hong Kong, People's Republic of China

14.1 Introduction

The use of organoselenium reagents in organic synthesis has a long and rich history, since such compounds can participate in a vast array of electrophilic, nucleophilic and radical reactions.[1–6] In many applications of these reagents the organoselenium group is removed from the final target molecule by an oxidation/elimination reaction sequence, carbon–selenium bond homolysis or nucleophilic displacement, and such processes have been applied in the field of solid-phase synthesis. When used in this context, the role of the organoselenium group that is temporarily attached to the synthesis substrate is that of a linker group. This review summarizes the literature regarding the use of such selenium-based linker groups in solid-phase organic synthesis,[7–9] and includes the limited number of reports regarding related tellurium-based linkers. Because the methodologies described in this review are generally not polymer-support dependent, the discussion of the actual polymers used and how the linker groups are attached to them is kept to a minimum, and it is organized into three sections: (i) the selenium- and tellurium-based linker reagents that have been used and their syntheses; (ii) methods for the attachment of synthesis substrates to these linker reagents; and (iii) methods for synthesis product cleavage from the linker groups. Apologies are made in advance for any errors, omissions or misrepresentations.

14.2 Selenium- and tellurium-based linker group reagents and their syntheses

The various polymer-supported selenium and tellurium linker group reagents (**A–Q**) that have been used to immobilize synthesis substrates for polymer-supported synthesis are shown in Figure 14.1. The first

Linker Strategies in Solid-Phase Organic Synthesis Edited by Peter Scott
© 2009 John Wiley & Sons, Ltd

392 *Multifunctional Linker Units for Diversity-Oriented Synthesis*

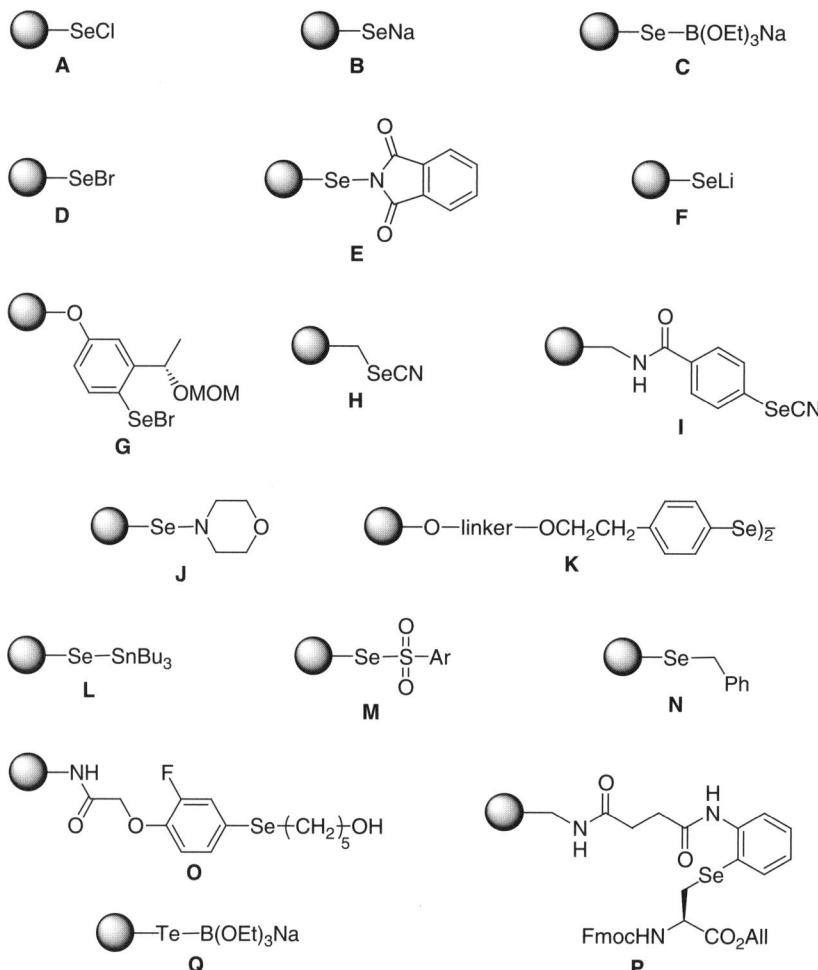

Figure 14.1 Polymer-supported selenium and tellurium linker group reagents

report of a polymer-supported selenium reagent and its use as a linker group was described by Heitz et al.[10] In this work 4-chlorostyrene (**1**) was converted via the corresponding Grignard reagent into 4-vinylbenzeneselenol (**2**), which was copolymerized with divinylbenzene under radical conditions to afford cross-linked polystyrene **3** (Scheme 14.1). Heterogeneous selenol **3** was then reacted with either sulfuryl chloride to afford **A**, or sodium borohydride to form **B**. As will be discussed later, **A** and **B** were used as electrophilic and nucleophilic regents, respectively, for the attachment of synthesis substrates.

After this initial report, the field of polymer-supported selenium-based linker group reagents fell dormant for over two decades, until the late 1990s when the concept of combinatorial chemistry reignited interest in this subject. At this time the Ruhland and Nicolaou groups virtually simultaneously reported the synthesis and use of such materials starting from preformed polystyrene beads. In the manuscript by Ruhland et al.,[11] it was reported that the procedure described by Heitz and coworkers for the synthesis of polymer-bound selenium was not successful in their hands and so they used the procedure outlined in Scheme 14.2 to prepare reagent **C** from a commercially available polymeric starting material. Brominated cross-linked

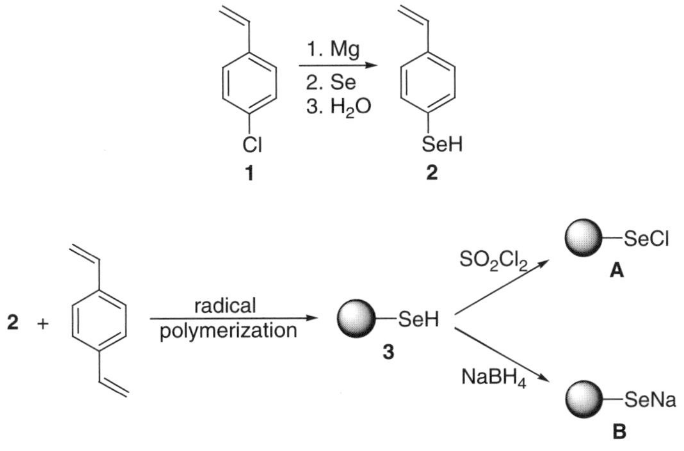

Scheme 14.1 Synthesis of selenium-based reagents **A** and **B**

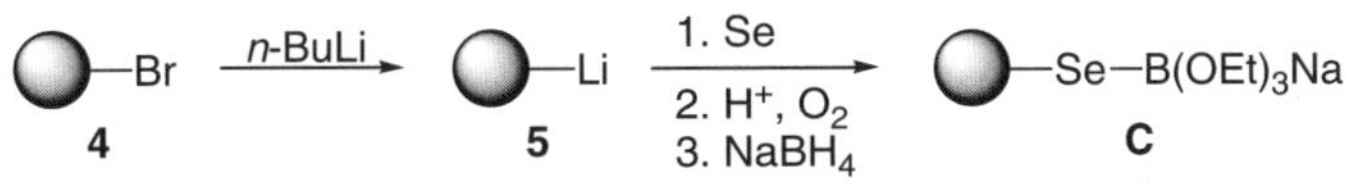

Scheme 14.2 Synthesis of nucleophilic reagent **C**

polystyrene (**4**) was treated with butyl lithium to generate **5**, which was reacted sequentially with selenium, aqueous acid in the presence of oxygen, and sodium borohydride to afford nucleophilic reagent **C**.

At almost the same time, Nicolaou and coworkers reported the synthesis of polystyrene-supported reagents **D**–**F**.[12] Treatment of unfunctionalized cross-linked polystyrene beads (**6**) with *n*-butyl lithium in the presence of tetramethylethylenediamine, followed by dimethyl diselenide afforded heterogeneous methyl selenide **7** (Scheme 14.3). Resin **7** was then treated with bromine to afford **D**, which was subsequently converted into both **E** and **F** by reaction with potassium phthalimide and lithium borohydride, respectively. The use of the phenylselenylbromide linker group reagent **D** with amphiphilic Argogel, a heterogeneous poly(ethylene glycol) grafted polystyrene material, as the resin support has also been reported.[13]

Additionally, a chiral analogue of both **A** and **D** was subsequently described by Uehlin and Wirth.[14, 15] Chiral phenol **8** was prepared and attached to a number of different alkyl/benzyl bromine functionalized polymer supports to afford **9** (Scheme 14.4). Reaction of **9** with bromine afforded chiral linker group reagent **G**, which was used in asymmetric solid-phase ether synthesis (*vide infra*).

Fujita *et al.* have reported the use of polymer-supported selenocyanates in solid-phase organic synthesis.[16, 17] Linker group reagent **H** was prepared by treatment of Merrifield resin (**10**) with potassium selenocyanate (Scheme 14.5). Quantitative formation of **H** was monitored using infrared (IR) spectroscopy by the appearance of a cyano group stretching vibration at 2149 cm^{-1}. They also reported the synthesis of polymer-supported arylselenocyanate **I** by the attachment of carboxylic acid **11** to a polystyrene resin functionalized with aminomethyl groups.

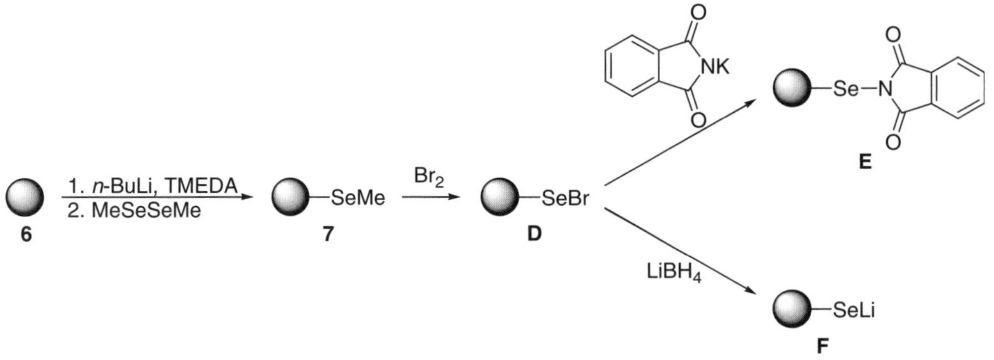

Scheme 14.3 Synthesis of polymer-supported selenyl bromide **D**, phthalimide **E** and nucleophilic reagent **F**

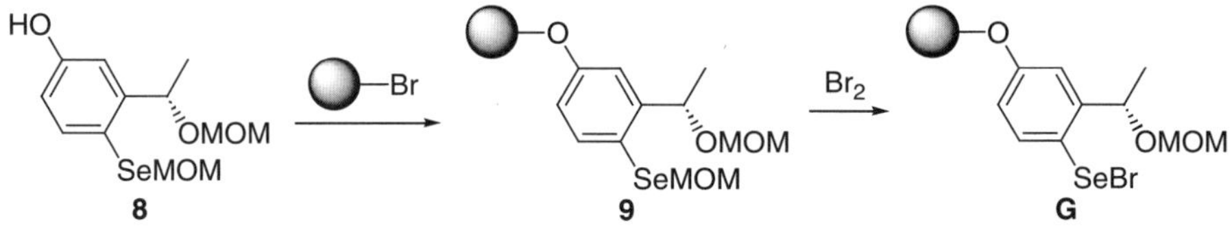

Scheme 14.4 Synthesis of chiral linker group reagent **G**

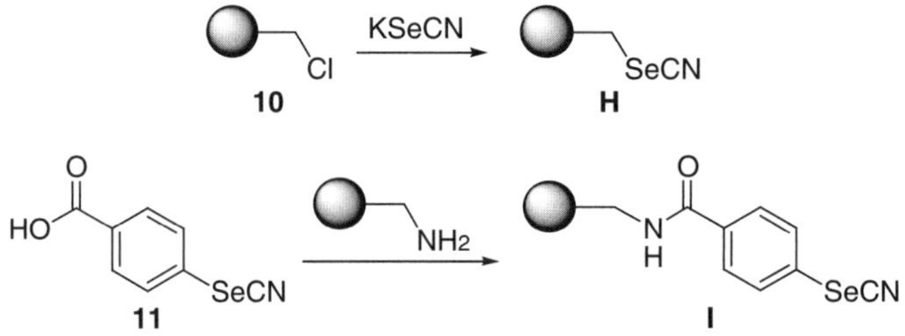

Scheme 14.5 Synthesis of polymer-supported selenocyanate reagents **H** and **I**

Sheng and Huang have reported another electrophilic selenium-based linker reagent, **J**, which is prepared by treatment of **D** with morpholine (Scheme 14.6).[18–20] Reagent **J** was found to be effective for the immobilization of aldehyde synthesis substrates (*vide infra*).

Later, Kulkarni and coworkers reported the synthesis of diphenyl diselenide resin **K** from 2-(4-bromophenyl)ethanol (**12**) (Scheme 14.7).[21] Alcohol **12** was converted stepwise into diaryl diselenide **13**, which was anchored to cross-linked polystyrene by either a trityl linker group or Ellman's dihydropyran linker. It is worth noting that the use of a linker group to attach a second linker group to a polymer allows for the selective cleavage of the primary group that is attached directly to the polymer, so that the cleaved product contains the synthesis substrate attached to the secondary linker group. Spectroscopic analysis of this cleaved product can allow for this strategy to be used as a method to determine the efficiency of the reaction for substrate attachment to the selenium group. In order to attach synthesis substrates

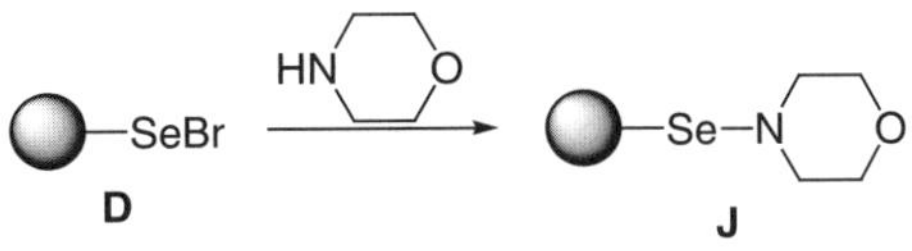

Scheme 14.6 Synthesis of reagent **J**

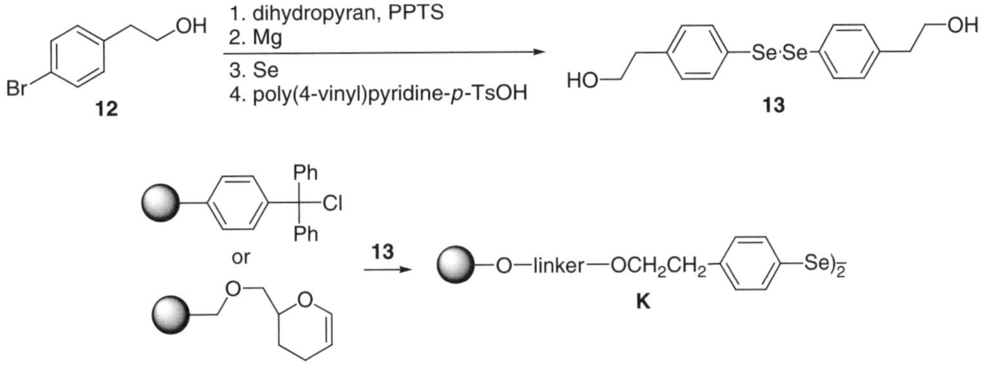

Scheme 14.7 Synthesis of diphenyl diselenide polymer **K**

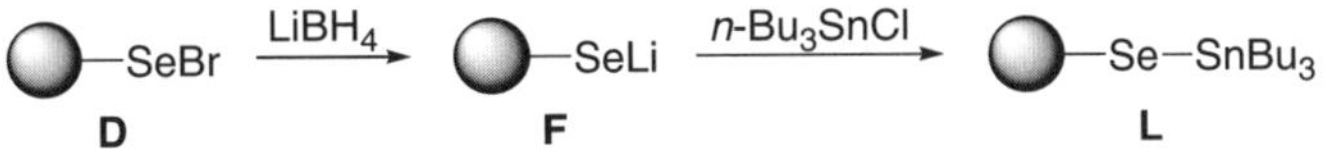

Scheme 14.8 Synthesis of reagent **L**

to **K**, it was first treated with sodium borohydride to render it nucleophilic (*vide infra*). Engman *et al.* subsequently showed that **K** can be prepared without the use of an intermediary linker group, where the diaryl diselenide groups are attached directly to the polymer backbone.[22] This strategy of direct attachment of the linker groups to the polymer presumably allows this reagent to be useful in a wider range of reaction conditions.

Nicolaou and coworkers also reported moderately air stable polymer-supported arylselenol equivalent **L** for use in supported glycoside synthesis.[23] This reagent was prepared from **D** by sequentially treating it with lithium borohydride, to form **F**, and chlorotributyltin (Scheme 14.8).

Linker reagent groups **M** (Ar = Ph or *p*-Tol) were designed by Huang *et al.* for the immobilization of alkene substrates via radical addition reactions.[24–26] They were prepared by treatment of either **A** or **D** with the appropriate sodium arylsulfinate in *N*,*N*-dimethylformamide (DMF) at room temperature (Scheme 14.9). The exact structure of the aryl group of **M** was found not to be significant.

The first example of a selenium-based linker reagent group where the synthesis substrate is not attached via a bond to selenium was reported by Huang and Xu.[27] Reaction of **F** with benzylselenium bromide in tetrahydrofuran (THF) at low temperature afforded **N** in good yield (Scheme 14.10). Attempts to synthesize **N** by treatment of **F** with dibenzyl diselenide were unsuccessful, as a mixture of products was formed. With this linker reagent, synthesis substrate attachment was at the selenium activated benzylic position rather than at selenium (*vide infra*).

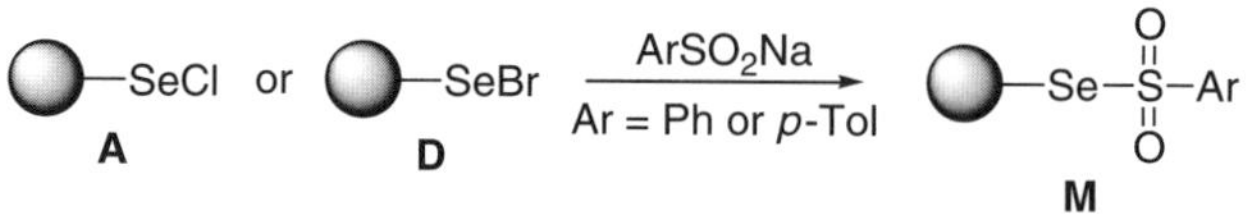

Scheme 14.9 Synthesis of reagent **M**

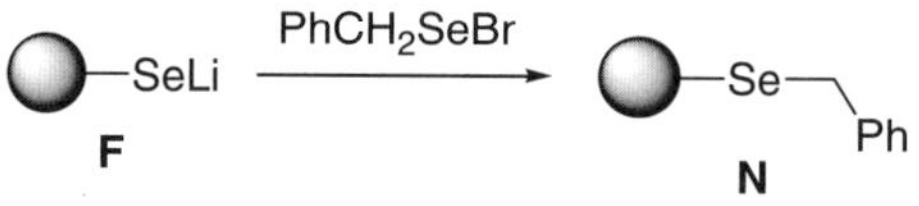

Scheme 14.10 Synthesis of reagent **N**

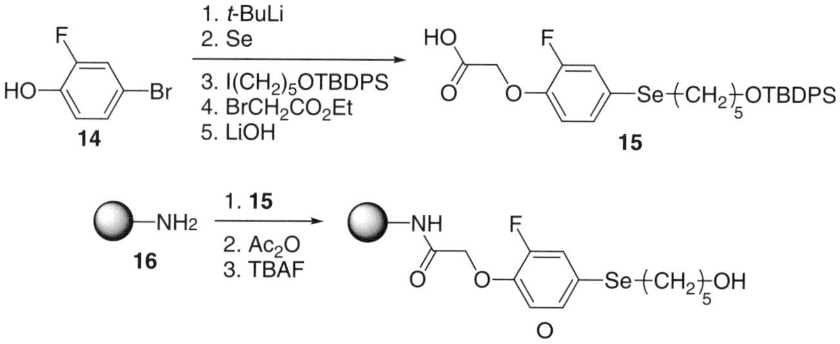

Scheme 14.11 Synthesis of reagent **O**

Pentenyl glycosides are useful reagents in oligosaccharide synthesis and fluorinated reagent **O**, another example where synthesis substrate attachment is not at the selenium atom, has been introduced for their polymer-supported preparation.[28] Resin **O** was synthesized from 4-bromo-2-fluorophenol (**14**) by way of carboxylic acid **15**, as outlined in Scheme 14.11. After treatment of amino group functionalized ArgoGel (**16**) with **15**, any residual unreacted amine groups were functionalized with acetic anhydride. This was followed by deprotection of the hydroxyl group to afford **O**. It is noteworthy that the incorporation of the fluorine atom in the linker group allows for facile reaction monitoring by gel-phase ^{19}F NMR analysis.

The final example of the use of a selenium-based linker group where the synthesis substrate is not attached at the selenium atom, is the synthesis of dehydropeptides using reagent **P**.[29] Nakamura and coworkers prepared **P** from protected serine **17** as shown in Scheme 14.12. The resin used for attachment of carboxylic acid **18** was glycine-loaded Wang resin **19**. The preparation of a second generation reagent which does not contain the succinate bridge between the polymer and the arylselenyl group was subsequently reported.[30]

Lastly, since the homolysis of alkyl tellurides was previously reported to be more facile than for alkyl selenides, Ruhland *et al.* synthesized linker group reagent **Q**, the tellurium-based analogue of **A**, using a similar reaction sequence and described its use (Scheme 14.13).[31, 32]

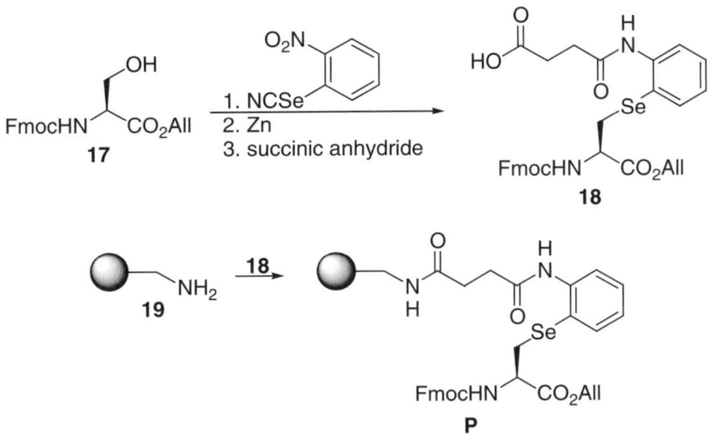

Scheme 14.12 Synthesis of amino acid-based reagent **P**

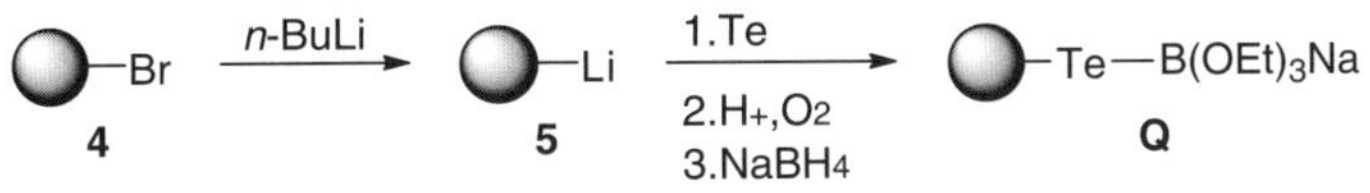

Scheme 14.13 Synthesis of tellurium-based linker group reagent **Q**

Typical Experimental Procedures

Synthesis of linker reagent D[12]

*Polystyrene beads were reacted with n-BuLi (24 equiv.) in the presence of TMEDA in cyclohexane for four hours at 65 °C. After filtration and washing with THF, dimethyl diselenide (2 mmol) was added at 0 °C in THF and reacted for 30 minutes. At this time the polymer beads were filtered and washed to afford polymer-supported methyl selenide 7. Resin 7 was swollen in chloroform and bromine (0.9 equiv.) was added at 0 °C. The suspension was stirred for 10 minutes and then filtered, washed with warm ethanol and dried in vacuum to afford resin **D**.*

Synthesis of linker reagent F[12]

*Resin **D** was swollen in THF and LiBH$_4$ (2 equiv.) was added. The suspension was stirred for one hour at room temperature. At this time the polymer was filtered, washed, and dried in vacuum to afford resin **F**.*

Synthesis of linker reagent K[21]

*A polystyrene solid-phase synthesis resin with either a trityl linker group or Ellman's dihydropyran linker group was swollen in 1,2-dichloroethane and pyridinium p-toluenesulfonate (2 equiv.) and diselenide diol 13 (5 equiv.) was added. The reaction mixture was refluxed for 24 hours and then the polymer was filtered, washed sequentially with DMF, methanol, dichloromethane (DCM), and dried in vacuum to afford resin **K**.*

Synthesis of linker reagent M[24]

*Resin **A** was swollen in DMF and sodium benzenethiosulfonate (4 equiv.) was added. The reaction mixture was stirred for six hours at room temperature and then the polymer was filtered, washed and dried in vacuum to afford resin **M**.*

14.3 Selenium-based linker group attachment methods

Rather than presenting each reported example individually, this section is organized according to where and how the synthesis substrate is attached to the linker group reagent. This format is designed to present related reactions together, so as to facilitate a rapid and coherent survey of the literature. As mentioned previously, attachment of a synthesis substrate to a polymeric selenium-based linker reagent can occur either at selenium or another position of the linker group, and when attachment is at selenium the selenium atom may react as an electrophile, nucleophile or radical. Separate sections are dedicated to each of these attachment methods.

14.3.1 Electrophilic attachment at selenium

Of the linker group reagents presented in Figure 14.1, **A**, **D**, **E**, and **G–J** have been used as electrophilic reagents. Examples of the use of these reagents with representative samples of the types of polymer-supported compounds synthesized are shown in Table 14.1. As is readily evident, when a selenium-based linker reagent group is used as an electrophile, the synthesis substrate attached is generally an alkene and the selenium atom attaches at this group in a manner to form the most stable possible carbocation (Entries 2–7, 9–11 and 13–17). In most of these examples, the synthesis substrate being attached contains a nucleophilic group that adds to the other alkene position. When the substrate contains no nucleophilic group, the bromine atom from the linker group reagent serves this role (Entries 14 and 16), or a separate alcohol or amine nucleophile can be added (Entries 2, 3 and 15). In reactions that do not involve addition of the selenium group to an alkene, Entries 1 and 8 illustrate how **A** and **J** can be attached at a position adjacent to ketone or aldehyde groups, respectively. The reaction shown in Entry 12 is part of a process for the synthesis of polymer-supported vinyl selenides, which in turn can be converted into various carbonyl compounds. Entry 17 presents a process that is catalytic in **D** and stoichiometric in N-chlorosuccinimide that results in the ultimate formation of an allylic chloride functionalized product.

14.3.2 Nucleophilic attachment at selenium

Linker group reagents **B**, **C**, **F**, **K**, **L** and **Q** have been used as nucleophilic reagents for the attachment of electrophilic synthesis substrates and examples of their use are summarized in Table 14.2. In most cases, the electrophilic synthesis substrate is an alkyl halide (Entries 1, 2, 5–9, 11 and 13), although epoxides (Entries 3 and 12) and allylic acetates (Entry 10) have also been used. As mentioned previously, reduction of the diselenide group of **K** using of sodium borohydride is required before reaction with an electrophile is possible (Entry 2). Entry 4 shows how **L** can react with a trichloroacetimidate to form a selenoglycoside in the presence of a Lewis acid. Comparison of Entry 1 with Entry 5, Entry 3 with Entry 12, Entry 6 with Entry 11, and Entry 8 with Entry 9 demonstrates that the identity of the cationic counter ion of the linker group reagent is not important and that **B**, **C** and **F** are functionally equivalent. Entry 10 illustrates how **F** can participate in an S_N2' reaction for substrate loading. Tellurium-based reagent **Q** was found to exhibit reactivity similar to selenium-based analogue **C** (Entries 1 and 14).

Table 14.1 Selenium-based linker reagent group attachment at selenium using electrophilic reactions

Entry	Linker Group Reagent	Substrate(s)	Product(s)	Reference(s)
1	●—SeCl **A**	4-methylcyclohexanone	●—C$_6$H$_4$—Se-(2-oxo-4-methylcyclohexyl)	[10]
2	●—SeBr **D**	R_1CH=CHR_2, ROH	●—Se—CH(R_1)—CH(R_2)—OR	[12, 45]
3	●—Se—N(phthalimide) **E**	R_1CH=CHR_2, ROH	●—Se—CH(R_1)—CH(R_2)—OR	[12]
4	●—SeBr **D**	R_1,R_2-substituted acrylic acid (COOH)	●—Se—γ-butyrolactone (R_1, R_2)	[13]
5	●—O—C$_6$H$_3$(CH(CH$_3$)OMOM)—SeBr **G**	Ph-C(=CH$_2$)-CH$_2$-CH$_2$-OH	●—O—aryl—tetrahydrofuran—CH$_2$—Se (Ph)	[14, 15]
6	●—CH$_2$—SeCN **H**	R_1,R_2-substituted acrylic acid (COOH)	●—Se—γ-butyrolactone (R_1, R_2)	[16]
7	●—CH$_2$—NH—C(O)—C$_6$H$_4$—SeCN **I**	R_1,R_2-substituted acrylic acid (COOH)	●—Se—γ-butyrolactone (R_1, R_2)	[17]
8	●—Se—N(morpholine) **J**	$R_1$$R_2$CH—CHO	●—Se—C(R_1)(R_2)—CHO	[18–20]
9	●—SeBr **D**	prenyl-substituted cyclohexanone with OAc, C(O)R_3, R_1, R_2	●—Se-bicyclic product with R_1, R_2, C(O)R_3	[33]

(*continued overleaf*)

Table 14.1 (continued)

#	Linker	Substrate	Product	Ref.
10	D—SeBr	aromatic with R_1, R_2, R_3, R_4(OH), R_5, R_6 on allyl side chain	chroman with Se-linker, R_1–R_6 substituents	[34, 35, 39–42, 54, 58]
11	D—SeBr	aromatic with R_1, R_2, R_3, R_4(NH$_2$), allyl	indoline with Se-linker, R_1–R_4	[38, 44]
12	D—SeBr	$RHC=PPh_3$	D—SeCR=PPh$_3$	[46]
13	D—SeBr	1,3-dicarbonyl with R_1, R_2 and allyl bearing R_3, R_4, R_5	dihydrofuran with Se-linker, R_1–R_5	[53]
14	D—SeBr	$R_1R_2C=CHR_3$ (H)	D—Se–CHR$_1$–CR$_2$R$_3$Br	[59–61]
15	D—SeBr	R_1CH=CH–COOMe, R_2NH_2	D—Se–CH(COOMe)–CHR$_1$–NHR$_2$	[63]
16	D—SeBr	$CH_2=CH_2$	D—Se–CH$_2$CH$_2$Br	[69]
17	D—SeBr (cat.) + NCS	prenyl-OAc, n = 2 or 3	Cl-substituted prenyl-OAc, n = 2 or 3	[73]

Table 14.2 Selenium and tellurium linker reagent group attachment at selenium using nucleophilic reactions

Entry	Linker Group Reagent	Substrate(s)	Product(s)	Reference(s)
1	●—Se—B(OEt)$_3$Na **C**	RCl	●—SeR	[11]
2	●—O—linker—O—CH$_2$CH$_2$—C$_6$H$_4$—Se$_{1/2}$ **K** + NaBH$_4$	Br(CH$_2$)$_6$—OTBDPS	●—O—linker—O—CH$_2$CH$_2$—C$_6$H$_4$—Se(CH$_2$)$_6$OTBDPS	[21]
3	●—SeNa **B**	epoxide—R	●—Se—CH$_2$CH(OH)R	[22]
4	●—Se—SnBu$_3$ **L** + BF$_3$-EtO$_2$	Cl$_3$C(HN)CO—O-pyranose(AcO, OAc, OPMB)	●—Se-pyranose(AcO, OAc, OPMB)	[23]
5	●—SeNa **B**	RCl, RBr or RI	●—SeR	[36, 43, 47, 48, 55, 62, 65, 66]
6	●—SeNa **B**	BrCH$_2$SO$_2$Ar	●—SeCH$_2$SO$_2$Ar	[49]
7	●—SeNa **B**	Br-CH$_2$-C≡CH	●—Se-CH$_2$-C≡CH	[50, 56, 57]
8	●—SeLi **F**	ClCH$_2$PO(OEt)$_2$	●—SeCH$_2$PO(OEt)$_2$	[51, 67]
9	●—SeNa **B**	ClCH$_2$PO(OEt)$_2$	●—SeCH$_2$PO(OEt)$_2$	[52]

(continued overleaf)

402   *Multifunctional Linker Units for Diversity-Oriented Synthesis*

Table 14.2 (continued)

10	⬤—SeLi **F**	R—C(OAc)(COOMe)=CH₂	⬤—Se—C(R)=C(H)—CH₂—CO₂Me	[64]
11	⬤—SeLi **F**	BrCH$_2$SO$_2$Ar	⬤—SeCH$_2$SO$_2$Ar	[68]
12	⬤—SeLi **F**	epoxide-R	⬤—Se—CH$_2$—CH(OH)—R	[70, 72]
13	⬤—SeLi **F**	RBr	⬤—SeR	[71]
14	⬤—Te—B(OEt)$_3$Na **Q**	RCl	⬤—TeR	[31, 32]

14.3.3 Radical attachment at selenium

The reported examples of the use of a radical reaction for the attachment of synthesis substrates to reagent **M** are listed in Table 14.3. Such reactions are much less common than the nucleophilic and electrophilic reactions described previously and require the use of radical chain initiator, generally 2,2-azobisisobutyronitrile (AIBN), for the addition to the carbon–carbon multiple bonds of the substrates to proceed efficiently. Entry 3 illustrates how a reaction cascade can occur with a non-conjugated diene synthesis substrate.

14.3.4 Attachment at other positions

The examples of the use of linker group reagents **N–P** to attach synthesis substrates at positions other than selenium of the linker reagent groups are shown in Table 14.4. Linker group reagent **N** can be deprotonated at the activated benzylic position so that it becomes nucleophilic enough to react with an electrophilic epoxide synthesis substrate (Entry 1). In chemistry similar to that used for the loading of **L**, reaction of a trichloroacidimidate with linker group reagent **O** in the presence of a Lewis acid results in formation of a polymer-supported glycoside (Entry 2). Linker group reagent **P** is an orthogonally protected polymer-supported amino acid derivative where attachment is at the side-chain. Selective deprotection of either the acid or amine group allows for standard solid-phase peptide synthesis to be preformed (Entry 3).

Table 14.3 Selenium-based linker reagent group attachment at selenium using radical reactions

Entry	Linker group reagent	Substrate(s)	Product(s)	Reference
1	⬤—Se—S(=O)₂—Ar **M** + AIBN	R₁–CH=CH–R₂	⬤—Se–CH(R₁)–CH(R₂)–SO₂Ar	[24]
2	⬤—Se—S(=O)₂—Ar **M** + AIBN	R–C≡C–H	⬤—Se–C(R)=C(H)–SO₂Ar	[25]
3	⬤—Se—S(=O)₂—Ar **M** + AIBN	CH₂=CH–CH₂–C(CO₂Et)₂–CH₂–CH=CH₂	⬤—Se–CH₂–[cyclopentane with C(CO₂Et)₂]–CH₂–SO₂Ar	[26]

14.4 Selenium-based linker group cleavage methods

As with the methods for synthesis substrate attachment to the selenium-based linker group reagents, the methods for synthesis product cleavage from the linker group are limited to a small group of reactions, with an oxidation/elimination reaction sequence that results in the formation of a carbon–carbon multiple bond being the most prevalent strategy. Oxidation at selenium is most commonly performed using hydrogen peroxide, though *t*-butylperoxide and *m*-chloroperbenzoic acid have also been used. Another strategy employed is an alkylation/nucleophilic displacement reaction sequence, where selenium is alkylated and a polymer-supported selenide is displaced by a nucleophile. Additionally, homolysis under neutral reaction conditions of the selenium–substrate bond has also been reported. Examples of each of these cleavage methodologies are presented in separate groups, as are the other miscellaneous techniques reported in the literature.

14.4.1 Oxidative cleavage

In the vast majority of the examples of the use of selenium-based linkers in organic synthesis, the cleavage of the synthesis product from the linker group is initiated by oxidation at selenium. The oxidized products of such reactions are reactive and readily participate in elimination and, occasionally, substitution reactions.

Cleavage processes using hydrogen peroxide for selenium oxidation are summarized in Table 14.5, with the yields reported being generally for the overall cleavage reaction sequence. As can be seen, a

Table 14.4 Selenium-based linker reagent group attachment not at selenium

Entry	Linker group reagent	Substrate(s)	Product(s)	Reference(s)
1	●—Se—CH₂Ph **N** + LDA	epoxide R₁—CH—CH—R₂	●—SeCHPh R₁—CHCH(OH)R₂	[27]
2	●—NH—C(O)—CH₂—O—(2-F-C₆H₃)—Se—(CH₂)₅—OH **O** + TMSOTf	Cl₃C-imidate sugar with p-FBzO, OBzp-F, Php-F	●—NH—C(O)—CH₂—O—aryl—Se—(CH₂)₅—O—sugar (p-FBzO, OBzp-F, Php-F)	[28]
3	●—CH₂—NH—C(O)—CH₂CH₂—C(O)—NH—aryl—Se—CH₂—CH(NHFmoc)—CO₂All **P**	standard deprotection and solid-phase peptide synthesis	●—Gly—...—cyclic peptide with Se linkage, C₃H₆Ph	[29–30]

large variety of alkene products can be prepared by such cleavage processes, including dehydropeptides (Entry 7) and a protected version of the important antibiotic natural product vancomycin (Entry 43). This last example is significant in that it shows how a selenium-based linker group can be cleaved to afford a readily cleavable allyl ester protecting group. Entry 42 illustrates how even carbon–carbon triple bonds can be formed using this methodology.

As before with hydrogen peroxide, cleavage reactions effected by selenium oxidation using *t*-butylhydroperoxide result in the formation of an alkene group in the cleaved product and published examples are listed in Table 14.6. The cleavage reaction presented in Entry 1 results in the synthesis of an *n*-pentenyl glycoside, which can be used in glycoslyation reactions. Entry 2 is another example of the formation of a dehydropeptide and the reported yield is for the overall synthesis performed.

The use of *m*-chloroperbenzoic acid for selenium oxidation is summarized in Table 14.7. Interestingly, when an unprotected hydroxyl group is located at an appropriate position with respect to the selenium-based linker group, elimination to form an alkene can occur, but when this hydroxyl group is protected as a silyl ether, no carbon–carbon double bond is formed (Entry 2 versus Entry 3). Furthermore, in both of these examples, the adjacent hydroxyethyl ether is converted to the corresponding cyclic acetal. Entry 4 shows that an intramolecular nucleophilic displacement of the oxidized selenium-based linker group can occur.

Table 14.5 Selenium-based linker group oxidative cleavage using hydrogen peroxide

Entry	Polymer attached substrate(s)	Product(s)	Yield(s) (%)	Reference(s)
1	⬤–O–linker–O–CH₂CH₂–(C₆H₄)–Se(CH₂)₆OTBDPS	CH₂=CH(CH₂)₃OTBDPS	60	[21]
2	⬤–SeCHPh(CH₂R)	Ph–CH=CH–R	7 examples 71–91	[27]
3	⬤–SeCHPh, R₁–CHCH(OH)R₂	Ph–C(R₁)=CH–CH(OH)R₂	5 examples 77–88	[27]
4	⬤–Se-tetrahydroquinoline-CN	dihydroquinoline-CN	95	[44]
5	⬤–Se-indoline (R₁, R₂, R₃, R₄; N-CO₂Et)	isopropenyl indoline (R₁, R₂, R₃, R₄; N-CO₂Et)	5 examples 75–95	[44]
6	⬤–SeCH₂CH₂O–Ar(R)	H₂C=HC–O–Ar(R)	8 examples 92–96	[69]
7	⬤–Se–CH(R₁)CH₂–OCHR₂R₃	R₁–C(=CH₂)–OCHR₂R₃	10 examples 74–90	[70]
8	⬤–Se–C(CH₃)(EWG)–CH(OH)R	R–CH(OH)–C(=CH₂)–EWG	16 examples 83–91	[71]
9	⬤–NHC(O)–(C₆H₄)–Se–peptide (MeOOC-Ala-Phe-Asp(OtBu)-Xaa-Gly-Arg(Pbf)-NHAc); Pbf = O₂S-(dihydrobenzofuran)	MeOOC-Ala-Phe-Asp(OtBu)-dehydroAla-Gly-Arg(Pbf)-NHAc	77	[30]
10	⬤–Se–CH₂CH₂–CONR₁R₂	CH₂=CH–C(O)NR₂R₁	9 examples 85–93	[65]

(continued overleaf)

Table 14.5 (continued)

11	●—SeCHCONR₂R₃ with CH₂R₁	R₁-CH=CH-C(O)NR₂R₃	10 examples 84–91	[66]
12	●—Se— with isoxazoline (R₂, R₃) and isoxazole (R₁) substituents	vinyl-linked isoxazoline/isoxazole	11 examples 56–76	[50]
13	●—Se— dihydrofuran with R₁, R₂, C=O	vinyl dihydrofuran with R₁, R₂, C=O	85	[53]
14	●—Se— bicyclic furan with R₁, R₂, C=O	bicyclic furan with R₁, R₂, C=O	2 examples 80–82	[53]
15	●—Se— with isoxazoline (R₃, R₂) and triazole (R₁)	vinyl-linked isoxazoline/triazole	7 examples 43–66	[56]
16	●—Se— with isoxazoline (R₃, R₂) and isoxazole (R₁)	vinyl-linked isoxazoline/isoxazole	6 examples 57–77	[56]
17	●—Se— with isoxazoline (R₃, R₂) and oxadiazole (R₁)	vinyl-linked isoxazoline/oxadiazole	19 examples 48–72	[56]
18	●—Se— with R₂ and triazole-N-R₁	vinyl triazole with R₂ and N-R₁	5 examples 71–82	[57]

19	⬤–Se–CH₂CH₂–[triazole with N-N, R₁, R₂]	vinyl–[triazole with N-N, R₁, R₂]	26 examples 49–69	[62]
20	⬤–Se–CH₂CH₂–[1,3,4-oxadiazole]–R	vinyl–[1,3,4-oxadiazole]–R	6 examples 57–67	[62]
21	⬤–Se–CH₂CH₂–[1,2,4-oxadiazole with R]	vinyl–[1,2,4-oxadiazole with R]	8 examples 53–62	[62]
22	⬤–C₆H₄–Se–(4-methylcyclohexanone)	4-methylcyclohex-2-enone	91	[10]
23	⬤–Se–(2-formyl-4-R-cyclohexanone)	2-formyl-4-R-cyclohex-2-enone	4 examples 85–90	[20]
24	⬤–Se–(γ-butyrolactone, Ph)	Ph–butenolide	59	[13, 17]
25	⬤–Se–(tetrahydropyran, R)	dihydropyran, R	5 examples 41–83	[13]
26	⬤–Se–(bicyclic diketone lactone with iPr)	(bicyclic enone diketone lactone with iPr)	90	[33]

(*continued overleaf*)

Table 14.5 (continued)

#	Substrate	Product	Examples (yield %)	Ref.
27	Polymer-Se-CH₂-chromane with R₁, R₂, R₃, R₄	Chromene with R₁, R₂, R₃, R₄	49 examples 41–97	[34, 39–42]
28	Polymer-Se-cyclohexane-OR	Cyclohexenyl-OR	8 examples 75–85	[45]
29	Polymer-Se-γ-butyrolactone with R₁, R₂, R₃	Butenolide with R₁, R₂, R₃	13 examples 76–90	[48]
30	Polymer-Se-flavanone with R₁–R₅	Flavone with R₁–R₅	40 examples 23–92	[54, 58]
31	Polymer-Se-isoxazoline-R	Isoxazole-R	32 examples 78–88	[59]
32	Polymer-Se-isoxazoline with COOEt, R	Isoxazole with COOEt, R	5 examples 80–89	[61]
33	Polymer-Se-triazoline with COOEt, N-R	Triazole with COOEt, N-R	5 examples 61–79	[61]

34	●—Se, X, EtOOC, X = CH₂, O	EtOOC, X	2 examples 88–92	[61]
35	●—Se, EtOOC, (methylcyclohexene)	EtOOC—⟨⟩—CH₃	79	[61]
36	●—Se, barbiturate with R₁, R₂, R₃	pyrimidinedione with R₁, R₂, R₃	11 examples 65–83	[63]
37	●—Se, R₁, R₂, SO₂Ar	R₂, R₁, H, SO₂Ar	19 examples 70–92	[24]
38	●—Se—⟨cyclopentane⟩—SO₂Ar, EtO₂C, CO₂Et	=CH₂ cyclopentane—SO₂Ar, EtO₂C, CO₂Et	8 examples 44–58	[26]
39	●—SeCHSO₂Ar, CH₂CH(OH)R	R, OH, SO₂Ar	3 examples 79–85	[49]
40	●—SeCHSO₂Ar, CH₂R	R, SO₂Ar	24 examples 73–93	[49, 52, 68]
41	●—SeCHPO(OEt)₂, CH₂R	R, PO(OEt)₂	17 examples 62–91	[52, 67]
42	●—Se, H, R, SO₂Ar	RS≡O₂Ar	13 examples 69–90	[25]
43	●—Se(CH₂)₃OC(O)—protected vancomycin	allyl-O-C(O)—protected vancomycin	90	[36, 43]

410   Multifunctional Linker Units for Diversity-Oriented Synthesis

Table 14.6 *Selenium-based linker group oxidative cleavage using* tert-*butylhydroperoxide*

Entry	Polymer Attached Substrate	Product	Yield (%)	Reference
1			92	[28]
2			10	[29]

Typical Experimental Procedures
Oxidative Cleavage of the Synthesis Product[21]

The synthesis product functionalized resin was suspended in THF and 30% hydrogen peroxide (11 equiv.) was added. The reaction mixture was stirred at room temperature for 30 minutes and then the polymer was filtered and rinsed with THF. The filtrate was washed with saturated sodium bicarbonate and extracted with diethyl ether. The combined organic layers were collected, dried over anhydrous sodium sulfate, and concentrated to afford the crude product.

14.4.2 Nucleophilic displacement cleavage

The previously described oxidation/elimination cleavage reaction sequences resulted in the formation of carbon–carbon π-bond. An alternative cleavage strategy that is similar to the previously reported solid-phase synthesis of various selenides,[74, 75] that does not result in such bond formation is a alkylation/nucleophilic displacement reaction sequence, in which the carbon–selenium bond attaching the synthesis product to the linker is replaced by a carbon–nucleophile bond. Entries 1–5 of Table 14.8 illustrate how this can be achieved. In these examples, the alkylating reagent is iodomethane and the resulting iodide ion serves as the nucleophile. Entry 6 shows how both bromine and chlorine can participate in similar reactions, and Entry 7 illustrates a related cleavage process in which an acylselenide undergoes nucleophilic acyl substitution when reacted with a copper acetylide. As mentioned previously Table 14.7, Entry 4 represents another example where nucleophilic displacement of the selenium-based linker group occurs.

Table 14.7 Selenium-based linker group oxidative cleavage using meta-chloroperbenzoic acid

Entry	Polymer Attached Substrate(s)	Product(s)	Yield(s)	Reference(s)
1	⬤–Se, Ph, lactone	Ph-furanone	46	[16]
2	⬤–Se, HO-CH₂CH₂-O-pyranose-OPMB, OH	dioxolane-pyranone-OPMB	81	[23, 37]
3	⬤–Se, HO-CH₂CH₂-O-pyranose-OPMB, OTBS	dioxolane-pyranone-OPMB, OTBS	85	[23, 37]
4	⬤–Se, OCONHBz, R_1, R_2	BzN-oxazolidinone, R_1, R_2	8 examples 75–85	[72]

Typical Experimental Procedures

Nucleophilic Substitution Cleavage of the Synthesis Product[53]

The synthesis product functionalized resin was suspended in dry DMF, and sodium iodide (6 equiv.) and iodomethane (CH_3I) (16 equiv.) were added under nitrogen. The reaction mixture was stirred at 75 °C for 18 hours and then the polymer was filtered and rinsed with DCM. The filtrate was washed sequentially with saturated sodium bicarbonate, saturated sodium thiosulfate ($Na_2S_2O_3$) and water. The combined organic layers were collected, dried over anhydrous sodium sulfate (Na_2SO_4) and concentrated to afford the crude product.

14.4.3 Homolytic cleavage

The final general method used for the cleavage of selenium- and tellurium-based linker groups is homolytic cleavage using radical reaction conditions and examples of this strategy are presented in Table 14.9. These reactions generally require the use of a radical chain initiator such as AIBN and a donor reagent such as tributyltinhydride (Entries 1–6 and 9). In Entry 7 the hydrogen atom source was tris(trimethylsilyl)silane. As mentioned previously, the original idea behind using tellurium analogues of the previously studied selenium-based linker groups is that the former were presumed to undergo homolysis more readily than the later. However, comparison of the results shown in Entries 1–4 show that this was not observed. Why this is in fact not the case is not understood at the present. Finally, Entry 8 shows how carbon

Table 14.8 Selenium-based linker group cleavage by nucleophilic displacement of selenium

Entry	Polymer attached substrate(s)	Product(s)	Yield(s) (%)	Reference
1	(Se-furanone with R1, R2) + MeI	(I-furanone with R1, R2)	7 examples 76–89	[53]
2	(Se-furan with R2, R3, OR1) + MeI	(I-furan with R2, R3, OR1)	9 examples 72–86	[53]
3	(Se-isoxazoline with R) + MeI	(I-isoxazoline with R)	7 examples 71–87	[55]
4	(Se-isoxazoline with R) + MeI	(I-isoxazoline with R)	7 examples 78–85	[59]
5	(Se-CH2-C(CO2Me)=CHR) + MeI	(I-CH2-C(CO2Me)=CHR)	10 examples 78–95	[64]
6	R_1R_2CCHO with Se + Br_2 or $SOCl_2$	R_1R_2C(X)CHO	11 examples 70–92	[19]
7	—SeCOR$_1$ + R$_2$≡—Cu	R$_2$—≡—C(O)R$_1$	9 examples 73–92	[47]

Selenium- and Tellurium-Based Linker Units 413

Table 14.9 Selenium- and tellurium-based linker group cleavage by homolysis

Entry	Polymer Attached Substrate (s)	Product(s)	Yield(s) (%)	Reference(s)
1	⬤–Se–CH₂CH₂–O–CH₂CH₂–O–C₆H₄–R	Et–O–CH₂CH₂–O–CH₂CH₂–O–C₆H₄–R	3 examples 78–88	[11]
2	⬤–Se–(CH₂)₅–O–C₆H₄–R	pentyl–O–C₆H₄–R	3 examples 80–87	[11]
3	⬤–Te–CH₂CH₂–O–CH₂CH₂–O–C₆H₄–R	Et–O–CH₂CH₂–O–CH₂CH₂–O–C₆H₄–R	3 examples 73–83	[31, 32]
4	⬤–Te–(CH₂)₅–O–C₆H₄–R	pentyl–O–C₆H₄–R	3 examples 73–86	[31, 32]
5	⬤–O–aryl with CH(CH₃)OMOM and Se–CH₂–(2-Ph-tetrahydrofuran)	Ph-tetrahydrofuran (2-methyl-2-Ph)	58%	[15]
6	⬤–O–linker–O–CH₂CH₂–C₆H₄–Se(CH₂)₆OTBDPS	hexyl–OTBDPS	65%	[21]
7	⬤–Se–CH(Bn)–CH(OH)–CH₃ + (Me₃Si)₃SiH	Bn–CH₂–CH(OH)–CH₃	97%	[22]
8	⬤–Se–CH₂–CH(Bn)–O–CH=CH–COOEt + CO	Bn-substituted γ-butyrolactone with CH₂COOEt	55%	[22]

(*continued overleaf*)

Table 14.9 (continued)

#	Substrate	Product	Yield	Ref.
9	Se-linked indoline with N-C(O)NR₂, R₁ substituent	2-methyl indoline with N-C(O)NR₂, R₁ substituent	6 examples 14 – 34%	[38, 44]
10	Se-linked sugar with RO, OBn, OBn, OBn	Sugar with RO, OBn, OBn, OBn	9 examples 23 – 96%	[12]
11	Se-linked cyclooctanol (HO)	Cyclooctanol (HO)	80%	[12]
12	Se-linked pyranose with BzO-CH₂CH₂-O, OPMB, OTBS	Pyranose with OBz, OPMB, OTBS	95%	[23, 37]
13	Se-linked bicyclic ketoester with OMe	Bicyclic ketoester with OMe	85%	[33]
14	Se-linked bicyclic ketoester with OEt + n-Bu₃SnCH₂CH=CH₂	Allyl-bicyclic ketoester with OEt	37%	[33]

Table 14.10 Miscellaneous methods for cleavage of selenium-based linker groups

Entry	Polymer Attached Substrate	Reagent(s)	Products(s)	Yields (s) (%)	Reference
1	⬤—Se—CHR$_1$—CHR$_2$(OH) (with H's shown)	SOCl$_2$ + Et$_3$N	R$_1$CH=CHR$_1$ (alkene)	11 examples 76–85	[18]
2	⬤—SeCR$_1$=CHR$_2$	TFA	R$_1$C(O)CH$_2$R$_2$	7 examples 40–82	[46]
3	⬤—SeCH=CHR	HBr	HC(O)CH$_2$R	6 examples 72–78	[46]

monoxide can be incorporated into the product under such radical cleavage reaction conditions and Entry 10 illustrates the use of allyltributyltin as the donor reagent.

Typical Experimental Procedures

Homolytic Cleavage of the Synthesis Product[11]

The synthesis product functionalized resin was suspended in benzene, and n-Bu$_3$SnH (5 equiv.) and AIBN (0.25 equiv.) were added. The reaction mixture was refluxed for three hours. After cooling to room temperature 2 N sodium hydroxide was added and the reaction mixture was stirred for three hours more at room temperature. The polymer was filtered and the filtrate was dried over anhydrous sodium sulfate and concentrated to afford the crude product.

14.4.4 Miscellaneous cleavage methods

Finally, the other reported methods used for selenium-based linker group cleavage are listed in Table 14.10. It has been found that when there is an α-hydroxyl group to the selenide moiety, eliminative cleavage forming a carbon–carbon double bond can occur with treatment by thionyl chloride and triethylamine (Entry 1). In these reactions, a mixture of alkene isomers is formed, with *E*-stereochemistry being favored. The vinyl selenide products formed by the transylidation reaction shown in Table 14.1, Entry 12 can be cleaved using trifluoroacetic acid to afford the corresponding ketone in the case of dialkyl substituted substrates (Entry 2), or they can be reacted to hydrogen bromide to afford the corresponding aldehyde in the case of monoalkyl substituted substrates (Entry 3).

14.5 Conclusions

In summary, the use of selenium- and tellurium-based linker groups in polymer-supported organic synthesis has mirrored the use of organoselenium reagents in traditional organic synthesis. The compatibility of selenides to a wide variety of reducing and nucleophilic reagents makes them appealing linker groups for supported synthesis. Furthermore, their utility is enhanced because, depending upon the cleavage reaction conditions employed, the synthesis products can either contain a group amenable to further functionalization, such as an alkene or alkyl halide, or can bear no trace of the position of linker group attachment. It is anticipated that as organoselenium chemistry further evolves, that not only will the utility of the linker groups presented here expand, but that new and currently unimagined strategies for using selenium- and tellurium-based linker group strategies will emerge.

References

[1] Paulmeier, C. (ed.), *Selenium Reagents and Intermediates in Organic Synthesis*, Pergamon Press, Oxford, **1986**.
[2] Liotta, D. (ed.), *Organoselenium Chemistry*, John Wiley & Sons, Inc., New York, **1987**.
[3] Krief, A., and Hevesi, L. (eds), *Organoselenium Chemistry I*, Springer-Verlag, Berlin, Germany, **1988**.
[4] Back, T. G. (ed.), *Organoselenium Chemistry: A Practical Approach*, Oxford University Press, Oxford, **1999**.
[5] Wirth, T (ed.), *Organoselenium Chemistry: Modern Developments in Organic Synthesis*, Springer, Berlin, Germany, **2000**.
[6] Wirth, T.; *Tetrahedron* **1999**, *55*, 1.
[7] Zaragoza, F.; *Angew. Chem. Int. Ed.* **2000**, *39*, 2077.
[8] McAllister, L. A., McCormick, R. A., and Procter, D. J.; *Tetrahedron* **2005**, *61*, 11527.
[9] Kan, J. T. W., and Toy, P. H.; *J. Sulfur Chem.* **2005**, *26*, 509.
[10] Michels, R., Kato, M., and Heitz, W.; *Makromol. Chem.* **1976**, *177*, 2311.
[11] Ruhland, T., Andersen, K., and Pedersen, H.; *J. Org. Chem.* **1998**, *63*, 9204.
[12] Nicolaou, K. C., Pastor, J., Barluenga, S., and Winssinger, N.; *Chem. Commun.* **1998**, 1947.
[13] Fujita, K.-I., Hashimoto, S., Oishi, A., and Taguchi, Y.; *Tetrahedron Lett.* **2003**, *44*, 3793.
[14] Uehlin, L., and Wirth, T.; *Chimia* **2001**, *55*, 65.
[15] Uehlin, L., and Wirth, T.; *Org. Lett.* **2001**, *3*, 2931.
[16] Fujita, K.-I., Watanabe, K., Oishi, A., *et al.*; *Synlett* **1999**, 1760.
[17] Fujita, K.-I., Taka, H., Oishi, A., *et al.*; *Synlett* **2000**, 1509.
[18] Sheng, S. R., and Huang, X.; *J. Chin. Chem. Soc.* **2003**, *50*, 893.
[19] Sheng, S. R., and Huang, X.; *Org. Prep. Proc. Int.* **2003**, *35*, 383.
[20] Liu, X. L., Wang, X. C., Sheng, S. R., and Huang, X.; *Chin. Chem. Lett.* **2004**, *15*, 1009.
[21] Li, Z., Kulkarni, B. A., and Ganesan, A.; *Biotechnol. Bioeng.* **2000**, *71*, 104.
[22] Berlin, S., Ericsson, C., and Engman, L.; *J. Org. Chem.* **2003**, *68*, 8386.
[23] Nicolaou, K. C., Fylaktakidou, K. C., Mitchell, H. J., *et al.*; *Chem. Eur. J.* **2000**, *6*, 3166.
[24] Qian, H., and Huang, X.; *Synlett* **2001**, 21913.
[25] Qian, H., and Huang, X.; *Tetrahedron Lett.* **2002**, *43*, 1059.
[26] Qian, H., and Huang, X.; *J. Comb. Chem.* **2003**, *5*, 569.
[27] Huang, X., and Xu, W.; *Tetrahedron Lett.* **2002**, *43*, 5495.
[28] Mogemark, M., Gustafsson, L., Bengtsson, C., *et al.*; *Org. Lett.* **2004**, *6*, 4885.
[29] Horikawa, E., Kodaka, M., Nakahara, Y., *et al.*; *Tetrahedron Lett.* **2001**, *42*, 8337.
[30] Nakamura, K., Ohnishi, Y., Horikawa, E., *et al.*; *Tetrahedron Lett.* **2003**, *44*, 5445.
[31] Ruhland, T., Torang, J., Pedersen, H., *et al.*; *Synthesis* **2004**, 2323.
[32] Ruhland, T., Torang, J., Pedersen, H., *et al.*; *Synthesis* **2005**, 1635.
[33] Nicolaou, K. C., Pfefferkorn, J. A., Cao, G.-Q., *et al.*; *Org. Lett.* **1999**, *1*, 807.
[34] Nicolaou, K. C., Pfefferkorn, J. A., and Cao, G.-Q.; *Angew. Chem. Int. Ed.* **2000**, *39*, 734.
[35] Nicolaou, K. C., Cao, G.-Q., and Pfefferkorn, J. A.; *Angew. Chem. Int. Ed.* **2000**, *39*, 739.
[36] Nicolaou, K. C., Winssinger, N., Hughes, R., *et al.*; *Angew. Chem. Int. Ed.* **2000**, *39*, 1084.
[37] Nicolaou, K. C., Mitchell, H. J., Fylaktakidou, K. C., *et al.*; *Angew. Chem. Int. Ed.* **2000**, *39*, 1089.
[38] Nicolaou, K. C., Roecker, A. J., Pfefferkorn, J. A., and Cao, G.-Q.; *J. Am. Chem. Soc.* **2000**, *122*, 2966.
[39] Nicolaou, K. C., Pfefferkorn, J. A., Roecker, A. J., *et al.*; *J. Am. Chem. Soc.* **2000**, *122*, 9939.

[40] Nicolaou, K. C., Pfefferkorn, J. A., Mitchell, H. J., *et al.*; *J. Am. Chem. Soc.* **2000**, *122*, 9954.
[41] Nicolaou, K. C., Pfefferkorn, J. A., Barluenga, S., *et al.*; *J. Am. Chem. Soc.* **2000**, *122*, 9968.
[42] Nicolaou, K. C., Pfefferkorn, J. A., Schuler, F., *et al.*; *Chem. Biol.* **2000**, *7*, 979.
[43] Nicolaou, K. C., Cho, S. Y., Hughes, R., *et al.*; *Chem. Eur. J.* **2001**, *7*, 3798.
[44] Nicolaou, K. C., Roecker, A. J., Hughes, R., *et al.*; *Bioorg. Med. Chem.* **2003**, *11*, 465.
[45] Sheng, S. R., and Huang, X.; *J. Chem. Res.* **2002**, 184.
[46] Huang, X., and Sheng, S. R.; *Tetrahedron Lett.* **2001**, *42*, 9035.
[47] Qian, H., Shao, L.-X., and Huang, X.; *Synlett* **2001**, 1571.
[48] Huang, X., and Sheng, S. R.; *J. Comb. Chem.* **2003**, *5*, 273.
[49] Xu, W.-M., Wu, L. L., and Huang, X; *Chin. Chem. Lett.* **2004**, *15*, 1047.
[50] Huang, X., and Xu, W.-M.; *Org. Lett.* **2003**, *5*, 4649.
[51] Sheng, S. R., Liu, X.-L., Zhou, W., and Huang, X.; *Chin. Chem. Lett.* **2005**, *16*, 285.
[52] Xu, W. M., Tang, E., and Huang, X.; *Synthesis* **2004**, 2094.
[53] Tang, E., Huang, X., and Xu, W.-M.; *Tetrahedron* **2004**, *60*, 9963.
[54] Huang, X., Tang, E., Xu, W.-M., and Cao, J.; *J. Comb. Chem.* **2005**, *7*, 802.
[55] Xu, W.-M., Wang, Y.-G., Miao, M.-Z., and Huang, X.; *Synthesis* **2005**, 2143.
[56] Xu, W.-M., Huang, X., and Tang, E.; *J. Comb. Chem.* **2005**, *7*, 726.
[57] Xu, W.-M., Wu, L. L., and Huang, X.; *Chin. Chem. Lett.* **2006**, *17*, 603.
[58] Cao, J., Tang, E., Huang, X., *et al.*; *Chin. Chem. Lett.* **2006**, *17*, 857.
[59] Sheng, S. R., Xin, Q., Liu, X.-L., *et al.*; *Synthesis* **2006**, 2293.
[60] Huang, X., and Wang, Y.-G.; *J. Comb. Chem.* **2007**, *9*, 121.
[61] Wang, Y.-G., Xu, W.-M., and Huang, X.; *Synthesis* **2007**, 28.
[62] Wang, Y.-G., Xu, W.-M., and Huang, X.; *J. Comb. Chem.* **2007**, *9*, 513.
[63] Huang, X., and Cao, J.; *Synthesis* **2007**, 2947.
[64] Sheng, S. R., Hu, M. G., Xin, Q., *et al.*; *Chin. Chem. Lett.* **2007**, *18*, 377.
[65] Sheng, S. R., Wang, X.-C., Liu, X.-L., and Song, C.-S.; *Synth. Commun.* **2003**, *33*, 2867.
[66] Liu, X.-L., Wang, X.-C., and Sheng, S. R.; *J. Chin. Chem. Soc.* **2004**, *51*, 1303.
[67] Sheng, S. R., Zhou, W., Liu, X.-L., and Song, C.-S.; *Synth. Commun.* **2004**, *34*, 1011.
[68] Sheng, S. R., Zhou, W., Liu, X.-L., and Song, C.-S.; *Chin. Chem. Lett.* **2005**, *16*, 583.
[69] Sheng, S. R., Liu, X.-L., Wang, X.-C., *et al.*; *Synthesis* **2004**, 2833.
[70] Liu, X.-L., Sheng, S. R., Wang, Q.-Y., *et al.*; *J. Chem. Res.* **2006**, 118.
[71] Huang, B., Sheng, S. R., Huang, Y.-X., *et al.*; *J. Chem. Res.* **2006**, 623.
[72] Hu, Q.-S., Sheng, S. R., Lin, S.-Y., *et al.*; *J. Chem. Res.* **2007**, 74.
[73] Barrero, F., Qurlez der Moral, J. F., Herrador, M. M., *et al.*; *J. Org. Chem.* **2006**, *71*, 5811.
[74] Weber, J. V., Faller, P., Kirsch, G., and Schneider, M.; *Synthesis* **1984**, 1044.
[75] Yanada, K., Fujita, T., and Yanada, R., *Synlett* **1998**, 971.

15

Linker Units Cleaved by Radical Processes: Cleavage of Carbon-Sulfur, -Selenium, -Tellurium, -Oxygen, -Nitrogen and -Carbon Linkers

Giuditta Guazzelli, Marc Miller and David J. Procter

The School of Chemistry, University of Manchester, United Kingdom

15.1 Introduction

A wide variety of radical transformations are available to the synthetic chemist ranging from functional group interconversions to asymmetric reactions and sequential cyclization reactions during which molecules can be built-up rapidly. Modern radical processes are often complimentary to more traditional polar processes and are characterized by attractive features, such as the mild, neutral reaction conditions involved and the compatibility of the processes with functional groups present in the substrates.

With the advent of solid-phase synthesis, and its application in the high-throughput synthesis of 'small molecule' libraries, solution-phase radical processes have been adapted for the manipulation of substrates attached to polymer supports. In recent years, significant progress has been made and both inter- and intramolecular radical reactions are nearing routine status as tools for solid-phase synthesis. The use of radical chemistry in solid-phase synthesis has recently been reviewed.[1]

The choice of the linker unit is crucial to the success of any solid-phase synthesis. One of the key requirements of a linker is that cleavage of the unit must be possible under conditions that do not affect the valuable functionality installed in the molecule under construction. Given the mild, neutral reaction conditions involved in many radical reactions and the functional group tolerance often observed, it is perhaps not surprising that chemists have turned to the development of linkers that can be cleaved by radical processes.

Linker Strategies in Solid-Phase Organic Synthesis   Edited by Peter Scott
© 2009 John Wiley & Sons, Ltd

This review describes the use of radical cleavage processes using 'traditional' radical reagents such as tributyltin hydride, and linker scission using electron-transfer reagents such as the lanthanide reagent samarium(II) iodide (SmI_2). The use of electron-transfer[2] in solid-phase synthesis and the role of lanthanide reagents[3] in the field are the subject of recent reviews. This review also covers linker systems that exploit *cyclative–capture* and *cleavage–cyclization* strategies. These innovative approaches shorten phase-tag assisted syntheses by obtaining a higher synthetic return from the linker formation and cleavage steps. The use of linkers designed to undergo cleavage upon photolysis is covered in Chapter 5. Sulfur, selenium and tellurium linkers cleaved by radical processes are discussed in this review but the reader is directed to a recent review[4] and to Chapters 13 and 14 for full discussions of these classes of linker. Representative experimental procedures for the construction and cleavage of selected linker units are also included.

15.2 Linkers cleaved using tin hydride, alkyltin and silicon hydride reagents

15.2.1 Oxygen-based linkers

Nicolaou has reported a sulfonate linker for use in solid-phase library synthesis.[5] The linker is constructed by reaction of epoxides, formed *in situ* by epoxidation of alkenes with dimethyldioxirane, with a polystyrene-derived sulfonic acid resin, followed by oxidation with the Dess–Martin periodinane (DMP), to give immobilized α-sulfonyloxy ketones such as resin **2**. The sulfonate linker can be cleaved using a variety of nucleophilic approaches that result in the introduction of diversity. The linker can also be cleaved under radical conditions, for example, treatment of **2** with allyltributyltin gives **3**, the product of cleavage–carbon–carbon bond formation (Scheme 15.1).[5]

Typical Experimental Procedures

Preparation of α-sulfonyloxy ketone resin 2

To a solution of epoxide (2.5 equiv.) in dichloromethane (DCM) (0.1 M) was added the polystyrene-derived sulfonic acid resin (1.0 equiv.). After four hours at 25 °C, aqueous saturated sodium bicarbonate (6.0 equiv.) and DMP (2.0 equiv.) were added and the reaction mixture was stirred for an additional 12 hours at 25 °C. Filtration, washing alternately with water (3 ×) and methanol (3 ×), followed by alternate washing with water (3 ×), methanol (3 ×), DCM (CH_2Cl_2) (3 ×) and, finally, by alternate washes with methanol (3 ×), DCM (3 ×), then drying in vacuo furnished the corresponding α-sulfonyloxy ketone resin.

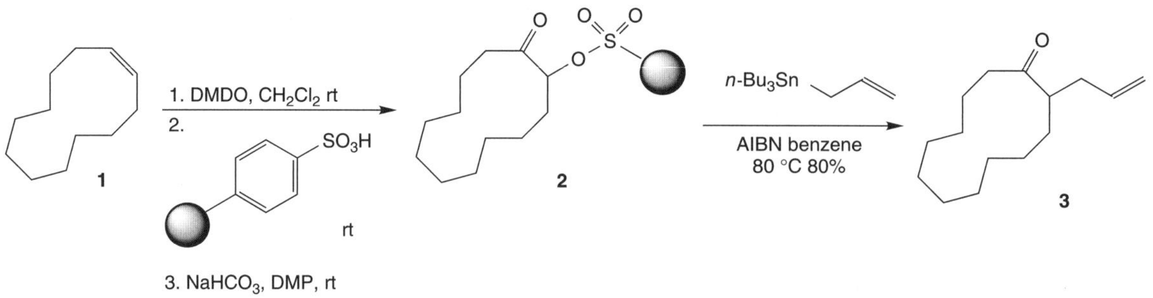

Scheme 15.1 *Nicolaou's sulfonate linker*

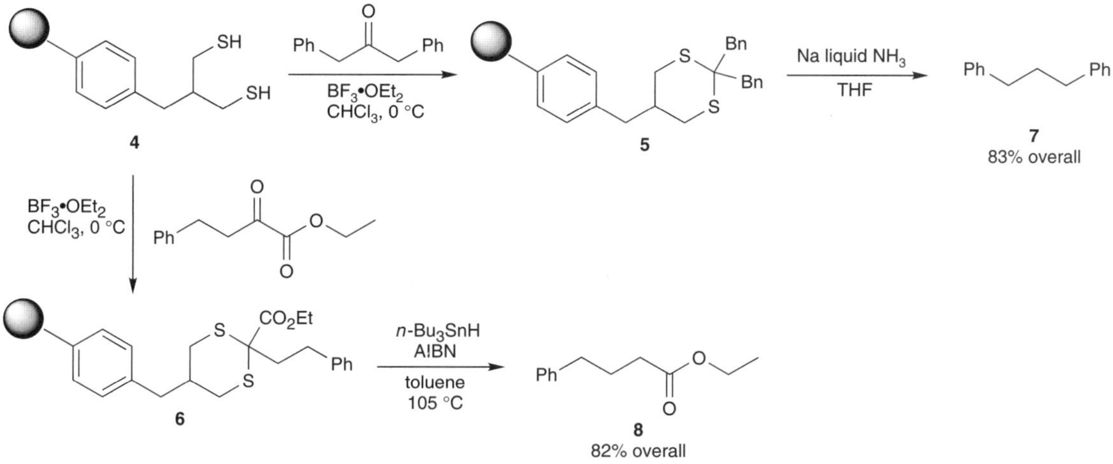

Scheme 15.2 Bertini's 1,3-dithiane linker

Cleavage of α-sulfonyloxy ketone resin 2

To a suspension of resin 2 (25 mg, 0.025 mmol) in benzene (0.5 ml) was added AIBN (1 mg, 0.005 mmol) followed by allyltri(n-butyl)tin (42 mg, 0.125 mmol) and the reaction mixture was heated at 80 °C. After 24 hours the reaction mixture was filtered, diluted with diethyl ether (3 ml), washed with saturated aqueous sodium bicarbonate (2 × 2 ml) and brine (2 ml), dried (magnesium sulfate) and concentrated in vacuo. The residue was purified by column chromatography on silica gel (eluting with hexanes:diethyl ether, 3:1) to furnish α-allyl ketone 3 (5.0 mg, 90%).

15.2.2 Sulfur-based linkers

Bertini has reported preliminary studies on a 1,3-dithiane linker that can be cleaved by radical reduction with tributyltin hydride and by reductive electron transfer using sodium in ammonia.[6] A full discussion of linkers cleaved by reductive electron transfer can be found in Section 15.4. Dithiol resin **4** was used to immobilize ketones giving 1,3-dithiane resins, such as **5** and **6**. Reductive radical cleavage using the complimentary conditions gave the expected products **7** and **8**, respectively, in good yield (Scheme 15.2).[6]

Winssinger has used a sulfur linker for the solid-phase synthesis of the natural products radicicol and pochonin C.[7] Although the sulfide linkage was cleaved by oxidation to the sulfoxide and elimination, radical cleavage of the linker was used to assess the efficiency of the steps on solid phase, for example, treatment of amides **9** and **10** with tributyltin hydride, under microwave conditions, gave the expected products, **11** and **12**, respectively. Although no yields were given for the cleavage, the efficiency of the Pummerer allylation to give **10** can be conveniently assessed using the radical cleavage process (Scheme 15.3).[7]

15.2.3 Selenium-based linkers

Nicolaou has done much to popularize the use of selenium linkers in solid-phase synthesis and was amongst the first to prepare a series of selenium resins for linker construction.[8] Treatment of polystyrene with *n*-butyllithium followed by dimethyldiselenide gave selenide resin **13**. Treatment of **13** with bromine gave selenenyl bromide resin **14**, that was reduced with lithium borohydride to give lithium selenide resin **15** (Scheme 15.4).[8]

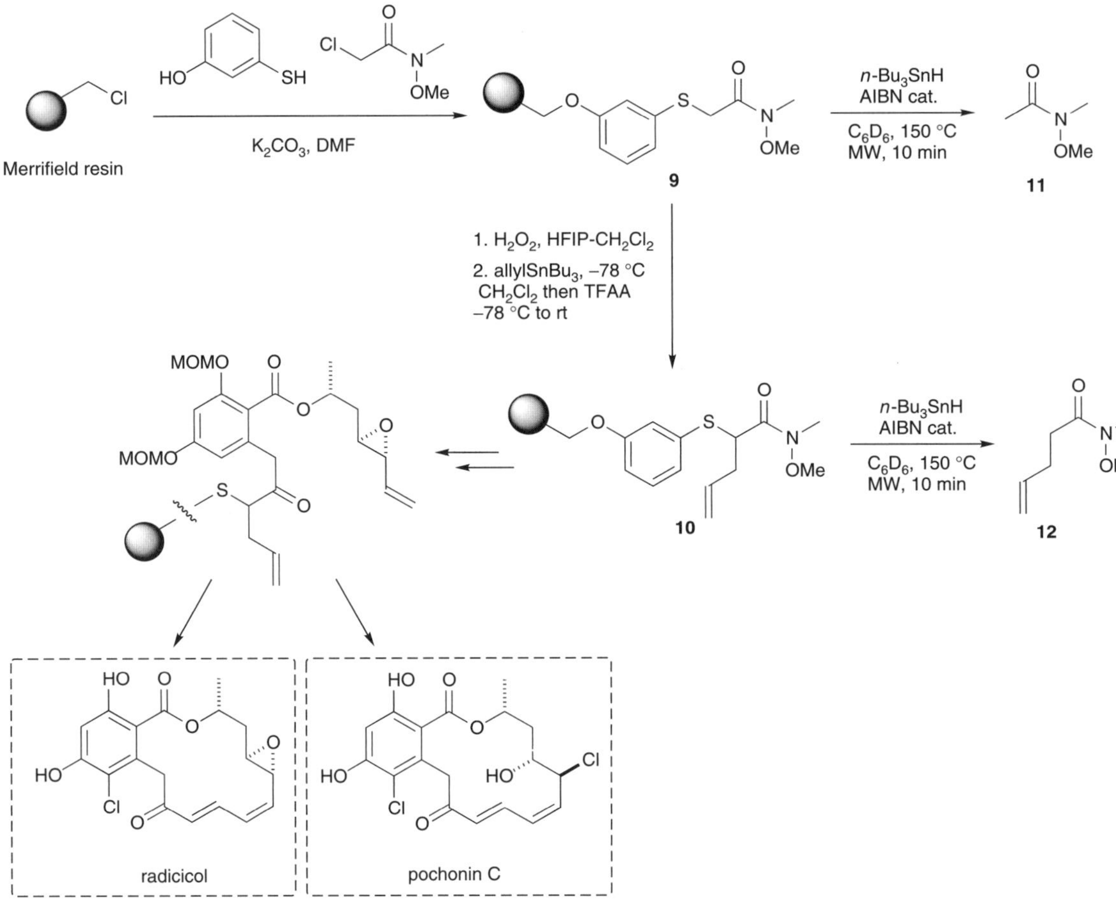

Scheme 15.3 Winssinger's cleavage of an α-arylsulfanyl carbonyl linker using tributyltin hydride

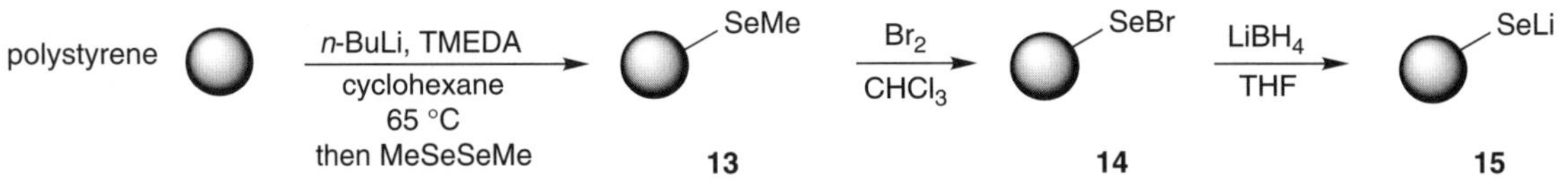

Scheme 15.4 Synthesis of selenium resins by Nicolaou

With convenient access to selenium resins **14** and **15**, Nicolaou explored the utility of selenium linker systems. Alkyl iodide **16** was efficiently immobilized using lithium selenide resin **15** (Scheme 15.4).[8] The selenide linkage in **17** could be cleaved in one of two ways: radical cleavage with tributyltin hydride resulted in release of alkane **18** in 89% overall yield, whereas oxidation of the selenide to the corresponding selenoxide and eliminative cleavage gave **19** in 78% overall yield (Scheme 15.5).[8]

Concurrently with Nicolaou, Ruhland developed a selenium linker that also allows the formation of aliphatic carbon–hydrogen bonds upon radical cleavage with tributyltin hydride (Scheme 15.6).[9] Sodium

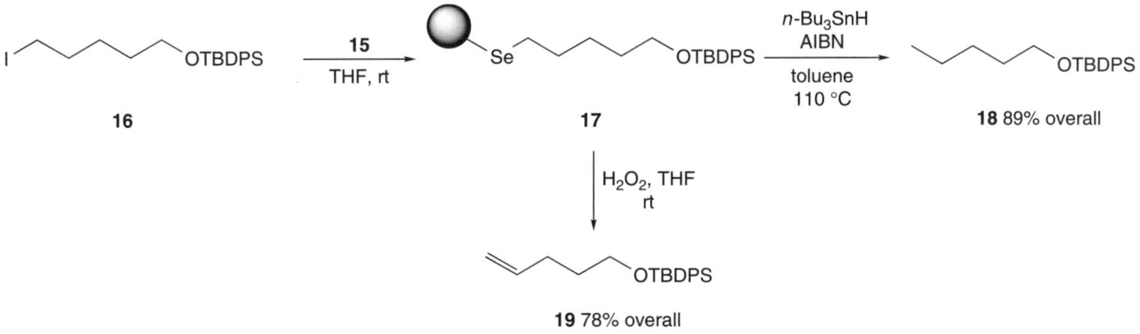

Scheme 15.5 Nicolaou's early studies on selenium linkers

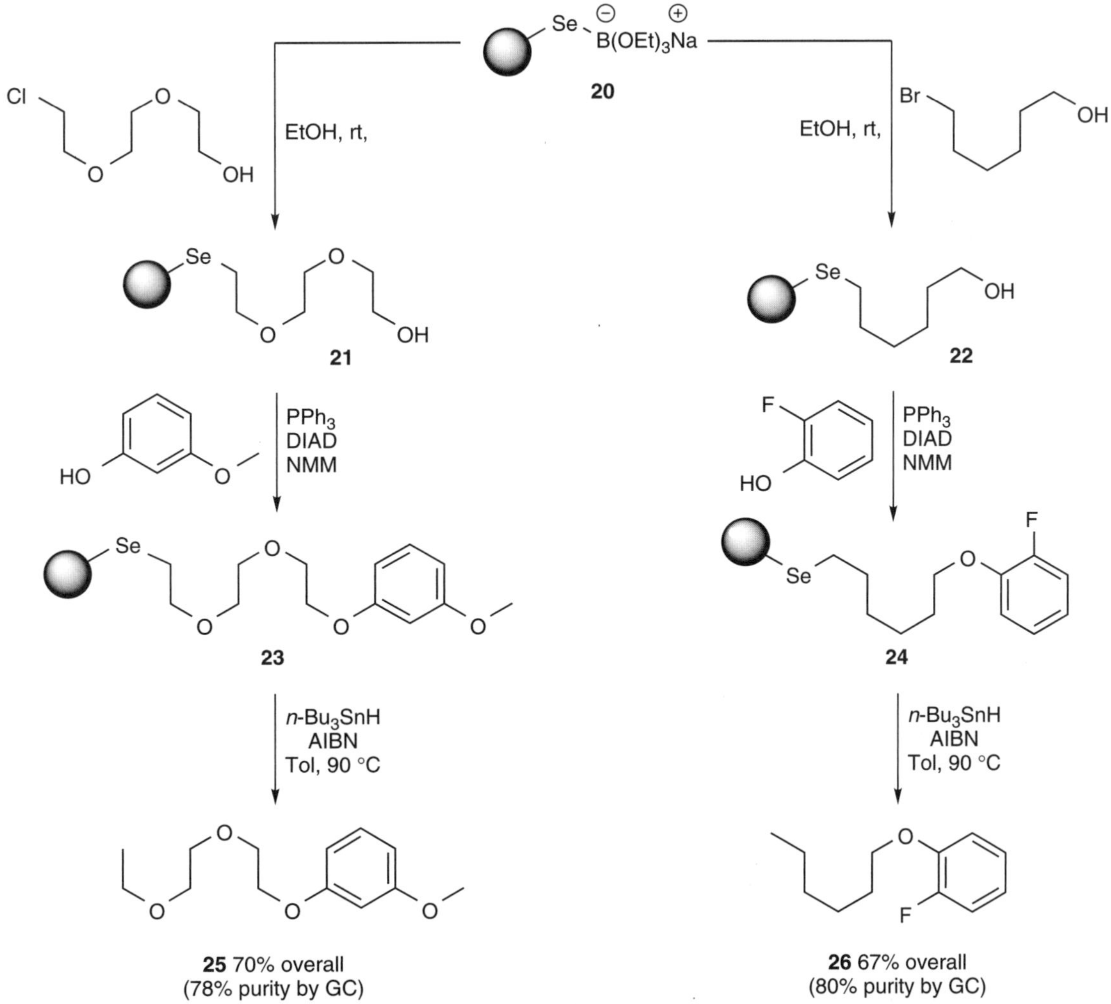

Scheme 15.6 Ruhland's early studies on selenium linkers

424 Multifunctional Linker Units for Diversity-Oriented Synthesis

seleno(triethylborate) resin **20** was prepared from bromopolystyrene resin by lithium–bromine exchange, treatment of the resulting lithiated resin with selenium powder and, finally, treatment with sodium borohydride in ethanol, thus reducing any diselenide present and eliminating selenium–selenium cross-linking. Selenolate resin **20** underwent alkylation upon treatment with alkyl chlorides and bromides to give products such as **21** and **22**. The resin-bound alcohols then underwent Mitsunobu reaction with phenols to give **23** and **24**. Reductive cleavage was then carried out using tributyltin hydride. Cleavage of the aliphatic carbon–selenium bond occurs preferentially to the aromatic carbon–selenium bond, to give aryl ethers **25** and **26** in good yields and high purity by GC (Scheme 15.6). The selenide linker system was used to prepare a 2 × 3 array of alkyl aryl ethers in yields between 57 and 83% and GC purities between 78 and 88%.[9]

Building on his initial work in the area,[8] Nicolaou applied polystyrene-based selenenyl bromide resin **14** in the cyclative capture of substituted cyclohexane-based enol acetates, such as **27**, to give immobilized, substituted [3.3.1] bicyclic systems **28** (Scheme 15.7).[10]

Methods for the radical cleavage of the selenium linker were investigated and are shown in Scheme 15.8.[10] Reductive cleavage to give **30** was achieved by treatment with tributyltin hydride and radical cleavage using allyltributyltin resulted in cleavage of the selenium link and introduction of an allyl group at the former point of attachment, to give **31**. Although low yielding, compared to the other cleavage methods, this strategy, in principle, allows further diversification to be introduced via carbon–carbon bond formation in the cleavage step (Scheme 15.8).[10]

Nicolaou has also exploited a selenium linker, cleaved under radical conditions, in a cyclative capture strategy for the solid-phase synthesis of indolines (Scheme 15.9).[11, 12] Treatment of *o*-allylanilines with selenenyl bromide resin in the presence of a Lewis acid results in cyclative capture to give immobilized indoline **33**. Further functionalization was then carried out prior to cleavage: Acylation of nitrogen with

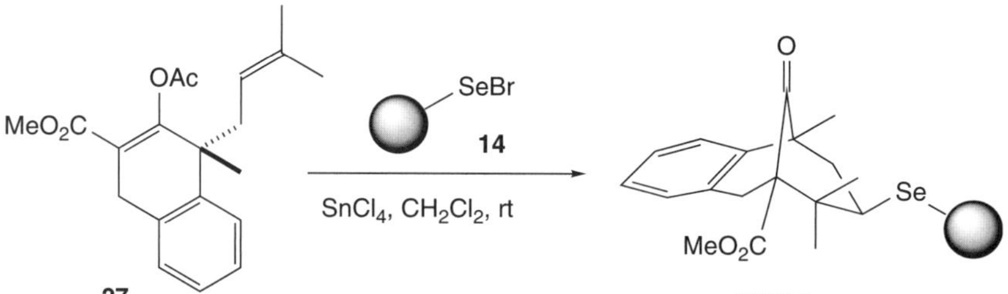

Scheme 15.7   Nicolaou's cyclative–capture using a selenenyl bromide resin

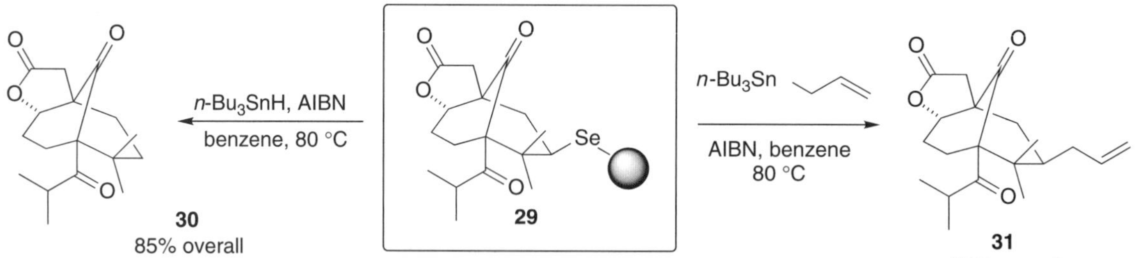

Scheme 15.8   Nicolaou's multifunctional cleavage of a selenium linker

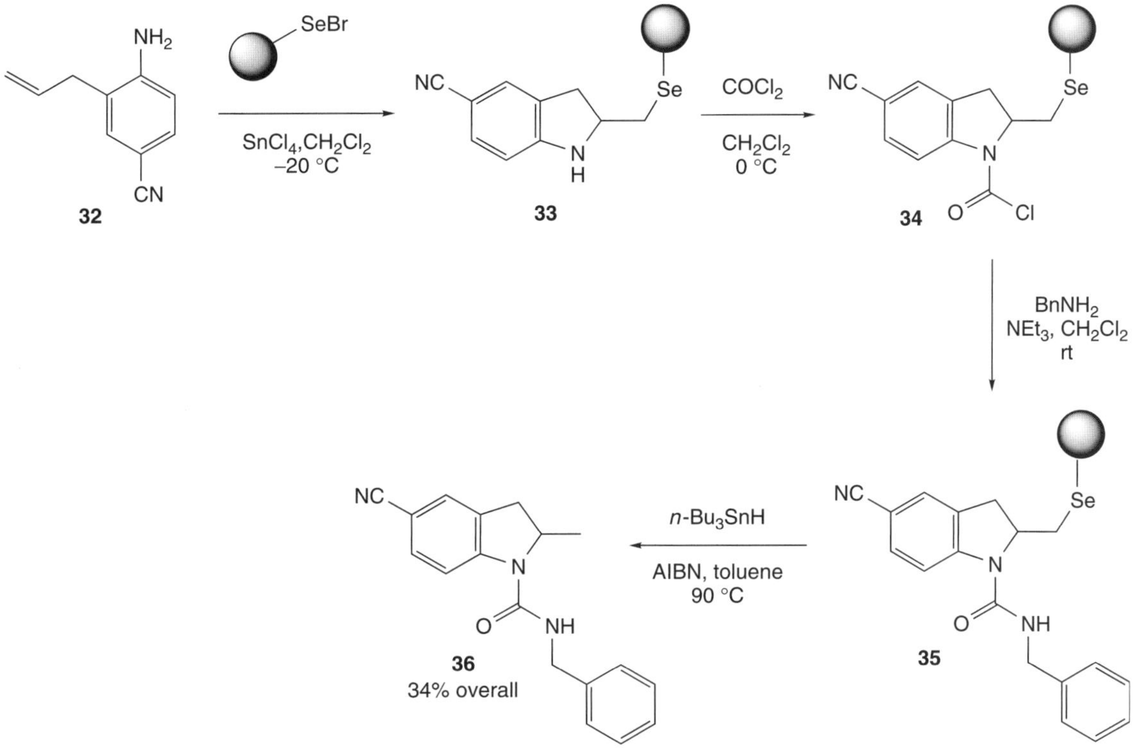

Scheme 15.9 Nicolaou's cyclative–capture approach to indolines

phosgene gave **34** and further reaction with amines gave **35**. Radical cleavage with tributyltin hydride gave functionalised 1-methylindoline **36** in good yield (Scheme 15.9). A range of 1-methylindolines was synthesized using this approach in overall yields between 14 and 34%.[11, 12]

Typical Experimental Procedures

Preparation of selenium linker in 33

To a suspension of selenium bromide resin (approximately 200 mg of 0.75 mmol/g, 1.0 equiv.) in DCM (4 ml) at −20 °C was added a solution of the o-allyl aniline 32 (4.0 equiv.) in DCM (1 ml) followed by addition of tin)IV) chloride (1.0 M solution in DCM, 5.5 equiv.). The reaction mixture was slowly stirred at −20 °C for one hour then quenched by addition of triethylamine (0.5 ml). The resulting suspension was poured into a fritted funnel and the resin was washed with DCM (4 × 15 ml), methanol (4 × 15 ml), and diethyl ether (2 × 15 ml).

Cleavage of the selenium linker in 35

The resin 35 (approximately 200 mg, 1.0 equiv.) was suspended in toluene (2 ml) and n-Bu₃SnH (4.0 equiv.) and 2,2-azobisisobutyronitrile (AIBN) (1.3 equiv.) were added; the reaction mixture was heated at 90 °C and stirred for two hours. After cooling to 25 °C, the resin was removed by filtration, washed with DCM

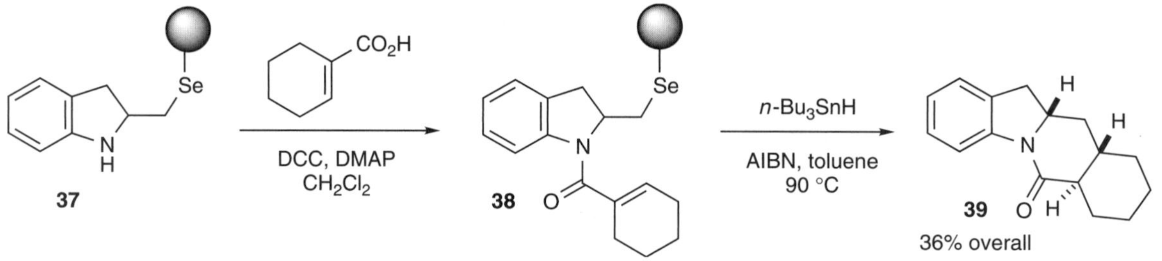

Scheme 15.10 Nicolaou's cleavage–cyclization using a selenium linker

(4 × 15 ml) and the resulting filtrate was concentrated to afford crude **36** which was purified by column chromatography.

Further complexity can be introduced in the cleavage step when a suitably positioned radical acceptor is present.[11, 12] Resin-bound indoline **37** was coupled with 1-cyclohexenecarboxylic acid to form **38** (Scheme 15.10). Cleavage of the selenium linker with tributyltin hydride gave a carbon-centred radical that underwent 6-*endo*-trig cyclization onto the tethered double bond to give **39** thus illustrating the feasibility of accessing polycycles using this methodology (Scheme 15.10).[11, 12] This was the first example of radical cleavage–cyclization.

In an attempt to introduce more diversity upon cleavage, exposure to a radical initiator in the absence of a radical quench led to cleavage and radical rearrangement allowing access to 2-methyl indoles, for example, treatment of **40** with AIBN gave primary radical **41** initially upon cleavage that rearranged to give 2-methyl indole **42** in good overall yield (Scheme 15.11).[11]

Nicolaou has used a selenium linker in the synthesis of 2-deoxyglycosides (Scheme 15.12).[8] Treatment of tri-*O*-benzylglucal **43** with **14** in the presence of an alcohol such as **44** resulted in glycosylation and immobilization of the product through a selenide linkage at the 2-position of the sugar ring. Radical cleavage of the carbon–selenium bond using tributyltin hydride released the 2-deoxyglycoside **45** in 61% yield, as an 8:1 mixture of α and β anomers (Scheme 15.12).[8]

Nicolaou has exploited this approach in the solid-phase synthesis of 2-deoxyglycosides, 2-deoxyorthoesters and 2,3-allyl orthoesters.[13] Resin-bound tributyltin selenolate **46** was conveniently

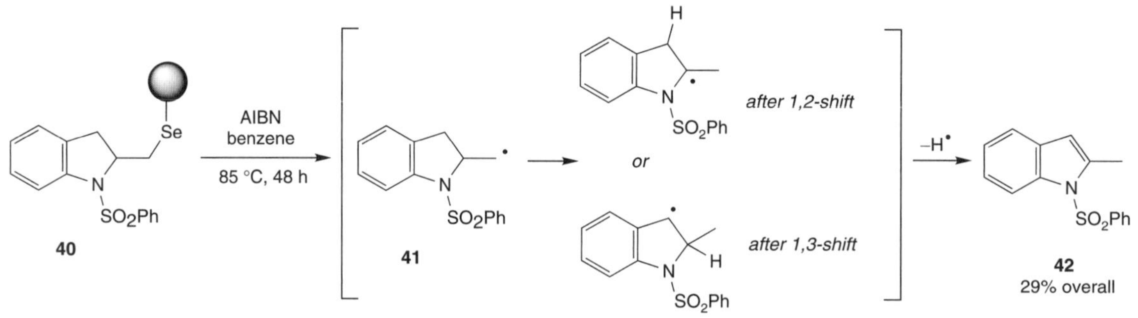

Scheme 15.11 Nicolaou's cleavage–radical rearrangement using a selenium linker

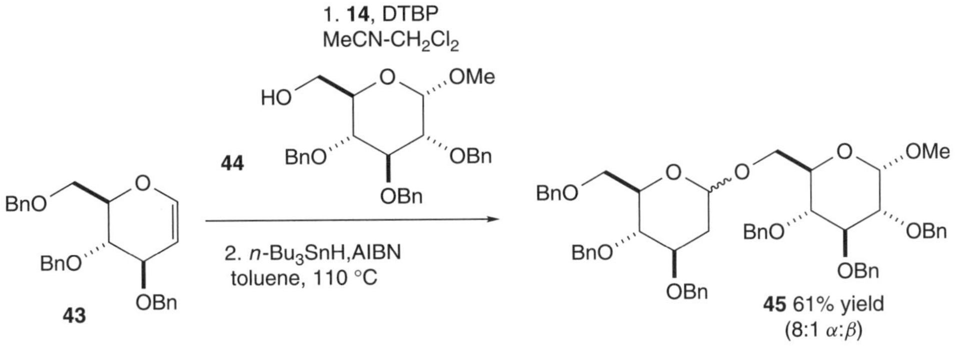

Scheme 15.12 *Nicolaou's 2-deoxyglycoside synthesis using a selenium linker*

synthesized in two steps from a selenenyl bromide resin. Treatment of **46** with trichloroacetimidate donor **47** gave resin-bound selenoglycoside **48** (Scheme 15.13). Removal of the C2 and C3 acetate protecting groups and re-protection of the C3 hydroxyl gave **49** ready for a 1,2-selenomigration reaction. Treatment of **49** with diethylaminosulfur trifluoride (DAST) promoted 1,2-migration of the selenium link to resin in a stereospecific manner to give **50**, now linked to resin at the C2 position of the sugar, whilst simultaneously generating an anomeric fluoride leaving group. Glycosyl fluoride donor **50** then underwent glycosylation with alcohols or sugars, for example; exposure to Lewis acid and monoprotected diol **51** selectively gave the α-glycoside **52**.[13] Radical cleavage of the selenium linker with tributyltin hydride gave the desired 2-deoxyglycoside **53** in excellent yield. A 3 × 3 library of 2-deoxyglycosides was synthesized using this approach in overall yields between 11% and 32% (Scheme 15.13).

Nicolaou has applied this methodology in synthetic studies towards the antibiotic everinomicin 13,384-1.[14] Glycosylation of immobilized fluoride donors such as **54** (Scheme 15.14) with sugar acceptors allowed the synthesis of disaccharides such as **55**, attached to solid support through a selenium linker. As illustrated in previous model studies, radical cleavage allowed access to 2-deoxy glycosides such as **56** (Scheme 15.14).[14]

Building on Nicolaou's studies, Engman has reported the first example of the homolytic cleavage of a selenide linker followed by carbonylation of the resultant radical and cyclization (Scheme 15.15).[15] In Engman's studies, tris(trimethylsilyl)silane (TTMSS) was employed, in place of tributyltin hydride, for the cleavage of the selenium linker. The loading of precursor **57** was determined by radical cleavage with TTMSS and AIBN to give **58** in excellent yield. Treatment of substrate **57** with TTMSS and AIBN under an atmosphere of carbon monoxide gave lactone **59** in good overall yield and with good diastereoselectivity (Scheme 15.15).[15]

Typical Experimental Procedures
Preparation of the selenium linker en route to 59

To a suspension of polymer-supported diaryl diselenide (300 mg, 0.675 mmol) in ethanol (5 ml) was added sodium borohydrate (51 mg, 1.35 mmol). The suspension was gently stirred for six hours after which time benzyloxirane (450 mg, 3.38 mmol) was added and the mixture stirred overnight. An aqueous solution of hydrochloric acid (2 M, 2 ml) was added, and the mixture was stirred for another two hours. The filtered resin was washed successively with water, ethanol, diethyl ether and ethanol.

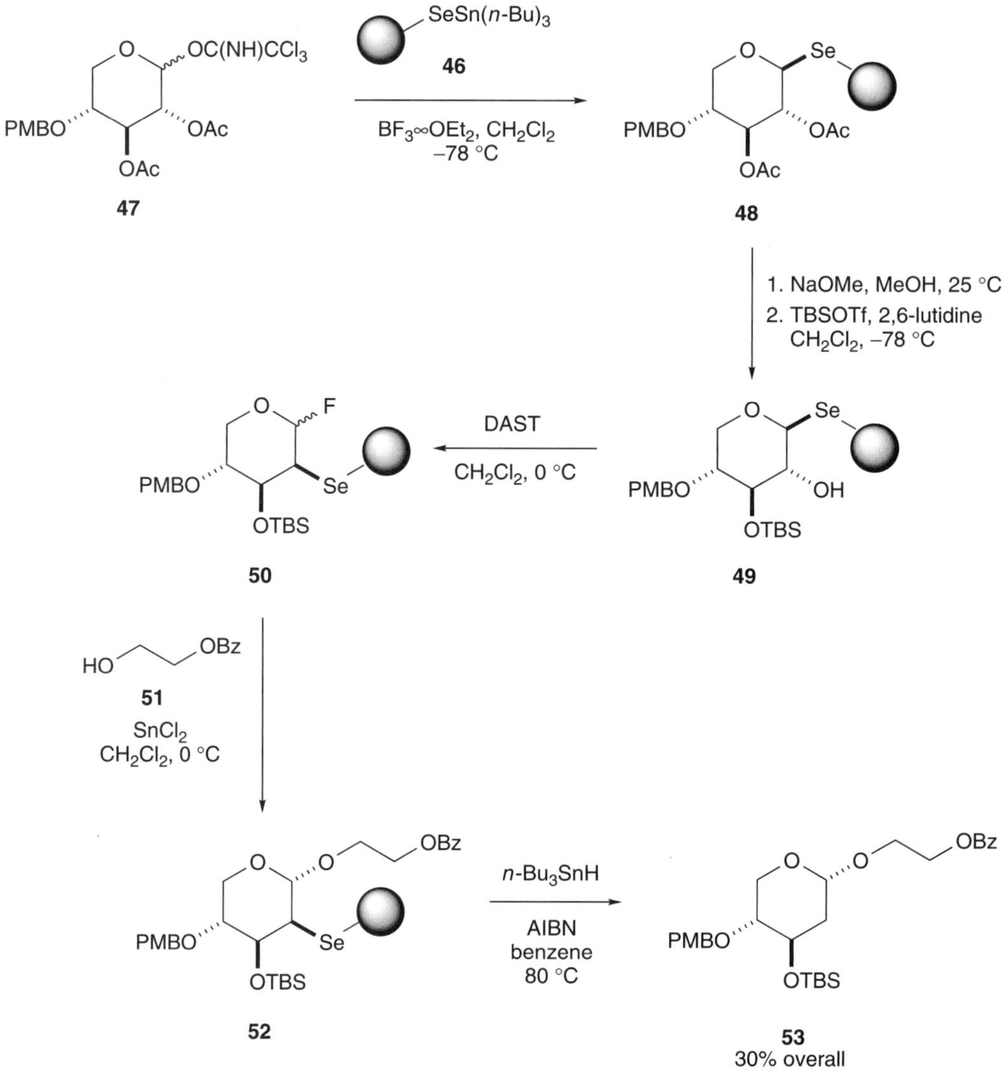

Scheme 15.13 *Nicolaou's synthesis of a library of 2-deoxyglycosides using a selenium linker*

Cleavage of the selenium linker in 57

*To resin **57** in benzene (114 ml) were added AIBN (33 mg, 0.2 mmol) and TTMSS (354 μl, 1.14 mmol, 0.01 M). The suspension was pressurized with carbon monoxide at 80 atm and heated to 80 °C for 12 hours. After filtration and washing with diethyl ether, the combined filtrate was concentrated and purified by column chromatography to afford **59** (97 mg, 55% over three steps) as a 9:1 mixture of cis and trans isomers.*

Wirth has reported the first polymer-bound, enantiomerically pure, electrophilic selenium reagents for asymmetric additions of nucleophiles to alkenes and has used radical processes to cleave the selenium linker and to release enantiomerically enriched products from the support (Scheme 15.16).[16]

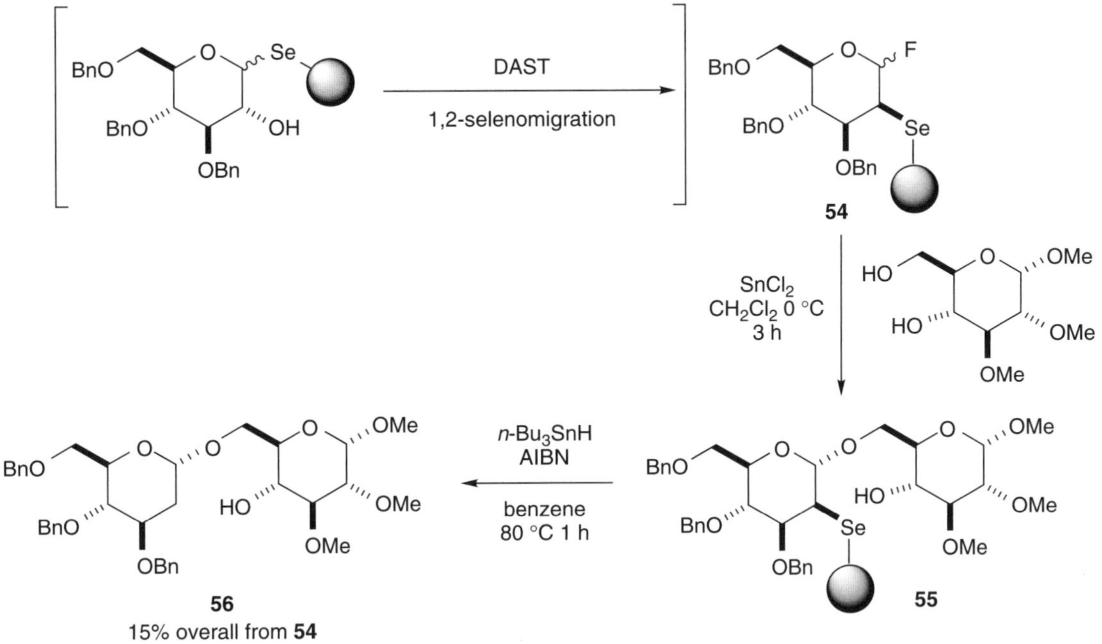

Scheme 15.14 Nicolaou's studies towards the antibiotic everinomicin 13,384-1 using a selenium linker

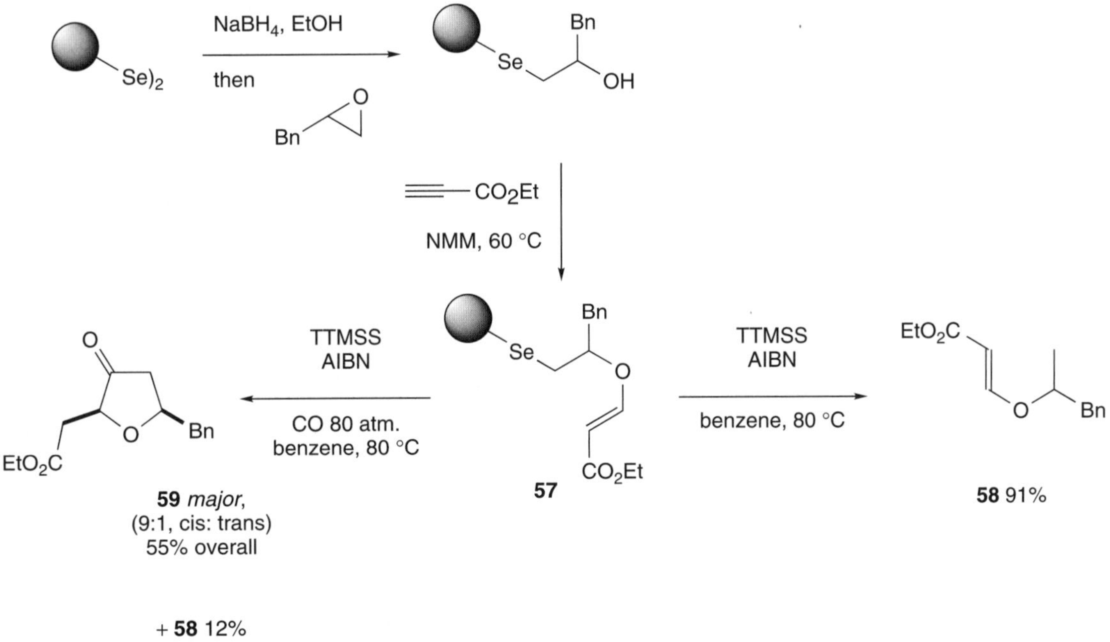

Scheme 15.15 Engman's cleavage–carbonylation–cyclization using a selenium linker

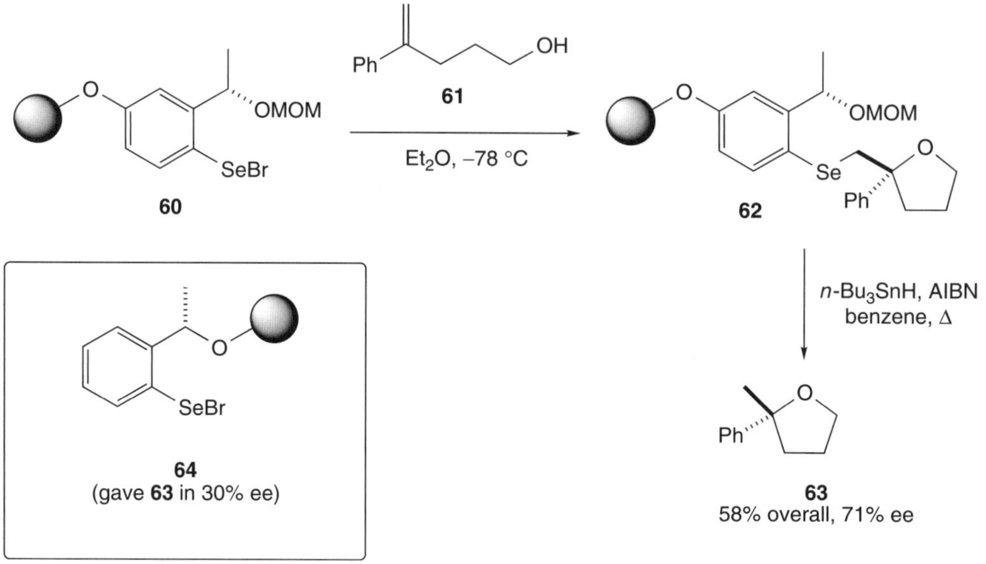

Scheme 15.16 Wirth's asymmetric cyclative–capture using an enantiomerically enriched selenenyl bromide resin

Chiral selenenyl bromides were prepared and immobilized on polystyrene, Tentagel and mesoporous silica supports. Polystyrene-based selenenyl bromide resin **60** was found to be most effective in asymmetric selenenylation reactions. Treatment of unsaturated alcohol **61** with chiral resin **60** led to diastereoselective seleniranium ion formation and intramolecular, nucleophilic addition of the tethered alcohol, to give diastereomerically-enriched tetrahydrofuran **62** immobilized via a selenium linker. This is the first example of *asymmetric* cyclative capture. Radical cleavage using tributyltin hydride gave tetrahydrofuran **63** in 58% yield and 71% ee (Scheme 15.16).[16] Alternative chiral selenenyl bromide **64**, where immobilization is achieved through the side chain hydroxyl group, was found to give lower selectivities (Scheme 15.16).[17]

15.2.4 Tellurium-based linkers

Ruhland has reported the first use of a tellurium-based linker.[18] A library of simple aryl alkyl ethers was prepared using polystyrene-bound telluride complex **65**, analogous to selenolate resin **20**. As the homolysis of carbon–tellurium bonds is known to occur at lower temperatures than the homolysis of the corresponding selenides, it was proposed that tellurium linkers might prove valuable for the solid-phase synthesis of temperature-sensitive targets. A library of simple aryl alkyl ethers was prepared using polystyrene-bound telluride complex **65** (Scheme 15.17), prepared from bromopolystyrene in five steps.[18] Alkylation of **65** with two alkyl halides and subsequent Mitsunobu coupling with three phenols gave ethers **68** and **69** ready for radical cleavage from the polymer support. The cleavage reaction was studied at 60 °C and 90 °C. Surprisingly, the yields of products **70** and **71** obtained using the tellurium linker system were found to be consistently lower than those obtained using an analogous selenium linker system.[18] The authors speculate that the presence of traces of elemental tellurium in the resin may be interfering with the radical chain process (Scheme 15.17).

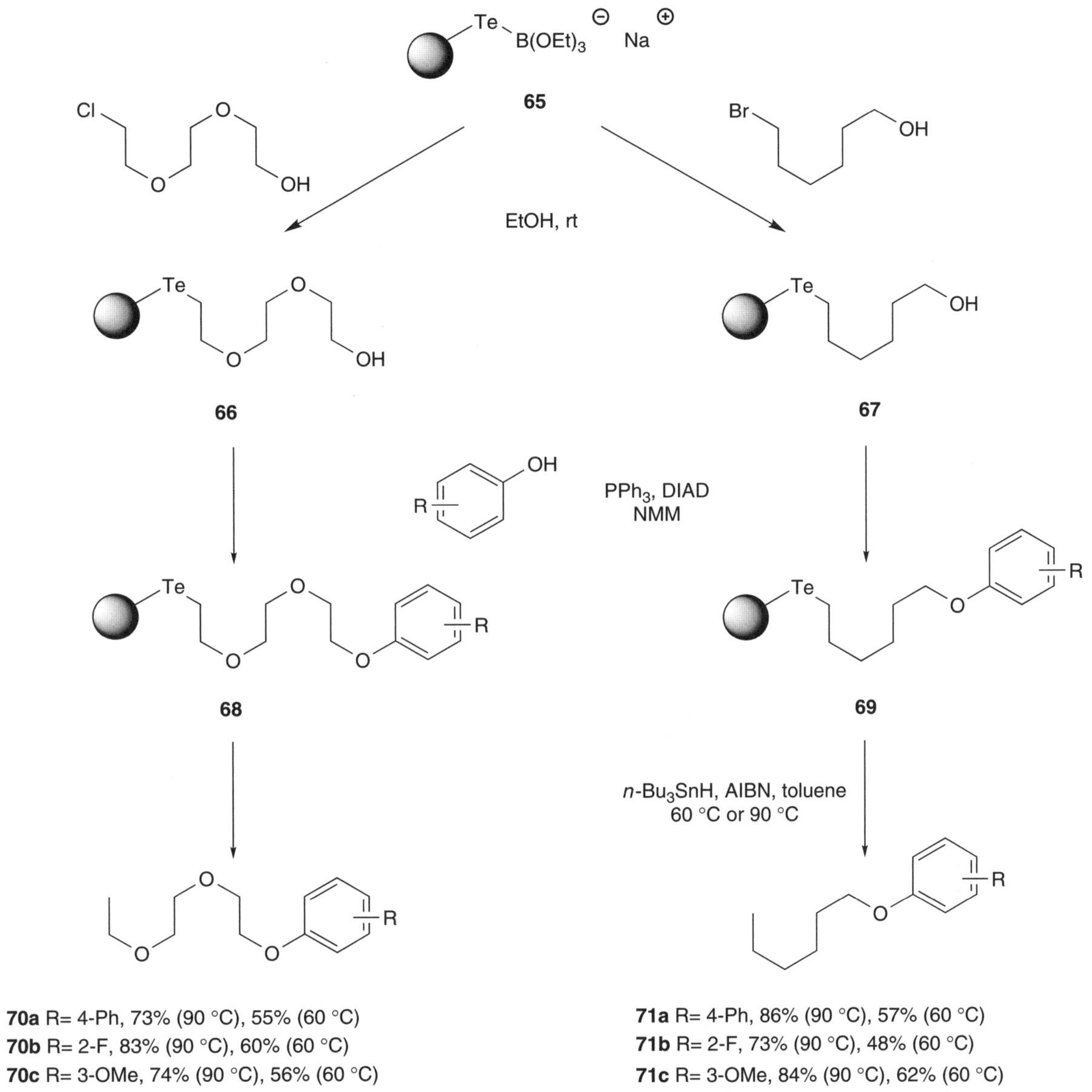

Te
B(OEt)3
Na
65
Cl
O
O
OH
Br
OH
EtOH, rt
Te
O
O
OH
66
Te
OH
67
OH
R
PPh3, DIAD
NMM
Te
O
O
O
R
68
Te
O
R
69
n-Bu3SnH, AIBN, toluene
60 °C or 90 °C
O
O
O
R
O
R
70a R= 4-Ph, 73% (90 °C), 55% (60 °C)
70b R= 2-F, 83% (90 °C), 60% (60 °C)
70c R= 3-OMe, 74% (90 °C), 56% (60 °C)
71a R= 4-Ph, 86% (90 °C), 57% (60 °C)
71b R= 2-F, 73% (90 °C), 48% (60 °C)
71c R= 3-OMe, 84% (90 °C), 62% (60 °C)

Scheme 15.17 *Ruhland's tellurium linker*

15.3 Linkers cleaved by oxidative electron-transfer

15.3.1 Ether and amine linkers cleaved by oxidative electron transfer

A range of benzyl amine, aniline, benzyl alcohol and phenol-based linkers have been developed that can be cleaved by oxidative electron transfer. The most well known of these is the Wang linker that is more usually cleaved using TFA.

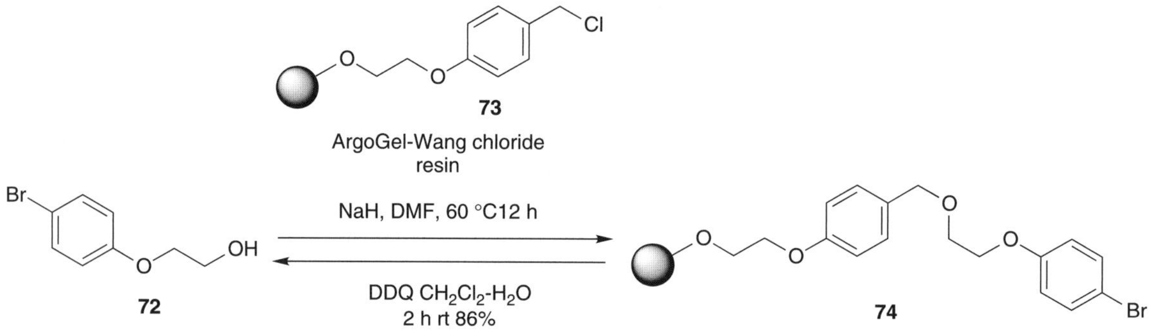

432 *Multifunctional Linker Units for Diversity-Oriented Synthesis*

Scheme 15.18 Porco's ether linker

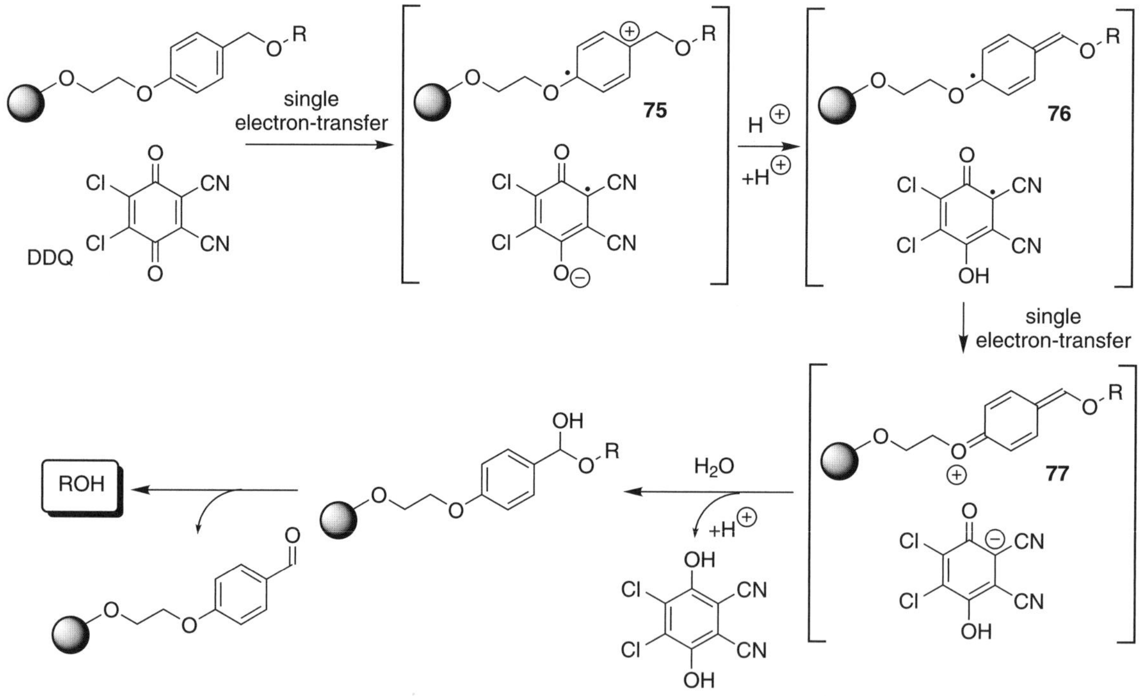

Scheme 15.19 Mechanism of ether linker cleavage using DDQ

Linker Units Cleaved by Radical Processes 433

Porco has reported a Wang-derived ether linker that is cleaved by oxidation with 2,3-dichloro-5,6-dicyanobenzoquinone (DDQ).[19] The linker was formed by etherification of alcohols, such as **72**, with ArgoGel-Wang resin **73**. The resulting resin-bound ethers, such as **74**, were then cleaved using DDQ to return the alcohols in high yield, after removal of excess reagent and reduced reagent by filtration through ion exchange resins (Scheme 15.18).[19]

DDQ is a facile one-electron oxidant and the mechanism of linker cleavage is likely to be similar to that proposed for the deprotection of *para*-methoxybenzyl (PMB) and meta-dimethoxybenzyl (DMB) ethers.[20] One-electron oxidation gives radical cation **75** that is then deprotonated to give radical **76**. A second one-electron oxidation then gives oxonium ion **77** that is hydrolysed, thus releasing the alcohol from the support (Scheme 15.19).

The oxidative radical cleavage of the Wang linker using DDQ has been employed by Oh in a synthesis of 1,3-oxazolidines, for example, treatment of **79** with DDQ gave **80** in good overall yield (Scheme 15.20).[21]

Waldmann has carried out the asymmetric solid-phase synthesis of 6,6-spiroketals using a Wang linker that is cleaved by oxidative electron transfer using DDQ.[22] Aldehyde **81** was prepared from Wang resin and underwent an asymmetric boron mediated aldol reaction to give immobilized adduct **82**, after protection of the secondary hydroxyl. A second asymmetric aldol reaction then gave **83** after protection. Oxidative scission of the Wang linker and simultaneous removal of the PMB protecting group using DDQ resulted in cleavage spiroketalisation to give **84** in good overall yield (Scheme 15.21).[22]

Waldmann has used this asymmetric solid-phase approach, in conjunction with the oxidative cleavage of the Wang linker, to prepare a collection of spiroketals that contains phosphatase inhibitors and compounds that modulate the formation of the tubulin cytoskeleton.[23]

Waldmann has recently used the same linker system and cleavage method in a solid-phase approach to polyoxygenated natural products displaying a broad range of biological activity.[24] Immobilization of enantiomerically-enriched alcohol **85** using Wang-trichloroacetimidate resin **86** and ozonolysis gave **87**.

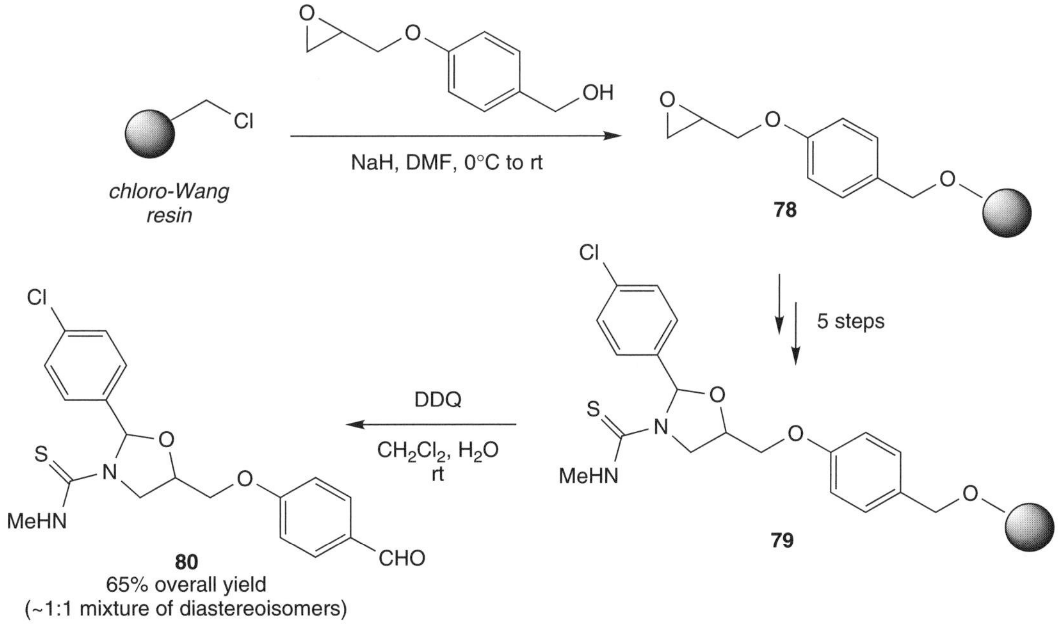

Scheme 15.20 *Oh's synthesis of 1,3-oxazolidines using an ether linker*

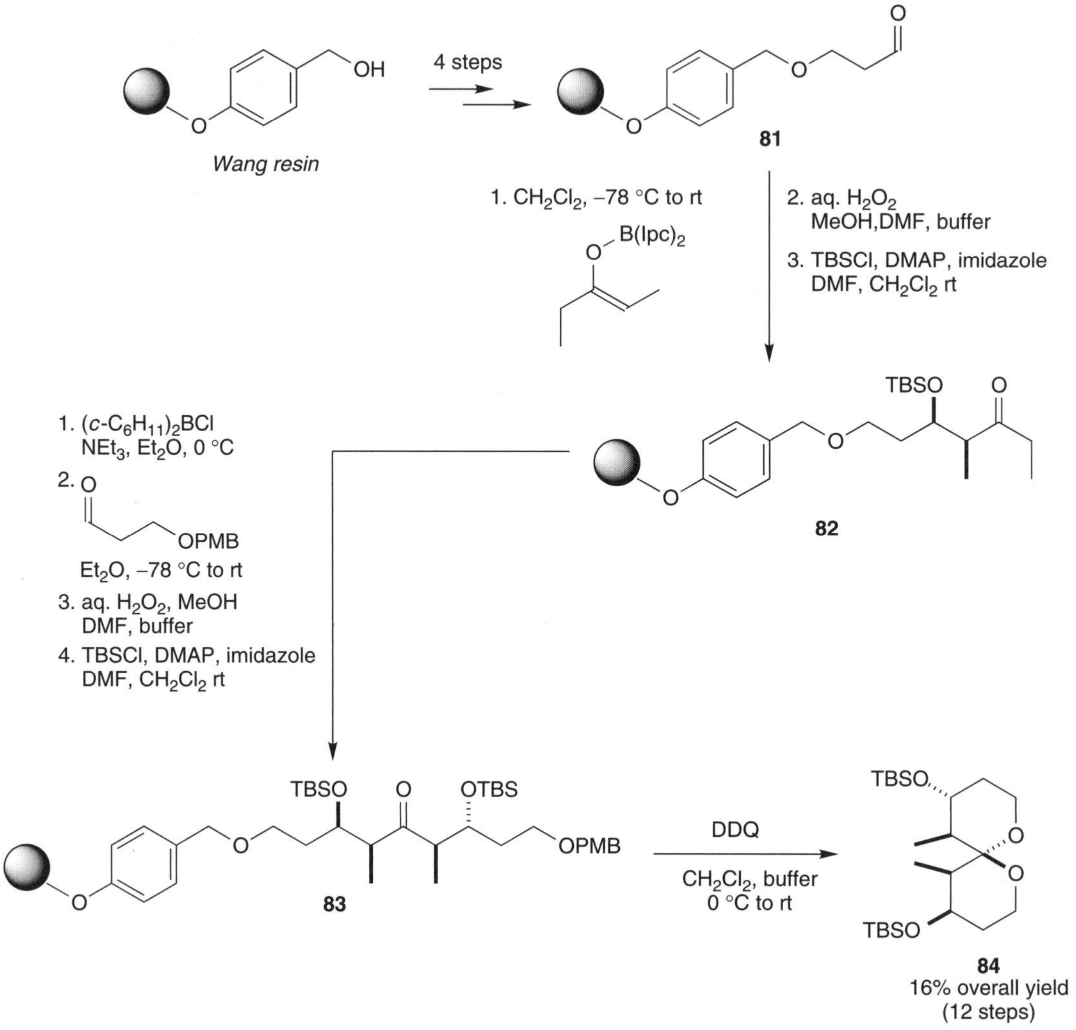

Scheme 15.21 Waldmann's synthesis of 6,6-spiroketals using an ether linker

Brown's asymmetric allylation and protection of the resultant secondary hydroxyl gave **88**. Repeating the ozonolysis and allylation steps then gave **89**. Subsequent formation of the corresponding acrylate ester and RCM gave lactone **90**. Oxidative radical cleavage of the Wang linker using DDQ then gave **91** in 9% overall yield after nine steps (Scheme 15.22).[24]

Typical Experimental Procedures
Cleavage of resin 90

*Resin **90** (250 mg, loading 0.4 mmol/g) was swollen in a mixture of DCM and pH 7 phosphate buffer (0.5 ml); recrystallized DDQ (180 mg, 0.8 mmol) was added at 0 °C. The mixture was shaken for 10 hours while allowing the temperature to rise to room temperature. The resin was then filtered off and washed with*

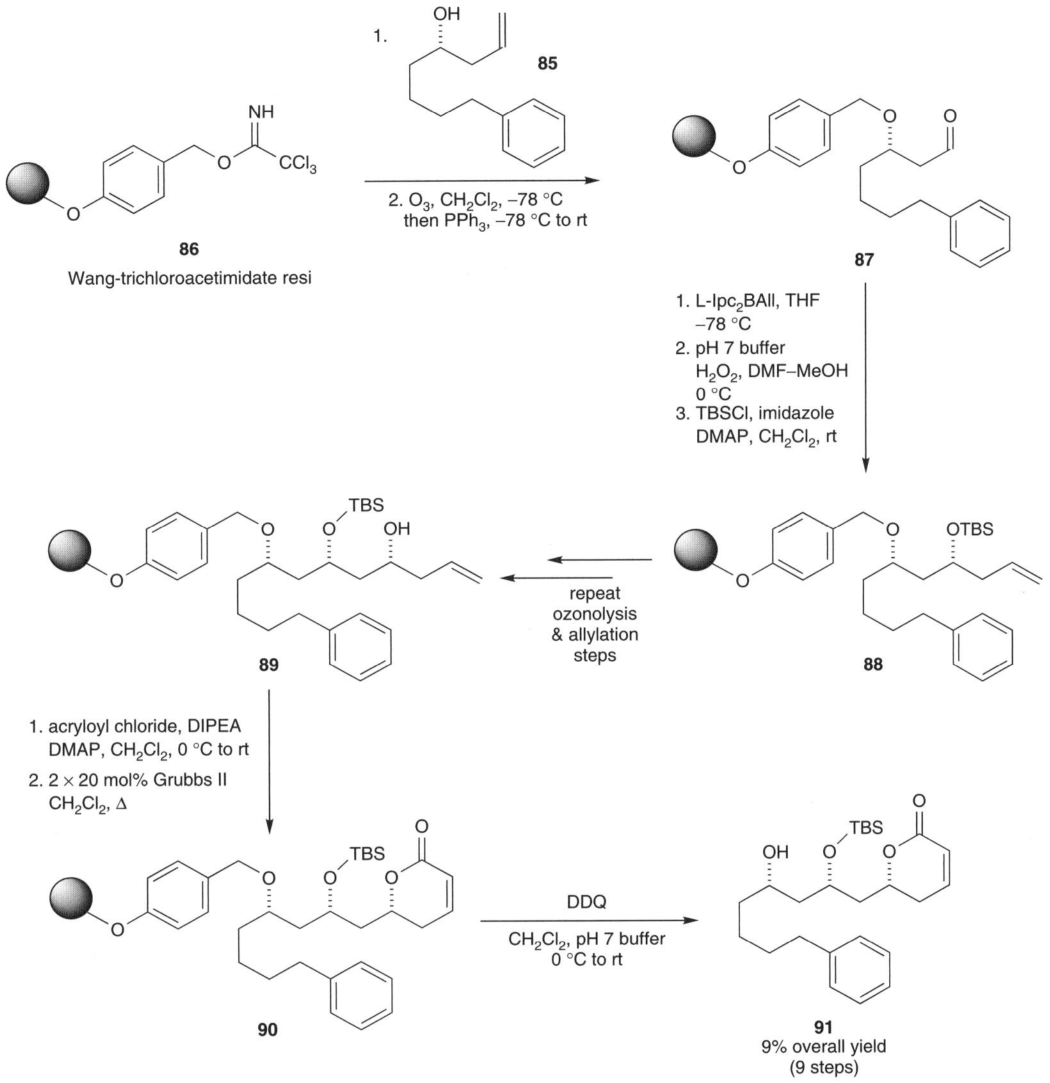

Scheme 15.22 Waldmann's synthesis of polyoxygenated natural products using an ether linker

DCM. The filtrate was washed with a saturated aqueous solution of sodium bicarbonate (3 × 15 ml) and brine and was then dried over sodium sulfate. After removal of solvent under reduced pressure, lactone **91** was obtained in 80–90% purity. Purification by silica-gel column chromatography gave **91** (10 mg, 9.3% over nine steps).

Kobayashi has prepared 4-benzyloxybenzylamine (BOBA) resin **92** from Merrifield resin.[25] Imines derived from **92** undergo Lewis acid catalyzed Mannich reactions with silylenolethers to give adducts such as **94**. Cleavage of the benzylamine linker was achieved by oxidation with DDQ to give β-aminoester products, such as **95**, in good overall yields (Scheme 15.23).[25]

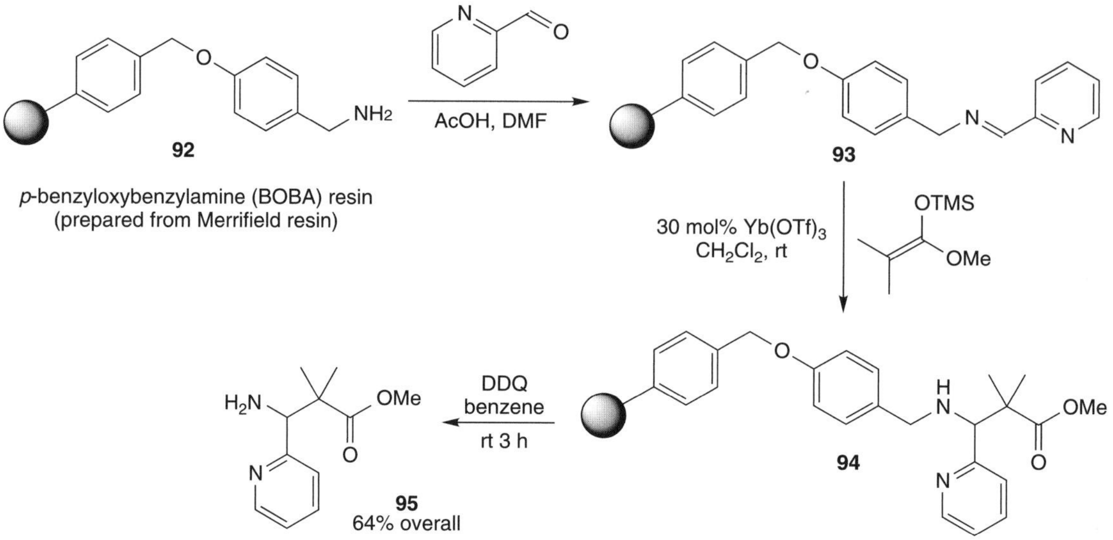

436 Multifunctional Linker Units for Diversity-Oriented Synthesis

Scheme 15.23 *Kobayashi's amine linker cleaved using DDQ*

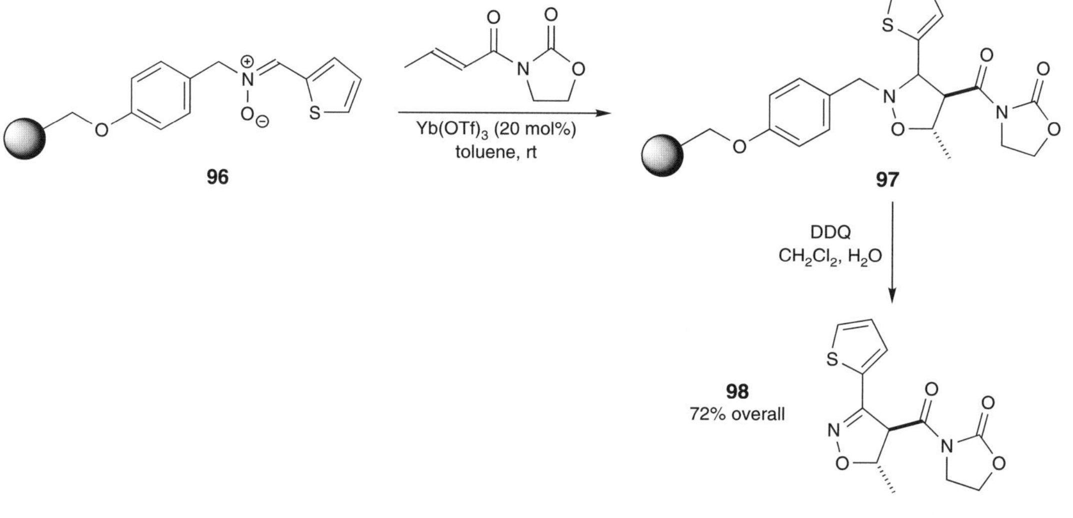

Scheme 15.24 *Kobayashi's synthesis of isoxazolines using an amine linker cleaved using DDQ*

Kobayashi has used the linker system to immobilize nitrones in a solid-phase cycloaddition approach to isoxazolines.[26] Ytterbium(III) triflate catalyzed 1,3-dipolar cycloadditions of nitrones, such as **96**, with oxazolidinone dipolarophiles gave isoxazoline products **98** in high yield and as single diastereoisomers after oxidative radical cleavage from the polystyrene support using DDQ (Scheme 15.24).[26]

Kobayashi has also employed the 4-benzyloxybenzylamine (BOBA) resin in a method for reductive amination on solid phase.[27] Imine formation using **92**, followed by reduction gives secondary amines

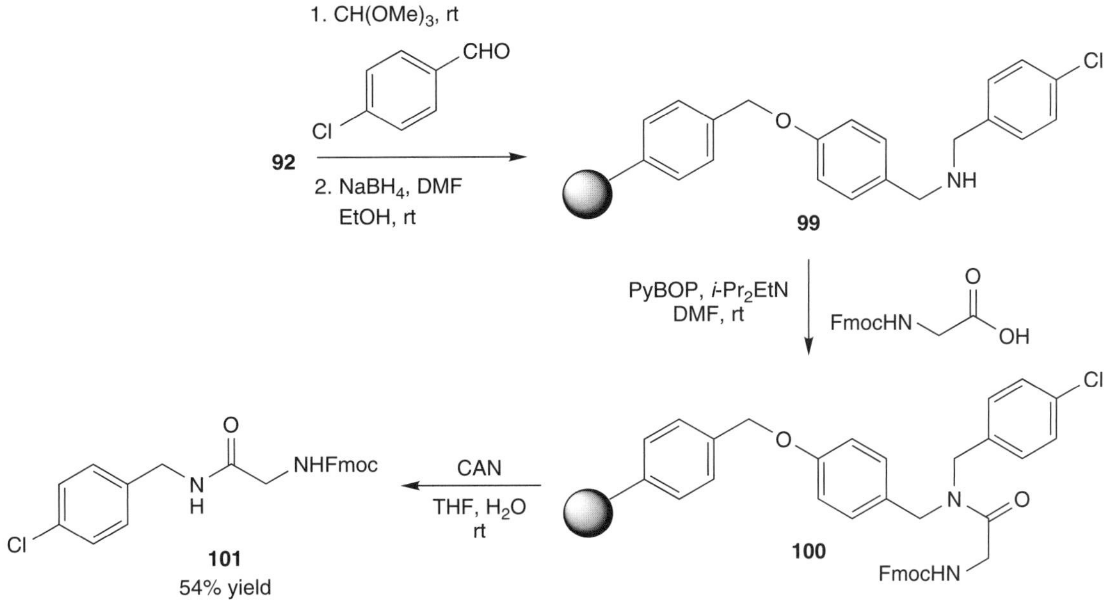

Scheme 15.25 *Kobayashi's synthesis of amides using an amine linker cleaved using CAN*

such as **99** that are then acylated to give amides **100**. Oxidative electron-transfer using cerium(IV) ammonium nitrate (CAN), cleaves the linker and releases amide **101** from the support in good overall yield (Scheme 15.25).[27] Cerium(IV) compounds are the most stable tetrapositive lanthanide species and are versatile one-electron oxidizing agents, making them a useful alternative to DDQ.

Typical Experimental Procedures
Preparation of BOBA resin 92

To 4-hydroxybenzamide (5.60 g, 40.8 mmol) in DMF (65 ml) was added sodium hydroxide (1.80 g, 45 mmol) at room temperature. The mixture was stirred for one hour at 90 °C before the addition of chloromethylated polystyrene (1% cross linked, 15 g, 1.0 mmol/g). The mixture was then stirred for an additional three hours at 90 °C. The resultant resin was washed with water, THF and diethyl ether and dried under reduced pressure. The amide resin was then treated with $BH_3 \cdot SMe_2$ (7.5 ml) in dimethoxyethane (DME) (300 ml). The mixture was stirred at reflux for 20 hours and then quenched by the addition of water. The resin was treated with aqueous 10% sodium hydroxide (50 ml) in THF (200 ml) for 10 hours at room temperature and the resultant resin washed with water, THF and diethyl ether. Drying under reduced pressure afforded BOBA resin **92** (0.80 mmol/g).

Cleavage of the BOBA linker in 100

To a suspension of resin **100** in THF (2 ml) was added water (0.2 ml) and CAN (5 equiv.). The mixture was stirred for five hours at room temperature; DCM and water were then added. The aqueous layer was extracted with DCM. The combined organic layers were washed with saturated aqueous sodium chloride and

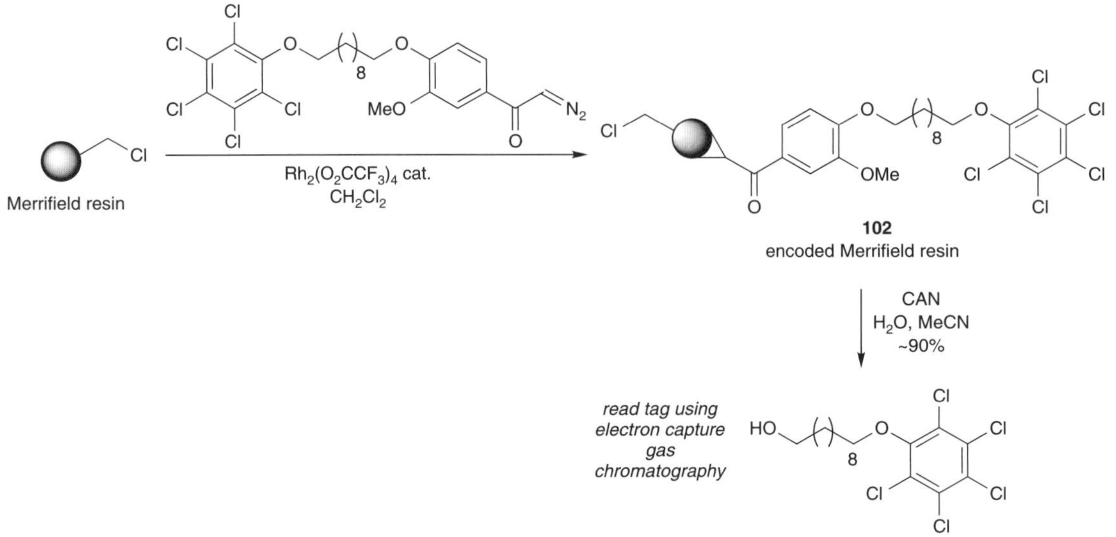

438   *Multifunctional Linker Units for Diversity-Oriented Synthesis*

Scheme 15.26   *Still's use of CAN to cleave molecular tags*

*dried over sodium sulfate. After removal of the solvents, the crude product was purified using preparative TLC to afford amide **101** in 54% overall yield.*

Still used CAN to oxidatively cleave molecular tags from a polymer support in his seminal work on the development of encoding methods for combinatorial chemistry.[28] Acylcarbenes bearing tags were used to encode Merrifield resin by reaction at unfunctionalized sites on the polymer backbone. The resulting encoded resin **102** can then be used for library synthesis. The tag can be read by electron capture gas chromatography after efficient cleavage of the tag using CAN. The mechanism of cleavage using CAN is likely to be very similar to that using DDQ (Scheme 15.19) proceeding through radical cation and radical intermediates (Scheme 15.26).[28]

Fukase reported preliminary studies on the use of a *p*-glutarylaminophenyl glycoside linker in carbohydrate synthesis. For example, treatment of glucose derivative **103**, attached to an aminomethylated polystrene resin, with ceric ammonium nitrate (CAN) released **104** from the support in 70% yield (Scheme 15.27).[29]

Fukase and Kusumoto have employed a related acylaminobenzyl linker in oligosaccharide synthesis using a macroporous aminomethylated polystyrene resin.[30] In this study the acylaminobenzyl linker is attached to the 2-hydroxyl rather than to the anomeric hydroxyl (Scheme 15.28).[30] The linker is constructed using glucose derivative **105**, bearing a nitrobenzyl group. Reduction of the nitro group and reaction with glutaric acid gives **106**, which is then coupled to aminomethylated macroporous polystyrene resin to give **107**. Fukase and Kusumoto found the use of a macroporous polystyrene support was particularly suitable for α-selective glycosylations using diethyl ether as solvent. To illustrate the utility of the acylaminobenzyl linker for oligosaccharide synthesis, **107** was coupled twice with thioglycosyl donor **108**, followed by oxidative cleavage using DDQ to give trisaccharide **109** in 50% overall yield with excellent α-selectivity at the two glycosidic linkages (Scheme 15.28).[30]

Balasubramanian has reported a related benzyloxyaniline linker **110** for the synthesis of amides and lactams that is cleaved by oxidative electron transfer.[31] The linker has been used for the synthesis of

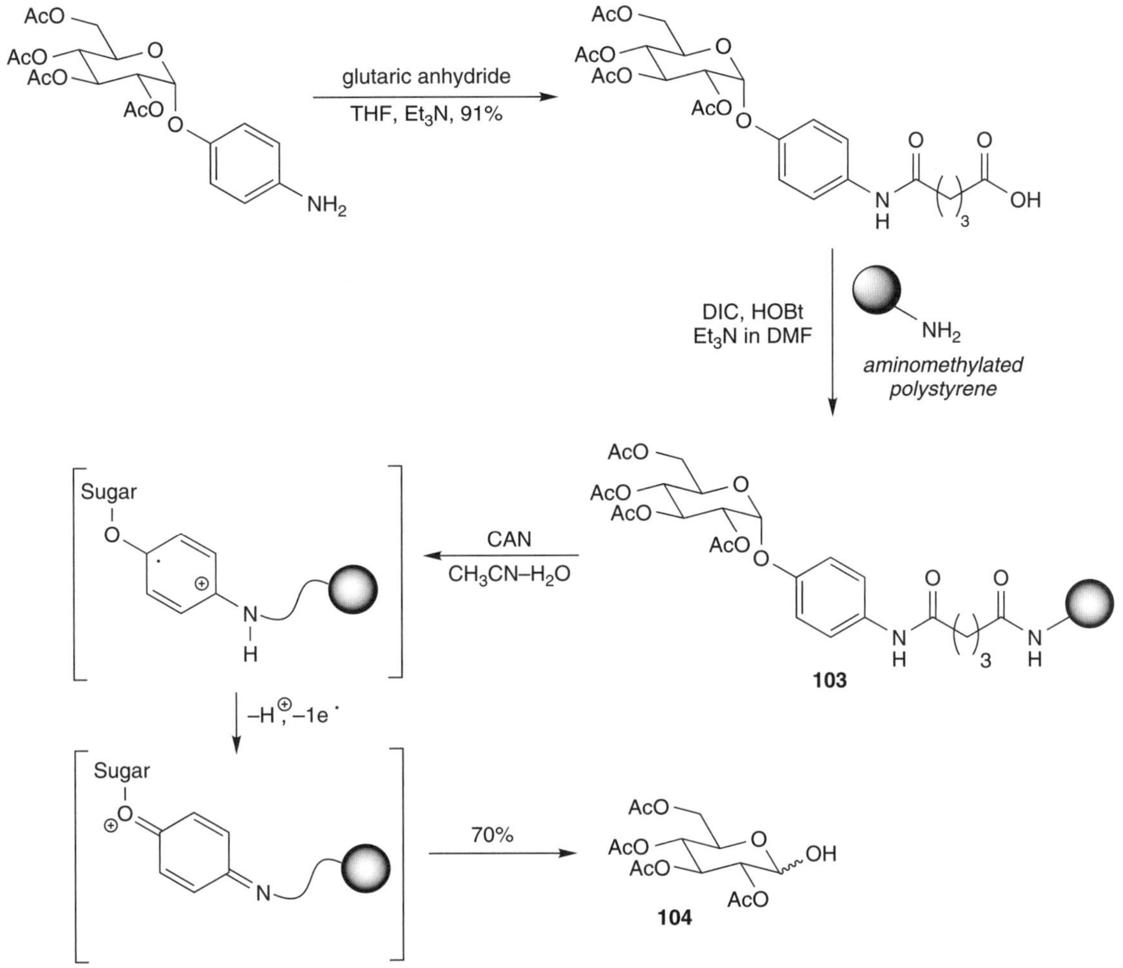

Scheme 15.27 *Fukase's ether linker for carbohydrate synthesis*

β-lactams such as **113** using the Staudinger reaction of immobilized imines **111** with a ketene. Radical cleavage of the linker by two single electron transfers gave β-lactams, such as **113**, in good yield and purity (Scheme 15.29).[31]

The linker can also be used for the synthesis of secondary amides.[31] Reductive amination of aldehydes using resin **110** gave immobilized amines, such as **114**. Acylation and cleavage using CAN gave amide products, such as **116**, in excellent yield and purity (Scheme 15.30).[31] In contrast to the BOBA linker reported by Kobayashi,[25] the benzyloxyaniline linker is stable to acid.

15.3.2 A homobenzylic ether linker cleaved by oxidative electron transfer

Floreancig has described a new solid-phase linker system cleaved by oxidative electron transfer using CAN.[32] The cleavage process is based on the oxidative fragmentation of homobenzylic ethers. On treatment with CAN, vinyl acetate **117**, immobilized on a soluble oligonorbornene scaffold prepared

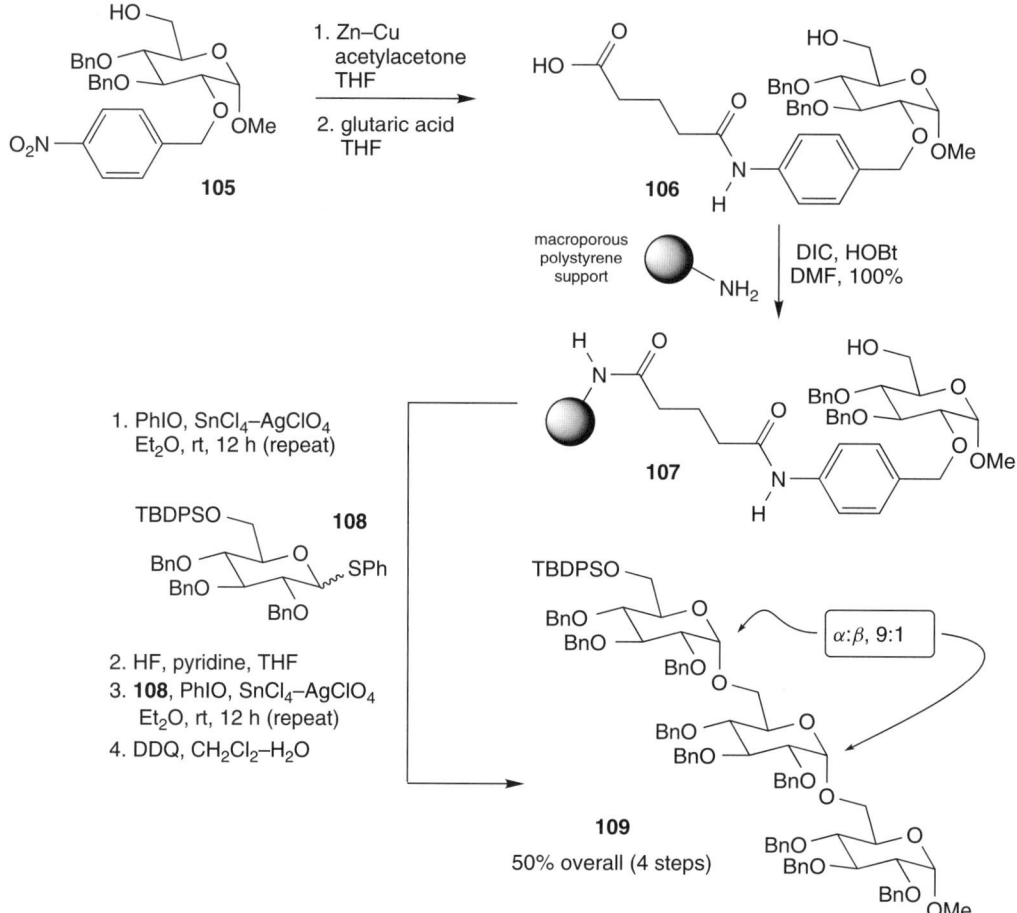

Scheme 15.28 Fukase and Kusumoto's oligosaccharide synthesis using an ether linker cleaved using DDQ

by ring-opening metathesis polymerisation (ROMP), undergoes a cleavage–cyclization process to give pyranone **118** as a single diastereoisomer (Scheme 15.31).[32]

Cleavage occurs via radical fragmentation and formation of the benzylic radical **119** and oxonium ion **120** (Scheme 15.32).[32] The latter then undergoes cyclization to give the observed product **118**.

Floreancig has also shown that the linker strategy is suitable for the synthesis of simpler ketones and aldehydes.[32] Treatment of immobilized substrates **121** and **122** with an alternative reagent system for oxidative electron transfer, gave aldehyde **123** and ketone **124**, respectively, in excellent yield (Scheme 15.33). In the absence of an appended nucleophile, the intermediate oxonium ions, analogous to **120**, break down to give the corresponding carbonyl compound.[32]

15.3.3 A sulfur linker cleaved by oxidative electron transfer with CAN

Procter has developed HASC (α-Heteroatom Substituted Carbonyl) linkers for solid-phase and fluorous synthesis. These linkers are typically cleaved by radical processes initiated by reductive electron transfer

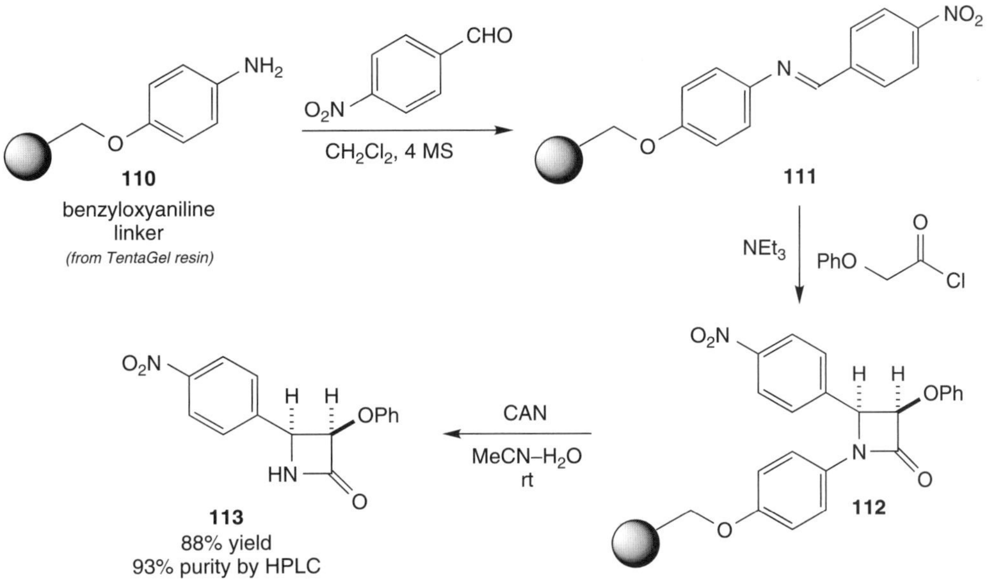

Scheme 15.29 Balasubramanian's β-lactam synthesis using a linker cleaved using CAN

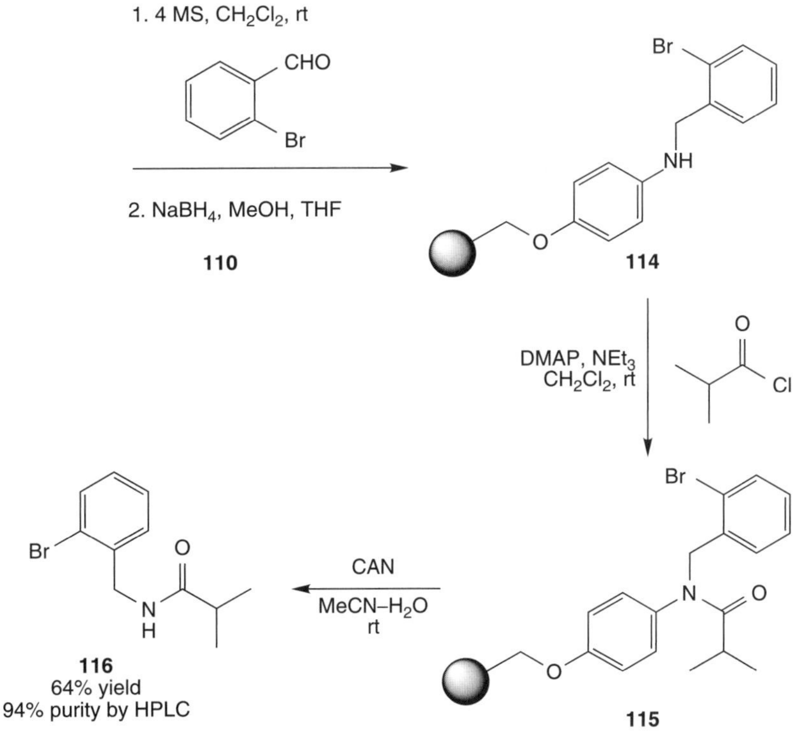

Scheme 15.30 Balasubramanian's amide synthesis using a linker cleaved using CAN

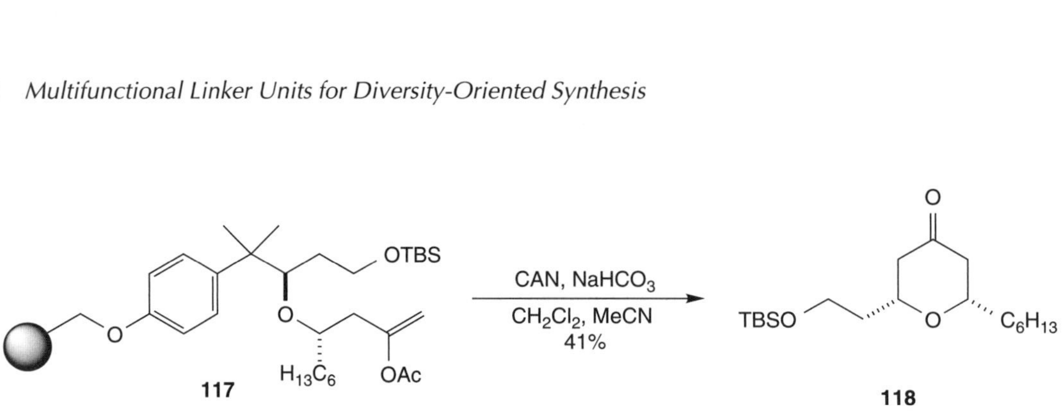

Scheme 15.31 Floreancig's homobenzylic ether linker cleaved using CAN

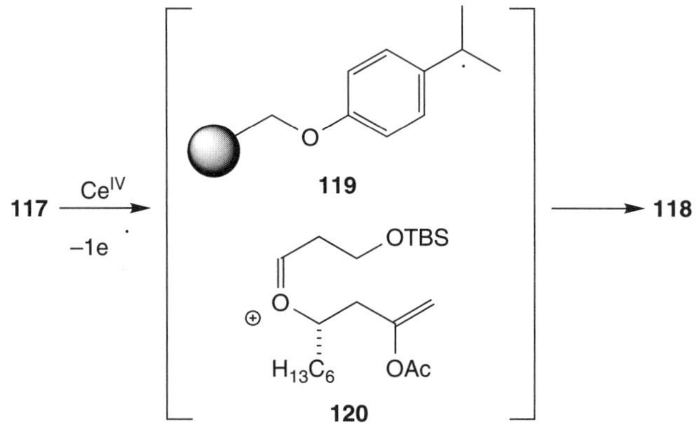

Scheme 15.32 Mechanism of homobenzylic ether linker cleavage

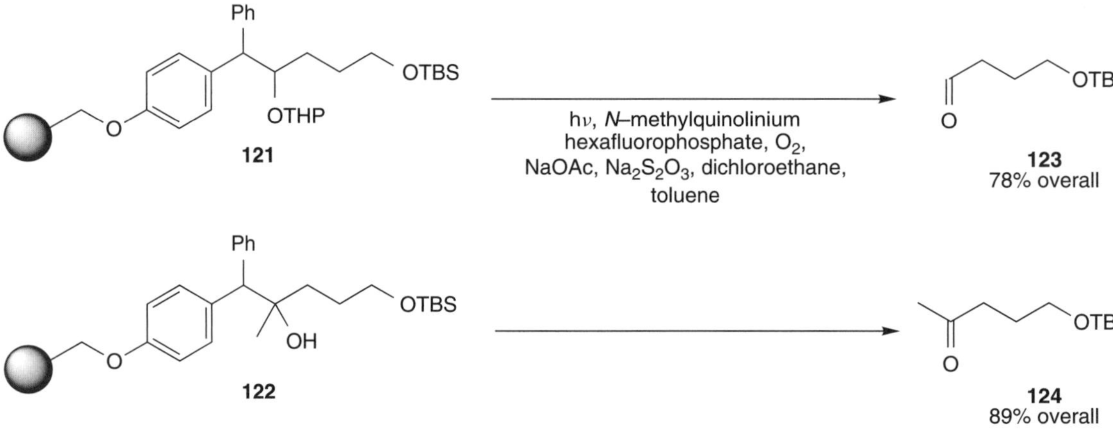

Scheme 15.33 Floreancig's synthesis of aldehydes and ketones using a homobenzylic ether linker

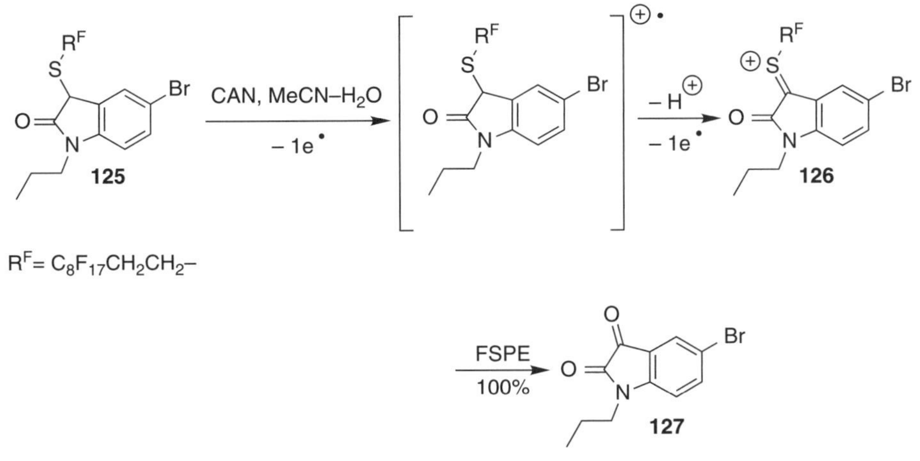

Scheme 15.34 Procter's oxidative cleavage of an α-heteroatom substituted carbonyl (HASC) linker

and are discussed in detail in Section 15.4. In some cases, however, oxidative radical processes can be used to trigger the cleavage of sulfur HASC linker, for example treatment of fluorous-tagged oxindole **125** with CAN results in two, one-electron oxidations and the formation of thionium ion **126** that is then hydrolyzed *in situ*.[33] Purification using fluorous solid-phase extraction (FSPE) gave isatin **127** in quantitative yield (Scheme 15.34).

The oxidative radical cleavage can be used in combination with known reactions of isatins to introduce additional diversity (Scheme 15.35).[33]

15.3.4 Safety-catch linkers cleaved by oxidative electron transfer

Li has reported a family of phenanthridine linkers that are cleaved using CAN; for example, phenanthridine resin **130** was prepared by the coupling of hydroxymethylpolystyrene with acid **128**, followed by reduction (Scheme 15.36).[34, 35]

The linker is stable to a range of conditions until oxidation with CAN activates the linker to hydrolysis. The efficacy of the linker system was evaluated in the synthesis of amide **133**. Acylation with 4-nitrophenylacetic acid, enolate alkylation, nitro group reduction, acylation and *N*-alkylation gave resin **132**. Oxidative cleavage using CAN released **133** from the support in excellent yield and purity (Scheme 15.37).[34, 35]

Typical Experimental Procedures
Preparation of phenanthridine linker 130

*Hydroxylmethyl polystyrene (100 mg, 0.16 mmol) was suspended in dry DMF (1.6 ml) and treated with acid **128** (0.135 g, 0.48 mmol), diisopropylcarbodiimide (DIC) (0.075 ml, 0.48 mmol) and 4-(N,N-dimethyl)aminopyridine (DMAP) (7.8 mg, 0.064 mmol) under an argon atmosphere. The mixture was agitated for 12 hours at room temperature, filtered and washed with DMF (2 × 2 ml), MeOH (2 × 2 ml) and DCM (2 × 2 ml). The resulting resin **129** was dried in vacuo. Sodium borohydride (24 mg, 0.63 mmol) in ethanol and THF at room temperature was added and the mixture shaken for four hours. Resin **130** was then collected by filtration and washed with DMF (2 × 2 ml), methanol (2 × 2 ml) and DCM (2 × 2 ml) before drying by flushing with nitrogen.*

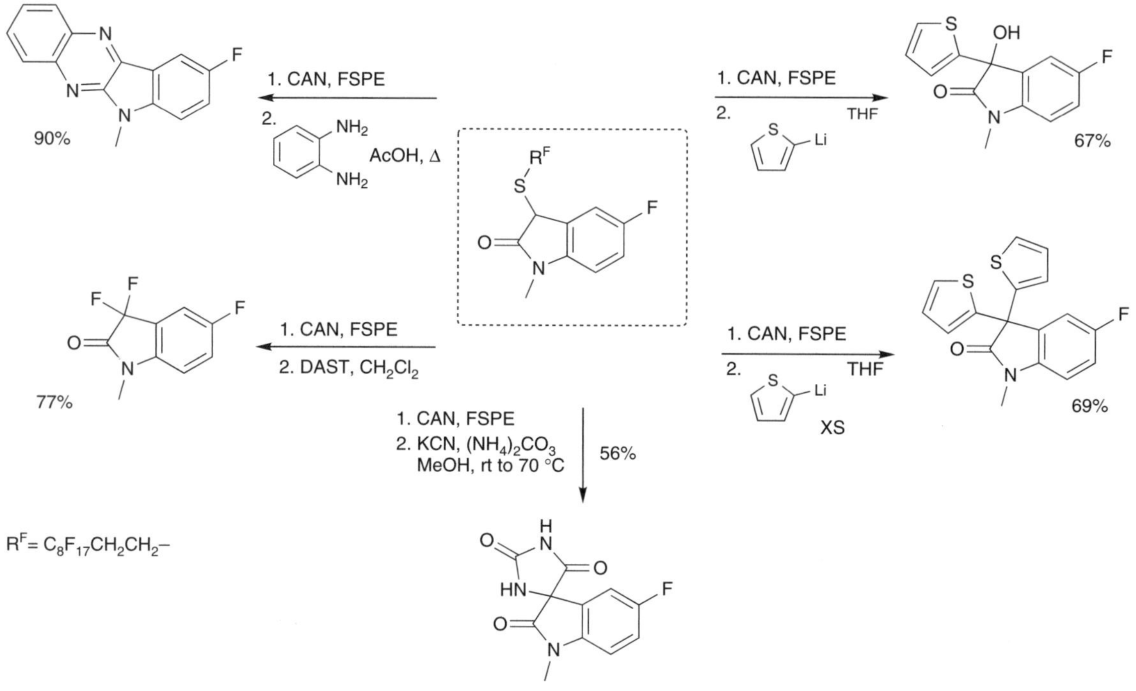

Scheme 15.35 Procter's library synthesis using the oxidative cleavage of a HASC linker

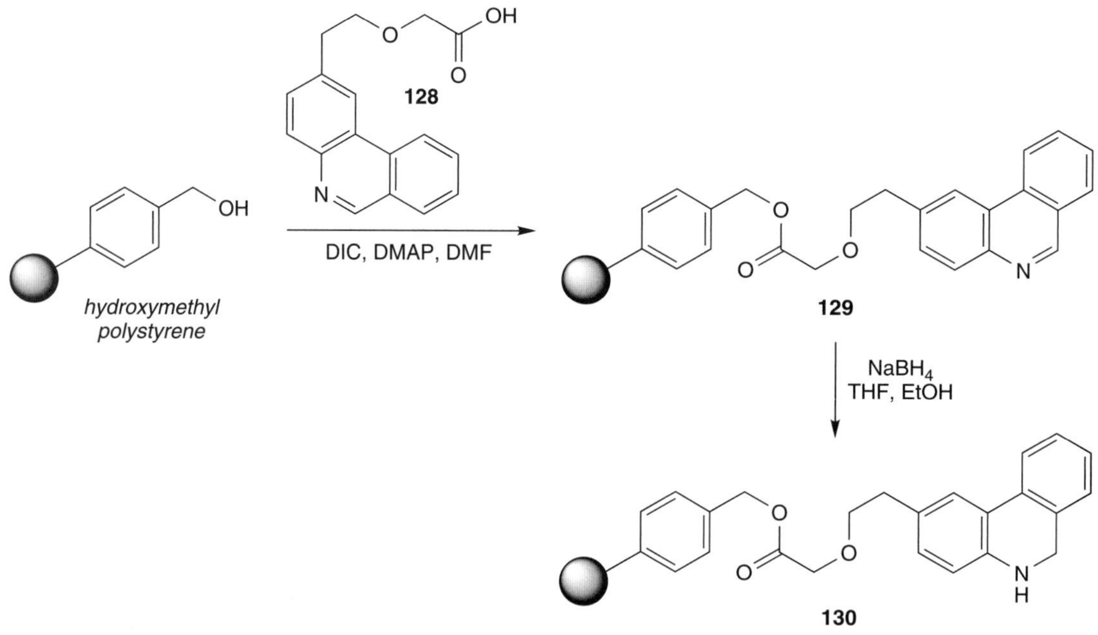

$R^F = C_8F_{17}CH_2CH_2-$

Scheme 15.36 Li's phenanthridine linker cleaved using CAN

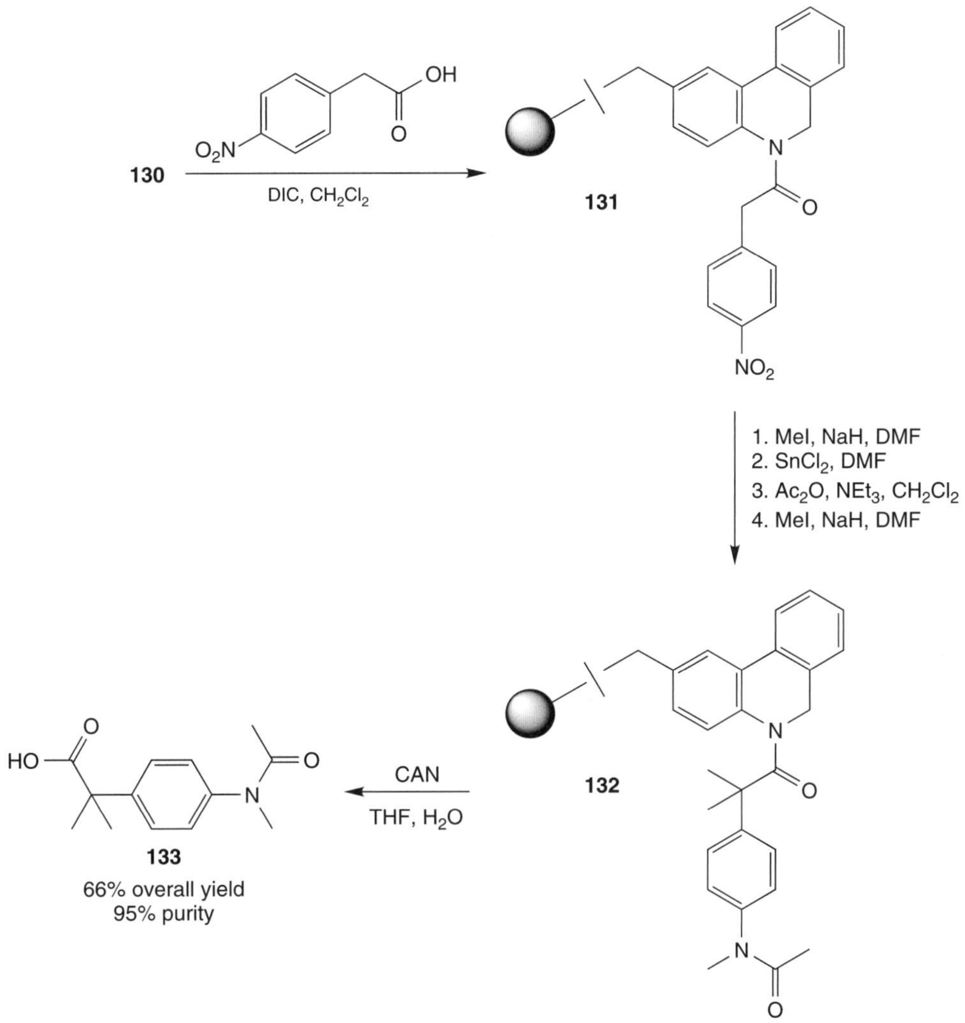

Scheme 15.37 Li's amide synthesis using a phenanthridine linker cleaved using CAN

Cleavage of the linker in 132

*To resin **132** (0.102 mmol) in THF (0.64 ml) were added water (0.2 ml) and CAN (67 mg, 0.122 mmol). The mixture was stirred for five minutes at room temperature and the resin removed by filtration. The filtrate was concentrated in vacuo and diluted with ethyl acetate (30 ml). The organic phase was washed with aqueous 10% hydrochloric acid (10 ml) and saturated aqueous sodium chloride (10 ml). The organic layer was dried (sodium sulfate), filtered and concentrated to afford **133** in 66% overall yield.*

Li also employed the phenanthridine linker for the synthesis of an *N*-acylated dipeptide **136**.[34] Coupling of **130** with Fmoc protected alanine and deprotection gave **134**, which was then coupled with Boc protected phenylalanine. Deprotection and *N*-acylation then gave resin **135**. Rapid oxidative cleavage of the linker

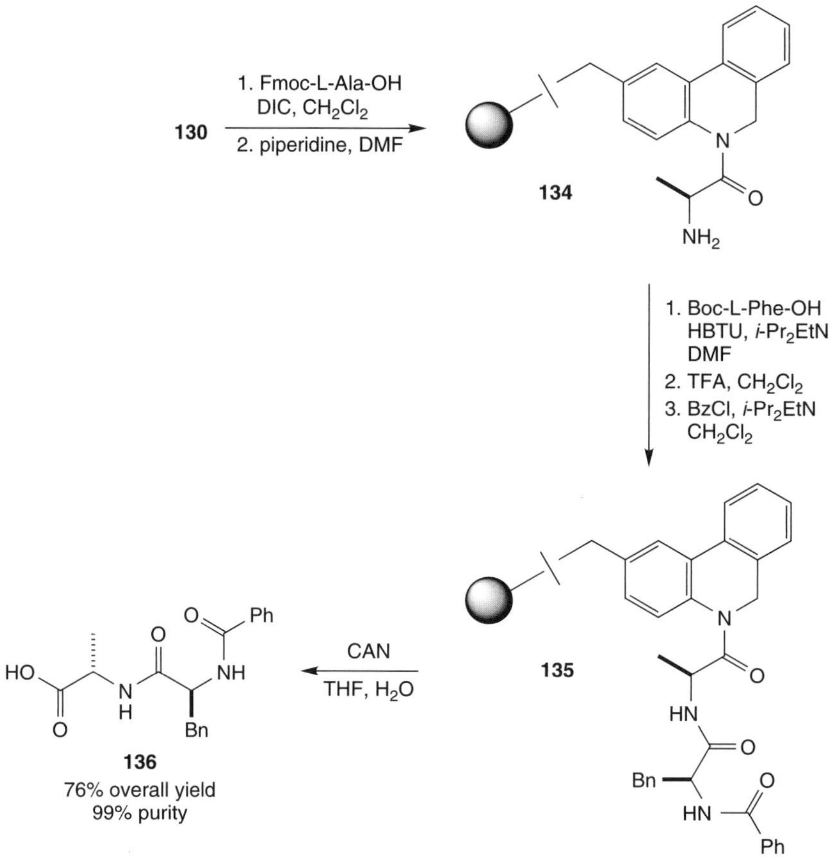

446 *Multifunctional Linker Units for Diversity-Oriented Synthesis*

Scheme 15.38 Li's dipeptide synthesis using a phenanthridine linker cleaved using CAN

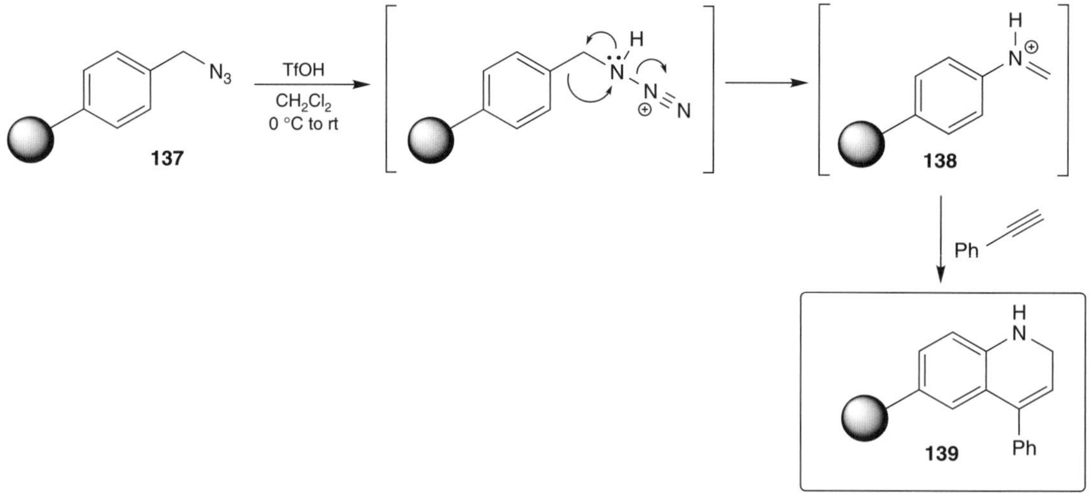

Scheme 15.39 Wagner and Mioskowski's safety catch linker based on a 4-phenyl-1,2-dihydroquinoline resin

using CAN (approximately five minutes) released **136** from the support in excellent yield and purity. Li's phenanthridine linkers are stable to acidic and basic conditions and are clearly orthogonal to Fmoc and Boc protection (Scheme 15.38).[34, 35]

Wagner and Mioskowski have reported a similar safety-catch linker based on the use of 4-phenyl-1,2-dihydroquinoline resin **139**.[36] The resin is prepared from Merrifield-derived azide **137** by an acid promoted Schmidt rearrangement to give **139** after aza-Diels–Alder reaction of the resultant supported iminium ion with phenylacetylene (Scheme 15.39).[36]

Acylation of **139** gave **140** that was stable under acidic, basic and reducing conditions. The linker was activated for cleavage by oxidation using CAN to give quinolinium ion **142**, via radical cation **141**. Hydrolysis gave 4-fluorobenzoic acid in 62% isolated yield (Scheme 15.40).[36] A similar radical mechanism is presumably involved in the activation and cleavage of Li's phenanthridine linkers.[34]

15.4 Linkers cleaved by reductive electron-transfer

The propensity of lanthanide(II) complexes to revert to the more stable lanthanide(III) oxidation state makes these species useful reducing agents. The single electron-transfer reagent samarium(II) iodide has been used extensively to mediate radical and organometallic reactions in solution and has now begun to find application in solid-phase organic synthesis.[3] The first use of samarium(II) iodide in solid-phase chemistry was reported by Armstrong in 1997.[37, 38] An efficient synthesis of benzofuran derivatives on solid support was achieved by samarium(II) iodide mediated cyclization of unsaturated aryl iodides

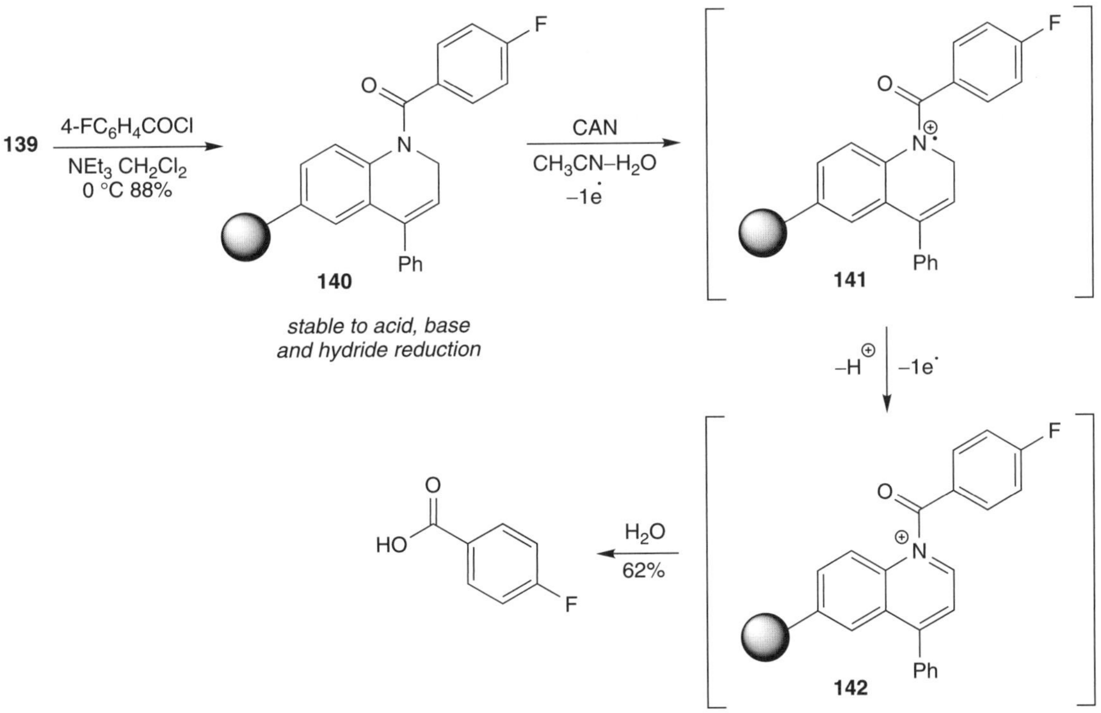

Scheme 15.40 Cleavage of Wagner and Mioskowski's safety-catch linker with CAN

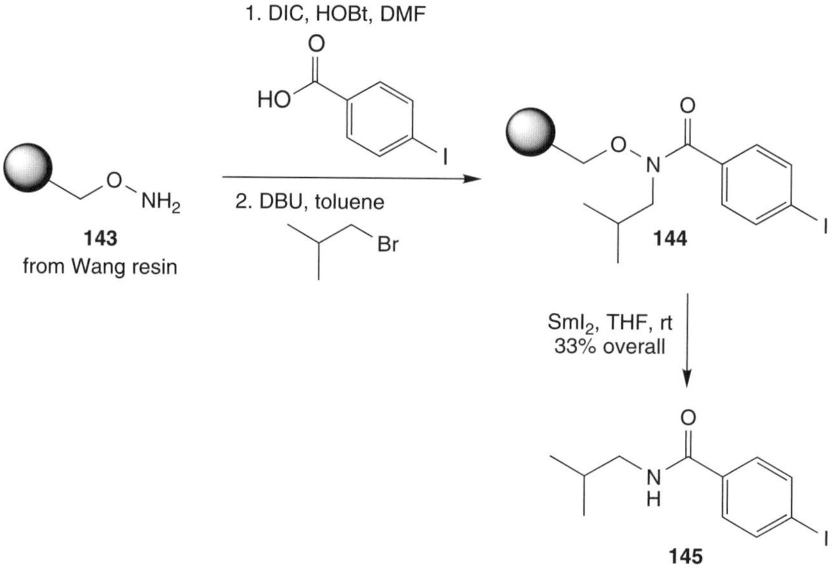

Scheme 15.41 Abell's synthesis of amides using an N–O linker

attached to a Rink resin. Benzofuran products were obtained in good yield after conventional hydrolytic cleavage of the products from the support. Armstrong's pioneering work was crucial in that it illustrated the compatibility of samarium(II) reagents with the common classes of polymer support. The mild, neutral electron-transfer conditions associated with the use of samarium(II) iodide make the reagent ideal for the selective cleavage of linkers in solid-phase organic synthesis.

15.4.1 N–O linkers cleaved using samarium(II) iodide

The first linker strategies involving samarium(II) iodide were based on the reduction of heteroatom–heteroatom bonds, more specifically nitrogen–oxygen bonds. Abell has developed a linker strategy for the synthesis of amides and ureas based on the reduction of nitrogen–oxygen bonds by samarium(II) iodide.[39] Wang-based hydroxylamine resin **143** was used to prepare a range of immobilized ureas and amides, such as **144**. Traceless radical cleavage of the linker gave amide **145** in good overall yield and in high purity (Scheme 15.41).[39]

Typical Experimental Procedures
Loading of Wang-based hydroxylamine resin 143

Hydroxylamine resin **143** (1.00 g, 1.17 mmol) was suspended in DMF (10 ml) and 4-iodobenzoic acid (0.94 g, 3.78 mmol), diisopropylcarbodiimide (0.58 ml, 3.78 mmol) and N-hydroxybenzotriazole hydrate (0.59 g, 3.78 mmol) added. The suspension was shaken at 25 °C for 16 hours. The resultant resin was washed with methanol, ethyl acetate and DCM (5 × 25 ml of each) and then dried under high vacuum for 16 hours. The resin was then suspended in a solution of DBU (0.9 ml, 6.3 mmol) in toluene (15 ml), and isobutyl bromide (1.33 g, 12.2 mmol) was added. The suspension was shaken at 25 °C for 48 hours. Resin **144** was washed and dried as described above.

Cleavage of linker in resin 144

*The resin **144** (1.21 g, 1.05 mmol) was pre-swollen with THF (2.1 ml), and samarium(II) iodide (0.1 M in THF, 21.0 ml, 2.10 mmol) was added. The reaction suspension was shaken at 25 °C for three hours. The resin was filtered off and rinsed with DCM (5 × 10 ml) and the cleavage solution and washings were collected. The filtrate was evaporated to give a dark yellow residue, which was re-dissolved in a solution of Et2O (diethyl ether) acetate (25 ml), 1 M hydrochloric acid (20 ml), and 10% aqueous sodium thiosulfate (5 ml). The mixture was transferred to a separating funnel and shaken until it became colourless. The organic layer was collected and the aqueous layer was extracted with diethyl ether (2 × 20 ml). The combined organic layers were washed with brine (2 × 20 ml) and dried over magnesium sulfate. The solid obtained after evaporation was re-dissolved in the minimum amount of DCM (approximately 0.3–0.5 ml) and filtered through a short pad of silica (eluting with 20% ethyl acetate in hexanes). The filtrate was collected and evaporated to afford 4-iodo-N-isobutylbenzamide **145** (49 mg, 33%).*

A related linker strategy has been used by Taddei in a solid-phase approach to β-lactams.[40] Treatment of immobilized substrate **146** with samarium(II) iodide resulted in smooth reduction and release of **147** from the support, further illustrating the suitability of the electron-transfer cleavage approach using samarium(II) iodide for the synthesis of highly-functionalized targets (Scheme 15.42).[40]

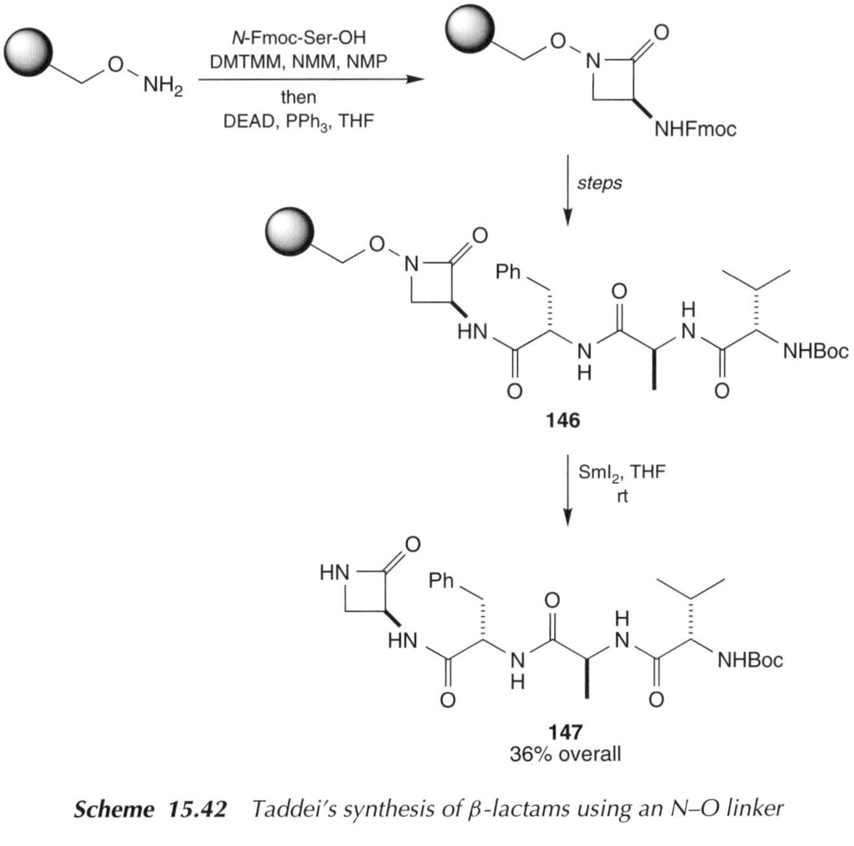

Scheme 15.42 *Taddei's synthesis of β-lactams using an N–O linker*

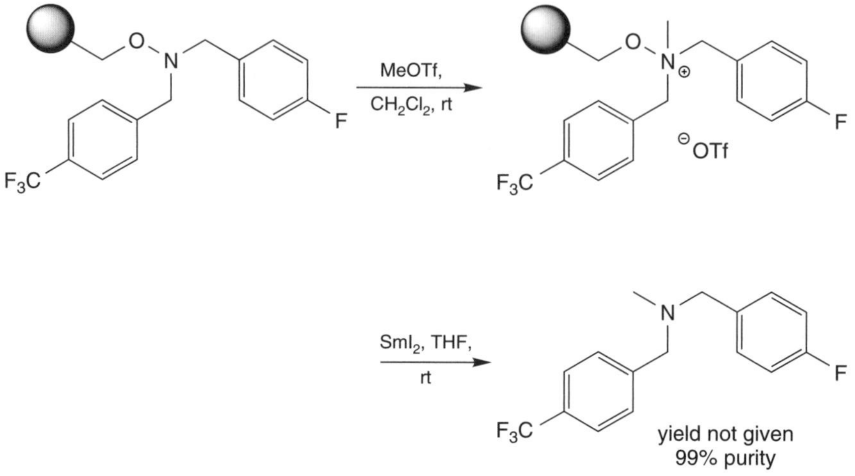

450 *Multifunctional Linker Units for Diversity-Oriented Synthesis*

Scheme 15.43 *Andersson's synthesis of amines using an N–O linker*

A similar N–O linker, cleaved by a radical process mediated by samarium(II) iodide, has been used by Andersson in a solid-phase approach to tertiary amines (Scheme 15.43).[41]

15.4.2 Sulfonamide linkers cleaved by reductive electron-transfer

A sulfonamide linker has been developed for the solid-phase synthesis of secondary amides.[42] Immobilization of primary amines using sulfonyl chloride resin **148** gave sulfonamide resins **149** that were then acylated on nitrogen. The sulfonamide linker was cleaved by electron transfer from a low-valent titanium complex to the sulfonamide carbonyl and radical fragmentation to release secondary amide products, such as **151**, in good yield (Scheme 15.44).[42]

Pilard has reported the first example of solid-phase synthesis where the support is a platininum electrode with a conducting thiophene monomer **152** electrodeposited on the surface.[43, 44] A sulfonamide linker was used for the synthesis of amine **156**. Reaction of sulfonyl chloride modified electrode **153** with 1-azidopropylamine, gave sulfonamide **154**. Azide reduction and amide formation then gave **155**. The sulfonamide linker underwent cathodic electrochemical cleavage to release amine **156** in 60% overall yield and in high purity (Scheme 15.45).[44] FT–IR on a small area of the conducting polymer was used to monitor the progress of each step.

15.4.3 Ether linkers cleaved using samarium(II) iodide

Procter has developed HASC linkers for the synthesis of carbonyl compounds (see also Sections 15.3.3 and 15.4.4).[45] The reductive cleavage of HASC linkers is based on the well established reduction of α-heteroatom-substituted carbonyl compounds to the parent ketone, ester or amide using samarium(II) iodide. Lactone **157**, immobilized using an oxygen HASC linker, was converted to a range of ketones and amides, including **158** and **159** (Scheme 15.46). Treatment with samarium(II) iodide released cyclopropyl ketone **160** and morpholine amide **161** from the support in good overall yield.[45, 46]

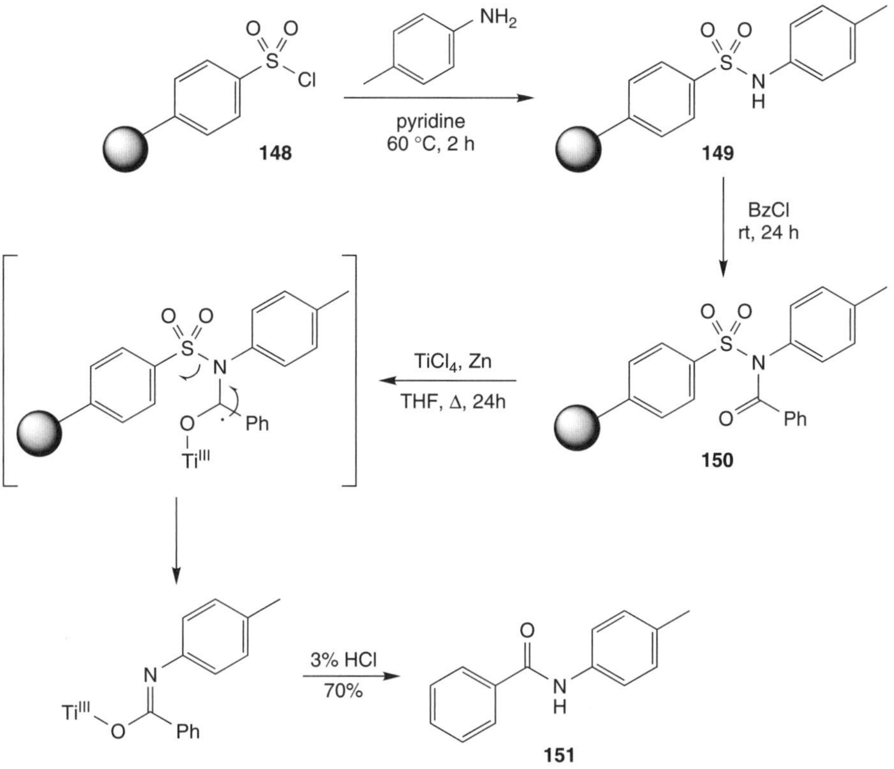

Scheme 15.44 Huang's sulfonamide linker

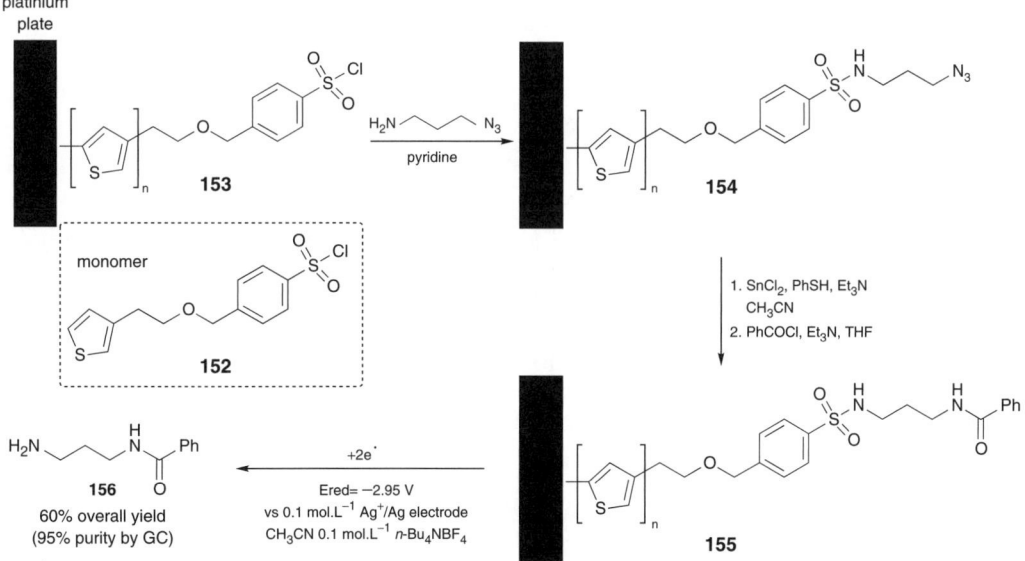

Scheme 15.45 Pilard's sulfonamide linker to a platinum electrode

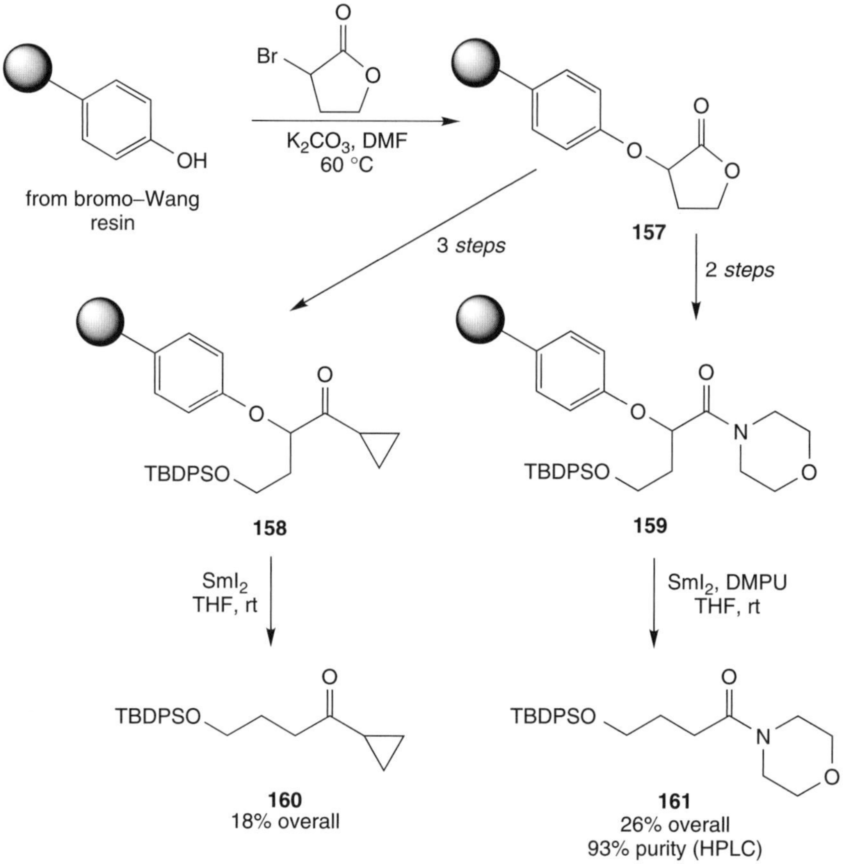

Scheme 15.46 Procter's oxygen HASC linker cleaved using samarium(II) iodide

Typical Experimental Procedures

Preparation of resin 157

*Phenol resin (1.31 g, 1.26 mmol) was swollen in DMF (30 ml). α-Bromo-γ-butyrolactone acid (1.67 ml, 20.2 mmol) was then added followed by potassium carbonate (1.78 g, 12.6 mmol) and the reaction mixture stirred slowly at 60 °C for 24 hours. The resin was then collected by filtration, washed with THF (30 ml), THF:water (2:1) (3 × 30 ml), THF:water (1:1) (3 × 30 ml), THF:water (1:2) (3 × 30 ml), THF (2 × 30 ml), then alternate washings with DCM (3 × 30 ml) and methanol (3 × 30 ml), finishing with THF (2 × 30 ml). The resin was then left to dry for 10 minutes under water pump pressure before being dried for six hours under high vacuum to give lactone resin **157**.*

Cleavage of the linker in 159

*DMPU (967 μl, 8.00 mmol) was added to a suspension of **159** (0.498 mmol) in THF (3 ml). Samarium(II) iodide (0.1 M in THF, 25.0 ml, 2.50 mmol) was added drop-wise and the mixture stirred slowly for 12 hours.*

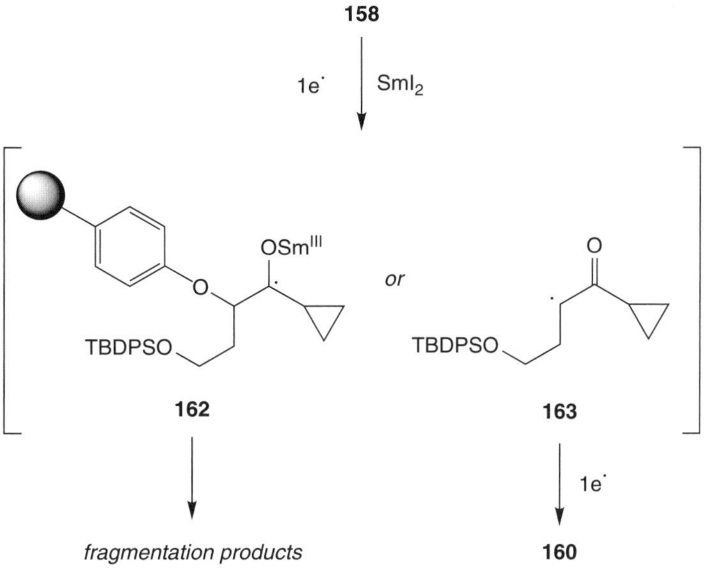

Linker Units Cleaved by Radical Processes 453

Scheme 15.47 *Mechanism of cleavage of Procter's ether HASC linker*

*The resin was then removed by filtration and washed with THF (100 ml). The filtrate was then concentrated in vacuo to give the crude product as a yellow oil. Filtration through a short pad of silica gel (eluting with 50% ethyl acetate/petroleum ether) gave **161** (54 mg, 0.131 mmol, 26% for four steps from the phenol resin starting material).*

Immobilized cyclopropyl ketone **158** was prepared to probe the mechanism of the cleavage reaction.[46] Isolation of **160** where the cyclopropyl ring was intact suggests that cleavage proceeds via formation of radical **163** rather than ketyl radical anion **162**, formed by single electron transfer to the ketone carbonyl, as cyclopropylmethyl radical anions are known to undergo facile fragmentation (Scheme 15.47).[46]

15.4.4 Alkyl and aryl sulfide/sulfone linkers cleaved by reductive electron-transfer

Janda has played an important role in pioneering the use of sulfide linkers in solid–phase synthesis and has developed several linkers for the traceless, radical cleavage of aliphatic molecules from a soluble polymer support.[47] CH$_3$O–PEG acid soluble polymer **164** was coupled with amino thiol **165** to form thiol resin **166**, which was then treated with alkyl halides, such as **167**, to form a sulfide linkage. The sulfur linker was cleaved by a radical process using Bu$_3$SnH and AIBN, thus releasing the desired amide **169** in moderate yield. In this case, more efficient cleavage could be achieved using hydrogen and Raney nickel (Scheme 15.48). The use of soluble polymers allowed easy isolation and purification of the released product; precipitation of desulfurized CH$_3$O–PEG resin with ether and concentration of the filtrate gave the product in high purity.[47, 48]

The strong reducing conditions required for the cleavage step limit the functional groups that can be present in the molecule being prepared. This problem was addressed by oxidation of the linking sulfur atom to the sulfone and the use of a more selective reducing agent. Reductive electron-transfer using sodium–mercury was employed in an improved cleavage strategy (Scheme 15.49).[49] After attachment

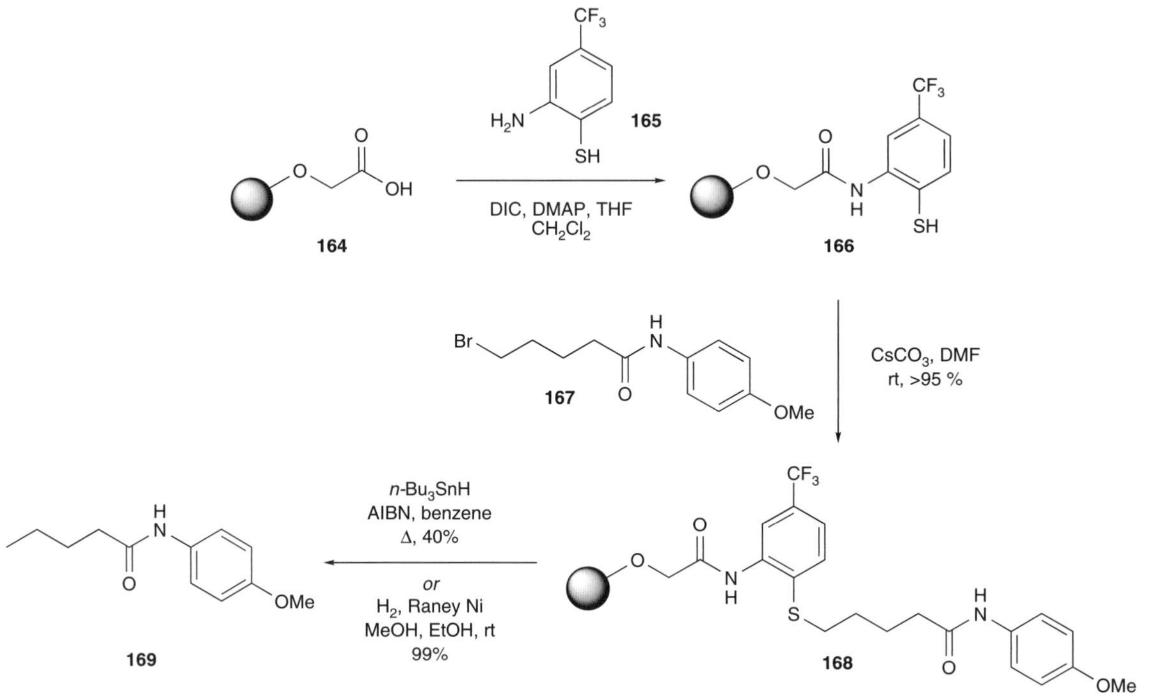

Scheme 15.48 Janda's early sulfide linker

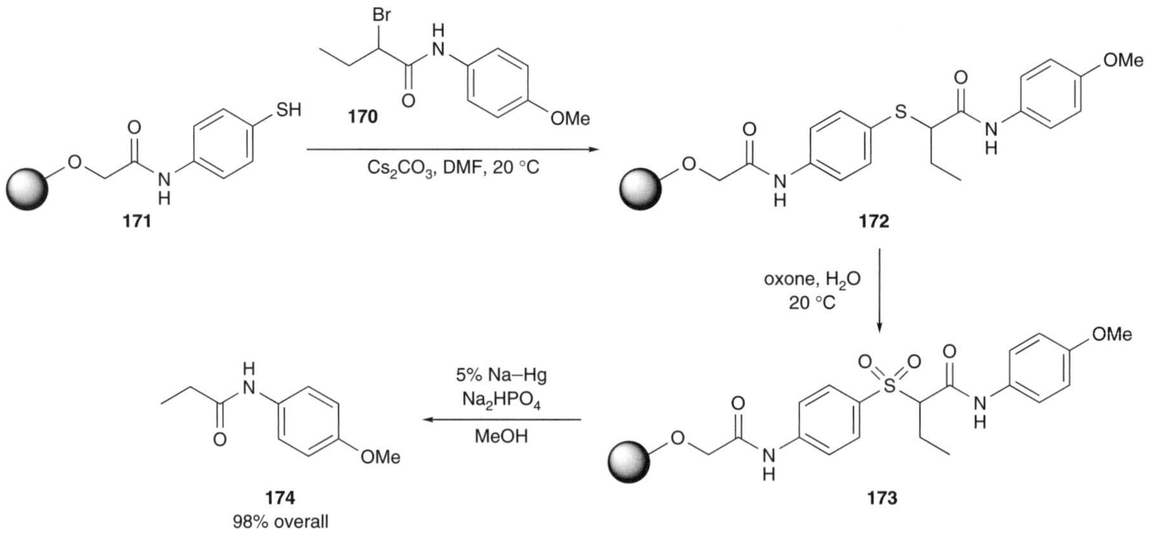

Scheme 15.49 Janda's early sulfone linker

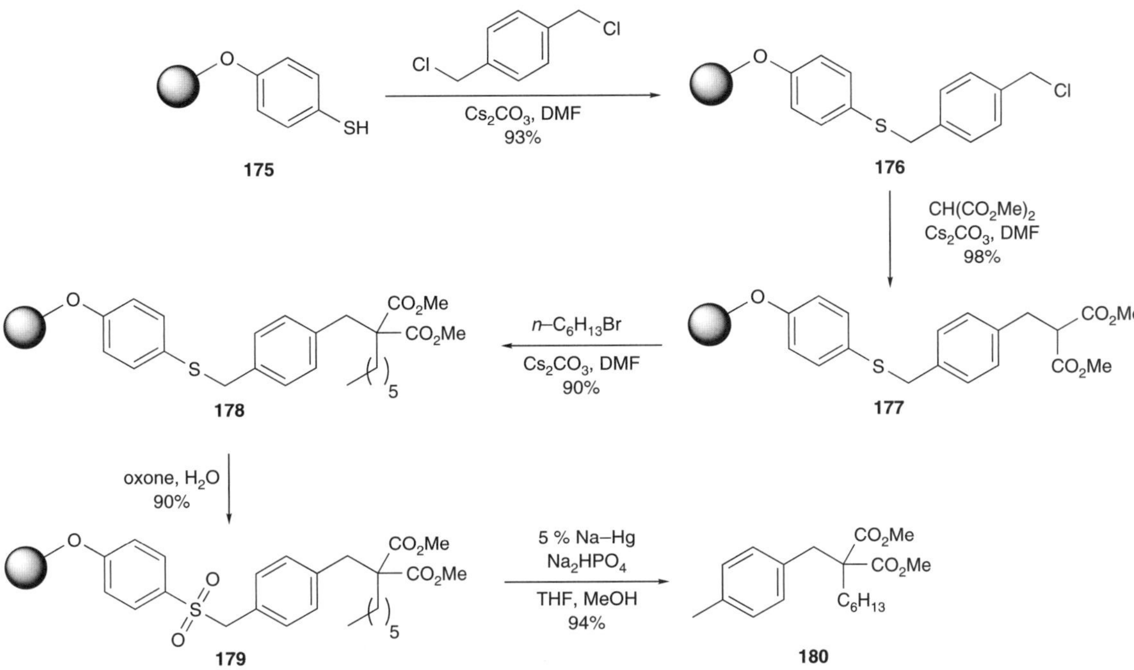

Scheme 15.50 Janda's malonate synthesis using an early sulfone linker

of the substrate **170** to Janda's second generation thiol resin **171**, the sulfide link in **172** was oxidized to the corresponding sulfone **173** using oxone. Reductive, radical cleavage of the linker gave product **174** in high yield (Scheme 15.49).[49, 50]

The amide functionality in Janda's sulfide linker systems renders them incompatible with reagents such as lithium aluminium hydride and strong base or acid. To increase the generality of these linkers, a third generation thiol **175** was synthesized and used to prepare a more robust, ether-based linker system. To demonstrate the utility of Janda's latest sulfide linker, a solid-phase synthesis of alkylated malonate derivatives was undertaken (Scheme 15.50).[51] Alkylation of thiol **175** with a dihaloalkane to form **176** was followed by addition of a malonate anion. The resulting polymer-supported malonate **177** was alkylated once more to furnish **178** before oxidation to the sulfone **179** and radical cleavage of the linker. Malonate derivative **180** was isolated in excellent yield (Scheme 15.50).[51] This linker system has also been applied in the synthesis of 3,5-pyrazolidinediones.[52]

De Clercq has used a sulfone linker in a solid-phase Julia-type olefination process.[53] Olefins were released from α-benzoyloxy sulfone resins, such as **181**, upon reduction with an electron-transfer reagent and elimination of the sulfone linkage (Scheme 15.51). Samarium(II) iodide proved to be the most suitable reagent for the process and was used with the promoters hexamethylphosphoric triamide (HMPA) and 1,3-dimethyl-3,4,5,6-tetrahydro-2(1H)-pyrimidinone (DMPU).[53] The stereoselectivity of the olefination was found to be strongly dependant upon the additive, DMPU giving rise to higher selectivities for the E-alkene than HMPA (Scheme 15.51).

Procter has developed a sulfur HASC linker for the solid-phase synthesis of N-heterocycles. A route to oxindoles has been developed that employs the first Pummerer cyclizations carried out on solid phase to construct the heterocyclic ring system.[54] The linker was formed by alkylation of thiol resin **183**, conveniently prepared from Merrifield resin, with α-bromoamide **182**. Selective oxidation of the resultant

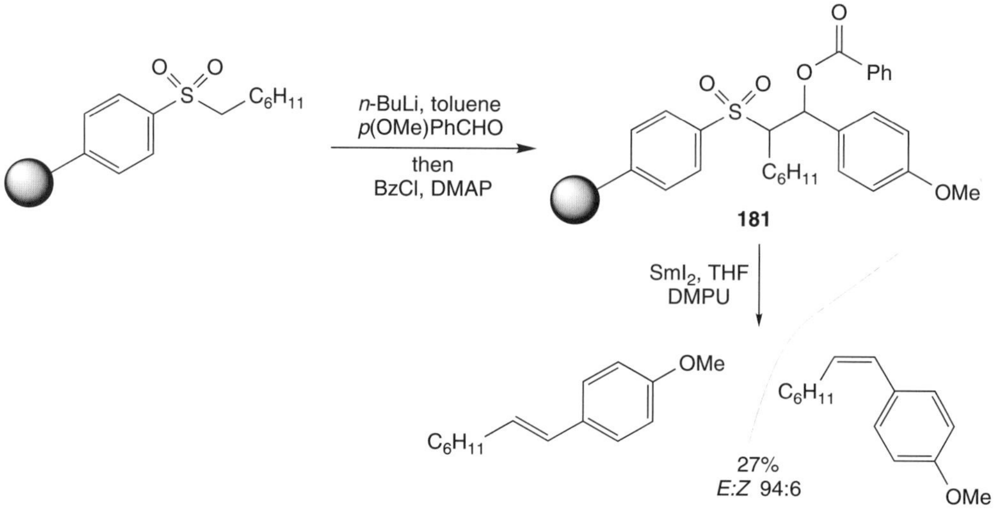

Scheme 15.51 De Clercq's alkene synthesis using a sulfone linker

immobilized sulfide to the sulfoxide **185** and Pummerer cyclization using trifluoroacetic anhydride (TFAA) gave oxindole **186** (Scheme 15.52). Radical cleavage of the sulfide linkage in **186** with samarium(II) iodide gave oxindole **187** in good overall yield. As the sulfide linkage remains intact during the Pummerer cyclization, the sulfur atom can be used in further elaborations of the heterocyclic framework. Oxidation to the sulfone facilitates alkylation to give **188** and samarium(II) iodide radical cleavage provides **189** in 30% overall (Scheme 15.52).[54, 55]

Procter has also used the HASC sulfur linker in a route to tetrahydroquinolines that involves a microwave-assisted Heck reaction followed by a Michael cyclization.[56] Again, the sulfone oxidation state of the linker is exploited to carry out carbon–carbon bond forming reactions in the construction of the target. At the end of the sequence, tetrahydroquinolines such as **194** are released from the polymer support by reduction of intermediate sulfones **193** with samarium(II) iodide. In this case lithium chloride is used to increase the reducing ability of the lanthanide reagent (Scheme 15.53).[55, 56]

In the same study, Procter illustrated the feasibility of a cyclative–cleavage strategy using the HASC linker family.[56] Treatment of immobilized sulfone **195** with samarium(II) iodide results in cleavage and cyclization to give tetrahydroquinoline **196**. This cyclization proceeds by a mechanism involving either a radical or a samarium enolate, formed by reduction of the initial radical (Scheme 15.54).[55, 56]

Procter has also shown the sulfide HASC linker to be effective in fluorous synthesis, where the polymer phase tag is replaced by a perfluoroalkyl group, or fluorous tag.[57]

In phase tag assisted synthesis, extra steps to introduce and remove the tag are unavoidable. As previously discussed, it is desirable to gain more synthetic value from these steps by using the construction of the linker and/or its cleavage to trigger other reactions that result in the construction of valuable motifs such as heterocyclic systems. Examples of both cyclative–capture and cleavage–cyclization have been noted elsewhere in this review. In an extension of their approach to *N*-heterocycles using a Pummerer cyclization (Section 15.4.4, Scheme 15.52),[54] Procter has developed a cyclative–capture process that allows the phase tag to be introduced, the HASC linker formed and a heterocyclic scaffold to be constructed. The cyclative–capture approach is based on the addition of the fluorous thiol **197** to glyoxamides such as **198** (Scheme 15.55). FSPE can then be used for rapid purification of the tagged heterocycles. The sulfide

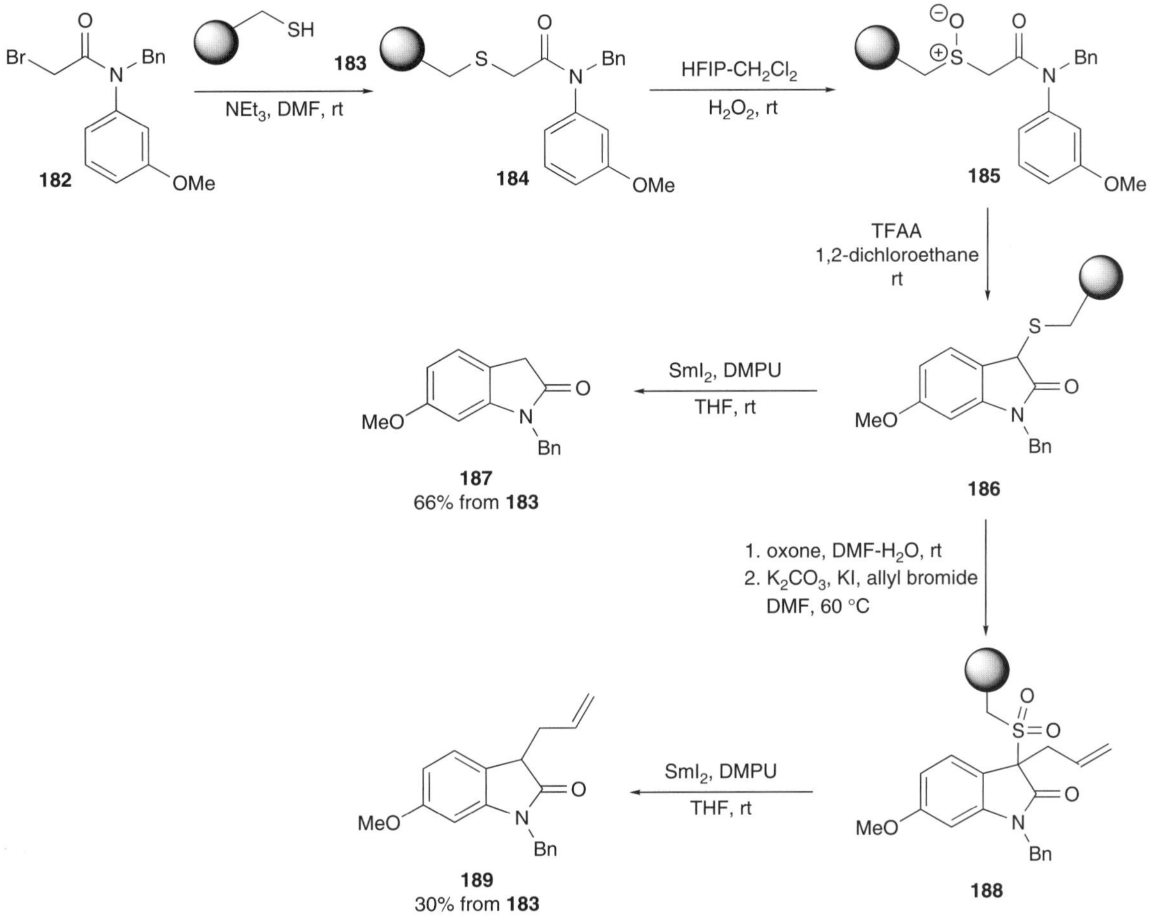

Scheme 15.52 *Procter's oxindole synthesis using a sulfur HASC linker*

linkage in **199** was shown to be compatible with a variety of transformations including palladium catalyzed cross-couplings. After elaboration of the heterocyclic framework, the linkage in **201** was cleaved using samarium(II) iodide to give product oxindoles, such as **202**, in high yield (Scheme 15.55).[57]

Typical Experimental Procedures

Cyclative–capture using fluorous thiol 197

*To the crude glyoxamide **198** (4.10 mmol, 1 equiv.) in DCM (20 ml) at room temperature was added fluorous thiol **197** (1.77 mmol, 0.7 equiv.). After 18 hours, TFAA (2.27 mmol, 9 equiv.) was added and, after a further one hour, $BF_3 \cdot OEt_2$ (12.7 mmol, 5 equiv.) was added. After a further 60 minutes, the reaction was quenched with aqueous saturated sodium bicarbonate. The organic layer was separated and washed with aqueous saturated sodium bicarbonate, dried (magnesium sulfate) and concentrated in vacuo. The crude product was purified using fluorous silica gel (eluting with 80% acetonitrile/water then acetonitrile) to give **199** as an oil (85% over two steps).*

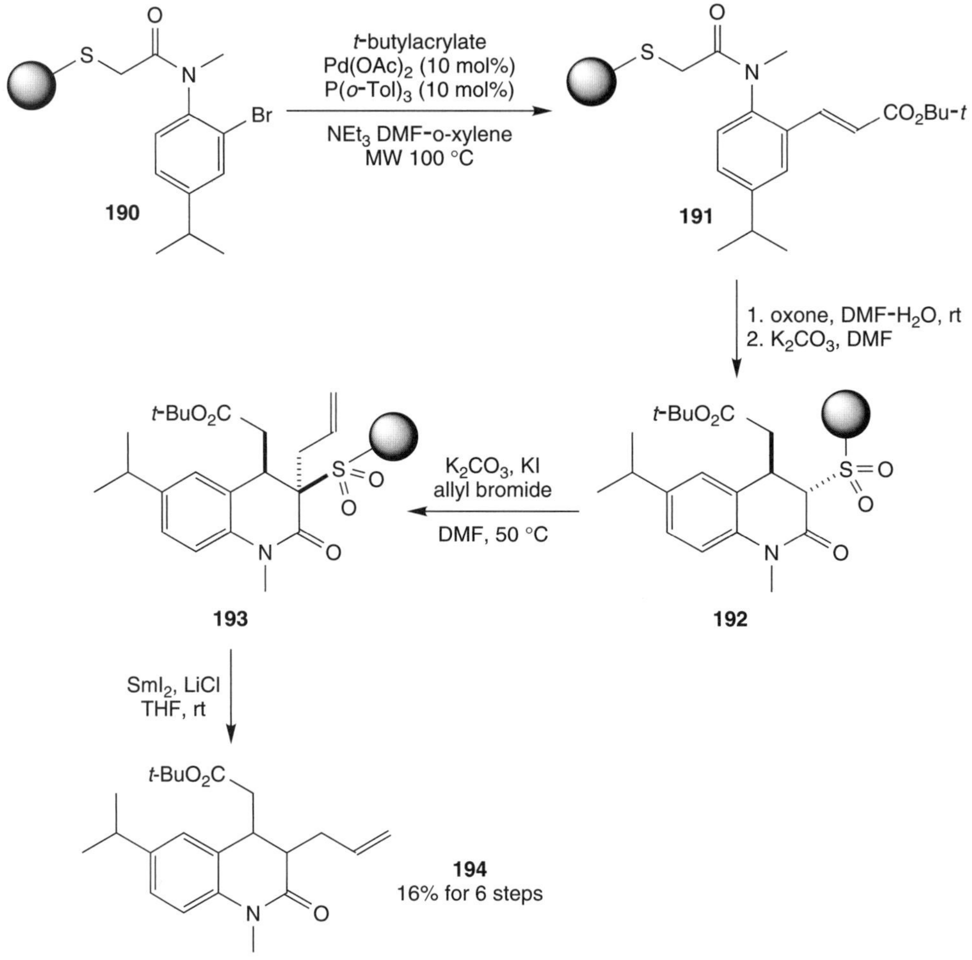

Scheme 15.53 Procter's tetrahydroquinoline synthesis using a sulfur HASC linker

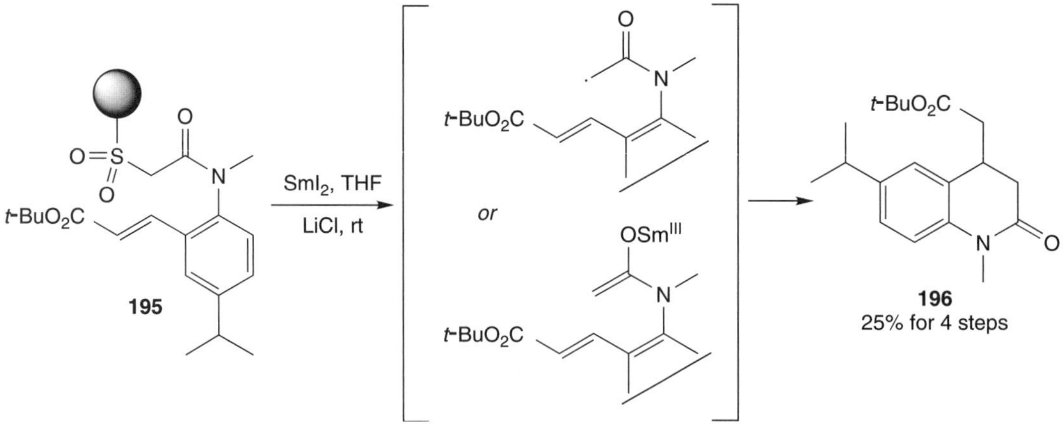

Scheme 15.54 Procter's cleavage–cyclization using a sulfur HASC linker

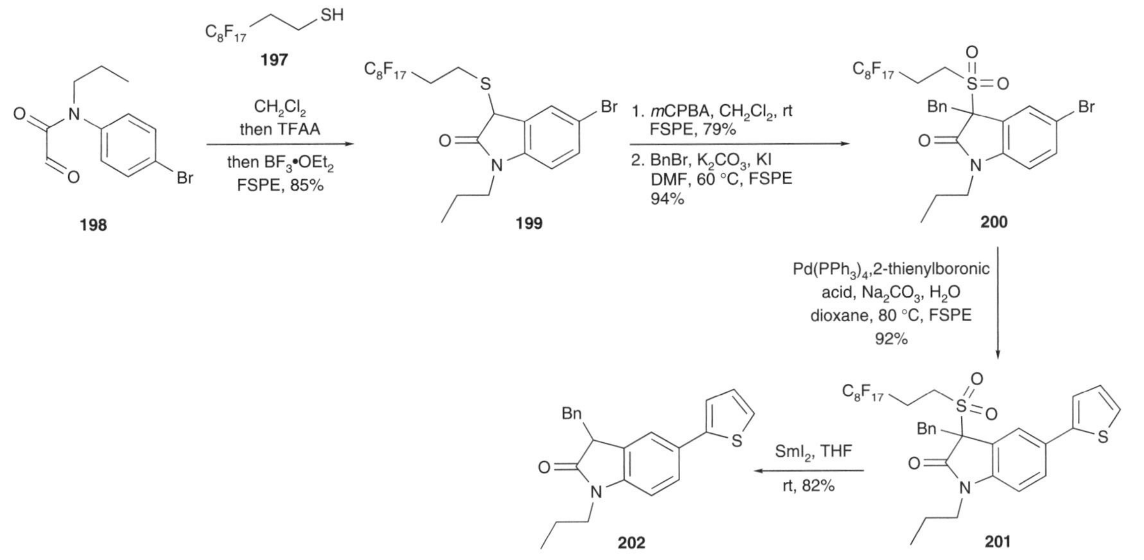

Scheme 15.55 Procter's cyclative–capture to construct oxindoles

Cleavage of the sulfur linker in 201

*To a solution of **201** (0.16 mmol, 1 equiv.) in THF (5 ml) was added samarium(II) iodide (0.1 M in THF, 0.35 mmol, 2.2 equiv.) at room temperature. After 18 hours, aqueous saturated sodium bicarbonate was added, the organic layer separated and washed with aqueous saturated sodium bicarbonate, dried (magnesium sulfate) and concentrated in vacuo. The crude product was purified by filtration through a short pad of silica (eluting with 30% ethyl acetate/petroleum ether (40–60 °C)) to give **202** (82%).*

Variation of the glyoxamide substrate allows access to other N-heterocycles;[57] for example, cyclative–capture of glyoxamide **203** with fluorous thiol **197** gave tagged tetrahydrobenzazepinone **204**. Modification of the heterocyclic framework by alkylation and cleavage of the HASC linker in **205** using samarium(II) iodide gave **206** in good yield (Scheme 15.56).[57, 58]

Procter has carried out preliminary investigations into the development of sequential cleavage–cyclization processes where the intermediates or products formed upon scission of the sulfur linker are exploited in the formation of cyclic motifs.[58] It was found that varying the order of addition in the samarium(II) iodide mediated cleavage of the HASC linker in oxindole **207** gave two very different cyclization products. Slow addition of samarium(II) iodide to the substrate gave spirocycle **208** resulting from alkylation of the intermediate samarium enolate, whilst slow addition of the substrate to samarium(II) iodide, gave indolocarbazole **209**, that may result from the Barbier cyclization of organosamarium intermediate **210**. In both processes, anionic intermediates are formed by reduction of initially-formed radicals. Thus, order of addition has been used to introduce diversity in the radical cleavage of a sulfur HASC linker (Scheme 15.57).[58]

Procter has also developed cleavage–cyclization processes, compatible with the HASC linker system, to form heterocyclic rings.[59] For example, heterocycles can be formed by a carbon–carbon bond forming process triggered by the removal of the fluorous tag. Treatment of vinyl sulfone **211** with samarium(II) iodide results in radical cleavage of the linker and addition of the resultant reactive intermediate to the electron-deficient alkene, thus generating a quaternary centre where the tag had been located (Scheme 15.58).[59]

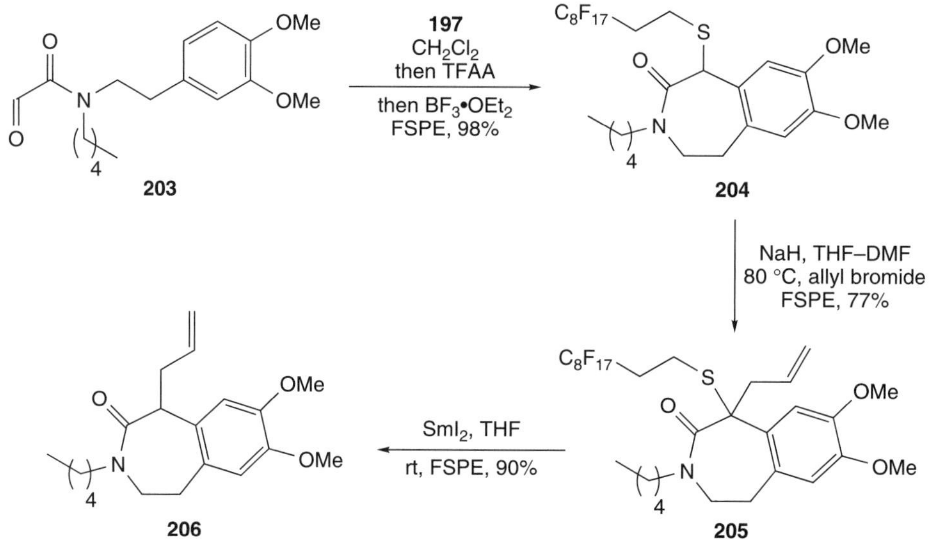

Scheme 15.56 Procter's cyclative–capture to construct benzazepinones

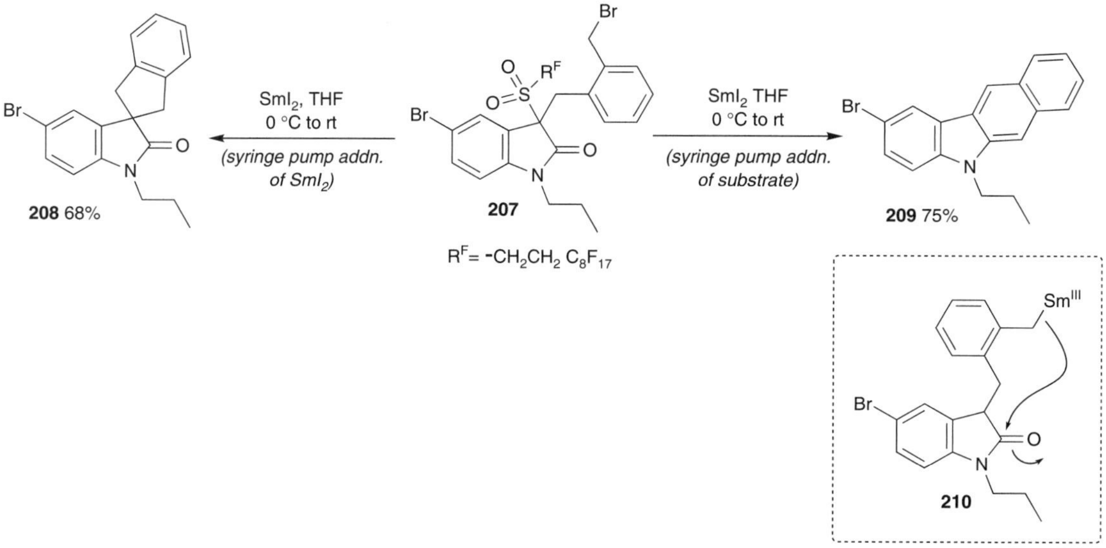

Scheme 15.57 Procter's cleavage–cyclization using a sulfur HASC linker

In the same study, Procter investigated a cleavage–cyclization process that constructs a heterocyclic ring through the formation of carbon–heteroatom bonds.[59] Treatment of **213** with samarium(II) iodide over 48 hours results in sequential removal of the fluorous tag and reduction of the aryl nitro group. Acid mediated cyclization of the resultant aniline **214**, gives indoloquinoline **215** in moderate overall yield (Scheme 15.59).[59] Interestingly, interrupting the samarium(II) iodide reduction of **213** after 3.5 hours

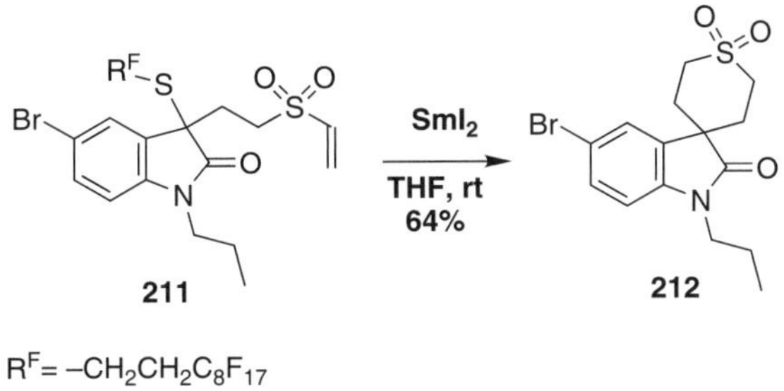

Scheme 15.58 Procter's cleavage–cyclization to form spirooxindoles

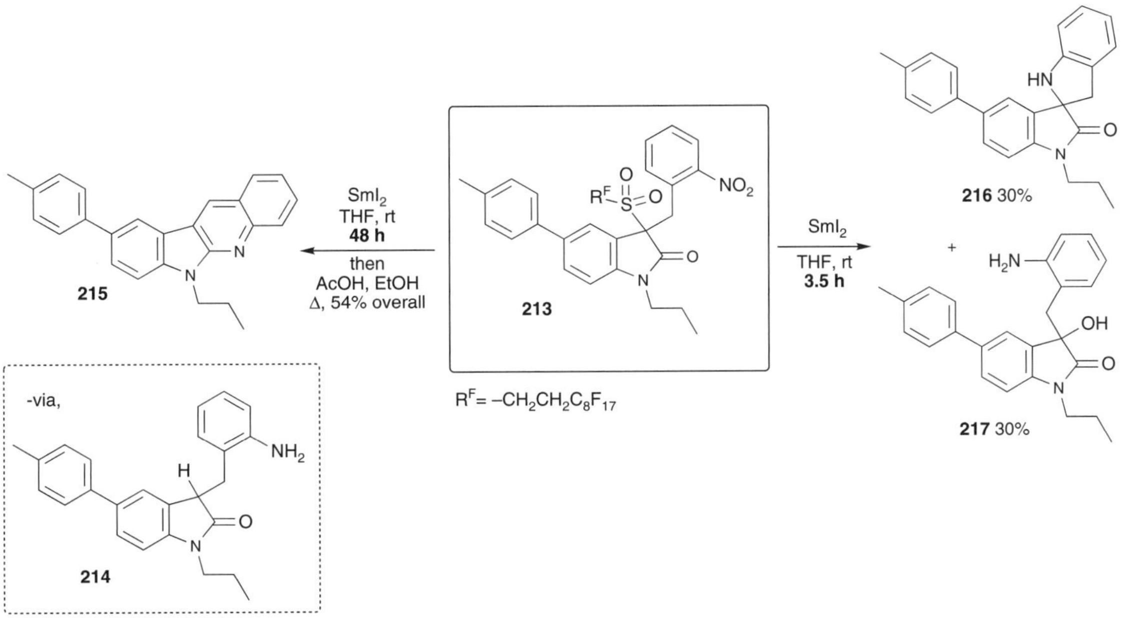

Scheme 15.59 Procter's cleavage–cyclization to form indoloquinolines

allowed two intermediates to be isolated, giving an insight into the mechanism of the sequential, radical reduction: azaspirocycle **216** and tertiary alcohol **217** were obtained in a 1:1 ratio and in 60% yield. Both intermediates are reduced further by samarium(II) iodide to give **214**. Azaspirocycle **216** appears to arise from attack of a Sm(III)–enolate intermediate, formed by the reduction of a radical intermediate initially formed upon cleavage of the linker, at the nitrogen of the nitro group, while tertiary alcohol **217** appears to be formed by attack of the Sm(III)–enolate intermediate at the oxygen of the nitro group. In this example, variation of reaction time with samarium(II) iodide can be used to introduce diversity in radical cleavage–cyclization reactions using the HASC linker (Scheme 15.59).[59]

A portfolio of cleavage–cyclization processes, compatible with the HASC linker system, will allow product scaffolds to be built in a traceless fashion while introducing diversity during the removal of the phase tag.

462 *Multifunctional Linker Units for Diversity-Oriented Synthesis*

15.5 Radical processes that indirectly trigger linker cleavage

15.5.1 Nitro group reduction as a trigger for linker cleavage

As seen in Scheme 15.59, samarium(II) iodide is also a useful reagent for the reductive manipulation of functional groups. Ito has employed samarium(II) iodide to reduce a nitro group in a solid-phase carbohydrate synthesis.[60] Treatment of immobilized monosaccharide **218** with samarium(II) iodide reduces the nitro group to initially give radical anion **219** that is then reduced further to give a mixture of the corresponding hydroxylamine and aniline **220**. Cyclative–cleavage then releases **221** from the support in good yield (Scheme 15.60).[60]

15.5.2 Radical carbon–carbon bond formation as a trigger for linker cleavage

Procter has reported the intermolecular radical coupling of aldehydes and ketones with α,β-unsaturated esters, immobilized using an ephedrine chiral linker, to give enantiomerically enriched γ-butyrolactones.[61] Treatment of acrylate and crotonate ephedrine resins **222** and **223** with cyclohexanecarboxaldehyde, employing samarium(II) iodide in THF with *t*-butanol as a proton source, gave lactones **224** and **225**, respectively, in moderate yield and good to high enantiomeric excess (Scheme 15.61). The process can be considered as an example of an asymmetric catch–release process, where a substrate immobilized using a chiral linker captures a reactive intermediate, in this case a ketyl-radical anion, from solution.[61, 62] The chiral linker controls the asymmetry of the capture step and leads to a diastereomeric, resin-bound

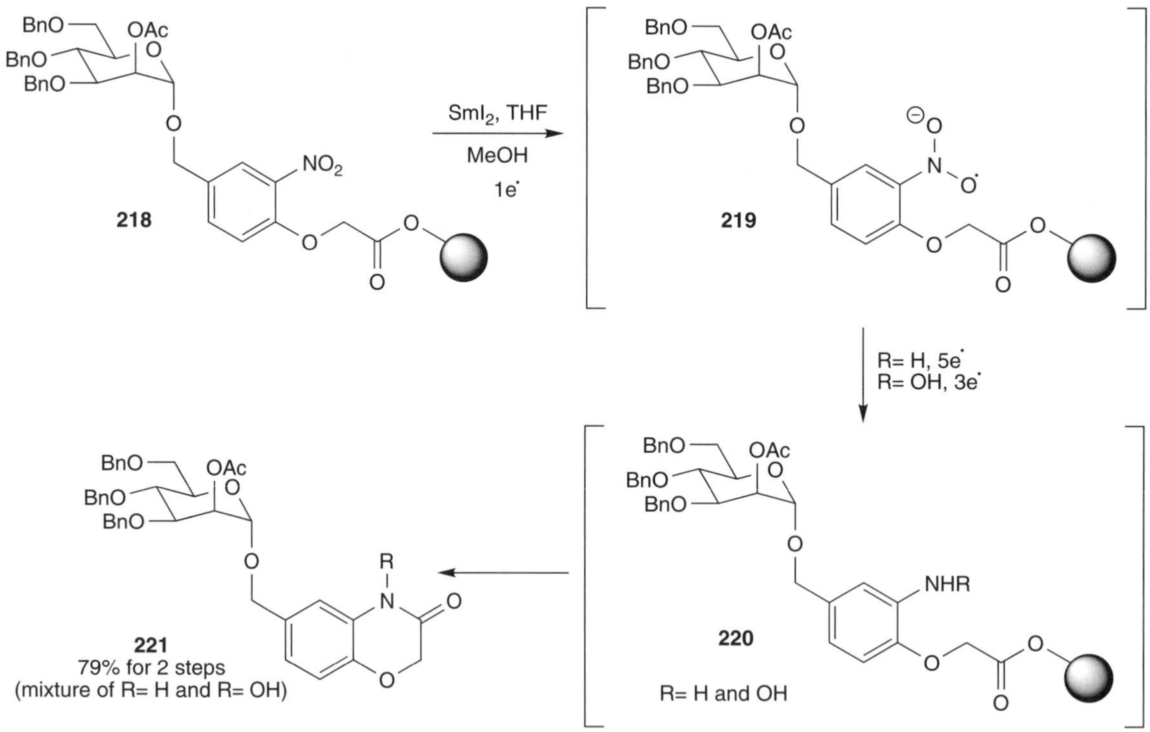

Scheme 15.60 Ito's linker cleavage triggered by nitro group reduction

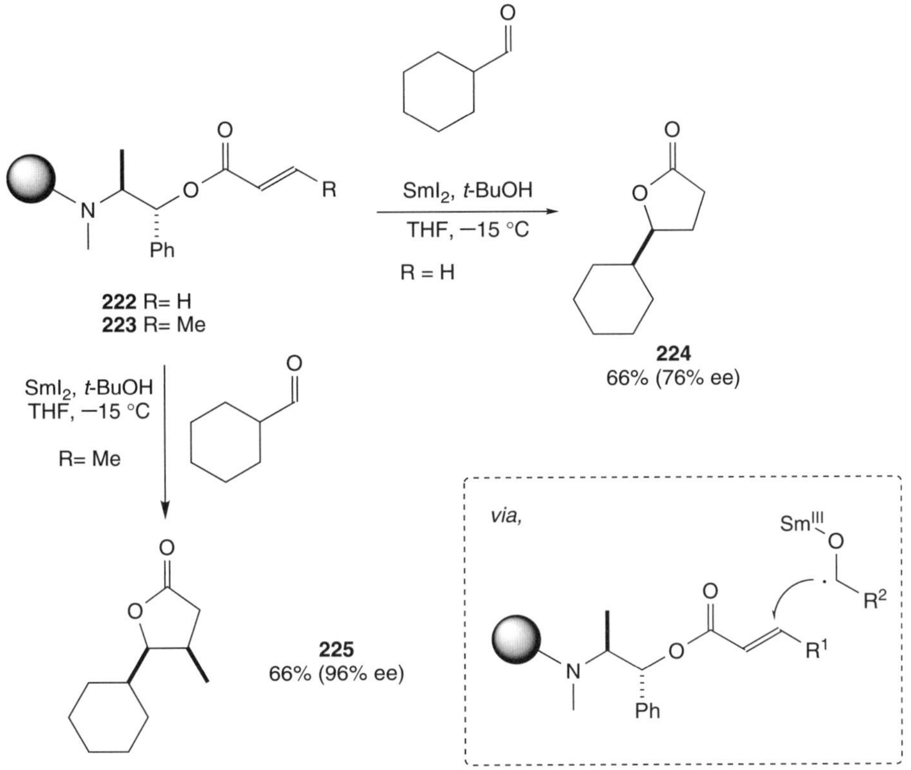

Scheme 15.61 Procter's asymmetric catch–release using an ephedrine chiral linker

intermediate that undergoes lactonization resulting in cleavage of the ephedrine linker and the release of a non-racemic product (Scheme 15.61).

Typical Experimental Procedures

Preparation of lactone 225

A suspension of (1R,2S)-N-Wang bound ephedrinyl crotonate resin **223** (0.49 mmol, 3.2 equiv.) in THF (5 ml) under argon was gently stirred for 15 minutes prior to the addition of cyclohexane carboxaldehyde (18.4 µl, 0.15 mmol, 1 equiv.) and t-butanol (2 equiv.). The resultant suspension was then allowed to stir for another 60 minutes at room temperature before being cooled to −15 °C. A pre-cooled samarium(II) iodide solution (0.1 M in THF, 5.5 equiv.) was then added and the dark blue solution was allowed to stir at −15 °C until TLC analysis showed that the aldehyde had been consumed. The reaction was then allowed to warm to room temperature over 5–6 hours. The resin was removed by filtration and was washed with THF. The filtrate was then concentrated (approximately 30 ml) and washed with aqueous saturated sodium chloride (4 ml). The aqueous layer was separated and washed with diethyl ether (3 × 15 ml). The combined organic layers were then dried (sodium sulfate) and concentrated in vacuo. The crude product was purified by column chromatography on silica (eluting with 10% ethyl acetate/petroleum ether (40–60 °C)) to give **225** (18.2 mg, 66%) as a pale yellow solid.

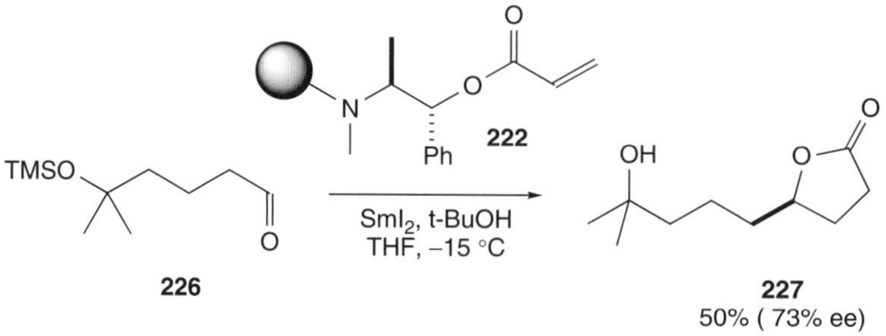

Scheme 15.62 Procter's synthesis of a DNA-binding metabolite using asymmetric catch–release

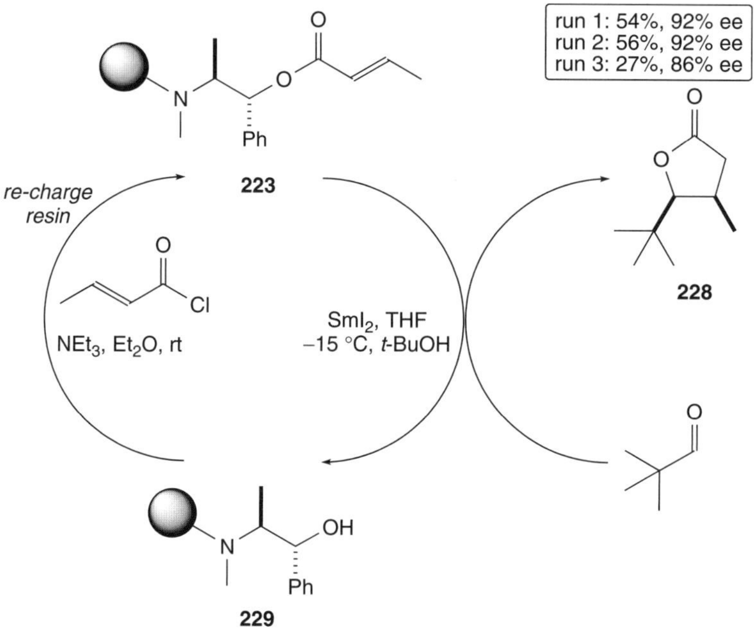

Scheme 15.63 Resin recycling in Procter's asymmetric catch–release approach

Procter has used the approach in a short, asymmetric synthesis of γ-butyrolactone **227**, a moderate DNA-binding metabolite isolated from Streptomyces GT61115, by reaction of aldehyde **226** with acrylate immobilized using an ephedrine linker (Scheme 15.62).[61, 62]

In further studies, the feasibility of recycling the chiral ephedrine resin has been investigated.[62] Employing crotonate resin **223** and 2,2-dimethylpropanal gave lactone **228** in 54% yield and 92% *ee* (Scheme 15.63). Recovery of the ephedrine resin **229** and re-esterification with crotonyl chloride gave recycled **223**. Re-treatment with 2,2-dimethylpropanal gave **228** in virtually identical yield and enantiomeric excess. Recovery and re-use for a third time, however, led to a substantially lower yield although the enantioselectivity of the process was still high (86% ee) (Scheme 15.63).[62]

15.6 Conclusions

There are now many examples of linker systems that are cleaved by radical processes. Many of these linkers were originally designed for traceless solid-phase synthesis, that is a carbon–hydrogen bond is formed on cleavage of the linker therefore leaving no residual functional groups as a consequence of the chosen linking strategy. In more recent times, however, many groups have recognized that the reactive intermediates formed on radical cleavage of a linker are rich in synthetic potential and can be exploited in both intermolecular and intramolecular reactions, thus allowing diversity to be increased in the final step of a solid-phase synthesis. These cleavage–cyclization and cleavage–functionalization strategies are fast becoming a requirement for new linker systems.

Radical cleavage of linkers has been carried out using a variety of reagents, ranging from 'classical' reagents for radical chemistry, such as tributyltin hydride, to relative newcomers to the field of solid-phase synthesis, samarium(II) iodide and CAN. The future promises increasingly powerful linker systems that operate in a multifunctional sense, for example, linkers that can be cleaved as part of complex, radical reaction sequences, that can enable a variety of bond-forming reactions to be carried out on immobilized intermediates, and that can control relative and absolute stereochemistry. As the search for new linkers with widespread applicability continues, synthetically powerful radical cleavage approaches that also display high functional group compatibility and selectivity will play an important role.

References

[1] McGhee, A. M., and Procter, D. J.; *Top. Curr. Chem.* **2006**, *264*, 93.
[2] Mentel, M., and Breinbauer, R.; *Eur. J. Org. Chem.* **2007**, 4283.
[3] Sloan, L. A., and Procter, D. J.; *Chem. Soc. Rev.* **2006**, *35*, 1221.
[4] McAllister, L. A., McCormick, R. A., and Procter, D. J.; *Tetrahedron* **2005**, *61*, 11527.
[5] Nicolaou, K. C., Montagnon, T., Ulven, T., et al.; *J. Am. Chem. Soc.* **2002**, *124*, 5718.
[6] Bertini, V., Lucchesini, F., Pocci, M., et al.; *Synlett* **2003**, 1201.
[7] Barluenga, S., Moulin, E., Lopez, P., and Winssinger, N.; *Chem. Eur. J.* **2005**, *11*, 4935.
[8] Nicolaou, K. C., Pastor, J., Barluenga, S., and Winssinger, N.; *Chem. Commun.* **1998**, 1947.
[9] Ruhland, T., Andersen, K., and Pedersen, H. J.; *J. Org. Chem.* **1998**, *63*, 9204.
[10] Nicolaou, K. C., Pfefferkorn, J. A., Cao, G., et al.; *Org. Lett.* **1999**, *1*, 807.
[11] Nicolaou, K. C., Roecker, A. J., Hughes, R., et al.; *Biorg. Med. Chem.* **2003**, *11*, 465.
[12] Nicolaou, K. C., Roecker, A. J., Pfefferkorn, J. A., and Cao, G.; *J. Am. Chem. Soc.* **2000**, *122*, 2966.
[13] Nicolaou, K. C., Mitchell, H. J., Fylaktakidou, K. C., et al.; *Angew. Chem. Int. Ed.* **2000**, *39*, 1089.
[14] Nicolaou, K. C., Fylaktakidou, K. C., Mitchell, H. J., et al.; *Chem. Eur. J.* **2000**, *6*, 3166.
[15] Berlin, S., Ericsson, C., and Engman, L.; *J. Org. Chem.* **2003**, *68*, 8386.
[16] Uehlin, L., and Wirth, T.; *Org. Lett.* **2001**, *3*, 2931.
[17] Uehlin, L., and Wirth, T.; *Chimia* **2001**, *55*, 65.
[18] Ruhland, T., Torang, J., Pedersen, H., et al.; *Synthesis* **2004**, 2323.
[19] Deegan, T. L., Gooding, O. W., Baudart, S., and Porco, J. A. Jr.; *Tetrahedron Lett.* **1997**, *38*, 4973.
[20] Kocienski, P. J., *Protecting groups*. 2000: Thieme.
[21] Oh, H. S., Hahn, H.-G., Cheon, S. H., and Ha, D.-C.; *Tetrahedron Lett.* **2000**, *41*, 5069.
[22] Barun, O., Sommer, S., and Waldmann, H.; *Angew. Chem. Int. Ed.* **2004**, *43*, 3195.
[23] Barun, O., Kumar, K., Sommer, S., et al.; *Eur. J. Org. Chem.* **2005**, 4773.
[24] Umarye, J. D., Leßmann, T., Garcra, A. B., et al.; *Chem. Eur. J.* **2007**, *13*, 3305.
[25] Kobayashi, S., and Aoki, Y.; *Tetrahedron Lett.* **1998**, *39*, 7345.
[26] Kobayashi, S., and Akiyama, R.; *Tetrahedron Lett.* **1998**, *39*, 9211.

[27] Aoki, Y., and Kobayashi, S.; *J. Comb. Chem.* **1999**, *1*, 371.
[28] Nestler, H. P., Bartlett, P. A., and Still, W. C.; *J. Org. Chem.* **1994**, *59*, 4723.
[29] Fukase, K., Egusa, K., Nakai, Y., and Kusumoto, S.; *Mol. Divers.* **1996**, *2*, 182.
[30] Fukase, K., Nakai, Y., Egusa, K., *et al.*; *Synlett* **1999**, 1074.
[31] Gordon, K. H., and Balasubramanian, S.; *Org. Lett.* **2001**, *3*, 53.
[32] Liu, H., Wan, S., and Floreancig, P. E.; *J. Org. Chem.* **2005**, *70*, 3814.
[33] McCormick, R. A., James, K. M., Willetts, N., and Procter, D. J.; *QSAR Comb. Sci.* **2006**, *25*, 709.
[34] Li, W.-R., Lin, Y.-S., and Hsu, N.-M.; *J. Comb. Chem.* **2001**, *3*, 634.
[35] Li, W.-R., Hsu, N.-M., Chou, H.-H., *et al.*; *Chem. Commun.* **2000**, 401.
[36] Arseniyadis, S., Wagner, A., and Mioskowski, C.; *Tetrahedron Lett.* **2004**, *45*, 2251.
[37] Du, X., and Armstrong, R. W.; *J. Org. Chem.* **1997**, *62*, 5678.
[38] Du, X., and Armstrong, R. W.; *Tetrahedron Lett.* **1998**, *39*, 2281.
[39] Myers, R. M., Langston, S. P., Conway, S. P., and Abell, C.; *Org. Lett.* **2000**, *2*, 1349.
[40] Meloni, M. M., and Taddei, M.; *Org. Lett.* **2001**, *3*, 337.
[41] Gustafsson, Olsson, R., and Andersson, C-M.; *Tetrahedron Lett.* **2001**, *42*, 133.
[42] Luo, J., and Huang, W.; *Mol. Divers.* **2003**, *6*, 33.
[43] Pilard, J. F., Marchand, G., and Simonet, J.; *Tetrahedron* **1998**, *54*, 9401.
[44] Marchand, G., Pilard, J. F., and Simonet, J.; *Tetrahedron Lett.* **2000**, *41*, 883.
[45] McKerlie, F., Procter, D. J., and Wynne, G.; *Chem. Commun.* **2002**, 584.
[46] McKerlie, F., Procter, D. J., Rudkin, I. M., and Wynne, G.; *Org. Biomol. Chem.* **2005**, *3*, 2805.
[47] Jung, K. W., Zhao, X., and Janda, K. D.; *Tetrahedron Lett.* **1996**, *37*, 6491.
[48] Jung, K. W., Zhao, X., and Janda, K. D.; *Tetrahedron* **1997**, *53*, 6645.
[49] Zhao, X., Jung, K. W., and Janda, K. D.; *Tetrahedron Lett.* **1997**, *38*, 977.
[50] Zhao, X., and Janda, K. D.; *Biorg. Med. Chem. Lett.* **1998**, *8*, 2439.
[51] Zhao, X., and Janda, K. D.; *Tetrahedron Lett.* **1997**, *38*, 5437.
[52] Zhao, X., Metz, W. A., Sieber, F., and Janda, K. D.; *Tetrahedron Lett.* **1998**, *39*, 8433.
[53] D'herde, J. N. P., and De Clercq, P. J.; *Tetrahedron Lett.* **2003**, *44*, 6657.
[54] McAllister, L. A., Brand, S., de Gentile, R., and Procter, D. J.; *Chem. Commun.* **2003**, 2380.
[55] McAllister, L. A., Turner, K. L., Brand, S., *et al.*; *J. Org. Chem.* **2006**, *71*, 6497.
[56] Turner, K. L., Baker, T. M., Islam, S., *et al.*; *Org. Lett.* **2006**, *8*, 329.
[57] McAllister, L. A., McCormick, R. A., Brand, S., and Procter, D. J.; *Angew. Chem. Int. Ed.* **2005**, *44*, 452.
[58] McAllister, L. A., McCormick, R. A., James, K. M., *et al.*; *Chem. Eur. J.* **2007**, *13*, 1032.
[59] James, K. M., Willetts, N., Procter, J.; *Org. Lett.* **2008**, *10*, 1203.
[60] Manabe, S., Nakahara, Y., and Ito, Y; *Synlett* **2000**, 1241.
[61] Kerrigan, N. J., Hutchison, P. C., Heightman, T. D., and Procter, D. J.; *Chem. Commun.* **2003**, 1402.
[62] Kerrigan, N. J., Hutchison, P. C., Heightman, T. D., and Procter, D. J.; *Org. Biomol. Chem.* **2004**, *2*, 2476.

16
Silicon and Germanium Linker Units

Alan C. Spivey[1] and Christopher M. Diaper[2]

[1]*Department of Chemistry Imperial College, United Kingdom*
[2]*NAEJA Pharmaceutical Inc., Canada*

16.1 Introduction

Silyl ethers have been widely used in solution phase as hydroxyl protecting groups since their introduction in the early 1970s.[1] This period also witnessed work in a number of laboratories directed towards the transposition of certain concepts of solution-phase protecting group chemistry to the development of linkers for solid-phase chemistry.[2] Inevitably, silyl ethers were included in this effort and the direct functionalization of polystyrene with silicon,[3] and its use as a 'polymer-anchored organosilyl protecting group' in solid-phase synthesis soon followed.[4]

Over the next twenty years, a number of different resin-bound silyl ethers appeared in the literature, mirroring the development of solution-phase silyl ether protecting groups. In both cases, the relative stability of the silyl ether linker/protecting group can be controlled by altering the size and stereoelectronic characteristics of the silicon substituents: an increase in the steric demand around silicon generally results in greater stability towards both acidic and basic conditions whereas electron withdrawing groups promote hydrolysis under basic conditions.[4] Silyl linkers can also be used to anchor other functional groups, such as thiols, amines, and carboxylic acids, to resins by using linkers that are cleaved by fragmentation. Silyl linkers have also been developed for the immobilization of aromatic nuclei as arylsilanes; cleavage is then possible either in a traceless fashion by *ipso*-protodesilylation with acid or with concomitant introduction of a diverse range of functionality via *ipso*-substitution with a variety of other electrophiles (e.g. halonium ions). In this context, germanium-based linkers have also been introduced which can also be cleaved tracelessly or with concomitant diversification. Arylgermanes are more susceptible towards cleavage by acids and electrophiles but display enhanced stability towards bases and nucleophiles compared to their

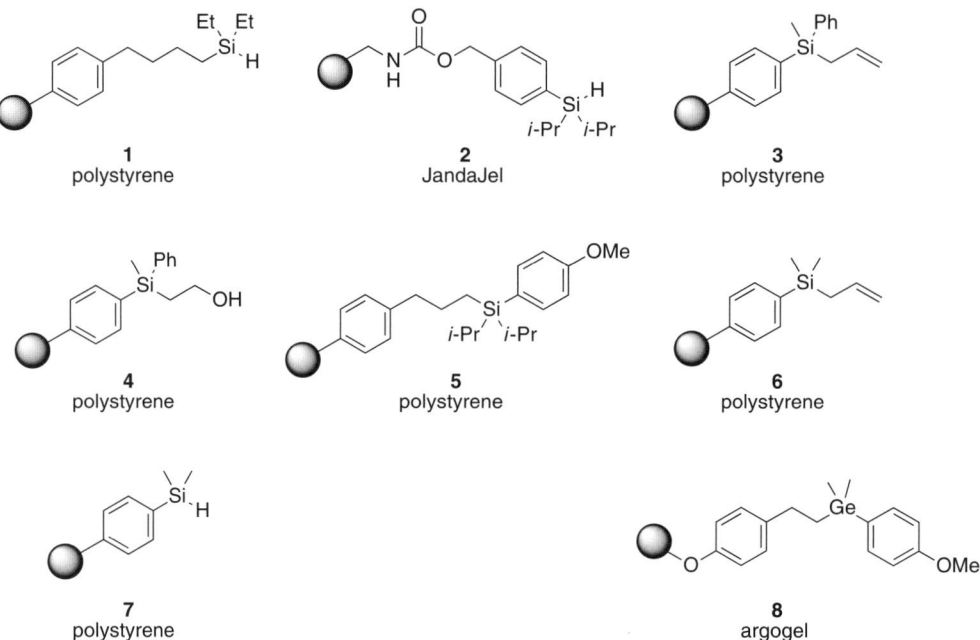

Figure 16.1 *Commercially available silicon and germanium based resins suppliers, Biotage (**1**), Aldrich (**1–5**), Novabiochem (**5–7**) and AFChemPharm (**8**)*

silicon counterparts: this distinctive reactivity profile complements the silicon linkers.[5] Collectively, with the choice of different solid supports,[6] the silicon/germanium-based linkers developed to date provide a spectrum of reactivity, which can be exquisitely tailored to the requirements of a specific library synthesis.

Recently, the commercialization of resins based on both silicon and germanium has provided easy access to appropriately functionalized supports for the non-specialist and has promoted their wider use (Figure 16.1).

This chapter is designed to provide a comprehensive overview of the field and also serve as a resource for potential practitioners, to facilitate identification of the most suitable linker for a given synthetic application. As such, all of the silyl and germyl linkers described to date have been collected into tables for easy reference.

16.2 Silicon-based linkers

16.2.1 The preparation of silyl resins

There are three general approaches for the introduction of silicon onto a solid support: direct functionalization; addition of a preformed silyl linker to the resin via a suitable functional handle; or incorporation of a silyl monomer in the polymerization process.

16.2.1.1 Direct silylation of resins

The direct functionalization of polystyrene is the simplest method of incorporating silicon onto a polystyrene support. This is generally accomplished by a two-step process whereby lithiation of

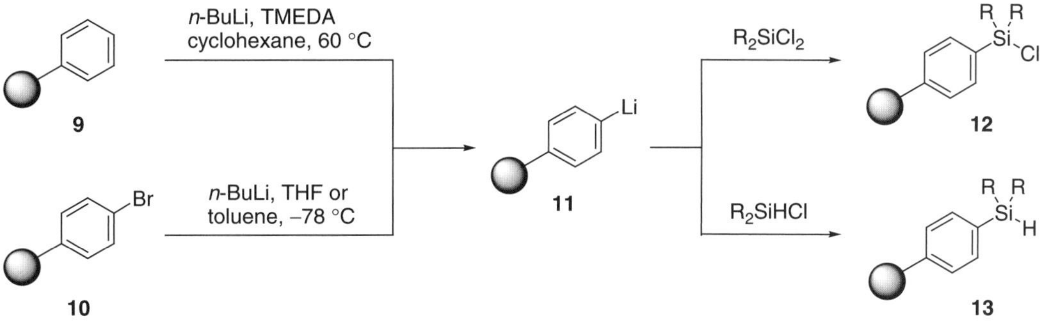

Scheme 16.1 *Silylation of lithiated polystyrene*

commercial cross-linked polystyrene **9**[3, 7] or *para*-bromopolystyrene **10**[8, 9] is followed by reaction with a dialkyldichlorosilane to generate the active polymer bound silyl chloride **12** (Scheme 16.1).[3] The incorporation of silicon can be verified by the appearance of diagnostic silicon–aryl (1108 and 1430 cm^{-1}),[10] or silicon–methyl (835 and 1250 cm^{-1})[10] bands in the IR spectrum of the silylated polymer. The absolute silicon content can be determined by elemental analysis[11] or Inductively Coupled Plasma-Optical Emission Spectroscopy (ICP-OES).[12] However, this may not provide an accurate picture of the number of silicon functional groups available to react, so an effective loading level is generally determined by chemical analysis: the silyl chloride is hydrolyzed to the corresponding silanol, which can then be quantified by acid–base titration.[10]

This procedure for resin preparation can be problematic. The use of highly activated, unhindered silanes, such as dimethyldichlorosilane, can result in unwanted cross-linking reactions, and the resulting resins can be contaminated with lithium salts, which in some cases cannot be removed completely, even by extensive washing, due to concomitant degradation of the Si–Cl functionality.[13] Hydrolysis of the silicon–chlorine bonds also makes resin **12** unsuitable for long term storage under atmospheric conditions.[10] These factors are widely believed to be the cause of the variable loading capacities and inconsistent results that have been reported across different batches of resin.[14] To circumvent these problems, lithiated polystyrene can be quenched with dialkylchlorosilanes to give the stable silanes **13**,[15] which are chemically stable and can be activated immediately prior to use (*vide infra*).

Although widely used, silyl polystyrenes have one further drawback: the Si–C$_{aryl}$ bond is incompatible with reactions that require strongly acidic or electrophilic conditions (*vide infra*). This limitation can be circumvented by the use of alkyl spacers, which are introduced between the polystyrene backbone and the reactive silane. Due to the lability of benzylic silanes (e.g. to oxidation), two or more carbon atoms are required for a stable alkyl tether.[16] One direct approach is the hydrosilylation of (vinyl)polystyrene **14** (commercially available as Amberlite XAD-2 and XAD-4) with dialkylchlorosilanes using dicobalt octacarbonyl as a catalyst to give resins **15** (Scheme 16.2).[17] An alternate two-step sequence has been employed on a commercial scale to generate a four-carbon tether:[13, 18] hydrosilylation of homoallylic polystyrene **17** (synthesized by the reaction of Merrifield resin **16** with allylmagnesium chloride) with dialkylsilanes gives resins **18** (Method A, Scheme 16.2). Similarly, treatment with a dichlorosilane followed by alkylation of the reactive silicon–chlorine bonds using Grignard or organolithium reagents can also be used to prepare resins **18** (Method B, Scheme 16.2). The method of choice is usually dictated by availability and cost of the silane. If dialkylarylsilanes are used in the hydrosilylation, a stable arylsilane resin **19** is produced (Method C, Scheme 16.2).[19] Hydrosilylation is also compatible with PEG-based supports such as ArgoGel allyl resin.[13]

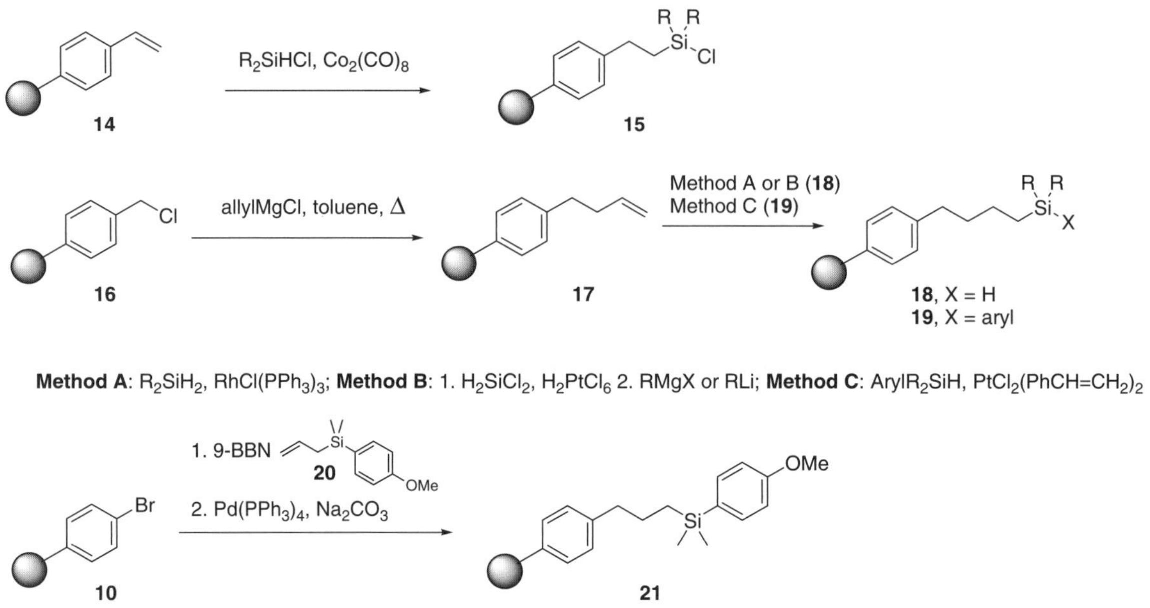

470 Multifunctional Linker Units for Diversity-Oriented Synthesis

Scheme 16.2 *The synthesis of two-, three- and four-carbon tethers*

A longer sequence has been used by several groups in which allylsilane **20** is attached to commercial *p*-bromopolystyrene **10** (or 3-iodobenzamidomethyl polystyrene)[20] by a one-pot hydroboration–Suzuki coupling process that furnishes an arylsilane attached via a three-carbon tether **21** (Scheme 16.2).[21]

These hydrosilylation and Suzuki-based methods are particularly suited to the direct preparation of resin bound simple arylsilanes as shelf stable precursors to reactive chloro- and bromosilanes. A typical such aryl group is a *p*-anisole[21] unit, which allows facile activation by *ipso*-chloro- or *ipso*-bromodesilylation using hydrochloric acid or hydrobromic acid, respectively (*vide infra*). Alternatively, a functionalized aromatic library progenitor can be introduced directly, that is a preformed handle approach.[19, 22]

16.2.1.2 Attachment of preformed linkers to resins

Although the silylated polystyrenes and alkyl chain tethered silyl polystyrenes are the most commonly used silicon-based solid supports for library synthesis, different linker architectures and solid supports, such as PEG functionalized polymers[6] or CPG (Figure 16.2),[23] are sometimes advantageous for a particular synthesis. These systems are usually constructed by attaching a preformed silyl linker to the solid support via a suitable handle such as an amide. This method is often used during the development and evaluation of a new linker or library design, as the new chemistry can be performed initially in solution, allowing convenient analysis via standard procedures without recourse to specialist techniques or equipment such as MAS-NMR. In more complex linkers, it allows the linker to be purified and characterized fully prior to attachment, thereby ensuring that it is free of impurities resulting from the synthetic pathway which might interfere with library generation or cleavage.

The successful loading of a preformed linker can be confirmed by ^{1}H, ^{13}C or ^{29}Si MAS-NMR;[24, 25] loading levels are typically estimated chemically by cleaving the library progenitor from the resin-bound linker followed by GC[24] or HPLC quantification of the amount of material released.

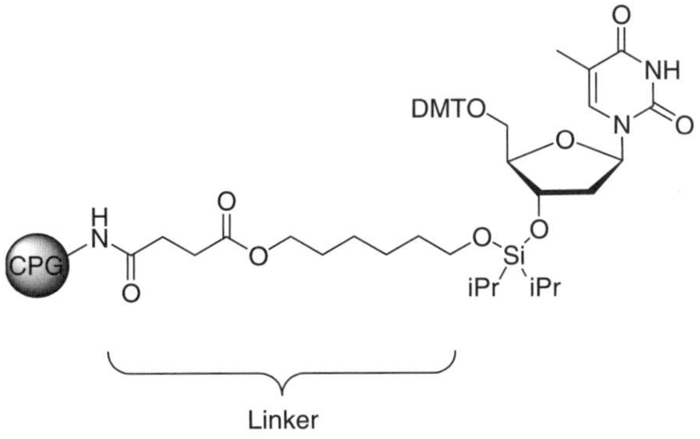

Figure 16.2 A preformed silylether linker attached via an amide to controlled pore glass (CPG) for use in automated oligonucleotide synthesis[23]

The stability of the functional handle between the linker and solid support is often checked using solution phase models. Typically, benzyl groups are used to mimic the attachment of linkers via ether,[24] silyl ether[26] or amide[27] linkages to Merrifield resins, whereas an ethoxyethyl ether has been used to mimic ether linkages to PEG-derived polymers.[28] This practice also applies to directly silylated polystyrenes: the stability of the aryl silane bond is sometimes checked in solution using the corresponding phenyl silyl ether protecting group.[29]

16.2.1.3 *Polymerization of silylated monomers*

Instead of modifying preformed polystyrene, it is also possible to introduce silyl-functionalized styrene monomers directly into the polymerization process. This has been studied in some detail using arylsilane monomers such as **22**,[30] **23**[31] and **24**[32] under anionic living polymerization conditions to give block co-polymers (Figure 16.3). Monomers with alkyl tethers such as **25** can be polymerized under free radical conditions with divinylbenzene (DVB) to give cross-linked silylated DVB–polystyrenes suitable for solid-phase organic synthesis.[31] One disadvantage of these polymers is that many of the functional groups introduced during the polymerization process are trapped in the hydrophobic core of cross-linked matrix, resulting in low loading levels of available silyl functionality.[14] A conceptually different approach is realized in the 'Rasta silane' resins, where a preformed cross-linked DVB–polystyene polymer is functionalized with 2,2,6,6-tetramethylpiperidine-1-oxyl (TEMPO) and then subjected to living free radical polymerization with monomer **23** (Figure 16.3).[14] This produces a structure in which the silicon functionality is appended to linear chains that extend from the main polymer core, represented in cartoon form by **27**. The functionalized chains are readily accessible to the solvent and therefore improve the effective loading of the silyl groups relative to standard DVB–polystyrene supports. As an alternative to polystyrene supports, PEG-based polymers incorporating silylethyl vinyl ethers **26** can be produced by photocationic suspension copolymerization.[33]

16.2.2 Activation of Si–H and Si–Aryl resins for substrate attachment

As indicated above, activated silicon containing polymers such as the silyl chloride resins are too reactive for long term storage due to hydrolysis,[10] resulting in the formation of the corresponding silanol. Inert silicon functionality such as silanes, for example PS-DES **1** (Scheme 16.3)[18] or aryl silanes **21** and **29** (Scheme 16.4),[21] are therefore required when resins are intended for long term storage prior to use.

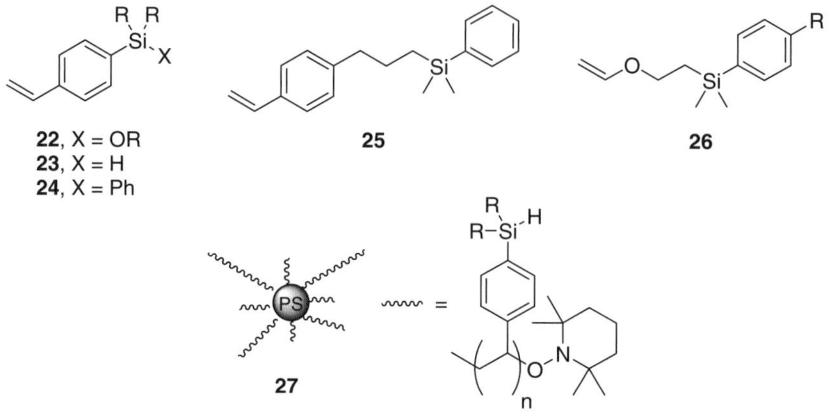

472 *Multifunctional Linker Units for Diversity-Oriented Synthesis*

22, X = OR
23, X = H
24, X = Ph

25

26

27

Figure 16.3 Top: Monomers used to produce silylated polystrenes. Bottom: Cartoon representation of 'Rasta silane' resins

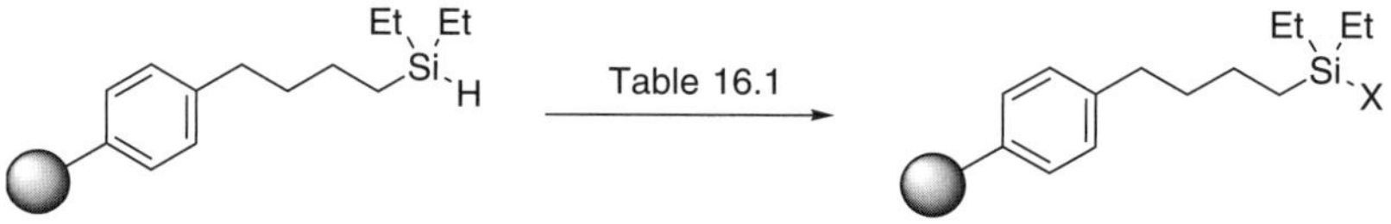

Scheme 16.3 Activation of PS–DES resin

21 R^1 = alkyl, R^2 = OMe
29 R^1 = methyl, R^2 = H

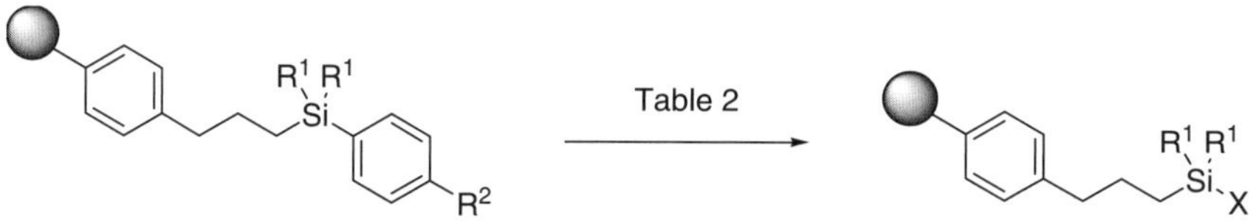

Scheme 16.4 Activation of arylsilane resins

16.2.2.1 Activation of Si–H resins

A number of procedures have been developed for activating PS-DES resin (Table 16.1), thereby allowing substrate loading under standard conditions. The application of these conditions to other, non-aliphatic silyl linkers must take into consideration the other functionality present. For example, strongly electrophilic or acidic conditions will result in the cleavage of silicon from directly silylated polystyenes such as **7** (Figure 16.1), due to the facile cleavage of the Si–C$_{aryl}$ bond.[34]

A convenient chemical method for determining silicon–hydrogen loading of PS–DES is the conversion of the silane to the corresponding bromide using trityl bromide. This generates triphenylmethane, which can be quantified using GC.[18] The activation of PS–DES resin is also easily monitored using the diagnostic

Table 16.1 Activation of PS-DES resin

Reagents and conditions	X	Analysis method	Ref.
1,3-dichloro-5,5-dimethylhydantoin, dichloromethane (CH$_2$Cl$_2$), rt, 90 min (trichloroisocyanuric acid, trityl chloride, or N-chlorosuccinimide (NCS) can also be used as chlorinating agents)	Cl	IR: loss of Si–H 2100 cm^{-1} Chemical: Hydrolysis of Si–Cl to Si–OH followed by acid–base titration.	[10, 13, 18]
CF$_3$SO$_3$H, DCM, rt, 5 min or 1. DIPEA, 2.5% TMSCl/DCM. 2. CF$_3$SO$_3$H, DCM, rt, 10 min.	OTf	^{19}F MAS NMR: −79.1 ppm (C$_6$D$_6$, CCl$_3$F internal standard). ^{29}Si MAS NMR: 44.2 ppm (C$_6$D$_6$).	[35, 36]
1. 1,3-dichloro-5,5-dimethyl hydantoin, DCM, rt. 2. TMSCN, 80 °C.	CN	IR: Si–CN 2185 cm^{-1}.	[37]
1. 1,3-dichloro-5,5-dimethyl hydantoin, DCM, rt. 2. TMSN$_3$, 80 °C	N$_3$	IR: Si–N$_3$ 2134 cm^{-1}.	[37]

signals of the silicon–hydrogen bond in the IR: there is a distinctive silicon–hydrogen stretch over the 2000–2200 cm^{-1} range.[18] Any subsequent hydrolysis of the Si–X bond to the corresponding silanol can also be monitored by IR (absorption range 3200–3700 cm^{-1}).[13] The disappearance of the silicon–hydrogen bond can also be monitored using ^{1}H-MAS-NMR.[37]

Typical Experimental Procedures

PS–DES Chloride[18]

PS–DES resin **1** (100 mg, 0.075 mmol) was added to a solution of 1,3-dichloro-5,5 dimethylhydantoin in DCM (0.3 M, 0.8 ml, 0.225 mmol) at room temperature under an atmosphere of argon. After 1.5 hours, the mixture was filtered and washed with DCM (3 × 3 ml) and anhydrous tetrahydrofuran (THF) (2 × 3 ml). The resin was used immediately after washing.
NB: The concentration of the chlorinating agent should be 0.3 M. It is important to use this concentration for the complete chlorination of the silane.

PS–DES triflate[35]

PS–DES resin **1** (500 mg, 0.36 mmol) was washed with DCM (3 × 10 ml) under an inert atmosphere. DCM (5 ml) was added to the resin followed by a solution of trifluoromethanesulfonic acid (96 µl, 1.08 mmol) in DCM (5 ml). After shaking for five minutes, the solution was drained and the resin was washed with DCM (3 × 10 ml) to give PS–DES triflate as a light tan resin.

PS–DES cyanide[37]

Trimethylsilyl cyanide (TMSCN) (45 ml, 0.34 mol) was added as a single portion to PS–DES chloride (5 g, 4.8 mmol) under an atmosphere of nitrogen. The mixture was gently stirred at 80 °C for 36 hours. The resulting mixture was cooled to room temperature, the resin isolated by filtration under a positive pressure of nitrogen and washed with DCM (20 × 20 ml) to give PS–DES cyanide.

PS–DES azide[37]

Trimethylsilyl azide (TMSN₃) (40 ml, 0.30 mol) was added as a single portion to PS–DES chloride (10 g, 6.8 mmol) under an atmosphere of nitrogen. The mixture was gently stirred at 80 °C for 72 hours. The resulting mixture was cooled to room temperature, the resin isolated by filtration under a positive pressure of nitrogen and washed with DCM (20 × 20 ml) to give PS–DES cyanide.

16.2.2.2 Activation of Si–Aryl resins

The Si–C$_{aryl}$ bond of the carbon tethered arylsilanes **21** and **29** (Scheme 16.4) are easily cleaved under electrophilic conditions to generate the activated resins **30** via an electrophilic aromatic *ipso*-substitution process (Table 16.2). In the case of resin **21**, the loading level is determined chemically by treating the resin with bromine to generate 4-bromoanisole, which is then quantified.[21] The loss of the aromatic residue can also be monitored on the resin by IR: a diagnostic shift of the silicon–methyl band is observed with the concomitant loss of the silicon–aryl bands (Table 16.2).[31] In the case of resin **29**, the silicon–methyl chemical shift (−2.93 ppm) in the ^{13}C NMR is sensitive to modifications at silicon and so monitoring this peak is an alternate method for following the progress of these reactions (Table 16.2).[31]

Typical Experimental Procedures

Resin 30 ($R^1 = CH_3$, $X = Cl$)[21]

*A 100 ml resin reactor equipped with a magnetic stirrer and septa fitted on the sidearm and ground glass joint was charged with resin **21**($R^1 = CH_3$, 0.88 g, 1.09 mmol) and anhydrous DCM (30 ml). The suspension was stirred gently and anhydrous hydrogen chloride was sparged for 30 seconds. The mixture was stirred at room temperature for 1.5 hours and then a nitrogen needle was inserted through the septum on the ground glass joint. The solvent was drained and the resin rinsed with DCM (5 × 20 ml), diethyl ether (2 × 20 ml), DCM (3 × 20 ml), diethyl ether (2 × 20 ml), THF (3 × 20 ml), and diethyl ether (2 × 20 ml). The resin was used immediately after washing.*

Table 16.2 Activation of arylsilane resins

Reagents and conditions	R^1	R^2	X	Analysis method	Ref.
HCl, DCM, rt.	Me	OMe	Cl	Chemical	[21]
SOCl₂, cat. H₂O, 40 °C, 70 h.	Me	H	Cl	^{13}C NMR: SiMe₂Cl 1.83 ppm IR: Si–Cl 467 cm^{-1}; loss of Si–Aryl 1426 and 1113 cm^{-1}; SiMe₂Ph → SiMe₂Cl, 881 to 846 cm^{-1}	[31]
TFA, TFAA, 35 °C, 60–70 h.	Me	H	COCF₃	^{13}C NMR: SiMe₂TFA−2.1 ppm IR: loss of Si–Aryl 1426 and 1113 cm^{-1}; Si–TFA 1768, 1220 and 1169 cm^{-1}	[31]
CF₃SO₃H, DCM, rt.	*i*-Pr	OMe	OTf	^{29}Si MAS-NMR: 44.0 ppm (C₆D₆)	[38]
Pr₃SiCl, imidazole, DCM, rt, 2 h.	*i*-Pr	OMe	Cl	not reported	[39]

Resin 30 ($R^1 = CH_3$, $X = CF_3CO_2$)[31]

Trifluoroacetic acid (TFA) (4 g) and TFAA (2 g) were added to resin **29** (2.0 g) and the mixture was stirred for 60–70 hours at 35 °C. The resin was then filtered, washed with carbon tetrachloride under an atmosphere of nitrogen and dried in vacuo at 40 °C for 24 hours.

Resin 30 ($R^1 = i$-Pr, $X = OTf$)[38]

A 10 ml polypropylene PD-10 column fitted with a Teflon stopcock was charged with resin **21**($R^1 = i$-Pr, 1.43 mequiv. of Si/g) that had been dried under high vacuum for 12 hours. The resin was then swollen in DCM (10 ml/g of resin) under a nitrogen atmosphere for 30 minutes. The solvent was then drained under a positive nitrogen pressure and 4% TfOH in DCM (6 equiv. relative to Si) was added by syringe. The resin turned bright red/orange on exposure to the acid. The mixture was gently agitated for 30 minutes and then washed with DCM (×2) before use.

16.2.3 Silyl ether linkers

As explained in the introduction, the first and probably still most popular use of silyl linkers is for the attachment of alcohol functional groups to a solid support as silyl ethers in a manner closely analogous to using a silyl protecting group.

16.2.3.1 Loading the alcohol

Typically, alcoholic substrates are loaded onto the activated silyl chloride,[10, 18] silyl trifluoroacetate[31] or triflate[14, 38] resins **31** using an amine base: the addition of DMAP is often required when using silyl chlorides[7, 14, 24] or trifluoroacetates.[31] It is also possible to load alcohols directly onto silanes **32** using three different methods: alcoholysis with catalytic tetrabutylammonium fluoride (TBAF);[13] alcoholysis with Wilkinson's catalyst;[40] or hydrosilylation using aldehyde or ketone substrates (Scheme 16.5).[40]

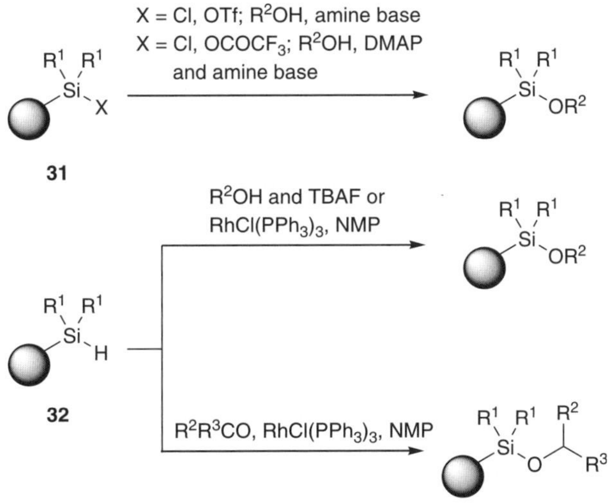

Scheme 16.5 The loading of alcohols onto silylated resins

Typical Experimental Procedures

Loading alcohols onto PS–DES chloride[18]

The alcohol (3 equiv.) and imidazole (3.5 equiv.) were added to a suspension of the resin **31**($R^1 = CH_4$; $X = Cl$ 1 equiv.) in DCM under an atmosphere of argon at room temperature. After four hours the mixture was washed with N,N-dimethylformamide (DMF) (×2), 1:1 DMF/water (×2), 1:1 THF/water (×2), and finally THF (×2).

Loading alcohols onto TFA resin 31[31]

The resin **31**($R^1 = CH_3$; $X = CO_2CF_3$, 2.4 mmol) was added to a solution of the alcohol (3.2 mmol) and triethylamine (3.2 mmol) in DCM (5 ml). The reaction mixture was then heated at 40 °C for 48 hours. The polymer was then filtered and washed with DCM until the alcohol was completely removed. The resin was then dried in vacuo at 40 °C.

Loading alcohols onto triflate resin 31[38]

2,6-Lutidine (0.27 ml, 8 equiv. relative to Si) was added to the resin **31**($R^1 = i$-Pr; $X = OTf$) and left for 15 minutes. An azeotropically dried solution of the alcohol in 2,6-lutidine (1.0 M, 2 equiv.) was added resulting in a colorless resin. The resin was then gently agitated for 10 hours under a nitrogen atmosphere. The beads were then drained, exposed to the atmosphere and washed with DCM (2 × 3 ml, 45 minutes), THF (2 × 3 ml, 30 minutes), 3:1 THF/IPA (2 × 3 ml, 30 minutes), 3:1 THF/water (2 × 3 ml, 30 minutes), 3:1 THF/IPA (2 × 3 ml, 30 minutes), DMF (2 × 3 ml, 30 minutes) and THF (2 × 3 ml, 30 minutes). The resin was air-dried for three hours and then placed under high vacuum for 24 hours.

Loading alcohols directly onto PS–DES Using TBAF[13]

A solution of TBAF in THF (2 mol%) was added to a suspension of PS–DES **1** (800 mg), the alcohol (1.5 equiv.) and N-methylpyrrolidone (NMP) (6.4 ml) under an atmosphere of argon. The reaction mixture was then agitated for 6–8 hours. The mixture was then filtered and washed with toluene (×2), methanol and THF (×2). The resin was then dried in vacuo.

Loading alcohols directly onto PS–DES using rhodium catalysis[13]

The alcohol (0.28 mmol) was added to a suspension of PS–DES resin **1** (200 mg) and $Rh_2(pfb)_4$ (1.7 mg) in DCM at room temperature under an atmosphere of argon. After three hours the reaction mixture was filtered and washed with DCM (×3), toluene (×2), 1:1 THF/water (×2) and THF (×3). The resin was then dried in vacuo.

Loading via carbonyl hydrosilylation of PS–DES Resin[40]

A mixture of the aldehyde (0.75 mmol), PS–DES resin **1** (0.375 mmol), $RhCl(PPh_3)_3$ (4 mol%) and NMP (5 ml) were heated at 60 °C for two hours under an atmosphere of argon. The resin was then filtered and the resin washed with NMP (3 × 5 ml), DCM (3 × 5 ml) and THF (3 × 5 ml).

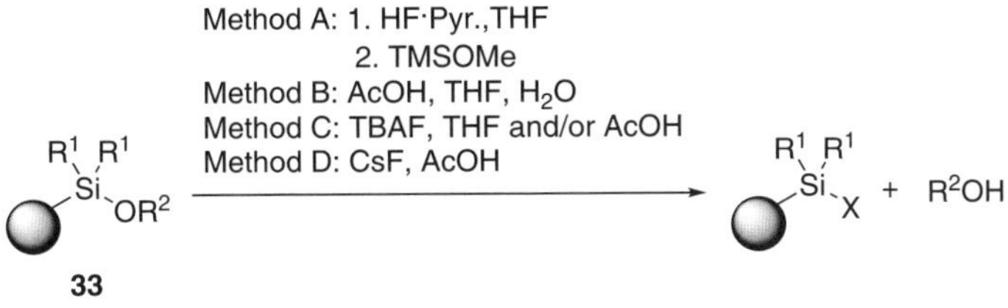

Scheme 16.6 The cleavage of alcohols from silyl ether resins

16.2.3.2 Cleavage of the alcohol

The release of alcohols from silyl ether resins **33** is usually achieved under mild conditions by treatment with either fluoride ion or hydrofluoric acid (Scheme 16.6). Cleavage with HF·pyridine is carried out in THF:[18] excess HF can be simply removed after cleavage by the scavenging with TMSOCH$_3$, which reacts to form trimethylfluorosilane (FSiMe$_3$) and methanol which are both volatile and can be removed under vacuum.[40] Acidic cleavage with either acetic acid[18] or trifluoroacetic acid[34] can also be used in some cases. The reagent of choice for basic cleavage of alcohols is usually TBAF.[41] An alternate source of fluoride ion is cesium fluoride:[42] the addition of 18-crown-6 is often required however to solubilize the metal cation and enable efficient cleavage.[43, 44]

Typical Experimental Procedures

Method A: use of HF·Pyr[40]

A solution of HF·pyridine complex in THF (0.4 M, 5 ml) was added to the resin (0.375 mmol) and the mixture was agitated for two hours at room temperature. methoxytrimethylsilane (TMSOCH$_3$) (approximately 2 equiv., 0.5 ml) was added and the mixture agitated for two hours. The resin was removed by filtration and washed with THF. The filtrate and washings were combined and concentrated in vacuo to give the alcohol.

Method B: use of acetic acid[18]

The resin (approximately 200 mg) was heated with 6:6:1 acetic acid/THF/water at 50 °C for four hours. The mixture was filtered and the filtrate was concentrated in vacuo to give the alcohol.

Method C: use of TBAF[41]

A solution of TBAF in THF (1 M, 150 μl) was added to a suspension of resin (45–65 mg) swollen in THF (0.7 ml). The resulting mixture was vortexed using a Vibrax shaker at 25 °C for four hours. The mixture was filtered and the resin washed once with THF (1 ml) and diethyl ether (2 × 1 ml). The filtrate was then washed with aqueous sodium bicarbonate (3 ml) and the aqueous phase was frozen in 14 ml polypropylene tubes using an ethanol/dry ice bath. The organic phase was decanted and the aqueous phase re-extracted with diethyl ether (2 ml). The organic layer was washed successively with water and brine using the freezing/decanting procedure. The organic layer was then evaporated under a stream of nitrogen followed by drying overnight in vacuo to give the alcohol.

Method D: use of cesium fluoride[42]

The resin (385 mg) was stirred with cesium fluoride (152 mg, 0.2 mmol) and acetic acid (0.3 ml, 1.0 mmol) in DMF (5 ml) overnight. The mixture was filtered and the resin was washed with DMF, DCM and diethyl ether. The combined filtrate was concentrated in vacuo and the residue was extracted with trichloromethane. The extract was washed with water and brine, dried (sodium sulfate) and concentrated in vacuo. The crude product was further purified by chromatography as required.

16.2.3.3 Resin recycling

A number of studies have focused on methods of regenerating active silyl functionality after the cleavage of alcohols from silyl ethers **34**. Acidic cleavage using trifluoroacetic acid in methanol results in the formation of resin bound methyl silyl ethers **35** (X = OCH_3),[31] which can then be reactivated with either thionyl chloride or trifluoroacetic anhydride (Scheme 16.7, Method A). The recycled resin can then be reloaded with alcoholic substrates using standard conditions (for example Scheme 16.5). This sequence requires the use of alkyl tethered resins, as the silylated polystyrenes suffer from competitive *ipso*-substitution of the polystyrene Si–C_{aryl} bond, resulting in leaching of the silyl groups from the polymeric backbone and, hence, loss in activity.[34] A more general procedure involves cleavage using HF: the resulting silyl fluorides **35** (X = F) can be converted into the corresponding chlorides **35** (X = Cl) using boron trichloride (Scheme 16.7, Method B).[8, 17]

Typical Experimental Procedures
Method A: use of acetic acid–methanol/TFAA[31]

Step 1, Cleavage: a mixture of resin **34** (R^1 = CH_3, n = 2, 1.1 g), acetic acid (0.15 g) and methanol (4 ml) was stirred at room temperature for 36 hours. The mixture was filtered and the resin washed repeatedly with DCM and methanol. The combined filtrate was concentrated in vacuo to give the alcohol.

Step 2, Regeneration: the recovered polymer was heated with TFAA (4 ml) at 35 °C for 49 hours. The mixture was filtered and the resin was washed with anhydrous carbon tetrachloride under nitrogen and then dried in vacuo at 40 °C to give the resin **35** (X = trifluoroacetate (OTFA)).

Method B: use of HF/BCl₃[17]

Step 1, Cleavage: the resin **34** (R^1 = CH_3, n = 2, 25 mmol) was suspended in a mixture of 52% HF (10 ml, 270 mmol) and acetonitrile (50 ml) at room temperature for three hours. The suspension was then filtered and the filtrate was concentrated in vacuo.

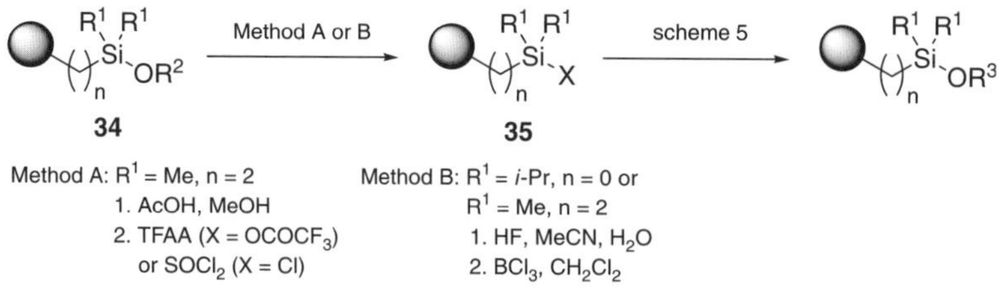

Scheme 16.7 *Regeneration of silyl ether resins*

Step 2, Regeneration: the resin (1.29 mmol) was added to a polypropylene tube in an ice bath and purged with nitrogen. A solution of boron trichloride (BCl₃) in DCM (1 M, 5 ml, 5 mmol) was added slowly and the mixture was then stirred for three hours. The ice bath was then removed and nitrogen was passed over the suspension until all the solvent had been evaporated. The remaining beads were rinsed with DCM under nitrogen and then dried in vacuo for 12 hours to give light brown beads.

16.2.3.4 Linker designs

The various silyl ether linkers developed to date are listed in Table 16.3. The table provides a concise overview of the practical use and reported applications of each linker. Only the loading and cleavage conditions specifically reported in the original literature are given.

16.2.4 Fragmentation-based silyl linkers

16.2.4.1 Linkers for alcohols, thiols and carboxylic acids

The direct attachment of amines or carboxylic acids to silicon generally results in a covalent linkage that is too labile to be of practical use in solid-phase synthesis,[4] although the high reactivity of the nitrogen–silicon bond has been used for solid-phase synthesis of pyrrolidines.[64] This situation has led to the development of linkers based on the well established 2-(trialkylsilyl)ethyl ether **36**, 2-(trimethylsilyl)ethoxymethyl (SEM) ether **37** and 2-(trimethylsilyl)ethyl ester (TMSE) protecting groups (Table 16.4). These resins accommodate a greater range of substrates than the silyl ethers, including alcohols, thiols, amines and carboxylic acids. The ethyl bridge between the silicon and heteroatom is sufficiently robust to survive a range of chemistry, but remains susceptible towards cleavage under mild conditions via either acidolysis or fluoridolysis (Scheme 16.8). These conditions trigger the fragmentation of the 2-(trialkylsilyl)ethyl unit, releasing the substrate, ethylene and, in the case of SEM derivatives **37**, formaldehyde. Another variation of this general concept are silicon linkers such as **38**, which rely on a 1,6-elimination process for cleavage (Scheme 16.8).

16.2.4.2 Linkers for amines–resins for peptide synthesis

The first indirect silicon linkers were developed in the late 1980s for Solid-Phase Peptide Synthesis (SPPS). They were designed to be orthogonal to standard peptide protecting groups,[65] opening up the possibility of cleaving peptides from the solid support with TFA or TBAF without concomitant side chain deprotection thereby allowing the synthesis of protected peptide fragments.

An additional benefit of the mild cleavage conditions is that some problematic side reactions, which occur during cleavage of certain peptides from conventional resins, can be avoided. One example is the use of the Silyl Amide Linker (SAL)[66]: cleavage with a cocktail of TFA, 1,2-ethanedithiol (EDT), phenol and thioanisole (90:5:3:2) prevents the alkylation of tryptophan residues which is observed upon cleavage of some other linkers.[67, 68]

Although cleavage with TBAF is relatively mild it can still present some problems in peptide synthesis. The Silyl ACid linker (SAC) (Table 16.4) was designed to suppress the formation of diketopiperazines (DKPs) during the release of C-terminal proline peptide chains from the solid support by using TBAF;[69] however, the basic nature of TBAF in solvents such as DMF, NMP or THF often results in the conversion of β-ester protected aspartyl residues into succinimides. The basicity of TBAF can be reduced by buffering with weak acids,[69] but this can result in slow and/or incomplete cleavage from the resin. In contrast to the SAC linker where the carbon–oxygen bond is benzylic, the (2-phenyl-2-trimethylsilyl) ethyl linker (PTMSEL) instead places the carbon–silicon bond at this position (Table 16.4).[70] This makes the

Table 16.3 Linkers for silyl ether resins

Resin structure, solid support and synthesis	Loading conditions	Cleavage conditions	Comments
PS–DES resin.[45] Hydrosilylation of allylpolystyrene[18] (structure: polystyrene–CH$_2$CH$_2$CH$_2$–Si(Et)(Et)–OR)	**Si–H:** 1° ROH: ROH, Rh$_2$(pfb)$_4$, DCM.[18] 1° and 2° ROH: ROH/aldehyde/ketone, RhCl(PPh$_3$)$_2$, NMP, 60°C.[40] ROH, TBAF, THF.[13] **Si–Cl:** 1° and 2° ROH: ROH, imidazole, DCM.[18] **Si–OTf:** Silyloxydiene: α,β-unsaturated ketone, (i-Pr)$_2$NEt, DCM.[35] Silyl ketene acetal: ester, Et$_3$N, DCM, rt, 2 h.[36] **Si–CN:** Cyanohydrins: aldehyde, ZnI$_2$, DCM, rt.[37]	1° and 2° ROH: HF-Pyr., THF.[18, 37, 46, 47] HF-Pyr., THF, TMSOMe.[13, 40, 48] 1° ROH: AcOH, THF, H$_2$O, 50°C, 4–8 h.[18, 49] TBAF, THF, rt.[41] 2° ROH: AcOH, THF, H$_2$O, 60–80°C, 8–12 h.[18] Silyl Enol Ether: TFA.[35] Cyanohydrins: HCl, DCM, EtOH.[37]	Commercially available in Si–H form. Used to synthesize: silyl enol ethers[35] α-hydroxy esters[37] oligosaccharides[49] vitamin D$_3$[46] Libraries: hydroxysteroids[47] 1,3-oxazolidines[41]
Addition of R$_2^1$SiCl$_2$ to lithiated polystyrene. R^1 = CH$_3$,[3, 10] Bu,[50] i-Pr,[7, 8, 51] Ph[7, 10] (structure: polystyrene–Ph–Si(R^1)(R^1)–OR^2)	R^{2}OH, imidazole, DCM, rt.[50] R^1 = i-Pr; Si-Cl: *1° R^{2}OH*: R^{2}OH, (i-Pr)$_2$NEt, DMAP, DCM, 3 d.[52] R^1 = Ph; Si-Cl: *1° R^{2}OH*: R^{2}OH, (i-Pr)$_2$NEt, DMAP, DCM, 2 d.[7] R^{2}OH, (i-Pr)$_2$NEt, DCM, 2 d.[10]	**R^1 = Bu:** 2° R^{2}OH: HF-Pyr., THF, rt.[50] R^1 = i-Pr: 1° R^{2}OH: TBAF, AcOH, THF, 40°C, 18 h.[7, 52] **R^1 = Ph:** 1° R^{2}OH: TBAF, DCM, rt, 5 h.[10] TBAF, AcOH, THF, 40°C, 18 h.[7]	R^1 = CH$_3$: Commercially available in Si–H form.[53] R^1 = i-Pr: Used for synthesis of: prostaglandin library.[50] Oligosaccharides.[7, 52, 54] Diels–Alder reactions of amino furanones.[51] Recycled using HF/BCl$_3$.[8]
(structure: polystyrene–Ph–C(i-Pr)(i-Pr)–N(cyclohexyl)–Si(i-Pr)(i-Pr)–OR)	R^{2}OH, Pyr., DCM, 2 d.[10] **Si–H:** 1° ROH:	1° and 2° ROH:	High loading levels.

Rasta Silanes.[14]		aldehyde, RhCl(PPh$_3$)$_2$, NMP, 60 °C, 8 h.	HF·Pyr., THF, TMSOMe.
Living free radical polymerization: polystyrene.		**Si–Cl**: 1° and 2° ROH: ROH, imidazole, DMAP, DCM, rt, 24 h. **Si-OTf**: 1° and 2° ROH: ROH, Pyr, DCM, rt, 3 h. **Si–Cl**: 1° ROH: ROH, Et$_3$N, DMAP, DCM.[31] **Si–OCOCF$_3$**: 1° ROH: ROH, Et$_3$N, DMAP, DCM.[31]	1° ROH: TBAF 2° ROH: TFA, DCM. 1° ROH: AcOH, MeOH, rt, 36 h[31] TFA, MeOH.[34]	Recycled using TFA then SOCl$_2$ or TFAA.[31, 34]
Copolymerization: polystyrene[31]	Si(iPr)(iPr)OR on polystyrene		
Suzuki coupling with 4-bromopolystyrene.[55]	Si(iPr)(iPr)OR on polystyrene	**Si–OTf**: 1° and 2° ROH: ROH, 2,6-lutidine, DCM.[38] **Si–Cl**: 1° ROH: ROH, 2,6-lutidine, DCM.[39] ROH, DMAP, DCM, DMF.[56]	1° and 2° ROH: HF·Pyr., TMSOMe, THF.[38, 57]	Commercially available in Si-4-PhOCH$_3$ form.[53, 55] Used for synthesis of: galanthamine analogues,[57] biaryls,[58] benzofurans[59] and the attachment of alcohols to glass slides for use in microarrays.[60]
Hydrosilylation of vinylpolystyrene.[17]	Si(iPr)(iPr)OR on polystyrene	**Si–Cl**: 1° ROH: ROH, Pyr., DCM.	1° ROH: HCl (aq), dioxane, 30 min. HF (aq), 30 min. TBAF>18 h.	Recycled using HF then BCl$_3$.
Preformed linker attached to AP–CPG silica[54, 55] or (aminomethyl)polystyrene.[55]	amide-Ph-Si(iPr)(iPr)OR on support	Preformed linker (AP–CPG silica).[54, 55]	2° ROH: 1 M TBAF, THF, rt.	Commercially available in Si–H form (polystyrene support).[55] Oligonucleotide synthesis.

(continued overleaf)

482 *Multifunctional Linker Units for Diversity-Oriented Synthesis*

Table 16.3 (continued)

Resin structure, solid support and synthesis	Loading conditions	Cleavage conditions	Comments
Preformed linker attached to HMP resin.[42, 44, 61]	Preformed linker.	1° ROH: TBAF, AcOH, THF.[61] CsF, AcOH.[42] CsF, 18-crown-6, AcOH, THF.[44]	Glycopeptide synthesis.
Preformed linker attached to Merrifield resin.[62]	Preformed linker.	1° ROH: TBAF, AcOH, THF 40°C, 14 h.	Polyketide library.
Addition of (CH₃)₂SiCl₂ or Ph₂SiCl₂ to modified Merrifield resin (R¹ = CH₃ or Ph).[24, 43]	R^1 = **Ph, Si–Cl**: 1° and 2° ROH: ROH, Et₃N, DMAP, DCM. R^1 = **CH₃, Si–Cl**: 1°, 2°, and 3° ROH: ROH, Et₃N, DMAP, DCM.	1°, 2° and 3° ROH: TBAF, THF, rt. 1°, 2°, and 3° ROH: KF, 18-crown-6, THF, rt. 1° ROH: 30% TFA in DCM. 2° ROH: HF·Pyr, rt.	R^1 = **Ph, Si–Cl**: Shelf stable for >1 year. R^1 = **CH₃**: Cleavage with fluoride generates volatile (CH₃)₂SiF₂ which is removed *in vacuo*.
Preformed linker added to CPG silica.[25]	Preformed linker.	2° ROH: 1 M TBAF, THF, rt.	Oligonucleotide synthesis.
Preformed linker added to CPG silica.[63]	Preformed linker via treatment of R₃Si–H with 1,3-dichloro-5,5-dimethylhydantoin to give R₃SiCl then ROH, imidazole, DCM, rt.	2° ROH: 0.2 M Et₃N· 3HF in THF, rt, 4 h.	Oligonucleotide synthesis.

Silicon and Germanium Linker Units 483

Table 16.4 Silyl linkers cleaved by fragmentation

Resin structure, solid support and synthesis	Loading conditions	Cleavage conditions	Comments
BMPSE linker ($R^1 = CH_3$).[29] Support: Polystyrene.	Loading onto alcohol resin: $R^1 = CH_3$: R^2COCl, Et_3N, DMAP, DCM.[29] $R^1 = Ph$: R^2CO_2H, DIPEA, HATU, HOBt, DMF.[9]	$R^1 = Me$: 20% TFA, DCM.[29] $R^1 = Ph$: 30% AcOH, MeOH.[9]	$R^1 = CH_3$: Synthesis of isoxazoline/isoxazole library. $R^1 = Ph$: Commercially available in $Si(CH_2)_2OH$ form.[55] Synthesis of tryprostatin B.[9]
Support: Aminomethyl polystyrene.[72]	Preformed linker with acetylated glycoside.	$BF_3 \cdot OEt_2$, Ac_2O, toluene, 23 h.	Carbohydrate synthesis. Anomeric centre ~15:1, β/α after cleavage.
DSEM[73] Support: Polystyrene.	Loading onto DSEM-Cl: X = OR, R^1R^2 NH, or RCO_2 DIPEA, DCE or DCM, rt. X = amide or nitrogen heterocycle BEMP, THF, rt.	X = OR, R^1R^2 NH, or RCO_2 5–50% TFA in DCM.	DSEM-Cl (X = Cl) stable at 0°C for 6 months. Resin-bound precursor commercially available.[54] Dimethyl analogue: unstable to acidic conditions. Di-isopropyl analogue: low loading.
	Preformed linker with Fmoc protected residue.	SAL:	Used for peptide synthesis.

(continued overleaf)

484 *Multifunctional Linker Units for Diversity-Oriented Synthesis*

Table 16.4 (continued)

Resin structure, solid support and synthesis	Loading conditions	Cleavage conditions	Comments
SAL (X = NHR)[66]		TFA, EDT, phenol, thioanisole (90:5:3:2).	SAL: Inert towards TBAF.
SAC (X = O₂CR)[69]		SAC:	Elimination assisted by β-effect at benzylic position under acidic conditions.[66]
Support: AMPS.		TBAF in DMF, THF, or DCM.	Self scavenging for TFA conjugate base.[66]
		BTMDF, DMF.	SAC: Suppresses diketopiperazine (C-terminal proline) formation.
		1% TFA, DCM.	TBAF cleavage causes succinimide formation with β-ester protected Asp's.
(structure with TMS, phenyl ether linker) PTMSEL[70, 71] Support: AMPS or NovaSyn TG resin.	Preformed linker with Fmoc protected residue.	TBAF · 3H₂O, DCM.	Used for glycopeptide[71] and lipopeptide[74] synthesis. Cleavage with almost neutral TBAF · 3H₂O, in DCM suppresses side reactions involving Asp residues.
(structure with Si-Ph, silyl linker) Support: AMPS[27]	Loading onto alcoholic resin: DIC, HOBt, DMAP.	TBAF · 3H₂O, DCM	Preformed linker which is loaded with substrate on resin.
(structure with TMS, Bn, XR) SEM Linker[75] Support: Aminomethyl polystyrene.	Loading onto SEM-Cl: X = O: ROH, DIPEA, DCM, rt. X = S: RSH, t-BuOK, t-BuOH, DMF, rt. Loading onto SEM-SePh:	XR = 1° and 2° OR: TBAF, TMU, rt and sonicate, or heating: conventional (100°C, 1 h) or microwave (100 W, 1 h). XR = 1° SR: TBAF, CsF, DMPU, sonicate.	SEM-Cl stable for prolonged periods. Steroid library synthesis.

Silicon and Germanium Linker Units

Structure	Conditions	Notes
Pbs[76, 77] Support: Aminomethyl polystyrene (AMPS).	X = O: ROH, NIS, TfOH, DCM, dioxane. Preformed linker with Fmoc protected residue.	TBAF, DMF, PhSH. TFA, DMS, PhSH, DCM. Used for peptide synthesis. thiophenol required to scavenge quinine methide generated by linker fragmentation.
Support: AMPS.[65, 78]	Preformed linker with Fmoc protected residue.	TBAF in DMF, MeCN or DCM, 5 min. Used for peptide synthesis. Linker labile in acidic conditions: stability can be improved using pyridine in place of the aryl ring.[79]
Support: Modified Merrifield or formyl polystyrene[80]	Loading onto alcoholic resin: Acids (X = alkyl, aryl): Amines (X = NHR): DIC, pyr., DMAP, HOBt, rt, 18 h. 1, CDI, pyr., DCM, rt, 1 h. 2, RNH₂, pyr., DMF, rt, 18 h. Alcohols (X = OR): 1, CDI, pyr., DCM, rt, 1 h. 2, DBU, DCM, rt, 15 min.	Acids (X = alkyl, aryl): TFA, DCM, rt. TBAF, DMF, rt. CsF, DMF, rt. Amines (X = NHR) CsF, DMF, rt. TBAF, DMF, rt. Alcohols (X = OR). TBAF, DMF, rt. Preformed linker which is loaded with substrate on resin.

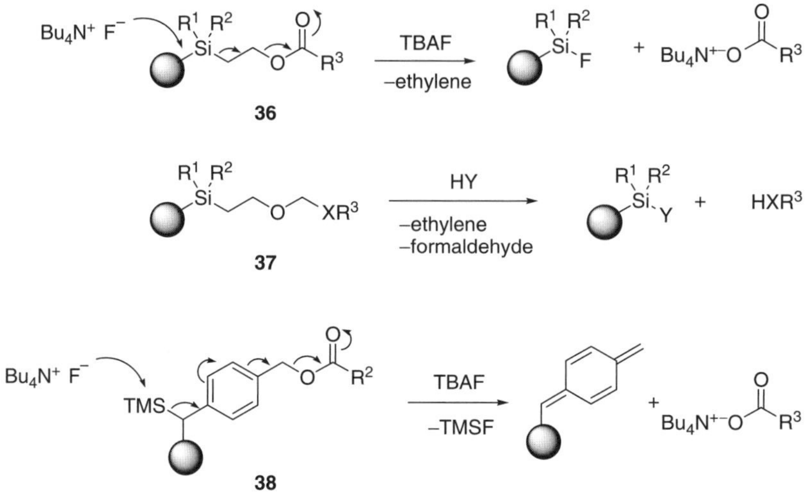

Scheme 16.8 Mechanisms of elimination for indirect silicon linkers

carbon–silicon bond more sensitive towards fluoridolysis, permitting cleavage with TBAF · 3H$_2$O in dichloromethane: this reagent/solvent system is considered to be almost neutral in character.[71] This reduction in basicity is sufficient to suppress the aspartyl residue reactions.[70]

The *O*-silyl carbamate-based triisopropylsilyloxycarbonyl (Tsoc) linker **39** can also be used to attach peptides via the *N*-terminus to the solid phase. Following SPPS the peptide is cleaved from the resin by treatment with hydrogen fluoride, which results in initial release of the *N*-terminal carbamic acid derivative of the peptide which spontaneously looses carbon dioxide to give the desired *N*-unprotected peptide **42** (Scheme 16.9).[8] This resin can be recycled by treatment with boron trichloride to regenerate the silyl chloride **39**.

The various fragmentation linkers developed to date are listed in Table 16.4. The table provides a concise overview of the practical use and reported applications of each linker. Only the loading and cleavage conditions specifically reported in the original literature are given.

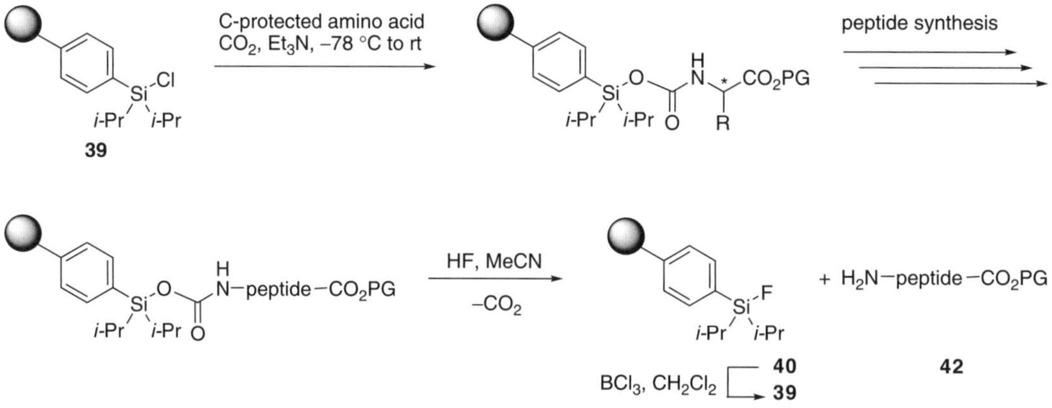

Scheme 16.9 Peptide synthesis and recycling of the Tsoc linker

16.2.5 Traceless/diversity silyl linkers

As indicated in the introduction, silyl linkers are also widely used to immobilize aromatic rings to the solid phase as arylsilanes. Cleavage can then be effected by *ipso*-substitution of the silyl group with an electrophile. Owing to their electropositive character, group 14 elements such as silicon and germanium exert a strong β-effect that selectively stabilizes the Wheland intermediate for this mode of *ipso*-substitution.

16.2.5.1 Loading the aromatic

Although the majority of reports using traceless/diversity arylsilane linkers have adopted a preformed linker approach, there are methods available for the attachment of aromatic library precursors directly to commercial resins such as PS–DES (Figure 16.1). The most direct method is palladium(0) mediated coupling of an aryl bromide **50** to the resin-bound silane **49** (Scheme 16.10).[81] This is advantageous because of the broad range of synthetically useful functional groups that are tolerated by this method. An alternate protocol involves the reaction of a silyl chloride resin **51** with an aryl- or heteroaryl lithium or Grignard species **52**.[13, 18, 82] Any chlorosilane groups remaining following arylation can be quenched by methanolysis.[82]

Typical Experimental Procedures

Loading aromatics using palladium coupling of aryl halides[81]

A mixture of the aryl bromide **50** *(4.43 mmol), PS–DES (8.0 g, 1.45 mmol/g), potassium acetate (KOAc) (1.28 g, 13.0 mmol) and NMP (50 ml) in a three-necked round bottom flask was degassed by sparging with argon. Pd$_2$(dba)$_3$·CHCl$_3$ 400 mg, 0.40 mmol) and P(o-tol)$_3$ (450 mg, 1.5 mmol) were added and then sparging with argon was continued for a further 15 minutes. The mixture was then stirred at 120–125 °C for 20 hours under an atmosphere of argon. The mixture was then allowed to cool to room temperature and filtered. The resin was washed with DMF (×2), 1 M hydrochloric acid in THF (×3), methanol (×3) and DCM (×3). The resin was then dried in vacuo.*

Loading aromatics using organolithium reagents[18]

A solution of the organolithium **52** *(5 equiv.) in THF (5 ml) was added to a suspension of PS–DES chloride* **51** *(~ 500 mg) in THF (5 ml) at − 78 °C. The reaction mixture was then allowed to warm to room*

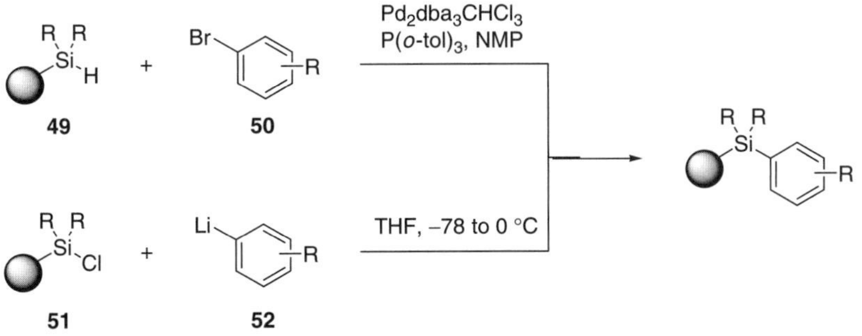

Scheme 16.10 Loading of aromatic substrates onto silyl linkers

488 Multifunctional Linker Units for Diversity-Oriented Synthesis

temperature in four hours. The mixture was then filtered and the resin washed with THF (3 × 7 ml), 1:1 THF/water (3 × 7 ml), THF (3 × 7 ml) and DCM (3 × 7 ml). The resin was then dried in vacuo for 24 hours.

16.2.5.2 Traceless cleavage for aromatics

Although the reactivity of the Si–C$_{aryl}$ bond towards strongly acidic or electrophilic conditions can be problematic in the context of silyl ether linkers,[34] it is central to the design of traceless linkers[83] for libraries containing aryl substituents **43**. Under acidic conditions, *ipso*-substitution of the Si–C$_{aryl}$ bond by a proton results in the introduction of a hydrogen atom at the former site of attachment of the linker to the substrate via a Wheland intermediate (Scheme 16.11). As indicated above, this mode of attack results from the well documented silicon β-effect.[84]

As expected from this mechanism, electron rich aromatics are more reactive towards *ipso*-substitution than electron poor substrates. Typically, electron rich aromatics can be cleaved with trifluoroacetic acid (either neat, as a vapor, or as a solution in dichloromethane),[18, 85] whereas electron poor substrates can be resistant to acidic cleavage, necessitating the use of hydrofluoric acid.[86] In some cases, silyl linkers are orthogonal to more acid sensitive protecting groups such as *N*-Boc,[87] permitting solid phase synthesis of phenylalanine containing cyclic peptides such as sansalvamide A[20, 81] via a different linker attachment site from standard linkers; namely, the 4-position of the aromatic ring of a phenylalanine residue.[22]

It is also often the case that the rate of substrate release from the resin can be in the order of hours. This can be used for the partial release of material by exposure to trifluoroacetic acid for a limited time period. This is useful in single-bead combinatorial screening, where a portion of the material is cleaved and assayed while the remaining material remains on the bead for subsequent characterization and identification of active compounds.[85]

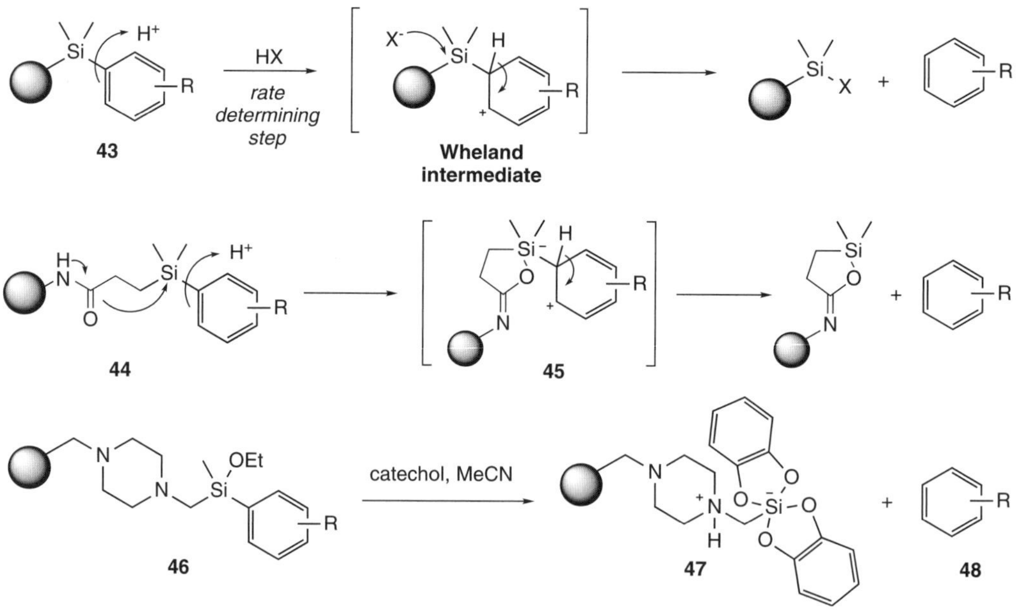

Scheme 16.11 Traceless cleavage of aromatics using silicon linkers

One approach to increasing the reactivity towards acidic cleavage is the introduction of β-amide into the linker design (**44**, Scheme 16.11).[85, 88–90] This has a pronounced influence on the cleavage rate of aromatic substrates.[85, 88] The observed reaction rate increase is attributed to intramolecular coordination of the amide carbonyl to the Lewis acidic silicon. From a mechanistic standpoint, this enhances the rate of the rate-determining *ipso*-protonation step by increasing the nucleophilicity of the *ipso*-position and stabilizing the resulting Wheland intermediate **45** (Scheme 16.11).[88]

Another design which facilitates the cleavage of aromatic substrates from the solid support is the silyl ether linker **46**: cleavage can be achieved using very mild conditions using catechol: this forms a resin bound zwitterionic λ^5 Si-silicate **47** with concomitant release of the desired aromatic compound **48** (Scheme 16.11).[25]

Silyl linkers can also be susceptible towards fluoridolysis with either tetrabutylammonium fluoride[18] or cesium fluoride.[26] This provides a method for the traceless cleavage of aromatics under basic conditions; however, it is limited to electron poor aromatic substrates. Increasing the electron density of aromatic substituents reduces the sensitivity of Si–C$_{aryl}$ bonds towards basic fluorodolysis.[91]

Typical Experimental Procedures
Traceless cleavage using TFA[18]

A mixture of resin loaded PS–DES (100 mg) in a solution of TFA and DCM (1:1, 3 ml) was stirred at room temperature for three hours. The resin was filtered and washed with DCM (3 × 2 ml). The combined filtrate was washed with saturated aqueous sodium bicarbonate, dried and concentrated in vacuo.

Traceless cleavage using HF[87]

Caution: *HF is extremely toxic. Its fumes are severely irritating and extremely destructive to the respiratory system. These reactions should only be performed with the proper equipment and training.*
Resin 53 (Table 16.5) was transferred to a perfluorinated plastic reaction vessel. Anhydrous HF gas was condensed into the reaction vessel and allowed to react (under some pressure since HF boils at 19.5 °C) for 12 hours. The HF was then removed with a flow of nitrogen through two sequential potassium hydroxide traps. The resin was washed with 4:1 DCM/methanol (5 × 2 min). The combined filtrates were concentrated in vacuo and further purified by chromatography.

Traceless cleavage using TBAF[21]

A mixture of loaded PS–DES (5 g) in a solution of TBAF in THF (1 M, 4 ml) was stirred gently for two hours. The mixture was then filtered and the resin washed with THF (4 × 2 ml), methanol (3 × 2 ml), THF (4 × 3 ml) and methanol (3 × 3 ml). The combined filtrate was concentrated in vacuo that was subjected to extractive work up with 0.1 N sodium hydroxide and DCM. The organic phases were combined, dried and concentrated in vacuo to give a residue that was then taken up in DCM (4 ml). This solution was extracted with 0.1 N sodium hydroxide (5 × 1 ml) and then concentrated in vacuo. Further purification was carried out by chromatography.

16.2.5.3 Cleavage with introduction of diversity for aromatics

In addition to their role in traceless cleavage strategies, arylsilanes can also undergo *ipso*-substitution with other electrophiles, making them useful as diversity linkers.[92] Although a broad range of electrophilic

490   Multifunctional Linker Units for Diversity-Oriented Synthesis

Table 16.6 Arylsilane linkers that allow traceless cleavage

Resin structure, solid support and synthesis	Loading conditions	Cleavage conditions	Comments
53 Support: (aminomethyl)polystyrene[86, 87]	Preformed linker	Traceless: HF	1,4-benzodiazepine library. 3-carbon analogue decomposes with HF to give silylated products.[87]
54 Support: polystyrene[21, 22, 99, 100] or (3-iodobenzamidomethyl)polystyrene.[20, 100]	Preformed linker[22, 100, 104] Si–Cl: ArLi, THF[21]	TBAF, THF[21] Traceless: TFA, DCM[22, 99, 100, 104] TFA, thioanisole, DCM, rt.[20] Diversity: X = Br: Br$_2$, DCM, rt.[20, 22, 99, 100] Diversity: X = I: ICl, DCM, rt.[22, 99]	Synthesis of 2-methoxyanilines.[104] 1,4-benzodiazepine library.[21] Peptide synthesis,[22, 99] e.g. sansalvamide A.[20]
Support: BTCore EM[105] TentaGel S NH$_2$ resin.[106]	Preformed linker.[105, 106]	Traceless: TFA, DCM, rt.[105, 106]	Indole library[106] Oxygen required in alkyl chain to improve solvation cf. carbon analogue.[105]

Silicon and Germanium Linker Units 491

Support/Linker	Loading	Cleavage	Notes
PS–DES resin.[81, 82]	Si–H: ArBr, P(o-tol)$_3$, Pd$_2$(dba)$_3$·CHCl$_3$, NMP.[81] Si–Cl: ArLi, THF;[13, 18, 82] residual SiCl consumed by methanolysis.[82] Preformed linker.	Traceless: Electron rich aryl: TFA, DCM.[13, 18, 81] Electron poor aryl: TBAF.[18]	Pyridine[82] and quinoline[13] synthesis. Cyclic peptides, e.g. sansalvamide A.[81]
Support: (hydroxymethyl)polystyrene[26, 102, 103]	Preformed linker.	Traceless: TBAF, THF[102] or DMF.[103] CsF, DMF, 60 °C[26]	Synthesis of benzopyran-4-ones[26] and oligothiophenes.[102] Poor swelling for basic amines used as solvent, e.g. piperidine or pyridine.[26] Sensitive to acid: cleavage with TFA gives silanols.[103] TBAF gives silanols with resin bound thiophenes[102] or benzopyrans.[26]
Support: isocyanatepolystyrene[107]		Traceless: TBAF, DMF, 60 °C.	

(continued overleaf)

Table 16.5 (continued)

Resin structure, solid support and synthesis	Loading conditions	Cleavage conditions	Comments
Support: Merrifield[101, 108] or Wang resin[19]	Preformed linker.	Traceless: neat TFA: electron rich aromatics.[108] CsF, DMF, H_2O, 100 °C: electron poor substrates.[108] HF.[108] Diversity: X = Br: Br_2, DCM.[19, 101]	Benzazepinone library synthesis.[101] H_2O and Me_2S scavengers required for TFA cleavage of furans.[108]
Support: Wang resin.[91]	Preformed linker.	Traceless: TFA, Me_2S, DCM. Diversity: X = Br: Br_2, pyr., DCM. X = I: ICl, DCM, rt.	Biphenyl library. Undesired cleavage of Wang resin when cleaving with TFA.
Support: Benzhydrylamine (BHA) resin,[85][85, 90] argoGel amine resin,[89] and (aminomethyl)polystyrene.[88]	Preformed linker.	Traceless: TFA vapor.[85] 30–50% TFA in DCM. Diversity: X = I: ICl, DCM, rt.[89]	Tamoxifen library.[89] Anchiomeric assistance from the amide increases reactivity towards acids.[88]

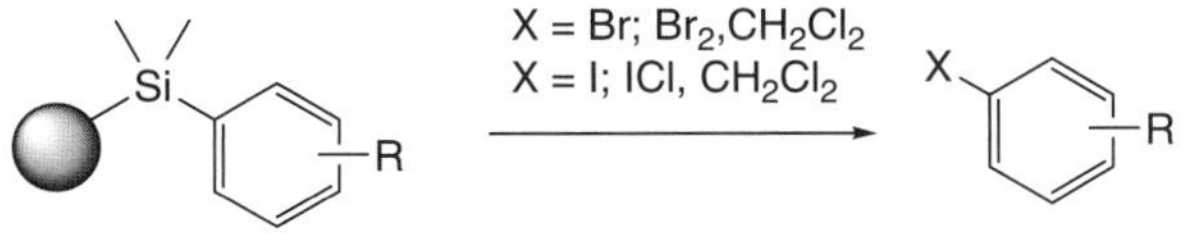

Scheme 16.12 Halogenation of arylsilane linkers

ipso-substitutions of arylsilanes are known in the solution phase,[93–98] only brominations[22, 91,99–102] and iodinations[22, 89, 91, 99] have successfully cleaved silyl linkers with the introduction of the corresponding halogen at the point of attachment (Scheme 16.12). Selective *ipso*-substitution during cleavage is strongly dependant on the electronic properties of the aromatic substrate. A lack of selectivity has been reported during the bromination of electron rich aromatics, resulting in the isolation of dibromoarenes.[91, 103]

Typical Experimental Procedures

Incorporation of bromine during cleavage[100]

Bromine (0.1 ml) was added to a suspension of resin **54** (Table 16.5) (~150 mg) in DCM and the mixture was stirred for 10 minutes at room temperature. The mixture was filtered and the filtrate was concentrated in vacuo to give the aryl bromide.

Incorporation of iodine during cleavage[22]

Iodine chloride (ICl) (30 mg) was added to a suspension of resin **54** (Table 16.5) (240 mg) in DCM (6 ml) at room temperature. After five minutes the mixture was filtered and the resin washed with DCM (3 ml). The combined filtrates were concentrated to give the aryl iodide.

16.2.5.4 Linker designs

The various arylsilane traceless linkers developed to date are listed in Table 16.5. The table provides a concise overview of the practical use and reported applications of each linker. Only the loading and cleavage conditions specifically reported in the original literature are given.

16.2.5.5 Diversity linkers for allyl groups

Although the majority of silicon traceless linkers are designed for aromatic substrates, the use of the allylsilyl linker **55** in combination with ruthenium-based metathesis methodology has broadened the scope of substrates to include terminal alkenes **57**[12] and 1,3-dienes **58**.[109] Alternatively, allylsilane **55** can be reacted with aldehydes in the presence of Lewis acids to produce homoallylic alcohols **59**,[110] homoallylic ethers **56**[12] or tetrahydropyrans **60** (Scheme 16.13).[111] Different oxygen heterocycles have also been synthesized using alternate resin-bound allylsilanes. Chiral allylsilane **61** can be used in the asymmetric synthesis of cycloheptenes **62**, whereas the cyclic allylsilyl ether **63** has been used as a cyclo-release linker to give tetrahydropyrans and tetrahydrofurans such as **64** via a ring closure mechanism (Scheme 16.13).[112, 113]

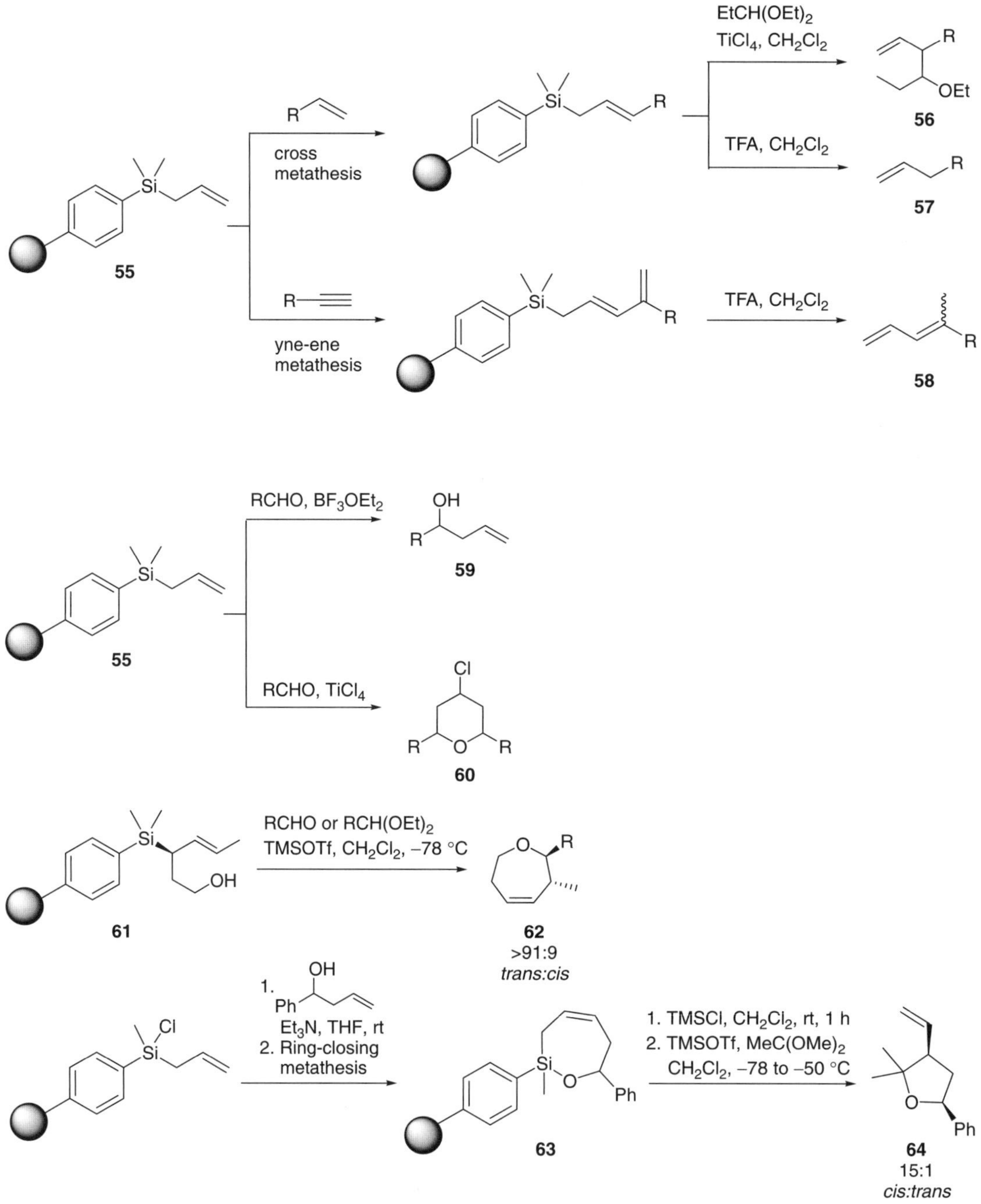

Scheme 16.13 Polymer-bound allylsilanes

16.3 Germanium-based linkers

Despite the popularity and widespread use of arylsilanes as traceless/diversity linkers for SPOS of aryl-containing libraries there are some significant limitations associated with their use. Firstly, electron deficient arylsilanes can be very reluctant to cleave by *ipso*-electrophilic desilylation and, therefore, require cleavage conditions that are sufficiently harsh to endanger other functionality present in the library. Secondly, arylsilanes in general and particularly electron deficient ones are labile to cleavage by strong nucleophiles.

The necessity for harsh electrophilic cleavage conditions reflects the relatively modest β-effect by which the Si–C$_{aryl}$ bond stabilizes the Wheland intermediate during cleavage (cf. Scheme 16.10). This also accounts for the relatively narrow repertoire of electrophiles, in addition to the proton, that can be used to generate diversity concomitant with cleavage. The first arylgermane linker to be developed, linker **66**, was introduced by Ellman in direct response to this limitation (Scheme 16.14).[87]

Arylgermanes display increased reactivity towards electrophiles due to the more pronounced β-effect exerted by germanium (cf. Scheme 16.10).[84] This increased reactivity of germanium towards cleavage by acid was demonstrated by the direct comparison of the germanium linker-containing system **66** with the analogous silicon linker-containing system **65**. Cleavage of electron poor benzodiazepines required hydrofluoric acid in the case of the silicon-based system **65**, whereas trifluoroacetic acid was a sufficiently strong acid to efficiently cleave the more reactive Ge–C$_{aryl}$ bond in **66** (Scheme 16.14). This greater reactivity towards electrophiles has also been exploited by Spivey to allow a greater variety of cleavage protocols for arylgermane linkers as compared to arylsilane linkers (*vide infra*).

Arylgermanes also display markedly greater stability towards basic conditions than arylsilanes due to what is known as the 'scandide contraction effect'.[114, 115] Thus, although germanium is below silicon in Group 14 of the Periodic Table it has only a marginally larger atomic radius and displays slightly greater electronegativity because its full shell of d-electrons are rather inefficient at shielding the nuclear charge. As a result, nucleophile attack at a germanium center is less favorable as compared to attack at a corresponding silicon center. This difference in behavior towards nucleophiles gives germanium-based linkers some unique advantages for certain applications and notably allows orthogonal cleavage of arylsilanes in the presence of arylgermanes using fluoride (*vide infra*).

It is worth noting that arylstannanes are significantly more reactive than arylgermanes towards electrophiles and have found use in a linker context for specialist applications, such as the preparation of radio-labeled aryl iodides for medical diagnosis by Positron Emission Tomography (PET).[116, 117] However, the extreme lability of the Sn–C$_{aryl}$ bond towards adventitious electrophiles during synthesis and also the greater lability of arylstannanes towards nucleophiles as compared to even arylsilanes precludes their use for most conventional applications (Chapter 17).

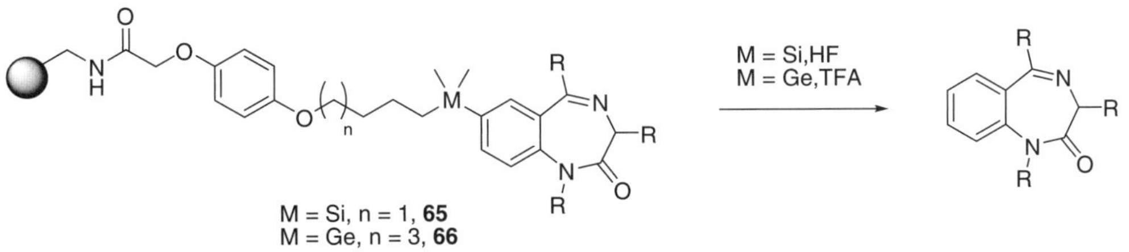

Scheme 16.14 *Comparison of germanium and silicon linkers*

16.3.1 The preparation of germyl resins

Ellman prepared resin **66** (Scheme 16.14) from commercially available aminomethylpolystyrene using a preformed handle strategy. The arylgermane component was prepared via a multi-step synthesis using commercially available but expensive dimethyldichlorogermane as the source of germanium (Scheme 16.15).

Mochida developed resin bound diethylgermanes **67** and **68** for use as a solid-phase hydride source for radical reactions. Two methods were used for germanium immobilization: addition of diethyldichlorogermane (Et_2GeCl_2) to lithiated PS–DVB resin (Method A, giving resin **67**) and incorporation of a germanium-derived monomer into a radical polymerisation reaction (Method B, giving resin **68**). The sources of germanium were the appropriate low molecular weight partially ethylated germyl chlorides (Scheme 16.16).[118]

Subsequent to these studies, Spivey et al. developed the synthesis of a trimethylgermyl resin **71** using a preformed handle method in which the less expensive germanium(IV) chloride ($GeCl_4$) was the source of germanium (Scheme 16.17).[119] Activation of this resin for aryl attachment was accomplished using tin(IV) chloride ($SnCl_4$) (*vide infra*). This method has also been used for the synthesis of a solid-supported dimethylgermane for use in radical synthesis.[120]

A more versatile variant on this approach was subsequently disclosed in which the germanium center was decorated with two 'spectator' groups and a *p*-anisyl residue, which could be readily activated by *ipso*-protodegermylation using hydrochloric acid or hydrobromic acid (*vide infra*) (Scheme 16.17).

This latter route offers an attractive method by which to tune the reactivity of the eventual Ge–C$_{aryl}$ bond linking the library aryl progenitor to the resin. In contrast to silicon, increasing the steric bulk of alkyl substituents around germanium does not strongly affect the cleavage rate by *ipso*-substitution by

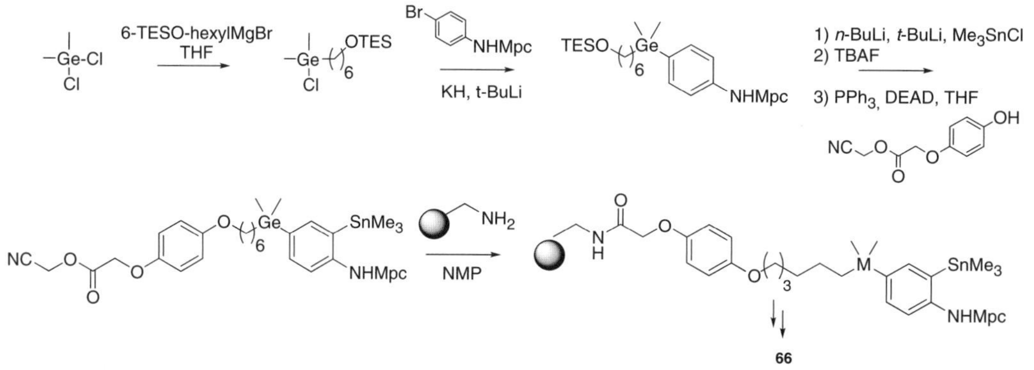

Scheme 16.15 *Ellman's arylgermane linker synthesis*

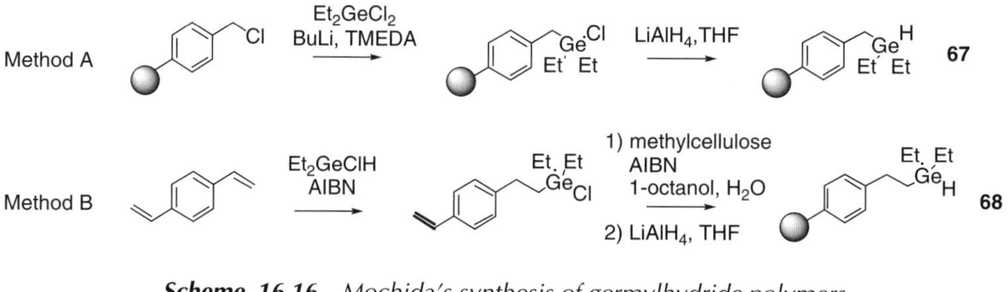

Scheme 16.16 *Mochida's synthesis of germylhydride polymers*

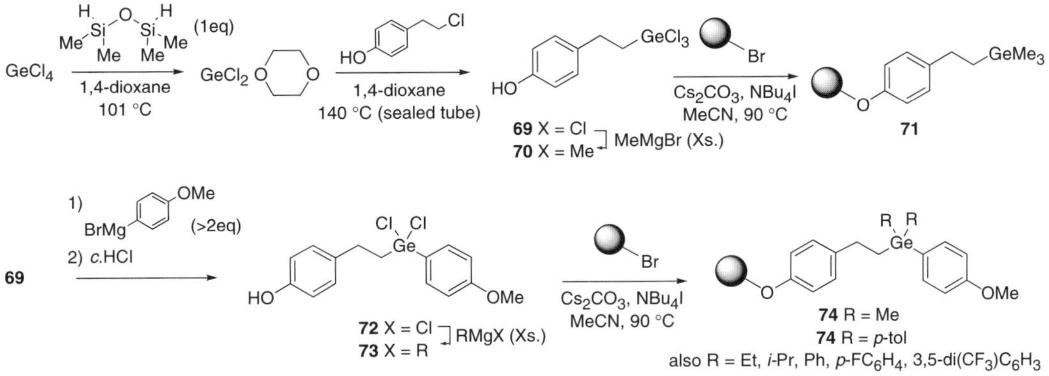

Scheme 16.17 *Spivey's organogermane linker synthesis*

acids. However, aryl spectator ligands such as *para*-tolyl groups (e.g. in **75**, Scheme 16.17) increase the stability of other more reactive aryl units such as anisole towards acidolysis.[114]

16.3.2 Activation of Ge–Methyl and Ge–Aryl resins for substrate attachment

Although germyl chlorides can be stored for prolonged periods under nitrogen in the dark, it is more convenient for long-term storage to use the above-described fully alkylated/arylated organogermanium linkers, which are stable to atmospheric conditions and can be conveniently activated immediately prior to use.[119]

Mono-chlorodemethylation of the trimethyl germane **71** requires the use of strongly Lewis acidic conditions and this does limit the choice of solid support compatible with this linker. Although it works smoothly with hydroxyethylpolystyrene (HEPS) and Quadragel resins, it is not compatible with ArgoGel resin. The activation of the germanium can be conveniently monitored by IR spectroscopy germanium–methyl bonds absorb around $600\,\text{cm}^{-1}$ and they are sufficiently sensitive to changes in the substitution patterns at germanium to be of diagnostic use (Scheme 16.18).[119]

Activation of the *p*-anisylgermane linkers, can be accomplished in a fashion closely analogous to that used for activating the corresponding arylsilane linkers. For example, resin **75** can be activated as either

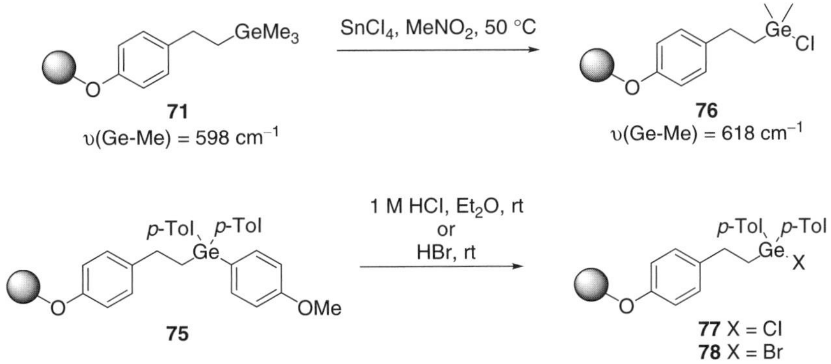

Scheme 16.18 *Activation of shelf stable germanium resins **71** and **75***

498 Multifunctional Linker Units for Diversity-Oriented Synthesis

the germyl chloride or germyl bromide by brief treatment with hydrochloric acid or hydrobromic acid, respectively. *p*-Anisole is sufficiently reactive that it can be selectively removed in the presence of the *p*-tolyl germanium substituents.[5] Although phenyl spectator groups can be used in place of *p*-tolyl groups, the methyl substituents of both anisole and toluene provide convenient diagnostic NMR signals which can be used to monitor these reaction by ^{1}H MAS-NMR (Scheme 16.18).[5]

Typical Experimental Procedures
Activation of trimethyl linker to give dimethylgermylchloride[119]

Tin(IV) chloride (153 µl, 1.30 mmol) was added drop-wise to a suspension of trimethylgermyl resin 71 (260 mg, 1.0 mmolg^{-1}, 0.26 mmol) in nitromethane (5 ml) at room temperature and then the reaction mixture heated at 50 °C for 20 hours. The resin was separated by filtration, washed with THF (3 × 30 ml), diethyl ether (3 × 30 ml), DCM (3 × 30 ml) and dried under high vacuum for 48 hours to give a free flowing dark brown resin 76 (230 mg, 0.8 mmolg^{-1}).

Activation of p-anisyl linker to give germyl chloride[5]

A solution of hydrochloric acid in diethyl ether (1 M, 65 ml, 65 mmol) was added to a suspension of resin 75 (6.8 mmol) in DCM (50 ml) at room temperature. The resulting mixture was stirred for 16 hours and then filtered to give resin 77 as brown granules.

16.3.3 Traceless/diversity germyl linkers

16.3.3.1 Loading aromatics

Once activated, aromatic library progenitors can be loaded onto germyl chloride/bromide linkers **77** and **78** as their Grignard or organolithium derivatives (Scheme 16.19).[5, 114, 119, 121] The efficiency of this process is determined by the choice of solid support: both Quadragel and HypoGel are suitable solid supports, but Merrifield resin sometimes results in low loading levels, presumably due to its poor swelling properties in ether or tetrahydrofuran.[5]

Typical Experimental Procedures
Loading aromatics using organometallic reagents[5]

A preformed degassed solution of the Grignard or organolithium reagent (2.2 mmol) in THF (4 ml) was added to a degassed suspension of the resin 77 (0.42 mmol) in THF (10 ml) at −50 °C. The mixture was stirred for one hour at −40 °C, warmed to room temperature and stirred for a further one hour. The reaction was quenched by addition of saturated aqueous ammonium chloride (50 ml). The mixture was filtered and the resin washed with DMF (3 × 50 ml), 1:1 THF/water (3 × 50 ml), THF (3 × 50 ml) and methanol (3 × 50 ml). The resin was then dried in vacuo overnight to give the resin 79 as yellow/orange granules.

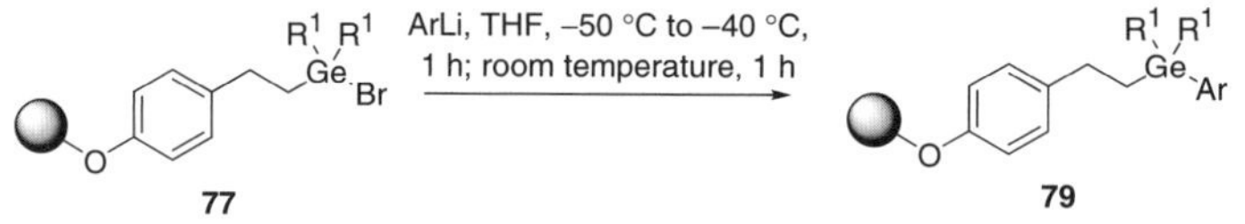

***Scheme 16.19** Loading of germanium linkers using organometallics*

16.3.3.2 Traceless cleavage of aromatics

The difference in reactivity between silicon and germanium towards electrophiles and nucleophiles has been used synthetically in the development of orthogonal protection strategies for dialkynes[122] and aromatics[28] in solution. This concept has been adapted to the solid phase using the germanium linker **80** to perform the iterative synthesis of conjugated oligomers (Scheme 16.20).[5, 114]

In this case, the silicon *tert*-butyldimethylsilyl (TBDMS) protecting group can be removed from resin **82** via basic fluoridolysis without affecting the germanium linker. The final resin-bound oligothiophenes **82** are sufficiently electron rich that both the germanium and silicon substituents are removed on exposure to trifluoroacetic acid.

Typical Experimental Procedures

Traceless cleavage using TFA[123]

*A suspension of resin **82** (27 mg) in freshly distilled TFA (2 ml) was stirred for 16 hours. The mixture was then filtered and washed extensively with DCM. The combined filtrates were concentrated in vacuo to give the target compound.*

16.3.3.3 Cleavage with introduction of diversity for aromatics

The greater propensity for germanium derivatives to participate in electrophilic substitution reactions as compared to their silicon congeners is important in the context of their development as diversity linkers. As observed previously with silicon linkers, the introduction of bromine[87, 124] and iodine[124] at the point of attachment is easily achieved during cleavage of aromatic substrates from germanium linkers **83** (Scheme 16.21). Moreover, the scope of this chemistry has been extended to include chlorine[123, 124] and fluorine substitution.[125] Significantly, conditions which are compatible with the radioisotope source [^{123}I]-sodium iodide and potentially [^{18}F]-fluorine or [^{18}F]-methoxymethyl fluoride are also compatible with the linker **83**. The late stage of isotope incorporation, combined with the rapid and simple isolation of radio-labeled pharmaceuticals after cleavage has great potential in the field of PET and Single Photon Emission Computed Tomography (SPECT) imaging (Scheme 16.21; see also Chapter 22).[125]

Although, as indicated above, tin linkers have also been used in this capacity (Chapter 17), the reduced stability of organostannes compared to their corresponding germanium analogues limits the scope of compatible chemistry for SPOS.[126, 127]

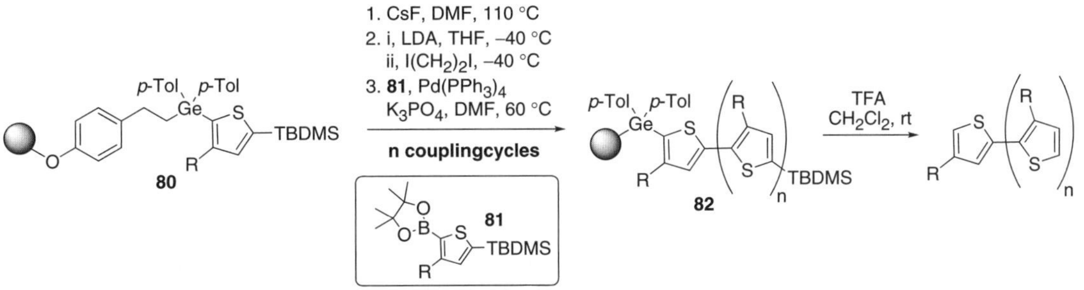

Scheme 16.20 *Solid-phase iterative synthesis of oligothiophenes using the orthogonality of silicon and germanium towards fluoridolysis*

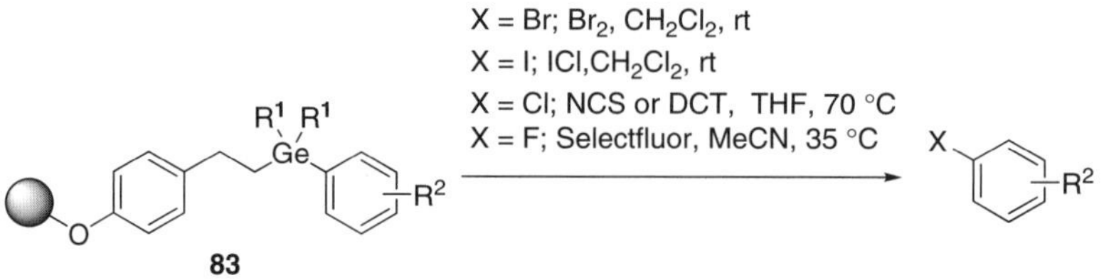

Scheme 16.21 Diversity cleavage using a germanium-based linkers

Typical Experimental Procedures
Incorporation of bromine during cleavage[123]

A solution of bromine (4 mg, 26 μmol) in DCM (0.25 ml) was added drop-wise to a suspension of resin **77** ($R^1 = CH_3$; R^2 = 4-anisole, 35 mg) in DCM (2 ml) at room temperature. The mixture was stirred for 40 minutes. The mixture was filtered and the resin washed extensively with DCM. The combined filtrates were concentrated in vacuo to give the aryl bromide.

Incorporation of iodine during cleavage[123]

A solution of iodine chloride (3 mg, 18 μmol) in DCM (1 ml) was added drop-wise to a suspension of resin **77** ($R^1 = CH_3$; R^2 = 4-anisole, 19 mg) in DCM (1 ml) at room temperature. The mixture was stirred for 40 minutes. The mixture was filtered and the resin washed extensively with DCM. The combined filtrates were washed with 1 M sodium thiosulfate ($Na_2S_2O_3$), dried (magnesium sulfate) and concentrated in vacuo to give the aryl iodide.

Incorporation of chlorine during cleavage[123]

NCS (11 mg, 79 μmol) was added as a single portion to a suspension of resin **77** ($R^1 = CH_3$; R^2 = 4-anisole, 37 mg) in THF (3 ml). The mixture was then heated at 70 °C for 16 hours. The mixture was filtered and the resin was washed extensively with DCM. The combined filtrates were concentrated in vacuo and the residue purified by passing through a plug of silica eluting with DCM to give the aryl chloride.

16.4 Conclusions

Just as silyl ether protecting groups occupy a central position in the repertoire of protecting groups for the temporary protection of alcohol functions during complex organic synthesis, silyl ether linkers have proved to be of great utility for the SPOS of libraries of molecules for property screening. Their utility arises from the straightforward and high yielding alcohol loading protocols that have been developed, the good stability profile of the resulting silicon–oxygen linkages during ensuing library elaboration and the range of facile and high yielding protocols available for release of the final library members from the solid support. The approach has been successful on a wide range of scales and to prepare a wide variety of structurally diverse compounds including heterocycles and oligosaccharides.

The commercial availability, at relatively low cost, of several of these systems has encouraged application-led uptake of the methodology. The use of silicon-based linkers for the attachment of several other functional groups has also been accomplished and their use for 'traceless' immobilisation of aromatic rings (i.e. as arylsilanes) has proved of special utility due in part to the prevalence of aromatic and heteroaromatic ring systems in compounds of interest as potential drugs, agrochemicals and new electronic materials. Traceless cleavage is achieved by electrophilic *ipso*-demetalation with acids, a mechanism which offers the opportunity also to demetalate with alternative electrophiles such that cleavage is accompanied by the introduction of a functional handle at the position of previous attachment to the resin. This has been achieved for arylsilanes with iodonium and bromonium ions to release aryl iodides and bromides respectively but has been exploited more extensively for arylgermane-based linkers.

Aryl germanium containing linkers, although more expensive than their silicon counterparts, display enhanced stability towards bases and nucleophiles and slightly enhanced lability towards electrophiles. Their deployment therefore allows for a wider variety of chemical procedures during library elaboration and then enables more facile cleavage of electron deficient aryl members within a library and also cleavage with a wider variety of electrophiles. The opportunities that this can afford continue to be developed; preliminary studies on solution phase models have demonstrated that cleavage to directly liberate radiolabelled [^{18}F]-fluoroaryls as PET imaging probes is possible[5] and that cleavage with concomitant carbon–carbon bond formation is possible using Friedel–Crafts acylation giving aryl ketones.[128] Moreover, model studies using solution-phase models and a fluorous tag have shown that suitably functionalized aryl germanes can be cleaved by treatment with BCl$_3$-propylene oxide followed by work-up with pinacol to give boronic esters,[5] and also by a germyl-Stille type cross-coupling process to allow biaryl formation.[129, 130] These latter processes in particular, significantly extend the range of diversification achievable upon library cleavage and should allow the methodology to be used in diverse range of applications. These developments are eagerly awaited.

References

[1] Corey, E. J., and Venkates. A; *J. Am. Chem. Soc*. **1972**, *94*, 6190.
[2] Frechet, J. M. J.; *Tetrahedron* **1981**, *37*, 663.
[3] Farrall, M. J., and Frechet, J. M. J.; *J. Org. Chem*. **1976**, *41*, 3877.
[4] Kocienski, P. J., *Protecting Groups*, 2nd edn, Georg Thieme Verlag, Stuttgart, Germany, **2000**.
[5] Turner, D. J., Anemian, R., Mackie, P. R., *et al.*; *Org. Biomol. Chem*. **2007**, *5*, 1752.
[6] Labadie, J. W.; *Curr. Opin. Chem. Bio*. **1998**, *2*, 346.
[7] Randolph, J. T., McClure, K. F., and Danishefsky, S. J.; *J. Am. Chem. Soc*. **1995**, *117*, 5712.
[8] Lipshutz, B. H., and Shin, Y. J.; *Tetrahedron Lett*. **2001**, *42*, 5629.
[9] Wang, B. B., Chen, L., and Kim, K. J.; *Tetrahedron Lett*. **2001**, *42*, 1463.
[10] Chan, T. H., Huang, W. O.; *J. Chem. Soc. Chem. Comm*. **1985**, 909.
[11] Fauvel, A., Deleuze, H., and Landais, Y.; *Eur. J. Org. Chem*. **2005**, 3900.
[12] Schuster, M., Lucas, N., and Blechert, S.; *Chem. Comm*. **1997**, 823.
[13] Hu, Y., Labadie, J. W., Porco, J. A., and Trost, B. M.; Compositions for organic synthesis on solid phase and methods of using the same, US Patent 2000/6147159, **1999**.
[14] Lindsley, C. W., Hodges, J. C., Filzen, G. F., *et al.*; *J. Comb. Chem*. **2000**, *2*, 550.
[15] Whitlock, H. W., and Maxson, K. K.; Silicon-containing solid support linker, US Patent 1999/5859277, **1999**.
[16] Darling, G. D., and Frechet, J. M. J.; *J. Org. Chem*. **1986**, *51*, 2270.
[17] Stranix, B. R., Liu, H. Q., and Darling, G. D.; *J. Org. Chem*. **1997**, *62*, 6183.
[18] Hu, Y. H., Porco, J. A., Labadie, J. W., *et al.*; *J. Org. Chem*. **1998**, *63*, 4518.
[19] Chenera, B., Elliott, J., Moore, M., and Weinstock, J.; Compounds and methods, US Patent 1998/5773512, **1998**.
[20] Lee, Y., and Silverman, R. B.; *Org. Lett*. **2000**, *2*, 3743.
[21] Woolard, F. X., Paetsch, J., Ellman, J. A.; *J. Org. Chem*. **1997**, *62*, 6102.
[22] Lee, Y., and Silverman, R. B.; *J. Am. Chem. Soc*. **1999**, *121*, 8407.

[23] Routledge, A., Wallis, M. P., Ross, K. C., and Fraser, W.; *Bioorg. Med. Chem. Lett.* **1995**, *5*, 2059.
[24] Meloni, M. M., White, P. D., Armour, D., and Brown, R. C. D.; *Tetrahedron* **2007**, *63*, 299.
[25] Tacke, R., Ulmer, B., Wagner, B., and Arlt, M.; *Organometallics* **2000**, *19*, 5297.
[26] Harikrishnan, L. S., and Showalter, H. D. H.; *Tetrahedron* **2000**, *56*, 515.
[27] Iyer, P., and Ghosh, S. K.; *Tetrahedron Lett.* **2002**, *43*, 9437.
[28] Spivey, A. C., Turner, D. J., Turner, M. L., and Yeates, S.; *Org. Lett.* **2002**, *4*, 1899.
[29] Alonso, C., Nantz, M. H., and Kurth, M. J.; *Tetrahedron Lett.* **2000**, *41*, 5617.
[30] Hirao, A., Hatayama, T., and Nakahama, S.; *Macromolecules* **1987**, *20*, 1505.
[31] Stover, H. D. H., Lu, P. Z., and Frechet, J. M. J.; *Polymer Bulletin* **1991**, *25*, 575.
[32] Stranix, B. R., Gao, J. P., Barghi, R., *et al.*; *J. Org. Chem.* **1997**, *62*, 8987.
[33] Porres, L., Deleuze, H., and Landais, Y.; *J. Chem. Soc., Perkin Trans. 1* **2002**, 2198.
[34] Frechet, J. M. J., Darling, G. D., Itsuno, S., *et al.*; *Pure & Appl. Chem.* **1988**, *60*, 353.
[35] Smith, E. M.; *Tetrahedron Lett.* **1999**, *40*, 3285.
[36] Hu, Y., and Porco, J. A.; *Tetrahedron Lett.* **1999**, *40*, 3289.
[37] Missio, A., Marchioro, C., Rossi, T., *et al.*; *Biotech. Bioeng.* **2000**, *71*, 38.
[38] Tallarico, J. A., Depew, K. M., Pelish, H. E., *et al.*; *J. Comb. Chem.* **2001**, *3*, 312.
[39] Liao, Y., Fathi, R., Reitman, M., *et al.*; *Tetrahedron Lett.* **2001**, *42*, 1815.
[40] Hu, Y. H., and Porco, J. A.; *Tetrahedron Lett.* **1998**, *39*, 2711.
[41] Tremblay, M. R., Wentworth, P., Lee, G. E., and Janda, K. D.; *J. Comb. Chem* **2000**, *2*, 698.
[42] Ishii, A., Hojo, H., Kobayashi, A., *et al.*; *Tetrahedron* **2000**, *56*, 6235.
[43] Meloni, M. M., Brown, R. C. D., White, P. D., and Armour, D.; *Tetrahedron Lett.* **2002**, *43*, 6023.
[44] Nakamura, K., Hanai, N., Kanno, M., *et al.*; *Tetrahedron Lett.* **1999**, *40*, 515.
[45] Commercially available from Biotage and Aldrich.
[46] Doi, T., Hijikuro, I., and Takahashi, T.; *J. Am. Chem. Soc.* **1999**, *121*, 6749.
[47] Maltais, R., Tremblay, M. R., and Poirier, D.; *J. Comb. Chem* **2000**, *2*, 604.
[48] Dragoli, D. R., Thompson, L. A., O'Brien, J., and Ellman, J. A.; *J. Comb. Chem.* **1999**, *1*, 534.
[49] Doi, T., Sugiki, M., Yamada, H., *et al.*; *Tetrahedron Lett.* **1999**, *40*, 2141.
[50] Thompson, L. A., Moore, F. L., Moon, Y. C., and Ellman, J. A.; *J. Org. Chem.* **1998**, *63*, 2066.
[51] Schlessinger, R. H., and Bergstrom, C. P.; *Tetrahedron Lett.* **1996**, *37*, 2133.
[52] Zheng, C., Seeberger, P. H., and Danishefsky, S. J.; *J. Org. Chem.* **1998**, *63*, 1126.
[53] Commercially available from Novabiochem.
[54] Seeberger, P. H., Beebe, X., Sukenick, G. D., *et al*; *Ang. Chem. Int. Ed.* **1997**, *36*, 491.
[55] Commercially available from Aldrich.
[56] Kubota, H., Lim, J., Depew, K. M., and Schreiber, S. L.; *Chem. & Biol.* **2002**, *9*, 265.
[57] Pelish, H. E., Westwood, N. J., Feng, Y., *et al.*; *J. Am. Chem. Soc.* **2001**, *123*, 6740.
[58] Spring, D. R., Krishnan, S., Blackwell, H. E., and Schreiber, S. L.; *J. Am. Chem. Soc.* **2002**, *124*, 1354.
[59] Liao, Y., Fathi, R., and Yang, Z.; *Org. Lett.* **2003**, *5*, 909.
[60] Hergenrother, P. J., Depew, K. M., and Schreiber, S. L.; *J. Am. Chem. Soc.* **2000**, *122*, 7849.
[61] Nakamura, K., Ishii, A., Ito, Y., and Nakahara, Y.; *Tetrahedron* **1999**, *55*, 11253.
[62] Reggelin, M., Brenig, V., vWelcker, R.; *Tetrahedron Lett.* **1998**, *39*, 4801.
[63] Ohkubo, A., Kasuya, R., Aoki, K., *et al.*; *Bioorg. Med. Chem.* **2008**, *16*, 5345.
[64] Komatsu, M., Okada, H., Akaki, T., *et al.*; *Org. Lett.* **2002**, *4*, 3505.
[65] Ramage, R., Barron, C. A., Bielecki, S., and Thomas, D. W.; *Tetrahedron Lett.* **1987**, *28*, 4105.
[66] Chao, H. G., Bernatowicz, M. S., and Matsueda, G. R.; *J. Org. Chem.* **1993**, *58*, 2640.
[67] Albericio, F., Kneib-Cordonier, N., Biancalana, S., *et al.*; *J. Org. Chem.* **1990**, *55*, 3730.
[68] Atherton, E., Cameron, L. R., and Sheppard, R. C.; *Tetrahedron* **1988**, *44*, 843.
[69] Chao, H. G., Bernatowicz, M. S., Reiss, P. D., *et al*,; *J. Am. Chem. Soc.* **1994**, *116*, 1746.
[70] Wagner, M., and Kunz, H.; *Ang. Chem. Int. Ed.* **2002**, *41*, 317.
[71] Wagner, M., Dziadek, S., and Kunz, H.; *Chem. Eur. J.* **2003**, *9*, 6018.
[72] Weigelt, D., and Magnusson, G.; *Tetrahedron Lett.* **1998**, *39*, 2839.
[73] Kim, K., and Wang, B. B.; *Chem. Comm.* **2001**, 2268.

[74] Lumbierres, M., Palomo, J. M., Kragol, G., and Waldmann, H.; *Tetrahedron Lett*. **2006**, *47*, 2671.
[75] Koot, W. J.; *J. Comb. Chem* **1999**, *1*, 467.
[76] Mullen, D. G., and Barany, G.; *Tetrahedron Lett*. **1987**, *28*, 491.
[77] Mullen, D. G., and Barany, G.; *J. Org. Chem*. **1988**, *53*, 5240.
[78] Ramage, R., Barron, C. A., Bielecki, S., *et al*.; *Tetrahedron* **1992**, *48*, 499.
[79] Ramage, R., Andrews, M. J. I., Raphy, J., and Wang, P.; *Tetrahedron Lett*. **2004**, *45*, 2403.
[80] Routledge, A., Stock, H. T., Flitsch, S. L., and Turner, N. J.; *Tetrahedron Lett*. **1997**, *38*, 8287.
[81] Liu, S. X., Feng, X., Geng, Y. L., *et al*.; *Chinese J. Org. Chem*. **2005**, *25*, 604.
[82] Pierrat, P., Gros, P. C., and Fort, Y.; *J. Comb. Chem*. **2005**, *7*, 879.
[83] Comely, A. C., Gibson, S. E.; *Ang. Chem. Int. Ed*. **2001**, *40*, 1012.
[84] Lambert, J. B., Zhao, Y., Emblidge, R. W., *et al*.; *Acc. Chem. Res*. **1999**, *32*, 183.
[85] Newlander, K. A., Chenera, B., Veber, D. F., *et al*.; *J. Org. Chem*. **1997**, *62*, 6726.
[86] Plunkett, M. J., and Ellman, J. A.; *J. Org. Chem*. **1995**, *60*, 6006.
[87] Plunkett, M. J., and Ellman, J. A.; *J. Org. Chem*. **1997**, *62*, 2885.
[88] Hone, N. D., Davies, S. G., Devereux, N. J., *et al*.; *Tetrahedron Lett*. **1998**, *39*, 897.
[89] Brown, S. D., and Armstrong, R. W.; *J. Org. Chem*. **1997**, *62*, 7076.
[90] Newlander, K. A., and Moore, M. L.; Silyl linker for solid phase organic synthesis of aryl-containing molecules, US 2000/6127489, **1997**.
[91] Han, Y. X., Walker, S. D., and Young, R. N.; *Tetrahedron Lett*. **1996**, *37*, 2703.
[92] Scott, P. J. H., and Steel, P. G.; *Eur. J. Org. Chem*. **2006**, 2251.
[93] Bennetau, B., Rajarison, F., Dunogues, J., and Babin, P.; *Tetrahedron* **1993**, *49*, 10843.
[94] Eaborn, C.; *J. Organomet. Chem*. **1975**, *100*, 43.
[95] Lothian, A. P., and Ramsden, C. A.; *Synlett* **1993**, 753.
[96] Mandal, S. K., and Sarkar, A.; *J. Org. Chem*. **1998**, *63*, 1901.
[97] Prouilhaccros, S., Babin, P., Bennetau, B., and Dunogues, J.; *Bull. Soc. Chim. Fr*. **1995**, *132*, 513.
[98] Wuts, P. G. M., and Wilson, K. E.; *Synthesis* **1998**, 1593.
[99] Lee, Y., and Silverman, R. B.; *Org. Lett*. **2000**, *2*, 303.
[100] Lee, Y., and Silverman, R. B.; *Tetrahedron* **2001**, *57*, 5339.
[101] Chenera, B., Elliott, J., Moore, M., and Weinstock, J., *Compounds and Methods*. **1995**.
[102] Briehn, C. A., Kirschbaum, T., vBauerle, P.; *J. Org. Chem*. **2000**, *65*, 352.
[103] Boehm, T. L., and Showalter, H. D. H.; *J. Org. Chem*. **1996**, *61*, 6498.
[104] Curtet, S., and Langlois, M.; *Tetrahedron Lett*. **1999**, *40*, 8563.
[105] Mun, H. S., and Jeong, J. H.; *Archives of Pharmacal Research* **2004**, *27*, 371.
[106] Mun, H. S., Ham, W. H., and Jeong, J. H.; *J. Comb. Chem*. **2005**, *7*, 130.
[107] Cereda, E., Pellegrini, C. M., Quai, M., and Barbaglia, W.; New polymers based on N-carbamyl-N'-dimethylsilyl methyl-piperazine traceless linkers for the solid phase synthesis of phenyl based libraries, US 2004/0186243, **2002**.
[108] Chenera, B., Finkelstein, J. A., and Veber, D. F.; *J. Am. Chem. Soc*. **1995**, *117*, 11999.
[109] Schuster, M., and Blechert, S.; *Tetrahedron Lett*. **1998**, *39*, 2295.
[110] Suginome, M., Iwanami, T., and Ito, Y.; *J. Am. Chem. Soc*. **2001**, *123*, 4356.
[111] Reginato, G., and Taddei, M.; *Il Farmaco* **2002**, *57*, 373.
[112] Cossy, J., Meyer, C., Bhatnagar, N., *et al*.; Synthesis of novel heterocyclic molecules oxygenated by metathesis reaction on solid support using novel silylated linkers, Patent WO 2002/051883, **2002**.
[113] Meyer, C., and Cossy, J.; *Tetrahedron Lett*. **1997**, *38*, 7861.
[114] Spivey, A. C., Turner, D. J., Turner, M. L., and Yeates, S.; *Synlett* **2004**, 111.
[115] Pyykko, P.; *Int. J. Quant. Chem*. **2001**, *85*, 18.
[116] Culbert, P. A., and Hunter, D. H.; *Reactive Polymers* **1993**, *19*, 247.
[117] Hunter, D. H., and Zhu, X.; *J. Labelled Compd. Radiopharm*. **1999**, *42*, 653.
[118] Mochida, K., Sugimoto, H., and Yokoyama, Y.; *Polyhedron* **1997**, *16*, 1767.
[119] Spivey, A. C., Srikaran, R., Diaper, C. M., and Turner, D. J.; *Org. & Biomol. Chem*. **2003**, *1*, 1638.
[120] Bowman, W. R., Krintel, S. L., and Schilling, M. B.; *Synlett* **2004**, 1215.

[121] Spivey, A. C., Turner, D. J., Cupertino, D. C., *et al.*; *Preparation of a conjugated molecule and materials for use therein*, Patent WO 2003/089499, **2003**.
[122] Ernst, A., and Vasella, A.; *Helv. Chim. Acta* **1996**, *79*, 1279.
[123] Spivey, A. C., Diaper, C. M., Adams, H., and Rudge, A. J.; *J. Org. Chem.* **2000**, *65*, 5253.
[124] Spivey, A. C., Diaper, C. M., and Rudge, A. J.; *Chem. Comm.* **1999**, 835.
[125] Spivey, A. C., Martin, L. J., Noban, C., *et al.*; *J. Labelled Compd. Radiopharm.* **2007**, *50*, 281.
[126] Hunter, D. H., and McRoberts, C.; *Organometallics* **1999**, *18*, 5577.
[127] Hunter, D. H., and Zhu, X. Z.; *J. Labelled Compd. Radiopharm.* **1999**, *42*, 653.
[128] Spivey, A. C., Gripton, C. J. G., Noban, C., and Parr, N. J.; *Synlett* **2005**, 2167.
[129] Spivey, A. C., Tseng, C. C., Hannah, J. P., *et al.*; *Chem. Comm.* **2007**, 2926.
[130] Spivey, A. C., Gripton, C. J. G., Hannah, J. P., *et al.*; *Appl. Organomet. Chem.* **2007**, *21*, 572.

17
Boron and Stannane Linker Units

Peter J. H. Scott

Department of Radiology, University of Michigan Medical School, USA

17.1 Introduction

The process of generating carbon–carbon bonds in organic synthesis was revolutionized by the discovery of palladium (and nickel) catalyzed cross-coupling reactions.[1] In the three decades following the introduction of, for example, the original Stille[2] and Suzuki–Miyaura[3] reactions, a sophisticated array of synthetic procedures have been developed in which an electrophilic component (for example: aryl halides, aryl and vinyl triflates, aryl diazonium species, aryl sulfonyl chlorides) is cross-coupled with a nucleophile (for example: formic acid, organoborons, organostannanes, organozincs, Grignard reagents, alkenes, alkynes, amines) in the presence of a palladium catalyst, as shown in Figure 17.1.

The enormous range of possible nucleophilic and electrophilic components that can be employed in these transition metal mediated cross-coupling reactions, many of which are now commercially available, has led to their extensive application in combinatorial library preparation and solid-phase synthesis and such applications have been recently reviewed.[4, 5] A number of solid-phase strategies have been developed, such as the simple application of these reactions to build up molecules on solid supports. For example, a polymer-bound electrophile, such as aryl halide **1**, may react with a solution-phase nucleophile, such as an organostannane **2**. The opposite approach is also viable, and resin-bound organostannanes **3** have been shown to be reactive towards solution-phase aryl halides **4** (Scheme 17.1).[6]

Other solid-phase techniques include the use of resin-bound palladium catalysts, which are attractive as they allow for simple removal of the catalyst following reaction.[7–12] The use of polymer-supported palladium facilitates not only purification of such cross-coupling reactions but also recycling of the catalyst. For example, Steel and Teasdale demonstrated that polymer-supported catalyst **5** could be recycled more than 14 times without any noticeable deterioration in the yield of simple Suzuki and Heck reactions (Scheme 17.2).[7]

Linker Strategies in Solid-Phase Organic Synthesis   Edited by Peter Scott
© 2009 John Wiley & Sons, Ltd

506 Multifunctional Linker Units for Diversity-Oriented Synthesis

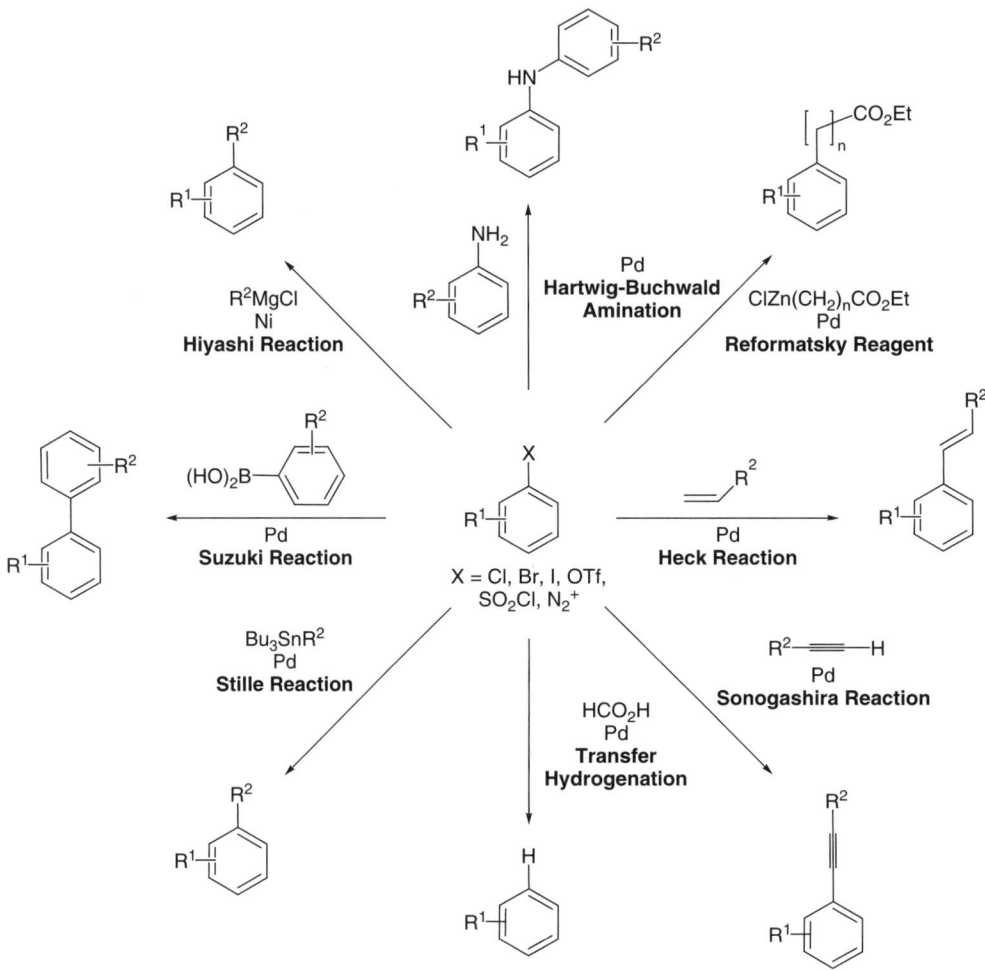

Figure 17.1 Palladium catalyzed cross-coupling reactions

Whilst the above examples demonstrate the simplest cases of solid-supported transition metal promoted cross-coupling reactions, more sophisticated examples where substrates for these reactions form an integral part of a linker unit have also been shown. A number of resin-bound electrophiles have been developed, such as the triazene linker units (Chapter 9), phosphonate linker units (Chapter 12) and triflate linker units (Chapter 13). Traceless cleavage from such linker units can be achieved using palladium mediated transfer hydrogenation, whilst multifunctional cleavage is possible by treatment with any of the large range of commercially available nucleophiles described above and shown in Figure 17.1.

The converse approach is also viable in which cleavage of substrates from resin-bound pro-nucleophilic linker units is realized by treatment with electrophiles. Such nucleophilic linker approaches have concentrated on the organostannane and organoboron linker units; these are discussed below in Sections 17.2 and 17.3, respectively. Cleavage from resin-bound pro-nucleophilic components occurs on treatment with electrophiles such as aryl halides or aryl and vinyl triflates in the presence of palladium. The ready availability of large numbers of commercial halides and triflates is an obvious advantage to using

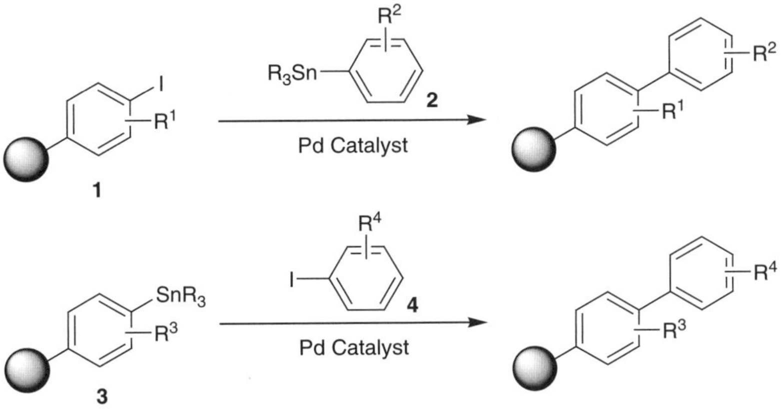

Scheme 17.1 Solid-phase synthesis using palladium

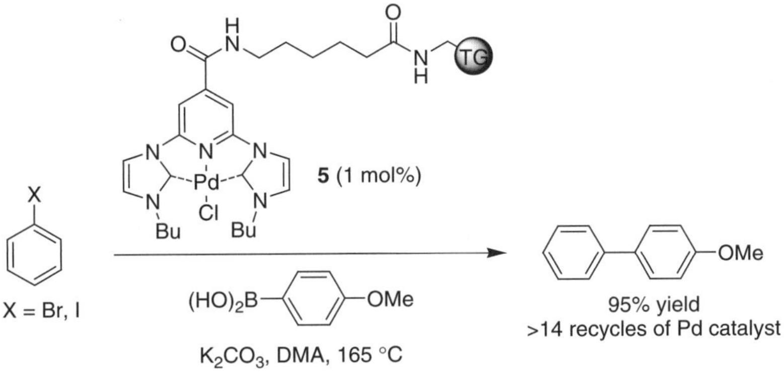

Scheme 17.2 Solid-supported palladium catalysts

organoboron and organostannane linker units, offering great scope for the simple introduction of diversity into molecules and associated compound libraries during the cleavage step in SPOS.

17.2 Organostannane linker units

17.2.1 Introduction

Stille first reported the cross-coupling reaction between organostannanes and aryl halides in 1978,[2] and it has since grown into one of the most important and exploited carbon–carbon bond forming reactions in the organic chemists arsenal.[13–20] Polymer-supported organotin compounds were introduced by Kuhn in 1994 and are particularly attractive reagents for synthetic chemists because following the Stille cleavage reaction all trialkyltin by-products remain attached to the polymer.[21] This allows for their simple and efficient removal by filtration and overcomes purification problems that have traditionally limited the use of organotin species in synthesis.

17.2.2 Organostannane linker units

Polymer-supported tin chloride or the corresponding tin hydride have both been used as organostannane linker units and are prepared as shown in Scheme 17.3. The tin chloride linker can be prepared from commercially available Merrifield resin (**6**).[21–26] Initial oxidation provides a resin-bound aldehyde, which can be converted to a polystyrene vinyl resin (**7**) using traditional Wittig chemistry. This vinyl resin is then reacted with di-*n*-butyltin chlorohydride (nBu_2SnHCl) (generated *in situ* from equal amounts of nBu_2SnCl_2 and nBu_2SnH_2) in the presence of catalytic 2,2-azobisisobutyronitrile (AIBN) to give polymer-supported tin chloride (**8**). Loading of the resin has been calculated to be 90% by mass gain of the polymer at this stage.[22] Reduction of polymer-supported tin chloride **8** with lithium borohydride provides the related tin hydride linker unit **9**.

Alkyne-containing substrates have been loaded onto the tin chloride linker unit using a Grignard reaction (Scheme 17.4) or onto the tin hydride linker via a hydrostannation reaction (mediated by AIBN or tetrakis(triphenylphosphine) palladium ($Pd(PPh_3)_4$)) (Scheme 17.5). In the former approach, the tin chloride resin is suspended in ether under an argon atmosphere, cooled to 0 °C and treated with the five equivalents of the Grignard reagent over one hour. The reaction is then warmed to room temperature and stirred for 15 hours, after which time filtration and washing provides resin-bound organostannanes.[21] For example, reaction of (2-phenylethynyl)magnesium chloride (**10**) with tin chloride resin **8** provided polymer-supported organostannane **11**.

In the latter case of the tin hydride linker, terminal alkynes were loaded via a hydrostannation reaction mediated by either a palladium catalyst (tin hydride resin should be added slowly to a mixture of the alkyne and palladium catalyst to prevent distannane formation) or AIBN. For example, the resin-bound tin hydride linker (**9**) was suspended in dry toluene under argon and treated with two equivalents of 1-ethynylbenzene (**12**) and 0.08 equivalents of AIBN. The mixture was stirred at 60 °C for 20 hours, adding an additional 0.08 equivalents of AIBN every 2.5 hours for the first 10 hours. After this time, filtration of the resin, washing and drying provided the resin-bound organostannane **13**. Multifunctional cleavage of substrates

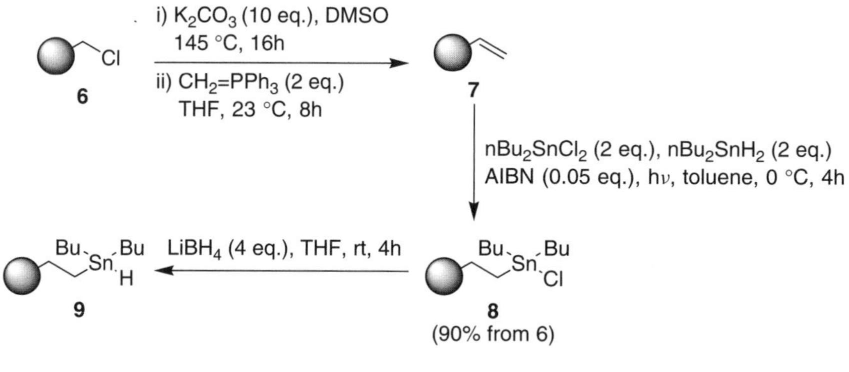

Scheme 17.3 Organostannane linker units

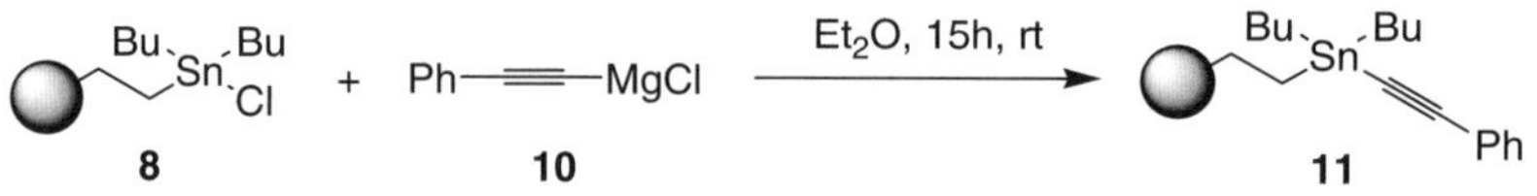

Scheme 17.4 Loading substrates onto stannane linkers

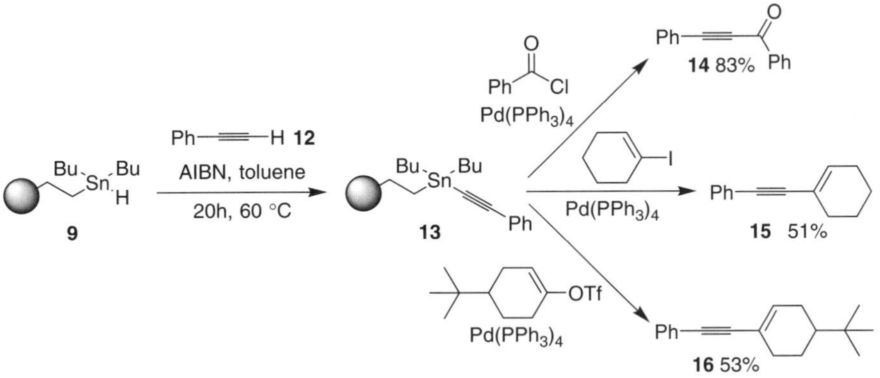

Scheme 17.5 Multifunctional cleavage strategies using organostannane linker units

can be achieved via simple treatment with iodine to cleave the corresponding iodides. This approach has found extensive application in solid-supported ^{123}I and ^{131}I-labelled radiopharmaceutical synthesis and is discussed in more detail in Chapter 22.[27–29] Other diversity can be introduced by cleavage using Stille cross-coupling reactions with acid chlorides, vinyl halides or vinyl triflates to provide a range of substituted alkynes (**14**–**16**) in moderate to high yields (Scheme 17.5).

Loading substrates onto the tin chloride linker described above is not confined to the limited availability of Grignard reagents. An alternative approach is to convert substrates to lithium reagents, which also react with polymer-supported tin chloride, and the potential of such an approach has been elegantly demonstrated by Nicolaou in his synthesis of the natural product (*S*)-zearalenone (**23**, Scheme 17.6).[22] Presumably, a range of other common nucleophilic substrates can also be loaded onto the tin chloride linker unit, but to date organostannane linkers have seen limited use and their full potential remains to be realized. Nicolaou loaded **17** and after deprotection and oxidation obtained exclusively the *E*-isomer of resin-bound vinyl aldehyde **18**. Interestingly, attempts to load the analogous acetylenic Weinreb amide onto the tin hydride linker unit and eliminate the need for deprotection and oxidation were hampered by the 50:50 mixture of *E* and *Z* alkenes resulting from the hydrostannation reaction. Addition of Grignard reagent **19** to resin-bound aldehyde **18** followed by Corey/Kim oxidation gave **20**. Deprotection of the silyl group and coupling with carboxylic acid **21** then provided the resin-bound precursor to the natural product **22**. Cleavage was then carried out by treatment with Pd(PPh$_3$)$_4$ to provide (*S*)-zearalenone **23**. The Stille cleavage reaction with the aryl iodide is an intramolecular process and so, in this case, the organostannane linker unit is bifunctional as it is also functioning as a cyclo-release linker unit (Chapter 4), but the opportunity for diversity cleavage also exists.

Typical Experimental Procedures

Preparation of polymer-supported tin chloride 8 (scheme 17.3)[22]

Commercially available Merrifield resin **6** *(1.0 equiv.) was oxidized to the corresponding aldehyde by suspending in dimethylsulfoxide(DMSO) and treating with potassium carbonate (10.0 equiv.). The mixture was heated at 145 °C for 15 hours after which time it was filtered and excess potassium carbonate was washed away. The resin was then suspended in tetrahydrofuran (THF) and treated with methylenetriphenylphosphorane for eight hours at room temperature, after which time it was filtered and washed to provide polystyrene vinyl resin* **7**. *Vinyl resin* **7** *was then suspended in toluene, cooled to 0 °C and treated with nBu$_2$SnCl$_2$*

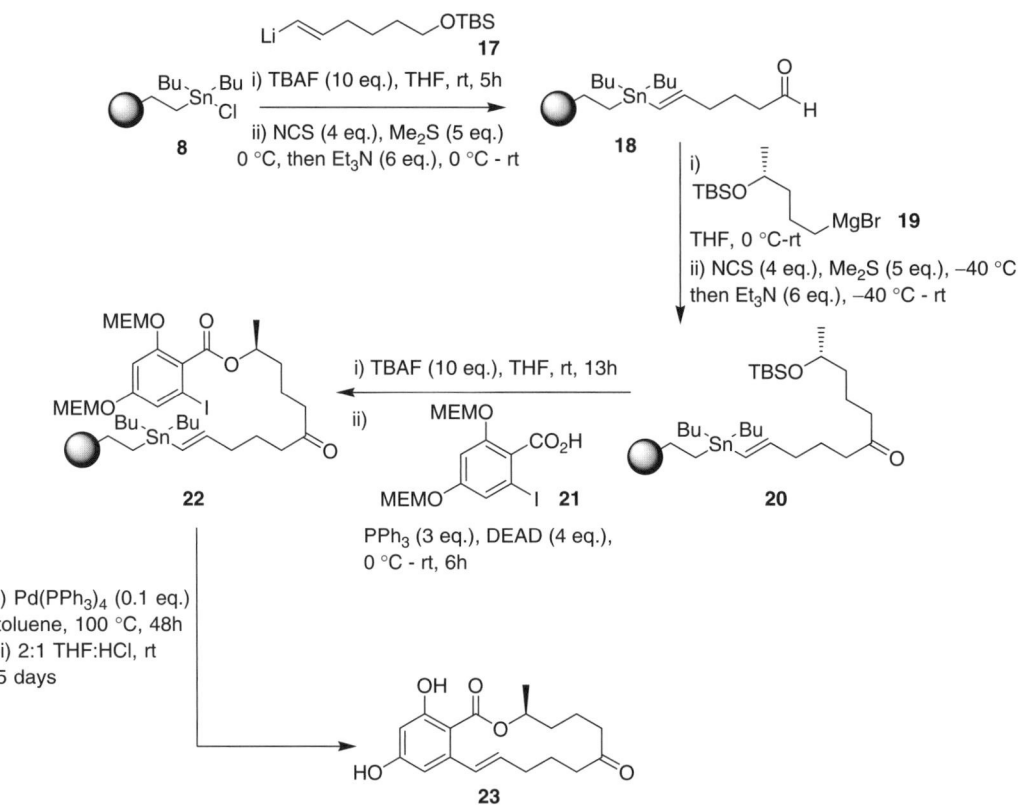

Scheme 17.6 Solid-phase synthesis of (s)-zearalenone

(2 equiv.), nBu₂SnH₂ (2 equiv.) and AIBN (0.05 equiv.). The reaction was stirred for four hours after which time it was filtered wand washed to provide resin-bound tin chloride (90% from 6).

Loading of substrates via a Grignard Reaction (Scheme 17.4)[21]

Dry tin chloride resin 8 was swollen under an argon atmosphere in dry ether at 0 °C for 0.5 hours. After this time, the Grignard reagent (5 equiv.) was slowly added in dry ether over one hour whilst maintaining slow stirring. After 15 hours at room temperature, the resin was filtered and washed with ether and water to provide resin-bound organostannane species.

Preparation of polymer-supported tin hydride 9 (scheme 17.3)[22]

Resin-bound tin chloride (8) prepared as described above was suspended in THF and treated with lithium borohydride (4.0 equiv.). The reaction was shaken for four hours at room temperature after which time the resin was filtered, washed and dried to provide resin-bound tin hydride 9.

Loading of alkynes via a hydrostannation reaction (scheme 17.5)[21]

The tin hydride resin 9 was suspended in dry toluene under argon and treated with alkyne (2 equiv.) and AIBN (0.08 equiv.). The mixture was heated to 60 °C and stirred slowly for 10 hours during which time additional AIBN was added to the reaction mixture every 2.5 hours (4 × 0.08 equiv.). The reaction was then stirred at 60 °C for an additional 10 hours after which time it was filtered, washed with dry toluene followed by ether and dried in vacuo to provide the resin-bound Stille reagent.

Typical cleavage procedure via a Stille reaction (scheme 17.5)[21]

Diphenyl-2-propyne-1-one 14

To phenylethyltin resin **13** (1.33 g, 1.5 mmol/g, 2.0 mmol of tin) in dry toluene (8 ml) under argon was added Pd(PPh$_3$)$_4$ (0.023 g, 0.02 mmol) and benzoyl chloride (0.28 g, 2.0 mmol). The mixture was slowly stirred and heated to 80 °C for one hour after which time the GC yield of diphenyl-2-propyne-1-one **14** was 83%.

17.3 Organoboron linker units

17.3.1 Introduction

Despite the advantages of using resin-bound organotin species, the toxicity of tin compounds creates a general reluctance to use Stille reactions when alternative methods are available. In particular, the availability of less toxic organoboronates has led to the Suzuki–Miyaura reaction becoming the method of choice for such transition metal catalyzed cross-coupling reactions. First reported in 1981, the cross-coupling reaction between an organoboronate and an aryl halide has the further advantage of water soluble by-products making purification operationally simpler than the analogous solution-phase Stille reactions.[3] Building on this initial work, extensive investigation has led to the development of many well established Suzuki reaction protocols which are beyond the scope of the present review. Readers are referred to the numerous excellent reviews of Suzuki reactions and the extensive references therein for further information.[13, 14, 30–39]

As with the Stille reaction described above, the Suzuki reaction has also evolved into a solid phase version and a number of immobilized organoboron-based linkers were reported by a number of groups in 1999.[40–42] For example, Burgess prepared boron linker **26** by first loading the desired boronic acid onto diol **24** in solution phase. Deprotection of the resulting ester provided acid **25** which was loaded onto Rink amide resin using standard peptide coupling techniques to give linker **26** (Scheme 17.7).

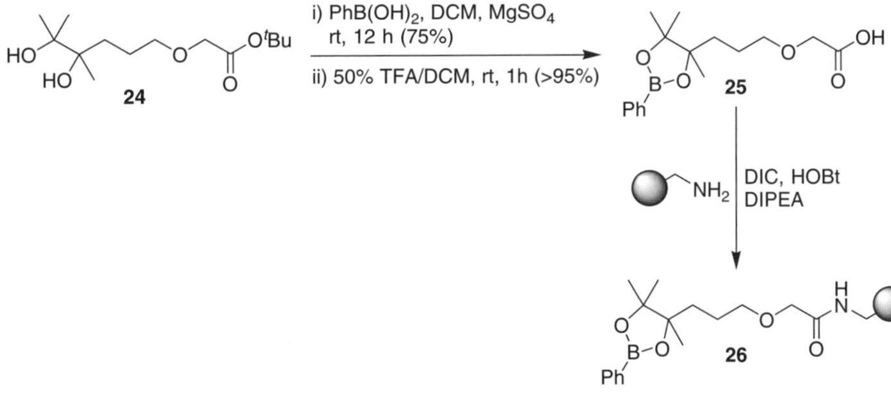

Scheme 17.7 Early boronate ester linker unit

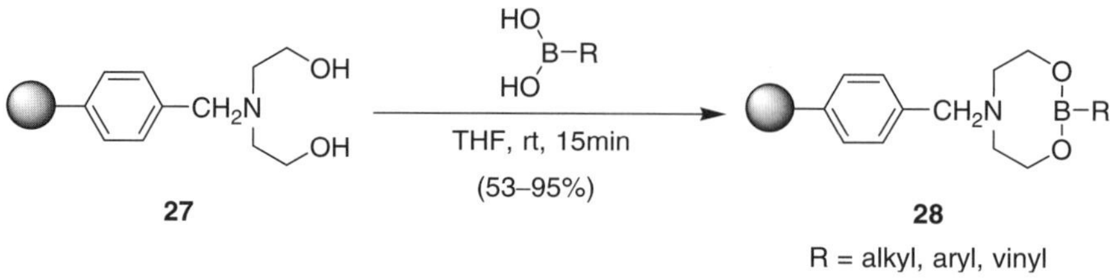

Scheme 17.8 Linker unit for boronic acids

A drawback of this approach is that the diversity resulting from the boronic acid is introduced early in solution phase rather than loading directly onto the resin. However, this can be avoided by loading boronic acids directly onto resin-bound diols, such as **27** as shown by both the Hall and Carboni groups (Scheme 17.8).[41, 42]

Typical Experimental Procedures

Loading of boronic acids onto resin-bound diols (scheme 17.8)[41, 42]

Resin bound diol (0.5 g, 1 mmol/g, 0.5 mmol) was suspended in THF (5 ml). The boronic acid (1.5 mmol) was added and the resulting suspension was stirred for 16 hours under reflux. After this time the reaction was cooled and the resin-bound boronic acid was filtered, washed successively with THF (3 × 5 ml), dichloromethane (DCM) (3 × 5 ml), diethyl ether (3 × 5 ml) and dried under vacuum at 60 °C for three hours.

17.3.2 Diversity cleavage through suzuki–miyaura reactions

Burgess and Carboni have both demonstrated cleavage of aromatic groups from organoboron linker units using Suzuki chemistry to afford biphenyls (Table 17.1).[40, 43] A number of palladium catalyst and base combinations have been used to achieve cleavage, the most effective of which appears to be potassium phosphate and either $PdCl_2$binap or $PdCl_2$dppf in DMF. These conditions provide the desired biphenyls in 45–75% yield. An advantage of this solid-phase approach to preparing biphenyls is that the pseudo-high-dilution conditions of solid-phase chemistry minimize the homo-coupling of boronic acids that is a commonly observed side reaction that reduces the efficiency of solution-phase Suzuki reactions.

Typical Experimental Procedures

Typical cleavage procedure via a suzuki reaction (table 17.1)[43]

The supported arylboronic acid (0.4 mmol) was suspended in DMF (8 ml). Aryl halide (2 mmol), $PdCl_2$ (dppf) (0.02 mmol) and a solution of potassium phosphate (2 M, 1.2 mmol) were added and the mixture was degassed and heated at 60 °C under argon for 24 hours. The reaction was then cooled to room temperature and the resin filtered and washed with DMF (2 × 5 mL), DCM (3 × 5 ml) and THF (3 × 5 ml). The filtrate was collected, partially concentrated, diluted with ethyl acetate (3 × 10 ml) and washed with water (10 ml). The aqueous layer was extracted with further ethyl acetate (3 × 10 ml) and the combined organic fractions were dried (magnesium sulfate), filtered and concentrated in vacuo. Purification by flash chromatography provided biphenyls.

Table 17.1 Multifunctional cleavage via Suzuki–Miyaura reactions

Entry	Resin	Aryl Iodide	Product	Ref.
1	Resin-CH2-boronate-Ph	4-iodoanisole	4-methoxybiphenyl, 67%[43] / 85% with PdCl$_2$binap[40]	[40, 43]
2	Resin-CH2-boronate-C6H4-3-NO2	iodobenzene	3-nitrobiphenyl, 75%	[43]
3	Resin-CH2-boronate-C6H4-3-NHCOEt	4-iodoanisole	4'-methoxy-3-NHCOEt-biphenyl, 75%	[43]
4	Resin-CH2-boronate-C6H4-4-CH2NHBn	4-iodoanisole	4'-methoxy-4-CH2NHBn-biphenyl, 53%	[43]
5	Resin-CH2-boronate-C6H4-4-C(O)N(Bn)CH(iPr)C(O)NHBn	4-iodoanisole	4'-methoxybiphenyl-4-C(O)N(Bn)CH(iPr)C(O)NHBn, 45%	[43]

17.3.3 Alternative cleavage strategies from organoboron linkers

The Suzuki reaction is large enough in scope to justify development of multifunctional organoboron linker units alone. However, given the increasing role of boronic acids in synthesis, it is not surprising that other cleavage strategies are also possible. Boronic acids have been shown to participate in a range of cross-coupling reactions such as rhodium(I) mediated addition to aldehydes,[44, 45] α,β-unsaturated carbonyls[46–50] and alkenes;[50, 51] palladium mediated reactions with sulfonium salts,[52, 53] thioesters[54–56] and thioalkynes;[57] nickel catalyzed reactions with allylic species;[58–60] copper catalyzed functionalisation of N–H and O–H bonds[61–76] and thiols;[77] and mercury(II) acetate/lead(IV) acetate couplings with active methylene species.[78] In addition, boronic acids have been shown to participate in multicomponent reactions. The Petasis reaction (a borono-Mannich variant between aldehyde, amine and boronic acid) allows for the generation of amines containing three points of diversity.[79, 80] Whilst cleavage from boron-based linkers using some of these newly developed reactions is highlighted below, many have yet to be adapted into SPOS cleavage protocols. Such strategies, particularly multicomponent reactions, would greatly enhance the role of boron linkers in library preparation and diversity oriented synthesis.

Traceless cleavage can be realized by treating resin-bound organoborons with Tollen's reagent (ammonical silver nitrate solution) and Carreaux and Carboni employed this strategy in the preparation of α-(acylamino) amides **29** as shown in Scheme 17.9.[81]

An alternative (and also traceless) cleavage strategy allows for detachment of boronic acids (or the corresponding boronote esters) from polymer supports as shown in Scheme 17.10. Such cleavage has been investigated by Carboni and Deluze[41] and in more detail by Hall using his N,N-diethanolaminomethyl polystyrene (DEAM–PS) support.[42, 82, 83] Treatment with water in THF cleaves the boronic acids **30**, whilst the analogous reaction with methanol in THF provides the corresponding methyl boronate esters **31**.

Alternative multifunctional cleavage strategies are also possible. Carboni has recently reported cleavage from organoboronate linker units by rhodium (I)-catalysed 1,2-addition to aldehydes to provide libraries

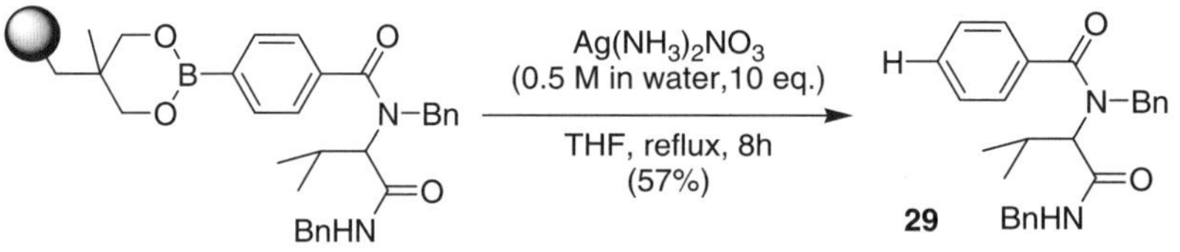

Scheme 17.9 *Traceless cleavage with Tollen's reagent*

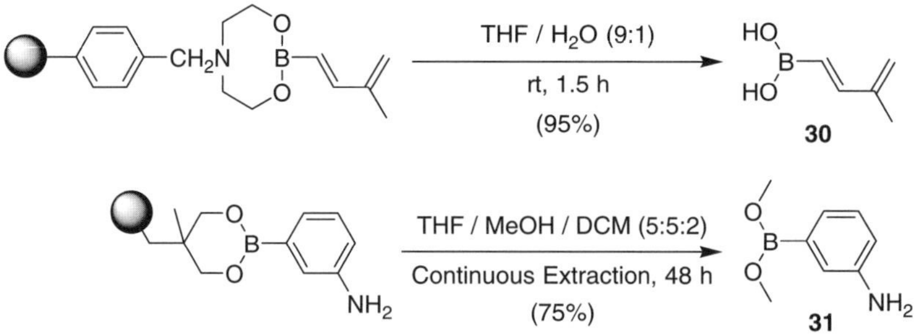

Scheme 17.10 *Cleavage of boronic acids and boronate esters*

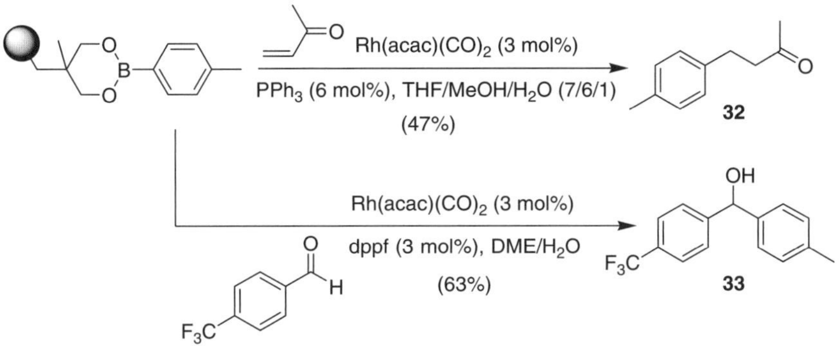

Scheme 17.11 Rhodium catalyzed multifunctional cleavage reactions

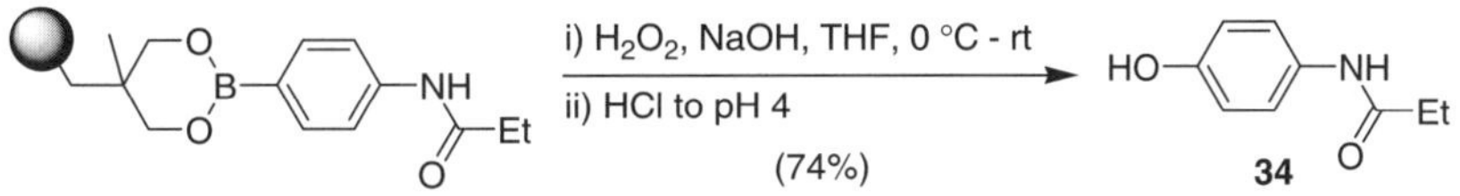

Scheme 17.12 Cleavage of phenols

of secondary alcohols **32** and 1,4-addition to α,β-usaturated ketones to give ketones **33** as alternative ways of introducing diversity into target molecules (Scheme 17.11). Given the large numbers of aldehydes and ketones that are commercially available (in addition to the extensive substrates for Suzuki reactions described above), the boronate multifunctional linkers should grow into powerful tools for diversity oriented synthesis in the future.

Finally, oxidative cleavage of aryl groups attached through boronate linkers as the corresponding phenols **34** is also possible (Scheme 17.12).[43] Carboni and Deluze demonstrated that cleavage using a mixture of hydrogen peroxide and sodium hydroxide, standard oxidation conditions compatible with many common functional groups,[84] gave phenols in reasonable yields (63–74%).

Typical Experimental Procedures

Traceless cleavage using aqueous silver diamine nitrate (scheme 17.9)[81]

To the resin bound boronic acid suspended in THF was added a solution of aqueous silver diamine nitrate $(Ag(NH_3)_2NO_3)$ (0.5 M, 10 equiv.). The reaction was then heated at 75 °C for eight hours after which time the resin was filtered. The filtrate was concentrated, redissolved in ethyl acetate and then washed with water. The layers were separated and the organic fraction was dried (magnesium sulfate) and evaporated to provide products (35–95%, >85 % purity). Note: after washing the resin (water, THF and diethyl ether), re-use was possible without apparent loss of activity.

Rhodium(I) mediated cleavage reactions (scheme 17.11)[43]
p-tolyl(4-(trifluoromethyl)phenyl)methanol 33

A suspension of 4-(trifluoromethyl)benzaldehyde (0.3 mmol), resin-bound boronic acid (0.3 mmol), [Rh(acac)(CO)₂] (0.009 mmol) and 1,1′-bis(diphenylphosphanyl)ferrocene (dppf) (0.009 mmol) in dimethoxyethane (DME)/water (5:1, 6 ml) was heated at 80 °C for 24 hours. After this time the resin was filtered and washed with DCM (3 × 5 ml) and THF (3 × 5 ml). The layers were separated and the aq. layer was extracted with EtOAc (3 × 5 ml). The combined organic fractions were dried (magnesium sulfate), filtered and concentrated. The resulting solid was re-dissolved in toluene and p-toluenesulfonic acid (TsOH) (0.01 equiv.) and diol resin (0.3 mmol, scavenger for unreacted aldehyde) were added. The mixture was heated at reflux for 24 hours, after which time the reaction was filtered (fritted glass funnel) and the resin was successively washed with toluene (10 ml), DCM (10 ml) and THF (10 ml). The combined organic fractions were concentrated and the resulting solid was extracted with DCM. Filtration through silica and concentration gave p-tolyl(4-(trifluoromethyl)phenyl)methanol (63% yield, >95% GC purity).

Oxidative cleavage of phenols (scheme 17.12)[41]
N-(4-Hydroxyphenyl)propionamide 34

To a suspension of resin-bound boronic acid (500 mg, 1 mmol/g, 0.5 mmol) in THF (5 ml) cooled to 0 °C was added 35% hydrogen peroxide (1.5 equiv.) and 3 M sodium hydroxide (1 equiv.). The reaction was maintained at 0 °C and stirred for one hour, after which time it was warmed to room temperature and stirred for an additional two hours. The suspension was treated with 1 M aqueous hydrochloric acid until pH 4, filtered and washed successively with THF (3 × 5 ml), DCM (3 × 5 ml) and diethyl ether (3 × 5 ml). The filtrate was partially evaporated and the product was extracted with ethyl acetate, dried (magnesium sulfate), filtered and concentrated to provide N-(4-hydroxyphenyl)propionamide (**34**, 74% yield).

17.4 Conclusion

The spectacular array of palladium mediated cross-coupling reactions that can be used to cleave substrates from the boron and stannane linker units makes them among the most powerful multifunctional linkers developed to date. The additional simplification of purification when compared to the analogous solution phase reactions further enhances their potential. In addition to model cleavage reactions, proof of concept has been demonstrated by Nicolaou who developed a solid-phase synthesis of the natural product (*S*)-zearalenone using a stannane linker unit.

References

[1] Hassan J, Sévignon M, Gozzi C, *et al.*; *Chem. Rev.* **2002**, *102*, 1359.
[2] Milstein, D., and Stille, J. K.; *J. Am. Chem. Soc.* **1978**, *100*, 3637.
[3] Miyauri, N., Yanagi, T., and Suzuki, A.; *Synth. Commun.* **1981**, *11*, 513.
[4] Bräse, S., Kirchoff, J. H., and Kobberlin, J.; *Tetrahedron* **2003**, *59*, 885.
[5] Todd, M. H., and Abell, C.; in *Solid-Phase Organic Synthesis* (Ed. K. Burgess), John Wiley and Sons, Inc., New York, **2000**, 25.
[6] Brody, M. S., and Finn, M. G.; *Tetrahedron Lett.* **1999**, *40*, 415.
[7] Steel, P. G., and Teasdale, C. W. T.; *Tetrahedron Lett.* **2004**, *45*, 8977.
[8] Cornils, B., and Hermann, W. A.; *J. Catal.* **2003**, *216*, 23.
[9] McNamara, C. A., Dixon, M. J., and Bradley M.; *Chem. Rev.* **2002**, *102*, 3275.
[10] Clapham, B., Reger, T. S., and Janda, K. D.; *Tetrahedron* **2001**, *57*, 4637.
[11] de Miguel, Y. R., Brule, E., and Margue, R. G.; *J. Chem. Soc., Perkin Trans. 1* **2001**, 3085.
[12] Herrmann, W. A., Ofele, K., Von Preysing, D., and Schneider, S. K.; *J. Organomet. Chem.* **2003**, *687*, 229.
[13] Lunxiang, Y., and Liebscher, J.; *Chem. Rev.* **2007**, *107*, 133.
[14] Stanforth, S.; *Tetrahedron* **1998**, *54*, 263.

[15] Mentzel, U. V., Tanner, D., and Tonder, J. E.; *Dansk Kemi* **2006**, *87*, 29.
[16] De Souza, M. V. N.; *Curr. Org. Synth*. **2006**, *3*, 313.
[17] Espinet, P., and Echavarren, A. M.; *Angew. Chem. Int. Ed*. **2004**, *43*, 4704.
[18] Kosugi, M., and Fugami, K.; in *Handbook of Organopalladium Chemistry for Organic Synthesis* (Ed. N. Ei-Ichi), John Wiley & Sons, Inc., Hoboken, **2002**, *1*, 263.
[19] Farina, V., Krishnamurthy, V., and Scott, W. J.; *The Stille Reaction*, John Wiley and Sons, Inc, New York, **1997**.
[20] Farina, V., and Roth, G. P.; *Adv. Met. Org. Chem*. **1996**, *5*, 1.
[21] Kuhn, H., and Neumann, W. P.; *Synlett* **1994**, 123.
[22] Nicolaou, K. C., Winssinger, N., Pastor, J., and Murphy, F.; *Angew. Chem. Int. Ed*. **1998**, *37*, 2534.
[23] Neumann, W. P., and Pedain, J.; *Tetrahedron Lett*. **1964**, *36*, 2461.
[24] Bokelmann, C., Neumann, W. P., and Peterseim, M.; *J. Chem. Soc., Perkin Trans. 1* **1992**, 3165.
[25] Neumann, W. P., and Peterseim, M.; *React. Polym*. **1993**, *20*, 189.
[26] Gerigk, U., Gerlach, M., Neumann, W. P., *et al*.; *Synthesis* **1990**,
[27] Hunter, D. H., and Zhu, X.; *J. Labelled Cpd. Radiopharm*. **1999**, *42*, 653.
[28] Kabalka, G. W., Namboodiri, V., Akula, M. R.; *J. Labelled Cpd. Radiopharm*. **2001**, *44*, 921.
[29] Culbert, P. A., and Hunter, D. H.; *React. Polym*. **1993**, *19*, 247.
[30] Suzuki, A.; *Handbook of Organopalladium Chemistry for Organic Synthesis* (Ed. N. Ei-Ichi), John Wiley & Sons, Inc., Hoboken, **2002**, *1*, 249.
[31] Miyauri, N., and Suzuki, A.; *Chem. Rev*. **1995**, *95*, 2457.
[32] Doucet, H.; *Eur. J. Org. Chem*. **2008**, 2013.
[33] Phan, N. T. S., Van Der Sluys, M., and Jones, C. W.; *Adv. Synth. Cat*. **2006**, *348*, 609.
[34] Nicolaou, K. C., Bulger, P. G., and Sarlah, D.; *Angew. Chem. Int. Ed*. **2005**, *44*, 4442.
[35] Bellina, F., Carpita, A., and Rossi, R.; *Synthesis* **2004**, 2419.
[36] Suzuki, A.; in *Modern Arene Chemistry* (Ed. D. Astruc), Wiley-VCH Verlag GmbH, Weinheim, Germany, **2002**, 53.
[37] Herrmann, W.; in *Applied Homogenous Catalysis with Organometallic Compounds*, 2nd edn (Eds. B. Cornils and W. A. Hermann), Wiley-VCH Verlag GmbH, Weinheim, Germany, **2002**, 591.
[38] Chemler, S. R., Trauner, D., and Danishefsky, S. J.; *Angew. Chem. Int. Ed*. **2001**, *40*, 4544.
[39] Suzuki, A.; *J. Organomet. Chem*. **1999**, *576*, 147.
[40] Li, W., and Burgess, K.; *Tetrahedron Lett*. **1999**, *40*, 6527.
[41] Carboni, B., Pourbaix, C., Carreaux, F., *et al*.; *Tetrahedron Lett*. **1999**, *40*, 7979.
[42] Hall, D. G., Tailor, J., and Gravel, M.; *Angew. Chem. Int. Ed*. **1999**, *38*, 3064.
[43] Pourbaix, C., Carreaux, F., and Carboni, B.; *Org. Lett*. **2001**, *3*, 803.
[44] Ueda, M., and Miyaura, N.; *J. Org. Chem*. **2000**, *65*, 4450.
[45] Sakai, M., Ueda, M., and Miyaura, N.; *Angew. Chem. Int. Ed*. **1998**, *37*, 3279.
[46] Sakuma, S., Sakai, M., Itooka, R., and Miyaura, N.; *J. Org. Chem*., **2000**, *65*, 5951.
[47] Sakai, M., Hayashi, H., and Miyaura, N.; *Organometallics* **1997**, *16*, 4229.
[48] Takaya, Y., Ogasawara, M., and Hayashi, H.; *Tetrahedron Lett*. **1998**, *39*, 8479.
[49] Takaya, Y., Ogasawara, M., and Hayashi, T.; *J. Am. Chem. Soc*. **1998**, *120*, 5579.
[50] Yoshida, K., and Hayashi, T.; in *Boronic Acids* (Ed. D. G. Hall), Wiley-VCH Verlag GmbH, Weinheim, Germany, **2005**, 171.
[51] Lautens, M., Roy, A., Fukuoka, K., and Martin-Matute, B.; *J. Am. Chem. Soc*. **2001**, *123*, 5358.
[52] Zhang, S., Marshall, D., and Liebeskind, L. S.; *J. Org. Chem*. **1999**, *64*, 2796.
[53] Srogl, J., Allred, G. D., and Liebeskind, L. S.; *J. Am. Chem. Soc*. **1997**, *119*, 12376.
[54] Liebeskind, L. S., and Srogl, J.; *J. Am. Chem. Soc*. **2000**, *122*, 12260.
[55] Zeysing, B., Gosch, C., and Terfort, A.; *Org. Lett*. **2000**, *2*, 1843.
[56] Savarin, C., Srogl, J., and Liebeskind, L. S.; *Org. Lett*. **2000**, *2*, 3229.
[57] Savarin, C., Srogl, J., and Liebeskind, L. S.; *Org. Lett*. **2001**, *3*, 91.
[58] Kobayashi, Y., and Ikeda, E.; *J. Chem. Soc., Chem. Commun*. **1994**, 1789.
[59] Mizojiri, R., and Kobayashi, Y.; *j. Chem. Soc., Perkin Trans. 1* **1995**, 2073.

[60] Trost, B. M., and Spagnol, M. D.; *J. Chem. Soc., Perkin Trans. 1* **1995**, 2083.
[61] Collman, J. P., Zhong, M., Zeng, L., and Costanzo, S.; *J. Org. Chem.*, **2001**, *66*, 1528.
[62] Chan, D. M. T., and Lam, P. Y. S.; in *Boronic Acids* (Ed. D. G. Hall), Wiley-VCH Verlag GmbH, Weinheim, Germany, **2005**, 205.
[63] Jung, M. E., and Lazarova, T. I.; *J. Org. Chem.* **1999**, *64*, 2976.
[64] Chan, D. M. T., Monaco, K. L., Wang, R. P., and Winters, M. P.; *Tetrahedron Lett.* **1998**, *39*, 2933.
[65] Evans, D. A., Katz, J. L., and West, T. R.; *Tetrahedron Lett.* **1998**, *39*, 2937.
[66] Lam, P. Y. S., Clark, C. G., Saubern, S., et al.; *Tetrahedron Lett.* **1998**, *39*, 2941.
[67] Cundy, D. J., and Forsyth, S. A.; *Tetrahedron Lett.* **1998**, *39*, 7979.
[68] Combs, A. P., Saubern, S., Rafalski, M., and Lam, P. Y. S.; *Tetrahedron Lett.* **1999**, *40*, 1623.
[69] Mederski, W. W. K. R., Lefort, M., Germann, M., and Kux, D.; *Tetrahedron* **1999**, *55*, 12757.
[70] Lam, P. Y. S., Vincent, G., Clark, C. G., et al.; *Tetrahedron Lett.* **2001**, *42*, 3415.
[71] Collot, V., Bovy, P. R., and Rault, S.; *Tetrahedron Lett.* **2000**, *41*, 9053.
[72] Collman, J. P., and Zhong, M.; *Org. Lett.* **2000**, *2*, 1233.
[73] Antilla, J. C., and Buchwald, S. L.; *Org. Lett.* **2001**, *3*, 2077.
[74] Lam, P. Y. S., Clark, C. G., Saubern, S., et al.; *Synlett* **2000**, 674.
[75] Petrassj, H. M., Sharpless, K. B., and Kelly, J. W.; *Org. Lett.* **2001**, *3*, 139.
[76] Desicco, C. P., Song, Y., and Evans, D. A.; *Org. Lett.* **2001**, *3*, 1029.
[77] Herradura, P. S., Pendola, K. A., and Guy, R. K.; *Org. Lett.* **2000**, *2*, 2019.
[78] Morgan, J., and Pinhey, J. T.; *J. Chem. Soc., Perkin Trans. 1* **1990**, 715.
[79] Petasis, N. A., and Akritopoulou, I.; *Tetrahedron Lett.* **1993**, *34*, 583.
[80] Petasis, N. A., and Zavialov, I. A.; *J. Am. Chem. Soc.* **1998**, *120*, 11798.
[81] Pourbaix, C., Carreaux, F., Carboni, B., and Deleuze, H.; *Chem. Commun.* **2000**, 1275.
[82] Gravel, M., and Hall, D. G.; *Spec. Chem. Mag.* **2003**, *23*, 31.
[83] Gravel, M., Thompson, K. A., Zak, M., et al.; *J. Org. Chem.*, **2002**, *67*, 3.
[84] Simon, J., Salzbrunn, S., Prakash, G. K. S., et al.; *J. Org. Chem.*, **2001**, *66*, 633.

18
Bismuth Linker Units

Peter J. H. Scott

Department of Radiology, University of Michigan Medical School, USA

18.1 Introduction

The reactivity of triaryl bismuth(III) species is a fairly new area of discovery in organic chemistry and, consequently, their use as synthetic tools to date has been somewhat limited, although applications (particularly in arylation reactions) have been the subject of several major review articles.[1–5] This range of possible synthetic applications using triaryl bismuthanes makes them ideally suited as potential linker scaffolds. Unlike the related palladium mediated cross-coupling reactions, which frequently require development of very specific reaction conditions for each individual substrate, reactions involving organobismuth species appear far more versatile. This potential has been recognized and exploited by Ruhland who has developed the only two bismuth-based linker units reported to date.[6, 7]

18.2 Bismuth linker units

The first linker was a triaryl bismuth linker **5** whilst the second was a triaryl bismuth diacetate species **6**. Preparation of resin-bound bismuthanes is shown in Scheme 18.1. Initially, 4-iodophenol **1** was coupled to commercially available chloromethyl polystyrene (loading 2.38 mmol/g, cross-linked with 1–2% DVB, mesh 200). The resulting aryliodide **2** was converted to the corresponding resin-bound Grignard reagent **3** using isopropylmagnesium bromide. The choice of bismuth electrophile to react with the Grignard is critical for successful preparation of the linker unit. A number of possibilites were considered including bismuth(III) chloride ($BiCl_3$), diarylbismuth chloride (Ar_2BiCl) and diarylbismuth triflate

Linker Strategies in Solid-Phase Organic Synthesis Edited by Peter Scott
© 2009 John Wiley & Sons, Ltd

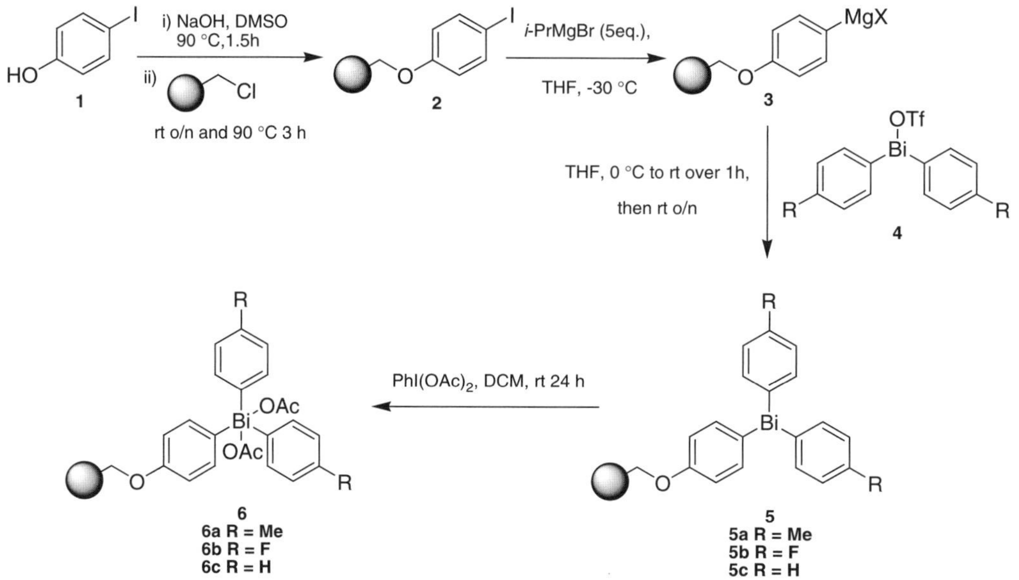

Scheme 18.1 Preparation of triarylbismuthane and triarylbismuthane diacetate linker units

(Ar$_2$BiOTf). Bismuth(III) chloride was abandoned because of the possibility of the intermediate resin-bound dichloroarylbismuth undergoing reaction with additional equivalents of Grignard reagents. Furthermore, it is known that scrambling of R-groups via ligand–ligand exchange can occur and this would lead to loss of the bismuth from the resin.[7] Similarly, Ar$_2$BiCl is problematic as it can undergo dismutation reactions in solution. Therefore, Ruhland settled upon reaction of Ar$_2$BiOTf. 2 HMPA complexes **4** (prepared by treatment of the analogous triaryl bismuth with trimethylsilyl trifluoromethanesulfonate (TMSOTf) (1 equiv.) and hexamethylphosphoric triamide (HMPA) (2 equiv.)) with Grignard reagents to generate solid-supported bismuthanes. These bismuth complexes are readily accessible, air stable and soluble in most organic solvents.

It is important to note that the resin-bound triarylbismuth species obtained are unsymmetrical. Previous reports have shown that the most electron-deficient aryl groups are transferred in arylation reactions using unsymmetrically substituted triaryl bismuthanes, and so the phenoxy spacer unit incorporated by loading 4-iodophenol onto the resin is essential to ensure selective reaction for a single arylbismuth group.[7, 8] Loading of the bismuth species was monitored by high-resolution magic angle spinning (HR MAS) ^{1}H NMR. For example, generation of linker unit **5a** was confirmed by the presence of an Ar–CH_3 signal at 2.15 ppm in the HR MAS ^{1}H NMR spectrum. These parent linker units were converted into the corresponding triaryl bismuth(V) diacetate linkers by simple treatment with diacetoxy iodobenzene (Scheme 18.1). Quantitative oxidation was confirmed by the appearance of an acetoxy CH_3CO$_2$–signal at 1.60 ppm in the HR MAS ^{1}H NMR spectrum.

A range of cleavage strategies from these resin-bound bismuthanes have been explored. Cleavage from the triaryl bismuth linkers (**5**) is illustrated in Scheme 18.2. Traceless cleavage of substrates was realized by treatment with 50% trifluoroacetic acid (TFA) in dichloromethane (DCM) to provide substituted benzenes (**7**). Multifunctional cleavage has also been demonstrated by treating with imides, carbamate or amides in the presence of 1.5 equivalents of copper(II) acetate and a base (pyridine or triethylamine) to provide a

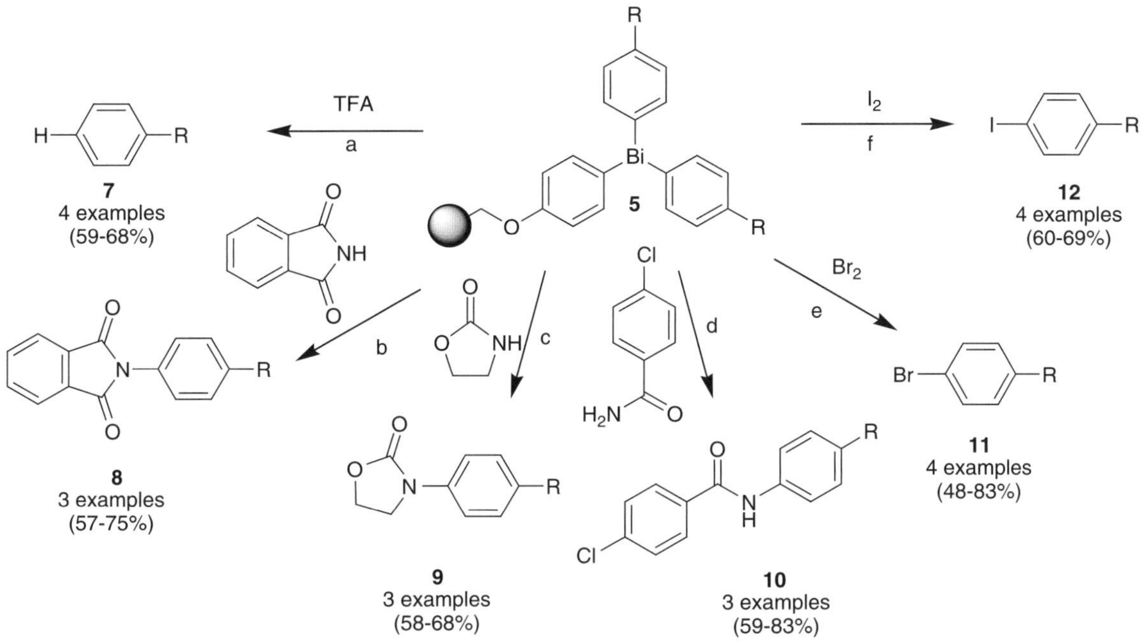

Scheme 18.2 *Multifunctional cleavage from triarylbismuthane linker units (a) TFA/DCM (1:10, rt, 1 h; (b) Phthalimide, Cu(OAc)$_2$ (1.5 equiv.), pyridine, DCM, 40 °C, 48 h; (c) 2-oxazolidone, Cu(OAc)$_2$ (1.5 equiv.), Et$_3$N, DCM, 40 °C, 48 h; (d) 4-chlorobenzamide, Cu(OAc)$_2$ (1.5 equiv.), pyridine, DCM, 40 °C, 48 h; (e) Br$_2$ (2 equiv.), DCE, 60 °C, 12 h; (f) I$_2$ (2 equiv.), THF, 60 °C, 12 h*

variety of substituted N-phenylphthalimides (**8**), N-phenyl-oxazolidin-2-ones (**9**) and N-phenylbenzamides (**10**), respectively. The base is essential to prevent unwanted protonation, which would result in cleavage of bismuth from the resin. N-arylated products were obtained in 57–83% yield following chromatography. These yields are lower than the analogous solution-phase reactions and this was suspected to be due to the poor solubility of copper(II) acetate in DCM or the fact that solution phase reactions were carried out using symmetrical triaryl bismuth species. Therefore, more soluble sources of copper (copper(II) pivaloate and copper(II) triflate) were investigated as a solution to the problem but actually resulted in diminished yields, suggesting that the latter hypothesis is correct. In further work, Ruhland reported halogen based cleavage strategies.[6] Treatment of polymer-supported triaryl bismuthanes with bromine or iodine at 60 °C gave bromo- (**11**) and iodo-substituted (**12**) biphenyls respectively.

The main benefit of using the triarylbismuth diacetate linker **6** over its unoxidized counterpart is that the enhanced reactivity of the diacetate derivative allows for cleavage to be achieved using 10% copper(II). The greatly reduced levels of copper by-products requiring removal during work-up simplifies purification. Cleavage from triarylbismuthane diacetate linkers, has been demonstrated using a range of nucleophiles, resulting in the formation of carbon–nitrogen (**13-16**), carbon–oxygen (**17**) and carbon–carbon (**18**) bonds in target molecules (Scheme 18.3). For example, cleavage in the presence of anisole gave the corresponding N-phenylanisole (**16**) whilst cleavage with phenols resulted in biphenyl ethers (**17**). Interestingly, in the absence of copper, *ortho*-arylation is the preferred cleavage pathway. Treatment with β-naphthol and 2-tert-butyl-1,1,3,3-tetramethylguanidine (TMG) as base gave the C-arylated derivative (**18**).

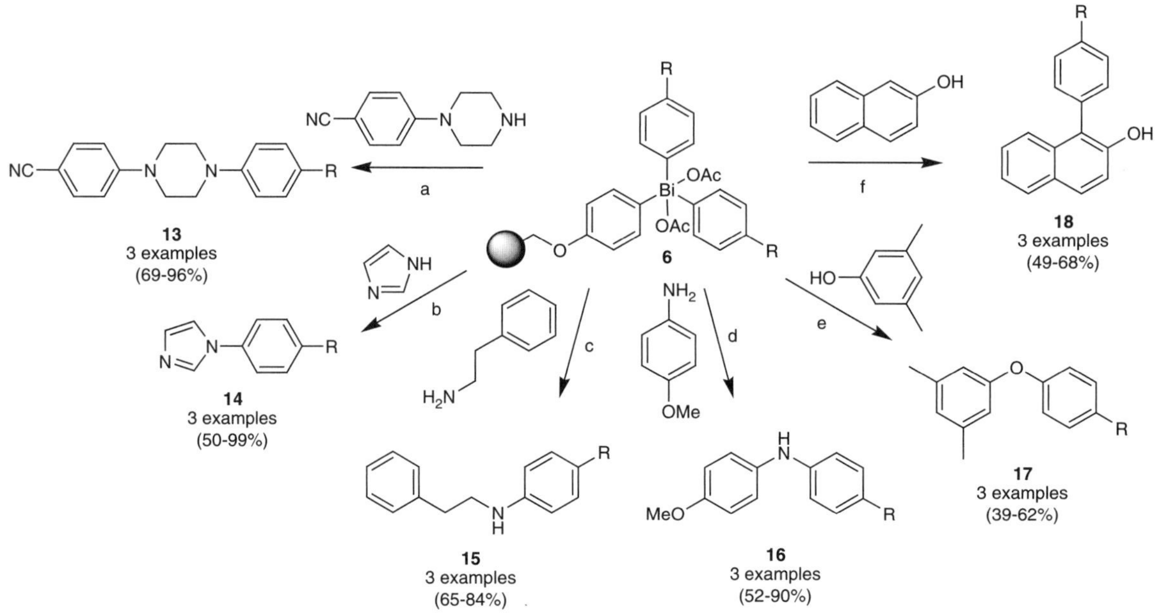

Scheme 18.3 Multifunctional cleavage from triarylbismuthane diacetate linker units (a) 2 equiv. 4-(piperazin-1-yl)benzonitrile, 10% Cu(OAc)$_2$, THF/Et$_3$N (25% v/v), rt, 24 h; (b) 2 equiv. imidazole, 10% Cu(OAc)$_2$, THF, 50 °C, 24 h; (c) 1.5 equiv. 2-phenylethanamine, 10% Cu(OAc)$_2$, THF, rt, 24 h; (d) 1.5 equiv. anisole, 10% Cu(OPiv)$_2$, THF, rt, 24 h; (e) 1.5 equiv. 3,5-dimethylphenol, 10% Cu(OPiv)$_2$, DCM, rt, 24 h; (f) 1.5 equiv. 2-naphthol, 1.2 equiv. TMG, THF, rt, 24 h

Typical Experimental Procedures

Preparation of triaryl bismuth linker (Scheme 18.1)[7]

To a suspension of 4-iodophenoxymethyl polystyrene **2** (1.5 g 2.3 mmol) in THF (5 ml) at −30 °C was added iPrMgBr (18.2 ml, 11.6 mmol). The reaction was stirred for three hours at −30 °C after which time the resin-bound Grignard reagent **3** was filtered and washed with THF (10 ml). The resin was then re-suspended in THF (7 ml) and warmed to 0 °C. A solution of Ph$_2$BiOTf·2HMPA complex **4** (4.7 mmol in 5 ml THF) was added and the reaction was warmed to room temperature overnight. The triarylbismuthane resin was filtered, washed with THF (3 × 10 ml), methanol (3 × 10 ml), THF (3 × 10 ml) and DCM (3 × 10 ml), and dried overnight in vacuo at 40 °C to provide a light yellow resin **5** (1.9 g, 1.2 mmol/g).

Typical cleavage procedure from triaryl bismuth linker (Scheme 18.2)[7]
3-Bromobiphenyl 11

To a suspension of resin **3** (0.50 g, 0.80 mmol) in dichloroethane (10 ml) was added bromine (0.50 g, 3.2 mmol). After stirring at 60 °C for 12 hours, the reaction was filtered. The resin was extracted with DCM (2 × 5 ml) and THF (2 × 5 ml). To the combined organic phases was added aqueous sodium thiosulfate (1 M, 5 ml). The organic solvents were removed in vacuo and to the residue were added water (20 ml) and ethyl acetate (20 ml). The phases were separated and the aqueous phase extracted with further ethyl acetate (2 × 20 ml). The combined organic phases were washed with brine (20 ml) and dried (magnesium *sulfate). The crude product was concentrated in vacuo and purified by column chromatography (heptane) to yield 113 mg (62%) of **11** as an oil.*

Preparation of triaryl bismuth diacetate linker preparation (Scheme 18.1)[7]

*To a suspension of resin **5** (200 mg, 0.23 mmol) in DCM (4 ml) was added diacetoxy iodobenzene (151 mg, 0.47 mmol) and the reaction mixture was stirred for 24 hours at room temperature. The resin was isolated by filtration, washed with DCM (3 × 5 ml), THF (3 × 5 ml), DCM (3 × 5 ml) and dried overnight in vacuo at 40 °C to provide the product as a yellow resin **6** (223 mg, 1.0 mmol/g) was obtained.*

Typical cleavage procedure from triaryl bismuth diacetate linker (Scheme 18.3)[7]
1,3-Dimethyl-5-phenoxy-benzene 17

*To a suspension of resin **6** (300 mg, 0.27 mmol) in DCM (5 ml) were added 3,5-dimethylphenol (122 mg, 0.41 mmol) and copper(II) pivaloate (7 mg, 0.03 mmol), the mixture was stirred for 24 hours at 25 °C. The reaction mixture was filtered and the resin was washed with DCM (3 × 5 ml). The combined filtrates were concentrated in vacuo. Flash chromatography of the residue (eluent: heptane) yielded 1,3-Dimethyl-5-phenoxy-benzene **17** (23 mg, 43%).*

18.3 Conclusions

The range of cleavage strategies possible for substrates attached to supports via bismuth linkers allows for easy generation of diverse compound libraries in a simple fashion. This is exemplified by the fact that ten or so distinct families of compounds have been generated so far from only two known linker units. In addition, the very general reaction conditions which do not require time-consuming development (when compared to their highly substrate specific palladium counterparts) simplifies the process of library generation.

References

[1] Barton, D. H. R., and Finet, J.-P.; *Pure Appl. Chem.* **1987**, *59*, 937.
[2] Finet, J. P.; *Chem. Rev.* **1989**, *7*, 1487.
[3] Abramovitch, R. A., Barton, D. H. R., and Finet, J. P.; *Tetrahedron* **1988**, *44*, 3039.
[4] Elliott, G. I., and Konopelski, J. P.; *Tetrahedron* **2001**, *57*, 5683.
[5] Suzuki, H., Ikegami, T., and Matano, Y.; *Synthesis* **1997**, 249.
[6] Rasmussen, L. K., Begtrup, M., and Ruhland, T.; *J. Org. Chem.* **2006**, *71*, 1230.
[7] Rasmussen, L. K., Begtrup, M., and Ruhland, T.; *J. Org. Chem.* **2004**, *69*, 6890.
[8] Barton, D. H. R., Bhatnagar, N. Y., Finet, J. P., and Motherwell, W. B.; *Tetrahedron* **1986**, *42*, 3111.

19
Transition Metal Carbonyl Linker Units

Susan E. Gibson and Amol A. Walke

Department of Chemistry, Imperial College London, United Kingdom

19.1 Introduction

Groups made up of unsaturated carbon atoms, such as arenes, alkynes and alkenes, are ubiquitous in organic chemistry. These groups are readily attached to transition metals by a bonding mode described by the Dewar–Chatt model, in which donation of π electrons from the unsaturated organic group into empty d_σ orbitals on the metal is accompanied by donation from filled metal d_π orbitals into empty π^* orbitals on the organic partner. This type of interaction is found in arene–chromium complexes **1**, alkyne–cobalt complexes **2** and alkene–manganese complexes **3** (Figure 19.1). Removal of the unsaturated organic species from metal complexes of this type may often be achieved without alteration of the organic molecule.

In the context of solid-phase organic synthesis, it is possible to envisage binding organic molecules containing unsaturated functional groups to a solid support via a linker composed of a transition metal species and one of its ligands (Figure 19.2). This chapter describes how this concept has been realised and used for arenes, and summarises work on other transition metal-based systems that also have the potential to deliver linkers that attach unsaturated organic molecules to solid supports via non-covalent bonding.

19.2 Chromium carbonyl linker units

The principle of attaching an arene to a solid support via a chromium carbonyl linker unit is depicted in Scheme 19.1. Manipulation of the arene substituent(s) on the solid support (R → R') is followed by release of the modified arene. One of the apparent advantages of this approach is that the functional group used to attach the substrate to the solid support is unaffected by the sequence, that is there is no trace whatsoever of the linker in the final product. Moreover, the point of attachment of the substrate molecule is an arene ring, a functional group present in many potential substrates for solid-phase chemistry.

Linker Strategies in Solid-Phase Organic Synthesis   Edited by Peter Scott
© 2009 John Wiley & Sons, Ltd

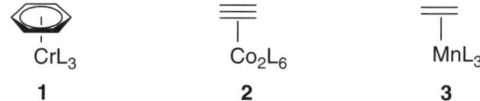

Figure 19.1 Examples of transition metal complexes of arenes, alkynes and alkenes

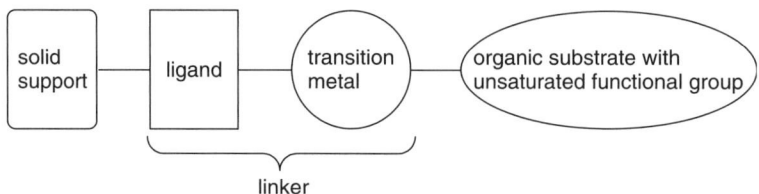

Figure 19.2 A transition metal/ligand linker

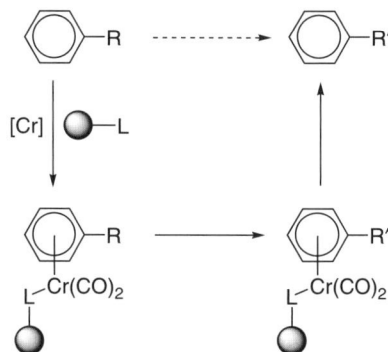

Scheme 19.1 The principle of a chromium carbonyl linker for arenes*

The first demonstration of this concept was executed in our laboratories.[1, 2] The arene 4-(4-methoxyphenyl)butan-2-one was heated with hexacarbonylchromium(0) to give the arene chromium complex **4** as yellow crystals in 71% yield (Scheme 19.2). Complex **4** was then irradiated in the presence of commercially available polymer-bound triphenylphosphine[†] in tetrahydrofuran (THF). Filtration, washing and drying of the polymer gave a brown powder that was characterised as predominantly **5** on the basis of its IR and ^{31}P NMR spectra. The IR spectrum contained strong absorptions at 1870 and 1802 cm^{-1}, which are characteristic of an [(arene)(CO)$_2$(PPh$_3$)Cr(0)] complex, and an absorption at 1710 cm^{-1} corresponding to the ketone. The ^{31}P NMR spectrum of **5** exhibited a resonance at δ 90, a value typical of an [(arene)(CO)$_2$(PPh$_3$)Cr(0)]complex. The ^{31}P NMR spectrum revealed that the polymer coverage with the arene chromium dicarbonyl complex was about 40%, the remainder of the phosphorus sites being polymer–PPh$_2$Cr(CO)$_5$ (20%), polymer–P(O)Ph$_2$ (10%) and polymer–PPh$_2$ (30%). Polymer **5** was reduced with lithium aluminium hydride to give, after work-up, a dark yellow powder that did not absorb at 1710 cm^{-1}. The ^{31}P NMR spectrum of the reduced material revealed that the loading of **6** on the polymer was 25%, indicating some loss of material from the polymer, perhaps as a result of complex

---

*○ represents polystyrene throughout this chapter.
[†]Polymer-bound triphenylphosphine describes a diphenylphosphine polystyrene polymer cross-linked with 2% divinylbenzene.

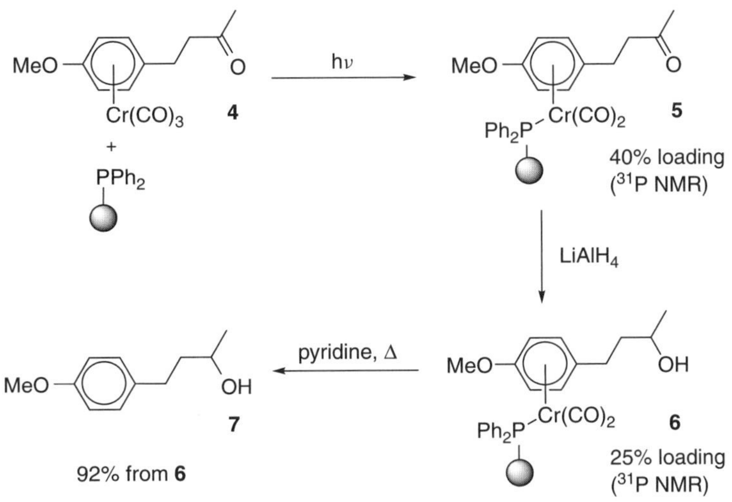

Scheme 19.2 A chromium carbonyl linker for arenes based on a phosphine ligand

decomposition resulting from nucleophilic addition at the metal carbonyl ligands. The reduction product was released from the polymer by heating in pyridine for two hours. Filtration, washing the polymer with THF and diethyl ether, and concentration of the washings gave alcohol **7** in ≥ 95% purity. The yield for the release of **7** from the polymer was 92%.

The above experiments demonstrated for the first time that an organic molecule containing an aromatic ring can be attached to a solid support via a transition metal/ligand linker, manipulated and subsequently released (Semmelhack had previously discussed this concept, but not provided a demonstration).[3] Each step in the sequence proceeded in tolerably good yield (unoptimised). Particularly encouraging at this stage, however, was the usefulness of IR and ^{31}P NMR spectroscopy for obtaining direct information about the structure and loading of polymer-bond complexes **5** and **6**.

Rigby subsequently introduced a new method for the attachment of an arene chromium carbonyl complex to a polymer support.[4] Performing the photolysis step in the presence of cyclo-octene generated a reactive intermediate **8** that cleanly added to polymer-bound benzyldiphenylphosphine. Under optimised conditions a range of complexes were generated in good yields (84–95%) (Scheme 19.3).

Grignard addition to ketones and esters (76–82%),[4] acetylation of primary alcohols (80–86%),[4] addition of organolithium reagents to esters (83–89%),[5] generation and alkylation of enolate anions (95–97%)[5] and substitution of alcohols via the Mitsunobu reaction (69–79%)[5] were all demonstrated to proceed well. Decomplexation was effected either by gently warming in acetonitrile or by oxidation with iodine. Overall efficiency was examined and it was found that optimisation of the polymer loading step had lead to overall yields of 61–78% for attachment of the arene complex, chemical transformation and removal of the product from the resin.

Rigby demonstrated that the methodology is amenable to the simultaneous incorporation of several starting materials on the solid support.[4] Irradiation of a mixture of four chromium complexes with excess cyclo-octene in toluene followed by treatment with the resin-bound phosphine afforded polymer **9** loaded with four different substrates (Figure 19.3). Reaction of this with excess ethylmagnesium bromide and subsequent acetonitrile-assisted cleavage of the linkers furnished a mixture of tertiary alcohols in the expected ratio of 1:2:1 in an overall yield of 84% based on **9** and 75% based on the mixture of chromium complexes.

Maiorana replaced the phosphine ligand in the linker unit with an isonitrile ligand[6] by introducing the isonitrile polymer **10** (Scheme 19.4), which was synthesised from hydroxymethyl polystyrene in four steps.

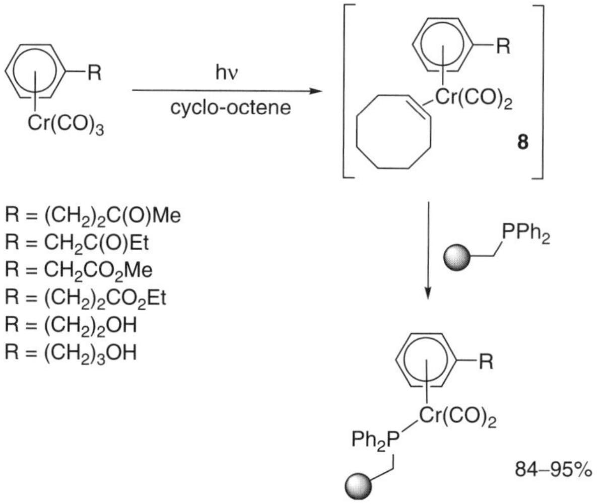

528 Multifunctional Linker Units for Diversity-Oriented Synthesis

Scheme 19.3 *Photolysis in the presence of cyclo-octene leads to efficient loading*

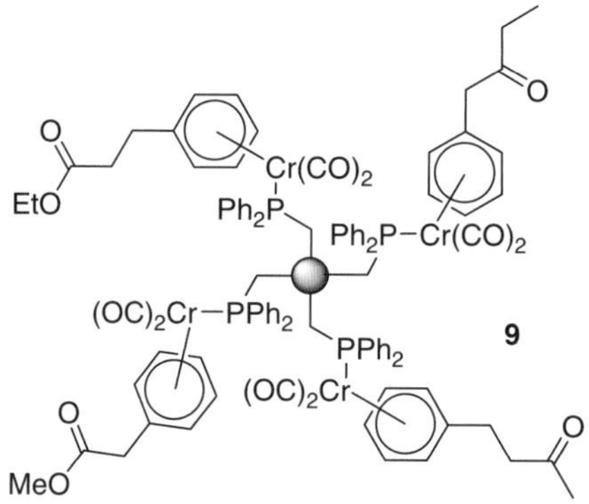

Figure 19.3 *Polymer loaded with several substrates*

The chromium complex of fluorobenzene was attached to the resin by photochemical irradiation to give **11**. A preliminary evaluation of the reactivity of these resins was carried out using the nitrogen nucleophiles pyrrolidine, piperidine and the *N*-anion of indole. After cleavage with iodine or by prolonged exposure to air and sunlight, the nucleophilic substitution products **12** were isolated in an average of 80% yield.

In an extension of this study,[7] chlorobenzenes were successfully introduced, an expanded range of nucleophiles were used in the nucleophilic aromatic substitution reaction and a reductive amination was performed. The usefulness of the IR active isonitrile group for resin analysis was noted. It is also of note that this study focussed on exploiting an aspect of the extensive chemistry of arene chromium carbonyl complexes, that is facile nucleophilic substitution, rather than carrying out chemistry on a more remote functional group.

Applications of chromium carbonyl linkers have started to appear. Jones has discovered that upon cycloaromatisation the enediyne **13** induces degradation of the human oestrogen receptor α(hERα), a receptor of importance in endocrinological pathways related to cancer. Enediyne **13** was identified from

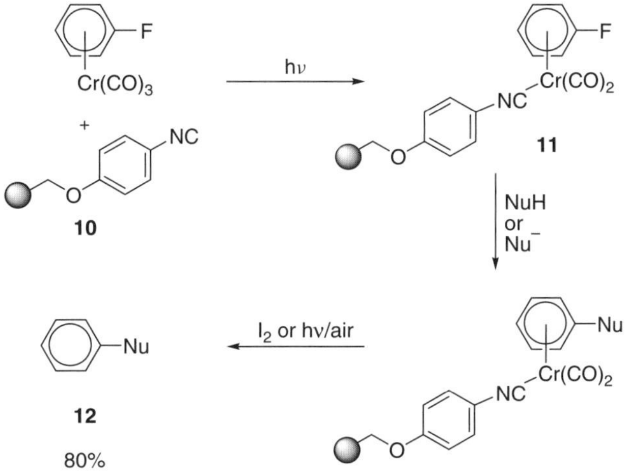

Scheme 19.4 *A chromium carbonyl linker for arenes based on an isonitrile ligand*

a screen of arenes **15** on the basis that as cycloaromatisation takes place via a late-stage transition state, these arenes bear a close structural relationship to the active diyl radicals **14** (Figure 19.4).[8]

In preparation for further screens, Jones developed a solid-phase synthesis of **20**, the arene that led to the identification of **13**.[8] Alcohol **16** was converted to its chromium carbonyl complex **17** and subjected to photolytic ligand exchange with polymer-bound triphenylphosphine to yield adduct **18** (Scheme 19.5). Esterification with **19**, deprotection and removal from the polymer gave a sample of **20**, identical with that prepared using the solution-phase method. This technology facilitates the rapid assembly of diverse libraries of enediyne mimics.

Elsewhere, it has been demonstrated that the amino acid phenylalanine can be immobilised via its arene, manipulated on the polymer and, subsequently, released.[2] Resin **21** was prepared from the corresponding Fmoc–Phe-O*t*Bu chromium complex by photolysis. Analysis by ^{31}P NMR spectroscopy revealed a loading of 70%. Amine deprotection of **21** to give **22** was effected by piperidine and then **22** was reacted with Fmoc-Val-OH under standard peptide coupling conditions to give **23** (Scheme 19.6). Oxidative cleavage of the chromium(0) linker effected release of the protected dipeptide Fmoc-Val-Phe-*t*Bu **24** in 81% overall yield from **21**.

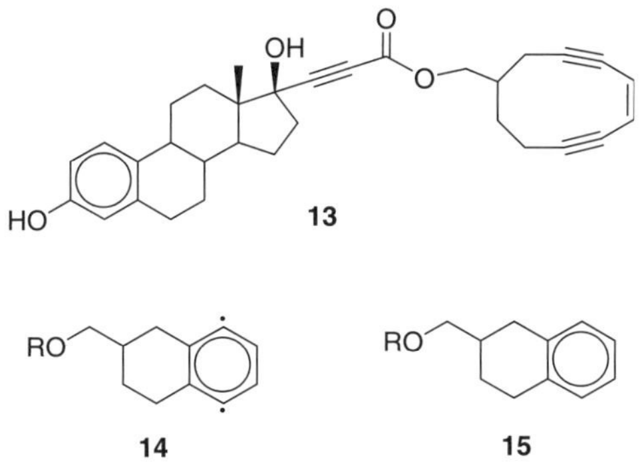

Figure 19.4 *A screen of arenes 15 led to the identification of the protein degrading enediyne 13*

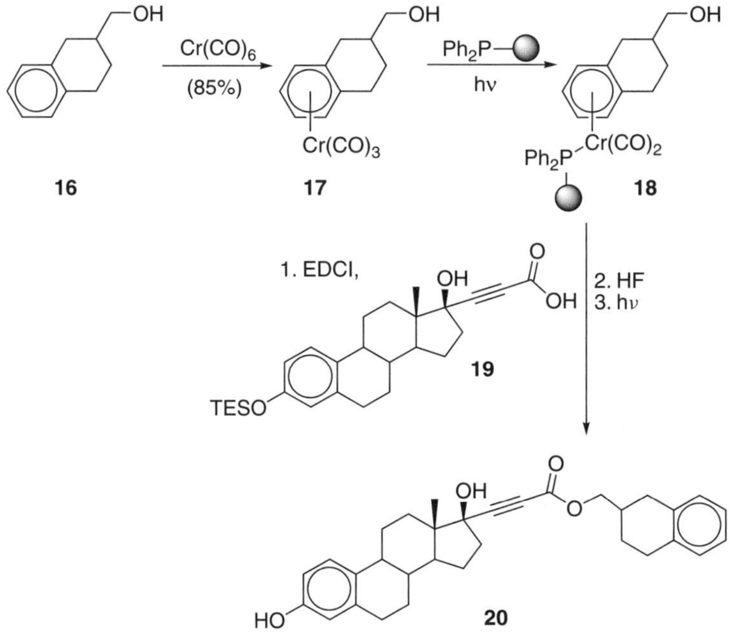

Scheme 19.5 A solid-phase synthesis of the enedyne mimic **20**

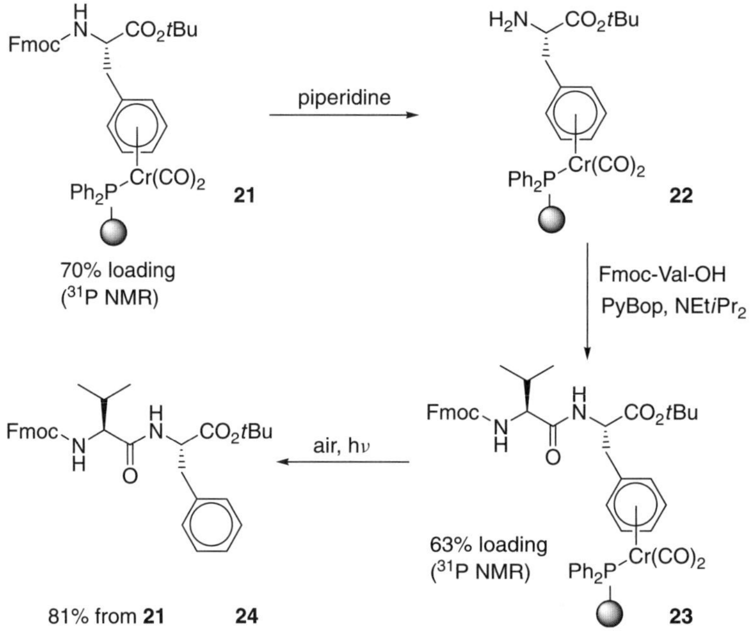

Scheme 19.6 Phenylalanine bound via its arene

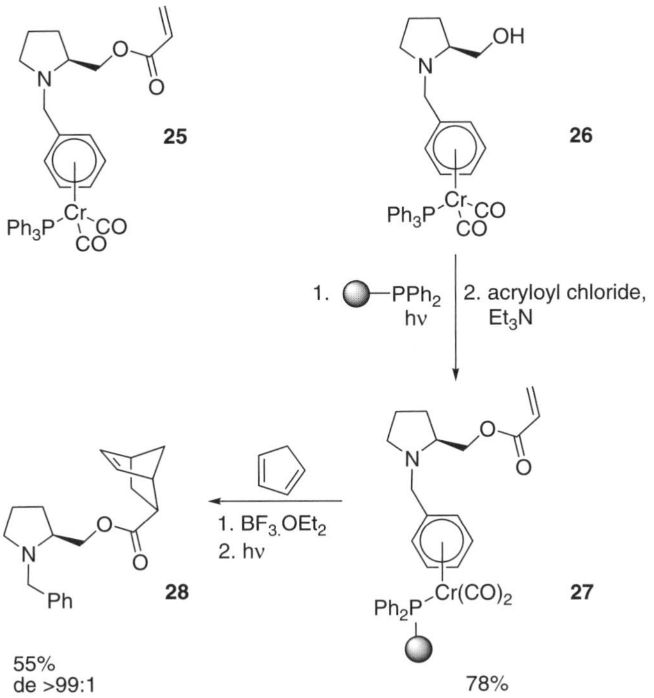

Scheme 19.7 Preparation of a screen to identify selective cycloadditions

Jones has reported that the proline-containing acrylate **25** reacts with cyclopentadiene to give their cycloaddition adduct in 99% yield and 98% *de*. To facilitate a screening programme to identify equally selective partners, Jones developed a solid-phase version of the reaction.[9, 10] Irradiation of **26** in the presence of polymer-bound triphenylphosphine and reaction with acryloyl chloride resulted in the formation of **27** (Scheme 19.7). Cycloaddition proceeded with essentially complete diastereoselectivity, and the adduct **28** was removed from the support by air/light oxidation.

To summarise this section, the principle of using chromium carbonyl linker units to attach an arene to a solid support has been demonstrated using a phosphine or an isonitrile ligand to connect the chromium to the solid support. The methods of substrate attachment, reaction types that have been shown to be compatible with the linker to date and methods of product detachment are summarised in Scheme 19.8. For the latter, air and light is a very mild method, but prolonged exposure may be necessary, particularly if sunlight intensity is weak. Ligand displacement methods (pyridine or acetonitrile) and iodine oxidation are generally faster but the former requires heating and the latter may be incompatible with functional groups in the product.

Typical Experimental Procedures

Typical loading of an arene chromium complex onto a phosphine polymer (scheme 19.6)

To a suspension of polymer-bound triphenylphosphine (450 mg, 0.72 mmol) at ambient temperature in anhydrous THF (200 cm³) left under constant nitrogen agitation for 20 minutes was added tricarbonyl(Fmoc-Phe-OtBu)chromium(0) (500 mg, 0.86 mmol). The yellow mixture was subjected to periodic irradiation (4 × 10

532 Multifunctional Linker Units for Diversity-Oriented Synthesis

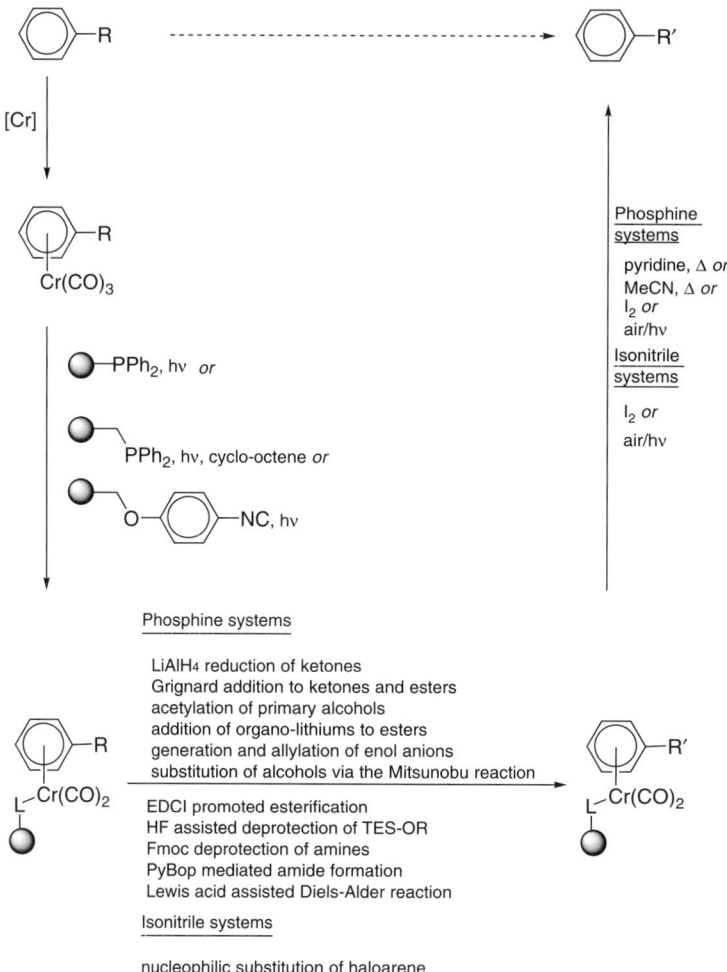

Scheme 19.8 *A summary of the methods used for chromium carbonyl linkers*

min) with a 125 W mercury vapour lamp over a 48 hour period. The resulting deep-red beads were filtered, washed thoroughly with alternate aliquots of THF and diethyl ether, and dried in vacuo to afford the resin **21** *(732 mg, 0.69 mmol [Fmoc-Phe-OtBu]/g).*

Typical air/light cleavage procedure for polymer-bound arenes (scheme 19.6)

Dicarbonyl(Fmoc-Val-Phe-OtBu)(polymer–PPh$_2$)chromium(0) **23** *(0.60 mmol/g, 63% loading, 100 mg, 0.06 mmol) was ground to a fine yellow powder and suspended in DCM (15 cm^3) in a 25 cm^3 round-bottomed flask equipped with a condenser and calcium(II) chloride (CaCl$_2$) drying tube, and stirred at ambient temperature in air under white light (100 W) for 48 hours. The resulting fine brown suspension was filtered through Celite and the polymeric residue was washed with dichloromethane (DCM). The combined filtrate and washings were concentrated at reduced pressure to afford the dipeptide* **24** *(29 mg, 0.054 mmol, 90% from* **23***).*

Typical ligand displacement procedure for polymer-bound arenes (scheme 19.2)

[4-(4-Methoxyphenyl)butan-2-ol]dicarbonyl(triphenylphosphine polymer)chromium(0) **6** (0.72 mmol/g, 25% loading, 500 mg, 0.36 mmol) was added to pyridine (10 cm^3), the mixture degassed and subsequently refluxed under a nitrogen atmosphere for two hours, during which time the solution became bright red. The mixture was allowed to cool and the solid to settle out. Supernatant liquid was removed via a cannula and the residual solid was washed successively with THF (2 × 20 cm^3) and diethyl ether (20 cm^3). The organic washings were reduced in vacuo and filtered through a short pad of silica (diethyl ether). Diethyl ether was removed in vacuo to afford the title compound **7** as a colourless oil of ≥ 95% purity (60 mg, 0.33 mmol, 92% from **6**).

Typical iodine cleavage procedure for polymer-bound arenes (scheme 19.4)

Product **12** was released from the solid support by swelling the resin in DCM (10 cm^3) and treating it with a 0.1 M DCM solution of iodine (1.1 equiv.). After stirring for four hours at room temperature, the brown polymer was filtered and washed with DCM (8 × 5 cm^3). The combined organic phases were washed with an aqueous solution of sodium thiosulfate and then dried over sodium sulfate and evaporated. The residue was taken up with diethyl ether (5 cm^3) and filtered over a pad of celite and evaporated. The crude mixture contained, as indicated by NMR spectroscopy, compound **12**, a minor unidentified hydrocarbon and aromatic contaminants, possibly due to fragmentation of the polystyrene resin. A filtration through a short pad of silica provided analytically pure **12**.

19.3 Cobalt carbonyl linker units

Cobalt carbonyl complexes of alkynes are well established as useful reagents for organic synthesis. Given their ease of formation and their tolerance of diverse reaction conditions, it was reasoned that cobalt carbonyls could be viewed as potential linker units for the immobilisation of alkynes to a solid support (Scheme 19.9).[11, 12]

5-Hexyn-1-ol **29** was loaded onto a solid support by a 'direct' method and an 'indirect' method (Scheme 19.10). In the 'direct' method, polymer-bound triphenylphosphine was reacted with octacarbonyldicobalt(0) to generate a cobalt carbonyl coated resin **30**. Treatment of **30** with 5-hexyn-1-ol **29** generated the purple resin **31**, which was characterised by IR and ^{31}P NMR spectroscopy (both mono and bisphosphine substituted alkyne complexes were formed). In the 'indirect' method, 5-hexyn-1-ol **29** was

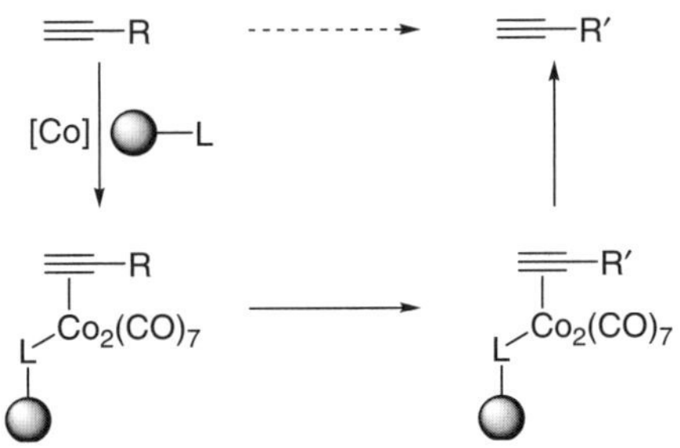

Scheme 19.9 *The principle of a cobalt carbonyl linker for alkynes*

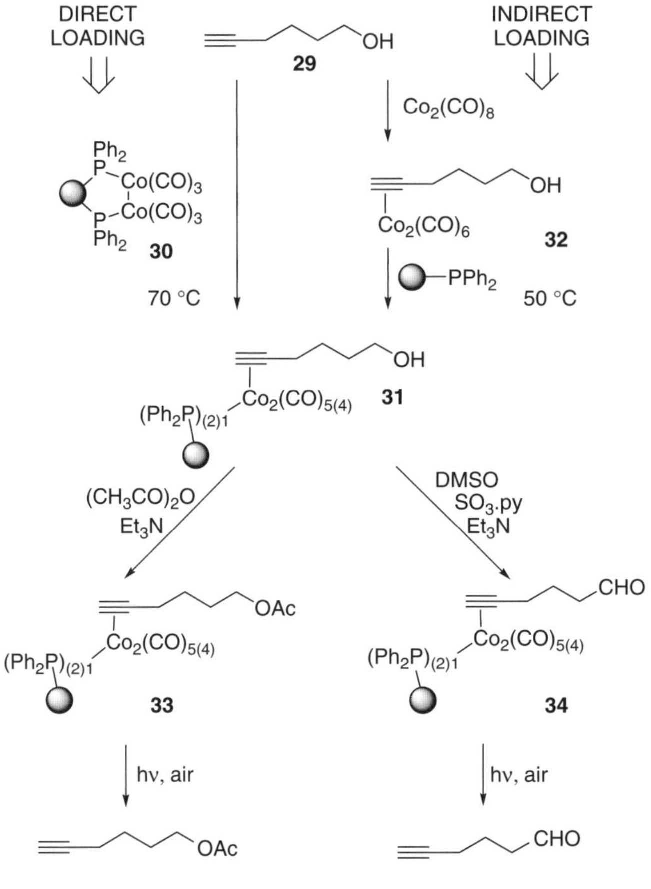

534 *Multifunctional Linker Units for Diversity-Oriented Synthesis*

Scheme 19.10 *A cobalt linker for alkynes*

treated with octacarbonyldicobalt(0) to provide cobalt alkyne complex **32**. This was then loaded onto the polymer by gentle warming in THF to provide resin **31**. The linker was shown to be compatible with acetylation of the primary alcohol of 5-hexyn-1-ol to give ester **33** and oxidation of the alcohol to the aldehyde **34**. The products **35** and **36** were liberated from the polymer by prolonged exposure to air and light.

Typical Experimental Procedures

Loading of an alkyne cobalt complex 32 onto a phosphine polymer (scheme 19.10)

*Polymer-bound triphenyl phosphine (1 g, 1.6 mmol P) was suspended at ambient temperature in anhydrous THF (5 cm^3) and after 20 minutes under constant nitrogen agitation, a solution of **32** (1.2 g, 3.2 mmol) in THF (5 cm^3) was added. The mixture was heated at 50 °C under constant nitrogen agitation for four hours The resulting deep purple beads were filtered, washed with alternate aliquots (20 cm^3) of THF and diethyl ether until the filtrate became colourless, and dried in vacuo to afford resin **31** (1.52 g, 0.94 ± 0.02 mmol[hex-5-yn-1-ol]/g).*

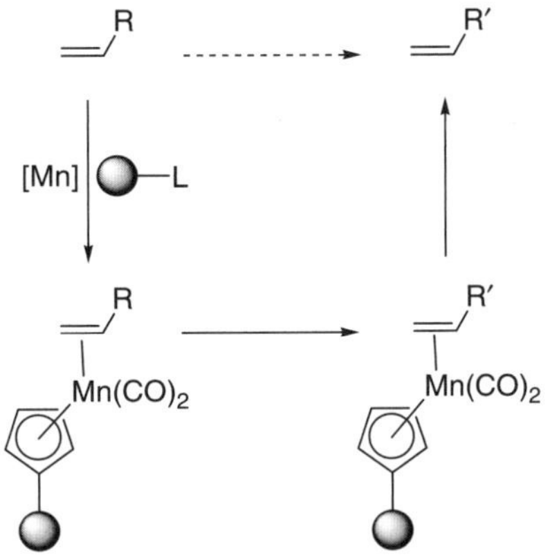

Scheme 19.11 *The principle of a manganese carbonyl linker for alkenes*

19.4 Manganese carbonyl linker units

Among the various metal carbonyl compounds that might be used to form the basis of a linker unit between alkenes and a solid support, the cyclopentadienylmanganese dicarbonyl unit stands out as a stable system with demonstrated compatibility towards a variety of bond-forming conditions (Scheme 19.11).[13]

Lepore has reported the addition of a cyclopentadienyl–manganese tricarbonyl unit onto aminomethyl polystyrene via an amide linker **37** (Scheme 19.12). Attachment of several alkenes via photolysis in

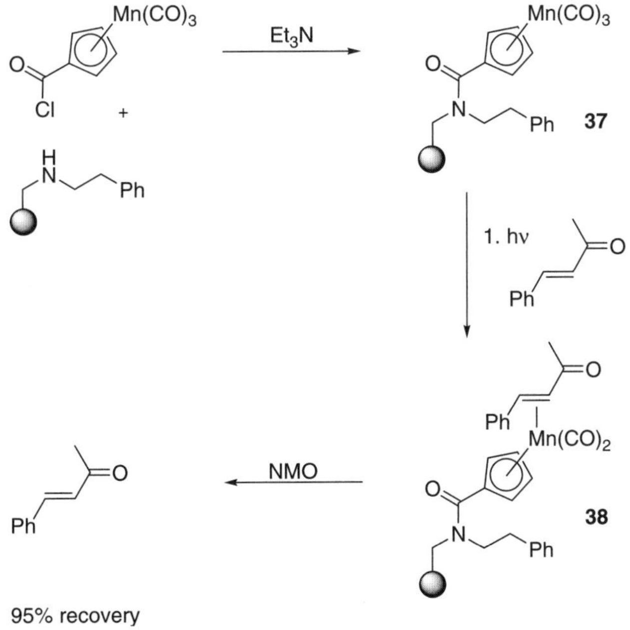

Scheme 19.12 *Towards a manganese carbonyl linker for alkenes based on a cyclopentadienyl ligand*

THF was demonstrated, giving resins such as **38**, was their removal using the mild oxidant *N*-methylmorpholine *N*-oxide (NMO). To date, modification of the alkene on the solid support has not been reported.

Typical Experimental Procedures

Loading of an alkene onto polymer 37 (scheme 19.12)

Resin 37 (500 mg, 1.0 mmol/g) was added to a microfilter funnel (25 cm^3). The funnel was sealed and flushed with argon followed by the addition of THF (8 cm^3). The funnel was vibrated while irradiating with UV (254 nm) for eight hours. Alkene was then added (1 mmol) and the funnel was vibrated at 50 °C for 15 hours. The resin suspension was filtered and the resin was washed with THF and methanol and dried in vacuo for five hours to give 38 as a yellow coloured resin.

19.5 Conclusion

The principle of attaching unsaturated organic molecules to a solid support via a transition metal/ligand linker has been established. In the case of the chromium carbonyl phosphine linker for arenes, the linker has been demonstrated to be compatible with a good range of reaction conditions.

Finally, it should be noted that immobilisation of chromium Fischer carbene complexes by loading them onto a polymer support,[14] or by incorporating them into a silica matrix via sol–gel processing,[15] has been reported. The reactivity of the immobilised carbenes has been probed in each case, and it is anticipated that their removal from their respective supports should be relatively straightforward.

References

[1] Gibson, S. E., Hales, N. J., and Peplow, M. A.; *Tetrahedron Lett*. **1999**, *40*, 1417.
[2] Comely, A. C., Gibson, S. E., Hales, N. J., and Peplow, M. A.; *J. Chem. Soc. Perkin Trans. 1* **2001**, 2526.
[3] Semmelhack, M. F., Hilt, G., and Colley, J. H.; *Tetrahedron Lett*. **1998**, *39*, 7683.
[4] Rigby, J. H., and Kondratenko, M. A.; *Org. Lett*. **2001**, *3*, 3683.
[5] Rigby, J. H., and Kondratenko, M. A.; *Bioorg. Med. Chem. Lett*. **2002**, *12*, 1829.
[6] Maiorana, S., Baldoli, C., Licandro, E., *et al.*; *Tetrahedron Lett*. **2000**, *41*, 7271.
[7] Baldoli, C., Maiorana, S., Licandro, E., *et al.*; *J. Comb. Chem*. **2003**, *5*, 809.
[8] Jones, G. B., Wright, J. M., Hynd, G., *et al.*; *Org. Lett*. **2000**, *2*, 1863.
[9] Xie, L., and Jones, G. B.; *Tetrahedron Lett*. **2005**, *46*, 3579.
[10] Li, J., Xie, L., Guzel, M., *et al.*; *J. Organometal. Chem*. **2007**, *692*, 5459.
[11] Comely, A. C., Gibson, S. E., and Hales, N. J.; *Chem. Commun*. **1999**, 2075.
[12] Comely, A. C., Gibson, S. E., Hales, N. J., *et al.*; *Org. Biomol. Chem*. **2003**, *1*, 1959.
[13] Zhang, Z., and Lepore, S. D.; *Tetrahedron Lett*. **2002**, *43*, 7357.
[14] Maiorana, S., Seneci, P., Rossi, T., *et al.*; *Tetrahedron Lett*. **1999**, *40*, 3635.
[15] Klapdohr, S., Dötz, K. H., Assenmacher, W., *et al.*; *Chem. Eur. J*. **2000**, *6*, 3006.

20
Linkers Releasing Olefins or Cycloolefins by Ring Closing Metathesis

Jan H. van Maarseveen

*Van't Hoff Institute for Molecular Sciences, University of Amsterdam,
The Netherlands*

20.1 Introduction

After the discovery of Grubbs I (**GI**) catalyst in the beginning of the 1990s, olefin metathesis became right from the start a popular reaction in many synthesis laboratories (Scheme 20.1).[1] Olefin metathesis is a fascinating and Nobel prize winning reaction that inspires many synthetic chemists to develop creative synthetic approaches to complex molecules.[2-4] Ring closing metathesis (RCM) provides facile access to cycloolefins of various ring sizes, while cross metathesis (CM) is a powerful method for the synthesis of internal alkenes. Recently, even more powerful precatalysts were developed, of which the Grubbs II[5] (**GII**) and Grubbs–Hoveyda[6] (**GH**) catalysts are the cream of the crop with respect to functional group tolerance, stability and reactivity.

Concurrently with the discovery of synthetically useful metathesis catalysts, the field of combinatorial synthesis of drug-like molecules via solid-phase synthesis emerged. In these first solid-phase syntheses of small organic molecules, mostly resins and linkers were used, especially designed for peptide synthesis resulting in molecules still containing a carboxylic acid or amide group to which they were anchored to the solid support. Because highly polar carboxylic acid or amide moieties are detrimental for the pharmacodynamics and pharmacokinetics of many biologically active molecules, so-called traceless linkers were sought. It is no surprise that medicinal chemists soon realized that solid-supported RCM and also CM enable a traceless approach to cyclic and linear olefins.[7] Also, it was realized at that time that an alkene linker is very robust and allows a wide variety of conditions for construction of the final RCM or

---

Linker Strategies in Solid-Phase Organic Synthesis   Edited by Peter Scott
© 2009 John Wiley & Sons, Ltd

538 *Multifunctional Linker Units for Diversity-Oriented Synthesis*

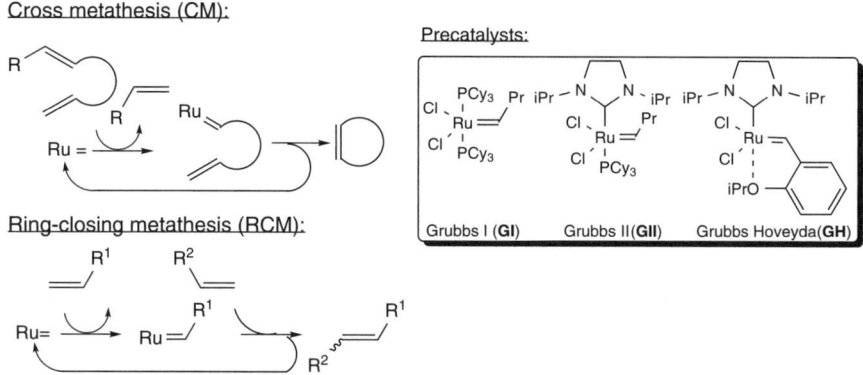

Scheme 20.1 *Cross-metathesis (CM) and ring-closing metathesis (RCM)*

CM precursors. In addition, the current metathesis catalysts are compatible with many different functional groups, making the combination of metathesis and solid-phase chemistry a powerful combination for traceless diversity oriented synthesis. This chapter describes all solid-supported approaches published so far whereby cyclic or acyclic olefins are released from the solid support using olefin metathesis. It has become evident over the last decade that solid-supported olefin metathesis is a very powerful method to prepare olefins and cycloolefins in a parallel fashion. It is, therefore, surprising that still relatively few publications have appeared on approaches relying on metathesis as the final cleavage reaction.

In Scheme 20.2 the key strategies of using olefin metathesis to release cyclic (**I**) or acyclic (**II, III**) olefins are depicted. Method I produces cycloolefins using RCM in a cyclization–cleavage fashion. If the synthesis of the RCM precursor ends with the terminal olefin, deletion sequences lacking the final olefin will remain on the resin ensuring a high product purity. As will be discussed below, an intrinsic feature of method I is the fact that after the RCM reaction the catalytically active ruthenium–alkylidene (or ruthenium–benzylidine) species is trapped on the solid support. If the cyclization precursor is linked via the fragment in between two alkenes (method II), RCM liberates an alkene. This is a very powerful approach as the catalyst is released in solution after each cycle. Finally, mono-alkenes may be released by simple CM (method III), which is a useful procedure to prepare asymmetric 1,2-disubstituted alkenes.

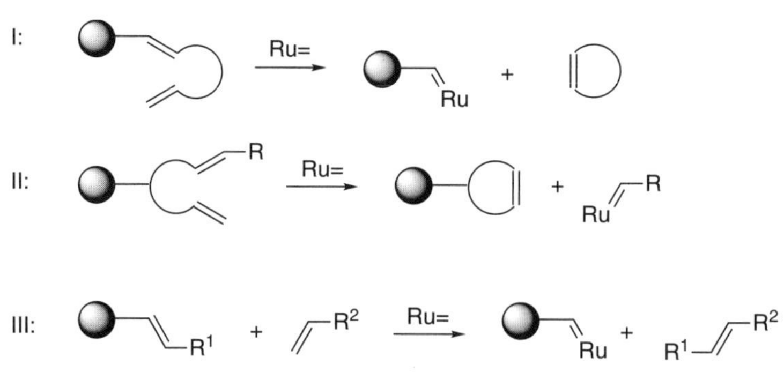

Scheme 20.2 *Olefin metathesis mediated cleavage strategies*

In the three sections below, all approaches published so far will be discussed according to their classification as depicted in Scheme 20.2.

20.2 Cycloolefins via method I

The first proof-of-principle was shown by us in an approach to seven-membered lactams (Scheme 20.3).[8] In the first attempt only trace amounts of the cycloolefin were released, probably caused by the fact that the active catalyst resides on the resin inaccessible by other substrates. It was reasoned that catalyst recycling may be accomplished by adding a monoolefin liberating the catalytic ruthenium–alkylidene complex via a CM process. Indeed, in the presence of 1-octene the isolated yield of cycoolefin was increased to 54%.

In a more detailed follow-up study aimed at allylglycine-derived azacycles it was found that styrene is the best additive (Scheme 20.4).[9] In the absence of styrene, acceptable yields were obtained also but the reaction did not go to completion. Cleavage of the residual material from the resin only revealed the starting diene and residual allylic ether ruling out inter-resin CM. The beneficial role of styrene was ascribed to the greater stability of a ruthenium–benzylidene species as compared to a ruthenium–alkylidene species, allowing more turnovers, lower catalyst loadings and higher yields.[10] Using this optimized procedure several six- to eight-membered lactams were made from α-alkenyl amino acids. Also, acrylamide-type linkers were used successfully providing six- and seven-membered α,β-unsaturated lactams.

Typical Experimental Procedures

Procedure for the synthesis of the α,β-unsaturated carboxylic acid chloride resin (Scheme 20.4)

A suspension of carboxyl polystyrene (1.89 g, 1.2 mmol), cesium carbonate (Cs_2CO_3) (1.0 g, 3.07 mmol), potassium iodide (25 mg, 0.15 mmol), 1,5-Diazabicyclo[4.3.0]non-5-ene (DBN) (19 μl, 0.15 mmol) and 1,1-diethoxy-4-chlorobutane (0.83 g, 4.6 mmol) in N-methylpyrrolidone (NMP) (20 ml) was swirled for 72 hours at 70 °C. The reaction mixture was filtered, washed (dichloromethane (DCM), methanol, DCM, methanol, DCM, diethyl ether, DCM, diethyl ether, DCM) and dried in a vacuum stove (2.05 g, 1.15 mmol) in 96% yield. IR v_{max} (neat) cm^{-1} 2980, 1720. The acetal resin was suspended in NMP/acetone (150 ml, 2:1) together with p-toluenesulfonic acid (pTsOH) (20 mg, 0.11 mmol) and swirled for three hours at 50 °C. The

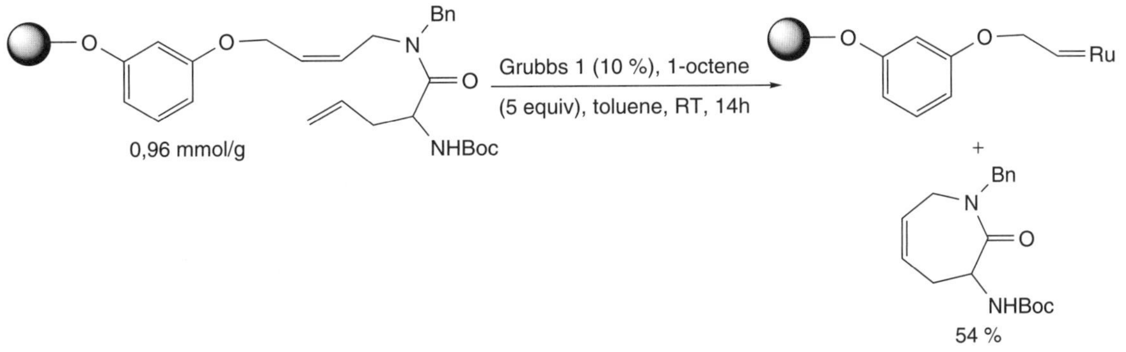

Scheme 20.3 *First cyclization–cleavage RCM and catalyst recycling via 1-octene addition*

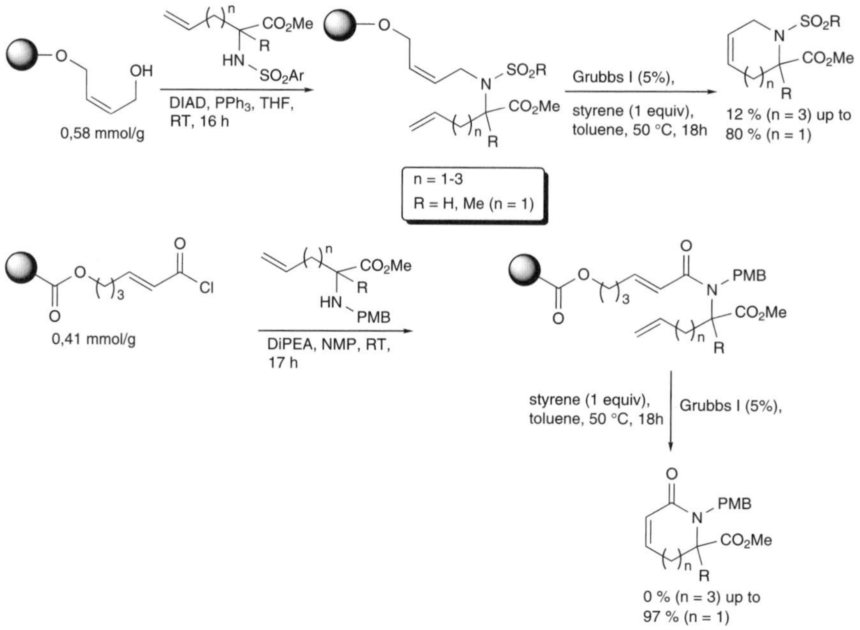

Scheme 20.4 Synthesis of six- to eight-membered azacycles

resin was filtered, washed as above, and dried in a vacuum stove. The resin was subjected to this procedure three times to yield the aldehyde resin in quantitative yield. IR v_{max} (neat) cm^{-1} 2720, 1720.

To a suspension of the aldehyde resin in toluene (70 ml) was added butyl (triphenylphosphoranylidene)acetate (1.2 3g, 3.27 mmol) followed by swirling for four hours at 50 °C. The resin was filtered, washed and dried in a vacuum stove to provide the α,β-unsaturated ester resin in 100% yield. IR v_{max} (neat) cm^{-1} 2980, 1720. Removal of the tBu-group was accomplished by suspension of the resin in DCM/TFA (30 ml, 4:1). After swirling for two hours at room temperature the resin was filtered, washed and dried in a vacuum stove to afford α,β-unsaturated carboxylic acid resin in 94% yield. IR v_{max} (neat) cm^{-1} 3600–2400, 1725, 1720, 1695. The carboxylic acid resin was suspended in 1,2-dichloroethane (45 ml), and oxalyl chloride (0.47 ml, 5.41 mmol) together with DMF (one drop) was added followed by swirling for two hours at room temperature. The resin was filtered off and washed successively with DCM, diethyl ether, DCM, diethyl ether, DCM, and dried in a vacuum to give the α,β-unsaturated carboxylic acid chloride resin (2.62 g, 1.08 mmol, 94% overall yield over five steps) with a loading of 0.41 mmol/g. IR v_{max} (neat) cm^{-1} 1760, 1720.

Shortly after our approach Piscopio described the synthesis of similar compounds that were applied as Freidinger-lactam type β-turn mimics (Scheme 20.5).[11–13] The synthesis of the RCM precursor was performed via a linear sequence relying on the Fukuyama–Mitsunobu reaction as the key step on a styrene-type linker. The final cyclization–cleavage step was performed using only 0.05 equivalent of Grubbs I precatalyst, liberating the cycloolefins in yields of 15–36%. A very efficient route towards these β-turn mimics was developed via the Ugi 4-component reaction to give the RCM precursor sequence in a single step, followed by RCM to give the final lactams in impressive overall yields up to 62%. It is likely that the use of a styrene-type linker is very beneficial for the efficiency of the process because after the RCM step the catalyst resides as a stable ruthenium–benzylidine species on the resin.

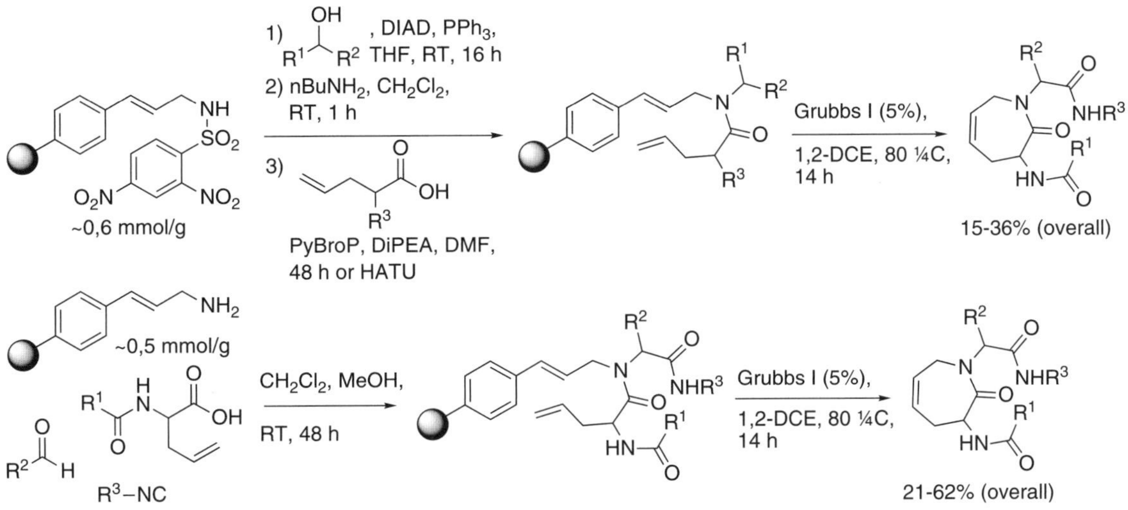

Scheme 20.5 Synthesis of 5-6-dehydroazepines on styrene-type linkers

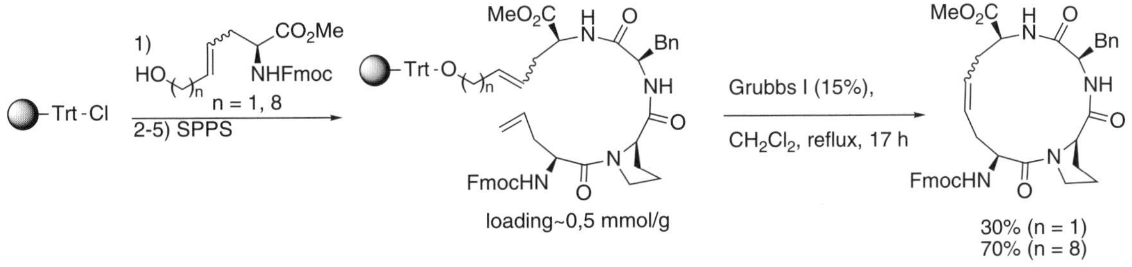

Scheme 20.6 Pseudo cyclic tetrapeptide synthesis using a tethered linker to facilitate catalyst recycling

An approach to cyclic pseudo tetrapeptides was published by the Blechert group (Scheme 20.6).[14] The terminal olefin moiety of Fmoc-AllGly-OMe was functionalized with alcohol tethers via CM. The alcohols were loaded onto trityl-polystyrene followed by build-up of the peptide chain via standard Fmoc solid-phase peptide synthesis (SPPS) ending with Fmoc-AllGly as the N-terminal residue. Cleavage occurred in low yield for the short spacer (n = 1) but was quite successful for the longer spacer (n = 8), which was ascribed to the greater mobility of the immobilized ruthenium–alkylidine species facilitating the on-resin transfer to another RCM precursor.[15] Yields up to 70% of the cyclic pseudo tetrapeptides as E/Z-mixtures were obtained using 0.15 equivalent of Grubbs I catalyst in refluxing DCM. Replacement of proline by linear amino acids was detrimental for the outcome of the reaction in terms of reaction rates and yields underscoring the necessity of a correct conformation of the cyclization precursor.

An impressive application of cyclization–cleavage RCM was published soon thereafter by Nicolaou, who reported the total synthesis of epothilone A and a library of more than 50 analogs using cyclization–cleavage RCM as the key step (Scheme 20.7).[16, 17] Starting from low-loaded Merrifield resin (0.3 mmol/g) the RCM precursor was obtained as a mixture of diastereomers after a nine-step sequence in good overall yields. After the addition of substoechiometric amounts of Grubbs I precatalyst,

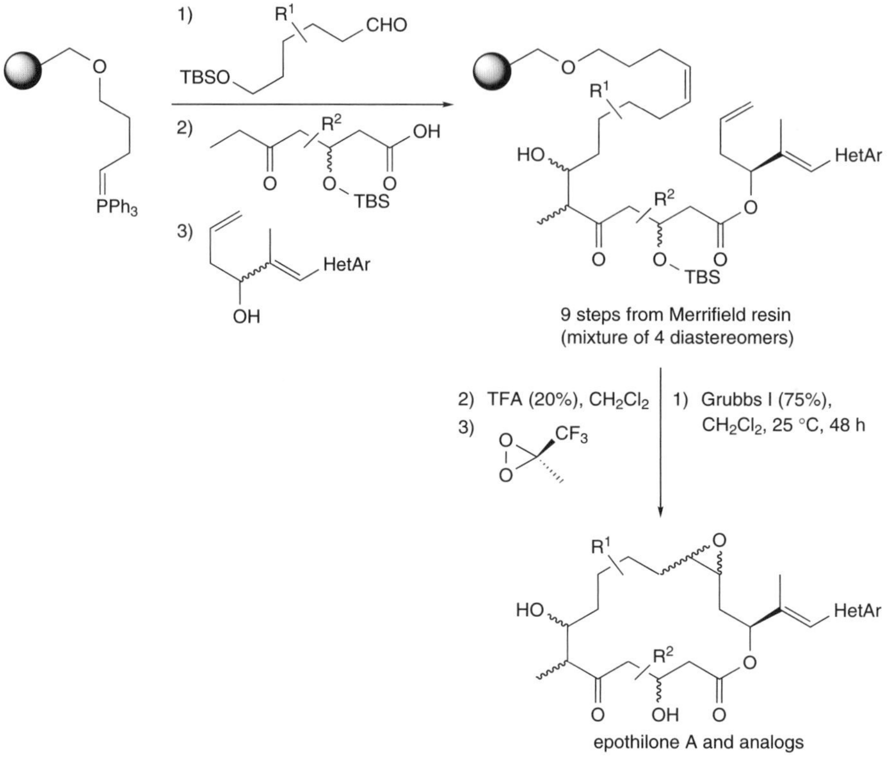

Scheme 20.7 A solid-supported cyclization–cleavage approach towards epothilone A and analogs

a mixture of four diastereomeric macrocycloolefins (including E/Z-stereoisomers) that could be separated by chromatography was obtained. The correct precursor within the library was further elaborated into epothilone A in two solution phase steps, thus showing that the total synthesis of natural products by solid-phase chemistry is not restricted just to biopolymers. The library members were subjected to a tubulin assembly assay and cytotoxicity assays with human ovarian and breast cancer cells. This study revealed some highly potent members together with a wealth of structure–activity relationship information.

The Maier group published both the solution and solid-phase synthesis of 12-membered lactams starting from Merrifield resin (1.5 mmol/g) etherified with an allylic alcohol-type linker containing an additional C3 spacer (Scheme 20.8).[18] The synthesis started by $BF_3 \cdot Et_2O$ mediated loading of the supported trichloroacetimidate with an asymmetric aldol reaction derived chiral secondary alcohol, followed by basic auxiliary removal. Reaction of the resulting carboxylic acid with an amine containing the terminal alkene moiety gave the RCM precursor. The final RCM reaction required no less than 0.7 equivalent of Grubbs I catalyst to obtain the tetrafunctionalized macrocycle in 30% isolated yield.

Carbohydrates were used as scaffolds for combinatorial decoration by the Overkleeft group (Scheme 20.9).[19] Rink resin (0.55 mmol/g) was loaded with a D-(+)-mannitol derived pyran scaffold via a C9 tether. The hydroxyl group was reacted with isocyanates, followed by Staudinger reduction of the azides and acylation of the resulting amines with acid chlorides or chloroformates. Final cleavage was achieved using Grubbs II catalyst furnishing a 3 × 3 set of cis-fused pyranofurans in overall yields of

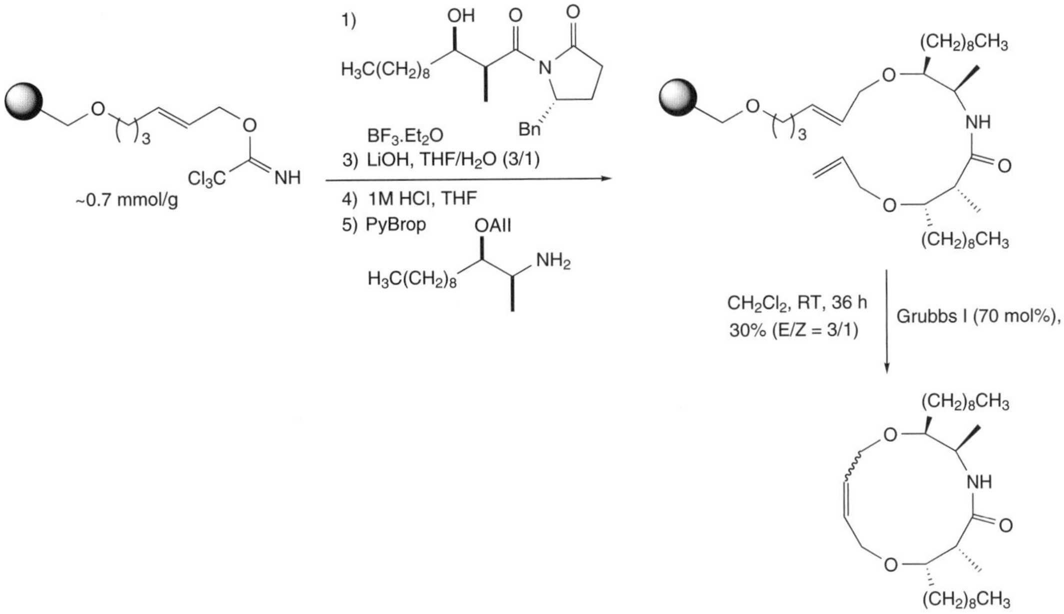

Scheme 20.8 A solid-phase synthesis of 12-membered lactams

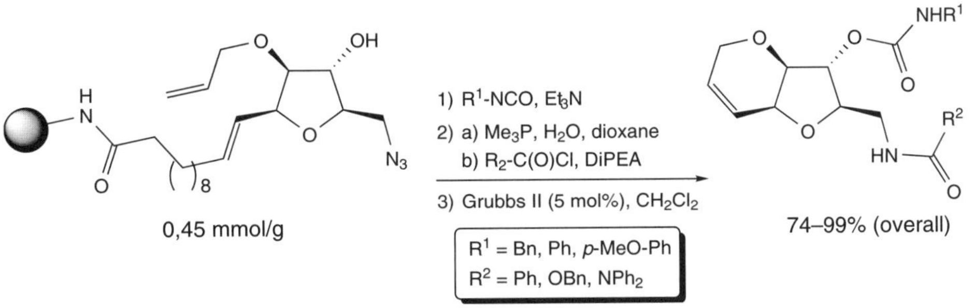

Scheme 20.9 Synthesis of bisfunctionalized pyranofurans by using a tethered linker to facilitate catalyst recycling

74–99% pointing to near quantitative yields for each step. The fact that only 5 mol% of the ruthenium catalyst was required underscores the beneficial effect of the C9 spacer, as was found also by Blechert and coworkers in their approach to cyclic pseudotetrapeptides (Scheme 20.6).

Typical Experimental Procedures
Protocol for the RCM cyclization–cleavage reaction (Scheme 20.9)

The RCM precursor resin(0.11 g, 0.05 mmol) was coevaporated with dichloroethane and dispersed in DCM (1 ml) under an argon atmosphere. Grubbs II catalyst(2 mg, 2.5 µmol, 5 mol%) was added and the suspension was refluxed for 16 hours. The resin was filtered off and washed with DCM. After concentration of the

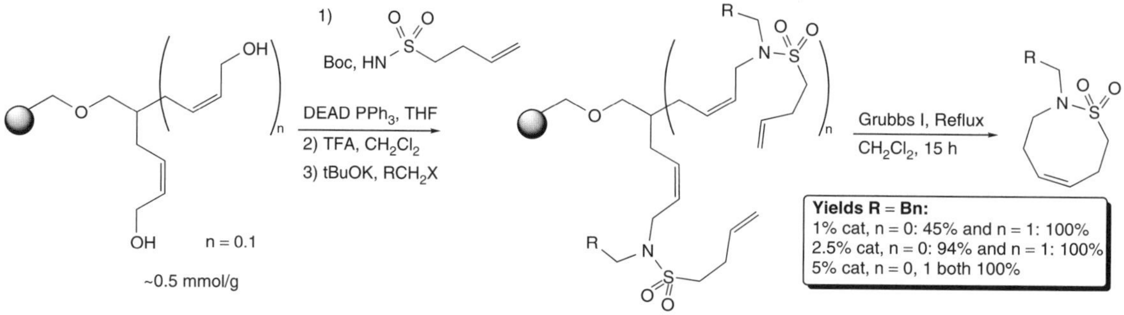

Scheme 20.10 Application of a double-armed olefin linker to enable catalyst recycling

mother liquor, the residue was taken up in 15% ethyl acetate in toluene and purified by filtration over a silica plug using the same eluent yielding the homogeneous pyranofuran.

Instead of using tethered linkers, the intrinsic problem of catalyst recovery was cleverly solved by Brown and coworkers using a double-armed olefin linker in an approach to seven-membered sulfonamides (Scheme 20.10).[20] After the first RCM reaction from the double-armed linker, the trapped ruthenium–alkylidene reacts further with the aptly positioned second arm with concomitant release of the other RCM precursor, which then cyclizes in solution regenerating the ruthenium–alkylidene species. Synthesis of the RCM precursor was accomplished by a Fukuyama–Mitsunobu reaction from N-Boc-3-butene-1-sulfonamide followed by Boc-removal and N-alkylation. For the RCM reaction only small amounts of Grubbs I catalyst were required to obtain the cyclic sulfonamides in very high yields. With catalyst loadings up to 5% the reaction worked equally well on a single-armed linker. However, with very low loadings (1%) the double-armed linker was required for obtaining high yields underscoring the efficacy of this strategy.

Typical Experimental Procedures

Procedure for the synthesis of the double-armed linker (Scheme 20.10)

To a solution of the bis-THP protected starting alcohol (490 mg, 1.38 mmol) in DMF (10 ml) was added sodium hydride (60 mg, 1.5 mmol). When the gas evolution ceased, the solution was added drop-wise to Merrifield resin (200 mg, loading 2.3 mmol Cl g^{-1}, 0.5 mmol) swollen in THF (5 ml). The reaction mixture was stirred at 60 °C for 15 hours. Excess sodium hydride was carefully quenched by drop-wise addition of water. The resin was collected by filtration and washed with DMF (10 ml), water (10 ml), DMF (10 ml) and DCM (10 ml). The resin was then dried in vacuo at 50 °C for five hours. IRv_{max} (neat) cm^{-1} 2921, 1601. Removal of the THP protective groups was accomplished by suspension of the resin (325 mg) in methanol (5 ml) followed by the addition of p-TSA (360 mg, 1.90 mmol). The mixture was stirred at room temperature for 15 hours. The resin was collected by filtration, washed with DCM (10 ml), methanol (10 ml), DCM (10 ml) and dried in vacuo at 50 °C for five hours. IRv_{max} (neat) cm^{-1} 3334 (br), 2920, 1601. The loading of the resin was estimated to be 0.97 mmol OH g^{-1}.

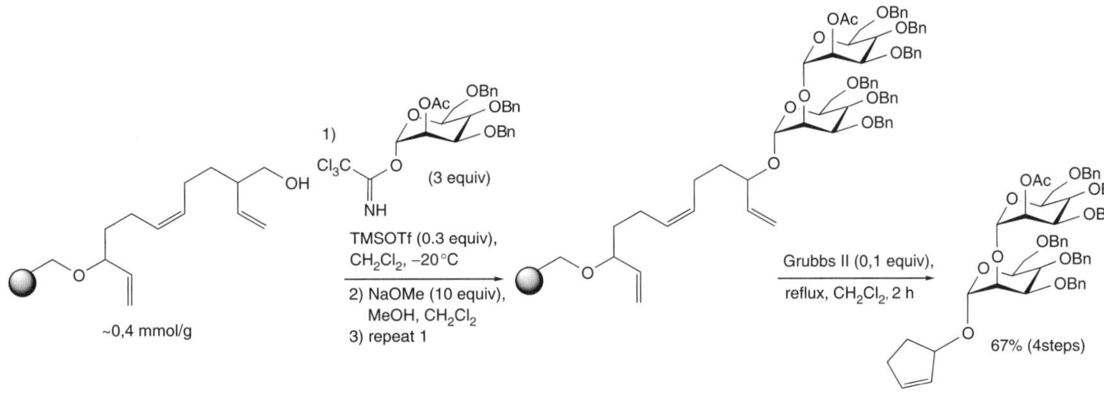

Scheme 20.11 A double-armed linker approach towards saccharides

At the same time, a similar strategy to release cycloolefins and regenerate the catalyst in the same step was published by Overkleeft and coworkers (Scheme 20.11).[21] A double-armed mono-hydroxyl-functionalized linker was loaded with the trichloroacetimidate of a suitably protected mannose followed, after selective O-deprotection, by another glycosylation. The addition of Grubbs II precatalyst (10%) induced the tandem cyclization–cleavage RCM and a concomitant catalyst regeneration process to fashion the disaccharide in an overall yield of 67%. Removal of the remaining cyclopentenyl moiety could be accomplished after Wilkinson's catalyst induced alkene migration followed by the addition of iodine to cleave the resulting vinyl ether providing the anomeric free sugar (not shown).

20.3 Terminal olefins via route II

Besides cycloolefins, linear olefins are useful molecules to be prepared via traceless solid phase RCM. An early contribution from the Blechert group describes a very elegant approach to a series of 4-functionalized styrenes (Scheme 20.12).[22] A styrene bearing an aldehyde was decorated by reduction/esterification, reductive amination/Boc-installation or the Wittig reaction. The addition of Grubbs I catalyst (two or three portions of 3%) liberated the 4-functionalized styrene's in acceptable yields.

Inspired by the results from the Blechert group a similar approach was published by Schmidt and coworkers to prepare oligosaccharides (Scheme 20.13).[23, 24] A protected lactosyl trichloroacetimidate was reacted with a solid-supported allylic alcohol followed by attachment of two additional lactose fragments arriving at a branched hexasaccharide. Final detachment by the Grubbs I precatalyst (12%) gave the

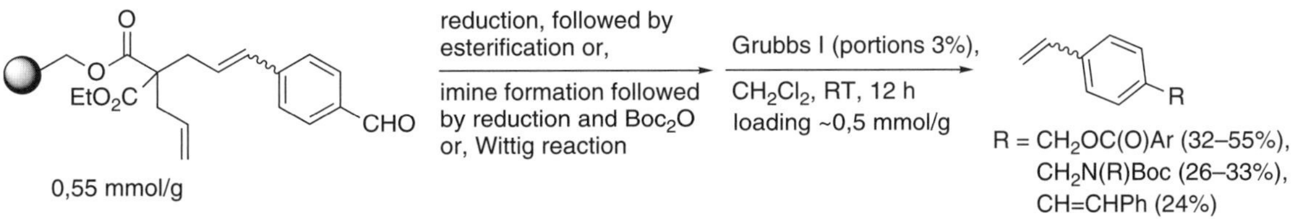

Scheme 20.12 An RCM approach to functionalized styrenes

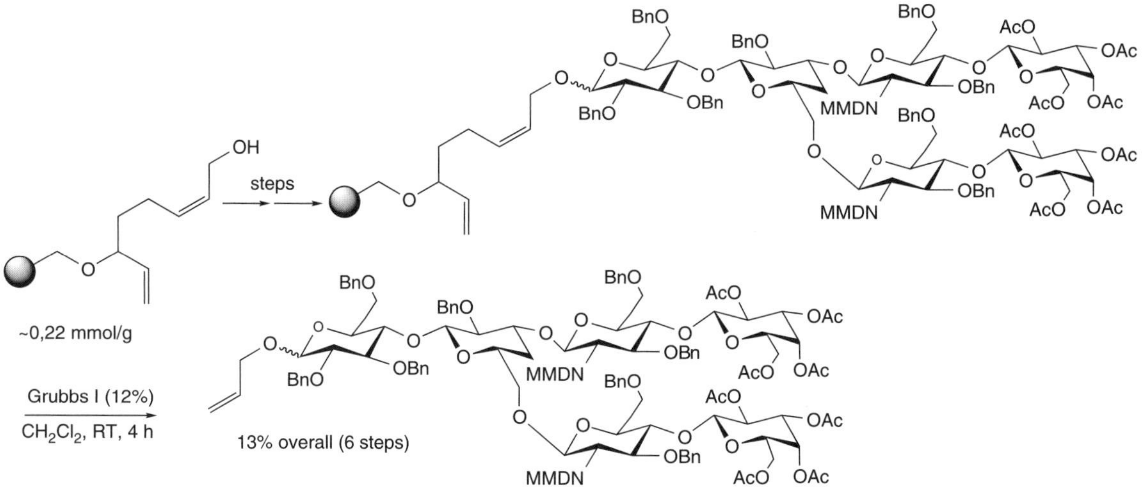

Scheme 20.13 Hexasaccharide synthesis on a allyl ether releasing RCM linker

anomeric allyl saccharide (1:1 α/β ratio) in an overall yield of 13% over six steps. It should be reported that both the glycosylation and RCM protocols were performed twice.

An impressive approach towards several analogs of the potent protein phosphatase inhibitor dysidiolide, using a similar linker liberating alkenes via RCM, was shown by the Waldmann group (Scheme 20.14).[25, 26] Besides 6-epi-dysidiolide, eight close analogs were made starting from an aldehyde-loaded resin in a 10-step sequence. Key on-resin steps in the synthesis were Wittig, Diels–Alder, acetal hydrolysis, 3-furyllithium addition and singlet oxygen oxidation reactions underscoring the robustness of an alkene linker. The final cleavage reaction was conducted using Grubbs I catalyst (0.1 equiv.) to liberate the final products in a very satisfying 14% overall yield.

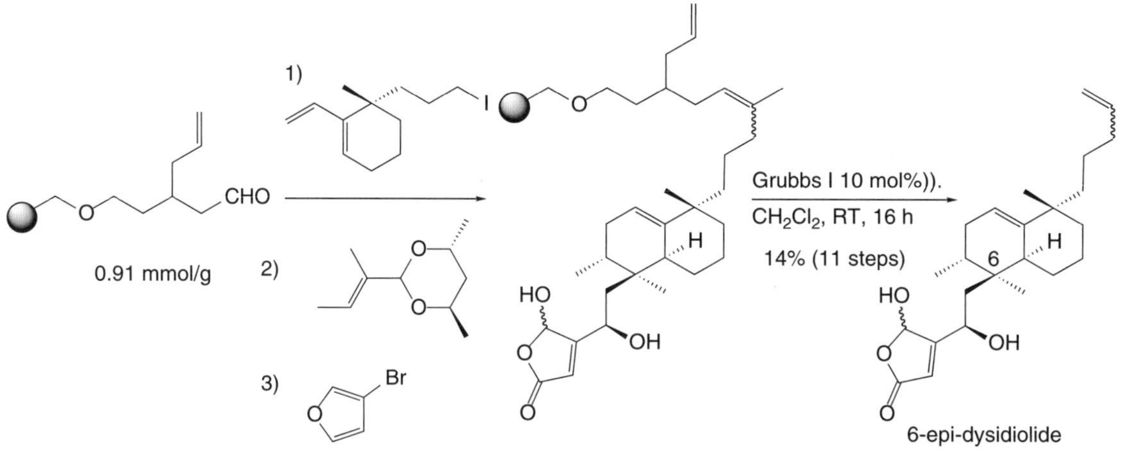

Scheme 20.14 Solid-phase synthesis of 6-epi-dysidiolide and analogs on an olefin releasing RCM linker

Typical Experimental Procedures
Synthesis of the aldehyde resin (Scheme 20.14)

A dispersion of sodium hydride in mineral oil (60%, 912 mg, 22.8 mmol) was added to a stirred solution of the mono-THP protected alcohol(5.2 g, 22.8 mmol) in dry DMF (50 ml) at 0 °C. After two hours Merrifield resin (5.51 g, 6.06 mmol, c = 1.1 mmol) and tetrabutylammonium iodide (215 mg, 0.61 mmol) were added and the mixture was shaken at room temperature for 20 hours. The reaction was quenched by addition of water (20 ml) and the resin was filtered followed by washings with DMF/water (1:1, 3 times), DMF (3 times), THF (3 times) and DCM (3 times), and dried to constant weight in vacuo to yield 6.79 g (6.06 mmol, c = 0.91 mmol/g, quantitative) of the THP protected alcohol yellowish resin. The loading and the yield were calculated on the basis of the weight difference. IR v_{max} (neat) cm^{-1} 1639.

For removal of the THP protective group a suspension of the obtained resin (6.72 g, 6.05 mmol) and pyridinium-p-toluene sulfonate (3.04 g, 12.1 mmol) in 1,2-dichlorethane/ethanol (1:1, 60 ml) was refluxed for 20 hours. After cooling to room temperature, the resin was filtered, washed with THF (3 times) and DCM (3 times), and dried in vacuo to afford the alcohol resin in a yield of 6.21 g (5.95 mmol, c = 0.96 mmol/g, 98%.) as a yellowish resin. The loading and yield were determined by the Fmoc-method. IR v_{max} (neat) cm^{-1} 3459, 1639.

Oxidation of the primary alcohol to the aldehyde was accomplished by adding a solution of 2-iodoxybenzoic acid (IBX) (7.7 g, 27.5 mmol) in DMSO (40 ml) to a suspension of the alcohol resin (5.6 g, 5.5 mmol) in THF (40 ml) at room temperature. The mixture was shaken for eight hours and then filtered. The resin was washed with DMSO/THF (1:1, 3 times), THF (3 times) and DCM (3 times), and dried in vacuo to yield 5.6 g (5.08 mmol, c = 0.91 mmol/g, 92%) of the yellowish aldehyde resin. The loading and yield were determined by the DNPH method (see supporting info ref. 25). IR v_{max} (neat) cm^{-1} 2718, 1724, 1639.

20.4 Terminal and internal olefins via route III

The Seeberger group developed a solid-supported route towards oligosaccharides using an 4-octene-1,8-diol based linker that was etherified to Merrifield resin (Scheme 20.15).[27, 28] Coupling of the first sugar and elongation of the oligosaccharide were performed using orthogonally protected glucose trichloroacetimidates or phosphates (not shown) as donors. Treatment of the supported oligosaccharides with Grubbs I catalyst in the presence of an ethylene atmosphere liberated the final sugars via CM in acceptable overall yields (average 84–95% yield for each coupling step). A later contribution of the same group showed that by using a phosphine-free catalyst even azide-containing sugars may be prepared, although 1-pentene had to be used for the final cleavage CM reaction instead of ethylene.[27]

Typical Experimental Procedures
Synthesis of the starting alcohol resin (Scheme 20.15)

8-(4,4'-dimethoxytrityl)-4-(Z)-octenol (see supporting info ref. 26) (1.49 g, 3.17 mmol, 3.3 equiv.) was dissolved in DMF (10 ml) and after cooling to 0 °C sodium hydride (60% dispersion in mineral oil, 0.13 g, 3.17 mmol) was added followed by stirring for one hour. Merrifield resin (1% cross-linked, 0.80 g, 0.96 mmol, 1.0 equiv.) was added along with tetrabutylammonium iodide (36 mg, 96 μmol, 0.1 equiv.). After shaking for one hour at 0 °C, the reaction mixture was warmed to room temperature and shaken for 12 hours. Capping of the unreacted sites was accomplished with methanol (0.10 ml) and sodium hydride (60% dispersion in

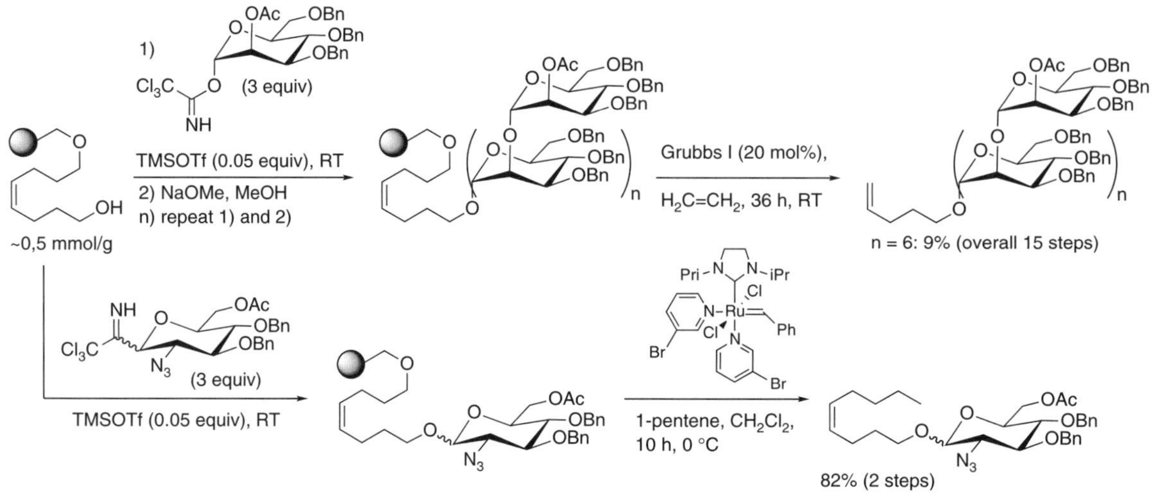

Scheme 20.15 Solid-phase saccharide synthesis on an olefin linker and final detachment by CM

mineral oil, 0.10 g) for four hours. Methanol (5 ml) was added and the resin was washed subsequently with portions (10 ml) of methanol/DMF (1:1), DMF, THF (three times) and DCM (three times). Drying in vacuo over phosphorus pentoxide (P_2O_5) afforded 1.08 g of the DMT protected alcohol resin. The loading (0.55 mmol/g) was determined by the DMT-cation assay (see supporting info ref. 27). Removal of the DMT protective group was effected by by washing the resin with 3% dichloroacetic acid/DCM (3 × 20 ml) followed by further washing with DCM (3 × 20 ml), 1% triehtylamine/DCM, DCM and drying in vacuo to afford 0.95 g of the alcohol resin (0.62 mmol/g)

20.5 Conclusion

Ten years after the first publication on the use of olefin metathesis as a product releasing reaction it is clear that great advances have been made. Key elements for a successful approach are the type of olefin linker that one may choose in combination with the introduction of a short spacer between the polymer matrix and linking olefin, and the type of catalyst. Many different reactions have been carried out for the synthesis of the cleavage precursors on several types of olefin linkers using a wide range of conditions, such as strong acids and bases but also oxidations/reductions and transition metal catalyzed transformations, indicating the versatility of the cleavage-by-metathesis approach. Therefore, many more applications on the use of olefin metathesis to release linear or cyclic olefins in solid-phase diversity-oriented synthetic approaches may be expected in the near future.

References

[1] Grubbs, R. H., Miller, S. J., and Fu, G. C.; *Accounts of Chemical Research* **1995**, *28*, 446.
[2] Grubbs, R. H.; *Advanced Synthesis & Catalysis* **2007**, *349*, 34.
[3] Schrock, R. R.; *Advanced Synthesis & Catalysis* **2007**, *349*, 41.
[4] Chauvin, Y.; *Advanced Synthesis & Catalysis* **2007**, *349*, 27.
[5] Scholl, M., Ding, S., Lee, C. W., and Grubbs, R. H.; *Organic Letters* **1999**, *1*, 953.

[6] Garber, S. B., Kingsbury, J. S., Gray, B. L., and Hoveyda, A. H.; *J. Amer. Chem. Soc.* **2000**, *122*, 8168.
[7] Piscopio, A. D., and Robinson, J. E.; *Curr. Opin. Chem. Biol.* **2004**, *8*, 245.
[8] van Maarseveen, J. H., den Hartog, J. A. J., Engelen, V., *et al.*; *Tetrahedron Lett.* **1996**, *37*, 8249.
[9] Veerman, J. J. N., van Maarseveen, J. H., Visser, *et al.*; *Eur. J. Org. Chem.* **1998**, 2583.
[10] Brittain, D. E. A., Gray, B. L., and Schreiber, S. L.; *Chemistry-A European Journal* **2005**, *11*, 5086.
[11] Piscopio, A. D., Miller, J. F., and Koch, K.; *Tetrahedron Lett.* **1997**, *38*, 7143.
[12] Piscopio, A. D., Miller, J. F., and Koch, K.; *Tetrahedron* **1999**, *55*, 8189.
[13] Piscopio, A. D., Miller, J. F., and Koch, K.; *Tetrahedron Lett.* **1998**, *39*, 2667.
[14] Pernerstorfer, J., Schuster, M., and Blechert, S.; *Chem. Comm.* **1997**, 1949.
[15] Garner, A. L., and Koide, K.; *Org. Lett.* **2007**, *9*, 5235.
[16] Nicolaou, K. C., Winssinger, N., Pastor, J., *et al.*; *Nature* **1997**, *387*, 268.
[17] Nicolaou, K. C., Winssinger, N., Pastor, J., and Murphy, F.; *Angew. Chem. Int. Ed.* **1998**, *37*, 2534.
[18] Sasmal, S., Geyer, A., and Maier, M. E.; *J. Org. Chem.* **2002**, *67*, 6260.
[19] Timmer, M. S. M., Verdoes, M., Sliedregt, L. A. J. M., *et al.*; *J. Org. Chem.* **2003**, *68*, 9406.
[20] Moriggi, J. D., Brown, L. J., Castro, J. L., and Brown, R. C. D.; *Org. Biomol. Chem.y* **2004**, *2*, 835.
[21] Timmer, M. S. M., Codee, J. D. C., Overkleeft, *et al.*; *Synlett* **2004**, 2155.
[22] Peters, J. U., and Blechert, S.; *Synlett* **1997**, 348.
[23] Knerr, L., and Schmidt, R. R.; *Synlett* **1999**, 1802.
[24] Knerr, L., and Schmidt, R. R.; *Eur. J. Org. Chem.* **2000**, 2803.
[25] Brohm, D., Metzger, S., Bhargava, A., *et al.*; *Angew. Chem. Int. Ed.* **2001**, *41*, 307.
[26] Brohm, D., Philippe, N., Metzger, S., *et al.*; *J. Amer. Chem. Soc.* **2002**, *124*, 13171.
[27] Andrade, R. B., Plante, O. J., Melean, L. G., and Seeberger, P. H.; *Organic Letters* **1999**, *1*, 1811.
[28] Kanemitsu, T., and Seeberger, P. H.; *Organic Letters* **2003**, *5*, 4541.

Part IV

Alternative Linker Strategies

21
Fluorous Linker Units

Wei Zhang

Department of Chemistry, University of Massachusetts, USA

21.1 Introduction

21.1.1 Polymer versus fluorous linkers

Linkers and associated techniques play key roles in solid-phase organic synthesis. The development of fluorous linkers provides a solution-phase alternative for parallel and high-throughput synthesis.[1] Perfluorocarbon chains (Rf, also called fluorous tags) are lipophobic and hydrophobic. This special property promises the easy separation of fluorous molecules from the mixture containing non-fluorous organic compounds.[2, 3] The utility of solid-supported linkers and fluorous linkers is similar in concept, but very different in practice. Fluorous linkers are real molecules; they can be soluble in common organic solvents; they can be analyzed by LC-MS and NMR; and they are relied on selective fluorous methods for separations.[4]

Polymer linkers have bifunctional moieties to connect the reactive functional group and the polymer support. Since the fluorous tag can be connected to the linker through a carbon–carbon bond, fluorous linkers have mono- and bifunctional versions. For example, the bifunctional fluorous PMB-OH linker is structurally similar to the Wang resin, except the support is a fluorous tag instead of a polymer bead. The fluorous tag of the monofunctional fluorous PMB-OH linker is directly attached to the reactive group through the carbon–carbon bond (Scheme 21.1).

21.1.2 Different types of fluorous linkers

Fluorous linkers can be classified into protective, displaceable and safety-catch linkers. Fluorous protective linkers are derivatives of common protecting groups such as tri-*iso*-propylsilyl (TIPS), *tert*-butoxycarbonyl

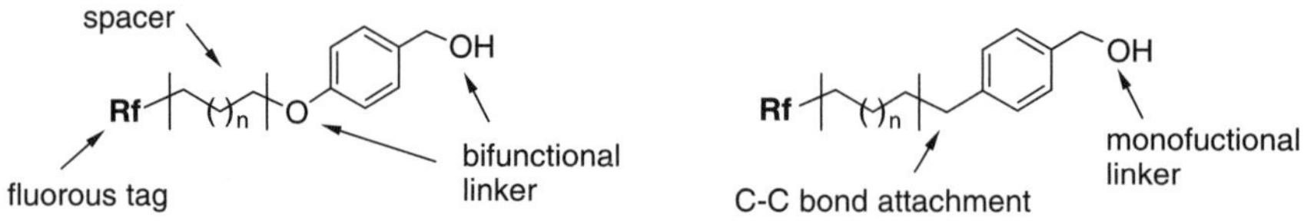

554 Alternative Linker Strategies

Scheme 21.1 *Fluorous bifunctional (left) and monofunctional (right) PMB-type linkers*

(Boc), 9-fluorenylmethoxycarbonyl (Fmoc), *para*-methoxybenzyl (PMB) and benzyloxycarbonyl (Cbz) (Table 21.1). They are used to protect amino, hydroxyl, carboxyl and other functional groups. The protected functional groups can be regenerated after the linker cleavage. Reaction conditions for attachment and cleavage of fluorous linkers are similar to those used for corresponding normal protecting groups.

Table 21.1 *Fluorous protective linkers*

Linker	Structure	Ref	Linker	Structure	Ref
Boc	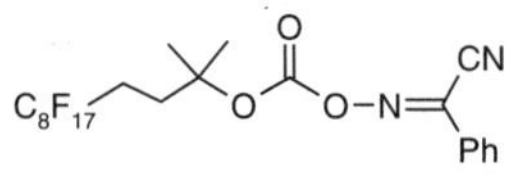 C_8F_{17} ...	[5-7]	PMB	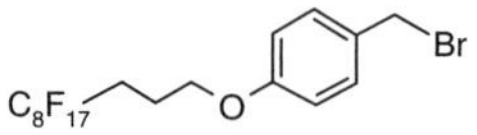 C_8F_{17} ... Br	[18,19]
Cbz	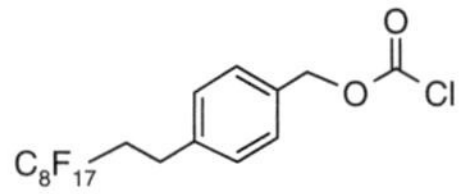 C_8F_{17} ... Cl	[8]		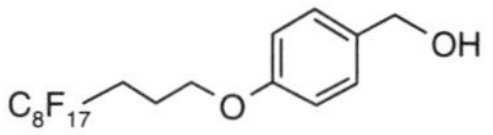 C_8F_{17} ... OH	[20]
	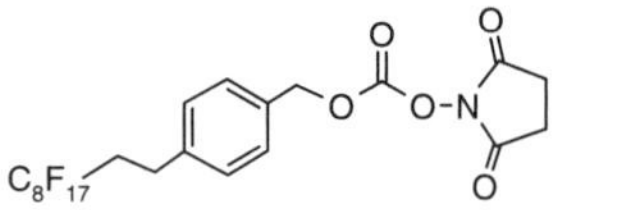 C_8F_{17} ...	[9]	DMTC	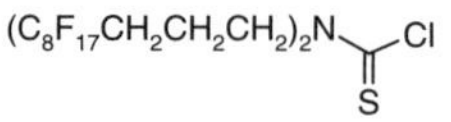 $(C_8F_{17}CH_2CH_2CH_2)_2N-C(=S)Cl$	[21]
Fmoc	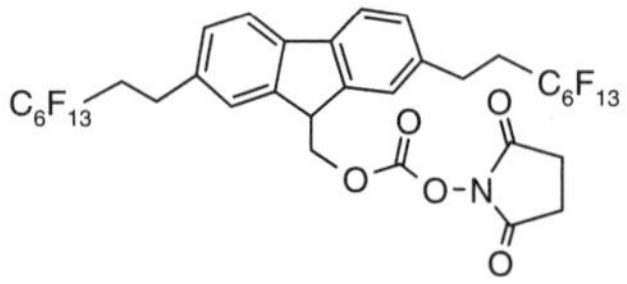 C_6F_{13} ... C_6F_{13}	[10]	Arylsulfonyl	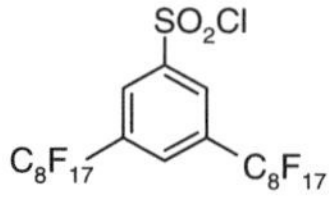 SO_2Cl / C_8F_{17} ... C_8F_{17}	[22]
			Silyl	$(C_6F_{13}CH_2CH_2)_3SiBr$	[23]
Msc	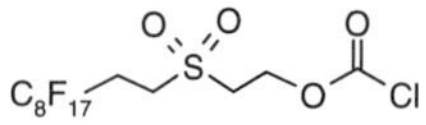 C_8F_{17} ... Cl	[11]		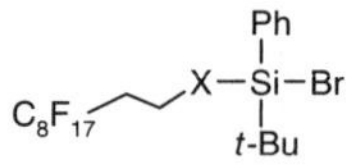 C_8F_{17}–X–Si(Ph)(t-Bu)–Br, X = O or nothing	[24-26]
Froc	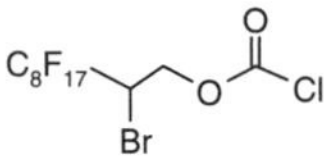 C_8F_{17} ... Cl (Br)	[12]			
Teoc	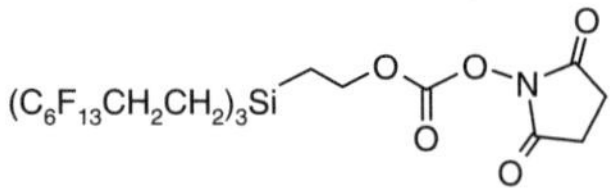 $(C_6F_{13}CH_2CH_2)_3Si$ ...	[13]	TIPS	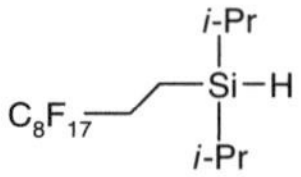 C_8F_{17}–Si(i-Pr)$_2$–H	[27-29]

(continued overleaf)

Fluorous Linker Units 555

Table 21.1 (continued)

Linker	Structure	Ref	Linker	Structure	Ref
Benzaldehyde	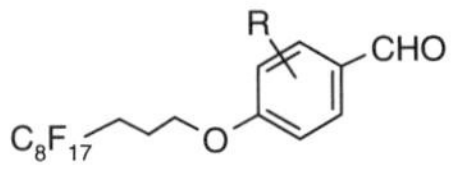 C_8F_{17}–O–C6H3(R)–CHO	[14,15]	Alkoxyethyl	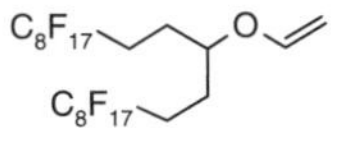 (C_8F_{17})$_2$CH–O–CH=CH$_2$	[30]
Trityl	C_8F_{17}–C6H4–C(Cl)(C6H4R)$_2$	[16,17]	MOM	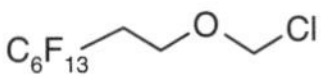 C_6F_{13}–CH$_2$CH$_2$–O–CH$_2$Cl	[31,32]
	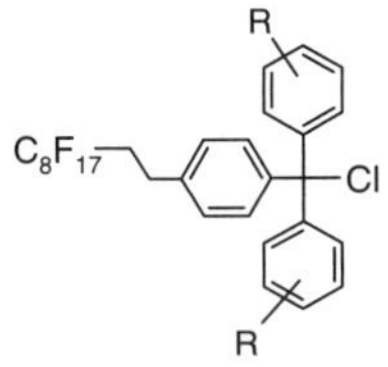		THP	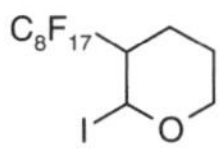 C_8F_{17}–(tetrahydropyran)–I	[33]
Bfp	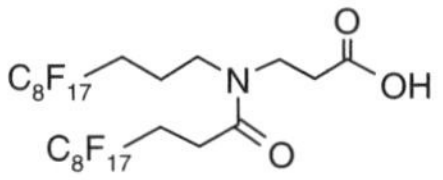 (C_8F_{17}CH$_2$CH$_2$CH$_2$)$_2$N–CH$_2$C(O)OH (with acyl)	[34]	TMSE	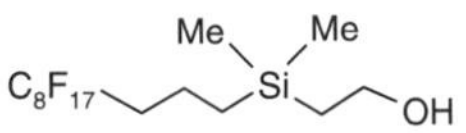 C_8F_{17}–(CH$_2$)$_3$–Si(Me)$_2$–CH$_2$CH$_2$OH	[38]
TfBz	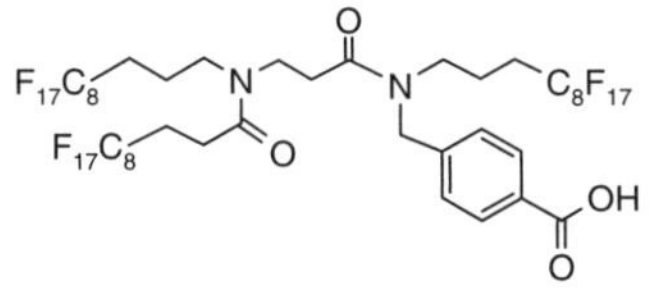 (F$_{17}$C$_8$–)$_2$N–…–N(–C$_8$F$_{17}$)–CH$_2$–C6H4–COOH	[35]	t-Butyl	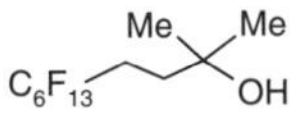 C_6F_{13}–CH$_2$CH$_2$–C(Me)$_2$OH	[39]
Phenole	(C_8F_{17}–CH$_2$CH$_2$CH$_2$O)$_3$CH–CH$_2$–O–C6H4–OH	[36]		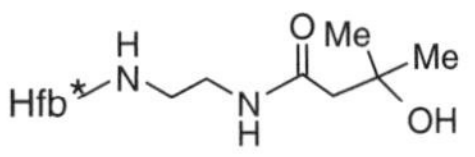 Hfb*–NH–CH$_2$CH$_2$–NH–C(O)–C(Me)$_2$OH	[37]
	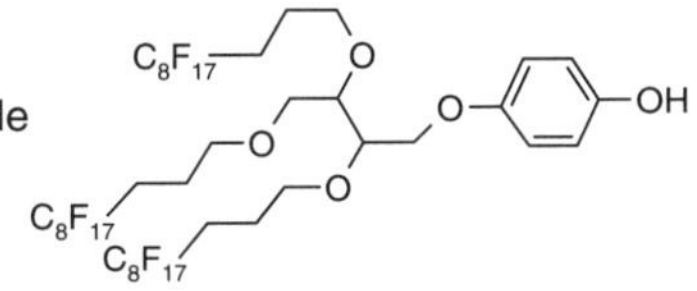		Diol	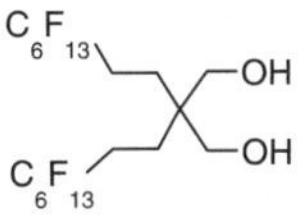 (C_6F_{13}–)$_2$C(CH$_2$OH)$_2$	[40]
Rink	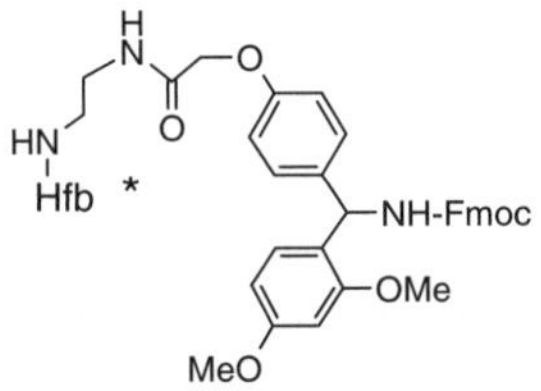 Hfb*–NH–CH$_2$CH$_2$–NH–C(O)CH$_2$–O–C6H3(OMe)–CH(NH-Fmoc)–C6H4–OMe	[37]	Benzylamine	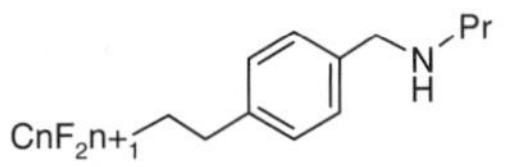 CnF$_{2n+1}$–CH$_2$CH$_2$–C6H4–CH$_2$–NH–Pr	[41]

*Hfb ia a heavy fluorous tag containing six C_8F_{17} groups

Fluorous displaceable linkers convert the original functional group to a new group after the linker cleavage (Table 21.2). The linker cleavage can be accomplished by a displacement reaction to introduce a new diversity point or by a cyclization reaction to form a new ring. The displaceable linker can also be cleaved in a traceless fashion to form a carbon–hydrogen bond. Fluorous sulfonyl, thiol and alcohol displaceable linkers have been used in the diversity-oriented synthesis (DOS) of library scaffolds.

556 *Alternative Linker Strategies*

Table 21.2 Fluorous displaceable linkers

Linker	Structure	Ref	Linker	Structure	Ref
Benzophenone amino	C_8F_{17}-C$_6$H$_4$-C(=NH)-C$_6$H$_5$	[42]	Alcohol	C_8F_{17}-(CH$_2$)$_3$-OH	[45]
Sulfonyl	$C_8F_{17}O_2SOF$	[43]	Thiophenol	C_8F_{17}-CH$_2$CH$_2$-C(O)-N(Me)-C$_6$H$_4$-SH	[46]
FluoMar	C_8F_{17}-S-C$_6$H$_4$-CH$_2$OH	[44]	Alkylstannyl	$(C_8F_{17}CH_2CH_2)_3$Sn-Br	[47]

Table 21.3 Safety-catch linkers

Linker	Structure	Ref	Linker	Structure	Ref
Alkylthio	C_8F_{17}-CH$_2$CH$_2$-SH	[48–50]	Arylgermyl	C_8F_{17}-CH$_2$CH$_2$-Ge(2-Nap)$_2$-Br	[51]

Fluorous safety-catch linkers are stable under common reaction conditions (Table 21.3). They need to be activated before the cleavage. Alkylthio and arylgemanyl are the only two fluorous safety-catch linkers that have been developed so far. They need to be activated by oxidation and photolysis, respectively.

Typical Experimental Procedures

Attachment and cleavage of fluorous TIPS[27]

CF_3SO_3H (16.4 mmol) is added to (3,3,4,4,5,5,6,6,7,7,8,8,9,9,10,10,10-heptadecafluorodecyl)diisopropylsilane ($C_8F_{17}CH_2CH_2(i\text{-}Pr)_2SiH$) (21.4 mmol) at 0 °C. The mixture is stirred at 25 °C for 15 hours. A solution of the alcohol (16.4 mmol) and 2,6-lutidine (32.8 mmol) in dry dichloromethane (CH_2Cl_2) (40 ml) is added. After two hours, the mixture is quenched with aqueous ammonium chloride and extracted with dichloromethane (DCM) and ether. The concentrated crude product is purified by F-SPE to give protected product. To a solution of F-TIPS attached compounds (0.01 mmol) in tetrahydrofuran (THF) (0.2 ml) is added 10 drops of HF-pyridine at 25 °C, and then it is heated at 60 °C for 1–10 hours. The reaction mixture is diluted with ethyl acetate, washed with aqueous sodium bicarbonate. The concentrated organic layer is subjected to F-SPE to give deprotected product.

Attachment and cleavage of F-Cbz[9]

To a solution of L-phenylalanine (2.16 mmol) is added triethylamine (2.16 mmol) and F-Cbz-OSu to pH 2 with aqueous hydrochloric acidl and extracted with ethyl acetate. The organic layer was extracted with water *and dried. The concentrated crude product is purified by recrystallization or by F-SPE to give protected product in 67%. The fluorous linker can be cleaved under standard hydrogenation conditions and purified by F-SPE.*

21.1.3 Methods for separation of fluorous linker attached substrates

Fluorous liquid–liquid extraction (F-LLE), fluorous solid-phase extraction (F-SPE) and fluorous high-performance chromatography (F-HPLC) are three primary methods for fluorous separations. F-LLE is for the separation of 'heavy fluorous' molecules that contain long or many perfluorocarbon chains.[52] Both the reaction and purification of heavy fluorous molecules may require fluorous solvents. Perfluorohexanes (FC-72), perfluoromethylcyclohexane (PFMC), perfluorobutylmethylether (HFE-7100), perfluorobutylethylether and $C_3F_7CF(OC_2H_5)CF(CF_3)_2$ (HFE-7500) are some of the common fluorous solvents for F-LLE.

Fluorous silica gel-based F-SPE has been developed for the separation of 'light fluorous' molecules that contain short and few perfluorocarbon chains, such as C_6F_{13} and C_8F_{17}.[53] Light fluorous molecules have reasonable solubility in common organic solvents. They have a low partition coefficient in fluorous solvents and are not suitable for F-LLE, but can be efficiently retained on fluorous silica gel when elute with fluorophobic solvents such as 80:20 methanol:water.

The HPLC-grade fluorous silica gel is able to separate a mixture of fluorous compounds based on their different fluorine contents.[54] The mobile phase is a gradient of methanol-water or acetonitrile-water, which is similar to that used for reverse-phase HPLC. Preparative F-HPLC is an important technique for fluorous mixture synthesis (FMS).[55]

Typical Experimental Procedures

Typical F-SPE procedures

A concentrated reaction mixture containing the fluorous molecules is loaded onto an F-SPE cartridge. The suggested mass loading is between 5 and 10%. On a SPE vacuum manifold, the cartridge is first washed with fluorophobic 80:20 methanol:water for 4–8 column volume to collect the non-fluorous organic compounds, and then washed with fluorophilic methanol or acetone to collect the fluorous component. The cartridge can be regenerated by washing thoroughly with acetone or THF for reuse.

21.2 Fluorous linkers for synthesis of small molecules

Over the last decade, a wide range of fluorous protective, displaceable and safety-catch linkers have been developed for solution-phase parallel and high-throughput synthesis of small molecules.[56, 57] Only a few of representative examples are described in this section.

21.2.1 Synthesis of heterocyclic compounds

The Curran group first introduced the fluorous Boc and demonstrated its utility in the synthesis of a small amide library of isonipecotic acid derivatives.[5] Zhang and Tempest used the fluorous Boc to improve the efficiency of Ugi/de-Boc/cyclization in the synthesis of quinoxalinone **3** and benzimidazole **5** analogs (Scheme 21.2).[6] The Ugi reactions involving F-Boc protected aniline derivative **1** were carried out under microwave conditions to form **2** and **4**, respectively. Excess aldehydes and unreacted acids were readily removed by F-SPE. The combination of microwave and F-SPE techniques was also applied to the de-Boc/cyclization step for facile synthesis and the separation of final products.

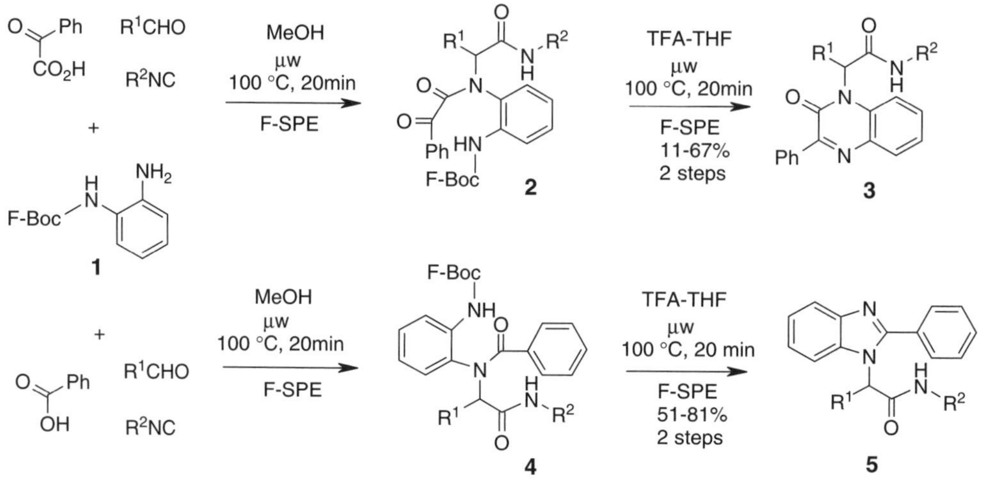

Scheme 21.2 *Microwave-assisted fluorous synthesis of quinoxalinones 3 and benzimidazoles 5*

Typical Experimental Procedures

Attachment and cleavage of fluorous Boc[5]

A mixture of F-BOC-ON reagent (1.3 mmol), isonipecotic acid (1.2 mmol) and triethylamine (20 mmol) is stirred in dry DCM (10.0 ml) at 22 °C for two hours. The reaction mixture is concentrated to <0.3 ml and then loaded on to a cartridge for F-SPE purification. Collection and then concentration of the fluorophilic wash fraction provides protected amino acid in 96% yield. F-Boc protected amine (0.05 mmol) is stirred with 1:3 concentrated hydrochloric acid:methanol (1.0 ml) at 65 °C for 16 hours. The mixture is concentrated to <0.1 ml and then purified by F-SPE to give deprotected product in typical yields of 90–100%.

The Firestine group applied F-Boc in the synthesis of polyamide **9**, a minor groove binding agent related to distamycin which has high affinity to specific DNA sequences.[7] The target heterocycle-conjugated polyamide has a hairpin-like structure (Scheme 21.3). The nitro group of F-Boc protected dipeptide **6** was reduced and then underwent sequential amide coupling reactions to afford polyamide **7**. The final hairpin polyamide **9** was prepared by the coupling of fluorous fragment **7** with non-fluorous fragment **8** followed by F-Boc cleavage.

Schwinn and Bannwarth applied F-Cbz protected aniline derivatives for synthesis of quinazoline-2,4-diones **12** (Scheme 21.4).[8b] Amide coupling of fluorous protected acids **10** followed by cyclative deprotection of **11** led to the formation of desired products.

Perfluorooctanesulfonyl attached benzaldehydes have been used as displaceable linkers for multicomponent reactions (MCRs). In the fluorous Groebke reaction, a reaction of fluorous benzaldehyde **13**, isonitrile and 2-aminopyridines afforded imidazo[1,2+a]pyridine **14**.[43e] The isolated condensation product was then subjected to a Suzuki-type reaction to form compound **15** (Scheme 21.5). This general protocol has been used for parallel synthesis of an 80-member compound library.

Fluorous benzaldehyde **16** has been used in a three-component [3 + 2] cycloaddition with a slightly excessive amount of an amine and an activated alkene (Scheme 21.6).[43f] The fluorous linker of the cycloaddtion product **17** was removed by the Suzuki coupling reaction to afford biaryl-substituted bicyclic proline **18**.

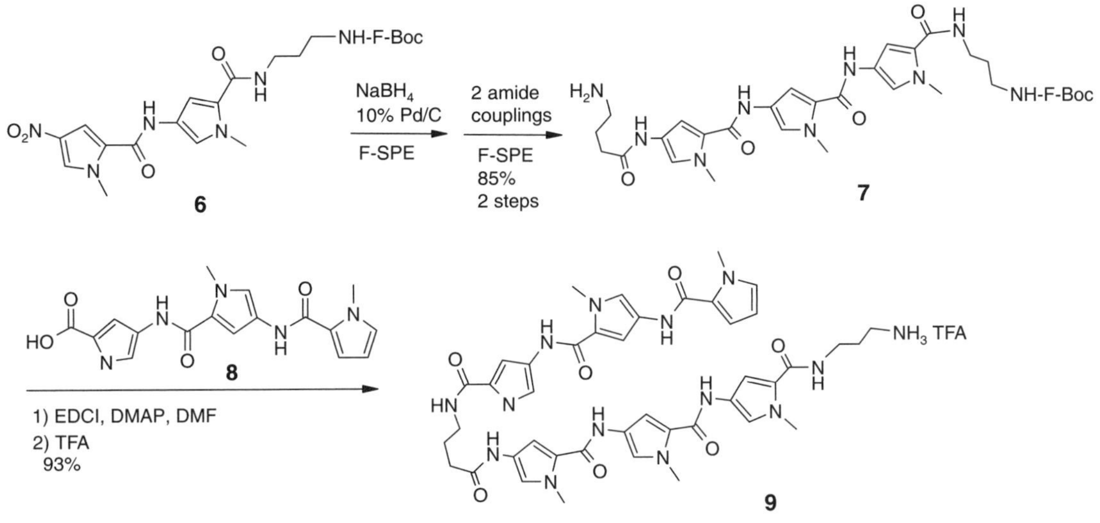

Scheme 21.3 F-Boc-facilitated synthesis of hairpin polyamide

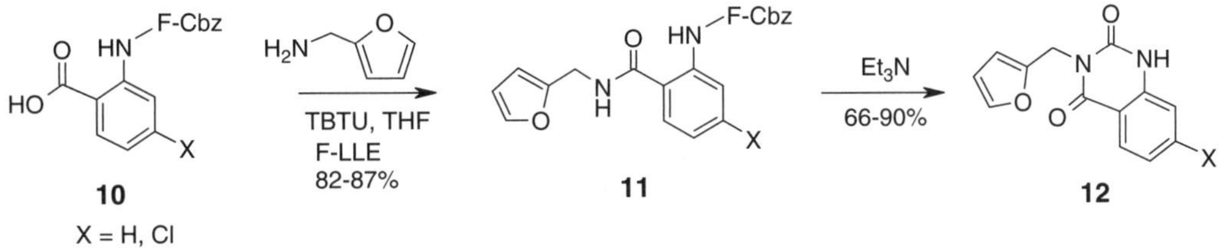

Scheme 21.4 F-Cbz facilitated synthesis of quinazoline-2,4-diones

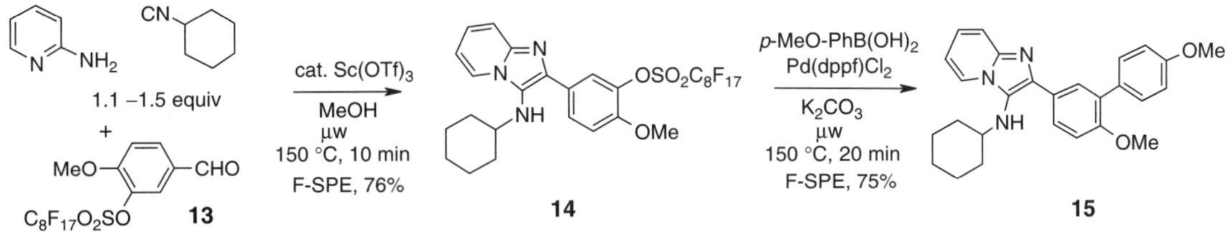

Scheme 21.5 Fluorous benzaldehyde facilitated synthesis of imidazo[1,2+a]pyridine

The Zhang group reported the synthesis of an *N*-alkylated dihydropteridinone library using fluorous alcohol attached amino acids as the starting materials (Scheme 21.7).[45b] The first chlorine of 4,6-dichloro-5-nitropyrimidine was displaced with fluorous amino acids **19**. The second chlorine was displaced with secondary amines to form compound **20**. The reduction of the nitro group of **21** followed by cyclization reaction and selective *N*-alkylation reaction gave *N*-alkylated dihydropteridinones **23**.

Sun and coworkers employed the fluorous alcohol linker in the synthesis of a hydantion-fused tetrahydro-β-carboline library (Scheme 21.8).[45e] The fluorous alcohol attached Boc-*L*-tryptpohan **24** was subjected to the Pictect–Spengler reaction with aldehydes to form the tetrahydro-β-carboline compounds **25**. This compound was then treated with an isocyanate, which led to the formation of the final products **26**.

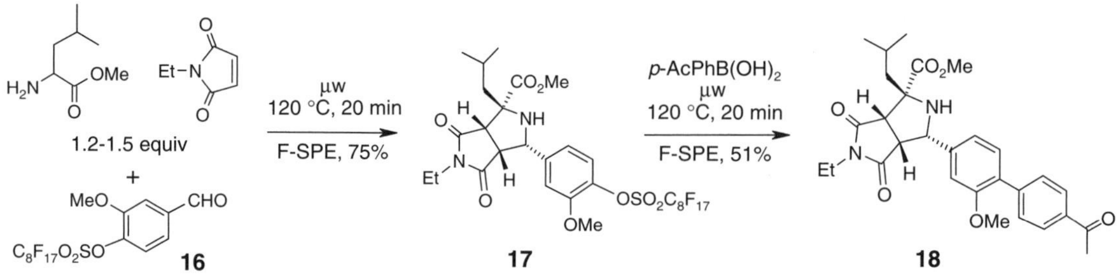

Scheme 21.6 Synthesis of a biaryl-substituted heterocyclic compound

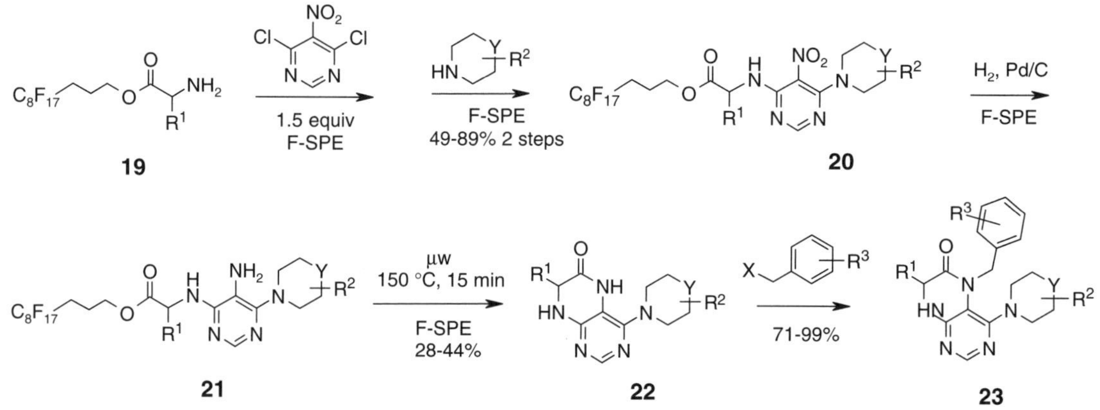

Scheme 21.7 Synthesis of N-alkylated dihydropteridinones

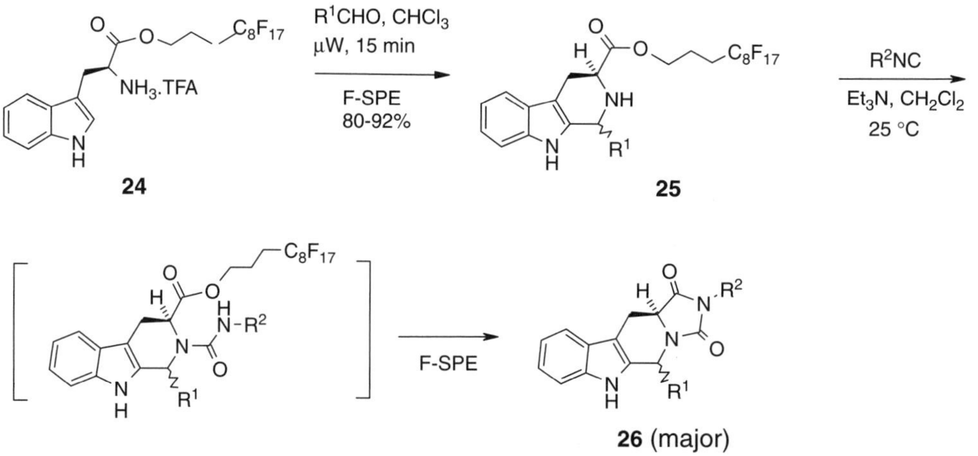

Scheme 21.8 Synthesis of hydantion-fused tetrahydro-β-carbolines

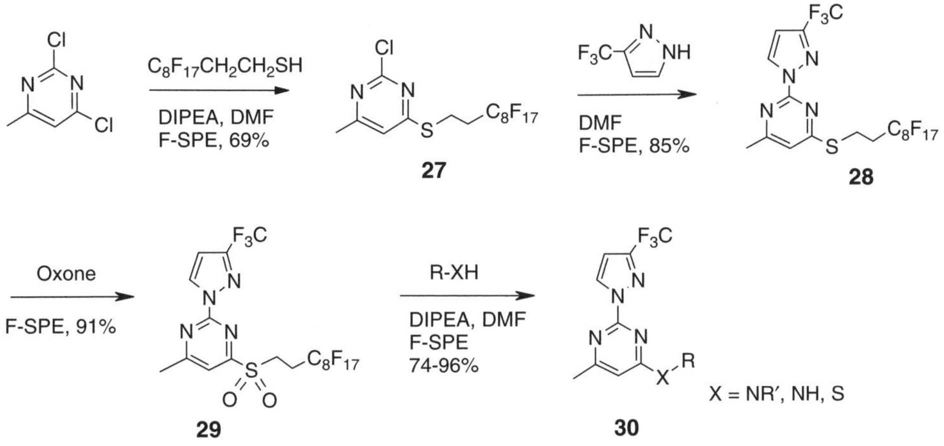

Scheme 21.9 Synthesis of 2,4-disubstituted pyrimidines

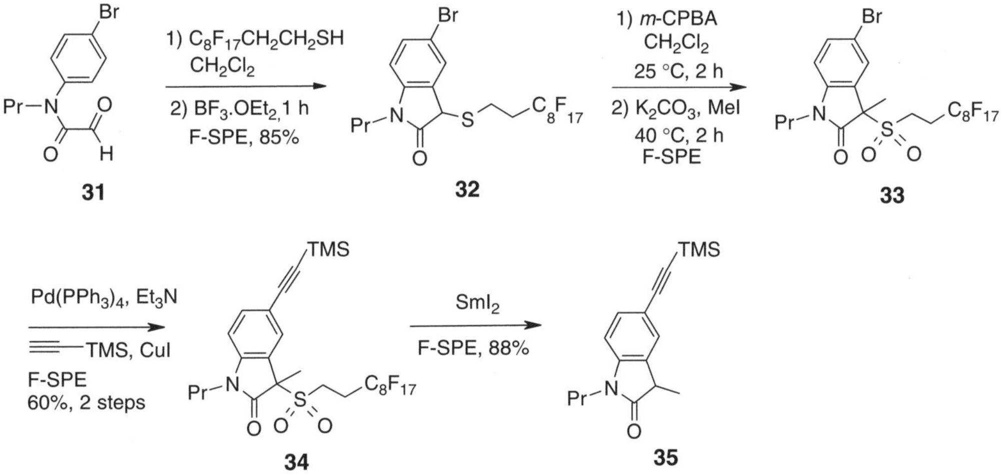

Scheme 21.10 Fluorous Pummerer cyclization and post-cyclization modifications

Fluorous thiol has been employed as a safety-catch linker for 2,4-dichloropyrimidine (Scheme 21.9).[48] The attached substrate **27** was displaced with 3-(trifluoromethyl)pyrazole to give **28**. The thiol tag was activated by oxidation with Oxone to form sulfone **29** and then displaced by a set of nucleophiles to afford disubstituted pyrimidines **30**.

The Procter group employed the fluorous thiol to catch Pummerer cyclization products.[49] The attached compound **32** was oxidized to sulfone, followed by methylation, palladium catalyzed coupling reaction and traceless tag cleavage to afford the substituted indolinone compound **35** (Scheme 21.10).

The Spivey group introduced arylgermanyl linkers to catch arylhalides through aryllithium or Grignard reactions.[51] Fluorous arylgermanyl compound **37** was activated by photolysis and then coupled with aryl bromides to form biaryl product **38**. Reduction with tin(II) chloride and amide coupling afforded fungicide Boscalid **39** (Scheme 21.11).

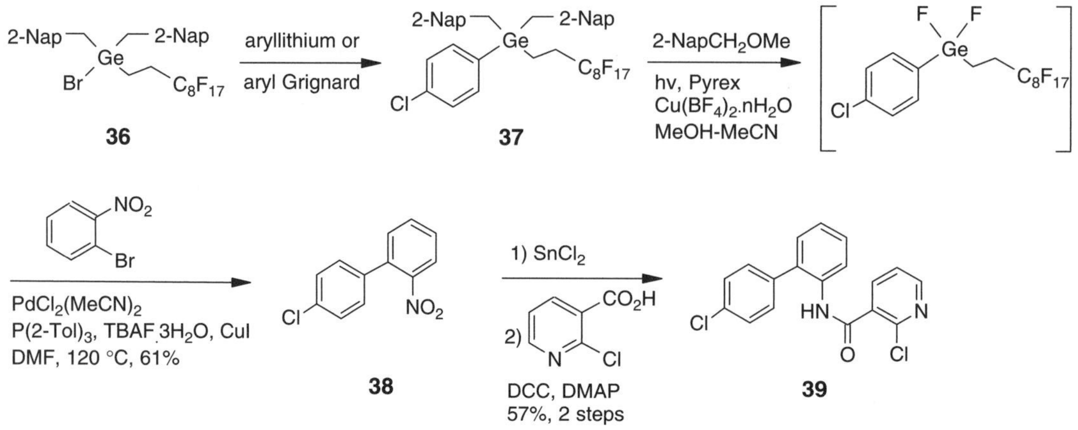

562   Alternative Linker Strategies

Scheme 21.11 *Fluorous arylgermanyl linker for the synthesis of Boscalid*

21.2.2 Synthesis of natural product analogs

Bistratamides were isolated from ascidians *Lissoclinum bistratum*. They have a range of biological activities. The Takeuchi group developed a heavy fluorous 2-[tri(perfluorodecyl)silyl]ethoxycarbonyl (F-Teoc) linker and applied it in the synthesis of bistratamide H (Scheme 21.12).[13] The F-Teoc has three C_8F_{17} chains to ensure the attached compounds have good partition coefficiency for F-LLE. The coupling of F-Teoc attached thiazole amino acid **40** with unprotected thiazole amino acid gave thiazole dipeptide ester **41**. It was hydrolyzed and then coupled with the oxazole amino acid methyl ester to give tripeptide methyl

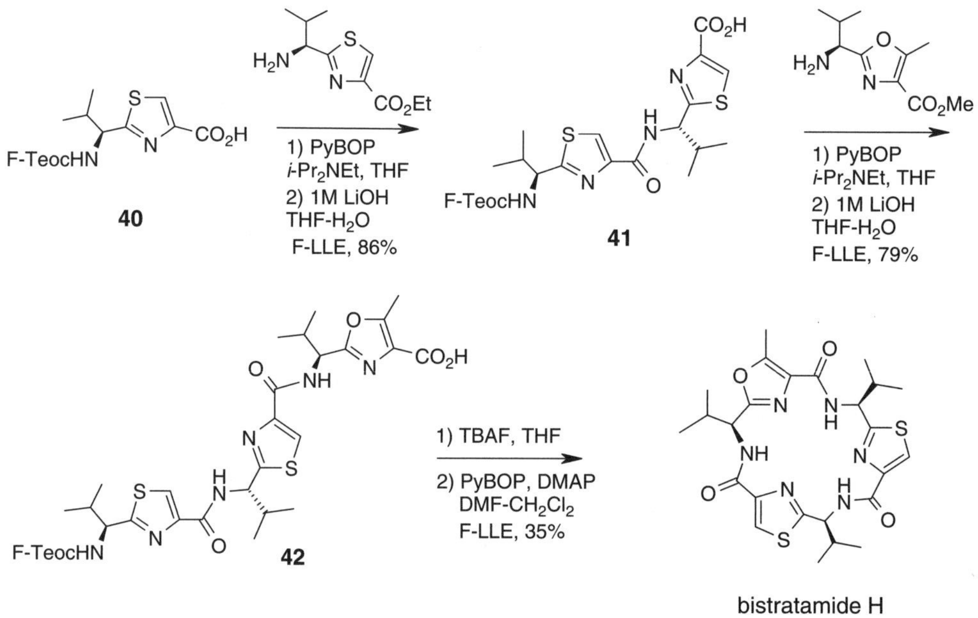

Scheme 21.12 *F-Teoc facilitated synthesis of bistratamide H*

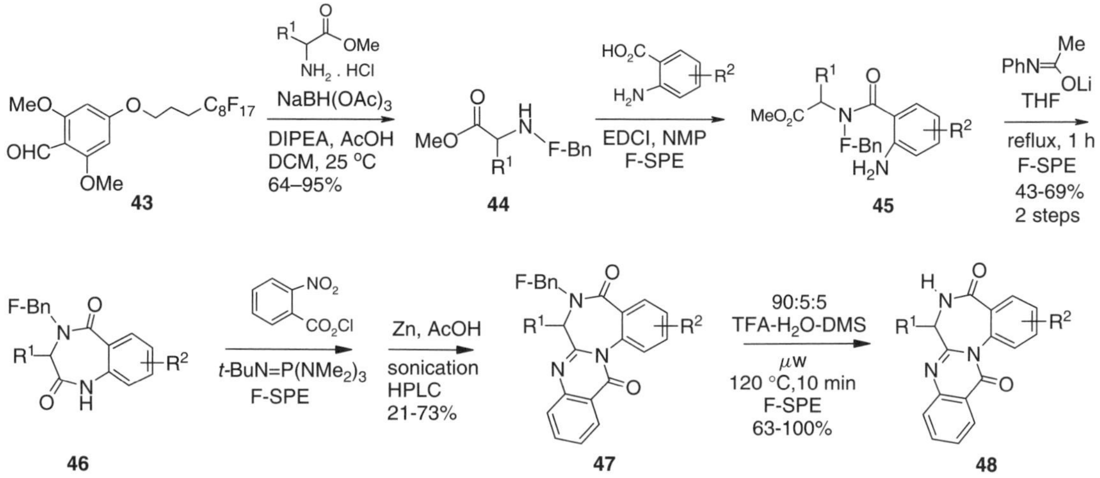

Scheme 21.13 Synthesis of sclerotigenin analogs

acid **42**. Cleavage of the F-Teoc protecting group with tetrabutylammonium fluoride (TBAF) followed by the intramolecular coupling reaction and preparative TLC gave bistratamide H.

The Zhang group reported the fluorous synthesis of benzodiazepine-quinazolinone sclerotigenin analogs **48** (Scheme 21.13).[15] Fluorous benzaldehyde **43** was attached to amino esters by reductive amination. The key intermediate benzodiazepinedinon **46** was prepared by base-promoted cyclization. N-Acylation, nitro group reduction and then cyclization led to the formation of attached benzodiazepine-quinazolinone **47**. The linker cleavage afforded final products **48**.

The Kondo group used fluorous sulfonyl linker in the synthesis of yuehchukene (Scheme 21.14).[22] This linker has no acidic protons and was thus stable for the lithiation reaction. The linker attached indole **49** was treated with mesityllithium and then reacted with a monoterpenoid aldehyde to give alcohol **50**. Oxidization of **50** with 2-iodoxybenzoic acid (IBX) followed by boron trifluoride etherate (BF$_3$·OEt$_2$) catalyzed cyclization gave the *cis*-hexahydroindeno[2,1+*b*]indol-6-one **51**. The ketone was stereoselectivly reduced to alcohol using DIBAH and then condensed with indole to afford attached product **52**. The fluorous linker was removed by treatment with TBAF to afford yuehchukene.

Winssinger and coworkers developed a fluorous PMB-trichloroacetimidate linker and used it in the total synthesis of Radicicol A (Scheme 21.15).[58] Cross-metathesis of the linker with the vinyl borolane afforded *trans*-vinyl borolane **53**. Treatment of **53** with *t*-BuLi followed by the addition of a *tert*-butyldiphenylsilyl (TBDPS) protected aldehyde led to the formation of product **54**. Iodo exchange of the silyl group followed by alkylation with an aromatic fragment gave **56**. The removal of the selenide and then the F-PMB groups afforded a hydroxyl acid. This compound underwent macrocyclization followed by the cleavage of ethoxymethyl (EOM) and acetonide groups to afford radicicol A.

21.2.3 Fluorous mixture synthesis

Fluorous mixture synthesis (FMS) is a solution-phase synthetic technology for making individual pure compound analogs and libraries. This technology was introduced by the Curran group in 2001 and has been applied to the synthesis of natural product enantiomers, diastereomers and drug-like compounds.[55] FMS employs a mixture of substrates attached to different lengths of fluorous tags as the starting material. The mixture can be treated as a single compound and subject to one-pot or split-parallel synthesis.

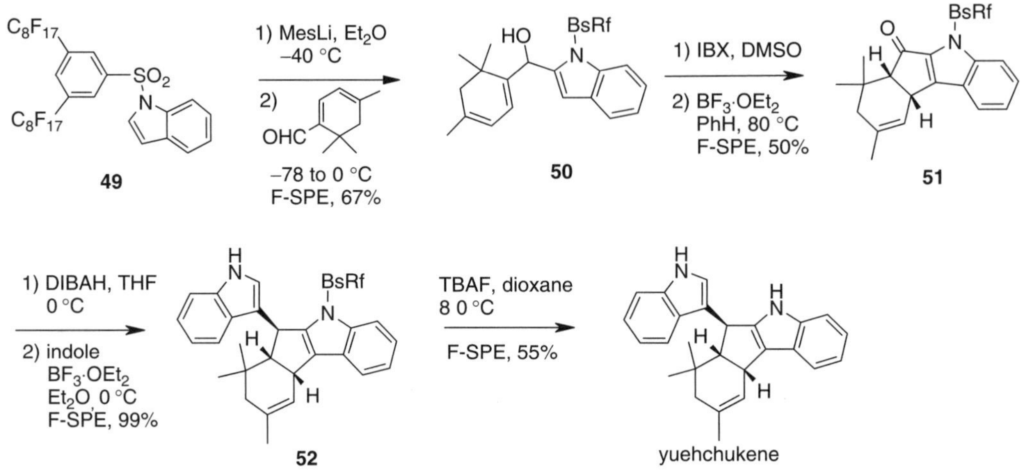

Scheme 21.14 Fluorous arylsulfonyl linker facilitated synthesis of yuehchukene

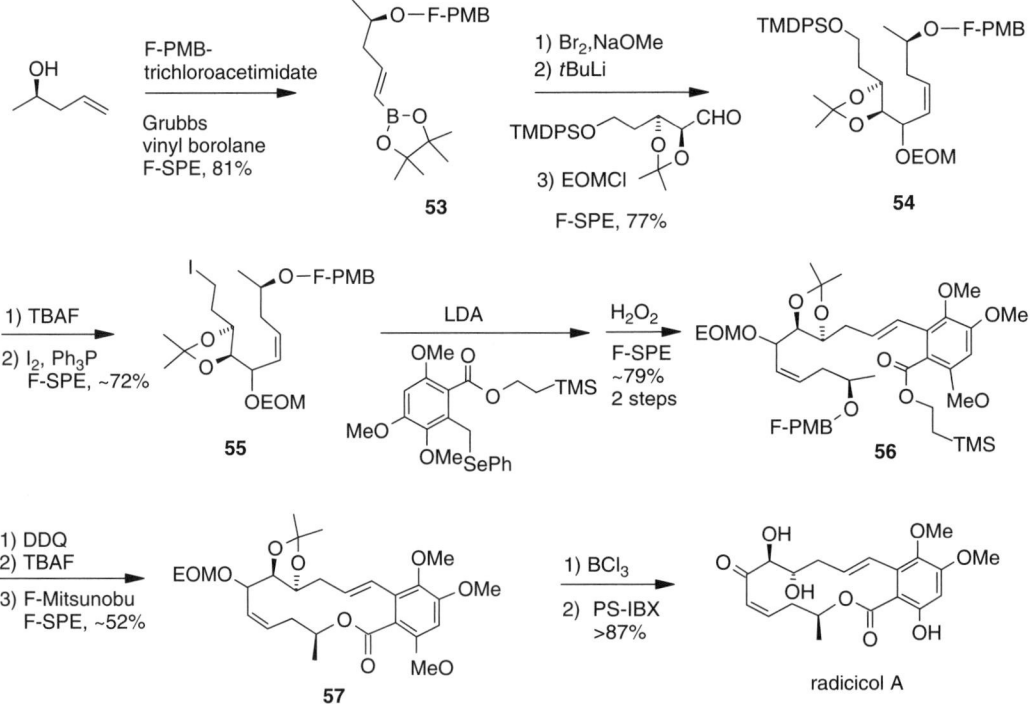

Scheme 21.15 F-PMB facilitated synthesis of radicicol A

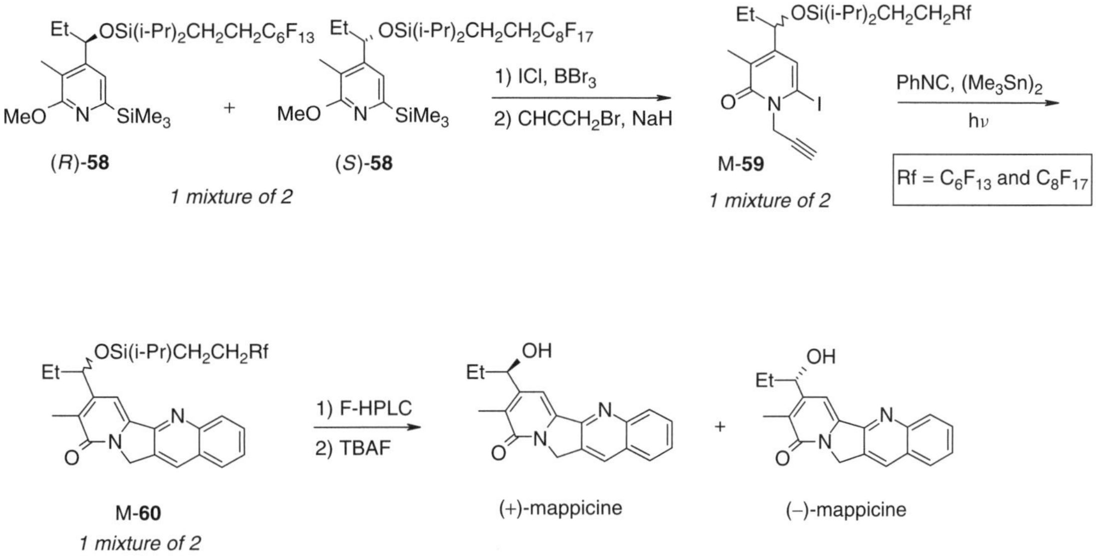

Scheme 21.16 *FMS of enantiopure (+)-and (–)-mappicines*

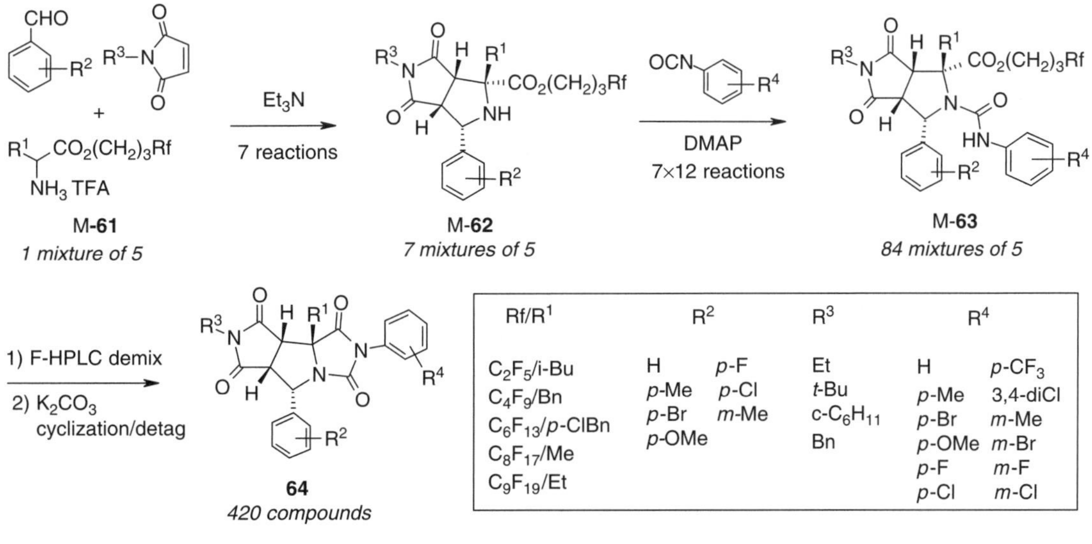

Rf/R¹	R²		R³	R⁴	
C_2F_5/i-Bu	H	*p*-F	Et	H	*p*-CF_3
C_4F_9/Bn	*p*-Me	*p*-Cl	*t*-Bu	*p*-Me	3,4-diCl
C_6F_{13}/*p*-ClBn	*p*-Br	*m*-Me	c-C_6H_{11}	*p*-Br	*m*-Me
C_8F_{17}/Me	*p*-OMe		Bn	*p*-OMe	*m*-Br
C_9F_{19}/Et				*p*-F	*m*-F
				p-Cl	*m*-Cl

Scheme 21.17 *Five-component FMS of a 420-member fused-hydantoin library*

The fluorous product mixtures are demixed by F-HPLC, followed by linker cleavage to release the final products.

Mappicine and mappicine ketone are active on the herpes viruse (HSV) and the human cytomegalovirus (HCMV). Fluorous TIPS-type linkers with C_6F_{13} and C_8F_{17} tags have been used in the quasiracemic synthesis of enantiomers of mappicine (Scheme 21.16).[29] Enantiomeric (*R*)- and (*S*)-alcohols were attached to C_6F_{13}- and C_8F_{17}-silanes, respectively, to form quasienantiomers (R)-**58** and (S)-**58**. A mixture of the two compounds was subjected to TMS group exchange with iodine chloride, demethylation with boron tribromide and N-propargylation to form M-**59**. The radical cycloaddition of M-**59** with phenyl isonitrile provided the quasiracemic mixture M-**60**. F-PHLC separation of this mixture followed by the linker cleavage afforded enantiopure (+)-mappicine and (−)-mappicine. Extension of this protocol for the synthesis of a 560-member mappicine library has been accomplished by the Zhang and Curran groups in a seven-component FMS.[27]

Another example of FMS is related to the preparation of a 420-member tricyclic compound library (Scheme 21.17).[59] Five α-amino acids bearing different R^1 groups were paired with five perfluoroalkyl alcohols in such: C_2F_5/i-Bu, C_4F_9/Bn, C_6F_{13}/p-ClBn, C_8F_{17}/Me, C_9F_{19}/Et. An equal molar mixture of five fluorous aminoester M-**61** was split into seven portions for 1,3-dipolar cycloaddition reactions with one of the seven benzaldehydes and one of the four maleimides. The resulting seven mixtures of M-**62** were each split into twelve portions and reacted with one of the twelve phenylisocyanates to form 84 mixtures of M-**63**. F-HPLC demixing followed by parallel detagging produced a 420-member library of **64**.

Fluorous linkers such as TIPS, Boc, Fmoc, PMB, Cbz and alcohols have been used for FMS. Some natural product and drug-like scaffolds prepared by FMS are listed in Table 21.4.

21.3 Fluorous linkers for synthesis of biomolecules

A wide range of fluorous linkers has been employed in the synthesis of biomolecules.[67] Only select examples are presented in this section.

21.3.1 Synthesis of peptides

The van Boom group made the first effort to introduce fluorous F-Cbz-type linkers in Fmoc-based solid-phase peptide synthesis (Scheme 21.18).[8a] In this work, the unreacted free amine was capped with an acetyl group after each amide coupling step. At the end of the coupling sequence, deprotection of the final Fmoc group followed by the attachment of F-CbzCl gave the desired fluorous product. The desired fluorous product and acetyl-capped byproducts were cleaved from the resin and separated by F-HPLC. This work demonstrated that peptides up to 22 amino acids could be synthesized. Since F-Cbz was found partially deprotected during the acidic resin cleavage, Overkleeft and coworkers later on developed fluorous methylsufonylethoxycarbonyl (F-Msc) as a base-labile linker for amine group protection.[11]

The Fustero group developed an easily cleavable F-TMSE linker and applied it in the synthesis of tripeptide **69** (Scheme 21.19).[38] The linker attachment of **65** was carried out under a standard esterification condition with diisopropylcarbodiimide (DIC) and N-hydroxybenzotriazole (HOBt). The linker cleavage of **68** was performed with the treatment of TBAF.

β-Peptides may significantly differ from α-peptides in structure, folding pattern, and potential biological activities. The Nelson and Curran groups developed a new strategy using iterative coupling and reduction of β-azido acids **71** to prepare β-peptides (Scheme 21.20).[20] Fluorous PMB attached azido-esters **72** facilitated the purification of intermediates in the synthesis of a tri-β-peptide library **74**.

21.3.2 Synthesis of oligosaccharides

Significant progress has been made in F-LLE-based 'heavy fluorous' synthesis of oligosaccharides. Mizuno, Inazu and their coworkers developed a series of carbonyl linkers with fluorous tags containing 34 to 102 fluorine atoms. Highlighted in Scheme 21.21 is the synthesis of disaccharide **78** from three Bfp attached mannose derivative **76**.[34] The triphenylmethyl (Trt) group of **76** was selectively removed by

Table 21.4 Fluorous linkers and the scaffolds made by FMS

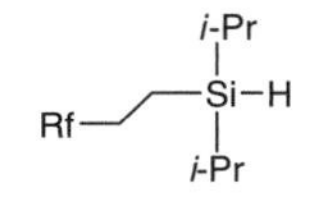

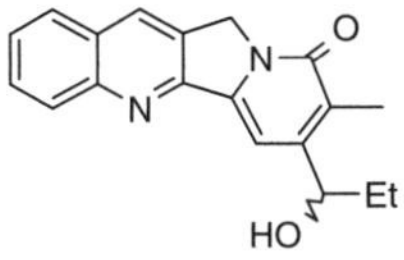

(+)- and (-)-mappicines
(ref. 29)

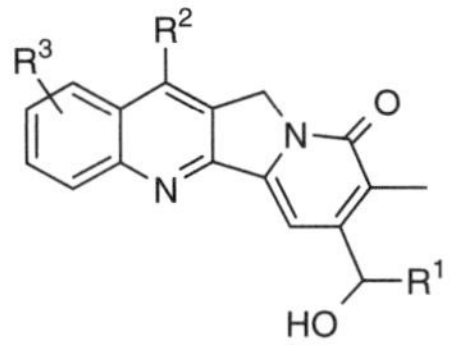

560 mappcine analogs
(ref. 27)

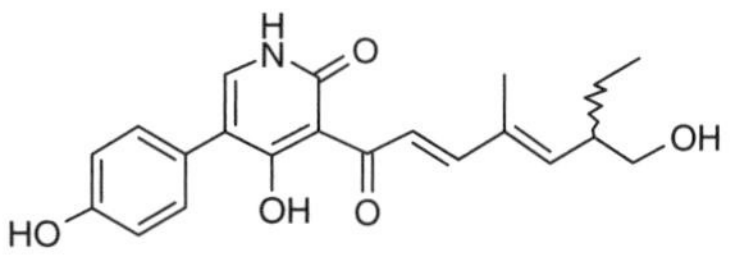

(S)- and (R)-pyridovericin
(ref. 29)

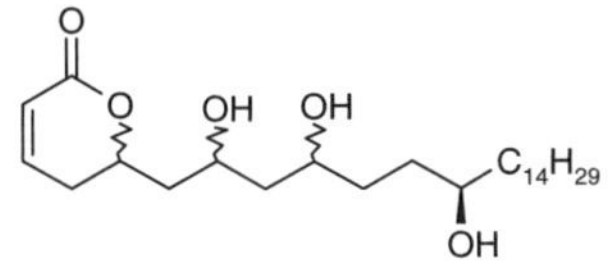

8 diastereomers of passifloricin A
(ref. 60)

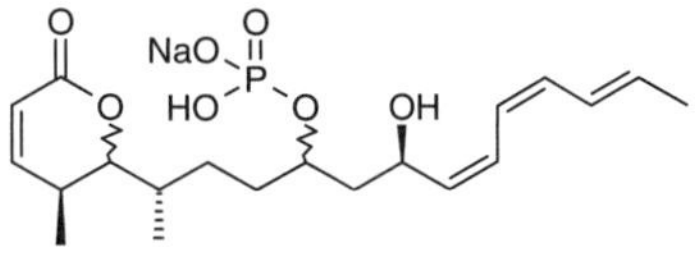

(+)-cytostatin and 3 diastereomers
(ref. 61)

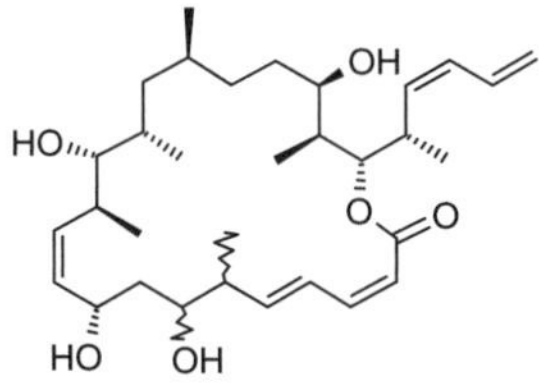

3 diastereomers of dictyostatin
(ref. 62)

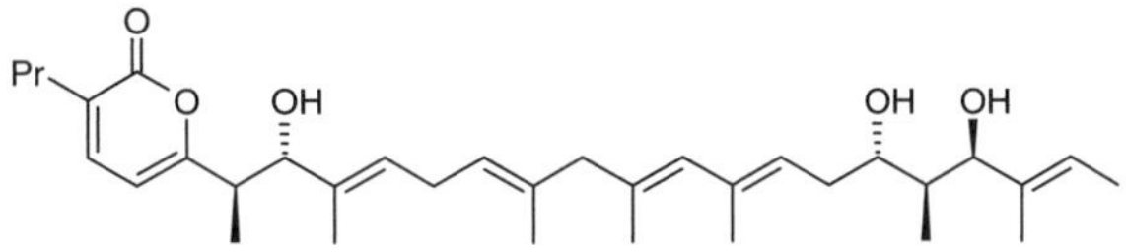

4 diastereomers of langunapyrone B
(ref. 63)

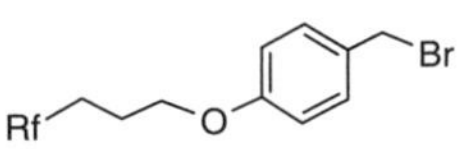

F-PMB-Br

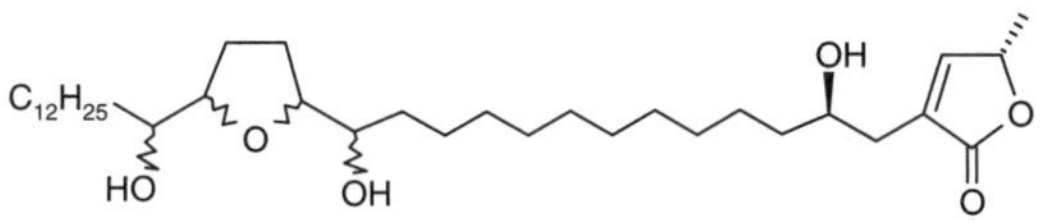

16 murisoline stereoisomers
(ref. 18)

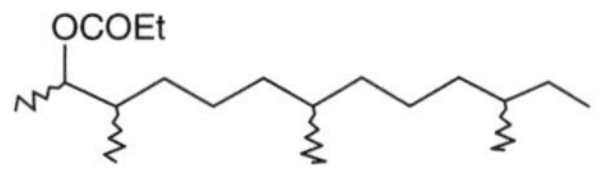

16 stereoisomers of pinesaw fly sex pheromone
(ref. 64)

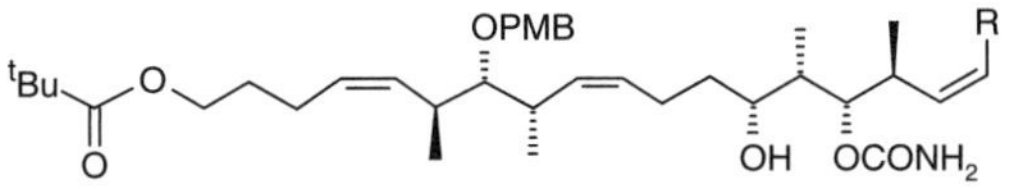

4 truncated discodermolide analogs
(ef. 19)

(continued overleaf)

Table 21.4 (continued)

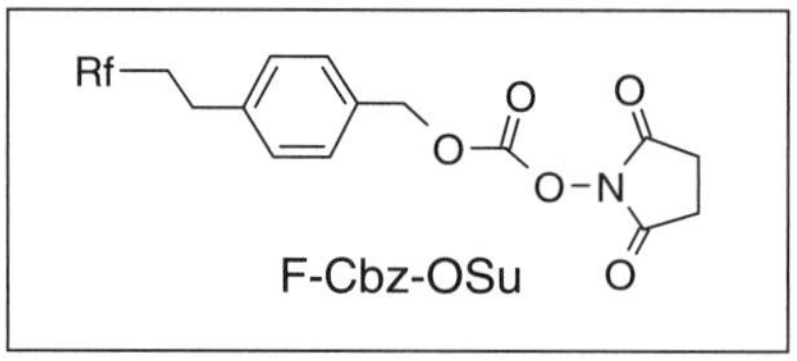

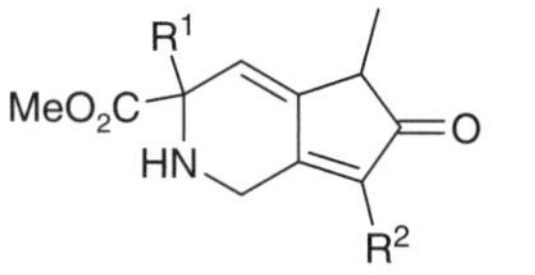

F-Cbz-OSu

16 tecomanine-type analogs
(ref. 65)

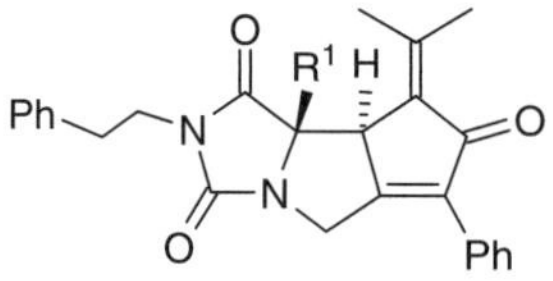

6 compounds
(ref. 66)

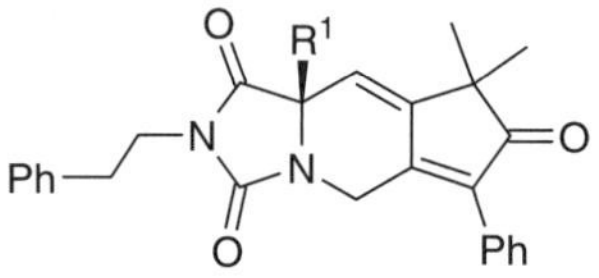

6 compounds
(ref. 66)

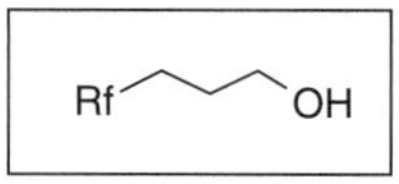

Rf∼∼∼OH

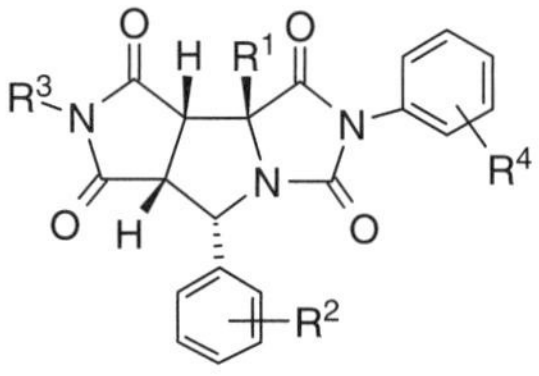

420 compounds
(ref. 59)

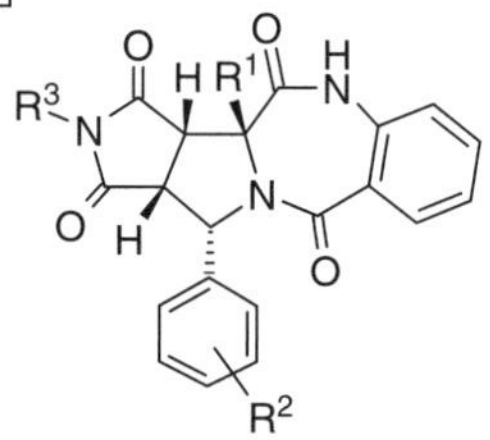

60 compounds
(ref. 59)

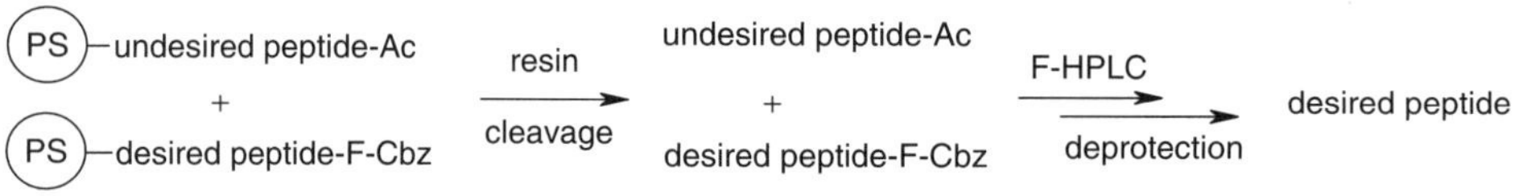

PS—undesired peptide-Ac   →   undesired peptide-Ac   →   desired peptide
     +   resin   +   F-HPLC
PS—desired peptide-F-Cbz cleavage desired peptide-F-Cbz deprotection

Scheme 21.18 *F-Cbz for purification of synthetic peptides*

the treatment of 10-camphorsulfonic acid (CSA). The deprotected hydroxyl group was coupled with a galactose derivative to give attached disaccharide **77**. Deprotection of both the acetyl and Bfp groups followed by FC-72/methanol extraction gave disaccharide **78** in the methanol layer. The protecting group was recovered from the FC-72 layer as a methyl ester.

The Hfb linker has a total of 102 fluorine atoms. The single Hfb linker has been proven to have enough fluorophilicity for F-LLE in the synthesis of trisaccharide **82** (Scheme 21.22).[68]

Manzoni and Castelli modified the structure of 2,2,2-trichloroethoxycarbonyl (Troc) and developed the 2-bromo-3,3,4,4,5,5,6,6,7,7,8,8,9,9,10,10,10-heptadecafluorodecyl chloroformate (Froc) linker for the protection of the amino group in carbohydrate synthesis (Scheme 21.23).[12] Froc protected glucosamine **83** was peracetylated to form **84**. It was then deacetylated at the anomeric position to and then re-protected with the texyldimethylsilyl group. Removal of the remaining acetyl groups followed by the treatment of the crude product with benzaldehyde dimethyl acetal afforded the fluorous glycosyl acceptor **85**. Glycosylation followed by the cleavage of Froc afforded the disaccharide **86**.

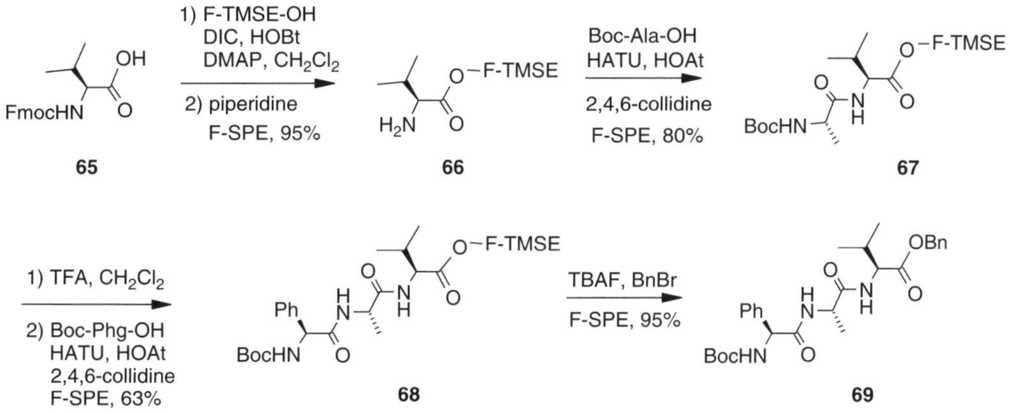

Scheme 21.19 F-TMSE facilitated synthesis of peptides

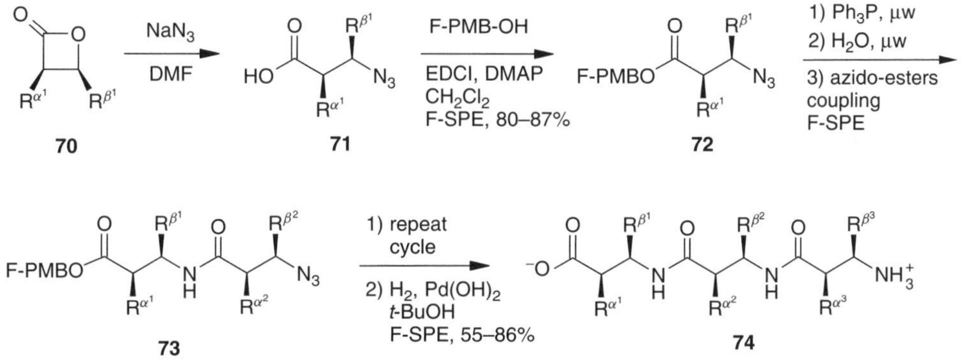

Scheme 21.20 Fluorous synthesis of tri-β-peptides

Fluorous PMB aldehyde has been used as an acetal protecting group in the regioselective synthesis of carbohydrates (Scheme 21.24).[69] The benzaldehyde was first converted to dimethyl acetal through reacting with trimethylorthoformate. The methyl α-D-glucopyranoside was added to the reaction mixture to form benzylidene acetal **87**. The acylation followed by the selective ring opening of the acetal group afforded compound **89**. The glycosidation followed by linker cleavage afforded disaccharide **91**.

21.3.3 Synthesis of glycopeptides

There is only one literature report on the fluorous synthesis of glycopeptides (Scheme 21.25).[70] The Mizuno group employed a heavy fluorous PMB attached leucine to prepare tripeptide **92**. It was then coupled with Fmoc-Asn(GlcNAc)-OMpt Fmoc-Asn(GlcNAc)-OH which has no protection of the sugar hydroxy groups. The glycopeptide intermediate **93** underwent de-Fmoc reaction and was then coupled with Boc-Ile-OMpt to afford desired glycopeptide **94**.

21.3.4 Synthesis of oligonucleotides

Tripathi and coworkers used fluorous silyl protected phosphoramidite **95** for the synthesis of oligonucleotides.[26] The fluorous monomer was incorporated in the last synthesis cycle. After the

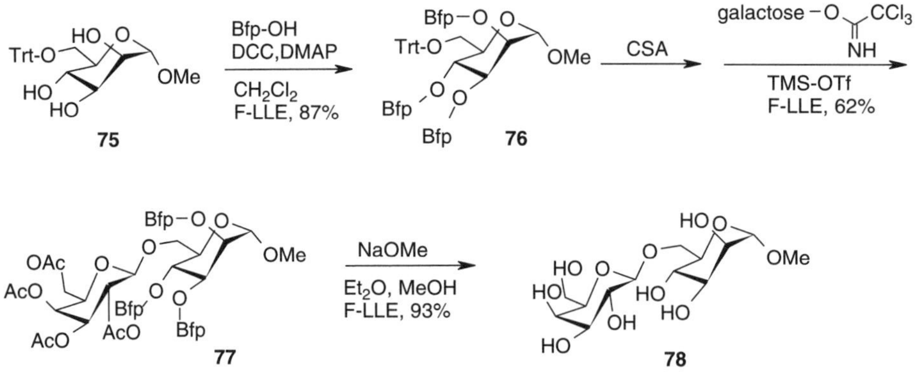

570 *Alternative Linker Strategies*

Scheme 21.21 *Bfp facilitated synthesis of a disaccharide*

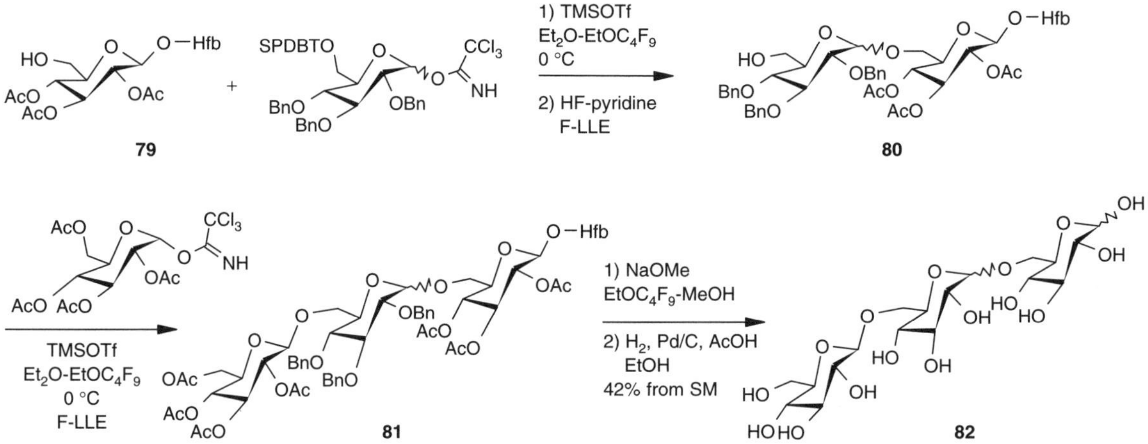

Scheme 21.22 *Hfb facilitated synthesis of a trisaccharide*

linker cleavage, the product was isolated by F-LLE and then by HPLC. The Pearson group employed fluorous dimethoxytrityl (FDMT attached nucleoside phosphoramidite **96** in the solid-phase synthesis of oligonucleotides (Scheme 21.26).[16] F-SPE of the resin-cleaved mixture retained the fluorous component and removed the failure sequences. On-column TFA detritylation followed by elution with aqueous acetonitrile afforded the clean products. Oligonucleotides up to 100-mers have been prepared by this protocol. Beller and Bannwarth used fluorous trityl linkers attached nucleoside phosphoramidites **97** for the synthesis of oligonucleotide and F-SPE for product purification.[17]

21.4 Other applications of fluorous linkers

The unique lipophobic and hydrophobic properties of fluorous linkers are not only limited to the separation of synthetic molecules. They have been applied to biotechnology related fields, including proteomics, microarray and enzymatic synthesis.[71]

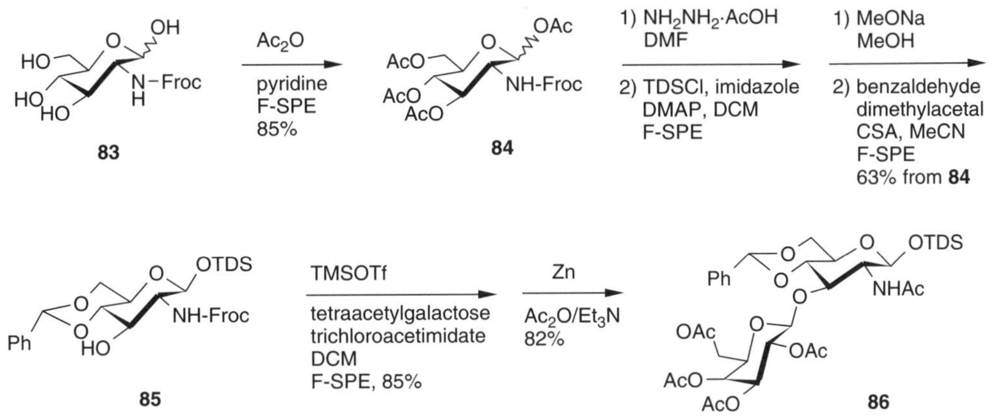

Scheme 21.23 Froc facilitated synthesis of a disaccharide

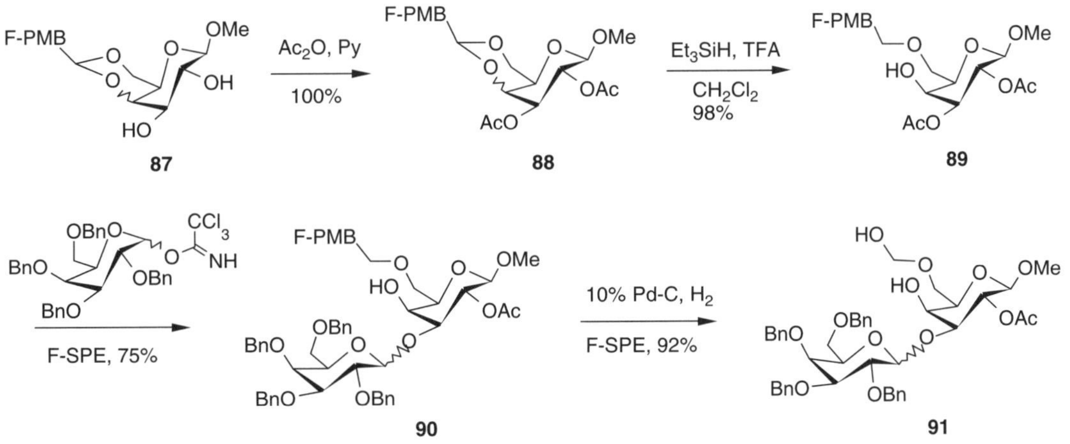

Scheme 21.24 F-PMB facilitated synthesis of a disaccharide

21.4.1 Isolation of proteomics samples

Separation of targeted peptides from a mixture of biological samples is a challenging task in proteomics research. The Peters group developed a new strategy for peptide enrichment by using fluorous linkers to attach desired peptides.[72a] In a preliminary study, the phosphorylated peptides were attached by fluorous thiol through β-elimination/Michael addition and then isolated by F-SPE from a mixture of digested yeast protein. The enriched fluorous peptides were analyzed by MALDI-MS or ESI-MS. Compared to the traditional biotin approach, the fluorous method avoids the problems of fragmentation during MS/MS and reduces the cost on the reagents. The same research laboratory later on introduced a modified fluorous protocol for the enrichment and subsequent direct MS characterization of the targeted peptide.[72b] The desired components were first selectively attached to fluorous linkers; the entire sample was then applied to a fluorous porous silicon surface. The unlabeled sample components were removed by simple surface washes and the retained sample fraction was directly analyzed by desorption/ionization on silicon mass spectrometry (DIOS-MS).

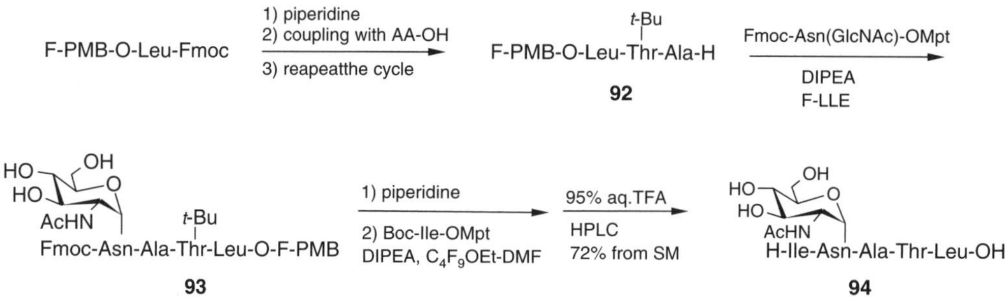

572 *Alternative Linker Strategies*

Scheme 21.25 *F-PMB facilitated synthesis of a glycopeptide*

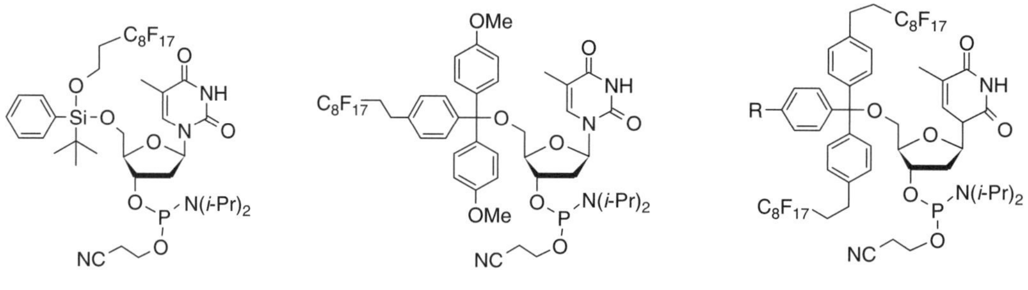

Scheme 21.26 *Fluorous nucleoside phosphoramidites for oligonucleotide synthesis*

21.4.2 Microarray screening

Fluorous linkers can be employed to synthesize small molecules and biomolecules and then immobilize the attached substrates on the fluorous slides for microarray screening. Both cleavable and permanent linkers can be used for microarray applications.

The Pohl group first applied the fluorous immobilization technique for the microarray of carbohydrates (Scheme 21.27). A series of fluorous mono- and disaccharides was synthesized by the glycosylation of corresponding glycosyl trichloroacetimidates attached to fluorous alcohol linkers. The attached saccharides such as **98** were spotted on a glass microscope slide whose surface had been pretreated with a fluorous silane. The fluorophilicity is strong enough to hold the substrates on the slide for screening with FITC-labeled lectins.[73, 74]

Fluorous microarray of small molecules has also been reported. The Spring group prepared fluorous biotin and incubated it on the slide with a Cy5 labeled avidin. The fluorous immobilized substrates withstood the protein incubation and washes. The microarray of fluorous biotin such as **99** (Figure 21.1), untagged biotin and other fluorous small molecules were tested.[75] The fluorescence was only detected with the fluorous biotin. It was also found that the fluorous immobilization was reversible. The slides could be washed and reused up to five times without any change in performance.

The Schreiber group reported the fluorous small molecule microarray for screening against *histone deacetylases* (HDACs) inhibitors.[76] An array of 20 compounds including known active such as **100** and inactives were spotted on the slides (Figure 21.2). These arrays were then exposed to three different purified HDACs. Their work was validated by comparing the screening results with those using other assay methods.

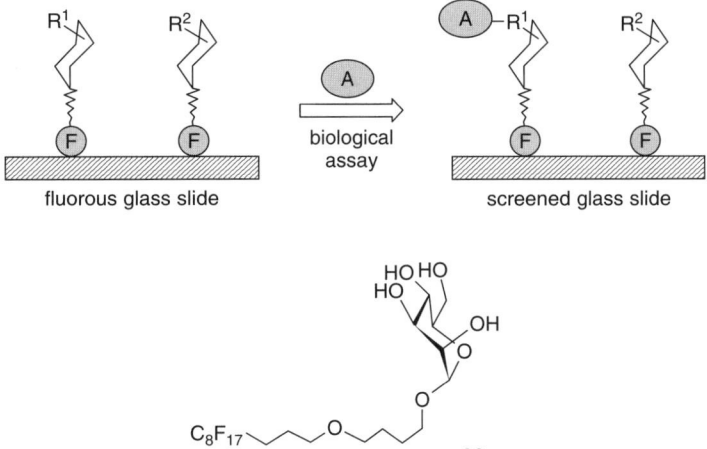

Scheme 21.27 Microarry screening of fluorous carbohydrates

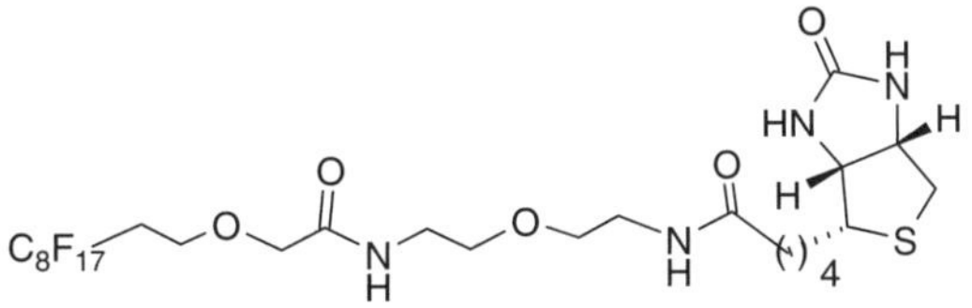

Figure 21.1 Flourous biotin

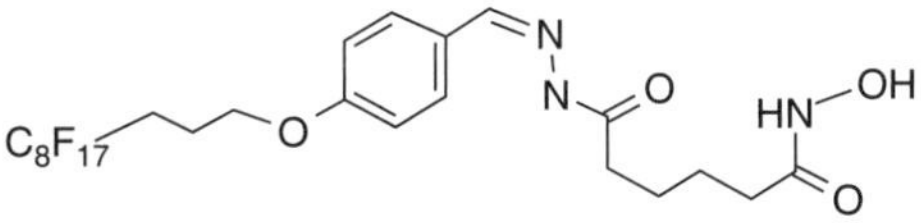

Figure 21.2 Fluorous histone deacetylases (HDACs) inhibitors

21.4.3 Enzymatic synthesis

Ikeda and coworkers developed 2-(perfluorohexyl)ethoxymethyl chloride as a fluorous version of methyloxymethyl (MOM)-type linker for the chemoenzymatic synthesis of a potent sialidase inhibitor 2-deoxy-2,3-didehydrosialic acid, (Scheme 21.28).[31] F-MOM was selectively attached to the primary hydroxyl function of N-acetyl-D-mannosamine to form **101**. After the removal of acetyl and the glycosidic benzyl groups, attached molecule **102** was used for Neu5Ac aldolase catalyzed enzymatic reaction. Sequential esterification, acetyl chloride treatment and linker cleavage afforded desired product **105**.

The Hatanaka group reported another example of fluorous enzymatic reactions. Fluorous alcohol attached lactoside primer **106** was introduced into muse B16 cells for an *in vivo* glycosylation to produce a GM3-type

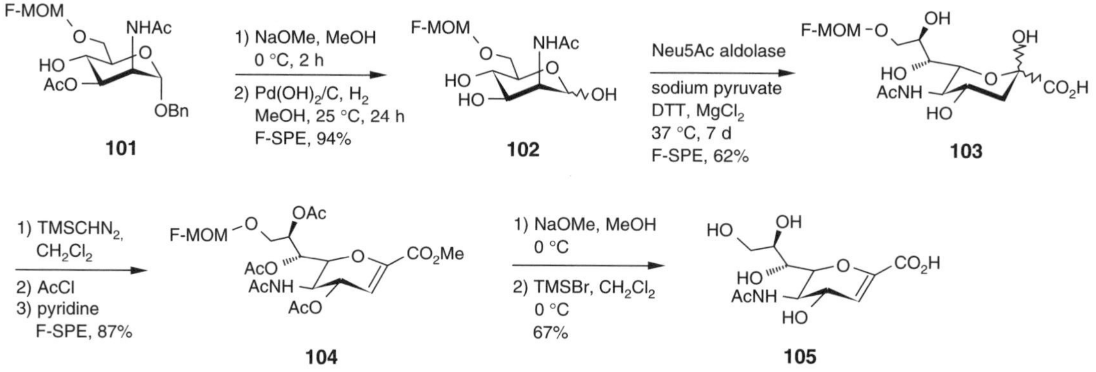

Scheme 21.28 F-MOM facilitated synthesis of sialidase inhibitor 105

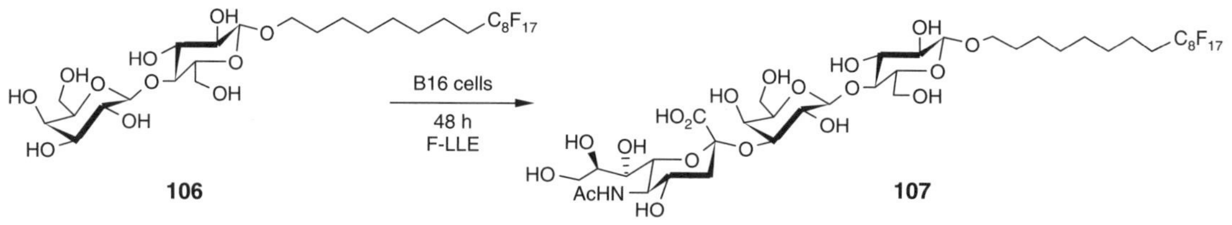

Scheme 21.29 Fluorous enzymatic glycosylation for a trisaccharine

oligosaccharide **107** (Scheme 21.29).[77] The fluorous link was found to be non-cytotoxic and had increased hydrophobicity and membrane permeability of the primer. This unique cellular enzymatic approach is simple and also environmentally friendly, since much less reaction solvent is involved in the production process.

21.5 Conclusion

Complementary to polymer linkers for solid-phase synthesis, fluorous linkers have been developed for solution-phase synthesis of small molecules and biomolecules. The reaction and separation processes are integrated to improve synthetic efficiency. Fluorous proteomics, microarray and biocatalysis have extended the applications of fluorous linkers in biological areas.

References

[1] a) Zhang, W.; in *The Handbook of Fluorous Chemistry* (eds Gladysz, J. A., Curran, D. P., and Horvath, I. T.); Wiley-VCH Verlag GmbH, Weinheim, Germany, **2004**, 222–236. b) Zhang, W.; *Curr. Opin. Drug Discov. Develop*. **2004**, *7*, 784. c) Zhang, W.; *Chem. Rev*. **2009**, *109*, 749.
[2] *The Handbook of Fluorous Chemistry* (eds Gladysz, J. A., Curran, D. P., and Horvath, I. T.); Wiley-VCH Verlag GmbH, Weinheim, Germany, **2004**.
[3] Special issues on fluorous chemistry: a) Gladysz, J. A., and Curran, D. P.; *Tetrahedron*, **2002**, *58*, 3823. b) Zhang, W.; *QSAR Comb. Sci*. **2006**, *25*, 679.

[4] Zhang, W.; *Tetrahedron* **2003**, *59*, 4475.
[5] Luo, Z., Williams, J., Read, R. W., and Curran, D. P.; *J. Org. Chem.* **2001**, *66*, 4261.
[6] Zhang, W., and Tempest, P.; *Tetrahedron Lett.* **2004**, *45*, 6757.
[7] Mamidyala, S. K., and Firestine, S. M.; *Tetrahedron Lett.* **2006**, *47*, 7431.
[8] a) Filippov, D. V., van Zoelen, D. J., Oldfield, S. P., *et al.*; *Tetrahedron Lett.* **2002**, *43*, 7809. b) Schwinn, D., and Bannwarth, W.; *Helv. Chim. Acta* **2002**, *85*, 255.
[9] Curran, D. P., Amatore, M., Campbell, M., *et al.*; *J. Org. Chem.* **2003**, *68*, 4643.
[10] Matsugi, M., Yamanaka, K., Inomata, I., *et al.*; *QSAR Comb. Sci.* **2006**, *25*, 713.
[11] de Visser, P. C., van Helden, M., Filippov, D. V., *et al.*; *Tetrahedron Lett.* **2003**, *44*, 9013.
[12] Manzoni, L., and Castelli, R.; *Org. Lett.* **2006**, *8*, 955.
[13] Nakamura, Y., and Takeuchi, S.; *QSAR Comb. Sci.* **2006**, *25*, 703.
[14] Villard, A.-L., Warrington, B. H., and Ladlow, M. L.; *J. Comb. Chem.* **2004**, *6*, 611.
[15] Zhang, W., Williams, J. P., Lu, Y., *et al.*; *Tetrahedron Lett.* **2007**, *48*, 563.
[16] Pearson, W. H., Berry, D. A., Stoy, P., *et al.*; *J. Org. Chem.* **2005**, *70*, 7114.
[17] Beller, C., and Bannwarth, W.; *Helv. Chim. Acta* **2005**, *88*, 171.
[18] Zhang, Q. S., Lu, H. J., Richard, C., and Curran, D. P.; *J. Am. Chem. Soc.* **2004**, *126*, 36.
[19] Curran, D. P., and Furukawa, T.; *Org. Lett.* **2002**, *4*, 2233.
[20] Wang, X., Nelson, S. G., and Curran, D. P.; *Tetrahedron* **2007**, *63*, 6141.
[21] Kojima, M., Nakamura, Y., Ishikawa, T., and Takeuchi, S.; *Tetrahedron Lett.* **2006**, *47*, 6309.
[22] Naka, H., Akagi, Y., Yamada, K., *et al.*; *Eur. J. Org. Chem.* **2007**, 4635.
[23] Studer, A., and Curran, D. P.; *Tetrahedron* **1997**, *53*, 6681.
[24] Rover, S., and Wipf, P.; *Tetrahedron Lett.* **1999**, *40*, 5667.
[25] Wipf, P., Reeves, J. T., and Day, B. W.; *Curr. Pharm. Design.* **2004**, *10*, 1417.
[26] Tripathi, S., Misra, K., and Sanghvi, Y. S.; *Org. Prep. Proc.* **2005**, *37*, 257.
[27] Zhang, W., Luo, Z., Chen, H.-T., and Curran, D. P.; *J. Am. Chem. Soc.* **2002**, *124*, 10443.
[28] Palmacci, E. R., Hewitt, M. C., and Seeberger, P. H.; *Angew. Chem. Int. Ed.* **2001**, *40*, 4433.
[29] Zhang, Q. S., Rivkin, A., and Curran, D. P.; *J. Am. Chem. Soc.* **2002**, *124*, 5774.
[30] Wipf, P., and Reeves, J.; *Tetrahedron Lett.* **1999**, *40*, 5139.
[31] Ikeda, K., Mori, H., and Sato, M.; *Chem. Commun.* **2006**, 3093.
[32] Curran, D. P., and Ogoe, C.; *QSAR Comb. Sci.* **2006**, *25*, 732.
[33] Wipf, P., and Reeves, J. T.; *Tetrahedron Lett.* **1999**, *40*, 4649.
[34] Miura, T., Goto, K., Waragai, H., *et al.*; *J. Org. Chem.* **2004**, *69*, 5348.
[35] Miura, T., Satoh, A., Goto, K., *et al.*; *Tetrahedron Asym.* **2005**, *16*, 3.
[36] Goto, K., and Mizuno, M.; *Tetrahedron Lett.* **2007**, *48*, 5605.
[37] Mizuno, M., Goto, K., Miura, T., *et al.*; *Chem. Commun.*, **2003**, 972.
[38] Fustero, S., Sancho, A. G., Chiva, G., *et al.*; *J. Org. Chem.* **2006**, *71*, 3299.
[39] Pardo, J., Cobas, A., Guitian, E., and Castedo, L.; *Org. Lett.* **2001**, *3*, 3711.
[40] a) Read, R. W., and Zhang, C.; *Tetrahedron Lett.* **2003**, *44*, 7045. b) Huang, Y., and Qing, F.-L.; *Tetrahedron* **2004**, *60*, 8341.
[41] Jian, H., and Tour, J. M.; *J. Org. Chem.* **2005**, *70*, 3396.
[42] Cioffi, C. L., Berlin, M. L., and Herr, R. J.; *Synlett* **2004**, *5*, 841.
[43] a) Zhang, W., Chen, C. H.-T., Lu, Y., and Nagashima, T.; *Org. Lett.* **2004**, *6*, 1473. b) Zhang, W., Lu, Y., and Chen, C. H.-T.; *Mol. Diversity* **2003**, *7*, 199. c) Zhang, W., and Nagashima, T.; *J. Fluorine Chem.* **2006**, *127*, 588. d) Zhang, W., Nagashima, T., Lu, Y., and Chen, C. H.-T.; *Tetrahedron Lett.* **2004**, *45*, 4611. e) Lu, Y., and Zhang, W.; *QSAR Comb. Sci.* **2004**, *23*, 827. f) Zhang, W., and Chen, C. H.-T.; *Tetrahedron Lett.* **2005**, *46*, 1807.
[44] Chen, C. H.-T., and Zhang, W.; *Org. Lett.* **2003**, *5*, 1015.
[45] a) Zhang, W., and Lu, Y.; *Org. Lett.* **2003**, *5*, 2555. b) Nagashima, T., and Zhang, W.; *J. Comb. Chem.* **2004**, *6*, 942. c) Zhang, W., Lu, Y., Chen, C. H.-T., et al.; *Eur. J. Org. Chem.* **2006**, 2055. d) Zhang, W., and Lu, Y., Geib, S.; *Org. Lett.* **2005**, *7*, 2269. e) Lin, M. J., Zhang, W., and Sun, C.-M.; *J. Comb. Chem.* **2007**, *9*, 951.

[46] Jing, Y., and Huang, X.; *Tetrahedron Lett.* **2004**, *45*, 4615.
[47] Donovan, A., Forbes, J., Dorff, P., et al.; *J. Am. Chem. Soc.* **2006**, *128*, 3536.
[48] Zhang, W.; *Org. Lett.* **2003**, *5*, 1011.
[49] a) McAllister, L. A., McCormick, R. A., Brand, S., and Procter, D. J.; *Angew. Chem. Int. Ed.* **2005**, *44*, 452. b) McCormick, R. A., James, K. M., Willetts, N., and Procter, D. J.; *QSAR Comb. Sci.* **2006**, *25*, 709. c) McAllister, L. A., McCormick, R. A., James, K. M., et al.; *Chem. Eur. J.* **2007**, *13*, 1032. d) James, K. M., Willetts, N., and Procter, D. J.; *Org. Lett.* **2008**, *10*, 1203. e) Ovens, C., Martin, N. G., and Procter, D. J.; *Org. Lett.* **2008**, *10*, 1441.
[50] Horhant, D., Lamer, A.-C. L., Boustie, J., et al.; *Tetrahedron Lett.* **2007**, *48*, 6031.
[51] Spivey, A. C., Tseng, C.-C., Hannah, J. P., et al.; *Chem. Commun.* **2007**, 2926.
[52] Chu, Q., Yu, M. S., and Curran, D. P.; *Tetrahedron* **2007**, *63*, 9890.
[53] Zhang, W., and Curran, D. P.; *Tetrahedron* **2006**, *62*, 11837.
[54] Zhang, W. *J. Fluorine Chem.* **2008**, *129*, 910.
[55] a) Luo, Z. Y., Zhang, Q. S., Oderaotoshi, Y., and Curran, D. P.; *Science* **2001**, *291*, 1766. b) Zhang, W. *Arkivoc* **2004** (i), *101*.
[56] Zhang, W., *Chem. Rev.* **2004**, *104*, 2531.
[57] Zhang, W.; *Comb. Chem. High Throuput Screening* **2007**, *10*, 219.
[58] Dakas, P.-Y., Barluenga, S., Totzke, F., et al.; *Angew. Chem. Int. Ed.* **2007**, *46*, 6899.
[59] Zhang, W., Lu, Y., Chen, C. H.-T., et al.; *J. Comb. Chem.* **2006**, *8*, 687.
[60] Curran, D. P., Moura-Letts, G., and Pohlman, M.; *Angew. Chem. Int. Ed.* **2006**, *45*, 2423.
[61] Jung, W.-H., Guyenne, S., Riesco-Fagundo, C., et al.; *Angew. Chem. Int. Ed.* **2008**, *47*, 1130.
[62] Fukui, Y., Brückner, A. M., Shin, Y., et a.; *Org. Lett.* **2006**, *8*, 301.
[63] Yang, F.; and Curran, D. P.; *J. Am. Chem. Soc.* **2006**, *128*, 14200.
[64] a) Dandapani, S., Jeske, M., and Curran, D. P.; *Proc. Natl. Acad. Sci. U.S.A.* **2004**, *101*, 12008. b) Dandapani, S., Jeske, M., and Curran, D. P.; *J. Org. Chem.* **2005**, *70*, 9447.
[65] Manku, S., and Curran, D. P.; *J. Comb. Chem.* **2005**, *7*, 63.
[66] Manku, S., and Curran, D. P.; *J. Org. Chem.* **2005**, *70*, 4470.
[67] Zhang, W.; in *Current Fluoroorganic Chemistry. New Synthetic Directions, Technologies, Materials and Biological Applications* (eds Soloshonok, V. A., Mikami, K., Yamazaki, T., et al.); Oxford University Press **2006**, 207–220.
[68] Goto, K., Miura, T., Hosaka, D., et al.; *Tetrahedron* **2004**, *60*, 8845.
[69] Kojima, M., Nakamura, Y., and Takeuchi, S.; *Tetrahedron Lett.* **2007**, *48*, 4431.
[70] Mizuno, M., Goto, K., and Miura, T.; *Chem. Lett.* **2005**, *34*, 426.
[71] Zhang, W., Cai, C., *Chem. Commun.*. **2008**, 5686.
[72] Brittain, S. M., Ficarro, S. B., Brock, A., and Peters, E. C. et al.; *Nat. Biotech.* **2005**, *23*, 463. b) Go E. P., Uritboonthai W., Apon J. V., *J. Proteome. Research* **2007**, *6*, 1492.
[73] a) Ko, K.-S., Jaipuri, F. A., and Pohl, N. L.; *J. Am. Chem. Soc.* **2005**, *127*, 13162. b) Mamidyala, S. K., Ko, K.-S., Jaipuri, F. A., et al.; *J. Fluorine Chem.* **2006**, *127*, 571.
[74] Jaipuri, F. A., Collet, B. Y. M., and Pohl, N. L.; *Angew. Chem. Int. Ed.* **2008**, *47*, 1707.
[75] Nicholson, R. L., Ladlow, M. L., and Spring, D. R.; *Chem. Commun.* **2007**, 3906.
[76] Vegas, A. J., Bradner, J. E., Tang, W., et al.; *Angew. Chem. Int. Ed.* **2007**, *46*, 7960.
[77] a) Kasuya, M. C. Z., Cusi, R., Ishihara, O., et al.; *Biochem. Biophy. Res. Commun.* **2004**, *316*, 599. b) Kasuya, M. C. Z., Ito, A., Cusi, R., et al.; *Chem. Lett.* **2005**, *34*, 856.

22
Solid-Phase Radiochemistry

Brian G. Hockley, Peter J. H. Scott and Michael R. Kilbourn

Department of Radiology, University of Michigan Medical School, USA

22.1 Introduction

Radiochemistry is the process by which molecules are labeled (or tagged) with a radioactive isotope. The molecules in question can range from the small and simple (small molecules, peptides, amino acids, etc.) to the large and complex (taxol, large peptides, proteins, anti-bodies, etc.) and the choice of radioisotope to label with depends very much upon the goal in question, as discussed below. Due to high levels of radiation involved, synthesis of radiopharmaceuticals is not a trivial process and, in the last two or three decades, many ingenious radiochemical reactions and automated hardware solutions have been developed to simplify the process. A detailed discussion of such is beyond the scope of this chapter, but the subject is receiving much attention in recent review articles in the mainstream organic chemistry literature.[1–4] As more and more traditional organic chemists venture into the radiochemistry arena, a fact supported by this recent flurry of review articles, more adventurous organic chemistry techniques are being adapted for radiochemical synthesis.

One such technology is solid-phase organic synthesis (SPOS), which has facilitated the rapid creation of diverse libraries of compounds appropriate for high-throughput screening in a number of industrial and medical applications. Building on this, SPOS is beginning to impact radiochemistry in two main areas. Firstly, the ease of purification after cleavage of substrates from the resin bound linker units is particularly attractive because one major challenge facing radiochemistry research groups is the initial synthesis and purification of a precursor molecule that will be labeled with a radioactive isotope. SPOS addresses this issue very neatly, allowing the synthesis of libraries of precursor candidates with controlled and systematic diversity, which could then be examined for their utility as potential biomarkers. Secondly, SPOS could also play a very revolutionary role in the radiolabeling process itself because the another major challenge

Linker Strategies in Solid-Phase Organic Synthesis Edited by Peter Scott
© 2009 John Wiley & Sons, Ltd

in the radiochemistry laboratory is the purification and formulation of the final radioactive biomarker. These purifications are not always trivial and can be very time consuming if multiple semi-preparative HPLC purifications and solid-phase extractions are required. When using short-lived isotopes such as carbon-11 and fluorine-18, whose half-lives are about 20 minutes and 110 minutes, respectively, these extended reaction and purification times can render an otherwise promising biomarker unusable. Most of the chemistry and purification must also be performed remotely from outside of lead-lined hot cells to protect the operator from excessive radiation exposure. Solid-phase radiochemistry (SPRC), the application of SPOS to traditional radiolabeling could very effectively address this issue allowing the rapid labeling and pre-clinical screening of the above-mentioned libraries of potential biomarkers.

Although radiochemistry is simply the chemical addition of radioactive isotopes to an otherwise useful molecule to create a radioactive biomarker or radiotracer, the utility of the technique can have a sweeping impact on health care and research. These biomarkers are used *in vivo* to generate images using scanners that are sensitive to the particle emissions from the decay of the radioactive isotopes. There are two major techniques that utilize radioactively tagged molecular probes, Positron Emission Tomography (PET) and Single Photon Emission Computed Tomography (SPECT). The production of specific isotopes used in PET and SPECT imaging is beyond the scope of this review but the process is well established and there are many excellent publications on the topic which have been thoroughly reviewed in the Handbook of Radiopharmaceuticals.[5, 6]

Molecular imaging with PET and SPECT agents allows researchers and physicians to collect dynamic, functional and quantitative information about specific biochemical pathways. The PET and SPECT imaging modalities are used routinely by large pharmaceutical companies to rapidly screen potential drug candidates for efficacy and specificity. Molecular imaging is used in medicine, pre-clinical medical research to examine many biological processes such as receptor binding, gene expression, cellular metabolism and structural features and abnormalities[7, 8]. Radiolabeled biologically active compounds are also used extensively as radiotherapeutic treatment agents in clinical oncology. In addition to oncology,[9-15] the use of radiolabeled biomarkers and bioactive compounds is also very common in neurology[16-27] and cardiology[28-32] and is increasingly used in other fields of medicine and research such as gene therapy and stem cell research.

Despite the relative popularity of molecular imaging and radiotherapy, to date there has been limited use of SPOS in the routine preparation of radiolabeled biomarkers and it is a field still in its infancy. This brief review chapter is intended to introduce the field and encourage further development of solid-supported synthetic strategies in radiochemistry.

22.2 Solid-phase surrogates in radiochemistry

22.2.1 Thin film radiochemistry

Solid-supported radiochemistry has been practiced in small academic research laboratories and large corporate drug discovery facilities alike for decades.[33-37] Much of that work incorporated surrogates to solid-phase synthesis such as solid-supported, thin film or captive-solvent chemistry. The thin film approach had a similar impact on radiopharmaceutical development as it did on other aspects of chemical synthesis. Adaptation of traditional solution-phase radiochemistry to solid-supported methods was particularly straightforward with carbon-11 labeled radiopharmaceuticals because the radioactive isotope is most often delivered as gaseous methyl iodide or methyl triflate. Improvements in reaction times and purification were especially valuable in radiochemistry applications, where the use of short half-life isotopes such as carbon-11 with its 20.4 minute half-life were very common. Although captive solvent or thin film radiochemistry is not truly a solid-phase-synthetic technique, its impact cannot be ignored and thus it has been included in this chapter as a pseudo-solid-phase method.

Reactions with carbon-11 are also routinely performed using disposable solid-phase extraction media (SPE). The array of Sep-Pak style stationary phases now available has further expanded the use of thin film or captive solvent radiochemistry. There are many examples of this application but two very well know and thoroughly tested examples of this are the production [^{11}C]choline (Scheme 22.1) and [^{11}C]methionine. Both radiotracers are widely used in oncology imaging. The precursor molecule is loaded in solution onto a Sep-Pak that serves not only as the solid support but also as the purification media. This method is simple, reproducible, clean, disposable, minimizes transfer losses and shortens process times.[38–42]

The immobilization of one or more reagents on a solid support in combination with a thin film was illustrated by Jewet et al. in the synthesis of (+)-[^{11}C]dihydrotetrabenazine ((+)-[^{11}C]DTBZ) (Scheme 22.2), a high affinity ligand for the vesicular monoamine transporter (VMAT2).[43] In this example bulk alumina was swollen in aqueous potassium hydroxide, similar to the longstanding use of KF alumina which is common in organic synthesis. The phenolic precursor in solution was then applied as a film to a small column of this basic alumina effectively providing the solid-supported, reaction-ready precursor prior to the addition of the radioactive [^{11}C]methyl group. Gaseous [^{11}C]methyl triflate was directed across the alumina at a controlled flow rate and reaction occurred within seconds. The alumina was then used as a small normal phase chromatography column to purify the product from precursor and radioactive contaminants.

The most common current application of thin film or captive-solvent radiochemistry relies solely on the advantages of a thin film reaction. This technique is now widely used in radiopharmaceutical research as well as being used in many pharmaceutical clinical trials. The precursor molecule to the target radiopharmaceutical and the reaction solvent are applied in a thin coating to a column of beads or length of tubing.[33, 34, 44–46] Gaseous [^{11}C]methyl iodide or [^{11}C]methyl triflate is simply passed through the tubing or across the beads at very slow flow rates. Owing to the reaction rates of methyl triflate and methyl iodide this technique can be very effective and very rapid. The decreased reaction times coupled with increased yields has allowed for continuous growth of carbon-11 radiopharmaceutical use worldwide and has encouraged corporate expansion into the field of radiopharmacy. The use of an unadulterated HPLC injector loop

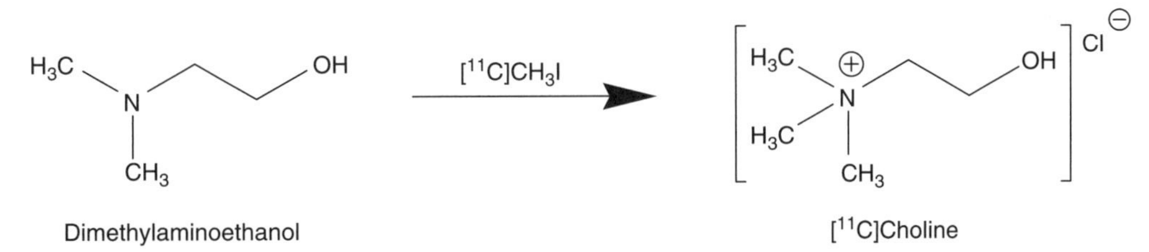

Scheme 22.1 Simple synthesis of [^{11}C]Choline

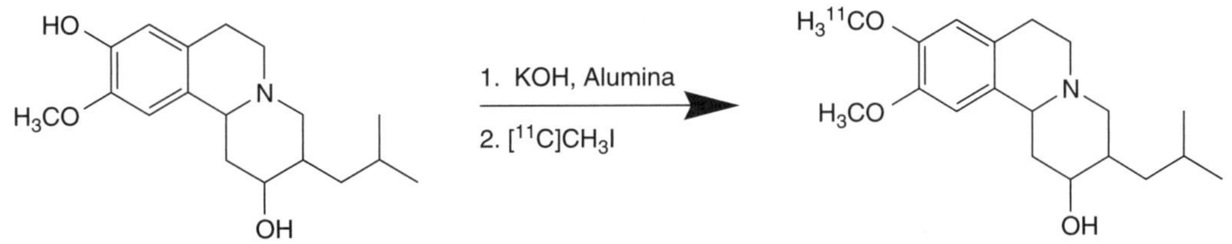

Scheme 22.2 Simple synthesis of [^{11}C]DTBZ

580 Alternative Linker Strategies

as the surface for the thin film reaction was a noteworthy development. In addition to the simplicity of preparation, the advantage in this case is that the reaction mixture is ready for HPLC purification without requiring fluid transfer using additional solvents which may adversely affect the purification. The reaction conditions and optimization of this technique are outlined elsewhere[42, 44–48] and in the patent held by Wilson *et al.*[47] The work done by Wilson and other academic researchers before him in combination with growth in the use of PET radiopharmaceuticals paved the way for commercialization of the injector loop technique.

22.2.2 Germanium solid-phase surrogates

Radiochemistry applications routinely require the use of low molecular weight volatile starting materials or intermediates, which present very unique problems related to containment of said radioactive reagents and can severely decrease the yield of solution phase reactions. Germanium-based linker strategies have been demonstrated to eliminate volatile radioactive intermediates via a solid-phase-surrogate.[49] The Spivey group employed a solution-phase germanium analogue of an initial reagent which was then reacted with solid carbon dioxide to give a non-volatile intermediate that could then be carried forward without concern for loss of end-product potency (Scheme 22.3) (see Chapter 16 for the preparation of germanium-based linker units).

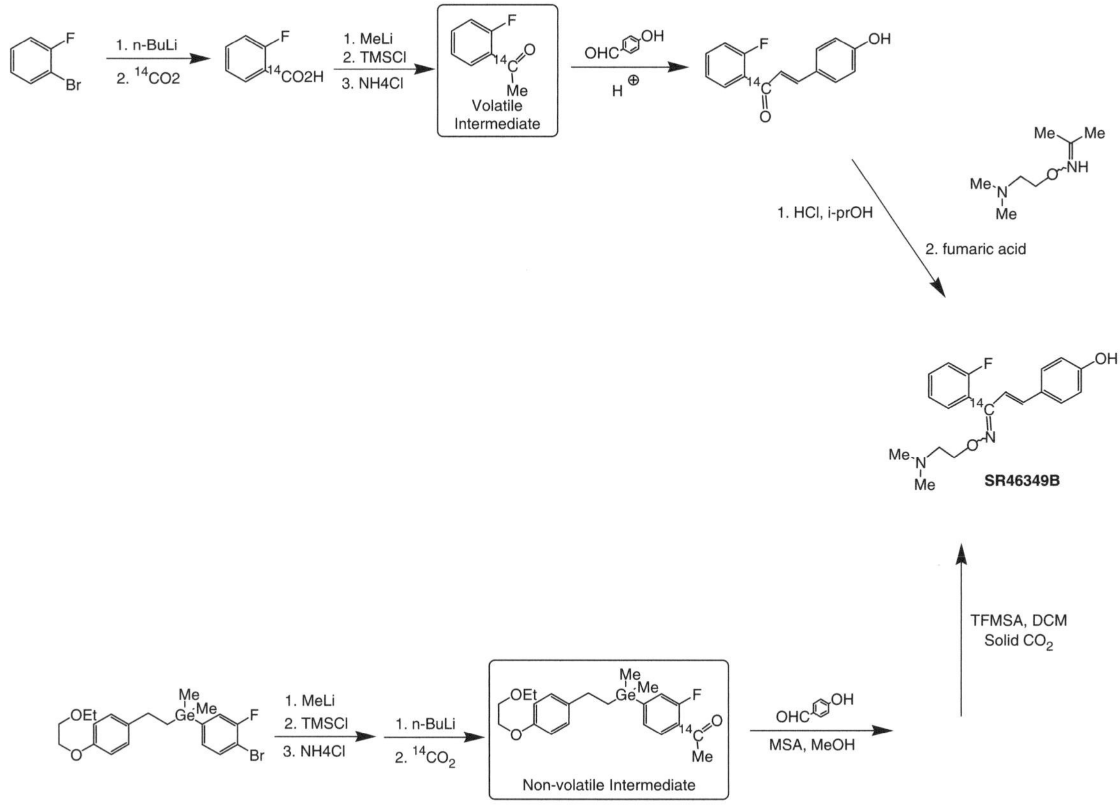

Scheme 22.3 *Germanium-based solid-phase surrogate for the purposes of devolatilization*

The Spivey group developed the solution-phase ethoxyethanol functionalized analogue of a commercially available resin-bound linker in order to address detrimental volatility issues discovered during synthetic development of the Sanofi–Aventis radioligand SR46349B. Following the synthesis of the germanium-bound product, traceless cleavage to the final product using 50% trifluormethanesulfonic acid in dichloromethane (DCM) proceeded smoothly in yields of 54%. Although the above carboxylation step in the presence of the germanium group is very efficient when a vast excess of carbon dioxide is used, when using one equivalent or less of carbon dioxide, which is a much better approximation of actual radiochemistry conditions, yields have been poor. This work was largely conceptual but following optimization the principal established could be readily applicable to routine radiochemistry reactions where intermediates are often lost due to their volatility.

22.3 Solid-phase radiochemistry

22.3.1 Stannane linkers in radiochemistry

As noted previously, the application of solid-phase synthetic techniques to radiochemistry is uncommon but those techniques offer very attractive options at both the reaction and purification stages of the radiolabeling process. This strategy is demonstrated in the work of Hunter and Culbert, who used tin linker units in the production of radiohalogenated benzylguanidines (Scheme 22.4) and amphetamines, for use as diagnostic

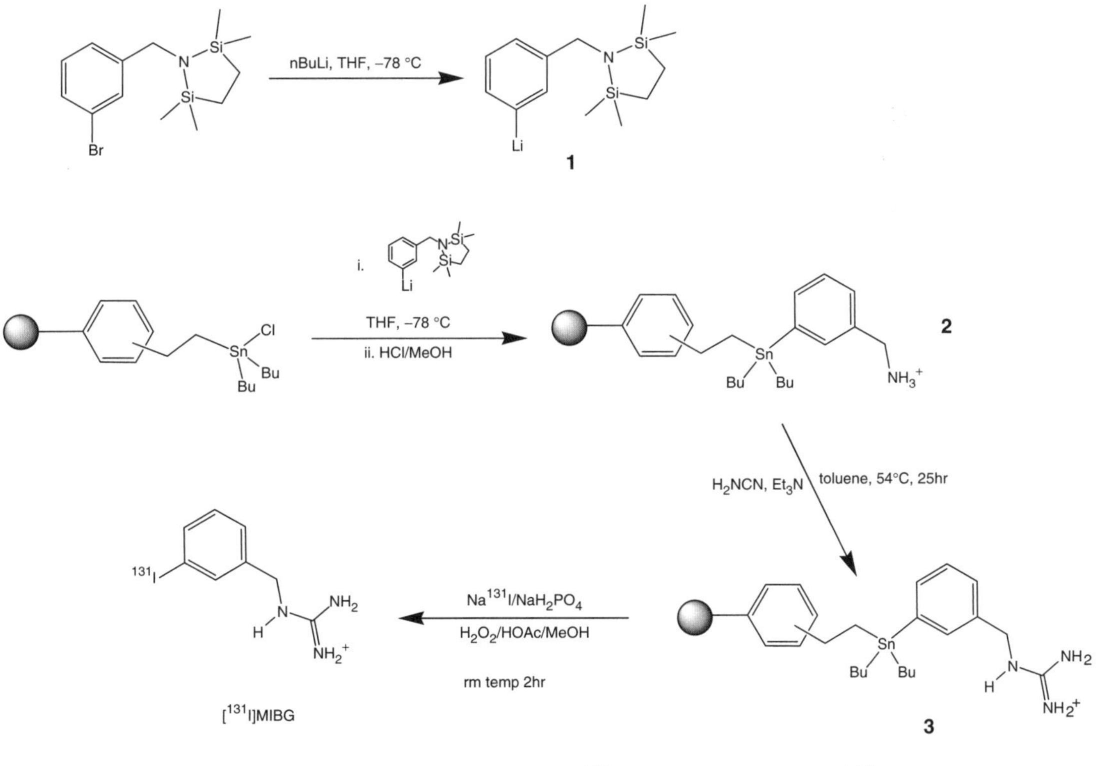

Scheme 22.4 Synthesis of meta-[^{131}I]iodobenzylguanidine[47]

agents in cardiology and oncology as well as for therapeutic agents,[50, 51] and later by Kabalka,[52] who applied the technique to the radioiodination of the azo dye Congo Red, a stain with high affinity for some forms of β-amyloid. Reaction Scheme 22.4 outlines the path used by Hunter in the synthesis of [^{131}I]iodobenzylguanidine.

The resin-bound guanidinium unit **3** was created in two steps starting with a lithium derivatized benzylamine **1** and a tin chloride resin. The newly formed resin bound benzylammonium compound **2** was treated with cyanamide and triethylamine to yield the polymer-bound tin linked benzylguanidine **3**. Cleavage of this linker was achieved by radioiodination under oxidative conditions using a buffered, labeling grade solution of sodium[^{131}I]iodide to give the desired iodobenzylguanidine ([^{131}I]MIBG). Unreacted resin-bound precursor was simply filtered off and excess [^{131}I]iodide could be removed by simple HPLC or use of an anion scavenger resin filter. Radioiodinated amphetamines have also been produced by Culbert and Hunter using the same organostannane linker units.[50] For additional detailed information regarding organostannane linkers and their preparation see Chapter 17.

Typical Experimental Procedures

Polymer-supported 3-benzylammonium chloride 2

*Under an argon atmosphere a solution of freshly distilled THF and 76 mmol of the silyl-dervatized bromobenzyl amine was cooled to $-78\,°C$ and one equivalent of n-butyllithium was added slowly. The tin chloride resin was added via powder addition to the flask and the reaction mixture was stirred for seven hours maintaining the $-78\,°C$ condition and for an additional one hour after rising to room temperature over two hours. The reaction was quenched with methanol (10 ml) and acidified with hydrochloric acid to give a pH of 4–5 and allowed to stir overnight. The polymer was filtered out and it was washed with 50% water methanol, methanol (3 × 100 ml) and 95% ethanol followed by vacuum drying at room temperature produced to yield 84 wt-% of **2**.*

Polymer-supported benzylguanadinium 3

*Under argon 20.0 g of **2**, 15.1 g (360 mmol) of cyanamide, 100 µl (0.72 mmol) triethylamine were combined in 250 ml of toluene and heated for 25 hours at 54 °C. The polymer was removed by filtration and washed with acetonitrile (4 × 100 ml), methanol (4 × 100 ml) and acetonitrile (4 × 100 ml) and dried at room temperature under vacuum which gave 20.7 g of polymer **3**.*

Radioiodination of polymer 3

*In a 2 ml vial 0.5 mg of **3**, 300 µl of methanol and 100 µl of 0.1 M potassium dihydrogen orthophosphate were combined with 50 µl (45.5 MBq) of a radiolabeling grade Na^{131}I solution, 450 µl of distilled water, 100 µl of a solution of acetic acid (700 mM) and hydrogen peroxide (600 mM). The mixture was shaken for two hours and 200 µl of a 0.1 M sodium metabisulfate was added. The polymer was removed by filtration and the filtrate was analyzed for [^{131}I] MIBG. Labeling efficiency of this reaction was 97.6%.*

22.3.2 Germanium linkers in radiochemistry

Radiohalogenation through germanium linkers that can be prepared from readily available commercial resins is also very feasible (see Chapter 16 for a further discussion of germanium-based linker units). These germanium linkers have been used here to conceptually establish synthetic routes to clinically relevant

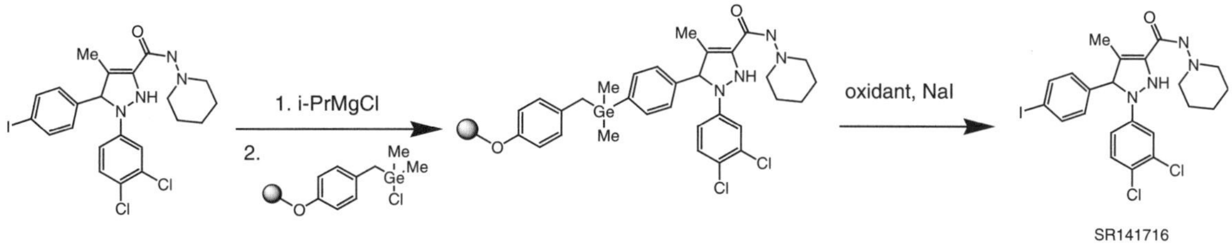

Scheme 22.5 *Conceptual synthetic route to iodinated SPECT radioligand*

PET and SPECT agents by the Spivey group [49] (Scheme 22.5). The germanium linked precursor unit was prepared from a Grignard intermediate of the SR141716 precursor and cleaved by treatment with the oxidant dichloramine-T and sodium iodide to produce a stable iodinated analogue of the radiopharmaceutical SR141716. Despite degradation under the strong oxidative conditions in initial experiments optimization of this method of production of radiopharmaceuticals is ongoing and unpublished but presents another solid-phase radiochemistry example for the use of solid-phase linkers and cleavage strategies.

Typical Experimental Procedures

Cleavage of the germanium linker was accomplished with dichloramine-T and aqueous sodium iodide solution (isotopically stable) and shaken at room temperature for two hours. The polymer was filtered off, unreacted iodine was removed by an ion exchange Sep-Pak and the product mixture was analyzed.

22.3.3 Fluorous linkers in radiochemistry

The most common clinical biomarker in use today is the radiohalogenated [^{18}F]FDG (2-[^{18}F]fluoro-2-deoxyglucose). FDG is used daily as a diagnostic tool in neurology, cardiology and oncology to measure cellular glucose utilization. Although this radiotracer is almost exclusively made in solution using [^{18}F]fluoride and a phase transfer agent on commercially available synthesis modules, solid-phase synthetic routes to FDG are also beginning to be explored. To date, two strategies have been reported – the first reported by Kilbourn[36] and later Iwata[53] traps fluoride on a polymer support using a basic counter ion. In this case, the linker is not intrinsic to the synthesis but rather the polymer support functions as a phase transfer catalyst and eliminates the need for toxic Kryptofix[2.2.2] traditionally employed in solution-phase FDG syntheses. The FDG precursor is then added in solution and radiofluorination and deprotection steps produce [^{18}F]FDG (Scheme 22.6).

The second approach offers a more traditional SPS route to [^{18}F]FDG, whereby the FDG precursor is loaded onto a solid support and treated with [^{18}F]fluoride. In this case the choice of linker unit is

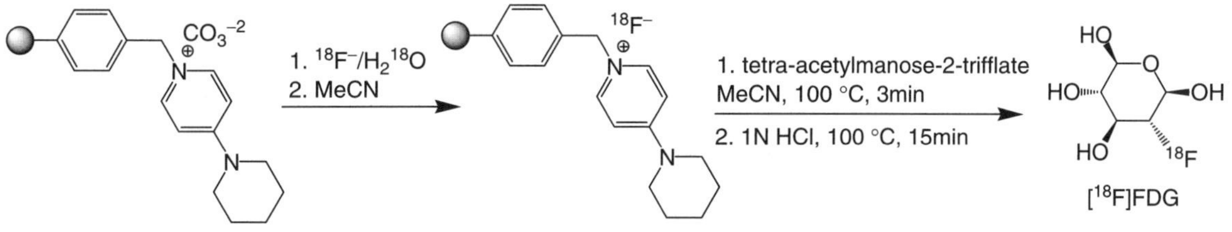

Scheme 22.6 *Polymer-supported synthesis of [^{18}F]FDG*

critical and Brown has successfully established a high yielding solid-phase route to [^{18}F]FDG using a perfluorsulfonate linker[54, 55]. The perfluorosulfonate linker was loaded onto polystyrene and employed to achieve an average of 73% (decay corrected) yield of [^{18}F]FDG (albeit with radiochemical purities that rival the common solution phase reaction results) (Schemes 22.7 and 22.8). Initial experiments using a perfluoroalkylsulfonyl fluoride linker resulted in excessive elimination of [^{19}F]fluoride *in situ*, which would be detrimental to potency of the final radiolabeled [^{18}F]FDG. Additionally, the unreacted resin-bound sulfonyl fluoride reacted with the precursor decreasing the yield of reaction. After much experimentation a conjugate of the precursor and a linker unit was produced prior to attachment to an amino functionalized resin (Scheme 22.7).

As illustrated, the precursor and new sulfonylfluoride linker unit was assembled in solution before loading on to an amino functionalized resin. This approach was designed to specifically minimize the elimination products mentioned above and to avoid side reactions of the previously described sulfonyl fluoride resin-bound linker. The addition of the iodo functionalized sulfonyl fluoride unit using sodium hexamethyldisilazide (NaHMDS) produced yields of 77% but when attempted with sodium hydride (NaH) the yields and reproducibility were very poor. Enoic acids of various chain lengths were used but the 5-iodooctafluoro-3-oxapentanesulfonyl fluoride was experimentally determined to be the most promising candidate. Deiodination of the newly formed iodic acid was carried out in refluxing ether at 80 °C for three hours under acidic conditions in the presence of zinc (6 equiv.). Despite the acidic conditions the FDG protecting groups remained in place during this deiodination (59% yield). Loading this linker on to an aminomethyl resin was accomplished in 90% yield and was then carried forward to fluorination experiments.

Experimental fluorinations were performed using a vast excess of ^{19}F$^-$ to illuminate the identity of potential impurities that may be encountered in future radiochemical reactions. The desired product was

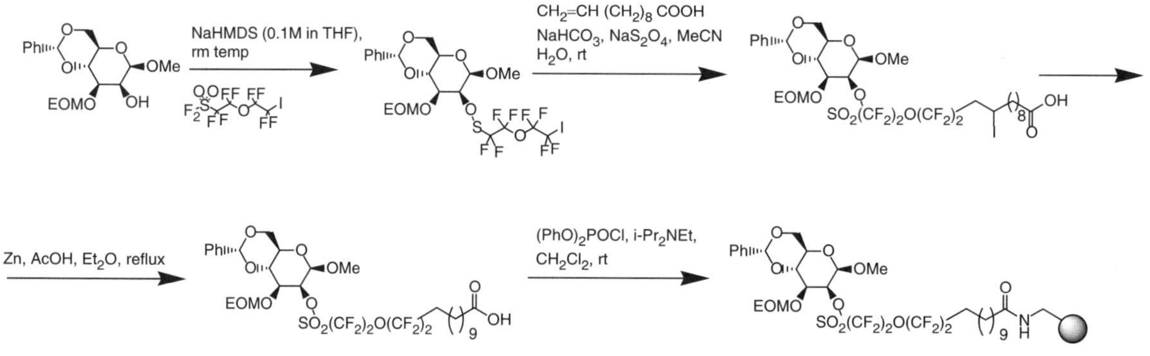

Scheme 22.7 *Preparation of solid-supported FDG precursor*[50, 51]

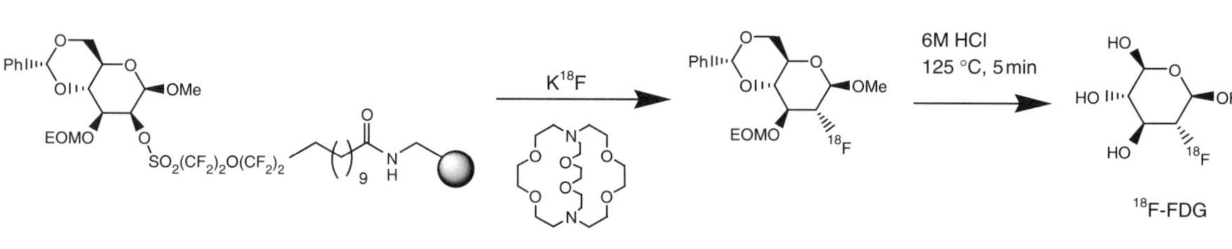

Scheme 22.8 *Synthesis of [^{18}F]FDG*

produced in 64% yield with three impurities, none of which were fluorinated. Further experiments limiting [^{19}F]fluoride, which more closely approximate actual radiochemical conditions, were performed. The incorporation results obtained were similarly promising. Additionally the resin could be used up to four times with acceptable yields and purities.

Low level radiochemical reactions (5–10 mCi) with [^{18}F]KF and kryptofix[2.2.2] using the resin bound precursor at 86 °C for four minutes resulted in comparable yields to the typical solution phase reaction. Yields of the fluorination were 80–85% by reverse phase HPLC and deprotection yields were 91–97% by ion exchange HPLC. High level reactions (5–6 Ci) produced incorporation yields of 68–77%.

The success of the above solid-phase radiofluorination is very significant to the radiochemistry and molecular imaging community because the use [^{18}F]fluoride as a radiolabel is increasing annually in commercial production operations and medical research. The fact that the above fluorous linked resin could be reused with acceptable results demonstrates the viability of the development of a radiochemistry kit-style reaction resins for use in routine clinical/research production. To date, efforts have primarily focused upon preparing FDG using SPRC due to its widespread clinical use. However, the reported techniques are equally applicable to the preparation of other [^{18}F]labeled radiopharmaceuticals. For example, Tang et al. recently reported a similar solid-phase synthesis of S-(2-[^{18}F]fluoroethyl)-L-methionine.[56] In addition, the automation and parallel synthesis strategies already well established in traditional SPOS could be adapted to enhance and streamline the radiopharmaceutical development process, perhaps allowing for hot-library synthesis in the future.

Typical Experimental Procedures

Cleavage of perfluorosulfonyl linker and deprotection to [^{18}F]FDG

An aqueous solution of ^{18}F fluoride ion is combined with potassium carbonate (8 mg in 300 µl) and Kryptofix [2.2.2.] (22 mg in 300 µl of acetonitrile) and heated to 125 °C under vacuum for up to 40 minutes to remove the water. The K[2.2.2.]potassium fluoride mixture was redissolved in acetonitrile and combined with the resin and heated at 86 °C for four minutes and then cooled to room temperature. The polymer was then filtered off and the filtrate was diluted with 1 ml of water and passed through a C18 Sep-Pak and then through an alumina Sep-Pak. The resulting solution was dried in a sealed vial under nitrogen flow and heated at 100 °C for 10 minutes. Hydrolysis in 0.5 ml of 6 M hydrochloric acid at 120 °C for five minutes. The resulting mixture was purified by HPLC and analyzed. Incorporation of fluoride ion was typically 73–91% with nearly quantitative yields from the deprotection step.

22.4 Conclusions and perspectives

The steadily increasing clinical demand for radiolabeled biomarkers and mounting regulatory pressures necessitate the development of high yield, high quality and reproducible production strategies. Despite limited use, solid-phase radiochemistry (SPRC) has been an effective tool to achieving those goals while producing product purities at least equal to those obtained by solution-phase reactions in a shorter time. The variety of sophisticated multifunctional linker units developed to date presents many attractive options to radiochemistry research and pharmaceutical drug development groups. The initial experiments with some of those linkers described here establish the foundation for future innovation and perhaps the development of fully disposable 'kit-style' production and screening products. Further investigation, capital investment and significant refinement of this approach are required. Acknowledging that work has yet to be done, solid-phase radiochemistry (SPRC) potentially represents the future of molecular imaging research from lead development to routine clinical production.

References

[1] Miller, P. W., Long, N. J., Vilar, R., and Gee, A. D.; *Angew. Chem. Int. Ed.* **2008**, *47*, 8998.
[2] Cai, L., Lu, S., and Pike, V.; *Eur. J. Org. Chem.* **2008**, 2843.
[3] Schirrmacher, R., Wängler, C., and Schirrmacher, E.; *Mini Rev. Org. Chem.* **2007**, *4*, 317.
[4] Wuest, F., Berndt, M., and Kniess, T.; *Ernst Schering Res. Found. Workshop* **2007**, *62*, 183.
[5] Schleyer, D. J., Production of Radionuclides in Accelerators, in *Handbook of Radiopharmaceuticals, Radiochemistry and Applications* (eds Welch, M. J., and Redvanly, C.S.), **2003**, John Wiley & Sons Ltd, Chichester
[6] Mausner, L., and Mirzadeh, S., Reactor Production of Radionuclides, in *Handbook of Radiopharmaceuticals, Radiochemistry and Applications* (eds Welch, M. J., and Redvanly, C.S.), **2003**, John Wiley & Sons Ltd, Chichester
[7] Jones, T.; *Eur. J. Nuc. Med.* **1996**, *23*, 207.
[8] Wang, J., and Maurer, L.; *Current topics in medicinal chemistry* **2005**, *5*, 1053.
[9] Mercer, J.; *J. Pharm. Pharm. Sci.* **2007**, *10*, 180.
[10] Orichi, N., Higuchi, T., Ishikita, T., *et al.*; *Cancer Science* **2006**, *97*, 1291.
[11] Didier, L.; *J. Fluor. Chem.* **2006**, *127*, 1488.
[12] *International congress series 1264* **2004**, 158.
[13] Scott, A.; *Positron emission tomography* **2005**, 311.
[14] Kubota, K.; *Ann. Nuc. Med.* **2001**, *15*, 471.
[15] Gambhir, S.; *Nat. Rev. Cancer* **2002**, *2*, 683.
[16] Hilker, R., Thomas, A. V., Klein, J. C., *et al.*; *Neurology* **2007**, *65*, 1716.
[17] Wu, C., Pike, V., and Wang, Y.; *Curr. Top. Dev. Biol.* **2005**, *70*, 171.
[18] Natsume, J., Kumakura, Y., Bernasconi, N., *et al.*; *Neurology* **2003**, *60*, 756.
[19] Toczek, M. T., Carson, R. E., Lang, L., *et al.*; *Neurology* **2003**, *60*, 749.
[20] Burn, D., and O'Brien, J.; *Movement Disorders* **2003**, *18*, S88.
[21] Zhang, W., Kung, M., Oya, S., *et al.*; *Nucl. Med. Biol.* **2007**, *34*, 89.
[22] Seneca, N., Cai, L., Liow, J.-S., *et al.*; *Nucl. Med. Biol.* **2007**, *34*, 681.
[23] Kung, H.; *Int. Congress Series 1264* **2004**, 3.
[24] Brooks, D.; *Mol. Imag. Biol.* **2007**, *9*, 217.
[25] Nordberg, A.; *Lancet Neurology* **2004**, *3*, 519.
[26] Sanchez-Pernaute, R., Brownell, A.-L., Jenkins, B., and Isacson, O.; *Tox. Appl. Pharm.* **2005**, *207*, S251.
[27] Troiano, A., and Stoessl, J., *Positron emission tomography in Parkinson's disease: Cerebral activation studies and neurochemical and receptor research*, in *Bioimaging* in *Neurodegeneration* (eds P. Broderick, D. Rahni, and E. Kolodny) **2005**, Humana Press Inc., Totowa, NJ.
[28] Schwaiger, M., Ziegler, S., and Nekolla, S. G.; *J. Nucl. Med.* **2005**, *46*, 1664.
[29] Knuuit, J., and Bengal, F.; *Heart* **2008**, *94*, 360.
[30] Giovanni, L.; *Eur. J. Nuc. Med. Mol. Imaging* **2006**, *33*, 621.
[31] Amol, T., Ayse, M., Abass, A., and Luis, A.; *Radiologic clinics of North America* **2005**, *43*, 107.
[32] Maisey, M.; *Nuc. Med. Comm.* **2000**, *21*, 234.
[33] Jewett, D., Mangner, T., and Watkins, L.; *Proc. Am. Chem. Soc. Int. Symp* **1991**, 387.
[34] Jewett, D., Ehrenkaufer, R., and Ram, S.; *Int. J. Appl. Rad. Isot* **1985**, *36*, 672.
[35] Hamacher, K., Coenen, H., and Stocklin, G.; *J. Nucl. Med.* **1986**, *27*, 235.
[36] Toorongian, S., Mulholland, G., Jewett, D., *et al.*; *Nucl. Med. Biol.* **1990**, *17*, 273.
[37] Mulholland, G., Mangner, T., Jewett, D., and Kilbourn, M.; *J. Labelled Cmpd. Radiopharm.* **1989**, *26*, 378.
[38] Mizuno, K., Yamazaki, S., Pascali, C., and Ido, T.; *Appl. Rad. Isotop.* **1993**, *44*, 788.
[39] Mitterhauser, M., Wadsak, W., Krcal, A., *et al.*; *Applied Radiation and Isotopes* **2005**, *62*, 441.
[40] Pascali, C., Bogni, A., Iwata, R., *et al.*; *J. Labelled Cmpd. Radiopharm.* **1999**, *42*, 715.
[41] Jinming, Z., Jiahe, T., Wushang, W., and Baoli, L.; *J. Radioanal. Nucl. Chem.* **2006**, *276*, 665.
[42] Wilson, A., DaSilva, J., and Houle, S.; *J. Labelled Cmpd. Radiopharm.* **1995**, *38*, 149.
[43] Jewett, D., Kilbourn, M. R., and Lee, L.; *Nucl. Med. Biol.* **1997**, *24*, 197.
[44] Wilson, A., Garcia, A., Chestakova, A., *et al.*; *J. Laballed Cpd. Radiopharm.* **2004**, *47*, 679.

[45] Wilson, A., Garcia, A., Jin, L., and Houle, S.; *Nucl. Med. Biol.* **2000**, *27*, 529.
[46] Ono, M., Wilson, A., Nobrega, J., *et al.*; *Nuclear Medicine and Biology* **2003**, *30*, 565.
[47] Wilson, A., Garcia, A., Jin, L., and Houle, S., *Method for synthesis of radiolabeled compounds*. **2004**: US 424001110; 530402000.
[48] Iwata, R., Pascali, C., Bogni, A., *et al.*; *Appl. Rad. Isotop.* **2001**, *55*, 17.
[49] Spivey, A. C., Laetitia, M. J., Noban, C., *et al.*; *J. Labelled Cmpd. Radiopharm.* **2007**, *50*, 281.
[50] Culbert, P., and Hunter, D.; *Reactive Polymers* **1993**, *19*, 247.
[51] Hunter, D., and Zhu, X.; *J. Labelled Cmpd. Radiopharm.* **1999**, *42*, 653.
[52] Kabalka, G., Namboodiri, V., and Akula, M.; *J. Labelled Cmpd. Radiopharm.* **2001**, *44*, 921.
[53] Ohsaki, K., Endo, Y., Yamazaki, S., *et al.*; *Appl. Rad. Isotop.* **1998**, *49*, 373.
[54] Brown, L., Bouvet, D., Champion, S., *et al.*; *Angew. Chem* **2007**, *119*, 959.
[55] Brown, L., Ma, N., Bouvet, D., *et al.*; *Org. Biomol. Chem.* **2009**,
[56] Tang, G., Wang, M., Tang, X., *et al.*; *Nucl. Med. Biol.* **2003**, *30*, 509.

Part V

Linker Selection Tables

23
Linker Selection Tables

Peter J. H. Scott

Department of Radiology, University of Michigan Medical School, USA

23.1 Introduction

It is always difficult deciding how best to classify linker units. In the hope of offering a practical guide to aid readers in selecting an appropriate linker unit for their synthesis, the tables below catalogue the linker units described in this volume according to the substrate liberated upon cleavage and conditions used to achieve such cleavage.

Linker Strategies in Solid-Phase Organic Synthesis   Edited by Peter Scott
© 2009 John Wiley & Sons, Ltd

592 Linker Selection Tables

23.2 Linkers for alcohols, phenols and diols

	Linker	Cleavage conditions	Product	Type	Chapter
1		10% TFA	R¹-OH	Acid	2
2		5% TFA	R¹-OH	Acid	2
3		TFA–water (95:5)	R¹-OH	Acid	2
4		2% TFA or formic acid	R¹-OH	Acid	2
5		Acids, TBAF	R¹-OH	Acid, Traceless	2, 16
6		Acids, TBAF	R¹-OH	Acid, Traceless	2, 16

#	Structure	Conditions	Product	Type	Ref
7		HF•pyridine, TBAF	R^1-OH	Acid, Traceless	2, 16
8		HF•pyridine, TBAF, AcOH-THF-H$_2$O (6/6/1)	R^1-OH	Acid, Traceless	2, 16
9		HF•pyridine, TBAF	R^1-OH	Acid, Traceless	2, 16
10		95% TFA–water	R^1 with OH, OH	Acid	2
11		15% TFA, 1% H$_2$O/DCM or AcOH 40 °C	R^1 with OH, OH	Acid	2
12		NaOMe/MeOH	R^1OH	Base	3
13		R^2MgCl	R^1, R^2, R with HO	Base, Multifunctional	3

594 Linker Selection Tables

	Linker	Cleavage conditions	Product	Type	Chapter
14		DIBAL-H		Base	3
15		hv, 350 nm		Photolabile	5
16		hv, 300 nm	R^1-OH	Photolabile	5
17		50% TFA/DCM		Safety Catch	6
18		0.1 M Et$_3$N/NH$_3$		Safety Catch	6

Linker Selection Tables 595

#	Structure	Cleavage	Product	Type	Ref
19		NaOH	ROH	Base, Safety Catch	3, 6
20		$Na_2S_2O_4$		Safety Catch	6
21		$NaBH_4$, Et_3N, EtOH		Safety Catch, Multifunctional	6, 13
22		Et_3N, DCM	ArOH	Safety Catch	6
23		Penicillin G Amidase	ROH	Enzyme	7
24		Penicillin G Amidase	ROH	Enzyme	7
25		5% TFA		Multifunctional	9
26		10% TFA/THF/H_2O m-CPBA, $NaBH_4$, m-CPBA, or BH_3	Aldehydes, ketones, nitriles, alcohols, acids, amines	Multifunctional	10

596 Linker Selection Tables

	Linker	Cleavage conditions	Product	Type	Chapter
27		Et₃N, EtOH, NaBH₄		Multifunctional	13
28		Piperazine	R-OH	Multifunctional	13
29		LiBH₄		Multifunctional	13
30		LiBH₄		Multifunctional	13
31		PhMgBr		Multifunctional	13
32		DDQ		Multifunctional	15
33		DDQ		Multifunctional	15
34		CAN		Multifunctional	15
35		CAN		Multifunctional	15

Linker Selection Tables 597

#	Linker	Conditions	Product	Type	Ref
35	R¹,R¹ Si-OR² (resin)	Method A: 1. HF·Pyr., THF; 2. TMSOMe. Method B: AcOH, THF, H₂O. Method C: TBAF, THF and/or AcOH. Method D: CsF, AcOH	R^2-OH	Multifunctional	16
36	allyl-dimethylsilyl-aryl (resin)	RCHO, BF₃·OEt₂	homoallyl alcohol	Multifunctional	16
37	boronate-tolyl (resin)	4-CF₃-benzaldehyde	diarylmethanol (CF₃)	Multifunctional	17
38	boronate-aryl-NHC(O)Et (resin)	Rh(acac)(CO)₂ (3 mol%), dppf (3 mol%), DME/H₂O	4-hydroxy-aryl-NHC(O)Et	Multifunctional	17
39	OSi(i-Pr)CH₂CH₂Rf camptothecin derivative	i) H₂O₂, NaOH; ii) HCl to pH 4	hydroxymethyl camptothecin	Fluorous	21
40	Bfp-protected sugar	TBAF	deprotected sugar	Fluorous	21
41	F-PMB protected sugar	NaOMe, EtOH/MeOH	deprotected sugar	Fluorous	21
42	F-MOM protected sialic acid methyl ester	10% Pd/C, H₂		Fluorous	21
		i) NaOMe, MeOH; ii) TMSBr, DCM	sialic acid		

598 Linker Selection Tables

23.3 Linkers for carboxylic acids, esters and related compounds

	Linker	Cleavage conditions	Product	Type	Chapter
1	4-(hydroxymethyl)phenyl resin (Wang-type, benzyl ester)	HF, 0 °C, (30–60 min)	$R^1\text{COOH}$	Acid	2
2	4-(hydroxymethyl)phenoxymethyl resin (Wang)	>20% TFA	$R^1\text{COOH}$	Acid	2
3	4-(4-hydroxymethylphenoxy)alkanoyl-PAL type	>20% TFA	$R^1\text{COOH}$	Acid	2
4	2,4-dimethoxy-4'-(hydroxymethyl)phenoxymethyl (HMPB)	0.1% TFA	$R^1\text{COOH}$	Acid	2
5	2-methoxy-4-(hydroxymethyl)phenoxymethyl	0.1–1% TFA	$R^1\text{COOH}$	Acid	2
6	trialkoxydiphenylmethyl (Rink acid)	0.1% TFA	$R^1\text{COOH}$	Acid	2
7	4-(hydroxymethyl)phenylacetamidomethyl (PAM)	HF, 0 °C, (30–60 min)	$R^1\text{COOH}$	Acid	2

#	Structure	Cleavage conditions	Product	Type	Ref
8	(4-hydroxymethyl-3-methoxyphenoxy)butyric acid linker	0.1–1% TFA	R¹COOH	Acid	2
9	Trityl-type linker (X = H or Cl)	1% TFA, 5% triisopropylsilane	R¹COOH	Acid	2
10	Silyl ethyl linker	30% AcOH/MeOH, 60 °C	R¹COOH	Acid, Traceless	2
11	Trichloro racemic linker	20% TFA	R¹COOH	Acid	2
12	Benzyl ether linker	NaOR² (R² = H, CH₃)	R¹CO-OR²	Base	3
13	Indole-2-carboxylate linker	NaOMe, MeOH	Methyl indole-2-carboxylate derivative	Base	3

600 Linker Selection Tables

	Linker	Cleavage conditions	Product	Type	Chapter
14		NH_2NH_2		Base	3
15		5% KCN, MeOH		Base	3
16		LAH	Boc-Phe-Val-Ala-H	Base, Traceless	3
17		NuH Nu: MeOH, py; H_2O, py; EtOH, DBU etc.	Fmoc–Peptide–OH	Base	3
18		KOH		Base, Asymmetric	3
19		Toluene, heat		Base, Cyclative, Multifunctional	3, 4

Linker Selection Tables 601

#	Structure	Conditions	Product	Type	Ref
20		$R^2\text{-}NH_2$		Cyclative	4
21		i) AcOH, NaBH$_3$CN; ii) iPrNEt$_2$, chlorobenzene		Cyclative	4
22		Toluene, 90 °C		Cyclative	4
23		1% Bu$_2$SnO chlorobenzene		Cyclative	4
24		$h\nu$, 350 nm	pGlu-His-Trp-Ser-Tyr-Gly-Leu-Arg-Pro-Gly-OH	Photolabile	5
25		$h\nu$, 350 nm	HO-Gly-Tyr-Ser-N-Boc	Photolabile	5
26		$h\nu$, 350 nm		Photolabile, Safety Catch	5, 6
27		$h\nu$, 350 nm	Peptide	Photolabile	5
28		$h\nu$, 350 nm	HO-Gly-Phe-Phe-Z-Lys(Z)	Photolabile	5

	Linker	Cleavage conditions	Product	Type	Chapter
29		hv, 350 nm		Photolabile	5
30		hv, 350 nm		Photolabile	5
31		hv, 300–340 nm		Photolabile	5
32		i) activation with CH$_2$N$_2$; ii) cleavage with NaOH		Safety Catch	6
33		LiOH, H$_2$O, H$_2$O$_2$, THF or NaOMe, MeOH/THF	R^4 = H or CH$_3$	Safety Catch	6
34		LiOH or NaOH		Safety Catch	6

Linker Selection Tables 603

35	[structure]	i) oxidative activation; ii) H₂O cleavage	$R^1\text{-COOH}$	Safety Catch	6
36	[structure]	NaOH, MeOH, dioxane or MeOH, THF, cat. NaNH₂	$R^1\text{-COOH}$ or $R^1\text{-COOMe}$	Safety Catch	6
37	[structure]	5% TFA in dioxane, microwave, 50 °C	Peptide-COOH	Safety Catch	6
38	[structure]	i) TFA activation ii) cleavage with pH 8 phosphate buffer	$R^1\text{-COOH}$	Safety Catch	6
39	[structure]	pH 7.5 phosphate buffer	peptide-COOH	Safety Catch	6
40	[structure]	5% TFA/DCM/Anisole	R-COOH	Safety Catch, Multifunctional	6, 19

604 Linker Selection Tables

	Linker	Cleavage conditions	Product	Type	Chapter
41		Phosphodiesterase, pH 5.7	HO-Gly-Ala-Leu-Gly-Pro-Ala-Fmoc	Enzyme	7
42		Chymotrypsin, H$_2$O, pH 7.0		Enzyme	7
43		Bovine trypsin	Peptide nucleic acid	Enzyme	7
44	Linker = HMPA, HMPB, HOA	Chymotrypsin, 0.1 M potassium phosphate buffer, pH 8		Enzyme	7
45				Multifunctional	9
46				Multifunctional	9
47				Multifunctional	9
48		10% TFA/THF/H$_2$O, H$_2$O$_2$		Multifunctional	10

#	Structure	Cleavage	Product	Type	Ref
49	(sulfone-linked glycine derivative with NHPh)	4 M NaOH followed by HCl	glycine NHPh derivative	Multifunctional	13
50	(dihydroacridine-linked N-acetyl-N-methyl aniline isobutyric acid)	CAN	N-acetyl-N-methylanilino isobutyric acid	Multifunctional	15
51	(quinolone-linked 4-fluorobenzoyl)	CAN	4-fluorobenzoic acid	Multifunctional	15
52	(silyl-linked TMSE ester)	TBAF	carboxylic acid R³COOH	Multifunctional	16
53	(F-TMSE ester of Boc-Phe-Ala-Val dipeptide)	TBAF, BnBr	OBn ester dipeptide	Fluorous	21
54	(F-PMBO tripeptide with azide)	H₂, Pd(OH)₂ t-BuOH	free amine tripeptide	Fluorous	21

23.4 Linkers for aldehydes, ketones and related carbonyl compounds

	Linker	Cleavage conditions	Product	Type	Chapter
1		Dowex1 × 2–100/ wet acetone		Acid	2
2		THF, H$_2$O, HCHO		Acid	2
3		THF/H$_2$O		Acid	2
4		THF-H$_2$O-CH$_3$CHO-TFE		Acid	2
5		AcOH, NaOAc		Acid, Asymmetric	2

#	Structure	Reagent	Product	Type
6	(lactone with iodoethyl group)	I₂ (aq), THF	(resin-bound pyrrolidinium-furan intermediate with OBn)	Acid, Asymmetric — 2
7	R³-CH(COR¹)(COR²)	R³-CH₂-C(O)R², LDA, THF	(benzotriazole-acyl resin)	Base — 3
8	PhCH₂CH₂-CHO	LAH	(resin Weinreb amide with PhCH₂CH₂)	Base, Traceless — 3
9	PhCH₂CH₂-C(O)R¹	R¹MgCl	(resin Weinreb amide with PhCH₂CH₂)	Base, Multifunctional — 3
10	(succinimide with R₁, R₃, N-R₂)	Bu₄NOH	(Wang-type resin with amide R₁, R₂, R₃)	Cyclative — 4
11	(benzothiazinone-dioxide with R₁, R₂, R₃)	NaH, DMF	(resin-bound benzoate-sulfonamide with R₁, R₂, R₃)	Cyclative — 4

608 Linker Selection Tables

	Linker	Cleavage conditions	Product	Type	Chapter
12	(structure)	Et$_3$N, EtOH	(structure)	Safety Catch, Multifunctional	6, 13
13	(structure)	Me$_3$SiCl, DCM	(structure)	Multifunctional	9
14	(structure)	Me$_3$SiCl, DCM	(structure)	Multifunctional	9
15	(structure)	AcOH, HCl	(structure)	Multifunctional	10
16	(structure)	TFA, H$_2$O, MeCHO, CF$_3$CH$_2$OH	(structure)	Multifunctional	10
17	(structure)	10% TFA/THF/H$_2$O	(structure)	Multifunctional	10
18	(structure)	10% TFA in wet THF	(structure)	Multifunctional	10

Linker Selection Tables 609

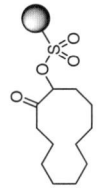

610 Linker Selection Tables

	Linker	Cleavage conditions	Product	Type	Chapter
26	(Se-CH(Bn)-O-CH=CH-COOEt) + CO	CO	lactone with Bn and COOEt	Multifunctional, Cyclative	14
27	—SeCR₁=CHR₂	TFA	R₁C(O)CH₂R₂	Multifunctional	14
28	—SeCH=CHR	HBr	RCH₂CHO	Multifunctional	14
29	Ph-C(OH)(Ar-OCH₂-resin)-CH₂CH₂CH₂-OTBS with methyl	hν, N-methylquinolinium hexafluorophosphate, O₂, NaOAc, Na₂S₂O₃	methyl ketone with OTBS chain	Multifunctional	15
30	Ph-CH(Ar-OCH₂-resin)-CH(OTHP)-CH₂CH₂CH₂-OTBS	hν, N-methylquinolinium hexafluorophosphate, O₂, NaOAc, Na₂S₂O₃	aldehyde with OTBS chain	Multifunctional	15
31	oxazolidine-thione with 4-chlorophenyl, N-MeHN-C(S)-, and CH₂-O-C₆H₄-OCH₂-resin	DDQ	oxazolidine-thione with CHO-aryl group	Multifunctional	15
32	resin-CH₂CH₂CH₂-Sn(Bu)₂-C≡C-Ph	PhC(O)Cl, Pd(PPh₃)₄	Ph-C(O)-C≡C-Ph	Multifunctional	17

23.5 Linkers for amides, ureas and related compounds

	Linker	Cleavage conditions	Product	Type	Chapter
1	(benzhydrylamine linker on resin, Ph-CH(NHC(O)R¹)-C₆H₄-resin)	HF, 0 °C, (30–60 min)	R¹-C(O)-NH₂	Acid	2
2	(4-methylbenzhydrylamine linker, 4-MeC₆H₄-CH(NHC(O)R¹)-C₆H₄-resin)	HF, 0 °C, (30 min) TFA (24 h)	R¹-C(O)-NH₂	Acid	2
3	(Rink amide linker, 2,4-dimethoxybenzhydrylamine attached via phenoxyacetamide to benzyl resin)	>20% TFA	R¹-C(O)-NH₂	Acid	2
4	(xanthenylamine linker on resin)	2% TFA	R¹-C(O)-NH₂	Acid	2

612 Linker Selection Tables

	Linker	Cleavage conditions	Product	Type	Chapter
5	(xanthene-based linker structure)	2% TFA	$R^1\text{-C(O)-NH}_2$	Acid	2
6	(dimethoxytrityl-based linker structure)	R_1 = H: TFA-DCM (50:50), thioanisol as a scavenger	$R^1\text{-C(O)-NH}_2$	Acid	2
7	(NHBoc α-substituted acrylate linker)	H_2O, 1 N HCl/AcOH, 50 °C	$H_2N\text{-CH(R)-C(O)-NH}_2$ with NHBoc	Acid	2
8	(tetrafluorophenyl ester linker)	R^2R^3 NH	$R^1\text{-C(O)-N}R^2R^3$	Base, Multifunctional	3
9	(tetrafluorophenyl sulfonate linker)	R^2R^3 NH	$R^1\text{-S(O)}_2\text{-N}R^2R^3$	Base, Multifunctional	3
10	(chlorotriazine linker)	R^1 NH_2	$R\text{-C(O)-NH-}R^1$	Base	3

Linker Selection Tables   613

614 Linker Selection Tables

	Linker	Cleavage conditions	Product	Type	Chapter
19		NH$_2$OH		Base	3
20		R^{1}COCl, MeCN		Base	3
21		pyrrolidine, LiClO$_4$		Base, Cyclative	3, 4
22		KOEt, EtOH		Base, Cyclative	3, 4
23		Et$_3$N		Base, Cyclative	3, 4
24		MeONa, MeOH		Base, Cyclative	3, 4
25		K$_2$CO$_3$, DMF		Base, Cyclative, Multifunctional	3, 4

Linker Selection Tables 615

#	Loaded Structure	Conditions	Product	Type	Refs
26		t-BuONa, THF		Base, Cyclative, Multifunctional	3, 4
27		2% Et₃N/DMF		Safety Catch, Cyclative	4, 6
28		i) ICH₂CN, DIPEA N-methylpyrrolidine; ii) 5% TFA/DCM, 2 h; 3 equiv. iPr₂NEt, THF, overnight		Cyclative	4
29		1 equiv. Ag(OCOCF₃) 2 equiv. iPr₂NEt, THF		Cyclative	4
30		Et₃N, THF/DMF 4:1 μW		Cyclative	4
31		R²-NH₂, Et₃N		Cyclative	4

616 Linker Selection Tables

	Linker	Cleavage conditions	Product	Type	Chapter
32		DMF, MW		Cyclative	4
33		$R^2\text{-}NH_2$		Cyclative	4
34		$R^2\text{-}NH_2$		Cyclative	4
35		$R^3\text{-}NH_2$, AcOH, $NaBH_3CN$		Cyclative	4
36		20% Piperidine/DCM		Cyclative	4
37		i) 25% TFA/DCM; ii) AcOH/toluene		Cyclative	4
38		i) $R_3\text{-}HN\text{-}R_4$, DIC; ii) 10% AcOH/DCM		Base, Cyclative	2, 4

Linker Selection Tables

#	Resin-bound structure	Conditions	Product	Type	Ref
39	(structure)	20% Et₃N/CHCl₃	(structure)	Base, Cyclative	2, 4
40	(structure)	DMF, 140 °C	(structure)	Cyclative	4
41	(structure)	lithiated oxazolidinone	(structure)	Cyclative	4
42	(structure)	hν, 350 nm	Amido Peptide	Photolabile	5
43	(structure)	hν, 365 nm	(structure)	Photolabile	5

618 Linker Selection Tables

	Linker	Cleavage conditions	Product	Type	Chapter
44		i) 10% AcCl in MeOH; ii) hν, 366 nm MeOH		Photolabile	5
45		hν, 354 nm MeOH		Photolabile	5
46		hν, 365 nm DMSO		Photolabile	5
47		hν, > 290 nm R^1R^2 NH		Photolabile	5
48		i) activation with CH_2N_2; ii) cleavage with R^2-NH_2		Safety Catch	6
49		i) activation with H_2O_2; ii) cleavage with NH_2-$CH(R^3)$-CO_2Na		Safety Catch	6

#	Structure	Cleavage conditions	Product	Type	Ref
50		NH$_3$	R^1-C(O)-NH$_2$	Safety Catch	6
51		i) oxidative activation; ii) R^2-NH$_2$ cleavage	R^1-C(O)-NHR^2	Safety Catch	6
52		R^2-NH$_2$ cleavage	R^1-C(O)-NHR^2	Safety Catch	6
53		R^2-NH$_2$ cleavage	R^1-C(O)-NHR^2	Safety Catch	6
54		i) acid activation; ii) R^2-NH$_2$ cleavage	R^1-C(O)-NHR^2	Safety Catch	6
55		TFA	R-C(O)-NH$_2$	Acid, Safety Catch	1, 6

620 Linker Selection Tables

	Linker	Cleavage conditions	Product	Type	Chapter
56		50% TFA/DCM		Acid, Safety Catch	1, 6
57		R^3-NH_2		Safety Catch	6
58		Lipase or esterase		Enzyme	7
59		i) R^2COCl, THF, Et$_3$N; ii) Me$_3$SiCl, DCM		Multifunctional	9
60		Me$_3$SiCl, DCM		Traceless, Multifunctional	9
61		10% TFA/DCM		Multifunctional	9

62		i) SCN(CH$_2$)$_6$CH$_3$, DMF; ii) 5% TFA, DCM		Multifunctional	9
63		i) OCN(CH$_2$)$_3$CH$_3$, DMF; ii) 5% TFA, DCM		Multifunctional	9
64		10% TFA, DCM		Multifunctional	9
65		i) BH$_3$·THF, THF; ii) HCl (aq); iii) R$_3$COCl, Et$_3$N, DMAP		Multifunctional	10
66		R^3–NH–R^4 (5 equiv.) chlorobenzene		Multifunctional	11
67		TFA		Multifunctional	13
68		mCPBA		Multifunctional	14
69		CAN		Multifunctional	15
70		CAN		Multifunctional	15
71		SmI$_2$		Multifunctional, Traceless	15

622 Linker Selection Tables

	Linker	Cleavage conditions	Product	Type	Chapter
72		SmI$_2$		Multifunctional, Traceless	15
73		i) TiCl$_4$, Zn; ii) 3% HCl		Multifunctional, Traceless	15
74		i) TBAF; ii) PyBOP, DMAP DMF-DCM		Fluorous, Cyclative	21
75		K$_2$CO$_3$		Fluorous, Cyclative	21
76		TFA–H$_2$O–DMS, microwave		Fluorous, Traceless	21
77		Et$_3$N		Fluorous	21
78		microwave		Fluorous	21

23.6 Linkers for amines

	Linker	Cleavage conditions	Product	Type	Chapter
1		TFA–DCM (50:50). Addition of scavenger is recommended	R^1-NH_2	Acid	2
2		TFA–DCM (50:50) in the presence of scavenger (i.e. Et_3SiH)	R^1-NH_2	Acid	2
3		TFA–water (95:5)	R^1-NH_2	Acid	2
4		TFA in the presence of scavenger (e.g. Et_3SiH)	R^1-NH_2	Acid	2
5		0.67 N HCl	R^1-NH_2	Acid	2

624 Linker Selection Tables

	Linker	Cleavage conditions	Product	Type	Chapter
6		10–50% TFA	$R^1\text{-}NH_2$	Acid	2
7		10–50% TFA	$R^1\text{-}NH_2$	Acid	2
8		20–25% TFA	$R^1\text{-}NH_2$	Acid	2
9		25% TFA	$R^1\text{-}NH_2$	Acid	2
10		R^1R^2 NH, Pd Catalyst		Base, Multifunctional	3
11		NaOMe/MeOH		Base	3
12		$MeNH_2$, MeOH, $CHCl_3$		Base	3

#	Structure	Conditions	Product	Type	Refs
13	(resin)-CH2O-CH2CH2-[2,5-dimethylpyrrole-N-R1]	NH2OH	R1NH2	Base	3
14	(resin)-CO2CH2CH2-N+(R1)(R2)(R3)	DIPEA	R1R2R3N	Base, Safety Catch	2, 6
15	(resin)-NHC(O)CH2-N+Me2-CH2CH2-NHC(O)-NHR	DIPEA	CH2=C(R2)C(O)NHR1	Base, Safety Catch	2, 6
16	(resin)-CH2-S-CH(R2)-C(O)NHR1	i) mCPBA activation; ii) DBU cleavage	CH2=C(R2)C(O)NHR1	Base, Safety Catch, Multifunctional	2, 6, 13
17	(resin)-OCH2CH(R1)-NHC(O)-Ar(R2) with OTs	Et3N	2-aryl-4-R1-oxazoline	Base, Cyclative, Multifunctional	3, 4
18	(resin)-C6H4-CH2-N(Me)-CH=C(NC)-C(O)R1	R2-NH2, n-BuOH	1-R2-imidazole-4-C(O)R1	Base, Cyclative, Multifunctional	3, 4
19	(resin)-piperazine-CH=C(C(O)R1)(C(O)YEt) with amidinium NO3	EtONa, EtOH, THF	pyrimidine derivative	Base, Cyclative, Multifunctional	3, 4

Y = NEt2, O.

626 Linker Selection Tables

	Linker	Cleavage conditions	Product	Type	Chapter
20	Cellulose-NH linker, Y=NH, NR³	NH₂–XH	Pyrazole product, Y=NH, NR³; X=O, NR⁴	Cyclative	4
21	Dimethoxy aniline imine linker	Piperidine, EtOH	Dimethoxy quinoline product	Cyclative	4
22	o-Nitrobenzyl photolabile linker	hν	4-hydroxybenzyl ester product	Photolabile	5
23	TentaGel-PEG linker with Bu^tO-Leu	Lipase RB001-05 pH 5.8, buffer	H₂N-Leu-O^tBu	Enzyme	7
24	Glycan-Phe linker	A− Chymotrypsin, Tris-HCl buffer, pH 7.8, 40 °C	Glycan product	Enzyme	7

Linker Selection Tables

#	Structure	Conditions	Product	Type	Ref
25		hν, 350 nm		Photolabile, Multifunctional	9
26		5% TFA/DCM		Acid, Multifunctional	1, 9
27		i) NaH, Br-(CH₂)₃-CH=CH₂; ii) 5% TFA		Multifunctional	9
28		BH₃, THF, reflux		Multifunctional	10
29		mCPBA		Multifunctional	10
30		R⁴MgCl (5 equiv.)		Multifunctional	11
31		NaBH₄ (20 equiv.)		Multifunctional	11
32		BnZnBr		Multifunctional	11
33		HCl		Multifunctional	13
34		DIPEA		Multifunctional	13
35		BnNH₂		Multifunctional	13

628 Linker Selection Tables

	Linker	Cleavage conditions	Product	Type	Chapter
36		$R^2\ NH_2$		Multifunctional	13
37		SmI_2		Multifunctional, Traceless	15
38		DDQ		Multifunctional, Traceless	15
39		HF	H_2N–peptide–CO_2PG	Multifunctional	16
40		TBAF, dioxane		Fluorous	21
41		TFA		Fluorous	21
42		R-XH, DIPEA, DMF	X = NH, NR, S	Fluorous	21

23.7 Linkers thiols, thioethers and disulfides

	Linker	Cleavage conditions	Product	Type	Chapter
1	(structure)	$Me\overset{\oplus}{-}\underset{Me}{\overset{S-S}{}}Me \; BF_4^{\ominus}$	(structure)	Multifunctional	13
2	(structure)	NCS, DMS	(structure)	Multifunctional, Cyclative	13
3	(structure)	R-XH, DIPEA, DMF	(structure) X = NH, NR, S	Fluorous	21

23.8 Linkers for sugars

	Linker	Cleavage conditions	Product	Type	Chapter
1	(structure)	NBS	(structure)	Multifunctional	13
2	(structure)	NBS, MeOH	(structure)	Multifunctional	13
3	(structure)	NaNu	(structure) Nu = N_3, I, OAc	Multifunctional	13

	Linker	Cleavage conditions	Product	Type	Chapter
4		$n\text{-}Bu_3SnCH_2CH=CH_2$		Multifunctional	14
5		$n\text{-}Bu_3SnCH_2CH=CH_2$		Multifunctional	14
6		$n\text{-}Bu_3SnCH_2CH=CH_2$		Multifunctional	14
7		CAN		Multifunctional	15
8		Grubbs (I) Catalyst, 1-Pentene		Multifunctional	4, 20
9		NaOMe, EtOH/MeOH		Fluorous	21
10		10% Pd/C, H_2		Fluorous	21
11		i) NaOMe, MeOH; ii) TMSBr, DCM		Fluorous	21

23.9 Linkers liberating alkyl groups

	Linker	Cleavage conditions	Product	Type	Chapter
1		DBU		Safety Catch, Cyclative, Multifunctional	4, 6, 13
2		hν, 350 nm, Bu$_3$SnH		Photolabile	5
3		hν, 350 nm		Photolabile	5
4		Heat	H—R	Traceless	12
5		Me$_3$SiCl, DCM		Multifunctional	9
6		Me$_3$SiX, DCM		Multifunctional	9
7		LAH or NaBH$_4$		Multifunctional	10
		KCN		Multifunctional	10

632 Linker Selection Tables

	Linker	Cleavage conditions	Product	Type	Chapter
8		Heat	H–R	Traceless	12
9		Bu$_3$SnH or Raney Ni, H$_2$		Multifunctional	13, 15
10		SmI$_2$, DMPU		Multifunctional	13, 15
11		SmI$_2$, DMPU		Multifunctional	13, 15
12		SmI$_2$, LiCl		Multifunctional	13, 15
13		SmI$_2$		Multifunctional, Fluorous	13, 21
14		SmI$_2$		Multifunctional, Fluorous	13, 21
15		NaI, MeI		Multifunctional	13

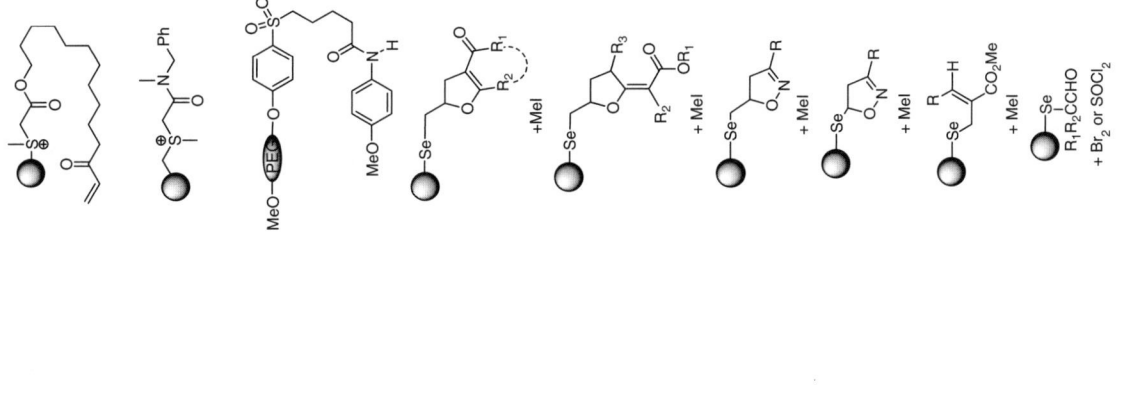

634 Linker Selection Tables

	Linker	Cleavage conditions	Product	Type	Chapter
25		AIBN, Bu$_3$SnH		Multifunctional	14, 15
26		AIBN, Bu$_3$SnH		Multifunctional	14, 15
27		AIBN, Bu$_3$SnH		Multifunctional	14, 15
28		AIBN, Bu$_3$SnH		Multifunctional	14, 15
29		AIBN, Bu$_3$SnH		Multifunctional	14, 15
30		AIBN, Bu$_3$SnH		Multifunctional	14, 15
31		AIBN, Bu$_3$SnH		Multifunctional	14, 15
32		(Me$_3$Si)$_3$SiH		Multifunctional	14, 15
33		SmI$_2$		Multifunctional	15

	Linker	Cleavage conditions	Product	Type	Chapter
34	(structure with CO₂Et, S, S, Ph on resin)	AIBN, Bu₃SnH	Ph-CH₂CH₂CH₂-C(=O)-OEt	Multifunctional	13, 15
35	(structure with Bn, S, S on resin)	Na/NH₃ (l)	Ph-CH₂CH₂CH₂-Bn	Multifunctional	13, 15
36	(TMS-alkynyl indolinone with C₈F₁₇ sulfonyl)	SmI₂	(TMS-alkynyl N-Pr indolinone)	Fluorous, Traceless	21

23.10 Linkers for alkenes, alkynes and related compounds

	Linker	Cleavage conditions	Product	Type	Chapter
1	(enaminone-piperazine linker with OTBS, R¹, R²)	MeOH, reflux	(chromone with R¹, R²)	Base, Cyclative	3, 4
2	(propargyl ether with Co₂(CO)₆, R¹, R², R³)	i) NuH; ii) (NH₄)₂Ce(NO₃)₆	(propargyl Nu, R¹, R², R³)	Base, Multifunctional	2
3	(sulfonylhydrazone R₁, R₂)	NaOCH₂CH₂OH, ethylene glycol	(alkene R₁, R₂)	Multifunctional	10

636 Linker Selection Tables

	Linker	Cleavage conditions	Product	Type	Chapter
4	Ph₃P-R on resin	R¹C(O)R² ketone	R¹R²C=CR (alkene)	Multifunctional	12
5	N⁺Me₃ benzyl phosphonate (O=P(OR)₂-CHX) on resin	R¹C(O)R² ketone	R¹R²C=CHX	Multifunctional	12
6	Pentafluorophenoxy phosphonate with CO₂Et, fluorinated on resin	RCHO, NaH	R-CH=CH-CO₂Et	Multifunctional	12
7	Phosphonate-tethered enone/ester on resin	K₂CO₃, 18-Crown-6	macrocyclic lactone	Multifunctional	12
8	4-MeO-phenyl-Cr(CO)₂-Ph₂P on resin, with OH side chain	Pyridine, heat	R¹R²C=CR	Traceless	19

9	(structure)	Dioxane, heat	(structure)	Multifunctional	13
10	(structure)	Benzene reflux	(structure)	Multifunctional	13
	(structure)	TBAF	(structure)	Multifunctional	13
11	(structure)	DBU	(structure)	Base, Multifunctional	13
12	(structure)	10% NaOH	(structure)	Base, Multifunctional	13
13	(structure)	Swern Oxidation	(structure)	Multifunctional	13
14	(structure)	Swern Oxidation	(structure)	Multifunctional	13
15	(structure)	Swern or SO$_3$/pyridine/Et$_3$N or TPAP/NMO	(structure)	Multifunctional	13
16	(structure)	SmI$_2$, DMPU	(structure)	Multifunctional	13

638 Linker Selection Tables

	Linker	Cleavage conditions	Product	Type	Chapter
17		iPrMgCl, CuI, THF		Multifunctional	13
18		EtO$_2$C—$\diagup$—CO$_2$Et, Na$^+$, Pd(PPh$_3$)$_4$, THF		Multifunctional	13
19		MeO$_2$C—$\diagup$—CO$_2$Me, Pd(PPh$_3$)$_4$, THF		Multifunctional	13
20		H$_2$O$_2$		Multifunctional	14
21		H$_2$O$_2$		Multifunctional	14
22		H$_2$O$_2$		Multifunctional	14
23		H$_2$O$_2$		Multifunctional	14
24		H$_2$O$_2$		Multifunctional	14
25		H$_2$O$_2$		Multifunctional	14
26		H$_2$O$_2$		Multifunctional	14
27		H$_2$O$_2$		Multifunctional	14

Linker Selection Tables   639

#	Structure	Cleavage	Product	Type	Ref
28		H₂O₂		Multifunctional	14
29		H₂O₂		Multifunctional	14
30		H₂O₂		Multifunctional	14
31		H₂O₂		Multifunctional	14
32		H₂O₂		Multifunctional	14
33		H₂O₂		Multifunctional	14
34		H₂O₂		Multifunctional	14
35		H₂O₂		Multifunctional	14

640 Linker Selection Tables

	Linker	Cleavage conditions	Product	Type	Chapter
36		H_2O_2		Multifunctional	14
37		H_2O_2		Multifunctional	14
38		H_2O_2		Multifunctional	14
39		H_2O_2		Multifunctional	14
40		H_2O_2		Multifunctional	14
41		H_2O_2		Multifunctional	14
42		H_2O_2		Multifunctional	14
43		H_2O_2		Multifunctional	14
44		H_2O_2		Multifunctional	14
45		H_2O_2		Multifunctional	14

46		H₂O₂	Multifunctional	14
47		H₂O₂	Multifunctional	14
48		H₂O₂	Multifunctional	14
49		H₂O₂	Multifunctional	14
50		H₂O₂	Multifunctional	14
51		H₂O₂	Multifunctional	14
52		H₂O₂	Multifunctional	14
53		H₂O₂	Multifunctional	14
54		H₂O₂	Multifunctional	14
55		H₂O₂	Multifunctional	14

642   Linker Selection Tables

	Linker	Cleavage conditions	Product	Type	Chapter
56		H_2O_2		Multifunctional	14
57		H_2O_2		Multifunctional	14
58		H_2O_2		Multifunctional	14
59		H_2O_2		Multifunctional	14
60		H_2O_2		Multifunctional	14
61		H_2O_2		Multifunctional	14
62		H_2O_2		Multifunctional	14
63		t-BuOOH		Multifunctional	14
64		t-BuOOH		Multifunctional	14

#	Structure	Reagents	Type	Ref
65	(selenide-bound γ-phenyl butyrolactone)	mCPBA	Multifunctional	14
66	(bis-selenide pyranose with OPMB)	mCPBA	Multifunctional	14
67	SeCOR₁	R₂—≡—Cu	Multifunctional	14
68	(β-hydroxy selenide)	SOCl₂ + Et₃N	Multifunctional	14
69	(allyl silane linker)	EtCH(OEt)₂, TiCl₄	Multifunctional	16
70	(diene silane linker)	TFA	Multifunctional	16
71	(alkynyl stannane linker)	TFA	Multifunctional	16
72	(resin-bound iodo vinyl stannane with MEMO groups)	i) Pd(PPh₃)₄; ii) 2:1 THF:HCl, rt, 5 days	Cyclative, Multifunctional	17
73	(Mn(CO)₂ complex linker)	NMO	Multifunctional, Traceless	19

644 Linker Selection Tables

Linker	Cleavage conditions	Product	Type	Chapter
74	hν, air		Multifunctional, Traceless	19
75	Grubbs (II) Catalyst		Cyclative, Multifunctional	4, 20
76	Grubbs (I) Catalyst		Cyclative, Multifunctional	4, 20
77	Grubbs (I) Catalyst		Cyclative, Multifunctional	4, 20
78	Grubbs (I) Catalyst, 1-Pentene		Multifunctional	4, 20

23.11 Linkers for aryl compounds

Linker	Cleavage conditions	Product	Type	Chapter
1	R^1COCl, Et_3N		Base, Cyclative, Multifunctional	3, 4
2	$Cu(OAc)_2$, Py, MeOH		Safety Catch, Traceless	6

#	Structure	Conditions	Product	Type	Ref
3	(resin-SO2-pyrimidine with Ph, amide R1,R2,R3)	i) mCPBA activation; ii) NuH	(pyrimidine-Nu, Nu = RNH, N3)	Safety Catch	6
4	(resin-SO2-CH2-C6H4-O-triazine-NHAr1, NR1R2)	R1R2 NH Cleavage	(triazine with HN-Ar1, HN-Ar2, NR1R2)	Safety Catch	6
5	(resin-S+(Et)-CH2-Ar-R)	ArB(OH)2, PdCl2(dppf), K2CO3, THF	(Ar-CH2-C6H4-R)	Safety Catch, Multifunctional	6, 13
6	(resin-CH2-N=N-C6H4-R1, Ph)	HSiCl3, DCM	(C6H5-R1)	Traceless, Multifunctional	9
7	(resin-CH2-N=N-C6H4-R1, Ph)	NaN3	(N3-C6H4-R1)	Traceless, Multifunctional	9
8	(resin-CH2-N=N-C6H4-R1, Ph)	i) TFA, MeOH; ii) Pd, =R2	(R2-CH=CH-C6H4-R1)	Multifunctional	9
9	(resin-CH2-N=N-C6H4-2-Br, Ph)	i) TFA, DCM, Me3SiN3; ii) H2O; iii) Norbornene	(norbornene-fused triazoline with 2-bromophenyl)	Cyclative, Multifunctional	9

646 Linker Selection Tables

	Linker	Cleavage conditions	Product	Type	Chapter
10		HY, acetone/H$_2$O		Cyclative, Multifunctional	9
11		5% TFA/DCM		Cyclative, Multifunctional	9
12		5% TFA/DCM		Cyclative, Multifunctional	9
13		5% TFA/DCM		Cyclative, Multifunctional	9
14		TFA, DCM		Cyclative, Multifunctional	9
15		HCl/THF		Traceless, Multifunctional	9
16		CH$_3$I		Multifunctional	9

Linker Selection Tables   647

#	Resin-bound substrate	Conditions	Product	Type	Ref
17	(resin-CH₂-N-piperazine-N=N-C₆H₄-CH=CH-CO₂t-Bu)	i) DIBAL-H, DCM; ii) AD-Mix β, acetone/H₂O, methansulfonamide; iii) THF/conc. HCl	Ph-CH(OH)-CH(OH)-CO₂t-Bu	Multifunctional	9
18	(resin-CH₂-N-piperazine-N=N-C₆H₄-CH=CH-CO₂t-Bu)	i) DIBAL-H, DCM; ii) Ac₂O, Et₃N, DCM; iii) [Pd(dba)2], PPh₃, Bn₂NH, THF; iv) THF/conc. HCl	Ph-CH=CH-CH₂-NBn₂	Multifunctional	9
19	(resin-CH₂-N-piperazine-N=N-C₆H₄-CH=CH-CO₂t-Bu)	i) DIBAL, DCM; ii) CPD; iii) THF/conc. HCl	norbornene-Ph, CO₂t-Bu	Multifunctional	9
20	(resin-CH₂-N-piperazine-N=N-C₆H₄-cyclohexene-Ph, MeO₂C, cyclopropane)	CPD, TFA, Pd(OAc)₂, MeOH	cyclopentene-C₆H₄-cyclohexene-Ph, MeO₂C, cyclopropane	Multifunctional	9

	Linker	Cleavage conditions	Product	Type	Chapter
21	(polymer-bead)-O-CH₂CH₂-N(Et)-N=... linker with C₁₂H₂₅ and SiMe₃ alkyne substituents on aryl, n=16	CH₃I	C₁₂H₂₅ and SiMe₃-alkynyl aryl iodide, n=16	Multifunctional	9
		NaOMe, MeOH	4-MeO-C₆H₄-C(O)-NH-(2-methylphenyl)	Traceless	12
22	(bead)-CH₂-PPh₂⁺ Br⁻ on benzyl, ortho to NH-C(O)-C₆H₄-OMe	i) Methyl-4-formylbenzoate, NaOMe, MeOH; ii) Girard's reagent	stilbene with CO₂Me and ortho NH-C(O)-C₆H₄-OMe	Multifunctional	12
		i) Toluene, DMF, distil; ii) KOtBu, reflux	2-(4-methoxyphenyl)indole	Multifunctional	12
23	(bead)-C₆H₄-O-P(=O)(OPh)-O-(N-Boc azepine enol)	Pd(PPh₃)₄, ArB(OH)₂, Na₂CO₃, DME/H₂O/EtOH	N-Boc-2-Ar-azepine	Multifunctional	12

24	(structure)	(structure)	Multifunctional	13
25	(structure)	(structure)	Multifunctional	13
26	(structure)	(structure)	Multifunctional	13
27	(structure)	(structure)	Multifunctional	13

650 Linker Selection Tables

Linker	Cleavage conditions	Product	Type	Chapter
28			Multifunctional	13
29			Multifunctional	13
30			Multifunctional	13

Linker Selection Tables 651

31	(resin-OSO2-Ar-R1)	Me2NH, malonate (EtO2C-CH2-CO2Et); Et3N, HCO2H, Pd(OAc)2, dppp, DMF	pyrrole with EtO2C, fused ring	Multifunctional	13
		B(OH)2-C6H4-Me	biphenyl-H-R1	Multifunctional	13
32	(resin-S-O-C6H4-NHAc)	NiCl2(PCy3)2, K3PO4, PCy3, dioxane	4-NHAc-biphenyl-4'-Me	Multifunctional	13
		Me-B(OH)2, Pd(OAc)2, K3PO4, XPHOS, THF, MW	4-Me-C6H4-NHAc	Multifunctional	13
		cycloheptanone; Pd(OAc)2, Cs2CO3, XPHOS, t-BuOH, MW	2-(4-NHAc-phenyl)cycloheptanone	Multifunctional	13
		PhNHMe; Pd(OAc)2, Cs2CO3, XPHOS, t-BuOH, MW	Ph-N(Me)-C6H4-NHAc	Multifunctional	13
33		R3-C6H4-MgBr	biphenyl-R1/R3	Multifunctional	13
34		Pd(OAc)2, dppp DMF, Et3N, formic acid	diphenylmethyl-piperazine-CH2-tolyl-H	Multifunctional	13

652 Linker Selection Tables

	Linker	Cleavage conditions	Product	Type	Chapter
35	(perfluoroalkyl sulfonate linker with R¹-aryl, amide-tethered to resin)	R^2–C₆H₄–B(OH)₂, PdCl₂(dppf), Et₃N, DMF	4-R¹-4′-R²-biphenyl	Multifunctional	13
36	(4-cyanophenyl pentafluorobenzoate sulfonate linker on resin)	Pd(OAc)₂, dppp, formic acid, Et₃N	4′-H-4-CN-biphenyl	Multifunctional	13
		C₆H₁₃–ZnI, Ni(PPh₃)₂Cl₂, PPh₃, LiCl, THF	4′-C₆H₁₃-4-CN-biphenyl	Multifunctional	13
		R–C₆H₄–B(OH)₂, PdCl₂(dppf), K₂CO₃	R-substituted 4-CN-terphenyl	Multifunctional	13
37	(aryl-silyl piperazinyl-methyl linker on resin)	Catechol	R–C₆H₅	Multifunctional, Traceless	16
38	(aryl-dimethylsilyl linker on resin)	X = Br; Br₂, DCM X = I; ICl, DCM	X–C₆H₄–R	Multifunctional	16
39	(aryl-germyl linker on phenoxy resin)	X = Br; Br₂, DCM X = I; ICl, DCM X = Cl; NCS, THF X = F; Selectfluor, MeCN	X–C₆H₄–R²	Multifunctional	16
40	(bis-thienyl-germyl TBDMS linker on resin)	TFA	oligothiophene product	Multifunctional, Traceless	16

41	(boronate on resin, R¹-aryl)	PdCl₂dppf (3 mol%) K₃PO₄ (2M, 3 equiv.) DMF, 60 °C, 24 h + R²-aryl-I	biaryl R¹/R²	Multifunctional	17
42	(boronate-benzamide-N-Bn-aa-NHBn on resin)	Ag(NH₃)₂NO₃ (0.5 M in water, 10 equiv.)	benzamide-N-Bn-aa-NHBn (H)	Traceless	17
43	(boronate-m-aniline on resin)	THF–MeOH–DCM (continuous extraction)	m-aminophenyl boronic acid dimethyl ester	Multifunctional	17
44	(boronate-p-tolyl on resin)	methyl vinyl ketone, Rh(acac)(CO)₂ (3 mol%), PPh₃ (6 mol%)	4-(p-tolyl)butan-2-one	Multifunctional	17
45	(triarylbismuth on resin)	TFA	Ar–H	Multifunctional	18
		phthalimide	Ar–N-phthalimide	Multifunctional	18
		Br₂	Ar–Br	Multifunctional	18
		I₂	Ar–I	Multifunctional	18

654 Linker Selection Tables

Linker	Cleavage conditions	Product	Type	Chapter
46	oxazolidinone NH	N-aryl oxazolidinone	Multifunctional	18
	4-chlorobenzamide	4-chloro-N-aryl benzamide	Multifunctional	18
	1-(4-cyanophenyl)piperazine	1,4-diaryl piperazine	Multifunctional	18
	imidazole	N-aryl imidazole	Multifunctional	18
	phenethylamine	N-aryl phenethylamine	Multifunctional	18
	4-methoxyaniline	diarylamine	Multifunctional	18
	3,5-dimethylphenol	diaryl ether	Multifunctional	18
	2-naphthol	1-aryl-2-naphthol	Multifunctional	18

Linker Selection Tables 655

#	Structure	Conditions	Product	Notes	Ref
47	(resin-bound Cr(CO)₂ arene complex with MeO, OH side chain, Ph₂P)	pyridine, Δ	MeO-phenyl-CH₂CH₂CH(OH)CH₃	Traceless	19
48	(Cr(CO)₂ arene-NC-phenyl-O-resin, Nu)	I₂ or hν/air	Ph-Nu	Traceless	19
49	(imidazopyridine with OSO₂C₈F₁₇, OMe, NH-cyclohexyl)	MeO-C₆H₄-B(OH)₂, Pd(dppf)Cl₂, K₂CO₃, microwave	(biaryl imidazopyridine with OMe, OMe, NH-cyclohexyl)	Fluorous, Multifunctional	21
50	(F-Boc-NH-benzamide-benzimidazole, R¹, NHR²)	TFA–THF, microwave	(benzimidazole-phenyl, R¹, NHR²)	Fluorous, Multifunctional, Cyclative	21
51	(ArGeF₂C₈F₁₇ with Cl-phenyl)	o-O₂N-C₆H₄-Br, PdCl₂(MeCN)₂, P(2-Tol)₃, TBAF·3H₂O, CuI, DMF	(2-nitro-4'-chlorobiphenyl)	Fluorous, Multifunctional	21

Index

ABB' three-component coupling reaction 256–7
acetal linkers 41, 57–60
acetic acid 477, 478
acid chlorides 509
acidic cleavage 14, 15, 31–70, 385
acid labile linkers
    alcohol linkers 41–51
    amine linkers 51–7
    carbonyl linkers 60–4
    carboxamide linkers 38–41
    carboxylic acid linkers 31–8
    in asymmetric synthesis 64–70
    ketal linkers 57–60
acrylamides 28, 207–8
α-(acylamino) amides 514
acylaminobenzyl linker 438
acylbenzotriazoles 317, 325
C-acylation, of ketones 110–11
8-acyl-8-azabicyclo[3.2.1]oct-3-en-2-ones 112, 113
acyl group-based linkers 223, 225
acylhydrazoles *see* hydrazides
acyl hydrazones 312
acylsulfonamides 196
1,2-additions
    to aldehydes 514–15
    to hydrazones 308, 309, 310
1,4-additions
    to α,β-unsaturated ketones 515
air/light cleavage 531, 532, 534
alanines 183, 184, 285
alcohol linkers 41–51, 555
alcohol resins 547–8
alcohols
    acetylation of 527
    cleavage from lithium borohydride 385–7
    cleavage from organoboronate linkers 514–15
    cleavage from polymer-supported esters 115–16
    cleavage from silyl ether resins 477–8
    cleavage from sulfoxide linkers 355–8
    enzymatic cleavage of 222–7
    immobilization of 41, 44–50
    linker selection tables for 592–7
    loading onto silyl ether resins 475–6
    NPPOC as protecting group 177
    photolabile linkers for 185–6
    safety-catch linkers for 212, 216
    silyl linkers for 46–50, 479
    substitution via Mitsunobu reaction 527
alcoxides 295
aldehyde resins 546–7
aldehydes
    1,2-additions to 514–15
    asymmetric alkylation of 311–12
    carbonyl linkers for 60, 63
    for loading amines 56
    homobenzylic ether linkers for 440, 442
    hydrazone linkers for 310–11
    immobilization of 57, 307–8
    linker selection tables for 606–10
    preparation of 356–8
    reaction with sulfur ylides 218
    reductive alkylation of 141
    safety-catch linkers for 216
    solid-phase Mannich reaction of 21–2
    Weinreb amide-based linkers for 113–14
    Wittig reaction of 308
aldol reactions
    asymmetric 48, 50
    of nortropinone 295, 296
aldol reduction 113
alkaloids 112
alkanesulfonate ester linkers
    cleavage by cross-coupling reactions 376–8
    cleavage by nucleophilic substitution 373–6
alkene–manganese complexes 525, 526
alkenes; *see also* olefins, cycloolefins
    asymmetric 1,2-disubstituted 538
    catalytic asymmetric cyclopropanations of 69–70
    for introducing diversity 362, 364
    in copper(I) mediated coupling reactions 270

Linker Strategies in Solid-Phase Organic Synthesis   Edited by Peter Scott
© 2009 John Wiley & Sons, Ltd

658 Index

alkenes; *see also* olefins, cycloolefins (*continued*)
  linker selection tables for 635–44
  manganese carbonyl linkers for 535–6
  sulfone linkers for 455, 456
  synthesis of 332–3; *see also* Wittig reaction
alkenyloxiranes 70
alkoxyamines 100
p-alkoxylbenzyl ether linkers 50–1, 52
alkoxyprolines 96, 97
alkylation
  asymmetric 311–12
  microwave-assisted 117
  of aldehydes 141, 311
  of amines 116–17, 208
  of ketones 311–12
  of thioethers 218–19
  reductive 141
N-alkylation 100, 200–1, 204
  activation of carbonyl group 196
  activation of sulfonamide linker group 15
O-alkylation 295
alkyldiisopropylsilyl linker 47
O-alkyl glycosides 58
alkyl groups, linkers liberating 631–5
alkyl halides 286, 347
O-alkyl hydroxylamines 99–101
alkyl iodides 347
alkyl sulfonates 287–90
2-alkylthio bis-benzimidazoles 111–112
alkyne-cobalt complexes 215–16, 525, 526
alkynes
  cobalt carbonyl linkers for 533–4
  coupling with aryl iodides 296–7
  desilylation of 298
  linker selection tables for 635–44
  loading onto organostannane linkers 509–11
  solid-phase Mannich reaction of 21–2
alkynol 117–18
allyl bromides 95
allyl esters 286
2,3-allyl orthoesters 426
allyl phenyl ethers 91, 92
allylsilanes 493–4
allyltributyltin 420, 424
amides 14
  benzyloxyaniline linker for 438–9
  cleavage by aminolysis 86–90, 95–6
  cleavage by oxidative electron transfer 437–9, 443
  cleavage by reduction of N–O bonds 448–9
  esters for preparation of 86, 88–90, 95–6
  HASC linkers for 450

  Lewis acid assisted cleavage of 90–1
  linker selection tables for 611–22
  nitroveratryl-based linkers for 167
  phenanthridine linkers for 443–5
  safety-catch linkers for 196, 199, 200, 203–4, 443
  sulfonamide linkers for 450
  T2 linkers for 282–3, 286–7
amide T2 resins 286–7
amido acids 32
amidomethyl ketone 61
amidopeptides 161
aminals 322, 323, 324
amine hydrochlorides 57
amine linkers 51–7, 435–7
amines
  aldehyde linkers for loading 58
  alkylation of 116–17, 208
  BAL linker for 54
  carbamate linkers for immobilization of 57, 58
  chiral 308
  cleavage by elimination 362–3
  cleavage with 98
  containing three points of diversity 514
  deprotection of 217
  diazonium T2* resins as scavengers for 292
  hydrazone linkers for 308–9
  linker selection tables for 623–8
  Mannich reaction of 21–2
  monomethylation of 116
  N–O linker for 450
  nucleophilic cleavage with 86–8, 90–2, 94–8, 370–2
  optically active 57
  phenylfluorenyl linkers for 52, 55
  protection of 51–2, 107
  safety-catch linkers for 207, 208–9, 215
  silyl linkers for 479
  sulfonamide linkers for 450, 451
  T2 linkers for 282, 283, 286, 287
  transformation of primary into secondary and tertiary 116–17
  triazene linkers for 188
  trityl linkers for 51, 55
amine T2 resins 286, 287
amino acids 6–7
  acylation of 146, 147
  asymmetric synthesis of 67, 68
  N^α-Boc protected 37
  Fmoc-protected 35, 37
  generation of 223–4
  nucleophilic cleavage of 348
  optically active 69

sensitive 198
   unnatural 67, 68–9
2-aminoacrylate linker 38, 39, 40
amino alcohols 225, 253
aminolysis, cleavage by 86–101
(R)-1-amino-2-methoxymethylpyrrolidine (RAMP) 308
(S)-1-amino-2-methoxymethylpyrrolidine (SAMP) 308, 309, 312
aminomethyl-3,5-dimethoxyphenoxy linker see PAL
aminomethyl polystyrene (AM-PS) 320
3-amino-3-(2-nitrophenyl)propionyl (ANP) linker 162–3
β-aminophenyloxycarbonyl-2,3-diaminopropionic acid 199
amino-polystyrene resin 293, 295
aminopyrimido[4,5-d]pyrimidines 346
2-aminothiophenol 374
ortho-aminothylphosphonium salts 336
6-amino-1,3,5-triazine-2,4-diones 120–1
amphetamines 581–2
AM-PS see aminomethyl polystyrene
amylose 307
aniline cellulose 124
anilines 57, 292
p-anisole 470
p-anisylgermane linkers 497–8
ANP linker see 3-amino-3-(2-nitrophenyl)propionyl linker
antibacterial compounds 252
aqueous buffers 208–12
arene–chromium complexes 525, 526
arenes 118
   chromium carbonyl linkers for 525–33
   directed ortho-metalation (DoM) of 126
ArgoGel resin 29, 30
aromatics 377
   cleavage with introduction of diversity 489–93, 499–500
   loading onto germyl linkers 498–500
   loading onto silyl linkers 487–8
   traceless cleavage using germyl linkers 499
   traceless cleavage using silyl linkers 488–9
aromatization 199–200
aryl alkyl ethers 430
arylation 17
aryl compounds, linker selection tables for 644–55
aryl enamines 338
aryl ethers 424
arylgermanes 467–8, 495, 496, 501, 561
aryl groups 515
aryl halides 487, 511

aryl hydrazides 118–19, 201
arylhydrazine linkers 200
aryl hydrazines 118, 270
aryl iodides
   coupling with terminal alkynes 296–7
   in Stille cleavage reaction 509
   radio-labelled 495
aryloxime adducts 62
arylsilanes 467, 469–70, 471, 472
   activation of 474–5
   limitations of use as linkers 495
arylsilane traceless/diversity linkers 487–93
arylstannanes 495
arylsulfide-based safety-catch linkers 216
aryl sulfonate esters 376
aryloxime resins 62
asymmetric alkylation 311–12
asymmetric catch–release process 462
asymmetric cyclative capture 430
asymmetric synthesis
   hydrazone linkers for 308–12
   of amino acids 67, 68
   of carboxylic acids 67, 68
   using acid (or electrophile) cleavable linkers 64–70
automated synthesis 104, 268, 283, 304–5, 326
azacycles 539, 540
azides 268–9, 272, 278
azido anilins 276
azolides 110–11

Baeyer–Villiger oxidation 163
BAL (backbone amide linker) 54, 56
Barbier cyclization 459
base labile linkers 79
bases
   cleavage mediated by 14–15, 16, 79–81
   epoxide opening induced by 186
   stability of silyl ethers against 45–6
Bayliss–Hillman adducts 144
benzaldehydes 183, 288, 291
   fluorous 558, 559
benzazepines 246
benzazepinones 459, 460
benzene-1,2-diamine 374
benzenes 378, 520
benzenesulfonyloxy linkers 376
benzhydrylamine (BHA) linker 38–40
benzimidazoles 111–12, 558–9
benzimidazolyl benzimidazolones 111–112
1,4-benzodiazepine-2,5-diones 197
benzodiazepines 3, 4, 16, 246, 365

1,4-benzodiazepin-5-ones 275–6
benzo[d][1,3]dioxoles 385
benzofurans 125–6, 447–8
benzofuranyl 213
benzofuroxane 268
benzoin-based linkers 180–4, 212–13
benzoins 180
benzophenone imines 67–9
benzophenone oxime linker 38
benzopyran motif 246–7
benzopyrans 83
benzopyrones 127–8
benzo[1--3]thiadiazoles 279, 281
benzothiazepinones 144, 145
benzothiazinones 147
benzotriazinones 270, 272
benzotriazole azolides 325, 326, 328
benzotriazole linkers 110, 317–29
benzotriazoles 317–18
   amide tethered 320–2, 323, 325
   carbon–carbon tethered 318–19, 325, 327
   ester tethered 322, 325
   ether tethered 319–20, 321, 323–5, 327–8
   in Mannich-type reaction 317, 318, 322–5
   synthesis of 270, 271
   synthesis of library 276–7
benzoxazoles 206
benzylamine linkers 435
benzyl bromides 95
N- benzyl deprotection 217–18
benzylguanidines 581
benzyl hydrazide 200
benzylhydrazine 307
N-benzyl-1-hydroxybenzotriazole-6-sulfonamide 88, 89
benzylic alcohol linker 14
O-benzylidene acetals 41
benzyloxyaniline linker 438
4-benzyloxybenzylamine (BOBA) 435, 436–8
benzyloxycarbonyl (Cbz) 32, 554
benzyl thioethers 218
BHA linker see benzhydrylamine linker
biaryl methanes 352, 354
biaryls 263, 264, 270
Biginelli condensation 140
biologically-oriented synthesis (BIOS) 247–8
biomarkers 578
biomolecules 241, 244, 245, 246
   fluorous linkers for synthesis of 566–70
BIOS see biologically-oriented synthesis
biotin 572, 573

biphenyl ethers 521
biphenyls 215, 377–8, 379, 381, 512, 521
bismuth(III) chloride 519–20
bismuth linkers 519–23
bistratamide H 562–3
1,4-bis(4-vinylphenoxy)butane 28
BOBA see 4-benzyloxybenzylamine
Boc see tert-butoxycarbonyl
N-Boc activated safety-catch linkers 197–8
N-Boc-glycine-phenacyl resin 176
Boc-gly resin 158
Boc strategy 32–7, 39, 40
boronate esters
   cleavage of 514
   in cross-coupling reactions 513
   in multicomponent reactions 513
   loading onto resin-bound diols 512
boron linkers see organoboron linkers
Borsche's conditions 147
Boscalid 561–2
bovine pancreatic ribonuclease A 34
bovine trypsin 230, 233
Bradley safety-catch linker 208
bradykinin 32
branching pathways 247, 248
branching strategy 246–7
Bredereck's reagent 125
bromine 493, 499, 500
3-bromobiphenyl 522
2-bromo-3,3,4,4,5,5,6,6,7,7,8,8,9,9,10,10, 10-heptadecafluorodecyl chloroformate (Froc) 568, 571
bromomethyl 158
bromomethylbenzotriazoles 319, 320
Brønstead acids 45–6
'build/couple/pair' strategy 249
tert-butanesulfinamide linker 57
tert-butoxycarbonyl (Boc) 32, 196, 197, 553–4, 557–9;
   see also Boc strategy
N-tert-butoxycarbonyl-7-(2'-nitrophenyl)caprolactam 165
tert-butylcyclohexanone hydrazones 310
tert-butyldiphenylsilyl (TBDPS) linker 47, 48
tert-butylhydroperoxide 404, 410
tert-butylketones 184
butyrolactones 140–2, 143
γ-butyrolactones 462, 464
   optically active 65, 67

Cadiot–Chodkiewicz-type reaction 295–6
CAN see cerium(IV) ammonium nitrate

Index 661

captive solvent radiochemistry 578–9
carbamates 57, 58, 95, 203
carbazate-derived hydrazone linkers 305–6
carbazates 305–6, 312–13
carbohydrates
  as scaffolds 542
    microarray screening of 572, 573
    synthesis of 438, 439, 462, 569
carbon-11 578–9
carbon–carbon bond formation see C–C bond formation
carbonic acid esters 179–80, 293, 295
carbonyl activation
  by inductive effect 196–9
  by negative mesomeric effect 199–202
  by positive mesomeric effect 202–5
carbonylating reagents 57
carbonyl compounds
  in Wittig reaction 333–4
  linker selection tables for 606–10
carbonyl imidazole 37
carbonyl linkers 60–4
carboxamide linkers 38–41, 45
carboxamides 38–40
carboxylic acid chloride resin 539–40
carboxylic acid linkers 31–8, 45, 57
carboxylic acids
  asymmetric synthesis of 67, 68
  benzoin-based linker for 212
  α-chiral 67, 68
  immobilization of 32, 36–7
  linker selection tables for 598–605
  pivaloyl-based photolabile linkers for 185
  safety-catch linkers for 196, 199, 212, 215
  silyl linkers for 479
carboxy linkers 101
β-casomorphin 225
catalytic asymmetric reactions 69–70
catch-and-release strategy 65, 118, 336
  asymmetric 462–3
catechol 489
catechol-derived safety-catch linker 214–15
catechol ester linkers 139, 140
cathepsin B 307
Cbz see benzyloxycarbonyl
C–C bond formation 117, 218, 146–8
  radical 462–4
  triple 21–2, 404
cellulose 169–70
ceramide glycanase 230, 232
cerium(IV) ammonium nitrate (CAN) 86, 437–47
cesium carboxylate 34

cesium fluoride 477, 478, 489
chemical descriptors 243–4
chemical genetics 241–2
chemical space 243–4
ChemMatrix resin 29, 30
chemo-enzymatic synthesis 228, 229, 231, 232
chiral amines 308
chiral auxiliaries 64, 311–12
chiral centres 244
chiral hydrazines 308–10
chiral linkers
  ephedrine 462–4
  hydrazone 308–12
chlorine 499, 500
chlorodimethyl(phenyl)silane linker 46
chlorohydrins 70, 71
chloromethyl ketones
  immobilization and functionalization of 61, 63, 305–7
  loading on carbazate resin 312–13
chloromethyl linker 32–5
chloromethyl polystyrene (CM-PS) 319
m-chloroperbenzoic acid 404, 411
chlorotriazine 88–9
chlorotrityl linker 32, 36, 51
[^{11}C]choline 579
chromium carbonyl linkers 525–33
  based on isonitrile ligand 527–8, 529
  based on phosphine ligand 526–7
α-chymotrypsin 228–30, 231, 234–5, 236–7
cinnamates 144
cinnamonic acid ester 266
cinnolines 270
CLEAR resin see cross-linked ethoxylate acrylate resin
cleavage see also cyclative cleavage; diversity cleavage; multifunctional cleavage; radical cleavage; traceless cleavage
  acidic 14, 15, 31–70, 385
  base-mediated 14–15, 16, 79–81
  by air/light 531, 532, 534
  by aminolysis 86–101
  by cross-coupling reactions 278–9, 281, 352–4, 372–3, 376–8, 509–13
  by elimination reactions 350–1, 362–70, 479, 486
  by epoxidation 352
  by esterification 80–6
  by Grignard reagents 115–16, 372, 377, 378
  by hydrazinolysis 101–5
  by hydroxylamines 105–9
  by iodine 528, 533
  by radical processes 345, 419–65

cleavage *see also* cyclative cleavage; diversity
    cleavage; multifunctional cleavage; radical
    cleavage; traceless cleavage (*continued*)
  by ring closing metathesis 537–48
  by saponification or trans-esterification 80–6
  electrophilic 17, 27–71, 285
  enzymatic 222–37
  halogen based 521
  homolytic 411–15, 427
  Lewis acid assisted 90, 91
  nucleophilic 14, 15, 79–129, 347–50, 370–3,
    383–5, 410–12, 521, 528
  oxidative 403–10, 411, 515, 516
  palladium assisted 91–2
  photololytic 151–91, 286
  rhodium(I) mediated 514–15
  using Pummerer rearrangement 355–8
  with amines 98
  with Raney nickel 346–7
cleavage–carbonylation–cyclization strategy 427, 429
cleavage–cross coupling strategy 269, 278, 280, 281
cleavage–cyclization strategy 420, 426
  using HASC linker system 459–61
cleavage–radical rearrangement strategy 426
CLEPSER *see* cross-linked polystyrene-ethyleneglycol
    acrylate resin
CM-PS *see* chloromethyl polystyrene
C–N bond formation 137–45
cobalt carbonyl linkers 533–4
C–O bond formation 145–6
colagenase 225, 228
combinatorial techniques 18, 20, 27, 78
compound libraries 7, 8, 77, 78, 79, 257–8
  combinatorial 244
  natural product-like 83, 246–7
  skeletally diverse 242, 243, 244, 248–57
Congo Red 582
controlled pore glass (CPG) 470, 471
Cope rearrangement 141
copolymers 10
copper(I) mediated coupling reactions 270
Corey/Kim oxidation 509
CPG *see* controlled pore glass
CR *see* cross metathesis
cross-coupling reactions *see* nickel-catalysed
    cross-coupling reactions; palladium catalysed
    cross-coupling reactions
cross-linked ethoxylate acrylate resin (CLEAR) 29, 30
cross-linked polystyrene-ethyleneglycol acrylate resin
    (CLEPSER) 28, 29

cross metathesis (CR) 537–8, 547
4-cyanobiphenyl 382
4-cyano-3-fluorobenzamide linker 62
cyanophosphoranes 339–40
cyclative capture 420, 424–5
  asymmetric 430
  using HASC linker 456–7, 459
cyclative cleavage 127, 136–48, 197, 217, 277
  base catalysed 121, 122
  nucleophilic 119
  of aromatic quinolines 147, 148
  of benzotriazinones 270–1
  of cyclic depsipeptides 145–6
  of diketopiperazines 143–4
  of peptides 138–9
  of ureas 140, 141
  using HASC linker 456, 458
  via intramolecular cyclopropanation 351–2
  via ring-closing metathesis 148–9, 538–45
cyclative release approach 125
cyclic lactones 218
cyclic peptides 45, 135–9, 188, 197–8, 214–15,
    488
cyclic pseudo tetrapeptides 541, 543
cyclization–cleavage strategy *see* cyclative cleavage
cycloadditions
  1,3-dipolar 92, 112, 249, 257, 436
  [2+2] 170
  [3+2] 558
  [4+2] 84
cyclobutanols 372
cyclocleavage *see* cyclative cleavage
cyclodepsipeptides 138–9
cycloheptenes 272–3, 493
cycloheptenones 295
cycloisomerisation reactions 255
cyclo-octene 527, 528
cycloolefins 539–45; *see also* olefins
cyclooxygenase-1 inhibitors 243
cyclopentadienyl ligand 535
cyclopentenones 367
cyclopropanations
  cyclative cleavage by 351–2
  of alkenes 69–70
  of lactones 218
cyclopropanes 69, 352, 354
cyclopropyl ketone 450, 453
cyclo-release linkers 80, 352, 509
cysteine-based linkers 208
cysteine protease inhibitors 305, 306

Dankward's approach 106
DCC see dicyclohexylcarbodiimide
Dde see
    N-1-(4,4-dimethyl-2,6-dioxocyclohexylidene)ethyl
DDQ see 2,3-dichloro-5,6-dicyanobenzoquinone
DEAD see diethyl azocarboxylate
DEAM–PS see N,N-diethanolaminomethyl polystyrene
decapeptides 230
dehydration-activated linkers 202
dehydroalanine derivatives 208
5,6-dehydroazepines 541
dehydropeptides 396, 404
demethoxyfumitremorgin C 142, 143
densely-functionalized molecules see pluripotent
    molecules
2-deoxy-2,3-didehydrosialic acid 573
2-deoxyglycosides 426–7, 428
deoxyvasicinone 276
depsipeptides 145–6, 204
Dess–Martin periodinane (DMP) 50, 295, 298–9, 374,
    420
devolatilization 580–1
Dewar–Chatt model 525
dialkoxyalkylsulfinylbenzhydrylamine (DSA) linker
    203
dialkylhydrazines 310
dialkylhydrazones 303, 311, 312
diamino alcohols 41, 44
diaminomethylaminopyridine (DMAP) 37
2,4-diaminothiazoles 350–1
diarylbismuth chloride 519–20
diarylbismuth triflate 519–20
diarylethers 271
diazoacetic esters 296, 298
diazo compounds 270
diazoimide resins 93
diazonium salt linker 57
diazonium T2 resins 282, 286, 287; see also T2 linker
    units
diazonium T2* resins 287–8, 290–3; see also T2 linker
    units
N,N-dibenzyl-N'-aryltriazene 268
dibenzyl-type T1 resins see T1 linkers
dibromoarenes 483
DIC see diisopropylcarbodiimide
2,3-dichloro-5,6-dicyanobenzoquinone (DDQ) 432–8
dicyclohexylcarbodiimide (DCC) 37
Diels–Alder reactions 60, 84, 222, 260, 447
N,N-diethanolaminomethyl polystyrene (DEAM–PS)
    514

diethyl azocarboxylate (DEAD) 37
diethyl(2-(3-benzyloxycyclobutylidene)ethyl)
    methylmalonate 372–3
diethylgermanes 496
diethyl zinc 288, 291
dihydropteridinones 559, 560
dihydropyridones 80–1, 140
1,2-dihydroquinoline 199
[^{11}C]dihydrotetrabenazine ([^{11}C]DTBZ) 579
1,4-diiodooctafluorobutane 379
diisopropylcarbodiimide (DIC) 37
diisopropyldialkoxysilyl (RO-Si-OR) linker 48
diisopropylphenylsilyl linker 46, 67
diketones 104, 107–8, 110
diketopiperazines, 37, 127, 142–4, 169, 197, 211, 479
2,6-dimethoxy-4-alkoxybezaldehyde linker see BAL
4,5-dimethoxy-2-nitrobenzyl see ortho-nitroveratryl
    group
N-1-(4,4-dimethyl-2,6-dioxocyclohexylidene)ethyl
    (Dde) 101–2, 103
4-{N-[1-(4,4-dimethyl-2,6-dioxocyclohexylidene)
    -3-methylbutyl]amino}benzyl ester (O Dmab)
    101
dimethylgermanes 496
dimethylgermylchloride 498
4-(2,5-dimethyl-4-methylsulfinylphenyl)
    -4-hydroxybutanoic acid (DSB) linker 202–3
1,3-dimethyl-5-phenoxy-benzene 521
2,5-dimethylpyrroles 107, 109
Dimroth cyclization 296, 299
dinucleoside monophosphates 362
diols 57
    immobilization of 41
    linker selection tables for 592–7
    synthesis of library 113, 114
dioxolanes 41
dioxothiones 84–5
diphenyl diselenide resin 394, 395
2,2-diphenyl-2-hydroxyethyl ester 202
diphenylmethanol linkers 45
2-(diphenylmethylsilyl)ethoxymethyl chloride
    (DSEM-Cl) 37
diphenyl-2-propyne-1-one 511
directed ortho-metalation (DoM) 126
directing metalation groups (DMGs) 126
disulfides 348
    linker selection tables for 629
dithianes 57, 184, 421
dithioacetal-based linkers 188, 191
diversity18, 245–6; see also skeletal diversity

diversity cleavage
    oxidative of cyanophosphoranes 339–40
    through Horner–Wadsworth–Emmons reaction 336–8
    through palladium catalysed cross-coupling reactions 338–9, 379, 381
    through Wittig reaction 332–6
    using germanium-based linkers 500–1
    using organoboron linkers 512–13
    using silicon-based linkers 489–93
    using sulfide linkers 347
    using sulfone linkers 370–3
    using tetrafluoroarylsulfonyl linkers 381–3
    via nucleophilic substitution 370–6
diversity elements 6
diversity/cyclative cleavable linker 362
diversity-based linker groups 13, 14, 258–60, 493–4
diversity-building cleavage reaction 95
diversity linkers 17, 258–60
diversity-oriented synthesis (DOS) 241–2, 244–5
    based on privileged scaffolds 246–8
    compared with target-oriented synthesis 245
    from simple starting materials 248
    generating skeletal diversity 248–57
    linker strategies 258–60
'diversomer' technology 3–4, 5
divinylbenzene (DVB) 28
O Dmab see 4-{N-[1-(4,4-dimethyl-2,6-dioxocyclohexelidene)-3-methylbutyl]amino} benzyl ester
DMAP see diaminomethylaminopyridine
DMGs see directing metalation groups
DMP see Dess–Martin periodinane
DoM see directed ortho-metalation
DOS see diversity-oriented synthesis
double-armed linkers 544–5
Dpr(phoc) safety-catch linker 199
drug-like molecules 537
DSA linker see dialkoxyalkylsulfinylbenzhydrylamine linker
DSB linker see 4-(2,5-dimethyl-4-methylsulfinylphenyl)-4-hydroxybutanoic acid linker
DSEM-Cl see 2-(diphenylmethylsilyl)ethoxymethyl chloride
[^{11}C]DTBZ see [^{11}C]dihydrotetrabenazine
DVB see divinylbenzene
DVB/polystyrene 47, 471
dysidiolide 546

electrophilic approach 259–60
electrophilic cleavage 17, 27–71, 285

electrophilic linkers 17, 30–70
    resins for use with 28–30
Electrospray Ionization-Mass Spectrometry (ESI-MS) 162
elemental analysis 469
elimination reactions
    cleavage of silicon linkers via 479, 486
    multifunctional cleavage via 350–1, 362–70
β-elimination reactions, 207–8
Ellman's reagent 344
enamine 147, 362
enaminodienones 125
enaminoesters 81
enaminones 123–4, 127
endo linkers 225–37
enedyine 528–9
enedyine mimics 529, 530
enol acetates 424
enolates 325, 326, 328, 527
enol phosphonates 338–9
enones 367
enyne metathesis reaction 255
enzymatic synthesis 573–4
enzyme-cleavable linkers 221–2
    endo 225–37
    exo 222–5
enzyme recognition motif 222, 225, 234
ephedrine chiral linkers 462–4
epothilone A 541–2
epoxides 186–7, 352
erythromycin 253
electrophilic cleavage, from T2 resin 285
ESI-MS see Electrospray Ionization-Mass Spectrometry
esterase 223
esterification 35, 37–8, 296, 299–300
    cleavage by 80–6
esters
    aminolysis of 38, 86, 88–90
    as linkers 38, 90
    diversity synthesis of 336
    for preparation of alcohols 115–16
    for preparation of amides 86, 88–90, 95–6
    linker selection tables for 598–605
    safety-catch linkers for 200
ethanols 374
4-(ethenylphenyl)diphenyl methanol 44
ether linkers 433, 450–3
ethyl acetoacetate 374
3,4-ethylenedioxythiophene 56
ethylene glycol linker 61, 64
ethyl 4-nitrocinnamate 337, 338

Evans–Tishchenko reaction 50
everinomicin 13, 384–1, 427, 429
exo linkers 222–5

farnesoid X receptor (FXR) 83–4
F-Boc see fluorous Boc
F-Cbz see fluorous benzyloxycarbonyl
[^{18}F]FDG see (2-[^{18}F]fluoro-2-deoxyglucose)
FDMT see fluorous dimethoxytrityl
F-HPLC see fluorous high-performance liquid chromatography
F-LLE see fluorous liquid–liquid extraction
9-fluorenylmethoxycarbonyl (Fmoc) 35, 196, 554; see also Fmoc strategy
9-fluorenylmethoxycarbonyl solid-phase technique see Fmoc strategy
fluoridolysis 489, 499
4-fluorobenzoic acid 447
2-fluorobenzonitrile linker 62, 65
(2-[^{18}F]fluoro-2-deoxyglucose) ([^{18}F]FDG) 583–5
fluorohydrocarbons 12
fluorous alcohol linkers 559
fluorous benzaldehyde 558, 559
fluorous benzyloxycarbonyl (F-Cbz) 556–7, 558, 559
　for synthesis of peptides 566, 568
fluorous Boc (F-Boc) 557–9
fluorous dimethoxytrityl (FDMT) 570
fluorous high-performance liquid chromatography (F-HPLC) 557, 570
fluorous linkers 12
　applications in proteomics 571
　for microarray applications 572–3
　for synthesis of biomolecules 566–70
　for synthesis of small molecules 557–66
　in enzymatic synthesis 573–4
　in radiochemistry 583–5
　methods for separation of substrates 557
　synthesis of natural product analogs 562–3
　types of 553–7
　versus polymer linkers 553
fluorous liquid–liquid extraction (F-LLE) 557, 562
　in synthesis of oligonucleotides 570
　in synthesis of oligosaccharides 566–9
fluorous mixture synthesis (FMS) 557, 563–6, 567–8
fluorous para-methoxybenzyl (F-PMB) 563, 564, 566; see also F-PMB-trichloroacetimidate linker
　in synthesis of carbohydrates 569, 571
　in synthesis of glycopeptides 569, 572
fluorous solid-phase extraction (F-SPE) 443, 557
　in synthesis of oligonucleotides 570
fluorous solvents 557

fluorous supports 12
fluorous tag 249, 456, 459, 460, 553
fluorous thiols 456, 457, 561
fluorous 2-(trimethylsilyl)ethyl (F-TMSE) 566, 569
fluorous 2-[tri(perfluorodecyl)silyl]ethoxycarbonyl (F-Teoc) 562–3
Fmoc see 9-fluorenylmethoxycarbonyl
Fmoc strategy 35–7, 40
F-MOM see 2-(perfluorohexyl)ethoxymethyl chloride
FMS see fluorous mixture synthesis
focused-library synthesis 244
folding pathways 248, 249 256–7
formates 44, 45
formic acid 376, 378, 379, 381, 382
four-carbon tether 469, 470
F-PMB see fluorous para-methoxybenzyl
F-PMB-trichloroacetimidate linker 563, 564
Frank safety-catch linker 210
Friedel–Crafts reactions 152, 173, 174, 161
Froc see 2-bromo-3,3,4,4,5,5,6,6,7,7,8,8,9,9,10,10,10-heptadecafluorodecyl chloroformate
F-SPE see fluorous solid-phase extraction
F-Teoc see fluorous 2-[tri(perfluorodecyl)silyl]ethoxycarbonyl
F-TMSE see fluorous 2-(trimethylsilyl)ethyl
Fukuyama method 116
Fukuyama–Mitsunobu reaction 540, 544
functional groups 6, 10–11, 13–14
　pluripotent 248–53
FXR see farnesoid X receptor

galactosylamine auxiliary 67, 68
ganglioside 230, 232
gemmacin 252–3
germanium-based linkers 17, 467–8, 495
　activation of Ge–Methyl and Ge–Aryl resins 497–500
　compared to silicon-based linkers 495
　in radiochemistry 582–3
　preparation of germyl resins 496–7
　traceless/diversity germyl linkers 497–8
germanium-based solid-phase surrogates 580–1
germyl bromides 498
germyl chlorides 496, 497–8
germyl linkers 497–8
germyl resins 496–7
Geysen safety-catch linker 208–9
GlcNAc-polymer linker system 228
glucan structures 103
p-glutarylaminophenyl glycoside linker 438
N-glycan motif 235

glycinates 67–9
glycine-derived resins 152, 158
glycine esters 296
glycopeptides 228–9, 235–7, 569
gold-mediated alkyne addition reaction 255
Gomberg–Bachmann radical reaction 270
GPCRs *see* G-protein coupled receptors
G-protein coupled receptors (GPCRs) 21, 22
Grignard reactions 295–6, 508, 510
Grignard reagents 115–16, 372, 377, 378, 509
Groebke reaction 558
Grubbs catalysts 537
Grubbs–Hoveyda catalysts 537
guanidine 109, 121, 123–4, 144, 287, 289

hairpin polyamide 558, 559
HAL linker *see* hypersensitive acid labile linker
halogens 17, 521
Hartwig–Buchwald amination reaction 276, 277
HASC linker *see* α-Hetero-Atom Substituted Carbonyl linker
HDACs inhibitors *see* histone deacetylases inhibitors
HDAR *see* hetero Diels–Alder reactions
'heavy fluorous' molecules 557
Heck cross-coupling reactions 222, 456
    of *p*-iodobenzaldehyde 307–8
    using T1 resin 266, 273–4, 278
Heck–Heck–Diels–Alder–Heck reaction sequence 279, 281–2
hemiacetal linker 59–60
hetero Diels–Alder reactions (HDAR) 84–5
α-Hetero-Atom Substituted Carbonyl (HASC) linker 346, 440–3, 444, 450
    ether 450–3
    oxygen 450, 452
    reductive cleavage of 450–3
    sulfide 456
    sulfur 443, 455–7
heteroatoms 6
heterocycles
    cleavage from sulfone linkers 364–5, 366–7
    cyclative cleavage strategy 119–20, 459–60
    five-membered 139–142
    fluorous linkers for 557–62
    seven-membered 142–5
    six-membered 84, 142–5
    synthesis using nucleophilic substitution 205, 374, 376
    synthesis using polymer-bound enaminones 124
    synthesis using T1 resins 267

N-heterocycles 455–7, 459
hexapeptides 45
4′-hexylbiphenyl-4-carbonitrile 383
5-hexyn-1-ol 533–4
HF·pyridine complex 477
high-performance liquid chromatography (HPLC) 579–80
high-throughput purification methods 98
high-throughput synthesis 249, 258
histone deacetylases (HDACs) inhibitors 572, 573
HMBA *see* hydroxymethylbenzoic acid
HMB-PS *see* hydroxymethylpolystyrene
HMPA (hydroxymethylphenoxyacetic acid) *see* Wang linker
HMPB ((4-(4-hydroxymethyl-3-methoxyphenoxy)-butyric acid) linker 33, 35
HOA *see* hydroxy octanoic acid
HOBt *see* 1-hydroxybenzotriazole
Hoffmann elimination 15, 207
homoallylic alcohols 493
homoallylic ethers 493
homobenzylic ether linker 439–40, 442
homolytic cleavage 411–15, 427
Horner–Wadsworth–Emmons (HWE) reaction 249, 332, 336–8
HPLC *see* high-performance liquid chromatography
HTPM linker *see* hydroxytetrachlorodiphenylmethyl linker
human cancer cell line 247
human oestrogen receptor α(hERα) 528
HWE *see* Horner–Wadsworth–Emmons reaction
hydantoins 4, 5, 121, 122, 139–40, 565, 566
hydrazide-based linkers 121, 122, 200–2
hydrazides 196, 303
hydrazinecarboxylate linker 61, 63
hydrazine-derived resins 307, 313
hydrazines 308–311, 313
hydrazinolysis 101–5
hydrazone linkers 303–8, 313–14
    chiral 308–12
    for aldehydes 310–11
    for amines 308–9
    for asymmetric synthesis 308–12
hydroboration–Suzuki coupling process 470
hydrocarbons 12
'hydrogen active' components 21
hydrogen bromide 34
hydrogen fluoride 34, 486
hydrogen peroxide 403, 404, 405–9
hydrosilanes 46

hydrostannation reaction 508, 510–11
hydroxamic acids 105–7
hydroxyamides 120
4-hydroxybenzyl alcohol linker 35; see also Wang linker
1-hydroxybenzotriazole (HOBt) 317, 318
N-hydroxybenzotriazole polystyrene (PS-HOBt) linker 90
4-(hydroxydiphenylmethyl)benzoic acid 44
4-(4-(1-hydroxyethyl)-2-methoxy-5-nitrophenoxy)butanoic acid 173
hydroxyketones 285
hydroxylamines
    alkyl 99–100
    cleavage by 105–9
    Wang-based resin 448–9
hydroxymethylbenzoic acid (HMBA) 101, 234–5
hydroxymethyl linker see Merrifield linker, hydroxymethyl
(4-(4-hydroxymethyl-3-methoxyphenoxy)-butyric acid) linker see HMPB linker and SASRIN
hydroxymethylphenoxyacetic acid (HMPA) see Wang linker
hydroxymethylpolystyrene (HMB-PS) 319
hydroxy octanoic acid (HOA) 235
(R)-2-hydroxy-2-phenylpropanoic acid 65, 66
N-(4-hydroxyphenyl)propionamide 516
N-hydroxyphthalimide 100
hydroxytetrachlorodiphenylmethyl (HTPM) linker 32, 34, 37
    for immobilization of amines 57
    for immobilization of alcohols and phenols 45
4-hydroxy-2,3,5,6-tetrafluorobenzamidomethyl polystyrene (PS-TFP) linker 90
hypersensitive acid labile (HAL) linker 101
HypoGel resin 29, 30

ICP-OES see Inductively Coupled Plasma-Optical Emission Spectroscopy
IDU see 5-iodo-2′-deoxyuridine
imidazo[1,2-a]pyridine-8-carboxamides 98
imidazo[1,2+a]pyridines 558, 559
imidazo[1,2-a]pyridines 364, 365
imidazoles 121–3, 306, 367–8
imidazolidinones 249, 252
imines 65, 66, 98
indenones 355
indole amides 200
indole-based linkers 203–4
indole-like scaffolds 257

indole ring formation 200
indoles 81–3
indolidones 561
indolines 424
indolocarbazoles 459
indoloquinolines 460–1
Inductively Coupled Plasma-Optical Emission Spectroscopy (ICP-OES) 469
injector loop technique 579–80
iodides 95, 509
iodine 493, 499, 500
    cleavage with 528, 533
iodine chloride 493
[^{131}I]iodobenzylguanidyne 581–2
5-iodo-2′-deoxyuridine (IDU) 85–6
4-iodo-N-isobutylbenzamide 449
iron tricarbonyl 90, 91
isatins 443
isoindolinones 141, 142
isomunchones 92
isonitrile ligand 527–8, 529
isoxazoles 61–2, 64, 124
isoxazolines 365, 436
isoxazolocyclobutanones 367, 368

JandaJel resin 28, 29
JandaJel-trityl linker 44
Jones' reagent 60
Julia–Lythgoe olefination see Julia olefination
Julia olefination 369–70, 455

kahalalide A 138
Kenner linkers 138–9, 196–7
ketal linkers 57–60
keto amides 339
keto esters 172, 339, 340
ketones
    C-acylation of 111
    1,4-additions to 515
    alkylation of 311–12
    carbonyl linkers for 60
    cleavage from hydrazine-based resin 307, 313
    cyclopropyl 450, 453
    homobenzylic ether linker for 440, 442
    immobilization of 57, 61
    linker selection tables for 606–10
    α-sulfonated 374–5
    thioester linkers for 385
    α,β-unsaturated 374, 515
    Weinreb amide linker for 113–14
ketopiperazines 197, 200–1
kryptofix[2.2.2] 583, 585

668 Index

Lacey–Dieckmann cyclization 146, 147
lactams 163, 217, 539, 542, 543
β-lactams 170, 173, 270–1, 439, 441, 449
lactol 253, 255
lactones 427, 434–5
   asymmetric catch–release approach 462–4
   cyclic 218
   saponification of 163
δ-lactones 60
lactonization reaction 213–14, 255
lactosyl ceramide 230
Lanterns 29, 30, 45, 47
lanthanide(II) complexes 447
LCAA-CPG *see* Long Chain Alkyl Amine Controlled Pore Glass
leaving group linkers 110
LEDs *see* light-emitting diodes
Lemieux oxidation 350
Lemieux-type glycosylation 347
Leu-enkephalin 225
levulinoyl ester 103–4
Lewis acids 90, 91, 187
libraries *see* compound libraries
'libraries from libraries' approach 7; *see also* branching strategy
ligand displacement procedure 531, 533
ligands
   for proteins 6–7
   triazene 288–90, 291
light-emitting diodes (LEDs) 164
'light fluorous' molecules 557
linker groups (linker units) 5, 12–17
   as protecting groups 109–110
   choice of 27–8, 30–1
   definition of and classification 78–9
linker selection tables
   alcohols 592–7
   aldehydes 606–10
   alkenes 635–44
   alkynes 635–44
   amides 611–22
   amines 623–8
   aryl compounds 644–5
   carbonyl compounds 606–10
   carboxylic acids 598–605
   diols 592–7
   disulfides 629
   esters 598–605
   ketones 606–10
   linkers liberating alkyl groups 631–5
   phenols 592–7
   sugars 629–30
   thioethers 629
   thiols 629
   ureas 611–22
lipases 223, 226
liquid crystals 98
liquid–liquid extraction 12, 98
lithium borohydride 385–7
lithium phenyl sulfinate 358, 359–60
Long Chain Alkyl Amine Controlled Pore Glass (LCAA-CPG) 155, 166
Lyttle safety-catch linker 210–11

macrolactone 337
malonate derivatives 455
maltoheptaose 307
manganese carbonyl linkers 535–6
Mannich reaction 21–2, 435
   benzotriazole-mediated 317, 318, 322–5
   using benzotriazones 322–5, 328
mappicines 565
Marshall linker 95–8, 198
Marshall resin 96–7
Marshall-type safety-catch linker 206
MBHA linker *see* 4-methylbenzhydrylamine linker
MCR4 *see* melanocortin subtype-4 receptor
MCRs *see* multicomponent reactions
MDDR *see* MDL Drug Data Repository
MDL Drug Data Repository (MDDR) 252
meclizine 379, 380–1
melanocortin subtype-4 receptor (MCR4) 170
mercaptomethyl ketones 307
Merrifield, Bruce 3
Merrifield linker 38
   chloromethyl 32–5
   hydroxymethyl 33, 35
   benzyldehyde 41
Merrifield resin 79
mesomeric effect 199–205
metal triazenes 288–91
methanolic potassium carbonate 374
[^{11}C]methionine 579
4-methoxybenzhydrylamine (MOBHA) linker 40
p-methoxybenzyl 554
p-methoxyphenacyl 176–80, 187, 189
3″-methoxy-[1,1′,4′,1″]terphenyl-4-carbonitrile 383
methyl 4-(4-acetyl-2-methoxy-5-nitrophenoxy)butanoate 172
methyl 4-(4-acetyl-2-methoxyphenoxy)butanoate 172
4-methylbenzhydrylamine (MBHA) linker 38, 39, 40, 61

methyl indole carboxylates 82, 83
2-methyl indoles 426
methyl iodide 298
methyl silyl ethers 478
methyl sulfone 358
methyl thioether 218
methyl triflate 184
(*S*)-methylcyclohexanone 65
1-methylindoline 425
Michael addition 15, 207, 218, 362
Michael addition–elimination 218
Michael cyclization 456
microarray screening 572–3
microsporin 139
microwave-assisted reactions 204
   alkylation reaction 117
   Nicholas reaction 118
   pyrrole ring formation 109
   synthesis of benzofurans 125–6
   synthesis of dihydropyrimidinones 140
   synthesis of diketones 108
   synthesis of diketopiperazines 170
   synthesis of hydantoins 139–140
   synthesis of imidazoles 121, 123
   synthesis of isoxazoles 124
   synthesis of peptide acids 204
   synthesis of pyrazoles 124
mimics 139, 540, 529, 530
Mitsunobu conditions 147, 177
Mitsunobu reactions 222, 430
   N-alkylation 100, 200–1, 204
   coupling of alcohols with phenols 424
   esterification of carboxylic acids 37
   forming ether bonds 91–2
   substitution of alcohols 527
MOBHA linker *see* 4-methoxybenzhydrylamine linker
MOE *see* molecular operating environment
molecular diversity spectrum 245, 246
molecular imaging 579
molecular operating environment (MOE) 252
molecular tags 438
molybdenum 372
monomethoxypolyethylene glycol 322
monomethylation 116
morpholine 374
morpholine amide 450
motilin antagonists 139
multi-cleavable resins 155
multicomponent reactions (MCRs) 249, 253, 513, 514, 558; *see also* Ugi multicomponent reactions
multi-detachable linkers 152–3

multidimensional descriptor space *see* chemical space
multifunctional cleavage
   from bismuth linkers 520–21, 522
   from sulfamate linkers 383
   from sulfide linkers 344
   from thioester linkers 385
   of hydrazone linkers 311
   using organostannane linkers 508–9
   using Pummerer rearrangement 355–8
   via elimination reactions 350–1, 362–70
   via nucleophilic substitution 347–50, 370–3
multiple cleavable linkers 211–12

naltrindoles 204, 205
natural product-like libraries 83, 246–7
natural products 244, 246–7, 433–5
natural products analogs 562–3
Negishi reaction 381–2, 383
neo-glycoconjugate 228, 231
Nicholas reaction 117–18
nickel-catalysed cross-coupling reactions 377
nitriles 311–12
3-nitro-4-aminomethylbenzamidomethyl PS-TTEGDA resin 160
nitroarenes 276
nitrobenzyl alcohol 204–5
nitrobenzyl alcohol derivatives 152, 153
ortho-nitrobenzylamino group-based linkers 158–61
ortho-nitrobenzyloxy group-based linkers 151–8
   α-substituted 161–5
3-nitro-4-bromomethylbenzamide PS-TTEGDA resin 160
nitro-esters 172
nitro group reduction 462
nitroindoline-based linkers 187, 189
p-nitromandelic acid 204, 205
nitrophenol resin 86
nitrophenyl resin 88
6-(2′-nitrophenyl)-6-(*N-tert*-butoxycarbonyl)amino caproic acid 165
7-(2′-nitrophenyl)caprolactam 165
p-nitrophenyl carbonate resin 120–1
2-(2′-nitrophenyl)cyclohexanone 165
3′-nitrophenylpropyoxycarbonyl (NPPOC) 177
ortho-nitroveratryl group 165–73
N–O linkers 448–50
nortropinone 295, 296
NovaGel resin 29, 30
NPPOC *see* 3′-nitrophenylpropyoxycarbonyl
nucleophilic cleavage 14, 15, 79–129, 347–50, 370–3, 383–5, 410–12, 521, 528; *see also* nucleophilic substitution

nucleophilic displacement cleavage *see* nucleophilic substitution
nucleophilic labile linkers 79–129
nucleophilic substitution
 alkanesulfonate ester linkers 373–6
 diversity cleavage via 370–6
 heterocycle synthesis 205, 374, 376
 multifunctional cleavage via 347–50, 370–3
 selenium-based linkers 410–11, 412
 sulfone linkers 370–3
nucleoside phosphoramidite 570, 572

4-octene-1,8-diol based linker 547
olefination; *see also* Wittig reaction; Horner–Wadsworth–Emmons reaction
 diversity cleavage by 332–8
 Julia-type 369–70, 455
olefins 545–7; *see also* alkenes, cycloolefins
oligonucleotides
 automated synthesis of 471
 fluorous linkers for synthesis of 569–70
 light-driven synthesis of 177
 photolabile linkers for 155, 166
 preparation of 210
oligopeptides 31–5
oligo(1,4-phenylene-ethylene)s 293, 294
oligo(phenylentriacetylene)s 296
oligosaccharides 46, 48
 acetal linkers for 58–9
 acylaminobenzyl linker for 438, 440
 ANP-based linker for 162, 163
 cleavage by nucleophilic substitution 373, 374
 Dde-based linkers for 102–3
 fluorous enzymatic synthesis of 573–4
 fluorous linkers for 566–9
 4-octene-1,8-diol based linker for 547–8
 photolabile linkers for 155, 156, 157
 safety-catch linkers for 215
 synthesis using RCM linker 545–6
oligothiophenes 499
'one-synthesis/one-skeleton' approach 244
optically active amines 57
optically active linkers 65
organic substrate 5
organoboronates 511
organoboron linkers 511–16
organocuprates 372
organolithium reagents 487, 527
organostannane linkers 507–11, 581–2
organostannanes 505–8

organotin compounds 507, 511
(1S,16S)-3-oxa-bicyclo[14.1.0]heptadecane-2,15-dione 352
oxacillin 253
oxazines 104–5, 120
oxazoles 367, 368, 369
1,3-oxazolidines 433
oxazolidinones 119–20
oxazolines 67, 120
oxidation
 Bayer–Villiger 163
 Corey/Kim 509
 Lemieux 350
 of arylsulfide for Pummerer rearrangement 216–17
 of sulfides 358–9, 360
 Swern 367
 with Dess–Martin periodinane 295, 296, 298–9
oxidation/elimination cleavage strategy 364–5, 367, 369, 403, 410
oxidative aromatization 199–200
oxidative electron transfer
 cleavage of ether and amine linkers 431–9
 cleavage of homobenzylic ether linker 439–40
 cleavage of safety-catch linkers 443–7
 cleavage of sulfur HASC linker 440–3, 444
oximes 60
oxindole quinazolines 348–9, 350
oxindoles
 cyclative–capture approach synthesis 456–7, 459
 fluorous-tagged 443
 nucleophilic cleavage with 348, 350
 reductive traceless cleavage of 346, 361
 synthesis using HASC linker 455–6, 457, 459
oxygen-based linkers 420–1

pairing reactions 255–6
PAL *see* peptide amide linker
palladium-catalysed cross-coupling reactions 86, 270, 505–6; *see also* cross-coupling reactions, Heck cross-coupling reactions, Sonogashira cross-coupling reactions, Stille cross-coupling reactions, Suzuki cross-coupling reactions
 azide functionality in 273–4
 cleavage by 278–9, 281, 338–9, 352–4, 372–3, 377
 functionalization/cleavage by 218
 in loading aromatics onto silyl linkers 487–8
palladium catalysts 91–2
 in hydrostannation reaction 508
 polymer-suported triazenes as 290, 291
 resin-bound 505, 507
PAM linker *see* phenylacetamide linker

PA polymers *see* polyacrylamide polymers
parallel synthesis 116
Passerini reaction 156
Pauson–Khand reaction 255
PCA *see* principle component analysis
PEG *see* polyethylene glycol
PEG-5000 supported ketal linker 61
PEGA resin *see* polyethylene glycol-polyacrylamide resin
penicillin G acylase 222
penicillin G amidase 224–5, 226–7
pentyl glycoside 396, 404
Pepsyn K 225
peptide acids 80, 199, 204; *see also* peptides
peptide aldehydes 60, 113–14, 115, 304–5
peptide amide linker (PAL) 54, 56
peptide amides 199, 202
peptide carboxamides *see* carboxamides
peptide esters 37
peptide hydrazides 303
peptide hydroxamic acids 105–6
peptide nucleic acid (PNA) 230, 233
peptides 3–4, 6, 31–3, 99, 160–1
   aryl hydrazide linkers for 118, 119
   Boc strategy 32, 35–7
   cleavage via elimination 362
   cyclative cleavage of 138–9
   cyclic 45, 188, 197–8, 135–7, 214–15, 488
   enrichment of 571
   enzymatic synthesis of 225
   fluorous linkers for 566, 568
   Fmoc strategy 35–7, 234
   Houghten syntheses of 8, 9
   long 177
   o-nitroveratryl-based linkers for 170
   photolabile linkers for 152, 154, 161
   polyacrylamide polymers for synthesis of 28
   redox-sensitive linker for 213–14
   safety-catch linkers for 208–10
   silyl linkers for 479, 486
   tripodal 169
β-peptides, 566, 569
peptidomimetics 51–2, 60
peptidyl ketones 60, 61, 305, 306
peptidyl privileged structures 215
perfluoralkanesulfonate esters 379, 380
perfluoralkanesulfonyl (PFS) linkers 378–81
perfluorinated hydrocarbons 12
perfluorocarbon chains *see* fluorous tags
2-(perfluorohexyl)ethoxymethyl chloride (F-MOM) 573, 574

PET *see* Positron Emission Tomography
Petasis reaction 253, 255, 513
PFS linkers *see* perfluoralkanesulfonyl linkers
phakellistatin 138
phase tag assisted synthesis 456
phase transfer catalysis (PTC) 69
phenacyl group-based linkers 173–6
phenacyl resins 175–6
phenanthridine linkers 443–7
phenol derivatives 217
phenols 292
   cleavage of 515, 516
   colorimetric test for 96
   immobilization of 45
   linker selection tables for 592–7
   loading 379–80
phenol–sulfide linker *see* Marshall linker
phenylacetamide (PAM) linker 33, 35
phenylacetylene oligomers 263
phenylalanine 488, 529, 530
N-phenylanisole 521
N-phenylbenzamide 521
4-phenyl-1,2-dihydroquinoline 446, 447
phenylene ethylene oligomers 297
9-phenylfluoren-9-yl group 52
phenylfluorenyl linkers 52, 55
phenylglyoxylate 65
2-phenyl-2-hexene 335
phenylhydrazide linkers 200
N-phenyl-oxazolidin-2-ones 521
N-phenylphthalimides 521
phenyltriazene linkers 188
2-phenyl-2-trimethylsilyl) ethyl linker (PTMSEL) 479
phosphine ligands 526–7
phosphinic esters 287, 290
phosphodiesterase 225, 228
phosphodiester group 225
phosphonate-based linker groups 17
phosphonates 332, 336
phosphonium salts 332, 333, 335, 336
phosphoranes 333–4
phosphoric esters 287, 290
phosphorus linkers 331–41
photolabile linkers 151
   based on benzoin group 180–4
   based on *ortho*-nitrobenzylamino function 158–61
   based on *ortho*-nitrobenzyloxy function 151–8
   based on *ortho*-nitroveratryl group 165–73
   based on *para*-methoxyphenacyl group 176–80
   based on phenacyl group 173–6
   based on pivaloyl group 184–7

photolabile linkers (*continued*)
  based on α-substituted*ortho*-nitrobenzyl group 161–5
  traceless 187
photolysis 187, 527
photolytic cleavage 161, 286
phthalides 126–7, 145, 146
phthalimides 140, 141
Pictet–Spengler reaction 140–2, 224, 559
piperazine amide 57
piperazines 142
piperazinyl-type T1 resins *see* T1 linkers
piperidine derivatives 67
piperidones 362, 364
pivaloyl-based linkers 184–7
platinum electrode supports 450, 451
pluripotent molecules 248, 253–6
pluripotent functional groups 248–53
PMB *see para*-methoxybenzyl
PMB-OH linker 553
Pnm linker *see p*-nitromandelic acid
pochonin C 421
POE 6000 (linear polyethyleneglycol) 222, 224
polyacrylamide (PA) resins 28, 29
polycyclic scaffolds 255–6
polyethylene glycol (PEG) 10–11, 30, 61, 174, 222, 224
poly(ethylene glycol) acrylamide (PEGA) resins 307
polyethylene glycol-polyacrylamide (PEGA) resins 30, 225
polyethylene glycol-derived polymers 153–5, 470, 471
polymer-supported reagents 65
polymyxins 138
polynucleotides 27
polypeptides 202
polysaccharides 27
poly(styrene-oxyethylene) graft copolymers 30
polystyrene (PS) resins 10–11, 28–30, 344–5
polystyrenes 468–70, 472
polystyrene sulfonium resin 344–5
polystyrene thiol resin 344, 345
poly(triacetylene)-derived oligomers 295–6, 297
Positron Emission Tomography (PET) 495, 499, 579, 583
potassium cyanide 111, 112
potassium iodide 113
preformed handle strategy 496
principle component analysis (PCA) 243, 252
privileged scaffolds 246–8
proline 558
1,3-propanedithiol linker 349
propargyl alcohol 215

propargyl derived safety-catch linkers 215, 216
protecting groups 6, 27–8, 553–4
  deprotection of 152
  for alcohols 37, 41, 45–6, 47, 177
  for amines 37, 51, 52, 107
  for amino acids 32, 35, 37
  for carboxylic acids 37
  linkers as 79
proteins, ligands for 6–7
protein structure similarity clustering (PSSC) 247
proteomics 571
ipso-protodesilylation 467
PS-DES resins 471–6
pseudo tetrapeptides 541, 543
PS-HOBt linker *see N*-hydroxybenzotriazole polystyrene linker
PS resins *see* polystyrene resins
PSSC *see* protein structure similarity clustering
PS-TFP linker *see* 4-hydroxy-2,3,5,6-tetrafluorobenzamidomethyl polystyrene linker
PTC *see* phase transfer catalysis
pteridinediones 370–1
PTMSEL *see* (2-phenyl-2-trimethylsilyl) ethyl linker
Pummerer cyclization 346, 455–6, 561
Pummerer rearrangement 216, 355–8
purification 6, 577
purines 206, 370
pyranofurans 542–3, 544
pyranones 440
pyrazoles 94, 104, 105, 124, 365
3,5-pyrazolidinediones 312, 455
pyrazolines 365
pyrazolones 121, 122, 312
pyridazines 365
pyridazinones 140
N-pyridin-2-ylmethylbenzamide 86, 87
pyridones 365
2-pyridylthiocarbonate linker 50
pyrimidine nucleosides 85–6, 87
pyrimidines 123–4, 205–6, 365, 370, 561
pyrimidinones 144, 145
pyrrole derivatives 109
pyrroles 374
(3*R*)-pyrrolidinol 309
pyrrolidinones 65, 140–2, 143
pyrrolo-benzodiazepinones 144, 145

quinazoline-2,4-diones 121, 558, 559
quinazolinones 144
quinolines 147, 148

quinone-based linkers 213
quinoxalinones 558–9

radical cleavage strategy 345
  by nitro group reduction 462
  by oxidative electron-transfer 431–47
  by radical carbon–carbon bond formation 462–4
  by reductive electron transfer 447–61
  using reagents 420–31
radical processes 419–20
radicicol 421
Radicicol A 563, 564
radiochemistry *see* solid-phase radiochemistry
radioiodination 582
radiolabeling 577–8
radiopharmaceuticals 577
RAM linker *see* Rink linker, amide
RAMP *see* (*R*)-1-amino-2-methoxymethylpyrrolidine
Raney nickel cleavage strategy 346–7
Rapp resin *see* TentaGel resin
'Rasta silane' resins 471, 472
RCM *see* ring-closing metathesis
reagent-based approach *see* branching pathways
redox-sensitive linkers 213–14
reductive amination 56, 277
reductive aromatization 213–14
reductive electron transfer 421, 447–8
  cleavage of ether linkers 450–3
  cleavage of N–O linkers 448–50
  cleavage of sulfonamide linkers 450
  of alkyl and aryl sulfide/sulfone linkers 453–61
Reformatsky reagent 324
regenerative Michael acceptor (REM)
  linker 207
  resin 14–15
REM *see* regenerative Michael acceptor
resins (solid supports) 5, 10–11
rhodium catalysts 257, 476, 515–16
Richter-type cleavage protocol 270
ring closing metathesis (RCM)
  cyclative cleavage via 148–9
  in releasing olefins and cycloolefins 537–47
Rink linker 14–15, 32, 34–6, 38–40, 45
RO5 *see* rule of five
rod-shaped compounds 98–9
RO-Si-OR linker *see* diisopropyldialkoxysilyl linker
rule of five (RO5) 244
ruthenium–alkylidene species 536, 537, 539
ruthenium–benzylidine species 536, 539
ruthenium catalysts 290, 291

SAC *see* Silyl Acid Linker
saccharides
  double-armed linker approach 545
  linker selection tables for 629–30
  nucleophilic cleavage of 347, 348, 350
safety-catch amide linker (SCAL) 202, 203
'safety-catch' concept 195–6
safety catch linkers 15–16, 80, 137, 195–6, 258–9
  activated by formation of alkyne-cobalt complex 215–16
  activated by indole ring formation 200
  activated by intramolecular H-bonding 214–15
  activated by oxidation of arylsulfide for Pummerer rearrangement 216–17
  activated by oxidative aromatization 199–200
  activated by oxidative *N*-benzyl deprotection 217–18
  activated by reductive aromatization 213–14
  activated by thioether alkylation 218–19
  benzyhydryl-based 202–3
  N-Boc activated 197–8
  cleaved by oxidative electron transfer 443–7
  dehydration-activated 202
  Dpr(phoc) 199
  enzyme-cleavable 222, 223
  fluorous 556
  for release in aqueous buffers 208–12
  fragmentation by β-elimination 207–8
  indole-based 203–4
  Kenner-type 138–8, 196–7
  p-metoxyphenacyl group-based 177, 179
  nitrobenzyl alcohol-based 204–5
  nucleophilic aromatic substitution 205–6
  photochemical activation 212–13
  sulfide/sulfone-based 198
SAL *see* Silyl Amide Linker
samarium (II) iodide 369, 447–8
  cleavage of HASC linkers 346–7, 450–3, 456, 459–61
  cleavage of N–O linkers 448–50
  cleavage of sulfone linkers 360–1, 455
  in nitro group reduction 482
  in radical carbon–carbon bond formation 462
SAMP *see* (*S*)-1-amino-2-methoxymethylpyrrolidine
Sanofi–Aventis radioligand SR46349B 581
sansalvamide A 488
saponification 80–6, 163
SAR *see* structure–activity relationship
sarcodictyin analogs 59–60
SASRIN (Super Acid Sensitive ResIN) 32, 33, 35, 101; *see also* HMPB

scaffolds
   ABB' MCR for synthesis of 256
   indole-like 257
   polycyclic 255–6
   privileged 246–8
SCAL *see* safety-catch amide linker
'scandide contraction effect' 495
Schmidt glycosylations 59
Schmidt rearrangement 57, 163, 447
sclerotigenin analogs 563
SCONP *see* structural classification of natural products
SCX extraction *see* strong cationic exchanger extraction
selenium-based linker groups
   attachment methods 398–403
   cleavage methods 403–16
   radical cleavage of 421–30
   reagents 391–8
selenocyanates 393, 394
SEM *see* 2-(trimethylsilyl)ethoxymethyl
SEM-Cl *see* 2-(trimethylsilyl)ethoxymethyl chloride
semicarbazide linker 60, 61, 63
semicarbazide resins 307, 313
semicarbazones 60, 61, 304–5
serine protease analogues 169
serratamolide 139
Sharpless asymmetric dihydroxylation (AD) reaction 249
sialidase inhibitor 573, 574
Sieber linker 38, 39, 40
silanes 469, 471
silicon-based linkers 67, 101, 467–8, 500–1
   activation of Si–Aryl resins 474–5
   activation of Si–H resins 471–4
   cleavage via 1,3-elimination 479, 486
   compared to germanium-based linkers 495
   fragmentation-based silyl linkers 479–86
   indirect 479, 486
   preparation of silyl resins 468–71
   silyl ether linkers 475–9
   traceless/diversity silyl linkers 16, 487–94
silicon-based resins 468
silver diamine nitrate 515
Silyl Acid Linker (SAC) 479
Silyl Amide Linker (SAL) 479
silyl chloride 46, 47, 469, 471, 486
silyl enol ethers 21
silyl ether linkers 500
   cleavage of alcohols from resins 477–8
   linker designs 479, 480–2
   loading alcohols onto resins 475–6
   recycling of resins 478–9

silyl ethers 45–6, 467
silyl linkers 467
   for alcohols 46–50
   for amines 479
   fragmentation-based 479–86
   traceless/diversity 487–94
silyl resins 468–71
silyltriflate linker 47, 48
Single Photon Emission Computed Tomography (SPECT) 499, 579, 583
skeletal diversity 246, 247, 248–57; *see also* diversity
skeletally diverse libraries 242, 243, 244, 248–57
small molecules 241–4, 247; *see also* skeletally diverse libraries
   biologically active 241, 243, 244, 247
   fluorous linkers for 557–66
   microarray sceening of 572
sol–gel processing 536
solid-phase asymmetric synthesis (SPAS) 308–12; *see also* asymmetric synthesis
solid-phase extraction (SPE) 207, 579
solid-phase organic synthesis (SPOS) 7, 17–18, 20
   apparatus for conducting 9
   in generation of DOS libraries 257–60
   versus solution-phase synthesis 4–5, 6, 77–8
solid-phase radiochemistry (SPRC) 577–8
   fluorous linkers in 583–5
   germanium linkers in 582–3
   stannane linkers in 581–2
solid-phase surrogates
   germanium-based 580–1
   thin film radiochemistry 578–80
solid supports 5, 10–12
somatostatin mimics 139
Sonogashira cross-coupling reactions 118, 127, 222
   using T1 resins 274–5, 278
   using T2* resins 290
   using triazene linker 295, 297
spacers 31
SPAS *see* solid-phase asymmetric synthesis
SPE *see* solid-phase extraction
SPECT *see* Single Photon Emission Computed Tomography
spiroketals 50, 61, 64, 433, 434
spirooxindoles 8–9, 461
spiropyrans 128–9
'split and pool' techniques 8–9, 31
SPOCC resin 230, 234
SPOS *see* solid-phase organic synthesis
SPRC *see* solid-phase radiochemistry
stannane linkers *see* organostannane linkers

starting material *see* organic substrate
Stille cross-coupling reactions 509, 511
streptogramins 138
strong cationic exchanger (SCX) extraction 116
structural classification of natural products (SCONP) 247
structure–activity relationship (SAR) 18
styrenes 539, 540, 545
ipso-substitution 467
substrate-based approach *see* folding pathways
subtilisin 233–4
succinate linker 128
sugars 225; *see also* saccharides; oligosaccharides
sulfamate linkers 383–5
sulfide (thioether) linkers 344–5
   cleavage via elimination reactions 350–1
   cleavage via nucleophlic substitution 347–50
   cleavage via reductive electron transfer 453–61
   reductive traceless cleavage 345–7
   safety-catch 198
sulfides 358–9, 360
sulfonamide linkers 15–16, 355, 356
   aliphatic 196–7
   aryl 196
   cleaveage via reductive electron transfer 450, 451
   Kenner linker 138–9
sulphonamides
   library synthesis 86, 88
   preparation of 116, 197
   safety-catch linkers for release of 203
sulfonate ester linkers 373
   alkanesulfonate ester linkers 373–8
   perfluoralkanesulfonyl (PFS) linkers 378–81
   tetrafluoroarylsulfonyl linkers 381–3
sulfonate linkers 420
sulfone linkers 358–60
   cleavage via elimination reactions 362–70
   cleavage via nucleophilic substitution 370–3
   cleavage via reductive electron transfer 453–61
   reductive traceless cleavage 360–1
   safety-catch 198
sulfonic acids 288
sulfonic esters 287, 290
sulfonium linkers 351–1
sulfonylhydrazone linkers 303–4
α-sulfonyloxy ketone resin 420–1
2-sulfonylpyrimidine linker 205
sulfoxide linkers 354–5
   cleavage using Pummerer rearrangement 355–8
   traceless cleavage from 355
sulfoximines 70, 71

sulfur-based linkers 343–87, 421
sulfur ylides 218
Suzuki cross-coupling 110, 222, 318, 512, 558
   diversity cleavage of enol phosphonates 338
   in synthesis of benzotriazinones 270–1
   one pot process with hydroboration 470
   using organoboron linkers 511
   using T1 resin 273–4, 278
   using T2* resins 290
Suzuki–Miyaura coupling reaction 62, 339, 511, 512–13
Swern oxidation 367
SynPhase Lantern 29, 30, 45

T1 linker units 264–5
   dibenzyl-type T1 resins 264–5, 266–78
   piperazinyl-type T1 resins 264–5, 278–82
   resin synthesis and washing procedure 265–6
T2 linker units
   T2 linkers 282–7
   T2* linkers 287–93
   T2* scavenger resins 292–3
tags *see* molecular tags
target-oriented synthesis (TOS) 245
TBAF *see* tetrabutylammonium fluoride
T-BAL 56
TBDPS linker *see* *tert*-butyldiphenylsilyl linker
Tebbe reagent 62
tellurium-based linker groups
   attachement methods 398
   cleavage methods 411
   radical cleavage of 430–1
   reagents 391–2, 396–7
TEMP *see* 2,2,6,6-tetramethylpiperidine-1-oxyl
TentaGel 10–11, 29, 30, 108, 223, 225, 226
tethered linkers 542–3
tethers 469, 470, 317, 318
tetrabutylammonium fluoride (TBAF) 475, 476, 477, 479, 489
tetraethyleneglycol diacrylate (TTEGDA) 159, 160
tetrafluorophenol resins 86, 88
tetrahydro-β-carbolines 224, 559, 560
tetrahydrofurans 430, 493
tetrahydropyrans 493
tetrahydropyranyl (THP) linker 45
tetrahydroquinolines (THQs) 324–5, 456, 458
2,2,6,6-tetramethylpiperidine-1-oxyl (TEMPO) 471
tetramic acids 146, 147
tetrapeptides 139
TFA *see* trifluoroacetic acid
thiazoles 367, 368

thin film radiochemistry 578–80
thioacetals 349
thioacetamide 374
thioesters 105, 106, 385–6
thioether linkers *see* sulfide linkers
thioethers
    alkylation of 218–19
    linker selection tables for 629
thiohydroxamic acid-derived linkers 187, 188
thiols
    fluorous 561
    linker selection tables for 629
    silyl linkers for 479
thiophenol 374
thiophenol esters 95–6
2-thiopyrimidine-based linker 205–6
Thorpe–Ingold effect 217
THP linker *see* tetrahydropyranyl linker
THQs *see* tetrahydroquinolines
tin chloride 508–9
tin hydride 187, 508, 510
tin linkers 499
TIPS *see* tri-*iso*-propylsilyl
TMSE *see* 2-(trimethylsilyl)ethyl
Tollen's reagent 514
toluene 187
toluenesulfonamide 324
p-toluenesulfonic acid (PPTS) 45
p-tolyl(4-(trifluoromethyl)phenyl)methanol 515–16
TOS *see* target-oriented synthesis
tosylamine 324
tosylates 120
traceless cleavage 378
    from bismuth linkers 520, 522
    from organoboron linkers 514
    from perfluoralkanesulfonyl (PFS) linkers 379–81
    from sulfide linkers 345–7
    from sulfone linkers 360–1
    from sulfoxide linkers 355
    from tetrafluoroarylsulfonyl linkers 381–2
    of aromatics 488–9, 499, 501
    of piperazinyl-type T1 resin 280
    using palladium-mediated reactions 376–7, 379, 380–1
    using silicon-based linker group 16
    using silver diamine nitrate 515
    using trichlorosilane 267–8, 277–8
traceless linkers 13, 14, 80, 258
    aryl hydrazides as 118
    germyl 497–8
    photolabile 187

    silicon-based 16, 487–94
traditional (classic) linker groups 13–14, 258, 259
transition metal carbonyl linkers
    chromium 525–33
    cobalt 533–4
    manganese 535–6
transition metal/ligand linkers 526–7
2-(trialkylsilyl)ethanol linker 37
2-(trialkylsilyl)ethyl ether 479
triaryl bismuthanes 519–20, 521, 522
triazene ligands 288–90, 291
triazene linkers 188, 190, 259, 263, 293–300; *see also* T1 linker units; T2 linker units
triazene T2 resins 282, 287, 288
triazenes 57, 263, 265
triazines 206, 370
triazoles 94, 296, 299
triazolines 268
$4H-^{[1-3]}$-triazolo[5,1-c]$^{[1,4]}$benzothiazine 280, 282
tributyltin hydride 420–31
trichloroacetimidate linker 50
2,2,2-trichloroethoxycarbonyl (Troc) 568
trichlorosilane 267–8, 277–8
trifluoroacetic acid (TFA) 34, 477, 489
trifluoromethanesulfonic acid 34
    fluorous 556, 565
tri-*iso*-propylsilyl (TIPS) 553
triisopropylsilyloxycarbonyl (Tsoc) linker 486
trimethylgermyl resin 496
1,3,3-trimethyl-indoline-5-yl-succinic amide Wang ester 128–9
2-(trimethylsilyl)ethoxymethyl (SEM) 479
2-(trimethylsilyl)ethoxymethyl chloride (SEM-Cl) 37
2-(trimethylsilyl)ethyl (TMSE) 37, 479
tripetides 162
triphenylmethyl (trityl) group 51; *see also* trityl linkers
triphenyl phosphine 37
triphenylphosphonium salts 335
triphenylsilyl linker 46
tripodal linkers 169, 170
tris(trimethylsilyl)silane (TTMSS) 427
trityl chloridere resin 383–5
trityl linkers 101
    for amines 51–2, 55
    for immobilization of alcohols 44
    for immobilization of carboxylic acids 32, 36–7
    for synthesis of peptide amides 40
Troc *see* 2,2,2-trichloroethoxycarbonyl
tropane derivatives 113
tropane ring system 111
tropinones 112–13

Tsoc linker *see* triisopropylsilyloxycarbonyl linker
TTEGDA *see* tetraethyleneglycol diacrylate
TTMSS *see* tris(trimethylsilyl)silane
tyrocidine antibiotics 139

Ugi multicomponent reactions 142, 169, 170, 171, 197, 198, 206, 540, 557
Ullman reaction 263, 271–3, 275
Upjohn's conditions 61
ureas 14, 94–5, 198, 325–6, 328–9
   formation and cyclative cleavage 140, 141
   linker selection tables for 611–22
   preparation of T2 resins 286–7
   synthesis using N–O linker 448
   synthesis using T2 linkers 282–4, 286–7

valinomycin 135, 136
vancomycin 263, 264, 404
vinyl ethers 61, 62, 64, 92
vinyl halides 509
vinylic sulfonamidopeptides 170
vinyl sulfones 208, 358
vinyl triflates 509

Wang linker 15, 38, 41, 50, 79, 101, 234
   carbamate 57
   cleaved by oxidative electron transfer 431–4
   2,4-dimethoxy substituted 35
   2,6-dimethoxy-substituted 32, 35
   lactoside 58, 59
   modified 32, 33, 35
Wang resin 14, 15
Weinreb amide 509
Weinreb amide-based linkers 113, 114–15
Wheland intermediate 487, 488, 489, 495
Wittig reaction 362
   aza- 275
   carbonyl compounds in 333–4
   diversity cleavage through 332–6
   of aldehydes 308

9-xanthenyl linker *see* Sieber linker

yuehchukene 563, 564

(*S*)-zearalenone 509, 510